Organic Electronic Spectral Data

Volume VIII 1966

Organic Electronic Spectral Data

Volume VIII 1966

JOHN P. PHILLIPS, HENRY FEUER
& B. S. THYAGARAJAN

EDITORS

CONTRIBUTORS

J. C. Craig
H. Feuer
L. D. Freedman
K. Genest
M. K. Hrenoff
P. M. Laughton
C. M. Martini
F. C. Nachod
J. P. Phillips
B. S. Thyagarajan
O. H. Wheeler

INTERSCIENCE PUBLISHERS

a Division of JOHN WILEY & SONS
New York • London • Sydney • Toronto

Library of Congress Catalog Card Number: 60-16428

ISBN 0-471-68799-5

Printed in the United States of America

10 9 8 7 6 5 4 3 2 1

INTRODUCTION TO THE SERIES

In 1956 a cooperative effort to abstract and publish in formula order all the ultraviolet-visible spectra of organic compounds presented in the journal literature was organized through the enterprise and leadership of M.J. Kamlet and H.E. Ungnade. Organic Electronic Spectral Data was incorporated in 1957 to create a formal structure for the venture, and coverage of the literature from 1946 onward was then carried out by chemists with special interests in spectrophotometry through a page by page search of the major chemical journals. After the first two volumes (covering the literature from 1946 through 1955) were produced, a regular schedule of one volume for each subsequent period of two years was instituted. With this volume an annual schedule is inaugurated. Eight volumes have now been published.

Altogether, more than fifty chemists have searched a group of journals totalling more than a hundred titles during the course of this sustained project. Additions and subtractions from both the list of contributors and of journals have occurred from time to time, and it is estimated that the effort to cover all the literature containing spectra may not be more than 95% successful. However, the total collection is by far the largest ever assembled, amounting to well over 170,000 spectra through the volumes so far published.

Volume IX (with the same editors) is in preparation.

PREFACE

Processing of the data provided by the contributors to Volume VIII was performed at the University of Louisville. The editors are grateful to Marion E. Hill of the Stanford Research Institute for help with the initial stages of this volume.

John P. Phillips
Henry Feuer
B.S. Thyagarajan

ORGANIZATION AND USE OF THE DATA

The data in this volume were abstracted from the journals listed in the reference section at the end. Although a few exceptions were made, the data generally had to satisfy the following requirements: the compound had to be pure enough for satisfactory elemental analysis and for a definite empirical formula; solvent and phase had to be given; and sufficient data to calculate molar absorptivities had to be available. Later it was decided to include spectra even if solvent was not mentioned.

All entries in the compilation are organized according to the molecular formula index system used by Chemical Abstracts. Most of the compound names have been made to conform with the Chemical Abstracts system of nomenclature.

Solvent or phase appears in the second column of the data lists, often abbreviated according to standard practice; there is a key to abbreviations on the next page.

The numerical data in the third column present wavelength values in millimicrons (or nanometers) for all maxima, shoulders and inflections, with the logarithms of the corresponding molar absorptivities in parentheses. Shoulders and inflections are marked with a letter s. In spectra with considerable fine structure in the bands a main maximum is listed and labelled with a letter f. Numerical values reported by authors are given to the nearest 0.1 millimicron for wavelength (if possible) and 0.01 unit for the logarithm of the molar absorptivity. Spectra that change with time are labelled "changing" and temperatures are sometimes indicated if unusual.

The reference column contains the code number of the journal, the page number of the paper, and in the last two digits the year. A letter is added for journals with more than one volume in a year. The complete list of all articles with authors is found in the References at the end of the volume.

Several journals that were abstracted for previous volumes in this series have been omitted and a few new ones have been added. In a very few instances a code number for an omitted journal has been assigned to a new one in this volume.

ABBREVIATIONS

s	shoulder or inflection
f	fine structure
n.s.g.	no solvent given in original reference
C_5H_5N	pyridine
C_6H_6	benzene
$C_6H_{11}Me$	methylcyclohexane
C_6H_{12}	cyclohexane
DMF	dimethylformamide
DMSO	dimethylsulfoxide
THF	tetrahydrofuran

Other solvent abbreviations generally follow the practice of Chemical Abstracts.

Underlined data were estimated from graphs.

JOURNALS ABSTRACTED

Journal	No.
Acta Chem. Scand.	1
Indian J. Chem.	2
Anal. Chem.	3
J. Heterocyclic Chem.	4
Ann. Chem. Liebigs	5
Ann. Chim. (Paris)	6
Ann. chim. (Rome)	7
Appl. Spectroscopy	9
Arch. Biochem. Biophys.	10
Arkiv Kemi	11
Australian J. Chem.	12
Steroids	13
Bol. inst. quim. univ. na. auton Mex.	16
Boll. sci. fac. chim. ind. Bologna	17
Bull. Chem. Soc. Japan	18
Bull. Acad. Polon. Sci.	19
Bull. soc. chim. Belges	20
Bull. soc. chim. France	22
Can. J. Chem.	23
Chem. Ber.	24
Chem. and Ind. (London)	25
Chimia	27
Compt. rend.	28
Discussions Faraday Soc.	29
Doklady Akad. Nauk S.S.S.R.	30
Experientia	31
Gazz. chim. ital.	32
Helv. Chim. Acta	33
J. Chem. Eng. Data	34
J. Am. Chem. Soc.	35
J. Pharm. Sci.	36
J. Biol. Chem.	37
J. Chem. Phys.	38
J. Chem. Soc.	39
J. Chim. phys.	41
J. Optical Soc. Am.	43
J. Org. Chem.	44
J. Phys. Chem.	46
Monatsh. Chem.	49
Nature	50
Naturwiss.	51
Proc. Royal Soc.	52
Rec. trav. chim.	54
Roczniki Chem.	56
Science	57
Spectrochim. Acta	59
Trans. Faraday Soc.	60
Ber. Bunsengesellschaft Phys. Chem.	61
Z. physik. Chem. (Frankfurt)	62
Zhur. Obshchei Khim.	65
J. Structural Chem.	67
Biochemistry	69
Izvest. Akad. Nauk S.S.S.R.	70
Biochem. Biophys. Acta	71
Coll. Czech. Chem. Comm.	73
Chemical Communications	77
Tetrahedron	78
Revue Romaine Chim.	80
J. Mol. Spectr.	82
Arch. Pharm.	83
Talanta	86
J. Med. Chem.	87
Tetrahedron Letters	88
Angew. Chem.	89
J. Inorg. Nucl. Chem.	90
J. Applied Chem. S.S.S.R.	93
Chem. Pharm. Bull. (Japan)	94
J. Pharm. Soc. Japan	95
The Analyst	96
Z. Chemie	97
Theor. Exptl. Chem.	99
Lloydia	100
J. Organometallic Chem.	101
Phytochemistry	102

Organic Electronic Spectral Data

Volume VIII 1966

Compound	Solvent	λ max(log ε)	Ref.
CH_4N_2S			
Thiourea	H_2O	198(3.3),236(3.3)	52-0122-66
$CN_2O_6^-$			
Trinitromethyl anion	H_2O	355(4.16)	70-0615-66
COS			
Carbonyl sulfide	gas	222.5(--)	35-2412-66
$C_2F_6N_2O$			
N-Nitrosobis(trifluoromethyl)amine	gas	386(1.56),402(1.67), 422(1.60)	39-0933-66A
$C_2H_2Cl_2N_2O_4$			
Ethane, 1,2-dichloro-1,1-dinitro-	MeCN	277(1.86)	44-0369-66
$C_2H_2N_2S$			
1,2,5-Thiadiazole	n.s.g.	254(3.94)	89-0514-66
$C_2H_3ClN_2O_5$			
Ethanol, 2-chloro-2,2-dinitro-	MeCN	276(1.91)	44-0369-66
$C_2H_3N_3O_7$			
Ethanol, 2,2,2-trinitro-	MeCN	277(2.02)	44-0369-66
$C_2H_3N_3S$			
1,2,3-Thiadiazole, 5-amino-	MeOH	224(3.37),295(3.91)	24-1618-66
$C_2H_3O^+$			
Methyloxocarbonium ion	$H_2SO_4-SO_3$	less than 215 nm	35-1488-66
$C_2H_4N_2O_2$			
Ethylnitrosolic acid	ether	262(4.00),732(1.52)	22-1658-66
anion	n.s.g.	336(4.40),505(1.56), 652(1.68)	22-1658-66
$C_2H_4N_2O_4$			
Carbamic acid, nitro-, methyl ester	hexane	210(3.85)	44-3038-66
	EtOH	217(3.93),257(3.45)	44-3038-66
$C_2H_4N_4O_2$			
Formamide, 1,1'-azobis-	pH 5	290(4.3)	22-1658-66
	aq. DMF	245(3.32),420(1.73)	22-1658-66
ion I	n.s.g.	347(4.4)	22-1658-66
ion II	n.s.g.	360(4.5)	22-1658-66
Formamide, 1-(formylazo)-, oxime	acid	270(4.0),425(1.78)	22-1658-66
ion	n.s.g.	345(4.3)	22-1658-66
$C_2H_4O_2$			
Acetic acid	H_2O	203(1.58)	23-1961-66
C_2H_5NOS			
Thiocarbamic acid, methyl ester	C_6H_{12}	242(4.0),290(2.0)	60-0788-66
iodine complex	C_6H_{12}	304(4.4)	60-0788-66
C_2H_6ClN			
Dimethylamine, N-chloro-	C_6H_{12}	275(2.7)	24-1357-66
	H_2SO_4	320(1.9)	24-1357-66

Compound	Solvent	λ_{max}(log ϵ)	Ref.
$C_2H_6N_2S$			
Dimethylamine, N-(thionitroso)-	C_6H_{12}	306(4.08),587(1.44), 705(0.18)	35-3842-66
	EtOH	306(4.03),533(1.24), 680(0.00)	35-3842-66
	CCl_4	309(4.09),576(1.58)	35-3842-66
Thiourea, N-methyl-, iodine complex	$CHCl_3$	300.5(4.64)	60-0029-66
C_2H_6S			
Methyl sulfide	gas	195f(4.0),203(3.7), 220f(2.8)	38-1367-66B
	hexane	191(3.40),200(3.38), 214(2.80),233s(1.9)	38-1367-66B
	H_2O	194(3.17),210(3.08), 234s(2.0)	38-1367-66B
	MeOH	199(3.24),214(2.94), 233s(1.9)	38-1367-66B
	EtOH	199(3.29),212(2.95), 233s(2.0)	38-1367-66B
	EtOH	201(3.32)	39-0239-66A
$C_2H_6S_2$			
Methyl disulfide	hexane	195(3.84),253(2.62)	38-1367-66B
	H_2O	189(3.73),251(2.52)	38-1367-66B
	MeOH	193(3.72),253(2.52)	38-1367-66B
	EtOH	195(3.73),253(2.59)	38-1367-66B
	EtOH	201(3.34),254(2.60)	39-0239-66A
tetracyanoethylene complex	CH_2Cl_2	425(3.33)	35-0204-66
$C_2H_6S_3$			
Methyl trisulfide	hexane	200s(--),212(4.08), 253s(2.61),285s(2.5)	38-1367-66B
	H_2O	203s(3.83),251s(2.53), 290s(2.5)	38-1367-66B
	MeOH	198s(--),211(3.97), 253s(2.58),290s(2.5)	38-1367-66B
	EtOH	211(3.98),253s(2.59), 290s(2.5)	38-1367-66B
$C_2H_6S_4$			
Methyl tetrasulfide	C_6H_{12}	200(--),207(4.27), 258s(3.30),302(3.27)	38-1367-66B
	MeOH	207(4.17),252s(3.34), 300(3.30)	38-1367-66B
	EtOH	207(4.11),258s(3.32), 300(3.28)	38-1367-66B
$C_2H_7BN_4$			
Cycotetrazenoborane, 2,5-dimethyl-	gas	198(3.80)	39-0580-66A

Compound	Solvent	λ_{max}(log ϵ)	Ref.
$C_3Cl_3N_3$			
Cyanuric chloride	EtOH	221(2.33)	80-0139-66
$C_3F_9N_3O$			
Hydrazine, nitrosotris(trifluoro-methyl)-	gas	215(3.11),360s(1.40), 371(1.60),387(1.74), 405(1.70)	39-1236-66C
	gas	374(1.60),388(1.71), 406(1.68)	39-0933-66A
$C_3F_9N_3O_2$			
Hydrazine, nitrotris(trifluoro-methyl)-	gas	230s(1.87),270s(1.18)	39-1236-66C
$C_3H_2Br_2N_2$			
Pyrazole, 3(5),4-dibromo-	EtOH	223.5(3.58)	22-3744-66
$C_3H_2Cl_2N_4$			
s-Triazine, 2-amino-4,6-dichloro-	EtOH	262(3.30)	80-0139-66
$C_3H_2IN_3O_2$			
Imidazole, 2-iodo-4(or 5)-nitro-	6M H_2SO_4	307(3.81)	4-0454-66
	pH 3.1	240s(3.66),325(3.83)	4-0454-66
	pH 13	273(3.53),363(3.97)	4-0454-66
	MeOH	235s(3.67),314(3.90)	4-0454-66
$C_3H_2N_4S_2$			
s-Triazolo[3,4-b][1,3,4]thiadiazole-6-thiol	MeOH	285(4.06)	44-3528-66
anion	n.s.g.	287(4.09)	44-3528-66
$C_3H_3ClO_2$			
Acrylic acid, 3-chloro-, cis	EtOH	213(4.00)	44-0646-66
C_3H_3FO			
Acrolein, 2-fluoro-	dioxan	220(--)	88-5379-66
C_3H_3N			
Acrylonitrile, radical anion	H_2O	255(2.83)	77-0498-66
$C_3H_3NOS_2$			
Rhodanine	EtOH	253(4.15),295(4.24)	95-0095-66
C_3H_3NS			
Thiazole	C_6H_{12}	232(3.49)	22-2857-66
$C_3H_3NS_2$			
4-Thiazoline-2-thione	acid	235s(--),310(4.14)	39-0092-66B
	1.82N HCl	235s(3.23),313(4.08)	59-2005-66
	H_2O	310(4.14)	39-0092-66B
	H_2O	200(3.96),235s(3.23), 313(4.08)	59-2005-66
	base	290(3.99)	39-0092-66B
	1.92N NaOH	294(3.94)	59-2005-66
	EtOH	321(4.13)	59-2005-66
$C_3H_3N_3O_2$			
Pyrazole, 4-nitro-	0.05N HCl	274(3.91)	54-1191-66
	H_2O	274(3.92)	54-1191-66
	0.05N NaOH	319(4.06)	54-1191-66

Compound	Solvent	$\lambda_{max}(\log \epsilon)$	Ref.
$C_3H_3N_3O_4$			
Diazoacetic acid, α-nitro-, methyl ester	MeOH	290(4.24)	88-5821-66
$C_3H_3N_5$			
5H-s-Triazolo[5,1-c]-s-triazole	MeOH	227(3.64)	25-2168-66
$C_3H_3N_5S$			
s-Triazolo[3,4-b][1,3,4]thiadiazole, 6-amino-	MeOH	238(3.79)	44-3528-66
	acid	243(3.88)	44-3528-66
5H-s-Triazolo[5,1-c]-s-triazole-3-thiol	MeOH	233(3.73),283(4.14)	25-2168-66
$C_3H_4ClNO_2$			
Pyruvoyl chloride, 1-oxime	pH 6	229(3.93)	28C-0985-66A
	base	282(4.20)	28C-0985-66A
$C_3H_4ClN_5$			
s-Triazine, 2,4-diamino-6-chloro-	EtOH	256(3.51)	80-0139-66
$C_3H_4N_2$			
Imidazole, HI salt	C_6H_{12}-2%BuOH	220(4.1)(anom.)	10-0332-66D
	H_2O	225(4.3)(anom.)	10-0332-66D
$C_3H_4N_2O$			
Isoxazole, 3-amino-	EtOH	228.5(3.45)	94-1277-66
$C_3H_4N_2OS$			
1,3,4-Oxadiazoline-5-thione, 2-methyl-	EtOH	262(4.15)	1-0057-66
$C_3H_4N_2O_4$			
2-Oxazolidinone, 3-nitro-	EtOH	227(3.72),290(1.96)	44-3038-66
$C_3H_4N_2S$			
1,2,5-Thiadiazole, 3-methyl-	n.s.g.	259(3.94)	89-0514-66
$C_3H_4N_2S_2$			
1,3,4-Thiadiazoline-5-thione, 2-methyl-	EtOH	309(4.14)	1-0057-66
$C_3H_4N_4O$			
1,2,4-Triazin-5-one, 3-amino-2,5-dihydro- (or azaisocytosine)	pH -1.1	257(3.70)	73-1864-66
	pH 5.1	247(3.71)	73-1864-66
	pH 11.4	286(3.80)	73-1864-66
	EtOH	248(3.75),293s(2.71)	73-1864-66
$C_3H_4N_6$			
5H-s-Triazolo[5,1-c]-s-triazole, 3-amino-	pH 1	245(4.12)	25-2168-66
	MeOH	239(4.13)	25-2168-66
$C_3H_4N_6S$			
7H-s-Triazolo[5,1-c]-s-triazole-3-thiol, 7-amino-	H_2O	230(3.70),278(3.95)	25-2168-66
C_3H_4O			
Acrolein	isooctane	203(4.03),335(1.30)	99-0469-66
	H_2O	210(3.99),315(1.43)	99-0469-66
	EtOH	207(4.08),328(1.33)	99-0469-66
Cyclopropanone	CH_2Cl_2	310(1.36),330s(--)	35-3672-66

Compound	Solvent	λmax(log ε)	Ref.
$C_3H_5ClN_2O_2$			
Glyoxime, chloromethyl-	pH 1	233(4.15)	28C-0985-66A
$C_3H_5ClN_2O_4$			
Propane, 1-chloro-2,2-dinitro-	MeCN	277.5(1.75)	44-0369-66
$C_3H_5ClN_2O_5$			
Ether, 2-chloro-2,2-dinitro-ethyl methyl	MeCN	275(1.92)	44-0369-66
C_3H_5ClPd			
Palladium, chloro-π-allyl-	EtOH	237(3.90),324(3.25)	101-0578-66B
	$CHCl_3$	245(4.00),325(3.06)	101-0578-66B
	DMSO	258(3.95),316(3.35)	101-0578-66B
$C_3H_5Cl_2NO_2$			
Carbamic acid, N,N-dichloro-, ethyl ester	CCl_4	315(2.46)	88-6157-66
C_3H_5HgI			
Allylmercuric iodide	isooctane	233(4.13)	44-4290-66
	MeOH	246(4.21)	44-4290-66
	tert-BuOH	245(4.15)	44-4290-66
	MeCN	246(4.23)	44-4290-66
	ether	230(4.20)	44-4290-66
	4% MeOH	248(4.20)	44-4290-66
C_3H_5I			
Propene, 1-iodo-, cis	n.s.g.	220(3.9),247(2.7)	44-1852-66
Propene, 1-iodo-, trans	n.s.g.	225(3.9),252(2.7)	44-1852-66
$C_3H_5IN_2$			
Imidazolium iodide	H_2O	225(4.3)(anom.)	10-0332-66D
$C_3H_5IN_2O_2$			
Glyoxime, iodomethyl-	pH 1	232(4.08)	28C-0985-66A
$C_3H_5KN_2O_5$			
Potassium, (2-methoxy-1,1-dinitroethyl)-	pH 13	364(4.25)	44-0369-66
C_3H_5NO			
Acrylamide, radical anion	H_2O	275(4.02)	77-0498-66
radical adduct	H_2O	370(2.95)	77-0498-66
$C_3H_5NOS_2$			
2-Thiazolidinethione, 4-hydroxy-	acid	245(3.89),275(4.20)	39-0092-66B
$C_3H_5NO_2$			
Pyruvaldehyde, 1-oxime	pH 6	225(4.01)	28C-0985-66A
	base	276(4.23)	28C-0985-66A
$C_3H_5NS_2$			
2-Thiazolidinethione	C_6H_{12}	280(4.2),339(2.1)	60-0788-66
	1.82N HCl	285(4.14)	59-2005-66
	acid	270(4.10)	39-0092-66B
	H_2O	208(3.72),270(4.18)	59-2005-66
	H_2O	254s(--),270(4.10)	39-0092-66B
	1.92N NaOH	240(4.04)	59-2005-66
	base	240(4.08)	39-0092-66B

Compound	Solvent	λ_{max}(log ε)	Ref.
2-Thiazolidinethione (cont.)	EtOH	204(3.72),252(3.88), 274(4.17),329(2.04)	59-2005-66
iodine complex	C_6H_{12}	308(--)	60-0788-66
$C_3H_5N_3O$			
Isoxazole, 3,5-diamino-	EtOH	245.5(4.19)	94-1277-66
$C_3H_5N_3O_2$			
1,2,4-Triazoline-3,5-dione, 4-methyl-	CH_2Cl_2	538(2.18)	44-3444-66
$C_3H_5N_5$			
s-Tetrazine, 3-amino-6-methyl-	dioxan	530(2.87)	88-5369-66
$C_3H_5N_7$			
7H-s-Triazolo[5,1-c]-s-triazole, 3,7-diamino-	pH 1	247(3.75)	25-2168-66
	H_2O	243(3.51)	25-2168-66
$C_3H_6N_2$			
1-Pyrazoline	MeOH	315(2.65)	35-3959-66
$C_3H_6N_2O_3$			
Acetamide, N-methyl-N-nitro-	hexane	239(3.88)	44-3038-66
Pyruvhydroxamic acid, oxime	pH 1	235(4.03)	28C-0985-66A
monoanion	n.s.g.	275(4.04)	28C-0985-66A
$C_3H_6N_2O_4$			
Carbamic acid, methylnitro-,	hexane	228(3.77),288(2.30)	44-3038-66
methyl ester	EtOH	232(3.78),288(2.42)	44-3038-66
Propane, 1,3-dinitro-	85%EtOH-NaOH	247(4.22)(anom.)	78-0925-66
$C_3H_6N_2O_5$			
Propanol, 2,2-dinitro-	MeCN	277(1.75)	44-0369-66
$C_3H_6N_4O_2$			
Formamidine, N,N'-bisformamido-	EtOH	245(4.09)	44-3442-66
$C_3H_6N_4O_4$			
2,3-Diaza-1-butene, 1,1-dinitro-3-methyl-	EtOH	326(4.06)	44-0923-66
$C_3H_6N_4S$			
s-Triazole, 5-aminomethyl-3-mercapto-, HBr salt	H_2O	243.4(4.00)	39-2038-66C
$C_3H_6N_8$			
7H-s-Triazolo[5,1-c]-s-triazole, 3,6,7-triamino-	pH 1	231(4.06)	25-2168-66
C_3H_7NS			
Thioacetamide, N-methyl-	C_6H_{12}	265(4.1),343s(1.5)	60-0788-66
	isooctane	264(4.04),360(1.42)	1-0689-66
	EtOH	261(4.13),321(1.69)	1-0689-66
iodine complex	C_6H_{12}	296(4.5)	60-0788-66
$C_3H_8N_2S$			
Thiourea, N,N'-dimethyl-, iodine complex	$CHCl_3$	302(4.61)	60-0029-66

Compound	Solvent	$\lambda_{max}(\log \epsilon)$	Ref.
C_3H_9B			
Borane, trimethyl-	gas	175(2.9),195-225f(1.7)	46-4097-66
C_3H_9N			
Iscpropylamine, iodine complex	benzene	380(3.43)	35-3905-66
	heptane	248(4.36),412(3.18)	35-3905-66
	dioxan	253(4.55),375(3.30)	35-3905-66
C_3H_9NO			
Trimethylamine, N-oxide	MeCN	198s(3.3)	82-0226-66B
C_3H_9OP			
Phosphine, trimethoxy-, trpylium perchlorate	MeCN	267(3.39)	27-0172-66

Compound	Solvent	λ_{max}(log ε)	Ref.
$C_4Cl_2F_4$			
1,3-Butadiene, 1,4-dichloro-tetrafluoro-, cis-cis	hexane	225.0(3.56)	78-1755-66
cis-trans	hexane	229.0(4.05)	78-1755-66
trans-trans	hexane	231.0(4.20)	78-1755-66
C_4F_4O			
Furan, tetrafluoro-	hexane	232(2.85)	77-0048-66
$C_4H_2BrCl_2N_3$			
Pyrimidine, 2-amino-5-bromo-4,6-dichloro-	EtOH	240(4.23),316(3.40)	78-2401-66
$C_4H_2Br_2N_2O_4$			
1,3-Butadiene, 1,4-dibromo-1,4-dinitro-	MeOH	339(4.00)	23-2115-66
$C_4H_2Cl_2N_2O_4$			
1,3-Butadiene, 1,4-dichloro-1,4-dinitro-	MeOH	330(4.11)	23-2115-66
$C_4H_2Cl_3N_4O_3P$			
Phosphoramidic dichloride, (4-chloro-5-nitro-2-pyrimidinyl)-	MeOH	224(4.04),324(4.16)	87-0121-66
$C_4H_2F_3N_3O_2$			
as-Triazine-3,5(2H,4H)-dione, 6-(trifluoromethyl)-	pH 1	262(3.78)	87-0715-66
	pH 1	263(3.81)	87-0876-66
	pH 7.6	258(3.72)	87-0876-66
	pH 12.6	292(3.84)	87-0876-66
	N NaOH	292(3.85)	87-0715-66
$C_4H_2O_4$			
Cyclobutenedione, 1,2-dihydroxy- (or squaric acid)	H_2O	269(4.43)	35-1533-66
$C_4H_3BrN_2O_2$			
Uracil, 5-bromo-	H_2O	276(3.86)	50-1222-66B
$C_4H_3Cl_2N_3$			
Pyrimidine, 2-amino-4,6-dichloro-	EtOH	296(3.96)	78-2401-66
$C_4H_3FN_2$			
Pyrazine, fluoro-	EtOH	261(3.72),265(3.72), 295s(--)	4-0435-66
$C_4H_3NO_2$			
Maleimide	EtOH	216(4.18),223s(4.11)	44-1311-66
$C_4H_3NO_2S$			
2H-1,3-Thiazine-2,4(3H)-dione	EtOH	223(3.89),270(3.80)	88-3225-66
$C_4H_3NO_2S_2$			
Thiazoline-4,5-dione, 2-(methylthio)-	dioxan	420(1.7)	24-3572-66
$C_4H_3NO_3$			
4-Isoxazolecarboxylic acid	EtOH	206(3.81)	32-0443-66
$C_4H_3N_3O$			
4-Oxazolecarbonitrile, 5-amino-	EtOH	244(4.18)	35-3829-66

Compound	Solvent	λ_{max}(log ϵ)	Ref.
$C_4H_3N_5$			
8-Azapurine	pH 0	248(3.91)	39-0427-66B
	pH 3.4	263(3.87)	39-0427-66B
	pH 7.0	268(3.89)	39-0427-66B
	dioxan	264(3.90)	39-0433-66B
Tetrazolo[1,5-a]pyrazine	H_2O	203(4.08),270(3.65)	4-0435-66
	pH 13	211(3.66),271(3.65)	4-0435-66
$C_4H_3N_5O$			
8-Azapurin-2-ol	pH -3.87	244(3.58),315(3.57)	39-0427-66B
	pH 2.5	209(3.73),241(3.82)	39-0427-66B
	pH 8.3	210(4.17),276s(3.66), 311(3.83)	39-0427-66B
	pH 12.0	213(4.33),267(3.64), 320(3.75)	39-0427-66B
$C_4H_3N_5S$			
8-Azapurine-2-thiol	pH 2.5	228(4.13),267(3.93), 297(3.71)	39-0427-66B
	pH 7.2	229(4.19),265(3.91), 290s(3.81)	39-0427-66B
$C_4H_4BrClN_2$			
Pyrazole, 4-bromo-5(3)-chloro-3(5)-methyl-	EtOH	226(3.52)	22-3744-66
$C_4H_4BrN_3$			
Pyrimidine, 2-amino-5-bromo-	EtOH	238(4.27),316(3.48)	78-2401-66
$C_4H_4Br_2N_2$			
Pyrazole, 4,5(3)-dibromo-3(5)-methyl-	EtOH	225(3.62)	22-3744-66
$C_4H_4ClN_3$			
Pyrimidine, 2-amino-4-chloro-	EtOH	229(4.16),296(3.59)	95-0952-66
Pyrimidine, 2-amino-5-chloro-	EtOH	238(4.32),316(3.60)	78-2401-66
Pyrimidine, 4-amino-2-chloro-	EtOH	235(4.05),274(3.65)	95-0952-66
$C_4H_4ClN_5O_2$			
Pyrimidine, 2,4-diamino-6-chloro-5-nitro-	pH 1	261(3.84),338(4.13)	87-0573-66
	pH 11	235(4.16),267(3.71), 340(4.35)	87-0573-66
$C_4H_4FN_3O_2$			
6-Azauracil, 5-(fluoromethyl)-	EtOH	265(3.78)	87-0419-66
$C_4H_4F_3N_3O_3$			
Pyruvic acid, trifluoro-, semicarbazone	H_2O	250(4.22)	87-0715-66
$C_4H_4IN_3O_2$			
Imidazole, 2-iodo-1-methyl-4-nitro-	6M H_2SO_4	308(3.79)	4-0454-66
	pH 6.95	243s(3.65),327(3.89)	4-0454-66
	MeOH	240s(3.70),315(3.88)	4-0454-66
Imidazole, 2-iodo-1-methyl-5-nitro-	6M H_2SO_4	307(3.75)	4-0454-66
	pH 6.95	262(3.59),331(3.83)	4-0454-66
	MeOH	258(3.61),323(3.84)	4-0454-66
$C_4H_4N_2O$			
4-Pyrimidinol	H_2O	223(3.79),240(3.61), 260(3.52)	44-0175-66

Compound	Solvent	λ_{max}(log ϵ)	Ref.
4-Pyrimidinol (cont.)	25% EtOH	222(3.80),240(3.57), 262(3.52)	44-0175-66
	50% EtOH	222(3.82),241(3.49), 265(3.51)	44-0175-66
	75% EtOH	221(3.82),241(3.39), 268(3.50)	44-0175-66
	EtOH	221(3.82),271(3.49)	44-0175-66
$C_4H_4N_2O_2$			
Cyclobutenedione, diamino-	n.s.g.	275(4.42),363(3.04)	35-1533-66
4-Isoxazolecarboxaldehyde, oxime	EtOH	221(3.83)	32-0443-66
Pyridazinedione	EtOH	209(4.13),225s(--)	44-1311-66
4,6-Pyrimidinediol	0.01M H_2SO_4	242(3.80)	39-0565-66B
	pH 4.8	252(3.86)	39-0565-66B
	pH 12	252(3.79)	39-0565-66B
	H_2O	240(4.0)	30-0635-66A
Pyrrole, 2-nitro-	hexane	232(3.46),320(4.12)	78-0057-66
	MeOH	231(3.61),335(4.23)	78-0057-66
Pyrrole, 3-nitro-	hexane	247(3.18),289(2.99)	78-0057-66
	MeOH	268(3.86),315(3.73)	78-0057-66
$C_4H_4N_2O_2S$			
Uracil, 1-hydroxy-2-thio-	pH 1	217(4.12),264(4.08), 297(3.98)	89-0511-66
	pH 13	244(4.41),274(4.15), 320(3.87)	89-0511-66
$C_4H_4N_2O_3S$			
Pyrazinesulfonic acid, sodium salt	H_2O	204(3.87),267(3.91), 274s(--),309(2.92)	4-0435-66
$C_4H_4N_2O_4$			
1,3-Butadiene, 1,4-dinitro-	MeOH	281(4.27)	23-2115-66
$C_4H_4N_4$			
Imidazole-4-carbonitrile, 5-amino-	EtOH	246(4.04)	35-3829-66
$C_4H_4N_4S$			
(Diaminomethylene)sulfonium dicyanomethylide	EtOH	277(2.86)	44-3731-66
5-Thiazolecarbonitrile, 2,4-diamino-	EtOH	224(4.23),302(3.99)	44-3731-66
$C_4H_4N_4S_2$			
s-Triazolo[3,4-b][1,3,4]thiadiazole, 6-(methylthio)-	MeOH	259(4.11)	44-3528-66
s-Triazolo[3,4-b][1,3,4]thiadiazole-6-thiol, 3-methyl-	MeOH	290(4.18)	44-3528-66
anion	MeOH	288(4.18)	44-3528-66
$C_4H_4N_6$			
8-Azapurine, 2-amino-	pH 0	227(3.98),235s(3.98), 326(2.53)	39-0427-66B
	pH 4.5	217(4.38),237s(3.68), 311(3.84)	39-0427-66B
	pH 8.5	218(4.35),273s(3.67), 315(3.72)	39-0427-66B
	dioxan	223(4.15),311(3.88)	39-0433-66B
$C_4H_4N_6O$			
8-Azapurine, 6-hydroxyamino-	N HCl	273(4.09)	69-3057-66

Compound	Solvent	λ_{max}(log ϵ)	Ref.
8-Azapurine, 6-hydroxyamino- (cont.)	pH 6.7	279(4.12)	69-3057-66
	N NaOH	275(3.96),286s(3.93), 316(4.00)	69-3057-66
8-Azapurin-6-ol, 2-amino-	pH 3.8	247(4.05),266(3.83)	39-0427-66B
	pH 8.75	214(4.34),244(3.76), 278(3.79)	39-0427-66B
C_4H_4O			
3-Butyn-2-one	EtOH	212(3.3)	88-5811-66
Furan, tetracyanoethylene complex	$CHCl_3$	450(3.04)	60-0018-66
$C_4H_4O_3$			
Fumaraldehydic acid	C_6H_{12}	216(4.26),356(1.32)	5-0042-66G
Malealdehydic acid	MeOH	209(3.65)	5-0042-66G
C_4H_4S			
Thiophene, tetracyanoethylene complex	$CHCl_3$	399(4.16)	60-0018-66
$C_4H_4S_3$			
Ethylenetrithiocarbonate	C_6H_{12}	315(4.2),467(1.9)	60-0788-66
iodine complex	C_6H_{12}	460(--)	60-0788-66
C_4H_4Te			
Tellurophene	C_6H_{12}	280(3.62)	89-0896-66
C_4H_5Br			
1,2-Butadiene, 1-bromo-	EtOH	201(3.72),215s(3.54)	39-1223-66C
$C_4H_5BrN_2$			
Pyrazole, 3(5)-bromo-4-methyl-	EtOH	224(3.62)	22-3744-66
Pyrazole, 4-bromo-1-methyl-	EtOH	228.5(3.58)	22-3744-66
Pyrazole, 4-bromo-3(5)-methyl-	EtOH	224.5(3.55)	22-3744-66
Pyrazole, 5(3)-bromo-3(5)-methyl-	EtOH	216(3.66)	22-3744-66
$C_4H_5BrN_2O_6$			
Ethanol, 2-bromo-2,2-dinitro-, acetate	MeCN	280s(2.09)	44-0369-66
$C_4H_5ClN_2$			
Pyrazole, 4-chloro-3(5)-methyl-	EtOH	223(3.53)	22-3744-66
$C_4H_5ClN_2O_6$			
Ethanol, 2-chloro-2,2-dinitro-, acetate	MeCN	277(1.89)	44-0369-66
$C_4H_5ClO_2$			
Acrylic acid, 3-chloro-, methyl ester, cis	EtOH	216(4.34)	44-0646-66
C_4H_5N			
Pyrrole, tetracyanoethylene complex	$CHCl_3$	530(--)	60-0018-66
C_4H_5NOS			
Thiosuccinimide	EtOH	269(4.32),392(1.32)	1-0689-66
	heptane + 0.1% CH_2Cl_2	265(4.26)	1-0689-66
	25% CH_2Cl_2	398.5(1.30)	1-0689-66
$C_4H_5NOS_2$			
Rhodanine, N-methyl-	EtOH	259(4.2),294(4.21)	95-0095-66

Compound	Solvent	λ_{max}(log ϵ)	Ref.
1,3-Thiazin-4-one, 2,3,5,6-tetrahydro-2-thioxo-	EtOH	258(4.12),309(4.14)	95-0095-66
$C_4H_5NS_2$			
Succinimide, dithio-	EtOH	237(3.56),321(4.57), 402(2.13)	1-0689-66
	heptane + 0.06%CH_2Cl_2	234(3.69),315(4.55)	1-0689-66
	12.5%CH_2Cl_2	406(2.20)	1-0689-66
2H-1,3-Thiazine-2-thione, 3,6-dihydro-	acid	298(4.03),317s(3.89)	39-0092-66B
	H_2O	256(3.37),298(4.03), 315s(3.88)	39-0092-66B
	base	291(3.78),316(3.75)	39-0092-66B
4-Thiazoline-2-thione, 3-methyl-	acid	237s(--),305(4.16)	39-0092-66B
	H_2O	305(4.16)	39-0092-66B
	base	305(4.16)	39-0092-66B
4-Thiazoline-2-thione, 4-methyl-	acid	236(3.42),313(4.16)	39-0092-66B
	H_2O	313(4.16)	39-0092-66B
	base	258(--),295(3.99)	39-0092-66B
$C_4H_5N_3O$			
3(2H)-Pyridazinone, 4-amino-	EtOH	290(4.04)	94-1065-66
3(2H)-Pyridazinone, 5-amino-	EtOH	274(3.72)	94-1065-66
$C_4H_5N_3O_2$			
Pyrazole, 1-methyl-4-nitro-	H_2O	281(3.96)	54-1191-66
Pyrazole, 3(5)-methyl-4-nitro-	0.05N HCl	283(3.91)	54-1191-66
	H_2O	282(3.91)	54-1191-66
	0.05N NaOH	322(4.08)	54-1191-66
s-Triazine-2,4-diol, 6-methyl-	pH 10	250(3.89)	39-0056-66C
$C_4H_5N_3O_3$			
Imidazole, 2-methoxy-4(or 5)-nitro-	6M H_2SO_4	225s(3.54),307(3.90)	4-0454-66
	pH 3.02	237(3.57),330(4.12)	4-0454-66
	pH 13	237(3.60),285(3.53), 366(4.15)	4-0454-66
	MeOH	233(3.58),330(4.07)	4-0454-66
$C_4H_5N_5$			
8-Azapurine, 1,6-dihydro-	pH 3.4	262(3.70)	39-0427-66B
	pH 7.28	277(3.78)	39-0427-66B
$C_4H_5N_5S$			
8-Azapurine-2-thiol, 1,6-dihydro-	pH 5.2	265(4.27)	39-0427-66B
	pH 9.8	244(4.14),276(4.00)	39-0427-66B
s-Triazolo[3,4-b][1,3,4]thiadiazole, 6-amino-3-methyl-	MeOH	238(3.83)	44-3528-66
cation	MeOH	241(3.90)	44-3528-66
C_4H_6			
Butadiene, tetracyanoethylene complex	$CHCl_3$	425(--)	60-0018-66
$C_4H_6Cl_2N_4$			
Diimide, bis(N-chloroacetimidoyl)-	hexane	273(4.26)	88-3537-66
$C_4H_6FN_3O_2S$			
Pyruvic acid, fluoro-, 3-thiosemicarbazone	EtOH	230(3.84),301(4.13)	87-0419-66

Compound	Solvent	λ_{max}(log ϵ)	Ref.
$C_4H_6N_2$			
1-Pyrazoline, 4-methylene-	hexane	324(2.70),329(2.70)	35-2587-66
$C_4H_6N_2O$			
Butyraldehyde, α-diazo-	n.s.g.	249(4.00),284(3.81), 383(1.51)	88-1109-66
3-Pyrazolin-5-one, 3-methyl-	EtOH	219(3.52),245(3.45)	94-0770-66
$C_4H_6N_2OS$			
1,3,4-Oxadiazole, 2-methyl-5-(methylthio)-	EtOH	219.5(3.92)	1-0057-66
1,3,4-Oxadiazoline-5-thione, 2,4-dimethyl-	EtOH	263(4.20)	1-0057-66
$C_4H_6N_2O_2$			
Diimide, diacetyl-	dioxan	267s(2.48)	24-2039-66
$C_4H_6N_2O_4$			
Formic acid, azodi-, dimethyl ester	dioxan	405(1.75)	89-0372-66
2H-1,3-Oxazin-2-one, tetrahydro-3-nitro-	EtOH	233(3.76),290(2.28)	44-3038-66
$C_4H_6N_2S_2$			
1,3,4-Thiadiazole, 2-methyl-5-(methylthio)-	EtOH	269(3.94)	1-0057-66
1,3,4-Thiadiazoline-5-thione, 2,4-dimethyl-	EtOH	306.5(4.04)	1-0057-66
$C_4H_6N_4$			
s-Tetrazine, 3,6-dimethyl-	EtOH	274(3.56),538(2.75)	88-5067-66
$C_4H_6N_4O$			
Imidazole-4-carboxamide, 5-amino-	pH 1.4	266(3.97)	69-2082-66
	pH 5.8	266(4.05)	69-2082-66
	pH 11.5	277(4.09)	69-2082-66
1,2,4-Triazin-5-one, 3-amino-2,5-dihydro-6-methyl-	pH -1.1	210(4.00),257(3.70)	73-1864-66
	pH 5.1	243(3.78)	73-1864-66
	pH 11.4	285(3.67)	73-1864-66
	EtOH	246(3.78),291s(2.66)	73-1864-66
1,2,4-Triazin-5-one, 3-amino-4,5-dihydro-4-methyl-	pH -1.1	257(3.71)	73-1864-66
	pH 9.3	219(3.90),297(3.82)	73-1864-66
	EtOH	220(3.82),301(3.80)	73-1864-66
$C_4H_6N_4OS$			
s-Triazole-5-thiol, 4-formamido-3-methyl-	MeOH	257(4.23)	44-3528-66
C_4H_6O			
3-Buten-2-one	isooctane	205(4.00),331(1.38)	99-0469-66
	H_2O	213(4.04),308(1.50)	99-0469-66
	EtOH	209(4.02),319(1.44)	99-0469-66
Crotonaldehyde, trans	isooctane	213(4.21),329(1.41)	99-0469-66
	H_2O	223(4.27),303(1.58)	99-0469-66
	EtOH	220(4.18),317(1.45)	99-0469-66
Methacrolein	isooctane	215(4.02),329(1.42)	99-0469-66
	H_2O	221(4.00),312(1.51)	99-0469-66
	EtOH	218(4.13),323(1.50)	99-0469-66
$C_4H_6O_2S$			
Acrylic acid, β-methylthio-, trans	n.s.g.	270(4.19)	12-1259-66

Compound	Solvent	$\lambda_{max}(\log \epsilon)$	Ref.
C_4H_7ClPd			
Palladium, chloro-π-crotyl-	EtOH	232(3.40),328(3.27)	101-0578-66B
	$CHCl_3$	244(4.05),327(3.18)	101-0578-66B
	DMSO	258(4.00),309(3.56)	101-0578-66B
Palladium, chloro-π-β-methallyl-	EtOH	229(3.91),327(3.25)	101-0578-66B
	$CHCl_3$	241(3.98),327(2.99)	101-0578-66B
	DMSO	258(3.90),309(3.47)	101-0578-66B
C_4H_7NO			
Methacrylamide, radical anion	H_2O	275(3.62)	77-0498-66
radical adduct	H_2O	370(2.93)	77-0498-66
C_4H_7NOS			
Thioacetamide, N-acetyl-	heptane	279(4.36),425(1.58)	1-0689-66
	EtOH	282(4.30),429(1.51)	1-0689-66
$C_4H_7NOS_2$			
2H-1,3-Thiazine-2-thione, tetrahydro-4-hydroxy-	acid	241(3.89),286(4.06)	39-0092-66B
	H_2O	241(3.84),286(4.06)	39-0092-66B
	base	239(3.88),256(3.85)	39-0092-66B
2-Thiazolidinethione, 4-hydroxy-3-methyl-	acid	250s(--),270(4.19)	39-0092-66B
	H_2O	251s(--),270(4.19)	39-0092-66B
	base	267(4.20)	39-0092-66B
2-Thiazolidinethione, 4-hydroxy-4-methyl-	H_2O	245(3.84),275(4.19)	39-0092-66B
	base	248(4.02)	39-0092-66B
C_4H_7NS			
2-Pyrrolidinethione	EtOH	267(4.16),319(1.82)	1-0689-66
	heptane + 0.1% CH_2Cl_2	270(4.17)	1-0689-66
	20% CH_2Cl_2	335(1.67)	1-0689-66
$C_4H_7NS_2$			
2H-1,3-Thiazine-2-thione, tetrahydro-	acid	242(3.87),280(4.10)	39-0092-66B
	H_2O	242(3.83),280(4.10)	39-0092-66B
	base	242(3.90),279(3.92)	39-0092-66B
2-Thiazolidinethione, 3-methyl-	acid	265(4.22)	39-0092-66B
	H_2O	265(4.22)	39-0092-66B
	base	265(4.22)	39-0092-66B
$C_4H_7N_3S$			
1,3,4-Triazoline-5-thione, 1,2-dimethyl-	EtOH	253(4.19)	1-0057-66
$C_4H_7N_5$			
1,2,4-Triazine, 3,5-diamino-6-methyl-	EtOH-HCl	252(3.81)	4-0188-66
	EtOH-NaOH	221(4.00),302(3.70)	4-0188-66
1,3,5-Triazine, 2,4-diamino-6-methyl-	MeOH	255(3.49)	78-0157-66
$C_4H_7N_5O$			
1,3,5-Triazine-2-methanol, 4,6-diamino-	H_2O	255(3.55)	39-0056-66C
$C_4H_7N_5S$			
1,3,5-Triazine, 2,4-diamino-6-mercaptomethyl-, hydrate	pH 5	258(3.40)	39-0056-66C
	pH 8	254(3.48)	39-0056-66C
$C_4H_8N_2$			
1-Pyrazoline, 3-methyl-	MeOH	319(2.58)	35-3959-66

Compound	Solvent	$\lambda_{max}(\log \epsilon)$	Ref.
1-Pyrazoline, 4-methyl-	MeOH	318(2.49)	35-3959-66
2-Pyrazoline, 4-methyl-	MeOH	230(3.60)	35-3959-66
$C_4H_8N_2O$			
5-Pyrimidinol, 1,4,5,6-tetrahydro-	MeOH	208(3.83)	44-3838-66
hydrochloride	MeOH	208(3.84)	44-3838-66
$C_4H_8N_2O_5$			
Propane, 1-methoxy-2,2-dinitro-	MeCN	275s(1.89)	44-0369-66
$C_4H_8N_4$			
1,2,4,5-Tetrazine, 1,6-dihydro-3,6-dimethyl-	EtOH	310(3.48),426(2.62)	88-5067-66
$C_4H_8N_4S$			
s-Triazole, 5-aminomethyl-3-(methylthio)-, hydrobromide	H_2O	232(3.53)	39-2038-66C
dihydrobromide	H_2O	234.6s(3.45)	39-2038-66C
C_4H_9I			
tert-Butyl iodide	C_6H_{12}	269(2.77)	3-0612-66
C_4H_9N			
Methylamine, N-isopropylidene-	heptane	181(3.78)	88-0507-66
	C_6H_{12}	244(2.20)	88-0507-66
	EtOH	231(2.20)	88-0507-66
$C_4H_{10}N_2S$			
Thiourea, N-isopropyl-N'-methyl-, iodine complex	$CHCl_3$	304(4.60)	60-0029-66
$C_4H_{10}OS$			
Ethyl sulfoxide	EtOH	201(3.30),214s(3.08)	39-0239-66A
$C_4H_{10}S$			
Ethyl sulfide	EtOH	202(3.32)	39-0239-66A
$C_4H_{10}S_2$			
Ethyl disulfide	EtOH	201(3.35),254(2.60)	39-0239-66A
$C_4H_{11}N$			
Diethylamine, iodine complex	heptane	260(4.40),410(--)	35-3905-66
	dioxan	263(4.46),385(3.34)	35-3905-66
$C_4H_{12}Cl_2Si_2$			
Disilane, 1,2-dichloro-1,1,2,2-tetramethyl-	C_6H_{12}	204.0(3.51)	101-0392-66A
$C_4H_{12}GeS_4$			
Germane, tetrakis(methylthio)-	hexane	204(3.86),248(3.81)	39-0242-66A
$C_4H_{12}N_4$			
Tetrazene, tetramethyl-	C_6H_{12}	278.0(3.94)	39-0813-66C
$C_4H_{12}SSi$			
Silane, trimethyl(methylthio)-	hexane	205(3.48),225(2.15)	39-0242-66A
$C_4H_{12}S_2Si$			
Silane, dimethylbis(methylthio)-	hexane	203(3.56),225(3.11)	39-0242-66A

Compound	Solvent	λ_{max}(log ϵ)	Ref.
$C_4H_{12}S_3Si$			
Silane, methyltris(methylthio)-	hexane	205(3.63),226(3.34)	39-0242-66A
$C_4H_{12}S_4Si$			
Tetramethyl orthothiosilicate	hexane	203(3.67),227(3.61)	39-0242-66A
$C_4H_{13}ClSi_2$			
Disilane, 1-chloro-1,1,2,2-tetra-methyl-	C_6H_{12}	194.0(3.44)	101-0392-66A
$C_4H_{14}B_{10}N_2$			
Acetonitrile, compound with $B_{10}H_8$	MeCN	239(4.69)	35-0935-66
$C_4H_{14}Si_2$			
Disilane, 1,1,2,2-tetramethyl-	C_6H_{12}	198.0(3.31)	101-0392-66A
$C_4H_{16}Br_2N_5O_3Rh$			
Rhodium, trans-dibromobisethylene-diamine, nitrate	n.s.g.	276(3.24),425(2.04)	90-1429-66
$C_4H_{16}Cl_3CoN_4O_4$			
Cobalt, dichlorobisethylenediamine, perchlorate, cis	DMF	540(2.04)	12-0949-66
trans	DMF	390(1.70),460(1.41), 620(1.60)	12-0949-66
$C_4H_{20}B_{10}N_6$			
Tetramethylammonium octahydroazido-(dinitrogen)decaborate(1-)	MeCN	218(4.15),284(4.34)	35-0935-66
$C_4H_{24}B_{10}N_4O_2$			
Ammonium octahydrobis(acetamidato)-decaborate(2-)	H_2O	230(4.28),260s(4.06)	35-0935-66

Compound	Solvent	$\lambda_{max}(\log \epsilon)$	Ref.
$C_5Cl_4N_2$			
Cyclopentadiene, tetrachlorodiazo-	EtOH	307(4.40)	44-0768-66
C_5HCl_4NO			
Pyridine, 2,3,4,5-tetrachloro-, N-oxide	EtOH	238(4.38),282(4.10)	24-0698-66
$C_5H_2ClN_7$			
Purine, 6-azido-2-chloro-	pH 1	217(4.19),250(3.85), 287(4.19)	44-2210-66
	pH 7	220(4.24),289(4.14)	44-2210-66
	pH 13	228(4.29),297(4.03)	44-2210-66
$C_5H_2Cl_2N_2O_2$			
Pyridine, 2,4-dichloro-3-nitro-	pH 1,7,13	264(3.29),274s(3.28)	87-0105-66
$C_5H_2N_{10}$			
Purine, 2,6-diazido-, tetrazolo tautomer	pH 1	245(4.35),297(4.10), 305(4.05)	44-2210-66
	pH 7	243(4.31),300(4.05)	44-2210-66
	pH 13	245s(4.31),306(3.99)	44-2210-66
$C_5H_3BrN_4O_2$			
2,8-Purinediol, 6-bromo-	pH 1	218(4.37),305(3.90)	73-1824-66
	pH 13	218(4.27),316(4.06)	73-1824-66
$C_5H_3Br_2NO$			
2-Pyridinol, 3,5-dibromo-	H_o -7.0	216(4.23),235s(3.78), 313(3.79)	39-0996-66B
2-Pyridone, 3,5-dibromo-	pH 5	216(4.27),236s(3.80), 323(3.80)	39-0991-66B
$C_5H_3ClN_4$			
Imidazo[4,5-c]pyridazine, 6-chloro-	EtOH	210(4.36),261(3.92), 285(3.89),315s(3.27)	4-0218-66
Purine, 6-chloro-	pH 5	265(3.96)	39-0010-66C
$C_5H_3ClN_8$			
Pyrimidine, 4-azido-5-(azidomethyl)-2-chloro-	EtOH	209s(4.00),241(4.22), 290(3.82)	5-0119-66B
$C_5H_3Cl_2IN_2$			
Pyrimidine, 2,4-dichloro-5-(iodomethyl)-	EtOH	206(4.07),232(4.22), 272(3.79)	5-0119-66B
$C_5H_3Cl_2NO$			
2-Pyridinol, 3,5-dichloro-	H_o -7.0	235(3.75),310(3.78)	39-0996-66B
2-Pyridone, 3,5-dichloro-	pH 5	238(3.83),320(3.89)	39-0991-66B
$C_5H_3Cl_3N_2$			
Pyrimidine, 2,4-dichloro-5-(chloromethyl)-	EtOH	222(4.01),263(3.57)	5-0119-66B
$C_5H_3F_3N_2O_2$			
Uracil, 5-(trifluoromethyl)-	pH 1	257(3.91)	87-0876-66
	pH 9.3	281(3.96)	87-0876-66
	pH 13.4	273(3.90)	87-0876-66

Compound	Solvent	λ_{max}(log ε)	Ref.
$C_5H_3I_2NO$			
2-Pyridinol, 3,5-diiodo-	H_o -7.0	233(4.18),244s(4.03), 325(3.81)	39-0996-66B
$C_5H_3N_7O$			
6-Purinol, 2-azido-, tetrazolo tautomer	pH 1	261(3.62),296(3.95)	44-2210-66
	pH 7	266(3.61),302(4.01)	44-2210-66
	pH 13	267(3.61),315(3.92)	44-2210-66
$C_5H_3N_{11}$			
Pyrimidine, 2,4-diazido-5-(azido-methyl)-	EtOH	208(3.79),240(4.18), 292(3.75)	5-0119-66B
C_5H_4BrNO			
Pyridine, 3-bromo-, N-oxide	H_2O	197(4.08),222(4.31), 260(4.08)	39-1058-66B
	EtOH	209s(4.00),226(4.25), 272(4.11)	39-1058-66B
Pyridine, 4-bromo-, N-oxide	H_2O	208(4.16),269(4.15)	39-1058-66B
2-Pyridinol, 3-bromo-	H_o -7.0	226(3.47),290(3.90)	39-0996-66B
2-Pyridinol, 5-bromo-	H_o -3.6	225(3.86),296(3.75)	39-0996-66B
2-Pyridone, 3-bromo-	pH 5	230(3.66),306(3.87)	39-0991-66B
2-Pyridone, 5-bromo-	pH 5	233(3.98),310(3.66)	39-0991-66B
C_5H_4ClNO			
Pyridine, 4-chloro-, N-oxide	H_2O	209(4.28),265(4.29)	39-1058-66B
2-Pyridinol, 3-chloro-	H_o -7.0	220(3.49),287(3.90)	39-0996-66B
2-Pyridinol, 5-chloro-	H_o -3.6	223(3.83),294(3.80)	39-0996-66B
2-Pyridone, 3-chloro-	pH 5	229(3.72),303(3.84)	39-0991-66B
2-Pyridone, 5-chloro-	pH 5	233(4.00),310(3.70)	39-0991-66B
$C_5H_4ClN_3O_2$			
Pyridine, 2-amino-4-chloro-3-nitro-	pH 1	220s(--),265s(--), 327(3.51),352(3.52)	87-0354-66
	pH 7	226(4.13),250s(--), 385(3.52)	87-0354-66
	pH 13	226(4.14),250s(--), 385(3.52)	87-0354-66
Pyridine, 4-amino-2-chloro-3-nitro-	pH 1	239(4.15),260s(--), 340(3.23)	87-0105-66
	pH 7	239(4.16),278(3.23), 362(3.30)	87-0105-66
	pH 13	238(4.16),278(3.23), 362(3.30)	87-0105-66
$C_5H_4ClN_5$			
s-Triazolo[4,3-b]pyridazine, 8-amino-6-chloro-	EtOH	212(4.24),230s(3.84), 298(4.02)	4-0218-66
s-Triazolo[4,3-b]pyridazine, 8-amino-7-chloro-	EtOH	203(4.3),227(3.94), 255(3.74),263(3.73), 300(3.99)	4-0218-66
$C_5H_4Cl_2N_2O$			
3(2H)-Pyridazinone, 4,5-dichloro-2-methyl-	EtOH	300(3.46)	95-1099-66
$C_5H_4Cl_2N_4O_{11}$			
Ethanol, 2-chloro-2,2-dinitro-, carbonate	MeCN	276(2.19)	44-0369-66

Compound	Solvent	λ_{max}(log ε)	Ref.
$C_5H_4Cl_2O_2$			
2-Cyclopenten-1-one, 2,5-dichloro-3-hydroxy-	H_2O	272(4.07)	12-2107-66
$C_5H_4F_2N_2O_2$			
Uracil, 5-(difluoromethyl)-	pH 1	263(3.87)	87-0876-66
	pH 7	265(3.87)	87-0876-66
$C_5H_4F_7NO$			
Butyrimidic acid, heptafluoro-, methyl ester	EtOH	220s(2.1),265(1.2)	65-0862-66
C_5H_4INO			
2-Pyridinol, 5-iodo-	H_0 -3.6	230(4.09),305(3.63)	39-0996-66B
2-Pyridone, 5-iodo-	pH 5	235(4.24),316(3.63)	39-0991-66B
$C_5H_4N_2O$			
Fumaronitrile, methoxy-	ether	242(4.00)	24-2572-66
Maleonitrile, methoxy-	ether	250(4.08)	24-2572-66
$C_5H_4N_2O_3$			
Pyridine, 2-nitro-, N-oxide	H_2O	210(4.13),245(4.04)	39-1058-66B
Pyridine, 3-nitro-, N-oxide	H_2O	242(4.23)	39-1058-66B
Pyridine, 4-nitro-, N-oxide	H_2O	227(3.93),315(4.11)	39-1058-66B
	EtOH	232(3.96),330(4.13)	39-1058-66B
2-Pyridinol, 3-nitro-	H_0 -9.4	240(3.54),303(3.87)	39-0996-66B
2-Pyridinol, 5-nitro-	H_0 -4.9	290(4.00),334s(3.59)	39-0996-66B
2-Pyridone, 3-nitro-	pH 5	257(3.40),362(3.85)	39-0991-66B
2-Pyridone, 5-nitro-	pH 5	211(3.93),301(4.03), 314s(4.02)	39-0991-66B
5-Uracilcarboxaldehyde	pH 1	231(3.99),277(4.06)	88-5253-66
	pH 13	253(3.94),300(4.07)	88-5253-66
$C_5H_4N_4O$			
Hypoxanthine	2N HCl	247(4.04)	50-1129-66A
	neutral	249(4.02)	44-1411-66
	pH 7	249(4.02)	50-1129-66A
	base	258(4.05)	44-1411-66
Pyrazolo[3,2-c]-as-triazin-4-ol	EtOH	256(3.79),289(2.94), 332(3.60)	39-1127-66C
$C_5H_4N_4O_2$			
Xanthine	pH 8.0	200(4.4),270(4.1)	37-3468-66
$C_5H_4N_4O_3$			
Xanthine, 7-N-oxide	pH 3-4	272(4.01)	44-0178-66
	pH 7	212(4.3),272(3.78)	44-0178-66
	pH 10-12	224(4.38),297(3.93)	44-0178-66
$C_5H_4N_6O$			
Imidazole-4-carboxaldehyde, 4-(5-tetrazolyl)-	pH 1	247(4.19)	44-2210-66
	pH 7	245(4.17)	44-2210-66
	pH 13	237(4.16)	44-2210-66
$C_5H_4N_8$			
Purine, 2-amino-6-azido-	pH 1	270(3.95),292(3.99)	44-2210-66
	pH 7	269(3.89),301(3.92)	44-2210-66
	pH 13	275(3.81),317(3.85)	44-2210-66
Purine, 6-amino-2-azido-, tetrazolo tautomer	pH 1	233(4.29),271(4.06), 311s(3.18),321s(3.04)	44-2210-66

Compound	Solvent	$\lambda_{max}(\log \epsilon)$	Ref.
Purine, 6-amino-2-azido-, tetrazolo tautomer (cont.)	pH 7	230(4.41),267(3.95), 310(3.55),321s(3.44)	44-2210-66
	pH 13	234(4.45),271(3.91), 324(3.47)	44-2210-66
$C_5H_4O_3$			
1,3,4-Cyclopentanetrione	EtOH-NaOH	226(4.12),313(4.13)	99-0230-66
$C_5H_4O_4$			
Cyclobutanedione, 1-hydroxy-2-methoxy-	H_2O	245(4.36),265(4.34)	35-1533-66
$C_5H_5BrClN_3$			
Pyrimidine, 2-amino-5-bromo-4-chloro-6-methyl-	EtOH	310(3.56)	78-2401-66
$C_5H_5BrN_2O$			
Pyrimidine, 5-bromo-2-methoxy-	pH 7	218(4.11),286(3.52)	12-2321-66
Pyrimidine, 5-bromo-4-methoxy-	pH 7	226(3.83),262(3.62)	12-2321-66
4-Pyrimidinol, 5-bromo-6-methyl-	EtOH	235(3.46),282(3.53)	78-2401-66
$C_5H_5BrN_2S$			
Pyrimidine, 5-bromo-2-(methylthio)-	pH -3	269(4.33),336(3.57)	12-2321-66
	pH 4	261(4.32),302s(3.27)	12-2321-66
Pyrimidine, 5-bromo-4-(methylthio)-	pH -1	214(3.92),238(3.74), 311(4.23)	12-2321-66
	pH 5	256(4.00),288(3.92)	12-2321-66
C_5H_5BrO			
2-Cyclopenten-1-one, 2-bromo-	EtOH	238(3.93)	88-3737-66
$C_5H_5BrO_2$			
2-Cyclopenten-1-one, 3-bromo-2-hydroxy-	hexane	257.0(4.15)	39-0075-66B
$C_5H_5ClN_2O$			
4-Pyrimidinol, 5-chloro-6-methyl-	EtOH	232(3.45),279(3.54)	78-2401-66
$C_5H_5ClN_2O_2$			
6(1H)-Pyridazinone, 4-chloro-3-hydroxy-1-methyl-	EtOH	218(4.32),328(3.46)	95-0487-66
6(1H)-Pyridazinone, 5-chloro-3-hydroxy-1-methyl-	EtOH	214(4.23),330(3.48)	95-0487-66
$C_5H_5Cl_2N_3$			
Pyrimidine, 2-amino-4,5-dichloro-6-methyl-	EtOH	310(3.48)	78-2401-66
Pyrimidine, 2,4-dichloro-5-(methylamino)-	MeOH	258(4.19)	87-0121-66
$C_5H_5IN_2O$			
4-Pyrimidinol, 5-iodo-6-methyl-	EtOH	238(2.75),290(2.96)	78-2401-66
C_5H_5NO			
Pyridine, N-oxide	EtOH	213(4.22),265(4.11)	24-0698-66
2-Pyridinol	H_o -4.9	209(3.51),277(3.82)	39-0996-66B
2-Pyridone	pH 5	224(3.85),294(3.75)	39-0991-66B

Compound	Solvent	λ_{max}(log ε)	Ref.
C_5H_5NOS			
2-Pyridinethiol, 1-oxide, ferric complex	$CHCl_3$	550f(3.53)	96-0098-66
chromic complex	$CHCl_3$	650(2.34)	96-0098-66
(also other metal complexes)			
1-Thiauracil, 6-methyl-	EtOH	222(3.84),268(3.81)	88-3225-66
$C_5H_5NO_2$			
Glutaconimide	2N H_2SO_4	218(3.76),295(4.04)	39-0562-66B
	pH 2.6	224(3.80),306(3.84)	39-0562-66B
	pH 6.8	235(3.95),319(4.18)	39-0562-66B
2,3-Pyridinediol	H_o -7.0	222(3.45),290(3.92)	39-0996-66B
2,6-Pyridinediol	H_o -1.76	217(3.68),292(4.05)	39-0996-66B
2-Pyridone, 3-hydroxy-	pH 5	234(3.61),297(3.87)	39-0991-66B
	pH 12	255(3.81),310(3.96)	39-0991-66B
2-Pyridone, 6-hydroxy-	pH 2.8	225(3.83),306(3.89)	39-0991-66B
	pH 7	234(3.92),322(4.17)	39-0991-66B
$C_5H_5NO_2S$			
1-Thiathymine	EtOH	224(3.86),272(3.80)	88-3225-66
$C_5H_5NO_2S_2$			
2-Thioacetic acid, S-(2-thiazolyl)-	acid	246s(--),276(3.79)	39-0092-66B
	H_2O	275(3.84)	39-0092-66B
	base	275(3.86)	39-0092-66B
$C_5H_5NO_3$			
5-Isoxazoleacetic acid	H_2O	218(3.8)	94-0089-66
$C_5H_5N_3$			
5-Pyrimidinecarbonitrile, 2-amino-	pH -0.4	246(4.32),307(3.36)	39-0164-66C
	pH 4.0	254(4.41),294(3.52)	39-0164-66C
$C_5H_5N_3O$			
Oxazole-4-carbonitrile, 5-amino-2-methyl-	EtOH	244(4.13)	35-3829-66
$C_5H_5N_3O_2$			
2-Pyrazinecarboxylic acid, 3-amino-	pH 7.0	243(4.03),340(3.76)	39-1112-66C
$C_5H_5N_3O_2S$			
Pyrimidine, 2-(methylthio)-5-nitro-	pH 3.0	219(3.73),325(4.15)	12-2321-66
$C_5H_5N_3O_3$			
Pyrimidine, 2-methoxy-5-nitro-	pH 7	271(4.12)	12-2321-66
$C_5H_5N_5$			
Adenine	pH 2	263(4.12)	69-2082-66
	pH 6.5	260(4.2),268s(4.1)	28C-0018-66A
	pH 11	266(4.2),277(3.9)	28C-0018-66A
	pH 13	269(4.09)	69-2082-66
8-Azapurine, 2-methyl-	pH 0.8	247(3.96)	39-0427-66B
	pH 4.14	267(3.88)	39-0427-66B
	pH 7.5	273(3.93)	39-0427-66B
8-Azapurine, 6-methyl-	pH -1.5	260(3.85)	39-0427-66B
	pH 2.9	260(3.89)	39-0427-66B
	pH 7.3	267(3.91)	39-0427-66B
8-Azapurine, 9-methyl-	pH -2.0	263(3.73)	39-0427-66B
	pH 2.5	264(3.88)	39-0427-66B
	dioxan	265(3.89)	39-0433-66B

Compound	Solvent	λ_{max}(log ε)	Ref.
Pyrazolo[3,2-c]-as-triazine, 4-amino-	EtOH	224(4.41),290(3.74), 349(3.85)	39-1127-66C
s-Triazolo[4,3-a]pyrazine, 3-amino-	H_2O	228(4.19),335(3.38)	39-2038-66C
s-Triazolo[4,3-b]pyridazine, 8-amino-	EtOH	205(4.29),232(4.15), 300(4.20)	4-0218-66
s-Triazolo[2,3-a]pyrimidine, 2-amino-	MeOH	222(4.43),287s(3.56), 324(3.88)	39-2031-66C
$C_5H_5N_5O$			
8-Azapurin-2-ol, 6-methyl-	pH -1.7	251(3.47),311(3.81)	39-0427-66B
	pH 2.9	216(4.20),239s(3.39), 318(3.69)	39-0427-66B
	pH 8.0	210(4.15),276s(3.68), 307(3.89)	39-0427-66B
	pH 12.9	215(4.33),263(3.63), 316(3.80)	39-0427-66B
Purine, 6-hydroxyamino-	pH 6.7	268(4.07)	69-3057-66
$C_5H_5N_5O_2$			
Guanine 7-N-oxide	pH 1	245(3.89),267(3.98)	44-0178-66
	pH 9	226(4.48),254(3.72), 292(3.82)	44-0178-66
	pH 12	227(4.49),283(3.99)	44-0178-66
2-Purinol, 6-hydroxyamino-	N HCl	294(4.08)	69-3057-66
	pH 6.7	272(4.12)	69-3057-66
	N NaOH	275(4.00)	69-3057-66
$C_5H_5N_5O_5$			
4(3H)-Pyrimidinone, 2-amino-6-(methylamino)-5-nitro-	pH -2.7	225(4.23),248s(3.70), 322(3.93)	24-2984-66
	pH 5.0	230s(4.19),290s(3.70), 333(4.17)	24-2984-66
	pH 11.0	230(3.96),260s(3.57), 344(4.22)	24-2984-66
$C_5H_5N_5S$			
8-Azapurine, 2-(methylthio)-	pH -2.4	249(4.29),264s(4.04), 321(3.35)	39-0427-66B
	pH 2.5	234(4.24),250s(3.89), 315(3.87)	39-0427-66B
	pH 7.5	226s(4.10),240(4.25), 313(3.80)	39-0427-66B
	dioxan	235(4.31),312(3.99)	39-0433-66B
8-Azapurine-2-thiol, 6-methyl-	pH 2.5	223(4.17),292(3.91)	39-0427-66B
	pH 7.2	232(4.21),288(3.86)	39-0427-66B
6(1H)-Purinethione, 8-amino-	pH 1	238(4.14),332(4.27)	44-1417-66
	pH 7	248(4.27),333(4.38)	44-1417-66
	pH 13	241(4.26),314(4.30)	44-1417-66
Thiazolo[5,4-d]pyrimidine, 2,7-diamino-	pH 1	225(4.26),280(4.13), 310s(3.82)	44-1417-66
	pH 7	220(4.36),274(4.14)	44-1417-66
	pH 13	220(4.33),274(4.11)	44-1417-66
C_5H_6			
1-Buten-3-yne, 2-methyl-	EtOH	222(4.04),236(3.99)	39-1976-66C
$C_5H_6BrN_3$			
Pyrimidine, 4-amino-5-bromo-6-methyl-	EtOH	240(4.04),279(3.62)	78-2401-66

Compound	Solvent	λ_{max}(log ε)	Ref.
$C_5H_6Br_2N_2$			
Pyrazole, 3,4-dibromo-1,5-dimethyl-	EtOH	228(3.62)	22-0293-66
Pyrazole, 4,5-dibromo-1,3-dimethyl-	EtOH	238(3.70)	22-0293-66
$C_5H_6ClN_3$			
Pyrimidine, 2-amino-4-chloro-6-methyl-	EtOH	295(3.50)	78-2401-66
	EtOH	230(4.14),293(3.70)	95-0952-66
Pyrimidine, 4-amino-2-chloro-5-methyl-	EtOH	235(3.96),276(3.79)	95-0952-66
Pyrimidine, 4-amino-2-chloro-6-methyl-	EtOH	237(4.00),271(3.66)	95-0952-66
$C_5H_6ClN_3O$			
4-Amino-6-chloro-3-hydroxy-1-methylpyridazinium hydroxide, inner salt	MeOH	235(4.35),313(4.08)	94-1090-66
3(2H)-Pyridazinone, 4-amino-6-chloro-2-methyl-	MeOH	298(4.01)	94-1065-66
$C_5H_6Cl_2N_4O_{10}$			
Bis(2-chloro-2,2-dinitroethyl)formal	MeCN	278(2.17)	44-0369-66
$C_5H_6Cl_2O$			
Cyclopentanone, 2,2-dichloro-	C_6H_{12}	304(1.53),311s(1.52)	28C-0924-66A
Cyclopentanone, 2,5-dichloro-, cis	C_6H_{12}	320(1.57)	28C-0924-66A
Cyclopentanone, 2,5-dichloro-, trans	C_6H_{12}	316(1.3)	28C-0924-66A
$C_5H_6FN_3OS$			
as-Triazin-5(2H)-one, 6-(fluoromethyl)-3-(methylthio)-	EtOH	235(4.26)	87-0419-66
$C_5H_6N_2$			
Pyridine, 3-amino-	pH 10	288(3.49)	35-0354-66
Pyridine, 4-amino-	pH 10	265s(3.34)	35-0354-66
$C_5H_6N_2O$			
Pyrazine, 2-methoxy-	MeOH	211(3.96),276(3.62), 288(3.59)	24-0542-66
4-Pyridazinone, 1-methyl-	EtOH	269(4.17)	83-0883-66
Pyrimidine, 2-methoxy-	pH 7.0	264(3.68)	12-1487-66
4-Pyrimidinol, 6-methyl-	EtOH	269(3.44)	78-2401-66
$C_5H_6N_2OS$			
Pyrazine, 1,2-dihydro-3-methyl-2-thiono-, 4-oxide	pH 3	220(4.03),272(4.37), 395(3.83)	39-0495-66C
	pH 9	211(4.12),259(4.27), 365(3.71)	39-0495-66C
$C_5H_6N_2OS_2$			
2-Thioacetamide, S-(2-thiazolyl)-	acid	245s(--),275(3.78)	39-0092-66B
	H_2O	273(3.78)	39-0092-66B
	base	273(3.78)	39-0092-66B
$C_5H_6N_2O_2$			
Pyrazole-1-acetic acid	0.05N HCl	219(3.69)	54-1194-66
	H_2O	219(3.71)	54-1194-66
	0.05N NaOH	219(3.71)	54-1194-66
Pyrazole-4-carboxylic acid, 1-methyl-	MeOH	227(3.91)	18-0253-66
6(1H)-Pyridazinone, 3-hydroxy-4-methyl-	EtOH	304(3.42)	49-0644-66
Pyridine, 4-hydroxyamino-, 1-oxide	EtOH	290(4.31)	88-4729-66

Compound	Solvent	λ_{max}(log ε)	Ref.
4,6-Pyrimidinediol, 1-methyl-	EtOH	259(3.43)	30-0144-66*
	DMSO	273(3.42)	30-0144-66*
2-Pyrimidinol, 4-methoxy-	pH 1	271(3.67)	44-1163-66
	pH 7	268(3.69)	44-1163-66
	pH 13	218(3.97),278(3.76)	44-1163-66
6-Pyrimidinol, 4-methoxy-	20N H_2SO_4	241(3.91)	39-0565-66B
	pH 7	233(3.56)	39-0565-66B
	pH 12	230(3.58)	39-0565-66B
4(3H)-Pyrimidinone, 6-hydroxy- 3-methyl-	pH 2	250(3.8)	30-0635-66A
	EtOH	259(3.43)	30-0635-66A
Thymine	pH 5.43	265.5(3.98)	24-2391-66
	pH 9.85	267(3.85)	24-2391-66
	pH 12-13	291(3.80)	24-2391-66
	pH 14	282(3.81)	24-2391-66
Uracil, 6-methyl-	EtOH	259(4.07)	70-1171-66
$C_5H_6N_2O_2S$			
Uracil, 5-mercaptomethyl-	pH 1	266(3.92)	87-0097-66
	pH 6.9	267(3.90)	87-0097-66
	pH 12	287(3.75)	87-0097-66
$C_5H_6N_2O_2S_2$			
5-Thiazolidineacetamide, 4-oxo- 2-thioxo-	EtOH	253(3.99),296(4.22)	95-0101-66
$C_5H_6N_2O_3$			
Isoxazole-5-acetic acid, α-amino- (or 3-deoxyibotenic acid)	H_2O	217(3.72)	94-0089-66
6(1H)-Pyridazinone, 3,5-dihydroxy- 1-methyl-	EtOH	213(4.20),233s(--), 296(3.54)	95-0487-66
Uracil, 5-(hydroxymethyl)-	10N HCl	264(3.93)	87-0097-66
	5N HCl	262(3.91)	87-0097-66
	pH 1	261(3.91)	87-0097-66
	pH 6.9	261(3.91)	87-0097-66
	pH 12	286(3.87)	87-0097-66
$C_5H_6N_2O_4$			
Malonic acid, diazo-, dimethyl ester	CH_2Cl_2	251(3.92)	24-3128-66
$C_5H_6N_2S$			
Pyrazine-2(1H)-thione, 3-methyl-	pH 3	228(3.63),275(4.06), 380(3.87)	39-0495-66C
	pH 9	230(3.87),265(4.03), 343(3.77)	39-0495-66C
Pyrimidine, 2-(methylthio)-	pH -2.8	214(3.48),255(4.10), 313(3.51)	12-1487-66
	pH 7	250(4.11),285s(3.25)	12-1487-66
Pyrimidine, 4-(methylthio)-	pH 0	221(3.47),302(4.23)	12-1487-66
	pH 7	257(3.85),279(3.91)	12-1487-66
$C_5H_6N_4O$			
Pyrimidine, 2-amino-5-carbamoyl-	pH -1.0	244(4.31),305(3.53)	39-0164-66C
	pH 5.0	255(4.32),290s(3.58)	39-0164-66C
$C_5H_6N_4O_2$			
2-Pyrimidinol, 4-amino-5-formamido-	pH 1.3	213(4.09),282(3.98)	39-0438-66B
	pH 7.0	215s(4.17),272(3.76)	39-0438-66B
	pH 13.1	287(3.89)	39-0438-66B

Compound	Solvent	λ_{max}(log ϵ)	Ref.
$C_5H_6N_4O_3$			
Uracil, 6-amino-5-formamido-	H_2O	265(4.28)	95-0854-66
$C_5H_6N_4O_4$			
4(3H)-Pyrimidinone, 2-amino-6-hydroxy-	pH 1.0	240(3.65),317(3.93)	24-2997-66
1-methyl-5-nitro-	pH 8.0	226(4.07),332(3.92)	24-2997-66
4(3H)-Pyrimidinone, 2-amino-6-methoxy-5-nitro-	pH 3.0	223(4.03),250s(3.55), 322(3.98)	24-2997-66
	pH 9.0	230(3.95),288s(3.50), 343(3.92)	24-2997-66
Uracil, 4-(methylamino)-5-nitro-	pH 3.0	227(4.32),323(4.12)	24-2984-66
	pH 8.0	286s(3.60),332(4.19)	24-2984-66
$C_5H_6N_4S_2$			
s-Triazolo[3,4-b][1,3,4]thiadiazole, 3-methyl-6-(methylthio)-	MeOH	259(4.06)	44-3528-66
s-Triazolo[3,4-b][1,3,4]thiadiazole-6-thiol, 3-ethyl-	MeOH	287(4.04)	44-3528-66
anion	MeOH	291(3.98)	44-3528-66
$C_5H_6N_6$			
8-Azapurine, 2-amino-6-methyl-	pH -0.26	222(4.52),241s(3.56), 318(3.68)	39-0427-66B
	pH 4.5	217(4.39),236s(3.70), 306(3.87)	39-0427-66B
	pH 9.0	218(4.35),270(3.64), 309(3.77)	39-0427-66B
$C_5H_6N_6O$			
Purine, 2-amino-6-hydroxyamino-	N HCl	254(4.06),283(3.99), 310s(3.71)	69-3057-66
	pH 6.7	253(3.92),282(4.04)	69-3057-66
C_5H_6O			
2-Cyclopenten-1-one	EtOH	218(4.09)	39-0164-66B
	n.s.g.	217.0(4.06)	39-0073-66B
$C_5H_6O_3$			
Fumaraldehydic acid, methyl ester	C_6H_{12}	216(4.26),356(1.32)	5-0042-66G
	MeOH	211(4.11),348(1.81)	5-0042-66G
Malealdehydic acid, methyl ester	C_6H_{12}	218(4.07),369(1.40)	5-0042-66G
C_5H_7Br			
1,2-Pentadiene, 1-bromo-	EtOH	205(3.85),215s(3.74)	39-1223-66C
$C_5H_7BrN_2$			
Pyrazole, 4-bromo-1,3-dimethyl-	EtOH	232(3.67)	22-0293-66
Pyrazole, 4-bromo-1,5-dimethyl-	EtOH	229(3.54)	22-0293-66
Pyrazole, 4-bromo-3,5-dimethyl-	EtOH	224.5(3.61)	22-3744-66
$C_5H_7BrO_2$			
2,4-Pentanedione, 3-bromo-, palladium complex	EtOH	332(3.84)	30-0103-66D
$C_5H_7ClN_2$			
Pyrazole, 4-chloro-1,3-dimethyl-	EtOH	230.5(3.73)	22-3744-66
Pyrazole, 4-chloro-3(5)-ethyl-	EtOH	223(3.54)	22-3744-66

Compound	Solvent	λ_{max}(log ϵ)	Ref.
$C_5H_7ClN_2O_7$			
2-Chloro-2,2-dinitroethyl ethyl carbonate	MeCN	275(1.90)	44-0369-66
$C_5H_7ClO_2$			
2,4-Pentanedione, 3-chloro-, palladium complex	EtOH	210(4.79),228(4.34), 255s(3.98),344(3.83)	30-0103-66D
$C_5H_7ClO_2S$			
Acrylic acid, 3-chloro-2-(methylthio)-, methyl ester	ether	286(3.57)	24-1558-66
C_5H_7NOS			
Thioglutarimide	EtOH	279(4.31),414(1.40)	1-0689-66
	heptane + 0.05%CH_2Cl_2	276(4.25)	1-0689-66
	20%CH_2Cl_2	417(1.43)	1-0689-66
Thiosuccinimide, N-methyl-	heptane	269(4.36),395(1.52)	1-0689-66
	EtOH	271(4.35),386(1.48)	1-0689-66
$C_5H_7NOS_2$			
1,3-Thiazine-2-thione, 2,3,5,6-tetrahydro-N-methyl-4-oxo-	EtOH	266(4.12),310(4.01)	95-0095-66
$C_5H_7NO_2S_2$			
2-Thioacetic acid, S-(4,5-dihydro-2-thiazolyl)-	acid	254(4.07)	39-0092-66B
	H_2O	235s(3.76),253(3.79)	39-0092-66B
	base	231(3.88),253s(--)	39-0092-66B
$C_5H_7NS_2$			
Dithioglutarimide	EtOH	237(3.84),336(4.51), 422(2.23)	1-0689-66
	heptane + 0.05%CH_2Cl_2	237(3.85),331(4.47)	1-0689-66
	2% CH_2Cl_2	427(2.24)	1-0689-66
	10%CH_2Cl_2	475.5(1.86)	1-0689-66
Dithiosuccinimide, N-methyl-	heptane	236(3.59),320(4.62), 324(4.62),395(2.21), 410(2.26)	1-0689-66
	EtOH	238(3.62),322(4.61), 395s(2.26),403(2.27)	1-0689-66
Δ^4-1,3-Thiazine-2-thione, 3-methyl-	acid	298(4.04),314s(3.86)	39-0092-66B
	H_2O	298(4.04),314s(3.86)	39-0092-66B
	base	298(4.04),314s(3.86)	39-0092-66B
Δ^4-1,3-Thiazine-2-thione, 4-methyl-	acid	234s(3.51),294(4.02), 313s(3.77)	39-0092-66B
	H_2O	298(4.05),315s(3.88)	39-0092-66B
	base	284(3.76),313(3.78)	39-0092-66B
$C_5H_7N_3$			
Pyridine, 1-amino-1,2-dihydro-2-imino-	MeOH	231(3.93),306(3.75)	44-0260-66
Pyrimidine, 4-amino-2-methyl- (cation)	H_2O	244(4.12)	94-1321-66
	HOAc	250(4.08)	94-1321-66
	MeOH	245(4.15)	94-1321-66
	EtOH	247(4.15)	94-1321-66
	iso-PrOH	248(4.17)	94-1321-66
	MeCN	241(4.10)	94-1321-66
Pyrimidine, 4-amino-6-methyl-	EtOH	269(3.62)	78-2401-66
s-Triazine, 2,4-dimethyl-	C_6H_{12}	267(2.93)	54-1101-66

Compound	Solvent	$\lambda_{max}(\log \epsilon)$	Ref.
$C_5H_7N_3O$			
4-Amino-3-hydroxy-1-methylpyridazinium hydroxide, inner salt	MeOH	229(4.20),311(4.13)	94-1090-66
Pyridazine, 4-amino-3-methoxy-	EtOH	252(3.96),275(3.89)	94-1065-66
3(2H)-Pyridazinone, 4-amino-2-methyl-	MeOH	295(4.04)	94-1090-66
	EtOH	295(4.04)	94-1065-66
3(2H)-Pyridazinone, 4-amino-6-methyl-	EtOH	291(4.02)	94-1065-66
3(2H)-Pyridazinone, 5-amino-2-methyl-	EtOH	277.5(3.79)	94-1065-66
$C_5H_7N_3O_2$			
Pyrazole, 3,5-dimethyl-4-nitro-	0.05N HCl	284(3.93)	54-1191-66
	H_2O	284(3.94)	54-1191-66
	0.05N NaOH	321(4.10)	54-1191-66
	EtOH	222(3.48),277(3.95)	22-3744-66
Uracil, 5-(methylamino)-	pH 1	258(3.88)	87-0121-66
	pH 11	227(3.94),289(3.65)	87-0121-66
$C_5H_7N_3O_2S$			
1,2,3-Thiadiazole-4-carboxylic acid, 5-amino-, ethyl ester	MeOH	263(4.05)	24-1618-66
$C_5H_7N_3O_3$			
Imidazole, 2-methoxy-1-methyl-4-nitro-	6M H_2SO_4	230s(3.52),294(3.84)	4-0454-66
	pH 7.10	240s(3.52),320(3.87)	4-0454-66
	NaOH	235s(3.56),307(3.83)	4-0454-66
Imidazole, 2-methoxy-1-methyl-5-nitro-	6M H_2SO_4	225s(3.56),305(3.96)	4-0454-66
	pH 6.90	248(3.63),339(4.09)	4-0454-66
	NaOH	247(3.61),327(4.06)	4-0454-66
$C_5H_7N_3O_5$			
2-Isoxazoline, 5,5-dimethyl-3,4-dinitro-	EtOH	268(3.77)	44-0394-66
$C_5H_7N_3O_9$			
Ethyl trinitroethyl carbonate	MeCN	276(1.97)	44-0369-66
$C_5H_7N_5$			
8-Azapurine, 1,6-dihydro-2-methyl-	pH 4.0	259(3.79)	39-0427-66B
	pH 7.81	276(3.75)	39-0427-66B
	pH 11.6	275(3.79)	39-0427-66B
$C_5H_7N_5O$			
2-Pyrazinecarbohydrazide, 3-amino-	pH -2.0	243(4.10),357(3.90)	39-1112-66C
	pH 2.1	247(4.08),354(3.84)	39-1112-66C
	pH 6.9	248(4.06),352(3.82)	39-1112-66C
Pyrimidine, 4-amino-5-(methylnitrosamino)-	pH 1.5	228(4.17)	39-0427-66B
	pH 6.0	239(4.08),270s(3.68), 358(2.26)	39-0427-66B
$C_5H_7N_5O_2$			
Pyrimidine, 2-amino-4-(methylamino)-5-nitro-	MeOH	259(3.86),293(3.69), 356(4.10)	87-0121-66
Pyrimidine, 4-amino-2-(methylamino)-5-nitro-	MeOH	265(3.71),350(4.41)	87-0121-66
$C_5H_7N_5O_3$			
4(3H)-Pyrimidinone, 2,6-diamino-3-methyl-5-nitro-	pH -0.89	223(4.36),245s(3.66), 315(4.07)	24-2984-66

Compound	Solvent	λ_{max}(log ε)	Ref.
4(3H)-Pyrimidinone, 2,6-diamino-3-methyl-5-nitro- (cont.)	pH 5.0	230(4.20),329(4.11)	24-2984-66
$C_5H_7N_5S$			
s-Triazolo[3,4-b][1,3,4]thiadiazole, 6-amino-3-ethyl-	MeOH	238(3.80)	44-3528-66
cation	MeOH	242(3.69)	44-3528-66
$C_5H_8FN_3O_2S$			
Pyruvic acid, fluoro-, 3-methyliso-thiosemicarbazone	EtOH	262(4.16)	87-0419-66
$C_5H_8N_2$			
Pyrazole, 1,3-dimethyl-	EtOH	221(3.66)	22-0293-66
Pyrazole, 1,4-dimethyl-	EtOH	226(3.62)	22-3744-66
Pyrazole, 1,5-dimethyl-	EtOH	217(3.62)	22-0293-66
Pyrazole, 1-ethyl-	EtOH	217.5(3.66)	22-3744-66
$C_5H_8N_2O$			
2(1H)-Pyrazinone, 1-methyl-	MeOH	224(3.91),318(3.70)	24-0542-66
$C_5H_8N_2OS_2$			
2-Thioacetamide, S-(4,5-dihydro-2-thiazolyl)-	acid	252(3.95)	39-0092-66B
	H_2O	227s(3.76),248s(3.44)	39-0092-66B
$C_5H_8N_2O_3$			
Urea, acetoacetyl-	EtOH	260(3.56)	70-1171-66
	hexanol	260(2.18)	70-1171-66
$C_5H_8N_2O_4$			
2-Butene, 3-methyl-1,2-dinitro-	EtOH	251(3.75)	44-0394-66
Cyclopropane, 3,3-dimethyl-1,2-di-nitro-, trans	MeOH	270(2.13)	44-0394-66
2-Oxazolidinone, 4,4-dimethyl-3-nitro-	EtOH	228(3.62),290(2.18)	44-3038-66
$C_5H_8N_2O_5$			
Oxetane, 3,3-bis(nitromethyl)- (all spectra anom.)	pH 11	246(3.82),257s(3.76)	78-0925-66
	pH 12	246(4.20),257(4.20)	78-0925-66
	pH 13	246(4.29),257(4.28)	78-0925-66
	0.00015N NaOH	246(3.61)	78-0925-66
$C_5H_8N_2O_6$			
Propanol, 2,2-dinitro-, acetate	MeCN	277(1.75)	44-0369-66
$C_5H_8N_4O$			
4-Pyrimidinol, 5,6-diamino-2-methyl-	pH 6.5	214(4.01),283(3.94)	39-1112-66C
	pH 13.0	274(3.93)	39-1112-66C
1,2,4-Triazin-5-one, 3-amino-2,5-dihydro-2,6-dimethyl-	pH -1.1	212(3.99),265(3.76)	73-1864-66
	pH 5.1	248(3.73)	73-1864-66
	EtOH	257(3.73)	73-1864-66
1,2,4-Triazin-5-one, 3-amino-4,5-dihydro-4,6-dimethyl-	pH -1.1	211(4.02),259(3.76)	73-1864-66
	pH 7.0	222(3.91),296(3.85)	73-1864-66
	EtOH	222(3.90),299(3.89)	73-1864-66
1,2,4-Triazin-5-one, 2,5-dihydro-6-methyl-3-(methylamino)-	pH -1.1	214(4.10),262(3.61)	73-1864-66
	pH 5.6	213(4.22),243s(3.80), 280s(3.10)	73-1864-66
	pH 11.4	294(3.64)	73-1864-66
	EtOH	214(4.25),241s(3.88), 292s(2.94)	73-1864-66

Compound	Solvent	λ max(log ε)	Ref.
$C_5H_8N_6S$			
4(1H)-Pyrimidinethione, 6-amino-5-guanidino-	pH 1	227(4.23),297(3.71),364(4.23)	44-1417-66
	pH 7	226(4.19),339(4.05),368(4.12)	44-1417-66
C_5H_8O			
Cyclopropanone, 2,2-dimethyl-	CH_2Cl_2	340.0(1.60)	35-2880-66
$C_5H_8O_3$			
Acrylic acid, 3-methoxy-, methyl ester, cis	ether	225(4.13)	24-0450-66
trans	EtOH	227(4.26)	44-0646-66
	ether	229(4.06)	24-0450-66
Formic acid, 2,3-epoxy-2-butyl ester	n.s.g.	284(1.33)	33-1002-66
$C_5H_8O_3S$			
Thiophene, 4,5-dihydro-3-methoxy-, 1,1-dioxide	H_2O	219.5(3.14)	73-3744-66
C_5H_8S			
4H-Thiopyran, 2,3-dihydro-	MeCN	226(3.77),249(3.47)	44-2333-66
$C_5H_8S_2$			
1,4-Dithiin, 5,6-dihydro-2-methyl-	EtOH	278(3.48)	44-0586-66
$C_5H_9ClN_2O_4$			
2,3-Dihydro-1,4-diazepinium perchlorate	MeOH	339(4.05)	39-0093-66C
$C_5H_9NOS_2$			
1,3-Thiazine-2-thione, tetrahydro-4-hydroxy-3-methyl-	acid	244(4.02),281(4.08)	39-0092-66B
	H_2O	244(3.99),281(4.08)	39-0092-66B
	base	247(3.96),276(4.08)	39-0092-66B
1,3-Thiazine-2-thione, tetrahydro-4-hydroxy-4-methyl-	acid	241(3.90),285(4.07)	39-0092-66B
	H_2O	241(3.85),285(4.07)	39-0092-66B
	base	243(3.76),287(3.83)	39-0092-66B
$C_5H_9NO_3$			
2-Propanol, 2-nitroso-, acetate	C_6H_{12}	680(1.27)	39-0441-66B
C_5H_9NS			
Thiopiperidone	EtOH	276(4.11)	1-0689-66
	heptane + 0.05%CH_2Cl_2	281(4.08)	1-0689-66
	10%CH_2Cl_2	340(1.73)	1-0689-66
Thiopyrrolidone, N-methyl-	EtOH	265(4.18),320(1.83)	1-0689-66
	heptane + 0.05%CH_2Cl_2	269(4.17)	1-0689-66
	10%CH_2Cl_2	347.5(1.56)	1-0689-66
$C_5H_9NS_2$			
1,3-Thiazine-2-thione, tetrahydro-3-methyl-	acid	250(3.97),277(4.13)	39-0092-66B
	H_2O	250(3.96),277(4.13)	39-0092-66B
	base	250(3.95),277(4.13)	39-0092-66B
1,3-Thiazine-2-thione, tetrahydro-4-methyl-	acid	244(3.91),280(4.11)	39-0092-66B
	H_2O	245(3.86),280(4.11)	39-0092-66B
	base	243(3.93),279(3.96)	39-0092-66B

Compound	Solvent	$\lambda_{max}(\log \epsilon)$	Ref.
$C_5H_9N_3S$			
1,3,4-Triazole, 1,2-dimethyl-5-(methylthio)-	EtOH	211(3.89)	1-0057-66
1,3,4-Triazoline-5-thione, 1,2,4-trimethyl-	EtOH	253(4.18)	1-0057-66
$C_5H_9N_5$			
Pyrimidine, 2,4-diamino-5-(methylamino)-	pH 1	224(4.23),292s(3.51)	87-0121-66
	pH 11	240(3.96),302(3.73)	87-0121-66
Pyrimidine, 2,5-diamino-4-(methylamino)-	MeOH-H_2SO_4	235(4.24),302(3.72)	87-0121-66
Pyrimidine, 4,5-diamino-2-(methylamino)-	MeOH-H_2SO_4	234(4.30),297s(3.08)	87-0121-66
1,3,5-Triazine, 2,4-diamino-6-ethyl-	MeOH	255(3.47)	78-0157-66
$C_5H_9N_5O$			
1,3,5-Triazine, 2,4-diamino-6-methoxymethyl-	MeOH	257(3.53)	78-0157-66
$C_5H_9N_5S$			
1,3,5-Triazine, 2,4-diamino-6-(methylthiomethyl)-	pH 5.2	258(3.60)	39-0056-66C
$C_5H_{10}N_2$			
1-Pyrazoline, 3,3-dimethyl-	MeOH	321(2.45)	35-3959-66
1-Pyrazoline, 3,5-dimethyl-, cis	MeOH	323(2.48)	35-3959-66
1-Pyrazoline, 3,5-dimethyl-, trans	MeOH	323(2.48)	35-3959-66
1-Pyrazoline, 4,4-dimethyl-	MeOH	323(2.51)	35-3959-66
2-Pyrazoline, 1,5-dimethyl-	EtOH	242(3.47)	4-0413-66
2-Pyrazoline, 4,4-dimethyl-	MeOH	228(3.63)	35-3959-66
$C_5H_{10}N_2O$			
5-Pyrimidinol, 1,4,5,6-tetrahydro-2-methyl-	MeOH	207(3.85)	44-3838-66
$C_5H_{10}N_2O_3$			
Propionamide, N-ethyl-N-nitro-	hexane	240(3.83)	44-3038-66
$C_5H_{10}N_2O_4$			
Propane, 2-methyl-1-nitro-2-(nitromethyl)-	85% EtOH-NaOH	238.5(4.10)(anom.)	78-0925-66
$C_5H_{10}N_4O_2$			
Formamidine, N,N'-diacetamido-	EtOH	242(4.15)	44-3442-66
$C_5H_{10}N_6O_4$			
2-Tetrazene, 1,1,4-trimethyl-4-(2,2-dinitrovinyl)-	EtOH	249(3.84),389(4.36)	44-0923-66
$C_5H_{10}O$			
2-Pentanone	C_6H_{12}	280(1.26)	44-4237-66
	EtOH	278(1.30)	44-4237-66
$C_5H_{11}I$			
Butane, 2-iodo-2-methyl-	C_6H_{12}	267(2.78)	3-0612-66
$C_5H_{11}N_7O$			
Pyruvaldehyde, 1-guanylhydrazone 2-semicarbazone	pH 1	283(4.53)	87-0609-66
	pH 11	294(4.41)	87-0609-66

Compound	Solvent	$\lambda_{max}(\log \epsilon)$	Ref.
$C_5H_{11}N_7S$			
Pyruvaldehyde, 1-guanylhydrazone	pH 1	308(4.68)	87-0609-66
2-thiosemicarbazone	pH 11	332(4.65)	87-0609-66
Pyruvaldehyde, 2-guanylhydrazone	pH 1	305(4.70)	87-0609-66
1-thiosemicarbazone	pH 11	335(4.64)	87-0609-66
$C_5H_{12}N_2S$			
Thiourea, N,N-diethyl-, iodine complex	$CHCl_3$	304(4.59)	60-0029-66
Thiourea, tetramethyl-, iodine complex	$CHCl_3$	334(4.56)	60-0029-66
$C_5H_{12}OSSi$			
Silane, thionoacetoxytrimethyl-	hexane	207(3.44),245(3.66), 389(1.06)	44-3439-66
$C_5H_{12}S$			
Sulfide, tert-butyl methyl	hexane	212.0(3.36)	39-0242-66A
$C_5H_{12}S_2$			
Acetone, dimethyl mercaptole	hexane	202(3.59),235(2.84)	39-0242-66A
$C_5H_{12}S_3$			
Orthoacetic acid, trithio-, trimethyl ester	hexane	202(3.65),235(3.08)	39-0242-66A
$C_5H_{12}S_4$			
Orthocarbonic acid, tetrathio-, tetramethyl ester	hexane	204(3.67),245(3.40)	39-0242-66A

Compound	Solvent	λ_{max}(log ε)	Ref.
$C_6Cl_4O_2$			
1,2-Cyclobutanedione, 3,4-bis-(dichloromethylene)-	C_6H_{12}	234(4.17),280(3.46), 383(4.06),403(4.06), 555(2.60)	18-0160-66
$C_6F_{10}N_4$			
1,2,4,5-Tetrazine, 6-(heptafluoro-propyl)-3-(trifluoromethyl)-	isooctane	252(3.23),533(2.70)	44-0781-66
$C_6H_2Br_2N_4$			
Pyrazino[2,3-d]pyridazine, 5,8-dibromo-	EtOH	205(4.00),247(4.15), 291(3.75)	4-0512-66
$C_6H_2Cl_2N_4$			
Pyrazino[2,3-d]pyridazine, 5,8-dichloro-	EtOH	203(3.83),236(4.10), 290(3.39)	4-0512-66
$C_6H_2F_{10}N_4$			
1,2,4,5-Tetrazine, 6-(heptafluoro-propyl)-3-(trifluoromethyl)-1,2-dihydro-	iso-PrOH	227(3.61)	44-0781-66
$C_6H_2N_4$			
Pyrazine-2,3-dicarbonitrile	EtOH	207s(3.67),229(3.98), 240s(3.81),275(3.74), 281s(3.71)	4-0512-66
$C_6H_3BCl_2F_4NS_2$			
4,6-Dichloro-1,3,2-benzothiazolium tetrafluoroborate	CF_3COOH	367(4.20),461(3.63)	4-0518-66
$C_6H_3BrO_2$			
p-Benzoquinone, bromo-	EtOH	256(4.06)	23-2867-66
$C_6H_3Br_2ClO_2$			
2-Cyclohexene-1,4-dione, 5,6-dibromo-2-chloro-	CCl_4	271(3.88),365(2.18)	12-0617-66
$C_6H_3Br_2ClO_3S$			
Benzenesulfonyl chloride, 3,5-dibromo-4-hydroxy-	n.s.g.	265(3.92)	44-2672-66
$C_6H_3Br_3O_2$			
2-Cyclohexene-1,4-dione, 2,5,6-tribromo-	CCl_4	286(3.90),366(2.28)	12-0617-66
$C_6H_3ClO_2$			
p-Benzoquinone, chloro-	EtOH	250(4.10)	23-2867-66
$C_6H_3Cl_2N_3$			
1,3,5-Triazaindene, 2,6-dichloro-	pH -1.0	217(4.73),220s(4.71), 257(3.65),281(3.65), 288s(3.56)	39-0285-66B
	pH 4.0	210(4.64),240s(3.67), 245(3.70),251s(3.66), 268s(3.58),273(3.65), 280(3.60),287s(3.11)	39-0285-66B
	pH 9.5	217(4.67),256s(3.55), 270s(3.70),274(3.71), 280s(3.67),290(3.47)	39-0285-66B

Compound	Solvent	$\lambda_{max}(\log \epsilon)$	Ref.
$C_6H_3Cl_3O_3S$			
Benzenesulfonyl chloride, 3,5-dichloro-4-hydroxy-	n.s.g.	264(3.95)	44-2672-66
$C_6H_3F_3N_4O$			
2-Purinol, 8-(trifluoromethyl)-	pH 2.0	213(4.31),281(3.77), 316(3.75)	39-0438-66B
	pH 8.0	215(4.42),271(3.78), 322(3.73)	39-0438-66B
	pH 13.0	218(4.39),263(3.64), 316(3.92)	39-0438-66B
$C_6H_3F_6N_3$			
2-Pyrazoline-3-carbonitrile, 5,5-bis(trifluoromethyl)-	EtOH	264(4.03)	35-3617-66
$C_6H_3N_3O_4$			
Benzofurazan, 5(6)-nitro-, oxide	EtOH	231(4.02),261(4.13), 387(3.75)	95-0766-66
Benzofuroxan, 5-nitro-	benzene	391(3.65)	39-0489-66B
	EtOH	304(3.30),391(3.70)	39-0489-66B
	DMF	391(3.67)	39-0489-66B
	$HCONH_2$	391(3.67)	39-0489-66B
$C_6H_3N_3O_6$			
Benzene, 1,3,5-trinitro-	C_6H_{12}	221(4.4)	49-1365-66
$C_6H_3N_5O$			
3,4-Pyrazoledicarbonitrile, 5-formamido-	EtOH	216(4.16),245(4.06)	44-0342-66
4(5H)-Pyrazolo[3,4-d]pyrimidinone, 3-cyano-	EtOH	225(3.99),232(4.09), 261(4.15)	44-0342-66
$C_6H_3N_5O_4$			
Benzene, 1-azido-2,4-dinitro-	benzene	391(2.00)	39-0489-66B
	EtOH	304(4.06),391(2.95)	39-0489-66B
	EtOH	227s(4.08),260(4.00), 298(3.99)	95-0766-66
	DMF	391(2.63)	39-0489-66B
	$HCONH_2$	391(2.78)	39-0489-66B
$C_6H_4BClF_4NS_2$			
6-Chloro-1,3,2-benzothiazathiolium tetrafluoroborate	CF_3COOH	366(4.18),419(3.70)	4-0518-66
$C_6H_4BF_4NS_2$			
1,3,2-Benzothiazathiolium tetrafluoroborate	CF_3COOH	344(4.15),423(3.28)	4-0518-66
$C_6H_4Br_2O_2$			
2-Cyclohexene-1,4-dione, 5,6-dibromo-	hexane	220(4.02)	23-2867-66
$C_6H_4ClNO_3$			
Phenol, 2-chloro-6-nitro-	C_6H_{12}	218(4.06),279(3.74), 353(3.42)	23-0961-66
	MeOH	277.5(3.65)	23-0961-66
	EtOH	217(4.12),354(3.30)	23-0961-66
$C_6H_4ClNO_4S_2$			
1,3,2-Benzothiazathiolium perchlorate	CF_3COOH	344(4.15),424(3.26)	4-0518-66

Compound	Solvent	λ_{max}(log ε)	Ref.
$C_6H_4ClN_3$			
3H-Imidazo[4,5-b]pyridine, 7-chloro-	pH 1	243s(--),250s(--), 259s(--),273s(3.92), 280(3.97)	87-0354-66
	pH 7	251(3.77),277(3.94), 282s(--)	87-0354-66
	pH 13	287(4.03)	87-0354-66
1,3,5-Triazaindene, 4-chloro-	pH 0.0	208s(4.46),241(3.44), 250s(3.48),270(3.79), 275s(3.77),285s(3.39)	39-0285-66B
	pH 6.0	208s(4.46),245(3.67), 251(3.69),259s(3.70), 266(3.78),273(3.71), 285s(2.67)	39-0285-66B
	pH 12.1	212(4.63),254s(3.55), 266s(3.70),273s(3.75), 279(3.76),286s(3.63)	39-0285-66B
1,3,5-Triazaindene, 6-chloro-	pH 0.5	233(3.56),241s(3.48), 274(3.60)	39-0285-66B
	pH 6.5	245(3.58),252s(3.53), 272(3.55),278(3.50), 288s(3.00)	39-0285-66B
	pH 12.4	216(4.62),256s(3.47), 276(3.59)	39-0285-66B
s-Triazolo[1,5-a]pyridine, 2-chloro-	MeOH	258(3.61),273s(3.49), 283(3.27)	44-0265-66
$C_6H_4ClN_3O$			
1,3,5-Triazainden-2-ol, 6-chloro-	pH -0.5	225(4.70),258(3.68), 285(3.77)	39-0285-66B
	pH 6.0	210(4.67),233(3.79), 278(3.70),284s(3.61)	39-0285-66B
	pH 12.0	218(4.59),260(3.83), 283(3.85),289s(3.80)	39-0285-66B
$C_6H_4ClN_3S$			
1,3,5-Triazaindene-2-thiol, 6-chloro-	pH -1.5	230(3.90),258(4.60), 324(4.23)	39-0285-66B
	pH 4.3	217(4.12),247(4.44), 296s(4.32),305(4.41)	39-0285-66B
	pH 10.6	238(4.43),274s(3.90), 304(4.28)	39-0285-66B
	pH 14.5	241(4.55),307(4.22)	39-0285-66B
$C_6H_4ClNaO_3S$			
Benzenesulfonic acid, p-chloro-, sodium salt	H_2O	224(4.2),264(2.5)	65-2135-66
$C_6H_4Cl_2O_2$			
2-Cyclohexene-1,4-dione, 5,6-dichloro-	hexane	230(3.99)	23-2867-66
$C_6H_4Cl_2O_4$			
2-Cyclopentene-1-carboxylic acid, 3,5-dichloro-4-oxo-2-hydroxy-	H_2O	276(3.94)	12-2107-66
$C_6H_4Cl_3F_3O_3$			
Acetoacetic acid, 4,4,4-trichloro-, 2,2,2-trifluoroethyl ester	hexane	252(3.33)	44-3369-66
	EtOH	252(2.89)	44-3369-66

Compound	Solvent	$\lambda_{max}(\log \epsilon)$	Ref.
$C_6H_4FNO_3$			
Phenol, 3-fluoro-6-nitro-	C_6H_{12}	211(4.17),278(3.82), 334(3.61)	23-0961-66
	MeOH	273(3.72)	23-0961-66
	EtOH	210(4.15),334(3.51)	23-0961-66
$C_6H_4F_3N_5$			
Purine, 2-amino-8-(trifluoromethyl)-	pH 0.0	221(4.59),241s(3.65), 318(3.64)	39-0438-66B
	pH 4.37	217(4.40),281(3.48), 312(3.80)	39-0438-66B
	pH 9.0	220(4.37),270(3.58), 308(3.84)	39-0438-66B
$C_6H_4F_{10}N_4O$			
Butyrhydrazide, N^2-(α-hydrazono-perfluoroethyl)perfluoro-	iso-PrOH	263(4.02)	44-0781-66
Propionhydrazide, N^2-(α-hydrazono-perfluoropropyl)perfluoro-	iso-PrOH	265(4.07)	44-0781-66
C_6H_4KN			
Cyclopentadienecarbonitrile, potassium derivative	H_2O	264(4.20)	35-3046-66
$C_6H_4N_2$			
Mucononitrile, cis-cis	EtOH	259.5(1.48)	39-0385-66C
cis-trans	EtOH	259(1.52)	39-0385-66C
trans-trans	EtOH	259(1.57)	39-0385-66C
$C_6H_4N_2O_2$			
Benzofuroxan	xylene	359(3.83)	39-0489-66B
	DMF	359(3.86)	39-0489-66B
3,4'-Biisoxazole	EtOH	222(3.87)	32-0443-66
5,5'-Biisoxazole	EtOH	260(4.25)	32-0443-66
$C_6H_4N_2O_4$			
Benzene, m-dinitro-	C_6H_{12}	228(4.3)	49-1365-66
	MeOH-MeONa	520(1.56)	39-0498-66B
(and other solvent mixtures not listed)	64% MeOH-DMSO-MeONa	520(1.36)	39-0498-66B
Benzene, p-dinitro-	C_6H_{12}	255(4.3)	49-1365-66
$C_6H_4N_2O_5$			
Phenol, 2,4-dinitro-	C_6H_{12}	256(4.19),335(3.53), 390s(3.80)	19-0023-66
	MeOH	212(4.09),257(4.06), 292(3.99),340s(3.51)	19-0023-66
	20% MeOH	212(4.01),261(4.09), 294(3.97),350s(3.50)	19-0023-66
Phenol, 2,6-dinitro-	C_6H_{12}	228(3.98),260s(3.79), 343(3.67)	19-0023-66
	20% MeOH	250(3.95),347(3.67)	19-0023-66
$C_6H_4N_4$			
Pyrazino[2,3-d]pyridazine	EtOH	219(4.12),284(3.12), 295(2.99)	4-0512-66
$C_6H_4N_4O$			
2-Pteridinol	pH 11.8	224(4.31),265(3.82), 375(3.83)	12-0337-66

Compound	Solvent	λ max (log ε)	Ref.
2-Pteridinol (cont.)	pH 12.0	230(3.96),267(3.83), 313(3.76)	12-0337-66
	pH 15.0	266(3.95),339(3.81)	12-0337-66
6-Pteridinol	pH 11.3	236(3.39),294(4.03)	12-0337-66
	pH 12.0	222(4.27),256(3.88), 358(3.74)	12-0337-66
	pH 15.0	303(3.94)	12-0337-66
$C_6H_4N_4O_2$			
Benzene, 1-azido-2-nitro-	xylene	359(2.89)	39-0489-66B
	DMF	359(3.18)	39-0489-66B
Pyrazino[2,3-d]pyridazine-5,8-diol	EtOH	206(4.12),265(4.01)	4-0512-66
$C_6H_4N_4O_3$			
Lumazine, 3-hydroxy-	EtOH	234(4.09),322(3.78)	4-0224-66
Pyrazolo[4,3-d]pyrimid-7-one, 3-carboxy-	pH 1	279(3.90)	4-0110-66
	pH 11	291(3.90)	4-0110-66
	N NaOH	293s(3.96),300(3.99), 312s(3.80)	4-0110-66
$C_6H_4N_4O_6$			
Aniline, 2,4,6-trinitro-	EtOH-15% HOAc	240(4.21),317(4.08), 410(3.91)	65-1034-66
$C_6H_4N_6$			
Pyrazolo[3,4-d]pyrimidine-3-carbonitrile, 4-amino-	EtOH	233(3.98),282(4.00)	44-0342-66
$C_6H_4O_2S$			
2,3-Thiophenedicarboxaldehyde	EtOH	225(3.61),269(4.02)	44-3592-66
$C_6H_4O_4$			
p-Benzoquinone, 2,5-dihydroxy-	EtOH	288(4.37),405(2.47)	12-0169-66
C_6H_5BO			
Boronic anhydride, phenyl-	C_6H_{12}	268(2.82),274(2.90), 280(2.80)	39-0566-66C
	EtOH	262(2.56),267(2.64), 274(2.54)	39-0566-66C
$C_6H_5BrN_2$			
Aniline, 2,4-dibromo-	2N HCl	272(2.50)	28C-0362-66A
	pH 9.2	299(3.34)	28C-0362-66A
Aniline, 2,5-dibromo-	2N HCl	272(2.60)	28C-0362-66A
	pH 9.2	295(4.48)	28C-0362-66A
$C_6H_5BrO_2S$			
Benzenesulfinic acid, p-bromo-	H_2O	230(4.2),250s(3.8), 275s(2.9)	18-1788-66
anion	H_2O	230(4.2),275(3.2)	18-1788-66
$C_6H_5BrO_3$			
2-Furoic acid, 5-bromo-3-methyl-	EtOH	260(4.17)	39-0976-66C
$C_6H_5Br_2NO$			
Pyridine, 3,5-dibromo-2-methoxy-	H_o -3.6	217(4.19),236s(3.75), 305(3.83)	39-0996-66B
	MeOH	227(3.98),294(3.73)	39-0991-66B

Compound	Solvent	λ max(log ε)	Ref.
C_6H_5ClFN			
Aniline, 3-chloro-4-fluoro-	2N HCl	265(2.95)	28C-0362-66A
	pH 9.2	293(3.37)	28C-0362-66A
$C_6H_5ClN_2O_2$			
Aniline, 4-chloro-2-nitro-	5N HCl	262(3.74)	28C-0362-66A
	pH 9.2	421(3.66)	28C-0362-66A
$C_6H_5ClN_2O_2S$			
Pyridine, 2-chloro-4-(methylthio)-3-nitro-	pH 1, 7	220s(--),263(4.06)	87-0354-66
	pH 13	263(4.08)	87-0354-66
$C_6H_5ClN_2O_3$			
Pyridine, 2-chloro-4-methoxy-3-nitro-	pH 1,7,13	215s(4.06),250(3.15)	87-0354-66
$C_6H_5ClN_4$			
Purine, 6-chloro-3-methyl-	pH 5	278(3.85)	39-0010-66C
C_6H_5ClO			
Phenol, 2-chloro-	H_2O	273(3.35)	22-3328-66
	base	293.5(3.63)	22-3328-66
Phenol, 3-chloro-	H_2O	273.5(3.33)	22-3328-66
	base	291.5(3.53)	22-3328-66
$C_6H_5ClO_2S$			
Benzenesulfinic acid, p-chloro-	H_2O	225(4.3),245(4.0), 270(3.2)	18-1788-66
anion	H_2O	225(4.2),270(3.2)	18-1788-66
$C_6H_5ClO_3S$			
Benzenesulfonic acid, 4-chloro-, sodium salt	H_2O	224(4.2),264(2.5)	65-2141-66
$C_6H_5Cl_2N$			
Aniline, 2,3-dichloro-	2N HCl	268(2.48)	28C-0362-66A
	pH 9.2	291(3.36)	28C-0362-66A
Aniline, 2,4-dichloro-	2N HCl	268(2.48)	28C-0362-66A
	pH 9.2	296(3.33)	28C-0362-66A
Aniline, 3,4-dichloro-	2N HCl	270(2.51)	28C-0362-66A
	pH 9.2	296(3.24)	28C-0362-66A
Aniline, 3,5-dichloro-	2N HCl	270(2.54)	28C-0362-66A
	pH 9.2	293(3.24)	28C-0362-66A
$C_6H_5Cl_2NO$			
Pyridine, 3,5-dichloro-2-methoxy-	MeOH	229(3.95),292(3.69)	39-0991-66B
$C_6H_5FN_2O_2$			
Aniline, 4-fluoro-3-nitro-	5N HCl	250(3.83)	28C-0362-66A
	pH 9.2	357(3.23)	28C-0362-66A
$C_6H_5F_3N_4O_2$			
2-Pyrimidinol, 4-amino-5-(trifluoroacetamido)-	pH 1	211(4.04),282(3.95)	39-0438-66B
	pH 5.4	230s(3.97),272(3.73)	39-0438-66B
	pH 10.0	290(3.75)	39-0438-66B
C_6H_5N			
Cyclopentadienecarbonitrile	H_2O	268(3.90)	35-3046-66

Compound	Solvent	λ_{max}(log ε)	Ref.
C_6H_5NO			
Benzene, nitroso-	EtOH	217(3.79),281(3.99), 304(3.86)	35-5010-66
C_6H_5NOS			
4H-Thieno[3,2-b]pyrrol-5(6H)-one	EtOH	223(4.07),281(3.32)	70-1549-66
$C_6H_5NO_2$			
Benzene, nitro-	C_6H_{12}	252(3.93),285s(3.1), 350s(2.1)	49-1365-66
	hexane	250(3.93),290s(3.1), 350s(2.0)	49-1365-66
	heptane	252(4.00),280s(3.0), 330s(2.1)	49-1365-66
	acid	275.0(3.97)	61-0862-66
electron complex	pH 13	285.0(4.15)	61-0862-66
Phenol, p-nitroso-	EtOH	236(3.42),302(4.22), 708(0.90)	12-0841-66
	ether	229(3.46),290(4.20), 727(0.89)	12-0841-66
	dioxan	292(4.22),735(0.90)	12-0841-66
$C_6H_5NO_2S$			
Benzenethiol, o-nitro-	hexane	225(4.36),240(--), 258(4.00),344(3.78)	65-1577-66
	EtOH	239(4.11),271(3.58), 357(3.43)	65-1577-66
	ether	238(4.17),269(3.62), 350(3.55)	65-1577-66
	dioxan	239(4.14),270(3.64), 352(3.45)	65-1577-66
Benzenethiol, p-nitro-	hexane	224(3.80),305(4.08), 346(2.75)	65-1577-66
	EtOH	221(4.07),322(4.38)	65-1577-66
	ether	218(3.87),313(4.11)	65-1577-66
	dioxan	318(3.94)	65-1577-66
$C_6H_5NO_3$			
Phenol, 2-nitro-	C_6H_{12}	230s(3.55),272(3.87), 347(3.57)	19-0023-66
	MeOH	212(4.10),230s(3.49), 273(3.69),346(3.51)	19-0023-66
	20% MeOH	210(4.07),230s(3.49), 278(3.78),350(3.49)	19-0023-66
Phenol, 3-nitro-	hexane	222s(4.04),258(3.80), 313(3.34)	19-0023-66
	MeOH	212(4.08),230(3.95), 269(3.73),332(3.33)	19-0023-66
	20% MeOH	210(4.08),229(3.86), 274(3.76),332(3.30)	19-0023-66
Phenol, 4-nitro-	hexane	219(3.98),226s(3.90), 285(4.01),340s(2.55)	19-0023-66
	MeOH	227(3.84),312(4.03)	19-0023-66
	20% MeOH	227(3.78),317(4.00)	19-0023-66
Picolinic acid, 3-hydroxy-, hydrochloride	EtOH	224(3.86),304(3.89)	44-0636-66
$C_6H_5NO_4$			
Catechol, 3-nitro-	C_6H_{12}	240(3.79),292(3.87), 365(3.25)	23-0961-66

Compound	Solvent	$\lambda_{max}(\log \epsilon)$	Ref.
Catechol, 3-nitro- (cont.)	EtOH	242(3.23),296(3.27), 367(2.64)	23-0961-66
$C_6H_5NO_4S$			
Benzenesulfonic acid, p-nitro-	H_2O	265f(4.0)	18-1788-66
anion	H_2O	270f(4.0)	18-1788-66
$C_6H_5N_3$			
1,3,5-Triazaindene	pH 4.0	248s(3.57),262(3.64), 267s(3.63)	39-0285-66B
	pH 8.5	241(3.58),247(3.60), 257(3.56),263(3.60), 269(3.54),273(3.06)	39-0285-66B
	pH 13.0	268(3.70),274s(3.65)	39-0285-66B
s-Triazolo[1,5-a]pyridine	MeOH	258(3.61),273s(3.49), 282s(3.27)	44-0265-66
s-Triazolo[4,3-a]pyridine	MeOH	266(3.9),280(3.9)	24-2593-66
$C_6H_5N_3O$			
1,3,5-Triazainden-2-ol	pH 3.9	215(4.59),254s(3.58), 276(3.88)	39-0285-66B
	pH 8.12	224s(3.88),270(3.75), 275s(3.73),286s(3.34)	39-0285-66B
	pH 12.4	256(3.76),277(3.86), 280s(3.84)	39-0285-66B
s-Triazolo[4,3-a]pyridin-3-ol	MeOH	260(3.8),268(3.8), 313(3.6)	24-2593-66
$C_6H_5N_3O_2$			
Uracil-5-carbonitrile, 6-methyl-	0.08N HCl	216(4.05),271(4.07)	25-0381-66
	0.08N NaOH	239(4.09),283(4.11)	25-0381-66
$C_6H_5N_3O_2S$			
Uracil, 5-thiocyanomethyl-	pH 1	260(3.95)	87-0097-66
	pH 6.9	260(3.95)	87-0097-66
	pH 11	278(3.90)	87-0097-66
$C_6H_5N_3O_3$			
Uracil-5-carbonitrile, 1-hydroxy-6-methyl-	0.08N HCl	223(3.92),284(4.06)	25-0381-66
	0.08N NaOH	257(3.82),320(3.97)	25-0381-66
$C_6H_5N_3O_4$			
Aniline, 2,4-dinitro-	EtOH	225(4.00),257(3.96), 335(4.14),375s(3.85)	65-1034-66
Aniline, 2,6-dinitro-	EtOH	250(4.04),422(3.98)	65-1034-66
Aniline, 3,5-dinitro-	5N HCl	232(4.23)	28C-0362-66A
	pH 9.2	376(3.29)	28C-0362-66A
$C_6H_5N_3S$			
1,3,5-Triazaindene-2-thiol	pH 2.7	247(4.46),314(4.26)	39-0285-66B
	pH 6.78	216(4.11),243(4.30), 292s(4.24),297(4.28), 315s(3.81)	39-0285-66B
	pH 11.0	224s(4.33),230s(4.34), 266s(3.84),296(4.29)	39-0285-66B
$C_6H_5N_5$			
Pteridine, 2-amino-	pH 7.0	225(4.39),259(3.81), 370(3.82)	39-1117-66C

Compound	Solvent	λ_{max}(log ε)	Ref.
$C_6H_5N_5O$			
Pteridin-7-one, 2-aminodihydro-	pH -0.89	222(4.46),258(4.11), 322(4.01)	24-0536-66
	pH 4.0	233(3.88),290(3.69), 341(4.22)	24-0536-66
	pH 10.0	225(4.54),269(3.74), 341(4.22)	24-0536-66
$C_6H_5N_5O_2$			
Pyrazolo[3,2-c]-as-triazine-3-carboxylic acid, 4-amino-	EtOH	232(4.24),305(4.02), 347(3.90)	39-1127-66C
Pyrimido[5,4-e]-as-triazine-5,7-dione, tetrahydro-6-methyl-	pH 1	232(4.24),272(3.33), 333(3.73)	44-0900-66
	pH 11	253(4.31),388(3.62)	44-0900-66
	EtOH	234(4.20),260s(3.59), 334(3.31)	44-0900-66
$C_6H_5N_5O_4$			
s-Triazolo[4,3-c]pyrimidine, 1,5,6,7-tetrahydro-6-methyl-8-nitro-5,7-dioxo-	pH 1	249(4.24),328(4.22)	44-0900-66
	pH 11	352(4.18)	44-0900-66
C_6H_6			
Benzene, tetracyanoethylene complex	$CHCl_3$	391(3.46)	60-0018-66
Fulvene	C_6H_{12}	236(4.04),241(4.10), 244(4.10),258(3.60), 362(2.33)	33-1278-66
	EtOH	242(4.06),360(2.35)	33-1278-66
$C_6H_6BrD_3O$			
Cyclohexanone-2,6,6-d_3, 2-bromo-	C_6H_{12}	312(2.00)	23-0759-66
C_6H_6BrN			
Aniline, m-bromo-	2N HCl	265(2.42)	28C-0362-66A
	pH 9.2	286(3.26)	28C-0362-66A
Aniline, o-bromo-	2N HCl	262(2.36)	28C-0362-66A
	pH 9.2	287(3.35)	28C-0362-66A
Aniline, p-bromo-	2N HCl	265(2.42)	28C-0362-66A
	pH 9.2	291(3.15)	28C-0362-66A
C_6H_6BrNO			
Pyridine, 3-bromo-2-methoxy-	pH 5	220(3.78),280(3.71)	39-0991-66B
	MeOH-HCl	290(3.88)	39-0996-66B
Pyridine, 5-bromo-2-methoxy-	H_o -3.6	225(3.93),300(3.78)	39-0996-66B
	pH 5	222(4.01),284(3.54)	39-0991-66B
$C_6H_6Br_2O_2$			
1,3-Cyclohexanedione, 2,4-dibromo-	MeOH-HCl	282(4.05)	20-0391-66
	MeOH-NaOH	312(4.12)	20-0391-66
$C_6H_6ClD_3O$			
Cyclohexanone-2,6,6-d_3, 2-chloro-	C_6H_{12}	303(1.54)	23-0759-66
C_6H_6ClN			
Aniline, m-chloro-	2N HCl	264(2.40)	28C-0362-66A
	pH 9.2	286(3.21)	28C-0362-66A
Aniline, o-chloro-	2N HCl	262(2.35)	28C-0362-66A
	pH 9.2	286(3.32)	28C-0362-66A
Aniline, p-chloro-	2N HCl	264(2.30)	28C-0362-66A
	pH 9.2	290(3.14)	28C-0362-66A

Compound	Solvent	$\lambda_{max}(\log \epsilon)$	Ref.
C_6H_6ClNO			
Pyridine, 3-chloro-2-methoxy-	pH 5	220(3.82),278(3.69)	39-0991-66B
	MeOH-HCl	223(3.69),288(3.86)	39-0996-66B
Pyridine, 5-chloro-2-methoxy-	H_o -3.6	221(3.89),298(3.81)	39-0996-66B
	pH 5	223(4.00),284(3.55)	39-0991-66B
$C_6H_6Cl_2N_2S$			
Pyrimidine, 2,4-dichloro-5-methyl-mercaptomethyl-	EtOH	221(3.99),263(3.66)	5-0119-66B
$C_6H_6Cl_2O$			
Cyclopentanone, 2-(dichloro-methylene)-	EtOH	256(4.14)	77-0567-66
4H-Pyran, 4-(dichloromethyl)-	EtOH	none	24-2351-66
$C_6H_6D_4O$			
Cyclohexanone-2,2,6,6-d_4	C_6H_{12}	291(1.18)	23-0759-66
C_6H_6FN			
Aniline, m-fluoro-	2N HCl	259(2.91)	28C-0362-66A
	pH 9.2	278(3.18)	28C-0362-66A
Aniline, o-fluoro-	2N HCl	259(2.90)	28C-0362-66A
	pH 9.2	277(3.22)	28C-0362-66A
Aniline, p-fluoro-	2N HCl	260(2.79)	28C-0362-66A
	pH 9.2	287(3.25)	28C-0362-66A
$C_6H_6FN_3O_2$			
Cytosine, N-acetyl-5-fluoro-	pH 7.05	216(4.16),240(4.02), 305(3.79)	87-0566-66
C_6H_6IN			
Aniline, m-iodo-	pH 9.2	289(3.33)	28C-0362-66A
Aniline, o-iodo-	pH 9.2	289(3.38)	28C-0362-66A
Aniline, p-iodo-	pH 9.2	287(3.17)	28C-0362-66A
C_6H_6INO			
Pyridine, 5-iodo-2-methoxy-	H_o -1.38	231(4.12),307(3.68)	39-0996-66B
	pH 7	231(4.16),290(3.47)	39-0991-66B
$C_6H_6N_2$			
Malononitrile, isopropylidene-	EtOH	232(4.08)	44-2784-66
$C_6H_6N_2O_2$			
Aniline, m-nitro-	5N HCl	258(3.89)	28C-0362-66A
	pH 9.2	359(3.16)	28C-0362-66A
Aniline, o-nitro-	5N HCl	266(3.86)	28C-0362-66A
	H_2O	280(3.73),409(3.66)	35-3140-66
	pH 9.2	410(3.65)	28C-0362-66A
	tert-BuOH	275(3.71),405(3.74)	35-3140-66
Aniline, p-nitro-	5N HCl	255(3.91)	28C-0362-66A
	H_2O	381.5(4.09)	35-3140-66
	pH 9.2	379(4.11)	28C-0362-66A
	tert-BuOH	378.5(4.18)	35-3140-66
$C_6H_6N_2O_3$			
Pyridazinedione, N-acetyl-	n.s.g.	216(4.16),285(3.61)	44-1311-66
$C_6H_6N_2O_3S$			
Uracil, 5-mercapto-, S-acetyl-	EtOH	271(3.77)	88-1759-66

Compound	Solvent	λ_{max}(log ε)	Ref.
$C_6H_6N_2O_4$			
Furoxan, diacetyl-	MeOH	275(3.62)	88-1727-66
3,4-Pyrazoledicarboxylic acid,	EtOH	229(3.95)	44-2491-66
1-methyl-	dioxan	237(3.98)	44-2491-66
4,5-Pyrazoledicarboxylic acid, 1-methyl-	EtOH	239(4.00)	44-2491-66
$C_6H_6N_2O_5S$			
Benzenesulfonic acid,	H_2O	366.8(4.15)	35-3140-66
2-amino-5-nitro-	tert-BuOH	372.5(4.21)	35-3140-66
$C_6H_6N_4$			
Purine, 3-methyl-	pH 1	275(3.99)	4-0241-66
	pH 11	276(3.84)	4-0241-66
	MeOH	277(3.84)	4-0241-66
Pyrazolo[3,2-c]-as-triazine, 4-methyl-	EtOH	226(4.50),287(3.28), 295s(3.15),350(3.41)	39-1127-66C
5-Pyrimidinecarbonitrile, 4-amino-	H_2O	247(4.14),279(3.64)	94-1321-66
2-methyl-, cation	MeOH	249(4.16),279(3.63)	94-1321-66
	EtOH	249(4.16),279(3.63)	94-1321-66
	iso-PrOH	250(4.16),283(3.61)	94-1321-66
	HOAc	250(4.12),275(3.64)	94-1321-66
	MeCN	248(4.16),276s(3.69)	94-1321-66
5-Pyrimidinecarbonitrile, 1,2-dihydro-2-imino-1-methyl-	pH 4.9	246(4.31),270s(3.67), 308(3.56)	39-0164-66C
5-Pyrimidinecarbonitrile,	pH -1.2	254(4.41),318(3.47)	39-0164-66C
2-(methylamino)-	pH 5.0	263(4.48),303(3.54)	39-0164-66C
s-Triazolo[1,5-a]pyridine, 2-amino-	MeOH	212(4.39),273s(3.54), 277(3.56),292(3.51), 299(3.55),304(3.50), 313(3.42)	44-0265-66
$C_6H_6N_4O$			
Purine, 6-methoxy-	pH 5	253(3.99)	39-0010-66C
2-Purinol, 6-methyl-	pH 0.0	259(3.64),318(3.91)	39-0438-66B
	pH 5.6	213(4.53),236(3.56), 312(3.82)	39-0438-66B
	pH 10.7	215(4.38),269(3.68), 312(3.78)	39-0438-66B
Pyrazolo[3,2-c]-as-triazin-4-ol, 3-methyl-	EtOH	235s(3.85),260(3.96), 332(3.67)	39-1127-66C
$C_6H_6N_4OS$			
2-Purinol, 8-(methylthio)-	pH 0.2	212(4.18),282(3.84), 342(4.31)	39-0438-66B
	pH 5.0	217(4.23),240(4.19), 287(3.66),326(4.00)	39-0438-66B
	pH 10.0	220(4.23),234(4.23), 284(4.06),326(3.92)	39-0438-66B
	pH 14.4	228(4.33),266(3.73), 323(4.15)	39-0438-66B
$C_6H_6N_4O_2$			
1H-Pyrazolo[4,3-d]pyrimidine-5,7(4H,6H)-dione, 1-methyl-	EtOH	287(3.75)	44-2491-66
2H-Pyrazolo[3,4-d]pyrimidine-4,6(5H,7H)-dione, 2-methyl-	EtOH	260(3.79)	44-2491-66
Uracil-5-carbonitrile,	0.08N HCl	223(3.97),281(4.08)	25-0381-66
1-amino-6-methyl-	0.08N NaOH	225(4.04),279(3.93)	25-0381-66
Xanthine, 3-methyl-	H_2O	273(4.04)	95-0854-66

Compound	Solvent	$\lambda_{max}(\log \epsilon)$	Ref.
$C_6H_6N_4O_2S$			
Purine, 8-(methylsulfonyl)-	pH 2.65	215(4.39),271(4.01)	39-0438-66B
	pH 7.5	279(4.13)	39-0438-66B
$C_6H_6N_4O_3$			
1H-Pyrazolo[4,3-d]pyrimidine-5,7-(4H,6H)-dione, 6-hydroxy-1-methyl-	EtOH	243(3.90),284(3.79)	44-2491-66
2H-Pyrazolo[3,4-d]pyrimidine-4,6-(5H,7H)-dione, 5-hydroxy-2-methyl-	EtOH	254(3.87)	44-2491-66
2H-Pyrazolo[4,3-d]pyrimidine-5,7-(4H,6H)-dione, 6-hydroxy-2-methyl-	EtOH	242(3.76),282(3.68)	44-2491-66
$C_6H_6N_4O_5$			
Imidazo[1,2-c]pyrimidine-5,7-dione, 1,2,3,5,6,7-hexahydro-3-hydroxy-8-nitro-	pH 5.0	227(4.29),311(4.12)	24-2984-66
	pH 11.0	234(4.11),328(4.14)	24-2984-66
$C_6H_6N_5NaO_3S$			
4-Pteridinesulfonic acid, 2-amino-3,4-dihydro-, sodium salt	pH 7.0	285(3.91),339(3.87)	39-1117-66C
$C_6H_6N_6$			
Pyrazino[2,3-d]pyridazine, 5,8-diamino-	EtOH	220(4.16),273(4.05)	4-0512-66
$C_6H_6N_6S$			
2H-Thiazolo[3,4,5-gh]purine, 2,2-diamino-	pH 1	218(4.47),264(4.18)	44-1417-66
	pH 7	217(4.34),282(4.12), 292(4.18)	44-1417-66
	pH 13	237(4.07),297(4.44)	44-1417-66
$C_6H_6O_2S$			
Benzenesulfinic acid	H_2O	215s(3.8),237(3.4), 265(2.8)	18-1788-66
anion	H_2O	215(3.9),265(2.9)	18-1788-66
$C_6H_6O_3$			
2-Furoic acid, methyl ester	EtOH	252(4.13)	39-0976-66C
$C_6H_6O_4$			
2,4-Furandione, 3-acetyltetrahydro-	EtOH-NaOH	230(4.10),262(4.14)	99-0230-66
2-Furoic acid, 4,5-dihydro-5-oxo-, methyl ester	ether	232(3.99)	89-0579-66
3-Hexenedioic acid, 3-hydroxy-, γ-lactone	pH 8.0	210(3.3)	37-3776-66
Muconic acid	pH 7.0	260(4.2)	37-3776-66
C_6H_6S			
Benzenethiol	hexane	236(3.90),275(2.92), 286(2.76)	65-1577-66
	EtOH	238(3.77),265(--), 277(2.82)	65-1577-66
	EtOH	235(3.95),279(2.75)	22-0228-66
	ether	237(3.84),282(2.77), 285(2.55)	65-1577-66
	dioxan	241(3.82),286(2.80)	65-1577-66
$C_6H_6S_2$			
Thieno[3,4-b]thiophene, 4,6-dihydro-	EtOH	233(3.62)	44-3363-66
	EtOH	233(3.72)	44-3592-66

Compound	Solvent	λ_{max}(log ε)	Ref.
$C_6H_7BrN_2O$			
2-Pyrimidinol, 5-bromo-4,6-dimethyl-	H_2O	312(3.52)	78-2401-66
$C_6H_7BrN_2OS$			
4-Pyrimidinol, 5-bromo-2-(ethylthio)-	EtOH	245(3.46),302(3.68)	78-2401-66
4-Pyrimidinol, 5-bromo-6-methyl-2-(methylthio)-	EtOH	245(3.39),299(3.51)	78-2401-66
$C_6H_7BrN_2O_2$			
3(5)-Pyrazolecarboxylic acid, 4-bromo-, ethyl ester	EtOH	220(3.76),244s(3.62)	22-3744-66
3-Pyrazolecarboxylic acid, 4-bromo-1,5-dimethyl-	EtOH	222(3.81),247s(3.54)	22-0293-66
5-Pyrazolecarboxylic acid, 4-bromo-1,3-dimethyl-	EtOH	233(3.88),261s(3.73)	22-0293-66
Uracil, 5-bromo-1,3-dimethyl-	H_2O	283(3.93)	50-1222-66B
$C_6H_7Br_3N_2$			
2,6-Dibromo-2,3-dihydro-1H-pyrazolo[1,2-a]pyrazol-4-ium bromide	n.s.g.	244(3.50)	35-5588-66
C_6H_7Cl			
2-Hexen-4-yne, 1-chloro-, cis	n.s.g.	231.0(4.15)	39-0578-66C
2-Hexen-4-yne, 1-chloro-, trans	n.s.g.	232.0(4.21)	39-0578-66C
$C_6H_7ClN_4$			
Acetaldehyde, (6-chloro-3-pyridazinyl)hydrazone	EtOH	270(4.34)	78-2073-66
C_6H_7ClO			
2-Cyclohexen-1-one, 2-chloro-	n.s.g.	243.0(3.83)	39-0073-66B
2-Cyclohexen-1-one, 3-chloro-	n.s.g.	231.5(4.17)	39-0073-66B
$C_6H_7Cl_2N_3$			
Pyrimidine, 2,4-dichloro-5-(dimethylamino)-	MeOH	260(4.15)	87-0121-66
$C_6H_7Cl_3O_3$			
Acetoacetic acid, 4,4,4-trichloro-, ethyl ester	hexane	248(3.42)	44-3369-66
	EtOH	246(2.95)	44-3369-66
C_6H_7N			
Aniline	2N HCl	253(2.23)	28C-0362-66A
	pH 5	230(3.94),280(3.22)	28C-0885-66C
	pH 9.2	280(3.16)	28C-0362-66A
	pH 10	278(3.17)	35-0354-66
mercuric chloride complex (2:1)	EtOH	234(4.50),285(3.84)	90-0147-66
4-Picoline	EtOH-HCl	253.5(3.59)	44-0399-66
	EtOH-NaOH	256(3.26)	44-0399-66
C_6H_7NO			
2-Hexenenitrile, 4,5-epoxy-	EtOH	219(4.11)	70-0646-66
Phenol, m-amino-	2N HCl	270(3.28)	28C-0362-66A
	pH 7.3	281(3.29)	28C-0362-66A
Phenol, o-amino-	2N HCl	270(3.30)	28C-0362-66A
	pH 7.3	282(3.43)	28C-0362-66A
Phenol, p-amino-	2N HCl	272(3.16)	28C-0362-66A
	pH 8.4	297(3.33)	28C-0362-66A
Pyridine, 2-methoxy-	H_o -3.6	279(3.85)	39-0996-66B
	pH 7	212s(3.7),269(3.53)	39-0991-66B

Compound	Solvent	λ_{max}(log ε)	Ref.
2-Pyridinol, 3-methyl-	H_o -3.6	215(3.48),279(3.87)	39-0996-66B
2-Pyridinol, 4-methyl-	H_o -3.6	211(3.56),274(3.82)	39-0996-66B
2-Pyridinol, 5-methyl-	H_o -3.6	215(3.70),285(3.86)	39-0996-66B
2-Pyridinol, 6-methyl-	H_o -3.6	214(3.60),284(3.96)	39-0996-66B
2-Pyridone, 3-methyl-	pH 7	226(3.75),294(3.83)	39-0991-66B
2-Pyridone, 4-methyl-	pH 7	226(3.69),290(3.76)	39-0991-66B
2-Pyridone, 5-methyl-	pH 7	227(3.91),302(3.76)	39-0991-66B
2-Pyridone, 6-methyl-	pH 7	225(3.86),301(3.89)	39-0991-66B
Pyrrole-2-carboxaldehyde, 5-methyl-	MeOH	248(3.49),298(4.23)	88-1347-66
$C_6H_7NO_2$			
Isoxazolin-5-one, 4,4-ethylene-3-methyl-	MeOH	219(3.52)	24-2962-66
Maleimide, N-ethyl-	EtOH	217(4.10),224(4.03), 310(2.70)	44-1311-66
2,4-Pentanedione, 3-cyano-, palladium complex	EtOH	223(4.27),255s(4.19), 325(3.89)	30-0103-66D
3-Pyridinol, 2-methoxy-	H_o -1.76	229(3.60),292(3.92)	39-0996-66B
	pH 5	221(3.76),280(3.76)	39-0991-66B
	pH 12	242(3.95),296(3.83)	39-0991-66B
2-Pyridone, 6-hydroxy-1-methyl-	2N H_2SO_4	219(3.69),295(3.95)	39-0562-66B
	pH 2.6	208(4.00),230(3.73), 308(3.51)	39-0562-66B
	pH 6.8	238(3.93),320(4.10)	39-0562-66B
2-Pyridone, 6-methoxy-	2N H_2SO_4	217(3.76),295(4.05)	39-0562-66B
	pH 6.8	224(3.83),305(4.00)	39-0562-66B
	N NaOH	230(3.87),294(3.86)	39-0562-66B
$C_6H_7NO_2S_2$			
2-Thioacetic acid, S-(4-methyl-2-thiazolyl)-	acid	248(3.40),288(3.83)	39-0092-66B
	H_2O	279(3.83)	39-0092-66B
	base	280(3.84)	39-0092-66B
$C_6H_7NO_3$			
4-Isoxazolemethanol, acetate	EtOH	214(3.36)	32-0443-66
Nicotinic acid, 1,4,5,6-tetrahydro-6-oxo-	N HCl	277(4.15)	37-1807-66
	N NaOH	268(4.15)	37-1807-66
	MeOH	275(4.17)	37-1807-66
$C_6H_7NO_3S_2$			
5-Thiazolidineacetic acid, N-methyl-4-oxo-2-thioxo-	EtOH	260(4.13),295(4.22)	95-0095-66
$C_6H_7NO_4$			
1-Cyclopentene-1-carboxylic acid, 2-nitro-	EtOH	220(3.71),265(3.62)	39-0255-66C
	EtOH-NH_3	235(3.63),291(3.72)	39-0255-66C
$C_6H_7N_3O$			
Isoniazid	EtOH	264(3.64)	39-2121-66C
Malononitrile, (aminoethoxymethylene)-	H_2O	251(4.23)	24-2302-66
	MeCN	255(4.27)	24-2302-66
	dioxan	259(4.28)	24-2302-66
4-Oxazolecarbonitrile, 5-amino-2-ethyl-	EtOH	245(4.21)	35-3829-66
5-Pyrimidinecarboxaldehyde, 4-amino-2-methyl-, cation	H_2O	249(4.05),289(3.63)	94-1321-66
	MeOH	246(4.10)	94-1321-66
	EtOH	249(4.07)	94-1321-66
	HOAc	251(4.06),287(3.63)	94-1321-66
	MeCN	249(4.08),285(3.67)	94-1321-66

Compound	Solvent	λ_{max}(log ε)	Ref.
$C_6H_7N_3O_2$			
2-Pyrazinecarboxylic acid, 3-amino-5-methyl-	pH 7.0	246(4.01),342(3.87)	39-1112-66C
2-Pyrazinecarboxylic acid, 3-amino-6-methyl-	pH 7.0	245(4.06),347(3.77)	39-1112-66C
5-Pyrimidinecarboxylic acid,	H_2O	246(4.15),280(3.65)	94-1321-66
4-amino-2-methyl-, cation	HOAc	248(4.13),280(3.62)	94-1321-66
5-Pyrimidinecarboxylic acid, 2-(methylamino)-	pH 10.6	260(4.38),303(3.51)	39-0164-66C
$C_6H_7N_3O_2S$			
Pyridine, 2-amino-4-(methylthio)-3-nitro-	pH 1	209(4.15),263(4.28), 352(4.00)	87-0354-66
	pH 7	221(4.29),243(4.17), 265s(--),354(3.76)	87-0354-66
	pH 13	224(4.22),242(4.18), 265s(--),354(3.77)	87-0354-66
Uracil, 5-(S-dithiocarbamyl)methyl-	pH 1	247(3.03),268(4.16)	87-0097-66
	pH 6.9	248(4.01),267(4.13)	87-0097-66
$C_6H_7N_3S$			
Thiopheno[2,3-d]pyrimidine, 2-amino-5,6-dihydro-	pH 1	217(4.39),240(3.89), 276(3.91),316(4.02)	24-0872-66
	MeOH	213(4.32),236(4.23), 316(4.01)	24-0872-66
$C_6H_7N_5$			
8-Azapurine, 6,9-dimethyl-	pH -1.5	265(3.79)	39-0427-66B
	pH 4.0	261(3.93)	39-0427-66B
Pteridine, 2-amino-3,4-dihydro-	pH 6	245(3.65),308(3.89)	39-1117-66C
	pH 10	282(3.89),339(3.89)	39-1117-66C
Purine, 2-amino-9-methyl-	pH 1	248(3.56),314(3.60)	87-0373-66
	pH 12	242(3.74),302(3.90)	87-0373-66
$C_6H_7N_5O$			
Isoguanine, 9-methyl-	pH 1	235(3.87),282(3.90)	95-0649-66
4-Pteridinol, 2-amino-3,4-dihydro-	pH 8.5	230(3.78),266(4.14), 328(3.91)	39-1117-66C
4-Pteridinol, 2-amino-7,8-dihydro-,	pH 1	256(4.11),314(3.75)	39-0285-66C
as bisulfite salt	pH 13	284(3.98),330(3.86)	39-0285-66C
Purine, 6-hydroxyamino-9-methyl-	N HCl	268(4.18)	69-3057-66
	pH 6.7	268(4.12)	69-3057-66
	N NaOH	253(3.95),296(3.49)	69-3057-66
s-Triazolo[2,3-a]pyrimidin-7-ol, 2-amino-5-methyl-	MeOH	225(4.45),271(3.92)	39-2031-66C
s-Triazolo[4,3-a]pyrimidin-5-ol, 3-amino-7-methyl-	MeOH	228(4.33),248s(3.76), 320(3.41)	39-2031-66C
$C_6H_7N_5O_2$			
s-Triazolo[4,3-c]pyrimidine-5,7-di-	pH 1	264(4.05)	44-0900-66
one, 8-amino-1,5,6,7-tetrahydro-6-methyl-	pH 11	281(4.08)	44-0900-66
s-Triazolo[2,3-a]pyrimidin-7-ol, 2-amino-5-methoxy-	MeOH	221(4.36),243s(3.66), 291(3.86)	39-2031-66C
$C_6H_7N_5O_2S$			
Purine, 2-amino-8-(methylsulfonyl)-	pH 0	225(4.60),326(3.74)	39-0438-66B
	pH 3.85	223(4.45),281(3.68), 322(3.90)	39-0438-66B
	pH 8.0	223(4.43),273(3.68), 318(3.98)	39-0438-66B

Compound	Solvent	λ max(log ε)	Ref.
$C_6H_7N_5O_3S$			
4-Pteridinesulfonic acid, 2-amino-3,4-dihydro-, sodium salt	pH 7.0	285(3.91),339(3.87)	39-1117-66C
$C_6H_7N_5O_4$			
Acetaldehyde, [(2-amino-1,6-dihydro-5-nitro-6-oxo-4-pyrimidinyl)-amino]-, hydrochloride	pH -0.89	226(4.26),245(3.87), 316(4.07)	24-2984-66
	pH 6.0	233(4.17),285s(3.61), 327(4.13)	24-2984-66
	pH 11.0	234(3.99),343(4.18)	24-2984-66
$C_6H_7N_5O_5$			
Pyrimidine, 2-amino-4-(dimethyl-amino)-dihydro-5-nitro-6-oxo-	pH -2.7	235(4.20),265(3.87), 322(3.68),346s(3.63)	24-2984-66
	pH 5.0	224(4.30),242s(4.17), 304s(3.75),350(3.89)	24-2984-66
	pH 11.0	223(4.17),360(4.02)	24-2984-66
Pyrimidine, 2-amino-dihydro-4-(methyl-amino)-1-methyl-5-nitro-6-oxo-	pH -2.7	228(4.37),250s(3.72), 324(4.06)	24-2984-66
	pH 6.0	284s(3.63),335(4.16)	24-2984-66
Uracil, 6-(2-formylhydrazino)-3-methyl-5-nitro-	pH 1	227(4.34),318(4.24)	44-0900-66
	pH 11	230(4.31),260(4.33), 320(4.17)	44-0900-66
$C_6H_7N_5S$			
8-Azapurine, 6-methyl-2-(methylthio)-	pH -2.7	245(4.20),269(4.06), 318(3.47)	39-0427-66B
	pH 2.7	234(4.23),252s(3.88), 308(3.87)	39-0427-66B
	pH 8.0	240(4.27),307(3.81)	39-0427-66B
Purine, 2-amino-8-(methylthio)-	pH 2.0	222(4.25),243(4.26), 288(3.87),326(3.98)	39-0438-66B
	pH 6.5	224(4.25),319(4.12)	39-0438-66B
	pH 11.0	231(4.31),277s(3.60), 317(4.15)	39-0438-66B
C_6H_8			
3-Hexen-1-yne	EtOH	223(4.11)	39-1976-66C
2-Penten-4-yne, 2-methyl-	EtOH	220(4.00)	39-1615-66C
3-Penten-1-yne, 3-methyl-	EtOH	222(3.95)	39-1976-66C
$C_6H_8BrN_3$			
Pyrimidine, 2-amino-5-bromo-4,6-dimethyl-	EtOH	304(3.43)	78-2401-66
Pyrimidine, 4-amino-5-bromo-2,6-dimethyl-	EtOH	279(3.51)	78-2401-66
$C_6H_8BrN_3O$			
Pyrimidine, 2-amino-5-bromo-4-methoxy-	EtOH	289(3.56)	78-2401-66
$C_6H_8Br_2N_2$			
1H-Pyrazolo[1,2-a]pyrazol-4-ium bromide, 2-bromo-2,3-dihydro-	n.s.g.	224(3.65)	35-5588-66
$C_6H_8ClNO_5$			
1-Methoxypyridinium perchlorate	H_2O	217(3.56),258(3.72)	39-0870-66B

Compound	Solvent	$\lambda_{max}(\log \epsilon)$	Ref.
$C_6H_8ClN_3$			
Pyrimidine, 2-amino-4-chloro-5,6-dimethyl-	EtOH	230(4.16),300(3.67)	95-0952-66
Pyrimidine, 2-amino-5-chloro-4,6-dimethyl-	EtOH	304(3.27)	78-2401-66
Pyrimidine, 4-amino-2-chloro-5,6-dimethyl-	EtOH	237(3.92),274(3.81)	95-0952-66
Pyrimidine, 4-amino-5-chloro-2,6-dimethyl-	EtOH	278(3.89)	78-2401-66
$C_6H_8ClN_3O$			
Pyrimidine, 2-amino-5-chloro-4-methoxy-6-methyl-	EtOH	289(3.76)	78-2401-66
$C_6H_8Cl_2O$			
3-Pentanone, 2-(dichloromethylene)-	EtOH	244(3.54)	77-0567-66
$C_6H_8IN_3$			
Pyrimidine, 2-amino-5-iodo-4,6-dimethyl-	EtOH	305(3.22)	78-2401-66
$C_6H_8N_2O$			
5-Pyrazolone, 4,4-ethylene-3-methyl-	MeOH	252(3.69)	24-2962-66
Pyrimidine, 2-ethoxy-	pH 7.0	266(3.62)	12-1487-66
Pyrimidine, 2-methoxy-4-methyl-	pH 7	264(3.73)	12-2321-66
2-Pyrimidinol, 4,6-dimethyl-	H_2O	296(3.02)	78-2401-66
$C_6H_8N_2OS$			
2(1H)-Pyrazinethione, 3,6-dimethyl-, 4-oxide	pH 3	224(4.00),270(4.36),405(3.91)	39-0495-66C
	pH 9	214(4.12),255(4.31),365(3.83)	39-0495-66C
4-Pyrimidinol, 2-(ethylthio)-	EtOH	230(3.88),281(3.82)	78-2401-66
4-Pyrimidinol, 6-methyl-2-(methylthio)-	EtOH	232(3.63),284(3.47)	78-2401-66
$C_6H_8N_2OS_2$			
2-Thioacetamide, S-(1,3-thiazin-2-yl)-	H_2O	263(3.82),305(3.47)	39-0092-66B
	base	263(3.82),300(3.51)	39-0092-66B
2-Thioacetamide, S-(4-methyl-2-thiazolyl)-	acid	285(3.76)	39-0092-66B
	H_2O	278(3.75)	39-0092-66B
	base	278(3.76)	39-0092-66B
$C_6H_8N_2O_2$			
4-Hydroxy-1,3-dimethyl-6-oxopyrimidinium hydroxide, inner salt	pH 5.84	253.5(3.73)	30-0144-66*
	7.4N HCl	244.5(3.70)	30-0144-66*
Muconamide, cis-cis	EtOH	259(1.43)	39-0385-66C
Muconamide, cis-trans	EtOH	258(1.44)	39-0385-66C
2-Picoline, 4-hydroxyamino-, 1-oxide	EtOH	285(4.32)	88-4729-66
3-Picoline, 4-hydroxyamino-, 1-oxide	EtOH	289(4.24)	88-4729-66
3(5)-Pyrazolecarboxylic acid, ethyl ester	EtOH	218(3.95)	22-3744-66
3-Pyrazolecarboxylic acid, 1,5-dimethyl-	EtOH	223(3.93)	22-0293-66
5-Pyrazolecarboxylic acid, 1,3-dimethyl-	EtOH	225(4.03),247s(3.73)	22-0293-66
Pyridazine, 3,6-dimethoxy-	EtOH	288(3.26)	95-1099-66
3,6-Pyridazinedione, 1,2-dihydro-1,2-dimethyl-	EtOH	336(3.50)	95-1099-66
	n.s.g.	210(4.20),340(3.42)	44-1311-66

Compound	Solvent	$\lambda_{max}(\log \epsilon)$	Ref.
$C_6H_8N_4O_2$			
2-Pyrimidinol, 4-amino-5-formamido-6-methyl-	pH 1.6	211(4.10),281(4.05)	39-0438-66B
	pH 7.5	211s(4.23),272(3.84)	39-0438-66B
	pH 13.0	223s(4.04),283(3.91)	39-0438-66B
$C_6H_8N_4O_2S$			
Uracil, 5-(S-thioureido)methyl-	pH 1	265s(3.97)	87-0097-66
	pH 6.9	261(3.98)	87-0097-66
$C_6H_8N_4O_3$			
4-Pyridazineacetic acid, 1,6-dihydro-3-hydroxy-6-oxo-, hydrazide	DMF	320(3.46)	49-1494-66
Uracil, 5-acetamido-6-amino-	H_2O	265(4.29)	95-0854-66
$C_6H_8N_4O_4$			
3,3'-Azobis(2-oxazolidinone), cis	MeCN	217(3.78),290(3.58)	35-1079-66
3,3'-Azobis(2-oxazolidinone), trans	MeCN	252s(4.00),272(4.25)	35-1079-66
4-Pyrimidinone, 2-amino-dihydro-6-methoxy-5-nitro-1-methyl-	pH 7.0	226(4.07),260s(3.50),328(3.94)	24-2997-66
Uracil, 4-(methylamino)-1-methyl-5-nitro-	pH 3.0	229(4.33),324(4.12)	24-2984-66
	pH 8.0	336(4.21)	24-2984-66
Uracil, 4-(methylamino)-3-methyl-5-nitro-	pH 6.0	237(4.21),260s(3.84),331(3.84)	24-2984-66
	pH 10.0	237(4.06),340(3.95)	24-2984-66
	pH 14.0	222(4.11),335(4.05)	24-2984-66
$C_6H_8N_6$			
s-Triazolo[4,3-b]pyridazine, 6-hydrazino-3-methyl-	EtOH	228(4.32),286(3.68)	78-2073-66
C_6H_8O			
2-Cyclohexen-1-one	C_6H_{12}	220.5(3.84)	39-0164-66B
	EtOH	225(4.06)	39-0164-66B
	n.s.g.	225.0(4.14)	39-0073-66B
2-Cyclopenten-1-one, 2-methyl-	n.s.g.	226.0(3.93)	39-0073-66B
2-Cyclopenten-1-one, 3-methyl-	n.s.g.	226.0(4.14)	39-0073-66B
$C_6H_8O_2$			
1,3-Cyclopentanedione, 4-methyl-	MeOH-acid	243(4.06)	20-0349-66
	MeOH-base	260(4.29)	20-0349-66
$C_6H_8O_2S$			
Thiophene, 2,3-bis(hydroxymethyl)-	EtOH	237(2.92),241(3.69),246(3.63),251s(3.43),260(2.91),264(2.86),267(2.81),269(2.94)	44-3592-66
$C_6H_8O_3$			
1,3-Cyclopentanedione, 4-hydroxy-5-methyl-	acid	252(4.29)	39-2308-66C
	base	272(4.46)	39-2308-66C
2-Cyclopenten-1-one, 2,3-dihydroxy-4-methyl-	H_2O	273(4.31)	5-0062-66G
2(5H)-Furanone, 5-ethoxy-	MeOH	208(3.48)	5-0042-66G
3(2H)-Furanone, 3-hydroxy-2,5-dimethyl-	pH 1	286(3.83)	33-0053-66
	H_2O	286(3.84)	33-0053-66
	MeOH	289(3.85)	33-0053-66
	ether	288(3.83)	33-0053-66

Compound	Solvent	λ_{max}(log ε)	Ref.
3(2H)-Pyridazinone, 6-methoxy-2-methyl-	EtOH	312(3.41)	95-1099-66
Pyrimidine, 4,6-dimethoxy-	20N H_2SO_4	247(3.89)	39-0565-66B
	pH 7	240(4.38)	39-0565-66B
4(1H)-Pyrimidinone, 6-methoxy-1-methyl-	20N H_2SO_4	252(3.31),278(3.09)	39-0565-66B
	pH 7	253(3.34),285(3.06)	39-0565-66B
4(3H)-Pyrimidinone, 6-methoxy-3-methyl-	20N H_2SO_4	241(3.81)	39-0565-66B
	7.4N HCl	240(3.75)	30-0635-66A
	pH 7	259(3.60)	39-0565-66B
	H_2O	259(3.47)	30-0635-66A
	EtOH	269(3.44)	30-0635-66A
	DMSO	273(3.42)	30-0635-66A
Thymine, 1-methyl-	pH 5.9	273(4.05)	24-2391-66
	pH 10.15	272(3.98)	24-2391-66
	pH 11.5-13	271(3.90)	24-2391-66
Thymine, 3-methyl-	pH 5.9	264.5(3.91)	24-2391-66
	pH 10.33	271(3.80)	24-2391-66
	pH 13	221(3.94),290(4.06)	24-2391-66
$C_6H_8N_2O_2S$			
Uracil, 5-methylthiomethyl-	pH 1	265(3.89)	87-0097-66
	pH 6.9	265(3.90)	87-0097-66
	pH 12	283(3.88)	87-0097-66
$C_6H_8N_2O_2S_2$			
5-Thiazolidineacetamide, 3-methyl-4-oxo-2-thioxo-	EtOH	260(4.15),295(4.24)	95-0101-66
$C_6H_8N_2O_3$			
Acetoacetic acid, α-diazo-, ethyl ester	CH_2Cl_2	257(3.90)	24-3128-66
6(1H)-Pyridazinone, 1,3-dimethoxy-	EtOH	313(3.458)	95-0082-66
$C_6H_8N_2S$			
2(1H)-Pyrazinethione, 3,6-dimethyl-	pH 3	270(4.04),385(3.94)	39-0495-66C
	pH 9	235(3.90),263(4.01),345(3.85)	39-0495-66C
Pyrimidine, 2-(ethylthio)-	pH -3.0	215(3.61),258(4.15),316(3.59)	12-1487-66
	pH 7.0	251(4.14),285s(3.30)	12-1487-66
Pyrimidine, 4-(ethylthio)-	pH 0.0	219(3.73),303(4.27)	12-1487-66
	pH 7	257(3.89),280(3.96)	12-1487-66
Pyrimidine, 4-methyl-2-(methylthio)-	pH 0	215(3.8),253(4.15),304(3.66)	12-2321-66
	pH 7	210(3.6),250(4.14),280s(3.40)	12-2321-66
$C_6H_8N_4O$			
Pyrimidine-5-carboxaldehyde, 4-amino-2,3-dihydro-2-imino-3-methyl-	pH 6.0	267(4.05),303(4.28)	39-0164-66C
	pH 10.6	243(3.95),314(4.10)	39-0164-66C
Pyrimidine-5-carboxamide, 4-amino-2-methyl-, cation	H_2O	247(4.13),272s(3.62)	94-1321-66
	MeOH	247(4.14),272s(3.63)	94-1321-66
	EtOH	248(4.15),272s(3.63)	94-1321-66
	iso-PrOH	249(4.14),274s(3.62)	94-1321-66
	HOAc	249(4.12),278(3.62)	94-1321-66
	MeCN	246(4.13),273s(3.64)	94-1321-66
Pyrimidine-5-carboxamide, 1,2-dihydro-2-imino-1-methyl-	pH 4.9	244(4.28),303(3.66)	39-0164-66C
	pH 11.0	266(4.36),315(3.09)	39-0164-66C
Pyrimidine-5-carboxamide, 2-(methylamino)-	pH -0.5	253(4.33),317(3.56)	39-0164-66C
	pH 5.0	266(4.39),305s(3.62)	39-0164-66C

Compound	Solvent	$\lambda_{max}(\log \epsilon)$	Ref.
$C_6H_8O_4$			
1,3-Cyclopentanedione, 4,5-di-hydroxy-5-methyl-	acid	247.5(4.17)	39-2308-66C
	base	264(4.37)	39-2308-66C
$C_6H_8O_6$			
Ascorbic acid	neutral	243(4.0)	95-0376-66
anion	n.s.g.	265(4.2)	95-0376-66
C_6H_9Br			
1,2-Hexadiene, 1-bromo-	EtOH	205(3.90),217s(3.76)	39-1223-66C
3,5-Hexadiene, 1-bromo-	C_6H_{12}	226.5(4.41)	35-1732-66
1,2-Pentadiene, 1-bromo-3-methyl-	EtOH	201(3.85),220s(3.79)	39-1223-66C
$C_6H_9BrN_2$			
Pyrazole, 4-bromo-1,3,5-trimethyl-	EtOH	229(3.79)	22-3744-66
C_6H_9BrO			
Cyclohexanone, 2-bromo-	C_6H_{12}	312(2.00)	23-0759-66
$C_6H_9BrO_2$			
2-Butenoic acid, 2-bromo-3-methyl-, methyl ester	EtOH	259.0(3.72)	39-0075-66B
C_6H_9Cl			
1,2-Hexadiene, 1-chloro-	EtOH	204(3.52)	39-1223-66C
$C_6H_9ClIN_3$			
3,4-Diamino-2-chloro-1-methylpyri-dinium iodide	pH -2.2	220(4.20),226s(4.09), 270(4.34)	39-0285-66B
	pH 5.0	220s(4.21),236(4.41), 295(3.93)	39-0285-66B
$C_6H_9ClN_4$			
Pyrimidine, 5-amino-2-chloro-6-(methylamino)-4-methyl-	pH 1.0	217(4.21),295(4.08)	39-0226-66C
	pH 7.0	264(3.91),288(3.93)	39-0226-66C
C_6H_9ClO			
Cyclohexanone, 2-chloro-	C_6H_{12}	304(1.54)	23-0759-66
C_6H_9I			
1,2-Hexadiene, 1-iodo-	EtOH	207(4.08),239(3.60)	39-1223-66C
C_6H_9NO			
1-Pyrroline, 5,5-dimethyl-, 1-oxide	H_2O	226(3.88)	39-0412-66C
	acid	226(3.88)	39-0412-66C
	base	226(3.88)	39-0412-66C
	EtOH	234(3.88)	39-0412-66C
C_6H_9NOS			
Thioglutarimide, N-methyl-	EtOH	279(4.32),417(1.48)	1-0689-66
	heptane + 0.025% CH_2Cl_2	278(4.31)	1-0689-66
	10% CH_2Cl_2	424(1.51)	1-0689-66
$C_6H_9NOS_2$			
1,3-Thiazine-2-thione, 2,3,5,6-tetra-hydro-N-ethyl-4-oxo-	EtOH	268(4.16),313(4.11)	95-0095-66
$C_6H_9NO_2$			
β-Aziridineacrylic acid, methyl ester	ether	250(4.15)	24-0450-66

Compound	Solvent	$\lambda_{max}(\log \epsilon)$	Ref.
$C_6H_9NO_3$			
2-Isoxazolin-5-one, 4-(2-hydroxyethyl)-3-methyl-	MeOH	257.5(4.00)	24-2962-66
$C_6H_9NO_5$			
Cyclopentanecarboxylic acid, 1-hydroxy-2-nitro-	EtOH	274(1.62)	39-0255-66C
$C_6H_9NS_2$			
Dithioglutarimide, N-methyl-	EtOH	235(3.62),334(4.52), 421(2.40),435(2.41)	1-0689-66
	heptane + $0.05\%CH_2Cl_2$	234(3.65),333(4.54)	1-0689-66
	$2\%CH_2Cl_2$	423(2.40),443(4.54)	1-0689-66
	$10\%CH_2Cl_2$	500s(1.60)	1-0689-66
$C_6H_9N_3$			
Pyridine, 1-amino-1,2-dihydro-2-imino-3-methyl-	MeOH	233(4.00),314(3.78)	44-0260-66
Pyridine, 1-amino-1,2-dihydro-2-imino-4-methyl-	MeOH	233(3.78),301(3.69)	44-0260-66
Pyridine, 1-amino-1,2-dihydro-2-imino-5-methyl-	MeOH	233(3.82),313(3.50)	44-0260-66
Pyridine, 1-amino-1,2-dihydro-2-imino-6-methyl-	MeOH	233(3.95),311(3.91)	44-0260-66
Pyrimidine, 2-amino-4,6-dimethyl-	EtOH	290(3.27)	78-2401-66
Pyrimidine, 4-amino-2,6-dimethyl-	pH 5.0	248(4.06)	39-0226-66C
	pH 9.5	233(4.01),266(3.68)	39-0226-66C
	EtOH	270(3.42)	78-2401-66
s-Triazine, 2-ethyl-4-methyl-	C_6H_{12}	255(2.88)	54-1101-66
$C_6H_9N_3O$			
4-Amino-3-hydroxy-1,6-dimethylpyridazinium hydroxide, inner salt	MeOH	230(4.19),307(4.03)	94-1090-66
Pyridazine, 4-amino-3-methoxy-6-methyl-	EtOH	251(3.90),278(3.86)	94-1065-66
3(2H)-Pyridazinone, 4-amino-2,6-dimethyl-	MeOH	294.5(4.01)	94-1090-66
	EtOH	294.5(4.01)	94-1065-66
Pyrimidine, 2-amino-4-methoxy-6-methyl-	EtOH	276(3.56)	78-2401-66
Pyrimidine, 2-(2-hydroxyethylamino)-	pH 0.5	229(4.23),315(3.52)	39-0226-66C
	pH 7.0	234(4.23),305(3.42)	39-0226-66C
2(1H)-Pyrimidinone, 4-(methylamino)-1-methyl-	pH 1	216(4.03),284(4.12)	65-2098-66
	pH 13	273(3.97)	65-2098-66
	EtOH	272(4.19)	65-2098-66
$C_6H_9N_3OS$			
4-Pyrimidinol, 2-amino-5-(2-mercaptoethyl)-	pH 13.5	233(4.02),279(3.83)	24-0872-66
	DMSO	296(4.02)	24-0872-66
$C_6H_9N_3O_2$			
4-Amino-3-hydroxy-6-methoxy-1-methylpyridazinium hydroxide, inner salt	MeOH	231(4.30),276(4.75), 304(3.98)	94-1090-66
Crotonic acid, β-azido-, ethyl ester	EtOH	266(4.24)	44-3907-66
Histidine, as HI salt	MeOH	none	10-0332-66D
Pyrazole, 1,3,5-trimethyl-4-nitro-	H_2O	290(3.97)	54-1191-66
3(2H)-Pyridazinone, 4-amino-6-methoxy-2-methyl-	MeOH	289.5(3.89)	94-1090-66
Uracil, 5-(dimethylamino)-	pH 1	259(3.79)	87-0121-66
	pH 11	230(3.79),291(3.60)	87-0121-66

Compound	Solvent	$\lambda_{max}(\log \epsilon)$	Ref.
$C_6H_9N_5O$			
2-Pyrazinecarboxylic acid, 3-amino-5-methyl-, hydrazide	pH -2	247(4.10),360(4.04)	39-1112-66C
	pH 2.1	251(4.02),353(3.94)	39-1112-66C
	pH 6.7	251(4.06),350(3.93)	39-1112-66C
2-Pyrazinecarboxylic acid, 3-amino-6-methyl-, hydrazide	pH -2	247(4.11),368(3.87)	39-1112-66C
	pH 2.0	250(4.06),361(3.81)	39-1112-66C
	pH 6.9	250(4.09),360(3.80)	39-1112-66C
$C_6H_9N_5O_2$			
Pyrimidine, 2-amino-4-(dimethyl-amino)-5-nitro-	MeOH	270(4.08),363(3.83)	87-0121-66
Pyrimidine, 4-amino-2-(dimethyl-amino)-5-nitro-	MeOH	272(3.65),359(4.33)	87-0121-66
$C_6H_9N_5O_3$			
Pyrimidine, 2-aminodihydro-4-(methyl-amino)-1-methyl-5-nitro-6-oxo-	pH -2.7	228(4.37),250s(3.72), 324(4.06)	24-3022-66
	pH 6	284s(3.63),335(4.16)	24-3022-66
$C_6H_{10}CuN_3O_4$			
Diglycylglycine, copper complex	H_2O	660(1.88)	37-1439-66
$C_6H_{10}IN_3$			
3,4-Diamino-1-methylpyridinium iodide	pH -2.0	267(4.24)	39-0285-66B
	pH 5.0	228(4.30),294(3.97)	39-0285-66B
$C_6H_{10}N_2$			
Pyrazole, 3,4,5-trimethyl-	EtOH	222(3.73)	22-2845-66
	EtOH	222(3.73)	22-2971-66
$C_6H_{10}N_2O$			
Nicotinamide, 1,4,5,6-tetrahydro-	EtOH	287(4.29)	44-2487-66
$C_6H_{10}N_2OS_2$			
2-Thioacetamide, S-(4,5-dihydro-2-thiazinyl)-	acid	238(3.69),260s(--)	39-0092-66B
	H_2O	225(3.76),237(3.73)	39-0092-66B
$C_6H_{10}N_2O_2$			
2-Pyrazolin-5-one, 4-(2-hydroxy-ethyl)-3-methyl-	MeOH	248.5(3.65)	24-2962-66
4,6(1H,5H)-Pyrimidinedione, dihydro-1,3-dimethyl-	7.4N HCl	244.5(3.70)	30-0635-66A
	pH 5.84	253.5(3.73)	30-0635-66A
$C_6H_{10}N_2O_2S_2$			
2-Thioacetamide, S-(4,5-dihydro-4-hydroxythiazin-2-yl)-	acid	233(3.59),265s(--)	39-0092-66B
	H_2O	225(3.78),236(3.77)	39-0092-66B
$C_6H_{10}N_2O_3$			
Urea, 3-(methoxycrotonoyl)-	EtOH	253(4.21)	70-1171-66
isomer	EtOH	246.5(4.32)	70-1171-66
$C_6H_{10}N_2O_4$			
Formic acid, azodi-, diethyl ester	dioxan	405(1.67)	89-0372-66
$C_6H_{10}N_4$			
1,2,4,5-Tetrazine, 3,6-diethyl-	EtOH	273(3.50),540(2.70)	88-5067-66
$C_6H_{10}N_4O$			
1,2,4-Triazin-5-one, 2,5-dihydro-3-(dimethylamino)-2-methyl-	pH -1.1	221(4.19),270(3.57)	73-1864-66

Compound	Solvent	λ_{max}(log ε)	Ref.
1,2,4-Triazin-5-one, 2,5-dihydro-3-(dimethylamino)-2-methyl- (cont.)	pH 4.0	223(4.28),247s(3.84),283s(2.92)	73-1864-66
	pH 11.4	232(4.15),303(3.56)	73-1864-66
	EtOH	220(4.27),247s(3.80),285(2.66)	73-1864-66
1,2,4-Triazin-5-one, 2,5-dihydro-3-(methylamino)-2,6-dimethyl-	pH -1.1	221(4.12),274(3.74)	73-1864-66
	pH 7.0	218s(4.19),245s(3.86)	73-1864-66
	EtOH	250(3.85)	73-1864-66
1,2,4-Triazin-5-one, 2,3,4,5-tetrahydro-3-imino-2,4,6-trimethyl-	pH 0.2	211(3.91),268(3.71)	73-1864-66
	pH 7.0	217(4.05),262(3.57),307s(3.2)	73-1864-66
	EtOH	216(3.88),266(3.58),308s(2.4)	73-1864-66
$C_6H_{10}N_4O_4$			
Uracil, 4-(dimethylamino)-5-nitro-	pH 2.0	238(4.23),258s(3.99),347(3.68)	24-2984-66
	pH 8.0	223(4.32),303s(3.70),350(3.97)	24-2984-66
	pH 14.0	270(3.60),352(4.13)	24-2984-66
$C_6H_{10}O$			
Cyclohexanone	C_6H_{12}	293(1.18)	23-0759-66
3-Penten-2-one, 4-methyl-	C_6H_{12}	232(4.08),328(1.62)	88-5081-66
	isooctane	232(4.09),329(1.61)	99-0469-66
	H_2O	244(4.06),294s(2.06)	99-0469-66
	EtOH	237(4.08),313(1.78)	88-5081-66
	EtOH	238(4.07),316(1.79)	99-0469-66
4-Penten-2-one, 4-methyl-	isooctane	283(1.86)	88-3671-66
$C_6H_{10}OS_2$			
m-Oxathiane-2-thione, 5,5-dimethyl-	MeOH	225(3.72),295(4.08)	50-0965-66C
$C_6H_{10}O_2$			
2,4-Pentanedione, 3-methyl-, palladium derivative	EtOH	211(4.81),227(4.37),251s(4.15),344(3.83)	30-0103-66D
$C_6H_{10}O_2S$			
1,3-Dioxolane-2-thione, 4,4-dimethyl-	MeOH	244(4.16)	50-0965-66C
$C_6H_{10}O_2S_2$			
Propionic acid, β-(1,2-dithiolan-4-yl)-	$CHCl_3$	333(2.16)	11-0263-66
$C_6H_{10}O_3$			
Crotonic acid, 2-ethoxy-	MeOH	219(3.98)	5-0053-66I
Formic acid, 2,3-epoxy-2-pentyl ester	n.s.g.	286(1.44)	33-1002-66
$C_6H_{10}S$			
2H-Thiopyran, 3,4-dihydro-6-methyl-	MeCN	228(3.60),247(3.37)	44-2333-66
$C_6H_{10}S_2$			
5H-Dithiepin, 6,7-dihydro-2-methyl-	EtOH	285(3.68)	44-0586-66
$C_6H_{11}^+$			
Cyclopropyldimethylcarbonium ion	FSO_3H-SbF_5	289(4.03)	35-1488-66

Compound	Solvent	$\lambda_{max}(\log \epsilon)$	Ref.
$C_6H_{11}BrN_2S$			
1,2,3,5,6,7-Hexahydroimidazo[2,1-b]-[1,3]thiazinium bromide, mercuric bromide complex	EtOH	227(4.18)	39-1226-66C
$C_6H_{11}NO$			
2-Piperidone, 1-methyl-	MeOH	204(3.86)	24-0724-66
$C_6H_{11}NOS_2$			
1,2-Dithiolane, 4-(acetamidomethyl)-	EtOH	205(3.59),331(2.11)	11-0135-66
$C_6H_{11}NO_2$			
2-Aziridinecarboxylic acid, isopropyl ester	C_6H_{12}	188.5(3.68)	33-0359-66
$C_6H_{11}NO_3$			
Pentanoic acid, 4-methyl-4-nitroso-	EtOH	295(3.67)	39-0412-66C
$C_6H_{11}NS$			
Thiopiperidone, N-methyl-	heptane	276(4.14),362(1.51)	1-0689-66
	EtOH	272(4.23),320s(1.87)	1-0689-66
$C_6H_{11}N_3O$			
2-Pyrazolin-5-one, 4-(3-aminopropyl)-	EtOH	237(3.76)	44-2487-66
$C_6H_{11}N_3O_2$			
1,2,4-Triazoline-3,5-dione, 4-butyl-	C_6H_{12}	544(2.32)	44-3444-66
$C_6H_{11}N_4O_3PS_2$			
O,O-Dimethyl-S-(4-amino-6-hydroxy-1,3,5-triazin-2-yl) methyl phosphorodithioate	pH 2	250(3.88)	39-0056-66C
	pH 10	256(3.61)	39-0056-66C
$C_6H_{11}N_5$			
Pyrimidine, 5-amino-2,4-bis-(methylamino)-	pH 11	238(4.10),332(3.83)	87-0121-66
	MeOH-acid	238(4.36),304(3.83)	87-0121-66
Pyrimidine, 2,4-diamino-5-(dimethylamino)-	pH 1	224(4.20),280s(3.52)	87-0121-66
	pH 11	240(3.93),298(3.77)	87-0121-66
Pyrimidine, 2,5-diamino-4-(dimethylamino)-	MeOH-acid	247(4.13),316(3.75)	87-0121-66
Pyrimidine, 4,5-diamino-2-(dimethylamino)-	MeOH-acid	240(4.15),315s(3.57)	87-0121-66
1,3,5-Triazine, 2,4-diamino-6-isopropyl-	MeOH	255(3.55)	78-0157-66
C_6H_{12}			
2-Butene, 2,3-dimethyl-	gas	178f(4.0),218f(3.0)	38-1803-66A
$C_6H_{12}ClNO$			
Acetamide, N-tert-butyl-N-chloro-	isooctane	256(2.61)	35-3051-66
$C_6H_{12}N_2$			
1-Pyrazoline, 3,3,5-trimethyl-	MeOH	325(2.57)	35-3959-66
$C_6H_{12}N_2O_4$			
Carbamic acid, butylnitro-, methyl ester	hexane	233(3.73),290(2.53)	44-3038-66
	EtOH	236(3.71),290(2.53)	44-3038-66
Carbamic acid, sec-butylnitro-, methyl ester	hexane	236(3.58),290(2.48)	44-3038-66
	EtOH	239(3.54),290(2.53)	44-3038-66

Compound	Solvent	$\lambda_{max}(\log \epsilon)$	Ref.
Carbamic acid, tert-butylnitro-, methyl ester	hexane	238s(3.20),330(1.82)	44-3038-66
	EtOH	235(3.30),330(1.70)	44-3038-66
$C_6H_{12}N_2S$			
Thiourea, N-tert-butyl-N'-methyl-, iodine complex	$CHCl_3$	306.5(4.62)	60-0029-66
$C_6H_{12}N_3O_6PS$			
O,O-Dimethyl-S-(4,6-dihydroxy-1,3,5-triazin-2-yl) methyl phosphorothiolate, hydrate	pH 10	255(3.91)	39-0056-66C
$C_6H_{12}N_4$			
1,2,4,5-Tetrazine, 3,6-diethyl-1,6-dihydro-	EtOH	310(3.48),426(2.61)	88-5067-66
$C_6H_{12}N_4O_2$			
Formamide, 1,1'-azobis[N,N-dimethyl-	MeOH	293(3.38)	24-2039-66
	dioxan	260s(3.20),435(1.59)	24-2039-66
	dioxan	260s(3.20),435(1.64)	89-0372-66
$C_6H_{12}O_2S_2Si$			
Silane, di(thionoacetoxy)dimethyl-	hexane	222(3.60),243(3.64), 389(1.34)	44-3439-66
$C_6H_{13}I$			
Pentane, 2-iodo-2-methyl-	C_6H_{12}	268(2.79)	3-0612-66
$C_6H_{13}NO_2$			
Hexyl nitrite	n.s.g.	356f(2.0)	93-2413-66
$C_6H_{13}N_4O_5PS$			
O,O-Dimethyl S-(4-amino-6-hydroxy-1,3,5-triazin-2-yl) methyl phosphorothiolate	pH 5	250(3.60)	39-0056-66C
	pH 10	256(3.56)	39-0056-66C
$C_6H_{13}O_4P$			
Phosphonic acid, acetyl-, diethyl ester	C_6H_{12}	334(1.72)	44-1304-66
$C_6H_{14}OS$			
Isopropyl sulfoxide	EtOH	202(3.30),218(2.90)	39-0239-66A
Propyl sulfoxide	EtOH	201(3.28),214(3.08)	39-0239-66A
$C_6H_{14}OSi$			
2-Propanone, 1-trimethylsilyl-	C_6H_{12}	283(1.91)	44-4237-66
	EtOH	276(2.06)	44-4237-66
$C_6H_{14}S$			
Isopropyl sulfide	EtOH	203(3.32),209(3.34)	39-0239-66A
Propyl sulfide	EtOH	202(3.32)	39-0239-66A
$C_6H_{14}S_2$			
Isopropyl disulfide	EtOH	201(3.46),245(2.62)	39-0239-66A
Propyl disulfide	EtOH	201(3.40),252(2.60)	39-0239-66A
$C_6H_{16}NOP$			
2-Picolylphosphine oxide	N H_2SO_4	218(4.28),262(4.05)	39-0631-66B
	pH 7.00	217(4.69),262(3.82)	39-0631-66B

Compound	Solvent	$\lambda_{max}(\log \epsilon)$	Ref.
3-Picolylphosphine oxide	N H_2SO_4	222(4.34),264(3.87)	39-0631-66B
	pH 7.00	218(4.18),264(3.57)	39-0631-66B
4-Picolylphosphine oxide	N H_2SO_4	222(4.35),256(3.88)	39-0631-66B
	pH 7.00	224(4.71),259(3.54)	39-0631-66B
$C_6H_{18}Cl_2Si_3$			
Trisilane, dichlorohexamethyl-	C_6H_{12}	219.0(3.94)	101-0392-66A
$C_6H_{18}Ge_2S$			
Digermthiane, hexamethyl-	hexane	213.0(3.72)	39-0242-66A
$C_6H_{18}SSi_2$			
Disilthiane, hexamethyl-	hexane	202.5(3.46)	39-0242-66A
$C_6H_{19}ClSi_3$			
Trisilane, chlorohexamethyl-	C_6H_{12}	217.5(3.85)	101-0392-66A
$C_6H_{20}Si_3$			
Trisilane, hexamethyl-	C_6H_{12}	218.0(3.83)	101-0392-66A

Compound	Solvent	λ_{max}(log ε)	Ref.
$C_7F_5O^+$			
Pentafluorophenyloxocarbonium ion	H_2SO_4-SO_3	258(4.15),309(3.35)	35-1488-66
$C_7H_2Br_3O^+$			
2,4,6-Tribromophenyloxocarbonium ion	H_2SO_4-SO_3	315(4.46),371(3.89)	35-1488-66
$C_7H_2Cl_3NO$			
Anthranil, 5,6,7-trichloro-	EtOH	266(3.11),277(3.32), 288(3.46),301(3.57), 325(3.73)	4-0422-66
$C_7H_3ClN_2O_3$			
Anthranil, 5-chloro-6-nitro-	EtOH	257(3.74),286(3.40), 301(3.42),325(3.49)	4-0422-66
$C_7H_3ClN_4$			
Indazole, 4-chloro-3-diazo-	n.s.g.	364(3.85)	5-0017-66G
Indazole, 5-chloro-3-diazo-	n.s.g.	354(3.85)	5-0017-66G
Indazole, 6-chloro-3-diazo-	n.s.g.	360(3.85)	5-0017-66G
$C_7H_3Cl_2NO$			
Anthranil, 5,6-dichloro-	EtOH	263(3.12),276(3.36), 285(3.48),298(3.59), 316(3.64)	4-0422-66
$C_7H_3Cl_5O$			
Cyclobutene, 1-chloro-3,4-bis(di- chloromethylene)-2-methoxy-	EtOH	233s(4.45),238(4.48), 277(4.00)	18-0160-66
$C_7H_3KN_2$			
Cyclopentadiene-1,3-dicarbonitrile, potassium derivative	H_2O	218(--),274(4.32)	35-3046-66
$C_7H_3N_7$			
Tetrazole, 2-methyl-5-tricyanovinyl-	MeCN	233(3.78),297(4.21)	44-3849-66
$C_7H_4ClN_3O_2$			
Benzimidazole, 2-chloro- 5(or 6)-nitro-	EtOH	237(4.25),260s(3.92), 305(3.89)	4-0107-66
$C_7H_4Cl_2N_2$			
Benzimidazole, 2,5(or 6)-dichloro-	EtOH	208(4.57),245(3.74), 288(3.86)	4-0107-66
$C_7H_4Cl_2O_2$			
p-Benzoquinone, 2,3-dichloro- 5-methyl-	n.s.g.	205(3.77),250(4.06), 356(2.00)	16-0054-66
Tropolone, 3,6-dichloro-	MeOH	252(4.39),336(3.98), 377(3.75),388(3.75), 397s(3.68),428(3.31)	18-1310-66
$C_7H_4Cl_4O_2$			
Catechol, tetrachloro-, methyl ether	EtOH	299(3.30)	78-0745-66
	EtOH-base	316(3.68)	78-0745-66
$C_7H_4CrO_3S$			
Chromium, tricarbonyl(thiophene)-	C_6H_{12}	226(4.49),260s(4.13), 320(3.82),410(3.83), 520s(3.00)	24-1732-66

Compound	Solvent	$\lambda_{max}(\log \epsilon)$	Ref.
$C_7H_4CrO_3Se$			
Chromium, tricarbonyl(selenophene)-	C_6H_{12}	225(4.44),260s(4.10), 375(3.86),420s(3.74), 532s(2.90)	24-1732-66
$C_7H_4FO^+$			
2-Fluorophenyloxocarbonium ion	$H_2SO_4-SO_3$	249(4.09),297(3.20)	35-1488-66
3-Fluorophenyloxocarbonium ion	$H_2SO_4-SO_3$	262(4.15),317(3.47)	35-1488-66
4-Fluorophenyloxocarbonium ion	$H_2SO_4-SO_3$	270(4.22),294(3.50)	35-1488-66
$C_7H_4F_3NO_2$			
Toluene, α,α,α-trifluoro-2-nitro-	C_6H_{12}	235s(3.68),272s(3.15), 335s(2.48)	23-0961-66
	EtOH	244s(3.57),275s(--), 335s(2.7)	23-0961-66
$C_7H_4F_3NO_3$			
Phenol, 2-nitro-3-(trifluoromethyl)-	C_6H_{12}	230s(3.66),259(3.60), 338(3.54)	23-0961-66
	H_2O	235s(--),283(3.23)	23-0961-66
	N NaOH	235(4.04),294(3.40), 407(3.04)	23-0961-66
	EtOH	286(3.40)	23-0961-66
	ether	286(3.43)	23-0961-66
	$CHCl_3$	267(3.52),343(3.41)	23-0961-66
Phenol, 4-nitro-3-(trifluoromethyl)-	C_6H_{12}	215(4.00),265(3.74)	23-0961-66
	EtOH	222(3.91),291(3.75), 375s(3.32)	23-0961-66
$C_7H_4F_3N_3O_3$			
Aniline, 2-nitro-6-nitroso-4-(trifluoromethyl)-	EtOH	231(4.38),434(3.88)	88-2307-66
	EtOH-base	278(3.97),309(4.00), 393(4.23)	88-2307-66
$C_7H_4N_2$			
Cyclopentadiene-1,2-dicarbonitrile	H_2O	290(3.59)	35-3046-66
Cyclopentadiene-1,3-dicarbonitrile	H_2O	279(4.07),286(4.06), 298(3.80)	35-3046-66
	MeCN	226(3.35),243(3.45), 278(4.07),292(3.71)	35-3046-66
$C_7H_4N_2O_3$			
Anthranil, 5-nitro-	n.s.g.	253(4.20),292(3.42), 304(3.43),343(3.56)	78-0049-66A
Anthranil, 6-nitro-	EtOH	256(4.25),345(3.38)	4-0422-66
Anthranil, 7-nitro-	n.s.g.	237s(--),364(3.88), 267s(--)	78-0049-66A
$C_7H_4N_4$			
Indazole, 3-diazo-	n.s.g.	356(3.85)	5-0017-66G
$C_7H_5ClN_2$			
Benzimidazole, 2-chloro-	EtOH	274(3.85),280(3.95)	4-0107-66
$C_7H_5ClN_2O_6S$			
6-Nitrothiazolo[3,2-a]pyridinium perchlorate	EtOH	213(3.89),329(3.08)	4-0027-66

Compound	Solvent	$\lambda_{max}(\log \varepsilon)$	Ref.
$C_7H_5ClN_4O$			
7(8H)-Pteridinone, 4-chloro-8-methyl-	pH 10.0	225(4.25),232s(4.19),242s(4.02),309(3.92),315s(3.91),330s(3.75)	39-1065-66C
C_7H_5ClOS			
Oxythiobenzoyl chloride, cis	n.s.g.	229(3.90),323(3.95)	23-0819-66
Oxythiobenzoyl chloride, trans	n.s.g.	226(3.90),328(4.15)	23-0819-66
$C_7H_5ClO_2$			
Tropone, 2-chloro-5-hydroxy-	pH 13	228(4.33),260s(3.52),381(4.37)	18-1310-66
	MeOH	220(4.27),350(4.14)	18-1310-66
$C_7H_5ClO_3$			
Tropolone, 3-chloro-6-hydroxy-	MeOH	255(4.47),340(3.81),367(3.93)	18-1310-66
$C_7H_5Cl_2NO_4S$			
6-Chlorothiazolo[3,2-a]pyridinium perchlorate	EtOH	257(4.20),303(3.65)	4-0027-66
8-Chlorothiazolo[3,2-a]pyridinium perchlorate	EtOH	249(4.07),298(3.67),318s(3.01)	4-0027-66
$C_7H_5F_3$			
Toluene, α,α,α-trifluoro-	hexane	250(--),256(--),260(--),266(--),292(1.45)	99-0201-66
	H_2SO_4	260(3.47),300(2.98)	99-0201-66
$C_7H_5F_3O$			
Phenol, 3-(trifluoromethyl)-	C_6H_{12}	214(3.73),275(3.35),282(3.31),286(3.26)	23-0961-66
	EtOH	219(3.79),278(3.35)	23-0961-66
$C_7H_5KN_2O_2$			
Phenylazo potassium carbonate	dioxan	265(4.02),418(1.99)	24-3337-66
$C_7H_5NO_4$			
Benzoic acid, p-nitro-	pH 5.3	269(4.0)	10-0129-66C
$C_7H_5NO_5$			
Acrylic acid, 3-(5-nitro-2-furyl)-	EtOH	236(4.09),273(3.83),344(4.24)	95-1187-66
$C_7H_5NS_2$			
Benzothiazoline-2-thione (also spectra in many other solvents)	hexane	231(4.15),240(4.13),250(3.92),282(3.54),298(3.77),324(4.40),329(4.41),333(4.40)	59-2005-66
$C_7H_5N_3$			
3,5-Pyridinedicarbonitrile, 1,4-dihydro-	n.s.g.	209(4.23),354(3.72)	39-1083-66B
$C_7H_5N_3O$			
Isoxazole, 4-(2-pyrazinyl)-	EtOH	234(4.01),286(3.91)	44-3845-66

Compound	Solvent	λ_{max}(log ε)	Ref.
$C_7H_5N_3O_6$			
Benzoic acid, 2-amino-3,5-dinitro-, anion	buffer	353(4.16)	35-0947-66
dianion	Me_4NOH	421(4.17)	35-0947-66
Benzoic acid, 4-amino-3,5-dinitro-, anion	buffer	436(3.95)	35-0947-66
dianion	Me_4NOH	554(3.94)	35-0947-66
$C_7H_5N_5O$			
3H-Imidazo[2,1-i]purin-8(7H)-one	pH 1	282(4.04)	69-2082-66
	pH 5.5	220(4.31),269(3.95), 303(4.16)	69-2082-66
	pH 13	306(4.19)	69-2082-66
$C_7H_5O^+$			
Phenyloxocarbonium ion	H_2SO_4-SO_3	259(4.23),308(3.57)	35-1488-66
$C_7H_6Br_2O_2$			
2-Cyclohexene-1,4-dione, 5,6-dibromo-2-methyl-	hexane	227(3.85),252(3.81)	23-2867-66
C_7H_6ClNO			
Tropone, 5-amino-2-chloro-	MeOH	230(4.28),260s(3.67), 383(4.32)	18-1310-66
$C_7H_6ClNO_4S$			
Thiazolo[3,2-a]pyridinium perchlorate	H_2O	208(4.00),224(4.08), 228s(4.02),295(4.11), 306(4.24)	4-0027-66
$C_7H_6ClN_3$			
1,3,5-Triazaindene, 6-chloro-1-methyl-	pH 0.46	211s(4.55),218s(4.24), 237(3.49),242(3.47), 252s(3.39),275(3.60), 282s(3.51)	39-0285-66B
	pH 5.0	210(4.66),252(3.60), 257(3.62),266(3.58), 272(3.59),279s(3.45)	39-0285-66B
1,3,5-Triazaindene, 6-chloro-3-methyl-	pH -4.0	211(4.59),214s(4.57), 240s(3.47),244(3.48), 283(3.69),290s(3.63)	39-0285-66B
	pH 0.5	236(3.54),241s(3.51), 250s(3.34),276(3.62), 283s(3.52),300s(2.55)	39-0285-66B
	pH 5.0	210(4.63),241(3.48), 249(3.50),255(3.51), 277s(3.58),284(3.63), 292s(3.53)	39-0285-66B
1,3,5-Triazaindene, 6-chloro-x-methyl-	pH -3.0	211(4.69),215(4.64), 241(3.55),245(3.55), 254s(3.26),282(3.69), 289s(3.64)	39-0285-66B
	pH 4.50	216(4.74),218s(4.70), 255(3.57),282(3.58)	39-0285-66B
	pH 9.0	226(4.65),266(3.67), 273s(3.62),301(3.52)	39-0285-66B
$C_7H_6ClN_3S$			
1,3,5-Triazaindene, 6-chloro-2-(methylthio)-	pH 0.0	243(4.46),293(4.12)	39-0285-66B

Compound	Solvent	$\lambda_{max}(\log \epsilon)$	Ref.
1,3,5-Triazaindene, 6-chloro-2-(methylthio)- (cont.)	pH 5.5	227(4.49),260(3.84), 286(4.08),292s(4.05)	39-0285-66B
	pH 11.0	229(4.55),291(4.08), 299s(3.96)	39-0285-66B
$C_7H_6ClN_5O$			
Adenine, N^6-chloroacetyl-	pH 1	274(4.11)	69-2082-66
	pH 7	278(4.05)	69-2082-66
	pH 13	276(3.94)	69-2082-66
$C_7H_6Cl_2F_4N_4O_2$			
s-Triazin-2-ol, 4-acetamido-6,6-bis-(chlorodifluoromethyl)-1,6-dihydro-	EtOH	235(3.70)	44-2568-66
$C_7H_6Cl_2O_2$			
2-Cyclohexene-1,4-dione, 5,6-dichloro-2-methyl-	hexane	247(4.08)	23-2867-66
$C_7H_6Cl_3NO_2$			
Pyridine, 2,3,5-trichloro-4,6-dimethoxy-	MeOH	240(3.8),283(3.7)	24-2818-66
Pyridine, 3,4,5-trichloro-2,6-dimethoxy-	MeOH	234(3.9),300(3.9)	24-2818-66
$C_7H_6Cl_4O$			
2-Oxatricyclo[5.1.0.0^{3,5}]octane, 4,4,8,8-tetrachloro-	EtOH	none	24-2351-66
$C_7H_6F_3N$			
Aniline, m-(trifluoromethyl)-	pH 10	288(3.19)	35-0354-66
Aniline, p-(trifluoromethyl)-	pH 10	275s(3.12)	35-0354-66
	EtOH	250(4.5),300s(3.6)	65-1248-66
$C_7H_6F_6N_2$			
3H-Pyrazole, 4,5-dimethyl-3,3-bis(trifluoromethyl)-	EtOH	273(3.45),358(2.51)	35-3617-66
$C_7H_6NO_7P$			
Benzoic acid, p-nitro-, anhydride with phosphoric acid	pH 5.3	264(4.1)	10-0129-66C
$C_7H_6N_2$			
Benzonitrile, m-amino-	2N HCl	271(2.93)	28C-0362-66A
	pH 9.2	308(3.43)	28C-0362-66A
Benzonitrile, o-amino-	2N HCl	271(3.06)	28C-0362-66A
	pH 9.2	315(3.57)	28C-0362-66A
Benzonitrile, p-amino-	pH 9.2	270(4.26)	28C-0362-66A
$C_7H_6N_2O$			
Nudiflorine	EtOH	206(4.27),254(4.32), 306(3.89)	78-1461-66
$C_7H_6N_2O_2$			
Malonaldehyde, 2-(2-pyrazinyl)-	EtOH	259(4.19),286s(3.92), 392(3.76)	44-3845-66
$C_7H_6N_2O_3$			
Benzofuroxan	xylene	311(3.80)	39-0489-66B
	DMF	310(3.82)	39-0489-66B

Compound	Solvent	λ_{max}(log ϵ)	Ref.
$C_7H_6N_2O_4$			
Benzoic acid, 2-amino-5-nitro-, anion	buffer	395(4.25)	35-0947-66
dianion	Me_4NOH	476(4.45)	35-0947-66
Benzoic acid, 4-amino-3-nitro-, anion	buffer	419(3.76)	35-0947-66
	Me_4NOH	520(3.92)	35-0947-66
$C_7H_6N_2O_5$			
Acrylamide, N-hydroxy-3-(5-nitro-2-furyl)-	EtOH	237(4.17),273(3.83),352(4.24)	95-1187-66
Anisole, 2,4-dinitro-	C_6H_{12}	240s(3.94),282(4.07),350s(2.66)	19-0023-66
	MeOH	215(4.13),252(3.85),291(4.04)	19-0023-66
	20% MeOH	214(4.11),258(3.95),300(4.03)	19-0023-66
$C_7H_6N_4$			
Pyrazino[2,3-d]pyridazine, 2-methyl-	EtOH	221(4.21),287(3.19),298(3.10)	4-0512-66
$C_7H_6N_4O$			
4-Pteridinol, 2-methyl-	pH 6.5	230(4.04),266(3.69),315(3.83)	39-1112-66C
	pH 11.3	243(4.28),335(3.78)	39-1112-66C
4-Pteridinol, 6-methyl-	pH 5.6	230(4.03),265(3.77),317(3.89)	39-1112-66C
	pH 10.9	244(4.28),337(3.83)	39-1112-66C
4-Pteridinol, 7-methyl-	pH 5.6	232(4.05),270(3.63),311(3.93)	39-1112-66C
	pH 10.9	244(4.24),331(3.83)	39-1112-66C
$C_7H_6N_4O_2$			
Benzotriazole, 1-methyl-7-nitro-	EtOH	519(3.71)	78-3351-66
Lumazine, 8-methyl-	pH -2.7	241(4.03),272(3.16),344(4.02)	24-3503-66
	pH 4	274(4.22),303(3.87)	24-3503-66
	pH 6	257(4.21),392(3.97)	24-3503-66
	pH 9	278(4.14),308(3.96)	24-3503-66
	pH 11.3	221(4.11),266(4.02),278(3.97),309(3.84),404(3.54)	24-3503-66
	pH 12	262(4.27),320(3.62),405(3.86)	24-3503-66
	pH 13	230(4.27),280(4.10),307(3.90),405(3.04)	24-3503-66
	pH 14.9	233(4.23),282(4.05),311(3.93)	24-3503-66
Pyrazolo[3,2-c]-as-triazine-3-carboxylic acid, 4-methyl-	EtOH	240(4.40),306(3.56),349(3.27)	39-1127-66C
$C_7H_6N_4O_3$			
Benzene, 1-azido-4-methoxy-2-nitro-	xylene	311(3.15)	39-0489-66B
	DMF	310(3.18)	39-0489-66B
C_7H_6O			
Allene, 1-(2-furyl)-	EtOH	254(4.18)	39-2283-66C
Benzaldehyde	EtOH	244(4.11),279(2.98)	49-0524-66
2-Heptene-4,6-diyn-1-ol, trans	EtOH	206(4.51),211(4.60),227(3.52),239(3.82),251(4.11),265(4.27),280(4.16)	39-2283-66C

Compound	Solvent	λ_{max}(log ε)	Ref.
1-Propyne, 1-(2-furyl)-	EtOH	248(4.20)	39-2283-66C
$C_7H_6O_2$			
Benzoic acid	pH 5.3	224(4.0),267(2.8)	10-0129-66C
$C_7H_6O_3$			
p-Benzoquinone, (hydroxymethyl)-	EtOH	247(4.38),432(1.32), 450s(--)	39-1627-66C
$C_7H_6O_3S_2$			
Thieno[3,4-b]thiophene-2-carboxylic acid, 4,6-dihydro-, 5-oxide	EtOH	250(3.97),280(3.92)	44-3363-66
$C_7H_6O_4$			
Protocatechuic acid	pH 7.0	250(3.9),290(3.6)	37-3776-66
$C_7H_6O_6$			
1,3-Butadiene-1,2,4-tricarboxylic acid	pH 7.0	260(3.9)	37-3776-66
1-Butene-1,3,4-tricarboxylic acid, 3-hydroxy-, γ-lactone	pH 8.0	230(3.61)	37-3776-66
$C_7H_6S_2$			
Benzoic acid, dithio-	C_6H_{12}	519s(1.23),537(1.85)	97-0318-66
$C_7H_7^+$			
Tropylium ion	MeCN	275(3.63)	27-0172-66
$C_7H_7BrN_4$			
Pyrazolo[3,2-c]-as-triazine, 8-bromo-3,4-dimethyl-	EtOH	235(4.48),293(3.28), 302(3.14),377(3.48)	39-1214-66C
$C_7H_7BrO_3$			
2-Furoic acid, 5-bromo-3-methyl-, methyl ester	EtOH	223s(3.38),262(4.19)	39-0976-66C
$C_7H_7ClN_2O$			
Aniline, m-chloro-N-methyl-N-nitroso-	H_2O	269.0(3.85)	39-0533-66B
Aniline, p-chloro-N-methyl-N-nitroso-	H_2O	273.0(3.90)	39-0533-66B
$C_7H_7ClN_4$			
Purine, 2-chloro-6,9-dimethyl-	pH -1.5	269(3.95),277s(3.86)	39-0226-66C
	pH 7.0	212(4.35),270(3.93)	39-0226-66C
Purine, 6-chloro-2-ethyl-	pH 1	268(3.95)	44-1417-66
	pH 7	270(3.94)	44-1417-66
	pH 13	276(3.94)	44-1417-66
s-Triazolo[2,3-a]pyrimidine, 2-chloro-5,7-dimethyl-	MeOH	214(4.30),276(3.74)	39-2031-66C
$C_7H_7ClN_4O$			
s-Triazolo[2,3-a]pyrimidine, 2-chloro-5-methoxy-7-methyl-	MeOH	216(4.22),272(3.70)	39-2031-66C
$C_7H_7Cl_3O_5$			
Succinaldehydic acid, 2-hydroxy-3-hydroxymethylene-, 2,2,2-trichloroethyl ester	EtOH-acid	247(4.28)	35-0852-66
	EtOH-base	269(4.44)	35-0852-66
C_7H_7N			
3H-Pyrrolizine	EtOH	218(3.40),290(3.79)	44-0870-66

Compound	Solvent	$\lambda_{max}(\log \epsilon)$	Ref.
C_7H_7NO			
Benzaldehyde, 2-amino-	1.6N HCl	246(4.06),284(3.20)	39-0956-66B
	H_2O	229(4.32),260(3.83), 365(3.59)	39-0956-66B
$C_7H_7NO_2$			
Phenol, 2-methyl-4-nitroso-	dioxan	295(4.30),731(0.00)	12-0841-66
	dioxan-NaOH	268(3.60),395(4.38)	12-0841-66
Phenol, 3-methyl-4-nitroso-	EtOH	300(4.27)	12-0841-66
	EtOH-NaOH	264(3.63),410(4.51)	12-0841-66
	dioxan	290(4.30),750(0.20)	12-0841-66
	dioxan-NaOH	271(3.60),395(4.40)	12-0841-66
Picolinic acid, 5-methyl-	MeOH	268(3.71)	32-0915-66
$C_7H_7NO_2S$			
Sulfide, methyl o-nitrophenyl	hexane	243(4.26),264(3.76), 360(3.56),388(3.13)	65-1577-66
	EtOH	244(4.20),270(3.58), 375(3.47)	65-1577-66
	ether	244(4.31),266(3.75), 367(3.58)	65-1577-66
	dioxan	245(4.23),266(3.75), 370(3.57)	65-1577-66
$C_7H_7NO_3$			
Anisole, o-nitro-	C_6H_{12}	251(3.52),308(3.40), 360s(2.44)	19-0023-66
	MeOH	212(4.17),230s(3.74), 258(3.55),323(3.41)	19-0023-66
	20% MeOH	230s(3.72),266(3.65), 336(3.48)	19-0023-66
Benzoic acid, 3-amino-5-hydroxy-	EtOH-HCl	230(4.63),298(4.44)	78-0171-66B
Nudifloric acid	EtOH	206(4.14),255(4.20), 300(3.72)	78-1461-66
$C_7H_7NO_3S$			
Ethylene, 1-(5-nitro-2-furyl)-2-(methylthio)-, cis	EtOH	267(4.05),385(4.20)	94-1300-66
trans	EtOH	272(4.16),404(4.31)	94-1300-66
$C_7H_7NO_6$			
3,5-Isoxazoledicarboxylic acid, 4-hydroxy-, dimethyl ester	EtOH	280(3.56),340(2.88)	95-0300-66
$C_7H_7NO_6S$			
Benzenesulfonic acid, 4-methoxy-2-nitro-, sodium salt	H_2O	224(4.3),262(3.6), 330(4.0)	65-2141-66
$C_7H_7N_3$			
Imidazo[4,5-b]pyridine, 3-methyl-	EtOH	250(3.61),278s(3.88), 282(3.91),288s(3.82)	39-0080-66C
1,3,5-Triazaindene, 1-methyl-	pH 3.9	209(4.67),255s(3.57), 263(3.63),267s(3.61)	39-0285-66B
	pH 8.5	251s(3.65),256(3.66), 263(3.64),271(3.48)	39-0285-66B
1,3,5-Triazaindene, 3-methyl-	pH 3.5	254s(3.68),285(3.78), 291s(3.72)	39-0285-66B
	pH 8.0	245s(3.58),250(3.60), 271s(3.66),275(3.71), 283(3.55)	39-0285-66B

Compound	Solvent	$\lambda_{max}(\log \epsilon)$	Ref.
s-Triazolo[4,3-a]pyridine, 3-methyl-	MeOH	268(3.9)	24-2593-66
	EtOH	215(4.06),268(3.56), 290(3.55)	30-0361-66D
$C_7H_7N_3O$			
Formamide, 1-phenylazo-	dioxan	281(4.01),426(2.09)	24-3337-66
	n.s.g.	282(4.01),425(2.09)	89-0372-66
s-Triazolo[1,5-a]pyridine, 1-methyl-2-oxo-	EtOH	270(3.79),306(3.59)	32-1020-66
s-Triazolo[4,3-a]pyrid-3-one, 2-methyl-	MeOH	260s(3.46),267(3.50), 277(3.37),329(3.44)	44-0265-66
$C_7H_7N_3OS$			
Formamide, 1-(phenylazo)thio-, S-oxide	10% EtOH	326(4.21),445(4.24)	88-3695-66
$C_7H_7N_3O_2$			
Uracil-5-carbonitrile, 1,6-dimethyl-	0.08N HCl	220(4.04),279(4.14)	25-0381-66
	0.08N NaOH	276(4.00)	25-0381-66
Uracil-5-carbonitrile, 6-ethyl-	0.08N HCl	217(4.05),272(4.07)	25-0381-66
	0.08N NaOH	240(4.08),285(4.11)	25-0381-66
$C_7H_7N_3O_5$			
Hydroxylamine, N-(5-methyl-2,4-dinitrophenyl)-	$CHCl_3$	272(4.09),338(3.97)	78-0995-66
$C_7H_7N_3S$			
3H-Imidazo[4,5-b]pyridine, 7-(methylthio)-	pH 1	226(3.92),289(4.17), 308(4.28)	87-0354-66
	pH 7	216(4.18),281(4.28), 286s(--)	87-0354-66
	pH 13	289(4.24),296s(--)	87-0354-66
Sulfide, methyl 3-s-triazolo[4,3-a]-pyridyl	MeOH	267s(3.69),272(3.70), 291s(3.60)	44-0265-66
1,3,5-Triazaindene, 2-(methylthio)-	pH 3.5	226s(4.29),236(4.40), 291(4.15)	39-0285-66B
	pH 7.96	231(4.33),287(4.11), 294s(4.07)	39-0285-66B
	pH 12.2	222(4.54),283(4.12)	39-0285-66B
1,3,5-Triazaindene-2-thiol, 1-methyl-	pH 2.0	231s(3.96),249(4.51), 316(4.26)	39-0285-66B
	pH 6.92	243(4.33),291s(4.25), 298(4.33),322s(3.41)	39-0285-66B
	pH 11.0	237(4.44),296(4.23)	39-0285-66B
1,3,5-Triazaindene-2-thiol, 3-methyl-	pH 2.0	227(3.98),249(4.51), 316(4.26)	39-0285-66B
	pH 6.98	226s(4.15),242(4.32), 291s(4.25),299(4.33), 320s(3.70)	39-0285-66B
	pH 11.5	230(4.43),276s(4.00), 299(4.34)	39-0285-66B
$C_7H_7N_5O$			
Adenine, N^6-acetyl-	pH 1	281(4.13)	69-2082-66
	pH 7.1	279(4.07)	69-2082-66
	pH 13	280(4.02)	69-2082-66
4(8H)-Pteridinone, 2-amino-8-methyl-	pH -0.89	226(4.48),261(4.13), 323(4.02)	24-0536-66
	pH 7.0	233s(3.86),291(3.61), 341(4.22)	24-0536-66

Compound	Solvent	λ max(log ε)	Ref.
as-Triazine, 3,5-diamino-6-(2-furyl)-	EtOH-HCl	217(4.29),258(3.98), 327(4.14)	4-0188-66
	EtOH-NaOH	265(4.02),354(4.19)	4-0188-66
$C_7H_7N_5O_2$			
Fervenulin	pH 1	237(4.23),275(3.29), 341(3.63)	44-0900-66
	pH 11	230s(3.69),275(3.28)	44-0900-66
	EtOH	238(4.27),275(3.20), 340(3.62)	44-0900-66
Toxoflavin	pH 1	258(4.26),394(3.65)	44-0900-66
	pH 11	320(3.81)	44-0900-66
	EtOH	258(4.35),391(3.74)	44-0900-66
$C_7H_7O_5P$			
Benzoic acid, anhydride with H_3PO_4	pH 5.3	231(4.1),274(3.1)	10-0129-66C
	pH 5.3	280(3.0)	31-0705-66
C_7H_8			
Norbornadiene	gas	182f(3.8),208f(3.4)	38-2664-66A
C_7H_8BrN			
Aniline, 4-bromo-2-methyl-	2N HCl	267(2.48)	28C-0362-66A
	pH 9.2	290(3.26)	28C-0362-66A
Aniline, 4-bromo-3-methyl-	2N HCl	266(2.60)	28C-0362-66A
	pH 9.2	290(3.16)	28C-0362-66A
$C_7H_8ClHgN_5$			
Purine, 9-(chloromercuri)-6-(dimethylamino)-	neutral	277(4.16)	44-1411-66
C_7H_8ClNO			
Aniline, 5-chloro-2-methoxy-	2N HCl	278(3.31)	28C-0362-66A
	pH 9.2	291(3.53)	28C-0362-66A
Pyridine, 4-chloro-3,5-dimethyl-, N-oxide	H_2O	215(3.38),265(3.20)	39-1058-66B
$C_7H_8Cl_2N_6S$			
Acetone thiosemicarbazone, S-(4,6-dichloro-1,3,5-triazin-2-yl)-	MeOH	302.0(4.11)	23-0829-66
$C_7H_8Cl_2O$			
Cyclohexanone, 2-(dichloromethylene)-	EtOH	248(3.72)	77-0567-66
$C_7H_8FN_3O_2$			
Cytosine, N -propionyl-5-fluoro-	EtOH	214(4.10),242(3.97), 309(3.68)	87-0566-66
$C_7H_8IN_3$			
1-Methyl-s-triazolo[4,3-a]pyridinium iodide	MeOH	270s(3.69),278(3.74)	44-0265-66
$C_7H_8N_2$			
Malononitrile, sec-butylidene-	C_6H_{12}	232(4.10)	35-1979-66
$C_7H_8N_2O$			
Aniline, N-methyl-N-nitroso-	H_2O	269.0(3.87)	39-0533-66B
Fumaronitrile, isopropoxy-	ether	247.5(4.11)	24-2572-66
Maleonitrile, isopropoxy-	ether	250.5(4.12)	24-2572-66
Urea, phenyl-	EtOH	237(4.25),272(3.04)	94-0001-66

Compound	Solvent	λ_{max}(log ε)	Ref.
$C_7H_8N_2O_2$			
Aniline, 2-methyl-4-nitro-	5N HCl	262(3.90)	28C-0362-66A
	H_2O	384.5(4.13)	35-3140-66
	pH 9.2	384(4.10)	28C-0362-66A
	tert-BuOH	374.5(4.20)	35-3140-66
Furfural acetylhydrazone	EtOH	301(4.48)	44-0052-66
Nudifloramide	H_2O	258(4.18),300(3.70)	78-1461-66
$C_7H_8N_2O_3$			
2,6-Lutidine, 3-nitro-, 1-oxide	H_2O	246(4.21)	39-0870-66B
2,6-Lutidine, 4-nitro-, 1-oxide	H_2O	318(4.02)	39-0870-66B
3,5-Lutidine, 4-nitro-, 1-oxide	H_2O	207(4.28),268(3.88)	39-1058-66B
1H-Pyrrolo[1,2-c]imidazole-1,3,5(2H,6H)-trione, 7,7a-dihydro-7a-methyl-	MeOH-HCl	207(4.02)	87-0142-66
	MeOH	207(4.02)	87-0142-66
	MeOH-KOH	223(4.11)	87-0142-66
$C_7H_8N_2O_3S$			
Uracil, 5-acetylthiomethyl-	pH 1	265(3.91)	87-0097-66
	pH 6.9	265(3.91)	87-0097-66
	pH 12	291(3.78)	87-0097-66
$C_7H_8N_2O_4$			
6(1H)-Pyridazinone, 1-acetoxy-3-methyl-	EtOH	313(3.441)	95-0082-66
$C_7H_8N_2O_4S$			
Methanesulfonamide, N-(4-nitrophenyl)-	EtOH	220s(3.87),307(4.07)	30-0114-66A
4-Thiazolecarboxylic acid, 2-(1-amino-2-carboxyethyl)-	2N HCl	237(3.78)	39-2115-66C
Uracil, 5-carboxymethylthiomethyl-	pH 1	265(4.03)	87-0097-66
	pH 6.9	265(3.89)	87-0097-66
	pH 11	283(3.88)	87-0097-66
$C_7H_8N_2S_2$			
Thiopheno[2,3-d]pyrimidine, 5,6-dihydro-2-mercapto-4-methyl-	pH 13.5	268(4.30),320(3.70)	24-0872-66
	MeOH	290(4.61),350(3.55)	24-0872-66
$C_7H_8N_4$			
Pyrazolo[3,2-c]-as-triazine, 3,4-dimethyl-	EtOH	228(4.55),289(3.35), 298s(3.23),357(3.45)	39-1127-66C
Pyrimidine-5-carbonitrile, 1,2-dihydro-1-methyl-2-methylimino-	pH 4.0	253(4.41),323(3.47)	39-0164-66C
$C_7H_8N_4O$			
Purine, 6-methoxy-3-methyl-	pH 5	270(3.90)	39-0010-66C
s-Triazolo[2,3-a]pyrazin-2-ol, 5,6-dimethyl-	EtOH	210(3.70),250(3.69), 320(3.68)	39-2038-66C
s-Triazolo[4,3-a]pyrazin-3-ol, 5,6-dimethyl-	EtOH	230(4.28),268(3.40), 288s(3.06),356(3.59)	39-2038-66C
s-Triazolo[2,3-a]pyrimidine, 5-methoxy-7-methyl-	MeOH	214(4.24),262(3.63)	39-2031-66C
s-Triazolo[2,3-a]pyrimidine, 7-methoxy-5-methyl-	MeOH	211(4.49),276(3.82)	39-2031-66C
$C_7H_8N_4OS$			
2-Hydroxy-7,9-dimethyl-6-thioxo-dihydropurinium betaine	pH 0	252(3.69),328(4.35)	24-0944-66
	pH 6	260(3.79),343(4.35), 352(4.37)	24-0944-66

Compound	Solvent	λ max(log ε)	Ref.
2-Mercapto-7,9-dimethyl-6-oxo-	pH -0.89	232(3.96),288(4.37)	24-0944-66
dihydropurinium betaine	pH 6	230s(4.00),289(4.33)	24-0944-66
$C_7H_8N_4O_2$			
Uracil-5-carbonitrile,	0.08N HCl	220(3.93),280(3.99)	25-0381-66
1-amino-6-ethyl-	0.08N NaOH	279(3.95)	25-0381-66
$C_7H_8N_4O_3$			
Xanthine, dimethyl-, 7-N-oxide	pH 2-3.5	237(3.92),287(4.09)	44-0178-66
	pH 7-12	242(3.99),294(4.07)	44-0178-66
	pH 14	243(3.71),287(3.98)	44-0178-66
$C_7H_8N_4O_4$			
Imidazo[1,2-c]pyrimidine-5,7-dione, 1,2,3,5,6,7-hexahydro-3-hydroxy-	pH 4.0	238(4.22),255s(4.02), 335(4.02)	24-2984-66
1-methyl-8-nitro-	pH 11.0	245(4.21),354(3.78)	24-2984-66
$C_7H_8N_4S$			
6-Mercapto-7,9-dimethyldihydro-	pH 1	224(3.94),329(4.25)	24-0944-66
purinium betaine	pH 9.2	229(4.14),326(4.23)	24-0944-66
6(1H)-Purinethione, 2-ethyl-	pH 1	227(3.97),329(4.24)	44-1417-66
	pH 7	227(4.00),326(4.28)	44-1417-66
	pH 13	231(4.17),309(4.22)	44-1417-66
$C_7H_8N_4S_2$			
2-Mercapto-7,9-dimethyl-6-thioxo-	pH -0.89	295(4.35),312s(4.29)	24-0944-66
dihydropurinium betaine	pH 3	257(3.98),298(4.40), 360(4.05)	24-0944-66
Purine, 2,8-bis(methylthio)-	pH -0.2	240(4.04),264(4.38), 326(4.03)	39-0438-66B
	pH 5.0	224(4.17),248(4.24), 319(4.20)	39-0438-66B
	pH 10.0	220(4.22),248(4.36), 318(4.21)	39-0438-66B
$C_7H_8N_6O_2$			
Pteridine, 2-amino-3,4-dihydro- 4-(nitromethyl)-	pH 5	240s(3.80),285s(3.74), 307(3.88)	39-1117-66C
	pH 9	278(3.93),335(3.88)	39-1117-66C
6-Pteridinecarboxamide, 2-amino-	pH 1	280(3.72),380(3.65)	39-0285-66C
7,8-dihydro-4-hydroxy-	pH 13	260(4.28),396(4.05)	39-0285-66C
C_7H_8O			
Anisole	C_6H_{12}	219(--),271(3.28)	30-0877-66F
	EtOH	219(3.90),271(3.24)	22-0228-66
AlBr$_3$ complex	C_6H_{12}	206(--),255(2.16)	30-0877-66F
p-Cresol	50% dioxan	278(3.28)	71-0550-66A
2,4-Pentadiyne, 1-ethoxy-	EtOH	228(3.04),240(3.11), 254(3.15)	22-3024-66
$C_7H_8O_2S$			
o-Toluenesulfinic acid	H_2O	220(3.8),240(3.5), 265(3.0)	18-1788-66
anion	H_2O	225(3.9),265(3.5)	18-1788-66
p-Toluenesulfinic acid	H_2O	223(3.8),240(3.5), 265(2.8)	18-1788-66
anion	H_2O	223(3.9),265(3.0)	18-1788-66

Compound	Solvent	$\lambda_{max}(\log \epsilon)$	Ref.
$C_7H_8O_3$			
1,3,4-Cyclopentanetrione,	acid	255(4.1)	39-2308-66C
2,5-dimethyl-	base	273(4.2)	39-2308-66C
2-Cyclopentene-1,4-dione, 2-methoxy-3-methyl-	MeOH	275(4.07)	20-0349-66
2-Furoic acid, 3-methyl-, methyl ester	EtOH	251(4.13)	39-0976-66C
2-Pyrone, 4-hydroxy-3,6-dimethyl-	EtOH	288(3.92)	35-0834-66
$C_7H_8O_3S$			
Benzenesulfinic acid, o-methoxy-	H_2O	222(3.8),245(3.3), 282(3.5)	18-1788-66
anion	H_2O	225(3.9),280(3.5)	18-1788-66
Benzenesulfinic acid, p-methoxy-	H_2O	235f(4.1),275(3.1)	18-1788-66
anion	H_2O	232(4.1),275(3.3)	18-1788-66
$C_7H_8O_4$			
Epoxydon	EtOH	240(3.71),334(1.65)	33-0204-66
	EtOH-NaOH	271(3.73)	33-0204-66
2,4-Hexadienoic acid, 5-formyl-	H_2O	325(4.43)	32-0915-66
2-hydroxy-	0.02N NaOH	380(4.62)	32-0915-66
	MeOH	291(4.13)	32-0915-66
$C_7H_8O_4S$			
Benzenesulfonic acid, 2-methoxy-, sodium salt	H_2O	222(3.9),278(3.5)	65-2141-66
Benzenesulfonic acid, 4-methoxy-, sodium salt	H_2O	232(4.2),270(3.0)	65-2141-66
2(5H)-Furanone, 4-acetylthio-methyl-3-hydroxy-	EtOH	240(4.16)	54-0347-66
C_7H_8S			
Sulfide, ethyl 1,3-pentadiynyl	EtOH	242(3.15),253(3.11), 268(3.08),284(2.92)	22-3024-66
Sulfide, methyl phenyl	EtOH	254(3.98),280s(3.00)	22-0228-66
$C_7H_9BrN_2O_2$			
Pyrazole-3(5)-carboxylic acid, 4-bromo-5(3)-methyl-, ethyl ester	EtOH	222(3.81),252s(3.59)	22-3744-66
Pyrimidine, 5-bromo-2,4-dimethoxy-6-methyl-	EtOH	270(4.06)	78-2401-66
$C_7H_9BrN_2S$			
2-Imidazolidinone, N-benzoyl-N'-(2-benzoylthioethyl)-, mercuric bromide salt	EtOH	213(4.19)	39-1226-66C
$C_7H_9BrN_4$			
3-Amino-1-methyl-s-triazolo[4,3-a]-pyridinium bromide	MeOH	229(4.21),267(3.86), 321(4.0)	44-0265-66
3-Amino-2-methyl-s-triazolo[4,3-a]-pyridinium bromide	MeOH	254s(3.44),264(3.52), 275(3.48),309(3.51)	44-0265-66
$C_7H_9BrO_5$			
Glutaric acid, 2-bromo-3-oxo-, dimethyl ester	EtOH	255(3.40)	95-0300-66
$C_7H_9ClN_2O_2$			
Pyrimidine, 5-chloro-2,4-dimethoxy-6-methyl-	EtOH	270(3.89)	78-2401-66

Compound	Solvent	$\lambda_{max}(\log \epsilon)$	Ref.
C_7H_9N			
Azepine, N-methyl-	hexane	236(3.71),249s(3.60), 258s(3.56),288s(3.29), 384s(1.81)	89-0839-66
m-Toluidine	2N HCl	260(2.50)	28C-0362-66A
	pH 9.2	283(3.16)	28C-0362-66A
o-Toluidine	2N HCl	259(2.41)	28C-0362-66A
	pH 9.2	280(3.23)	28C-0362-66A
p-Toluidine	2N HCl	261(2.30)	28C-0362-66A
	pH 9.2	287(3.13)	28C-0362-66A
C_7H_9NO			
m-Anisidine	2N HCl	269(3.27)	28C-0362-66A
	pH 9.2	282(3.28)	28C-0362-66A
o-Anisidine	2N HCl	268(3.31)	28C-0362-66A
	pH 9.2	282(3.40)	28C-0362-66A
mercuric chloride complex	EtOH	238(3.92),287(3.50)	90-0147-66
p-Anisidine	2N HCl	270(3.14)	28C-0362-66A
	pH 9.2	295(3.33)	28C-0362-66A
2,6-Lutidine, 1-oxide	H_2O	252(3.95)	39-0870-66B
3-Picoline, 6-methoxy-	H_o -3.6	215(3.68),290(3.83)	39-0996-66B
	pH 7	217(3.85),280(3.56)	39-0991-66B
Pyrrole, 4-acetyl-2-methyl-	EtOH	204(3.79),246(3.89), 270(3.72)	33-0349-66
C_7H_9NOS			
5-Pyrrolethiolcarboxylic acid, 2-methyl-, methyl ester	EtOH	283(4.34),312(3.67)	23-1007-66
4H-1,3-Thiazin-4-one, 2-(ethylthio)-6-methyl-	EtOH	254(4.34),284s(4.03)	88-3225-66
$C_7H_9NO_2$			
Pyridine, 2,6-dimethoxy-	2N H_2SO_4	216(3.64),294(4.08)	39-0562-66B
	pH 6.8	218(3.76),279(3.85)	39-0562-66B
	MeOH	278(4.00)	44-4054-66
2,6(1H,3H)-Pyridinedione, 3,3-dimethyl-	78% H_2SO_4	238(3.71)	39-0562-66B
	pH 6.8	210(4.23),230(3.69), 322(2.45)	39-0562-66B
	N NaOH	226(3.79),255(3.52)	39-0562-66B
2-Pyridone, 6-methoxy-1-methyl-	2N H_2SO_4	219(3.64),295(4.06)	39-0562-66B
	pH 6.8	226(3.82),306(4.05)	39-0562-66B
Pyrrole, 4-acetyl-2-(hydroxymethyl)-	EtOH	198(4.02),249(4.0), 270s(3.75)	33-0349-66
$C_7H_9NO_2S$			
2-Thiophenecarboxylic acid, 3-(2-aminoethyl)-	HCl	240(3.67),270(3.77)	4-0466-66
	NaOH	241(3.70),269(3.72)	4-0466-66
$C_7H_9NO_3$			
Nicotinic acid, 6-oxo-1,4,5,6-tetrahydro-, methyl ester	MeOH	277(4.17)	37-1807-66
2,3,6-Piperidinetrione, 5,5-dimethyl-	EtOH	262(3.56)	78-0015-66B
	EtOH-KOH	308(3.08)	78-0015-66B
$C_7H_9NO_3S_2$			
5-Thiazolidineacetic acid, N-ethyl-4-oxo-2-thioxo-	EtOH	262(4.14),298(4.24)	95-0095-66
5-Thiazolidineacetic acid, N-methyl-4-oxo-2-thioxo-, methyl ester	EtOH	260(4.12),294(4.20)	95-0095-66

Compound	Solvent	λ_{max}(log ϵ)	Ref.
$C_7H_9NO_4$			
Glutaconimide, 4,4-dimethoxy-	MeOH	218(3.79)	44-4056-66
$C_7H_9N_2O_3^+$			
1-Hydroxy-3-nitro-2,6-lutidinium	60% H_2SO_4	280(3.81)	39-0870-66B
1-Hydroxy-4-nitro-2,6-lutidinium	60% H_2SO_4	300(3.64)	39-0870-66B
$C_7H_9N_3$			
Pyrimidine, 2-allylamino-	pH 0.5	229(4.26),314(3.56)	39-0226-66C
	pH 7.0	234(4.26),305(3.43)	39-0226-66C
$C_7H_9N_3OS$			
5H-Thiopyrano[2,3-d]pyrimidin-2-ol, 4-amino-6,7-dihydro-	MeOH	221(4.21),294(4.07)	24-0872-66
$C_7H_9N_3O_2$			
2-Pyrazinecarboxylic acid, 3-amino-5,6-dimethyl-	pH 7.0	248(4.01),343(3.86)	39-1112-66C
$C_7H_9N_3O_3$			
2H-Pyrimido[4,5-e][1,3]oxazine-6,8-diol, 3,4-dihydro-3-methyl-	pH 1	276(4.16)	4-0115-66
	pH 11	240(3.94),301(4.06)	4-0115-66
$C_7H_9N_3S$			
Thiopheno[2,3-d]pyrimidine, 2-amino-5,6-dihydro-4-methyl-	pH 1	216(4.31),240(3.97), 273(3.98),318(4.11)	24-0872-66
	MeOH	212(4.27),235(4.33), 313(4.13)	24-0872-66
$C_7H_9N_3S_2$			
5H-Thiopyrano[2,3-d]pyrimidine, 4-amino-6,7-dihydro-2-mercapto-	pH 13.5	223(4.38),253(4.35), 300(3.81)	24-0872-66
	MeOH	247(4.25),273(4.41), 312(4.41)	24-0872-66
$C_7H_9N_5$			
Adenine, N,N-dimethyl-	neutral	275(4.24)	44-1411-66
anion	n.s.g.	281(4.25)	44-1411-66
s-Triazolo[4,3-a]pyrazine, 3-amino-5,6-dimethyl-	H_2O	230(4.26),261s(3.2), 339(3.54)	39-2038-66C
s-Triazolo[2,3-a]pyrimidine, 2-amino-5,7-dimethyl-	MeOH	224(4.29),281s(3.32), 312(3.62)	39-2031-66C
s-Triazolo[4,3-a]pyrimidine, 3-amino-5,7-dimethyl-	MeOH	228(4.39),259s(3.31), 324(3.27)	39-2031-66C
$C_7H_9N_5O$			
Adenine, 9-ethyl-, 1-oxide	pH 1	259(4.04)	25-1598-66
	pH 7	232(4.63),261(3.86)	25-1598-66
	pH 13	231(4.44),268(3.87)	25-1598-66
Isoguanine, 9-ethyl-	pH 1	231(3.80),282(4.10)	95-0649-66
	pH 13	246(3.75),285(4.00)	95-0649-66
Pteridine, 2-amino-3,4-dihydro-4-methoxy-	EtOH-HCl	235(4.01),285(3.89), 299s(3.84)	39-1117-66C
Pterin, 7,8-dihydro-6-methyl-, hydrochloride	pH 1	252(4.27),271s(3.88), 361(3.71)	24-3008-66
	pH 7.0	229(4.41),279(4.02), 324(3.75)	24-3008-66
	pH 14.0	231(4.17),282(3.90), 322(3.76)	24-3008-66

Compound	Solvent	$\lambda_{max}(\log \epsilon)$	Ref.
s-Triazolo[2,3-a]pyrimidine, 2-amino-5-methoxy-7-methyl-	MeOH	217(4.55),299(3.81)	39-2031-66C
s-Triazolo[2,3-a]pyrimidine, 2-amino-7-methoxy-5-methyl-	MeOH	229(4.21),261(3.72), 325(3.85)	39-2031-66C
$C_7H_9N_5O_2$			
Purine, 6-hydroxyamino-9-methoxymethyl-	N HCl	265(4.20)	69-3057-66
	pH 6.7	266(4.12)	69-3057-66
	N NaOH	252(4.01),295(3.47)	69-3057-66
Triazolo[2,3-a]pyrimidine, 2-amino-5,7-dimethoxy-	MeOH	228(4.33),255(3.64), 301(4.01)	39-2031-66C
s-Triazolo[4,3-a]pyrimidine, 3-amino-5,7-dimethoxy-	MeOH	223(4.40),263(3.37), 302(3.21)	39-2031-66C
$C_7H_9N_5O_4$			
Acetaldehyde, [(2-aminodihydro-1-methyl-5-nitro-6-oxo-4-pyrimidinyl)amino]-, hydrochloride	pH -0.89	228(4.30),248s(3.72), 312(4.08)	24-2984-66
	pH 6.0	230s(4.19),290s(3.67), 335(4.16)	24-2984-66
$C_7H_9N_5O_5$			
Pyrimidine, 2-aminodihydro-4-(dimethylamino)-1-methyl-5-nitro-6-oxo-	pH -2.7	237(4.27),267(3.96), 347(3.64)	24-2984-66
	pH 6.0	303s(3.80),356(4.00)	24-2984-66
$C_7H_9N_5S$			
2-Amino-6-mercapto-7,9-dimethylpurinium betaine	pH 4	254(3.90),287(3.75), 329(4.21)	24-0944-66
	pH 9.2	263(3.81),343s(4.28), 351(4.30)	24-0944-66
$C_7H_9N_5S_2$			
s-Triazolo[4,3-a]pyrimidine, 3-amino-5,7-bis(methylthio)-	MeOH	268(4.28),301(4.00), 309(3.97),385(3.57)	39-2031-66C
C_7H_{10}			
3-Penten-1-yne, 3-ethyl-	EtOH	222(4.14)	39-1976-66C
$C_7H_{10}BrN_3O$			
Pyrimidine, 2-amino-5-bromo-4-ethoxy-6-methyl-	EtOH	288(3.40)	78-2401-66
$C_7H_{10}ClN_3O_2S$			
6-Pyridazinesulfonamide, 3-chloro-N-propyl-	EtOH	264(4.18)	83-0648-66
$C_7H_{10}N_2O$			
2,1-Benzisoxazole, 3-amino-4,5,6,7-tetrahydro-	n.s.g.	243(3.87)	18-1125-66
Pyrimidine, 2-isopropoxy-	pH 7.0	268(3.55)	12-1487-66
Pyrimidine, 2-methoxy-4,6-dimethyl-	pH 10.2	264(3.82)	12-2321-66
Pyrimidine, 4-methoxy-2,5-dimethyl-	MeOH	218(3.81),255(3.61)	22-0527-66
Pyrimidine, 4-methoxy-2,6-dimethyl-	pH 7	217(3.68),249(3.56)	12-2321-66
Pyrimidine, 2-propoxy-	pH 7.0	266(3.63)	12-1487-66
4-Pyrimidone, 2,3,5-trimethyl	MeOH	226(4.05),276(4.06)	22-0527-66
HI salt	H_2O	220(4.30),271(3.79)	22-0527-66
2-Pyrrolidinone, 4-isopropylidene-5-imino-	EtOH	254(4.24)	22-3895-66

Compound	Solvent	λ_{max}(log ε)	Ref.
$C_7H_{10}N_2OS_2$			
4-Pyrimidinol, 2-mercapto-5-(2-mercaptoethyl)-6-methyl-	pH 13.5	232(4.30),262(4.27), 292s(4.03)	24-0872-66
	MeOH	217(4.12),280(4.32)	24-0872-66
2-Thioacetamide, S-(4-methyl-1,3-thiazin-2-yl)-	acid	237s(--),283(3.88)	39-0092-66B
	H_2O	262(3.79),310(3.43)	39-0092-66B
	base	262(3.79),310(3.43)	39-0092-66B
$C_7H_{10}N_2O_2$			
2,6-Lutidine, 4-hydroxyamino-, 1-oxide	n.s.g.	282(4.26)	88-4729-66
Pyrazole-1-acetic acid, ethyl ester	0.05N HCl	218(3.65)	54-1194-66
	H_2O	217(3.68)	54-1194-66
Pyrazole-1-acetic acid, 3,5-dimethyl-	0.05N HCl	226(3.89)	54-1194-66
	H_2O	224(3.83)	54-1194-66
	0.05N NaOH	223(3.77)	54-1194-66
Pyrazole-3(5)-carboxylic acid, 4-methyl-, ethyl ester	EtOH	222(3.91),240s(3.60)	22-3744-66
Pyrimidine, 2,4-dimethoxy-6-methyl-	EtOH	258(3.71)	78-2401-66
Thymine, 1,3-dimethyl-	pH 5-12	272(3.99)	24-2391-66
Thymine, 4-O-ethyl-	pH 5.9	275(3.75)	24-2391-66
	pH 11.52	222(4.13),284(3.80)	24-2391-66
	pH 13	223(4.15),286(3.85)	24-2391-66
	pH 14	226(4.17),286(3.85)	24-2391-66
$C_7H_{10}N_2O_2S$			
Uracil, 5-ethylthiomethyl-	pH 1	265(3.90)	87-0097-66
	pH 6.9	265(3.89)	87-0097-66
	pH 12	284(3.86)	87-0097-66
Uracil, 5-(2-mercaptoethyl)-6-methyl-	pH 13.5	222(4.05),280(3.75)	24-0872-66
	MeOH	267.5(3.96)	24-0872-66
$C_7H_{10}N_2O_2S_2$			
5-Thiazolidineacetamide, 3-ethyl-4-oxo-2-thioxo-	EtOH	262(4.12),296(4.22)	95-0101-66
$C_7H_{10}N_2O_3$			
6(1H)-Pyridazinone, 1-ethoxy-3-methoxy-	EtOH	314(3.436)	95-0082-66
6(1H)-Pyridazinone, 3-ethoxy-1-methoxy-	EtOH	315(3.406)	95-0082-66
$C_7H_{10}N_2O_4$			
Malonic acid, diazo-, diethyl ester	CH_2Cl_2	251(3.86)	24-3128-66
Urea, acetoacetyl-, enol acetate	EtOH	227.5(3.91)	70-1171-66
$C_7H_{10}N_2S$			
Pyrimidine, 2,4-dimethyl-6-(methylthio)-	pH 1	228(3.92),299(4.30)	12-2321-66
	pH 7	215(3.91),250(3.84), 280(4.01)	12-2321-66
Pyrimidine, 4,6-dimethyl-2-(methylthio)-	pH -1	217(3.67),253(4.06), 303(3.72)	12-2321-66
	pH 7	250(4.07),280s(3.48)	12-2321-66
Pyrimidine, 2-(isopropylthio)-	pH -2.5	219(3.41),258(4.13), 315(3.57)	12-1487-66
	pH 7.0	252(4.13),285s(3.30)	12-1487-66
$C_7H_{10}N_4$			
Pyrazolo[3,2-c]-as-triazine, 4,6-dihydro-3,4-dimethyl-	EtOH	237(3.67),273(3.74)	39-1127-66C

Compound	Solvent	$\lambda_{max}(\log \epsilon)$	Ref.
$C_7H_{10}N_4O$			
Pyrimidine-5-carboxaldehyde, 4-amino-	pH 4.0	269(4.05),308(4.32)	39-0164-66C
2,3-dihydro-3-methyl-2-methylimino-	pH 11.0	256(4.04),328(4.13)	39-0164-66C
$C_7H_{10}N_4O_3$			
Uracil, 5-acetamido-6-amino-1-methyl-	H_2O	268(4.18)	95-0854-66
$C_7H_{10}N_4O_4$			
2,6-Pyrimidinedione, tetrahydro-4-(methylamino)-1,3-dimethyl-5-nitro-	pH 7	240(4.20),260s(3.83),334(3.85)	24-3022-66
	pH 12	280s(3.44),336(4.05)	24-3022-66
$C_7H_{10}O$			
2-Cyclohepten-1-one	C_6H_{12}	223.5(3.79)	39-0164-66B
	EtOH	228(3.98)	39-0164-66B
2-Cyclopenten-1-one, 2,3-dimethyl-	n.s.g.	234.0(4.13)	39-0073-66B
2-Cyclopenten-1-one, 2,5-dimethyl-	EtOH	227(4.04)	70-2139-66
2-Cyclopenten-1-one, 2-ethyl-	n.s.g.	227.0(3.97)	39-0073-66B
2-Cyclopenten-1-one, 3-ethyl-	n.s.g.	227.0(4.05)	39-0073-66B
Cyclopropyl ketone (protonated)	FSO_3H-SbF_5	235(4.12)	35-1488-66
3,5-Heptadien-2-one	EtOH	271(4.35)	28C-0652-66A
2,4-Hexadienal, 5-methyl-	EtOH	285(4.46)	28C-0652-66A
4,5-Hexadien-2-one, 4-methyl-	hexane	197f(3.65),290(2.10)	28C-1601-66A
	EtOH	212f(3.08),285(2.18)	28C-1601-66A
$C_7H_{10}OS$			
2-Cyclohexen-1-one, 3-(methylthio)-	n.s.g.	287.0(4.35)	39-0073-66B
$C_7H_{10}OS_3$			
1,3-Benzodithiole-2-thione, 4,5,6,7-tetrahydro-, S-oxide	EtOH	284(3.62),349(4.03)	77-0016-66
$C_7H_{10}O_2$			
Furan, 3-acetyl-4,5-dihydro-	hexane	190(3.80),270(4.12)	78-0407-66
2-methyl-	EtOH	203(3.63),280(4.09)	78-0407-66
4,6-Heptadienoic acid	C_6H_{12}	225.4(4.40)	35-1732-66
3-Hepten-2-one, 5,6-epoxy-	EtOH	232(4.12)	70-0646-66
2,4-Hexadienal, 6-hydroxy-4-methyl-, all trans	pet ether	262(4.48)	33-0369-66
2,4-Pentadienoic acid, ethyl ester	EtOH	246.0(4.35)	39-1152-66C
$C_7H_{10}O_3$			
2-Cyclopenten-1-one, 4-hydroxy-3-methoxy-2-methyl-	MeOH	253(4.16)	20-0349-66
2-Cyclopenten-1-one, 5-hydroxy-3-methoxy-2-methyl-	MeOH	258(4.21)	20-0349-66
2,4,6-Heptanetrione	aq. NaOH	342(4.54)	99-0230-66
5-Hexenoic acid, 4-oxo-, methyl ester	n.s.g.	212(4.02)	70-0256-66
$C_7H_{10}O_4$			
Acrylic acid, 3-acetoxy-, ethyl ester, cis	n.s.g.	219(3.65)	16-0023-66
trans	n.s.g.	220(4.17)	16-0023-66
$C_7H_{10}O_5$			
Fumaric acid, methoxy-, dimethyl ester	ether	239(3.85)	24-0450-66
Glutaric acid, 3-oxo-, dimethyl ester	EtOH	240(3.50)	95-0300-66
Maleic acid, methoxy-, dimethyl ester	ether	222(4.04)	24-0450-66

Compound	Solvent	λ_{max}(log ϵ)	Ref.
$C_7H_{11}^+$			
Dicyclopropylcarbonium ion	FSO_3H-SbF_5	273(4.09)	35-1488-66
$C_7H_{11}Br$			
1,2-Pentadiene, 1-bromo-3-ethyl-	EtOH	206(3.89),220s(3.85)	39-1223-66C
1,3-Pentadiene, 3-bromo-2,4-dimethyl-	heptane	206s(3.67),250(2.98)	44-0081-66
$C_7H_{11}INOP$			
2-Picolylphosphine oxide, methiodide	N H_2SO_4	222(4.50),273(4.05)	39-0631-66B
3-Picolylphosphine oxide, methiodide	N H_2SO_4	222(4.59),266(3.84)	39-0631-66B
4-Picolylphosphine oxide, methiodide	N H_2SO_4	224(4.56),255(3.94)	39-0631-66B
$C_7H_{11}N$			
7-Norbornenylamine, anti	MeOH	187(4.04)	44-3941-66
7-Norbornenylamine, syn	MeOH	194(3.73)	44-3941-66
1-Penten-3-ynylamine, N,N-dimethyl-	EtOH	273(4.08)	70-1739-66
Pyrrole, 4-ethyl-2-methyl-	EtOH	202(3.75),279(2.27), 302(2.00)	33-0349-66
$C_7H_{11}NO_2$			
Propionitrile, 3-ethoxy-2-methoxy-methylene, cis	EtOH	232(4.17)	94-0238-66
trans	EtOH	229(4.12)	94-0238-66
$C_7H_{11}NO_2S_2$			
Carbamic acid, acetoacetyldithio-, ethyl ester	EtOH	261(4.10),309(4.17)	88-3225-66
$C_7H_{11}NO_3$			
1-Cyclopentanol, 1-nitroso-, acetate	C_6H_{12}	655(1.20)	39-0441-66B
2,4-Pentanedione, 3-acetamido-, palladium derivative	EtOH	209(4.83),232(4.34), 260s(4.02),334(3.86)	30-0103-66D
$C_7H_{11}NO_3S_2$			
1,2-Dithiolane, 4-(acetamido-carboxymethyl)-	EtOH-N HCl	330(2.12)	11-0135-66
$C_7H_{11}NO_4S$			
2(5H)-Furanone, 4-acetylthiomethyl-3-hydroxy-, ammonium salt	H_2O	238(4.05)	54-0347-66
$C_7H_{11}N_3$			
Pyrimidine, 4-(methylamino)-2,6-dimethyl-	pH 5.0	255(4.15)	39-0226-66C
	pH 10.0	242(4.07),275(3.65)	39-0226-66C
$C_7H_{11}N_3O$			
2,4-Pentadienal, 4-methyl-, semicarbazone	n.s.g.	290(4.63)	22-0717-66
Pyrimidine, 2-amino-4-ethoxy-6-methyl-	EtOH	276(3.56)	78-2401-66
s-Triazolo[4,3-a]pyridine, 5,6,7,8-tetrahydro-1-methyl-3-oxo-	EtOH	242(3.65)	7-0199-66
$C_7H_{11}N_3OS$			
Cytosine, 5-(2-mercaptoethyl)-6-methyl-	pH 13.5	236(4.05),278(3.89)	24-0872-66
	MeOH	229(3.57),289(3.51)	24-0872-66
$C_7H_{11}N_3OS_2$			
4-Pyrimidinol, 6-amino-2-mercapto-5-(3-mercaptopropyl)-	MeOH	206(4.40),287(4.44)	24-0872-66

Compound	Solvent	λ max(log ε)	Ref.
$C_7H_{11}N_3O_2S$			
2,6-Pyrimidinediol, 4-amino-5-(3-mercaptopropyl)-	pH 13.5	233(3.89),275(4.09)	24-0872-66
	MeOH	271(4.23)	24-0872-66
$C_7H_{11}N_5O$			
2-Pyrazinecarboxylic acid, 3-amino-5,6-dimethyl-, hydrazide	pH -2	251(4.12),371(4.05)	39-1112-66C
	pH 2.2	253(4.07),361(3.94)	39-1112-66C
	pH 6.7	253(4.08),355(3.91)	39-1112-66C
$C_7H_{11}N_5O_2$			
Pyrimidine, 2-(dimethylamino)-4-(methylamino)-5-nitro-	pH -0.89	243(4.28),265s(4.08), 331(3.98)	24-3022-66
	pH 7.0	223(4.22),268(3.55), 378(4.34)	24-3022-66
$C_7H_{12}N_2$			
Pyrazole, 1-tert-butyl-	EtOH	217(3.67)	22-3744-66
Pyrazole, 5-ethyl-3,4-dimethyl-	EtOH	223(3.72)	22-2971-66
3H-Pyrazole, 3,3,4,5-tetramethyl-	EtOH	269(3.84),339(2.42)	78-0463-66
1-Pyrazoline, 4,5,5-trimethyl-3-methylene-	EtOH	244(3.81),352(2.40)	78-0463-66
$C_7H_{12}N_2O$			
3H-Pyrazole, 3-methoxy-3,4,5-trimethyl-	EtOH	278(3.35),378(2.33)	78-0463-66
2(1H)-Pyrimidinone, 3,4-dihydro-4,4,6-trimethyl-	pH 5	246(3.28)	44-2700-66
$C_7H_{12}N_2O_2$			
3-Furamide, 2-amino-4,5-dihydro-4,5-dimethyl-, cis	EtOH	278(4.31)	44-4133-66
trans	EtOH	277(4.36)	44-4133-66
2-Heptenedioic acid diamide	n.s.g.	217(4.09)	44-1728-66
2-Pyrazolin-5-one, 4-(2-hydroxypropyl)-3-methyl-	MeOH	248(3.71)	24-2962-66
$C_7H_{12}N_4O$			
Δ^1-1,2,4-Triazolin-3-one, 4-isopropylimino-5,5-dimethyl-	EtOH	291(3.74)	88-5815-66
$C_7H_{12}N_4OS$			
4-Pyrimidinol, 2,6-diamino-5-(3-mercaptopropyl)-	pH 13.5	236(4.08),269(3.91)	24-0872-66
	MeOH	212(4.32),277(4.05)	24-0872-66
$C_7H_{12}N_4O_4$			
Uracil, dihydro-6-(dimethylamino)-1-methyl-5-nitro-	pH 3.0	241(4.25),260(4.00), 349(3.69)	24-2984-66
	pH 8.0	225(4.32),310s(3.72), 352(4.01)	24-2984-66
$C_7H_{12}O$			
4-Hexen-3-one, 5-methyl-	C_6H_{12}	232(4.04),325(1.58)	88-5081-66
	EtOH	236(4.02),314(1.74)	88-5081-66
3-Penten-2-one, 3,4-dimethyl-	isooctane	238(3.84)	22-0273-66
2-Propanone, 1-(1-methylcyclopropyl)-	C_6H_{12}	289(1.61)	88-5081-66
	EtOH	283.5(1.61)	88-5081-66
$C_7H_{12}O_2$			
1-Penten-3-one, 4-hydroxy-2,2-dimethyl-	EtOH	221(3.89)	88-6037-66

Compound	Solvent	$\lambda_{max}(\log \epsilon)$	Ref.
$C_7H_{12}O_2S_2$			
Butyric acid, γ-(1,2-dithiolan-4-yl)-	$CHCl_3$	333(2.17)	11-0263-66
$C_7H_{13}ClN_2O_4$			
2,3-Dihydro-1,4-dimethyl-1,4-diazepinium perchlorate	MeOH	340(4.23)	39-0093-66C
$C_7H_{13}N$			
Methylamine, N-cyclohexylidene-	heptane	179(3.98)	88-0507-66
	EtOH	235(2.30)	88-0507-66
$C_7H_{13}NO$			
1-Penten-3-one, 1-dimethylamino-	EtOH	303(5.40)	70-1739-66
2-Pyrrolidinone, 1-propyl-	MeOH	204(3.7)	24-0724-66
$C_7H_{13}NO_2$			
Acrylic acid, 3-(isopropylamino)-, methyl ester	EtOH	276(4.31)	24-2526-66
$C_7H_{13}N_3O$			
3-Butenal, 2,2-dimethyl-, semicarbazone	n.s.g.	230(4.13)	22-0734-66
2-Pentenal, 4-methyl-, semicarbazone	n.s.g.	263(4.42)	22-0717-66
3-Pentenal, 4-methyl-, semicarbazone	n.s.g.	231(4.12)	22-0717-66
$C_7H_{13}N_3O_4$			
L-Alanine, (glycylglycyl)-, copper complex	H_2O	555(2.19),580(1.92), 660(1.95)	37-1439-66
Glycine, (alanylglycyl)-, copper complex	H_2O	550(2.20),575(1.97), 655(1.91)	37-1439-66
Glycine, (glycylalanyl)-, copper complex	H_2O	550(2.19),575(1.97), 655(1.93)	37-1439-66
$C_7H_{13}N_5$			
Pyrimidine, 5-amino-2-(dimethylamino)-4-(methylamino)-, sulfate	pH 4	240(4.35),295(3.55)	24-3022-66
$C_7H_{14}N_2$			
2-Pyrazoline, 5-ethyl-1,4-dimethyl-	EtOH	243(3.52)	4-0413-66
1-Pyrazoline, 3,3,5,5-tetramethyl-	MeOH	324(2.20)	35-3959-66
$C_7H_{14}N_4O_2$			
Formamidine, N,N'-bis(propionamido)-	MeOH	245(4.17)	44-3442-66
$C_7H_{14}N_6OS_2$			
Butyraldehyde, 3-methoxy-2-oxo-, bis(thiosemicarbazone)	EtOH	233(3.96),271(3.87), 348(4.68)	87-0420-66
$C_7H_{14}O$			
2-Pentanone, 4,4-dimethyl-	C_6H_{12}	287(1.30)	44-4237-66
	EtOH	284(1.38)	44-4237-66
$C_7H_{15}ClN_2O_4$			
1,3-Bis(dimethylamino)propenium perchlorate	EtOH-100°K	315(4.72)	61-0052-66
phototropic form	EtOH-100°K	340(4.49)	61-0052-66
$C_7H_{15}I$			
Pentane, 3-ethyl-3-iodo-	C_6H_{12}	266(2.84)	3-0612-66

Compound	Solvent	$\lambda_{max}(\log \epsilon)$	Ref.
$C_7H_{15}N$			
Butylamine, N-isopropylidene-	C_6H_{12}	246(2.15)	88-0507-66
	heptane	179(3.95)	88-0507-66
	EtOH	232(2.30)	88-0507-66
$C_7H_{15}O_4P$			
Phosphonic acid, propionyl-, diethyl ester	C_6H_{12}	340(1.85)	44-1304-66
$C_7H_{16}MnN_5O_4$			
Manganese, hydroaquobiguanido-(acetylacetonato)-	MeOH	390(3.4),431s(3.2)	12-0881-66
$C_7H_{16}N_2S$			
Thiourea, diisopropyl-, bromine complex	$CHCl_3$	282(--)	60-0029-66
iodine complex	$CHCl_3$	305.5(4.59)	60-0029-66
ICl complex	$CHCl_3$	305(--)	60-0029-66
ICN complex	$CHCl_3$	307(--)	60-0029-66
$C_7H_{16}OSi$			
2-Butanone, 1-(trimethylsilyl)-	C_6H_{12}	279(1.97)	44-4237-66
	EtOH	276(2.08)	44-4237-66
2-Butanone, 4-(trimethylsilyl)-	C_6H_{12}	283(1.38)	44-4237-66
	EtOH	279(1.60)	44-4237-66

Compound	Solvent	$\lambda_{max}(\log \epsilon)$	Ref.
C_8ClF_5			
Benzene, (chloroethynyl)pentafluoro-	EtOH	230(4.18),240(4.35), 251(4.34)	39-0597-66C
C_8ClF_7			
Styrene, β-chloroheptafluoro-	hexane	243.0(3.94)	78-1755-66
$C_8Cl_2N_2O_2$			
1,4-Cyclohexadiene-1,2-dicarbonitrile, 4,5-dichloro-3,6-dioxo-, benzene complex	$CHCl_3$	427(3.40)	59-1869-66
biphenyl complex	$CHCl_3$	440(3.14),564(3.22)	59-1869-66
pyrone complex	$CHCl_3$	540(3.10)	59-1869-66
m-terphenyl complex	$CHCl_3$	574(2.95)	59-1869-66
toluene complex	$CHCl_3$	450(3.37)	59-1869-66
(also other complexes, not listed)			
$C_8Cl_3F_5$			
Styrene, α,β,β-trichloropentafluoro-	n.s.g.	249(3.66)	39-0597-66C
$C_8Cl_3F_5O$			
Acetophenone, 2,2,2-trichloro-2',3',4',5',6'-pentafluoro-	EtOH	211(3.86),270(3.03)	39-0597-66C
C_8F_{12}			
1,3-Cyclohexadiene, hexafluoro-1,4-bis(trifluoromethyl)-	hexane	249(3.76)	78-2555-66
1,4-Cyclohexadiene, hexafluoro-1,2-bis(trifluoromethyl)-	hexane	205(2.62)	78-2555-66
C_8F_{14}			
Cyclohexene, octafluoro-1,2-bis-(trifluoromethyl)-	hexane	200(2.42)	78-2555-66
$C_8F_{14}N_4$			
1,2,4,5-Tetrazine, 3,5-bis-(heptafluoropropyl)-	isooctane	252(3.22),525(2.72)	44-0781-66
$C_8HCl_2F_5$			
Styrene, β,β-dichloropentafluoro-	EtOH	239(4.10)	39-0597-66C
C_8HF_5			
Acetylene, pentafluorophenyl-	EtOH	222(4.26),234(4.24), 243(4.14)	39-0597-66C
$C_8HCl_2F_5O$			
Acetophenone, 2,2-dichloro-2',3',4',5',6'-pentafluoro-	EtOH	240(3.57),267(3.08)	39-0597-66C
$C_8H_2BrF_5O$			
Acetophenone, 2-bromo-2',3',4',5',6'-pentafluoro-	EtOH	236(3.78),272(3.03)	39-0597-66C
$C_8H_2Br_2N_4S_2$			
Dipyridazino[4,5-b:4,5-e]-1,4-dithiin, 1,6-dibromo-	EtOH	208(4.53),269(4.38), 300s(3.72)	4-0541-66
$C_8H_2Cl_2N_4S_2$			
Dipyridazino[4,5-b:4,5-e]-1,4-dithiin, 1,6-dichloroo-	EtOH	204(4.34),229(4.22), 264(4.43),296s(3.89)	4-0541-66

Compound	Solvent	λmax(log ε)	Ref.
$C_8H_2F_{14}N_4$			
1,2,4,5-Tetrazine, 3,6-bis(hepta-fluoropropyl)-1,2-dihydro-	iso-PrOH	232(3.54)	44-0781-66
$C_8H_2KN_3$			
1,2,3-Cyclopentadienetricarbo-nitrile, potassium deriv.	MeCN	220(4.35),288(4.27), 298(4.26)	35-3046-66
1,3,5-Cyclopentadienetricarbo-nitrile, potassium deriv.	MeCN	236(4.49),267(3.75)	35-3046-66
$C_8H_4BrN_3O_3$			
4-Cinnolinol, 3-bromo-6-nitro-	EtOH	210(4.22),238(4.16), 270s(4.00),279s(3.92), 324(3.93),370(4.11)	39-0470-66C
$C_8H_4ClN_3O_3$			
4-Cinnolinol, 6-chloro-3-nitro-	EtOH	222(4.42),343(4.11), 353s(4.05)	39-0470-66C
$C_8H_4Cl_4F_8N_4O_2$			
Dicyandiamide, dichlorotetra-fluoroacetone adduct	EtOH	214(4.17)	44-2568-66
$C_8H_4F_{14}N_4O$			
Hydrazine, 1,2-bis(heptafluoro-butyryl)-, hydrazone	iso-PrOH	272(4.05)	44-0781-66
$C_8H_4N_2OS$			
2-Benzothiazolecarbonitrile, 4-hydroxy-	EtOH	251(4.21),294(4.04), 353(3.54)	44-1484-66
2-Benzothiazolecarbonitrile, 5-hydroxy-	EtOH	227(4.28),287(4.02), 351(3.56)	44-1484-66
2-Benzothiazolecarbonitrile, 6-hydroxy-	EtOH	263(3.90),322(4.21)	35-2015-66
2-Benzothiazolecarbonitrile, 7-hydroxy-	EtOH	252(4.31),297(3.97), 347(3.37)	44-1484-66
$C_8H_4N_2O_2S$			
2-Benzothiazolecarbonitrile, 4,6-dihydroxy-	EtOH	265(4.07),316(4.07), 350(3.87)	44-1484-66
$C_8H_4N_2O_4$			
N,N'-Biisomaleimide	MeCN	292(4.26)	44-1311-66
N,N'-Bimaleimide	EtOH	215(4.28),253(4.22)	44-1311-66
Pyridazino[1,2-a]pyridazine-1,4,6,9-tetrone	dioxan	228(4.04),340(3.30)	44-1311-66
$C_8H_4N_2O_5$			
Furo[2,3-b]pyridine-2-carboxylic acid, 5-nitro-	5% EtOH	249(4.45)	4-0202-66
$C_8H_4N_2O_6$			
2,2'-Bifuran, 5,5'-dinitro-	$CHCl_3$	261s(3.72),377(4.43)	39-0976-66C
$C_8H_4N_4O_2S_2$			
Dipyridazino[4,5-b:4,5-e]-1,4-di-thiin-1,6(2H,7H)-dione	EtOH	214(4.57),294(3.64)	4-0541-66
$C_8H_4N_4O_4$			
1,2,4-Triazoline-3,5-dione, 4-(p-nitrophenyl)-	CH_2Cl_2	540(2.18)	44-3444-66

Compound	Solvent	λ max(log ε)	Ref.
$C_8H_4N_4S_4$			
Dipyridazino[4,5-b:4,5-e]-1,4-dithiin-1,6(2H,7H)-dithione	5% NaOH	218(--),250(--), 276(--),366(--)	4-0541-66
$C_8H_4O_2$			
Benzocyclobutenedione	90% EtOH	225(4.67),286(3.62), 292(3.79),301(3.82), 424(1.57)	12-1045-66
$C_8H_5BrN_2O$			
Acetophenone, 3'-bromo-2-diazo-	hexane	218(4.39),251(4.07), 289(4.05)	78-0209-66
	butanol	224(4.23),251(4.01), 296(4.09)	78-0209-66
Acetophenone, 4'-bromo-2-diazo-	hexane	214(4.17),262(4.29), 293s(4.06)	78-0209-66
	butanol	219(4.05),262(4.18), 301(4.15)	78-0209-66
$C_8H_5BrN_2O_4$			
Styrene, β-bromo-p,β-dinitro-	n.s.g.	320(4.34)	67-0042-66
$C_8H_5ClN_2O$			
Acetophenone, 3'-chloro-2-diazo-	hexane	212(4.41),250(4.05), 289(4.03)	78-0209-66
	butanol	210(4.17),250(4.03), 295(4.10)	78-0209-66
Acetophenone, 4'-chloro-2-diazo-	hexane	215(4.15),258(4.21), 294s(4.04)	78-0209-66
	butanol	216(4.06),258(4.13), 301(4.10)	78-0209-66
$C_8H_5ClN_2O_2$			
2-Quinoxalinol, 7-chloro-, 1-oxide	EtOH	213(4.38),238(4.36), 350(3.74)	30-0577-66C
$C_8H_5ClN_2S$			
1,3,4-Thiadiazole, 2-chloro-5-phenyl-	MeOH	277(4.12)	44-3528-66
$C_8H_5ClN_4O$			
3H-Indazole, 6-chloro-3-diazo-5-methoxy-	n.s.g.	354(3.90)	5-0017-66G
$C_8H_5FN_2$			
Quinoxaline, 2-fluoro-	EtOH	210(3.96),235(4.29), 238s(--),313(3.69), 325s(--)	4-0435-66
$C_8H_5FN_2O$			
Acetophenone, 2-diazo-3'-fluoro-	hexane	211(4.10),247(4.03), 285(4.00)	78-0209-66
	butanol	216(3.93),248(3.99), 298(4.04)	78-0209-66
Acetophenone, 2-diazo-4'-fluoro-	hexane	210(4.18),252(4.07), 290(3.95)	78-0209-66
	butanol	216(3.96),253(4.05), 297(4.09)	78-0209-66

Compound	Solvent	$\lambda_{max}(\log \epsilon)$	Ref.
$C_8H_5NO_2$			
Glyoxylonitrile, phenyl-, N-oxide	pH 4	276(4.13)	28C-0985-66A
$C_8H_5N_3O_2$			
1,2,4-Triazoline-3,5-dione, 4-phenyl-	CH_2Cl_2	545(2.13)	44-3444-66
$C_8H_5N_3O_3$			
Acetophenone, 2-diazo-2-nitro-	MeOH	260(4.08)	88-5821-66
Acetophenone, 2-diazo-3'-nitro-	hexane	224(3.75),275s(3.53), 295(3.42)	78-0209-66
	butanol	233(4.26),300(4.06)	78-0209-66
Acetophenone, 2-diazo-4'-nitro-	hexane	204(4.09),237(4.09), 263(4.17),300(4.01)	78-0209-66
	butanol	214(3.88),244s(4.09), 266(4.12),310(4.06)	78-0209-66
4-Cinnolinol, 6-nitro-	EtOH	236(4.14),258s(3.87), 266(3.86),275s(3.76), 322(3.85),365(3.99)	39-0470-66C
4-Cinnolinol, 8-nitro-	EtOH	230(4.16),255s(3.99), 280s(3.79),388(4.01)	39-0470-66C
$C_8H_5N_3O_3S$			
Thiazole, 2-[(5-nitrofurfurylidene)amino]-	EtOH	225(3.75),255(3.67), 305(3.87),378(4.04)	25-0031-66
$C_8H_5N_3O_4$			
Indole, 3,4-dinitro-	EtOH	231(4.06),291(3.88), 334(3.94)	44-0070-66
Indole, 3,5-dinitro-	EtOH	248(4.32),308(3.99), 346(3.99)	44-0070-66
Indole, 3,6-dinitro-	EtOH	227(3.99),295(4.21), 334(4.15)	44-0070-66
Indole, 3,7-dinitro-	EtOH	225(4.20),347(4.16)	44-0070-66
$C_8H_5N_3S$			
2-Benzothiazolecarbonitrile, 6-amino-	EtOH	241(3.83),277(3.80), 365(4.19)	35-2015-66
$C_8H_6BrNO_2$			
Styrene, β-bromo-β-nitro-	n.s.g.	322(4.13)	67-0042-66
C_8H_6ClN			
Indole, 2-chloro-	MeOH	215(4.70),266(3.94), 277(3.88),281(3.88), 288(3.79)	44-2627-66
$C_8H_6ClNO_2$			
Glyoxyloyl chloride, phenyl-, 1-oxime	pH 1	264(3.95)	28C-0985-66A
$C_8H_6Cl_3N_3O$			
Picolinic acid, (2,2,2-trichloroethylidene)hydrazide	EtOH	270(4.15)	95-0823-66
$C_8H_6N_2O$			
Acetophenone, 2-diazo-	hexane	211(4.13),251(4.11), 280s(4.01),290s(3.97)	78-0209-66
	C_6H_{12}	210(4.05),251(4.10), 280s(4.00),290s(3.96)	78-0209-66
	H_2O	200s(4.07),250(3.98), 297(4.18)	78-0209-66

Compound	Solvent	$\lambda_{max}(\log \epsilon)$	Ref.
Acetophenone, 2-diazo- (cont.)	EtOH	210(4.14),252(4.08), 297(4.10)	78-0209-66
	BuOH	210(3.79),252(4.08), 297(4.11)	78-0209-66
	$CHCl_3$	253(4.04),293(4.05)	78-0209-66
	MeCN	255(4.05),285s(4.03), 295(4.05)	78-0209-66
	dioxan	253(4.03),285s(3.98), 297(3.98)	78-0209-66
Cinnoline, 1-oxide	n.s.g.	230(4.4),303(3.7), 315(4.0),351(3.9), 367(3.9)	88-2899-66
Cinnoline, 2-oxide	n.s.g.	262(4.5),310(3.7), 352(3.7),361(3.7)	88-2899-66
Tropone, 2-amino-3-cyano-	MeOH	225s(4.12),248(4.30), 270(4.04),345(4.01), 406(4.06)	18-2538-66
Tropone, 2-amino-7-cyano-	MeOH	220(4.07),250(4.34), 273(4.13),283s(4.01), 349(3.99),424(4.31)	18-2538-66
$C_8H_6N_2OS$			
1,3,4-Oxadiazoline-5-thione, 2-phenyl-	EtOH	303(4.16)	1-0057-66
$C_8H_6N_2O_2$			
Benzyl cyanide, p-nitro-	EtOH-piperidine	530(1.44)	83-0131-66
	acetone-MeOH-KOH	528(4.60)	83-0131-66
+ formaldehyde	acetone-MeOH-KOH	541(4.26)	83-0131-66
+ benzaldehyde	EtOH-piperidine	550(1.80)	83-0131-66
+ m-nitrobenzaldehyde	EtOH-piperidine	550(2.14)	83-0131-66
+ o-nitrobenzaldehyde	EtOH-piperidine	550(2.04)	83-0131-66
Cinnoline, dioxide	n.s.g.	234(4.3),258(4.4), 273(4.6),340(3.8)	88-2899-66
Indole, 3-nitro-	EtOH	249(3.99),267(3.87), 273(3.85),351(3.99)	44-0070-66
Indole, 4-nitro-	EtOH	236(4.03),282s(3.38), 380(3.78)	24-0418-66
Indole, 5-nitro-	EtOH	265(4.25),324(3.92)	34-0418-66
Indole, 7-nitro-	EtOH	232(3.97),250(3.89), 365(3.86)	34-0418-66
2-Quinoxalinol, 1-oxide	EtOH	235(4.42),283(4.66), 350(3.72)	30-0577-66C
α-Tolunitrile, α-oximino-, N-oxide	pH 4	266(4.08)	28C-0985-66A
$C_8H_6N_2O_3$			
Anthranil, 3-methyl-5-nitro-	n.s.g.	256(4.27),298(3.56), 309(3.58),356(3.76)	78-0049-66A
$C_8H_6N_2O_4$			
Styrene, p,β-dinitro-	n.s.g.	298(4.60)	67-0042-66
$C_8H_6N_2S_2$			
1,3,4-Thiadiazoline-5-thione, 2-phenyl-	EtOH	337(4.26)	1-0057-66

Compound	Solvent	$\lambda_{max}(\log \epsilon)$	Ref.
$C_8H_6N_4O$			
3H-Indazole, 3-diazo-5-methoxy-	n.s.g.	352(3.80)	5-0017-66G
$C_8H_6N_4O_3$			
8(7H)-Pyrazino[2,3-d]pyridazinone, 5-acetyl-	EtOH	204(3.92),253(3.80), 317(3.18)	4-0512-66
$C_8H_6N_4O_8$			
Glycine, (trinitrophenyl)-	pH 7.4	345(4.18)	69-3385-66
$C_8H_6N_4S$			
2-Pyrimidinethiol, 5-(2-pyrazinyl)-	EtOH	255(3.66),340(4.37)	44-3845-66
C_8H_6O			
Benzofuran, tetracyanoethylene complex	$CHCl_3$	465(3.04)	60-0018-66
2,3-Octadiene-5,7-diyn-1-ol	EtOH	208(4.66),226(3.45), 237(3.60),249(3.93), 263(4.15),278(4.07)	39-0129-66C
3,5,7-Octatriyn-1-ol	EtOH	209.0(--)	39-0129-66C
$C_8H_6O_2$			
2,2'-Bifuran	EtOH	217(3.31),223(3.32), 256(3.96),269(4.22), 274(4.27),281(4.27), 287(4.18),295(4.09)	39-0976-66C
$C_8H_6O_5$			
2(5H)-Furanone, 5,5'-oxydi-	MeOH	209(3.83)	5-0042-66G
C_8H_6S			
Benzo[b]thiophene, tetracyanoethylene complex	$CHCl_3$	477(--),528(3.23)	60-0018-66
Benzo[c]thiophene	EtOH	272(3.34),278(3.34), 283(3.31),290(3.28), 305(3.16),313(3.01), 318(3.06),322(3.08), 328(3.13),333(3.15), 343(3.05),347(3.05)	88-1953-66
C_8H_7Br			
Styrene, m-bromo-	MeOH	248(4.14)	22-0043-66
Styrene, p-bromo-	MeOH	255(4.30)	22-0043-66
$C_8H_7BrN_2$			
Indazole, 6-bromo-3-methyl-	EtOH	255(4.64),295(3.73), 303(3.68)	78-3131-66
$C_8H_7BrN_2O$			
2-Azaquinolizinium bromide, 2-oxide	EtOH	288?(3.99),258s(3.63), 265s(3.59),275s(3.52), 308s(3.50),320(3.62), 339(3.66),360(3.32)	44-0941-66
$C_8H_7BrO_2$			
p-Benzoquinone, 5-bromo-2,3-dimethyl-	EtOH	275(4.13)	23-2867-66
C_8H_7Cl			
Styrene, m-chloro-	MeOH	247(4.08)	22-0043-66

Compound	Solvent	λ_{max}(log ϵ)	Ref.
Styrene, p-chloro-	EtOH	253(4.27)	22-0043-66
$C_8H_7ClN_2O_2$			
Glyoxime, 1-chloro-2-phenyl-	pH 1	232(4.10)	28C-0985-66A
$C_8H_7ClN_2O_6S$			
3-Methyl-6-nitrothiazolo[3,2-a]-pyridinium perchlorate	EtOH	214(4.01),246(4.12), 312(3.92)	4-0027-66
3-Methyl-8-nitrothiazolo[3,2-a]-pyridinium perchlorate	EtOH	222(3.87),267(3.72), 307(3.88),409(4.22)	4-0027-66
$C_8H_7ClN_4O$			
7(8H)-Pteridinone, 4-chloro-6,8-dimethyl-	pH 9.5	217(4.30),223s(4.28), 233s(4.16),243s(3.97), 304(4.07),312s(4.06), 325s(3.82)	39-1065-66C
	pH 14.5	258(3.96),286(3.84)	39-1065-66C
$C_8H_7ClN_6S$			
Purine, 2-chloro-6-(2-imidazolin-2-ylthio)-, hydrochloride	pH 1	291(4.09)	44-1417-66
	pH 7	294(3.99)	44-1417-66
	pH 13	299(4.03)	44-1417-66
Spiro[imidazolidine-2,5'-[5H]thiazolo[3,4,5-gh]purine, 2'-chloro-	pH 1	213(4.18),292(3.92), 327(3.73)	44-1417-66
	pH 7	220(4.18),312(3.96)	44-1417-66
	pH 13	227(4.15),309(4.04)	44-1417-66
$C_8H_7ClO_2$			
p-Benzoquinone, 5-chloro-2,3-dimethyl-	EtOH	267(4.12)	23-2867-66
Tropone, 2-chloro-5-methoxy-	MeOH	225(4.25),338(4.13)	18-1310-66
$C_8H_7Cl_2NO_4S$			
6-Chloro-3-methylthiazolo[3,2-a]pyridinium perchlorate	EtOH	221(4.20),245(4.20), 314s(4.07),323(4.16)	4-0027-66
8-Chloro-3-methylthiazolo[3,2-a]pyridinium perchlorate	EtOH	242(4.20),312s(4.06), 322(4.20)	4-0027-66
$C_8H_7Cl_3O$			
Bicyclo[3.1.0]hex-3-en-2-one, 6-methyl-6-(trichloromethyl)-	dioxan	226(3.86),334(1.46)	35-1825-66
C_8H_7N			
Indole, tetracyanoethylene complex	$CHCl_3$	543(3.48)	60-0018-66
$C_8H_7NO_2$			
Styrene, β-nitro-	n.s.g.	300(4.30)	67-0042-66
Styrene, m-nitro-	MeOH	240(4.29)	22-0043-66
Styrene, p-nitro-	EtOH	300(4.1)	65-1202-66
$C_8H_7NO_3$			
Phenol, o-(2-nitrovinyl)-	C_6H_{12}	236(3.95),290(4.10), 328(4.02)	19-0023-66
	20% MeOH	200(4.34),220s(3.84), 246(3.74),314(4.00), 356(3.98)	19-0023-66
Tropone, 2-amino-7-carboxy-	MeOH	249(4.25),272(4.00), 344(3.88),418(4.19)	18-2538-66

Compound	Solvent	$\lambda_{max}(\log \epsilon)$	Ref.
$C_8H_7NO_3S$			
1,3-Benzoxathian, 6-nitro-	MeOH	309(3.98)	35-5855-66
	MeOH-KOH	309(3.98)	35-5855-66
2H-1,2-Benzothiazin-4(3H)-one, 1,1-dioxide	EtOH	248(3.93),286(3.13), 297s(3.08)	44-0162-66
$C_8H_7NO_4$			
Benzoic acid, p-nitro-, methyl ester	EtOH	220(3.7),260(4.1)	65-1198-66
C_8H_7NS			
Isothiocyanic acid, benzyl ester	hexane	251.5(3.18)	88-5455-66
Thiocyanic acid, benzyl ester	hexane	260(2.67)	88-5455-66
$C_8H_7NS_2$			
Pyrrolo[2,1-b]thiazole-5-thiocarboxaldehyde, 6-methyl-	C_6H_{12}	407s(4.43),418(4.53), 427(4.55)	77-0401-66
$C_8H_7N_3$			
3,5-Pyridinedicarbonitrile, 1,4-dihydro-1-methyl-	n.s.g.	215(4.21),370(3.76)	39-1083-66B
3,5-Pyridinedicarbonitrile, 1,4-dihydro-4-methyl-	n.s.g.	210(4.23),234s(3.73), 347(3.83)	39-1083-66B
Quinazoline, 2-amino-	pH 0.0	229(4.32),245s(4.16), 247(4.17),284(3.64), 347(3.36)	39-0234-66C
	pH 9.0	232(4.47),247s(4.19), 258s(3.67),350(3.41)	39-0234-66C
$C_8H_7N_3O$			
Imidazo[4,5-b]pyridine, 2,3-(methyleneoxymethylene)-	EtOH	250(3.61),281s(3.90), 284(3.92),289s(3.83)	39-0080-66C
Imidazo[4,5-c]pyridine, 1,2-(methyleneoxymethylene)-	EtOH	253(3.6),259(3.61), 265(3.63),275(3.52)	39-0080-66C
1,2,4-Oxadiazole, 3-amino-5-phenyl-	EtOH	239(4.21)	39-1522-66C
1,2,4-Oxadiazole, 5-amino-3-phenyl-	EtOH	225(4.23)	39-1522-66C
v-Triazolo[3,4-a]pyridine, 3-acetyl-	MeOH	247s(3.58),254(3.65), 287(4.05),305(4.14)	24-2918-66
$C_8H_7N_3O_3$			
Benzofuroxan, 5-acetamido-	50% EtOH	250(4.2),330f(3.7), 395(3.6)	39-0971-66C
$C_8H_7N_3O_4$			
Formic acid, p-nitrophenylazo-, methyl ester	dioxan	283(4.24),430(2.28)	24-3337-66
1(2H)-Pyrimidineacetic acid, 5-cyano-3,4-dihydro-6-methyl-2,4-dioxo-	0.08N HCl	217(4.03),276(4.13)	25-0381-66
	0.08N NaOH	277(4.03)	25-0381-66
$C_8H_7N_3O_4S$			
Furfuryl alcohol, 5-nitro-α-(2-thiazolylamino)-	EtOH	229s(3.72),257(3.87), 307(4.05)	25-0031-66
$C_8H_7N_3S$			
4H-1,2,4-Triazole-3-thiol, 4-phenyl-	EtOH	269(3.90)	39-0001-66C
$C_8H_7N_5$			
Pyrimidine, 2-amino-5-(2-pyrazinyl)-	EtOH	236(3.75),266(4.21), 295(4.31)	44-3845-66

Compound	Solvent	λ_{max}(log ε)	Ref.
$C_8H_7N_5O_2$			
7(8H)-Pteridinone, 2-acetamido-	pH -0.89	232(4.46),281(3.97), 307(4.02)	24-0536-66
	pH 5.0	232s(3.95),280(4.21), 321(4.21)	24-0536-66
	pH 10.0	231(4.60),262s(3.76), 334(4.20)	24-0536-66
$C_8H_7O^+$			
p-Tolyloxocarbonium ion	H_2SO_4-SO_3	293.2(4.14)	35-1488-66
$C_8H_7O_2^+$			
p-Anisyloxocarbonium ion	H_2SO_4-SO_3	232(3.97),299(4.32)	35-1488-66
C_8H_8			
Semibullvalene	n.s.g.	230(3.39)	35-0183-66
Styrene	EtOH	248(4.14)	22-0043-66
tetracyanoethylene complex	$CHCl_3$	485(3.10)	60-0018-66
$C_8H_8Br_2O_2$			
2-Cyclohexene-1,4-dione, 5,6-dibromo-2,3-dimethyl-	hexane	222(3.77),267(3.92)	23-2867-66
C_8H_8ClN			
Methylamine, N-(p-chloro-benzylidene)-	C_6H_{12}	251(4.34)	35-2775-66
C_8H_8ClNO			
Tropone, 2-chloro-5-(methylamino)-	MeOH	230(4.29),365(3.75), 390(4.32)	18-1310-66
$C_8H_8ClNO_4S$			
3-Methylthiazolo[3,2-a]pyridinium perchlorate	H_2O	207s(3.99),233(4.10), 308s(4.10),312(4.20)	4-0027-66
$C_8H_8ClN_5$			
Pteridine, 7-chloro-2-(dimethylamino)-	MeOH	245(4.31),289(4.02), 408(3.87)	24-0536-66
$C_8H_8Cl_2N_2O$			
Pyrimidine, 5-[(allyloxy)methyl]-2,6-dichloro-	EtOH	218(3.97),262(3.56)	5-0119-66B
$C_8H_8Cl_2O_2$			
Benzene, 1,5-dichloro-2,4-dimethoxy-	$C_2H_4Cl_2$-SO_3	510(4.0)	65-1722-66
2-Cyclohexene-1,4-dione, 5,6-dichloro-2,3-dimethyl-	hexane	205(--),259(4.06)	23-2867-66
$C_8H_8Cl_2O_2S$			
Thiophene-2-carboxylic acid, 4,5-bis-(chloromethyl)-, methyl ester	EtOH	263(4.05)	44-3363-66
$C_8H_8F_8N_2O_2$			
Adipimidic acid, octafluoro-, dimethyl ester	EtOH	none	65-0862-66
$C_8H_8N_2$			
2-Azabicyclo[3.2.1]octa-3,6-diene, N-cyano-	isooctane	223(3.69),251(3.45)	44-1131-66
$\Delta^{1,\alpha}$-Cyclopentanemalononitrile	EtOH	238(4.04)	44-2784-66

Compound	Solvent	λ_{max}(log ε)	Ref.
Indazole, 3-methyl-	EtOH	255(3.42),290(3.58)	78-3131-66
$C_8H_8N_2O$			
Diimide, acetylphenyl-	dioxan	287(4.03),439(1.94)	24-3337-66
	n.s.g.	287(4.03),439(1.94)	89-0372-66
$C_8H_8N_2O_2$			
Benzaldehyde, 2-(methylnitrosamino)-	H_2O	241(4.22),287s(3.53)	39-0956-66B
Formic acid, phenylazo-,	dioxan	257(4.00),420(2.11)	24-3337-66
methyl ester	n.s.g.	257(4.00),420(2.12)	89-0372-66
$C_8H_8N_2O_4$			
1,1,2,2-Ethanetetracarboxylic 1,2;1,2-diimide, N,N-dimethyl-	MeCN	243(2.61),256(2.62), 265(2.59)	78-0279-66B
$C_8H_8N_4O$			
4-Pteridinol, 2,6-dimethyl-	pH 6.4	234(3.99),266(3.78), 320(3.79)	39-1112-66C
	pH 11.0	245(4.26),339(3.77)	39-1112-66C
4-Pteridinol, 2,7-dimethyl-	pH 6.4	233(4.07),270(3.70), 313(3.93)	39-1112-66C
	pH 11.0	245(4.29),331(3.86)	39-1112-66C
Pyrazolo[3,2-c]-as-triazine, 3-acetyl-4-methyl-	EtOH	250(4.36),315(3.75), 360s(3.38)	39-1127-66C
$C_8H_8N_4O_2$			
Lumazine, 3,8-dimethyl-	pH -1.9	241(4.03),285s(3.26), 343(4.03)	24-3503-66
	pH 4.0	277(4.16),308(3.75)	24-3503-66
	pH 7	259(4.32),394(3.97)	24-3503-66
	pH 13	230(4.30),282(4.13), 306(3.96)	24-3503-66
Lumazine, 8-ethyl-	pH -2.7	242(4.01),275(3.15), 344(4.01)	24-3503-66
	pH 6	258(4.17),394(3.94)	24-3503-66
	pH 13	230(4.16),266(4.00), 280(3.96),307(3.79), 405(3.41)	24-3503-66
7(8H)-Pteridinone, 2-hydroxy-6,8-dimethyl-	pH 5.0	218(4.41),258(4.11), 266s(4.08),328(4.06)	39-1065-66C
	pH 10.0	210(4.45),231(4.00), 266(3.61),290(3.70), 295s(3.69),345(4.24)	39-1065-66C
Pyrazino[2,3-d]pyridazine, 5,8-dimethoxy-	EtOH	205(3.92),255(4.28)	4-0512-66
$C_8H_8N_4O_3$			
Lumazine, 8-(2-hydroxyethyl)-	pH -2.7	242(4.04),272s(3.28), 347(4.00)	24-3503-66
	pH 6.0	258(4.18),398(3.97)	24-3503-66
	pH 13	231(4.31),280(4.11), 304(3.92),400(2.60)	24-3503-66
$C_8H_8N_4O_3S$			
Propionic acid, 3-(2-hydroxypurin-6-ylthio)-	pH 1	266(3.91),310(4.02)	73-1824-66
	pH 13	225(4.22),319(4.06)	73-1824-66
Propionic acid, 3-(8-hydroxypurin-6-ylthio)-	pH 1	226(4.11),296(4.23)	73-1824-66
	pH 13	223(4.18),302(4.23)	73-1824-66
Propionic acid, 3-(2-hydroxy-6-mertopurin-3-yl)-	pH 1	223(3.91),254(3.77), 339(4.37)	73-1824-66

Compound	Solvent	λ_{max}(log ε)	Ref.
Propionic acid, 3-(2-hydroxy-6-mercaptopurin-3-yl)- (cont.)	pH 13	227(4.25),257(3.88), 344(4.38)	73-1824-66
$C_8H_8N_4O_4$			
2-Imidazolidinone, 1-(5-nitrofurfurylideneamino)-	pH 7.4	270(4.1),385(4.25)	73-1824-66
$C_8H_8N_4O_4S$			
Propionic acid, 3-(2,8-dihydroxypurin-6-ylthio)-	pH 1	223(4.37),340(4.04)	73-1824-66
	pH 13	226(4.29),327(4.18)	73-1824-66
$C_8H_8N_4O_6$			
Aniline, N,N-dimethyl-2,4,6-trinitro-	EtOH	340s(3.98),370(4.08)	65-1034-66
Aniline, N-ethyl-2,4,6-trinitro-	EtOH	337(4.16),410(3.78)	65-1034-66
$C_8H_8N_4S$			
1,3,4-Thiadiazole, 2-hydrazino-5-phenyl-	MeOH	305(3.95)	44-3528-66
1,2,4-Triazole, 3-amino-5-mercapto-4-phenyl-	EtOH	265(3.93)	39-0001-66C
$C_8H_8N_6$			
Bis-s-triazolo[4,3-b:3',4'-f]pyridazine, 1,8-dimethyl-	EtOH	294(3.92)	78-2073-66
$C_8H_8N_6S$			
Purine, 6-(2-imidazolin-2-ylthio)-, hydrochloride	pH 1	280(4.16)	44-1417-66
	pH 7	282(4.07)	44-1417-66
	pH 13	291(4.04)	44-1417-66
Spiro[imidazolidine-2,5'[5H]thiazolo-[3,4,5-gh]purine]	pH 7	283(3.99),312(3.89)	44-1417-66
	pH 13	302(4.09)	44-1417-66
C_8H_8O			
Acetophenone (protonated)	FSO_3H-SbF_5	296(4.27),335(3.33)	35-1488-66
Benzofuran, 2,3-dihydro-	EtOH	283(3.59)	22-0228-66
Phenol, o-vinyl-	EtOH	245(3.85),249s(3.85), 274s(3.23),283(3.33), 294s(3.42),303(3.46), 318s(3.14)	1-1561-66
C_8H_8OS			
1,3-Benzoxathian	MeOH	273(3.23),281(3.26)	35-5855-66
$C_8H_8O_2$			
Acetic acid, phenyl ester	EtOH	260(2.48)	65-1198-66
Benzoic acid, methyl ester	EtOH	230(4.0),270(3.0), 280(3.0)	65-1198-66
$C_8H_8O_2S_2$			
Thieno[3,4-b]thiophene-2-carboxylic acid, 4,6-dihydro-, methyl ester	EtOH	249(3.96),290(3.85)	44-3363-66
$C_8H_8O_3$			
Acetic acid, p-hydroxyphenyl ester	EtOH	223(3.83),278(3.33)	65-1198-66
Acetic acid, p-hydroxyphenyl-	70% MeOH	225(3.8),278(3.2)	83-0303-66
p-Benzoquinone, (1-hydroxyethyl)-	EtOH	246(4.25),440(1.30)	39-1627-66C
p-Benzoquinone, (methoxymethyl)-	C_6H_{12}	244(4.31),421s(--), 440(1.30),455s(--)	39-1627-66C
Cyclohexen-1-ylacetic acid, 2-hydroxy-6-oxo-, lactone	EtOH	255(4.10),293s(--)	39-2100-66C

Compound	Solvent	λ_{max}(log ε)	Ref.
o-Vanillin	MeOH	230(4.02),265(3.98), 340(3.45)	65-1590-66
$C_8H_8O_3S$			
2-Thiophenecarboxaldehyde, 3-(1,3-dioxolan-2-yl)-	EtOH	247(3.16),253(3.81), 259(4.00),265(4.07), 267(4.11),291s(3.84)	44-3592-66
$C_8H_8O_4$			
Propionic acid, 3-α-furoyl-	EtOH	225(3.39),270(4.19)	12-0683-66
$C_8H_8O_4S$			
2-Pyrone, 3-(methylthio)-5-carbomethoxy-	ether	323(4.00)	24-1558-66
2,4-Thiophenedicarboxylic acid, dimethyl ester	ether	252(3.85)	24-1558-66
$C_8H_8O_4S_2$			
Thieno[3,4-b]thiophene-2-carboxylic acid, 4,6-dihydro-, methyl ester	EtOH	253(4.01),278(3.97)	44-3363-66
C_8H_8S			
Benzo[b]thiophene, 2,3-dihydro-	EtOH	250(3.94)	22-0228-66
C_8H_9Br			
Benzene, 2-bromo-1-ethyl-	EtOH	267(2.48),274(2.54)	22-1012-66
Benzene, 3-bromo-1-ethyl-	EtOH	261(2.40),268(2.50), 274(2.40)	22-1012-66
Benzene, 4-bromo-1-ethyl-	EtOH	262(2.52),269(2.61), 277(2.54)	22-1012-66
$C_8H_9BrO_2S$			
Benzenesulfinic acid, p-bromo-, ethyl ester	hexane	232(4.20),260(3.71)	18-1788-66
	MeOH	233(4.17),250(3.80)	18-1788-66
	ether	232(4.21),258(3.75)	18-1788-66
$C_8H_9ClN_2O$			
Pyridine, N-chlorocarbonyl-2-ethylamino-	heptane	273(3.33)	32-1020-66
$C_8H_9ClN_4O_4$			
2-Pyridinecarbamic acid, 4-amino-6-chloro-5-nitro-, ethyl ester	pH 1, 7	222(4.45),244s(4.18), 308(3.77)	44-1890-66
	pH 13	223(4.31),237s(4.27), 273(3.91),370(3.95)	44-1890-66
2-Pyridinecarbamic acid, 6-amino-4-chloro-5-nitro-, ethyl ester	pH 1	231(4.06),264(3.97), 295s(3.64),391(3.91)	44-1890-66
	pH 7	231(4.03),265(3.99), 295s(3.64),395(3.91)	44-1890-66
	pH 13	283(3.74),326(3.68), 400(4.19)	44-1890-66
$C_8H_9ClO_2S$			
Benzenesulfinic acid, p-chloro-, ethyl ester	hexane	227(4.14),260(3.58)	18-1788-66
	MeOH	228(4.13),250(3.73)	18-1788-66
$C_8H_9ClO_3S$			
Benzenesulfonyl chloride, 4-hydroxy-3,5-dimethyl-	n.s.g.	272(3.98)	44-2672-66

Compound	Solvent	λ max(log ε)	Ref.
C_8H_9N			
Aniline, m-vinyl-	H_2O	225(4.33)	22-0043-66
	acid	247(4.13)	22-0043-66
Methylamine, N-benzylidene-	C_6H_{12}	244(4.16)	35-2775-66
Pyridine, 4-cyclopropyl-	EtOH-HCl	256.5(4.11)	44-0399-66
	EtOH-NaOH	259(3.38)	44-0399-66
3H-Pyrrolizine, 1-methyl-	EtOH	287(3.79)	44-0870-66
C_8H_9NO			
Acetophenone, o-amino-	ether	252(3.87),362(3.80)	39-1433-66C
	MeOH	227(4.3),257(3.8), 360(3.8)	49-0409-66
Anthranilaldehyde, N-methyl-	2N HCl	247(4.05),283(3.20)	39-0956-66B
	H_2O	223(4.33),230s(4.31), 263(3.80),267s(3.77), 387(3.72)	39-0956-66B
1H-Pyrrolizin-1-one, 2,3-dihydro-7-methyl-	n.s.g.	288(4.22)	57-0583-66A
C_8H_9NOS			
3-Buten-2-one, 4-(2-thienyl)-, α-oxime	n.s.g.	306.5(4.48)	83-0626-66
3-Buten-2-one, 4-(2-thienyl)-, β-oxime	n.s.g.	313(4.31)	83-0626-66
4-Pyridinethiocarboxylic acid, 2-ethyl-	EtOH	278(3.72)	88-1145-66
$C_8H_9NO_2$			
Anthranilic acid, methyl ester	2N HCl	273(3.02)	28C-0362-66A
	pH 9.2	327(3.62)	28C-0362-66A
1H-Azepine-2,5-dione, 3,6-dimethyl-	EtOH	228(3.26),288(2.53)	78-1201-66
Benzaldehyde imine, 2-hydroxy-3-methoxy-	MeOH	237(4.61),261(4.25), 283(4.24),420(3.77)	65-1590-66
Benzene, 1-ethyl-2-nitro-	EtOH	256(3.76)	22-1012-66
Benzene, 1-ethyl-4-nitro-	EtOH	272(3.92)	22-1012-66
Benzoic acid, m-amino-, methyl ester	HCl	272(2.95)	28C-0362-66A
	pH 9.2	312(3.28)	28C-0362-66A
Benzoic acid, p-amino-, methyl ester	2N HCl	270(3.00)	28C-0362-66A
	pH 9.2	284(4.25)	28C-0362-66A
3-Buten-2-one, 4-(2-furyl)-, oxime	n.s.g.	296.5(4.49)	83-0626-66
isomer	n.s.g.	304.5(4.45)	83-0626-66
3,5-Hexadienoic acid, 6-cyano-, methyl ester	EtOH	254(4.39)	77-0141-66
Phenol, 2,3-dimethyl-4-nitroso-	EtOH	297(4.23)	12-0841-66
	EtOH-NaOH	262(3.70),418(4.36)	12-0841-66
	dioxan	290(4.26),750(-0.4)	12-0841-66
	dioxan-NaOH	264(3.68),408(4.29)	12-0841-66
Phenol, 2,5-dimethyl-4-nitroso-	EtOH	306(4.28),400(3.65)	12-0841-66
	EtOH-NaOH	260(3.59),400(4.49)	12-0841-66
	dioxan	295(4.32),750(-1)	12-0841-66
	dioxan-NaOH	264(3.70),390(4.38)	12-0841-66
Phenol, 2,6-dimethyl-4-nitroso-	EtOH	308(4.18),401(3.75)	12-0841-66
	EtOH-NaOH	261(3.70),400(4.32)	12-0841-66
	dioxan	302(4.30),750(-2)	12-0841-66
	dioxan-NaOH	264(3.70),396(4.38)	12-0841-66
Phenol, 3,5-dimethyl-4-nitroso-	EtOH	299(3.79),390(3.92)	12-0841-66
	EtOH-NaOH	266(3.08),390(4.20)	12-0841-66
	dioxan	289(4.32),773(0.52)	12-0841-66
	dioxan-NaOH	270(3.74),388(4.45)	12-0841-66
Pyridine, 4-(allyloxy)-, 1-oxide	H O	260(4.45)	24-0368-66
2(1H)-Pyridone, 1-(allyloxy)-	EtOH	229(3.78),301(3.73)	78-0025-66
Sorbic acid, 6-cyano-, methyl ester	EtOH	250(4.36)	77-0141-66

Compound	Solvent	$\lambda_{max}(\log \epsilon)$	Ref.
$C_8H_9NO_2S$			
Pyrrole-2-carbothioic acid, 4-formyl-5-methyl-, S-methyl ester	EtOH	229(4.30),273(4.04), 307(4.26)	23-1007-66
$C_8H_9NO_2S_2$			
4H-1,3-Thiazin-4-one, 5-acetyl-2-(ethylthio)-	EtOH	258(4.37),278s(4.24)	88-3225-66
$C_8H_9NO_3$			
Nudifloric acid, methyl ester	EtOH	205(4.16),264(4.20), 300(3.66)	78-1461-66
Phenol, 5,6-dimethyl-2-nitro-	C_6H_{12}	216(4.07),355(3.51)	23-0961-66
	EtOH	215(4.02),355(3.46)	23-0961-66
2(1H)-Pyridone, 4-acetoxy-6-methyl-	EtOH	229(3.82),300(3.85)	70-1546-66
2(1H)-Pyridone, 3-acetyl-4-hydroxy-6-methyl-	EtOH	230(4.06),266(3.41), 326(4.12)	70-0848-66
	EtOH	230(4.06),266(3.41) 325(4.12)	70-1546-66
$C_8H_9NO_3S$			
Anisole, α-(methylthio)-o-nitro-	MeOH	257(3.46),320(3.28)	35-5855-66
	MeOH-KOH	257(3.46),320(3.28)	35-5855-66
Anisole, α-(methylthio)-p-nitro-	MeOH	309(4.06)	35-5855-66
o-Cresol, α-(methylthio)-4-nitro-	MeOH	318(3.84)	35-5855-66
	MeOH-KOH	410(4.28)	35-5855-66
o-Cresol, α-(methylthio)-6-nitro-	MeOH	280(3.77),353(3.57)	35-5855-66
	MeOH-KOH	263(4.09),425(3.73)	35-5855-66
$C_8H_9NO_4$			
Benzene, 1,4-dimethoxy-2-nitro-	MeOH	357(3.39)	78-0093-66B
Benzoic acid, 4-amino-2,6-dihydroxy-, methyl ester	EtOH	235(3.92),292(4.24)	12-0275-66
Phenol, 2-ethoxy-5-nitro-	pH 13	227(3.98),264(4.06), 323(3.67),418(3.58)	94-0939-66
	EtOH	244(3.92),302(3.71), 344(3.81)	94-0939-66
$C_8H_9NO_4S$			
Benzenesulfinic acid, p-nitro-, ethyl ester	hexane	250(4.07),288(3.63)	18-1788-66
	MeOH	254s(4.03),280s(3.79)	18-1788-66
o-Cresol, α-(methylsulfinyl)-6-nitro-	MeOH	275(3.74),353(3.51)	35-5855-66
	MeOH-base	235(4.09),260(3.67)	35-5855-66
$C_8H_9N_3$			
Imidazo[1,2-a]pyridine, 5-amino-2-methyl-, hydrochloride	H_2O	225(4.48),313(4.05)	4-0093-66
s-Triazolo[4,3-a]pyridine, 3,5-dimethyl-	EtOH	215(4.06),273(3.66), 291(3.69)	30-0361-66D
s-Triazolo[4,3-a]pyridine, 3-ethyl-	MeOH	268(3.9)	24-2593-66
$C_8H_9N_3O$			
s-Triazolo[1,5-a]pyridine, 1-ethyl-2-oxo-	EtOH	270(3.86),306(3.67)	32-1020-66
s-Triazolo[4,3-a]pyridin-3(2H)-one, 1-ethyl-	EtOH	235(4.13),281(3.56), 342(3.53)	7-0190-66
$C_8H_9N_3O_2$			
Uracil, 5-cyano-6-ethyl-1-methyl-	0.08N HCl	221(4.02),280(4.13)	25-0381-66
	0.08N NaOH	278(3.99)	25-0381-66

Compound	Solvent	λ_{max}(log ε)	Ref.
$C_8H_9N_3O_4$			
Aniline, N-ethyl-2,4-dinitro-	EtOH	259(3.95),347(4.22), 410(3.81)	65-1048-66*
$C_8H_9N_3O_5$			
6-Azauracil, 1-(O^2,2'-cyclo-β-D-arabinofuranosyl)-	H_2O	254(3.877)	73-4002-66
$C_8H_9N_3S$			
1,3,5-Triazaindene, 1-methyl-2-(methylthio)-	pH 4.0	238(4.49),289(3.99)	39-0285-66B
	pH 9.0	219(4.43),278(4.03), 284s(4.00)	39-0285-66B
1,3,5-Triazaindene, 3-methyl-2-(methylthio)-	pH 4.0	229(4.24),294(4.37)	39-0285-66B
	pH 8.5	218(4.38),267(3.99), 275s(4.02),281(4.06), 289(3.96)	39-0285-66B
1,3,5-Triazaindene, 5-methyl-2-(methylthio)-	pH 2.0	238(4.42),295(4.17)	39-0285-66B
	pH 9.0	238(4.43),299(4.23)	39-0285-66B
$C_8H_9N_5$			
Pyrido[2,3-b]pyrazine, 6,8-diamino-2-methyl-	pH 1	221(4.57),334(4.20)	44-1890-66
	pH 7	220(4.53),255(3.97), 341(4.08)	44-1890-66
	pH 13	220(4.41),259(4.20), 354(3.98)	44-1890-66
$C_8H_9N_5O$			
7(8H)-Pteridinone, 2-(dimethylamino)-	pH -0.89	233(4.39),267(4.25), 343(3.81)	24-0536-66
	pH 5.0	223(4.17),245(4.17), 300(3.74),370(4.22)	24-0536-66
	pH 11.0	236(4.51),282(3.97), 364(4.16)	24-0536-66
$C_8H_9N_5O_2$			
Acetic acid, cyano[pyrazol-3(or 5)-ylazo]-, ethyl ester	EtOH	358(4.15)	39-1127-66C
Pyrazolo[3,2-c]-as-triazine-3-carboxylic acid, 4-amino-, ethyl ester	EtOH	234(4.14),310(4.05), 344(3.92)	39-1127-66C
$C_8H_9N_7S$			
Purine, 2-amino-6-(2-imidazolin-2-ylthio)-	pH 1	237s(4.15),329(3.94)	44-1417-66
	pH 7	324(3.98)	44-1417-66
	pH 13	321(4.00)	44-1417-66
C_8H_{10}			
Fulvene, 6,6-dimethyl-	C_6H_{12}	265(4.26),272(4.25), 359(2.55)	33-1278-66
	isóoctane	265(4.15),360(2.50)	44-4321-66
	EtOH	266(4.24),271(4.24), 357(2.58)	33-1278-66
$C_8H_{10}ClNO_2S$			
2-Thiophenecarboxylic acid, 3-(2-aminopropyl)-5-chloro-	HCl	255s(3.66),279(3.82)	4-0466-66
	NaOH	255s(3.67),277(3.80)	4-0466-66
$C_8H_{10}ClN_3O_3S$			
Morpholine, 4-[(6-chloro-3-pyridazinyl)sulfonyl]-	EtOH	274(4.20)	83-0648-66

Compound	Solvent	λ_{max}(log ε)	Ref.
$C_8H_{10}Cl_2N_2O_2$			
Pyrimidine, 2,4-dichloro-5-[(2-methoxyethoxy)methyl]-	EtOH	219(3.91),262(3.58)	5-0119-66B
$C_8H_{10}Cl_2N_2S$			
Pyrimidine, 2,4-dichloro-5-[(isopropylthio)methyl]-	EtOH	221(4.00),262(3.63)	5-0119-66B
$C_8H_{10}IN_3S$			
1-Methyl-3-(methylthio)-s-triazolo-[4,3-a]pyridinium iodide	MeOH	221(4.44),269(3.70),276(3.70),300(3.64)	44-0265-66
$C_8H_{10}NO_6P$			
Pyridoxal-5-phosphate	pH 7.1	330(3.38),388(3.70)	37-3424-66
$C_8H_{10}N_2$			
Malononitrile, (1,2-dimethyl-propylidene)-	EtOH	233(4.14)	44-2784-66
Malononitrile, (1-methylbutylidene)-	EtOH	234(4.12)	44-2784-66
$C_8H_{10}N_2O$			
Acetanilide, 2'-amino-	MeOH	234s(3.94),291(3.97)	78-1175-66
Acetanilide, 3'-amino-	MeOH	247s(4.08),296(3.40)	78-1175-66
Acetophenone, 2-amino-, oxime, syn, hydrochloride	EtOH	245.5(4.05)	24-1179-66
Acetophenone, 2'-amino-, oxime	MeOH	223(4.43),255s(3.74),322(3.52)	39-1433-66C
7-Azaindolin-6-ol, 4-methyl-	EtOH	260(3.8),356(4.1)	78-3233-66
	EtOH-dioxan(1:1)	260(3.8),355(4.1)	78-3233-66
Fumaronitrile, tert-butoxy-	ether	248.5(3.97)	24-2572-66
Maleonitrile, tert-butoxy-	ether	249(3.97)	24-2572-66
2-Pyrrolidinone, 5-pyrrol-2-yl-	EtOH-HCl	215(4.04)	78-0053-66
	EtOH	215(4.04)	78-0053-66
p-Toluidine, N-methyl-N-nitroso-	H_2O	274.0(3.89)	39-0533-66B
$C_8H_{10}N_2OS$			
p-Phenylenediamine, N,N-dimethyl-N'-sulfinyl-	C_6H_{12}	415(4.46)	35-3842-66
4(1H)-Pyrimidinone, 1-allyl-2-(methylthio)-	EtOH	234(4.46)	44-0406-66
4(1H)-Pyrimidinone, 3-allyl-2-(methylthio)-	EtOH	291(3.94)	44-0406-66
5H-Thiopyrano[2,3-d]pyrimidin-2-ol, 6,7-dihydro-4-methyl-	pH 13.5	226(4.12),248s(3.77),308(4.07)	24-0872-66
	MeOH	214(4.26),280(3.95),314(4.02)	24-0872-66
$C_8H_{10}N_2O_2$			
4,6-Pyrimidinediol, 1,2-tetra-methylene-	0.01M H_2SO_4	243(3.83)	39-0565-66B
	pH 4.8	255(3.97)	39-0565-66B
	pH 12	255(3.58)	39-0565-66B
2,4-Pyrimidinedione, 3-allyl-1-methyl-	EtOH	266(3.94)	44-0406-66
2-Pyrimidone, 4-(allyloxy)-1-methyl-	EtOH	276(3.79)	44-0406-66
$C_8H_{10}N_2O_2S$			
4,5-Thiazoledione, 2-piperidino-	dioxan	278(3.9),320s(3.7)	24-3572-66

Compound	Solvent	λ_{max}(log ε)	Ref.
$C_8H_{10}N_2O_4$			
Epoxydon, diazomethane adduct	EtOH	240(3.01),330(2.50)	33-0204-66
4-Pyridazineacetic acid, 1,6-dihydro-3-hydroxy-6-oxo-, ethyl ester	EtOH	310(3.46)	49-0644-66
6(1H)-Pyridazinone, 1-acetoxy-3-ethoxy-	EtOH	311(3.462)	95-0082-66
6(1H)-Pyridazinone, 1-acetoxy-3-methoxy-4-methyl-	EtOH	302(3.511)	95-0082-66
$C_8H_{10}N_2O_4S$			
4-Thiazoleacetic acid, 2-amino-5-carbomethoxy-, methyl ester	EtOH	300(4.23)	95-0300-66
$C_8H_{10}N_2O_5$			
4-Oxazoleacetic acid, 2-amino-5-carbomethoxy-, methyl ester	EtOH	280(4.185)	95-0300-66
$C_8H_{10}N_2S_2$			
5H-Thiopyrano[2,3-d]pyrimidine-2-thiol, 6,7-dihydro-4-methyl-	pH 13.5	269(4.43),318(3.69)	24-0872-66
	MeOH	210(3.94),289(4.50),353(3.53)	24-0872-66
$C_8H_{10}N_4OS$			
1,6-Dihydro-2-hydroxy-1,7,9-trimethyl-6-thioxopurinium betaine	pH 0	252(3.71),276(3.67),324(4.28)	24-0944-66
	pH 6	260(3.81),340s(4.27),348(4.28)	24-0944-66
1,6-Dihydro-2-mercapto-1,7,9-trimethyl-6-oxopurinium betaine	pH -0.89	241(4.00),286(4.31)	24-0944-66
	pH 6	235s(4.00),291(4.32)	24-0944-66
$C_8H_{10}N_4O_2$			
2-Imidazolidinone, 3-(4-cyano-2-oxobutylideneamino)-	H_2O	288(4.32)	10-0399-66A
Isocaffeine	neutral	239(3.88),267(3.95)	44-1411-66
1H-Pyrazolo[4,3-d]pyrimidine-5,7(4H,6H)-dione, 1,4,6-trimethyl-	EtOH	291(3.75)	44-2491-66
2H-Pyrazolo[3,4-d]pyrimidine-4,6(5H,7H)-dione, 2,5,7-trimethyl-	EtOH	240(3.73),264(3.85)	44-2491-66
2H-Pyrazolo[4,3-d]pyrimidine-5,7(4H,6H)-dione, 2,4,6-trimethyl-	MeOH	288(3.74)	44-2491-66
$C_8H_{10}N_4S_2$			
1,6-Dihydro-2-mercapto-1,7,9-trimethyl-6-thioxopurinium betaine	pH -0.89	296(4.35),340s(4.02)	24-0944-66
	pH 3	251(4.00),295(4.35),357(4.03)	24-0944-66
6-Mercapto-7,9-dimethyl-2-(methyl-	pH 1	227(4.12),245(4.11),268(3.99),338(4.21),350s(4.15)	24-0944-66
	pH 9.2	223(4.21),270(4.12),290s(3.92),332(4.20)	24-0944-66
$C_8H_{10}N_6$			
Acetaldehyde, (3-methyl-s-triazolo-[4,3-b]pyridazin-6-yl)hydrazone	EtOH	254(4.41),300(3.73)	78-2073-66
$C_8H_{10}N_6S$			
5H-Thiazolo[3,4,5-gh]purine, 5-amino-5-(ethylamino)-	pH 1	267(4.23)	44-1417-66
	pH 7	302(4.29)	44-1417-66
	pH 13	238(4.05),302(4.47)	44-1417-66

Compound	Solvent	$\lambda_{max}(\log \epsilon)$	Ref.
5H-Thiazolo[3,4,5-gh]purine, 5,5-diamino-2-ethyl-	pH 1	266(4.14)	44-1417-66
	pH 7	284(4.11)	44-1417-66
	pH 13	237(4.11),298(4.42)	44-1417-66
$C_8H_{10}O$			
Cyclobutanone, 2-cyclobutylidene-	EtOH	253(4.13)	25-1380-66
4-Octen-6-ynal	EtOH	227(4.00),323s(3.0)	28C-1433-66A
Phenetole	C_6H_{12}	220(--),272(3.30)	30-0877-66F
AlBr$_3$ complex	C_6H_{12}	256(2.21)	30-0877-66F
$C_8H_{10}OS$			
o-Cresol, α-(methylthio)-	MeOH	278(3.46)	35-5855-66
	MeOH-KOH	297(3.61)	35-5855-66
$C_8H_{10}O_2$			
Benzene, 1,4-dimethoxy-	MeOH	288(3.54)	78-0093-66B
1,4-Cyclohexadiene-1-carboxylic acid, 2-methyl-	EtOH	234(3.50)	22-2535-66
2(5H)-Furanone, 5-ethylidene-3,4-dimethyl-	n.s.g.	270(4.3)	54-0043-66
$C_8H_{10}O_2S$			
Benzenesulfinic acid, ethyl ester	hexane	215(3.96),257(3.46)	18-1788-66
	MeOH	216(3.89),246(3.52)	18-1788-66
$C_8H_{10}O_3$			
p-Benzoquinone, dimethyl acetal	MeOH	213(3.98)	44-4056-66
$C_8H_{10}O_4$			
Aconic acid, γ-propyl-	EtOH	216(4.03)	44-0616-66
$C_8H_{10}O_5$			
2,4-Hexadienedioic acid, 3-hydroxy-, dimethyl ester	ether	292(4.24)	24-2522-66
$C_8H_{10}O_6$			
Muconic acid, α,α'-dimethoxy-, trans-trans	EtOH	300(4.35)	33-0168-66
$C_8H_{10}S$			
Sulfide, ethyl phenyl	EtOH	255(3.93),278(3.04)	22-0228-66
$C_8H_{11}Br$			
2,4-Cyclooctadiene, 1-bromo-	C_6H_{12}	232(3.82)	88-0999-66
3-Hepten-5-yne, 1-bromo-4-methyl-	EtOH	230(4.04)	78-0443-66B
$C_8H_{11}BrN_2OS$			
Pyrimidine, 5-bromo-4-ethoxy-6-methyl-2-(methylthio)-	EtOH	261(3.87)	78-2401-66
$C_8H_{11}BrN_2O_2$			
3-Pyrazolecarboxylic acid, 4-bromo-1,5-dimethyl-, ethyl ester	EtOH	225(3.86),250s(3.57)	22-0293-66
5-Pyrazolecarboxylic acid, 4-bromo-1,3-dimethyl-, ethyl ester	EtOH	235(3.92),263(3.72)	22-0293-66
$C_8H_{11}Br_2CuNS$			
Copper, dibromo[2-[2-(methylthio)-ethyl]pyridine]-	$MeNO_2$	410(2.45),530s(1.60), 800(1.65)	12-1835-66

Compound	Solvent	λ_{max}(log ε)	Ref.
$C_8H_{11}Cl_2CuNS$			
Copper, dichloro[2-[2-(methylthio)-ethyl]pyridine]-	$MeNO_2$	785(2.18)	12-1835-66
$C_8H_{11}N$			
Aniline, 2,4-dimethyl-	2N HCl	262(2.46)	28C-0362-66A
	pH 9.2	286(3.28)	28C-0362-66A
Aniline, 2,6-dimethyl-	2N HCl	262(2.60)	28C-0362-66A
	pH 9.2	280(3.25)	28C-0362-66A
Aniline, o-ethyl-	2N HCl	259(2.40)	28C-0362-66A
	pH 9.2	280(3.26)	28C-0362-66A
Aniline, p-ethyl-	2N HCl	259(2.40)	28C-0362-66A
	pH 9.2	285(3.16)	28C-0362-66A
Ethylamine, α-phenyl-, mercuric chloride complex	EtOH	248(2.90),254(3.25), 260(3.20)	90-0147-66
$C_8H_{11}NO$			
Pyrrole-2-carboxaldehyde, 4-isopropyl-	EtOH	258(3.84),302(4.20)	23-1831-66
$C_8H_{11}NO_2$			
Anisole, 2,6-dimethyl-4-nitroso-	EtOH	229(3.81),294s(3.77), 323(3.97)	12-0841-66
	dioxan	229(3.81),296s(3.80), 323(3.94),747(1.66)	12-0841-66
Benzyl alcohol, α-(aminomethyl)-o-hydroxy-	EtOH	226(4.59),277(3.34)	6-0083-66
Pyrrole-3-carboxylic acid, 1,2-dimethyl-, methyl ester	EtOH	205(4.27),233(3.96), 252(3.88)	39-1950-66C
3-Pyrrolin-2-one, 5-(methoxymethylene)-3,4-dimethyl-	n.s.g.	302(4.34)	39-0040-66C
$C_8H_{11}NO_2S$			
2-Thiophenecarboxylic acid, 3-(2-aminopropyl)-	HCl	234(3.87),241(3.90), 270(4.04)	4-0466-66
	NaOH	234(3.88),242(3.90), 268(4.08)	4-0466-66
3-Thiophenecarboxylic acid, 2-(2-aminopropyl)-	HCl	211(4.42),246(4.04)	4-0466-66
	NaOH	220(3.98),246(4.09)	4-0466-66
$C_8H_{11}NO_3$			
4-Pentenal, 5-cyano-2,3-epoxy-, dimethyl acetal	EtOH	217(4.18)	70-0646-66
$C_8H_{11}NO_4$			
Fumaric acid, 1-aziridinyl-, dimethyl ester	ether	292(3.99)	24-0450-66
Maleic acid, 1-aziridinyl-, dimethyl ester	ether	254(4.27)	24-0450-66
$C_8H_{11}N_3O_2$			
5-Pyrimidinecarboxylic acid, 4-amino-2-methyl-, ethyl ester, cation	H_2O	245(4.15),282(3.66)	94-1321-66
	MeOH	246(4.16),281(3.64)	94-1321-66
	EtOH	246(4.16),283(3.62)	94-1321-66
	iso-PrOH	248(4.17),283(3.64)	94-1321-66
	HOAc	248(4.13),280(3.63)	94-1321-66
	MeCN	246(4.16),277(3.65)	94-1321-66
2H-Pyrimido[4,5-e]-1,3-oxazin-8-ol, 3,4-dihydro-3,6-dimethyl-	pH 1	270(4.13)	4-0115-66
	pH 11	270(4.12),294(4.13)	4-0115-66

Compound	Solvent	λ_{max}(log ϵ)	Ref.
$C_8H_{11}N_3O_5$			
as-Triazine-3,5(2H,4H)-dione, 2-(5-deoxy-β-D-ribofuranosyl)-	EtOH	265(3.68)	4-0226-66
$C_8H_{11}N_3O_6$			
6-Azauracil, 1-β-D-arabino- furanosyl-	H_2O	265(3.58)	73-4002-66
	pH 13	256(3.83)	73-4002-66
$C_8H_{11}N_3S$			
5H-Thiopyrano[2,3-d]pyrimidine, 2-amino-6,7-dihydro-4-methyl-	pH 1	219(4.42),235s(3.95), 281(4.03),316(4.18)	24-0872-66
	MeOH	212(4.21),234(4.19), 308(3.98)	24-0872-66
$C_8H_{11}N_5$			
Ammonium trimethylpurin-6-yl hydroxide, inner salt	pH 10	274(3.86)	87-0981-66
$C_8H_{11}N_5O$			
Adenine, 1-ethoxy-9-methyl-, HI salt	pH 1	260(4.08)	25-1598-66
	pH 13	257(4.09),264s(4.04)	25-1598-66
Isoguanine, 9-propyl-	pH 1	232(3.72),282(4.11)	95-0649-66
	pH 13	245(3.77),285(4.01)	95-0649-66
Pteridine, 2-amino-4-ethoxy-3,4-di- hydro-, hydrochloride	EtOH	232(3.96),286(3.89), 300s(3.82)	39-1117-66C
Pterin, 7,8-dihydro-6,7-dimethyl-, hydrochloride	pH 1.0	252(4.24),270s(3.89), 350(3.79)	24-3008-66
	pH 7.0	227(4.40),280(4.05), 318(3.79)	24-3008-66
	pH 14.0	231(4.16),281(3.92), 318(3.79)	24-3008-66
s-Triazolo[2,3-a]pyrimidine, 2-amino- 5-ethoxy-7-methyl-	MeOH	217(4.57),299(3.81)	39-2031-66C
$C_8H_{11}N_7O_2$			
Glycine, N-(5-amino-3-ethyl- 3H-v-triazolo[4,5-d]pyrim- idin-7-yl)-	pH 1	256(4.18),273(4.10)	87-0417-66
	pH 7	260s(--),289(4.10)	87-0417-.6
	pH 13	260s(--),289(4.10)	87-0417-66
$C_8H_{11}O_3PSe$			
Phosphonic acid, phenylseleno-, dimethyl ester	MeOH	248(3.39),271s(2.84)	65-0254-66
C_8H_{12}			
1-Buten-3-yne, 2-tert-butyl-	EtOH	210(3.86),218(4.00), 226(3.87)	39-1976-66C
$C_8H_{12}BrN_3$			
Pyrimidine, 5-bromo-2-(butylamino)-	pH 0.0	244(4.36),342(3.45)	12-2321-66
	pH 7.0	248(4.36),324(3.16)	12-2321-66
Pyrimidine, 5-bromo-2-(tert-butyl- amino)-	pH 0.0	245(4.35),342(3.51)	12-2321-66
	pH 7.0	249(4.35),324(3.37)	12-2321-66
Pyrimidine, 5-bromo-4-(butylamino)-	pH 2.0	266(4.16)	12-2321-66
	pH 7.0	249(4.10),288(3.61)	12-2321-66
Pyrimidine, 5-bromo-4-(tert-butyl- amino)-	pH 2.0	267(4.12)	12-2321-66
	pH 7.0	249(4.07),284(3.57)	12-2321-66
$C_8H_{12}ClNO_5$			
1-Methoxy-2,6-lutidinium perchlorate	H_2O	271(3.89)	39-0870-66B

Compound	Solvent	λ_{max}(log ε)	Ref.
$C_8H_{12}ClN_3O_2S$			
6-Pyridazinesulfonamide, N-butyl-3-chloro-	EtOH	264(3.50),319(2.95)	83-0648-66
$C_8H_{12}ClN_5O$			
Pteridine, 2-amino-4-ethoxy-3,4-dihydro-, hydrochloride	EtOH	232(3.96),286(3.89), 300s(3.82)	39-1117-66C
$C_8H_{12}Cl_2O$			
4-Heptanone, 3-(dichloromethylene)-	EtOH	244(3.52)	77-0567-66
$C_8H_{12}IN_5O$			
Adenine, 1-ethoxy-9-methyl-, hydroiodide	pH 1	260(4.08)	25-1598-66
	pH 13	257(4.09),264s(4.04)	25-1598-66
$C_8H_{12}NO_2$			
Norpseudopelletierin-N-oxyl	MeOH	445(0.81)	35-3180-66
	$CHCl_3$	462(0.93)	35-3180-66
$C_8H_{12}N_2O$			
5-Hepten-2-one, 1-diazo-6-methyl-	EtOH	246(3.95),267s(3.86)	22-3490-66
5-Hepten-2-one, 3-diazo-6-methyl-	EtOH	247(3.94),293(4.05)	22-3490-66
Pyrimidine, 2-butoxy-	pH 7.0	267.5(3.56)	12-1487-66
Pyrimidine, 2-sec-butoxy-	pH 7.0	267(3.59)	12-1487-66
Pyrimidine, 2-isobutoxy-	pH 7.0	266.5(3.62)	12-1487-66
2-Pyrrolidinone, 5-imino-4-isopropylidene-3-methyl-	EtOH	254(4.22)	22-3895-66
$C_8H_{12}N_2OS$			
Pyrimidine, 4-ethoxy-6-methyl-2-(methylthio)-	EtOH	258(3.94)	78-2401-66
4-Pyrimidinol, 5-(2-mercaptoethyl)-2,6-dimethyl-	pH 13.5	236(3.96),269(3.81)	24-0872-66
	MeOH	279(3.77)	24-0872-66
Thymine, 1-methyl-2-thioethyl-	pH 5.9-11.5	234.5(4.48)	24-2380-66
Thymine, 3-methyl-2-thioethyl-	pH 5.9-11.5	244(3.91),294(4.13)	24-2380-66
$C_8H_{12}N_2OS_2$			
4-Pyrimidinol, 2-mercapto-5-(3-mercaptopropyl)-6-methyl-	pH 13.5	226(3.94),279(3.83)	24-0872-66
	MeOH	281(4.23)	24-0872-66
$C_8H_{12}N_2O_2$			
Pyrazole-3-carboxylic acid, 1,5-dimethyl-, ethyl ester	EtOH	225(4.02)	22-0293-66
Pyrazole-3-carboxylic acid, 4,5-dimethyl-, ethyl ester	EtOH	226(3.95)	22-3744-66
Pyrazole-5-carboxylic acid, 1,3-dimethyl-, ethyl ester	EtOH	230(4.03),250s(3.73)	22-0293-66
Thymine, 4-ethyl-1-methyl-	pH 5-12	280(3.81)	24-2391-66
$C_8H_{12}N_2O_2S$			
Uracil, 5-(3-mercaptopropyl)-6-methyl-	pH 13.5	226(4.03),278(3.91)	24-0872-66
	MeOH	208(4.10),268(4.00)	24-0872-66
$C_8H_{12}N_2O_2S_2$			
5-Thiazolidineacetamide, N-ethyl-3-methyl-4-oxo-2-thioxo-	EtOH	260(4.14),295(4.23)	95-0101-66
5-Thiazolidineacetamide, 4-oxo-3-propyl-2-thioxo-	EtOH	263(4.14),297(4.25)	95-0101-66

Compound	Solvent	$\lambda_{max}(\log \epsilon)$	Ref.
$C_8H_{12}N_2O_3$			
Acetoacetic acid, 2-diazo-, tert-butyl ester	CH_2Cl_2	259(3.86)	24-3128-66
Acrylic acid, 3-amino-2-cyano-3-ethoxy-, ethyl ester	H_2O	256(4.38)	24-2302-66
	dioxan	263(4.28)	24-2302-66
	MeCN	260(4.32)	24-2302-66
6(1H)-Pyridazinone, 1,3-diethoxy-	EtOH	315(3.415)	95-0082-66
6(1H)-Pyridazinone, 1-ethoxy-3-methoxy-4-methyl-	EtOH	307(3.415)	95-0082-66
6(1H)-Pyridazinone, 3-ethoxy-1-methoxy-4-methyl-	EtOH	307(3.461)	95-0082-66
$C_8H_{12}N_2S_3$			
Pyrimidine, 2,4-bis(methylthio)-5-[(methylthio)methyl]-	EtOH	206(3.70),256(4.34), 307(3.74)	5-0119-66B
$C_8H_{12}N_4O_2$			
Pyrimidine, 2-(butylamino)-5-nitro-	pH -2.0	214(3.82),300(4.13)	12-2321-66
	pH 3.0	217(3.86),340(4.20)	12-2321-66
Pyrimidine, 2-(tert-butylamino)-5-nitro-	pH -2.0	215(3.81),300(4.18)	12-2321-66
	pH 3.0	218(3.87),344(4.24)	12-2321-66
$C_8H_{12}N_4O_3$			
Pyrimidine, 2-(dimethylamino)-4-ethoxy-5-nitro-	pH -0.89	278(4.08),306(3.93)	24-2997-66
	pH 5.0	287(3.56),360(4.25)	24-2997-66
$C_8H_{12}N_4O_5$			
6-Azacytosine, 1-β-D-arabinofuranosyl-	H_2O	264(3.899)	73-4002-66
$C_8H_{12}N_6O_6$			
Dehydroascorbic acid, disemicarbazone	pH 3.40	230(4.08),265(4.10)	22-0522-66
	pH 5.50	230(4.11),255(4.11)	22-0522-66
	pH 6.45	230(4.11),255(4.11)	22-0522-66
	pH 8.10	230(4.11),255(4.11)	22-0522-66
	pH 10.5	235(4.15),255(4.08)	22-0522-66
$C_8H_{12}O$			
Acetaldehyde, 2-cyclohexylidene-	EtOH	240(4.27)	22-1400-66
2-Cyclohepten-1-one, 3-methyl-	EtOH	236(4.06)	78-0105-66B
Cyclohexanone, 2-ethylidene-	n.s.g.	249.0(3.63)	39-0397-66C
2-Cycloocten-1-one	C_6H_{12}	224.5(3.83)	39-0164-66B
	EtOH	229.5(3.91)	39-0164-66B
3-Cycloocten-1-one	EtOH	286(2.06)	39-0164-66B
2-Cyclopenten-1-one, 2-ethyl-3-methyl-	n.s.g.	234.5(4.13)	39-0073-66B
2-Cyclopenten-1-one, 2-isopropyl-	n.s.g.	226.5(3.96)	39-0073-66B
2-Cyclopenten-1-one, 3-isopropyl-	n.s.g.	227.0(4.16)	39-0073-66B
3,5-Heptadien-2-one, 4-methyl-	EtOH	257(4.10)	78-0443-66B
3,5-Heptadien-2-one, 6-methyl-	EtOH	292(4.26)	22-3490-66
	EtOH	289(4.47)	28C-0652-66A
3-Hepten-5-yn-1-ol, 4-methyl-, cis	EtOH	225(4.08)	78-0443-66B
trans	EtOH	225(4.08)	78-0443-66B
3,7-Octadien-2-one	isooctane	216(4.16),324(1.53)	22-0287-66
3-Octyn-2-one	n.s.g.	220(3.4)	88-5811-66
Spiro[2.5]octan-4-one	isooctane	281(1.43),287(1.46), 295(1.43)	22-0121-66
$C_8H_{12}O_2$			
1,3-Cyclohexanedione, 5,5-dimethyl-	aq. NaOH	282(4.39)	30-0827-66B

Compound	Solvent	λ_{max}(log ε)	Ref.
1-Cyclohexene-1-carboxylic acid, 2-methyl-	EtOH	225(3.96)	22-2535-66
1-Cyclohexene-1-carboxylic acid, 6-methyl-	EtOH	214(3.82)	44-1472-66
2-Cyclohexen-1-one, 3-ethoxy-	EtOH	249(4.35)	22-1012-66
	n.s.g.	237.5(4.19)	39-0073-66B
2,4-Hexadienoic acid, ethyl ester	EtOH	258(4.45)	54-0929-66
$C_8H_{12}O_3$			
2-Cyclopenten-1-one, 3-ethoxy-2-hydroxy-5-methyl-	H_2O	266(4.24)	5-0062-66G
2(5H)-Furanone, 5-ethyl-5-hydroxy-3,4-dimethyl-	n.s.g.	213(4.1)	54-0043-66
2-Hexenoic acid, 4,5-epoxy-, ethyl ester	EtOH	217(4.18)	70-0646-66
$C_8H_{12}O_4$			
3(2H)-Furanone, 5-(1,2-dimethoxyethyl)-	H_2O	187(3.38),262(4.11)	88-1649-66
	pH 12	296(4.24)	88-1649-66
2,4-Pentanedione, 3-propionyloxy-, palladium derivative	EtOH	210(4.65),227(4.78), 250s(4.60),261s(4.65), 331(4.00)	30-0108-66D
$C_8H_{12}O_6$			
β-L-threo-Hex-4-enosiduronic acid, methyl 4-deoxy-, methyl ester	n.s.g.	238(3.81)	39-1549-66C
$C_8H_{12}S$			
Thiophene, 2,3-dihydro-, dimer	MeCN	247(3.84)	44-2333-66
$C_8H_{13}Br$			
1,2-Pentadiene, 1-bromo-3,4,4-trimethyl-	EtOH	201(3.95),224(3.82)	39-1223-66C
$C_8H_{13}ClO_2$			
2-Pentenoic acid, 2-chloro-4-methyl-, ethyl ester	MeOH	218(4.07)	5-0134-66C
$C_8H_{13}IN_2O$			
3,4-Dihydro-1,2,3,5-tetramethyl-4-oxopyrimidinium iodide	H_2O	202(4.11),223(4.26), 268(3.66)	22-0527-66
$C_8H_{13}NO$			
3-Buten-2-one, 4-(tert-butylimino)-	C_6H_{12}	214(4.02),253(4.19)	35-3169-66
Crotononitrile, β-butoxy-	MeOH	228(4.16)	35-3169-66
Crotononitrile, β-tert-butoxy-	EtOH	229(4.15)	35-3169-66
$C_8H_{13}NO_2$			
5-Heptenal, 6-methyl-2-oxo-, 1-oxime	EtOH	230(3.99)	22-3490-66
Nicotinic acid, 1,4,5,6-tetrahydro-, ethyl ester	EtOH	281(4.32)	44-2487-66
Propionitrile, 3-ethoxy-2-(ethoxymethylene)-, cis	EtOH	233(4.17)	94-0238-66
trans	EtOH	230(4.20)	94-0238-66
1-Pyrrolidineacrylic acid, methyl ester, trans	EtOH	228(3.30),283(4.48)	24-2526-66
$C_8H_{13}NO_3$			
Cyclohexanol, 1-nitroso-, acetate	C_6H_{12}	665(1.36)	39-0441-66B

Compound	Solvent	$\lambda_{max}(\log \epsilon)$	Ref.
2-Isoxazolin-5-one, 4-(2-hydroxy-2-methylpropyl)-3-methyl-	MeOH	257.5(4.03)	24-2962-66
$C_8H_{13}NO_4$			
Maleic acid, (dimethylamino)-, dimethyl ester	EtOH	225s(3.53),282(4.37)	24-2526-66
$C_8H_{13}NO_5$			
Adipic acid, α-oximino-	EtOH	218(3.91)	39-0255-66C
	EtOH-NaOH	245(3.58)	39-0255-66C
Pentanoic acid, 4-carbethoxy-4-nitroso-	EtOH	293(3.81)	39-0412-66C
$C_8H_{13}N_3$			
Pyrimidine, 2-(butylamino)-	pH 1	230(4.18),315(3.48)	12-1487-66
	pH 7.6	236(4.16),308(3.35)	12-1487-66
Pyrimidine, 2-(tert-butylamino)-	pH 1.8	231(4.23),320(3.55)	39-0226-66C
	pH 6.8	237(4.25),309(3.41)	39-0226-66C
Pyrimidine, 4-(butylamino)-	pH 2	256(4.24)	12-1487-66
	pH 12	244(4.20),279(3.56)	12-1487-66
Pyrimidine, 2-(dimethylamino)-4,6-dimethyl-	pH 3.5	238(4.18),319(3.64),	39-0226-66C
	pH 8.0	248(4.21),312(3.42)	39-0226-66C
Pyrimidine, 4-(dimethylamino)-2,6-dimethyl-	pH 5.5	264(4.13)	39-0226-66C
	pH 10.0	251(4.14),292s(3.61)	39-0226-66C
$C_8H_{13}N_3O$			
s-Triazolo[1,5-a]pyridine, 1-ethyl-5,6,7,8-tetrahydro-2-oxo-	EtOH	242(3.63)	32-1020-66
s-Triazolo[4,3-a]pyridine, 1-ethyl-5,6,7,8-tetrahydro-3-oxo-	EtOH	242(3.64)	7-0199-66
$C_8H_{13}N_3OS$			
4-Pyrimidinol, 2-amino-6-methyl-5-(3-mercaptopropyl)-	pH 13.5	235(4.13),278(3.92)	24-0872-66
	MeOH	228(3.92),290(3.83)	24-0872-66
$C_8H_{13}N_3O_2$			
Sorbaldehyde, 6-hydroxy-4-methyl-, semicarbazone, all trans	EtOH	293(4.71)	33-0369-66
1,2,4-Triazoline-3,5-dione, 4-cyclohexyl-	C_6H_{12}	545(2.31)	44-3444-66
$C_8H_{13}N_5O_2$			
Pyrimidine, 2-amino-4-(butylamino)-5-nitro-	MeOH	260(3.98),295(3.81), 358(4.15)	87-0121-66
Pyrimidine, 4-amino-2-(butylamino)-5-nitro-	MeOH	262(3.79),352(4.17)	87-0121-66
$C_8H_{13}N_5O_3$			
4(3H)-Pyrimidinone, 2,6-bis(methylamino)-5-nitro-	pH -3.8	245(4.04),306(4.14), 360s(3.40)	24-2984-66
	pH 6.0	311s(3.88),357(4.08)	24-2984-66
	pH 7.0	224(4.32),246s(4.17), 305(3.72),357(3.84)	24-2984-66
	pH 11.0	290s(3.71),367(4.23)	24-2984-66
C_8H_{14}			
1,3-Butadiene, 2-sec-butyl-	isooctane	219s(4.15),225(4.19), 232s(4.03)	32-0483-66
1,3-Heptadiene, 5-methyl-, trans	isooctane	220s(4.37),225(4.40), 232s(4.24)	32-0483-66

Compound	Solvent	λ_{max}(log ε)	Ref.
3,5-Octadiene, cis-trans	EtOH	232(4.12)	70-0468-66
3,5-Octadiene, trans-trans	EtOH	231(4.17)	70-0468-66
$C_8H_{14}ClNO_5$			
N-tert-Butyl-5-methylisoxazolium perchlorate	pH 1	234(3.95)	35-3169-66
$C_8H_{14}N_2$			
Pyrazole, 3,5-diethyl-4-methyl-	EtOH	224(3.74)	22-2971-66
$C_8H_{14}N_2O$			
3H-Pyrazole-5-methanol, α,α,3,3-tetramethyl-	hexane	245(3.26),356(2.32)	39-1719-66C
	EtOH	254(3.32),347(2.32)	39-1719-66C
$C_8H_{14}N_4$			
1,2,4,5-Tetrazine, 3,6-diisopropyl-	EtOH	273(3.48),545(2.67)	88-5067-66
1,2,4,5-Tetrazine, 3,6-dipropyl-	EtOH	275(3.47),542(2.67)	88-5067-66
$C_8H_{14}N_6O_2S_2$			
Butyraldehyde, 3-acetoxy-2-oxo-, bis(thiosemicarbazone)	EtOH	247(4.15),346(4.45)	87-0420-66
$C_8H_{14}O$			
2-Butanone, 1-(1-methylcyclopropyl)-	C_6H_{12}	288(1.52)	88-5081-66
	EtOH	285(1.61)	88-5081-66
3,5-Heptadien-1-ol, 4-methyl-, cis-cis	EtOH	224.5(3.77)	78-0443-66B
3-cis-5-trans	EtOH	234(4.18)	78-0443-66B
5-cis-3-trans	EtOH	230(4.15)	78-0443-66B
trans-trans	EtOH	233(4.31)	78-0443-66B
3-Hexenal, 4-ethyl-	n.s.g.	222(4.45)	22-0717-66
4-Hexen-3-one, 2,5-dimethyl-	C_6H_{12}	232(4.04),330(1.70)	88-5081-66
	EtOH	238(4.04),317(1.86)	88-5081-66
3-Hexen-2-one, 4-ethyl-	C_6H_{12}	233(4.00),320(1.56)	88-5081-66
	EtOH	238(3.95),310(1.75)	88-5081-66
2-Propanone, 1-(1-ethylcyclopropyl)-	C_6H_{12}	288.5(1.52)	88-5081-66
	EtOH	284(1.50)	88-5081-66
$C_8H_{14}O_2S_2$			
Isolipoic acid	$CHCl_3$	333.5(2.17)	11-0263-66
$C_8H_{14}O_3$			
Acrylic acid, β-methoxy-, tert-butyl ester	EtOH	227(4.08)	44-0646-66
Crotonic acid, 2-ethoxy-, ethyl ester, trans-cis	MeOH	222(3.94)	5-0053-66I
$C_8H_{14}O_3S_2$			
1,2-Dithiolane-4-valeric acid, 1-oxide	EtOH	243(2.94)	11-0263-66
$C_8H_{14}Pt$			
Platinum, cyclopentadienyltrimethyl-	dioxan	257(1.0)	101-0181-66A
$C_8H_{15}ClN_2O_4$			
2,3-Dihydro-1,4,6-trimethyl-1,4-diazepinium perchlorate	MeOH	362(4.19)	39-0093-66C
$C_8H_{15}NO$			
2-Butene, 2-morpholino-	ether	221(3.88)	22-2845-66

Compound	Solvent	$\lambda_{max}(\log \epsilon)$	Ref.
1-Pyrroline, 2-ethyl-5,5-dimethyl-, 1-oxide	EtOH	234(3.91)	39-0412-66C
$C_8H_{15}NO_2$			
Acrylic acid, 3-(butylamino)-, methyl ester	EtOH	276(4.31)	24-2526-66
Acrylic acid, 3-(diethylamino)-, methyl ester	ether	272(4.40)	24-0450-66
Crotonic acid, 3-(ethylamino)-, ethyl ester	MeOH	282(4.32)	35-2536-66
$C_8H_{15}NO_3S$			
3-Pentene-1-sulfonic acid, 2-(dimethylamino)-4-hydroxy-2-methyl-, sultone	MeOH	207(3.08),266(3.27)	5-0068-66I
$C_8H_{15}NO_5S_2$			
Thiazolidine-2-thione, 5-(D-arabino-tetrahydroxybutyl)-5-hydroxy-3-methyl-	MeOH	250s(3.92),273(4.22)	12-0445-66
$C_8H_{15}N_3O$			
3-Pentenal, 2,2-dimethyl-, semicarbazone	n.s.g.	230(4.15)	22-0734-66
$C_8H_{15}N_3O_3$			
Formic acid, [(diethylcarbamoyl)-azo]-, ethyl ester	MeOH	290(3.07)	24-2039-66
	dioxan	285(3.00)	24-2039-66
$C_8H_{15}N_5$			
Pyrimidine, 2,5-diamino-4-(butylamino)-	$MeOH-H_2SO_4$	237(4.25),304(3.77)	87-0121-66
$C_8H_{15}N_5O$			
4-Pyrimidinol, 2,5-diamino-6-(butylamino)-	EtOH	294(3.86),367s(3.27)	87-0977-66
$C_8H_{15}N_5S$			
4-Pyrimidinethiol, 2,5-diamino-6-(butylamino)-	pH 1	235(4.29),311(4.38)	87-0977-66
	pH 13	225(4.11),305(3.66)	87-0977-66
$C_8H_{16}N_4$			
1,2,4,5-Tetrazine, 1,6-dihydro-3,6-dipropyl-	EtOH	310(3.46),426(2.60)	88-5067-66
$C_8H_{17}I$			
Heptane, 4-iodo-4-methyl-	C_6H_{12}	269(2.79)	3-0612-66
Hexane, 4-iodo-2,4-dimethyl-	C_6H_{12}	267(2.80)	3-0612-66
$C_8H_{17}N_3O_2S$			
Pyruvaldehyde, diethyl acetal, thiosemicarbazone	pH 1	301(4.15)	87-0609-66
	pH 11	234(3.91),267(4.33)	87-0609-66
$C_8H_{18}N_2O_2$			
Hyponitrous acid, di-tert-butyl ester	n.s.g.	226(4.79)	88-6163-66
$C_8H_{18}OS$			
Butyl sulfoxide	EtOH	201(3.28),214s(3.08)	39-0239-66A
$C_8H_{18}O_5S$			
D-Galactose, diethyl dithioacetal	EtOH	211(2.87),264(1.60)	44-0509-66

Compound	Solvent	$\lambda_{max}(\log \epsilon)$	Ref.
$C_8H_{18}S$			
Butyl sulfide	EtOH	202(3.32)	39-0239-66A
tert-Butyl sulfide	hexane	215.5(3.53)	39-0242-66A
	EtOH	212(3.32)	39-0239-66A
$C_8H_{18}S_2$			
Butyl disulfide	EtOH	201(3.36),250(2.60)	39-0239-66A
tert-Butyl disulfide	EtOH	201(3.60),222s(2.66)	39-0239-66A
$C_8H_{20}NiO_4P_2S_4$			
Ethyl nickel phosphorodithioate	$CHCl_3$	283s(3.68),320(4.28), 383(2.91),525(1.96), 690(1.87)	80-0701-66
$C_8H_{24}Cl_2Si_4$			
Tetrasilane, 1,4-dichloro-octamethyl-	C_6H_{12}	235.0(4.24)	101-0392-66A
$C_8H_{25}ClSi_4$			
Tetrasilane, 1-chloro-1,1,2,2,3,3,4,4-octamethyl-	C_6H_{12}	235.0(4.13)	101-0392-66A
$C_8H_{26}Si_4$			
Tetrasilane, octamethyl-	C_6H_{12}	235.5(4.09)	101-0392-66A

Compound	Solvent	$\lambda_{max}(\log \epsilon)$	Ref.
C_9F_7N			
Isoquinoline, heptafluoro-	C_6H_{12}	213(4.62),260s(3.75), 272(3.82),282s(3.75), 323s(3.75),332(3.77)	39-2328-66C
Quinoline, heptafluoro-	C_6H_{12}	215s(4.28),227(4.45), 272(3.49),282(3.41)	39-2328-66C
	H_2SO_4	203(3.92),243(4.58), 314(3.77)	39-2328-66C
$C_9H_3Br_4NO_2$			
4H-3,1-Benzoxazin-4-one, 6,8-dibromo-2-(dibromomethyl)-	EtOH	244(4.38),266(3.96), 281(4.05),293(3.98), 324(3.60),335(3.54)	4-0422-66
$C_9H_3Cl_4NO_2$			
4H-3,1-Benzoxazin-4-one, 5,6,7,8-tetrachloro-2-methyl-	EtOH	244(4.53),264s(3.79), 273s(3.66),283(3.49), 332(3.34),345(3.33)	4-0422-66
$C_9H_4BrClN_4O_4$			
Pyrazole, 4-bromo-3-chloro-1-(2,4-dinitrophenyl)-	EtOH	315(4.10)	22-3744-66
$C_9H_4BrN_5O_6$			
Pyrazole, 4-bromo-1-picryl-	EtOH	320(3.75)	22-3744-66
$C_9H_4Br_2N_4O_4$			
Pyrazole, 3,4-dibromo-1-(2,4-dinitrophenyl)-	EtOH	320(4.12)	22-3744-66
$C_9H_5BrCrO_3$			
Chromium, (bromobenzene)-tricarbonyl-	C_6H_{12}	226(4.24),280(4.08), 318(4.05)	24-1732-66
$C_9H_5BrN_4O_4$			
Pyrazole, 4-bromo-1-(2,4-dinitro-phenyl)-	EtOH	314(4.10)	22-3744-66
$C_9H_5Br_2NO$			
8-Quinolinol, 5,7-dibromo-	N $HClO_4$	328(3.23),367(3.34)	39-0073-66A
	pH 13	343(3.62),368(3.58)	39-0073-66A
	$CHCl_3$	328(3.52)	39-0073-66A
copper chelate	pyridine	347(3.80),415(3.85)	39-0073-66A
	dioxan	419(3.88)	39-0073-66A
nickel chelate	pyridine	353(3.98),415(3.91)	39-0073-66A
$C_9H_5Br_2NO_2$			
4H-3,1-Benzoxazin-4-one, 6,8-dibromo-2-methyl-	EtOH	238(4.41),258(3.86), 267(3.89),278(3.79), 323(3.46),335(3.40)	4-0422-66
$C_9H_5Br_3N_2$			
Pyrazole, 4,5-dibromo-1-(p-bromophenyl)-	EtOH	252.5(4.22)	22-3744-66
$C_9H_5Br_4NO_3$			
Anthranilic acid, 3,5-dibromo-N-(dibromoacetyl)-	EtOH	219(4.43),257(3.9), 298(3.26)	4-0422-66

Compound	Solvent	λ_{max}(log ε)	Ref.
$C_9H_5ClCrO_3$			
Chromium, (chlorobenzene)-tricarbonyl-	C_6H_{12}	227(4.25),267(4.11), 318(4.12)	24-1732-66
C_9H_5ClINO			
8-Quinolinol, 5-chloro-7-iodo-	N $HClO_4$	332(3.42),368(3.43)	39-0073-66A
	pH 13	347(3.61),374(3.54)	39-0073-66A
	$CHCl_3$	328(3.51)	39-0073-66A
copper chelate	pyridine	350(3.89),415(3.92)	39-0073-66A
	dioxan	424(3.86)	39-0073-66A
nickel chelate	pyridine	357(3.97),416(3.87)	39-0073-66A
$C_9H_5ClN_2O_3$			
2-Quinoxalinecarboxylic acid, 7-chloro-3-hydroxy-	EtOH	213(4.59),234(4.34), 342(3.87)	30-0577-66C
$C_9H_5ClN_2O_4$			
2-Quinoxalinecarboxylic acid, 7-chloro-3-hydroxy-, 4-oxide	EtOH	213(4.32),268(4.28), 340(3.72),435(3.40)	30-0577-66C
$C_9H_5ClN_4O_4$			
Pyrazole, 3-chloro-1-(2,4-dinitrophenyl)-	EtOH	309(4.08)	22-3744-66
C_9H_5ClOS			
1-Thiochromone, 3-chloro-	EtOH	216(3.96),226(4.05), 250s(4.21),254(4.22), 282(3.53),292(3.49), 347(3.96)	12-1751-66
$C_9H_5Cl_2NO$			
8-Quinolinol, 5,7-dichloro-	N $HClO_4$	326(3.20),367(3.40)	39-0073-66A
	pH 13	342(3.58),369(3.56)	39-0073-66A
	$CHCl_3$	328(3.48)	39-0073-66A
copper chelate	pyridine	347(3.79),416(3.85)	39-0073-66A
	dioxan	421(3.78)	39-0073-66A
	$CHCl_3$	425(3.82)	39-0073-66A
nickel chelate	pyridine	353(3.92),415(3.87)	39-0073-66A
$C_9H_5Cl_2NOS_2$			
Rhodanine, N-(2,4-dichlorophenyl)-	EtOH	257(4.05),296(4.19)	95-0095-66
$C_9H_5Cl_3N_2$			
Quinoline, 7-amino-5,6,8-trichloro-	EtOH	257(4.29),288(3.21), 370(3.87)	95-0055-66
$C_9H_5CrFO_3$			
Chromium, tricarbonyl-(fluorobenzene)-	C_6H_{12}	226(4.30),257(4.09), 312(4.14)	24-1732-66
$C_9H_5CrIO_3$			
Chromium, tricarbonyl(iodobenzene)-	C_6H_{12}	228(4.26),304(4.16), 340s(3.88)	24-1732-66
$C_9H_5F_3N_2$			
Quinazoline, 2-(trifluoromethyl)-	pH -4.0	242(4.51),286(3.30), 334(3.23)	39-0234-66C
	pH 8.0	228(4.58),280(3.94), 316s(3.71)	39-0234-66C

Compound	Solvent	$\lambda_{max}(\log \epsilon)$	Ref.
$C_9H_5F_3N_2O$			
Acetophenone, 2-diazo-3'-(trifluoromethyl)-	hexane	213(4.12),228s(4.11),243(4.10),294(4.06)	78-0209-66
	butanol	220(3.90),245(3.98),301(4.05)	78-0209-66
$C_9H_5I_2NO$			
5-Isoquinolinol, 6,8-diiodo-	EtOH	220s(4.58),255(4.50),340(3.79)	87-0046-66
8-Isoquinolinol, 5,7-diiodo-, hydrochloride	iso-PrOH	225s(4.32),251(4.28),313(3.65),349(3.68)	87-0046-66
8-Quinolinol, 5,7-diiodo-	pH 13	350(3.67),370(3.64)	39-0073-66A
	$CHCl_3$	334(3.57)	39-0073-66A
copper chelate	pyridine	352(3.85),420(3.85)	39-0073-66A
	dioxan	427(3.86)	39-0073-66A
nickel chelate	pyridine	356(3.91),416(3.81)	39-0073-66A
$C_9H_5MnO_4$			
Manganese, tricarbonyl-(formylcyclopentadienyl)-	EtOH	213(4.35),284(3.22),336(3.19)	101-0357-66A
$C_9H_5NO_6$			
Furo[2,3-b]pyridine-2,5-dicarboxylic acid, 4-hydroxy-	H_2O	234(4.39)	4-0202-66
$C_9H_5N_3O_2$			
Indole-3-carbonitrile, 4-nitro-	EtOH	234(4.09),345(3.75)	44-0070-66
Indole-3-carbonitrile, 6-nitro-	EtOH	249s(4.23),258(4.32),261s(4.31),309(3.92),335s(3.83),354s(3.73)	44-0070-66
$C_9H_5N_3O_6$			
Indole-2-carboxylic acid, 3,4-dinitro-	EtOH	240(4.09),333(3.93)	44-0070-66
Indole-2-carboxylic acid, 3,5-dinitro-	EtOH	255(4.40),321(4.08),351s(4.01)	44-0070-66
Indole-2-carboxylic acid, 3,7-dinitro-	EtOH	229(4.20),347(4.12)	44-0070-66
$C_9H_5N_5O_2$			
1,2,4-Oxadiazole-3-carbonylazide, 5-phenyl-	EtOH	252(4.35)	39-1522-66C
$C_9H_5N_5O_6$			
Pyrazole, 1-picryl-	EtOH	305(3.81)	22-3744-66
$C_9H_6BrClN_2$			
Pyrazole, 4-bromo-5(3)-chloro-3(5)-phenyl-	EtOH	254(4.12)	22-3744-66
$C_9H_6BrClO_4S$			
6-Bromothiochromylium perchlorate	H_2SO_4	226s(4.16),270(4.52),354(3.90),385s(3.40)	17-0075-66
7-Bromothiochromylium perchlorate	H_2SO_4	270(4.53),348(3.77),402(3.79)	17-0075-66
C_9H_6BrNO			
8-Quinolinol, 5-bromo-	N $HClO_4$	312(3.04),376(3.34)	39-0073-66A
	pH 13	342(3.54),370(3.60)	39-0073-66A
	$CHCl_3$	332(3.59)	39-0073-66A

Compound	Solvent	λ_{max}(log ε)	Ref.
8-Quinolinol, 5-bromo-, copper chelate	$CHCl_3$	425(3.89)	39-0073-66A
	pyridine	346(3.72),422(3.89)	39-0073-66A
	dioxan	428(3.90)	39-0073-66A
nickel chelate	pyridine	353(3.86),421(3.93)	39-0073-66A
$C_9H_6BrN_3O_3$			
4(1H)-Cinnolinone, 3-bromo-1-methyl-6-nitro-	EtOH	210(4.22),240(4.13),259s(4.05),272s(4.00),282s(3.90),330(3.91),381(4.09)	39-0470-66C
$C_9H_6BrN_5O_6$			
2-Pyrazoline, 3-bromo-1-picryl-	$CHCl_3$	389(4.17)	22-0610-66
$C_9H_6ClI_2NO$			
8-Isoquinolinol, 5,7-diiodo-, hydrochloride	iso-PrOH	225s(4.32),251(4.28),313(3.65),349(3.68)	87-0046-66
C_9H_6ClNO			
8-Quinolinol, 5-chloro-	N $HClO_4$	312(3.00),376(3.34)	39-0073-66A
	pH 13	340(3.54),365(3.58)	39-0073-66A
	$CHCl_3$	332(3.53)	39-0073-66A
copper chelate	$CHCl_3$	425(3.86)	39-0073-66A
	pyridine	347(3.70),424(3.85)	39-0073-66A
	dioxan	428(3.90)	39-0073-66A
$C_9H_6ClNOS_2$			
Rhodanine, N-(p-chlorophenyl)-	EtOH	257(4.05),296(4.16)	95-0095-66
$C_9H_6ClNO_2$			
3-Isoxazolol, 5-(p-chlorophenyl)-	EtOH	266(4.32)	94-1277-66
$C_9H_6ClNO_3$			
Indole-2,3-dione, 6-chloro-7-methoxy-	EtOH	216(4.35),248(4.23),255(4.18),311(3.79)	44-1303-66
$C_9H_6ClNO_3S$			
Benzenesulfonyl chloride, m(p)-5-isoxazolyl-	CH_2Cl_2	269(4.20)	78-0321-66B
$C_9H_6ClNO_4$			
2,1-Benzisoxazole-3-carboxylic acid, 6-chloro-7-methoxy-, hydrochloride	EtOH	229(4.43)	44-1303-66
$C_9H_6ClN_3O_3$			
4(1H)-Cinnolinone, 6-chloro-1-methyl-3-nitro-	EtOH	224(4.36),350(4.13)	39-0470-66C
$C_9H_6ClN_5$			
5H-s-Triazolo[4,3-b]-s-triazole, 6-(p-chlorophenyl)-	EtOH	256(4.43)	4-0119-66
$C_9H_6ClN_5S$			
s-Triazolo[3,4-b][1,3,4]thiadiazole, 6-amino-3-(o-chlorophenyl)-	MeOH	249(4.12)	44-3528-66
cation	MeOH	255(4.10)	44-3528-66
$C_9H_6Cl_2N_2$			
Quinazoline, 2-(dichloromethyl)-	pH -3.0	243(4.56),278(3.52),318(3.34),330s(3.31)	39-0234-66C

Compound	Solvent	$\lambda_{max}(\log \epsilon)$	Ref.
Quinazoline, 2-(dichloromethyl)- (cont.)	pH 5.0	232(4.66),267(3.52), 306(3.34),320s(3.25)	39-0234-66C
$C_9H_6Cl_2O_4S$			
6-Chlorothiochromylium perchlorate	H_2SO_4	226s(4.19),266(4.56), 350(3.88),378s(3.45)	17-0075-66
7-Chlorothiochromylium perchlorate	H_2SO_4	266(4.57),344(3.75), 397(3.76)	17-0075-66
C_9H_6FNO			
8-Quinolinol, 5-fluoro-	N $HClO_4$	310(2.95),372(3.20)	39-0073-66A
	pH 13	340(3.54),365(3.43)	39-0073-66A
	$CHCl_3$	327(3.42)	39-0073-66A
copper chelate	$CHCl_3$	425(3.69)	39-0073-66A
	dioxan	425(3.67)	39-0073-66A
	pyridine	352(3.80),425(3.72)	39-0073-66A
nickel chelate	pyridine	357(3.93),425(3.78)	39-0073-66A
$C_9H_6F_6$			
1,3,5-Cycloheptatriene, 7,7-bis(trifluoromethyl)-	C_6H_{12}	277(3.50)	35-3617-66
	EtOH	276(3.51)	35-3617-66
	MeCN	276(3.50)	35-3617-66
$C_9H_6FeO_7$			
Iron, tricarbonyl(muconic acid)-	EtOH	206(4.27),229(4.29), 240s(4.20),303(3.54)	101-0370-66A
C_9H_6INO			
5-Isoquinolinol, 8-iodo-	MeOH-HCl	248(4.45),268(4.21), 304s(3.58),317s(3.62), 333(3.67),363s(3.54)	87-0046-66
	MeOH	249(4.61),305s(3.60), 321(3.69),331(3.72)	87-0046-66
8-Isoquinolinol, 5-iodo-	iso-PrOH-HCl	226s(4.32),243s(4.17), 260s(3.57),315(3.65), 341(3.72)	87-0046-66
	iso-PrOH	233s(4.20),242(4.21), 315(3.72),343(3.81)	87-0046-66
$C_9H_6N_2OS$			
2-Benzothiazolecarbonitrile, 4-methoxy-	EtOH	249(3.18),294(4.08), 345(3.60)	44-1484-66
2-Benzothiazolecarbonitrile, 5-methoxy-	EtOH	225(4.30),286(4.01), 340(3.29)	44-1484-66
2-Benzothiazolecarbonitrile, 7-methoxy-	EtOH	218(4.58),251(4.22), 294(3.88),339(3.27)	44-1484-66
$C_9H_6N_2O_2S$			
Isothiazole, 5-(p-nitrophenyl)-	C_6H_{12}	223(3.65),299(3.90)	78-2119-66
$C_9H_6N_2O_3$			
Indole-3-carboxaldehyde, 4-nitro-	EtOH	266(4.36),287(3.98), 355(3.64)	34-0418-66
Indole-3-carboxaldehyde, 5-nitro-	EtOH	236s(4.17),257(4.42), 262s(4.41),313(3.97)	34-0418-66
Indole-3-carboxaldehyde, 6-nitro-	EtOH	278(4.56),316s(4.07)	34-0418-66
Indole-3-carboxaldehyde, 7-nitro-	EtOH	222(4.28),268(4.15), 345(4.03)	34-0418-66
Isoxazole, 3-nitro-5-phenyl-	EtOH	250(4.3)	32-0454-66

Compound	Solvent	λ_{max}(log ε)	Ref.
2-Quinoxalinecarboxylic acid, 3-hydroxy-	EtOH	230(4.33),295(3.81), 345(3.78)	30-0577-66C
$C_9H_6N_2O_3S_2$			
Rhodanine, N-(p-nitrophenyl)-	EtOH	298(4.17)	95-0095-66
$C_9H_6N_2O_4$			
Indole-2-carboxylic acid, 3-nitro-	EtOH	256(4.09),264s(4.05), 354(3.91)	44-0070-66
3-Isoxazolol, 5-(p-nitrophenyl)-	EtOH	225(4.08),308(4.15)	94-1277-66
2-Quinoxalinecarboxylic acid, 3-hydroxy-, 4-oxide	EtOH	237(4.26),258(4.26), 320(3.69),410(3.43)	30-0577-66C
$C_9H_6N_4$			
4-Cycloheptimidazole-carbonitrile, 2-amino-	MeOH	263(4.46),302(3.93), 364(4.11),428(3.89)	18-2538-66
$C_9H_6N_4O_4$			
N-(1,6-Dihydro-2-hydroxy-5-nitro-6-oxo-4-pyrimidinyl)pyridinium betaine	pH -1.9	242(3.79),249(3.72), 255(3.69),313(4.02)	24-2997-66
	pH 2.0	255(3.97),340(3.94)	24-2997-66
$C_9H_6N_4S_2$			
s-Triazolo[3,4-b][1,3,4]thiadiazole-3-thiol, 6-phenyl-	MeOH	258(4.19)	44-3528-66
anion	MeOH	242(4.19),267(4.28)	44-3528-66
s-Triazolo[3,4-b][1,3,4]thiadiazole-6-thiol, 3-phenyl-	MeOH	275(4.26)	44-3528-66
anion	MeOH	273(4.17)	44-3528-66
$C_9H_6N_6O_2$			
5H-s-Triazolo[4,3-b]-s-triazole, 6-(p-nitrophenyl)-	EtOH	298(4.22)	4-0119-66
C_9H_6OS			
2-Penten-4-ynal, 5-(2-thienyl)-	ether	266(4.01),332(4.39), 341(4.38)	24-1642-66
$C_9H_6O_2$			
Cyclopropenone, hydroxy-phenyl-	H_2O	204(4.14),256(4.26), 265(4.26),275s(3.97)	35-3075-66
	1.25M H_2SO_4	202(4.27),248(4.26), 256(4.24),266s(4.01)	35-3075-66
	13.5M H_2SO_4	256(4.28)	35-3075-66
$C_9H_6O_2S$			
Benzo[c]thiophene-2-carboxylic acid	EtOH	237(4.27),286s(3.41), 299(3.59),313(3.66), 352(3.81)	88-1953-66
$C_9H_6O_4$			
[2,2'-Bifuran]-3-carboxylic acid	EtOH	204(4.12),232s(3.46), 308(4.20)	39-0976-66C
$C_9H_7BBrNO_3$			
5-Isoxazoleboronic acid, 4-bromo-3-phenyl-	EtOH	239(3.90)	101-0598-66B
5-Isoxazoleboronic acid, 3-(p-bromophenyl)-	EtOH	251(4.30)	101-0598-66B

Compound	Solvent	λ_{max}(log ϵ)	Ref.
$C_9H_7BrN_2$			
Pyrazole, 4-bromo-3(5)-phenyl-	EtOH	250(4.10)	22-3744-66
$C_9H_7BrN_2O$			
4-Cinnolinol, 3-bromo-8-methyl-	EtOH	214(4.38),235(4.09), 243(4.13),261(3.98), 271(3.95),293(3.59), 302(3.57),346(4.08), 362(3.97)	39-0470-66C
Isoxazole, 3-amino-5-(p-bromophenyl)-	EtOH	267(4.38)	94-1277-66
3-Pyrazolol, 4-bromo-1-phenyl-	EtOH	280(4.30)	44-1538-66
$C_9H_7BrN_4O_4$			
Pyrazoline, 3-bromo-1-(2,4-dinitrophenyl)-	$CHCl_3$	378(4.32)	22-0610-66
C_9H_7BrS			
2H-1-Benzothiopyran, 6-bromo-	EtOH	248(4.39),280s(3.50), 298s(3.33),336(3.08)	17-0075-66
2H-1-Benzothiopyran, 7-bromo-	EtOH	250(4.45),293(3.43), 302s(3.35),332(3.24)	17-0075-66
$C_9H_7Br_2NO_3$			
Anthranilic acid, N-acetyl-3,5-dibromo-	EtOH	219(4.45),251s(3.89), 302(3.01)	4-0422-66
$C_9H_7ClN_2$			
Isoquinoline, 1-amino-3-chloro-	EtOH	229(4.20),301(3.98), 338(3.74)	95-0544-66
Isoquinoline, 3-amino-1-chloro-	EtOH	236(4.66),287(3.99), 378(3.58)	95-0544-66
Pyrazole, 4-chloro-1-phenyl-	EtOH	262(4.15)	22-3744-66
Pyrazole, 5(3)-chloro-3(5)-phenyl-	EtOH	253.5(4.30)	22-3744-66
Quinazoline, 2-chloro-4-methyl-	pH -2.0	241(4.49),289(3.70), 340(3.30)	39-0234-66C
	pH 5.0	231(4.67),270(3.40), 314(3.45)	39-0234-66C
Quinazoline, 2-(chloromethyl)-	pH -1.0	209(4.16),270(3.87), 344s(2.27)	39-0234-66C
	pH 6.0	230(4.63),270(3.40), 308(3.33),318s(3.27)	39-0234-66C
$C_9H_7ClN_2O$			
Pyrazol-3-ol, 4-chloro-1-phenyl-	EtOH	279(4.30)	44-1538-66
C_9H_7ClOS			
2-Octene-4,6-diynal, 2-chloro-3-(methylthio)-, trans	ether	228(4.17),294(3.91), 314(3.96),341(4.03)	24-0138-66
$C_9H_7ClO_4$			
3(2H)-Benzofuranone, 7-chloro-4-hydroxy-6-methoxy-	pH 13	219(4.25),240(4.23), 284(4.25),355(3.79)	87-0242-66
	EtOH-HCl	210(4.30),236(4.25), 284(4.31),325(3.62)	87-0242-66
	EtOH	210(4.31),235(4.26), 283(4.33),325(3.62)	87-0242-66
$C_9H_7ClO_4S$			
1-Benzothiopyrylium perchlorate	HOAc	260(4.60),335(3.71), 385(3.58)	78-0007-66

Compound	Solvent	λ max(log ε)	Ref.
$C_9H_7ClO_4Se$			
1-Benzoselenopyrylium perchlorate	HOAc-1% $HClO_4$	260(4.56),348(3.69), 408(3.61)	20-0169-66
C_9H_7ClS			
2H-1-Benzothiopyran, 6-chloro-	EtOH	247(4.41),282(3.44), 298s(3.28),338(3.15)	17-0075-66
2H-1-Benzothiopyran, 7-chloro-	EtOH	250(4.44),293(3.38), 304s(3.28),332(3.19)	17-0075-66
$C_9H_7Cl_2NO$			
2-Indolinone, 3,5-dichloro-3-methyl-	MeOH	214(4.27),258(3.84)	44-2627-66
$C_9H_7F_7O_4$			
Malonic acid, [1,3,3,3-tetrafluoro-2-(trifluoromethyl)propylidene]-, dimethyl ester	hexane	221(3.33)	70-1000-66
	EtOH	224(3.35)	70-1000-66
C_9H_7N			
Isoquinoline	C_6H_{12}	218(4.66),267(3.74), 304(3.48),317(3.56)	39-2328-66C
Quinoline	C_6H_{12}	209s(4.46),222(4.51), 226(4.54),229(4.43), 270(3.70),300(3.47), 314(3.51)	39-2328-66C
	H_2SO_4	194(4.17),197s(4.27), 202(4.46),235(4.59), 238s(4.59),310s(4.00), 314(4.04)	39-2328-66C
C_9H_7NO			
Carbostyril	EtOH	250(4.08)	35-1205-66
Isocarbostyril	EtOH	232(4.02)	35-1205-66
8-Isoquinolinol	iso-PrOH-HCl	244(4.31),310(3.36), 335s(3.30),379(3.55)	87-0046-66
	iso-PrOH	233(4.38),294s(3.48), 332(3.72)	87-0046-66
Isoxazole, 5-phenyl-	EtOH	261(4.29)	78-0321-66B
Quinoline, N-oxide	MeCN	340(3.9)	23-0972-66
8-Quinolinol	C_6H_{12}	320(3.39)	101-0249-66B
	N $HClO_4$	318(3.23),357(3.23)	39-0073-66A
	pH 13	334(3.48),350(3.46)	39-0073-66A
	$CHCl_3$	306(3.43)	39-0073-66A
copper chelate	$CHCl_3$	410(3.74)	39-0073-66A
	pyridine	340(3.60),409(3.74)	39-0073-66A
	dioxan	411(3.75)	39-0073-66A
nickel chelate	pyridine	345(3.83),406(3.80)	39-0073-66A
$SnCl_4$ adduct	n.s.g.	385(3.30)	39-0544-66A
$TiCl_4$ adduct	n.s.g.	400(3.34)	39-0544-66A
(other metal complexes, not listed)			
$C_9H_7NOS_2$			
Rhodanine, N-phenyl-	EtOH	258(4.04),295(4.15)	95-0095-66
$C_9H_7NO_2$			
4H-1,3-Benzoxazin-4-one, 2-methyl-	EtOH	249(3.86),306(3.46), 315(3.43)	18-1942-66
3-Isoxazolol, 5-phenyl-	EtOH	260(4.28)	94-1277-66

Compound	Solvent	$\lambda_{max}(\log \epsilon)$	Ref.
$C_9H_7NO_2S$			
1,4-Benzothiazine-3-carboxylic acid	EtOH	208(4.23),252(4.24), 335(2.78)	32-1147-66
$C_9H_7NO_3$			
1-Indanone, 2-nitro-	EtOH-NaOH	263(3.78),371(4.37)	94-1408-66
C_9H_7NS			
Isothiazole, 3-phenyl-	C_6H_{12}	239s(3.82),270(4.18), 283s(3.99),291(3.81)	78-2119-66
Isothiazole, 4-phenyl-	C_6H_{12}	242(4.09),266(4.14)	78-2119-66
Isothiazole, 5-phenyl-	C_6H_{12}	266(3.90)	78-2119-66
8-Quinolinethiol	pH -1	242(4.22),316(3.63)	3-1702-66
	pH 3.2	252(4.16),278(4.25), 316(2.91),446(3.20)	3-1702-66
	pH 13	260(4.29),368(3.51)	3-1702-66
	EtOH	502(1.34)	3-1702-66
(also other solvents)	MeCN	550(0.90)	3-1702-66
$C_9H_7N_3O$			
2-Quinoxalinecarboxamide	EtOH	207(3.25),243(3.57), 316(2.83),324(2.83)	87-0266-66
$C_9H_7N_3O_2$			
1,3-Butanedione, 2-diazo-1-(2-pyridyl)-	CH_2Cl_2	241(4.20),287s(3.82)	24-3128-66
$C_9H_7N_3O_3$			
Cinnoline, 4-methoxy-6-nitro-	EtOH	218(4.38),235(4.18), 255(4.36),290(3.79), 299(3.76),349(3.67)	39-0470-66C
Cinnoline, 4-methoxy-8-nitro-	EtOH	221(4.32),293(3.71), 321(3.63),390(3.63)	39-0470-66C
4-Cinnolinol, 8-methyl-5-nitro-	EtOH	210(4.38),230(3.93), 257(3.83),266s(3.76), 290(3.51),300(3.57), 340(3.94),351s(3.90)	39-0470-66C
4-Cinnolinol, 8-methyl-6-nitro-	EtOH	240(4.19),273(3.88), 282s(3.83),360(4.03), 327(3.91),360(4.03)	39-0470-66C
4(1H)-Cinnolinone, 1-methyl-6-nitro-	EtOH	213(4.18),238(4.20), 261s(3.90),270(3.88), 279(3.80),328(3.99), 360(4.04)	39-0470-66C
4-Hydroxy-2-methyl-6-nitrocinnolinium hydroxide, inner salt	EtOH	210(4.30),257(4.16), 285s(3.79),322(3.49), 335(3.60),402(4.10)	39-0470-66C
4-Hydroxy-2-methyl-8-nitrocinnolinium hydroxide, inner salt	EtOH	210(4.47),253(3.86), 270s(3.63),313s(3.53), 327(3.64),358s(3.92), 377(4.03)	39-0470-66C
Indole-3-carboxaldehyde, 6-nitro-, oxime	EtOH	214(4.25),251s(4.04), 256(4.06),283(4.32), 323(3.87),382(3.74)	44-0070-66
Pyrazol-3-ol, 4-nitro-1-phenyl-	EtOH	238(4.15),270(3.88), 326(3.98)	44-1538-66
$C_9H_7N_5$			
5H-s-Triazolo[4,3-b]-s-triazole, 6-phenyl-	EtOH	249(4.34)	4-0119-66

Compound	Solvent	λ_{max}(log ε)	Ref.
$C_9H_7N_5O_3$			
N-(2-Amino-6-hydroxy-5-nitro-4-pyrimidinyl)pyridinium betaine	pH 0.0	258(3.93),341(3.92)	24-2997-66
	pH 6.0	255(3.93),352(3.84)	24-2997-66
$C_9H_7N_5O_6$			
2-Pyrazoline, 1-picryl-	$CHCl_3$	384(4.22)	22-0610-66
$C_9H_7N_5S$			
s-Triazolo[3,4-b][1,3,4]thiadiazole, 3-amino-6-phenyl-, anion	MeOH	240(4.26),267(4.29)	44-3528-66
s-Triazolo[3,4-b][1,3,4]thiadiazole, 6-amino-3-phenyl-	MeOH	257(4.34),269s(4.26), 284s(4.05)	44-3528-66
cation	MeOH	256(4.30),274s(4.20), 287s(3.08),312s(3.31)	44-3528-66
$C_9H_7O^+$			
Crotonaldehyde, carbonium ion	H_2SO_4-SO_3	326.5(4.47)	35-1488-66
C_9H_8			
Indene, tetracyanoethylene complex	$CHCl_3$	430(--),542(4.13)	60-0018-66
$C_9H_8BrN_3O$			
2-Pyrazoline, 3-(m-bromophenyl)-1-nitroso-	n.s.g.	298(4.34)	88-5525-66
2-Pyrazoline, 3-(p-bromophenyl)-1-nitroso-	n.s.g.	302(4.43)	88-5525-66
$C_9H_8BrN_3O_2$			
Pyrazoline, 3-bromo-1-(p-nitrophenyl)-	$CHCl_3$	390(4.39)	22-0610-66
C_9H_8ClN			
Indole, 3-chloro-2-methyl-	MeOH	223(4.45),274(3.93), 281(3.97),289(3.90)	44-2627-66
C_9H_8ClNO			
2-Indolinone, 3-chloro-3-methyl-	MeOH	207(4.42),253(3.77)	44-2627-66
2-Isoxazoline, 3-(p-chlorophenyl)-	n.s.g.	267(4.20)	88-5525-66
$C_9H_8ClNO_2$			
Tropone, 5-acetamido-2-chloro-	MeOH	223(4.23),248(4.15), 269s(3.91),350(4.17), 370s(4.06)	18-1310-66
$C_9H_8ClN_3O$			
2-Pyrazoline, 3-(p-chlorophenyl)-1-nitroso-	n.s.g.	300(4.38)	88-5525-66
$C_9H_8ClN_5$			
as-Triazine, 3,5-diamino-6-(p-chlorophenyl)-	EtOH-HCl	216s(4.18),315(3.79)	4-0188-66
	EtOH-NaOH	243(4.09),332(4.05)	4-0188-66
$C_9H_8Cl_2N_4O$			
Pyrazolo[3,2-c]-as-triazine, 6-(dichloroacetyl)-4,6-dihydro-3-methyl-4-methylene-	EtOH	227(4.10),298(4.26)	39-1127-66C
$C_9H_8Cl_4$			
Spiro[2.4]hepta-4,6-diene, 4,5,6,7-tetrachloro-1,2-dimethyl-, cis	EtOH	245(3.98),287(3.70)	44-0768-66

Compound	Solvent	$\lambda_{max}(\log \epsilon)$	Ref.
Spiro[2.4]hepta-4,6-diene, 4,5,6,7-tetrachloro-1,2-dimethyl-, trans	EtOH	244(3.84),286(3.44)	44-0768-66
$C_9H_8CrO_3Se$			
Chromium, tricarbonyl(2,5-dimethylselenophene)-	C_6H_{12}	226(4.47),392(3.86), 430s(3.76),521(2.97)	24-1732-66
$C_9H_8F_3^+$			
(Trifluoromethyl)methylphenylcarbonium ion	FSO_3H-SbF_5	347(3.53)	35-1488-66
$C_9H_8FeO_5$			
Iron, tricarbonyl(sorbic acid)-	EtOH	206(4.26),228(4.23), 240s(4.11),300(3.38)	101-0370-66A
$C_9H_8N_2$			
Cyclopropanecarbonitrile, 2-(4-pyridyl)-	EtOH-HCl	237(4.04),255s(3.87)	44-0399-66
	EtOH-NaOH	260(3.36)	44-0399-66
Isoquinoline, 1-amino-	EtOH	230(4.19),302(3.79), 331(3.69)	95-0544-66
Isoquinoline, 3-amino-	EtOH	234(4.69),284(3.81), 364(3.49)	95-0544-66
1,6-Naphthyridine, 2-methyl-	MeOH	205(4.78),225(4.56), 230s(4.44),298(3.53), 311(3.52)	44-3055-66
1,6-Naphthyridine, 4-methyl-	MeOH	203(4.81),208(4.53), 220s(4.44),254(3.48), 263s(3.45),275s(3.28), 295s(3.30),305(3.44), 317(3.43)	44-3055-66
Pyrazole, 4-phenyl-	EtOH	250.5(4.18)	22-3744-66
Quinoline, 8-amino-	H_2O	247(4.41),336(3.40)	1-1113-66
$C_9H_8N_2O$			
Acetophenone, 2-diazo-3'-methyl-	hexane	213(4.27),255(4.15), 286(4.06)	78-0209-66
	butanol	218(4.13),255(4.09), 293(4.17)	78-0209-66
Acetophenone, 2-diazo-4'-methyl-	hexane	214(4.21),260(4.25), 290s(4.06)	78-0209-66
	butanol	218(4.09),261(4.15), 300(4.21)	78-0209-66
Cinnoline, 4-methyl-, 1-oxide	EtOH	220(4.36),263(4.12), 315(3.58),360(3.75)	25-0157-66
Cinnoline, 4-methyl-, 2-oxide	EtOH	224(4.42),262(4.57), 357(3.54)	25-0157-66
4-Cinnolinol, 3-methyl-	EtOH	208(4.47),230s(4.10), 238(4.13),250(4.02), 282(3.56),292(3.54), 343(4.10),357(4.09)	39-0470-66C
4-Cinnolinol, 8-methyl-	EtOH	211(4.40),229(4.02), 236(3.99),258(3.90), 267(3.86),288(3.43), 297(3.42),340(4.04), 354s(3.95)	39-0470-66C
Imidazol-2-one, 4-phenyl-	EtOH	286(4.19)	44-3917-66
Isocarbostyril, 3-amino-	EtOH	230(4.36),299(4.25), 374(3.64)	95-0544-66
Pyrazol-3-ol, 1-phenyl-	EtOH	272(4.28)	44-1538-66

Compound	Solvent	$\lambda_{max}(\log \epsilon)$	Ref.
$C_9H_8N_2OS$			
Benzo[b]thiophene, 2-hydroxy-, diazomethane adduct	EtOH	220(4.70),250(4.85), 280s(4.23),290(4.17), 342(3.78),382(3.81)	12-1751-66
1,3,4-Oxadiazole, 5-(methylthio)- 2-phenyl-	EtOH	272(4.27)	1-0057-66
1,3,4-Oxadiazoline-5-thione, 4-methyl-2-phenyl-	EtOH	301.5(4.19)	1-0057-66
$C_9H_8N_2OS_2$			
Thiazolin-2-imine, 3-(thenoylmethyl)-	n.s.g.	259(4.23),288s(3.94)	20-0380-66
$C_9H_8N_2OSe$			
4-Selenazolinone, 2-amino-5-phenyl-	EtOH	225(4.40),264(3.6)	28C-0285-66A
$C_9H_8N_2O_2$			
Acetophenone, 2-diazo-4'-methoxy-	hexane	218(4.21),285(4.32)	78-0209-66
	butanol	219(4.15),305(4.43)	78-0209-66
Cinnoline, 4-methyl-, 1,2-dioxide	EtOH	236(4.42),275(4.60), 348(3.85)	25-0157-66
Indole, 5-methyl-7-nitro-	EtOH	233s(3.96),252s(3.83), 378(3.86)	34-0418-66
Indole, 7-methyl-4-nitro-	EtOH	238(4.00),379(3.84)	34-0418-66
Indole, 7-methyl-5-nitro-	EtOH	253s(4.17),268(4.23), 329(3.88)	34-0418-66
Pyrazolo[3,4-c]tropolone, 3'-methyl-	MeOH	226(4.28),255(4.38), 305(3.80),318(3.87), 379s(3.88),390(3.93)	18-0253-66
Pyrazolo[4,3-d]tropolone, 3'-methyl-	MeOH	228(4.27),280s(4.27), 287(4.31),296s(4.28), 398(3.61),435(3.55)	18-0253-66
$C_9H_8N_2O_2S$			
2-Benzothiazolecarboxamide, 4-methoxy-	EtOH	237(4.15),288(3.99), 331(3.52)	44-1484-66
Thiazolin-2-imine, 3-(furoylmethyl)-	iso-PrOH-acid	270.0(4.31)	20-0380-66
	iso-PrOH-base	267.0(4.27)	20-0380-66
$C_9H_8N_2O_3$			
Indole, 5-methoxy-7-nitro-	EtOH	230s(4.08),250s(3.78), 400(3.80)	34-0418-66
$C_9H_8N_2S_2$			
1,3,4-Thiadiazole, 5-(methylthio)- 2-phenyl-	EtOH	301(4.20)	1-0057-66
1,3,4-Thiadiazoline-5-thione, 4-methyl-2-phenyl-	EtOH	339(4.17)	1-0057-66
$C_9H_8N_2Se$			
2-Benzoselenazolinecarbonitrile, 3-methyl-	C_6H_{12}	225(3.96),252s(--), 304(3.00)	22-4028-66
$C_9H_8N_4O$			
as-Triazine, 3-amino-5-phenyl-, 1-oxide (unchanged in acid or base)	EtOH	227s(4.18),256(4.25), 280s(4.03),360(3.85)	44-3917-66
as-Triazine, 3-amino-5-phenyl-, 2-oxide	EtOH-HCl	229(4.19),291(3.96), 385(3.80)	44-3917-66

Compound	Solvent	λ_{max}(log ε)	Ref.
as-Triazine, 3-amino-5-phenyl-, 2-oxide (cont.)	EtOH	229(4.18),288(3.96), 398(3.98)	44-3917-66
	EtOH-NaOH	233(4.23),287(4.10), 410(3.74),450s(3.67)	44-3917-66
$C_9H_8N_4O_2$			
1,2,4-Oxadiazole, 3-phenyl-5-ureido-	EtOH	232(4.41)	39-1522-66C
1,2,4-Oxadiazole, 5-phenyl-3-ureido-	EtOH	239(4.26)	39-1522-66C
$C_9H_8N_4O_4$			
Pyrazoline, 1-(2,4-dinitrophenyl)-	$CHCl_3$	380(4.32)	22-0610-66
C_9H_8O			
Benzo[b]furan, 4-methyl-	EtOH	247(4.05),273(3.14), 283(3.15)	39-0734-66C
Inden-4-ol	n.s.g.	251(3.90),299(3.41), 305(3.41)	25-0340-66
Inden-7-ol	n.s.g.	249(3.94),288(3.26), 299(3.33)	25-0340-66
C_9H_8OS			
2-Octene-4,6-diynal, 3-(methylthio)-, cis	ether	216(4.35),222(4.37), 270(3.83),285(4.08), 303(4.20)	24-0138-66
$C_9H_8O_2$			
1,4-Benzodioxan, 2-methylene-	MeOH	278(3.35),283(3.30)	78-0931-66
1,4-Benzodioxin, 2-methyl-	MeOH	297(3.26)	78-0931-66
p-Benzoquinone, allyl-	C_6H_{12}	246(4.39),441(1.32)	39-1627-66C
p-Benzoquinone, isopropenyl-	C_6H_{12}	247(4.08),354(3.19), 440(1.87),468s(--)	39-1627-66C
Formic acid, styryl ester	n.s.g.	257(4.26),294(3.19)	33-1002-66
$C_9H_8O_2S$			
Isophthalaldehyde, 5-(methylthio)-	ether	244(4.31)	24-1558-66
Thiete, 3-phenyl-, 1,1-dioxide	EtOH	253(4.15)	87-0489-66
$C_9H_8O_3$			
7-Nonene-3,5-diynoic acid, 9-hydroxy-, trans	EtOH	213(4.70),227s(4.02), 239(4.00),252(4.16), 266(4.27),282(4.16)	39-0135-66C
C_9H_8S			
2H-1-Benzothiopyran	EtOH	242(4.36),275(3.41), 325(3.09)	78-0007-66
$C_9H_9BrN_2O$			
3-Methyl-2-azaquinolinium bromide, 2-oxide	EtOH	228(4.20),245(4.13), 268(4.02),276(3.99), 323(3.86),341s(3.84), 354(3.88),368(3.87)	44-0941-66
C_9H_9Cl			
Benzene, (1-chloropropenyl)-, cis	C_6H_{12}	246(3.55)	35-5555-66
Benzene, (1-chloropropenyl)-, trans	C_6H_{12}	249(3.77)	35-5555-66
Propene, 1-(p-chlorophenyl)-, cis	EtOH	247(4.23)	44-4043-66
Propene, 1-(p-chlorophenyl)-, trans	EtOH	255(4.34)	44-4043-66

Compound	Solvent	$\lambda_{max}(\log \varepsilon)$	Ref.
$C_9H_9ClN_4O$			
Pyrazolo[3,2-c]-as-triazine, 6-(chloroacetyl)-3-methyl-4-methylene-	EtOH	227(3.98),266s(4.14), 294(4.19)	39-1127-66C
$C_9H_9ClN_6O$			
Benzoic acid, p-chloro-, 2-(1-methyl-1H-tetrazol-5-yl)hydrazide	EtOH	238(4.15)	39-1202-66C
$C_9H_9ClO_3$			
Propionic acid, 3-(p-chlorophenyl)-3-hydroxy-	EtOH	220(4.07),253s(2.75), 259(2.83),266(2.86), 274(2.79)	22-3243-66
C_9H_9ClPd			
Palladium, chloro-π-cinnamyl-	$CHCl_3$	242(3.83),290(4.06), 352(3.39)	101-0578-66B
	DMSO	284(4.18),352(3.89)	101-0578-66B
$C_9H_9Cl_5Si$			
Silane, trimethyl(pentachlorophenyl)-	C_6H_{12}	216.5(4.84)	101-0102-66B
$C_9H_9CuN_3O_4$			
Diglycylglycine, copper complex	H_2O	555(2.17)	37-1439-66
$C_9H_9FN_4O$			
Pyrazolo[3,2-c]-as-triazine, 6-fluoroacetyl)-4,6-dihydro-3-methyl-4-methylene-	EtOH	228(3.96),266s(4.09), 286(4.13)	39-1127-66C
$C_9H_9FeNO_4$			
Iron, tricarbonyl(sorbamide)-	EtOH	207(4.34),234(4.19), 300(3.31)	101-0370-66A
$C_9H_9IN_2$			
N-Aminoquinolinium iodide	EtOH	318(3.86)	94-0512-66
$C_9H_9IN_4O$			
Pyrazolo[3,2-c]-as-triazine, 4,6-dihydro-6-(iodoacetyl)-3-methyl-4-methylene-	EtOH	224(4.03),302(4.20)	39-1127-66C
C_9H_9N			
1H-1-Pyrindine, 1-methyl-	MeOH-HI	249s(3.50),297(4.02)	88-2579-66
	C_6H_{12}	270(4.4),330f(3.9)	88-2579-66
Skatole	EtOH	224(4.58),283(3.78)	94-0555-66
	EtOH	224(4.52),283(3.76), 292(3.69)	94-0934-66
C_9H_9NO			
Acrylophenone, oxime	EtOH	222(4.21)	88-3477-66
Carbostyril, 2,3-dihydro-	EtOH	249(4.14),277s(3.43), 287s(3.07)	1-2467-66
Cinnamaldehyde, α-oxime	n.s.g.	285(4.66)	83-0626-66
Cinnamaldehyde, β-oxime	n.s.g.	289.5(4.50)	83-0626-66
2-Isoxazoline, 3-phenyl-	n.s.g.	262(4.06)	88-5525-66
$C_9H_9NO_2$			
1-Indolecarboxaldehyde, 2-hydroxy-	EtOH	248(4.13),278(3.61), 287(3.60)	88-4355-66

Compound	Solvent	$\lambda_{max}(\log \epsilon)$	Ref.
$C_9H_9NO_2S$			
Benzenesulfinic acid, p-cyano-, ethyl ester	hexane	230(4.18),275(3.50)	18-1788-66
	MeOH	231(4.24),270(3.64)	18-1788-66
2,1-Benzothiazine, N-methyl-, 2,2-dioxide	MeOH	222(4.44),268(3.95), 322(3.48)	44-3531-66
Sulfide, methyl p-nitrostyryl, 90% trans	EtOH	246(3.88),365(4.13)	35-5747-66
$C_9H_9NO_3$			
Anisole, o-(2-nitrovinyl)-	C_6H_{12}	236(4.00),292(4.10), 336(4.05)	19-0023-66
	MeOH	226s(3.83),238(3.86), 300(4.04),351(4.02)	19-0023-66
	20% MeOH	244s(3.79),313(4.01), 361(4.01)	19-0023-66
Tropone, 2-amino-7-carbomethoxy-	MeOH	240s(4.34),245(4.36), 265(4.01),274s(3.93), 344(3.99),409(4.11)	18-2538-66
$C_9H_9NO_3S$			
4H-1,2-Benzothiazin-4-one, 2,3-dihydro-2-methyl-, 1,1-dioxide	EtOH	248(3.93),286(3.11), 297s(3.08)	44-0162-66
$C_9H_9NO_5$			
Benzoic acid, 3-hydroxy-4-nitro-, ethyl ester	n.s.g.	272(3.88),353(3.46)	39-0424-66B
2-Furanacrylic acid, 5-nitro-, ethyl ester	EtOH	273(4.11),275(3.84), 344(4.24)	95-1187-66
Propionic acid, 3-(p-nitrophenoxy)-	EtOH	228(3.95),307(4.08)	94-0939-66
$C_9H_9N_3$			
3,5-Pyridinedicarbonitrile, 1,4-dihydro-1,4-dimethyl-	n.s.g.	213(4.34),363(3.79)	39-1083-66B
3,5-Pyridinedicarbonitrile, 1,4-dihydro-2,6-dimethyl-	n.s.g.	215(4.36),345(3.73)	39-1083-66B
6H-Pyrrolo[1',2":1,2]imidazo[4,5-c]-pyridine, 7,8-dihydro-	EtOH	252(3.53),259(3.54), 265(3.55),273(3.45)	39-0080-66C
6H-Pyrrolo[2',1':2,3]imidazo[4,5-b]-pyridine, 7,8-dihydro-	EtOH	248(3.60),282s(3.97), 284(3.98),291s(3.87)	39-0080-66C
Quinazoline, 2-amino-4-methyl-	pH 2.0	228(4.36),245(4.27), 280(3.34),342(3.25)	39-0234-66C
	pH 10.0	236(4.54),245s(4.35), 260(3.66),348(3.42)	39-0234-66C
Quinazoline, 2-(methylamino)-	pH 0.0	234(4.48),249s(4.25), 285(3.79),350(3.59)	39-0234-66C
	pH 9.0	239(4.55),252s(4.33), 363(3.47)	39-0234-66C
1H-1,2,3-Triazole, 5-methyl-1-phenyl-	EtOH	222(4.19)	44-1587-66
$C_9H_9N_3O$			
4-Naphthyridinol, 7-amino-2-methyl-	EtOH	250(4.42),320f(4.2), 340(4.08)	32-0103-66
2-Pyrazoline, 1-nitroso-3-phenyl-	n.s.g.	298(4.41)	88-5525-66
Pyrazol-3-ol, 4-amino-1-phenyl-	EtOH	297(4.08),400(2.38)	44-1538-66
v-Triazolo[3,4-a]pyridine, 3-acetyl-7-methyl-	MeOH	219(4.22),245(3.34), 252(3.25),309(4.24)	24-2918-66
$C_9H_9N_3O_2$			
Pyrazoline, 1-(p-nitrophenyl)-	$CHCl_3$	394.5(4.40)	22-0610-66

Compound	Solvent	λ_{max}(log ϵ)	Ref.
$C_9H_9N_3O_2S$			
1,3,4-Thiadiazole, 2-(N-methyl-acetamido)-5-(2-furyl)-	EtOH	306(4.31)	83-0921-66
$C_9H_9N_3O_3$			
1,2,4-Triazoline-3,5-dione, 4-(p-methoxyphenyl)-	CH_2Cl_2	546(2.18)	44-3444-66
$C_9H_9N_3O_6$			
Hydroxylamine, N-(5-methyl-2,4-di-nitrophenyl)-, O-acetate	$CHCl_3$	216(4.08),267(4.13), 322(3.97)	78-0995-66
$C_9H_9N_3S$			
1,3,4-Triazoline-5-thione, 1-methyl-2-phenyl-	EtOH	255(4.27),280s(3.90)	1-0057-66
$C_9H_9N_5$			
as-Triazine, 3,5-diamino-6-phenyl-	EtOH-HCl	237s(4.11),280s(3.77)	4-0188-66
	EtOH-NaOH	249(3.96),315(3.84)	4-0188-66
s-Triazine, 2-amino-4-anilino-	EtOH	261(4.09)	-0-0139-66
$C_9H_9N_5O_2$			
Pyrazolo[3,2-c]-as-triazine, 4-amino-, diacetyl derivative	EtOH	248(4.24),322(3.45)	39-1127-66C
C_9H_{10}			
Pentacyclo[4.3.0.0^{2,5}.0^{3,8}.0^{5,7}]-nonane (end absorption only)	EtOH	205(2.40)	88-3743-66
Styrene, m-methyl-	MeOH	250(4.10)	22-0043-66
Styrene, p-methyl-	MeOH	252(4.23)	22-0043-66
$C_9H_{10}BrClN_4O_2$			
Pyrazolo[3,2-c]-as-triazin-4-ol, 4-(bromomethyl)-6-(chloroacetyl)-4,6-dihydro-3-methyl-	EtOH	251(4.12)	39-1214-66C
$C_9H_{10}BrD_3O$			
2-Indanone-1,3,3-d_3, 1-bromo-hexahydro-, trans	C_6H_{12}	318s(2.06)	23-0759-66
$C_9H_{10}Br_2N_2$			
1-(3-Bromopropyl)-4-cyanopyridinium bromide	H_2O	228(4.10),276(3.64)	87-0638-66
$C_9H_{10}ClD_3O$			
2-Indanone-1,3,3-d_3, 1-chloro-hexahydro-, trans	C_6H_{12}	317(1.70)	23-0759-66
$C_9H_{10}ClN$			
1-Methylpyrindinium chloride	H_2O	233(4.07),275(4.10), 302(3.79)	88-2579-66
$C_9H_{10}ClNO_4S$			
2,3-Dimethylthiazolo[3,2-a]pyrid-inium perchlorate	H_2O	213(4.08),239(4.01), 305s(4.10),314(4.22)	4-0027-66
$C_9H_{10}ClN_3$			
Benzimidazole, 5(or 6)-chloro-2-(dimethylamino)-	EtOH	219(4.68),256(3.99), 298(4.16)	4-0107-66

Compound	Solvent	$\lambda_{max}(\log \epsilon)$	Ref.
$C_9H_{10}D_4O$			
2-Indanone-1,1,3,3-d_4, hexahydro-, trans	C_6H_{12}	301(1.48)	23-0759-66
$C_9H_{10}F_3N_3O_5$			
6-Aza-2'-deoxyuridine, 5-(trifluoromethyl)-	pH 1	268(3.81)	87-0715-66
	N NaOH	264(3.77)	87-0715-66
$C_9H_{10}IN$			
1-Methylpyrindinium iodide	H_2O	223(4.55),275(4.01), 302(3.75)	88-2579-66
	pH 12	257(4.39),321(3.90), 455(3.13)	88-2579-66
	aq. HI	276(4.01),301(3.77)	88-2579-66
$C_9H_{10}N_2$			
2-Cyclohexene-1-malononitrile	C_6H_{12}	210(1.76)(end abs.)	35-1979-66
Indazole, 3,6-dimethyl-	EtOH	215(4.57),267(3.60), 290(3.66),299(3.59)	78-3131-66
Malononitrile, cyclohexylidene-	EtOH	236(4.28)	44-2784-66
$C_9H_{10}N_2O$			
Indazole, 6-methoxy-3-methyl-	EtOH	216(4.45),285(3.83), 294(3.81)	78-3131-66
$C_9H_{10}N_2OS_2$			
Thiazolin-2-imine, 3-(2-hydroxy-2-thienylethyl)-	iso-PrOH-acid	236(4.08),255s(--)	20-0380-66
	iso-PrOH	237(4.11),259s(--)	20-0380-66
	iso-PrOH-base	236(4.13),264(3.88)	20-0380-66
$C_9H_{10}N_2O_2$			
Formic acid, (phenylazo)-, ethyl ester	n.s.g.	287(4.01),419(2.16)	89-0372-66
Indazole, 5,6-dimethoxy-	EtOH	215(4.20),267(3.69), 294(3.79),305(3.67)	78-3131-66
$C_9H_{10}N_2O_2S$			
Oxamide, N-(m-methoxyphenyl)-2-thio-	EtOH	340(3.98)	44-1484-66
$C_9H_{10}N_2O_4$			
Uracil, 1-(2-deoxy-3,5-epoxy-β-D-threo-pentofuranosyl)-	H_2O	262(3.98)	44-0205-66
2'-Uridenine, 2',3'-dideoxy-	H_2O	261(4.02)	44-0205-66
$C_9H_{10}N_2O_5$			
Uracil, 2,2'-anhydro-1-β-D-arabinofuranosyl-	pH 2	222(3.94),249(3.93)	69-2076-66
	pH 12	254(3.89)	69-2076-66
Uracil, 1-(5-deoxy-β-D-erythro-pent-4-enofuranosyl)-	MeOH	261(3.98)	35-5684-66
$C_9H_{10}N_2S$			
2-Thiazoline, 2-amino-3-phenyl-	EtOH	257(3.86)	94-1201-66
hydrobromide	EtOH	240s(3.88)	94-1201-66
2-Thiazoline, 2-anilino-	EtOH	257(4.08)	94-1201-66
hydrobromide	EtOH	245(4.04)	94-1201-66

Compound	Solvent	$\lambda_{max}(\log \epsilon)$	Ref.
$C_9H_{10}N_4O$			
4-Pteridinol, 2,6,7-trimethyl-	pH 6.5	233(4.06),267(3.81), 316(3.91)	39-1112-66C
	pH 11.3	245(4.32),332(3.85)	39-1112-66C
Pyrazolo[3,2-c]-as-triazine, 6-acetyl-4,6-dihydro-3-methyl-4-methylene-	EtOH	226(4.08),265s(4.14), 289(4.18)	39-1127-66C
$C_9H_{10}N_4O_2$			
Benzimidazole, 2-(dimethylamino)-5(or 6)-nitro-	pH 2	222(4.25),252(4.23)	4-0107-66
	EtOH	231(4.10),267(4.22), 338(3.93)	4-0107-66
Lumazine, 6,7,8-trimethyl-	pH -2.7	244(4.04),358(4.18)	24-3503-66
	pH 3	308(4.38),394(3.68)	24-3503-66
	pH 5.8	257(4.19),276(4.07), 403(4.17)	24-3503-66
	pH 7	256(4.17),275(4.05), 402(4.09)	24-3503-66
	pH 13	244(4.30),267s(3.85), 313(4.35),364(3.78)	24-3503-66
Pyrazolo[3,2-c]-as-triazine-3-carboxylic acid, 4-methyl-, ethyl ester	EtOH	242(4.50),306(3.70), 345s(3.37)	39-1127-66C
$C_9H_{10}N_4O_3$			
Formamide, N,N-dimethyl-1-[(p-nitrophenyl)azo]-	dioxan	284(4.14),448(2.24)	24-3337-66
Lumazine, 8-(2-hydroxyethyl)-3-methyl-	pH -2.7	240(4.28),346(4.02)	24-3503-66
	pH 7	261(4.29),400(3.97)	24-3503-66
	pH 13	233(4.30),280(4.06), 306(3.94)	24-3503-66
$C_9H_{10}N_4O_4$			
1,1-Hydrazinedicarboxylic acid, (dicyanomethylene)-, diethyl ester	MeCN	255(3.82),283(4.01)	35-1979-66
$C_9H_{10}N_4O_6S$			
Acetaldehyde, (methylsulfonyl)-, 2,4-dinitrophenylhydrazone	EtOH	219(4.14),348(4.36)	24-1580-66
$C_9H_{10}O$			
Benzofuran, 2,3-dihydro-2-methyl-	EtOH	225s(3.64),278(3.37), 282(3.37),288(3.26)	1-1561-66
	EtOH	282(3.49)	22-0228-66
Benzofuran, 2,3-dihydro-3-methyl-	EtOH	226s(3.65),276(3.41), 282(3.43),287s(3.26)	1-1561-66
Benzofuran, 2,3-dihydro-4-methyl-	EtOH	278(3.34),282(3.34)	39-0734-66C
4H-Benzopyran, 2,3-dihydro-	EtOH	276(3.32)	22-0228-66
Chroman	EtOH	273s(3.37),277(3.39), 283(3.33)	1-1561-66
2-Nonene-4,6-diyn-1-ol	EtOH	207(4.53),213(4.69), 228(3.46),240(3.80), 252(4.13),266(4.31), 282(4.20)	39-2283-66C
Oxirane, 3-methyl-2R-phenyl-, trans	EtOH	217(4.06),245s(2.00), 250s(2.15),256(2.27), 261(2.35),267(2.24), 270s(2.05)	44-3921-66
Oxirane, 3-methyl-2S-phenyl-, cis	EtOH	244(1.89),249s(2.08), 254(2.21),259(2.31), 266(2.17)	44-3921-66

Compound	Solvent	λ_{max}(log ε)	Ref.
Phenol, o-isopropenyl-	EtOH	237s(3.59),280(3.36), 305s(2.98)	1-1561-66
Styrene, p-methoxy-	MeOH	258(4.29)	22-0043-66
$C_9H_{10}OS$			
2-Octene-4,6-diyn-1-ol, 3-(methylthio)-, cis	ether	238(3.61),252(3.74), 268(3.90),284(3.92)	24-0138-66
trans	ether	239(3.65),253(3.82), 268(3.98),285(3.98)	24-0138-66
$C_9H_{10}O_2$			
1,4-Benzodioxan, 2-methyl-	MeOH	279(3.40),285(3.34)	78-0931-66
1,3-Benzodioxole, 2-ethyl-	MeOH	233(3.49),284(3.55)	78-0931-66
Tropolone, 5-ethyl-	isooctane	224(4.38),240s(4.28), 248s(4.16),311s(3.92), 323(3.96),358(3.76), 376(3.66)	31-0140-66
$C_9H_{10}O_3$			
Acetic acid, p-anisyl ester	EtOH	222(3.91),277(3.30)	65-1198-66
p-Benzoquinone, (1-hydroxy-1-methylethyl)-	C H	247(4.22),450(1.32)	39-1627-66C
Cyclohexanone, 2-methyl-6-oxalyl-, enol lactone	EtOH	290.0(4.36)	39-0764-66C
$C_9H_{10}O_4$			
2-Furanpropionic acid, 3-methyl-β-oxo-, methyl ester	EtOH	275.5(4.15)	39-0976-66C
$C_9H_{10}O_5S$			
2(5H)-Furanone, 3-acetoxy-4-(acetylthiomethyl)-	EtOH	216(4.13)	54-0347-66
$C_9H_{10}S$			
Benzothiopyran, 2,3-dihydro-	EtOH	260(4.03)	22-0228-66
Sulfide, methyl styryl, cis	EtOH	286(4.19)	35-5747-66
$C_9H_{11}^+$			
Dimethylphenylcarbonium ion	FSO_3H-SbF_5	326(4.04),390(3.15)	35-1488-66
$C_9H_{11}BrN_2O_3$			
Uracil, 5-bromo-1-(tetrahydro-pyran-2-yl)-	pH 1	277(4.04)	4-0005-66
	H_2O	277(4.02)	4-0005-66
	pH 11	232(3.82),274(3.87)	4-0005-66
$C_9H_{11}BrN_4O_2$			
Pyrazolo[3,2-c]-as-triazine, 6-acetyl-4-(bromomethyl)-4,6-di-hydro-4-hydroxy-3-methyl-	EtOH	249(4.16)	39-1214-66C
$C_9H_{11}ClN_4$			
9H-Purine, 6-chloro-2,9-diethyl-	pH 1	270(3.97)	44-1417-66
	pH 7	271(3.97)	44-1417-66
	pH 13	271(3.97)	44-1417-66
$C_9H_{11}ClN_4O_3$			
Acetoacetic acid, 2-chloro-2-(pyra-zol-3-ylazo)-, ethyl ester	EtOH	307(4.40)	39-1127-66C

Compound	Solvent	λ_{max}(log ε)	Ref.
$C_9H_{11}ClO_2$			
1,3-Cyclohexadiene-1-carboxylic acid, 3-chloro-5,5-dimethyl-	EtOH	295(3.76)	44-2073-66
$C_9H_{11}Cl_4N_3O$			
3(2H)-Pyridazinone, 2-[bis(2-chloroethyl)amino]methyl-4,5-dichloro-	EtOH	301(3.33)	95-1099-66
$C_9H_{11}IN_2O_3$			
Uracil, 5-iodo-1-(tetrahydropyran-2-yl)-	pH 1	288(3.85)	4-0005-66
	H_2O	283(3.90)	4-0005-66
	pH 11	230(4.04),283(3.76)	4-0005-66
$C_9H_{11}NO$			
1-Indanol, 2-amino-	EtOH	261(2.66),266(2.80), 273(2.70)	94-1408-66
Methylamine, N-(p-methoxybenzylidene)-	C_6H_{12}	260(4.31)	35-2775-66
$C_9H_{11}NO_2$			
Anthranilic acid, ethyl ester	2N HCl	272(3.04)	28C-0362-66A
	pH 9.2	327(3.60)	28C-0362-66A
1H-Azepine-2,5-dione, 3,4,6-trimethyl-	EtOH	228(3.04),288(2.52)	78-1201-66
Benzoic acid, p-amino-, ethyl ester	2N HCl	271(3.02)	28C-0362-66A
	pH 9.2	284(4.23)	28C-0362-66A
Benzoic acid, 4-amino-3-methyl-, methyl ester	2N HCl	270(3.11)	28C-0362-66A
	pH 9.2	281(4.18)	28C-0362-66A
Pyridine, 3,5-diacetyl-1,4-dihydro-	n.s.g.	232(4.10),265(3.72), 392(3.90)	39-1083-66
$C_9H_{11}NO_3$			
Ether, p-nitrophenyl propyl	EtOH	228(3.85),309(4.06)	94-0939-66
2(1H)-Pyridone, 4-acetoxy-1,6-dimethyl-	EtOH	230(3.79),303(3.86)	70-1546-66
2(1H)-Pyridone, 3-acetyl-4-hydroxy-1,6-dimethyl-	EtOH	232(3.97),269(3.51), 330(4.13)	70-1546-66
2(1H)-Pyridone, 4-hydroxy-6-methyl-3-propionyl-	EtOH	230(4.07),266(3.43), 324(4.12)	70-0848-66
Pyrrole, 2-acetoxymethyl-4-acetyl-	EtOH	194(4.04),245(3.97), 270(3.72)	33-0349-66
$C_9H_{11}NO_3S$			
3-Thiophenecarboxylic acid, 5-acetyl-2-amino-4-methyl-, methyl ester	MeOH	224(4.18),249(3.82), 347(4.19)	24-0094-66
$C_9H_{11}NO_4$			
Benzene, 1-ethoxy-2-methoxy-4-nitro-	EtOH	240(3.95),338(3.88)	94-0939-66
Benzene, 1-ethoxy-4-methoxy-2-nitro-	MeOH	357(3.39)	78-0093-66B
Benzene, 1-ethoxy-4-methoxy-3-nitro-	MeOH	357(3.39)	78-0093-66B
Ether, 2-hydroxypropyl p-nitrophenyl	EtOH	228(3.95),307(4.08)	94-0939-66
Oxazole-4-carboxylic acid, 2-acetonyl-, ethyl ester	n.s.g.	252(3.79),266s(3.78)	39-1653-66C
	EtOH-NaOH	302(4.20)	39-1653-66C
Phenol, 5-nitro-2-propoxy-	pH 13	227(3.98),264(4.07), 324(3.76),420(3.57)	94-0939-66
	EtOH	244(3.99),300(3.79), 343(3.86)	94-0939-66
5-Pyridineacetic acid, 3-hydroxy-4-(hydroxymethyl)-2-methyl-	pH 1	290(3.95)	4-0178-66
	pH 13	244(3.85),310(3.89)	4-0178-66

Compound	Solvent	λ max(log ε)	Ref.
$C_9H_{11}NO_5$			
Nicotinic acid, 1,2-dihydro-4,6-dihydroxy-1-methyl-2-oxo-, ethyl ester	EtOH	304(4.3)	78-0455-66
Succinamic acid, N-(2-hydroxy-3-methyl-4-oxo-2-cyclobuten-1-yl)-	MeOH-NaOH	251(4.25)	88-3031-66
$C_9H_{11}N_3$			
Benzimidazole, 2-(dimethylamino)-	EtOH	250(3.29),289(4.05)	4-0107-66
Fumaronitrile, piperidino-	ether	297.5(4.13)	24-2572-66
Maleonitrile, piperidino-	ether	304.5(4.13)	24-2572-66
s-Triazolo[4,3-b]pyridine, 3-propyl-	MeOH	268(3.9)	24-2593-66
$C_9H_{11}N_3O$			
Acetophenone, semicarbazone	EtOH	273(4.20)	22-0717-66
Formamide, N,N-dimethyl-1-(phenylazo)-	dioxan	284(4.03),434(2.11)	24-3337-66
	n.s.g.	284(4.03),434(2.11)	89-0372-66
Pyrazineacetaldehyde, α-[(dimethylamino)methylene]-	EtOH	287(4.42),330s(3.72)	44-3845-66
$C_9H_{11}N_5$			
Pteridine, 2-(propylamino)-	pH 1.0	235(4.12),288s(3.78), 304(3.91)	39-0226-66C
	pH 7	231(4.36),275(3.98), 393(3.80)	39-0226-66C
$C_9H_{11}N_5O$			
7(8H)-Pteridinone, 6,8-dimethyl-2-(methylamino)-	pH 0.5	230(4.40),264(4.29), 272(4.25),338(3.92)	39-1065-66C
	pH 5.0	218(4.34),241(4.12), 294(3.78),353(4.21)	39-1065-66C
7(8H)-Pteridinone, 6,8-dimethyl-4-(methylamino)-	pH 5.0	223(4.21),242(4.11), 255s(4.04),262s(3.92), 292s(3.69),297(3.71), 345(4.01)	39-1065-66C
$C_9H_{12}BN$			
Isoquinoline, 3,4-dihydro-, borane adduct	90% EtOH	277(3.87)	94-1382-66
$C_9H_{12}BrN_3O_2$			
Cytosine, 5-bromo-1-(tetrahydropyran-2-yl)-	pH 1	297(4.12)	4-0005-66
	H_2O	285(3.91)	4-0005-66
	pH 11	235(3.91),285(3.91)	4-0005-66
$C_9H_{12}Br_2N_2O$			
1-(3-Bromopropyl)-4-formylpyridinium bromide, oxime	H_2O	280(4.26)	87-0638-66
$C_9H_{12}Br_2O$			
2,5-Heptadien-4-one, 3,5-dibromo-2,6-dimethyl-	EtOH	282.0(3.74)	39-0075-66B
$C_9H_{12}Cl_2O$			
Cyclooctanone, 2-(dichloromethylene)-	EtOH	248(3.31)	77-0567-66
$C_9H_{12}FN_3O_2$			
Cytosine, N^4-pivaloyl-5-fluoro-	EtOH	214(4.08),242(3.83), 313(3.75)	87-0566-66

Compound	Solvent	λ_{max}(log ε)	Ref.
$C_9H_{12}FN_3O_4$			
Cytosine, 1-(2-deoxy-α-D-ribofuranosyl)-5-fluoro-	pH 1	291(4.08)	87-0566-66
$C_9H_{12}FN_3O_5$			
Cytosine, 1-β-D-arabinofuranosyl-5-fluoro-	N HCl	221(4.01),291(4.08)	87-0101-66
	pH 5-7	235(3.90),280(3.92)	87-0101-66
$C_9H_{12}IN_3O_2$			
Cytosine, 5-iodo-1-(tetrahydropyran-2-yl)-	pH 1	306(3.97)	4-0005-66
	H_2O	291(3.81)	4-0005-66
	pH 11	227(4.09),291(3.81)	4-0005-66
$C_9H_{12}I_2N_2O$			
1-(3-Iodopropyl)-4-formylpyridinium iodide, oxime	H_2O	224(4.23),279(4.24)	87-0638-66
$C_9H_{12}N_2$			
Malononitrile, (1,1,2-trimethylpropylidene)-	EtOH	238(4.10)	44-2784-66
$C_9H_{12}N_2O$			
Aniline, o-propionamido-	MeOH	234s(3.93),291(3.48)	78-1175-66
Cyclooct[c]isoxazole, 3-amino-4,5,8,9-tetrahydro-	n.s.g.	251(3.92)	18-1125-66
Urea, 1-ethyl-1-phenyl-	EtOH	236(3.54)	94-0001-66
Urea, 1-ethyl-3-phenyl-	EtOH	241(4.3),276(3.00)	94-0001-66
$C_9H_{12}N_2OS$			
Benzo[b]thiophene-3-carboxamide, 2-amino-4,5,6,7-tetrahydro-	MeOH	229(4.36),309(3.75)	24-0094-66
$C_9H_{12}N_2O_2$			
Alanine, p-aminophenyl-	3N HCl	256(2.68)	35-5010-66
	H_2O	283(3.11)	35-5010-66
Urea, (p-ethoxyphenyl)-	H_2O	236(4.14),281(3.25)	94-0001-66
$C_9H_{12}N_2O_2S$			
Uracil, 1-(tetrahydropyran-2-yl)-2-thio-	pH 1	246(3.56),329(4.38)	4-0005-66
	H_2O	245(3.46),328(4.35)	4-0005-66
	pH 11	234(3.66),314(4.33)	4-0005-66
$C_9H_{12}N_2O_3$			
Propionic acid, 2-hydrazino-3-(4-hydroxyphenyl)-	90% MeOH	207(4.25),226(4.42),279(3.59)	87-0439-66
Uracil, 1-(tetrahydropyran-2-yl)-	pH 1	259(4.11)	4-0005-66
	H_2O	259(4.04)	4-0005-66
	pH 11	230(3.77),259(3.99)	4-0005-66
$C_9H_{12}N_2O_4$			
6(1H)-Pyridazinone, 1-acetoxy-3-ethoxy-4-methyl-	EtOH	304(3.402)	95-0082-66
Uridine, 2',3'-dideoxy-	H_2O	263(3.99)	44-0205-66
$C_9H_{12}N_2O_4S_2$			
Uridine, 2,4-dithio-	pH 5.8	283(4.35),350s(4.11)	94-0666-66
	pH 9	280(4.23),320(4.39)	94-0666-66

Compound	Solvent	λ_{max}(log ϵ)	Ref.
$C_9H_{12}N_2O_5$			
Uracil, 1-(3-deoxy-β-D-threo-pentofuranosyl)-	EtOH	262(4.01)	44-0205-66
Uracil, 1-(3-deoxy-α-D-ribo-furanosyl)-	pH 1	205(3.95),263(4.00)	44-1163-66
	pH 7	205(3.97),264(4.02)	44-1163-66
	pH 13	213(4.05),263(3.90)	44-1163-66
Uridine, 3'-deoxy-	pH 1	212(3.96),263(4.03)	44-1163-66
	pH 7	212(3.97),264(4.03)	44-1163-66
	pH 13	263(3.88)	44-1163-66
$C_9H_{12}N_2O_5S$			
Uracil, 1-(4-thio-α-D-ribo-furanosyl)-	pH 2	266(4.09)	44-0813-66
	pH 12	266(3.93)	44-0813-66
β-isomer	pH 2	266(4.01)	44-0813-66
	pH 12	266(3.92)	44-0813-66
Uracil, 1-(5-thio-D-xylopyranosyl)-	pH 2	265(4.07)	44-0813-66
	pH 12	265(3.96)	44-0813-66
$C_9H_{12}N_2O_6$			
Uracil, 5-β-D-xylofuranosyl-	pH 7	262(3.83)	44-2215-66
	pH 12	289(3.82)	44-2215-66
Uridine, 2'-deoxy-5-hydroxy-	pH 1	281(3.92)	87-0366-66
	pH 12	304(3.83)	87-0366-66
$C_9H_{12}N_2S$			
Thiourea, N-benzyl-N'-methyl-, iodine complex	$CHCl_3$	305.5(4.58)	60-0029-66
$C_9H_{12}N_4O_2S$			
Theophylline, (2-hydroxyethyl)-6-thio-	EtOH	205(3.85),215(3.85), 265(3.36),335(4.36), 342(4.40)	22-3934-66
$C_9H_{12}N_4O_3$			
8H-Theophylline, 8-methoxy-8-methyl-	H_2O	257(3.74),305s(3.24)	5-0233-66C
$C_9H_{12}N_4O_5$			
Imidazo[1,2-c]pyrimidine-5,7(1H,6H)-dione, 3-ethoxy-2,3-dihydro-1-methyl-8-nitro-	pH 5.0	238(4.18),255s(3.99), 333(3.67)	24-2984-66
	pH 11.0	250(4.19),353(3.75)	24-2984-66
$C_9H_{12}N_4S$			
6(1H)-Purinethione, 2,9-diethyl-	pH 1	227(4.00),326(4.28)	44-1417-66
	pH 7	229(4.01),325(4.35)	44-1417-66
	pH 13	235(4.21),310(4.34)	44-1417-66
$C_9H_{12}O$			
Anisole, 2,6-dimethyl-	EtOH	268(2.52)	22-0228-66
Benzofuran, 4,5,6,7-tetra-hydro-3-methyl-	heptane	222.5(3.80)	39-0377-66C
2,4-Cyclopentadien-1-one, 3-tert-butyl-	isooctane	200(4.71),380(1.9)	35-3433-66
Dispiro[2.1.2.2]nonan-4-one	isooctane	214(2.83),283(1.64)	22-0116-66
4-Nonen-6-ynal	EtOH	227(4.12),323s(3.0)	28C-1433-66A
4-Octen-6-ynal, 3-methyl-	EtOH	227(4.07),323s(3.0)	28C-1433-66A
Phenol, 3-propyl-	H_2O	271(3.21)	22-3328-66
	base	288(3.46)	22-3328-66
Spiro[2.5]oct-5-en-4-one, 6-methyl-	isooctane	225(4.16),297(1.97), 328(1.85)	22-0116-66

Compound	Solvent	λ_{max}(log ε)	Ref.
$C_9H_{12}OS$			
1,3-Benzoxathian, 8-methyl-	MeOH	273(3.17),282(3.18)	35-5855-66
	MeOH-KOH	273(3.17),282(3.18)	35-5855-66
Phenol, 2-methyl-6-(methylthio-methyl)-	MeOH	276(3.30)	35-5855-66
	MeOH-KOH	240(3.53),283(3.30), 297(3.23)	35-5855-66
$C_9H_{12}O_2$			
Benzene, 1-ethoxy-4-methoxy-	MeOH	226(4.01),288(3.49)	78-0093-66B
Bicyclo[3.1.0]hexan-2-one, 3-(hy-droxymethylene)-6,6-dimethyl-	EtOH	273(4.01)	22-3490-66
Bicyclo[3.3.1]nonane-2,9-dione	MeOH	281(1.84)	88-4173-66
2,4-Heptadienoic acid, 4-hydroxy-2,3-dimethyl-, γ-lactone	n.s.g.	270(4.3)	54-0043-66
Indone, 3a,4,5,6,7,7a-hexahydro-3-hydroxy-	EtOH	245(3.76)	70-1167-66
Spiro[3.4]oct-1-ene-2-carboxylic acid	EtOH	221(4.03)	22-3947-66
$C_9H_{12}O_2S$			
p-Toluenesulfinic acid, ethyl ester	hexane	223(4.00),255(3.51)	18-1788-66
	MeOH	224(3.95),243(3.70)	18-1788-66
$C_9H_{12}O_3$			
1,3-Cyclohexanedione, 2-formyl-5,5-dimethyl-	EtOH-NaOH	258(4.27),272(4.19)	99-0230-66
5-Hepten-3-ynoic acid, 2-hydroxy-2,6-dimethyl-	EtOH	236(4.00)	39-1615-66
6-Heptynoic acid, 5-oxo-, ethyl ester	n.s.g.	217(3.85)	70-0256-66
$C_9H_{12}O_4$			
Cycloheptene-1,5-dicarboxylic acid	EtOH	223(3.72)	78-1521-66
4-Heptenoic acid, 2,3-epoxy-6-oxo-, ethyl ester	EtOH	230(4.06)	70-0646-66
Malonic acid, (3-methyl-2,4-penta-dienyl)-	EtOH	229(4.21)	33-0858-66
$C_9H_{12}O_5$			
2,4-Hexadienedioic acid, 3-methoxy-, dimethyl ester, cis	ether	281(4.15)	24-2572-66
trans	ether	280(4.02)	24-2572-66
$C_9H_{12}S$			
Sulfide, isopropyl phenyl	EtOH	258(3.75)	22-0228-66
$C_9H_{13}BrN_4O$			
Pyrimidine, 2-amino-5-bromo-4-morpholino-	EtOH	304(4.13)	78-2401-66
$C_9H_{13}BrO$			
2-Indanone, 1-bromohexahydro-, trans	C_6H_{12}	321(1.85)	23-0759-66
$C_9H_{13}BrO_2$			
2-Cyclopenten-1-one, 3-bromo-2-hydroxy-4,4,5,5-tetramethyl-	EtOH	262.0(4.12)	39-0075-66B
$C_9H_{13}ClO$			
2-Indanone, 1-chlorohexahydro-, trans	C_6H_{12}	322(1.70)	23-0759-66

Compound	Solvent	λ_{max}(log ε)	Ref.
$C_9H_{13}Cl_2N_3O_2$			
3(2H)-Pyridazinone, 2-[bis(2-chloroethyl)amino]methyl-6-hydroxy-	EtOH	322(3.28)	95-1099-66
$C_9H_{13}NO$			
2,5-Cyclohexadienecarboxamide, N,2-dimethyl-	EtOH	220(2.83)(end abs.)	44-3482-66
$C_9H_{13}NO_2$			
Ethylamine, 2-(p-hydroxyphenyl)-2-methoxy-	EtOH	226(4.07),277(3.25)	6-0083-66
2(1H)-Pyridone, 1-butoxy-	EtOH	227(3.81),300(3.75)	78-0025-66
Pyrrole-1-carboxylic acid, tert-butyl ester	EtOH	227(4.9)	44-0764-66
Pyrrole-3-carboxylic acid, 1,2,4-trimethyl-, methyl ester	EtOH	210(4.18),239(3.99), 258s(3.7)	39-1950-66C
Pyrrole-3-carboxylic acid, 1,2,5-trimethyl-, methyl ester	EtOH	205(4.32),235(3.93), 265(3.81)	39-1950-66C
$C_9H_{13}NO_2S$			
3-Thiophenecarboxylic acid, 2-amino-5-ethyl-, ethyl ester	MeOH	232(4.51),260(3.63), 304(3.76)	24-0094-66
$C_9H_{13}NO_3$			
Dimedon, carboxamido-	MeOH-HCl	258(4.25)	44-2429-66
	MeOH-NaOH	269(4.27)	44-2429-66
$C_9H_{13}NO_3S_2$			
5-Thiazolidineacetic acid, N-ethyl-4-oxo-2-thioxo-, ethyl ester	EtOH	261(4.12),295(4.19)	95-0095-66
5-Thiazolidineacetic acid, N-methyl-4-oxo-2-thioxo-, propyl ester	EtOH	260(4.16),294(4.21)	95-0095-66
5-Thiazolidineacetic acid, 4-oxo-N-propyl-2-thioxo-, methyl ester	EtOH	262(4.10),296(4.19)	95-0095-66
$C_9H_{13}NO_4$			
Pyridine, 2,3,5,6-tetramethoxy-	MeOH	229(4.08),305(3.98)	44-4054-66
$C_9H_{13}NS$			
Fulvene, 6-(dimethylamino)-6-(methylthio)-	EtOH	270(3.58),355(4.35)	24-3268-66
$C_9H_{13}N_3O_2$			
Cytosine, 1-(tetrahydropyran-2-yl)-	pH 1	276(4.13)	4-0005-66
	H_2O	268(3.96)	4-0005-66
	pH 11	235(3.89),268(3.96)	4-0005-66
Triazene, 1-(p-ethoxyphenyl)-3-hydroxy-3-methyl-	1% acetone	264(5.49)	65-2117-66
$C_9H_{13}N_3O_4$			
Cytidine, 3'-deoxy-	pH 1	216(3.88),282(4.12)	44-1163-66
	pH 7	215(3.88),232(3.83), 272(3.95)	44-1163-66
	pH 13	232(3.83),273(3.95)	44-1163-66
$C_9H_{13}N_3O_4S$			
Cytidine, 2-thio-	pH 1	230(4.21),276(4.24), 310s(3.73)	94-0666-66
	H_2O	247(4.35),270s(4.23)	94-0666-66

Compound	Solvent	λ_{max}(log ε)	Ref.
Cytosine, 1-(4-thio-α-D-ribofurano-syl)-, hydrochloride	pH 2	215(4.04),283(4.16)	44-0813-66
β-isomer	pH 2	214(4.02),282(4.16)	44-0813-66
Cytosine, 1-(5-thio-D-xylo-pyranosyl)-	pH 2	215(4.02),281(4.16)	44-0813-66
	pH 12	231(3.94),273(4.02)	44-0813-66
$C_9H_{13}N_3O_5$			
Cytidine	n.s.g.	270(4.0)	57-0379-66C
Cytosine, 1-β-D-arabinofuranosyl-	pH 2	280(4.13)	88-3499-66
Uridine, 5-amino-2'-deoxy-	pH 12	217(4.15),289(3.73)	87-0366-66
	pH 1	206(3.92),265(3.90)	87-0366-66
$C_9H_{13}N_5O$			
Adenine, N-ethoxy-9-ethyl-	pH 1	269(4.17)	25-1967-66
	H_2O	268(4.17)	25-1967-66
	pH 13	284(4.05)	25-1967-66
Isoguanine, 9-butyl-	pH 1	233(3.74),282(4.14)	95-0649-66
	pH 13	245(3.79),285(4.05)	95-0649-66
Pterin, 7,8-dihydro-6,7,7-trimethyl-	pH 1.0	252(4.25),270s(3.87), 350(3.83)	24-3008-66
	pH 7.0	227(4.41),278(4.06), 316(3.79)	24-3008-66
$C_9H_{13}N_5O_4$			
Imidazo[1,2-c]pyrimidin-7(1H)-one, 5-amino-3-ethoxy-2,3-dihydro-1-methyl-8-nitro-	pH 0	233(4.33),262(4.11), 325(3.71)	24-2984-66
	pH 6.0	246(4.29),352(3.76)	24-2984-66
	pH 14.0	230(4.18),280(3.93), 316(3.81)	24-2984-66
C_9H_{14}			
Cyclohexene, 1-isopropenyl-	EtOH	233(4.28)	44-3787-66
$C_9H_{14}IN$			
Trimethylphenylammonium iodide	CH_2Cl_2	239(--)	60-0286-66
	CCl_4	283(--)	60-0286-66
$C_9H_{14}IP$			
Trimethylphenylphosphonium iodide	CH_2Cl_2	233(--)	60-0286-66
	CCl_4	281(--)	60-0286-66
	dioxan	230(--)	60-0286-66
$C_9H_{14}N_2$			
Pyrazole, 5-cyclopropyl-3-ethyl-4-methyl-	EtOH	227(3.80)	22-2971-66
Pyridine, 1,2,3,4-tetrahydro-5-(1-pyrrolin-2-yl)-	EtOH	317(4.53)	44-2487-66
$C_9H_{14}N_2O$			
Cyclooct[c]isoxazole, 3-amino-4,5,6,7,8,9-hexahydro-	n.s.g.	252(3.99)	18-1125-66
5-Hepten-2-one, 1-diazo-3,6-dimethyl-	EtOH	250(3.88),268s(3.85)	22-3490-66
2-Pyrrolidinone, 4-(1-ethylpropyl-idene)-5-imino-	EtOH	256(4.24)	22-3895-66
2-Pyrrolidinone, 5-imino-4-iso-propylidene-3,3-dimethyl-	EtOH	255(4.30)	22-3895-66
$C_9H_{14}N_2OS$			
4(5H)-Pyrimidinone, 5-(3-mercapto-propyl)-2,6-dimethyl-	pH 13.5	236(4.02),268(3.85)	24-0872-66
	MeOH	232(3.78),272(3.77)	24-0872-66

Compound	Solvent	$\lambda_{max}(\log \epsilon)$	Ref.
$C_9H_{14}N_2O_2$			
5H-Furo[2,3-e]-1,4-diazepin-5-one, 1,2,3,4,6,7-hexahydro-6,7-dimethyl-, cis	EtOH	279(4.42)	44-4133-66
Pyrazole-1-acetic acid, 3,5-dimethyl-, ethyl ester	0.05N HCl	226(3.89)	54-1194-66
	H_2O	222(3.76)	54-1194-66
2(1H)-Pyrimidinone, acetyl-3,4-dihydro-4,4,6-trimethyl-	n.s.g.	235(3.47)	49-0036-66
Thymine, 2,4-diethyl-	pH 5-12	217(3.98),265(3.85)	24-2391-66
Uracil, 1-butyl-6-methyl-	EtOH	268(4.18)	70-1171-66
$C_9H_{14}N_2O_2S_2$			
5-Thiazolidineacetamide, 3-butyl-4-oxo-2-thioxo-	EtOH	263(4.12),297(4.26)	95-0101-66
5-Thiazolidineacetamide, N,3-diethyl-4-oxo-2-thioxo-	EtOH	262(4.03),296(4.14)	95-0101-66
$C_9H_{14}N_2O_3$			
6(1H)-Pyridazinone, 1,3-diethoxy-4-methyl-	EtOH	308(3.440)	95-0082-66
$C_9H_{14}N_2O_7$			
Uracil, 5-α-D-arabinitol-	pH 7	262(3.68)	44-2215-66
	pH 12	287(3.63)	44-2215-66
Uracil, 5-α-D-ribitol-	pH 7	262(3.75)	44-2215-66
	pH 12	287(3.57)	44-2215-66
$C_9H_{14}N_4O$			
Pyrimidine, 2-amino-6-methyl-4-morpholino-	EtOH	288(3.78)	78-2401-66
$C_9H_{14}N_4O_2$			
L-Ornithine, N^5-2-pyrimidinyl-	pH 2	230(4.25),315(3.54)	69-3454-66
$C_9H_{14}O$			
2-Cyclononen-1-one	C_6H_{12}	226(3.88)	39-0164-66B
	EtOH	233.5(3.86)	39-0164-66B
3-Cyclononen-1-one	EtOH	298(1.95)	39-0164-66B
Cyclooctanone, 3-methylene-	isooctane	285(1.54),293(1.58),301(1.57),312(1.43),323(1.10)	22-0121-66
2-Cyclopenten-1-one, 2-tert-butyl-	n.s.g.	226.5(3.94)	39-0073-66B
2-Cyclopenten-1-one, 3-tert-butyl-	n.s.g.	227.0(4.11)	39-0073-66B
2,5-Heptadien-4-one, 2,6-dimethyl-	EtOH	265(4.39)	28C-0652-66A
3,5-Heptadien-2-one, 4,6-dimethyl-	EtOH	227(3.83),286(4.21)	22-0689-66
4,6-Heptadien-2-one, 4,6-dimethyl-	EtOH	235(3.86)	22-0689-66
2-Indanone, hexahydro-, trans	C_6H_{12}	301(1.48)	23-0759-66
2-Methylcyclohexylideneacetaldehyde	EtOH	230(4.08)	22-1400-66
4-Methylcyclohexylideneacetaldehyde	EtOH	239.5(4.16)	22-1400-66
5,6-Octadien-2-one, 7-methyl-	hexane	210f(3.23)	28C-1601-66A
	EtOH	215f(3.11)	28C-1601-66A
Phorone	EtOH	265.0(4.29)	39-0075-66B
Propanal, 2-cyclohexylidene-	EtOH	249(4.09)	22-1400-66
2H-Pyran, 2,2,4,6-tetramethyl-	EtOH	282(3.59)	28C-0652-66A
Spiro[2.6]nonan-4-one	isooctane	276(1.44),291s(1.34),300(1.18)	22-0121-66
Spiro[2.5]octan-4-one, 6-methyl-	isooctane	291(1.49)	22-0116-66

Compound	Solvent	λ_{max}(log ε)	Ref.
$C_9H_{14}OS$			
Crotonaldehyde, 3-[(3-methyl-2-butenyl)thio]-	EtOH	299(4.12)	78-0259-66
$C_9H_{14}O_2$			
Cyclohexanone, 6-acetyl-2-methyl-	EtOH	289(3.95)	78-2039-66
2-Cyclohexen-1-one, 2-hydroxy-3,5,5-trimethyl-	EtOH	274(3.95)	35-2803-66
2-Cyclohexen-1-one, 2-methoxy-3,5-dimethyl-	n.s.g.	247.0(3.96)	39-0073-66B
2-Cyclopenten-1-one, 2-isobutoxy-	n.s.g.	248.0(3.80)	39-0073-66B
2-Cyclopenten-1-one, 3-isobutoxy-	n.s.g.	237.5(4.25)	39-0073-66B
Cyclopentylideneacetic acid, ethyl ester	EtOH	223(4.16)	54-0929-66
$C_9H_{14}O_2S$			
Crotonic acid, 3-[(3-methyl-2-butenyl)thio]-, cis	C_6H_{12}	221(3.28),289(3.93)	78-0259-66
trans	C_6H_{12}	216(3.66),278(4.21)	78-0259-66
$C_9H_{14}O_3$			
2-Heptenoic acid, 4,4-dihydroxy-2,3-dimethyl-, γ-lactone	n.s.g.	214(4.1)	54-0043-66
4-Heptenoic acid, 4-methyl-6-oxo-, methyl ester	EtOH	237(3.97)	44-1797-66
isomer	EtOH	235(4.06)	44-1797-66
2-Hexenoic acid, 3,4,4-trimethyl-5-oxo-, δ-lactol, cis	n.s.g.	222(4.02)	35-0618-66
$C_9H_{14}O_5$			
Maleic acid, isopropoxy-, dimethyl ester	ether	226(4.02)	24-0450-66
$C_9H_{14}O_6$			
Glutaconic acid, 4,4-dimethoxy-, dimethyl ester	MeOH	220(3.34)	44-4054-66
$C_9H_{15}Br$			
1,2-Pentadiene, 1-bromo-3-isopropyl-4-methyl-	EtOH	201(3.99),221s(3.84)	39-1223-66C
$C_9H_{15}N$			
1-Penten-3-ynylamine, N,N-diethyl-	EtOH	275(4.11)	70-1739-66
$C_9H_{15}NO$			
1-Azaspiro[4.5]decan-2-one	n.s.g.	208(3.8)	24-0724-66
$C_9H_{15}NO_2$			
Acrylic acid, piperidino-, methyl ester	ether	273(4.42)	24-0450-66
trans isomer	EtOH	282(4.51)	24-2526-66
1-Cyclopentene-1-carboxylic acid, 2-(methylamino)-, ethyl ester	EtOH	295(4.30)	78-2575-66
Indone, 3a,4,5,6,7,7a-hexahydro-3-hydroxy-, ammonium salt	EtOH	244(4.30)	70-1167-66
$C_9H_{15}NO_3$			
3-Furoic acid, 2-amino-4,5-dihydro-4,5-dimethyl-, ethyl ester, cis	EtOH	272(4.29)	44-4133-66
trans	EtOH	272(4.32)	44-4133-66

Compound	Solvent	λ_{max}(log ε)	Ref.
$C_9H_{15}NO_5$			
Adipic acid, α-oximino-, methyl ester	EtOH	220(3.93)	39-0255-66C
	EtOH-NaOH	270(4.08)	39-0255-66C
$C_9H_{15}N_2^+$			
Aniline, 3-trimethylammonium ion	pH 10	283(3.36)	35-0354-66
Aniline, 4-trimethylammonium ion	pH 10	280(3.16)	35-0354-66
$C_9H_{15}N_3$			
Pyrimidine, 1-butyl-1,2-dihydro-2-methylimino-	pH 8.0	230(4.2),320(3.7)	39-1165-66C
	pH 14.3	242(4.2),368(3.4)	39-1165-66C
Pyrimidine, 2-butylamino-4-methyl-	pH 2.0	230(4.22),313(3.58)	12-2321-66
	pH 7.0	237(4.22),305(3.50)	12-2321-66
Pyrimidine, 2-butylimino-1,2-dihydro-1-methyl-	pH 8.0	234(4.16),318(3.60)	39-1165-66C
	pH 14.3	244(4.17),368(3.36)	39-1165-66C
Pyrimidine, 2-(isopropylamino)-4,6-dimethyl-	pH 3.0	231(4.14),310(3.67)	39-0226-66C
	pH 8.0	239(4.07),300(3.46)	39-0226-66C
Pyrimidine, 2-(propylamino)-4,6-dimethyl-	pH 3.0	231(4.10),310(3.63)	39-0226-66C
	pH 8.0	238(4.21),303(3.61)	39-0226-66C
$C_9H_{15}N_3O$			
Bicyclo[3.1.0]hexan-2-one, 6,6-dimethyl-, semicarbazone	EtOH	238.5(4.29)	22-3490-66
Cyclohexylideneacetaldehyde, semicarbazone	EtOH	272(4.48)	22-1400-66
$C_9H_{15}N_3O_5$			
5,6-Dehydrocytidine	n.s.g.	243(4.0)	57-0379-66C
$C_9H_{15}N_5O_2$			
Formamide, N-[4-(ethoxyamidino)-3-ethylimidazol-5-yl]-	pH 1	253(3.90)	25-1967-66
	H_2O	218(4.07),247s(3.84)	25-1967-66
	pH 13	250(4.03)	25-1967-66
$C_9H_{16}N_2$			
Pyrazole, 4-ethyl-1-methyl-5-propyl-	EtOH	226(3.86)	4-0413-66
$C_9H_{16}N_2O_2$			
3-Furamide, tetrahydro-N,4,5-trimethyl-2-(methylimino)-, trans	EtOH	277(2.41)	44-4133-66
$C_9H_{16}N_2O_3$			
Urea, 1-acetoacetyl-3-butyl-	EtOH	263(3.99)	70-1171-66
$C_9H_{16}N_4O$			
Δ^1-1,2,4-Triazolin-3-one, 4-sec-butylideneamino-5-ethyl-5-methyl-	EtOH	295(3.73)	88-5815-66
$C_9H_{16}O$			
3,5-Heptadien-1-ol, 4,6-dimethyl-, 3-cis	EtOH	219.5(3.77)	78-0443-66B
3-trans	EtOH	232.5(4.00)	78-0443-66B
$C_9H_{16}O_3$			
Crotonic acid, 2-ethoxy-3-methyl-, ethyl ester	MeOH	229(4.00)	5-0053-66I
$C_9H_{16}O_4$			
Propionic acid, 3-ethoxy-2-methoxymethylene-, ethyl ester	EtOH	239(4.07)	94-0238-66

Compound	Solvent	$\lambda_{max}(\log \epsilon)$	Ref.
$C_9H_{16}Pt$			
Platinum, trimethyl(methyl-cyclopentadienyl)-	dioxan	256(0.9)	101-0181-66A
$C_9H_{17}ClN_2O_4$			
2,3-Dihydro-1,4,5,7-tetramethyl-1H-1,4-diazepinium perchlorate	MeOH	335(4.29)	39-0093-66C
[5-(Dimethylamino)-2,4-pentadienyl-idene]dimethylammonium perchlorate	EtOH-100°K	415(5.21)	61-0052-66
phototropic form	EtOH-100°K	242(4.13),442(4.92)	61-0052-66
$C_9H_{17}N$			
Butylamine, N-cyclopentylidene-	heptane	181(3.97)	88-0507-66
	C_6H_{12}	250(2.39)	88-0507-66
	EtOH	240(2.46)	88-0507-66
Cyclohexylamine, N-isopropylidene-	heptane	180(3.98)	88-0507-66
	C_6H_{12}	247(2.29)	88-0507-66
	EtOH	235(2.26)	88-0507-66
Pyridine, 2-butyl-3,4,5,6-tetrahydro-	heptane	178(3.92)	88-0507-66
	C_6H_{12}	252(2.47)	88-0507-66
	EtOH	235(2.48)	88-0507-66
$C_9H_{17}NO$			
2-Butene, 2-methyl-3-morpholino-	ether	221(3.88)	22-2845-66
1-Pentene, 2-morpholino-	ether	221(3.88)	22-2845-66
2-Pentene, 3-morpholino-	EtOH	221(3.88)	22-2845-66
1-Penten-3-one, 1-(diethylamino)-	EtOH	305(4.42)	70-1739-66
$C_9H_{17}NO_2$			
Crotonic acid, 3-(isopropylamino)-, ethyl ester	MeOH	285(4.29)	35-2536-66
Crotonic acid, 3-(propylamino)-, ethyl ester	MeOH	285(4.30)	35-2536-66
2-Pentenoic acid, 3-(ethylamino)-, ethyl ester	MeOH	283(4.28)	35-2536-66
$C_9H_{17}NO_3$			
Heptane, 2-acetoxy-2-nitroso-	C_6H_{12}	665(1.25)	39-0441-66B
Heptane, 4-acetoxy-4-nitroso-	C_6H_{12}	665(1.15)	39-0441-66B
$C_9H_{17}NO_5S_2$			
Thiazolidine-2-thione, 3-ethyl-5-hydroxy-5-(D-arabino-tetra-hydroxybutyl)-	MeOH	250s(3.95),273(4.23)	12-0445-66
$C_9H_{17}N_3O$			
3-Hexenal, 4-ethyl-, semicarbazone	n.s.g.	230(4.14)	22-0717-66
3-Pentenal, 2,2,4-trimethyl-, semicarbazone	n.s.g.	230(4.19)	22-0734-66
$C_9H_{17}N_3O_4$			
L-Alanine, alanylalanyl-, copper complex	H_2O	550(2.20),570(1.97), 655(1.93)	37-1439-66
$C_9H_{17}N_5O$			
Pyrimidine, 2,5-diamino-4-(butyl-amino)-6-methoxy-	EtOH	286(3.92),302(3.83)	87-0977-66

Compound	Solvent	λ_{max}(log ε)	Ref.
$C_9H_{18}BN$			
Borane, B-dimethylamino-B-butyl-B-(1-propynyl)-	C_6H_{12}	225(3.78)	22-3850-66
$C_9H_{18}ClNO$			
Valeramide, N-tert-butyl-N-chloro-	isooctane	256(2.67)	35-3051-66
$C_9H_{18}N_2$			
2-Pyrazoline, 4-ethyl-1-methyl-5-propyl-	EtOH	243(3.54)	4-0413-66
$C_9H_{18}N_2O$			
Piperidine, 2,2,6,6-tetramethyl-1-nitroso-	hexane	238(3.73),397(1.93)	39-0813-66C
$C_9H_{18}N_2O_4$			
Carbamic acid, (1,1-diethylpropyl)-nitro-, methyl ester	hexane	338(1.77)	44-3038-66
	EtOH	238(3.46),335(1.64)	44-3038-66
$C_9H_{18}N_2S$			
Thiourea, N,N'-di-tert-butyl-, iodine complex	$CHCl_3$	307.5(4.58)	60-0029-66
$C_9H_{18}N_6O_2S_2$			
Butyraldehyde, 3-(2-methoxyethoxy)-2-oxo-, bis(thiosemicarbazone)	EtOH	237(3.97),267(3.88), 347(4.68)	87-0420-66
$C_9H_{18}N_6O_4$			
2-Tetrazene, 1,4-diisopropyl-1-methyl-4-(2,2-dinitrovinyl)-	EtOH	252(3.84),391(4.36)	44-0923-66
$C_9H_{18}N_6S_2$			
Pyruvaldehyde, bis(4,4-dimethyl-thiosemicarbazone)	EtOH	232(4.19),328(4.39), 410(3.52)	87-0420-66
$C_9H_{18}O$			
3-Hexanone, 2,2,5-trimethyl-	EtOH	287.5(1.46)	22-3578-66
3-Pentanone, 2,2,4,4-tetramethyl-	hexane	296(1.32)	44-4237-66
$C_9H_{19}I$			
Octane, 4-iodo-4-methyl-	C_6H_{12}	269(2.82)	3-0612-66
$C_9H_{20}GeO$			
2-Propanone, (triethylgermyl)-	n.s.g.	283(2.08)	65-0158-66
$C_9H_{20}NO_4P$			
Phosphoramidic acid, formyl-, dibutyl ester	EtOH	221.3(2.06)	78-0987-66
	EtOH-KOH	224.4(2.53)	78-0987-66
$C_9H_{20}N_2$			
Pyrrolidine, 1-methyl-2-(3-methyl-aminopropyl)-	MeOH	205(3.3)	24-0737-66
$C_9H_{21}BF_4S_2$			
Bis(diethylsulfonium)methylide tetrafluoroborate	MeOH	215(4.00)	35-1560-66
$C_9H_{22}B_2F_8S_2$			
Methylenebis(diethylsulfonium) difluoroborate	MeOH	214(3.0)	35-1559-66
	MeOH-NaOMe	215(4.00)	35-1559-66

Compound	Solvent	λ_{max}(log ϵ)	Ref.
C_9N_6			
1,3-Cyclopentadiene-1,2,3,4-tetra-carbonitrile, 5-diazo-	MeCN	252(4.47),262(4.43), 331(4.17)	35-4055-66

Compound	Solvent	$\lambda_{max}(\log \epsilon)$	Ref.
$C_{10}Br_8$			
Fulvalene, octabromo-	n.s.g.	221(4.32),414(4.67), 638(2.37)	35-4541-66
$C_{10}Cl_8$			
Fulvalene, octachloro-	n.s.g.	208(4.45),390(4.61), 610(2.40)	35-4541-66
$C_{10}Cl_8N_2$			
Cyclopentadienone, tetrachloro-, azine	EtOH	305(4.59),316(4.71)	44-4260-66
$C_{10}F_{12}O_2S_4$			
p-Benzoquinone, tetrakis-(trifluoromethyl)-	CH_2Cl_2	330(3.82)	44-3671-66
$C_{10}F_{14}$			
Bi-1-cyclopenten-1-yl, tetradecafluoro-	EtOH	260.0(4.26)	78-1755-66
$C_{10}HF_7$			
Naphthalene, heptafluoro-	C_6H_{12}	218(4.62),270s(3.64), 278(3.68),285s(3.59), 320s(3.34),324(3.38)	39-2328-66C
$C_{10}H_2N_6$			
2,4-Cyclopentadiene-1,1,2,3,4-penta-carbonitrile, 5-amino-	MeCN	413(3.85)	35-4055-66
$C_{10}H_5BrCl_3NO$			
4(1H)-Quinolone, 6-bromo-2-(trichloromethyl)-	EtOH	244(4.51)	44-3369-66
4(1H)-Quinolone, 8-bromo-2-(trichloromethyl)-	EtOH	235(4.53)	44-3369-66
$C_{10}H_5ClN_2O$			
Quinoline-2-carbonitrile, 4-chloro-1,2-epoxy-	EtOH	242(4.40),325(3.58)	88-2145-66
$C_{10}H_5ClN_2S$			
Thiazolo[4,5-c]isoquinoline, 2-chloro-	EtOH	232s(4.39),249(4.51), 282s(3.72),292(3.77), 304(3.72),324(3.63), 337(3.70)	23-2465-66
Thiazolo[5,4-c]isoquinoline, 2-chloro-	EtOH	230s(4.64),235(4.69), 241s(4.52),274(3.68), 285(3.79),297(3.77), 321(3.75),330s(3.77), 336(3.85)	23-2473-66
$C_{10}H_5ClN_4O_2$			
Alloxazine, 8-chloro-	EtOH	220(4.66),247(4.67), 325(3.89),383(4.03)	30-1101-66F
$C_{10}H_5ClN_4O_3$			
Alloxazine, 7-chloro-, N^9-oxide	EtOH	220(4.46),305(4.58), 363(3.89),473(3.73)	30-0577-66C
Isoalloxazine, 8-chloro-10-hydroxy-	EtOH	220(4.46),280(4.58), 363(3.89),413(3.58)	30-1101-66F

Compound	Solvent	$\lambda_{max}(\log \epsilon)$	Ref.
$C_{10}H_5Cl_3$			
Azulene, 4,6,8-trichloro-	H_2O	246(4.32),285(4.54), 291(4.52),333(3.50), 346(3.58),540(2.85)	65-1728-66
$C_{10}H_5Cl_4NO$			
4(1H)-Quinolone, 6-chloro-2-(trichloromethyl)-	EtOH	242(4.50)	44-3369-66
4(1H)-Quinolone, 8-chloro-2-(trichloromethyl)-	EtOH	233(4.58)	44-3369-66
$C_{10}H_5N_3$			
α,α,α-Toluenetricarbonitrile	isooctane	248(2.2),255(2.4), 262(2.5),268(2.4)	44-0919-66
$C_{10}H_6BrClN$			
Dipyridazino[2,3-a:4,3-d]pyrrole, 5-bromo-8-chloro-3-methyl-	EtOH	265(4.40),293(4.21), 304(4.00)	1-2637-66
$C_{10}H_6BrClN_4O_4$			
Pyrazole, 4-bromo-3-chloro-1-(2,4-dinitrophenyl)-5-methyl-	EtOH	293(3.88)	22-3744-66
Pyrazole, 4-bromo-5-chloro-1-(2,4-dinitrophenyl)-3-methyl-	EtOH	286(3.96)	22-3744-66
$C_{10}H_6BrN_5O_6$			
Pyrazole, 4-bromo-3-methyl-1-picryl-	EtOH	338(3.90)	22-3744-66
$C_{10}H_6Br_2N_2O_4$			
Benzene, p-bis(2-bromo-2-nitrovinyl)-	n.s.g.	360(4.49)	67-0042-66
$C_{10}H_6Br_2N_4O_4$			
Pyrazole, 3,4-dibromo-5-methyl-1-(2,4-dinitrophenyl)-	EtOH	294(3.86)	22-3744-66
Pyrazole, 4,5-dibromo-3-methyl-1-(2,4-dinitrophenyl)-	EtOH	292(3.94)	22-3744-66
$C_{10}H_6Br_2O_2$			
1,4-Naphthoquinone, 2,3-dibromo-2,3-dihydro-	hexane	235(4.20),263(3.80), 308(2.32)	23-2867-66
$C_{10}H_6ClN_5O_4S$			
Alloxazine, 7-chloro-8-sulfamyl-	EtOH	222(4.54),255(4.57), 319(3.76),380(3.91)	30-1101-66F
$C_{10}H_6ClN_5O_5S$			
Isoalloxazine, 7-chloro-10-hydroxy-8-sulfamyl-	EtOH	222(4.22),287(4.42), 415(3.53)	30-1101-66F
$C_{10}H_6Cl_2N_2O_3$			
Pyrrole, 3-chloro-4-(3-chloro-6-hydroxy-2-nitrophenyl)-	EtOH	290(3.54)	94-1314-66
$C_{10}H_6Cl_2N_4O_4$			
Pyrazole, 3,4-dichloro-5-methyl-1-(2,4-dinitrophenyl)-	EtOH	291(3.95)	22-3744-66
$C_{10}H_6Cl_3NO$			
4(1H)-Quinolone, 2-(trichloromethyl)-	EtOH	234(4.46)	44-3369-66

Compound	Solvent	$\lambda_{max}(\log \epsilon)$	Ref.
$C_{10}H_6Cl_4O$			
4,7-Methanoinden-8-one, 4,5,6,7-tetrachloro-3a,4,7,7a-tetrahydro-	C_6H_{12}	204(4.09),246(3.14)	88-0201-66
$C_{10}H_6CrO_5$			
Chromium, tricarbonyl(benzoic acid)-	EtOH	319(3.6)	49-1029-66
$C_{10}H_6N_2O$			
Isoquinoline-1-carbonitrile, 1,2-epoxy-	EtOH	246(3.90),253(3.86), 345(3.68)	88-2145-66
laH-Oxazirino[2,3-a]quinoline-la-carbonitrile	EtOH	240(4.51),322(3.56)	88-2145-66
	EtOH	240(4.51),322(3.56)	94-0555-66
5-Quinolinecarbonitrile, 8-hydroxy-	N $HClO_4$	319(3.46),354(3.58)	39-0073-66A
	pH 13	363(3.94)	39-0073-66A
	$CHCl_3$	318(3.82)	39-0073-66A
copper chelate	pyridine	334(3.62),401(4.11)	39-0073-66A
	dioxan	405(4.17)	39-0073-66A
nickel chelate	pyridine	338(3.79),401(4.15)	39-0073-66A
$C_{10}H_6N_2OS$			
Thiazolo[4,5-c]isoquinolin-2-ol	EtOH	235s(4.36),247(4.45), 278(3.56),288(3.60), 301(3.55),350(3.66)	23-2465-66
Thiazolo[5,4-c]isoquinolin-2-ol	EtOH	240(4.51),261s(3.82), 293(3.58),304(3.59), 346(3.82)	23-2473-66
$C_{10}H_6N_2O_2$			
3-Isoquinolinecarbonitrile, 1,2-dihydro-2-hydroxy-1-oxo-	EtOH	262(3.57),277(3.46), 308(3.72),322(3.71), 335(3.60)	44-2090-66
$C_{10}H_6N_2O_7$			
Coumarin, 6,8-dinitro-7-hydroxy-4-methyl-	NH_3	402(4.16)	50-0298-66B
$C_{10}H_6N_2S$			
Thiazolo[4,5-c]isoquinoline	EtOH	237s(4.37),245(4.45), 276(3.70),286(3.69), 298(3.54),320(3.52), 334(3.56)	23-2465-66
Thiazolo[5,4-c]isoquinoline	EtOH	233(4.66),243s(4.42), 271(3.69),281(3.75), 293(3.69),319(3.64), 328s(3.64),333(3.72)	23-2473-66
$C_{10}H_6N_2S_2$			
Thiazolo[5,4-c]isoquinoline-2-thiol	EtOH	249(4.29),314(3.98), 320s(3.97),368(4.22)	23-2473-66
$C_{10}H_6N_4O_2$			
Alloxazine	EtOH	242(4.55),323(3.82), 380(3.82)	30-1101-66F
$C_{10}H_6N_4O_3$			
Alloxazine, N^9-oxide	pH 13	277(4.72),348(3.88), 453(3.94)	30-0577-66C
Isoalloxazine, 10-hydroxy-	EtOH	273(4.44),345(3.72), 415(3.64)	30-1101-66F

Compound	Solvent	$\lambda_{max}(\log \epsilon)$	Ref.
$C_{10}H_6O_2$			
2-Decene-4,6,8-triynoic acid	ether	243(--),253(--), 282(--),299(--), 319(--),342(--)	39-1617-66C
$C_{10}H_6O_2S_2$			
[2,2'-Bithiophene]-5,5'-dicarbox-aldehyde	ether	258(3.75),362(4.46)	24-0984-66
$C_{10}H_6O_3$			
1,4-Naphthoquinone, 2-hydroxy-	EtOH	232(4.14),270(4.33), 390(3.03)	23-1086-66
	EtOH-base	270(4.31),480(3.22)	23-1086-66
$C_{10}H_6O_4$			
[2,2'-Bifuran]-5,5'-dicarbox-aldehyde	$CHCl_3$	250(3.53),350(4.54), 368(4.51)	39-0976-66C
$C_{10}H_6O_8$			
1,4-Naphthoquinone, 2,3,5,6,7,8-hexahydroxy- (or spinochrome E)	EtOH	270(4.18),359(2.43), 450(3.40),477(3.38)	39-0426-66C
$C_{10}H_7BrN_2O_2$			
1,3-Butanedione, 1-(p-bromophenyl)-2-diazo-	CH_2Cl_2	250(4.34)	24-3128-66
$C_{10}H_7BrN_4O_4$			
Pyrazole, 3-bromo-1-(2,4-dinitro-phenyl)-4-methyl-	EtOH	328(4.15)	22-3744-66
Pyrazole, 4-bromo-1-(2,4-dinitro-phenyl)-3-methyl-	EtOH	326(4.08)	22-3744-66
Pyrazole, 4-bromo-1-(2,4-dinitro-phenyl)-5-methyl-	EtOH	286(3.88)	22-3744-66
$C_{10}H_7Br_2N_3O_2$			
Pyrazole, 4,5-dibromo-3-methyl-1-(p-nitrophenyl)-	EtOH	304(4.11)	22-3744-66
$C_{10}H_7ClN_2O_2$			
1,3-Butanedione, 1-(p-chloro-phenyl)-2-diazo-	CH_2Cl_2	248(4.34)	24-3128-66
$C_{10}H_7ClN_4$			
Dipyridazino[2,3-a:4,3-d]pyrrole, 8-chloro-3-methyl-	EtOH	259(4.47),289(4.21), 300(4.02)	1-2637-66
$C_{10}H_7ClN_4O_4$			
Pyrazole, 3-chloro-1-(2,4-dinitro-phenyl)-5-methyl-	EtOH	295(3.86)	22-3744-66
Pyrazole, 5-chloro-1-(2,4-dinitro-phenyl)-3-methyl-	EtOH	290(3.95)	22-3744-66
$C_{10}H_7ClO_2$			
Propiolic acid, (m-chlorophenyl)-, methyl ester	MeOH	253(4.20)	44-1135-66
Propiolic acid, (o-chlorophenyl)-, methyl ester	MeOH	260(4.15)	44-1135-66
Propiolic acid, (p-chlorophenyl)-, methyl ester	MeOH	264(4.29)	44-1135-66

Compound	Solvent	λ_{max}(log ε)	Ref.
$C_{10}H_7Cl_2NOS_2$			
1,3-Thiazin-4-one, 2,3,5,6-tetra-hydro-N-(2,4-dichlorophenyl)-2-thioxo-	EtOH	259(4.17),312(4.08)	95-0095-66
$C_{10}H_7F_3N_2$			
Quinazoline, 4-methyl-2-(trifluoromethyl)-	pH -3.0	242(4.58),282(3.41), 330(3.37)	39-0234-66C
	pH 6.0	228(4.64),280(3.40), 300(3.39),315s(3.31)	39-0234-66C
$C_{10}H_7F_6NO$			
2H-Pyran-5-carbonitrile, 3-methyl-6-[2,2,2-trifluoro-1-(trifluoro-methyl)ethyl]-	isooctane	298(3.64)	35-0487-66
$C_{10}H_7NO$			
Propiolonitrile, (p-methoxyphenyl)-	EtOH	289(4.42),294(4.45)	94-1277-66
$C_{10}H_7NO_2$			
Indole-3-glyoxylaldehyde, as bisulfite addition compound	H_2O	248(4.06),265(3.98), 314(4.09)	88-3445-66
Isomaleimide, N-phenyl-	dioxan	231(4.07),342(4.05)	44-1311-66
Maleimide, N-phenyl-	EtOH	220(4.26),320(2.70)	44-1311-66
5-Quinolinecarboxaldehyde,	N $HClO_4$	317(3.63),354(3.74)	39-0073-66A
8-hydroxy-	pH 13	327(3.76),380(4.34)	39-0073-66A
	$CHCl_3$	324(4.04)	39-0073-66A
copper chelate	$CHCl_3$	410(4.15)	39-0073-66A
	pyridine	333(4.14),405(4.39)	39-0073-66A
	dioxan	406(4.17)	39-0073-66A
nickel chelate	pyridine	338(4.10),401(4.33)	39-0073-66A
$C_{10}H_7NO_3$			
5-Quinolinecarboxylic acid,	N $HClO_4$	319(3.52),350(3.57)	39-0073-66A
8-hydroxy-	pH 13	332(3.64),360(3.79)	39-0073-66A
copper chelate	pyridine	332(3.66),401(4.13)	39-0073-66A
nickel chelate	pyridine	336(3.83),398(4.23)	39-0073-66A
$C_{10}H_7NO_4$			
Propiolic acid, (m-nitrophenyl)-, methyl ester	MeOH	246(4.38)	44-1135-66
Propiolic acid, (o-nitrophenyl)-, methyl ester	MeOH	234(4.21),262(3.88), 306(3.53)	44-1135-66
Propiolic acid, (p-nitrophenyl)-, methyl ester	MeOH	284(4.29)	44-1135-66
$C_{10}H_7NO_5$			
Coumarin, 7-hydroxy-4-methyl-	EtOAc	353(3.80)	50-0298-66B
6-nitro-	NH_3	408(3.97)	50-0298-66B
Coumarin, 7-hydroxy-4-methyl-	EtOAc	310(4.12)	50-0298-66B
8-nitro-	NH_3	353(4.23)	50-0298-66B
2-Isoindolineacetic acid,	EtOH	243s(3.97),339(3.81)	4-0328-66
4-hydroxy-1,3-dioxo-	dil. HCl	220(4.76),243s(3.96), 340(3.80)	4-0328-66
	dil. NaOH	276(4.17),396(2.62)	4-0328-66
2-Isoindolineacetic acid, 5-hydroxy-1,3-dioxo-	EtOH	232(4.72),288s(3.42), 330(3.51)	4-0328-66
	dil. HCl	233(3.36),280s(3.50), 333(4.60)	4-0328-66
	dil. NaOH	322(3.75),394(3.37)	4-0328-66

Compound	Solvent	λ_{max}(log ϵ)	Ref.
$C_{10}H_7NO_7$			
Phthalic anhydride, 3,4-dimethoxy-6-nitro-	ether	246(4.00),325(3.65)	44-1912-66
$C_{10}H_7NS_4$			
3H-1,2-Dithiole-3-thione, 4-(thiobenzamido)-	EtOH-2% DMF	271(4.33),310(4.22), 336s(4.05),413(4.09)	12-0503-66
$C_{10}H_7N_3$			
Imidazo[2,3-a]phthalazine	10% HCl	240(4.65),263(4.01), 273(4.01),283(3.95)	4-0381-66
	EtOH	245(4.59),252(4.62), 271(3.95),281(3.98), 292(3.83),333(3.48)	4-0381-66
2H-Pyrazolo[4,3-c]quinoline	EtOH	211(4.16),237(4.70), 249(4.22),264(3.78), 277(3.78),286(3.78), 305(3.16),320(2.70)	54-0681-66
$C_{10}H_7N_3O$			
Malononitrile, (aminophenoxymethylene)-	H_2O	255(4.36)	24-2307-66
	MeCN	259(4.32)	24-2307-66
	dioxan	262(4.28)	24-2307-66
Oxazole-4-carbonitrile, 5-amino-2-phenyl-	EtOH	227(4.24),258(4.09), 309(4.30)	35-3829-66
$C_{10}H_7N_3OS$			
2-Benzothiazolecarbonitrile, 6-acetamido-	EtOH	218(4.10),254(3.70), 323(3.95)	35-2015-66
$C_{10}H_7N_3O_2S$			
1H-Pyridazino[4,5-b][1,4]benzothiazine-1,4(10H)-dione, 2,3-dihydro-	EtOH	209(4.37),255(4.44), 328(3.60)	32-1147-66
$C_{10}H_7N_3O_4$			
1,3-Butanedione, 2-diazo-1-(o-nitrophenyl)-	CH_2Cl_2	234(4.36)	24-3128-66
1,3-Butanedione, 2-diazo-1-(p-nitrophenyl)-	CH_2Cl_2	255(4.29)	24-3128-66
$C_{10}H_7N_3S$			
10H-Dipyrido[2,3-b:2',3'-e]thiazine	EtOH	239(4.42),338(3.99)	87-0116-66
Thiazolo[4,5-c]isoquinoline, 2-amino-	EtOH	259(4.57),288s(3.66), 299(3.90),311(3.98), 360(3.48)	23-2465-66
Thiazolo[5,4-c]isoquinoline, 2-amino-	EtOH	222(4.47),230s(4.43), 238s(4.41),252(4.46), 303s(3.60),317s(3.66), 348(4.02)	23-2473-66
$C_{10}H_7N_5O_6$			
Pyrazole, 3-methyl-1-picryl-	EtOH	327(3.84)	22-3744-66
Pyrazole, 4-methyl-1-picryl-	EtOH	326(3.83)	22-3744-66
Uracil, 5-(2,4-dinitroanilino)-	pH 1	262(4.17),351(4.21)	87-0108-66
	pH 11	228(4.12),269(4.08), 362(4.21)	87-0108-66

Compound	Solvent	λ_{max}(log ε)	Ref.
$C_{10}H_8$			
Azulene	H_2O	235(4.27),274(3.72), 280(4.67),325(3.52), 339(3.64),573(2.55)	65-1728-66
Naphthalene, tetracyanoethylene complex	$CHCl_3$	430(--),550(3.20)	60-0018-66
$C_{10}H_8BrN_3O_2$			
Uracil, 5-(m-bromoanilino)-	pH 1	248(4.16)	87-0121-66
	pH 11	247(4.11),290(3.95)	87-0121-66
$C_{10}H_8BrClN_4$			
Dipyridazino[2,3-a:4,3-d]pyrrole, 5-bromo-8-chloro-6,7-dihydro-3-methyl-	EtOH	230(4.48),266(3.70), 352(3.63)	1-2637-66
$C_{10}H_8BrN_5O_2$			
Pyrimidine, 4-amino-2-(p-bromo-anilino)-5-nitro-	MeOH	250(4.24),364(4.40)	87-0121-66
$C_{10}H_8BrN_5O_6$			
Pyrazoline, 3-bromo-4-methyl-1-picryl-	$CHCl_3$	390(4.15)	22-0610-66
Pyrazoline, 3-bromo-5-methyl-1-picryl-	$CHCl_3$	395(4.16)	22-0610-66
$C_{10}H_8Br_2N_2O_2$			
8(1H)-Cycloheptapyrazolone, 4,6-di-bromo-7-methoxy-1-methyl-	MeOH	245s(4.27),264(4.34), 324(3.74),335s(3.70), 370(3.78)	18-0253-66
$C_{10}H_8ClNOS_2$			
1,3-Thiazin-4-one, 2,3,5,6-tetra-hydro-N-(p-chlorophenyl)-2-thioxo-	EtOH	262(4.17),314(4.16)	95-0095-66
$C_{10}H_8ClN_5$			
5H-s-Triazolo[4,3-b]-s-triazole, 6-(p-chlorophenyl)-3-methyl-	EtOH	257(4.43)	4-0119-66
$C_{10}H_8Cl_2N_2$			
Quinazoline, 2-(dichloromethyl)-4-methyl-	pH -2.0	244(4.32),278(3.21), 330(3.27)	39-0234-66C
	pH 7.0	231(4.38),264(3.30), 316(3.10)	39-0234-66C
$C_{10}H_8CrO_3$			
Chromium, tricarbonyl(toluene)-	n.s.g.	317(3.94)	101-0480-66A
2:1 trinitrobenzene complex	benzene	330(4.0)	101-0480-66A
3:1 trinitrobenzene complex	benzene	335(3.2)	101-0480-66A
$C_{10}H_8N_2$			
Indole-2-carbonitrile, 3-methyl-	EtOH	227(4.56),287(4.24)	94-0555-66
$C_{10}H_8N_2O$			
4(1H)-Pyridone, N-(4-pyridyl)-	H_2O	283(4.5)	24-0368-66
$C_{10}H_8N_2OS_2$			
5H-Thiazolo[4,5-g][1,4]benzothiazin-6(7H)-one, 2-methyl-	EtOH	260(4.5),290s(3.9), 330(3.5)	32-0941-66

Compound	Solvent	$\lambda_{max}(\log \epsilon)$	Ref.
4H-Thiazolo[5,4-e][1,4]benzothiazin-5(6H)-one, 2-methyl-	EtOH	250(4.4),270s(4.0), 310(3.8)	32-0941-66
6H-Thiazolo[5,4-g][1,4]benzothiazin-7(8H)-one, 2-methyl-	EtOH	220(4.2),260(4.4), 290s(3.7),330(3.5)	32-0941-66
7H-Thiazolo[5,4-f][1,4]benzothiazin-8(9H)-one, 2-methyl-	EtOH	220(4.1),260(4.4), 310(3.8)	32-0941-66
$C_{10}H_8N_2O_2$			
1,3-Butanedione, 2-diazo-1-phenyl-	CH_2Cl_2	245(4.34)	24-3128-66
1-Indolineglyoxylonitrile, 2-hydroxy-	EtOH	238(4.20)	88-4701-66
Isoxazole, 3-benzamido-	EtOH	237.5(4.20)	94-1277-66
Isoxazole, 5-benzamido-	EtOH	230(3.98),266(4.18)	94-1277-66
Naphtho[1,2-c]furazan, 4,5-dihydro-, 3-oxide	EtOH	236(3.85),289(3.76), 297s(3.74),303s(3.76)	39-0717-66C
1H-[1,2,4]Oxadiazolo[4,3-a]quinolin-1-one, 4,5-dihydro-	EtOH	236(4.09),267(3.04)	39-0717-66C
	EtOH-NaOH	266(4.28)	39-0717-66C
5-Quinolinecarboxaldehyde, 8-hydroxy-, oxime	N $HClO_4$	319(3.60),355(3.72)	39-0073-66A
	pH 13	380(3.81)	39-0073-66A
	$CHCl_3$	339(3.81)	39-0073-66A
copper chelate	pyridine	350(3.89),432(4.06)	39-0073-66A
	dioxan	428(3.98)	39-0073-66A
nickel chelate	pyridine	357(3.93),411(4.17)	39-0073-66A
Terephthalonitrile, 2,5-dimethoxy-	EtOH	222(4.54),240s(4.04), 245(4.10),252(4.20), 355(3.85)	44-3321-66
$C_{10}H_8N_2O_2S$			
2-Benzothiazolecarbonitrile, 4,6-dimethoxy-	EtOH	262(4.10),312(4.11), 338(3.93)	44-1484-66
$C_{10}H_8N_2O_3$			
Indole, 1-acetyl-5-nitro-	EtOH	232s(4.06),255(4.40), 263(4.40),297(4.00)	44-0070-66
Indole, 3-acetyl-4-nitro-	EtOH	227(4.20),274(4.02), 303s(3.74),347s(3.43)	44-0070-66
Indole, 3-acetyl-5-nitro-	EtOH	258s(4.26),267(4.31), 278s(4.26),315(3.93)	44-0070-66
Indole, 3-acetyl-6-nitro-	EtOH	279(4.38),318(3.89), 333s(3.86)	44-0070-66
3-Indolecarboxaldehyde, 5-methyl-7-nitro-	EtOH	224(4.28),269(4.11), 355(3.97)	34-0418-66
3-Indolecarboxaldehyde, 7-methyl-4-nitro-	EtOH	257(4.07),290(3.89), 361(4.18)	34-0418-66
3-Indolecarboxaldehyde, 7-methyl-5-nitro-	EtOH	235(3.95),366(4.11)	34-0418-66
2-Quinoxalinecarboxylic acid, 3-hydroxy-7-methyl-	EtOH	234(4.40),300(3.81), 357(3.72)	30-0577-66C
$C_{10}H_8N_2O_3S_2$			
1,3-Thiazin-4-one, 2,3,5,6-tetrahydro-N-(p-nitrophenyl)-2-thioxo-	EtOH	262(4.00),312(3.97)	95-0095-66
$C_{10}H_8N_2O_4$			
Benzene, p-bis(2-nitrovinyl)-	n.s.g.	345(4.51)	67-0042-66
3-Indolecarboxaldehyde, 5-methoxy-7-nitro-	EtOH	225(4.31),236s(4.15), 268(4.12),375(3.84)	34-0418-66
2-Quinoxalinecarboxylic acid, 3-hydroxy-7-methyl-, 4-oxide	EtOH	260(4.35),320(3.67), 375(3.43),420(3.48)	30-0577-66C

Compound	Solvent	$\lambda_{max}(\log \epsilon)$	Ref.
$C_{10}H_8N_2O_4S$			
Pyridazinedione, N-phenylsulfonyl-	n.s.g.	208(4.18),260(3.38), 267(3.41),274(3.04)	44-1311-66
$C_{10}H_8N_2O_5$			
Furo[2,3-b]pyridine-2-carboxylic acid, 5-nitro-, ethyl ester	5% EtOH	247.5(4.48)	4-0202-66
$C_{10}H_8N_4$			
Dipyridazino[2,3-a:4,3-d]pyrrole, 3-methyl-	EtOH	253(4.45),285(4.13), 296(3.97)	1-2637-66
$C_{10}H_8N_4O$			
2-Pyrazolin-5-one, 3-methyl-1-phenyl-4-diazo-	DMF	323(3.28)	5-0133-66I
$C_{10}H_8N_4OS_2$			
s-Triazolo[3,4-b][1,3,4]thiadiazole-6-thiol, 3-(p-methoxyphenyl)-	MeOH	287(4.12)	44-3528-66
anion	MeOH	287(4.08)	44-3528-66
$C_{10}H_8N_4O_4$			
1-(1,6-Dihydro-2-hydroxy-1-methyl-5-nitro-6-oxo-4-pyrimidinyl)-pyridinium hydroxide, inner salt	pH -1.9	240(3.75),250(3.75), 260s(3.71),313(3.87)	24-2997-66
	pH 2.0	256(3.96),342(3.91)	24-2997-66
1-(1,6-Dihydro-2-hydroxy-5-nitro-6-oxo-4-pyrimidinyl)-3-picolinium hydroxide, inner salt	pH -1.9	240s(3.71),270s(3.77), 313(3.91)	24-2997-66
	pH 2.0	263(3.94),342(3.94)	24-2997-66
1-(1,6-Dihydro-2-hydroxy-5-nitro-6-oxo-4-pyrimidinyl)-4-picolinium hydroxide, inner salt	pH -1.9	235s(4.00),258s(3.73), 312(3.93)	24-2997-66
	pH 2.0	232s(4.19),250s(4.00), 341(3.93)	24-2997-66
Pyrazole, 1-(2,4-dinitrophenyl)-3-methyl-	EtOH	322.5(4.03)	22-3744-66
Pyrazole, 1-(2,4-dinitrophenyl)-4-methyl-	EtOH	319(4.14)	22-3744-66
Pyrazole, 1-(2,4-dinitrophenyl)-5-methyl-	EtOH	290(3.78)	22-3744-66
$C_{10}H_8N_4S_2$			
s-Triazolo[3,4-b][1,3,4]thiadiazole, 3-(methylthio)-6-phenyl-	MeOH	271(4.51)	44-3528-66
s-Triazolo[3,4-b][1,3,4]thiadiazole, 6-(methylthio)-3-phenyl-, anion	MeOH	265(4.21)	44-3528-66
s-Triazolo[3,4-b][1,3,4]thiadiazole-6-thiol, 3-p-tolyl-	MeOH	275(4.33)	44-3528-66
$C_{10}H_8N_4S_4$			
p-Dithiino[2,3-d:5,6-d']dipyridazine, 1,6-bis(methylthio)-	EtOH	207(4.47),241(4.42), 248(4.43),330(3.40)	4-0541-66
$C_{10}H_8N_6O_2$			
5H-s-Triazolo[4,3-b]-s-triazole, 3-methyl-6-(p-nitrophenyl)-	EtOH	300(4.22)	4-0119-66
$C_{10}H_8O$			
3-Benzoxepin	EtOH	390(2.40),415(2.26), 445(1.68)	24-0634-66

Compound	Solvent	$\lambda_{max}(\log \epsilon)$	Ref.
2,8-Decadiene-4,6-diynal, 8-cis-2-trans	ether	236(4.30),247(4.37), 260(4.32),281(3.94), 299(4.15),318(4.26), 339(4.17)	24-3194-66
1-Naphthol	EtOH	220(4.60),230(4.55), 270s(3.50),280s(3.70), 290(3.72),302s(3.55), 322(3.55)	22-0228-66
2-Naphthol	EtOH	228(4.85),252s(3.48), 263(3.60),273(3.70), 285(3.55),320(3.25), 330(3.30)	22-0228-66
1,6-Oxido[10]annulene	C_6H_{12}	256(4.85),299(3.80), 394s(2.45)	33-2017-66
$C_{10}H_8O_2$			
Acetylene, m-acetoxyphenyl-	ether	226(4.44),269(4.45)	24-2822-66
Cyclopropenone, 2-methoxy-3-phenyl-	n.s.g.	207(4.12),255(4.17), 263(4.16)	35-3075-66
	18M H_2SO_4	248(4.35)	35-3075-66
2(5H)-Furanone, 4-phenyl-	EtOH	213(4.15),271(4.33)	39-1924-66C
3(2H)-Furanone, 5-phenyl-	EtOH	220(3.96),244(3.99), 306(4.27)	32-1073-66
	EtOH	220(3.96),244(3.99), 306(4.27)	88-0233-66
2-Hexen-4-ynal, 6-(2(5H)-furylidene)-	ether	365(4.47)	24-3544-66
Isocoumarin, 3-methyl-	EtOH	229(4.47),240(4.24), 255(3.87),263(3.93), 272(3.84),325(3.58)	12-1265-66
Matricaric acid, cis-cis	ether	244(4.34),257(4.28), 291s(4.02),309(4.12), 332(4.02)	24-2096-66
Pentacyclo[4.4.0.$0^{2,5}$.$0^{3,8}$.$0^{4,7}$]decane-9,10-dione	$CHCl_3$	490(1.20)	88-3743-66
Propiolic acid, phenyl, methyl ester	MeOH	258(4.20)	44-1135-66
$C_{10}H_8O_2S$			
Benzo[c]thiophene-2-carboxylic acid, methyl ester	EtOH	244(4.32),287s(3.51), 300(3.71),314(3.82), 356(3.88)	88-1953-66
2,4-Pentadienoic acid, 4-hydroxy-5-(5-methyl-2-thienyl)-, γ-lactone	ether	237(3.94),281(3.68), 371(4.46)	24-1226-66
2-Propyn-1-ol, 3,3'-(2,5-thiophenediyl)di-	ether	298(4.38),303(4.38), 311(4.33),316(4.30)	24-0984-66
$C_{10}H_8O_3$			
Benzocyclobutadienequinone, mono(ethylene ketal)	EtOH	241(4.20),279(3.62), 286(3.59)	44-1866-66
Chromone, 5-hydroxy-2-methyl-	MeOH	225(4.3),255(4.2), 325(3.8)	24-3842-66
Phthalic anhydride, 3,4-dimethyl-	ether	225(4.70),265(3.69), 314(3.68),326(3.71)	33-0858-66
Phthalic anhydride, 3,6-dimethyl-	ether	216(4.65),255(3.59), 308(3.62),320(3.62)	33-0858-66
Phthalic anhydride, 4,5-dimethyl-	ether	228(4.38),267(3.47), 291s(3.17),302(3.07)	33-0858-66

Compound	Solvent	λ_{max}(log ϵ)	Ref.
$C_{10}H_8O_3S$			
2-Naphthalenesulfonic acid, sodium salt	H_2O	265s(--),276(3.69), 286s(--),300s(2.86), 307(2.87),314(2.80), 322(2.92)	80-0263-66
$C_{10}H_8O_4$			
[2,2'-Bifuran]-3-carboxylic acid, methyl ester	EtOH	206(4.07),229(4.48), 310(4.15)	39-0976-66C
[2,2'-Bifuran]-3-carboxylic acid, 3'-methyl-	EtOH	202(4.14),224s(3.67), 296(4.04)	39-0976-66C
[2,2'-Bifuran]-3-carboxylic acid, 5'-methyl-	EtOH	207(4.16),236(3.57), 317(4.25)	39-0976-66C
Chromone, 3,5-dihydroxy-2-methyl-	MeOH	245(4.2),348(3.8)	24-3842-66
Isoscopoletin	EtOH	232(4.24),257(3.77), 261s(3.76),298(3.79), 351(4.04)	39-1805-66C
$C_{10}H_9BClF_4NO_3S$			
N-Methyl-5-(3'-and 4'-chlorosulfonyl-phenyl)isoxazolium tetrafluoro-borate	CH_2Cl_2	285(4.33)	78-0321-66B
$C_{10}H_9BrN_2$			
Pyrazole, 4-bromo-5(3)-methyl-3(5)-phenyl-	EtOH	250(4.11)	22-3744-66
$C_{10}H_9BrN_2O$			
3-Bromo-4-hydroxy-2,8-dimethyl-cinnolinium hydroxide, anhydro base	EtOH	216(4.45),241(3.89), 255(3.96),260s(3.95), 280(3.68),292(3.67), 321s(3.54),336(3.70), 354s(3.94),370(4.20), 390(4.23)	39-0470-66C
4(1H)-Cinnolinone, 3-bromo-1,8-dimethyl-	EtOH	220(4.30),239(4.05), 246(4.07),265(3.92), 275(3.92),298(3.57), 307(3.54),355(4.05), 367s(3.99)	39-0470-66C
$C_{10}H_9BrN_4$			
Dipyridazino[2,3-a:4,3-d]pyrrole, 5-bromo-6,7-dihydro-3-methyl-	EtOH	228(4.14),261(4.24), 349(3.68)	1-2637-66
$C_{10}H_9BrN_4O_4$			
Pyrazoline, 3-bromo-1-(2,4-dinitro-phenyl)-4-methyl-	$CHCl_3$	380(4.38)	22-0610-66
Pyrazoline, 3-bromo-1-(2,4-dinitro-phenyl)-5-methyl-	$CHCl_3$	386(4.31)	22-0610-66
$C_{10}H_9BrN_6O_2$			
Pyrimidine, 2,4-diamino-6-(p-bromo-anilino)-5-nitro-	pH 1	238(4.42),257s(4.37), 327(4.26)	87-0573-66
	pH 11	255(4.38),342(4.40)	87-0573-66
$C_{10}H_9BrN_6O_3$			
Pyrimidine, 2,4-diamino-6-(p-bromo-anilino)-5-nitro-, oxide	pH 1	242(4.40),265(4.38), 324(4.23)	87-0573-66
	pH 11	229(4.41),275(4.41), 325(4.20),374(3.92)	87-0573-66

Compound	Solvent	$\lambda_{max}(\log \epsilon)$	Ref.
$C_{10}H_9BrO$			
4,7-Methanoinden-1-one, 2-bromo-3a,4,7,7a-tetrahydro-	EtOH	246(3.78)	88-3737-66
$C_{10}H_9BrS$			
Benzo[b]thiophene, 5-bromo-2,3-dimethyl-	EtOH	240(4.51),244s(4.45),270(3.78),297(3.47),308(3.27)	22-3055-66
Benzo[b]thiophene, 6-bromo-2,3-dimethyl-	EtOH	240(4.52),244s(4.49),277(3.92),295s(3.69),305s(3.15)	22-3055-66
$C_{10}H_9Br_2NO_2$			
2-Indolinol, N-acetyl-5,7-dibromo-	EtOH	254(4.20),286(3.40),295(3.25)	1-2467-66
$C_{10}H_9ClN_2O_7S$			
2-Acetyl-3-methyl-6-nitrothiazolo-[3,2-a]pyridinium perchlorate	EtOH	249(3.99),324(3.93)	4-0027-66
$C_{10}H_9ClN_4$			
Dipyridazino[2,3-a:4,3-d]pyrrole, 8-chloro-6,7-dihydro-3-methyl-	EtOH	222(4.28),264(4.35),340(3.96)	1-2637-66
$C_{10}H_9ClN_4O_4$			
Pyrazoline, 3-chloro-5-methyl-1-(2,4-dinitrophenyl)-	$CHCl_3$	380(4.31)	22-0610-66
$C_{10}H_9ClO_4$			
Phthalide, 4-chloro-5,7-dimethoxy-	EtOH	227(4.53),258(4.13),298(3.80)	12-1265-66
$C_{10}H_9ClO_4S$			
3-Methyl-1-benzothiopyrylium perchlorate	HOAc	260(4.58),338(3.70),388(3.64)	78-0007-66
6-Methyl-1-benzothiopyrylium perchlorate	H_2SO_4	265(4.52),348(3.85),397(3.40)	17-0075-66
7-Methyl-1-benzothiopyrylium perchlorate	H_2SO_4	264(4.58),343(3.78),405(3.65)	17-0075-66
$C_{10}H_9ClO_4S_2$			
6-(Methylthio)-1-benzothiopyrylium perchlorate	H_2SO_4	232(4.32),241s(4.15),272s(3.91),289s(4.05),309(4.40),377(3.68),482(3.36)	17-0075-66
7-(Methylthio)-1-benzothiopyrylium perchlorate	H_2SO_4	238(4.44),304(4.33),346(3.51),486(4.06)	17-0075-66
4-p-Tolyl-1,2-dithiol-1-ium perchlorate	pH 1	249(4.06),361(2.88)	78-2119-66
$C_{10}H_9ClO_4Se$			
3-Methyl-1-benzoseleninium perchlorate	HOAc-1% $HClO_4$	262(4.39),351(3.46),410(3.34)	20-0169-66
6-Methyl-1-benzoseleninium perchlorate	HOAc-1% $HClO_4$	268(5.1),360(4.14),415(3.82)	20-0169-66
7-Methyl-1-benzoseleninium	HOAc-1% $HClO_4$	266(4.47),352(3.63),427(3.62)	20-0169-66

Compound	Solvent	$\lambda_{max}(\log \epsilon)$	Ref.
$C_{10}H_9ClO_5S$			
6-Methoxy-1-benzopyrylium perchlorate	H_2SO_4	235s(4.08),272(4.36), 356(3.87),405(3.28)	17-0075-66
7-Methoxy-1-benzopyrylium perchlorate	H_2SO_4	270s(4.38),274(4.40), 285s(3.59),343(3.63), 427(3.83)	17-0075-66
$C_{10}H_9ClO_5S_2$			
3-(p-Methoxyphenyl)-1,2-dithiol-1-ium perchlorate	pH 1	244(3.85),407(4.32)	78-2119-66
$C_{10}H_9ClS$			
Benzo[b]thiophene, 5-chloro-2,3-dimethyl-	EtOH	238(4.49),244(4.41), 270(3.78),297(3.43), 307(3.31)	22-3055-66
$C_{10}H_9IN_6O_2$			
Pyrimidine, 2,4-diamino-6-(p-iodo-anilino)-5-nitro-	pH 1	240(4.46),257s(4.42), 324(4.24)	87-0573-66
	pH 11	257(4.42),336(4.31)	87-0573-66
$C_{10}H_9MnO_5$			
Manganese, (3-acetyl-1-methyl-π-allyl)tetracarbonyl-	dioxan	215(4.34)	39-0194-66A
$C_{10}H_9N$			
1,6-Imino[10]annulene	C_6H_{12}	239(4.35),273(4.50), 312s(3.85),394f(2.65)	33-2017-66
Quinaldine	C_6H_{12}	269f(3.58)	5-0149-66H
	MeOH	272f(3.56)	5-0149-66H
	MeOH-H_2SO_4	318(3.95)	5-0149-66H
$C_{10}H_9NO$			
Isocarbostyril, 3-methyl-	EtOH	226(3.98),240(3.74), 247(3.60),278(3.74), 286(3.72),318(3.18), 332(3.35)	44-2090-66
Isoxazole, 3-p-tolyl-	EtOH	246(4.19)	101-0598-66B
Isoxazole, 5-p-tolyl-	EtOH	266(4.27)	78-0321-66B
Quinaldine, N-oxide	EtOH	319(3.99),329s(3.94)	1-2467-66
8-Quinolinol, 2-methyl-	N $HClO_4$	320(3.54),345(3.26)	39-0073-66A
	pH 13	334(3.49),355(3.43)	39-0073-66A
	$CHCl_3$	307(3.45)	39-0073-66A
copper chelate	pyridine	340(3.60),392(3.72)	39-0073-66A
	dioxan	399(3.71)	39-0073-66A
nickel chelate	pyridine	345(3.73),400(3.72)	39-0073-66A
8-Quinolinol, 5-methyl-	N $HClO_4$	323(3.00),333(3.26)	39-0073-66A
	pH 13	339(3.49),364(3.49)	39-0073-66A
	$CHCl_3$	330(3.46)	39-0073-66A
copper chelate	$CHCl_3$	429(3.78)	39-0073-66A
	pyridine	348(3.72),429(3.79)	39-0073-66A
nickel chelate	pyridine	355(3.85),425(3.80)	39-0073-66A
Spiro[cyclopropane-1,3'-indolin]-2'-one	EtOH	253(3.85),278s(3.24)	39-2245-66C
$C_{10}H_9NOS$			
Isothiazole, 3-(p-methoxyphenyl)-	C_6H_{12}	246s(3.81),280(4.25), 303s(3.88)	78-2119-66
Isothiazole, 5-(p-methoxyphenyl)-	C_6H_{12}	285(3.92)	78-2119-66

Compound	Solvent	$\lambda_{max}(\log \epsilon)$	Ref.
$C_{10}H_9NOS_2$			
Rhodanine, N-p-tolyl-	EtOH	258(3.98),295(4.16)	95-0095-66
1,3-Thiazin-4-one, 2,3,5,6-tetrahydro-N-phenyl-2-thioxo-	EtOH	264(4.10),314(4.09)	95-0095-66
$C_{10}H_9NO_2$			
Benzoic acid, o-(cyanomethyl)-, methyl ester	EtOH	229(3.81),275(3.00)	95-0544-66
Isocarbostyril, 2-hydroxy-3-methyl-	EtOH	225(4.13),241(3.76), 248(3.72),290(3.70), 318(3.40),328(3.24), 342(3.20)	44-2090-66
Isoxazole, 3-methoxy-5-phenyl-	EtOH	260(4.27)	32-0454-66
3-Isoxazolinone, 5-methyl-2-phenyl-	MeOH	262(4.01)	88-5451-66
2-Oxazolinone, 5-methyl-3-phenyl-	MeOH	247.5(3.99)	88-5451-66
$C_{10}H_9NO_3$			
Acetonitrile, [2-methoxy-4,5-(methylenedioxy)phenyl]-	EtOH	213(3.67),239(3.74), 301(3.84)	18-1525-66
3-Isoxazolol, 5-(p-methoxyphenyl)-	EtOH	273.5(4.37)	94-1277-66
$C_{10}H_9NO_3S$			
2-Benzothiazolecarboxylic acid, 4-methoxy-, methyl ester	EtOH	249(4.13),293(4.04), 341(3.57)	44-1484-66
$C_{10}H_9NO_4$			
Cinnamic acid, p-nitro-, methyl ester	EtOH	300(4.3)	65-1202-66
$C_{10}H_9NO_4S$			
N-Methyl-5-phenylisoxazolium 3'(4')-sulfonate (mixture)	pH 1	284(4.34)	78-0321-66B
$C_{10}H_9NO_5$			
Phthalic anhydride, 6-amino-3,4-dimethoxy-	ether	231(4.31),261(4.36), 392(3.79)	44-1912-66
$C_{10}H_9NS$			
3-Indolizinethiocarboxaldehyde, 2-methyl-	C_6H_{12}	420s(4.28),432(4.44), 442(4.60)	77-0401-66
Isothiazole, 4-p-tolyl-	C_6H_{12}	242(3.96),272(4.05)	78-2119-66
$C_{10}H_9N_3$			
1,4-Diazacycl[3.3.2]azine, 2,3-dimethyl-	EtOH	230(5.00),284(4.01), 292(4.02),320s(3.81), 337(4.01),343(4.23), 377(2.43)	88-3621-66
Imidazo[2,3-a]phthalazine, 1,2-dihydro-	10% HCl	215(4.61),237(4.16), 272(3.95),298(3.87)	4-0381-66
	EtOH	209(4.65),266(4.01), 274(4.19),335(3.81)	4-0381-66
s-Triazine, 4-methyl-2-phenyl-	50% EtOH	264(4.26)	54-1101-66
$C_{10}H_9N_3O_2S$			
1,2,3-Thiadiazole-5-carbamic acid, N-methyl-, phenyl ester	MeOH	227(3.63),270(4.06)	24-1618-66
Δ^3-1,2,3-Thiadiazoline-$\Delta^{5,N}$-carbamic acid, 2-methyl-, phenyl ester	MeOH	218(4.19),245(4.10), 337(4.09)	24-1618-66

Compound	Solvent	λ max(log ε)	Ref.
$C_{10}H_9N_3O_3$			
4-Hydroxy-2,8-dimethyl-5-nitrocinno-linium hydroxide, inner salt	EtOH	213(4.51),245(3.90), 276(3.48),287(3.62), 342s(3.94),359(4.15), 378(4.18)	39-0470-66C
4-Hydroxy-2,8-dimethyl-6-nitrocinno-linium hydroxide, inner salt	EtOH	209(4.24),256(4.18), 289s(3.76),328(3.49), 338(3.51),392s(4.08), 408(4.10)	39-0470-66C
Phthalhydrazide, 3-acetamido-	DMF	333(3.92)	18-0932-66
$C_{10}H_9N_5$			
5H-s-Triazolo[4,3-b]-s-triazole, 3-methyl-6-phenyl-	EtOH	248(4.35)	4-0119-66
5H-s-Triazolo[4,3-b]-s-triazole, 6-p-tolyl-	EtOH	253(4.39)	4-0119-66
$C_{10}H_9N_5O$			
Adenine, 9-(5-methyl-2-furyl)-	pH 1	252(4.36)	35-1549-66
	pH 11	251(4.28)	35-1549-66
	MeOH	249(4.30)	35-1549-66
4-Pyrimidinol, 2-amino-6-phenyl-	pH 1	238(4.14),266(4.03), 305(4.20)	4-0476-66
	pH 7	226(4.13),274s(4.05), 305(4.21)	4-0476-66
	pH 13	237(4.27),318(4.24)	4-0476-66
5H-s-Triazolo[4,3-b]-s-triazole, 6-(p-methoxyphenyl)-	EtOH	263(4.40)	4-0119-66
$C_{10}H_9N_5OS$			
s-Triazolo[3,4-b][1,3,4]thiadiazole, 6-amino-3-(p-methoxyphenyl)-	MeOH	274(4.50),292s(4.30)	44-3528-66
cation	MeOH	293(4.52),309(4.37)	44-3528-66
$C_{10}H_9N_5O_6$			
Pyrazoline, 3-methyl-1-picryl-	$CHCl_3$	394(4.30)	22-0610-66
Pyrazoline, 4-methyl-1-picryl-	$CHCl_3$	383(4.26)	22-0610-66
Pyrazoline, 5-methyl-1-picryl-	$CHCl_3$	394(4.28)	22-0610-66
$C_{10}H_9N_5S$			
s-Triazolo[3,4-b][1,3,4]thiadiazole, 6-amino-3-benzyl-	MeOH	236(3.99)	44-3528-66
cation	MeOH	243(4.06)	44-3528-66
s-Triazolo[3,4-b][1,3,4]thiadiazole, 6-amino-3-p-tolyl-	MeOH	267(4.39),272s(4.38), 287(4.11)	44-3528-66
cation	MeOH	246s(4.15),278(4.41)	44-3528-66
$C_{10}H_9N_7O_2$			
4,6-Pyrimidinediol, 5-(2-amino-3,4-dihydro-4-pteridinyl)-	pH 4	255(4.05),309(3.95)	39-1117-66C
	pH 12	256(3.95),283(3.92), 340(3.87)	39-1117-66C
$C_{10}H_9N_7O_2S$			
Barbituric acid, 5-(2-amino-3,4-di-hydro-4-pteridinyl)-2-thio-	pH 6	246s(4.15),265(4.28), 289(4.36),305s(4.28)	39-1117-66C
	pH 11	236(4.32),286(4.27), 340(4.02)	39-1117-66C

Compound	Solvent	$\lambda_{max}(\log \epsilon)$	Ref.
$C_{10}H_9N_7O_3$			
Barbituric acid, 5-(2-amino-3,4-dihydro-4-pteridinyl)-	pH 6	255(4.32),309(3.90)	39-1117-66C
	pH 11	258(4.33),282s(3.88),338(3.86)	39-1117-66C
$C_{10}H_{10}$			
1,3,5-Cyclooctatriene, 7,8-dimethylene-	hexane	228(4.23),272(3.96)	88-5961-66
	EtOH	269(3.53)	77-0508-66
Cyclopropane, benzylidene-	EtOH	245(4.09),254(4.17),263(3.99)	88-3267-66
3a,7a-Methanoindene	C_6H_{12}	270(3.32),258(3.30)	35-3461-66
$C_{10}H_{10}BNO_3$			
5-Isoxazoleboronic acid, 3-p-tolyl-	EtOH	248(4.16)	101-0598-66B
$C_{10}H_{10}BrClO_2$			
Propionic acid, β-bromo-β-(p-chlorophenyl)-α-methyl-	EtOH	232(4.02)	44-4043-66
$C_{10}H_{10}BrNO_2$			
2-Indolinol, N-acetyl-5-bromo-	EtOH	253(4.30),285(3.67),295(3.60)	1-2467-66
$C_{10}H_{10}BrN_3O$			
1,3,4-Oxadiazole, 5-(p-bromophenyl)-2-(dimethylamino)-	EtOH	299(4.35)	22-0153-66
$C_{10}H_{10}BrN_3O_2$			
Pyrazoline, 3-bromo-4-methyl-1-(p-nitrophenyl)-	$CHCl_3$	392(4.36)	22-0610-66
$C_{10}H_{10}BrN_5$			
Pyrimidine, 4,5-diamino-2-(p-bromoanilino)-	MeOH	277(4.39)	87-0121-66
$C_{10}H_{10}Br_2Ti$			
Titanium, dibromodicyclopentadienyl-	C_6H_{12}	290(3.93),327(3.74),427(3.44)	22-3548-66
$C_{10}H_{10}Br_2Zr$			
Zirconium, dibromodicyclopentadienyl-	C_6H_{12}	295(3.65),335(3.0)	22-3548-66
$C_{10}H_{10}ClNO_4S$			
2,3-Dihydro-1H-cyclopenta[4,5]thiazolo[3,2-a]pyridinium perchlorate	EtOH	212s(4.07),244(3.99),318(4.25)	4-0027-66
$C_{10}H_{10}ClNO_5$			
2-Methoxyisoquinolinium perchlorate	H_2O	272s(3.45),280(3.47),334(3.60)	39-0870-66B
$C_{10}H_{10}ClNO_5S$			
2-Acetyl-3-methylthiazolo[3,2-a]pyridinium perchlorate	EtOH	237(4.11),306s(4.03),314(4.16),327s(3.68)	4-0027-66
$C_{10}H_{10}ClN_3O$			
1,3,4-Oxadiazole, 5-(m-chlorophenyl)-2-(dimethylamino)-	EtOH	297(4.21)	22-0153-66
1,3,4-Oxadiazole, 5-(o-chlorophenyl)-2-(dimethylamino)-	EtOH	288(4.11)	22-0153-66

Compound	Solvent	$\lambda_{max}(\log \epsilon)$	Ref.
1,3,4-Oxadiazole, 5-(p-chlorophenyl)-2-(dimethylamino)-	EtOH	298(4.28)	22-0153-66
$C_{10}H_{10}ClN_5O_3$			
1-(2-Amino-1,6-dihydro-1-methyl-6-oxo-5-nitro-4-pyrimidinyl)-pyridinium chloride	pH 0.0	257(3.90),340(3.87)	24-2997-66
$C_{10}H_{10}Cl_2Hf$			
Hafnium, dichlorodicyclopentadienyl-	C_6H_{12}	260(3.62),300(3.01)	22-3548-66
$C_{10}H_{10}Cl_2N_4O_4$			
9H-Purine, 2,6-dichloro-9-β-D-ribofuranosyl-	pH 1	274(3.93)	44-3258-66
	pH 11	274(4.03)	44-3258-66
	EtOH	275(3.97)	44-3258-66
$C_{10}H_{10}Cl_2Ti$			
Titanium, dichlorodicyclopentadienyl-	C_6H_{12}	258(4.04),300(--),385(3.28)	22-3548-66
$C_{10}H_{10}Cl_2Zr$			
Zirconium, dichlorodicyclopentadienyl-	C_6H_{12}	285(3.54),330(2.96)	22-3548-66
$C_{10}H_{10}Cl_4$			
Spiro[2.4]hepta-4,6-diene, 4,5,6,7-tetrachloro-1-ethyl-2-methyl-, cis	EtOH	245(3.95),287(3.54)	44-0768-66
trans	EtOH	254(3.88),287(3.48)	44-0768-66
Spiro[2.4]hepta-4,6-diene, 4,5,6,7-tetrachloro-1,1,2-trimethyl-	EtOH	250(3.94),292(3.59)	44-0768-66
$C_{10}H_{10}FeO_4$			
Iron, dicarbonyl(carboxymethyl)-cyclopentadienyl-, methyl ester	C_6H_{12}	256(3.96),360(2.86)	101-0341-66A
Iron, tricarbonyl(sorbyl ketone)-	EtOH	202(4.41),225(4.27),250(4.16),310(3.53)	101-0370-66A
$C_{10}H_{10}FeO_5$			
Iron, tricarbonyl(sorbic acid)-, methyl ester	EtOH	202(4.31),228(4.19),240(4.11),300(3.27)	101-0370-66A
$C_{10}H_{10}INO$			
1-Methyl-5-hydroxyquinolinium iodide	0.05N HCl	226(4.32),254(4.63),307(3.20),318(3.24),375(3.50)	10-0195-66A
	0.05N NaOH	272(4.54),317(3.09),332(3.06),460(3.59)	10-0195-66A
1-Methyl-6-hydroxyquinolinium iodide	0.05N HCl	222(4.48),248(4.52),317(3.75),348(3.58)	10-0195-66A
	0.05N NaOH	268(4.55),317(3.43),325(3.44),408(3.61)	10-0195-66A
1-Methyl-7-hydroxyquinolinium iodide	0.05N HCl	227(4.40),248(4.47),307(3.40),353(3.83)	10-0195-66A
	0.05N NaOH	260(4.57),310(3.02),406(3.93)	10-0195-66A
1-Methyl-8-hydroxyquinolinium iodide	0.05N HCl	225(4.26),255(4.60),312(3.08),323(3.08),365(3.30)	10-0195-66A
	0.05N NaOH	273(4.56),332(3.06),345(3.11),445(3.19)	10-0195-66A

Compound	Solvent	λ_{max}(log ϵ)	Ref.
$C_{10}H_{10}I_2Zr$			
Zirconium, diiododicyclopentadienyl-	C_6H_{12}	295(--),360(3.84)	22-3548-66
$C_{10}H_{10}N_2$			
Pyrazole, 3(5)-methyl-4-phenyl-	EtOH	246.5(4.13)	22-3744-66
Pyrazole, 4-methyl-3(5)-phenyl-	EtOH	250(4.14)	22-3744-66
Quinazoline, 2-ethyl-	pH 0.0	261(3.97)	39-0234-66C
	pH 10.0	224(4.64),266(3.38), 310(3.38),320s(3.37)	39-0234-66C
$C_{10}H_{10}N_2O$			
Acetamide, N-benzyl-2-cyano-	EtOH	248(2.26),252(2.31), 259(2.37),264(2.24)	78-2575-66
Cinnoline, 4-methoxy-3-methyl-	EtOH	225(4.75),281(3.53), 292(3.53),326(3.61)	39-0470-66C
Cinnoline, 4-methoxy-8-methyl-	EtOH	219(4.67),286(3.68), 295(3.72),323(3.71)	39-0470-66C
4(1H)-Cinnolinone, 1,3-dimethyl-	EtOH	209(4.46),236s(4.11), 243(4.16),253(3.99), 262s(3.89),284(3.54), 295(3.56),305(4.12), 366(4.12)	39-0470-66C
4-Hydroxy-2,3-dimethylcinnolinium hydroxide, inner salt	EtOH	213(4.45),256(3.87), 267s(3.70),302s(3.48), 313s(3.46),325(3.63), 345(3.81),360(4.07), 377(4.08)	39-0470-66C
4-Hydroxy-2,8-dimethylcinnolinium hydroxide, inner salt	EtOH	213(4.43),234s(3.99), 241(4.03),248(4.00), 275(3.64),291(3.59), 331s(3.79),340s(3.91), 356(4.15),357(4.15)	39-0470-66C
Isocarbostyril, 3-amino-2-methyl-	EtOH	230(4.29),304(4.14), 372(3.57)	95-0544-66
Pyrazole, 3-methoxy-1-phenyl-	EtOH	272(4.24)	44-1538-66
Pyrazol-3-ol, 4-methyl-1-phenyl-	EtOH	278(4.37)	44-1538-66
Quinazoline, 2-methoxy-4-methyl-	pH 0.0	218(4.31),237(4.39), 290(3.66),330(3.31)	39-0234-66C
	pH 6.0	223s(4.45),226(4.51), 234s(4.37),255(3.51), 322(3.48),334s(3.33)	39-0234-66C
$C_{10}H_{10}N_2OSe$			
Selenazolidinone, 2-imino-3-methyl-5-phenyl-	EtOH	214s(4.3),244s(3.5)	28C-0285-66A
Selenazolinone, 2-(methylamino)-5-phenyl-	EtOH	229(4.38)	28C-0285-66A
$C_{10}H_{10}N_2O_2$			
Acetophenone, 2-diazo-4'-ethoxy-	hexane	210(4.23),218s(4.17), 287(4.31)	78-0209-66
	butanol	224(4.07),306(4.40)	78-0209-66
8(1H)-Cycloheptapyrazolone, 7-methoxy-1-methyl-	MeOH	230(4.56),255(4.41), 306(4.06),321(4.10), 380(3.91)	18-0253-66
3a,6a-Diazapentalene, 1,3-diacetyl-	n.s.g.	219(4.20),223(4.20), 247(3.80),262s(3.48), 322(4.30),396(4.50)	35-5588-66
Isoxazole, 3-amino-5-(p-methoxyphenyl)-	EtOH	272(4.40)	94-1277-66

Compound	Solvent	$\lambda_{max}(\log \epsilon)$	Ref.
1,2-Naphthoquinone, 3,4-dihydro-, dioxime	EtOH	222(3.87),267(3.89)	39-0717-66C
	EtOH-NaOH	240(4.07),313(3.90)	39-0717-66C
$C_{10}H_{10}N_2O_2S$			
Pyrazole, 1-p-toluenesulfonyl-	EtOH	240(4.25)	28C-0782-66A
$C_{10}H_{10}N_2O_3$			
3-Buten-2-one, 4-(o-nitrophenyl)-, α-oxime	n.s.g.	264(4.37),333(3.69)	83-0626-66
Formic acid, (benzoylazo)-, ethyl ester	MeOH	228(4.04)	24-2039-66
	dioxan	239s(4.14)	24-2039-66
	dioxan	239s(4.14),435(1.57)	89-0372-66
Furo[2,3-b]pyridine-2-carboxylic acid, 5-amino-, ethyl ester, hydrochloride	H_2O	272.5(4.26)	4-0202-66
1(2H)-Naphthalenone, 3,4-dihydro-2-nitro-, oxime	EtOH	258(4.08),292(3.08), 303(3.04)	39-0717-66C
	0.05N NaOH	240(4.04),299(3.98)	39-0717-66C
2,4(1H,3H)-Quinazolinedione, 3-(2-hydroxyethyl)-	EtOH	219(4.67),242(3.96), 309(3.52)	24-1532-66
Thymine, 1-(5-methyl-2-furyl)-	EtOH	212(4.02),264(3.92)	44-0205-66
$C_{10}H_{10}N_2O_3S$			
2-Benzothiazolecarboxamide, 4,6-dimethoxy-	EtOH	259(4.04),310(4.06)	44-1484-66
$C_{10}H_{10}N_2O_4$			
1-Indanone, 2-(hydroxymethyl)-2-nitro-, oxime	EtOH	258(4.11),286(3.59), 295(3.68),306(3.59)	94-1408-66
	EtOH-NaOH	249(4.08),308(4.00), 318(3.97)	94-1408-66
$C_{10}H_{10}N_2S$			
Quinazoline, 4-methyl-2-(methylthio)-	pH -1.0	237(4.09),249(4.22), 268(4.35),319(3.42), 353(3.67)	39-0234-66C
	pH 6.0	208(4.43),257(4.48), 338(3.45)	39-0234-66C
$C_{10}H_{10}N_2Se$			
2-Benzoselenazolinecarbonitrile, 2,3-dimethyl-	C_6H_{12}	226(4.49),263(3.63), 305(3.54)	22-4028-66
$C_{10}H_{10}N_4$			
Dipyridazino[2,3-a:4,3-d]pyrrole, 6,7-dihydro-3-methyl-	EtOH	223(4.29),259(4.28), 339(3.83)	1-2637-66
7H-Pyrrolo[2,3-c]pyridazine, 6-(2-cyanoethyl)-3-methyl-	EtOH	225(4.39),277(3.95), 327(3.72)	1-2637-66
$C_{10}H_{10}N_4O$			
Isocytosine, 5-anilino-	pH 1	235(3.98),272(4.05), 332s(3.67)	87-0108-66
	pH 13	240(4.07),280s(3.93)	87-0108-66
Propiophenone, 3-azido-3-(methylimino)-	EtOH	245(4.10),285(3.18), 348(3.78)	78-0415-66A
2-Pyrazolin-5-one, 3-methyl-4-(phenylazo)-	$CHCl_3$	410(4.16)	22-2990-66

Compound	Solvent	λmax(log ε)	Ref.
$C_{10}H_{10}N_4O_4$			
Pyrazoline, 1-(2,4-dinitrophenyl)-3-methyl-	$CHCl_3$	389(4.42)	22-0610-66
Pyrazoline, 1-(2,4-dinitrophenyl)-4-methyl-	$CHCl_3$	375.5(4.30)	22-0610-66
Pyrazoline, 1-(2,4-dinitrophenyl)-5-methyl-	$CHCl_3$	383(4.33)	22-0610-66
$C_{10}H_{10}N_4O_4S$			
Uracil, 5,5'-(thiodimethylene)di-	pH 1	264(3.86)	87-0097-66
	pH 11	283(4.18)	87-0097-66
$C_{10}H_{10}N_4O_4S_2$			
Uracil, 5,5'-(dithiodimethylene)di-	pH 11	282(4.15)	87-0097-66
$C_{10}H_{10}N_4O_5$			
Biacetyl, mono(dinitrophenyl-hydrazone)	EtOH-KOH	501(4.58)	96-0297-66
	$CHCl_3$	351(4.46)	96-0297-66
$C_{10}H_{10}N_4O_6S$			
Uracil, 5,5'-(sulfonyl-dimethylene)di-	pH 11	282(4.17)	87-0097-66
$C_{10}H_{10}N_4S$			
Benzamidine, N-(5-methyl-1,3,4-thiadiazol-2-yl)-	MeOH	235(3.99),302(4.26)	44-3528-66
$C_{10}H_{10}N_8$			
Glyoxal, bis(pyrazinylhydrazone)	dil. HCl	239(3.95),369(5.14)	39-2038-66C
$C_{10}H_{10}O$			
6(1H)-Azulenone, 2,3-dihydro-	EtOH	231(4.29),335(4.12), 336s(4.00)	44-1042-66
3,4-Decadiene-6,8-diyn-1-ol	EtOH	209(4.72),225(3.57), 237(3.80),249(4.05), 263(4.20),279(4.13)	39-0129-66C
	n.s.g.	208(4.75),225(3.59), 237(3.81),251(4.08), 264(4.20),280(4.11)	77-0585-66
4,5-Decadiene-7,9-diyn-1-ol	EtOH	207(4.65),224s(3.49), 237(3.69),249(3.96), 263(4.15),278(4.04)	39-0129-66C
	EtOH	225(3.48),237(3.72), 249(3.98),263(4.11), 278(4.05)	39-0129-66C
4,6,8-Decatriyn-1-ol	EtOH	209.5(5.17)	39-0129-66C
	ether	209.0(5.08)	39-0129-66C
Dicyclopentadienone, endo-	EtOH	226(3.87),323(1.56)	35-1059-66
Dicyclopentadienone, exo-	EtOH	217(3.87),248s(3.03), 323(1.68)	35-1059-66
1-Tetralone	EtOH	248(4.08),292(3.26)	22-1693-66
$C_{10}H_{10}OS$			
Thieno[3,4-d]oxepin, 1,3-dimethyl-	EtOH	247(4.57)	24-0634-66
Thiochromen, 6-methoxy-	EtOH	243(4.25),274s(3.50), 292s(3.23),345(3.14)	17-0075-66
Thiochromen, 7-methoxy-	EtOH	251(4.35),301(3.51), 317s(3.35)	17-0075-66

Compound	Solvent	λ_{max}(log ε)	Ref.
$C_{10}H_{10}O_2$			
2(3H)-Benzofuranone, 2,2-dimethyl-	EtOH	226s(3.70),251s(2.95), 272s(3.33),280(3.46), 283s(3.45),287s(3.38)	1-1561-66
2(3H)-Benzofuranone, 3,3-dimethyl-	EtOH	225s(2.59),262s(2.01), 269(2.13),276(2.10), 322(1.95)	1-1561-66
p-Benzoquinone, (3-butenyl)-	C_6H_{12}	245(4.31),433(1.30), 447s(--),457s(--)	39-1627-66C
Cinnamic acid, methyl ester	MeOH	216(4.20),221(4.14), 273(4.36)	33-1552-66
	$CHCl_3$	277(4.32)	33-1552-66
	CH_2Cl_2	276(4.34)	33-1552-66
	EtOH	280(4.5)	65-1202-66
4,5-Decadiene-7,9-diyne-1,3-diol	EtOH	208(4.71),225s(3.46), 237(3.68),250(3.98), 263(4.19),279(4.07)	39-0129-66C
5,7,9-Decatriyne-1,3-diol	EtOH	207.0(5.10)	39-0129-66C
1-Propen-2-ol, 1-phenyl-, formate ester	n.s.g.	241(4.17)	33-1002-66
Propiophenone, enol formate ester	n.s.g.	236(4.04)	33-1002-66
$C_{10}H_{10}O_3$			
Chromanone, 5-hydroxy-2-methyl-	MeOH	210(4.2),225s(4.1), 275(4.0),350(3.5)	24-3842-66
Ketone, o-carboxybenzyl methyl	EtOH	231(3.93),277(3.13)	12-1265-66
Mellein	MeOH	212(4.30),245(3.86), 312(3.62)	31-0209-66
7-Nonene-3,5-diynoic acid, 9-hydroxy-, methyl ester, trans	EtOH	214(4.13),226s(3.83), 239(3.93),252(4.12), 266(4.24),282(4.16)	39-0135-66C
1-Propanol, 1,2-epoxy-1-phenyl-, formate ester	n.s.g.	240(4.03),279(2.95)	33-1002-66
2-Propanol, 2,3-epoxy-3-phenyl-, formate ester	n.s.g.	242(3.30)	33-1002-66
Salicylic acid, 6-trans-propenyl-	MeOH	254(4.08),318(3.64)	31-0209-66
$C_{10}H_{10}O_4$			
p-Benzoquinone, (1-acetoxyethyl)-	EtOH	246(4.29),443(1.30)	39-1627-66C
Cyclohex-1-enylacetic acid, α-acetyl-2-hydroxy-6-oxo-, γ-lactone	EtOH	308(4.20)	39-2100-66C
Isocoumarin, 3,4-dihydro-6,8-dihydroxy-3-methyl-	EtOH	270(4.12),305(3.77)	39-0168-66C
Phthalide, 6,7-dimethoxy-	EtOH	303(3.68)	44-1912-66
Terephthalic acid, dimethyl ester	isooctane	240(4.29),284(3.23), 317(3.17)	51-0584-66
Tricyclo[3.2.1.0^{2,7}]oct-3-ene-1,8-dicarboxylic acid	EtOH	208(3.73)	88-1185-66
$C_{10}H_{10}O_5$			
Isovanillin, o-quinol acetate	MeOH	304(3.62)	49-0570-66
	DMF	298(3.59)	49-0570-66
7-Oxabicyclo[2.2.1]hepta-2,5-diene-2,3-dicarboxylic acid, dimethyl ester	ether	222s(3.63),288(3.03)	89-1039-66
4,5-Oxepinedicarboxylic acid, dimethyl ester	EtOH	269(3.32),304s(3.15)	89-1039-66
Vanillin, o-quinol acetate	MeOH	307(3.61)	49-0570-66
	dioxan	302(3.40)	49-0570-66
	DMF	295(3.57)	49-0570-66

Compound	Solvent	λ_{max}(log ε)	Ref.
$C_{10}H_{10}O_5S_3$			
3-(p-Methoxyphenyl)-1,2-dithiol-1-ium hydrogen sulfate	pH 1	244(3.88),407(4.30)	78-2119-66
$C_{10}H_{10}O_6$			
1,4-Naphthoquinone, 2,5,7,8-tetrahydroxy- (or mompain)	EtOH	228(4.43),272(4.06), 319(3.93),486(3.75), 518(3.80),554(3.63)	39-2184-66C
Phthalic acid, 3,4-dimethoxy-	EtOH	253(4.16),285s(3.57)	44-1912-66
$C_{10}H_{10}S$			
2H-1-Benzothiopyran, 2-methyl-	EtOH	242(4.35),280s(3.35), 324(3.13)	78-0007-66
2H-1-Benzothiopyran, 3-methyl-	EtOH	246(4.06),283s(3.76), 320(3.14)	78-0007-66
2H-1-Benzothiopyran, 4-methyl-	EtOH	244(4.33),272s(3.78), 323(3.12)	78-0007-66
2H-1-Benzothiopyran, 6-methyl-	EtOH	246(4.35),276(3.39), 296s(3.19),334(3.14)	17-0075-66
2H-1-Benzothiopyran, 7-methyl-	EtOH	248(4.42),290(3.36), 300s(3.27),328(3.15)	17-0075-66
$C_{10}H_{10}S_2$			
2H-1-Benzothiopyran, 6-(methylthio)-	EtOH	248(4.34),284(3.73), 305s(3.47),344(3.07)	17-0075-66
2H-1-Benzothiopyran, 7-(methylthio)-	EtOH	254(4.13),274(4.26), 307(3.77),312s(3.74), 330s(3.51)	17-0075-66
$C_{10}H_{11}^+$			
Cyclopropylphenylcarbonium ion	FSO_3H-SbF_5	343(4.26)	35-1488-66
$C_{10}H_{11}Br$			
2,6,8-Decatrien-4-yne, 1-bromo-	ether	301(4.37)	24-0142-66
$C_{10}H_{11}BrN_2O$			
1,3-Dimethyl-2-azaquinolizinium bromide, 2-oxide	EtOH	228(4.04),241(3.95), 252(3.94),272(3.81), 276(3.82),315s(3.61), 322(3.62),342(3.55), 358(3.50),370(3.60)	44-0941-66
$C_{10}H_{11}BrN_2O_8$			
β-D-Glucopyranuronic acid, 1-(5-bromo-1,2,3,4-tetrahydro-2,4-dioxo-1-pyrimidinyl)-1-deoxy-	H_2O	277(3.96)	94-1354-66
$C_{10}H_{11}BrO$			
Tropone, 2-bromo-4-isopropyl-	MeOH	225s(3.99),248(4.17), 323(3.80)	18-2538-66
Tropone, 2-bromo-6-isopropyl-	MeOH	237s(4.28),247(4.31), 320(3.83)	18-2538-66
$C_{10}H_{11}Cl$			
3a,7a-Methanoindan, 2-chloro-	C_6H_{12}	245(3.32),255(3.32), 273(3.36)	35-3461-66
$C_{10}H_{11}ClN_2O_2S$			
1,2,4-Benzothiadiazine, 6-chloro-2-ethyl-7-methyl-, 1,1-dioxide	EtOH	228(4.48),270(3.98), 279(3.93),301(3.65)	7-1083-66

Compound	Solvent	λ_{max}(log ε)	Ref.
$C_{10}H_{11}ClN_2O_3$			
Furo[2,3-b]pyridine-2-carboxylic acid, 5-amino-, ethyl ester, hydrochloride	H_2O	272.5(4.26)	4-0202-66
$C_{10}H_{11}ClN_4O$			
Purine, 6-chloro-9-(tetrahydro-pyran-2-yl)-	pH 7	264(3.94)	39-0692-66C
$C_{10}H_{11}ClN_4O_5$			
Inosine, 2-chloro-	pH 1	252(4.08)	44-3258-66
	pH 11	257(4.15)	44-3258-66
$C_{10}H_{11}ClO_4$			
Benzoic acid, 3-chloro-4,6-dihydroxy-2-methyl-, ethyl ester	EtOH	263(4.00),309(3.67)	12-1265-66
$C_{10}H_{11}ClS$			
Thiachroman, 6-chloro-4-methyl-	EtOH	262(4.25),295s(3.10)	78-0007-66
$C_{10}H_{11}F_3N_2O_5$			
Uridine, 2'-deoxy-5-(trifluoro-methyl)-	pH 1	262(4.01)	44-1163-66
	pH 7	262(3.99)	44-1163-66
	pH 11	260(3.84)	44-1163-66
$C_{10}H_{11}IN_2O_8$			
β-D-Glucopyranuronic acid, 1-deoxy-1-(1,2,3,4-tetrahydro-5-iodo-2,4-dioxopyrimidin-1-yl)-	H_2O	283(3.85)	94-1354-66
$C_{10}H_{11}N$			
Indole, 2,3-dimethyl-	EtOH	228(4.46),285(3.76), 292(3.69)	94-0934-66
Indole, 3-ethyl-	EtOH	224(4.52),284(3.85), 292(3.79)	94-0934-66
$C_{10}H_{11}NO$			
3-Buten-2-one, 4-phenyl-, α-oxime	n.s.g.	284(4.63)	83-0626-66
3-Buten-2-one, 4-phenyl-, β-oxime	n.s.g.	287.5(4.35)	83-0626-66
Carbostyril, 2,3-dihydro-N-methyl-	EtOH	250(4.01),275s(3.37), 283s(2.88)	1-2467-66
$C_{10}H_{11}NO_2$			
Acrylic acid, 3-anilino-, methyl ester, cis	CH_2Cl_2	291s(4.2),320(4.4)	24-2526-66
trans	dioxan	289(4.36),310(4.44)	24-2526-66
Cyclopropanecarboxylic acid, 2-(4-pyridyl)-, methyl ester, trans	EtOH-HCl	243(4.07),255s(4.01)	44-0399-66
	EtOH-NaOH	259(3.35)	44-0399-66
1-Indolinecarboxaldehyde, 2-hydroxy-5-methyl-	EtOH	253(4.16),286(3.58), 293(3.58)	88-4355-66
2-Indolinol, N-acetyl-	EtOH	247(4.15)	1-0262-66
	EtOH	248(4.14),279(3.55), 285(3.48)	1-2467-66
	EtOH	248(4.14),279(3.55), 285(3.48)	88-4355-66
	EtOH	247.5(4.23)	88-4701-66
2-Indolinone, 3-methoxy-3-methyl-	MeOH	210(4.34),253(3.81)	44-2627-66
$C_{10}H_{11}NO_3$			
Acetanilide, 4'-acetoxy-	EtOH	245(4.21),280s(3.00)	65-1198-66

Compound	Solvent	$\lambda_{max}(\log \epsilon)$	Ref.
1-Indolinecarboxaldehyde, 2-hydroxy-5-methoxy-	EtOH	257(4.12),294(3.58), 303s(3.51)	88-4355-66
$C_{10}H_{11}NO_3S$			
Thiophene, tetrahydro-2-(p-nitrophenoxy)-	MeOH	221(3.91),308(4.09)	35-5855-66
$C_{10}H_{11}NO_5$			
Benzoic acid, 4-acetamido-2,6-dihydroxy-, methyl ester	EtOH	218(4.39),281(4.42)	12-0275-66
Benzoic acid, 3-methoxy-4-nitro-, ethyl ester	n.s.g.	264(3.6),330(3.42)	39-0424-66B
Glycine, N-[(2,6-dimethyl-4-oxo-4H-pyran-3-yl)carbonyl]-	EtOH	233(4.35),266(4.13)	78-2003-66
$C_{10}H_{11}NO_6$			
Propionic acid, 3-(2-methoxy-4-nitrophenoxy)-	pH 13	229(3.94),264(3.89), 327(3.76)	94-0939-66
	EtOH	239(3.99),336(3.89)	94-0939-66
$C_{10}H_{11}N_2O_3P$			
1,3,2-Diazaphosphorin-4(1H)-one, 2,3-dihydro-6-methyl-2-phenoxy-, 2-oxide	EtOH	260(3.91)	78-2003-66
$C_{10}H_{11}N_3$			
Dipyrido[1,2-a:3',2'-d]imidazole, 6,7,8,9-tetrahydro-	EtOH	252(3.50),281s(4.05), 284(4.06),289s(3.98)	39-0080-66C
Dipyrido[1,2-a:3',4'-d]imidazole, 6,7,8,9-tetrahydro-	EtOH	253(3.58),261(3.58), 268(3.67),274(3.49)	39-0080-66C
3,5-Pyridinedicarbonitrile, 1,4-dihydro-1,2,6-trimethyl-	n.s.g.	218(4.38),250s(3.51), 348(3.72)	39-1083-66B
3,5-Pyridinedicarbonitrile, 1,4-dihydro-2,4,6-trimethyl-	n.s.g.	216(4.37),338(3.79)	39-1083-66B
Quinazoline, 2-(dimethylamino)-	pH 0.0	239(4.45),255s(4.24), 289(3.76),365(3.66)	39-0234-66C
	pH 9.0	244(4.49),260s(4.32), 375(3.46)	39-0234-66C
Quinazoline, 4-methyl-2-(methylamino)-	pH 2.0	234(4.43),247(4.28), 281(3.56),346(3.45)	39-0234-66C
	pH 10.0	241(4.57),250s(4.38), 266s(3.90),358(3.42)	39-0234-66C
$C_{10}H_{11}N_3O$			
Benzaldehyde, 4,4-dimethylsemicarbazone	EtOH	290(4.23)	22-0153-66
Phthalazine, 1-(2-hydroxyethylamino)-	10% HCl	214(4.51),243(4.19), 261(4.17),304(3.86), 316(3.79)	4-0381-66
	EtOH	208(4.70),227s(4.19), 256(3.77),313(3.95)	4-0381-66
1,2,4-Triazol-5-one, 2,3-dimethyl-4-phenyl-	EtOH	252(3.69)	7-0190-66
v-Triazolo[3,4-a]pyridine, 3-butyryl-	MeOH	249s(3.58),255(3.64), 287(4.04),305(4.11)	24-2918-66
$C_{10}H_{11}N_3OS$			
3-Pyrazolin-5-one, 4-[(o-aminophenyl)thio]-3-methyl-	EtOH	227(4.34),249(3.95), 310(3.52)	94-0770-66

Compound	Solvent	λ_{max}(log ϵ)	Ref.
$C_{10}H_{11}N_3O_2$			
β-Alanine, 2-diazo-N-phenyl-, methyl ester	EtOH	246(4.25),289(3.54)	24-0475-66
5-Amino-1-(carboxymethyl)-2-methyl-imidazo[1,2-a]pyridinium hydroxide	H_2O	228(4.45),320(4.17)	4-0093-66
hydrochloride	H_2O	223(4.45),320(4.17)	4-0093-66
Pyrazoline, 3-methyl-1-(p-nitrophenyl)-	$CHCl_3$	407(4.40)	22-0610-66
Pyrazoline, 4-methyl-1-(p-nitrophenyl)-	$CHCl_3$	393(4.38)	22-0610-66
Pyrazoline, 5-methyl-1-(p-nitrophenyl)-	$CHCl_3$	396(4.36)	22-0610-66
7H-Pyrrolo[2,3-c]pyridazine, 6-(2-carboxyethyl)-3-methyl-	EtOH	226(4.47),278(3.91), 326(3.73)	1-2637-66
$C_{10}H_{11}N_3O_2S$			
1,2,3-Triazole, 5-methyl-1-(p-tolylsulfonyl)-	EtOH	231(4.30)	44-1587-66
$C_{10}H_{11}N_3O_3$			
Quinazoline, 1,2-dihydro-4-methyl-2-(nitromethyl)-, 3-oxide	MeOH	234(4.06),295(3.94), 373(3.64)	39-1433-66C
$C_{10}H_{11}N_3O_4$			
Isoindoline-1,3-dione, 2,4-diamino-6,7-dimethoxy-	EtOH	230(4.24),267(4.13), 397(3.65)	44-1912-66
1,4-Phthalazinedione, 8-amino-2,3-dihydro-5,6-dimethoxy-	6N NH_3	322(3.87),355(3.91)	44-1912-66
$C_{10}H_{11}N_3S$			
1,3,4-Triazole, 1-methyl-5-(methylthio)-2-phenyl-	EtOH	254(4.06)	1-0057-66
1,3,4-Triazoline-5-thione, 1,4-dimethyl-2-phenyl-	EtOH	255(4.27),280s(3.91)	1-0057-66
$C_{10}H_{11}N_5$			
1,3,5-Triazine, 2,4-diamino-6-benzyl-	MeOH	258(3.55)	78-0157-66
$C_{10}H_{11}N_5O$			
Adenosine, 2',3',5'-trideoxy-2',3'-didehydro-	MeOH	259(4.17)	35-1549-66
as-Triazine, 3,5-diamino-6-(p-methoxyphenyl)-	EtOH-HCl	247(4.19),309(3.76)	4-0188-66
	EtOH-NaOH	252(4.10),326(3.95)	4-0188-66
$C_{10}H_{11}N_5O_2$			
Adenine, 9-(2-deoxy-3,5-epoxy-β-D-threopentofuranosyl)-	MeOH	259(4.18)	88-1343-66
Adenine, 9-(2,3-dideoxy-2-ene-β-D-glyceropentofuranosyl)-	MeOH	259(4.19)	88-1343-66
Adenosine, 2',3'-dideoxy-2',3'-didehydro-	MeOH	259.5(4.19)	35-1549-66
$C_{10}H_{11}N_5O_4$			
Adenine, 2',8-anhydro-8-hydroxy-β-D-arabinofuranosyl-	pH 1	260(4.03)	35-3165-66
	H_2O	260(4.04)	35-3165-66
	pH 14	260(4.03)	35-3165-66

Compound	Solvent	λ_{max}(log ε)	Ref.
$C_{10}H_{12}$			
Cyclopentadiene, 1-(2-cyclopenten-1-yl)-	EtOH	248(3.44)	89-0246-66
Cyclopentadiene, cyclopentylidene-	C_6H_{12}	268(4.20),276(4.30),284(4.24),293(3.90),366(2.55)	33-1278-66
	EtOH	276(4.25),282(4.22),363(2.58)	33-1278-66
1,3,5-Hexatriene, 3,4-divinyl-	EtOH	224(4.21),284(4.37)	88-2257-66
$C_{10}H_{12}BF_4NO_2S$			
p-Nitrostyryldimethylsulfonium fluoroborate, trans	EtOH	289(4.35)	35-5747-66
$C_{10}H_{12}ClNO_4$			
Benzene, 1-(3-chloropropoxy)-4-methoxy-2-nitro-	MeOH	357(3.36)	78-0093-66B
Benzene, 1-(3-chloropropoxy)-4-methoxy-3-nitro-	MeOH	357(3.36)	78-0093-66B
$C_{10}H_{12}ClN_3O_2$			
5-Amino-1-(carboxymethyl)-2-methylimidazo[1,2-a]pyridinium hydroxide, hydrochloride	H_2O	223(4.45),320(4.17)	4-0093-66
$C_{10}H_{12}Cl_2O$			
2,5-Cyclohexadien-1-one, 4-(dichloromethyl)-3,4,5-trimethyl-	EtOH	240.0(4.18)	39-0075-66B
$C_{10}H_{12}Cl_2O_2$			
Phenetole, 2,4-dichloro-5-ethoxy-	$C_2H_4Cl_2$-SO_3	510(4.0)	65-1722-66
$C_{10}H_{12}N_2$			
Indole, 5-(dimethylamino)-	EtOH	238(4.36)	23-0387-66
2-Pyrazoline, 3-methyl-1-phenyl-	EtOH	251(4.05)	78-2461-66
$C_{10}H_{12}N_2O$			
Biacetyl, phenylhydrazone, anti	$CHCl_3$	333(4.34)	22-2981-66
$C_{10}H_{12}N_2O_2$			
Acetamide, N,N'-m-phenylenebis-	MeOH	234(4.48),281s(3.11),292(3.04)	78-1175-66
Acetamide, N,N'-o-phenylenebis-	MeOH	239s(4.11)	78-1175-66
Acetamide, N,N'-p-phenylenebis-	MeOH	264(4.34)	78-1175-66
1H-Indazole, 5,6-dimethoxy-3-methyl-	EtOH	216(4.30),267(3.65),297(3.74),308(3.63)	78-3131-66
2,4-Pyrimidinedione, 1,3-diallyl-	EtOH	266(3.97)	44-0406-66
2(1H)-Pyrimidinone, 1-allyl-4-(allyloxy)-	EtOH	274(3.78)	44-0406-66
$C_{10}H_{12}N_2O_3S$			
Oxamide, N-(2,4-dimethoxyphenyl)-2-thio-	EtOH	233(4.06),252s(3.92),370(4.11)	44-1484-66
2'-[4-Thiothymidinene)-, 3'-deoxy-	EtOH	243(3.58),332(4.28)	44-0205-66
Thymine, 1-(2-deoxy-3,5-epoxy-β-D-threopentofuranosyl)-4-thio-	EtOH	243(3.57),332(4.26)	44-0205-66
$C_{10}H_{12}N_2O_4$			
2'-Thymidinene, 3'-deoxy-	H_2O	266(4.00)	44-0205-66

Compound	Solvent	$\lambda_{max}(\log \epsilon)$	Ref.
$C_{10}H_{12}N_2O_8$			
β-D-Glucopyranuronic acid, 1-deoxy-1-(1,2,3,4-tetrahydro-2,4-dioxo-1-pyrimidinyl)-	H_2O	260(4.09)	94-1354-66
$C_{10}H_{12}N_4O_2$			
Lumazine, 8-ethyl-6,7-dimethyl-	pH -2.7	246(4.05),358(4.19)	24-3503-66
	pH 5.8	258(4.17),277(4.08), 405(4.10)	24-3503-66
	pH 13	244(4.26),268s(3.94), 315(4.30),367(3.76)	24-3503-66
Lumazine, 3,6,7,8-tetramethyl-	pH -1.9	247(4.08),285s(3.06), 359(4.18)	24-3503-66
	pH 7	260(4.32),275s(3.90), 404(4.08)	24-3503-66
	pH 13	240(4.33),311(4.37), 362(3.86)	24-3503-66
Pyrazino[2,3-d]pyridazine, 5,8-diethoxy-	EtOH	206(3.94),257(4.30)	4-0512-66
Pyrazolo[3,2-c]-as-triazine-8-carboxylic acid, 3,4-dimethyl-, ethyl ester	EtOH	230(4.24),273(3.83), 322(3.89)	39-1127-66C
$C_{10}H_{12}N_4O_3$			
Lumazine, 8-(2-hydroxyethyl)-6,7-dimethyl-	pH -2.7	246(4.08),361(4.16)	24-3503-66
	pH 3	275(4.08),306(4.15)	24-3503-66
	pH 5.8	258(4.18),277(4.02), 408(4.08)	24-3503-66
	pH 13	230(4.26),280(4.01), 313(4.13),368s(3.41)	24-3503-66
Xanthine, 7-(tetrahydropyran-2-yl)-	pH 1	262(3.93)	44-2685-66
	pH 11	292.5(3.89)	44-2685-66
	EtOH	272.5(3.91)	44-2685-66
$C_{10}H_{12}N_4O_3S$			
6-Purinethiol, 2-hydroxy-3(9)-(2-carbethoxyethyl)-	pH 1	223(3.95),252(3.79), 338(4.38)	73-1824-66
	pH 13	227(4.27),256(3.92), 344(4.35)	73-1824-66
2-Purinol, 6-(4-carboxybutyl)thio-	pH 1	268(3.90),315(3.97)	73-1824-66
	pH 13	223(4.25),320(4.07)	73-1824-66
8-Purinol, 6-(4-carboxybutyl)thio-	pH 1	230(4.06),298(4.26)	73-1824-66
	pH 13	229(4.20),305(4.26)	73-1824-66
$C_{10}H_{12}N_4O_4$			
Isonebularine	N HCl	258(3.84)	88-0643-66
	H_2O	264(3.87)	88-0643-66
	pH 9.18	265(3.86)	88-0643-66
Nebularine	N HCl	262.8(3.77)	88-0643-66
	H_2O	262.5(3.85)	88-0643-66
	pH 9.18	263.0(3.85)	88-0643-66
Thymine dimer	pH 7	none	57-0379-66C
	pH 13	237(4.1)	57-0379-66C
$C_{10}H_{12}N_4O_4S$			
2,8-Purinediol, 6-(4-carboxybutyl)thio-	pH 1	223(4.34),338(4.07)	73-1824-66
	pH 3	229(4.12),330(4.15)	73-1824-66
$C_{10}H_{12}N_4O_5$			
Acetoin, 2,4-dinitrophenylhydrazone	$CHCl_3$	357(4.52)	96-0297-66

Compound	Solvent	$\lambda_{max}(\log \epsilon)$	Ref.
Hypoxanthine, 7-α-D-arabino-furanosyl-	pH 1	250(3.95)	44-1413-66
	pH 7	256(3.92)	44-1413-66
	pH 13	262(3.96)	44-1413-66
Hypoxanthine, 7-β-D-ribofuranosyl-	pH 1	249(3.95)	44-1413-66
	pH 7	251(3.93)	44-1413-66
	pH 13	261(3.93)	44-1413-66
Inosine	2N HCl	250(4.05)	50-1129-66A
	pH 7.0	249(4.09)	50-1129-66A
3-Isoinosine	pH 1	254(4.04)	35-0185-66
	pH 7	265(4.12)	35-0185-66
	pH 13	270(4.04)	35-0185-66
$C_{10}H_{12}N_4O_6$			
Aniline, N,N-diethyl-2,4,6-trinitro-	EtOH	340s(3.86),385(4.01)	65-1034-66
$C_{10}H_{12}N_6S$			
4(3H)-Pyrimidinethione, 6-amino-5-[(4,6-dimethyl-2-pyrimidinyl)-amino]-, dihydrochloride	pH 1	235(4.25),287(3.96), 300(3.97),385(4.22)	44-1417-66
	pH 7	242(4.19),272(4.02), 292s(3.95),372(4.15)	44-1417-66
	pH 13	244(4.27),333(4.08)	44-1417-66
$C_{10}H_{12}O$			
6(1H)-Azulenone, 2,3,5,7-tetrahydro-	EtOH	253s(--),314(--)	44-1042-66
6(1H)-Azulenone, 2,3,9,10-tetrahydro-	EtOH	238(--)	44-1042-66
Benzofuran, 2,3-dihydro-	EtOH	282(3.49)	22-0228-66
2,2-dimethyl-	EtOH	226s(3.70),251s(2.95), 272s(3.33),280(3.46), 283s(3.45),287s(3.38)	1-1561-66
Benzofuran, 2,3-dihydro-3,3-dimethyl-	EtOH	224(3.71),279(3.50), 285(3.43)	1-1561-66
1-Benzoxepin, 2,3,4,5-tetrahydro-	EtOH	267(2.83)	22-0228-66
8-Decene-4,6-diyn-1-ol, cis	ether	226(3.11),238(3.70), 251(4.07),265(4.26), 281(4.16)	24-0586-66
1H,4H-3a,6a-Ethenopentalen-1-one, 2,3,5,6-tetrahydro-	EtOH	307(2.00)	35-1330-66
7-Indanol, 4-methyl-	H_2O	273(3.00)	22-3328-66
	base	290(3.41)	22-3328-66
4,7-Methanoinden-1-one, 3a,4,5,6,7,7a-hexahydro-, endo	EtOH	228(3.90),323(1.58)	35-1059-66
exo	EtOH	226(3.92),321(1.79)	35-1059-66
Nezukone (or 4-isopropyltropone)	EtOH	230(4.47),313(4.17)	88-5875-66
$C_{10}H_{12}OS$			
Ether, (methylthiomethyl) 1-phenylvinyl	EtOH	232(3.98),278(3.00)	25-1963-66
$C_{10}H_{12}OS_2$			
1,3-Benzoxathian, 8-(methylthio-methyl)-	MeOH	227(3.36)	35-5855-66
	MeOH-KOH	227(3.36)	35-5855-66
$C_{10}H_{12}O_2$			
3H-2-Benzopyran-3-one, 5,6,7,8-tetra-hydro-1-methyl-	MeOH	211(3.86),311(3.87)	5-0074-66I
3-Carene-2,5-dione	pentane	240(4.135)	56-0463-66
1,5(6H)-Indandione, 7,7a-dihydro-7a-methyl-	EtOH	238(4.03)	88-6495-66
Mesitoic acid	EtOH	242s(3.5),275s(2.5)	28C-0369-66A

Compound	Solvent	$\lambda_{max}(\log \epsilon)$	Ref.
2(3H)-Naphthalenone, 4,4a,5,6-tetra-hydro-7-hydroxy-	EtOH	318(4.40)	88-0927-66
	NaOH	382(4.92)	88-0927-66
2,6-Octadien-4-ynal, 8-hydroxy-2,6-dimethyl-	pet ether	286s(4.21),298(4.32), 315(4.31)	33-0369-66
$C_{10}H_{12}O_2S$			
Benzo[c]thiophene-2-carboxylic acid, tetrahydro-, methyl ester	EtOH	263(4.0)	88-1953-66
$C_{10}H_{12}O_3$			
Phthalan, 4,5,6,7-tetrahydro-7,7-dimethyl-1,6-dioxo-	EtOH	215(3.98)	78-1659-66
Propionic acid, 3-(p-methoxyphenyl)-	50% dioxan	273(3.19)	71-0550-66A
$C_{10}H_{12}O_4$			
Benzoic acid, 2,6-dimethoxy-, methyl ester	EtOH	280(3.37)	12-0275-66
Butanoic acid, 3-(2-furyl)-2-methyl-	H_2O	226(3.46),276(4.19)	12-0683-66
	EtOH	224(3.42),270(4.16)	12-0683-66
2-Cyclopentenylidenepropionic acid, 2-hydroxy-4-oxo-, ethyl ester	EtOH	275.5(4.25)	35-3131-66
2-Furanpropionic acid, 5-methyl-β-oxo-, ethyl ester	EtOH	222(3.40),287(3.11)	39-0976-66C
o-Orsellinic acid, ethyl ester	EtOH	215(4.38),263(4.18), 299(3.75)	12-1265-66
$C_{10}H_{12}O_5$			
Benzoic acid, 3,4,5-trimethoxy-	EtOH	262(4.1),297s(3.5)	28C-0584-66A
1-Cyclopentene-1-carboxylic acid, 3,4-dihydroxy-5-(2-hydroxyethyl)-, δ-lactone, 3-acetate	n.s.g.	218(4.0)	88-1245-66
2-Octenoic acid, 6,7-epoxy-4,5-di-hydroxy-, δ-lactone, acetate	EtOH	204(4.08)	88-1969-66
$C_{10}H_{12}O_6$			
2,4-Hexadienedioic acid, 2-hydroxy-, dimethyl ester, acetate	ether	270(4.48)	24-2522-66
$C_{10}H_{12}S$			
1-Benzothiepin, 2,3,4,5-tetrahydro-	EtOH	262(3.82)	22-0228-66
Sulfide, cinnamyl methyl	EtOH	254(4.24),284s(3.37), 294(3.17)	44-1694-66
Thiachroman, 2-methyl-	EtOH	256(4.10),286s(3.07)	78-0007-66
Thiachroman, 3-methyl-	EtOH	257(4.07),296s(2.93)	78-0007-66
Thiachroman, 4-methyl-	EtOH	256(4.08),290s(3.00)	78-0007-66
$C_{10}H_{13}^+$			
Ethylmethylphenylcarbonium ion	FSO_3H-SbF_5	321(4.01),397(3.12)	35-1488-66
$C_{10}H_{13}BF_4S$			
Dimethylstyrylsulfonium fluoroborate, cis	EtOH	261(4.08)	35-5747-66
$C_{10}H_{13}ClN_4$			
Dipyridazino[2,3-a:4,3-d]pyrrole, 8-chloro-4,4a,5,5a,6,7-hexa-hydro-3-methyl-	EtOH	302(4.17)	1-2637-66
$C_{10}H_{13}ClO_2$			
Anisole, 4-(3-chloropropoxy)-	MeOH	225(3.95),287(3.35)	78-0093-66B

Compound	Solvent	λ_{max}(log ε)	Ref.
$C_{10}H_{13}IN_2O_3$			
2(1H)-Pyrimidinone, 5-iodo-4-methoxy-1-(tetrahydropyran-2-yl)-	MeOH	298(3.76)	4-0005-66
$C_{10}H_{13}N$			
1-Indanamine, 2-methyl-	EtOH	261(2.56),267(2.71), 273(2.63)	94-1408-66
Styrylamine, N,N-dimethyl-	EtOH	299(4.18)	35-5747-66
$C_{10}H_{13}NO$			
Azepino[3,2,1-hi]indole, 1,2,4,5,6,7-hexahydro-7-oxo-	EtOH	242(4.10),315(2.92), 376(3.48)	35-4061-66
Mesitamide	EtOH	240s(3.4),275s(2.4)	28C-0369-66A
$C_{10}H_{13}NO_2$			
1H-Azepine-2,5-dione, 6-isopropyl-3-methyl-	EtOH	228(3.16),288(2.47)	78-1201-66
1H-Azepine-2,5-dione, 3,4,6,7-tetramethyl-	EtOH	233(3.07),298(2.42)	78-1201-66
Benzoic acid, p-(dimethylamino)-, methyl ester	EtOH	257(4.18),305(3.38)	65-1198-66
Benzoquinone, tert-butyl-, oxime	EtOH-acid	303(4.28)	12-0617-66
	EtOH-base	397(4.46)	12-0617-66
Pyridine, 3,5-diacetyl-1,4-dihydro-1-methyl-	n.s.g.	233(4.07),274(3.75), 406(3.84)	39-1083-66B
Pyridine, 3,5-diacetyl-1,4-dihydro-4-methyl-	n.s.g.	234(4.10),263(3.68), 378(3.89)	39-1083-66B
2(1H)-Pyridone, 1-cyclopentyloxy-	EtOH	229(3.78),302(3.74)	78-0025-66
$C_{10}H_{13}NO_3$			
Ether, butyl p-nitrophenyl	EtOH	228(3.88),309(4.06)	94-0939-66
Glyoxal, (o-aminophenyl)-, 1-(dimethyl acetal)	MeOH	232(4.2),260(3.8), 267(3.8),378(3.7)	49-0409-66
Phenol, 3-tert-butyl-2-nitro-	C_6H_{12}	271(3.40),344(2.93)	23-0961-66
	EtOH	277(3.44),352(2.97)	23-0961-66
2(1H)-Pyridone, 3-butyryl-4-hydroxy-6-methyl-	EtOH	230(4.03),267(3.43), 325(4.09)	70-0848-66
$C_{10}H_{13}NO_3S_2$			
Phenol, 0,2-bis(methylthiomethyl)-p-nitro-	MeOH	309(4.01)	35-5855-66
	MeOH-KOH	309(4.01)	35-5855-66
Phenol, 0,6-bis(methylthiomethyl)-o-nitro-	MeOH	250(3.67)	35-5855-66
	MeOH-KOH	250(3.67)	35-5855-66
Phenol, 2,6-bis(methylthiomethyl)-p-nitro-	MeOH	317(3.86)	35-5855-66
	MeOH-KOH	422(4.32)	35-5855-66
$C_{10}H_{13}NO_4$			
Benzamide, 3,4,5-trimethoxy-	EtOH	262(4.0),292s(3.5)	28C-0584-66A
Benzene, 1-isopropoxy-4-methoxy-2-nitro-	MeOH	357(3.39)	78-0093-66B
Benzene, 1-isopropoxy-4-methoxy-3-nitro-	MeOH	357(3.39)	78-0093-66B
Benzene, 2-methoxy-4-nitro-1-propoxy-	EtOH	240(4.02),338(3.97)	94-0939-66
Benzene, 4-methoxy-2-nitro-1-propoxy-	MeOH	357(3.39)	78-0093-66B
Benzene, 4-methoxy-3-nitro-1-propoxy-	MeOH	357(3.39)	78-0093-66B
Benzoic acid, 4-amino-2,6-dimethoxy-, methyl ester	EtOH	213s(4.35),272(3.97)	12-0275-66

Compound	Solvent	λ_{max}(log ε)	Ref.
Muconic acid, α-aziridino-, dimethyl ester	ether	311(4.33)	24-0450-66
isomer	ether	324(4.16)	24-0450-66
Phenol, 2-butoxy-5-nitro-	pH 13	227(4.01),264(4.10), 324(3.76),418(3.72)	94-0939-66
	EtOH	244(4.01),302(3.81), 344(3.91)	94-0939-66
$C_{10}H_{13}NO_5$			
Anisole, 2-(2-hydroxypropoxy)-5-nitro-	EtOH	238(4.09),335(4.01)	94-0939-66
Nicotinic acid, 1-ethyl-1,2-dihydro-4,6-dihydroxy-2-oxo-, ethyl ester	EtOH	306(4.5)	78-0455-66
Pyrrole-2,5-dicarboxylic acid, 1-ethyl-4-hydroxy-, dimethyl ester	ether	274(4.30),300(3.99)	24-1558-66
$C_{10}H_{13}N_3O_4$			
Aniline, N,N-diethyl-2,4-dinitro-	EtOH	225(4.09),375(4.22)	65-1034-66
$C_{10}H_{13}N_3O_7$			
β-D-Glucopyranuronamide, 1-deoxy-1-(1,2,3,4-tetrahydro-2,4-dioxo-1-pyrimidinyl)-	H_2O	258.5(4.02)	94-1354-66
$C_{10}H_{13}N_4O_8P$			
Inosine-5'-monophosphate	2N HCl	252(4.06)	50-1129-66A
	pH 7.0	249(4.09)	50-1129-66A
$C_{10}H_{13}N_5$			
Pteridine, 2-(butylamino)-	pH 1.8	235(4.12),283s(3.74), 304(3.92)	39-0226-66C
	pH 6.0	231(4.36),275(3.98), 392(3.81)	39-0226-66C
Pteridine, 2-(sec-butylamino)-	pH 1.0	235(4.14),287s(3.79), 304(3.93)	39-0226-66C
	pH 7.0	232(4.36),276(4.00), 396(3.81)	39-0226-66C
Pteridine, 2-(tert-butylamino)-	pH 1.0	236(4.34),286s(3.78), 304(3.92)	39-0226-66C
	pH 7.0	232(4.35),276(4.02), 394(3.80)	39-0226-66C
Pteridine, 2-(diethylamino)-	pH 1.0	212(3.85),238(4.22), 287s(3.81),304(3.92)	39-0226-66C
	pH 7.0	238(4.34),283(4.05), 416(3.81)	39-0226-66C
$C_{10}H_{13}N_5O$			
Adenine, 9-(tetrahydropyran-2-yl)-	pH 7	261(4.15)	39-0692-66C
Adenosine, 2',3',5'-trideoxy-	MeOH	259.5(4.19)	69-0224-66
2-Indolinone, 3-(dihydrazinomethylene)-1-methyl-	EtOH	257(4.18),312(4.14)	95-1152-66
7(8H)-Pteridinone, 2-(dimethylamino)-6,8-dimethyl-	pH -0.89	236(4.37),268(4.36), 273s(4.34),345(3.79)	24-3022-66
	pH 6	248(4.22),297(3.81), 364(4.19)	24-3022-66
7(8H)-Pteridinone, 2-(dimethylamino)-8-ethyl-	pH -0.89	236(4.40),272(4.29), 345(3.78)	24-0536-66
	pH 5.0	224(4.29),244(4.17), 298(3.75),368(4.20)	24-0536-66

Compound	Solvent	λ_{max}(log ε)	Ref.
7(8H)-Pteridinone, 2-(dimethyl-amino)-8-ethyl- (cont.)	MeOH	223(4.14),242s(3.95), 300(3.55),364(4.07)	24-0536-66
Purine, 6-(4-hydroxy-3-methyl-but-2-enylamino)-, trans	pH 1	207(4.16),275(4.17)	39-0921-66C
	pH 7.2	212(4.23),270(4.21)	39-0921-66C
	pH 13	220(4.20),276(4.17)	39-0921-66C
$C_{10}H_{13}N_5O_2$			
Adenosine, 2',3'-dideoxy-	MeOH	259.5(4.16)	35-1549-66
	MeOH	259.5(4.16)	69-0224-66
Adenosine, 2',5'-dideoxy-	MeOH	259.5(4.16)	69-0224-66
7(8H)-Pteridinone, 2-(dimethylamino)-4-methoxy-8-methyl-	pH 1-13	240(4.2),292(3.7), 363(4.3)	33-1815-66
$C_{10}H_{13}N_5O_3$			
Adenine, 9-(2-deoxy-β-D-threo-pentofuranosyl)-	pH 1	258(4.15)	44-3263-66
	pH 7	259(4.18)	44-3263-66
	pH 13	259(4.20)	44-3263-66
Adenine, 9-(3-deoxy-β-D-threo-pentofuranosyl)-	pH 1	258(4.16)	44-3263-66
	pH 7	260(4.20)	44-3263-66
	pH 13	260(4.18)	44-3263-66
$C_{10}H_{13}N_5O_5$			
Guanine, 7-β-D-ribofuranosyl-	pH 1	252(3.94),275s(--)	94-1377-66
	H_2O	287(3.85)	94-1377-66
	pH 13	283.5(3.79)	94-1377-66
Purine, 6-(hydroxylamino)-9-β-D-ribo-furanosyl-	pH 1.4	265(4.25)	87-0143-66
	pH 6.7	265(4.16)	87-0143-66
	pH 12.2	252(--),310s(--)	87-0143-66
$C_{10}H_{13}N_7O_4$			
Aspartic acid, N-(5-amino-3-ethyl-3H-v-triazolo[4,5-d]pyrimidin-7-yl)-	pH 1	256(4.20),273(4.12)	87-0417-66
	pH 7	230s(4.22),265s(--), 291(4.11)	87-0417-66
	pH 13	230s(4.20),265s(--), 291(4.11)	87-0417-66
$C_{10}H_{13}N_7O_5$			
β-D-Glucopyranuronamide, 1-deoxy-1-(7-amino-3H-v-triazolo[4,5-d]-pyrimidin-3-yl)-	H_2O	280(4.08)	94-1360-66
$C_{10}H_{13}O_5P$			
Phosphonic acid, p-anisoyl-, dimethyl ester	C_6H_{12}	295(4.01)	44-1304-66
$C_{10}H_{14}$			
1,3,6-Cyclooctatriene, 5,8-dimethyl-	EtOH	265(2.30)	44-3958-66
Fulvene, 6,6-diethyl-	C_6H_{12}	269(4.27),274(4.26), 360(2.55)	33-1278-66
	EtOH	270(4.29),274(4.28), 359(2.56)	33-1278-66
Spiro[2.4]hepta-4,6-diene, 1,1,2-trimethyl-	pentane	234s(4.15),262(3.59)	44-3296-66
Tropilidene, 1,7,7-trimethyl-	EtOH	278(3.64)	35-2494-66
Tropilidene, 3,7,7-trimethyl-	EtOH	268(3.61)	35-2494-66
$C_{10}H_{14}BN$			
Isoquinoline, 3,4-dihydro-1-methyl-, borane adduct	90% EtOH	272(4.10)	94-1382-66

Compound	Solvent	$\lambda_{max}(\log \epsilon)$	Ref.
$C_{10}H_{14}BrClO$			
Camphor, 3-bromo-9-chloro-	C_6H_{12}	309(2.02),321(1.97), 336(1.59)	39-0885-66B
	EtOH	292(1.52)	39-0885-66B
$C_{10}H_{14}BrN_3O$			
Pyrimidine, 5-bromo-4,6-dimethyl-2-morpholino-	EtOH	255(4.03),314(3.08)	78-2401-66
$C_{10}H_{14}ClN_3O$			
Pyrimidine, 5-chloro-4,6-dimethyl-2-morpholino-	EtOH	252(4.23),313(3.11)	78-2401-66
$C_{10}H_{14}Cl_2N_6S$			
4(3H)-Pyrimidinethione, 6-amino-5-[(4,6-dimethyl-2-pyrimidinyl)-amino]-, dihydrochloride	pH 1	235(4.25),287(3.96), 300(3.97),385(4.22)	44-1417-66
	pH 7	242(4.19),272(4.02), 292s(3.95),372(4.15)	44-1417-66
	pH 13	244(4.27),333(4.08)	44-1417-66
$C_{10}H_{14}CoS_4$			
Cobalt, bis(2,4-pentane-dithionato)-	benzene	360s(4.18),415s(3.70), 462(3.51),572(3.60)	50-0522-66B
	MeOH	224(4.48),249(4.30)	50-0522-66B
	CCl_4	273(4.45),341(4.15)	50-0522-66B
$C_{10}H_{14}N_2$			
Benzopyrazole, 3-cyclopropyl-4,5,6,7-tetrahydro-	EtOH	228.5(3.82)	22-2971-66
Malononitrile, (1-ethylpentylidene)-	EtOH	236(4.09)	44-2784-66
Malononitrile, (1-isopropyl-2-methylpropylidene)-	EtOH	237.5(4.15)	44-2784-66
Malononitrile, (1-methylhexylidene)-	EtOH	235(3.85)	44-2784-66
$C_{10}H_{14}N_2O$			
Butyranilide, 2'-amino-	MeOH	234s(3.95),291(3.47)	78-1175-66
$C_{10}H_{14}N_2OS_2$			
Rhodanine, 3-methyl-5-(piperidino-methylene)-	dioxan	280(4.22),391(4.51)	24-0307-66
$C_{10}H_{14}N_2O_2$			
Pyridine, 2-(N-carbethoxy-N-ethyl-amino)-	EtOH	234(3.90),273(3.62)	32-1020-66
$C_{10}H_{14}N_2O_2S$			
Thiourea, (3,4-dimethoxybenzyl)-	EtOH	207(4.35),241(4.17), 278(3.46)	1-1778-66
$C_{10}H_{14}N_2O_3$			
2(1H)-Pyrimidinone, 4-methoxy-1-(tetrahydropyran-2-yl)-	MeOH	247(3.89)	4-0005-66
Thymine, 1-(tetrahydropyran-2-yl)-	pH 1	265(3.98)	4-0005-66
	pH 7	265(3.93)	4-0005-66
	pH 11	265(3.89),277(3.83)	4-0005-66
$C_{10}H_{14}N_2O_4$			
3,4-Pyrazoledicarboxylic acid, 1-methyl-, diethyl ester	EtOH	224(3.98)	44-2491-66

Compound	Solvent	λ_{max}(log ε)	Ref.
4,5-Pyrazoledicarboxylic acid, 1-methyl-, diethyl ester	EtOH	234(3.91)	44-2491-66
$C_{10}H_{14}N_2O_5$			
2(1H)-Pyrimidinone, 1-(3-deoxy-α-D-ribofuranosyl)-4-methoxy-	pH 1	210(4.06),276(3.83)	44-1163-66
	H_2O	204(4.25),275(3.84)	44-1163-66
	pH 13	275(3.85)	44-1163-66
2(1H)-Pyrimidinone, 1-(3-deoxy-β-D-ribofuranosyl)-4-methoxy-	pH 1	275(3.82)	44-1163-66
	H_2O	208(4.26),273(3.83)	44-1163-66
	pH 13	274(3.82)	44-1163-66
Thymidine	pH 7	210(4.0),267(4.0)	57-0379-66C
	pH 13	267(3.9)	57-0379-66C
Thymine, 1-(3-deoxy-α-D-ribofuranosyl)-	pH 1	269(4.00)	44-1163-66
	pH 7	269(4.00)	44-1163-66
	pH 13	268(3.88)	44-1163-66
Thymine, 1-(3-deoxy-β-D-ribofuranosyl)-	pH 1	269(3.98)	44-1163-66
	pH 7	269(3.98)	44-1163-66
	pH 13	268(3.85)	44-1163-66
$C_{10}H_{14}N_2O_5S$			
Thymine, 1-(4-thio-α-D-ribofuranosyl)-	pH 2	270.5(4.08)	44-0813-66
	pH 12	270.5(4.05)	44-0813-66
Thymine, 1-(4-thio-β-D-ribofuranosyl)-	pH 2	271(4.02)	44-0813-66
	pH 12	271(3.93)	44-0813-66
Thymine, 1-(5-thio-D-xylopyranosyl)-	pH 2	269(4.05)	44-0813-66
	pH 12	269(3.95)	44-0813-66
$C_{10}H_{14}N_2O_6$			
Pyrazine, 2-(β-D-glucopyranosyloxy)-	pH 7.0	204(4.02),273(3.72), 285(3.67)	24-0542-66
	MeOH	206(4.04),274(3.75), 290s(3.56)	24-0542-66
4-Pyridazinone, 1-(1-β-D-glucosyl)-	2N HCl	263(4.08)	83-0883-66
	H_2O	274(4.24)	83-0883-66
	EtOH	272(4.26)	83-0883-66
Thymine, 1-(β-D-arabinofuranosyl)-	pH 1	268(3.99)	44-1289-66
	pH 7	268(4.00)	44-1289-66
	pH 13	269(3.91)	44-1289-66
Uracil, 3-(4-deoxy-β-D-xylohexosyl)-	pH 7	260(4.00)	39-1549-66C
$C_{10}H_{14}N_4O_2S$			
Theophylline, 2'-hydroxypropyl-6-thio-	EtOH	204(3.85),217(3.88), 267(3.38),335(4.35), 345(4.35)	22-3934-66
$C_{10}H_{14}N_4O_3$			
Lumazine, 7,8-dihydro-8-(2-hydroxyethyl)-6,7-dimethyl-	pH 5	277(4.16),310(3.82)	24-3503-66
	pH 10	231(4.32),282(4.09), 315(3.81)	24-3503-66
Pyrazole-4-carboxylic acid, 3-[(1-methylacetonyl)azo]-, ethyl ester	EtOH	234s(3.84),314(4.25)	39-1127-66C
Theophylline, 7-hydroxy-8-isopropyl-	pH 1	274(4.03)	5-0142-66A
	pH 13	228(4.22),255s(3.90), 280(3.83)	5-0142-66A
$C_{10}H_{14}N_4O_4$			
Uracil, 6-(cyclohexylamino)-5-nitro-	pH 3	231(4.33),325(4.07)	24-3022-66
	pH 7	282s(3.46),334(4.13)	24-3022-66

Compound	Solvent	$\lambda_{max}(\log \epsilon)$	Ref.
$C_{10}H_{14}N_4O_6$			
β-D-Glucopyranuronamide, 1-deoxy-1-(4-amino-1,2-dihydro-2-oxo-1-pyrimidinyl)-	pH 1.2	276.5(4.08)	94-1354-66
	H_2O	239(3.90),269(3.92)	94-1354-66
	pH 12.6	237(3.89),269(3.91)	94-1354-66
$C_{10}H_{14}N_4S$			
Purine, 2-ethyl-6-(propylthio)-	pH 1	228(4.00),271(3.72), 311(4.26)	44-1417-66
	pH 7	225s(4.07),291(4.19)	44-1417-66
	pH 13	225(4.21),296(4.14)	44-1417-66
$C_{10}H_{14}N_5O_7P$			
Adenosine-5'-monophosphate	2N HCl	257.5(4.05)	50-1129-66A
	pH 7.0	259(4.16)	50-1129-66A
$C_{10}H_{14}N_6O_4$			
Purine, 2,6-diamino-9-β-D-ribofuranosyl-	pH 1	253(3.06),293(3.99)	94-1377-66
	pH 13	257(--),281(3.99)	94-1377-66
$C_{10}H_{14}N_6O_5$			
Purine, 2-amino-6-(hydroxyamino)-9-β-D-ribofuranosyl- (as hydrate)	N HCl	258(3.89),298(3.90)	69-3057-66
	pH 6.7	262s(3.89),280(3.99)	69-3057-66
	N NaOH	256(3.90),267s(3.88), 300(3.37)	69-3057-66
$C_{10}H_{14}N_{10}$			
Glyoxal, dipyrazinylhydrazone, hydrazine adduct	EtOH	241(4.00),318(4.69)	39-2038-66C
$C_{10}H_{14}NiO_2S_2$			
Nickel, bis(4-thiolopent-3-en-2-one)-	$CHCl_3$	490s(3.20),636(2.08)	12-1401-66
$C_{10}H_{14}NiO_4S_2$			
Nickel, bis(methyl thioloacetoacetate)-	$CHCl_3$	500(2.60),675(1.70)	12-1401-66
$C_{14}H_{10}O$			
Δ^3-Caren-10-al	MeOH	227(3.94),296(2.49)	78-0133-66
4-Caren-3-one	isooctane	260(3.74)	56-0463-66
Chroman, 5,6,7,8-tetrahydro-5-methylene-	EtOH	251(4.10)	44-2171-66
2,4-Cyclohexadien-1-one, 3,4,6,6-tetramethyl-	EtOH	316(3.36)	44-3736-66
Cyclopentanone, 2-cyclopentylidene-	EtOH	256.0(4.05)	39-0075-66B
1,4-Dioxaspiro[4.5]dec-7-ene, 7-vinyl-	EtOH	229(4.32)	65-0835-66
Dispiro[2.1.2.3]decan-4-one	isooctane	283(1.64)	22-0121-66
1H,4H-3a,6a-Ethanopentalen-1-one, tetrahydro-	EtOH	300(1.34)	35-1330-66
Ether, tert-butyl phenyl	EtOH	270(2.65)	22-0228-66
4-Nonen-6-ynal, 3-methyl-	EtOH	228(4.12),323s(3.0)	28C-1433-66A
3-Norcarene-3-carboxaldehyde, 7,7-dimethyl-	EtOH	263(4.07)	35-4113-66
Pericyclocamphanone	EtOH	282(1.48)	39-0885-66B
Phenol, 2-sec-butyl-	H_2O	270.5(3.30)	22-3328-66
	base	290.5(3.58)	22-3328-66
Phenol, 3-sec-butyl-	H_2O	270.5(3.20)	22-3328-66
	base	287.5(3.44)	22-3328-66
Phenol, 3-tert-butyl-	H_2O	272.5(3.22)	22-3328-66
	base	288.5(3.45)	22-3328-66

Compound	Solvent	λ max(log ε)	Ref.
Phenol, 4-sec-butyl-	H_2O	275.5(3.20)	22-3328-66
	base	293.5(3.40)	22-3328-66
Phenol, 3,5-diethyl-	H_2O	271(3.15)	22-3328-66
	base	289(3.45)	22-3328-66
Phenol, 4-ethyl-2,5-dimethyl-	H_2O	278.5(3.45)	22-3328-66
	base	296.5(3.58)	22-3328-66
Phenol, 5-isopropyl-3-methyl-	H_2O	272(3.11),277(3.07)	22-3328-66
	base	289(3.43)	22-3328-66
Piperitenone	EtOH	243(4.00),277(3.90)	44-2026-66
Spiro[2.6]nonan-4-one, 6-methylene-	isooctane	283(2.00),292(2.00), 299s(1.89),311(1.61), 325(1.11)	22-0121-66
Toluene, m-(α-hydroxyisopropyl)-	CCl_4	264(2.39),270(2.40)	78-0139-66
$C_{10}H_{14}OS$			
3,5-Heptadiyn-2-ol, 7-(ethylthio)-2-methyl-	EtOH	233(3.15),247(3.20), 261(3.08)	22-3024-66
Phenol, 3,6-dimethyl-2-(methylthiomethyl)-	MeOH	282(3.34)	35-5855-66
	MeOH-KOH	285(3.34),303(3.08)	35-5855-66
$C_{10}H_{14}OS_2$			
Phenol, 2,6-bis(methylthiomethyl)-	MeOH	283(3.46)	35-5855-66
	MeOH-KOH	307(3.74)	35-5855-66
$C_{10}H_{14}O_2$			
Benzene, 1-isopropoxy-4-methoxy-	MeOH	226(3.89),287(3.39)	78-0093-66B
Benzene, 1-methoxy-4-propoxy-	MeOH	225(3.96),287(3.43)	78-0093-66B
Bicyclo[3.3.1]non-3-en-2-one, 8-hydroxy-8-methyl-	EtOH	220(3.75)	39-0419-66C
isomer	EtOH	227(3.82)	39-0419-66C
Camphane-2,5-dione	EtOH	294(1.74)	39-0885-66B
1,3-Cyclohexadiene-1-carboxylic acid, 3,5,5-trimethyl-	EtOH	296(3.83)	44-2073-66
Fenchane-2,5-dione	EtOH	295(1.71)	39-0885-66B
3-Hepten-5-yn-1-ol, 4-methyl-, acetate	EtOH	232(4.08)	78-0443-66B
5(6H)-Indanone, 7,7a-dihydro-1-hydroxy-7a-methyl-	EtOH	240(4.13)	88-6495-66
Indone, 3a,4,5,6,7,7a-hexahydro-3-methoxy-	EtOH	240(4.04)	70-1167-66
2,4-Octadienoic acid, 4-hydroxy-2,3-dimethyl-, γ-lactone	n.s.g.	270(4.3)	54-0043-66
2,4,6-Octatrienal, 8-hydroxy-2,6-dimethyl-	pet ether	289(4.50),301(4.70), 314(4.67)	33-0369-66
Spiro[3.5]non-1-ene-2-carboxylic acid	EtOH	222(4.00)	22-3947-66
$C_{10}H_{14}O_3$			
1,3-Cyclohexanedione, 2-acetyl-5,5-dimethyl-	EtOH-NaOH	267(4.36),275s(4.30)	99-0230-66
2-Cyclohexene-1-carboxylic acid, 6-formyl-, ethyl ester	EtOH	285.0(1.57)	39-1152-66C
1-Cyclohexene-1-propionic acid, 2-methyl-6-oxo-	EtOH	242(4.11)	44-2192-66
$C_{10}H_{14}O_4$			
Aconic acid, γ-pentyl-	EtOH	217(4.01)	44-0616-66
$C_{10}H_{14}O_4Pd$			
Palladium, bis(2,4-pentanedionato)-	EtOH	209(4.63),225(4.55), 250(4.21),328(4.10)	30-0653-66*

Compound	Solvent	$\lambda_{max}(\log \epsilon)$	Ref.
$C_{10}H_{14}O_5$			
2-Hexenedioic acid, 4,5-epoxy-, diethyl ester	EtOH	213(4.16)	70-0646-66
Retronecic acid lactone	n.s.g.	222(4.04)	12-2127-66
$C_{10}H_{14}O_6$			
Muconic acid, α,α'-dimethoxy-, methyl ester, cis-cis	EtOH	315(4.29)	33-0168-66
cis-trans	EtOH	305(4.32)	33-0168-66
trans-trans	EtOH	305(4.38)	33-0168-66
$C_{10}H_{14}S$			
Sulfide, tert-butyl phenyl	EtOH	228(4.08),268(3.14)	22-0228-66
Sulfide, mesityl methyl	EtOH	223(4.03),268(3.36)	22-0228-66
$C_{10}H_{15}^+$			
Tricyclopropylcarbonium ion	FSO_3H-SbF_5	270(4.35)	35-1488-66
$C_{10}H_{15}BrO$			
Camphor, 3-bromo-, endo	C_6H_{12}	309(2.01)	39-0885-66B
	EtOH	307(2.03)	39-0885-66B
Camphor, 5-bromo-	EtOH	294(1.48)	39-0885-66B
2-Decalone, 3-bromo-, axial, trans	isooctane	310.5(2.02)	35-0334-66
	MeOH	307(2.02)	35-0334-66
equatorial	isooctane	286.8(1.45)	35-0334-66
	MeOH	283(1.50)	35-0334-66
$C_{10}H_{15}BrO_2$			
2-Cyclopenten-1-one, 3-bromo-2-methoxy-4,4,5,5-tetramethyl-	EtOH	261.0(4.07)	39-0075-66B
$C_{10}H_{15}ClO$			
Camphor, 9-chloro-	C_6H_{12}	292(1.52),301(1.49), 314(1.36)	39-0885-66B
	EtOH	291(1.61)	39-0885-66B
$C_{10}H_{15}FO_5$			
Fumaric acid, ethoxyfluoro-, diethyl ester	MeOH	236(3.92)	5-0134-66C
$C_{10}H_{15}Li_2N_3O_{11}P_2$			
Cytidine, N-methyl-, 5'-pyrophosphate, dilithium salt	pH 1	215(3.94),280(4.11)	39-0588-66C
$C_{10}H_{15}NO_3S_2$			
5-Thiazolidineacetic acid, N-butyl-4-oxo-2-thioxo-, methyl ester	EtOH	263(4.11),296(4.21)	95-0095-66
5-Thiazolidineacetic acid, N-ethyl-4-oxo-2-thioxo-, propyl ester	EtOH	261(4.12),296(4.21)	95-0095-66
5-Thiazolidineacetic acid, 4-oxo-N-propyl-2-thioxo-, ethyl ester	EtOH	262(4.10),295(4.19)	95-0095-66
$C_{10}H_{15}N_3$			
Pyrimidine, 1-allyl-1,2-dihydro-2-(propylimino)-	pH 8.0	232(4.24),320(3.69)	39-1165-66C
	pH 14.3	244(4.22),376(3.21)	39-1165-66C
Pyrimidine, 2-(allylimino)-1,2-dihydro-1-propyl-	pH 8.0	232(4.08),320(3.62)	39-1165-66C
	pH 14.3	246(4.19),368(3.42)	39-1165-66C
$C_{10}H_{15}N_3O$			
Apoverbenone, semicarbazone	EtOH	270(4.38)	22-1227-66

Compound	Solvent	λ_{max}(log ε)	Ref.
3-Hexen-5-ynal, 2,2,4-trimethyl-, semicarbazone	n.s.g.	226(4.24)	22-0734-66
Pyrimidine, 4,6-dimethyl-2-morpholino-	EtOH	247(4.15),300(3.04)	78-2401-66
$C_{10}H_{15}N_3O_2$			
Cytosine, 5-methyl-1-(tetra-hydropyran-2-yl)-	pH 1	284(4.20)	4-0005-66
	H_2O	275(4.00)	4-0005-66
	pH 11	235(3.66),275(4.02)	4-0005-66
2(1H)-Pyrimidinone, 4-(methylamino)-1-(tetrahydropyran-2-yl)-	pH 1	279(4.26)	4-0005-66
	H_2O	235(4.09),268(4.20)	4-0005-66
	pH 11	239(3.07),268(4.20)	4-0005-66
Triazene, 3-hydroxy-1-(p-methoxy-phenyl)-3-propyl-	1% acetone	294(4.20)	65-2117-66
Uracil, 5-(cyclohexylamino)-	pH 1	261(3.96)	87-0121-66
	pH 11	234(3.93),295(3.70)	87-0121-66
$C_{10}H_{15}N_3O_3$			
Uracil, 5-hydroxy-6-(piperidino-methyl)-	pH 1	283(3.79)	4-0115-66
	pH 11	241(3.79),310(3.75)	4-0115-66
$C_{10}H_{15}N_3O_4$			
Cytosine, 1-(3-deoxy-α-D-ribo-furanosyl)-5-methyl-	pH 1	210s(4.07),289(4.10)	44-1163-66
	pH 7	212(4.11),225s(3.94), 278(3.95)	44-1163-66
	pH 11	225s(3.97),278(3.94)	44-1163-66
Cytosine, 1-(3-deoxy-β-D-ribo-furanosyl)-5-methyl-	pH 1	214(4.04),288(4.10)	44-1163-66
	pH 7	212(4.10),225s(3.94), 278(3.93)	44-1163-66
	pH 11	225s(3.94),278(3.93)	44-1163-66
$C_{10}H_{15}N_3O_5$			
Uridine, 2'-deoxy-5-(methylamino)-	pH 1	204(3.97),266(3.97)	87-0366-66
	pH 5.8	203(3.99),237(3.86), 300(3.81)	87-0366-66
	pH 12	218(4.27),293(3.76)	87-0366-66
$C_{10}H_{15}N_5O$			
4(3H)-Pteridinone, 2-amino-7,8-dihydro-3,6,7,7-tetra-methyl-	pH 1.0	254(4.25),272s(3.76), 350(3.81)	24-3008-66
	pH 7.0	229(4.32),279(3.98), 316(3.71)	24-3008-66
$C_{10}H_{15}N_5O_2$			
Pyrimidine, 2-amino-4-(cyclohexyl-amino)-5-nitro-	MeOH	258(4.07),292(3.89), 358(4.31)	87-0121-66
Pyrimidine, 4-amino-2-(cyclohexyl-amino)-5-nitro-	MeOH	266(3.63),354(4.27)	87-0121-66
$C_{10}H_{15}N_5O_4$			
L-Histidine, glycylglycyl-, copper complex	H_2O	525(2.0)	37-0122-66
nickel complex	pH 11	425(2.15),480(1.79)	37-1072-66
Imidazo[1,2-c]pyrimidin-7(1H)-one, 3-ethoxy-2,3,5,6-tetrahydro-5-imino-1,6-dimethyl-	pH 2.0	234(4.21),262(3.97), 325(3.56)	24-2984-66
	pH 8.0	245(4.09),344(3.81)	24-2984-66
$C_{10}H_{15}O_3PSe$			
Ethyl phenyl phosphoroselenoate	MeOH	248(3.36),271s(2.85)	65-0254-66

Compound	Solvent	λ_{max}(log ε)	Ref.
$C_{10}H_{15}P$			
Phosphine, diethylphenyl-	EtOH	253(3.52),273(2.15)	77-0425-66
$C_{10}H_{16}BN$			
Borane, (diethylamino)- di-(1-propynyl)-	C_6H_{12}	220(4.30),245(4.23)	22-3850-66
	C_6H_{12}	220(4.30),245(4.23)	28C-0376-66A
$C_{10}H_{16}ClN_3$			
Pyrimidine, 2-(butylamino)-5-chloro- 4,6-dimethyl-	EtOH	246(4.36),317(3.42)	78-2401-66
$C_{10}H_{16}IN$			
Benzyltrimethylammonium iodide	CCl_4	284(--)	60-0286-66
	CH_2Cl_2	239(--)	60-0286-66
Ethylphenyldimethylammonium iodide	CCl_4	286(--)	60-0286-66
	CH_2Cl_2	242(--)	60-0286-66
$C_{10}H_{16}LiN_3O_{11}P_2$			
Cytidine, 3-methyl-, 5'-pyro- phosphate, lithium salt	pH 1	278(4.04)	39-0588-66C
	pH 7	277(4.04)	39-0588-66C
$C_{10}H_{16}NO_6P$			
Pyruvic acid, cyanophosphono-, triethyl ester	MeOH	258(3.90)	5-0134-66C
$C_{10}H_{16}N_2$			
3H-Azepine, 2-(diethylamino)-	n.s.g.	297(3.90)	78-0081-66
Pyrazole, 5-cyclobutyl-3-ethyl- 4-methyl-	EtOH	226(3.80)	22-2971-66
$C_{10}H_{16}N_2O$			
$\Delta^{2,3'}$-[Bipyrrolidin]-2'-one, 1,1'-dimethyl-	MeOH	218(4.0),296(4.6)	24-0724-66
hydrochloride	MeOH	204(4.1)	24-0724-66
2-Pyrrolidinone, 4-(α-ethylpropyli- dene)-3-methyl-5-imino-	EtOH	256(4.20)	22-3895-66
$C_{10}H_{16}N_2O_2S_2$			
5-Thiazolidineacetamide, N-ethyl- 4-oxo-3-propyl-2-thioxo-	EtOH	263(4.11),297(4.21)	95-0101-66
5-Thiazolidineacetamide, 3-ethyl- 4-oxo-N-propyl-2-thioxo-	EtOH	262(4.11),297(4.22)	95-0101-66
5-Thiazolidineacetamide, 4-oxo- 3-pentyl-2-thioxo-	EtOH	263(4.11),297(4.22)	95-0101-66
$C_{10}H_{16}N_2O_6$			
1H-Pyrimidin-2-one, 1-(β-D-arabino- furanosyl)-4-methoxy-5-methyl-	EtOH	243(2.94),284(3.82)	44-1289-66
$C_{10}H_{16}N_3O_8P$			
Cytidine, 3-methyl-, 2'(3')-phosphate	pH 1	276(4.06)	39-0588-66C
	pH 7	276(4.05)	39-0588-66C
Cytidine, N-methyl-, 5'-phosphate	pH 1	218(3.99),280(4.17)	39-0588-66C
$C_{10}H_{16}N_4O_6$			
Uracil, 4-(2,2-diethoxyethyl- amino)-5-nitro-	pH 2.0	228(4.49),250s(3.80), 322(4.10)	24-2984-66
	pH 7.0	230s(4.13),285s(3.58), 332(4.18)	24-2984-66

Compound	Solvent	$\lambda_{max}(\log \epsilon)$	Ref.
$C_{10}H_{16}O$			
Camphor	EtOH	291(1.51)	39-0885-66B
1,3-Cyclohexadiene, 5-(α-hydroxy-isopropyl)-1-methyl-	MeOH	264(3.74)	78-0139-66
1-Cyclohexene-1-carboxaldehyde, 3,5,5-trimethyl-	EtOH	232.0(4.13)	28C-0848-66A
Cyclononanone, 3-methylene-	isooctane	285(1.65),294(1.66), 303(1.62),313(1.46)	22-0121-66
Cyclooctylideneacetaldehyde	EtOH	238(4.08)	22-1400-66
3-Cyclopentene, 2-acetyl-1,2,3-tri-methyl-	C_6H_{12}	294(1.92),301(1.90), 322s(1.45)	22-0273-66
Epicamphor	EtOH	292(1.53)	39-0885-66B
4-Hexen-2-one, 3-isopropenyl-3-methyl-	MeOH	293(2.03)	22-0273-66
Nopinone	EtOH	236(3.90)	22-1227-66
3,7-Octadien-2-ol, 2-methyl-6-methylene-, trans	C_6H_{12}	225(4.30)	78-1929-66
3,7-Octadien-2-one, 3,4-dimethyl-, cis	C_6H_{12}	239(3.91),313(1.87)	22-0273-66
trans isomer	C_6H_{12}	240(3.93),308(1.90)	22-0273-66
4,7-Octadien-2-one, 3,4-dimethyl-	C_6H_{12}	289(2.27),297(2.27), 305(2.17),315(1.86)	22-0273-66
Spiro[2.7]decan-4-one	isooctane	285(1.30),294(1.23)	22-0121-66
Unidentified dihydrofuran from tiglaldehyde reduction	EtOH	216.5(3.56)	6-0179-66
$C_{10}H_{16}O_2$			
Alloocimene, diepoxide, trans-cis	heptane	205(4.18)	33-0617-66
trans-trans	heptane	205(4.18)	33-0617-66
1-Butanol, 3-(2-furyl)-2-methyl-	EtOH	214(4.01)	12-0683-66
Camphor, 5-hydroxy-	EtOH	292(1.51)	39-0885-66B
Cyclohexylideneacetic acid, ethyl ester	EtOH	222(4.19)	54-0929-66
Fenchol, 5-oxo-	EtOH	292(1.54)	39-0885-66B
3,5-Heptadien-1-ol, 4-methyl-, acetate, 3-cis-5-trans	EtOH	235.5(4.18)	78-0443-66B
trans-trans	EtOH	232(4.32)	78-0443-66B
$C_{10}H_{16}O_2S$			
1-Thiacycloheptane-4,5-dione, 3,3,6,6-tetramethyl-	EtOH	219(2.82),300(1.72), 333(1.62)	44-3954-66
4-Thia-2,6-octadienoic acid, 3,7-di-methyl-, methyl ester, cis	C_6H_{12}	284(4.09)	78-0259-66
trans	C_6H_{12}	211(3.65),274(4.22)	78-0259-66
$C_{10}H_{16}O_3$			
1,4-Butanediol, 4-(2-furyl)-2,3-dimethyl-	H_2O	215(3.96)	12-0683-66
2-Octenoic acid, 4,4-dihydroxy-2,3-dimethyl-, γ-lactone	n.s.g.	214(4.1)	54-0043-66
$C_{10}H_{16}O_5$			
Fumaric acid, ethoxy-, diethyl ester	MeOH	242(3.87)	5-0134-66C
Maleic acid, ethoxy-, diethyl ester	MeOH	228(3.99)	5-0134-66C
$C_{10}H_{16}O_5S$			
α-D-Xylose, 3-O-acetyl-5-deoxy-5-mer-capto-1,2-O-isopropylidene-	EtOH	230(4.03)	39-1287-66C
$C_{10}H_{16}O_6$			
Retronecic acid	n.s.g.	218(4.08)	12-2127-66

Compound	Solvent	$\lambda_{max}(\log \epsilon)$	Ref.
$C_{10}H_{16}S$			
Thiocamphor	C_6H_{12}	214(2.62),244s(4.1), 492(1.1)	60-0788-66
iodine complex	C_6H_{12}	305(--)	60-0788-66
$C_{10}H_{16}S_2$			
4H-Thiopyran, 2,3-dihydro-, dimer	MeCN	222(3.84),243(3.78)	44-2333-66
$C_{10}H_{16}S_3$			
Thiophene, 2,3-bis[(ethylthio)-methyl]-	EtOH	240.5(3.92)	44-3363-66
	EtOH	242.5(3.91)	44-3592-66
$C_{10}H_{17}BrO$			
Ether, 6-bromo-4-vinyl-4-hexenyl ethyl	ether	246(4.24)	33-0858-66
$C_{10}H_{17}NO$			
1-Azaspiro[4.5]decan-2-one, 1-methyl-	MeOH	208(3.9)	24-0724-66
1-Penten-3-one, 1-piperidino-	EtOH	307(4.41)	70-1739-66
$C_{10}H_{17}NO_2$			
Acrylic acid, 3-(cyclohexylamino)-, methyl ester, cis	CH_2Cl_2	281(4.3)	24-2526-66
trans isomer	CH_2Cl_2	270(4.4)	24-2526-66
2-Cyclohexenone, 2-(3-hydroxy-propyl)-3-methyl-, oxime	EtOH	240(4.43)	44-2171-66
$C_{10}H_{17}NO_4$			
Maleic acid, (diethylamino)-, dimethyl ester	ether	277(4.32)	24-0450-66
$C_{10}H_{17}NO_5$			
Adipic acid, α-oximino-, dimethyl ester	EtOH	220(3.90)	39-0255-66C
	EtOH-NaOH	270(4.14)	39-0255-66C
Azacyclohexadiene, pentamethoxy-	MeOH	213(3.04)	44-4054-66
$C_{10}H_{17}N_3$			
Pyrimidine, 2-(butylamino)-4,6-dimethyl-	pH 3.0	231(4.15),310(3.70)	12-2321-66
	pH 8.0	239(4.05),303(3.45)	12-2321-66
	pH 8.0	239(4.05),303(3.45)	39-0226-66C
	EtOH	239(3.72),312(3.05)	78-2401-66
Pyrimidine, 4-(butylamino)-2,6-dimethyl-	pH 5.0	258(4.21)	12-2321-66
	pH 10.0	244(4.03),275(3.61)	12-2321-66
	pH 10.0	244(4.03),275(3.61)	39-0226-66C
Pyrimidine, 2-(sec-butylamino)-4,6-dimethyl-	pH 3.0	232(4.12),310(3.64)	39-0226-66C
	pH 8.0	239(4.12),303(3.47)	39-0226-66C
Pyrimidine, 4-(sec-butylamino)-2,6-dimethyl-	pH 5.5	259(4.24)	39-0226-66C
	pH 10.0	245(4.15),276(3.72)	39-0226-66C
Pyrimidine, 2-(tert-butylamino)-4,6-dimethyl-	pH 3.5	232(4.17),310(3.76)	39-0226-66C
	pH 8.0	239(4.12),302(3.50)	39-0226-66C
Pyrimidine, 4-(tert-butylamino)-2,6-dimethyl-	pH 5.5	260(4.20)	39-0226-66C
	pH 10.0	244(4.02),276(3.71)	39-0226-66C
Pyrimidine, 2-(diethylamino)-4,6-dimethyl-	pH 3.5	238(4.18),319(3.64)	39-0226-66C
	pH 8.0	248(4.21),312(3.43)	39-0226-66C
Pyrimidine, 2-(isobutylamino)-4,6-dimethyl-	pH 3.0	232(4.18),310(3.67)	39-0226-66C
	pH 8.0	238(4.20),302(3.60)	39-0226-66C
Pyrimidine, 4-(isobutylamino)-2,6-dimethyl-	pH 5.0	258(4.19)	39-0226-66C
	pH 10.0	244(4.13),275(3.71)	39-0226-66C

Compound	Solvent	λ_{max}(log ϵ)	Ref.
$C_{10}H_{17}N_3O$			
2-Cyclohexylideneacetaldehyde, α-methyl-, semicarbazone	EtOH	275(4.39)	22-1400-66
3,5-Heptadien-2-one, 4,6-dimethyl-, semicarbazone	EtOH	292(4.38)	22-0689-66
4,6-Heptadien-2-one, 4,6-dimethyl-, semicarbazone	EtOH	230(4.15)	22-0689-66
3,5-Hexadienal, 2,2,4-trimethyl-, semicarbazone	n.s.g.	225(4.40)	22-0734-66
Nopinone, semicarbazone	EtOH	225(4.30)	22-1227-66
5,6-Octadien-2-one, 7-methyl-, semicarbazone	EtOH	228(4.17)	28C-1601-66A
$C_{10}H_{17}N_3O_3$			
Cyclohexaneacetic acid, 2-formyl-, semicarbazone	EtOH	230(4.14)	44-2695-66
$C_{10}H_{17}N_5$			
Pyrimidine, 2,4-diamino-5-(cyclohexylamino)-	pH 11	240(3.92),296(3.72)	87-0121-66
Pyrimidine, 2,5-diamino-4-(cyclohexylamino)-	MeOH-acid	235(4.41),300(3.91)	87-0121-66
Pyrimidine, 4,5-diamino-2-(cyclohexylamino)-	MeOH-acid	237(4.24),308s(3.23)	87-0121-66
$C_{10}H_{17}N_5O_5$			
Pyrimidine, 2-amino-4-(2,2-diethoxy-ethylamino)-dihydro-5-nitro-6-oxo-	pH -0.89	225(4.34),245s(3.84), 312(4.12)	24-2984-66
	pH 5.0	230s(4.18),285s(3.70), 332(4.16)	24-2984-66
	pH 11.0	232(3.97),342(4.24)	24-2984-66
$C_{10}H_{17}N_5O_6$			
Glycine, tetraglycyl-, copper complex	pH 11	510(2.17)	37-1072-66
nickel complex	pH 11	412(2.60)	37-1072-66
$C_{10}H_{18}N_2$			
$\Delta^{2,3'}$-Bipyrrolidine, 1,1'-dimethyl-	MeOH	207(4.1)	24-1923-66
Pyrazine, 2,3-dihydro-2,2,3,3,5,6-hexamethyl-	C_6H_{12}	231(3.98),358(3.08)	39-2300-66C
$C_{10}H_{18}N_2O$			
[2,3'-Bipyrrolidin]-2'-one, 1,1'-dimethyl-	MeOH	218(3.3)	24-1923-66
3H-Pyrazole-5-methanol, α,α-di-ethyl-3,3-dimethyl-	hexane	248(3.33),355(2.38)	39-1719-66C
	EtOH	254(3.37),346(2.36)	39-1719-66C
$C_{10}H_{18}N_2O_2$			
Pyrazine, 2,3-dihydrohexamethyl-, 1,4-dioxide	EtOH	218(3.73),347(4.09)	39-2300-66C
$C_{10}H_{18}O$			
5-Nonen-4-one, 6-methyl-	n.s.g.	241(1.77)	54-0753-66
7-Octen-4-ol, 2-methyl-6-methylene-	hexane	226(4.30)	78-1929-66
1-Octen-7-one, 2,6-dimethyl-	EtOH	283(1.51)	33-0617-66
2-Octen-7-one, 2,6-dimethyl-	EtOH	283(1.54)	33-0617-66
1,3-Pentadiene, 3-(3-ethoxypropyl)-	EtOH	229(4.31)	33-0858-66
$C_{10}H_{18}O_2$			
3,8-Decanedione	n.s.g.	278(1.78)	28C-1271-66A

Compound	Solvent	λ_{max}(log ϵ)	Ref.
Ether, ethyl 6-hydroxy-4-vinylhexyl	EtOH	230(4.30)	33-0858-66
2,5-Hexadienal, diethyl acetal, 2-cis	EtOH	230(3.31)	54-0117-66
2,4-Pentadien-1-ol, 3-(3-ethoxypropyl)-	EtOH	230(4.30)	33-0858-66
$C_{10}H_{18}O_3$			
2-Pentenoic acid, 2-ethoxy-4-methyl-, ethyl ester	MeOH	228(3.84)	5-0053-66I
$C_{10}H_{18}O_4$			
Propionic acid, 3-ethoxy-2-ethoxymethylene-, ethyl ester	EtOH	239(4.10)	94-0238-66
$C_{10}H_{19}N$			
Butylamine, N-cyclohexylidene-	heptane	180(3.98)	88-0507-66
	EtOH	238(2.36)	88-0507-66
$C_{10}H_{19}NO$			
2-Pyrrolidinone, 1-hexyl-	MeOH	208(3.8)	24-0724-66
$C_{10}H_{19}NO_2$			
Acrylic acid, 3-(diisopropylamino)-, methyl ester	EtOH	283(4.48)	24-2526-66
Crotonic acid, 3-(butylamino)-, ethyl ester	MeOH	285(4.33)	35-2536-66
$C_{10}H_{19}O_4P$			
Phosphonic acid, (cyclopentylcarbonyl)-, diethyl ester	C_6H_{12}	340(1.79)	44-1304-66
$C_{10}H_{20}N_2$			
3H-Azepine, 2-(diethylamino)-4,5,6,7-tetrahydro-	n.s.g.	224(3.60)	78-0081-66
$C_{10}H_{20}N_4$			
Piperidine, 1,1'-azodi-	MeOH	250s(--),280(3.97)	39-0813-66C
$C_{10}H_{20}O$			
3-Hexanone, 2,2,4,5-tetramethyl-	hexane	294(1.34)	22-3578-66
2-Octanone, 3,4-dimethyl-	isooctane	288(1.40)	22-0273-66
$C_{10}H_{20}O_2$			
4-Octene-2,7-diol, 2,6-dimethyl-	EtOH	239(4.31)	33-0617-66
$C_{10}H_{21}N$			
Nonane, 4-iodo-4-methyl-	C_6H_{12}	267(2.77)	3-0612-66
$C_{10}H_{22}OS$			
Pentyl sulfoxide	EtOH	201(3.28),214s(3.08)	39-0239-66A
$C_{10}H_{22}S$			
Pentyl sulfide	EtOH	202(3.43)	39-0239-66A
$C_{10}H_{22}S_2$			
Pentyl disulfide	EtOH	201(3.36),250(2.60)	39-0239-66A
$C_{10}H_{24}ClN_5O_4$			
Octamethylbiguanide perchlorate	MeOH-HCl	226(4.39)	87-0980-66
	MeOH	243(4.49)	87-0980-66

Compound	Solvent	$\lambda_{max}(\log \epsilon)$	Ref.
$C_{10}H_{28}N_6$ Pentaethylenehexamine, cobalt complex (also other metal complexes)	n.s.g.	345(2.23),488(2.35)	33-0625-66
$C_{10}H_{30}Cl_2Si_5$ Pentasilane, 1,5-dichloro-decamethyl-	C_6H_{12}	214(4.00),250(4.29)	101-0392-66A
$C_{10}H_{31}ClSi_5$ Pentasilane, 1-chloro-1,1,2,2,3,3,4,4,5,5-decamethyl-	C_6H_{12}	215(3.94),250(4.21)	101-0392-66A
$C_{10}H_{32}Si_5$ Pentasilane, decamethyl-	C_6H_{12}	214(3.97),249(4.13)	101-0392-66A
$C_{10}Mn_2O_{10}$ Manganese carbonyl	n.s.g.	276(5.20),342(4.18)	35-5124-66

Compound	Solvent	$\lambda_{max}(\log \epsilon)$	Ref.
$C_{11}H_3Cl_7N_2O$			
Pyrimidine, 2,4-dichloro-5-[(2,3,4,5,6-pentachlorophenoxy)methyl]-	EtOH	216(4.83)	5-0119-66B
$C_{11}H_5Cl_4N$			
Pyridine, 3,4,5,6-tetrachloro-2-phenyl-	MeOH	258(4.08),298(3.92)	24-2813-66
$C_{11}H_5F_6$			
Cyclopentene, heptafluoro-1-phenyl-	EtOH	251.3(4.23)	78-1755-66
$C_{11}H_5N_3$			
2,4-Quinolinedicarbonitrile	EtOH	325(3.77)	94-0512-66
$C_{11}H_6Cl_2N_4$			
s-Triazolo[4,3-b]pyridazine, 6-chloro-3-(p-chlorophenyl)-	EtOH	260(4.36),328(3.27)	78-2073-66
$C_{11}H_6F_{10}$			
1,3,5-Cycloheptatriene, 7,7-bis-(pentafluoroethyl)-	C_6H_{12}	292(3.37)	35-3617-66
	EtOH	293(3.36)	35-3617-66
	MeCN	293(3.35)	35-3617-66
$C_{11}H_6N_2O_2S_2$			
Benzothiazole, 2-(5-nitro-2-thienyl)-	EtOH	225s(4.21),268(3.95), 378(4.32)	87-0751-66
$C_{11}H_6N_2O_3S$			
Benzoxazole, 2-(5-nitro-2-thienyl)-	EtOH	249(3.98),262(4.01), 369(4.33)	87-0751-66
$C_{11}H_6O_4$			
Xanthotoxol	EtOH	224(4.59),250(4.44), 268(4.42),308(4.28)	93-0616-66
$C_{11}H_7BrOS$			
5H-Naphtho[1,8-bc]thiophen-5-one, 2-bromo-3,4-dihydro-	EtOH	227(3.72),257(3.87), 310(3.26)	12-1908-66
$C_{11}H_7BrO_2$			
1-Naphthoic acid, 2-bromo-	dioxan	275s(3.68),283(3.72), 293s(3.59),323(2.40)	78-1587-66
1-Naphthoic acid, 3-bromo-	dioxan	301s(3.70),335(3.04)	78-1587-66
1-Naphthoic acid, 4-bromo-	dioxan	305(3.97),328(2.95)	78-1587-66
1-Naphthoic acid, 5-bromo-	dioxan	305(3.90),328(2.70)	78-1587-66
1-Naphthoic acid, 6-bromo-	dioxan	297s(3.81),328(3.04)	78-1587-66
1-Naphthoic acid, 7-bromo-	dioxan	299s(3.81),331(3.42)	78-1587-66
1-Naphthoic acid, 8-bromo-	dioxan	283s(3.79),292(3.89), 303s(3.81)	78-1587-66
$C_{11}H_7ClFN_3O_2$			
Cytosine, N^4-(p-chlorobenzoyl)-5-fluoro-	EtOH	265(4.10),332(4.19)	87-0566-66
$C_{11}H_7ClN_2S$			
Imidazo[2,1-b]thiazole, 6-(p-chlorophenyl)-	n.s.g.	204(4.68),256(4.59)	22-1277-66

Compound	Solvent	λ_{max}(log ε)	Ref.
$C_{11}H_7ClN_4$			
s-Triazolo[4,3-b]pyridazine, 6-chloro-3-phenyl-	EtOH	252(4.35),339(3.29)	78-2073-66
$C_{11}H_7ClO_2$			
1-Naphthoic acid, 2-chloro-	dioxan	273s(3.69),282(3.75), 293s(3.63),322(2.30)	78-1587-66
1-Naphthoic acid, 3-chloro-	dioxan	298(3.73),335(3.11)	78-1587-66
1-Naphthoic acid, 4-chloro-	dioxan	304(3.94),327(2.95)	78-1587-66
1-Naphthoic acid, 5-chloro-	dioxan	303(3.85),327(2.70)	78-1587-66
1-Naphthoic acid, 6-chloro-	dioxan	296(3.81),328(3.11)	78-1587-66
1-Naphthoic acid, 7-chloro-	dioxan	298s(3.82),331(3.40)	78-1587-66
1-Naphthoic acid, 8-chloro-	dioxan	280s(3.78),290(3.85), 302s(3.74),326(2.00)	78-1587-66
$C_{11}H_7Cl_2N_3O_3$			
Pyrimidine, 2,6-dichloro-5-[(4-nitrophenoxy)methyl]-	EtOH	218(4.26),297(4.08)	5-0119-66B
$C_{11}H_7FN_2S$			
Imidazo[2,1-b]thiazole, 6-(p-fluorophenyl)-	n.s.g.	204(4.71),250(4.51)	22-1277-66
$C_{11}H_7FN_4O_4$			
Cytosine, 5-fluoro-N^4-(p-nitrobenzoyl)-	EtOH	270(4.27),344(4.25)	87-0566-66
$C_{11}H_7FO_2$			
1-Naphthoic acid, 2-fluoro-	dioxan	283(3.76),319(3.26)	78-1587-66
1-Naphthoic acid, 3-fluoro-	dioxan	296(3.79),330(3.16)	78-1587-66
1-Naphthoic acid, 4-fluoro-	dioxan	298(3.90),320(2.90)	78-1587-66
1-Naphthoic acid, 5-fluoro-	dioxan	300(3.77),325(2.60)	78-1587-66
1-Naphthoic acid, 7-fluoro-	dioxan	293s(3.78),328(3.54)	78-1587-66
$C_{11}H_7IO_2$			
1-Naphthoic acid, 2-iodo-	dioxan	275(3.80),284(3.81), 295s(3.67),326(2.40)	78-1587-66
1-Naphthoic acid, 3-iodo-	dioxan	303s(3.8),338(3.08)	78-1587-66
1-Naphthoic acid, 4-iodo-	dioxan	309(4.04),330(3.00)	78-1587-66
1-Naphthoic acid, 5-iodo-	dioxan	309(3.96),330(2.60)	78-1587-66
1-Naphthoic acid, 7-iodo-	dioxan	300s(3.76),333(3.36)	78-1587-66
1-Naphthoic acid, 8-iodo-	dioxan	285s(3.80),299(3.91), 313s(3.94)	78-1587-66
$C_{11}H_7I_2NO_2$			
5-Isoquinolinol, 6,8-diiodo-, acetate	iso-PrOH	224(4.38),250(4.50), 294(3.76),305(3.78), 320(3.69),333(3.71)	87-0046-66 87-0046-66
$C_{11}H_7NO_4$			
1-Naphthoic acid, 2-nitro-	dioxan	304(3.81),347(3.41)	78-1587-66
	base	309(3.88),363(3.51)	78-1587-66
1-Naphthoic acid, 3-nitro-	dioxan	308(3.86),351(3.40)	78-1587-66
	base	310(3.91),363(3.50)	78-1587-66
1-Naphthoic acid, 4-nitro-	dioxan	335(3.65)	78-1587-66
	base	353(3.72)	78-1587-66
1-Naphthoic acid, 5-nitro-	dioxan	310(3.80),328(2.70)	78-1587-66
	base	350(3.55)	78-1587-66
1-Naphthoic acid, 6-nitro-	dioxan	295s(3.90),345(3.32)	78-1587-66
	base	309(3.95),361(3.54)	78-1587-66

Compound	Solvent	λ_{max}(log ε)	Ref.
1-Naphthoic acid, 7-nitro-	dioxan	311(4.00),346(3.55)	78-1587-66
	base	312(3.92),359(3.54)	78-1587-66
1-Naphthoic acid, 8-nitro-	dioxan	281(3.73)	78-1587-66
	base	326(3.55)	78-1587-66
$C_{11}H_7NS_2$			
Acrylonitrile, 2,3-di-2-thienyl-, cis	n.s.g.	230(3.97),288(4.00), 331(4.08)	12-1243-66
Acrylonitrile, 2,3-di-2-thienyl-, trans	n.s.g.	233(3.96),286(3.70), 368(4.57)	12-1243-66
$C_{11}H_7N_3O$			
Pyrimidino[4,5-c]isocarbostyril, hydrochloride	dioxan	260(4.1),270(4.0), 300(4.1),310(4.2), 320(4.2)	32-1108-66
Quinaldamide, 4-cyano-	EtOH	323(3.82)	94-0512-66
$C_{11}H_7N_3O_2$			
Ketone, 2-furyl v-triazolo[1,5-a]-pyridin-3-yl	MeOH	222(4.18),282(3.96), 292(4.00),332(4.40)	24-2918-66
$C_{11}H_7N_3O_2S$			
Benzimidazole, 2-(5-nitro-2-thienyl)-	EtOH	252(3.90),270(3.90), 390(4.23)	87-0751-66
Benzothiazole, 2-(5-nitro-2-pyrrolyl)-	EtOH	224(4.28),277(3.95), 378(4.38)	87-0751-66
$C_{11}H_7N_3O_3$			
Benzoxazole, 2-(5-nitro-2-pyrrolyl)-	EtOH	247(4.03),264(4.06), 368(4.37)	87-0751-66
Oxazolo[5,4-d]pyrimidine-5,7(4H,6H)-dione, 2-phenyl-	EtOH	321(4.10)	50-0737-66C
s-Triazolo[4,3-a]pyridine-3-tetrolic acid, γ-oxo-, methyl ester	MeOH	268(3.95),312(3.60)	44-0265-66
$C_{11}H_8$			
1,3-Pentadiyne, 1-phenyl-	EtOH	222(4.81),232(3.40), 243(3.88),256(4.22), 271(4.40),287(4.27)	78-0867-66
1,4-Pentadiyne, 1-phenyl-	EtOH	238(4.30),249(4.31), 265(2.97),271(3.02), 278(2.86),284s(2.48)	78-0867-66
$C_{11}H_8BrNO$			
Aniline, p-bromo-N-furfurylidene-	EtOH	294(4.16),327(4.20)	12-1747-66
$C_{11}H_8BrNS$			
Aniline, p-bromo-N-thenylidene-	EtOH	273(4.05),331(4.20)	12-1747-66
$C_{11}H_8BrN_5O_6$			
Pyrazole, 4-bromo-3,5-dimethyl-1-picryl-	EtOH	329(3.65)	22-3744-66
$C_{11}H_8ClNO$			
Aniline, p-chloro-N-furfurylidene-	EtOH	293(4.19),325(4.23)	12-1747-66
$C_{11}H_8ClNS$			
Aniline, m-chloro-N-2-thenylidene-	EtOH	272(4.04),323(4.15)	12-1747-66
Aniline, p-chloro-N-2-thenylidene-	EtOH	272(4.03),330(4.16)	12-1747-66

Compound	Solvent	$\lambda_{max}(\log \epsilon)$	Ref.
$C_{11}H_8ClN_5$			
Adenine, N-(p-chlorophenyl)-	pH 1	280(3.84)	95-0649-66
	pH 6	293(3.98)	95-0649-66
	pH 13	295(4.01)	95-0649-66
Adenine, 9-(p-chlorophenyl)-	pH 1	261(4.08)	95-0649-66
	pH 13	265(4.11)	95-0649-66
s-Triazolo[4,3-b]pyridazine, 8-amino-6-chloro-3-phenyl-	EtOH	203(4.33),253(4.34),288(4.13),300s(4.03)	4-0218-66
s-Triazolo[4,3-b]pyridazine, 8-amino-7-chloro-3-phenyl-	EtOH	203(4.45),240(4.24),280(4.16),310(3.96)	4-0218-66
$C_{11}H_8Cl_2N_2O$			
Pyrimidine, 2,4-dichloro-5-(phenoxymethyl)-	EtOH	219(4.29),263(3.72)	5-0119-66B
$C_{11}H_8Cl_2N_2S$			
Pyridine, 2-[(o-aminophenyl)thio]-3,5-dichloro-	EtOH	251(4.18),309(3.80)	95-0050-66
$C_{11}H_8Cl_2N_4$			
Benzaldehyde, p-chloro-, (6-chloro-3-pyridazinyl)hydrazone	EtOH	240(4.10),323(4.55)	78-2073-66
$C_{11}H_8Cl_3NO$			
4(1H)-Quinolone, 6-methyl-2-(trichloromethyl)-	EtOH	238(4.44)	44-3369-66
$C_{11}H_8CrO_5$			
Chromium, tricarbonyl(m-toluic acid)-	EtOH	319(3.8)	49-1029-66
Chromium, tricarbonyl(o-toluic acid)-	EtOH	319(3.8)	49-1029-66
$C_{11}H_8FNO$			
Aniline, p-fluoro-N-furfurylidene-	EtOH	290(4.16),322(4.17)	12-1747-66
$C_{11}H_8FN_3O_2$			
Cytosine, N^4-benzoyl-5-fluoro-	EtOH	259(4.05),327(4.15)	87-0566-66
$C_{11}H_8F_3NO_4$			
Furo[2,3-b]pyridine-2-carboxylic acid, 4-hydroxy-6-(trifluoromethyl)-, ethyl ester	5% EtOH	327.5(4.26)	4-0202-66
$C_{11}H_8INO_2$			
5-Isoquinolinol, 8-iodo-, acetate	iso-PrOH	227(4.47),243(4.59),272(3.76),280(3.76),297s(3.60),310(3.43),325(3.52)	87-0046-66
8-Isoquinolinol, 5-iodo-, acetate	iso-PrOH	211(4.52),290s(3.76),300(3.80),323(3.70),333(3.73)	87-0046-66
$C_{11}H_8N_2O_2$			
Bicyclo[4.3.0]nona-2,4,7-triene-9,9-dicarbonitrile	C_6H_{12}	218(3.59),255(3.66),264(3.64),274s(3.38)	35-1979-66
Bicyclo[6.1.0]nona-2,4,6-triene-9,9-dicarbonitrile	C_6H_{12}	242(3.53)	35-1979-66
1,1-Cyclopropanedicarbonitrile, 2-phenyl-	C_6H_{12}	219(3.79),259(2.38),266(2.38),272(2.20)	35-1979-66
1H-Pyrrolo[2,3-f]isoquinoline	acid	245(4.01),287(4.40),355(3.92)	2-0118-66

Compound	Solvent	λ_{max}(log ϵ)	Ref.
1H-Pyrrolo[2,3-f]isoquinoline (cont.)	neutral	265(4.43),292s(3.77)	2-0118-66
Pyrrolo[1,2-a]quinoxaline	pH 1.0	225(4.52),240(4.44), 352(4.07)	39-0852-66C
	pH 6.9	224(4.44),247(4.39), 334(3.93)	39-0852-66C
Tricyclo[4.2.1.0^{2,5}]nona-3,7-diene-3,4-dicarbonitrile	EtOH	228(3.81),283(3.15)	35-4273-66
$C_{11}H_8N_2O$			
Isoquinoline-1-carbonitrile, 1,2-epoxy-3-methyl-	EtOH	248(3.97),255(3.93), 350(3.61)	88-2145-66
1aH-Oxazirino[2,3-a]quinoline-1a-carbonitrile, 3-methyl-	EtOH	239(4.42),315(3.50)	88-2145-66
	EtOH	239(4.42),315(3.58)	94-0555-66
Pyrrolo[1,2-a]quinoxalin-4(5H)-one	EtOH	227(4.45),251(4.16), 260(4.08),312(4.07), 325(4.00)	39-0852-66C
Quinaldonitrile, 3-hydroxy-4-methyl-	EtOH	236(4.79),295(3.95), 304(3.90),358(3.55), 420(3.41)	88-2145-66
	5% K_2CO_3	247(4.73),303(3.68), 314(3.68),410(3.85)	88-2145-66
$C_{11}H_8N_2OS$			
Thiazolo[4,5-c]isoquinoline, 2-methoxy-	EtOH	247(4.63),278(3.71), 289(3.79),301(3.73), 335(3.61),345(3.61)	23-2465-66
Thiazolo[5,4-c]isoquinoline, 2-methoxy-	EtOH	238(4.73),263s(3.52), 278s(3.59),288(3.69), 299(3.66),330s(3.80), 338(3.85)	23-2473-66
$C_{11}H_8N_2O_2$			
Quinaldonitrile, 3-hydroxy-4-methoxy-	EtOH	240(4.89),294(3.96), 303(3.92),360(3.68), 420(3.39)	88-2145-66
	5% K_2CO_3	248(4.65),302(3.40), 312(3.35),409(3.75)	88-2145-66
$C_{11}H_8N_2O_3$			
Aniline, N-furfurylidene-p-nitro-	EtOH	344(4.33)	12-1747-66
$C_{11}H_8N_2O_3S$			
2-Thiazoline-4-carboxylic acid, 2-(4-hydroxy-2-benzothiazolyl)-	EtOH	257(4.06),303(4.12)	44-1484-66
2-Thiazoline-4-carboxylic acid, 2-(5-hydroxy-2-benzothiazolyl)-	EtOH	229(4.29),294(4.17), 350(3.64)	44-1484-66
2-Thiazoline-4-carboxylic acid, 2-(6-hydroxy-2-benzothiazolyl)-	EtOH	269(3.85),330(4.26)	35-2015-66
2-Thiazoline-4-carboxylic acid, 2-(7-hydroxy-2-benzothiazolyl)-	EtOH	257(4.28),301(4.16), 365s(--)	44-1484-66
$C_{11}H_8N_2O_4S_2$			
2-Thiazoline-4-carboxylic acid, 2-(4,6-dihydroxy-2-benzothiazolyl)-	EtOH	269(4.04),324(4.17), 360s(--)	44-1484-66
$C_{11}H_8N_4$			
Purine, 8-phenyl-	pH 5	232(4.07),297(4.41)	39-0010-66C
$C_{11}H_8N_4O$			
Hypoxanthine, 8-phenyl-	pH 1	234(4.05),286(4.33)	4-0476-66

Compound	Solvent	λ_{max}(log ε)	Ref.
Hypoxanthine, 8-phenyl- (cont.)	pH 5	286(4.28)	39-0010-66C
	pH 7	235(4.07),286(4.33), 297s(4.28),312s(3.90)	4-0476-66
	pH 13	236(4.27),303(4.29)	4-0476-66
$C_{11}H_8N_4OS$			
2-Purinethiol, 6-hydroxy-8-phenyl-	pH 1	233s(4.14),296(4.28)	4-0476-66
	pH 7	223s(4.20),299(4.27)	4-0476-66
	pH 13	241(4.25),316(4.23)	4-0476-66
$C_{11}H_8N_4O_2$			
Alloxazine, 7-methyl-	EtOH	245(4.54),328(3.78), 387(3.78)	30-1101-66F
Benzimidazole, 2-(5-nitro-2-pyrrolyl)-	EtOH	235(4.08),287(3.94), 382(4.31)	87-0751-66
2,6-Purinediol, 8-phenyl-	2N HCl	236(3.99),296(4.27)	4-0476-66
	pH 13	233(4.30),241(4.30), 320(4.18)	4-0476-66
$C_{11}H_8N_4O_3$			
Alloxazine, 6-methyl-, N^9-oxide	pH 13	277(4.80),348(3.95), 455(3.97)	30-0577-66C
3H,10H-Benzo[g]pteridine-2,4-dione, 10-hydroxy-3-methyl-	EtOH	271(4.60),345(3.92), 415(3.70)	30-1101-66F
3H,10H-Benzo[g]pteridine-2,4-dione, 10-hydroxy-7-methyl-	EtOH	273(4.42),346(3.72), 405(3.56)	30-1101-66F
$C_{11}H_8N_4O_4$			
Cytosine, N^4-(p-nitrobenzoyl)-	EtOH	265(4.27),305(3.91)	44-4014-66
$C_{11}H_8N_4S$			
Purine-6-thiol, 8-phenyl-	pH 5	260(4.33),345(4.38)	39-0010-66C
s-Triazolo[4,3-b]pyridazine-6-thiol, 3-phenyl-	EtOH	238(4.22)	78-2073-66
$C_{11}H_8N_6O_3$			
6-Purinol, 2-amino-8-(p-nitrophenyl)-	pH 1	216(4.11),256(4.20), 345(3.83)	4-0476-66
	pH 7	233s(4.08),258(4.21), 357(3.76)	4-0476-66
	pH 13	251(4.23),263s(4.19), 402(3.81)	4-0476-66
$C_{11}H_8OS$			
3H-Cyclopenta[b][1]benzo[b]thiophen-3-one, 1,2-dihydro-	EtOH	236(4.17),247(4.19), 292(4.29),322(3.73)	12-1908-66
6H-Indeno[5,4-b]thiophen-6-one, 7,8-dihydro-	EtOH	240(4.56),272(3.86), 322(3.60)	12-1908-66
2H-Naphtho[1,8-bc]thiophene-2-ol	EtOH	244(3.91),321(3.47), 328(3.47)	12-1908-66
3H-Naphtho[1,8-bc]thiophen-3-one, 4,5-dihydro-	EtOH	248(3.78),319(3.71)	12-1908-66
Thiophene, 2-benzoyl-	MeOH	222s(3.84),260(4.11), 292(4.11),354s(2.35)	73-0835-66
Thiophene, 3-benzoyl-	MeOH	257(4.21),282s(3.33), 329(2.32)	73-0835-66
$C_{11}H_8OS_2$			
2-Propyn-1-ol, 3-[5-(2-thienyl)-2-thienyl]-	ether	233(3.55),334(4.28)	39-1101-66C

Compound	Solvent	λ_{max}(log ε)	Ref.
$C_{11}H_8O_2$			
Furan, 2-benzoyl-	MeOH	256s(4.09),288(4.36), 348s(2.52)	73-0835-66
1-Naphthoic acid	dioxan	298(3.82),325(2.00)	78-1587-66
$C_{11}H_8O_3$			
p-Benzoquinone, 2-(2-furyl)-5-methyl-	EtOH	257(4.23),300s(3.68), 430(3.63)	33-1794-66
p-Benzoquinone, 2-(5-methyl-2-furyl)-	EtOH	224(4.08),262(4.22), 465(3.71)	33-1794-66
$C_{11}H_8O_4$			
2(5H)-Furanone, 5-hydroxy-, benzoate	MeOH	230(4.19),275(3.04), 282(2.94)	5-0042-66G
Maleic anhydride, (p-methoxyphenyl)-	n.s.g.	227(4.00),308(4.02), 367(3.80)	39-1924-66C
1,4-Naphthoquinone, 2-hydroxy- 6-methoxy-	EtOH	278(4.208),294(4.192), 380(3.089)	23-1086-66
	EtOH-base	276(4.207),293(4.150), 470(3.310)	23-1086-66
1,4-Naphthoquinone, 2-hydroxy- 7-methoxy-	EtOH	264(4.382),272(4.411), 292(3.872),340(3.728), 450(2.827)	23-1086-66
	EtOH-base	264(4.390),272(4.413), 296(3.870),345(3.754), 465(3.492)	23-1086-66
$C_{11}H_8O_5$			
Phthalic anhydride, 4-(carboxy- methyl)-3-methyl-	ether	220(4.60),260(3.69), 300(3.55),310(3.58)	33-0858-66
$C_{11}H_8S$			
1,3-Pentadiyne, 1-(phenylthio)-	EtOH	248(4.14),261(3.94), 275(3.78),294(3.63)	22-3024-66
$C_{11}H_9BrClN_3$			
Pyrimidine, 5-bromo-4-(p-chloro- anilino)-6-methyl-	EtOH	288(4.35)	78-2401-66
$C_{11}H_9BrClN_3O$			
Pyrimidine, 2-amino-5-bromo- 4-(p-chlorophenoxy)-6-methyl-	EtOH	304(3.70)	78-2401-66
$C_{11}H_9BrN_2OS$			
Thiazolin-2-imine, 3-(m-bromo- benzoylmethyl)-	iso-PrOH- acid	248.2(4.26)	20-0380-66
$C_{11}H_9BrN_4O_4$			
Pyrazole, 4-bromo-3,5-dimethyl- 1-(2,4-dinitrophenyl)-	EtOH	315(3.80)	22-3744-66
Pyrazole, 4-bromo-3-ethyl- 1-(2,4-dinitrophenyl)-	EtOH	328(4.15)	22-3744-66
$C_{11}H_9ClN_2OS$			
Thiazolin-2-imine, 3-(m-chloro- benzoylmethyl)-	iso-PrOH- acid	247.2(4.29)	20-0380-66
Thiazolin-2-imine, 3-(p-chloro- benzoylmethyl)-	iso-PrOH- acid	256.5(4.40)	20-0380-66
	iso-PrOH- base	255.5(4.39)	20-0380-66

Compound	Solvent	λ_{max}(log ε)	Ref.
$C_{11}H_9ClN_4$			
Benzaldehyde, (6-chloro-3-pyridazinyl)hydrazone	EtOH	225(4.05),316(4.52)	78-2073-66
$C_{11}H_9ClN_4O$			
Benzoic acid, 2-(6-chloro-3-pyridazinyl)hydrazide	EtOH	274(4.25)	78-2073-66
$C_{11}H_9ClO$			
1-Naphthalenemethanol, 2-chloro-	dioxan	273(3.76),282(3.79), 293(3.65),323(2.48)	78-1587-66
1-Naphthalenemethanol, 8-chloro-	dioxan	280(3.77),290(3.83), 303s(3.68),325(2.65)	78-1587-66
$C_{11}H_9FN_2OS$			
Thiazolin-2-imine, 3-(p-fluorobenzoylmethyl)-	iso-PrOH-acid	249.0(4.33)	20-0380-66
	iso-PrOH-base	246.5(4.32)	20-0380-66
$C_{11}H_9FO$			
1-Naphthalenemethanol, 2-fluoro-	dioxan	272(3.81),277(3.81), 282(3.79),288s(3.73), 320(3.27)	78-1587-66
$C_{11}H_9N$			
Cyclobuta[b]quinoline, 1,2-dihydro-	C_6H_{12}	275f(3.65)	5-0149-66H
	MeOH	283f(3.58)	5-0149-66H
	MeOH-acid	319(4.04)	5-0149-66H
Pyridine, 2-phenyl-	MeOH	244(4.12),276(4.01)	24-2813-66
3H-Pyrrolo[1,2-a]indole	EtOH	215(4.05),265(4.25)	44-0870-66
$C_{11}H_9NO$			
Aniline, N-furfurylidene-	EtOH	290(4.04),320(4.03)	12-1747-66
	EtOH	230(3.8),290(4.2), 320(4.1)	24-3932-66
$C_{11}H_9NO_2$			
3,4-Benzazepine-2,5-dione, 6-methyl-	EtOH	207(3.19),233(3.08), 274(2.73)	78-1201-66
Benz[c]azepine-2,5-dione, 7-methyl-	n.s.g.	235(4.34),310(3.92)	88-2361-66
Maleimide, 3-methyl-4-phenyl-	MeOH	223(4.04),252s(--), 324(3.30)	44-0048-66
2,6-Pyridinediol, 5-phenyl-	n.s.g.	261(3.80),335(4.07)	22-2387-66
$C_{11}H_9NO_2S$			
Phthalimide, N-[2-(methylthio)-vinyl]-, cis	$CHCl_3$	270(4.02)	78-0033-66B
trans	$CHCl_3$	284(4.42)	78-0033-66B
$C_{11}H_9NO_3$			
1H-1-Benzazepine-4-carboxylic acid, 2,3-dihydro-2-oxo-	MeOH	232(4.53),273(3.94)	24-3070-66
$C_{11}H_9NO_4$			
2(5H)-Furanone, 5-hydroxy-, carbanilate	dioxan	243(3.94),265(3.91), 272(3.92),279(3.78)	5-0042-66G
Furo[3,2-d]oxazole-2,5(1H,3aH)-dione, dihydro-1-phenyl-	MeOH	234(4.01),267(3.74), 277(3.71),292(3.63)	5-0042-66G
	dioxan	265s(3.80),271(3.88), 278(3.77)	5-0042-66G

Compound	Solvent	λ_{max}(log ϵ)	Ref.
$C_{11}H_9NO_5$			
Coumarin, 7-hydroxy-	EtOAc	360(3.75)	50-0298-66B
4,8-dimethyl-6-nitro-	NH_3	420(3.90)	50-0298-66B
2-Isoindolineacetic acid,	dil. HCl	242s(3.89),342(3.72)	4-0328-66
4-methoxy-1,3-dioxo-	dil. NaOH	287(3.47)	4-0328-66
	EtOH	242s(3.91),336(3.72)	4-0328-66
2-Isoindolineacetic acid,	dil. HCl	235(4.60),330(3.42)	4-0328-66
5-methoxy-1,3-dioxo-	dil. NaOH	246s(3.97)	4-0328-66
	EtOH	232(4.69),284(3.25), 321(3.42)	4-0328-66
$C_{11}H_9NO_8$			
Phthalic anhydride, 4,5,6-trimethoxy-3-nitro-	ether	238(4.26),320(3.40)	44-1912-66
$C_{11}H_9NS$			
Aniline, N-2-thenylidene-	EtOH	270(3.99),324(4.08)	12-1747-66
$C_{11}H_9N_3O$			
2H-Pyrazolo[4,3-c]quinoline, 7-methoxy-	EtOH	213(4.34),245(4.72), 269(3.64),305(3.26), 321(3.34),335(3.34)	54-0681-66
$C_{11}H_9N_3O_2$			
Malononitrile, [amino-(o-methoxy-	H_2O	256(4.31)	24-2307-66
phenoxy)methylene]-	MeCN	259(4.35)	24-2307-66
	dioxan	261(4.29)	24-2307-66
$C_{11}H_9N_3O_2S$			
Pyridine, 4-[(o-aminophenyl)thio]-3-nitro-	EtOH	245(4.35),319(3.80)	95-0050-66
Δ^3-1,2,3-Thiadiazoline, 2-acetyl-5-benzoylimino-	MeOH	229(4.03),284(4.17), 372(3.32)	24-1618-66
$C_{11}H_9N_3O_2S_2$			
2-Thiazoline-4-carboxylic acid, 2-(6-amino-2-benzothiazolyl)-	EtOH	220(4.41),280(3.78), 363(4.19)	35-2015-66
$C_{11}H_9N_3O_3$			
Glycine, N-(2-quinoxalinylcarbonyl)-	EtOH	207(3.39),244(3.74), 32[illegible](3.00),327(3.00)	87-0266-66
$C_{11}H_9N_3O_4$			
Uracil, 1-(p-nitrobenzyl)-	pH 2.5	274(4.23)	39-1784-66C
$C_{11}H_9N_3O_6$			
Indole-2-carboxylic acid, 3,4-dinitro-, ethyl ester	EtOH	214(4.48),243(4.02), 324(3.95),357s(3.87)	44-0070-66
Indole-2-carboxylic acid, 3,5-dinitro-, ethyl ester	EtOH	252(4.24),323(4.01)	44-0070-66
Indole-2-carboxylic acid, 3,7-dinitro-, ethyl ester	EtOH	227s(4.24),348(4.13)	44-0070-66
$C_{11}H_9N_3S$			
10H-Dipyrido[2,3-b:2',3'-e]-[1,4]thiazine, 3-methyl-	EtOH	239(4.41),341(3.99)	87-0116-66
$C_{11}H_9N_5$			
8-Azapurine, 9-benzyl-	pH -2.3	262(3.71)	39-0427-66B

Compound	Solvent	λ_{max}(log ε)	Ref.
8-Azapurine, 9-benzyl- (cont.)	pH 2.5	263(3.88)	39-0427-66B
Purine, 3-methyl-8-(3-pyridyl)-	pH 8	228(4.24),311(4.41)	39-0010-66C
s-Triazolo[4,3-b]pyridazine, 6-amino-3-phenyl-	EtOH	266(4.34)	78-2073-66
s-Triazolo[4,3-b]pyridazine, 8-amino-3-phenyl-	EtOH	205(4.35),252(4.29), 294(4.21),306s(4.16)	4-0218-66
$C_{11}H_9N_5O$			
Isoguanine, 9-phenyl-	pH 1	282(4.58)	95-0649-66
	pH 13	285(4.17)	95-0649-66
2-Purinol, 6-anilino-	pH 1	283(4.87)	95-0649-66
	pH 13	285(4.76),303(4.28)	95-0649-66
$C_{11}H_9N_5OS$			
Xanthine, 3-methyl-8-(3-pyridyl)-2-thio-	pH 8	239(3.65),327(3.80)	39-0010-66C
$C_{11}H_9N_5O_6$			
Pyrazole, 3,4-dimethyl-1-picryl-	EtOH	345(3.91)	22-3744-66
Pyrazole, 3,5-dimethyl-1-picryl-	EtOH	330(3.63)	22-3744-66
$C_{11}H_9N_5S$			
8-Azapurine, 9-benzyl-6-mercapto-	H_2O	332(4.26)	89-0587-66
[1,2,3]Thiadiazolo[5,4-d]pyrimidine, 7-(benzylamino)-	H_2O	325(3.86)	89-0587-66
$C_{11}H_9N_5S_2$			
Xanthine, 3-methyl-8-(3-pyridyl)-2,6-dithio-	pH 8	267(4.41),303(4.29), 370(4.45)	39-0010-66C
$C_{11}H_{10}$			
1,6-Methanocyclodecapentaene	C_6H_{12}	254(4.90),298(3.80), 360f(2.25)	33-2017-66
$C_{11}H_{10}BrClN_4$			
Pyrimidine, 2-amino-5-bromo-4-(m-chloroanilino)-6-methyl-	EtOH	259(3.85),308(4.00)	78-2401-66
Pyrimidine, 2-amino-5-bromo-4-(o-chloroanilino)-6-methyl-	EtOH	260(4.06),310(4.20)	78-2401-66
Pyrimidine, 2-amino-5-bromo-4-(p-chloroanilino)-6-methyl-	EtOH	262(3.91),308(4.20)	78-2401-66
$C_{11}H_{10}BrN$			
Isoquinoline, 4-bromo-1-ethyl-	EtOH	256(3.82),262(3.82), 312(3.55),318(3.59)	22-1763-66
$C_{11}H_{10}BrN_3O_2$			
Pyrazole, 4-bromo-3,5-dimethyl-1-(p-nitrophenyl)-	EtOH	315(4.19)	22-3744-66
$C_{11}H_{10}Br_2N_2S$			
6H-Dipyrido[2,1-b:1',2'-e][1,3,5]thiadiazinediium dibromide	H_2O	235(3.95),290(4.13), 313(4.11)	77-0546-66
$C_{11}H_{10}ClN_3$			
Pyrimidine, 4-(p-chloroanilino)-6-methyl-	EtOH	292(4.30)	78-2401-66

Compound	Solvent	$\lambda_{max}(\log \epsilon)$	Ref.
$C_{11}H_{10}ClN_3O$			
Pyrimidine, 2-amino-4-(p-chloro-phenoxy)-6-methyl-	EtOH	280(3.81)	78-2401-66
$C_{11}H_{10}ClN_5$			
Benzaldehyde, (4-amino-6-chloro-3-pyridazinyl)hydrazone	EtOH	203(4.31),237(4.34), 344(4.32)	4-0218-66
$C_{11}H_{10}Cl_2N_4$			
Pyrimidine, 2-amino-5-chloro-4-(o-chloroanilino)-6-methyl-	EtOH	258(3.20),308(3.98)	78-2401-66
$C_{11}H_{10}Cl_4$			
Spiro[2,4-cyclopentadiene-1,7'-nor-carane], 2,3,4,5-tetrachloro-	EtOH	247(4.03),289(3.60)	44-0768-66
Spiro[2.4]hepta-1,4,6-triene, 4,5,6,7-tetrachloro-1,2-diethyl-	EtOH	239(3.76),284(3.53)	44-4260-66
$C_{11}H_{10}FeO_7$			
Iron, tricarbonyl(muconic acid)-, dimethyl ester	EtOH	202(4.39),228(4.40), 240s(4.31),303(3.52)	101-0370-66A
$C_{11}H_{10}MoO_5$			
Molybdenum, tricarbonyl(carboxy-methyl)cyclopentadienyl-, methyl ester	C_6H_{12}	254(4.06),310(3.38)	101-0341-66A
$C_{11}H_{10}N_2$			
3H-Pyrido[3,4-b]indole, 4,9-dihydro-	EtOH-HCl	246(4.04),359(4.34)	39-0425-66C
	EtOH	236(4.20),242(4.20), 318(4.16)	39-0425-66C
$C_{11}H_{10}N_2O$			
1-Acetamidoquinolinium hydroxide, inner salt	EtOH	322(3.82)	94-0512-66
5,6-Diazaspiro[2.4]hept-6-en-4-one, 7-phenyl-	MeOH	218(3.88),287(4.09)	24-2962-66
Quinoline, 7-acetyl-8-amino-	n.s.g.	245(4.18),285(4.51), 346(3.74),384(3.70)	24-3806-66
$C_{11}H_{10}N_2OS$			
Thiazolin-2-imine, 3-(benzoyl-methyl)-	iso-PrOH-acid	247.2(4.37)	20-0380-66
	iso-PrOH	243.8(4.33)	20-0380-66
	iso-PrOH-base	244.5(4.30)	20-0380-66
$C_{11}H_{10}N_2O_2$			
1,3-Butanedione, 2-diazo-1-p-tolyl-	CH_2Cl_2	248(4.32)	24-3128-66
2-Indolinone, 3-(2-oxazolidin-ylidene)-	EtOH	253(4.37),302(4.32)	95-1152-66
3-Pyrazolol, 4-acetyl-1-phenyl-	EtOH	243(4.11),300(4.20)	44-1538-66
3,6-Pyridazinedione, 1,2-dihydro-1-methyl-2-phenyl-	EtOH	337(3.52)	95-1082-66
6(1H)-Pyridazinone, 3-hydroxy-4-methyl-1-phenyl-	n.s.g.	324(3.66)	49-0644-66
3(2H)-Pyridazinone, 6-methoxy-2-phenyl-	EtOH	326(3.70)	95-1082-66

Compound	Solvent	λ max(log ε)	Ref.
$C_{11}H_{10}N_2O_2S$			
Thiazoline-4,5-dione, 2-ethyl-phenylamino-	dioxan	230s(4.1),280(4.0), 320s(3.7)	24-3572-66
$C_{11}H_{10}N_2O_2S_2$			
5-Thiazolidineacetamide, 4-oxo-3-phenyl-2-thioxo-	EtOH	258(4.00),297(4.18)	95-0101-66
Thiazolin-2-imine, N-acetyl-3-(2-thenoylmethyl)-	iso-PrOH-acid	269(4.21)	20-0380-66
	iso-PrOH	266(4.16)	20-0380-66
	iso-PrOH-base	262(4.09),294(4.29)	20-0380-66
$C_{11}H_{10}N_2O_3$			
1,3-Butanedione, 2-diazo-1-(p-methoxyphenyl)-	CH_2Cl_2	244(4.26),270s(4.15)	24-3128-66
4-Oxazoleacetic acid, 2-anilino-	EtOH	265(4.62)	95-0300-66
Pyrazole-5-carboxylic acid, 3-methoxy-1-phenyl-	EtOH	221(3.96),280(3.70)	95-0867-66
Pyrazoline-5-carboxylic acid, 2-methyl-3-oxo-1-phenyl-	EtOH	303(3.84)	95-0867-66
2-Quinoxalinecarboxylic acid, 3-hydroxy-6,7-dimethyl-	EtOH	235(4.38),320(3.91), 345(3.89)	30-0577-66C
$C_{11}H_{10}N_2O_3S$			
Thiazolin-2-imine, N-acetyl-3-(2-furoylmethyl)-	iso-PrOH-acid	274.8(4.41)	20-0380-66
$C_{11}H_{10}N_2O_4$			
2H-Cyclopenta[d]pyridazine-1,4-dicarboxylic acid, dimethyl ester	n.s.g.	216(4.35),291(4.33), 303s(4.24),339(3.38), 505(2.68)	88-4979-66
Indole-2-carboxylic acid, 3-nitro-, ethyl ester	EtOH	220s(4.22),257(4.08), 346(3.86)	44-0070-66
2-Quinoxalinecarboxylic acid, 3-hydroxy-6,7-dimethyl-, 4-oxide	EtOH	260(4.62),330(3.88), 395(3.79)	30-0577-66C
Troponitrile, acetyl-4-nitro-	MeOH-KOH	541(4.61)	83-0131-66
$C_{11}H_{10}N_2O_5$			
7,10-Diaza[3.3.3]propellan-3-one, 7,10-dimethyl-6,8,9,11-tetraoxo-	MeCN	246(2.73),259(2.73), 268(2.76),291s(1.04), 304s(0.90),314s(0.48)	78-0279-66B
$C_{11}H_{10}N_4OS_2$			
s-Triazolo[3,4-b][1,3,4]thiadiazole, 3-(p-methoxyphenyl)-6-(methylthio)-	MeOH	264(4.35),281(4.27)	44-3528-66
$C_{11}H_{10}N_4O_2$			
4-Pyridazinecarboxaldehyde, 1,6-dihydro-3-hydroxy-6-oxo-, 5-phenylhydrazone	THF	258(4.03),350(3.95)	49-1494-66
4-Pyrimidinol, 6-amino-5-benzamido-	pH 1	258(3.97)	4-0476-66
	pH 7	258(3.96)	4-0476-66
	pH 13	260s(3.84)	4-0476-66
as-Triazine, 3-acetamido-5-phenyl-, 2-oxide	EtOH	240(4.27),300(4.20), 385(4.18)	44-3917-66
$C_{11}H_{10}N_4O_2S$			
2-Pyrimidinethiol, 4-amino-5-benzamido-6-hydroxy-	pH 1	227(4.09),284(4.28)	4-0476-66
	pH 7	225(4.10),284(4.27)	4-0476-66

Compound	Solvent	$\lambda_{max}(\log \epsilon)$	Ref.
2-Pyrimidinethiol, 4-amino-5-benzamido-6-hydroxy- (cont.)	pH 13	237s(4.21),290(4.15)	4-0476-66
$C_{11}H_{10}N_4O_3$			
Cytosine, 1-(p-nitrobenzyl)-	pH 1	282(4.3)	39-1784-66C
	pH 7	276(--)	39-1784-66C
	pH 13	276(--)	39-1784-66C
2,4-Pyrimidinediol, 4-amino-5-benzamido-	pH 1	229(4.17),263(4.29)	4-0476-66
	pH 7	228(4.16),263(4.30)	4-0476-66
	pH 13	230s(4.16),266(4.22)	4-0476-66
$C_{11}H_{10}N_4O_4$			
Pyrazole, 3,4-dimethyl-1-(2,4-dinitrophenyl)-	EtOH	338(4.17)	22-3744-66
Pyrazole, 4,5-dimethyl-1-(2,4-dinitrophenyl)-	EtOH	301(3.88)	22-3744-66
Pyrazole, 3-ethyl-1-(2,4-dinitrophenyl-	EtOH	326(4.15)	22-3744-66
$C_{11}H_{10}N_4O_6$			
Inosine 2',3'-carbonate	pH 2	248(4.05)	69-2076-66
	pH 12	253(4.07)	69-2076-66
$C_{11}H_{10}N_4S_2$			
s-Triazolo[3,4-b][1,3,4]thiadiazole, 6-(methylthio)-3-p-tolyl-	MeOH	260(4.31)	44-3528-66
$C_{11}H_{10}N_6$			
Purine, 2,6-diamino-8-phenyl-	pH 1	227(4.32),315(4.39)	4-0476-66
	pH 7	228(4.36),314(4.37)	4-0476-66
	pH 13	242(4.37),321(4.33)	4-0476-66
s-Triazolo[4,3-b]pyridazine, 6-hydrazino-3-phenyl-	EtOH	270(4.27)	78-2073-66
$C_{11}H_{10}N_6O_2$			
5H-s-Triazolo[4,3-b]-s-triazole, 3-ethyl-6-(p-nitrophenyl)-	EtOH	298(4.22)	4-0119-66
$C_{11}H_{10}N_6O_2S$			
Sulfanilamide, N^1-6-purinyl-	pH 1.2	311(4.39)	87-0373-66
	pH 12.9	306(4.51)	87-0373-66
$C_{11}H_{10}N_6O_4$			
6-Pyrimidinol, 2,4-diamino-5-(p-nitrobenzamido)-	pH 1	264.5(4.46)	4-0476-66
	pH 7	267(4.39)	4-0476-66
	pH 13	262.5(4.31)	4-0476-66
$C_{11}H_{10}N_6S$			
Purine, 6-[(4,6-dimethylpyrimidin-2-yl)thio]-	pH 1	298(4.04+)	44-1417-66
	pH 7	295(4.03+)	44-1417-66
	pH 13	301(4.06+)	44-1417-66
Purine-6(1H)-thione, 7-(4,6-dimethylpyrimidin-2-yl)-	pH 1	227(4.19),321(4.33)	44-1417-66
	pH 7	233(4.26),315(4.26)	44-1417-66
	pH 13	232(4.35),310(4.31)	44-1417-66
Spiro[pyrimidine-2(1H),5'-[5H]-thiazolo[3,4,5-gh]purine, 4,6-dimethyl-	pH 1	215(4.29),221(4.29), 267(4.14),307(4.37), 348(3.64)	44-1417-66
	pH 7	239(4.05),290(4.41)	44-1417-66
	pH 13	220(4.14),318(4.24)	44-1417-66

Compound	Solvent	λ max(log ε)	Ref.
$C_{11}H_{10}O$			
1-Naphthalenemethanol	dioxan	272(3.74),282(3.80), 293(3.65),313(2.48)	78-1587-66
Tricyclo[4.4.1.0^{1,6}]undeca-3,7,9-trien-2-one	C_6H_{12}	215(4.06),274(3.42), 320(2.26)	89-0732-66
$C_{11}H_{10}OS_3$			
Carbonic acid, trithio-, cyclic ester with 1,2,3,4-tetrahydro-2,3-naphthalenedithiol, thiono-oxide	EtOH	283(3.69),347(4.05)	77-0016-66
$C_{11}H_{10}O_2$			
3-Butenoic acid, 2-methylene-3-phenyl-	EtOH	240(4.09)	88-1787-66
4,7-Ethenoindene-5,6-dione, 3a,4,5,6,7,7a-hexahydro-	EtOH	214(3.16),453(1.65)	77-0775-66
5,8-Methanonaphthalene-1,2-dione, 1,2,4a,5,8,8a-hexahydro-	EtOH	238(3.69),435(0.91)	77-0775-66
	n.s.g.	279(3.23)	88-6175-66
Propiolic acid, m-tolyl-, methyl ester	MeOH	253(4.18)	44-1135-66
Propiolic acid, o-tolyl-, methyl ester	MeOH	263(4.13)	44-1135-66
Propiolic acid, p-tolyl-, methyl ester	MeOH	266(3.94)	44-1135-66
4H-Pyran-4-one, 2,3-dihydro-6-phenyl-	EtOH	243(3.65),298(4.15)	32-1073-66
	EtOH	243(3.65),298(4.15)	88-0233-66
$C_{11}H_{10}O_2S$			
2,8-Decadiene-4,6-diynoic acid, 9-(methylthio)-, cis-trans	ether	259(4.17),288(4.28), 306(4.26),328(4.24), 352(4.27)	24-3437-66
trans-trans	ether	267(4.11),291(4.17), 309(4.15),330(4.14), 355(4.15)	24-3437-66
α-Pyrone, 6-[4-(methylthio)-1,2,3-pentatrienyl]-	ether	237(4.04),300(4.03), 418(4.40)	5-0149-66D
$C_{11}H_{10}O_3$			
Cinnamic acid, β-(hydroxymethyl)-p-methoxy-, γ-lactone	EtOH	223(4.16),296(4.40)	39-1924-66C
Propiolic acid, (m-methoxyphenyl)-, methyl ester	MeOH	263(4.11),300(3.62)	44-1135-66
Propiolic acid, (o-methoxyphenyl)-, methyl ester	MeOH	263(4.03),308(3.85)	44-1135-66
Propiolic acid, (p-methoxyphenyl)-, methyl ester	MeOH	284(4.29)	44-1135-66
$C_{11}H_{10}O_3S$			
2-Naphthalenesulfonic acid, 6-methyl-, sodium salt	H_2O	265s(3.61),277(3.68), 287s(3.57),308(2.81), 327(2.30)	80-0263-66
$C_{11}H_{10}O_4$			
[2,2'-Bifuran]-3-carboxylic acid, 5'-methyl-, methyl ester	EtOH	233s(3.63),320(4.26)	39-0976-66C
3-Chromancarboxaldehyde, 7-methoxy-4-oxo-	EtOH	230(4.06),250(3.87), 274(4.06),305(3.85), 355(3.98)	78-1027-66
Coumarin-4-acetic acid, 3,4-dihydro-	MeOH	260s(2.68),266(2.80), 273(2.76)	25-1344-66

Compound	Solvent	λ_{max}(log ε)	Ref.
$C_{11}H_{10}O_4S$			
α-Pyrone, 6-[4-(methylsulfonyl)-3-penten-1-ynyl]-	MeOH	246s(4.02),257(4.06),333(4.31)	5-0149-66D
$C_{11}H_{10}O_5$			
Phthalic anhydride, 4,6-dimethoxy-3-methyl-	EtOH	224(4.32),246(4.30),280s(3.51),346(3.84)	39-0126-66C
$C_{11}H_{10}S$			
Naphthalene, 2-(methylthio)-	EtOH	252(4.48),280(3.88)	22-0228-66
$C_{11}H_{11}BClF_4NO_3S$			
N-Ethyl-5-(3' and 4')-chlorosulfonylphenylisoxazolium fluoroborate	CH_2Cl_2	281(4.34)	78-0321-66B
$C_{11}H_{11}BrN_2OS$			
Thiazolin-2-imine, 3-[2-hydroxy-2-(m-bromophenyl)ethyl]-	iso-PrOH-acid	256.0(3.84)	20-0380-66
	iso-PrOH-base	264.0(3.87)	20-0380-66
$C_{11}H_{11}BrN_4$			
Pyrimidine, 2-amino-4-anilino-5-bromo-	EtOH	258(3.92),306(4.00)	78-2401-66
$C_{11}H_{11}BrS$			
Benzo[b]thiophene, 6-bromo-2,3,5-trimethyl-	EtOH	238(4.55),245s(4.51),277(3.92),295s(3.62),308(3.21)	22-3055-66
$C_{11}H_{11}ClN_2OS$			
Thiazolin-2-imine, 3-[2-hydroxy-2-(p-chlorophenyl)ethyl]-	iso-PrOH-acid	221(4.21),257(3.87)	20-0380-66
	iso-PrOH-base	219(4.26),264(3.94)	20-0380-66
$C_{11}H_{11}ClN_2O_6S$			
6,7,8,9-Tetrahydro-2-nitropyrido-[2,1-b]benzothiazolium perchlorate	EtOH	251(4.15),323(3.93)	4-0027-66
$C_{11}H_{11}ClN_4$			
Pyrimidine, 2-amino-4-(m-chloroanilino)-6-methyl-	EtOH	259(3.50),299(4.22)	78-2401-66
Pyrimidine, 2-amino-4-(o-chloroanilino)-6-methyl-	EtOH	258(3.81),298(4.34)	78-2401-66
Pyrimidine, 2-amino-4-(p-chloroanilino)-6-methyl-	EtOH	265(3.75),302(4.31)	78-2401-66
$C_{11}H_{11}ClN_6O_2$			
Hydrazine, N-acetyl-N'-(p-chlorobenzoyl)-N-(1-methyl-1H-tetrazol-5-yl)-	EtOH	241(4.27)	39-1202-66C
$C_{11}H_{11}ClO_2$			
1,4-Methanonaphthalene, 6-chloro-5,8-dihydroxy-7-methyl-	n.s.g.	204(4.43),300(3.60)	16-0054-66
$C_{11}H_{11}ClO_4$			
3(2H)-Benzofuranone, 7-chloro-4-ethoxy-6-methoxy-	EtOH	210(4.28),235(4.26),286(4.29),322(3.71)	87-0242-66

Compound	Solvent	$\lambda_{max}(\log \epsilon)$	Ref.
Phthalide, 4-(chloromethyl)-6,7-dimethoxy-	EtOH	313(3.69)	44-1912-66
$C_{11}H_{11}ClO_4S$			
2,4-Dimethyl-1-benzothiopyrylium perchlorate	HOAc	258(4.66),336(3.93), 369(3.77)	78-0007-66
3,4-Dimethyl-1-benzothiopyrylium perchlorate	HOAc	257(4.23),334(3.73), 384(4.10)	78-0007-66
$C_{11}H_{11}ClO_5S$			
7-Methoxy-4-methyl-1-benzothiopyrylium perchlorate	HOAc	271(4.53),339(3.70), 413(3.82)	78-0007-66
$C_{11}H_{11}ClS$			
Butadiene, 1-chloro-1-(methylthio)-2-phenyl-	EtOH	234(4.08),288(4.03)	44-1694-66
$C_{11}H_{11}FN_2OS$			
Thiazolin-2-imine, 3-[2-hydroxy-2-(p-fluorophenyl)ethyl]-	iso-PrOH-acid	256.2(3.86)	20-0380-66
	iso-PrOH-base	264.0(3.88)	20-0380-66
$C_{11}H_{11}IN_2O$			
N-Acetamidoquinolinium iodide	EtOH	318(3.85)	94-0512-66
$C_{11}H_{11}N$			
Cyclopent[b]indole, 1,2,3,8-tetrahydro-	EtOH	229(4.47),281(3.84)	94-0934-66
1,6-Methano[10]annulen-2-amine	C_6H_{12}	254(4.51),270s(4.32), 342(3.74),415(3.04)	89-0732-66
1,6-(N-Methylimino)[10]annulene	C_6H_{12}	253(4.75),293(3.95), 370s(2.55)	33-2017-66
Pyrrolo[3,2,1-hi]indole, 1,2-dihydro-4-methyl-	MeOH	223(4.19),283(3.66)	35-4061-66
1H-Pyrrolo[1,2-a]indole, 2,3-dihydro-	EtOH	229(4.23),283(3.81)	44-0870-66
Quinoline, 2,3-dimethyl-	C_6H_{12}	267f(3.64)	5-0149-66H
	MeOH-acid	320(4.04)	5-0149-66H
	MeOH	273f(3.62)	5-0149-66H
$C_{11}H_{11}NO$			
Indole-3-carboxaldehyde, 2,4-dimethyl-	EtOH	216(4.52),248(4.29), 268(4.02),313(4.07)	44-3321-66
2-Indolinone, 3-isopropylidene-	EtOH	216s(3.97),250(4.44), 253(4.44),259(4.52), 290(3.82),300s(3.71), 346(3.26)	44-0077-66
Isocarbostyril, 3-ethyl-	EtOH	226(4.15),239(3.92), 247(3.74),278(3.90), 286(3.84),317(3.51), 331(3.60),345(3.45)	44-2090-66
Isoquinoline, 1-methoxy-3-methyl-	EtOH	217(4.45),232(3.15), 237(3.67),263(3.70), 273(3.80),283(3.74), 303(3.38),310(3.52), 328(3.47)	44-2090-66
2-Naphthamide, 3,4-dihydro-	EtOH	225(4.30),287(4.18)	44-1372-66
Naphth[1,2-b]azet-2(1H)-one, 2a,3,4,8b-tetrahydro-	EtOH	263(2.48),271(2.48)	44-1372-66
3-Pyrrolin-2-one, 4-methyl-3-phenyl-	MeOH	230(4.02)	44-0048-66

Compound	Solvent	λ_{max}(log ϵ)	Ref.
Quinoline, 2,3-dimethyl-, N-oxide	EtOH	317(3.87),337s(3.76)	1-2467-66
Quinoline, 2,4-dimethyl-, N-oxide	EtOH	323(3.98),333s(3.93)	1-2467-66
Tropone, 2-cyano-4-isopropyl-	MeOH	245(4.31),305(3.82), 318(3.82),340(3.67)	18-2538-66
Tropone, 2-cyano-5-isopropyl-	MeOH	237(4.28),321(3.89), 340s(3.83)	18-2538-66
Tropone, 2-cyano-6-isopropyl-	MeOH	246(4.35),303(3.58), 315(3.59),343(3.60)	18-2538-66
$C_{11}H_{11}NOS_2$			
2-Indolinone, 3-(bismethylthio- methylene)-	EtOH	261(4.08),283(4.02), 366(4.22)	95-1152-66
$C_{11}H_{11}NOSe$			
Benzoselenazoline, 2-acetyl- methylene-3-methyl-	EtOH	355.0(4.56)	22-4028-66
$C_{11}H_{11}NO_2$			
Benz[f]azepinedione, dihydromethyl-	EtOH	229(4.27),265(3.88), 317(3.43)	88-2361-66
Indole-4,7-dione, 1-ethyl-2-methyl-	MeOH	230(4.19),254(4.07), 345(3.40),440(3.32)	35-0804-66
Isocarbostyril, 3-ethyl-2-hydroxy-	EtOH	225(4.22),241(3.94), 248(3.92),291(3.89), 318(3.63),327(3.66), 342(3.48)	44-2090-66
Isoxazole, 3-ethoxy-5-phenyl-	EtOH	258(4.12)	32-0454-66
Quinaldine, 6-methoxy-, N-oxide	EtOH	315(3.89),333s(3.63), 348s(3.49)	1-2467-66
Spiro[indoline-3,2'-oxiran]-2-one, 3',3'-dimethyl-	EtOH	218(3.45),248s(3.72), 260s(3.62),268s(3.43), 302(3.18)	44-0077-66
$C_{11}H_{11}NO_3$			
Isocarbostyril, 8-hydroxy- 7-methoxy-N-methyl-	EtOH	228(4.49),289(3.98), 298(3.83),353(3.76), 362s(--)	24-2703-66
$C_{11}H_{11}NO_4S$			
2H-1,2-Benzothiazine, 4-acetoxy- 2-methyl-, 1,1-dioxide	EtOH	282(3.89),304s(3.79)	44-0162-66
N-Ethyl-5-phenylisoxazolium- 3'-sulfonate	pH 1	283(4.35)	78-0321-66B
N-Ethyl-5-phenylisoxazolium- 4'-sulfonate (hydrate)	pH 1	288(4.45)	78-0321-66B
$C_{11}H_{11}NO_6$			
Phthalic anhydride, 3-amino- 4,5,6-trimethoxy-	ether	235(4.26),259(4.33), 298(3.38),390(3.84)	44-1912-66
$C_{11}H_{11}NO_9$			
Phthalic acid, 4,5,6-trimethoxy- 3-nitro-	EtOH	245s(3.88),295(3.18)	44-1912-66
$C_{11}H_{11}N_3$			
2-Benzimidazoleacetonitrile, α-ethyl-	$CHCl_3$	276(3.88),282(3.86)	22-3989-66
$C_{11}H_{11}N_3O$			
2-Indolinone, 3-(2-imidazol- idinylidene)-	EtOH	254(4.30),306(4.42)	95-1152-66

Compound	Solvent	λ_{max}(log ε)	Ref.
$C_{11}H_{11}N_3OS$			
Uracil, 2-thio-5-p-toluidino-	pH 5.0	235(4.09),315(4.22)	87-0108-66
	pH 10.7	238(4.15),298(4.17)	87-0108-66
$C_{11}H_{11}N_3O_2$			
Pyrazole, 4,5-dimethyl-1-(p-nitrophenyl)-	EtOH	313(4.09)	22-3744-66
Uracil, 5-m-toluidino-	pH 1	245(4.11),312(3.48)	87-0121-66
	pH 11	242(4.08),277(3.90)	87-0121-66
Uracil, 5-o-toluidino-	8N HCl	262.5(3.89)	87-0108-66
	pH 7	243(4.09),312(3.48)	87-0108-66
	pH 13	238(4.10),280(3.90)	87-0108-66
Uracil, 5-p-toluidino-	8N HCl	260(3.91)	87-0108-66
	pH 7	247(4.12),310s(3.54)	87-0108-66
	pH 11	238(4.10),282s(3.91)	87-0108-66
$C_{11}H_{11}N_3O_3$			
1,2,4-Oxadiazole-3-carbamic acid, 5-phenyl-, ethyl ester	EtOH	237(4.35)	39-1522-66C
Propionamide, N-(1,2,3,4-tetrahydro-1,4-dioxo-5-phthalazinyl)-	DMF	333(3.91)	18-0932-66
$C_{11}H_{11}N_3O_5$			
Formic acid, (benzoylazo)-, O-carboxy-oxime, dimethyl ester	$CHCl_3$	245(4.16),418(1.83)	88-0405-66
$C_{11}H_{11}N_3S_2$			
Uracil, 2,4-dithio-p-toluidino-	pH 8.6	248(3.86),308(4.07)	87-0108-66
	pH 13.0	283(3.98),358s(3.46)	87-0108-66
	EtOH	263(4.11),325(4.52), 415(3.68)	87-0108-66
$C_{11}H_{11}N_5$			
5H-s-Triazolo[4,3-b]-s-triazole, 3-ethyl-6-phenyl-	EtOH	248(4.34)	4-0119-66
5H-s-Triazolo[4,3-b]-s-triazole, 3-methyl-6-p-tolyl-	EtOH	253(4.39)	4-0119-66
$C_{11}H_{11}N_5O$			
Pyrimidine, 2,4-diamino-5-benzamido-	pH 1	230(4.36),275s(3.85)	4-0476-66
	pH 7	230(4.33),275s(3.81)	4-0476-66
	pH 13	225s(4.22),290(3.88)	4-0476-66
5H-s-Triazolo[4,3-b]-s-triazole, 6-(p-methoxyphenyl)-3-methyl-	EtOH	264(4.41)	4-0119-66
$C_{11}H_{11}N_5O_2$			
4-Pyrimidinol, 2,6-diamino-5-benzamido-	pH 1	236(4.13),265(4.30)	4-0476-66
	pH 7	232s(4.13),267(4.20)	4-0476-66
	pH 13	234s(4.16),263(4.05)	4-0476-66
$C_{11}H_{11}N_5O_5$			
Adenosine, 2',3'-cyclic carbonate	EtOH	258(4.13)	69-2076-66
$C_{11}H_{11}N_7$			
1H-Tetrazole, 5-(4-amino-1-benzylimidazol-5-yl)-	pH 1	243(3.88),267(3.88)	44-2210-66
	pH 7, 13	258(3.94)	44-2210-66
1H-Tetrazole, 5-(5-amino-1-benzylimidazol-5-yl)-	pH 1	246(3.99),267(4.06)	44-2210-66
	pH 7	246(4.09)	44-2210-66
	pH 13	245(4.11)	44-2210-66

Compound	Solvent	$\lambda_{max}(\log \epsilon)$	Ref.
$C_{11}H_{12}$			
Indene, 1,2-dimethyl-	EtOH	260(3.99)	11-0109-66
$C_{11}H_{12}BF_4NO$			
N-Ethyl-5-phenylisoxazolium tetrafluoroborate	CH_2Cl_2	295(4.32)	78-0415-66A
$C_{11}H_{12}BF_4NS$			
N-Ethyl-3-phenylisothiazolium tetrafluoroborate	pH 1	281(4.24)	78-2135-66
N-Ethyl-4-phenylisothiazolium tetrafluoroborate	pH 1	225(4.21),305(3.51)	78-2135-66
N-Ethyl-5-phenylisothiazolium tetrafluoroborate	pH 1	227s(3.62),260s(3.65), 306(4.14)	78-2135-66
$C_{11}H_{12}BF_4N_3O_2S$			
2-(N^{ω}-Carbethoxyazo)-3-methylbenzothiazolium tetrafluoroborate	MeCN	368(4.07)	5-0065-66J
$C_{11}H_{12}Br_3NO_3$			
Ether, 3-bromo-2,2-bis(bromomethyl)-propyl o-nitrophenyl	EtOH	213(4.22),263(3.67)	78-0199-66
$C_{11}H_{12}ClNOSn$			
Tin, chloro(8-quinolinolato)-trimethyl-	C_6H_{12}	397(3.16)	101-0249-66B
	pyridine	379(3.37)	101-0249-66B
	dioxan	378(3.34)	101-0249-66B
$C_{11}H_{12}ClNO_4S$			
6,7,8,9-Tetrahydropyrido[2,1-b]benzothiazolium perchlorate	EtOH	213(4.09),242(4.03), 306s(4.12),316(4.24)	4-0027-66
$C_{11}H_{12}ClN_3O_4$			
1H-Imidazo[4,5-c]pyridine, 4-chloro-1-α-D-ribofuranosyl-	pH 1	256(3.66),270(3.80)	69-0756-66
	pH 11	255(3.87),265(3.84), 273(3.76)	69-0756-66
1-β-D-isomer	pH 1	207(4.53),257(3.72), 265(3.74),273(3.72)	87-0105-66
	pH 1	253(3.75),266(3.76), 273(3.68)	69-0756-66
	pH 7	255(3.80),265(3.76), 273(3.65)	87-0105-66
	pH 11	252(3.85),265(3.77), 273(3.68)	69-0756-66
	pH 13	257(3.81),265(3.78), 273(3.67)	87-0105-66
3H-Imidazo[4,5-b]pyridine, 7-chloro-3-β-D-ribofuranosyl-	pH 1	253(3.76),274(3.77), 280s(--)	87-0354-66
	pH 7	255(3.77),277(3.76), 285s(--)	87-0354-66
	pH 13	257(3.76),278(3.78), 286s(--)	87-0354-66
$C_{11}H_{12}ClN_5$			
Pyrazino[2,3-d]pyridazine, 5-chloro-8-piperidino-	EtOH	211(4.19),220s(4.17), 248(4.06),270s(3.87)	4-0512-66
$C_{11}H_{12}ClN_5O_2$			
N-[2-(Dimethylamino)-5-nitro-4-pyrimidinyl]pyridinium chloride	pH 0.0	255(3.90),352(4.09)	24-2997-66

Compound	Solvent	λ_{max}(log ε)	Ref.
$C_{11}H_{12}Cl_4$			
Cyclohexane, (2,3,4,5-tetrachloro-2,4-cyclopentadien-1-yl)-	EtOH	299(3.25)	44-0768-66
Spiro[2.4]hepta-4,6-diene, 1-butyl-4,5,6,7-tetrachloro-	EtOH	241(3.90),287(3.53)	44-0768-66
Spiro[2.4]hepta-4,6-diene, 4,5,6,7-tetrachloro-1-methyl-2-propyl-, cis	EtOH	244(4.06),285(3.64)	44-0768-66
trans	EtOH	245(3.97),286(3.57)	44-0768-66
Spiro[2.4]hepta-4,6-diene, 4,5,6,7-tetrachloro-1,1,2,2-tetramethyl-	EtOH	254(4.03),296(3.24)	44-0768-66
$C_{11}H_{12}N_2$			
Benzimidazole, 2-(1-ethylvinyl)-	$CHCl_3$	293(4.15)	22-3989-66
Pyrazole, 4,5-dimethyl-3-phenyl-	EtOH	251(4.12)	22-2971-66
Pyrazole, 4,5(3)-dimethyl-3(5)-phenyl-	EtOH	251(4.12)	22-2845-66
	EtOH	251.5(4.12)	22-3744-66
5H-Pyrido[2,3-b]indole, 6,7,8,9-tetrahydro-	EtOH	232(4.38),295(3.92)	30-0361-66D
Quinazoline, 2-isopropyl-	pH 0.0	259(3.97)	39-0234-66C
	pH 10.0	224(4.62),265(3.38), 310(3.38),320s(3.27)	39-0234-66C
$C_{11}H_{12}N_2O$			
Indole-3-carboxaldehyde, 5-(dimethylamino)-	EtOH	240(4.29),282(4.06)	23-0387-66
2-Pyrazoline, N-acetyl-3-phenyl-	n.s.g.	286(4.36)	88-5525-66
$C_{11}H_{12}N_2OS$			
Thiazolin-2-imine, 3-(2-hydroxy-2-phenylethyl)-	iso-PrOH-acid	265.0(3.82)	20-0380-66
	iso-PrOH	263.5(3.83)	20-0380-66
	iso-PrOH-base	263.0(3.83)	20-0380-66
$C_{11}H_{12}N_2OSe$			
4-Selenazolidinone, 3-methyl-2-(methylimino)-5-phenyl-	EtOH	212s(4.3)	28C-0285-66A
4-Selenazolinone, 2-(dimethylamino)-5-phenyl-	EtOH	230(4.33),242(4.31)	28C-0285-66A
$C_{11}H_{12}N_2O_2$			
8(1H)-Cycloheptapyrazolone, 1,3-dimethyl-7-methoxy-	MeOH	230(4.37),250(4.37), 302(3.98),316(4.03), 370(3.94)	18-0253-66
2-Pyrazolin-5-one, 4-(2-hydroxyethyl)-3-phenyl-	MeOH	247(4.12)	24-2962-66
$C_{11}H_{12}N_2O_2S$			
Pyrazole, 3-methyl-1-(p-toluenesulfonyl)-	EtOH	243(4.20)	28C-0782-66A
3-Thiophenecarboxylic acid, 2-amino-5-(2-cyano-1-methylvinyl)-4-methyl-, methyl ester	MeOH	226(4.31),317s(3.90), 360(4.11)	24-0094-66
$C_{11}H_{12}N_2O_2S_2$			
Thiazolin-2-imine, N-acetyl-3-(2-hydroxy-2-thienylethyl)-	iso-PrOH-acid	234(4.01),287(3.89)	20-0380-66
	iso-PrOH	233(4.08),300(4.13)	20-0380-66
	iso-PrOH-base	236(4.06),300(4.13)	20-0380-66

Compound	Solvent	$\lambda_{max}(\log \epsilon)$	Ref.
$C_{11}H_{12}N_2O_3$			
Tryptophan, 6-hydroxy-	MeOH	266s(3.62),274(3.67), 296(3.71)	5-0157-66J
$C_{11}H_{12}N_2O_3S$			
Thiazolin-2-imine, N-acetyl- 3-(2-furyl-2-hydroxyethyl)-	iso-PrOH-acid	216(4.03),286(3.88)	20-0380-66
	iso-PrOH-base	216(4.09),299(4.15)	20-0380-66
$C_{11}H_{12}N_2O_4$			
7,10-Diaza[3.3.3]propellane, 7,10-dimethyl-6,8,9,11-tetraoxo-	MeCN	246(2.67),259(2.62), 268(2.63)	78-0279-66B
$C_{11}H_{12}N_4$			
Pyrazole, 3,5-dimethyl-4-(phenylazo)-	$CHCl_3$	330(4.26),410(2.98)	22-2990-66
Pyrimidine, 2-amino-4-anilino- 6-methyl-	EtOH	262(3.82),300(4.11)	78-2401-66
$C_{11}H_{12}N_4O$			
5-Pyrazolone, 1,3-dimethyl- 4-(phenylazo)-	$CHCl_3$	392(4.18)	22-2990-66
4-Pyrimidinol, 2-amino-5-m-toluidino-	pH 1.8	238(4.01),273(4.06), 332s(3.71)	87-0108-66
	pH 7.0	273(4.01)	87-0108-66
$C_{11}H_{12}N_4O_3S$			
Imidazole-5-carbonitrile, 4-amino-, p-toluenesulfonate	EtOH	220(4.41),246(4.09)	35-3829-66
$C_{11}H_{12}N_4O_4$			
Pyrazoline, 3,4-dimethyl- 1-(2,4-dinitrophenyl)-	$CHCl_3$	392(4.35)	22-0610-66
Pyrazoline, 3-ethyl- 1-(2,4-dinitrophenyl)-	$CHCl_3$	396(4.33)	22-0610-66
$C_{11}H_{12}N_4O_6S$			
2-Thiophenecarboxaldehyde, tetra- hydro-, 2,4-dinitrophenyl- hydrazone, 1,1-dioxide	EtOH	221(4.14),349(4.38)	24-1580-66
$C_{11}H_{12}N_6O$			
Benzamide, N-(2,4,6-triamino- 5-pyrimidinyl)-	pH 1	215(4.49),237s(4.20), 272(4.28)	4-0476-66
	pH 7	215(4.47),238s(4.17), 272(4.26)	4-0476-66
	pH 13	236s(4.21),267(4.09)	4-0476-66
$C_{11}H_{12}O$			
Chroman, 2-vinyl-	n.s.g.	269s(2.89),277(3.03), 284(3.00)	22-1974-66
1-Indanone, 2,3-dimethyl-	EtOH	245(4.08),291(3.41)	11-0109-66
5,6-Undecadiene-8,10-diyn-1-ol	EtOH	208(4.67),224(3.69), 237(3.84),249(3.92), 263(4.09),278(4.04)	39-0129-66C
	EtOH	225(3.57),236(3.79), 248(3.98),262(4.13), 277(4.07)	39-0129-66C
5,7,9-Undecatriyn-1-ol	EtOH	209.0(5.01)	39-0129-66C

Compound	Solvent	λ_{max}(log ϵ)	Ref.
$C_{11}H_{12}OS$			
1-Benzoxepin, 2,5-epithio-2,3,4,5-tetrahydro-9-methyl-	MeOH	275(3.26),285(3.30)	35-5855-66
$C_{11}H_{12}OS_2$			
2-Propenal, 3,3-bis(methylthio)-2-phenyl-	MeOH	257(3.68),326(4.04)	22-1582-66
$C_{11}H_{12}O_2$			
p-Benzoquinone, (3-methyl-2-butenyl)-	ether	243(4.28),315(2.87), 440(1.52)	24-0885-66
3-Butenoic acid, 2-methyl-3-phenyl-	EtOH	237.5(3.97)	35-1959-66
1,4-Methanonaphthalene, 5,8-dihydroxy-6-methyl-	n.s.g.	207(4.35),300(3.60)	16-0054-66
2,4-Pentanedione, 3-phenyl-	hexane	284(3.98)	18-0901-66
	EtOH	235s(--),286(3.95)	18-0901-66
	EtOH-NaOEt	306.5(4.26)	18-0901-66
palladium derivative	EtOH	208(4.82),235(4.66), 265s(4.35),340(3.99)	30-0103-66D
Tricyclo[4.2.1.0^{2,5}]nona-3,7-diene-3-carboxylic acid, methyl ester	EtOH	250(3.38)	35-4273-66
5,6-Undecadiene-8,10-diyne-1,4-diol	EtOH	208(4.56),225(3.62), 237(3.72),250(3.94), 264(4.12),279(4.03)	39-0129-66C
$C_{11}H_{12}O_2S$			
2-Thiopheneacrylic acid, 5-propenyl-, methyl ester	ether	227(4.14),345(4.42)	5-0149-66D
$C_{11}H_{12}O_3$			
Benzoic acid, o-acetonyl-, methyl ester	EtOH	231(3.99),274(3.12)	12-1265-66
Benzoic acid, o-(3-oxobutyl)-	EtOH	228(3.83),277(3.11)	44-2090-66
Isocoumarin, 3,4-dihydro-8-hydroxy-3,5-dimethyl-	MeOH	210(4.35),247(3.81), 323(3.61)	88-3723-66
Pentanoic acid, 2-(o-hydroxyphenyl)-4-hydroxy-, lactone	MeOH	274(3.40),280(3.36)	25-1344-66
	MeOH-NaOH	238(3.81),293(3.48)	25-1344-66
Tropone, 2-carboxy-4-isopropyl-	MeOH	238(4.27),319(3.90)	18-2538-66
Tropone, 2-carboxy-6-isopropyl-	MeOH	240(4.34),304s(3.74), 316(3.77)	18-2538-66
$C_{11}H_{12}O_4$			
Furo[2,3-b]benzofuran-4-ol, tetrahydro-6-methoxy-	EtOH	208(4.63),225(3.95), 270(2.85),278(2.74)	39-1308-66C
Isocoumarin, 3,4-dihydro-8-hydroxy-6-methoxy-3-methyl-	EtOH	267(4.21),301(3.97)	39-0168-66C
$C_{11}H_{12}O_5$			
2,4-Cyclohexadien-1-one, 4-acetyl-6-hydroxy-6-methoxy-, acetate	EtOH	301(3.49)	49-1384-66
Phthalide, 4,5,6-trimethoxy-	EtOH	256(3.93),296(3.61)	44-1912-66
$C_{11}H_{12}O_6$			
Epoxydon diacetate	EtOH	228(3.85),275s(2.3), 329(1.64)	33-0204-66
Phthalic acid, 4,6-dimethoxy-3-methyl-	EtOH	210(4.31),250(3.80), 298(3.67)	39-0126-66C

Compound	Solvent	λ max(log ε)	Ref.
$C_{11}H_{12}O_8$			
2,3,4-Furantricarboxylic acid, 5-methoxy-, trimethyl ester	EtOH	276(4.18)	88-1185-66
	ether	248s(4.04),273(4.19)	89-0579-66
$C_{11}H_{12}S$			
2H-1-Benzothiopyran, 2,4-dimethyl-	EtOH	244(4.29),275s(3.70), 322(3.05)	78-0007-66
$C_{11}H_{13}^+$			
Cyclopropylmethylphenylcarbonium ion	FSO_3H-SbF_5	316(4.18),404(3.33)	35-1488-66
$C_{11}H_{13}ClLiN_4O_7P$			
Inosine, 5'-chloromethyl-phosphonate, lithium salt	pH 2	249(4.01)	69-2076-66
	pH 12	253(4.07)	69-2076-66
$C_{11}H_{13}ClN_2O_2S$			
1,2,4-Benzothiadiazine-1,1-dioxide, 6-chloro-7-methyl-2-propyl-	EtOH	229(4.48),270(4.00), 279(3.95),300(3.66)	7-1083-66
$C_{11}H_{13}ClO_4$			
Benzoic acid, 3-chloro-6-hydroxy-4-methoxy-2-methyl-, ethyl ester	EtOH	260(4.02),306(3.68)	12-1265-66
$C_{11}H_{13}N$			
1-Aza-2,3-benzo-6-methylbicyclo-[2.2.1]heptene, exo-	ether	256(2.81),261(2.84), 267s(2.80)	5-0128-66D
Indole, 2,3,4-trimethyl-	EtOH	230(4.51),278s(3.84), 284(3.86),292s(3.77)	44-3321-66
3H-Indole, 2,3,3-trimethyl-	EtOH	258(3.80)	94-0934-66
$C_{11}H_{13}NO$			
Benzocyclobutene, O-methyl-1-acetyl-, oxime	EtOH	259(3.10),265(3.28), 272(3.26)	87-0656-66
3-Buten-2-one, 3-methyl-4-phenyl-, α-oxime	n.s.g.	269(4.47)	83-0626-66
3-Buten-2-one, 4-p-tolyl-, α-oxime	n.s.g.	281(4.35)	83-0626-66
β-oxime	n.s.g.	287(4.21)	83-0626-66
Carbostyril, 3,4-dihydro-5,8-dimethyl-	EtOH	253(4.03)	35-1205-66
Cinnamamide, α,β-dimethyl-, cis	EtOH	239(3.83)	35-1959-66
trans	EtOH	233(3.94)	35-1959-66
Isocarbostyril, 3,4-dihydro-5,8-dimethyl-	EtOH	238(3.82),290(3.23)	35-1205-66
1-Naphthonitrile, 1,2,3,4,6,7,8,8a-octahydro-6-oxo-	EtOH	235(4.21)	39-1866-66C
isomer	EtOH	235(4.21)	39-1866-66C
1-Penten-3-one, 1-phenyl-, α-oxime	n.s.g.	285(4.37)	83-0626-66
β-oxime	n.s.g.	287(4.30)	83-0626-66
2-Pyrrolidinone, 1-benzyl-	MeOH	217(4.3)	24-0724-66
2-Pyrrolidinone, 5-benzyl-	EtOH	243s(1.92),248s(2.09), 253(2.22),258(2.29), 264(2.18),268(2.05)	95-1213-66
$C_{11}H_{13}NO_2$			
Acetanilide, 2'-acetonyl-	EtOH	227s(3.79)	1-2467-66
Acrylic acid, α-cyano-β-cyclo-pentylidene-, ethyl ester	EtOH	235(4.11)	22-1033-66
Acrylic acid, 3-(N-methylanilino)-, methyl ester, trans	EtOH	220(3.83),297(4.47)	24-2526-66

Compound	Solvent	$\lambda_{max}(\log \epsilon)$	Ref.
3-Buten-2-one, 4-(p-nitrophenyl)-, α-oxime	n.s.g.	293(4.50)	83-0626-66
β-oxime	n.s.g.	298.5(4.43)	83-0626-66
Butyronitrile, 4-(p-methoxyphenoxy)-	MeOH	226(4.00),287(3.38)	78-0093-66B
Cinnamamide, 2-ethoxy-	MeOH	222(4.07),275(4.33)	5-0053-66I
2-Indolinol, N-acetyl-3-methyl-	EtOH	254(4.24),293(3.58), 303s(3.47)	1-2467-66
	EtOH	248(4.19),278(3.56), 285(3.51)	88-4355-66
$C_{11}H_{13}NO_3$			
Acetamide, N-(α-methoxyphenacyl)-	EtOH	253(4.32),274(4.09)	78-0441-66A
3-Aza[3.3.3]propellan-7-one, 3-methyl-2,4-dioxo-	MeCN	250(2.42),260s(2.33), 292s(1.48),305s(1.30)	78-0279-66B
2-Indolinol, N-acetyl-5-methoxy-	EtOH	254(4.24),293(3.58), 303s(3.47)	88-4355-66
$C_{11}H_{13}NO_4$			
1H-Azepine-3,6-dicarboxylic acid, 4,5-dihydro-2-(hydroxymethyl)-7-methyl-, γ-lactone, methyl ester	EtOH	226(4.21),315(4.19)	39-1075-66C
2-Carbomethoxypentylidenecyano-acetic acid, methyl ester	EtOH	237(4.07),355(2.45)	78-2575-66
2-Cyclobuten-1-one, 3-ethoxy-2-methyl-4-(2,5-dioxopyrrolidin-1-yl)-	MeOH	240(4.06)	88-3031-66
Terephthalic acid, 2-amino-3-methyl-, dimethyl ester	EtOH	228(4.44),255s(3.88), 359(3.70)	39-1950-66C
$C_{11}H_{13}NO_5$			
Pyrano[4,3-d][1,3]dioxin, 7-(ethyl-amino)-2,2-dimethyl-4,5-dioxo-	EtOH	330(4.19)	78-0455-66
4-Pyronecarboxylic acid amide, 2,6-dimethyl-N-(carboxy-methyl)-, methyl ester	EtOH	230(4.31),262(4.10)	78-2003-66
$C_{11}H_{13}NS$			
Indolizine, 1,2-dimethyl-3-thioacetyl-	C_6H_{12}	452(4.46)	77-0401-66
10bH-Thiazolo[2,3-a]isoquinoline, 2,3,5,6-tetrahydro-	C_6H_{12}	260(2.75),269(2.81), 276(2.78)	83-0846-66
	MeOH-HCl	267(2.88),274(2.85)	83-0846-66
	MeOH	261(2.76),268(2.82), 276(2.80)	83-0846-66
$C_{11}H_{13}N_3$			
3,5-Pyridinedicarbonitrile, 1,4-di-hydro-1,2,4,6-tetramethyl-	n.s.g.	219(4.38),245s(3.62), 342(3.73)	39-1083-66B
6H-Pyrido[3',2':4,5]imidazo[1,2-a]-azepine, 7,8,9,10-tetrahydro-	EtOH	252(3.68),281s(4.01), 285(4.02),290s(3.94)	39-0080-66C
6H-Pyrido[3',4':4,5]imidazo[1,2-a]-azepine, 7,8,9,10-tetrahydro-	EtOH	253(3.62),260(3.63), 266(3.64),273(3.57)	39-0080-66C
Quinazoline, 2-(dimethylamino)-4-methyl-	pH 2.0	239(4.45),252s(4.34), 282(3.67),355(3.64)	39-0234-66C
	pH 9.0	246(4.49),259s(4.34), 273(4.13),369(3.43)	39-0234-66C
$C_{11}H_{13}N_3O$			
1,3,4-Oxadiazole, 2-(dimethylamino)-5-m-tolyl-	EtOH	289(4.23)	22-0153-66

Compound	Solvent	λ_{max}(log ε)	Ref.
1,3,4-Oxadiazole, 2-(dimethylamino)-5-o-tolyl-	EtOH	284(4.20)	22-0153-66
1,3,4-Oxadiazole, 2-(dimethylamino)-5-p-tolyl-	EtOH	289(4.28)	22-0153-66
2-Pentanone, 4-anilino-3-diazo-	EtOH	243(4.35),280(3.86)	24-0475-66
1-Propanone, 2,2-dimethyl-1-v-triazolo[1,5-a]pyridin-3-yl	MeOH	248s(3.58),256(3.64), 288(4.04),307(4.13)	24-2918-66
$C_{11}H_{13}N_3O_2$			
p-Anisaldehyde, 4,4-dimethylsemicarbazone	EtOH	293(4.35)	22-0153-66
1,3,4-Oxadiazole, 5-m-anisyl-2-(dimethylamino)-	EtOH	292(4.22)	22-0153-66
1,3,4-Oxadiazole, 5-o-anisyl-2-(dimethylamino)-	EtOH	275(4.06),285(4.08), 307(4.12)	22-0153-66
$C_{11}H_{13}N_3O_2S$			
Acetamide, N-[5-(2-furyl)-1,3,4-thiadiazol-2-yl]-N-isopropyl-	EtOH	307(4.28)	83-0921-66
p-Toluenesulfonamide, N-(1-methylpyrazol-5-yl)-	MeOH	227(4.1),263s(3.0), 276(2.7)	24-0178-66
$C_{11}H_{13}N_3O_4$			
3H-Imidazo[4,5-b]pyridine, 3-β-D-ribofuranosyl-	pH 1	236(3.70),274(4.00), 281(3.94)	87-0354-66
	pH 7	244(3.71),277s(--), 281(3.93),287(3.82)	87-0354-66
	pH 13	246(3.68),277s(--), 281(3.91),286s(--)	87-0354-66
$C_{11}H_{13}N_3O_4S$			
1H-Imidazo[4,5-c]pyridine-4(5H)-thione, 1-β-D-ribofuranosyl-	pH 1	224(4.07),290s(--), 332(4.13)	87-0105-66
	pH 7	226(4.09),295s(--), 324(4.19)	87-0105-66
	pH 13	227(4.08),302(4.18)	87-0105-66
$C_{11}H_{13}N_3O_5$			
1,4-Phthalazinedione, 5-amino-2,3-dihydro-6,7,8-trimethoxy-	6N NH_3	230(4.45),318(3.85), 358(3.92)	44-1912-66
$C_{11}H_{13}N_3S$			
1,3,4-Thiadiazole, 2-(isopropylamino)-5-phenyl-	EtOH	228(3.73),313(4.07)	83-0921-66
$C_{11}H_{13}N_5$			
Pyrimidine, 2,4-diamino-5-m-toluidino-	pH 1	232(4.31),260s(4.04)	87-0121-66
	pH 11	240(4.16),278s(3.94)	87-0121-66
Pyrimidine, 2,4-diamino-5-p-toluidino-	12N HCl	225(3.93),278(3.82)	87-0108-66
	pH 1.0	234(3.98)	87-0108-66
	pH 10.0	240(3.88),280s(3.61)	87-0108-66
$C_{11}H_{13}N_5O_2$			
Pyrido[2,3-b]pyrazine-6-carbamic acid, 8-amino-2-methyl-, ethyl ester	pH 1	228(4.49),250s(4.10), 323(4.26)	44-1890-66
	pH 7	225(4.48),263(4.38), 331(4.01)	44-1890-66
	pH 13	226(4.42),263(4.35), 334(3.98)	44-1890-66

Compound	Solvent	λ max(log ε)	Ref.
$C_{11}H_{13}N_5O_5$			
Adenosine, 3'-formate	dioxan	259(4.22)	69-3638-66
$C_{11}H_{13}N_5O_7$			
Hypoxanthine, 9-(3-deoxy-3-nitro-β-D-glucopyranosyl)-	H_2O	247(4.05)	24-0575-66
$C_{11}H_{14}$			
Cyclohexane, (2,4-cyclopentadien-1-ylidene)-	C_6H_{12}	262(4.23),268(4.36), 277(4.35),287(4.07), 358(2.58)	33-1278-66
	EtOH	269(4.31),276(4.30), 355(2.54)	33-1278-66
Naphthalene, tetrahydro-1-methyl-	EtOH	254s(2.57),259s(2.63), 266(2.71),273(2.70), 282s(1.81),293s(1.62)	22-0645-66
Spiro[2,4-cyclopentadiene-1,7'-norcarane]	pentane	238(3.66),247s(--), 260s(3.46)	44-3296-66
$C_{11}H_{14}BN$			
Borane, (dimethylamino)phenyl-1-propynyl-	C_6H_{12}	225(4.23)	22-3850-66
	C_6H_{12}	225(4.23)	28C-0376-66A
$C_{11}H_{14}BrClN_4O_2$			
Pyrazolo[5,1-c]-as-triazine, 4-(bromomethyl)-6-(chloroacetyl)-4-ethoxy-4,6-dihydro-3-methyl-	EtOH	251(4.14)	39-1214-66C
$C_{11}H_{14}BrNS$			
10bH-Thiazolo[2,3-a]isoquinoline, 2,3,5,6-tetrahydro-, hydrobromide	H_2O	267(2.82),274(2.85)	83-0846-66
$C_{11}H_{14}ClNO_4$			
Benzene, 1-(4-chlorobutoxy)-4-methoxy-2-nitro-	MeOH	357(3.36)	78-0093-66B
Benzene, 1-(4-chlorobutoxy)-4-methoxy-3-nitro-	MeOH	357(3.37)	78-0093-66B
Benzene, 1-(3-chloro-2-methylpropoxy)-4-methoxy-2-nitro-	MeOH	357(3.33)	78-0093-66B
Benzene, 1-(3-chloro-2-methylpropoxy)-4-methoxy-3-nitro-	MeOH	357(3.34)	78-0093-66B
$C_{11}H_{14}ClNS$			
10bH-Thiazolo[2,3-a]isoquinoline, 2,3,5,6-tetrahydro-, HCl salt	H_2O	267(2.86),274(2.89)	83-0846-66
$C_{11}H_{14}ClN_3O$			
1-Acetonyl-5-amino-2-methylimidazo-[1,2-a]pyridinium chloride	H_2O	228(4.45),323(4.16)	4-0093-66
$C_{11}H_{14}ClN_5O_4$			
Adenine, 2-chloro-9-(5-deoxy-β-D-ribohexofuranosyl)-	pH 1	263(4.16)	87-0234-66
	pH 7	263(4.16)	87-0234-66
	pH 13	264(4.18)	87-0234-66
$C_{11}H_{14}ClO_4P$			
Phosphonic acid, (p-chlorobenzoyl)-, diethyl ester	C_6H_{12}	268(4.10)	44-1304-66

Compound	Solvent	λ_{max}(log ϵ)	Ref.
$C_{11}H_{14}N_2$			
Benzimidazole, 2-sec-butyl-	$CHCl_3$	276(3.83),282(3.82)	22-3989-66
Indole, 1-(dimethylaminomethyl)-	hexane	230(4.18),273(4.18)	24-0889-66
	EtOH	226(4.47),272(4.10)	24-0889-66
$C_{11}H_{14}N_2O$			
2-Benzimidazoleethanol, β-ethyl-	$CHCl_3$	275(3.89),282(3.90)	22-3989-66
Biacetyl, methylphenylhydrazone	$CHCl_3$	351(3.95)	22-2981-66
4-Isoquinolinecarboxamide, 1,2,3,4-tetrahydro-2-methyl-	EtOH	251(2.34),284(1.46), 290(1.46)	22-1763-66
Quinazoline, 1,2-dihydro-2,2,4-trimethyl-, 3-oxide	MeOH	236(4.29),293(3.90), 372(3.61)	39-1433-66
$C_{11}H_{14}N_2O_2$			
Pyruvic acid, ethyl ester, phenylhydrazone, anti	EtOH	323(4.34)	22-2981-66
syn	EtOH	351(4.23)	22-2981-66
L-Tryptophan, 4,7-dihydro-	pH 13	214(3.79)	35-3941-66
$C_{11}H_{14}N_2O_3$			
Butyric acid, 3-(3-phenylureido)-	EtOH	240(4.37),275(3.07)	95-0300-66
$C_{11}H_{14}N_2O_4$			
Glyoxylic acid, (phenylazo)-, methyl ester, dimethyl acetal	EtOH	275(3.8)	24-1899-66
$C_{11}H_{14}N_2O_8$			
β-D-Glucopyranuronic acid, 1-deoxy-1-(1,2,3,4-tetrahydro-2,4-dioxo-1-pyrimidinyl)-, methyl ester	MeOH	258.5(3.92)	94-1354-66
β-D-Glucopyranuronic acid, 1-deoxy-1-(1,2,3,4-tetrahydro-5-methyl-2,4-dioxo-1-pyrimidinyl)-	H_2O	265(4.08)	94-1354-66
$C_{11}H_{14}N_4O_3$			
Lumazine, 8-(2-hydroxyethyl)-3,6,7-trimethyl-	pH -2.7	248(4.12),361(4.15)	24-3503-66
	pH 7	262(4.32),408(4.09)	24-3503-66
	pH 12	232(4.27),282(3.99), 311(4.19),365(3.61)	24-3503-66
$C_{11}H_{14}N_4O_4$			
1H-Imidazo[4,5-c]pyridine, 4-amino-1-β-D-ribofuranosyl-	pH 1	262(4.01),274s(3.98)	69-0756-66
	pH 11	265(4.03)	69-0756-66
3,5,9-Triazatricyclo[5.3.0.0^{2,6}]-deca-4,8,10-trione, 1,3,7,9-tetramethyl-5-nitroso-	H_2O	264(3.75)	39-2239-66C
$C_{11}H_{14}N_4O_4S$			
2,8-Purinediol, 6-(4-carbomethoxybutyl)thio-	pH 1	222(4.30),339(4.03)	73-1824-66
	pH 13	228(4.21),326(4.14)	73-1824-66
6(1H)-Purinethione, 9-(5-deoxy-β-D-ribohexofuranosyl)-	pH 1	224(3.98),322(4.38)	87-0234-66
	pH 7	226(4.01),318(4.37)	87-0234-66
	pH 13	232(4.18),310(4.36)	87-0234-66
6-Purinethione, 9-(2-O-methyl-β-D-ribofuranosyl)-	pH 1	225(3.96),320(4.38)	35-3640-66
	pH 11	233(4.18),310(4.37)	35-3640-66
$C_{11}H_{14}N_4O_6$			
Inosine, 2-methoxy-	pH 1	249(3.95)	44-3258-66
	pH 11	259(4.03)	44-3258-66
	EtOH	247(4.02),257s(3.95)	44-3258-66

Compound	Solvent	λ_{max}(log ε)	Ref.
$C_{11}H_{14}N_6O_5$			
β-D-Glucopyranuronamide, 1-deoxy-1-(6-amino-9-purinyl)-	H_2O	259(4.20)	94-1360-66
$C_{11}H_{14}O$			
2(1H)-Azulenone, 4,5,6,7,8,8a-hexahydro-8-methylene-	EtOH	230(4.09)	88-3627-66
Benzaldehyde, 3,5-diethyl-	EtOH	255(4.17),295(3.36)	22-3238-66
Benzosuberone	EtOH	248(3.92),287(3.24),	22-1693-66
Bicyclo[3.3.]]nona-3,7-dien-2-one, 5,8-dimethyl-	EtOH	233(3.74)	39-0419-66C
Chroman, 2,2-dimethyl-	n.s.g.	269s(3.17),277(3.36), 284(3.34)	22-1974-66
Chroman, 2-ethyl-	n.s.g.	270s(3.22),277(3.37), 284(3.37)	22-1974-66
2(3H)-Naphthalenone, 4,4a,5,6-tetrahydro-4a-methyl-	EtOH	278(4.43)	44-3109-66
1-Naphthol, 5,6,7,8-tetrahydro-4-methyl-	EtOH	280(3.21)	44-0656-66
Tricyclo[4.4.0.0^{2,7}]dec-4-en-3-one, 6-methyl-	EtOH	222(3.48)	88-2043-66
$C_{11}H_{14}OS$			
Phenol, 2-methyl-6-(tetrahydro-2-thienyl)-	MeOH	278(3.32)	35-5855-66
	MeOH-KOH	283(3.38),297(3.30)	35-5855-66
Thiachroman, 7-methoxy-4-methyl-	EtOH	256(3.99),291(3.49), 299(3.42)	78-0007-66
Thiachroman, 8-methoxy-4-methyl-	EtOH	257(3.99),288(3.30)	78-0007-66
$C_{11}H_{14}O_2$			
2-[Bicyclo[2.2.1]heptane-1'-spiro-cyclobut-2'-ene-3'-carboxylic acid	EtOH	224.5(4.07)	22-3947-66
Chroman, 2-ethoxy-	n.s.g.	266s(3.06),274(3.23), 280(3.22)	22-1974-66
Eugenol, methyl-	isooctane	228(3.89),278(3.45)	102-0921-66
Hydroquinone, (3-methyl-2-butenyl)-	ether	209(4.08),295(3.61)	24-0885-66
$C_{11}H_{14}O_3$			
Hydroquinone, 2,5,6-trimethyl-, 1-acetate	EtOH	284(3.35)	73-4598-66
$C_{11}H_{14}O_4$			
Benzoic acid, 4,6-dihydroxy-2,3-dimethyl-, ethyl ester	EtOH	267(4.06),309(3.73)	39-0168-66C
$C_{11}H_{14}O_5$			
2-Cycloheptene-1,2-dicarboxylic acid, 7-oxo-, dimethyl ester, cis	C_6H_{12}	261(3.96)	35-4685-66
	EtOH	210(4.01),259(3.94)	35-4685-66
	MeCN	262(3.82)	35-4685-66
$C_{11}H_{14}S$			
Sulfide, 3-methyl-2-butenyl phenyl	EtOH	255(3.82)	44-1694-66
Thiachroman, 2,4-dimethyl-	EtOH	257(4.09),288s(2.99)	78-0007-66
Thiachroman, 3,4-dimethyl-	EtOH	255(3.95),283s(3.90)	78-0007-66
Thiachroman, 4,6-dimethyl-	EtOH	257(4.12),294(3.07)	78-0007-66
Thiachroman, 4,7-dimethyl-	EtOH	257(4.06),288(3.08)	78-0007-66
Thiachroman, 4,8-dimethyl-	EtOH	253(4.05),281s(3.10)	78-0007-66
Thiachroman, 4-ethyl-	EtOH	255(3.89),283s(3.84)	78-0007-66

Compound	Solvent	$\lambda_{max}(\log \epsilon)$	Ref.
$C_{11}H_{15}BF_4OS$			
(p-Methoxystyryl)dimethylsulfonium tetrafluoroborate, cis	EtOH	295(4.12)	35-5747-66
trans	EtOH	304(4.38)	35-5747-66
$C_{11}H_{15}Br$			
Naphthalene, 8a-(bromomethyl)-1,2,3,5,8,8a-hexahydro-	hexane	230(3.20)	78-0279-66B
$C_{11}H_{15}BrN_4O_2$			
Pyrazolo[3,2-c]-as-triazine, 6-acetyl-4-(bromomethyl)-4-ethoxy-4,6-dihydro-3-methyl-	EtOH	249(4.17)	39-1214-66C
$C_{11}H_{15}BrO_3$			
2-Cyclopenten-1-one, 2-acetoxy-3-bromo-4,4,5,5-tetramethyl-	EtOH	244.0(4.18)	39-0075-66B
$C_{11}H_{15}ClO_2$			
Benzene, 1-(4-chlorobutoxy)-4-methoxy-	MeOH	225(4.03),287(3.37)	78-0093-66B
Benzene, 1-(3-chloro-1-methylpropoxy)-4-methoxy-	MeOH	226(4.05),287(3.38)	78-0093-66B
$C_{11}H_{15}NOS$			
Fulvene, 6-(methylthio)-6-morpholino-	EtOH	278(3.66),354(4.43)	24-3268-66
$C_{11}H_{15}NO_2$			
3-Aza[3.3.3]propellane, 3-methyl-2,4-dioxo-	MeCN	250(2.21),259s(2.03)	78-0279-66B
Pyridine, 3,5-diacetyl-1,4-dihydro-1,4-dimethyl-	n.s.g.	236(4.25),272(3.94), 393(3.99)	39-1083-66B
Pyridine, 3,5-diacetyl-1,4-dihydro-2,6-dimethyl-	n.s.g.	251(4.09),277(3.95), 404(3.82)	39-1083-66B
$C_{11}H_{15}NO_2S$			
Styrylamine, N,N-dimethyl-β-(methylsulfonyl)-	EtOH	243(4.16)	87-0489-66
Vinylamine, 2-(benzylsulfonyl)-N,N-dimethyl-	EtOH	251(4.35)	87-0489-66
$C_{11}H_{15}NO_3$			
Ether, isopentyl p-nitrophenyl	EtOH	225(4.00),318(4.08)	94-0939-66
6-Isoquinolinol, 1,2,3,4-tetrahydro-7,8-dimethoxy-	pH 1	223s(3.92),273s(3.11), 279(3.15)	33-0403-66
	pH 13	240s(3.89),295(3.56)	33-0403-66
	iso-PrOH	230s(3.90),273s(3.20), 282(3.27)	33-0403-66
2-Naphthamide, 1,4,4a,5,6,7,8,8a-octahydro-3-hydroxy-1-oxo-	MeOH-HCl	259(4.22)	44-2429-66
	MeOH-NaOH	268(4.27)	44-2429-66
2(1H)-Pyridone, 4-hydroxy-3-iso-valeryl-6-methyl-	EtOH	230(4.05),267(3.43), 325(4.11)	70-0848-66
2(1H)-Pyridone, 4-hydroxy-6-methyl-3-valeryl-	EtOH	230(4.03),267(3.44), 327(4.13)	70-0848-66
$C_{11}H_{15}NO_4$			
Benzene, 1-butoxy-2-methoxy-4-nitro-	EtOH	240(3.98),338(3.90)	94-0939-66
Benzene, 1-butoxy-4-methoxy-2-nitro-	MeOH	357(3.39)	78-0093-66B
Benzene, 1-butoxy-4-methoxy-3-nitro-	MeOH	357(3.39)	78-0093-66B

Compound	Solvent	λ_{max}(log ε)	Ref.
Benzene, 1-isobutoxy-4-methoxy-2-nitro-	MeOH	357(3.39)	78-0093-66B
Benzene, 1-isobutoxy-4-methoxy-3-nitro-	MeOH	357(3.39)	78-0093-66B
Ether, 3-hydroxy-3-methylbutyl p-nitrophenyl	EtOH	228(3.94),308(4.06)	94-0939-66
3,5-Pyridinedicarboxylic acid, 1,4-dihydro-, diethyl ester	n.s.g.	222(4.05),244s(3.77), 376(3.93)	39-1083-66B
$C_{11}H_{15}NO_4S$			
1,4-Thiazine-3,6-dicarboxylic acid, 2,3-dihydro-2,2-dimethyl-, dimethyl ester	EtOH	280s(3.51),312(4.00)	88-1205-66
2,4-Thiophenedicarboxylic acid, 5-amino-3-methyl-, diethyl ester	MeOH	223(4.15),234(4.21), 317(4.20)	24-0094-66
$C_{11}H_{15}NO_5$			
Nicotinic acid, 1,2-dihydro-4,6-dihydroxy-2-oxo-1-methyl-, butyl ester	EtOH	302(4.32)	78-0455-66
Nicotinic acid, 1,2-dihydro-4,6-dihydroxy-2-oxo-1-methyl-, isobutyl ester	EtOH	306(4.3)	78-0455-66
Succinamic acid, N-(2-methoxy-3-methyl-4-oxo-2-cyclobuten-1-yl)-, methyl ester	MeOH	240(4.10)	88-3031-66
$C_{11}H_{15}NS$			
Fulvene, 6-(methylthio)-6-pyrrolidino-	EtOH	273(3.45),357(4.39)	24-3268-66
2(1H)-Isoquinolineethanethiol, 3,4-dihydro-, hydrochloride	MeOH	263(2.60),264(2.60), 271(2.56)	83-0846-66
$C_{11}H_{15}N_3$			
Benzimidazole, 2-(2-dimethylaminoethyl)-	EtOH	243(3.83),274(3.90), 281(3.96)	4-0278-66
Tetramethylammonium 1,2-dicyanocyclopentadienide	H_2O	267(4.11),283(4.16)	35-3046-66
$C_{11}H_{15}N_3O$			
Nezukone, semicarbazone	EtOH	325(4.55),385s(3.25)	88-5875-66
$C_{11}H_{15}N_5$			
Adenine, N-cyclohexyl-	pH 1	265(3.75)	95-0649-66
	pH 13	272(3.89)	95-0649-66
Adenine, 9-cyclohexyl-	pH 1	260(4.16)	95-0649-66
	pH 13	262(4.17)	95-0649-66
$C_{11}H_{15}N_5O$			
Isoguanine, 9-cyclohexyl-	pH 1	233(3.72),282(4.16)	95-0649-66
	pH 13	245(3.78),285(4.05)	95-0649-66
2-Purinol, 6-(cyclohexylamino)-	pH 1	285(4.17)	95-0649-66
	pH 13	285(4.13)	95-0649-66
$C_{11}H_{15}N_5O_2$			
Pyrido[2,3-b]pyrazine-6-carbamic acid, 8-amino-3,4-dihydro-2-methyl-, ethyl ester	pH 1	232(4.45),318(4.08)	44-1890-66
	pH 13	223(4.47),256s(4.19), 327(3.90)	44-1890-66

Compound	Solvent	$\lambda_{max}(\log \epsilon)$	Ref.
$C_{11}H_{15}N_5O_4$			
Adenine, 9-(4-deoxy-β-D-xylo-hexopyranosyl)-	pH 7	259(4.18)	39-1549-66C
Methyl 5-deoxy-5-(purin-6-yl)amino-β-D-ribofuranoside	EtOH-HCl	278(2.22)	4-0485-66
	EtOH	268(4.22)	4-0485-66
	EtOH-NaOH	275.5(4.23)	4-0485-66
6-Purinethione, 2'-O-methyl-2-amino-9-β-D-ribofuranosyl-	pH 1	342(4.14)	35-3640-66
$C_{11}H_{15}N_5O_5$			
Adenine, 9-β-D-allopyranosyl-	H_2O	259(4.13)	44-1503-66
Adenine, 9-α-D-altrofuranosyl-	H_2O	260(4.16)	44-1503-66
Adenine, 9-β-D-gulofuranosyl-	H_2O	259(4.16)	44-1503-66
Adenine, 9-β-L-gulofuranosyl-	H_2O	259(4.16)	44-1503-66
Adenine, 9-D-mannofuranosyl-	H_2O	259(4.17)	44-0339-66
Adenine, 9-α-D-talofuranosyl-	H_2O	257(4.16)	44-1503-66
Guanosine, 2'-O-methyl-	pH 1	255.5(4.03)	35-3640-66
	pH 11	258(3.99)	35-3640-66
2-Pyridinecarbamic acid, 6-(acetonyl-amino)-4-amino-5-nitro-, ethyl ester	pH 1	221(4.39),238(4.42), 268s(3.95),340(4.08)	44-1890-66
	pH 13	215(4.32),258s(3.95), 298s(3.80),348(4.09)	44-1890-66
$C_{11}H_{15}N_5O_7S$			
Guanosine, 8-(methylsulfonyl)-	pH 1	273(4.13),285s(--)	94-0046-66
	pH 7.1	272(4.07),285s(--)	94-0046-66
	pH 13	298(4.11)	94-0046-66
$C_{11}H_{15}O_4P$			
Phosphonic acid, benzoyl-, diethyl ester	C_6H_{12}	258(4.05)	44-1304-66
$C_{11}H_{16}$			
Fulvene, 6-tert-butyl-6-methyl-	C_6H_{12}	273(4.3),365(2.6)	33-0517-66
2-Norcarene, 7,7-dimethyl-3-vinyl-	EtOH	245(4.22)	35-4113-66
Spiro[2.4]hepta-4,6-diene, 1-butyl-	pentane	228(3.75),257s(3.40)	44-3296-66
Spiro[2.4]hepta-4,6-diene, 1-tert-butyl-	pentane	228(3.97),258s(3.49)	44-3296-66
Spiro[2.4]hepta-4,6-diene, 1,1,2,2-tetramethyl-	pentane	237(3.98),273s(3.26)	44-3296-66
$C_{11}H_{16}BF_4N$			
Trimethylstyrylammonium tetrafluoroborate	EtOH	248(4.20)	35-5747-66
$C_{11}H_{16}BNO_2$			
Isoquinoline, 3,4-dihydro-6,7-dimethoxy-, borane adduct	90% EtOH	242(4.31),290s(3.91), 299(3.98),332(4.01)	94-1382-66
$C_{11}H_{16}BrN_3$			
Pyrimidine, 5-bromo-4,6-dimethyl-2-piperidino-	EtOH	258(4.33),319(3.16)	78-2401-66
$C_{11}H_{16}ClN_3$			
Pyrimidine, 5-chloro-4,6-dimethyl-2-piperidino-	EtOH	256(4.13),320(3.12)	78-2401-66
$C_{11}H_{16}N_2$			
Benzopyrazole, 3-cyclobutyl-4,5,6,7-tetrahydro-	EtOH	227.5(3.83)	22-2971-66

Compound	Solvent	$\lambda_{max}(\log \epsilon)$	Ref.
Pyrazolidine, 3,3-dimethyl-1-phenyl-	EtOH	249(4.06)	78-2461-66
$C_{11}H_{16}N_2O$			
Anthranilamide, N-butyl-	pH 1.0	224(4.00)	35-4001-66
Anthranilamide, N-tert-butyl-	pH 1.0	223(3.98)	35-4001-66
Urea, 1,1-diethyl-3-phenyl-	EtOH	240(4.26),272(3.04)	94-0001-66
Urea, 1,3-diethyl-3-phenyl-	EtOH	243(3.56)	94-0001-66
$C_{11}H_{16}N_2O_2$			
L-Tryptophan, 4,5,6,7-tetrahydro-	H_2O	212(3.91)	35-3941-66
$C_{11}H_{16}N_2O_2S_2$			
5-Thiazolidineacetamide, 3-cyclohexyl-4-oxo-2-thioxo-	EtOH	265(4.11),298(4.21)	95-0101-66
$C_{11}H_{16}N_2O_5$			
2(1H)-Pyrimidinone, 1-(3-deoxy-α-D-ribofuranosyl)-4-methoxy-5-methyl-	H_2O	204(4.28),281(3.82)	44-1163-66
β-isomer	H_2O	203(4.27),215s(4.08), 280(3.82)	44-1163-66
$C_{11}H_{16}N_2O_6$			
Thymine, 3-(3-deoxy-β-D-xylo-hexopyranosyl)-	pH 7	265(3.98)	39-1549-66C
Thymine, 3-(4-deoxy-β-D-xylo-hexopyranosyl)-	pH 1	264.5(3.99)	39-1549-66C
	pH 7	265(3.99)	39-1549-66C
$C_{11}H_{16}N_2O_9S_2$			
Uridine, 2'-deoxy-, 3',5'-dimethanesulfonate	EtOH	258(3.98)	44-0205-66
$C_{11}H_{16}N_4O_3$			
Theophylline, 8-isopropyl-7-methoxy-	pH 1, 13	274(4.02)	5-0142-66A
$C_{11}H_{16}N_4O_4$			
Uracil, 6-(cyclohexylamino)-1-methyl-5-nitro-	pH 3	232(4.34),328(4.11)	24-3022-66
	pH 9	336(4.16)	24-3022-66
$C_{11}H_{16}N_6$			
Purine, 2-amino-6-(cyclohexylamino)-	pH 1	244(4.06),278(4.15)	95-0649-66
	pH 13	285(3.93)	95-0649-66
9H-Purine, 2,6-diamino-9-cyclohexyl-	pH 1	236(4.39),290(4.02)	95-0649-66
	pH 13	283(4.08)	95-0649-66
$C_{11}H_{16}N_6O_4$			
Purine, 2,6-diamino-9-(4-deoxy-β-D-xylohexosyl)-	pH 7	215(4.41),256(4.02), 280(4.04)	39-1549-66C
$C_{11}H_{16}O$			
Cyclohexanone, 3-methyl-6-(1-methyl-1,2-propadien-1-yl)-	hexane	204f(3.32),290(2.28)	28C-1601-66A
	EtOH	217f(3.08),290(2.41)	28C-1601-66A
Cyclohexanone, 2-(4-pentylidene)-	isooctane	237(4.26),330(1.67)	22-0287-66
2-Cyclohexen-1-one, 2-(3-butenyl)-3-methyl-	EtOH	244(4.09)	35-3408-66
2-Cyclohexen-1-one, 4-(2-methyl-2-butenyl)-	EtOH	223(4.00)	39-0419-66C
Cyclopentanone, 2-allyl-2-isopropenyl-	isooctane	292(1.06),302(1.15), 311(1.09),317(0.79)	22-0281-66

Compound	Solvent	$\lambda_{max}(\log \epsilon)$	Ref.
Cyclopentanone, 2-(1-methyl-4-pentenylidene)-, cis	isooctane	248(3.90),329(1.51), 341(1.60),356(1.62), 370(1.48),373s(1.11)	22-0281-66
isomer?	isooctane	249(3.88),327(1.62), 340(1.69),354(1.70), 369(1.56),373s(1.15)	22-0281-66
Jasmone, cis	EtOH	234(4.14)	44-0977-66
1(2H)-Naphthalenone, 3,5,6,7,8,8a-hexahydro-4-methyl-	EtOH	237(2.43)	44-3787-66
3,4-Nonadien-6-yn-1-ol, 8,8-dimethyl-	n.s.g.	219(4.17)	77-0585-66
Phenol, 5-ethyl-2-isopropyl-	H_2O	273(3.34)	22-3328-66
	base	291(3.62)	22-3328-66
Spiro[4.4]non-6-en-1-one, 6,9-dimethyl-	isooctane	300(1.69),310(1.71), 321(1.67)	22-0281-66
$C_{11}H_{16}OS$			
Phenol, 2,4,5-trimethyl-6-(methylthiomethyl)-	MeOH	287(3.38)	35-5855-66
	MeOH-KOH	287(3.36),312(2.26)	35-5855-66
$C_{11}H_{16}OS_2$			
Phenol, 2-methyl-4,6-bis-(methylthiomethyl)-	MeOH	285(3.34)	35-5855-66
	MeOH-KOH	256(3.99),305(3.58)	35-5855-66
$C_{11}H_{16}O_2$			
Benzene, 1-butoxy-4-methoxy-	MeOH	225(3.99),287(3.42)	78-0093-66B
Benzene, 1-isobutoxy-4-methoxy-	MeOH	226(4.09),287(3.51)	78-0093-66B
Bicyclo[3.3.1]non-3-en-2-one, 8-hydroxy-5,8-dimethyl-	EtOH	221(3.79)	39-0419-66C
isomer	EtOH	225(3.89)	39-0419-66C
1-Butanol, 3-(m-methoxyphenyl)-	n.s.g.	274(3.26),285(3.22)	78-1019-66
2,4-Heptadienoic acid, 3-ethyl-4-hydroxy-2,3-dimethyl-, γ-lactone	n.s.g.	282(4.3)	54-0043-66
2(3H)-Naphthalenone, 4,4a,5,6,7,8-hexahydro-1-hydroxy-4a-methyl-	EtOH	276(4.30)	23-1317-66
2,4-Nonadienoic acid, 4-hydroxy-2,3-dimethyl-, γ-lactone	n.s.g.	270(4.3)	54-0043-66
2,4-Octadienoic acid, 4-hydroxy-2,3,7-trimethyl-, γ-lactone	n.s.g.	270(4.3)	54-0043-66
$C_{11}H_{16}O_3$			
Bicyclo[3.1.0]hexane-3-carboxylic acid, 6,6-dimethyl-2-oxo-, ethyl ester	EtOH	272(3.41)	22-3490-66
Cycloheptanecarboxylic acid, 3-hydroxy-1,3,4-trimethyl-2-oxo-, lactone	EtOH	290.0(1.58)	39-0764-66C
2-Cyclopentene-1,4-dione, 3-iso-pentyl-2-methoxy-	MeOH	277(4.06)	20-0349-66
5-Hepten-3-ynoic acid, 2-hydroxy-2,6-dimethyl-, ethyl ester	EtOH	235(3.97)	39-1615-66C
$C_{11}H_{16}O_4$			
Aconic acid, γ-hexyl-	EtOH	218(4.01)	44-0616-66
2-Cyclohexene-1-carboxylic acid, 1-hydroxy-2,6,6-trimethyl-4-oxo-, methyl ester	n.s.g.	230(4.00)	28C-1725-66A
$C_{11}H_{17}Br$			
Methane, bromocyclocitrylidene-	MeOH	238.5(3.80)	24-0689-66

Compound	Solvent	λ_{max}(log ε)	Ref.
Naphthalene, 4a-(bromomethyl)-1,2,3,4,4a,5,6,7-octahydro-	hexane	230(3.08)	78-0279-66B
$C_{11}H_{17}Cl$			
Methane, chloro-β-cyclocitrylidene-, cis	MeOH	228s(3.54)	24-0689-66
trans	MeOH	238(3.77)	24-0689-66
$C_{11}H_{17}ClN_2O_2$			
3-Carbamoyl-1-(5-hydroxypentyl)-pyridinium chloride	pH 5.7	265(3.62)	5-0180-66J
$C_{11}H_{17}Li_2N_3O_{11}P_2$			
Cytidine, N,N-dimethyl-, 5'-pyrophosphate, dilithium salt	pH 1	218(3.84),287(4.16)	39-0588-66C
$C_{11}H_{17}NO$			
11-Azatricyclo[5.3.1.0^{2,6}]undecan-9-one, 11-methyl-	EtOH	251(3.10)	44-1042-66
$C_{11}H_{17}NOS$			
6-Thiacycloheptoxazole, 4,5,7,8-tetrahydro-4,4,8,8-tetramethyl-	EtOH	220(3.71)	44-3954-66
$C_{11}H_{17}NO_3$			
2,4-Pentanedione, 3-(2-oxocyclohexyl)-, oxime	EtOH	239.5(4.00)	18-1129-66
3-Pyridinebutyric acid, 1,4,5,6-tetrahydro-γ-oxo-, ethyl ester	EtOH	302(4.42)	44-2487-66
$C_{11}H_{17}NO_3S_2$			
5-Thiazolidineacetic acid, N-butyl-4-oxo-2-thioxo-, ethyl ester	EtOH	262(4.12),296(4.21)	95-0095-66
5-Thiazolidineacetic acid, N-ethyl-4-oxo-2-thioxo-, butyl ester	EtOH	260(4.09),294(4.17)	95-0095-66
5-Thiazolidineacetic acid, N-methyl-4-oxo-2-thioxo-, pentyl ester	EtOH	260(4.13),294(4.22)	95-0095-66
5-Thiazolidineacetic acid, 4-oxo-N-propyl-2-thioxo-, propyl ester	EtOH	262(4.06),296(4.16)	95-0095-66
$C_{11}H_{17}NO_4$			
Maleic acid, piperidino-, dimethyl ester	ether	278(4.31)	24-0450-66
$C_{11}H_{17}NO_6$			
Maleic acid, [(carboxymethyl)ethylamino]-, trimethyl ester	ether	273.5(4.24)	24-1558-66
$C_{11}H_{17}NO_6S$			
2-Hexenoic acid, 5-[(2-amino-2-carboxyethyl)thio]-3-methoxy-5-methyl-4-oxo-	H_2O	228(4.07)	39-1123-66C
$C_{11}H_{17}N_2O_5P$			
3-Carbamoyl-1-(5-hydroxypentyl)-pyridinium hydroxide, dihydrogen phosphate ester, inner salt	pH 5.7	265(3.66)	5-0180-66J
$C_{11}H_{17}N_3$			
Pyrimidine, 4,6-dimethyl-2-piperidino-	EtOH	250(3.83),308(2.84)	78-2401-66

Compound	Solvent	$\lambda_{max}(\log \epsilon)$	Ref.
$C_{11}H_{17}N_3O_2$			
Triazene, 1-(p-ethoxyphenyl)-3-hydroxy-3-propyl-	1% aq. acetone	294(3.94)	65-2117-66
$C_{11}H_{17}N_3O_5$			
Uridine, 2'-deoxy-5-(dimethylamino)-	pH 1	267(3.98)	87-0366-66
	pH 12	220(4.11),284(3.79)	87-0366-66
Uridine, 2'-deoxy-5-(ethylamino)-	pH 1	267(3.98)	87-0366-66
	pH 12	212(4.26),292(3.75)	87-0366-66
$C_{11}H_{17}N_5$			
Purine, 2-(butylamino)-6,9-dimethyl-	pH 1.5	230(4.60),250s(3.80),325(3.62)	39-0226-66C
	pH 7.0	225(4.46),250(3.91),256s(3.83),315(3.81)	39-0226-66C
$C_{11}H_{17}N_5O_8$			
4(3H)-Pyrimidinone, 2-amino-6-(β-D-glucopyranosylamino)-3-methyl-5-nitro-	pH -4.89	227(4.38),247s(3.80),317(4.02)	24-3022-66
	pH 1	235(4.16),285s(3.54),332(4.17)	24-3022-66
$C_{11}H_{18}$			
Methane, β-cyclocitrylidene-	MeOH	233(3.76)	24-0689-66
$C_{11}H_{18}BN$			
Borane, (tert-butylaminomethyl)-di(1-propynyl)-	C_6H_{12}	220(4.34),230(4.48)	22-3850-66
	C_6H_{12}	220(4.34),230(4.48)	28C-0376-66A
$C_{11}H_{18}N_2$			
Pyrazole, 5-cyclopentyl-3-ethyl-4-methyl-	EtOH	225(3.70)	22-2971-66
$C_{11}H_{18}N_2O$			
2-Pyrrolidinone, 4-(1-ethylpropylidene)-5-imino-3,3-dimethyl-	EtOH	256(4.22)	22-3895-66
$C_{11}H_{18}N_2O_2$			
Pyrazoline, 3,3,5,5-tetramethyl-4-(propionyloxymethylene)-	hexane	207(4.06),327(2.36)	39-0464-66C
	EtOH	325(2.24)	39-0464-66C
$C_{11}H_{18}N_2O_2S_2$			
5-Thiazolidineacetamide, N-butyl-3-ethyl-4-oxo-2-thioxo-	EtOH	262(4.10),296(4.22)	95-0101-66
5-Thiazolidineacetamide, 3-hexyl-4-oxo-2-thioxo-	EtOH	263(4.14),297(4.25)	95-0101-66
5-Thiazolidineacetamide, 3-isobutyl-N-ethyl-4-oxo-2-thioxo-	EtOH	263(4.08),297(4.19)	95-0101-66
5-Thiazolidineacetamide, 4-oxo-N,3-dipropyl-2-thioxo-	EtOH	263(4.06),297(4.17)	95-0101-66
$C_{11}H_{18}N_2O_3$			
Pyrimidine, 2,4-diethoxy-5-(ethoxymethyl)-	EtOH	218(4.16),262(3.97)	5-0119-66B
$C_{11}H_{18}N_2O_4$			
Thymine, 6-(α-acetoxyethyl)-5,6-dihydro-1,3-dimethyl-	n.s.g.	222(3.26)	57-0068-66A
isomer	n.s.g.	222(3.64)	57-0068-66A

Compound	Solvent	λ_{max}(log ϵ)	Ref.
$C_{11}H_{18}N_3O_8P$			
Cytidine, N,N-dimethyl-, 5'-phosphate	pH 1	218(3.94),287(4.19)	39-0588-66C
$C_{11}H_{18}N_4O_6$			
Uracil, 6-[methyl(2-diethoxyethyl)-amino]-5-nitro-, methylamino-acetate	pH 2.0	242(4.18),261s(4.05), 345(3.62)	24-2984-66
	pH 7.0	224(4.28),242s(4.12), 310s(3.74),352(3.92)	24-2984-66
	pH 14.0	226(4.11),355(4.12)	24-2984-66
$C_{11}H_{18}O$			
1-Buten-3-one, 1-(2-methyl-cyclohexyl)-	n.s.g.	229.0(3.99)	22-2212-66
Cyclohexylideneacetaldehyde, 2,2,6-trimethyl-	EtOH	237(4.00)	22-1400-66
6,7-Decadien-4-one, 6-methyl-	EtOH	216f(3.18),289(2.36)	28C-1601-66A
	hexane	210f(3.40),288(2.30)	28C-1601-66A
2H-Pyran, 2,6-diethyl-3,4-dihydro-2-methyl-	EtOH	252(4.25)	28C-0927-66A
2H-Pyran, 2-ethyl-6-ethylidene-3,5-dihydro-2,4-dimethyl-	EtOH	244(4.13)	28C-0927-66A
Spiro[4.4]nonanone, 6,9-dimethyl-	isooctane	284s(1.28),295(1.36), 307(1.42),319(1.36), 332(1.08)	22-0281-66
2-Undecen-4-yn-1-ol, trans	EtOH	227(4.17),236(4.10)	39-2283-66C
$C_{11}H_{18}O_2$			
Geranic acid, methyl ester, cis	EtOH	213(4.10)	39-2144-66C
Geranic acid, methyl ester, trans	EtOH	218(4.22)	39-2144-66C
3,5-Heptadien-1-ol, 4,6-dimethyl-, acetate	EtOH	227(3.92)	78-0443-66B
3-cis	EtOH	218.5(3.74)	78-0443-66B
3-trans	EtOH	232.5(3.95)	78-0443-66B
$C_{11}H_{18}O_3$			
2-Heptenoic acid, 5-ethyl-4,4-di-hydroxy-2,3-dimethyl-, γ-lactone	n.s.g.	208(4.1)	54-0043-66
2-Hexenoic acid, 3,4,4-trimethyl-5-oxo-, ethyl ester, trans	n.s.g.	222(4.10)	35-0618-66
2-Hexen-4-yn-1-ol, 6,6-diethoxy-3-methyl-	EtOH	225(4.19)	33-0369-66
2-Nonenoic acid, 4,4-dihydroxy-2,3-dimethyl-, γ-lactone	n.s.g.	213(4.1)	54-0043-66
2-Octenoic acid, 4,4-dihydroxy-2,3,7-trimethyl-, γ-lactone	n.s.g.	213(4.1)	54-0043-66
$C_{11}H_{18}O_4$			
Cyclohexanecarboxylic acid, 2-acetyl-2-hydroxy-1,3-dimethyl-	EtOH	290.0(1.59)	39-0764-66C
3-Hexenedioic acid, 3-methyl-, diethyl ester	EtOH	216(4.16)	39-2144-66C
Succinic acid, 2-heptylidene-	EtOH	216(3.08)	44-0616-66
$C_{11}H_{18}O_5$			
Butenedioic acid, 2-ethoxy-3-methyl-, diethyl ester	MeOH	242(3.91)	5-0053-66I
$C_{11}H_{19}Br$			
1,2-Hexadiene, 1-bromo-3-isobutyl-5-methyl-	EtOH	203(3.96),230(3.95)	39-1223-66C

Compound	Solvent	λ_{max}(log ε)	Ref.
1,2-Pentadiene, 1-bromo-3-tert-butyl-4,4-dimethyl-	EtOH	201(4.08),227(3.87)	39-1223-66C
$C_{11}H_{19}ClN_2O_4$			
7-(Dimethylamino)-2,4,6-heptatrien-ylidene-dimethylammonium perchlorate	EtOH-100°K	515(5.42)	61-0052-66
phototropic form	EtOH-100°K	266(--),544(--)	61-0052-66
$C_{11}H_{19}IN_2S_2$			
3-Ethyl-2-[(3-ethyl-2-thiazolin-ylidene)methyl]-2-thiazolinium iodide	MeOH	335(4.56)	9-0150-66
$C_{11}H_{19}NO_3S$			
3-Hexene-1-sulfonic acid, 2-pyrrolidino-5-methyl-4-hydroxy-, sultone	MeOH	208(3.18),269(2.92)	5-0068-66I
3-Pentene-1-sulfonic acid, 2-piperidino-3-methyl-4-hydroxy-, sultone	MeOH	210(3.39)	5-0068-66I
$C_{11}H_{19}N_3O$			
p-Menth-4(8)-en-9-ol, semicarbazone	EtOH	275(4.45)	44-3510-66
$C_{11}H_{19}N_3O_2$			
Urea, [3-(cyclohexylamino)-crotonoyl]-	EtOH	305(4.47)	70-1171-66
$C_{11}H_{19}N_5O_5$			
Acetaldehyde, [(2-amino-1,6-dihydro-1-methyl-5-nitro-6-oxo-4-pyrimidinyl)amino]-, diethyl ester	pH -0.89	228(4.34),250s(3.68), 312(4.10)	24-2984-66
	pH 7.0	230s(4.23),290s(3.70), 334(4.19)	24-2984-66
	pH 14.0	240s(3.81),346(4.31)	24-2984-66
Acetaldehyde, [(2-amino-1,6-dihydro-5-nitro-6-oxo-4-pyrimidinyl)-methylamino]-, diethyl ester	pH 6.0	224(4.33),246s(4.18), 297(3.76),355(3.85)	24-2984-66
	pH 14.0	223(4.28),364(4.01)	24-2984-66
$C_{11}H_{20}N_2O_4$			
1,2-Cyclohexanediol, 1-acetyl-, (ethoxycarbonyl)hydrazone, cis	EtOH	218(4.08)	70-2092-66
trans	EtOH	215(4.00)	70-2092-66
$C_{11}H_{20}O$			
Cyclopentanone, 2-(1-methylpentyl)-	isooctane	291(1.40),299(1.43), 307(1.36)	22-0281-66
$C_{11}H_{20}O_3$			
Acrylic acid, 3-tert-butoxy-, tert-butyl ester	EtOH	237(4.22)	44-0646-66
2,4-Hexadien-1-ol, 6,6-diethoxy-3-methyl-	EtOH	232(4.09)	33-0369-66
$C_{11}H_{20}O_5S$			
Methyl 4-acetylthio-4,6-dideoxy-2,3-di-O-methyl-β-D-glucoside	EtOH	231(3.60)	39-1291-66C
$C_{11}H_{21}ClN_2O_4$			
1-(3-Pyrrolidinylidene-1-propenyl)-pyrrolidinium perchlorate	EtOH-100°K	327(4.78)	61-0052-66
phototropic form	EtOH-100°K	350(4.62)	61-0052-66

Compound	Solvent	$\lambda_{max}(\log \epsilon)$	Ref.
$C_{11}H_{21}N$			
Butylamine, N-cycloheptylidene-	heptane	180(4.02)	88-0507-66
	C_6H_{12}	250(2.28)	88-0507-66
	EtOH	240(2.31)	88-0507-66
$C_{11}H_{21}NO$			
2-Heptene, 2-morpholino-	ether	221(3.88)	22-2845-66
$C_{11}H_{21}N_3O$			
2-Octen-7-one, 2,6-dimethyl-, semicarbazone	EtOH	226(4.13)	33-0617-66
$C_{11}H_{21}O_4P$			
Phosphonic acid, (cyclohexylcarbonyl)-, diethyl ester	C_6H_{12}	345(1.73)	44-1304-66
$C_{11}H_{22}N_2$			
2-Pyrazoline, 5-isobutyl-4-isopropyl-1-methyl-	EtOH	249(3.56)	4-0413-66
$C_{11}H_{22}N_2O_2$			
Δ^1-Pyrazine, tetrahydro-4-hydroxy-2,3,3,5,5,6,6-heptamethyl-, 1-oxide	EtOH	236.5(3.93)	39-2300-66C
$C_{11}H_{22}O$			
3-Heptanone, 5-ethyl-2,2-dimethyl-	hexane	289(1.47)	22-3578-66
3-Hexanone, 4-ethyl-2,2,5-trimethyl-	hexane	298(1.41)	22-3578-66
3-Nonanone, 5,7-dimethyl-	EtOH	279(1.51)	22-0689-66
	EtOH	280(1.50)	22-0689-66
2-Octanone, 6-ethyl-4-methyl-	EtOH	280(1.49)	22-0689-66
$C_{11}H_{24}N_2$			
Piperidine, 1-methyl-2-(4-methylaminobutyl)-	MeOH	204(3.4)	24-0737-66
$C_{11}H_{25}AgNPS_2$			
Silver, (N,N-diethyldithiocarbamato)(triethylphosphine)-	EtOH	257(4.17),297(3.60)	12-0555-66
$C_{11}H_{25}CuNPS_2$			
Copper, (N,N-diethyldithiocarbamato)(triethylphosphine)-	EtOH	259(3.94),305(3.40)	12-0555-66

Compound	Solvent	$\lambda_{max}(\log \epsilon)$	Ref.
$C_{12}Cl_8$			
Acenaphthylene, octachloro-	benzene	371(4.29)	88-2875-66
Biphenylene, octachloro-	dioxan	225(4.23),243(4.06), 268s(4.10),279(4.45), 290(5.15),354(3.05), 375(3.21),393(3.26), 428(2.91),452(2.87)	88-2947-66
$C_{12}Cl_{10}$			
Acenaphthene, decachloro-	C_6H_{12}	261(4.80)	88-2875-66
$C_{12}Cl_{12}$			
Tricyclo[4.2.0.0^{2,5}]octane, perchloro-3,4,7,8-tetramethylene-	C_6H_{12}	287(4.45),299(4.65), 312(4.43)	88-0489-66
$C_{12}F_{14}$			
Benzene, pentafluoro(nonafluoro-1-cyclohexen-1-yl)-	EtOH	227(3.20),264(3.14)	78-1755-66
$C_{12}F_{18}$			
Bi-1-cyclohex-1-enyl, octadecafluoro-	EtOH	280(4.13)	78-1755-66
$C_{12}H_4Cl_4N_2$			
Cyclobuta[b]quinoxaline, 1,2-bis-(dichloromethylene)-1,2-dihydro-	EtOH	271(4.37),279(4.37), 336(3.73),378(4.02), 401(4.09)	18-0160-66
$C_{12}H_4Cl_8$			
Tricyclo[4.2.0.0^{2,5}]octane, 1,2,5,6-tetrachloro-3,4,7,8-tetrakis-(chloromethylene)-	C_6H_{12}	278(4.26),290(4.04)	88-0489-66
$C_{12}H_5F_9$			
Benzene, (nonafluoro-1-cyclohexen-1-yl)-	EtOH	245.0(3.83)	78-1755-66
$C_{12}H_5N_3$			
Indene-1,1,3-tricarbonitrile	MeCN	224(4.51),274(3.97), 292(3.89)	35-3046-66
$C_{12}H_6Br_2N_2$			
Benzo[c]cinnoline, 1,7-dibromo-	EtOH	256(4.58)	39-1306-66C
	CCl_4	281(4.12),321(3.98), 332(3.99),360(3.22), 376(3.06)	39-1306-66C
Benzo[c]cinnoline, 2,7-dibromo-	EtOH	258(4.61)	39-1306-66C
	CCl_4	314(4.15),346(4.13), 356(3.57),372(3.51)	39-1306-66C
Benzo[c]cinnoline, 4,7-dibromo-	EtOH	257(4.48)	39-1306-66C
	CCl_4	280(4.04),319(4.01), 332(3.95),361(3.25), 377(3.06)	39-1306-66C
$C_{12}H_6Br_2N_2O$			
Benzo[c]cinnoline, 1,7-dibromo-, 5-oxide	CCl_4	304(4.04),316(4.05), 353(3.90),367s(3.87), 371s(3.86),391(3.61)	39-1306-66C
Benzo[c]cinnoline, 2,7-dibromo-, 5-oxide	CCl_4	301(4.13),313(4.15), 348(3.89),360(3.80), 378s(3.60),386s(3.55)	39-1306-66C

Compound	Solvent	$\lambda_{max}(\log \varepsilon)$	Ref.
$C_{12}H_6ClNOS$			
7-Phenothiazinone, 2-chloro-	MeOH	238(4.40),281(4.33), 290(4.32),365(4.00), 515(3.90)	95-0541-66
$C_{12}H_6Cl_2N_2$			
Nicotinonitrile, 2,6-dichloro- 5-phenyl-	n.s.g.	262(4.01),288s(3.72)	22-2387-66
$C_{12}H_6Cl_2N_2O_2S$			
Dibenzo[b,f][1,4,5]thiadiazepine, 3,8-dichloro-, 11,11-dioxide	EtOH	246(4.56),316(3.72), 430(2.78)	39-0255-66B
$C_{12}H_6Cl_2N_2O_3S$			
Dibenzo[b,f][1,4,5]thiadiazepine, 3,8-dichloro-, 5,11,11-trioxide	EtOH	244(4.48),330(3.74)	39-0255-66B
$C_{12}H_6Cl_4$			
Fulvene, 1,2,3,4-tetrachloro- 6-phenyl-	C_6H_{12}	242(4.15),333(4.36), 422s(2.61)	78-1275-66
	MeCN	240(4.06),336(4.20), 420s(2.67)	78-1275-66
$C_{12}H_7BrN_2$			
Benzo[c]cinnoline, 1-bromo-	EtOH	254(4.64)	39-1306-66C
	CCl_4	308(3.91),319(3.90), 337s(3.07),354(3.11), 370(3.02)	39-1306-66C
Benzo[c]cinnoline, 4-bromo-	EtOH	253(4.58)	39-1306-66C
	CCl_4	308(3.98),321(3.94), 339s(3.22),356(3.25), 371(3.17)	39-1306-66C
$C_{12}H_7BrN_2O$			
Benzo[c]cinnoline, 1-bromo-, 5-oxide	CCl_4	303(3.97),339(3.97), 352(3.96),370(3.81), 389(3.59)	39-1306-66C
Benzo[c]cinnoline, 1-bromo- 6-oxide	CCl_4	289s(4.11),303(4.05), 334s(4.00),339(4.01), 348(3.99),365(3.78), 385(3.54)	39-1306-66C
Benzo[c]cinnoline, 4-bromo-, 6-oxide	CCl_4	296(4.07),309(4.14), 346(3.94),350s(3.93), 356s(3.92),366s(3.82), 375s(3.65),386(3.58)	39-1306-66C
$C_{12}H_7ClN_2$			
Nicotinonitrile, 2-chloro-5-phenyl-	n.s.g.	263(4.20)	22-2387-66
$C_{12}H_7NO$			
4-Azafluorenone	MeOH	213s(4.18),227(4.21), 245(4.53),282s(4.04), 289(4.11),305s(3.53)	39-1121-66C
	MeOH-H_2SO_4	216(4.35),236(4.27), 300(4.30),320s(3.98)	39-1121-66C
$C_{12}H_7NOS$			
3H-Phenothiazin-3-one	MeOH	237(4.40),273(4.28), 288(4.24),368(4.04), 385s(--),505(3.93)	95-0541-66

Compound	Solvent	λ_{max}(log ε)	Ref.
$C_{12}H_7NO_2$			
1H-Pyrano[3,2-f]quinolin-1-one	EtOH	251(4.27),290(3.91), 314(3.76),328(3.80)	95-0114-66
$C_{12}H_7NO_3$			
p-Benzoquinone, 2-cyano- 3-(3-methyl-2-furyl)-	EtOH	222(4.22),253(4.10), 340(3.46),473(3.78)	33-1794-66
p-Benzoquinone, 2-cyano- 3-(5-methyl-2-furyl)-	EtOH	229(4.12),285(3.99), 362(3.55),493(3.86)	33-1794-66
3H-Phenoxazin-3-one, 8-hydroxy-	MeOH	259(4.24),352(4.10), 504(3.96)	24-1470-66
	MeOH-KOH	274(4.12),410(4.02), 610(3.62)	24-1470-66
$C_{12}H_8BClO_2$			
Benzo-1,3,2-dioxaborole, 2-(m-chlorophenyl)-	C_6H_{12}	272(4.06),278(4.03)	39-0314-66B
Benzo-1,3,2-dioxaborole, 2-(o-chlorophenyl)-	C_6H_{12}	273(4.02),281(3.95)	39-0314-66B
Benzo-1,3,2-dioxaborole, 2-(p-chlorophenyl)-	C_6H_{12}	272(4.19),278(4.18)	39-0314-66B
$C_{12}H_8Br_2O_3$			
Coumarin, 3-acetonyl-6,8-dibromo-	EtOH	210(4.35),225(4.24), 280(3.98),325(3.51)	25-2057-66
2(3H)-Furanone, 3-(3,5-dibromo- salicylidene)-5-methyl-	EtOH	212(4.07),222(4.08), 292(3.59),370(3.98)	25-2057-66
$C_{12}H_8Br_2O_4S_2$			
Disulfone, bis(m-bromophenyl)-	n.s.g.	235(4.18)	44-3418-66
Disulfone, bis(p-bromophenyl)-	n.s.g.	262(4.29)	44-3418-66
$C_{12}H_8ClN$			
Carbazole, 1-chloro-	EtOH	215(4.42),235(4.57), 239(4.58),259(4.24), 292(4.20),326(3.60), 339(3.61)	39-0521-66B
Carbazole, 2-chloro-	EtOH	213(4.48),236(4.71), 260(4.31),299(4.28), 325(3.63),338(3.49)	39-0521-66B
Carbazole, 3-chloro-	EtOH	236(4.60),247(4.38), 261(4.34),298(4.21), 332(3.53),346(3.48)	39-0521-66B
Carbazole, 4-chloro-	EtOH	216(4.43),239(4.63), 249(4.51),259(4.39), 292(4.20),326(3.58), 339(3.60)	39-0521-66B
$C_{12}H_8ClNO_4S_2$			
5H-Indeno[2,1-d]thiazolo[2,3-b]- thiazolium perchlorate	MeOH	213(4.38),236(4.06), 245(4.00),255(4.00), 320(4.19)	39-0686-66C
$C_{12}H_8ClNS$			
Phenothiazine, 2-chloro-	EtOH	256(4.71),320(3.69)	36-0144-66
$C_{12}H_8ClN_5$			
Imidazo[4,5-c]pyridazine, 3-(benzyli- deneamino)-6-chloro-	EtOH	216(4.38),286(4.37), 320s(4.29),286s(3.4)?	4-0218-66

Compound	Solvent	λ_{max}(log ε)	Ref.
$C_{12}H_8ClN_5O_2$			
Purine, 6-chloro-7-(p-nitrobenzyl)-	EtOH-HCl	268(4.23)	87-0576-66
	EtOH	268(4.23)	87-0576-66
	EtOH-NaOH	263(4.24)	87-0576-66
9H-Purine, 6-chloro-9-(p-nitrobenzyl)-	pH 1	268(4.27)	87-0576-66
	pH 7	268(4.31)	87-0576-66
	pH 13	268(4.31)	87-0576-66
$C_{12}H_8Cl_2OS$			
Sulfoxide, bis(m-chlorophenyl)	EtOH	202(4.65),237(4.27), 267(3.31),272(3.15)	39-0239-66A
Sulfoxide, bis(o-chlorophenyl)	EtOH	202(4.63),236(4.23), 271(3.49)	39-0239-66A
Sulfoxide, bis(p-chlorophenyl)	EtOH	202(4.64),244(4.38)	39-0239-66A
$C_{12}H_8Cl_2O_2S$			
Sulfone, bis(m-chlorophenyl)	EtOH	207(4.65),239(4.21), 271(3.34),277(3.47), 286(3.33)	39-0239-66A
Sulfone, bis(o-chlorophenyl)	EtOH	202(4.63),236(4.12), 276(3.47),284(3.42)	39-0239-66A
Sulfone, bis(p-chlorophenyl)	EtOH	202(4.63),226(4.19), 249(4.40),278(3.12)	39-0239-66A
$C_{12}H_8Cl_2S$			
Sulfide, bis(m-chlorophenyl)	EtOH	201(4.66),237(4.08), 254(4.16),279(3.86)	39-0239-66A
Sulfide, bis(o-chlorophenyl)	EtOH	201(4.65),248(4.13), 274(3.74)	39-0239-66A
Sulfide, bis(p-chlorophenyl)	EtOH	201(4.67),221(4.32), 256(4.21),280(3.97)	39-0239-66A
$C_{12}H_8Cl_4O$			
1,4:5,8-Dimethanonaphthalen-9-one, 5,6,7,8-tetrachloro-1,4,4a,5,8,8a-hexahydro-	C_6H_{12}	204(4.16)	88-0201-66
$C_{12}H_8Cl_6$			
Aldrin	hexane	210(3.75)	39-2026-66C
$C_{12}H_8Cl_6O$			
Dieldrin	hexane	215(3.76)	39-2026-66C
$C_{12}H_8FNO_2$			
Biphenyl, 4-fluoro-3'-nitro-	$C_6H_{11}Me$	248(4.28)	35-3318-66
$C_{12}H_8FN_3O_4$			
Cytosine, N^4-p-carboxybenzoyl-5-fluoro-	EtOH	267(4.35),332(4.42)	87-0566-66
$C_{12}H_8INO_2$			
Biphenyl, 4-iodo-3'-nitro-	$C_6H_{11}Me$	229(3.75),327(4.24)	35-3318-66
$C_{12}H_8N_2$			
Benzo[c]cinnoline	EtOH	253(4.61)	39-1306-66C
	CCl_4	297(3.90),308(3.85), 335(3.05),349(3.10), 364(3.08)	39-1306-66C

Compound	Solvent	$\lambda_{max}(\log \epsilon)$	Ref.
Benzo[f]quinazoline	EtOH-acid	208(4.46),229(4.42), 245s(4.18),253(4.32), 262(4.35),314s(3.51), 329(3.40),345(3.15)	44-2607-66
	EtOH	212(4.53),225s(4.53), 229(4.58),263(4.29), 300(3.79),328(3.42), 343(3.30)	44-2607-66
	EtOH-base	229(4.58),264(4.30), 304(3.79),329s(3.46), 344(3.32)	44-2607-66
Nicotinonitrile, 5-phenyl-	n.s.g.	257(4.09)	22-2387-66
1,5-Phenanthroline	heptane	215s(4.31),230(4.62), 262(4.27),290(3.90), 307(3.48),314(3.24), 322s(3.10),324s(3.18), 328(3.51),336(3.07), 339(3.13),344(3.61)	5-0001-66F
1,6-Phenanthroline	heptane	212(4.33),238(4.65), 265(4.10),298(3.27), 305(3.23),312(3.54), 318(3.44),326(3.88), 333(3.48),341(4.00)	5-0001-66F
1,7-Phenanthroline	heptane	228(4.67),232(4.75), 266(4.46),308(2.95), 314(2.75),322(2.84), 328(2.54),336(2.46), 344s(1.15)	5-0001-66F
1,8-Phenanthroline	heptane	204(4.57),214(4.26), 228(4.65),233(4.76), 270(4.42),282s(4.21), 308(3.01),314(2.88), 322(2.89),328(2.70), 338(2.64)	5-0001-66F
2,6-Phenanthroline	heptane	242(4.59),260s(4.09), 273s(3.96),284(3.95), 294(3.80),310(3.13), 316(2.89),324(3.31), 334(2.78),339(3.40)	5-0001-66F
2,7-Phenanthroline	heptane	215(4.38),234(4.62), 240(4.67),258s(3.95), 268(3.93),278(4.08), 290(4.03),308(2.94), 322(3.30),337(3.61), 348(3.47),353(3.71)	5-0001-66F
$C_{12}H_8N_2O$			
Benzo[c]cinnoline, oxide	CCl_4	288(4.10),300(4.10), 332(4.00),340(3.95), 347(4.00),356s(3.95), 365(3.80),374s(3.70), 383(3.95)	39-1306-66C
Benzo[f]quinazolin-1-ol	EtOH-acid	235(4.03),270(4.45), 312s(3.71),327s(3.61), 342(3.36)	44-2607-66
	EtOH	221s(4.08),243s(4.26), 255s(4.47),261(4.59), 270(4.55),287s(3.72), 298(3.72),307s(3.60), (continued next page)	44-2607-66

Compound	Solvent	λ_{max}(log ε)	Ref.
Benzo[f]quinazolin-1-ol (cont.)	EtOH	315(3.58),323(3.43), 329(3.62),338(3.34), 344(3.70)	
	EtOH-base	241s(4.18),247s(4.26), 265(4.60),290(3.94), 315(3.32),329(3.45), 344(3.48)	44-2607-66
Benzo[f]quinazolin-3-ol	EtOH-acid	211(4.31),232(4.76), 247s(4.11),255(4.08), 263(3.99),280(3.76), 285(3.73),292(3.91), 366(3.86)	44-2607-66
	EtOH	214(4.29),237(4.71), 246(4.64),263(3.89), 269(3.87),282(3.78), 292(3.68),321(3.51), 337s(3.34),364(3.26), 377(3.20)	44-2607-66
	EtOH-base	236(4.56),246(4.55), 262(4.37),270(4.35), 287(4.08),360(3.54), 370(3.54)	44-2607-66
Nicotinonitrile, 2-hydroxy-5-phenyl-	n.s.g.	258(4.33),357(3.88)	22-2387-66
2-Phenazinol	N HCl	217(4.38),263(4.79), 388(4.26)	69-0689-66
	N NaOH	229(4.43),275(4.72), 367(3.88)	69-0689-66
$C_{12}H_8N_2OS$			
Dibenzo[b,f][1,4,5]thiadiazepine, 5-oxide	EtOH	246(4.20),318(3.71)	39-0255-66B
$C_{12}H_8N_2OS_2$			
3H-1,2-Dithiole-Δ$^{3,\alpha}$-acetamide, α-cyano-5-phenyl-	EtOH	303(4.13),425(4.19)	44-3489-66
	EtOH-base	405(4.24)	44-3489-66
$C_{12}H_8N_2O_2$			
Benzo[f]quinazoline-1,3-diol	EtOH-acid	212(4.43),224(4.41), 232(4.44),241s(4.50), 246(4.54),252(4.57), 262s(4.28),299(3.70), 308(3.77),337(3.57), 350(3.54)	44-2607-66
	EtOH	225s(4.38),240s(4.52), 245(4.57),252(4.59), 262(4.30),282s(3.54), 294(3.67),307(3.74), 337(3.58),352(3.57)	44-2607-66
	EtOH-base	238s(4.35),254(4.59), 261(4.60),282s(3.90), 291s(4.35),303(3.62), 337s(3.51),351(3.63), 369s(3.45)	44-2607-66
3,4'-Biisoxazole, 5-phenyl-	EtOH	200(4.4),265(4.31)	32-0443-66
Nicotinonitrile, 2,6-dihydroxy-5-phenyl-	n.s.g.	280(4.07),348(4.26)	22-2387-66
piperidine salt	n.s.g.	280(4.05),348(4.24)	22-2387-66

Compound	Solvent	$\lambda_{max}(\log \epsilon)$	Ref.
1-Phenazinol, 5-oxide	toluene-EtOH-H_2O	355(3.43),363(3.45), 374(3.43),383(3.53), 448(3.44)	69-3824-66
1-Phenazinol, 10-oxide	EtOH	279(4.83),326(--), 334(--),368(--), 380(--),387(3.56), 468(3.33)	69-3824-66
$C_{12}H_8N_2O_2S$			
Dibenzo[b,f][1,4,5]thiadiazepine, 11,11-dioxide	EtOH	230(4.31),320(3.70), 436(2.76)	39-0255-66B
$C_{12}H_8N_2O_3$			
1-Phenazinol, 5,10-dioxide	toluene-EtOH-H_2O	374(3.49),382s(3.47), 502(3.72)	69-3824-66
$C_{12}H_8N_2O_3S$			
Dibenzo[b,f][1,4,5]thiadiazepine, 5,11,11-trioxide	EtOH	234(4.41),334(3.76)	39-0255-66B
$C_{12}H_8N_2O_4$			
Pyrazino[1,2-a]indole-1,3,4(2H)-tri-one, 6-hydroxy-2-methyl-	pH 13	251(4.37),297(4.27)	39-1799-66C
	MeOH	258(4.28),335(3.62), 400(3.82)	39-1799-66C
$C_{12}H_8N_2S$			
Benzo[f]quinazoline-1-thiol	EtOH-acid	212(4.63),223s(4.53), 270s(4.02),312(4.22), 378(3.93)	44-2607-66
	EtOH	217(4.64),253s(4.05), 260(4.05),270(3.97), 309(4.25),327s(3.94), 379(3.78)	44-2607-66
	EtOH-base	231(4.27),239(4.28), 247s(4.34),259(4.50), 291s(3.95),329s(3.49), 339s(3.42),345(3.43)	44-2607-66
Benzo[f]quinazoline-3-thiol	EtOH-acid	217(4.47),263(4.16), 268(4.16),315(4.55), 374(3.67)	44-2607-66
	EtOH	215(4.48),238(4.21), 272(4.54),295(4.24), 307(4.28),324s(3.89), 338s(3.60)	44-2607-66
	EtOH-base	220(4.38),233(4.35), 249(4.13),261(4.23), 297(4.54),317s(4.32), 375(3.36)	44-2607-66
Dibenzo[b,f][1,4,5]thiadiazepine	EtOH	246(4.11),313(3.62), 426(2.86)	39-0255-66B
$C_{12}H_8N_2S_2$			
2,5-Cyclohexadiene-$\Delta^{1,\alpha}$-malono-nitrile, 4-(1,3-dithiolan-2-ylidene)-	$CHCl_3$	512(4.61),550(4.69), 595(4.51)	89-0517-66
	CF_3COOH	500s(2.75),535(3.02), 576(3.09)	89-0517-66
$C_{12}H_8N_2S_3$			
3H-1,2-Dithiole-$\Delta^{3,\alpha}$-acetamide, α-cyano-5-phenyl-thio-	EtOH	252(4.59),297(4.29), 465(3.91)	44-3489-66

Compound	Solvent	λ_{max}(log ε)	Ref.
3H-1,2-Dithiole-$\Delta^{3,\alpha}$-acetamide, α-cyano-5-phenyl-thio- (cont.)	EtOH-base	405(4.15)	44-3489-66
$C_{12}H_8N_4O$			
v-Triazolo[3,4-a]pyridine, 3-(2-pyridoyl)-	MeOH	216s(4.19),269(3.79), 322(4.26)	24-2918-66
v-Triazolo[3,4-a]pyridine, 3-(3-pyridoyl)-	MeOH	222s(4.22),269(3.75), 279(3.75),293s(3.87), 322(4.29)	24-2918-66
$C_{12}H_8N_4O_2$			
Pyrazolo[3,2-c]-as-triazine-3-carboxylic acid, 4-phenyl-	EtOH	257(4.07),285(3.90), 345(3.94)	39-1127-66C
Pyrazolo[3,2-c]-as-triazin-4-ol, benzoate	EtOH	238(4.25),326(3.79)	39-1127-66C
$C_{12}H_8N_4O_5S$			
Lumazine, 3-[(phenylsulfonyl)oxy]-	EtOH	267(3.38),275(3.30), 318(3.88)	4-0224-66
$C_{12}H_8N_4O_6$			
Diphenylamine, 2,4,6-trinitro-	EtOH	235(4.29),365(4.16)	65-1034-66
$C_{12}H_8N_4O_9S$			
Sulfanilic acid, N-picryl-, anion	buffer	377(4.16)	35-0947-66
dianion	NaOH	438(4.33)	35-0947-66
$C_{12}H_8N_6$			
Bis-s-triazolo[4,3-b:3',4'-f]pyridazine, 1-phenyl-	EtOH	290(3.91)	78-2073-66
$C_{12}H_8O$			
Dibenzofuran, tetracyanoethylene complex	$CHCl_3$	498(4.22)	60-0018-66
$C_{12}H_8OS$			
5H-Naphtho[1,8-bc]thiophen-5-one, 2-methyl-	$HOAc-HClO_4$	255(3.79),347(3.76), 435(4.15)	12-1908-66
	EtOH	233(3.89),262(4.13), 270(4.09),296(3.45), 380(4.13)	12-1908-66
$C_{12}H_8O_2$			
2,4-Dodecadiene-6,8,10-triynoic acid, trans-trans	ether	250(4.21),265(4.41), 276(4.64),293(3.98), 311(4.27),333(4.41), 356(4.31)	39-1617-66C
2,4-Pentadiyn-1-ol, 5-benzoyl-	ether	266(4.13),278(4.22), 294(4.08)	24-2413-66
$C_{12}H_8O_3$			
4a,8a-Naphthalenedicarboxylic anhydride	MeOH	255(3.96),310(3.15)	89-0590-66
1-Naphthoic acid, 8-hydroxy-3-methoxy-, lactone	EtOH	234(3.54),245s(3.28), 265(2.78),277s(2.49), 325s(2.49),368(2.76)	39-0523-66C
1-Naphthoic acid, 8-hydroxy-6-methoxy-, lactone	EtOH	259(3.53),322s(2.56), 336(2.61),362(2.49)	39-0523-66C

Compound	Solvent	λ_{max}(log ε)	Ref.
Spiro[1,3-benzodioxole-2,1'-[2,5]-cyclohexadien-4'-one	C_6H_{12}	217(4.30),235s(3.73), 270s(3.60),274(3.66), 279(3.67),285(3.54)	25-1533-66
$C_{12}H_8O_3S$			
p-Benzoquinone, 2-acetyl-3-(2-thienyl)-	EtOH	246(4.24),430(3.55)	33-1794-66
$C_{12}H_8O_4$			
3-Benzo[c]furanacrylaldehyde, 3-methyl-4,7-dioxo-	ether	226(4.48),300s(4.07), 308(4.12),319s(4.02), 389(4.00)	33-1806-66
p-Benzoquinone, 2-acetyl-3-(2-furyl)-	EtOH	250(4.19),433(3.67)	33-1794-66
Xanthotoxin	EtOH	219(4.32),249(4.35), 300(4.06)	4-0042-66
$C_{12}H_8O_5$			
p-Benzoquinone, 2-carboxy-3-(5-methyl-2-furyl)-	EtOH	350(3.45),472(3.36)	33-1794-66
$C_{12}H_8S$			
Dibenzothiophene, tetracyanoethylene complex	$CHCl_3$	550(3.53)	60-0018-66
Naphtho[2,3-b]thiophene	MeOH	208(4.45),223(4.32), 253s(4.38),257(4.39), 266(4.43),271(4.53), 277(4.56),314s(3.78), 318(3.79),326s(3.71), 332(3.67),348(3.27)	35-4112-66
$C_{12}H_9BO_2$			
Benzo-1,3,2-dioxaborole, 2-phenyl-	C_6H_{12}	272(4.10),278(4.00)	39-0314-66B
$C_{12}H_9BrN_4O_6$			
Pyrazole-3-carboxylic acid, 4-bromo-1-(2,4-dinitrophenyl)-, ethyl ester	EtOH	302(4.00)	22-0619-66
$C_{12}H_9Br_2N$			
7-Bromo-6H-benzo[a]indolizinium bromide	H_2O	209(4.59),258(4.07), 312(4.10)	88-3341-66
$C_{12}H_9Br_2NO_3$			
Indole-3-carboxaldehyde, 5,6-dibromo-1-ethyl-4,7-dihydro-2-methyl-4,7-dioxo-	MeOH	231(4.26),301(4.18), 357(3.60),475(3.45)	35-0804-66
$C_{12}H_9Cl$			
Biphenyl, 4-chloro-	EtOH	255.0(4.37)	56-0429-66
$C_{12}H_9ClHgN_4O$			
Hypoxanthine, 1-benzyl-9-(chloromercuri)-	neutral	253.5(3.95)	44-1411-66
$C_{12}H_9ClHgN_4S$			
6(1H)-Purinethione, 1-benzyl-9-(chloromercuri)-	neutral	321(4.21)	44-1411-66
$C_{12}H_9ClN_2$			
Azobenzene, p-chloro-	$C_2H_4Cl_2$	325(4.39)	49-0171-66
	MeCN	325(4.39)	49-0171-66

Compound	Solvent	$\lambda_{max}(\log \epsilon)$	Ref.
Azobenzene, p-chloro-, $AlBr_3$ complex	$C_2H_4Cl_2$	455(4.53)	49-0171-66
	MeCN	425(4.53)	49-0171-66
$AlCl_3$ complex	$C_2H_4Cl_2$	455(4.53)	49-0171-66
	MeCN	425(4.53)	49-0171-66
$GaBr_3$ complex	$C_2H_4Cl_2$	455(4.53)	49-0171-66
	MeCN	425(4.53)	49-0171-66
$GaCl_3$ complex	$C_2H_4Cl_2$	455(4.53)	49-0171-66
	MeCN	425(4.53)	49-0171-66
$SbCl_5$ complex	$C_2H_4Cl_2$	455(4.53)	49-0171-66
	MeCN	425(4.53)	49-0171-66
$SnCl_4$ complex	$C_2H_4Cl_2$	455(4.53)	49-0171-66
	MeCN	425(4.53)	49-0171-66
$C_{12}H_9ClN_2O_3$			
4-Pyridazineacetic acid, 3-chloro-1,6-dihydro-6-oxo-1-phenyl-	THF	348(3.81)	49-0644-66
$C_{12}H_9ClN_2O_3S$			
Benzenesulfonic acid, p-[(p-chlorophenyl)azo]-	H_2O	327(4.37)	35-2240-66
	70.6% H_2SO_4	435(4.56)	35-2240-66
$C_{12}H_9ClN_4$			
Purine, 6-chloro-3-methyl-8-phenyl-	pH 8	234(3.64),321(3.86)	39-0010-66C
$C_{12}H_9ClN_4O$			
s-Triazolo[4,3-b]pyridazine, 6-chloro-3-(p-methoxyphenyl)-	EtOH	267(4.29),350(3.18)	78-2073-66
$C_{12}H_9ClOS_2$			
3-Butyn-2-ol, 1-chloro-4-[5-(2-thienyl)-2-thienyl]-	ether	245(3.96),331(4.40), 336(4.41)	39-1101-66C
$C_{12}H_9N$			
Carbazole, tetracyanoethylene complex	$CHCl_3$	605(3.46)	60-0018-66
Pyrido[2,1-a]isoindole	EtOH	255(4.61),308(3.65), 321s(3.62),334(3.76), 394(3.85)	88-3341-66
$C_{12}H_9NO$			
Δ^2(5H),α-Furanacetonitrile, 5-(2-hexen-4-ynylidene)-	ether	258(4.27),393(4.63)	24-3441-66
2-Furanacetonitrile, 5-(2,4-hexadiynyl)-	ether	220(3.96)	24-3441-66
Phenoxazine	EtOH	214(4.24),237(4.57), 240(4.60),318(3.88)	7-0182-66
$C_{12}H_9NO_2$			
Biphenyl, p-nitro-	EtOH	306.0(4.25)	56-0429-66
1H-Cyclopenta[b]quinoline-3,9(2H,4H)-dione	EtOH	215(4.01),247(4.25), 258(4.39),359(3.72), 375(3.67)	35-1049-66
1-Naphthoic acid, 8-amino-3-methoxy-, lactam	EtOH	231(3.59),253(3.28), 269s(2.57),280(2.57), 290(2.43),366(2.77)	39-0523-66C
1-Naphthoic acid, 8-amino-6-methoxy-, lactam	EtOH	258(3.54),264s(3.52), 311s(2.36),324(2.58), 339(2.69),376(2.46)	39-0523-66C
Nicotinic acid, 5-phenyl-	n.s.g.	254(4.09)	22-2387-66
Pyridine, 2-benzoyl-, N-oxide	MeOH	257(4.31)	44-2149-66

Compound	Solvent	λ_{max}(log ε)	Ref.
$C_{12}H_9NO_3$			
Nicotinic acid, 2-hydroxy-5-phenyl-	n.s.g.	258(4.28),352(3.73)	22-2387-66
$C_{12}H_9NS$			
Phenothiazine	EtOH	226s(4.06),254(4.64), 319(3.67)	7-0182-66
	EtOH	253(4.64),320(3.64)	36-0144-66
$C_{12}H_9NSe$			
Phenoselenazine	EtOH	226(4.07),231s(4.06), 257(4.51),318(3.65)	7-0182-66
$C_{12}H_9N_3$			
Benzo[f]quinazoline, 1-amino-	EtOH-acid	211(4.60),278(4.48), 303s(3.91),320s(3.72), 334(3.60),350(3.58)	44-2607-66
	EtOH	213s(4.50),233s(4.06), 245s(4.15),251s(4.21), 269(4.54),292s(4.02), 320(3.20),334(3.36), 349(3.40)	44-2607-66
	EtOH-base	234s(4.07),244s(4.11), 252s(4.20),269(4.54), 290s(4.01),319s(3.26), 333(3.38),348(3.39)	44-2607-66
Benzo[f]quinazoline, 3-amino-	EtOH-acid	215(4.39),237(4.76), 267(4.20),291(4.04), 331(3.78),364(3.82), 377(3.78)	44-2607-66
	EtOH	237(4.55),247(4.53), 264(4.42),272(4.43), 286s(4.17),360(3.49), 370(3.49)	44-2607-66
	EtOH-base	236(4.56),245s(4.48), 264(4.40),272(4.40), 286(4.15),360(3.51)	44-2607-66
$C_{12}H_9N_3O$			
Benzo[f]quinazolin-1-ol, 3-amino-	EtOH-acid	213(4.41),225s(4.27), 234s(4.30),255(4.57), 263(4.59),299s(3.64), 313(3.75),332(3.57), 342(3.51)	44-2607-66
	EtOH	213(4.35),237(4.19), 255s(4.56),263(4.66), 272s(4.43),284s(3.94), 295s(3.81),307(3.49), 333s(3.45),349(3.66), 364(3.60)	44-2607-66
	EtOH-base	230(4.20),263(4.70), 280s(4.07),291(4.03), 301s(3.48),348(3.52), 360(3.49)	44-2607-66
Benzo[f]quinazolin-3-ol, 1-amino-	EtOH-acid	213(4.42),229(4.51), 259(4.39),322(3.90), 352(3.72),365s(3.67)	44-2607-66
	EtOH	217(4.48),238(4.49), 255(4.57),293(3.78), 305(3.86),335s(3.40), 345(3.58),359(3.57)	44-2607-66

Compound	Solvent	λ_{max}(log ε)	Ref.
Benzo[f]quinazolin-3-ol, 1-amino- (cont.)	EtOH-base	237(4.47),255(4.52), 297s(3.78),305(3.83), 344(3.58),358(3.57)	44-2607-66
v-Triazolo[3,4-a]quinoline, 3-acetyl-	MeOH	228(4.19),253(4.23), 261s(4.06),281s(3.86), 290(4.08),301(4.19), 323(4.21),335(4.15)	24-2918-66
$C_{12}H_9N_3OS$			
Ketone, 7-methyl-v-triazolo[1,5-a]-pyridin-3-yl 2-thienyl	MeOH	221s(4.05),274s(3.96), 280(3.98),292(3.99), 340(4.39)	24-2918-66
Thiazolo[4,5-c]isoquinoline, 2-acetamido-	EtOH	258(4.42),266(4.44), 290s(3.86),302(3.97), 311(4.01),343(3.81)	23-2465-66
Thiazolo[5,4-c]isoquinoline, 2-acetamido-	EtOH	221(4.45),242(4.54), 249s(4.50),272s(3.90), 280(3.94),293(3.94), 305(3.87),333(4.07), 345(4.14)	23-2473-66
$C_{12}H_9N_3O_2$			
Azobenzene, p-nitro-	$C_2H_4Cl_2$	335(4.41)	49-0171-66
	MeCN	330(4.40)	49-0171-66
$AlBr_3$ complex	$C_2H_4Cl_2$	435(4.51)	49-0171-66
	MeCN	410(4.48)	49-0171-66
$AlCl_3$ complex	$C_2H_4Cl_2$	435(4.51)	49-0171-66
	MeCN	410(4.48)	49-0171-66
$GaBr_3$ complex	$C_2H_4Cl_2$	435(4.51)	49-0171-66
	MeCN	410(4.48)	49-0171-66
$GaCl_3$ complex	$C_2H_4Cl_2$	435(4.51)	49-0171-66
	MeCN	410(4.48)	49-0171-66
$SbCl_5$ complex	$C_2H_4Cl_2$	435(4.51)	49-0171-66
	MeCN	410(4.48)	49-0171-66
$SnCl_4$ complex	$C_2H_4Cl_2$	435(4.51)	49-0171-66
	MeCN	410(4.48)	49-0171-66
Ketone, 2-furyl 7-methyl-v-triazolo-[1,5-a]pyridin-3-yl	MeOH	226(4.12),281(4.00), 291(4.00),339(4.42)	24-2918-66
Uracil, 5-cyano-6-methyl-1-phenyl-	0.08N HCl	277(4.20)	25-0381-66
	0.08N NaOH	276(4.09)	25-0381-66
$C_{12}H_9N_3O_2S$			
Benzothiazole, 2-(1-methyl-5-nitro-2-pyrrolyl)-	EtOH	226(4.29),274(3.97), 374(4.38)	87-0751-66
1H-Thieno[3,2-c]pyrazole, 3-methyl-1-(p-nitrophenyl)-	EtOH	237(4.12),272s(--), 347(4.38)	39-1527-66C
$C_{12}H_9N_3O_3$			
Benzoxazole, 2-(1-methyl-5-nitro-2-pyrrolyl)-	EtOH	248s(4.04),263(4.09), 368(4.38)	87-0751-66
$C_{12}H_9N_3O_4$			
Diphenylamine, 2,2'-dinitro-	EtOH	225(4.18),264(4.22), 290(4.08),380(3.85), 420(3.97)	65-1034-66
Diphenylamine, 2,4-dinitro-	EtOH	232(4.15),257(4.05), 352(4.25),385(4.06)	65-1034-66

Compound	Solvent	λ_{max}(log ε)	Ref.
$C_{12}H_9N_3S$			
Benzo[f]quinazoline-3-thiol, 1-amino-	EtOH-acid	220s(4.36),258(4.14), 272(4.23),303(4.67), 380s(3.45)	44-2607-66
	EtOH	256s(4.28),264s(4.34), 274(4.49),301(4.54), 359(3.45)	44-2607-66
	EtOH-base	232(4.23),257s(4.31), 279(4.58),297s(4.42), 343s(3.48),357(3.48)	44-2607-66
$C_{12}H_9N_5O_2S$			
9H-Purine-6-thiol, 9-(p-nitrobenzyl)-	EtOH-N HCl	327(4.27)	87-0576-66
	EtOH	327(4.26)	87-0576-66
	EtOH-N NaOH	323(4.25)	87-0576-66
$C_{12}H_9N_5O_3$			
Hypoxanthine, 9-(p-nitrobenzyl)-	EtOH-N HCl	252(4.23)	87-0576-66
	EtOH	252(4.24)	87-0576-66
	EtOH-N NaOH	260(4.23)	87-0576-66
$C_{12}H_9N_7$			
7H-Tetrazolo[5,1-i]purine, 7-benzyl-	pH 7	253(3.74),262(3.79), 287(3.80)	44-2210-66
9H-Tetrazolo[5,1-i]purine, 9-benzyl-	pH 1	253(3.68),261(3.68), 288(3.93)	44-2210-66
	pH 7	253(3.68),261(3.68), 288(3.93)	44-2210-66
$C_{12}H_9O_3P$			
Phosphine, tri-2-furyl-	EtOH	243(4.34)	24-0712-66
$C_{12}H_9O_3PS$			
Phosphine sulfide, tri-2-furyl-	EtOH	241(4.48)	24-0712-66
$C_{12}H_9O_3PSe$			
Phosphine selenide, tri-2-furyl-	EtOH	239(4.38)	24-0712-66
$C_{12}H_9O_4P$			
Phosphine oxide, tri-2-furyl-	EtOH	245(4.59)	24-0712-66
$C_{12}H_{10}$			
Biphenyl	EtOH	250(3.82)	24-0712-66
	EtOH	247.5(4.29)	56-0429-66
tetracyanoethylene complex	$CHCl_3$	505(3.02)	60-0018-66
1,3-Hexadiyne, 1-phenyl-	EtOH	222(4.14),244(3.65), 257(3.87),271(4.03), 288(4.01)	70-0833-66
1,4-Pentadiyne, 1-p-tolyl-	EtOH	241(4.32),247(4.25), 252(4.32),270(3.33), 281(3.17),287(3.14)	78-0867-66
$C_{12}H_{10}BClN_2$			
Benzo-1,3,2-diazaboroline, 2-(m-chlorophenyl)-	C_6H_{12}	295(4.22),303(4.19), 317(3.75)	39-0314-66B
Benzo-1,3,2-diazaboroline, 2-(o-chlorophenyl)-	C_6H_{12}	297(4.16),308(4.11), 322(3.71)	39-0314-66B
Benzo-1,3,2-diazaboroline, 2-(p-chlorophenyl)-	C_6H_{12}	295(4.27),303(4.26), 316(3.82)	39-0314-66B

Compound	Solvent	λ_{max}(log ϵ)	Ref.
$C_{12}H_{10}BrN$			
6H-Pyrido[2,1-a]isoindolium bromide	H_2O	256(4.14),313(4.02)	88-3341-66
$C_{12}H_{10}ClN$			
4-Biphenylamine, 4'-chloro-	EtOH-HCl	256(4.36)	56-0429-66
	EtOH	288(4.34)	56-0429-66
$C_{12}H_{10}ClNO$			
3-Pyrrolin-2-one, 5-(chloromethylene)-4-methyl-3-phenyl-	MeOH	297(4.36)	44-0048-66
$C_{12}H_{10}ClNO_2$			
Cinnamic acid, p-chloro-α-cyano-, ethyl ester	EtOH	312(4.39)	22-1033-66
$C_{12}H_{10}ClNO_4S_2$			
2,3-Dihydro-5H-indeno[2,1-d]thiazolo-[2,3-b]thiazolium perchlorate	MeOH	243(3.53),270(3.47), 347(4.22)	39-0686-66C
$C_{12}H_{10}Cl_2N_6O_2S$			
Sulfanilamide, N^1-(2,6-dichloro-9-methyl-9H-purin-8-yl)-	pH 1.2	302(4.46)	87-0373-66
	pH 12.9	258(4.29),312(4.33)	87-0373-66
$C_{12}H_{10}Cl_2Se$			
Diphenylselenium dichloride	MeOH	230(3.35),238(3.25), 258(3.29),266(3.31), 272(3.21)	78-0653-66
$C_{12}H_{10}Cl_6$			
Aldrin, dihydro-	hexane	215(3.78)	39-2026-66C
$C_{12}H_{10}FN_3O_2$			
p-Toluamide, N-(5-fluoro-1,2-dihydro-2-oxo-4-pyrimidinyl)-	EtOH	265(4.10),325(4.23)	87-0566-66
$C_{12}H_{10}FN_3O_3$			
p-Anisamide, N-(5-fluoro-1,2-dihydro-2-oxo-4-pyrimidinyl)-	EtOH	286(4.04),332(4.23)	87-0566-66
$C_{12}H_{10}Fe$			
Ferrocene, ethynyl-	EtOH	265(3.52),445(2.06)	101-0173-66B
	$CHCl_3$	264(3.72),440(2.26)	101-0399-66B
$C_{12}H_{10}NNaO_4S$			
Carbazole-6(7)-sulfonic acid, 1,2,3,4-tetrahydro-4-oxo-, sodium salt	MeOH	220(4.54),240(4.35), 265(4.17),295(3.88)	5-0116-66F
$C_{12}H_{10}N_2$			
Azobenzene	dioxan	228(4.20),317(4.32), 438(2.65)	35-5010-66
anion	DMF	295(4.0),320s(3.9), 425(4.4)	77-0137-66
dianion	DMF	295(3.9)	77-0137-66
Azobenzene, trans	$C_2H_4Cl_2$	320(4.33)	49-0171-66
	MeCN	320(4.29)	49-0171-66
$AlBr_3$ complex	$C_2H_4Cl_2$	435(4.47)	49-0171-66
$AlCl_3$ complex	$C_2H_4Cl_2$	435(4.47)	49-0171-66
	MeCN	410(4.46)	49-0171-66

Compound	Solvent	$\lambda_{max}(\log \epsilon)$	Ref.
Azobenzene, trans, $GaBr_3$ complex	$C_2H_4Cl_2$	435(4.47)	49-0171-66
	MeCN	410(4.46)	49-0171-66
$GaCl_3$ complex	$C_2H_4Cl_2$	435(4.47)	49-0171-66
	MeCN	410(4.46)	49-0171-66
$SbCl_5$ complex	$C_2H_4Cl_2$	435(4.47)	49-0171-66
	MeCN	410(4.46)	49-0171-66
$SnCl_4$ complex	$C_2H_4Cl_2$	435(4.47)	49-0171-66
	MeCN	410(4.46)	49-0171-66
Pyridine, 2,2'-vinylenedi-	EtOH	216(4.04),263(4.14), 312(4.45)	19-0505-66
1H-Pyrrolo[2,3-f]isoquinoline, 2-methyl-	acid	245(3.96),290(4.37), 355(3.96),370(3.96)	2-0118-66
	neutral	270(4.56),305(3.80)	2-0118-66
1H-Pyrrolo[2,3-f]isoquinoline, 3-methyl-	acid	250(4.14),295(4.56), 368(4.01)	2-0118-66
	neutral	275(4.68),298s(3.84)	2-0118-66
$C_{12}H_{10}N_2O$			
Azobenzene, p-hydroxy-	MeOH	235(4.08),345(4.46), 425(3.07)	70-1414-66
	EtOH	236(4.1),350(4.4), 425s(3.2)	12-1887-66
	EtOH-NaOH	262(4.2),410(4.4)	12-1887-66
	$C_2H_4Cl_2$	237(4.20),342(4.50), 430(3.01)	70-1414-66
Azoxybenzene	EtOH	231(3.96),261(3.89), 322(4.19)	59-1701-66
Nicotinamide, 5-phenyl-	n.s.g.	253(4.10)	22-2387-66
2-Picoline, 3-(2-picolinoyl)-	MeOH	201(4.26),238(3.97), 272(3.95),340(2.76)	44-3206-66
Pyridine, 2,2'-vinylenedi-, N-oxide	EtOH	218(4.05),263(4.48), 319(4.51)	19-0505-66
Pyrrolo[1,2-a]quinoxalin-4(5H)-one, 5-methyl-	EtOH	228(4.44),260(4.11), 314(4.09),327(3.99)	39-0852-66C
$C_{12}H_{10}N_2O_2$			
4-Biphenylamine, 4'-nitro-	EtOH-HCl	222(4.07),299(4.22)	56-0429-66
	EtOH	252(4.10),378(4.16)	56-0429-66
	EtOH	250(4.08),380(4.14)	99-0204-66
	ether	250(4.23),380(4.35)	99-0204-66
Pyridine, 2,2'-vinylenedi-, N,N'-dioxide	EtOH	217(4.05),262(4.56), 322(4.31)	19-0505-66
$C_{12}H_{10}N_2O_3$			
1H-Pyrrolo[1,2-c]imidazole-1,3,5(2H,6H)-trione, 7,7a-dihydro-7a-phenyl-	MeOH-HCl	259(2.74),263(2.63)	87-0142-66
	MeOH	259(2.61),263(2.44)	87-0142-66
$C_{12}H_{10}N_2O_3S$			
Acetic acid, [(1,6-dihydro-6-oxo-1-phenyl-3-pyridazinyl)thio]-	EtOH	228(4.12),345(3.54)	49-0644-66
Benzenesulfonic acid, p-(phenylazo)-, anion	H_2O	320(4.37)	35-2240-66
	67.9% H_2SO_4	418(4.51)	35-2240-66
$C_{12}H_{10}N_2O_4$			
Cinnamic acid, α-cyano-m-nitro-, ethyl ester	EtOH	270s(4.26),289(4.25)	22-1033-66
Cinnamic acid, α-cyano-o-nitro-, ethyl ester	EtOH	250.5(4.16)	22-1033-66

Compound	Solvent	λmax(log ε)	Ref.
Cinnamic acid, α-cyano-p-nitro-, ethyl ester	EtOH	303(4.27)	22-1033-66
8,11-Diaza[4.3.3]propella-2,4-diene, 8,11-dimethyl-7,9,10,12-tetraoxo-	MeCN	243(3.14),252(3.28), 264s(3.26),270(3.26), 286s(3.04)	78-0279-66B
Orotic acid, 3-benzyl-	pH 1	283(3.85)	44-0201-66
	pH 7	279(3.87)	44-0201-66
	pH 13	300(3.87)	44-0201-66
4-Pyridazineacetic acid, 1,6-dihydro-3-hydroxy-6-oxo-1-phenyl-	EtOH	330(3.72)	49-0644-66
6(1H)-Pyridazinone, 1-benzoyloxy-3-methoxy-	EtOH	232(4.34),286(3.55), 310(3.60)	95-0082-66
$C_{12}H_{10}N_2O_4S$			
Benzenesulfonamide, N-(p-nitrophenyl)-	EtOH	227(4.08),372(4.51)	30-0114-66A
	$C_2H_4Cl_2$	226(4.22),350(4.52)	30-0114-66A
Benzenesulfonic acid, p-[(p-hydroxyphenyl)azo]-, anion	H_2O	350(4.41)	35-2240-66
	54% H_2SO_4	464(4.71)	35-2240-66
$C_{12}H_{10}N_2O_5S$			
Acetic acid, [(1,6-dihydro-6-oxo-1-phenyl-3-pyridazinyl)sulfonyl]-	EtOH	238(3.93),320(3.68)	49-0644-66
$C_{12}H_{10}N_2O_6$			
1,3-Cyclohexanedione, 2-(2,4-dinitrophenyl)-	EtOH-base	239(4.02),280(4.18), 430(3.54)	30-0827-66B
$C_{12}H_{10}N_2O_7$			
2-Isoxazoline-4-carboxylic acid, 3-[2-(5-nitro-2-furyl)vinyl]-5-oxo-, ethyl ester	EtOH	247(4.36),370(4.25)	95-1187-66
$C_{12}H_{10}N_2S$			
Picolinamide, N-methyl-thio-	heptane	235(4.26),288(4.14), 351(3.91)	65-1499-66
	EtOH	231(4.30),288(4.15), 338(3.91)	65-1499-66
	$CHCl_3$	290(4.15),344(3.90)	65-1499-66
	DMF	288(4.13),334(3.89)	65-1499-66
	H_2SO_4	266(4.07),333(3.88)	65-1499-66
Thieno[3,2-c]pyrazole, 3-methyl-1-phenyl-	EtOH	243(4.06),293(4.22)	39-1527-66C
$C_{12}H_{10}N_4$			
Benzo[f]quinazoline, 1,3-diamino-	EtOH-acid	217(4.53),237(4.38), 260(4.44),308(3.79), 341(3.54),356(3.51)	44-2607-66
	EtOH	214(4.37),218(4.38), 250s(4.36),266(4.64), 280s(4.19),292(4.04), 352(3.48),365(3.46)	44-2607-66
	EtOH-base	248s(4.32),266(4.62), 291s(4.00),353(3.46), 363(3.45)	44-2607-66
Purine, 2-methyl-8-phenyl-	pH 1	238(4.23),305(4.39)	4-0476-66
	pH 7	233(4.13),296s(4.41), 302(4.43),315s(4.18)	4-0476-66
	pH 13	233(4.28),308(4.42), 322s(4.22)	4-0476-66

Compound	Solvent	$\lambda_{max}(\log \epsilon)$	Ref.
Purine, 3-methyl-8-phenyl-, (as picrate)	pH 5	316(--)	39-0010-66C
$C_{12}H_{10}N_4O$			
Hypoxanthine, 1-benzyl-	neutral	251(3.96)	44-1411-66
anion	n.s.g.	261(3.99)	44-1411-66
Hypoxanthine, 3-benzyl-	neutral	265(4.14)	44-1411-66
anion	n.s.g.	264(4.02),276s(3.98)	44-1411-66
Hypoxanthine, 7-benzyl-	neutral	257(3.98)	44-1411-66
anion	n.s.g.	264(4.00)	44-1411-66
Hypoxanthine, 9-benzyl-	neutral	251(4.11)	44-1411-66
anion	n.s.g.	255(4.13)	44-1411-66
Hypoxanthine, 3-methyl-8-phenyl-	pH 5	232s(--),294(4.35)	39-0010-66C
$C_{12}H_{10}N_4OS$			
s-Triazolo[4,3-b]pyridazine, 6-(hydroxymethylthio)-3-phenyl-	EtOH	275(4.35)	78-2073-66
$C_{12}H_{10}N_4O_2$			
Alloxazine, 1,3-dimethyl-	EtOH	248(4.59),320(3.83), 380(3.84)	30-1101-66F
Alloxazine, 7,8-dimethyl-	EtOH	248(4.58),335(3.96), 385(3.93)	30-1101-66F
Benzimidazole, 2-(1-methyl-5-nitropyrrol-2-yl)-	EtOH	234(4.06),277(3.91), 376(4.28)	87-0751-66
$C_{12}H_{10}N_4O_3$			
Alloxazine, 1,3-dimethyl-, 10-oxide	EtOH	267(4.57),337(3.89), 400(3.69)	30-0577-66C
Alloxazine, 6,7-dimethyl-, N-oxide	pH 13	277(4.74),358(4.01), 450(3.95)	30-0577-66C
Isoalloxazine, 7,8-dimethyl-10-hydroxy-	EtOH	275(4.68),355(4.09), 413(3.73)	30-1101-66F
$C_{12}H_{10}N_4O_5S$			
1H-Pyrazolo[4,3-d]pyrimidine-5,7(4H,6H)-dione, 1-methyl-6-(p-tolylsulfonyloxy)-	EtOH	269s(3.74),276(3.80), 289(3.81)	44-2491-66
2H-Pyrazolo[3,4-d]pyrimidine-4,6(5H,7H)-dione, 2-methyl-5-(p-tolylsulfonyloxy)-	EtOH	262(3.95)	44-2491-66
$C_{12}H_{10}N_4O_6$			
Pyrazole-3-carboxylic acid, 1-(2,4-dinitrophenyl)-, ethyl ester	EtOH	292(4.06)	22-0619-66
$C_{12}H_{10}N_4S$			
Purine, 6-(methylthio)-8-phenyl-	pH 5	249(4.25),315(4.38)	39-0010-66C
2-Purinethiol, 6-methyl-8-phenyl-	2N HCl	257s(4.02),294(4.59), 384(3.97)	4-0476-66
	pH 13	259(4.47),345(4.16)	4-0476-66
6-Purinethiol, 2-methyl-8-phenyl-	pH 1	261(4.41),350(4.36)	4-0476-66
6-Purinethiol, 3-methyl-8-phenyl-	pH 5	261(4.27),361(4.43)	39-0010-66C
s-Triazolo[4,3-b]pyridazine, 6-(methylthio)-3-phenyl-	EtOH	228(4.18),268(4.29)	78-2073-66
$C_{12}H_{10}N_6$			
Benzaldehyde, s-triazolo[4,3-b]-pyridazin-6-ylhydrazone	EtOH	230(4.24),303(4.31)	78-2073-66

Compound	Solvent	$\lambda_{max}(\log \epsilon)$	Ref.
$C_{12}H_{10}N_6O_2$			
Adenine, 3-(p-nitrobenzyl)-	EtOH-N HCl	277(4.39)	87-0576-66
	EtOH	274(4.27)	87-0576-66
	EtOH-NaOH	270(4.38)	87-0576-66
Adenine, 7-(p-nitrobenzyl)-	EtOH-N HCl	273(4.30)	87-0576-66
	EtOH	269(4.19)	87-0576-66
	EtOH-NaOH	269(4.22)	87-0576-66
Adenine, 9-(p-nitrobenzyl)-	pH 1	264(4.34)	87-0576-66
	pH 7	265(4.32)	87-0576-66
	pH 13	264(4.34)	87-0576-66
Isoxanthopterin, 6-methyl-8-(2-pyridyl)-	pH 5.0	256(3.88),262(3.86), 294(4.06),342(4.16)	24-2997-66
	pH 11.0	257(4.19),285s(3.69), 358(4.17)	24-2997-66
$C_{12}H_{10}O$			
Phenol, p-phenyl-	EtOH	261.2(4.35)	56-0429-66
Phenyl ether	C_6H_{12}	205(--),225(4.05), 272(3.32)	30-0877-66F
	EtOH	224(4.05),271(3.31)	22-0228-66
$AlBr_3$ complex	C_6H_{12}	208(--),257(2.62)	30-0877-66F
$C_{12}H_{10}OS$			
1H-Naphtho[1,2-e][1,3]oxathiin	MeOH	232(4.88),266(3.63), 276(3.70),288(3.59), 316(3.24),332(3.35)	35-5855-66
	MeOH-KOH	232(4.88),266(3.63), 276(3.70),288(3.59), 316(3.24),332(3.35)	35-5855-66
4H-Naphtho[2,1-e][1,3]oxathiin	MeOH	216(4.73),235(4.58), 291(3.67),309(3.50), 323(3.41)	35-5855-66
	MeOH-KOH	216(4.73),235(4.58), 291(3.67),309(3.50), 323(3.41)	35-5855-66
5H-Naphtho[1,8-bc]thiophen-5-one, 3,4-dihydro-2-methyl-	EtOH	228(3.96),258(4.28), 315(3.50),342(3.50)	12-1908-66
Phenyl sulfoxide	EtOH	202(4.52),232(4.21), 266s(3.3)	39-0239-66A
$C_{12}H_{10}OS_2$			
[2,2'-Bithiophene]-5-(3-hydroxy-1-butynyl)-	ether	241(3.78),326(4.36), 333(4.37)	39-1101-66C
2-Propanone, 1-(4-phenyl-3H-1,2-dithiol-3-ylidene)-	EtOH	410(4.12),427(4.11)	44-3489-66
$C_{12}H_{10}O_2$			
2H-Benzocyclohepten-2-one, 3-methoxy-	n.s.g.	220(4.1),267(4.4), 302(4.3),347(4.2), 418s(3.6),434(3.6), 494(3.6)	77-0696-66
1-Naphthoic acid, 2-methyl-	dioxan	270s(3.66),282(3.74), 293s(3.63),321(2.30)	78-1587-66
1-Naphthoic acid, 3-methyl-	dioxan	299(3.84),330(2.00)	78-1587-66
1-Naphthoic acid, 4-methyl-	dioxan	300(3.96),325(2.93)	78-1587-66
1-Naphthoic acid, 5-methyl-	dioxan	303(3.82),326(2.70)	78-1587-66
1-Naphthoic acid, 6-methyl-	dioxan	301(3.82),329(2.88)	78-1587-66
1-Naphthoic acid, 7-methyl-	dioxan	301s(3.75),330(3.08)	78-1587-66
1-Naphthoic acid, 8-methyl-	dioxan	278s(3.74),288(3.83), 298s(3.79)	78-1587-66

Compound	Solvent	$\lambda_{max}(\log \epsilon)$	Ref.
$C_{12}H_{10}O_2S$			
Phenyl sulfone	EtOH	201(4.58),235(4.19), 260(3.24),266(3.33), 274(3.14)	39-0239-66A
$C_{12}H_{10}O_3$			
p-Benzoquinone, 2-methyl-5-(5-methyl-2-furyl)-	EtOH	264(4.19),300s(3.68), 466(3.67)	33-1794-66
2(3H)-Furanone, 3-acetyl-5-phenyl-	EtOH	254(3.786)	30-1343-66A
2(5H)-Furanone, 3-acetyl-5-phenyl-	EtOH	232(4.05),350(4.46)	30-1343-66A
2-Naphthaldehyde, 3-hydroxy-8-methoxy-	EtOH	223(4.48),257(4.48), 270(4.31),315(3.84), 323(3.88),398(3.39)	44-1747-66
2-Naphthoic acid, 1-hydroxy-, methyl ester	hexane	232(4.63),280(3.68), 290(3.65),303(3.45), 343(3.75),359(3.73)	59-1537-66
	EtOH	232(4.61),278(3.70), 287(3.67),301(3.46), 343(3.72),357(3.71)	59-1537-66
	ether	279(3.68),290(3.66), 302(3.46),343(3.75), 357(3.73)	59-1537-66
2-Naphthoic acid, 3-hydroxy-, methyl ester	hexane	240(4.69),277(3.76), 285(3.83),299(3.77), 371(3.42),385(3.35)	59-1537-66
	EtOH	241(4.68),277(3.75), 287(3.88),298(3.74), 368(3.39)	59-1537-66
	ether	280(3.71),287(3.84), 299(3.70),371(3.39)	59-1537-66
1-Naphthoic acid, 2-methoxy-	EtOH	273(3.49),281(3.60), 292(3.52),323(3.25), 335(3.32)	59-1537-66
	ether	270(3.55),280(3.66), 291(3.57),322(3.29), 335(3.33)	59-1537-66
2-Naphthoic acid, 1-methoxy-	hexane	237(4.69),272(3.69), 282(3.77),293(3.62), 321(3.14),336(3.29)	59-1537-66
	EtOH	231(4.56),273(3.71), 284(3.79),295(3.65), 323(3.24),336(3.26)	59-1537-66
	ether	274(3.71),283(3.77), 294(3.65),325(3.24), 337(3.29)	59-1537-66
2-Naphthoic acid, 3-methoxy-	hexane	235(3.60),270(3.71), 281(3.76),293(3.57), 334(3.18),349(3.22)	59-1537-66
	EtOH	234(4.70),271(3.37), 279(3.71),293(3.37), 344(3.19)	59-1537-66
	ether	269(3.70),279(3.69), 290(3.40),341(3.22)	59-1537-66
$C_{12}H_{10}O_4$			
Hydroquinone, 2-acetyl-3-(2-furyl)-	EtOH	216(4.14),271(3.89), 332(3.72)	33-1794-66
2-Naphthoic acid, 3-hydroxy-8-methoxy-	EtOH	227(4.53),235(4.61), 260(4.31),301(3.75), 309(3.67),370(3.46)	44-1747-66

Compound	Solvent	λ_{max}(log ϵ)	Ref.
1,4-Naphthoquinone, 2,6-dimethoxy-	EtOH	266(4.22),292(4.19), 378(3.52)	23-1086-66
1,4-Naphthoquinone, 2,7-dimethoxy-	EtOH	262(4.32),288(4.05), 334(3.41)	23-1086-66
1,4-Naphthoquinone, 6-hydroxy- 2(3)-methyl-5-methoxy- (or diomelquinone A)	EtOH	216(4.33),259(4.16), 308(3.86),544(3.55)	5-0172-66A
	EtOH-NaOH	202(4.63),226(4.33), 256(4.24),308(4.07), 544(3.75)	5-0172-66A
$C_{12}H_{10}O_4S_2$			
[2,2'-Bithiophene]-5,5'-dicarboxylic acid, dimethyl ester	ether	340(4.42)	24-0984-66
Phenyl disulfone	n.s.g.	237(4.30)	44-3418-66
$C_{12}H_{10}O_5$			
2-Benzopyran-1-one, 4-acetyl- 6,8-dihydroxy-5-methyl-	EtOH	237(4.36),337(3.72)	39-0126-66C
Isoscopoletin, O-acetate	EtOH	221(4.22),242(3.75), 252(3.62),285s(3.76), 294(3.87),328(4.15)	39-1805-66C
1,4-Naphthoquinone, 5-hydroxy- 2,7-dimethoxy-	$CHCl_3$	261(4.18),302(4.07), 435(3.63)	44-1496-66
Phthalic anhydride, 3-(2-carboxy- ethyl)-4-methyl-	ether	221(4.56),260(3.68), 301(3.50),311(3.51)	33-0858-66
Phthalic anhydride, 3-(2-carboxy- ethyl)-6-methyl-	ether	218(4.81),256(3.70), 307(3.71),317(3.72)	33-0858-66
Phthalic anhydride, 4-(2-carboxy- ethyl)-3-methyl-	ether	220(4.60),263(3.66), 300(3.54),310(3.57)	33-0858-66
$C_{12}H_{10}O_6$			
[2,2'-Bifuran]-5,5'-dicarboxylic acid, dimethyl ester	$CHCl_3$	240(3.51),317s(4.44), 326(4.55),342(4.45)	39-0976-66C
1,4-Naphthoquinone, 5,8-dihydroxy- 2,7-dimethoxy-	$CHCl_3$	285(3.94),308(3.98), 480(3.85),512(3.93), 550(3.74)	44-1496-66
$C_{12}H_{10}P$			
Diphenylphosphine tropylium (as perchlorate)	MeCN	262s(--),267(3.61), 273s(3.54)	27-0172-66
$C_{12}H_{10}S$			
2H-Naphtho[1,8-bc]thiophene, 2-methyl-	EtOH	215(4.18),233s(3.87), 246(3.85),319(3.45), 336(3.44)	12-1908-66
Phenyl sulfide	EtOH	206(4.39),250(4.07), 275(3.75)	22-0228-66
	EtOH	201(4.58),250(4.09), 275(3.75)	39-0239-66A
$C_{12}H_{10}S_2$			
Phenyl disulfide	MeOH	241(4.21)	35-5855-66
$C_{12}H_{10}S_3$			
2-Propanethione, 1-(4-phenyl- 3H-1,2-dithiol-3-ylidene)-	EtOH	252(4.65),484(3.80)	44-3489-66
$C_{12}H_{10}Se$			
Phenyl selenide	MeOH	235(3.92),256(4.02)	78-0653-66

Compound	Solvent	λ_{max}(log ϵ)	Ref.
$C_{12}H_{11}BN_2$			
Benzo-1,3,2-diazaboroline, 2-phenyl-	C_6H_{12}	290(4.18),297(4.22), 309(3.87)	39-0314-66B
$C_{12}H_{11}BO_2$			
2-Biphenylboronic acid	pet ether	246.0(4.15)	39-0566-66C
3-Biphenylboronic acid	pet ether	251.0(4.24)	39-0566-66C
4-Biphenylboronic acid	pet ether	260.0(4.41)	39-0566-66C
$C_{12}H_{11}BrO_3$			
1-Benzoxepin-3,5(2H,5H)-dione, 7-bromo-2,8-dimethyl-	EtOH	226(4.49),262(4.07), 303(4.07),328(4.20)	88-4993-66
$C_{12}H_{11}ClN_4O$			
p-Anisaldehyde, (6-chloro-3-pyridazinyl)hydrazone	EtOH	250(3.94),323(4.55)	78-2073-66
$C_{12}H_{11}ClN_6O_2S$			
Sulfanilamide, N^1-(2-chloro-7-methyl-6-purinyl)-	pH 1.2	290(4.30)	87-0373-66
	pH 12.9	256(4.18),292(4.34)	87-0373-66
Sulfanilamide, N^1-(2-chloro-9-methyl-6-purinyl)-	pH 1.2	281(4.20)	87-0373-66
	pH 12.9	258(4.20),289(4.41)	87-0373-66
$C_{12}H_{11}Cl_2N_3$			
Imidazole, 4-[2-[2,4-dichloro-benzylidene)amino]ethyl]-	EtOH	254(4.18)	44-2380-66
$C_{12}H_{11}F_3N_2O$			
Indole, 3-(trifluoroacetyl)-5-(dimethylamino)-	EtOH	240(4.26),300(4.00)	23-0387-66
$C_{12}H_{11}IN_2$			
Pyrrolo[1,2-a]quinoxaline methiodide	H_2O	227(4.41),244(4.32), 355(4.02)	39-0852-66C
$C_{12}H_{11}N$			
4-Biphenylamine	EtOH-HCl	250(4.26)	56-0429-66
	EtOH	278(4.27)	56-0429-66
	EtOH	280(4.28)	56-0429-66
1H-Cyclopenta[b]quinoline, 2,3-dihydro-	C_6H_{12}	273f(3.67)	5-0149-66H
	MeOH-acid	323(4.11)	5-0149-66H
	MeOH	280f(3.57)	5-0149-66H
Pyridine, 4-benzyl-	EtOH-HCl	254.5(3.67)	44-0399-66
	EtOH-NaOH	258(3.47)	44-0399-66
$C_{12}H_{11}NO$			
Benzylamine, N-furfurylidene-	EtOH	273(4.27)	12-1747-66
4-Biphenylamine, 4'-hydroxy-	EtOH-HCl	267(4.29)	56-0429-66
	EtOH	276(4.36)	56-0429-66
	EtOH-NaOH	293(4.42)	56-0429-66
Furanoquinoline, tetrahydro-4-methyl-	n.s.g.	271(3.76),310(3.81), 324(3.84)	70-0122-66
2-Naphthol, 1-(N-methylformimidoyl)-	CH_2Cl_2	402(3.85),421(3.85)	35-2407-66
p-Toluidine, N-furfurylidene-	EtOH	289(4.12),328(4.16)	12-1747-66
$C_{12}H_{11}NOS$			
p-Anisidine, N-2-thenylidene-	EtOH	272(3.99),296(4.01), 345(4.18)	12-1747-66

Compound	Solvent	$\lambda_{max}(\log \epsilon)$	Ref.
$C_{12}H_{11}NOS_2$			
1,3-Thiazin-4-one, 2,3,5,6-tetra-hydro-N-p-tolyl-2-thioxo-	EtOH	263(4.06),313(4.08)	95-0095-66
$C_{12}H_{11}NO_2$			
p-Anisidine, N-furfurylidene-	EtOH	288(4.13),340(4.23)	12-1747-66
Carbostyril, 1-(allyloxy)-	EtOH	229(4.58),245s(3.88), 271(3.83),277s(3.79), 316s(3.66),327(3.75), 342s(3.60)	78-0025-66
Cinnamic acid, α-cyano-, ethyl ester	EtOH	303(4.35)	22-1033-66
9H-Cyclopenta[b]quinolin-9-one, 1,2,3,4-tetrahydro-3-hydroxy-	EtOH	215(4.03),238(4.55), 242(4.52),321(3.76), 334(3.81)	35-1049-66
Naphtho[2,3-d]isoxazole, 5-methoxy-	EtOH	274(3.24),280(3.29)	44-1747-66
2-Naphthonitrile, 1,2,3,4-tetra-hydro-8-methoxy-3-oxo-	EtOH	272(3.92),278(3.88)	44-1747-66
$C_{12}H_{11}NO_2S$			
Benzenesulfonanilide	EtOH	220(4.13),265(3.41)	30-0114-66A
$C_{12}H_{11}NO_2S_2$			
4-Oxazolecarbothioic acid, 5-(methyl-thio)-2-phenyl-, S-methyl ester	EtOH	218(4.16),303(4.37)	12-0503-66
$C_{12}H_{11}NO_3$			
Cinnamic acid, α-cyano-p-hydroxy-, ethyl ester	EtOH	350.5(4.22)	22-1033-66
Indole-3-carboxaldehyde, 1-ethyl-4,7-dihydro-2-methyl-4,7-dioxo-	MeOH	246(4.28),256(4.27), 265(4.18),335(3.48), 430(3.48)	35-0804-66
$C_{12}H_{11}NO_3S_2$			
5-Thiazolidineacetic acid, N-benzyl-4-oxo-2-thioxo-	EtOH	261(4.05),297(4.15)	95-0095-66
$C_{12}H_{11}NO_5$			
2-Isoindolineacetic acid, 4-methoxy-1,3-dioxo-, methyl ester	dil. HCl	220(4.69),243s(3.88), 342(3.75)	4-0328-66
	dil. NaOH	289(3.54)	4-0328-66
	EtOH	218(4.84),242s(3.89), 338(3.76)	4-0328-66
2-Isoindolineacetic acid, 5-methoxy-1,3-dioxo-, methyl ester	dil. HCl	234(4.61),284(3.25), 332(3.44)	4-0328-66
	dil. NaOH	248s(3.97)	4-0328-66
	EtOH	232(4.69),284(3.25), 321(3.42)	4-0328-66
$C_{12}H_{11}NS$			
Benzylamine, N-2-thenylidene-	EtOH	255(4.06),283(3.95)	12-1747-66
m-Toluidine, N-2-thenylidene-	EtOH	270(4.12),307s(4.18), 328(4.20)	12-1747-66
p-Toluidine, N-2-thenylidene-	EtOH	271(4.01),306s(4.06), 333(4.14)	12-1747-66
$C_{12}H_{11}NS_4$			
Thiazole-4-carbodithioic acid, 5-(methylthio)-2-phenyl-, methyl ester	EtOH	217s(4.30),266(4.18), 324(4.40),396(4.04)	12-0503-66

Compound	Solvent	$\lambda_{max}(\log \epsilon)$	Ref.
$C_{12}H_{11}N_3O$			
Nicotinic acid, 5-phenyl-, hydrazide	n.s.g.	252(4.19)	22-2387-66
$C_{12}H_{11}N_3OS$			
8H-Oxazolo[5,4-e][1,4]diazepine-8-thione, 4,5,6,7-tetrahydro-2-phenyl-	EtOH	233(4.23),338(4.42)	12-0503-66
$C_{12}H_{11}N_3O_2$			
Indole-3-carboxylic acid, 2-(nitrosamino-1-propyl)-, lactam	MeOH	215(4.40),251(4.19),322(4.02)	5-0116-66F
6-Pyridazinone, 1-(p-aminophenyl)-, acetate	MeOH	208(--),238(--),319(--)	5-0042-66G
$C_{12}H_{11}N_3O_3$			
Alanine, N-(2-quinoxalinylcarbonyl)-	EtOH	206(2.93),244(3.30),317(2.57),328(2.58)	87-0266-66
$C_{12}H_{11}N_3O_3S$			
Cysteine, N-(2-quinoxalinylcarbonyl)-	EtOH	206(3.39),244(3.59),319(2.81),328(2.83)	87-0266-66
$C_{12}H_{11}N_3O_4$			
Phthalhydrazide, 3-acetamido-O-acetyl-	DMF	332(3.98)	18-0932-66
Serine, N-(2-quinoxalinylcarbonyl)-	EtOH	206(3.32),244(3.39),319(2.78),327(2.79)	87-0266-66
$C_{12}H_{11}N_3O_4S$			
Sulfanilamide, N-(p-nitrophenyl)-	EtOH	270(4.42),320(4.13)	30-0114-66A
$C_{12}H_{11}N_5$			
Adenine, N-p-tolyl-	pH 1	278(3.91)	95-0649-66
	pH 6	287(3.98)	95-0649-66
	pH 13	287(4.03)	95-0649-66
Adenine, 9-p-tolyl-	pH 1	260(4.17)	95-0649-66
	pH 13	263(4.19)	95-0649-66
Purine, 2-amino-6-methyl-8-phenyl-	pH 1	253(4.43),330(4.18)	4-0476-66
	pH 7	237(4.23),325(4.30)	4-0476-66
	pH 13	238(4.30),327(4.31)	4-0476-66
$C_{12}H_{11}N_5O$			
Isoguanine, 9-benzyl-	pH 1	231(3.82),282(4.18)	95-0649-66
	pH 13	248(3.82),285(4.08)	95-0649-66
2-Purinol, 6-(benzylamino)-	pH 1	288(4.18)	95-0649-66
	pH 13	286(4.14)	95-0649-66
$C_{12}H_{11}N_5O_2$			
Malononitrile, (s-triazolo[4,3-a]pyridin-3-ylcarbonyl)-, dimethyl acetal	MeOH	236(4.23),275s(3.82),283(3.83),295s(3.61)	44-0265-66
s-Triazolo[4,3-a]-s-triazine, 5,7-dimethoxy-3-phenyl-	EtOH	250(4.33)	49-1713-66
$C_{12}H_{11}N_5O_6$			
Pyrazole, 3,4,5-trimethyl-1-picryl-	EtOH	344(3.55)	22-3744-66
$C_{12}H_{11}N_7O$			
Formamide, N-(1-benzyl-5-tetrazol-5-ylimidazol-4-yl)-	pH 1	243(4.07)	44-2210-66
	pH 7	238(4.00)	44-2210-66

Compound	Solvent	$\lambda_{max}(\log \epsilon)$	Ref.
Formamide, N-(1-benzyl-5-tetrazol-5-ylimidazol-4-yl)- (cont.)	pH 13	244(4.02)	44-2210-66
$C_{12}H_{12}$			
Azulene, 1,4-dimethyl-	C_6H_{12}	240(4.57),242(4.34), 285(4.63),288(4.61), 302(3.78),332(3.45), 333(3.50),348(3.68), 365(3.46),555(2.43), 574(2.50),595(2.56), 625(2.49),652(2.47), 710(1.97)	24-2669-66
Azulene, 5,6-dimethyl-	EtOH	241(4.12),281(4.69), 287(4.66),299(3.91), 577(2.51),600(2.49), 639(2.50),658(2.26), 695(2.21)	44-3013-66
4a,8a-Ethanonaphthalene	C_6H_{12}	235(3.67),285(3.42)	88-0655-66
	EtOH	225s(3.59),233(3.62), 284(3.34)	88-2971-66
3-Hexen-1-yne, 6-phenyl-	n.s.g.	228(4.10),234s(4.10)	39-0578-66C
Naphthalene, 1,2-dimethyl-	EtOH	254s(3.24),265s(3.52), 275(3.70),280s(3.72), 284(3.73),291(3.61), 294s(3.56),307(2.88), 313s(2.62)	22-0645-66
Naphthalene, 1,3-dimethyl-	EtOH	253s(3.26),276s(3.69), 281(3.72),291s(3.61), 301s(2.99),307s(2.66), 321(2.63)	22-0645-66
Naphthalene, 1,4-dimethyl-	EtOH	255s(3.17),267s(3.51), 279s(3.73),288(3.66), 309s(3.07),316(2.82), 320(2.63)	22-0645-66
$C_{12}H_{12}BrN_3$			
Imidazole, 4-[2-(p-bromobenzylidene)amino]ethyl]-	EtOH	257(4.02)	44-2380-66
1H-Imidazo[4,5-c]pyridine, 4-(p-bromophenyl)-4,5,6,7-tetrahydro-	EtOH	218(4.22)	44-2380-66
$C_{12}H_{12}Br_2N_2S$			
6,7-Dihydrodipyrido[2,1-b:1',2'-f]-[1,3,6]thiadiazepinediium dibromide	H_2O	242(3.93),293(4.08), 323(4.18)	89-0605-66
$C_{12}H_{12}ClNO_4$			
Fumaric acid, (p-chloroanilino)-, dimethyl ester	EtOH	244(3.96),322(4.16)	24-2526-66
$C_{12}H_{12}ClNO_4S$			
5,6-Dihydro-3-methylthiazolo[2,3-a]-isoquinolinium perchlorate	EtOH	320(4.19)	4-0282-66
$C_{12}H_{12}ClN_3$			
Imidazole, 4-[2-(o-chlorobenzylidene)amino]ethyl]-	EtOH	247(4.13)	44-2380-66
Imidazole, 4-[2-(p-chlorobenzylidene)amino]ethyl]-	EtOH	253(4.28)	44-2380-66
1H-Imidazo[4,5-c]pyridine, 4-(p-chlorophenyl)-4,5,6,7-tetrahydro-	EtOH	217(4.24)	44-2380-66

Compound	Solvent	λ_{max}(log ε)	Ref.
$C_{12}H_{12}Cl_2N_2O$			
1-Pyrrolin-5-one, 3,3-dichloro-4-methyl-2-(methylamino)-3-phenyl-	EtOH	247.5(4.20)	28C-1276-66A
$C_{12}H_{12}INO_2$			
5-Acetoxy-1-methylquinolinium iodide	pH 7.0	237(4.39),273(3.27), 315(3.61)	10-0195-66A
6-Acetoxy-1-methylquinolinium iodide	pH 7.0	235(4.66),316(3.91)	10-0195-66A
7-Acetoxy-1-methylquinolinium iodide	pH 7.0	235(4.61),260(3.55), 316(3.87)	10-0195-66A
8-Acetoxy-1-methylquinolinium iodide	pH 7.0	240(4.56),273(3.06), 315(3.67)	10-0195-66A
$C_{12}H_{12}MoO_5$			
Tricarbonyl(carboxymethyl)cyclopentadienylmolybdenum ethyl ester	C_6H_{12}	310(3.56)	101-0341-66A
$C_{12}H_{12}N_2$			
Benzidine	EtOH	284(4.42)	56-0429-66
Hydrazobenzene	EtOH	246(4.32),290(3.59)	59-0399-66
Pyridine, 2-methyl-2',3'-methylenedi-	MeOH	263(3.90)	44-3206-66
9H-Pyrido[3,4-b]indole, 3,4-dihydro-1-methyl-	EtOH-acid	245(4.02),349(4.35)	94-0856-66
	EtOH	234(4.19),240(4.18), 315(4.18)	94-0856-66
$C_{12}H_{12}N_2O$			
Carbazol-4(1H)-one, 7-amino-2,3-dihydro-	0.2N HCl	217(4.75),239(4.33), 263(4.40),295s(4.18)	5-0116-66F
	MeOH	222(4.33),247(4.31), 283(4.19),314(3.82)	5-0116-66F
3,4-Diazabicyclo[4.2.0]oct-2-en-5-one, 2-phenyl-	MeOH	294(4.20)	24-1229-66
Indole-2-acetic acid, 3-(2-amino-1-ethyl)-, lactam	MeOH	220(5.11),275(4.31), 289s(4.15)	5-0116-66F
Indole-2-carboxylic acid, 3-(3-amino-1-propyl)-, lactam	MeOH	225(4.50),295(4.30)	5-0116-66F
Indole-3-carboxylic acid, 2-(3-amino-1-propyl)-, lactam	MeOH-HCl	213(4.20),241(4.12), 268(3.82),307(4.00)	5-0116-66F
	MeOH	216(4.78),231s(4.48), 256s(4.11),278(4.15), 286s(4.07)	5-0116-66F
5-Pyrazolinone, 4,4-ethylene-3-methyl-1-phenyl-	MeOH	208(4.28),242(4.06)	24-2962-66
3-Pyrazolol, 4-allyl-1-phenyl-	EtOH	279(4.30)	44-1538-66
$C_{12}H_{12}N_2OS$			
Propiophenone, 3-(2-thiazolylamino)-	MeOH	206(4.03),245(4.18)	23-1872-66
Thiazolin-2-imine, 3-(p-toluoylmethyl)-	iso-PrOH-acid	257.0(4.37)	20-0380-66
	iso-PrOH-base	255.0(4.38)	20-0380-66
$C_{12}H_{12}N_2O_2$			
2-Indolinone, 1-methyl-3-(2-oxazolidinylidene)-	EtOH	256(4.39),302(4.35)	95-1152-66
Pyrazole, 4-acetyl-3-methoxy-1-phenyl-	EtOH	230(4.15),296(4.22)	44-1538-66
Pyrazole-3(5)-carboxylic acid, 5(3)-phenyl-, ethyl ester	EtOH	228(4.25),250(4.25)	22-3744-66

Compound	Solvent	$\lambda_{max}(\log \epsilon)$	Ref.
$C_{12}H_{12}N_2O_2S$			
Sulfanilanilide	EtOH	225(4.03),270(4.30)	30-0114-66A
Thiazolin-2-imine, 3-(p-methoxy-benzoylmethyl)-	iso-PrOH-acid	221(4.13),279(4.31)	20-0380-66
	iso-PrOH-base	221(4.18),270(4.27)	20-0380-66
Uracil, 5-benzylthiomethyl-	pH 1	264(3.83)	87-0097-66
	pH 6.9	264(3.83)	87-0097-66
	pH 11	283(3.85)	87-0097-66
$C_{12}H_{12}N_2O_2S_2$			
5-Thiazolidineacetamide, 3-benzyl-4-oxo-2-thioxo-	EtOH	261(4.07),297(4.19)	95-0101-66
5-Thiazolidineacetamide, 4-oxo-2-thioxo-3-o-tolyl-	EtOH	258(4.11),297(4.20)	95-0101-66
Thiazolin-2-imine, N-propionyl-3-(2-thenoylmethyl)-	iso-PrOH-acid	265(4.21),288(4.26)	20-0380-66
$C_{12}H_{12}N_2O_3$			
Acrylic acid, 3-amino-2-cyano-3-phenoxy-, ethyl ester	H_2O	260(4.42)	24-2302-66
	dioxan	267(4.37)	24-2302-66
	MeCN	263(4.40)	24-2302-66
2,5-Piperazinedione, 3-(4-hydroxy-benzylidene)-6-methyl-	n.s.g.	226(4.1),319(4.2)	25-1300-66
3(2H)-Pyridazinone, 2-benzyloxy-6-methoxy-	EtOH	313(3.466)	95-0082-66
	EtOH	315(3.49)	95-1099-66
$C_{12}H_{12}N_2O_4$			
1,3-Butanedione, 2-diazo-1-(3,4-dimethoxyphenyl)-	CH_2Cl_2	238(4.35),306(3.91)	24-3128-66
8,11-Diaza[4.3.3]propell-3-ene, 8,11-dimethyl-7,9,10,12-tetraoxo-	MeCN	246(2.74),259(2.66), 268(2.69)	78-0279-66B
$C_{12}H_{12}N_4O_2$			
Imidazole, 4-[2-[(m-nitrobenzylidene)amino]ethyl]-	EtOH	234(4.39)	44-2380-66
Imidazole, 4-[2-[(p-nitrobenzylidene)amino]ethyl]-	EtOH	281(4.18)	44-2380-66
1H-Imidazo[4,5-c]pyridine, 4,5,6,7-tetrahydro-4-(m-nitrophenyl)-	EtOH	215(4.06)	44-2380-66
$C_{12}H_{12}N_4O_3$			
4-Pyridazineacetic acid, 1,6-dihydro-3-hydroxy-6-oxo-1-phenyl-, hydrazide	DMF	336(3.64)	49-1494-66
as-Triazine-3-carbamic acid, 5-phenyl-, ethyl ester, 2-oxide	EtOH	235(4.30),296(4.18), 384(4.20)	44-3917-66
$C_{12}H_{12}N_4O_4$			
2,5-Hexadienal, 2,4-dinitrophenyl-hydrazone, 2-trans	$CHCl_3$	374(4.47)	54-0117-66
2,4-Pentadienal, 4-methyl-, 2,4-dinitrophenylhydrazone	n.s.g.	265(4.24),300(4.10), 387(4.52)	22-0717-66
Pyrazole, 1-(2,4-dinitrophenyl)-5-ethyl-3-methyl-	EtOH	307(3.82)	28C-0782-66A
Pyrazole, 1-(2,4-dinitrophenyl)-3,4,5-trimethyl-	EtOH	322(3.83)	22-3744-66
Pyrrole , 1-(2,4-dinitroanilino)-2,5-dimethyl-	EtOH	220(4.13),328(4.05)	39-0341-66C

Compound	Solvent	λ_{max}(log ϵ)	Ref.
$C_{12}H_{12}N_4O_6$			
2-Pyrazoline-3-carboxylic acid, 1-(2,4-dinitrophenyl)-, ethyl ester	$CHCl_3$	387(4.42)	22-0619-66
$C_{12}H_{12}N_6$			
Adenine, 9-(p-aminobenzyl)-	pH 1	259(4.15)	87-0576-66
	pH 7	258(4.21)	87-0576-66
	pH 13	259(4.20)	87-0576-66
Purine, 2-amino-6-(benzylamino)-	pH 1	245(4.10),278(4.13)	95-0649-66
	pH 13	283(4.01)	95-0649-66
Purine, 2,6-diamino-9-benzyl-	pH 1	240(4.25),290(3.98)	95-0649-66
	pH 13	280(4.01)	95-0649-66
$C_{12}H_{12}N_6O_2S$			
Sulfanilamide, N^1-(7-methyl-6-purinyl)-	pH 1.2	293(4.36)	87-0373-66
	pH 12.9	256(4.25),287(4.39)	87-0373-66
Sulfanilamide, N^1-(9-methyl-2-purinyl)-	pH 1.2	232s(4.27),261s(3.99)	87-0373-66
	pH 12.9	227(4.44),255(4.30), 299s(3.88)	87-0373-66
Sulfanilamide, N^1-(9-methyl-6-purinyl)-	pH 1.2	287(4.38)	87-0373-66
	pH 12.9	258s(4.27),281(4.46)	87-0373-66
Sulfanilamide, N -(9-methyl-8-purinyl)-	pH 1.2	294(4.35)	87-0373-66
	pH 12.9	258(4.29),297(4.29)	87-0373-66
$C_{12}H_{12}N_6O_5$			
Pyrrolo[2,3-d]pyrimidine-2,4,6(3H)-trione, 5-(6-amino-1,2,3,4-tetrahydro-1-methyl-2,4-dioxo-5-pyrimidinyl)-5,7-dihydro-1-methyl-	MeOH	229(4.29),253(4.14), 267(4.22),325(3.75)	24-3524-66
$C_{12}H_{12}O$			
Benzobicyclo[3.2.1]oct-3-en-2-one	EtOH	248(4.00)	22-0147-66
Benzobicyclo[4.2.0]oct-3-en-2-one	EtOH	251(4.02)	22-0147-66
2-Cyclohexen-1-one, 3-phenyl-	EtOH	221(3.97),284(4.28)	22-1012-66
Dibenzofuran, 1,2,3,4-tetrahydro-	EtOH	251(4.03),277(3.51), 284(3.41)	88-5225-66
1-Naphthalenemethanol, 2-methyl-	dioxan	275(3.71),285(3.74), 295s(3.54),321(2.60)	78-1587-66
1-Naphthalenemethanol, 8-methyl-	dioxan	276(3.81),286(3.87), 298(3.73),321(2.40)	78-1587-66
$C_{12}H_{12}OS$			
1-Naphthol, 2-(methylthiomethyl)-	MeOH	214(4.63),237(4.64), 295(3.61)	35-5855-66
	MeOH-KOH	219(4.44),255(4.49), 343(3.96)	35-5855-66
2-Naphthol, 1-(methylthiomethyl)-	MeOH	231(4.79),281(3.72), 293(3.67),338(3.51)	35-5855-66
	MeOH-KOH	243(4.75),286(3.78), 357(3.61)	35-5855-66
$C_{12}H_{12}O_2$			
Cyclopropane, 1-acetyl-1-benzoyl-	hexane	214(3.67),244(4.09), 310(2.49)	44-0447-66
	EtOH	204(4.32),246(4.03), 287(2.46)	44-0447-66
3(2H)-Furanone, 2,2-dimethyl-5-phenyl-	EtOH	220(3.98),242(3.89), 305(4.23)	88-0233-66
	EtOH	220(3.98),242(3.89)	32-1073-66

Compound	Solvent	$\lambda_{max}(\log \epsilon)$	Ref.
1,4-Methanonaphthalene-5,6-dione, 1,4,4a,8a-tetrahydro-8-methyl-	n.s.g.	285(3.35)	88-6175-66
1(2H)-Naphthalenone, 3,4-dihydro-2-(methoxymethylene)-	n.s.g.	264(3.94),295(4.06)	88-2963-66
$C_{12}H_{12}O_2S$			
2,4-Decadiene-6,8-diynoic acid, 4-(methylthio)-, methyl ester, 2-cis-4-trans	hexane	203(4.35),279s(3.92), 327(4.01)	24-2096-66
4-cis-2-trans	ether	239(4.16),250(4.22), 268s(3.96),358(4.20)	24-2096-66
trans-trans	ether	243(4.16),296s(4.18), 312(4.25),332(4.23), 350s(4.16)	24-2096-66
2,4-Decadiene-6,8-diynoic acid, 5-(methylthio)-, methyl ester, trans	ether	238(4.21),296(4.11), 354(4.40)	24-0138-66
$C_{12}H_{12}O_3$			
1-Cyclobutanecarboxylic acid, 2-benzoyl-, cis	MeOH	243(4.10)	24-1229-66
	EtOH	244(4.11)	22-0147-66
2(1H)-Naphthalenone, 3,4-dihydro-3-(hydroxymethylene)-5-methoxy-	EtOH	205(3.97),280(3.60), 308(3.53)	44-1747-66
$C_{12}H_{12}O_4$			
Benzocyclobutadienequinone, bis-(ethylene ketal)	EtOH	256.5(3.08)	44-1866-66
2-Naphthoic acid, 1,2,3,4-tetra-hydro-8-methoxy-3-oxo-	EtOH	214(4.07),264(3.81), 279(3.59)	44-1747-66
2,4-Pentanedione, 3-(o-carboxy-phenyl)-	EtOH	224s(3.92),285(3.94)	12-1265-66
	EtOH-KOH	304(4.15)	12-1265-66
$C_{12}H_{12}O_4S$			
Acrylic acid, 3-hydroxy-2-(methyl-thio)-, methyl ester, benzoate	ether	238(4.28)	24-1558-66
2,4-Decadiene-6,8-diynoic acid, 4-(methylsulfonyl)-, methyl ester, 2-cis-4-trans	ether	314(4.16)	24-2096-66
4-cis-2-trans	ether	244(4.18),317(4.37), 333(4.42)	24-2096-66
di-cis	ether	211(4.38),248s(3.75), 265s(3.78),281s(3.98), 300(4.08)	24-2096-66
di-trans	ether	234(4.10),245(4.14), 315(4.32),333(4.34)	24-2096-66
2,4-Decadiene-6,8-diynoic acid, 5-(methylsulfonyl)-, methyl ester	ether	238(4.14),248(4.23), 328(4.35)	24-2096-66
$C_{12}H_{12}O_8$			
1,2,4-Benzenetricarboxylic acid, 3,5-dimethoxy-, 4-methyl ester	EtOH	245(3.89),302(3.47)	39-1608-66C
$C_{12}H_{13}BrS$			
Benzo[b]thiophene, 5-bromo-2,3,4,7-tetramethyl-	EtOH	238(4.56),240(4.48), 269(3.90),300(3.61), 312(3.70)	22-3055-66
Benzo[b]thiophene, 6-bromo-2,3,4,7-tetramethyl-	EtOH	238(4.43),244s(4.39), 269(3.96),277(3.97), 300(3.60),312(3.70)	22-3055-66

Compound	Solvent	$\lambda_{max}(\log \epsilon)$	Ref.
$C_{12}H_{13}ClN_4$			
Dipyridazino[2,3-a:4,3-d]pyrrole, 8-chloro-3-ethyl-6,7-dihydro-5-methyl-	EtOH	229(4.29),267(4.37), 338(3.83)	1-2637-66
$C_{12}H_{13}ClO_5$			
Phthalide, 7-(chloromethyl)-4,5,6-trimethoxy-	EtOH	258(3.92),298(3.60)	44-1912-66
$C_{12}H_{13}IN_2O$			
1-(N-Methylacetamido)quinolinium iodide	EtOH	318(3.91)	94-0512-66
$C_{12}H_{13}N$			
Carbazole, 1,2,3,4-tetrahydro-	EtOH	228(4.53),284(3.87), 292(3.81)	94-0934-66
$C_{12}H_{13}NO$			
Azepino[3,2,1-hi]indol-6(7H)-one, 1,2,4,5-tetrahydro-	MeOH	257(3.81),300s(3.18)	35-4061-66
Furfurylamine, 3-methyl-4-phenyl-, hydrochloride	EtOH	240(3.72)	44-0052-66
2-Indolinone, 3-isopropylidene-N-methyl-	EtOH	220s(3.97),252(4.47), 256(4.47),262(4.53), 280(3.69),291(3.77), 300s(3.69),346(3.08)	44-0077-66
Isocarbostyril, 3-isopropyl-	EtOH	226(4.23),239(4.03), 248(3.92),278(4.00), 287(3.99),319(3.64), 331(3.62),347(3.54)	44-2090-66
Isocarbostyril, 3-propyl-	EtOH	227(4.30),239(4.07), 248(3.97),280(4.05), 287(4.04),319(3.67), 332(3.76),347(3.59)	44-2090-66
Quinoline, 2,3,4-trimethyl-, N-oxide	EtOH	323(3.88),338s(3.81)	1-2467-66
$C_{12}H_{13}NOS$			
2-Propanone, [(3,4-dihydro-1-isoquinolyl)thio]-, HBr salt	EtOH	282(4.24)	4-0282-66
$C_{12}H_{13}NOS_2$			
2-Indolinone, 1-methyl-3-(bis-methylthiomethylene)-	EtOH	260(4.03),282(3.93), 366(4.13)	95-1152-66
$C_{12}H_{13}NO_2$			
Benzocyclobutene, 0,1-diacetyl-, oxime	EtOH	259(3.11),265(3.29), 271(3.28)	87-0656-66
Carbostyril, 1-propoxy-	EtOH	230(4.60),246s(3.92), 271(3.82),277s(3.79), 316s(3.67),328(3.74), 342s(3.60)	78-0025-66
Indole-3-carboxaldehyde, 1-ethyl-4-hydroxy-2-methyl-	MeOH	253(4.40),270(4.19), 345(3.94)	35-0804-66
Indole-4,5-dione, 1-ethyl-2,6-dimethyl-	MeOH	242(4.46),355(3.61), 475(3.15)	35-2536-66
	MeOH	243(4.45),355(3.61), 570(3.19)	44-1012-66
Indole-4,7-dione, 1-ethyl-2,6-dimethyl-	MeOH	232(4.18),254(4.07), 346(3.40),444(3.32)	44-1012-66

Compound	Solvent	$\lambda_{max}(\log \epsilon)$	Ref.
Isocarbostyril, 2-hydroxy-3-isopropyl-	EtOH	226(4.06),249(3.75), 287(3.74),291(3.72), 317(3.36),327(3.44), 343(3.24)	44-2090-66
Isocarbostyril, 2-hydroxy-3-propyl-	EtOH	226(4.31),242(3.98), 249(3.97),291(3.92), 327(3.72),343(3.57)	44-2090-66
Spiro[indoline-3,2'-oxiran]-2-one, 1,3',3'-trimethyl-	EtOH	219(4.47),242s(3.69), 252s(3.67),262s(3.62), 274s(3.45),302(3.11)	44-0077-66
$C_{12}H_{13}NO_2S$			
L-Proline, N-(thiobenzoyl)-	H_2O	278(3.96),340(2.58)	1-2781-66
	MeOH	280(3.72),367(2.44)	1-2781-66
	dioxan	282(3.97),385(2.46)	1-2781-66
$C_{12}H_{13}NO_2S_2$			
2-Indolinone, 5-methoxy-3-(bis-methylthiomethylene)-	EtOH	266(3.89),289(3.99), 366(4.08)	95-1152-66
$C_{12}H_{13}NO_3$			
1-Aza-8,9-benzocyclononene-2,7-dione, 3-hydroxy-, hemiketal	EtOH	250(4.03),281(3.34), 288(3.20)	35-1049-66
Indole-4,7-dione, 1-ethyl-3-(hydroxy-methyl)-2-methyl-	MeOH	232(4.30),257(4.13), 350(3.52),445(3.52)	35-0804-66
Indole-3-lactic acid, β-methyl-	EtOH	282(3.74)	30-0611-66A
Pyrazolin-5-one, 4-(2-hydroxyethyl)-3-methyl-1-phenyl-	MeOH	248(4.00)	24-2962-66
$C_{12}H_{13}NO_4$			
Alanine, N-carbobenzyloxy-dehydro-, methyl ester	80% MeOH	244(3.74)	44-3928-66
Fumaric acid, anilino-, dimethyl ester	EtOH	234(3.96),323(4.14)	24-2526-66
$C_{12}H_{13}NO_4S$			
N-Ethyl-5-p-tolylisoxazolium-3'-sulfonate	pH 1	296(4.40)	78-0321-66B
$C_{12}H_{13}N_2P$			
Phosphine, bis(2-cyanoethyl)phenyl-	MeOH	244(3.6)	51-0476-66
$C_{12}H_{13}N_3$			
Imidazo[2,3-a]phthalazine, 2-ethyl-2,3-dihydro-	10% HCl	216(4.63),238(4.18), 272(4.00),296(3.90)	4-0381-66
	EtOH	211(4.68),268(4.11), 276(4.26),336(3.95)	4-0381-66
Phenazine, 1-amino-6,7,8,9-tetrahydro-	EtOH	238(3.93),272(4.51), 326(3.42)	44-3384-66
$C_{12}H_{13}N_3O$			
Imidazo[2,1-b]quinazolin-5-one, 1-ethyl-	EtOH	232(4.59),267s(4.10), 274(4.16),331(3.50)	24-1532-66
2-Indolinone, 3-(2-imidazolidin-ylidene)-1-methyl-	EtOH	256(4.31),307(4.48)	95-1152-66
Phenol, p-[N-[2-imidazol-4-ylethyl]-formimidoyl]-	EtOH	214(4.34),271(4.29)	44-2380-66
s-Triazolo[4,3-a]pyridine, 5,6,7,8-tetrahydro-3-oxo-1-phenyl-	EtOH	268(3.81)	7-0199-66

Compound	Solvent	λ max(log ε)	Ref.
$C_{12}H_{13}N_3O_2$			
Hydrazine, 1-nitroso-1-(3-methyl-4-phenylfurfuryl)-	EtOH	249(4.49)	44-0052-66
Pyrazole, 1-(p-nitrophenyl)-3,4,5-trimethyl-	EtOH	325(4.13)	22-3744-66
Uracil, 6-methyl-5-p-toluidino-	9N HCl	258(3.95)	87-0108-66
	pH 7.0	246(4.21),288s(3.59)	87-0108-66
	pH 11.6	246(4.18),284(3.88)	87-0108-66
$C_{12}H_{13}N_3O_4$			
Succinic acid, 2-anilino-3-diazo-, dimethyl ester	EtOH	244(4.22),284s(3.40)	24-0475-66
$C_{12}H_{13}N_5$			
5H-s-Triazolo[3,4-b]-s-triazole, 3-ethyl-6-p-tolyl-	EtOH	253(4.39)	4-0119-66
$C_{12}H_{13}N_5O$			
Pyrimidine, 2,4-diamino-5-benzamido-6-methyl-	pH 1	230(4.38),268(3.90)	4-0476-66
	pH 7	228(4.27),281(3.94)	4-0476-66
	pH 13	228(4.23),283(3.92)	4-0476-66
5H-s-Triazolo[4,3-b]-s-triazole, 3-ethyl-6-(p-methoxyphenyl)-	EtOH	265(4.40)	4-0119-66
$C_{12}H_{13}N_5O_2$			
Benzaldehyde, (4,6-dimethoxy-s-triazin-2-yl)hydrazone	EtOH	300(4.49)	49-1713-66
$C_{12}H_{13}N_5O_4$			
Pyrazolo[3,2-c]-as-triazine-3-carboxylic acid, 4-amino-, ethyl ester, diacetyl derivative	EtOH	227(4.13),256(4.05), 332(3.81)	39-1127-66C
$C_{12}H_{14}$			
Fulvene, 6,6-dicyclopropyl-	isooctane	289(4.29),365(2.49)	44-4321-66
Indene, 1-ethyl-2-methyl-	EtOH	260(4.03)	11-0109-66
Indene, 2-ethyl-1-methyl-	EtOH	259.5(4.08)	11-0109-66
Indene, 1,2,3-trimethyl-	heptane	208(4.29),261(4.03)	44-0081-66
Naphthalene, 1,2-dihydro-2,4-dimethyl-	EtOH	223(4.28),259(3.98), 263s(3.96),283s(3.53), 294s(3.15)	22-0645-66
Naphthalene, 1,2-dihydro-3,4-dimethyl-	EtOH	223s(4.18),262(4.00), 268(3.99),285s(3.62), 296s(3.18)	22-0645-66
$C_{12}H_{14}BF_4NO$			
N-Ethyl-5-p-tolylisoxazolium tetrafluoroborate	CH_2Cl_2	309(4.40)	78-0321-66B
$C_{12}H_{14}BrClO_4$			
Benzoic acid, 2-(bromomethyl)-3-chloro-4,6-dimethoxy-, ethyl ester	EtOH	306(3.59)	12-1265-66
$C_{12}H_{14}BrNOS$			
2-Propanone, [(3,4-dihydro-1-isoquinolyl)thio]-, HBr salt	EtOH	282(4.24)	4-0282-66

Compound	Solvent	λ max(log ε)	Ref.
$C_{12}H_{14}BrN_3O_2$			
7H-Pyrrolo[2,3-c]pyridazine, 5-bromo-6-(2-carboxyethyl)-3-methyl-	EtOH	232(4.70),282(4.12), 331(3.77)	1-2637-66
$C_{12}H_{14}ClNO$			
Furfurylamine, 3-methyl-4-phenyl-, hydrochloride	EtOH	240(3.72)	44-0052-66
$C_{12}H_{14}Cl_2$			
Benzocyclobutene, 1,1-dichloro-tetramethyl-	C_6H_{12}	253(3.22),260(3.22), 265(3.26),276(3.43), 283(3.44),287(3.47)	44-2244-66
$C_{12}H_{14}N_2$			
α-Carboline, 9,10,11,12-tetrahydro-4-methyl-	EtOH	234(4.38),298(3.96)	30-0705-66
Pyrazole, 5-ethyl-4-methyl-3-phenyl-	EtOH	249(4.12)	22-2971-66
3H-Pyridazino[2,3-a]quinoline, 4,4a,5,6-tetrahydro-	EtOH	279(4.18),307(3.87)	95-0608-66
9H-Pyrido[2,3-b]indole, 5,6,7,8-tetrahydro-2-methyl-	EtOH	234(4.38),298(3.96)	30-0361-66D
Quinazoline, 2-tert-butyl-	pH 0.0	260(3.92)	39-0234-66C
	pH 10.0	224(4.61),265(3.42), 310(3.35),320s(3.24)	39-0234-66C
$C_{12}H_{14}N_2O$			
1H-1,2-Diazepin-6-ol, 6,7-dihydro-5-methyl-4-phenyl-	EtOH	303(3.72)	44-0052-66
Indole, 3-acetyl-5-(dimethylamino)-	EtOH	240(4.32),282(4.00)	23-0387-66
$C_{12}H_{14}N_2OS$			
L-Prolineamide, N-(thiobenzoyl)-	MeOH	283(4.01),368(2.37)	1-2781-66
	dioxan	288(4.27),383(2.48)	1-2781-66
$C_{12}H_{14}N_2O_2$			
8,11-Diaza[4.3.3]propellane, 8,11-dimethyl-7,9,10,12-tetraoxo-	MeCN	246(2.77),259(2.65), 268(2.67)	78-0279-66B
Indole-2-carboxylic acid, 3-(3-amino-1-propyl)-	MeOH	235(4.28),290(4.25)	5-0116-66F
$C_{12}H_{14}N_2O_2S$			
Pyrazole, 3,5-dimethyl-1-(p-toluenesulfonyl)-	EtOH	241(4.07)	28C-0782-66A
$C_{12}H_{14}N_2O_2S_2$			
Thiazolin-2-imine, N-propionyl-3-(2-hydroxy-2-thienylethyl)-	iso-PrOH-acid	233(4.08),286(3.92)	20-0380-66
$C_{12}H_{14}N_2O_4$			
Benzene, 1-(4-cyanobutoxy)-4-methoxy-2-nitro-	MeOH	357(3.38)	78-0093-66B
Benzene, 1-(4-cyanobutoxy)-4-methoxy-3-nitro-	MeOH	357(3.37)	78-0093-66B
$C_{12}H_{14}N_4$			
Pyrazole, 1,3,5-trimethyl-4-(phenylazo)-	$CHCl_3$	332(4.36),430(3.18)	22-2990-66

Compound	Solvent	λ_{max}(log ε)	Ref.
$C_{12}H_{14}N_4O$			
3-Pyrazolin-5-one, 1,2,3-tri-methyl-4-(phenylazo)-	$CHCl_3$	355(4.31),420(3.32)	22-2990-66
$C_{12}H_{14}N_4O_4$			
5,5'-Biuracil, 1,1',3,3'-tetra-methyl-	H_2O	272(3.18)	50-1222-66B
3-Butenal, 2,2-dimethyl-, 2,4-dinitrophenylhydrazone	n.s.g.	360(4.22)	22-0734-66
2-Pentenal, 4-methyl-, 2,4-dinitrophenylhydrazone	n.s.g.	245(4.19),373(4.46)	22-0717-66
3-Pentenal, 4-methyl-, 2,4-dinitrophenylhydrazone	n.s.g.	245s(4.06),360(4.33)	22-0717-66
2-Pyrazoline, 1-(2,4-dinitrophenyl)-3,5,5-trimethyl-	$CHCl_3$	402(4.31)	22-0610-66
$C_{12}H_{14}N_4O_6$			
Isovaleric acid, β-formyl-, 2,4-dinitrophenylhydrazone	EtOH	364(4.28)	78-0015-66B
$C_{12}H_{14}N_4O_8$			
Caproic acid, ε-picryl-	pH 7.4	348(4.20)	69-3385-66
$C_{12}H_{14}N_4S$			
2(1H)-Pyrazinethione, 3-[(3,6-di-methylpyrazinyl)methyl]-6-methyl?	pH 3	244s(3.70),280(4.20), 385(3.89)	39-0495-66C
	pH 9	235(3.88),278(4.14), 345(3.78)	39-0495-66C
$C_{12}H_{14}O$			
Benzocyclobutenone, tetramethyl-	EtOH	265(4.18),306(3.47)	44-2244-66
	H_2SO_4	320(4.26),380(3.31)	44-2244-66
Dibenzofuran, 1,2,3,4,4a,9b-hexa-hydro-	EtOH	267s(3.15),274(3.32), 280(3.35),285s(3.20)	1-1561-66
1-Indanone, 2-ethyl-3-methyl-	EtOH	245(4.06),290(3.38)	11-0109-66
1-Indanone, 3-ethyl-2-methyl-	EtOH	245(4.05),290(3.37)	11-0109-66
1-Indanone, 4,5,6-trimethyl-	MeOH	263(4.11),304(3.51)	35-2838-66
1-Indanone, 4,6,7-trimethyl-	MeOH	262(4.03),304(3.23)	35-2838-66
1-Indanone, 5,6,7-trimethyl-	MeOH	259(4.15),305(3.46)	35-2838-66
1(2H)-Naphthalenone, 3,4-dihydro-2,2-dimethyl-	MeOH	246(4.05),290(3.20)	35-5855-66
Phenol, o-(1-cyclohexenyl)-	EtOH	239s(3.76),275s(3.39), 281(3.45)	1-1561-66
Spiro[furan-2(3H),1'-indan], 4,5-dihydro-	MeOH	259(2.83),266(3.02), 272(3.07)	87-0719-66
$C_{12}H_{14}O_2$			
Bicyclo[4.2.1]nonatrien-9-ol, 9-methyl-, acetate	MeOH	219(3.50),256(3.57), 265(3.56)	35-3832-66
2-Butenoic acid, 3-phenyl-, ethyl ester, trans	EtOH	265(4.20)	88-1787-66
3-Butenoic acid, 3-phenyl-, ethyl ester	EtOH	243(4.03)	88-1787-66
6,9-Ethano-4-chromone, 5,6,9,10-tetrahydro-6-methyl-	EtOH	263(3.85)	39-2324-66C
2,4-Pentanedione, 3-benzyl-	hexane	262s(--),269s(--), 287(3.80)	18-0901-66
	EtOH	261s(--),269s(--), 289(3.58)	18-0901-66
	EtOH-NaOEt	309(4.29)	18-0901-66

Compound	Solvent	λ_{max}(log ε)	Ref.
$C_{12}H_{14}O_3$			
1-Azulenecarboxylic acid, 1,2,4,5-6,7,8,8a-octahydro-8-hydroxy-8-methyl-2-oxo-, γ-lactone	EtOH	246(4.00),271s(3.87)	88-3627-66
Benzoic acid, o-(3-oxopentyl)-	EtOH	228(3.81),277(3.02)	44-2090-66
1,3-Cyclohexanedione, 2-(3-oxo-1-cyclohexen-1-yl)-	EtOH	236s(4.06),268(4.14)	44-1050-66
	6N H_2SO_4	242s(4.09),273(4.19)	44-1050-66
	H_2O	239(4.07),284(4.21)	44-1050-66
	NaOH	238(4.04),286(4.34)	44-1050-66
2,6-Octadien-4-ynal, 8-acetoxy-2,6-dimethyl-	pet ether	283s(4.28),295(4.40), 313(4.37)	33-0369-66
$C_{12}H_{14}O_4$			
Benzene, 1,4-diacetyl-2,5-dimethoxy-	EtOH	225(4.25),261(4.00), 363(3.59)	44-3321-66
Furo[2,3-b]benzofuran, 2,3,3a,8a-tetrahydro-4,6-dimethoxy-	EtOH	207(4.63),271(2.92), 278(2.88),333s(3.86)	39-1308-66C
$C_{12}H_{14}O_4S$			
2,4,8-Decatrien-6-ynoic acid, 4-methylsulfonyl-, methyl ester, cis-cis-cis	ether	228(4.08),295s(4.09), 310(4.14),328s(4.10)	5-0149-66D
cis-trans-cis	ether	227(4.09),300s(4.18), 315(4.24),334s(4.23)	5-0149-66D
$C_{12}H_{14}S$			
Benzo[b]thiophene, 5-ethyl-2,3-dimethyl-	EtOH	235(4.62),265(3.85), 293(3.43),304(3.39)	22-3055-66
Benzo[b]thiophene, 6-ethyl-2,3-dimethyl-	EtOH	234(4.55),269(3.90), 291s(3.40),302(3.07)	22-3055-66
2H-1-Benzothiopyran, 2,2,4-trimethyl-	EtOH	240(4.26),298(3.05), 322(3.00)	78-0007-66
$C_{12}H_{15}^+$			
Cyclobutylmethylphenylcarbonium ion	FSO_3H-SbF_5	334(4.34),382(3.78)	35-1488-66
$C_{12}H_{15}ClN_2O$			
Furfurylhydrazine, 3-methyl-4-phenyl-, hydrochloride	EtOH	219(4.32)	44-0052-66
1(2H)-Pyridinecarbonyl chloride, 2-(cyclohexylimino)-	heptane	260(3.42),267(3.38)	32-1020-66
$C_{12}H_{15}ClO$			
Cyclopropenone, 2-(1-chloro-1-butenyl)-3-(1-pentenyl)-	n.s.g.	272(4.42),281(4.43)	88-3763-66
$C_{12}H_{15}ClO_4$			
Benzoic acid, 3-chloro-4,6-dimethoxy-2-methyl-, ethyl ester	EtOH	288(3.50)	12-1265-66
$C_{12}H_{15}Cl_2N_5O_2$			
Pyrazolo[5,1-c]-as-triazine-3-carboxylic acid, 4-[bis(2-chloroethyl)-amino]-, ethyl ester	EtOH	269(4.05),345(3.94)	39-1127-66C
$C_{12}H_{15}N$			
1-Aza-2,3-benzo-6,6-dimethylbicyclo[2.2.1]heptene	pentane	269(3.00),275(2.98)	5-0128-66D
3-Picoline, 1,2,5,6-tetrahydro-4-phenyl-	EtOH	237(3.87)	78-2745-66

Compound	Solvent	$\lambda_{max}(\log \epsilon)$	Ref.
$C_{12}H_{15}NO$			
3-Buten-2-one, 3-methyl-4-p-tolyl-, α-oxime	n.s.g.	273.5(4.36)	83-0626-66
Ketone, cyclohexyl 2-pyridyl, hydrochloride	MeOH	230(3.89),267(3.62)	44-2149-66
1-Penten-3-one, 1-p-tolyl-, α-oxime	n.s.g.	288.5(4.40)	83-0626-66
1-Penten-3-one, 1-p-tolyl-, β-oxime	n.s.g.	293.5(4.28)	83-0626-66
2-Piperidone, 6-methyl-1-phenyl-	MeOH	204(4.2),235(3.8)	24-0724-66
$C_{12}H_{15}NOSn$			
Tin, (8-quinolinolato)trimethyl-	C_6H_{12}	362(3.40)	101-0249-66B
$C_{12}H_{15}NO_2$			
Acrylic acid, α-cyano-β-cyclohexylidene-, ethyl ester	EtOH	236.5(4.07)	22-1033-66
8-Aza[4.3.3]propell-3-ene, 8-methyl-7,9-dioxo-	MeCN	248(2.33),258s(2.18)	78-0279-66B
2-Butanone, 3-(o-acetamidophenyl)-	EtOH	226(3.80)	1-2467-66
3-Buten-2-one, 4-(p-methoxyphenyl)-3-methyl-, α-oxime	n.s.g.	282.5(4.49)	83-0626-66
1-Penten-3-one, 1-(p-methoxyphenyl)-, α-oxime	n.s.g.	298.5(4.42)	83-0626-66
$C_{12}H_{15}NO_2S$			
L-Valine, N-(thiobenzoyl)-	H_2O	278s(3.80),364s(2.57)	1-2781-66
	MeOH	286(3.59),371(2.36)	1-2781-66
	dioxan	289(3.78),389(2.41)	1-2781-66
$C_{12}H_{15}NO_3$			
Benzoic acid, o-(1,1-dimethylacetonyl)-, oxime	EtOH	231(3.88)	44-2090-66
Hydrocotarnine	dioxan	210(4.3),298(3.2)	24-0273-66
$C_{12}H_{15}NO_4$			
3-Pyridineacetic acid, 5-carboxy-2,6-dimethyl-, dimethyl ester	EtOH	230(3.98),276(3.69), 281(3.64)	39-1075-66C
$C_{12}H_{15}NO_4S_2$			
[5-(2-Carboxy-3-oxo-2-butenyl)-4-hydroxy-3-methyl-4-thiazolin-2-ylidene]methylsulfonium hydroxide, inner salt	dioxan	256(4.27),339(3.73), 460(4.49)	24-0307-66
	CH_2Cl_2	257(4.19),335(3.66), 458(4.40)	24-0307-66
$C_{12}H_{15}NO_5$			
Benzoic acid, 4-acetamido-2,6-dimethoxy-, methyl ester	EtOH	217(4.47),260(4.17)	12-0275-66
$C_{12}H_{15}N_3$			
Pyrrolidine, 2,5-di-2-pyrrolyl-	EtOH	218(4.26)	78-0053-66
$C_{12}H_{15}N_3O$			
Phthalazine, 1-(1-hydroxy-2-butylamino)-	10% HCl	213(4.51),244(4.16), 262(4.16),305(3.85), 317(3.78)	4-0381-66
	EtOH	210(4.68),220s(4.05), 257(3.78),316(3.95)	4-0381-66
2-Pyrazolin-5-one, 4-(3-aminopropyl)-1-phenyl-	EtOH	252(4.17)	44-2487-66

Compound	Solvent	λ_{max}(log ϵ)	Ref.
$C_{12}H_{15}N_3OS$			
1,3,4-Thiadiazole, 2-(isopropyl-amino)-5-(p-methoxyphenyl)-	EtOH	253(3.59),314(4.29)	83-0921-66
$C_{12}H_{15}N_3O_3$			
Veratraldehyde, 4,4-dimethyl-semicarbazone	EtOH	292(4.31),303(4.33)	22-0153-66
$C_{12}H_{15}N_3O_4S$			
1H-Imidazo[4,5-c]pyridine, 4-(methyl-thio)-1-β-D-ribofuranosyl-	pH 1	222s(--),235s(--), 275s(--),300s(--), 307(4.23)	87-0105-66
	pH 7	216(4.23),284(4.13), 291s(--)	87-0105-66
	pH 13	284(4.14),291s(--)	87-0105-66
3H-Imidazo[4,5-b]pyridine, 7-(methyl-thio)-3-β-D-ribofuranosyl-	pH 1	225(4.03),280s(4.08), 288(4.11)	87-0354-66
	pH 7	218(4.17),282(4.27), 287s(4.26)	87-0354-66
	pH 13	282(4.28),287s(4.27)	87-0354-66
$C_{12}H_{15}N_5$			
Pteridine, 2-(allylamino)-4,6,7-trimethyl-	pH 1.0	224(4.24),235s(4.09), 360(3.93)	39-0226-66C
	pH 7.0	234(4.28),270(3.85), 380(3.77)	39-0226-66C
Pyrimidine, 2,4-diamino-5-(N,4-dimethylanilino)-	12N HCl	233(3.97),282(3.78)	87-0108-66
	pH 5.0	240(4.02)	87-0108-66
	pH 10.0	246(3.93),295s(3.59)	87-0108-66
$C_{12}H_{15}N_5OS$			
Adenine, 9-(5-S-ethyl-5-thio-2,3,5-trideoxy-2,3-didehydro-β-D-glyc-eropentofuranosyl)-	MeOH	259.5(4.17)	35-1549-66
$C_{12}H_{15}N_5O_4$			
Adenosine, 3'-O-acetyl-2'-deoxy-	MeOH	259(4.22)	69-0224-66
$C_{12}H_{15}N_5O_8$			
Lysine, picryl-	pH 7.4	348(4.19)	69-3385-66
$C_{12}H_{15}P$			
Phosphine, diethyl(phenylethynyl)-	EtOH	220(4.08),237(4.08), 247(4.11),258(4.08), 282(3.88)	22-1002-66
$C_{12}H_{16}$			
Biphenylene, 1,2,3,6,7,8,8a,8b-octahydro-	n.s.g.	237(4.176)	89-0846-66
1-Butene, 3,3-dimethyl-1-phenyl-, cis	EtOH	220(3.81)	35-0378-66
trans isomer	EtOH	251(4.26)	35-0378-66
Cycloheptane, (2,4-cyclopenta-dien-1-ylidene)-	C_6H_{12}	275(4.28),280(4.28), 364(2.44)	33-1278-66
	EtOH	276(4.26),362(2.51)	33-1278-66
Naphthalene, 1,2,3,4-tetrahydro-1,3-dimethyl-	EtOH	260s(2.63),266(2.74), 273(2.74),287s(1.69), 291s(1.37),297s(1.07)	22-0645-66
Naphthalene, 1,2,3,4-tetrahydro-1,4-dimethyl-	EtOH	259s(2.75),265(2.81), 273(2.77),295s(1.69)	22-0645-66

Compound	Solvent	$\lambda_{max}(\log \epsilon)$	Ref.
$C_{12}H_{16}ClN$			
3-Picoline, 1,2,5,6-tetrahydro-4-phenyl-, hydrochloride	EtOH	230(3.89)	78-2745-66
$C_{12}H_{16}ClNO$			
Ketone, cyclohexyl 2-pyridyl, hydrochloride	MeOH	230(3.89),267(3.62)	44-2149-66
$C_{12}H_{16}ClNO_4$			
Dimethyl(α-methylcinnamylidene)-ammonium perchlorate	MeCN	326(4.53)	24-2479-66
$C_{12}H_{16}ClN_3O_2$			
5-Amino-1-(carboxymethyl)-2-methyl-imidazo[1,2-a]pyridinium chloride, ethyl ester	H_2O	228(4.45),323(4.18)	4-0093-66
$C_{12}H_{16}CuN_6O_5$			
L-Histidine, glycylglycylglycyl-, copper complex	H_2O	550(2.05),590(1.81), 660(1.80),715(1.60)	37-0122-66
$C_{12}H_{16}N_2O$			
Tryptamine, 6-hydroxy-N,N-dimethyl-	MeOH	264s(3.53),275(3.63), 297(3.70)	5-0157-66J
$C_{12}H_{16}N_2OS$			
L-Valineamide, N-(thiobenzoyl)-	MeOH	288(3.89),388s(2.39)	1-2781-66
	dioxan	290(3.81),388(2.46)	1-2781-66
$C_{12}H_{16}N_2O_2$			
Nitrone, α-(diethylcarbamoyl)-N-phenyl-	EtOH	268(5.00)	44-1689-66
o-Phenylenediamine, N,N'-di-propionyl-	MeOH	240s(4.10)	78-1175-66
Pyruvic acid, ethyl ester, methylphenylhydrazone	EtOH	338(3.48)	22-2981-66
$C_{12}H_{16}N_2O_3$			
Anthranilic acid, N-(butyl-carbamoyl)-	pH 1.0	222(4.49),250(4.02), 313(3.58)	35-4001-66
Anthranilic acid, N-(tert-butyl-carbamoyl)-	pH 1	224(4.49),252(4.09), 314(3.62)	35-4001-66
D-Erythrose, 2,4-O-ethylidene-, phenylhydrazone	n.s.g.	280(4.34)	44-0223-66
$C_{12}H_{16}N_4O_3$			
Lumazine, 8-(3-hydroxypropyl)-3,6,7-trimethyl-	pH -1.9	248(4.04),360(4.11)	24-3503-66
	pH 5	262(4.27),276s(3.91), 408(4.06)	24-3503-66
	pH 13	238(4.20),312(4.18), 365(3.63)	24-3503-66
$C_{12}H_{16}N_4O_3S$			
Valeric acid, 5-[(2-hydroxy-purin-6-yl)thio]-, ethyl ester	pH 1	268(3.87),311(3.97)	73-1824-66
	pH 13	221(4.28),323(4.03)	73-1824-66
Valeric acid, 5-[(8-hydroxy-purin-6-yl)thio]-, ethyl ester	pH 1	227(4.08),297(4.23)	73-1824-66
	pH 13	230(4.16),304(4.26)	73-1824-66
$C_{12}H_{16}N_4O_4$			
Thymine, 1-methyl-, dimer	N NaOH	242(4.05)	39-1342-66C

Compound	Solvent	λ_{max}(log ε)	Ref.
Thymine, 1-methyl-, dimer, isomer B	H_2O	242(4.14)	39-1342-66C
$C_{12}H_{16}N_6NiO_5$			
L-Histidine, glycylglycylglycyl-, nickel complex	pH 11	430(2.18),480(1.81)	37-1072-66
$C_{12}H_{16}NiO_4S_2$			
Nickel, bis(ethyl thioloacetoacetate)	$CHCl_3$	513s(2.56),671(1.70)	12-1401-66
$C_{12}H_{16}O$			
Benzaldehyde, 3,5-diethyl-2-methyl-	EtOH	256(3.94),302(3.15)	22-3238-66
Benzocyclooctenone	EtOH	250(3.81),291(3.06)	22-1693-66
Inden-6(3aH)-one, 4,5,7,7a-tetrahydro-3a-methyl-3-vinyl-, cis	EtOH	234(4.12)	78-2021-66
Phenol, 2-cyclohexyl-	H_2O	271(3.34)	22-3328-66
	base	290.0(3.61)	22-3328-66
Phenol, 4-cyclohexyl-	H_2O	275(3.24)	22-3328-66
	base	293(3.42)	22-3328-66
$C_{12}H_{16}O_2$			
2H-Benzocycloheptene-2,5(3H)-dione, 4,4a,6,7,8,9-hexahydro-4a-methyl-	EtOH	238(4.09)	78-0949-66
2-Cyclohexen-1-one, 2-(3-oxocyclohexyl)-	EtOH	233(3.90),285(2.05)	44-0639-66
Cyclohexylidenecrotonic acid, γ-hydroxy-α,β-dimethyl-, γ-lactone	n.s.g.	282(4.3)	54-0043-66
2,4,6,8-Decatetraenal, 10-hydroxy-4,8-dimethyl-	pet ether	319(4.62),334(4.81), 351(4.78)	33-0369-66
6,9-Ethano-4-chromone, 5,6,7,8,9,10-hexahydro-6-methyl-	EtOH	263(3.85)	39-2324-66C
Indene, 1α-ethynyl-1β-hydroxy-5-oxo-octahydro-7a-methyl-, cis	EtOH	none	78-2021-66
$C_{12}H_{16}O_3$			
Butyric acid, 4-(6-hydroxy-2,5-xylyl)-	EtOH	280(3.09)	78-0949-66
Cinnamyl alcohol, 4-ethoxy-3-methoxy-	EtOH	212(4.28),262(4.13), 294s(3.72)	39-1775-66C
1,3-Cyclohexanedione, 2-(3-oxocyclohexyl)-	6N HCl	278(4.38)	44-1050-66
	2N NaOH	290(4.40)	44-1050-66
	EtOH	264(4.21)	44-1050-66
2,4,6-Octatrienoic acid, 2,6-dimethyl-8-oxo-, ethyl ester, trans	pet ether	316(4.68),330(4.61)	33-0369-66
Pericyclocamphanone, 6-acetoxy-	EtOH	278(1.64)	39-0885-66B
$C_{12}H_{16}O_4$			
2-Cyclopenten-1-ylidenepropionic acid, 2-ethoxy-4-oxo-, ethyl ester	EtOH	206(4.09),211(4.06), 270(4.22)	35-3131-66
2-Cyclopenten-1-ylidenepropionic acid, 4-ethoxy-2-oxo-, ethyl ester	EtOH	251(4.16),260(4.16)	35-3131-66
Hexanoic acid, 6-(3-oxocyclohexyl)-	EtOH	238(4.14)	44-1050-66
	5% $NaHCO_3$	242(4.16)	44-1050-66
	2N NaOH	397(4.7)	44-1050-66
$C_{12}H_{16}O_5$			
Benzoic acid, 3,4,5-trimethoxy-, ethyl ester	n.s.g.	214(4.69),266(4.29)	16-0034-66
3-Cycloocten-1-one, 2,3-dicarbomethoxy-, cis	C_6H_{12}	256(4.02)	35-4685-66
	EtOH	213(4.05),257(4.02)	35-4685-66
	MeCN	258(3.89)	35-4685-66

Compound	Solvent	λ_{max}(log ε)	Ref.
$C_{12}H_{16}O_7S$			
β-D-Mannose, 2,3-di-O-acetyl-4-(acetylthio)-1,6-anhydro-4-deoxy-	EtOH	228(3.58)	39-1291-66C
$C_{12}H_{16}S$			
Sulfide, cyclohexyl phenyl	EtOH	257(3.77)	22-0228-66
$C_{12}H_{17}BrO_3$			
Camphor, 6-acetoxy-5-bromo-	EtOH	295(1.54)	39-0885-66B
$C_{12}H_{17}Cl_2N$			
Butylamine, N-[4-(dichloromethyl)-4-methyl-2,5-cyclohexadien-1-ylidene]-	C_6H_{12}	239(4.27),307(2.54)	44-3178-66
Piperidine, 1-[6-(dichloromethylene)-1-cyclohexen-1-yl]-	isooctane	236(3.79)	77-0567-66
$C_{12}H_{17}N$			
Aniline, N-(1,2,2-trimethylpropylidene)-	C_6H_{12}	199(4.33),223(3.00), 275(3.36)	35-2775-66
Quinoline, 1,2,3,4-tetrahydro-2,2,4-trimethyl-	MeOH	247(3.98),300(3.39)	30-0341-66E
$C_{12}H_{17}NO$			
Morpholine, 4-[2-(1-cyclobuten-1-yl)-1-cyclobuten-1-yl]-	EtOH	287(3.69)	25-1380-66
$C_{12}H_{17}NO_2$			
8-Aza[4.3.3]propellane, 8-methyl-7,9-dioxo-	MeCN	248(2.31),258s(2.14)	78-0279-66B
Carbamic acid, N-methyl-, 2,7-dimethylocta-5,6-dien-3-yn-2-yl ester	n.s.g.	219(4.19)	25-1637-66
Pyridine, 3,5-diacetyl-1,4-dihydro-1,2,6-trimethyl-	n.s.g.	257(4.03),286(3.95), 377(3.81)	39-1083-66B
Pyridine, 3,5-diacetyl-1,4-dihydro-2,4,6-trimethyl-	n.s.g.	255(4.11),275(3.89), 379(3.79)	39-1083-66B
$C_{12}H_{17}NO_2S$			
3-Thiophenecarboxylic acid, 2-amino-4-cyclohexyl-, methyl ester	MeOH	226(4.34),301(3.81)	24-2714-66
$C_{12}H_{17}NO_2S_2$			
Pyrrole-2,4-dicarbothioic acid, 3,5-dimethyl-, S,S-diethyl ester	EtOH	237(4.26),308(4.38)	23-1007-66
$C_{12}H_{17}NO_3$			
Calycotomine, hydrochloride	n.s.g.	204s(4.67),232(3.89), 280(3.47),284(3.56)	78-1335-66
6-Isoquinolinol, 1,2,3,4-tetrahydro-7,8-dimethoxy-1-methyl-	pH 1	225s(3.95),274s(3.11), 280(3.15)	33-0403-66
	pH 13	237s(3.95),295(3.54)	33-0403-66
	iso-PrOH-HBr	230s(3.95),275s(3.11), 282(3.15)	33-0403-66
2(1H)-Pyridone, 3-hexanoyl-4-hydroxy-6-methyl-	EtOH	230(4.07),267(3.47), 326(4.13)	70-0848-66
$C_{12}H_{17}NO_3S$			
Pyrrole-2,4-dicarboxylic acid, 3,5-dimethyl-2-thio-, O,S-diethyl ester	EtOH	232(4.27),302(4.35)	23-1007-66

Compound	Solvent	$\lambda_{max}(\log \epsilon)$	Ref.
Pyrrole-2,4-dicarboxylic acid, 3,5-dimethyl-4-thio-, O,S-diethyl ester	EtOH	231(4.28),247(4.16), 286(4.14)	23-1007-66
$C_{12}H_{17}NO_4$			
Benzene, 1-(ethylpropoxy)-4-methoxy-3-nitro-	MeOH	357(3.38)	78-0093-66B
Benzene, 1-(1-methylbutoxy)-4-methoxy-2-nitro-	MeOH	357(3.39)	78-0093-66B
Benzene, 1-(1-methylbutoxy)-4-methoxy-3-nitro-	MeOH	357(3.38)	78-0093-66B
Benzene, 1-(3-methylbutoxy)-4-methoxy-2-nitro-	MeOH	357(3.38)	78-0093-66B
Benzene, 1-(3-methylbutoxy)-4-methoxy-3-nitro-	MeOH	357(3.38)	78-0093-66B
Benzene, 1-pentoxy-4-methoxy-2-nitro-	MeOH	357(3.38)	78-0093-66B
Benzene, 1-pentoxy-4-methoxy-3-nitro-	MeOH	357(3.38)	78-0093-66B
3,5-Pyridinedicarboxylic acid, 1,4-dihydro-1-methyl-, diethyl ester	n.s.g.	225(4.06),254(3.85), 387(3.89)	39-1083-66B
3,5-Pyridinedicarboxylic acid, 1,4-dihydro-4-methyl-, diethyl ester	n.s.g.	221(4.10),244s(3.75), 359(3.94)	39-1083-66B
$C_{12}H_{17}NO_5$			
2H-Azepine-3,6-dicarboxylic acid, 3,4-dihydro-2-hydroxy-2,7-dimethyl-, dimethyl ester	EtOH	286.5(4.25)	39-1075-66C
$C_{12}H_{17}NS$			
Fulvene, 6-(methylthio)-6-piperidino-	EtOH	276(3.59),358(4.38)	24-3268-66
$C_{12}H_{17}N_3$			
as-Triazine, 1,4,5,6-tetrahydro-1,5,5-trimethyl-3-phenyl-	MeOH	293(3.60)	87-0881-66
$C_{12}H_{17}N_3O$			
s-Triazolo[4,3-a]pyridine, 1-cyclohexyl-3-oxo-	EtOH	235(4.04),282(3.54), 346(3.51)	32-1020-66
$C_{12}H_{17}N_3O_6$			
Cytidine, N-methyl-, 5'-acetate	pH 1	217(3.98),280(4.13)	39-0588-66C
$C_{12}H_{17}N_5O_2S$			
Adenosine, 5'-S-ethyl-5'-thio-2',5'-dideoxy-	MeOH	259.5(4.13)	69-0224-66
Purine, 6-amino-9-(3-S-ethyl-3-thio-2,3-dideoxy-β-D-threo-pentofuranosyl)-	MeOH	259(4.19)	69-0224-66
$C_{12}H_{17}N_5O_3S$			
Adenine, 9-(3-S-ethyl-3-thio-β-D-arabinofuranosyl)-	pH 1	257(4.17)	44-3263-66
	pH 7	260(4.18)	44-3263-66
	pH 13	260(4.17)	44-3263-66
$C_{12}H_{17}N_5O_4$			
Pyrazolo[5,1-c]-as-triazine-3-carboxylic acid, 4-[bis(2-hydroxyethyl)amino]-, ethyl ester	EtOH	272(4.03),345(3.94)	39-1127-66C

Compound	Solvent	λ_{max}(log ε)	Ref.
$C_{12}H_{17}N_5O_4S$			
9H-Purine-6(1H)-thione, 2-(dimethyl-amino)-9-β-D-ribofuranosyl-	pH 1	277(4.17),359(4.31)	44-3258-66
	pH 11	265(4.28),282s(4.14), 327(4.18)	44-3258-66
$C_{12}H_{17}N_5O_5$			
Guanosine, N,N-dimethyl-	pH 1	264(4.11),293s(3.77)	44-3258-66
	pH 11	262(4.09),273s(4.03)	44-3258-66
$C_{12}H_{17}N_5O_6$			
Adenine, 9-β-D-glycero-D-gulo-heptofuranosyl-	H_2O	260(4.16)	44-1503-66
$C_{12}H_{17}N_7O_4$			
Aspartic acid, N-(5-amino-3-butyl-3H-v-triazolo[4,5-d]pyrimidin-7-yl)-	pH 1	256(4.21),274(4.13)	87-0417-66
	pH 7	218(4.26),230s(--), 265s(--),290(4.12)	87-0417-66
	pH 13	230s(--),265s(--), 290(4.12)	87-0417-66
$C_{12}H_{18}$			
1,4-Cyclohexadiene, 1,2,4,6,6-penta-methyl-3-methylene-	isooctane	253(4.29)	35-1005-66
Cyclopropane, triisopropylidene-	hexane	309.5(4.26)	35-0181-66
Naphthalene, 1,2,3,4,4a,8a-hexa-hydro-4a,8a-dimethyl-	EtOH	262(3.46)	44-1016-66
Prismane, hexamethyl-	isooctane	230(3.00)(end abs.)	35-5934-66
$C_{12}H_{18}BNO_2$			
Isoquinoline, 3,4-dihydro-6,7-di-methoxy-1-methyl-, borane adduct	90% EtOH	238(4.27),284s(3.93), 293(3.97),325(3.98)	94-1382-66
$C_{12}H_{18}Br_2O$			
A dibromo ketone	EtOH	294(2.04)	39-2095-66
$C_{12}H_{18}ClNO_3$			
Calycotomine, hydrochloride	n.s.g.	204s(4.67),232(3.89), 280(3.47),284(3.56)	78-1335-66
$C_{12}H_{18}N_2$			
Benzopyrazole, 3-cyclopentyl-4,5,6,7-tetrahydro-	EtOH	227(3.78)	22-2971-66
$C_{12}H_{18}N_2O$			
7-Azaindoline, 1-butyl-6-hydroxy-4-methyl-	EtOH	252(3.8),360(4.3)	78-3233-66
	75% EtOH	252(3.8),360(4.2)	78-3233-66
7(4H)-Benzimidazolone, 5,6-dihydro-2-isopropyl-5,5-dimethyl-	EtOH	272(4.27)	25-0731-66
$C_{12}H_{18}N_2O_8S_3$			
4-Thiothymidine, 3',5'-dimethane-sulfonate	EtOH	247(3.88),330(4.26)	44-0205-66
$C_{12}H_{18}N_4O$			
Semicarbazide, 1-cyclohexyl-1-(2-pyridyl)-	EtOH	248(4.12),292(3.70)	32-1020-66
$C_{12}H_{18}N_4O_9$			
Uracil, 6-(β-D-glucopyranosylamino)-1,3-dimethyl-5-nitro-, hydrate	pH 5	242(4.16),328(3.86)	24-3022-66

Compound	Solvent	λ_{max}(log ε)	Ref.
Uracil, 6-(β-D-glucopyranosylamino)-1,3-dimethyl-5-nitro-, hydrate (cont.)	pH 12	235s(4.14),285(3.43), 346(3.94)	24-3022-66
$C_{12}H_{18}N_4S$			
9H-Purine, 2,9-diethyl-6-(propylthio)-	pH 1	227(4.09),306(4.16)	44-1417-66
	pH 7	224(4.13),290(4.20)	44-1417-66
	pH 13	290(4.20)	44-1417-66
$C_{12}H_{18}O$			
Bicyclo[3.1.0]hex-3-en-2-one, 1,3,4,5,6,6-hexamethyl-	EtOH	235(3.80),274(3.51), 320(2.78)	35-1005-66
1-Cyclohexene, 1-acetyl-3-allyl-2-methyl-	EtOH	248(3.78)	22-0287-66
1-Cyclohexene-1-acrolein, 2,6,6-trimethyl-	n.s.g.	295(4.02)	78-0293-66
2-Cyclohexen-1-one, 4-methyl-4-(2-methyl-2-butenyl)-	EtOH	224(4.02)	39-0419-66C
Isoxylitone	MeOH	236(3.54),298(4.16)	44-2026-66
2(1H)-Naphthalenone, 4a,5,6,7,8,8a-hexahydro-4a,8a-dimethyl-	EtOH	232(3.88)	44-1016-66
2(3H)-Naphthalenone, 4,4a,5,6,7,8-hexahydro-4a,7-dimethyl-, trans	EtOH	238(4.16)	44-3109-66
Phenol, 2,5-dipropyl-	H_2O	274(3.34)	22-3328-66
	base	292(3.60)	22-3328-66
Phenol, 4-ethyl-5-isopropyl-2-methyl-	H_2O	278(3.31)	22-3328-66
	base	295.5(3.51)	22-3328-66
$C_{12}H_{18}OS$			
4-Thiaionone	C_6H_{12}	223(3.95),333(4.11)	78-0259-66
$C_{12}H_{18}O_2$			
Actinomycete metabolite	EtOH	225(3.52),301(3.91)	57-1372-66B
2(1H)-Azulenone, 4,5,6,7,8,8a-hexahydro-8-methoxy-8-methyl-	EtOH	238(4.10)	88-3627-66
Benzene, 1-(1-ethylpropoxy)-4-methoxy-	MeOH	226(3.98),287(3.59)	78-0093-66B
Benzene, 1-(1-methylbutoxy)-4-methoxy-	MeOH	226(3.91),287(3.40)	78-0093-66B
Benzene, 1-(3-methylbutoxy)-4-methoxy-	MeOH	226(3.89),287(3.36)	78-0093-66B
1,3-Cyclohexadiene, 5-(α-acetoxyisopropyl)-1-methyl-	MeOH	264(3.75)	78-0139-66
2,4-Decadienoic acid, 4-hydroxy-2,3-dimethyl-, γ-lactone	n.s.g.	270(4.3)	54-0043-66
2,4,6,8-Dodecatetraene-1,12-diol, all-trans	EtOH	274s(4.41),286(4.71), 298(4.88),312(4.84)	39-0135-66C
2(3H)-Naphthalenone, 4,4a,5,6,7,8-hexahydro-6-hydroxy-1,4a-dimethyl-	n.s.g.	248(4.19)	88-2615-66
Phthalide, 3-butyl-4,5,6,7-tetrahydro-, cis-anti	EtOH	none	39-1152-66C
trans	EtOH	220.0(4.01)	39-1152-66C
$C_{12}H_{18}O_2S_2$			
Xanthic acid, (5-oxo-2-bornyl)-, methyl ester	C_6H_{12}	214s(3.84),226(3.94), 278(4.10),356(1.63)	39-0885-66B
$C_{12}H_{18}O_2S_3$			
2-Thiophenecarboxylic acid, 4,5-bis-[(ethylthio)methyl]-, methyl ester	EtOH	257(3.87),289(3.95)	44-3363-66

Compound	Solvent	λ_{max}(log ϵ)	Ref.
$C_{12}H_{18}O_3$			
3-Bornanone, 6-hydroxy-, acetate	EtOH	293(1.49)	39-0885-66B
1-Cyclohexene-1-butyric acid, 2,5-dimethyl-6-oxo-	EtOH	244(4.05)	78-0949-66
2-Decenoic acid, 4,4-dihydroxy-2,3-dimethyl-, γ-lactone	n.s.g.	214(4.1)	54-0043-66
Fenchyl acetate, 5-oxo-	EtOH	292(1.52)	39-0885-66B
Inden-5-one, 1α-ethyl-2,3,5,6,7,7a-hexahydro-1β-hydroxy-7a-methyl-	EtOH	245(4.14)	78-2021-66
2,4,6-Octatrienal, 8,8-dimethoxy-3,7-dimethyl-, all-trans	pet ether	300s(4.46),313(4.62), 327(4.57)	33-0369-66
2,4,6-Octatrienoic acid, 8-hydroxy-2,6-dimethyl-, ethyl ester, trans	pet ether	295(4.57)	33-0369-66
$C_{12}H_{18}O_4$			
2,4-Octadienoic acid, 7-methoxy-3,7-dimethyl-6-oxo-, methyl ester	EtOH	284(4.40)	39-2166-66C
$C_{12}H_{19}BrO$			
5-Azulenone, 6α-bromooctahydro-3,8-dimethyl-	EtOH	288(1.87)	39-2095-66C
6β-isomer	EtOH	305(2.01)	39-2095-66C
	EtOH	301(1.63)	39-2095-66C
$C_{12}H_{19}ClO$			
5-Azulenone, 6α-chlorooctahydro-3,8-dimethyl-	EtOH	284.5(1.72)	39-2095-66C
$C_{12}H_{19}ClO_5$			
4-Methyl-2,6-dipropylpyrylium perchlorate (dihydrate)	H_2O	229(3.67),288(4.10)	78-0001-66
	CH_2Cl_2	291(4.10)	78-0001-66
$C_{12}H_{19}NO$			
Cyclohexene, 2-acetyl-1-pyrrolidino-	hexane	220(3.5),316(3.8)	5-0074-66I
$C_{12}H_{19}NO_2S_2$			
1αH,5αH-Tropan-3α-ol, 1,2-dithiolane-3-carboxylate (ester)	EtOH	277(2.54),320s(2.11)	88-6327-66
$C_{12}H_{19}NO_3S_2$			
5-Thiazolidineacetic acid, N-ethyl-4-oxo-2-thioxo-, pentyl ester	EtOH	261(4.11),296(4.20)	95-0095-66
5-Thiazolidineacetic acid, N-hexyl-4-oxo-2-thioxo-, methyl ester	EtOH	261(4.11),296(4.20)	95-0095-66
5-Thiazolidineacetic acid, 4-oxo-N-propyl-2-thioxo-, butyl ester	EtOH	262(4.10),296(4.20)	95-0095-66
$C_{12}H_{19}N_3O$			
2-Cyclohexen-1-one, 4-(2-methyl-2-butenyl)-, semicarbazone	EtOH	223(4.00)	39-0419-66C
s-Triazolo[1,5-a]pyridine, 1-cyclohexyl-4,6,7,8-tetrahydro-2-oxo-	EtOH	242(3.66)	32-1020-66
$C_{12}H_{19}N_3O_2$			
Bicyclo[2.2.2]oct-2-ene, 6-acetyl-1-methoxy-, semicarbazone	EtOH	228(4.19)	39-0419-66C
$C_{12}H_{19}N_5O_2$			
Pyrimidine, 4-(cyclohexylamino)-2-(dimethylamino)-5-nitro-	MeOH	228(3.97),270(3.51), 371(4.01)	24-3022-66

Compound	Solvent	λ_{max}(log ϵ)	Ref.
$C_{12}H_{19}N_5O_7$			
Pyrimidine, 4-(β-D-glucopyranosyl-amino)-2-(dimethylamino)-5-nitro-	pH -0.89	236(4.29),274(4.06), 328(3.93)	24-3022-66
	pH 5	226(4.27),273(3.71), 369(4.34)	24-3022-66
$C_{12}H_{19}N_6OP$			
Phosphonic diamide, P-(3-anilino-1H-1,2,4-triazol-1-yl)-N,N,N',N'-tetramethyl-	EtOH	245(4.19),262(4.15)	54-0429-66
Phosphonic diamide, P-(5-anilino-1H-1,2,4-triazol-1-yl)-N,N,N',N'-tetramethyl-	EtOH	261(4.40)	54-0429-66
$C_{12}H_{19}N_6O_7P$			
5'-Adenylic acid, 2-aminoethyl ester	n.s.g.	258.5(4.13)	33-0076-66
$C_{12}H_{19}O_3PSe$			
Isopropyl phenyl phosphoroselenoate	MeOH	248(3.38)	65-0254-66
Phenyl propyl phosphoroselenoate	MeOH	250(3.35),272s(2.85)	65-0254-66
$C_{12}H_{19}P$			
Phosphine, (1-cyclohexen-1-yl-ethynyl)diethyl-	EtOH	219(4.08),229(4.06), 240(4.02),252(3.95)	22-1002-66
$C_{12}H_{20}KO_8P$			
Potassium, [(1,2-dicarboxy-2-phos-phonovinyl)oxy]-, tetraethyl ester	MeOH	260(4.10)	5-0134-66C
$C_{12}H_{20}NO_3P$			
Phosphorimidic acid, phenyl-, triethyl ester	EtOH	238(4.11),277(3.03)	35-3781-66
$C_{12}H_{20}N_2$			
Pyrazole, 5-cyclohexyl-3-ethyl-4-methyl-	EtOH	224(3.76)	22-2971-66
$C_{12}H_{20}N_2O$			
[$\Delta^{2,3'}$-Bipiperidin]-2'-one, 1,1'-dimethyl-	MeOH	218(3.8),315(3.1)	24-0724-66
[$\Delta^{2,3'}$-Bipyrrolidin]-2'-one, 1,1',4,4'-tetramethyl-	MeOH	218(3.7),294(4.4)	24-0724-66
$C_{12}H_{20}N_2O_2$			
Aspergillic acid	EtOH	233(3.88),328(3.94)	44-4143-66
Neoaspergillic acid	EtOH	234(3.86),328(3.97)	44-4143-66
$C_{12}H_{20}N_4O_2$			
Piperidine, 1,1'-(azodicarbonyl)di-	MeOH	296(3.38)	24-2039-66
	dioxan	270s(3.18)	24-2039-66
	dioxan	270s(3.18),433(1.64)	89-0372-66
$C_{12}H_{20}O$			
5-Azulenone, octahydro-3,8-dimethyl-	EtOH	284.5(1.51)	39-2095-66C
1-Buten-3-one, (2,2-dimethyl-cyclohexyl)-	n.s.g.	229.0(4.10)	22-2212-66
2-Cyclododecen-1-one, cis	EtOH	228(3.72),320(1.67)	78-1207-66
2-Cyclododecen-1-one, trans	EtOH	230(4.05),321(2.24)	78-1207-66
3-Cyclododecen-1-one, cis	EtOH	289(1.95)	78-1207-66

Compound	Solvent	λ_{max}(log ε)	Ref.
2,4-Cyclohexadien-1-ol, 2,3,4,5,6,6-hexamethyl-	hexane	262(3.97)	35-1005-66
2-Cyclohexen-1-one, 2,3,4,5,6,6-hexamethyl-	EtOH	245(3.93)	35-1005-66
3-Cyclohexen-1-one, 2,3,4,5,6,6-hexamethyl-	EtOH	285(1.63)	35-1005-66
2-Propen-1-ol, 3-(2,6,6-trimethyl-1-cyclohexen-1-yl)-	n.s.g.	234(3.80)	78-0293-66
2H-Pyran, 2,2,6-triethyl-4-methyl-	EtOH	285(3.53)	28C-0652-66A
$C_{12}H_{20}O_2$			
5-Azulenone, octahydro-6-hydroxy-3,8-dimethyl-	EtOH	279(1.52)	39-2095-66C
	EtOH	282(1.61)	39-2095-66C
p-Menth-4(8)-en-9-oic acid, ethyl ester	EtOH	221(3.80)	44-3510-66
p-Menth-8(10)-en-9-oic acid, ethyl ester, cis	EtOH	209(3.83)	44-3510-66
trans	EtOH	209(3.88)	44-3510-66
$C_{12}H_{20}O_2S_2$			
Xanthic acid, (5-hydroxy-2-bornyl)-, methyl ester	C_6H_{12}	213s(3.82),227(3.92), 276(4.06),355(1.64)	39-0885-66B
$C_{12}H_{20}O_3$			
Cyclohexylideneacetic acid, α-ethoxy-, ethyl ester	MeOH	233(4.10)	5-0053-66I
2-Decenoic acid, 4,4-dihydroxy-2,3-dimethyl-,γ-lactone	n.s.g.	211(4.1)	54-0043-66
2-Hexenoic acid, 6-ethoxy-3-vinyl-, ethyl ester	EtOH	253(4.26)	33-0858-66
$C_{12}H_{20}O_4$			
Cyclohexanecarboxylic acid, 2-acetyl-2-hydroxy-1,3-dimethyl-, methyl ester	EtOH	300.0(1.59)	39-0764-66C
$C_{12}H_{20}O_6$			
Fumaric acid, diethoxy-, diethyl ester	MeOH	244(3.92)	5-0053-66I
	MeOH	244(3.92)	5-0134-66C
$C_{12}H_{21}Br$			
2,4-Dodecadiene, 1-bromo-	EtOH	228(4.38)	95-1051-66
$C_{12}H_{21}NO_3$			
Cyclohexanol, 4-tert-butyl-1-nitroso-, acetate	C_6H_{12}	673(1.22)	39-0441-66B
$C_{12}H_{21}N_5O_5$			
Pyrimidine, 5-amino-4-(β-D-glucopyranosylamino)-, hydrate	pH 4	241(4.41),310(3.56)	24-3022-66
	pH 11	241(4.28),316(3.68)	24-3022-66
$C_{12}H_{22}N_2$			
Piperidine, 1-methyl-3-(1,4,5,6-tetrahydro-1-methyl-2-pyridyl)-	MeOH	207(3.7)	24-1923-66
$C_{12}H_{22}O$			
6-Undecen-5-one, 7-methyl-	n.s.g.	242(1.95)	54-0753-66

Compound	Solvent	λ_{max}(log ε)	Ref.
$C_{12}H_{22}O_4S_4$			
Formic acid, dithiobis[thio-, O,O-bis(3-hydroxy-2,2-dimethylpropyl) ester	MeOH	240(4.30)	50-0965-66C
$C_{12}H_{23}N$			
1-Octen-3-one, 1-(diethylamino)-	EtOH	306(4.43)	70-1739-66
$C_{12}H_{23}NO$			
2-Pyrrolidinone, 1-octyl-	MeOH	207(3.7)	24-0724-66
$C_{12}H_{24}IN$			
Dicyclohexylammonium iodide	CH_2Cl_2	222(--)	60-0286-66
	THF	231(--)	60-0286-66
$C_{12}H_{24}N_2$			
2,3'-Bipiperidine, 1,1'-dimethyl-	MeOH	218(3.2)	24-1923-66
$C_{12}H_{24}O$			
3-Heptanone, 2,2,5,6,6-pentamethyl-	EtOH	288(1.50)	22-3578-66
$C_{12}H_{26}GeO$			
2-Propanone, (tripropylgermyl)-	n.s.g.	280(2.06)	65-0158-66
$C_{12}H_{36}Cl_2Si_6$			
Hexasilane, 1,6-dichloro-dodecamethyl-	C_6H_{12}	225(4.04),259(4.37)	101-0392-66A
$C_{12}H_{37}ClSi_6$			
Hexasilane, 1-chloro-dodecamethyl-	C_6H_{12}	225(4.08),258(4.34)	101-0392-66A
$C_{12}H_{38}Si_6$			
Hexasilane, dodecamethyl-	C_6H_{12}	225(4.05),258(4.29)	101-0392-66A

Compound	Solvent	$\lambda_{max}(\log \epsilon)$	Ref.
$C_{13}HF_{10}NO_3$			
Benzhydrol, decafluoro-, nitrate	EtOH	264(3.246)	70-0337-66
$C_{13}H_2F_6O_4$			
p-Benzoquinone, 2,2'-methylene-bis[3,5,6-trifluoro-	EtOH	270(4.32),396(3.12)	30-0598-66A
$C_{13}H_2F_8O_2$			
p-Benzoquinone, trifluoro-(2,3,4,5,6-pentafluorobenzyl)-	EtOH	278(4.00),400(2.75)	30-0598-66A
$C_{13}H_2F_{10}$			
Methane, bis(pentafluorophenyl)-	heptane	262(3.01)	30-0598-66A
$C_{13}H_5NO_4$			
Benz[cd]indole-4,5-dicarboxylic anhydride, 1,2-dihydro-2-oxo-	EtOH	223(4.66),269(4.19), 280s(4.11),334(3.43), 350(3.52),379(3.58)	39-1028-66C
$C_{13}H_6BrClN_4$			
Benzimidazo[1,2-c][1,2,3]benzotriazine, x-bromo-9(10)-chloro-	EtOH	229(4.51),288(4.58), 295(4.56),324s(3.98)	4-0444-66
$C_{13}H_6Br_2N_4$			
Benzimidazo[1,2-c][1,2,3]benzo-triazine, dibromo-	EtOH	230(4.48),289(4.57), 296(4.55),324s(4.00)	4-0444-66
$C_{13}H_6Cl_2N_4$			
Benzimidazo[1,2-c][1,2,3]benzo-triazine, 9,10-dichloro-	EtOH	229(4.54),287(4.61), 295(4.58),324(3.99)	4-0289-66
$C_{13}H_6Cl_6$			
2,5-Norbornadiene, 1,2,3,4,7,7-hexa-chloro-5-phenyl-	EtOH	223(4.06),254(3.95), 285(3.52)	44-1863-66
$C_{13}H_6N_2OS$			
3H-Phenothiazine-8-carbo-nitrile, 3-oxo-	MeOH	225(4.27),245s(--), 299(4.59),358(3.84), 377(3.71),488(3.87)	95-0541-66
$C_{13}H_7BrO$			
6H-Benz[cd]azulen-6-one, 9-bromo-	EtOH	232(4.33),271(3.90), 314(3.61),324(3.60), 380(3.60)	35-3950-66
8H-Benz[cd]azulen-8-one, 9-bromo-	EtOH	243(4.29),248s(4.19), 273(3.83),282(3.86), 354(3.34),371(3.45), 390(3.34)	35-3950-66
$C_{13}H_7Br_2NO_2S$			
Benzo[b]cyclohepta[e][1,4]thiazin-10(11H)-one, 7,9-dibromo-, 5-oxide	MeOH	236(4.27),268(4.30), 380(4.29)	18-1988-66
$C_{13}H_7Br_4NO_2$			
Salicylanilide, 2',3,4',5-tetra-bromo-	hexane	338(3.90)	59-0281-66
	THF	323(3.92)	59-0281-66
Salicylanilide, 3,3',4',5-tetra-bromo-	THF	332(4.09)	9-0281-66

Compound	Solvent	λ_{max}(log ε)	Ref.
$C_{13}H_7ClN_4$			
Benzimidazo[1,2-c][1,2,3]benzo-triazine, 2-chloro-	EtOH	233(4.40),290(4.56), 326s(3.83)	4-0289-66
Benzimidazo[1,2-c][1,2,3]benzo-triazine, 3-chloro-	EtOH	233(4.45),287(4.58), 293(4.58),327s(3.90)	4-0289-66
Benzimidazo[1,2-c][1,2,3]benzo-triazine, 10-chloro-	EtOH	228(4.41),283(4.55), 290(4.54),320s(3.88)	4-0289-66
$C_{13}H_7ClN_4O_8$			
Anthranilic acid, 5-chloro-N-picryl-, anion	buffer	400(4.23)	35-0947-66
dianion	Me_4NOH	441(4.34)	35-0947-66
$C_{13}H_7Cl_2N_3O_2$			
Benzimidazole, 5,6-dichloro-2-(o-nitrophenyl)-	EtOH	295(4.05)	4-0289-66
$C_{13}H_7Cl_3N_2$			
Benzimidazole, 5,6-dichloro-2-(o-chlorophenyl)-	EtOH	302(4.18)	4-0444-66
$C_{13}H_7FN_4$			
Benzimidazo[1,2-c][1,2,3]benzo-triazine, 3-fluoro-	EtOH	230(4.43),284(4.57), 289(4.56),324s(3.89)	4-0289-66
$C_{13}H_8BrClN_2$			
Benzimidazole, 2-(2-bromo-4-chlorophenyl)-	EtOH	285(4.07)	4-0444-66
Benzimidazole, 2-(o-bromophenyl)-5-chloro-	EtOH	240s(3.91),290(4.07)	4-0444-66
$C_{13}H_8BrClN_2O_6S$			
3-(p-Bromophenyl)-6-nitrothiazolo-[3,2-a]pyridinium perchlorate	EtOH	254(4.36),331(3.82)	4-0027-66
3-(p-Bromophenyl)-8-nitrothiazolo-[3,2-a]pyridinium perchlorate	EtOH	230(4.27),305(4.02), 413(4.28)	4-0027-66
$C_{13}H_8BrCl_2NO_4S$			
3-(p-Bromophenyl)-6-chlorothiazolo-[3,2-a]pyridinium perchlorate	EtOH	199(5.17),215s(4.76), 222s(4.72),244(4.82), 321(4.04)	4-0027-66
$C_{13}H_8BrNO_3S$			
Benzo[b]cyclohepta[e][1,4]thiazin-10(11H)-one, 9-bromo-, 5,5-dioxide	MeOH	233(4.28),366(4.18)	18-1988-66
$C_{13}H_8Br_2ClNO_2$			
Salicylanilide, 2',4'-dibromo-5-chloro-	hexane	330(3.97)	59-0281-66
	THF	312(3.86)	59-0281-66
$C_{13}H_8Br_3NO_2$			
Salicylanilide, 2',3,5-tribromo-	hexane	336(3.89)	59-0281-66
	THF	324(3.96)	59-0281-66
Salicylanilide, 3,3',5-tribromo-	hexane	336(3.86)	59-0281-66
	THF	331(4.03)	59-0281-66
Salicylanilide, 3,4',5-tribromo-	THF	331(4.06)	59-0281-66
$C_{13}H_8ClFN_2$			
Benzimidazole, 2-(2-chloro-5-fluorophenyl)-	EtOH	230s(4.07),290(4.15)	4-0444-66

Compound	Solvent	λ_{max}(log ε)	Ref.
$C_{13}H_8ClN_3$			
Quinoxaline, 2-chloro-3-(2-pyridyl)-	EtOH	244(4.67),328(4.11)	44-3384-66
$C_{13}H_8ClN_3O_2$			
Benzimidazole, 2-(5-chloro-2-nitrophenyl)-	EtOH	240s(3.99),273(4.11), 278(4.10)	4-0289-66
Benzimidazole, 5-chloro-2-(o-nitrophenyl)-	EtOH	243s(4.10),286(4.03)	4-0289-66
$C_{13}H_8ClN_3O_6$			
Anthranilic acid, N-(p-chlorophenyl)-3,5-dinitro-, anion	buffer	383(4.23)	35-0947-66
dianion	Me_4NOH	460(4.28)	35-0947-66
Benzoic acid, 4-(p-chloroanilino)-3,5-dinitro-, anion	buffer	415(3.76)	35-0947-66
dianion	Me_4NOH	570(3.83)	35-0947-66
$C_{13}H_8Cl_2N_2$			
Benzimidazole, 5-chloro-2-(o-chlorophenyl)-	EtOH	236s(3.94),295(4.14)	4-0444-66
Benzimidazole, 2-(2,4-dichlorophenyl)-	EtOH	290(4.26)	4-0444-66
Methane, diazobis(p-chlorophenyl)-	dioxan	294(4.39)	5-0001-66H
$C_{13}H_8Cl_2N_2O$			
Diimide, N-(p-chlorobenzoyl)-N'-(p-chlorophenyl)-	n.s.g.	298(4.14),443(2.13)	89-0372-66
$C_{13}H_8Cl_2O$			
Benzophenone, 4,4'-dichloro-, cation	H_2SO_4	378(4.58)	39-0650-66B
$C_{13}H_8FN_3O_2$			
Benzimidazole, 2-(5-fluoro-2-nitrophenyl)-	EtOH	243s(4.08),272(4.10), 279(4.10)	4-0289-66
$C_{13}H_8N_2O_3$			
1,2-Benzisoxazole, 3-(m-nitrophenyl)-	EtOH	228(4.35),287(3.89)	32-0559-66
$C_{13}H_8N_2O_4$			
3H-Phenoxazine-1-carboxylic acid, 2-amino-3-oxo-	EtOH	233(4.45),425(4.03), 442(4.02)	69-3824-66
$C_{13}H_8N_2O_4S$			
Benzo[b]cyclohepta[e][1,4]thiazin-10(11H)-one, 9-nitro-, 5-oxide	pH 13	227(4.29),400(3.81), 497(4.27)	18-1988-66
	MeOH	235(4.31),370(4.03), 427(4.18)	18-1988-66
$C_{13}H_8N_2O_5S$			
Benzo[b]cyclohepta[e][1,4]thiazin-10(11H)-one, 9-nitro-, 5,5-dioxide	MeOH	232(4.25),380(3.98), 420(4.05)	18-1988-66
$C_{13}H_8N_4$			
Benzimidazo[1,2-c][1,2,3]benzotriazine	EtOH	227(4.51),281(4.62), 287(4.60),320s(3.84)	4-0289-66
$C_{13}H_8N_4O_3$			
2(1H)-Quinoxalinone, 2-hydroxy-5-nitro-3-(2-pyridyl)-	EtOH	237(4.35),300(3.95), 363(4.05)	44-3384-66

Compound	Solvent	λ_{max}(log ε)	Ref.
$C_{13}H_8N_4O_4$			
Benzimidazole, 5-nitro-2-(o-nitrophenyl)-	EtOH	243(4.33),309(4.10)	4-0289-66
$C_{13}H_8N_4O_8$			
Anthranilic acid, N-picryl-, anion	buffer	395(4.20)	35-0947-66
dianion	Me_4NOH	436(4.37)	35-0947-66
Benzoic acid, m-(2,4,6-trinitro-anilino)-, anion	buffer	393(4.19)	35-0947-66
dianion	NaOH	438(4.34)	35-0947-66
Benzoic acid, p-(2,4,6-trinitro-anilino)-, anion	buffer	382(4.19)	35-0947-66
dianion	NaOH	442(4.36)	35-0947-66
$C_{13}H_8O$			
6H-Benz[cd]azulen-6-one	EtOH	226(4.37),266(3.85), 310s(3.44),324(3.47), 370(3.65)	35-3950-66
8H-Benz[cd]azulen-8-one	EtOH	243(4.18),248s(4.17), 275s(3.94),282(3.96), 320(3.94),350(3.11), 367(3.22),385(3.15)	35-3950-66
2-Heptene-4,6-diynal, 7-phenyl-	ether	253(4.40),265(4.37), 298(4.27),322(4.36), 336(4.27),346(4.25)	24-3201-66
$C_{13}H_8OS_2$			
Thioxanthene-9-thione, 10-oxide	C_6H_{12}	420(3.85)	88-0065-66
$C_{13}H_8O_2S$			
Thioxanthen-9-one, 10-oxide	C_6H_{12}	410(4.02)	88-0065-66
$C_{13}H_8O_4$			
Xanthen-9-one, 1,5-dihydroxy-	EtOH	252(4.62),318(3.92), 378(3.73)	88-6087-66
	EtOH-NaOH	252(4.62),318(3.82), 358(3.98),416(3.86)	88-6087-66
Xanthen-9-one, 1,7-dihydroxy-	EtOH	237(4.47),260(4.56), 287(3.83),385(3.88)	39-0430-66C
$C_{13}H_8O_5$			
Xanthen-9-one, 1,2,8-trihydroxy-	EtOH	241(4.42),265(4.56), 290(3.89),338(3.94)	39-2186-66C
	EtOH	241(4.42),265(4.56), 290(3.89),338(3.94)	78-1785-66
	EtOH-$AlCl_3$	243(4.41),268s(4.31), 286(4.44),315s(3.92), 367(3.94)	78-1785-66
	H_3BO_3-NaOAc	254(4.40),268(4.55), 278s(4.47),336(3.96)	78-1785-66
Xanthen-9-one, 1,3,5-trihydroxy-	EtOH	246(4.45),314(4.09), 350(3.87)	39-2186-66C
Xanthen-9-one, 1,3,6-trihydroxy-	n.s.g.	237(4.60),251(4.43), 287(4.02),313(4.38), 340s(--)	39-2186-66C
Xanthen-9-one, 1,5,6-trihydroxy-	EtOH	251(4.58),316(3.81), 332(4.18)	39-0430-66C
	EtOH	250(4.54),331(4.10)	39-2186-66C
Xanthen-9-one, 1,6,7-trihydroxy-	n.s.g.	224(4.39),253(4.41), 271(4.01),296(4.00), 373(4.01)	39-2186-66C

Compound	Solvent	λ_{max}(log ε)	Ref.
$C_{13}H_8O_6$			
Xanthen-9-one, 1,3,5,6-tetra-hydroxy-	MeOH	253(4.69),281(4.11), 326(4.33)	39-0178-66C
	MeOH	253(4.69),281(4.11), 326(4.33)	39-2186-66C
Xanthen-9-one, 1,3,6,8-tetra-hydroxy-	n.s.g.	252(4.45),271(3.77), 330(4.28)	39-2186-66C
$C_{13}H_9BO_4$			
Boric acid, cyclic o-phenylene ester, ester with salicylaldehyde	CH_2Cl_2	246(--),252s(--), 276(3.95),308(4.11), 347s(--),357(--), 364s(3.89)	80-1409-66
$C_{13}H_9BrClNO_2$			
Salicylanilide, 5-bromo-3'-chloro-	THF	322(3.81)	59-0281-66
$C_{13}H_9BrClNO_4S$			
3-(p-Bromophenyl)thiazolo[3,2-a]-pyridinium perchlorate	EtOH	234(4.38),312(4.08)	4-0027-66
$C_{13}H_9BrN_2$			
Benzimidazole, 2-(o-bromophenyl)-	EtOH	240s(3.86),282(4.06)	4-0444-66
$C_{13}H_9BrO_2$			
2,4-Hexadiynophenone, 2'-bromo-3'-methoxy-	ether	281(4.09),296(4.05)	24-2822-66
$C_{13}H_9Br_2N$			
Aniline, 2,4-dibromo-N-benzylidene-	hexane	262(4.35),325(3.85)	32-1423-66
	MeOH	261(4.34),319(3.90)	32-1423-66
Aniline, 2,6-dibromo-N-benzylidene-	hexane	257(4.35),320s(3.27)	32-1423-66
	MeOH	256(4.34),315s(3.32)	32-1423-66
$C_{13}H_9Br_2NO_2$			
Salicylanilide, 2',5-dibromo-	hexane	330(3.93)	59-0281-66
	THF	311(3.90)	59-0281-66
Salicylanilide, 3,5-dibromo-	hexane	333(3.88)	59-0281-66
	THF	330(4.05)	59-0281-66
Salicylanilide, 4',5-dibromo-	hexane	328(3.89)	59-0281-66
	THF	322(4.06)	59-0281-66
$C_{13}H_9ClN_2$			
Aniline, N-benzylidene-2,4-dichloro-	hexane	263(4.34),323(3.86)	32-1423-66
	MeOH	260(4.31),315(3.89)	32-1423-66
Aniline, N-benzylidene-2,6-dichloro-	hexane	254(4.45),321(3.42)	32-1423-66
	MeOH	255(4.42),312(3.41)	32-1423-66
Benzimidazole, 2-(o-chlorophenyl)-	EtOH	240s(3.97),287(4.10)	4-0444-66
Benzimidazole, 2-(p-chlorophenyl)-	EtOH	242(4.14),300s(4.39), 306(4.41),320s(4.17)	4-0444-66
$C_{13}H_9ClN_2O$			
Benzimidazole, 5-chloro-2-(o-hydroxyphenyl)-	EtOH	235(4.08),273(3.89), 288s(3.97),294(4.06), 319(4.34),332(4.30)	4-0444-66
Diimide, p-chlorophenyl-benzoyl-	n.s.g.	295(4.25),445(2.14)	89-0372-66

Compound	Solvent	λ_{max}(log ϵ)	Ref.
$C_{13}H_9ClN_2O_2$			
3-Nitrobenzo[c]quinolizinium chloride	EtOH	255(4.54),288(4.10), 310s(3.94),342s(3.66), 357(4.01),374(4.06)	44-2346-66
$C_{13}H_9ClN_2O_3$			
Pyrazino[1,2-a]indole-1,4-dione, 9(7)-chloro-2,3-dihydro-6-hydroxy-2-methyl-3-methylene-	pH 13	257(4.45),303(4.36)	39-1799-66C
	MeOH	226(4.45),232(4.32), 310(4.12),385(3.92)	39-1799-66C
$C_{13}H_9ClN_4O_4$			
Benzaldehyde, p-chloro-, 2,4-dinitrophenylhydrazone	EtOH	474(--)	65-1927-66
$C_{13}H_9ClO$			
Benzophenone, 3-chloro-, cation	H_2SO_4	340(4.36)	39-0650-66B
Benzophenone, 4-chloro-, cation	H_2SO_4	364(4.50)	39-0650-66B
Furan, 5-(chloromethyl)-2,5-dihydro-2-(2,4,6-octatriynylidene)-, cis	ether	268(4.27),280(4.36), 341(4.26),359(4.23)	24-1648-66
trans	ether	227(4.43),237(4.23), 270(4.44),282(4.56), 337(4.38),357(4.40), 379(4.17)	24-1648-66
2H-Pyran, 5-chloro-5,6-dihydro-2-(2,4,6-octatriynylidene)-, trans	ether	270(4.40),280(4.49), 342(4.37),358s(4.35)	24-1648-66
$C_{13}H_9ClO_2$			
2H-Pyran, 5-chloro-3,4-epoxy-5,6-dihydro-2-(2,4,6-octatriynylidene)-	EtOH	249(4.65),261(4.88), 282(4.08),300(4.39), 320(4.53),343(4.42)	24-1648-66
$C_{13}H_9Cl_2N$			
2-Chlorobenzo[c]quinolizinium chloride	EtOH	234(4.26),260(4.45), 300s(3.64),340(3.72), 357(4.08),374(4.19)	44-2346-66
$C_{13}H_9Cl_2NO_4$			
2-Chlorobenzo[c]quinolizinium perchlorate	EtOH	250s(--),265s(--), 280(4.14),330(3.55), 345(3.80),363(3.90)	44-2616-66
$C_{13}H_9Cl_2NO_4S$			
8-Chloro-3-phenylthiazolo[3,2-a]-pyridinium perchlorate	EtOH	238(4.26),321(4.08)	4-0027-66
$C_{13}H_9Cl_2N_3$			
Benzimidazole, 2-(o-aminophenyl)-5,6-dichloro-	EtOH	248(4.14),278(3.90), 296(3.99),307(4.10), 350(4.18)	4-0289-66
$C_{13}H_9F_2N$			
Aniline, N-benzylidene-2,4-difluoro-	hexane	262(4.26),314(3.88)	32-1423-66
	MeOH	262(4.26),310(3.96)	32-1423-66
$C_{13}H_9I_2N$			
Aniline, N-benzylidene-2,4-diiodo-	hexane	260(4.42),335(3.83)	32-1423-66
	MeOH	260(4.37),327(3.85)	32-1423-66
$C_{13}H_9N$			
Acridine	EtOH	355(3.9)	28C-0803-66A

Compound	Solvent	λ_{max}(log ε)	Ref.
Benz[f]isoquinoline	hexane	209(4.48),218(4.32), 242(4.71),248(4.76), 270(4.09),280(4.05), 292(4.12),303(2.80), 310(2.70),317(3.05), 324(2.88),328(2.98), 332(3.39),340(2.93), 344(3.02),348(3.53)	39-1078-66C
Benz[h]isoquinoline	hexane	219(4.27),246(4.66), 284(3.91),295(3.91), 313(3.03),320(2.82), 328(3.18),336(2.59), 344(3.20)	39-1078-66C
Benzo[f]quinoline (also inflections, not listed)	hexane	209(4.44),214(4.42), 233(4.55),240(4.51), 266(4.27),307(2.92), 314(3.26),321(3.10), 329(3.59),337(3.14), 345(3.72)	39-1078-66C
4-Biphenylcarbonitrile	EtOH	269.0(4.45)	56-0429-66
$C_{13}H_9NOS$			
Benzo[b]cyclohepta[e][1,4]thiazin-6(11H)-one	pH 1	235(4.28),292(4.41), 334(3.90),440(3.74)	18-1980-66
	MeOH	230(4.32),285(4.35), 320s(3.96),425(3.71)	18-1980-66
Benzo[b]cyclohepta[e][1,4]thiazin-8(11H)-one	pH 1	238(4.20),288(4.13), 358(3.95),465(3.46)	18-1980-66
	MeOH	225(4.24),275(4.25), 330(3.69),418(3.70)	18-1890-66
$C_{13}H_9NO_2$			
1H-Pyrano[3,2-f]quinolin-1-one, 3-methyl-	EtOH	251(4.29),290(3.91), 313(3.72),326(3.72)	95-0114-66
$C_{13}H_9NO_2S$			
Benzo[b]cyclohepta[e][1,4]thiazin-10(11H)-one, 5-oxide	pH 13	233(4.38),393(4.29), 466(4.15)	18-1988-66
	MeOH	234(4.42),365(4.22), 380(4.22)	18-1988-66
3H-Phenothiazin-3-one, 2-methoxy-	MeOH	243(4.56),280(4.24), 288(4.26),403(4.23), 492(3.89)	95-0541-66
$C_{13}H_9NO_3$			
p-Benzoquinone, 2-cyano-3-(3,4-dimethyl-2-furyl)-	EtOH	237(3.93),306(3.74), 370(3.28),514(3.88)	33-1794-66
p-Benzoquinone, 2-cyano-3-(3,5-dimethyl-2-furyl)-	EtOH	231(4.18),290s(3.88), 360s(3.16),508(3.79)	33-1794-66
1H-Pyrano[3,2-f]quinolin-1-one, 6-hydroxy-3-methyl-	EtOH	230(4.30),269(4.21), 307(3.95),333(3.80)	95-0114-66
$C_{13}H_9NO_3S$			
Benzo[b]cyclohepta[e][1,4]thiazin-10(11H)-one, 5,5-dioxide	pH 13	232(4.35),394(4.28), 459(4.15)	18-1988-66
	MeOH	232(4.44),358(4.20), 380(4.21)	18-1988-66

Compound	Solvent	λ_{max}(log ε)	Ref.
10-Sulfoacridizinium hydroxide, inner salt	H_2O	200s(4.18),228s(4.14), 236(4.19),254s(3.88), 330s(2.99),353(3.24), 376(3.20),396(3.13)	44-0565-66
$C_{13}H_9NO_4$			
Benzoic acid, p-nitro-, phenyl ester	EtOH	260(4.2)	65-1198-66
3H-Phenoxazin-3-one, 7-methoxy-, 10-oxide	EtOH	204(4.50),226(4.56), 494(4.18),525(4.21), 560s(3.98)	49-0129-66
$C_{13}H_9NO_5$			
Benzoic acid, p-nitro-, p-hydroxyphenyl ester	EtOH	260(4.1)	65-1198-66
$C_{13}H_9N_3$			
Quinoxaline, 2-(2-pyridyl)-	EtOH	250(4.43),275(4.16), 332(4.09)	44-3384-66
$C_{13}H_9N_3O$			
2-Quinoxalinol, 3-(2-pyridyl)-	EtOH	224(4.33),305(3.98), 364(3.93)	44-3384-66
v-Triazolo[3,4-a]pyridine, 3-benzoyl-	MeOH	217(4.25),256(3.95), 292s(4.03),319(4.29)	24-2918-66
$C_{13}H_9N_3O_2$			
Benzimidazole, 2-(o-nitrophenyl)-	EtOH	240s(4.06),272(4.02), 280(4.02)	4-0289-66
Methane, diazo(p-nitrophenyl)phenyl-	dioxan	392(4.08)	5-0001-66H
$C_{13}H_9N_3O_2S_2$			
[2,4'-Bithiazole]-4-carboxylic acid, 2'-(2-pyridyl)-, methyl ester	MeOH	216(4.41),230(4.35), 278(4.31),301(4.33)	39-1354-66C
	10N HCl	250(4.09),332(4.37)	39-1354-66C
$C_{13}H_9N_3O_3$			
Diimide, benzoyl(p-nitrophenyl)-	dioxan	285(4.39),458(2.15)	24-3337-66
$C_{13}H_9N_3O_6$			
Anthranilic acid, N-(2,4-dinitrophenyl)-, anion	buffer	383(4.32)	35-0947-66
dianion	Me_4NOH	425(4.32)	35-0947-66
Anthranilic acid, 3,5-dinitro-N-phenyl-, anion	buffer	383(4.23)	35-0947-66
dianion	Me_4NOH	458(4.28)	35-0947-66
Benzoic acid, m-(2,4-dinitroanilino)-, anion	buffer	368(4.27)	35-0947-66
dianion	Me_4NOH	430(4.29)	35-0947-66
Benzoic acid, p-(2,4-dinitroanilino)-, anion	buffer	373(4.27)	35-0947-66
dianion	Me_4NOH	435(4.30)	35-0947-66
$C_{13}H_9N_5O_6$			
Benzaldehyde, m-nitro-, 2,4-dinitrophenylhydrazone	EtOH	470(--)	65-1927-66
$C_{13}H_{10}$			
Fluorene, tetracyanoethylene complex	$CHCl_3$	419(--),567(3.30)	60-0018-66

Compound	Solvent	$\lambda_{max}(\log \epsilon)$	Ref.
$C_{13}H_{10}BrNO_2$			
Salicylanilide, 2'-bromo-	hexane	316(4.03)	59-0281-66
	THF	300(4.01)	59-0281-66
Salicylanilide, 4'-bromo-	hexane	315(4.03)	59-0281-66
	THF	309(4.14)	59-0281-66
Salicylanilide, 5-bromo-	THF	322(4.02)	59-0281-66
$C_{13}H_{10}BrNO_5$			
Nicotinic acid, 1-(m-bromophenyl)-1,2-dihydro-4,6-dihydroxy-2-oxo-, methyl ester	EtOH	305(4.40)	78-0455-66
Nicotinic acid, 1-(p-bromophenyl)-1,2-dihydro-4,6-dihydroxy-2-oxo-, methyl ester	EtOH	305(4.30)	78-0455-66
$C_{13}H_{10}Br_2N_2S$			
Carbanilide, 4,4'-dibromothio-, iodine complex	$CHCl_3$	318(--)	60-0029-66
$C_{13}H_{10}Cl^+$			
Chlorodiphenylcarbonium ion	FSO_3H-SbF_5	291(4.11),334(4.41), 434(4.40)	35-1488-66
$C_{13}H_{10}ClN$			
Benzo[c]quinolizinium chloride	EtOH	229(4.27),255(4.42), 280(3.98),295s(3.62), 332s(3.68),349(4.03), 365(4.16)	44-2346-66
Pyridine, 2-(o-chlorostyryl)-, cis	EtOH	285(4.03)	44-2346-66
$C_{13}H_{10}ClNO$			
Benzophenone, p-chloro-, oxime, cis	C_6H_{12}	235(4.32),258s(4.11)	35-2775-66
Benzophenone, p-chloro-, oxime, trans	C_6H_{12}	237(4.18),257(4.15)	35-2775-66
$C_{13}H_{10}ClNO_2$			
Nicotinic acid, 2-chloro-5-phenyl-, methyl ester	n.s.g.	260(4.24)	22-2387-66
$C_{13}H_{10}ClNO_4S$			
3-Phenylthiazolo[3,2-a]pyridinium perchlorate	H_2O	230(4.24),311(4.07)	4-0027-66
$C_{13}H_{10}ClNO_5$			
Nicotinic acid, 1-(m-chlorophenyl)-1,2-dihydro-4,6-dihydroxy-2-oxo-, methyl ester	EtOH	304(4.30)	78-0455-66
Nicotinic acid, 1-(p-chlorophenyl)-1,2-dihydro-4,6-dihydroxy-2-oxo-, methyl ester	EtOH	304(4.99)	78-0455-66
$C_{13}H_{10}ClN_3$			
Benzimidazole, 2-(2-amino-4-chlorophenyl)-	EtOH	257(4.20),291(4.07), 298(4.06),354(4.02)	4-0289-66
Benzimidazole, 2-(2-amino-5-chlorophenyl)-	EtOH	254(4.15),292(4.08), 298(4.12),344(4.16)	4-0289-66
Benzimidazole, 2-(o-aminophenyl)-5-chloro-	EtOH	251(4.16),277(3.93), 292(4.05),301(4.08), 344(4.12)	4-0289-66

Compound	Solvent	$\lambda_{max}(\log \epsilon)$	Ref.
$C_{13}H_{10}ClN_3O_2$			
Benzaldehyde, 4-chloro-, p-nitrophenylhydrazone	EtOH	524(--)	65-1927-66
$C_{13}H_{10}ClN_5$			
Pyrazino[2,3-d]pyridazine, 8-(benzylamino)-5-chloro-	EtOH	211(4.14),225s(4.10), 264(3.73)	4-0512-66
$C_{13}H_{10}ClN_5O$			
s-Triazolo[4,3-b]pyridazine, 8-acetamido-6-chloro-3-phenyl-	EtOH	207(4.36),264(4.38), 290s(4.18),315(3.94)	4-0218-66
$C_{13}H_{10}FN$			
Aniline, N-benzylidene-2-fluoro-	hexane	261(4.0),316(3.82)	32-1423-66
	MeOH	261(4.25),310(3.91)	32-1423-66
Aniline, N-benzylidene-4-fluoro-	hexane	263(4.26),314(3.93)	32-1423-66
	MeOH	263(4.22),310(3.98)	32-1423-66
$C_{13}H_{10}FN_3$			
Benzimidazole, 2-(2-amino-5-fluorophenyl)-	EtOH	247(4.09),290(4.13), 349(4.09)	4-0289-66
$C_{13}H_{10}N_2$			
Benzimidazole, 2-phenyl-	EtOH	240(4.09),294s(4.33), 301(4.36),315s(4.13)	4-0444-66
	$HOCH_2CH_2OMe$	303(4.38)	44-1498-66
4-Biphenylcarbonitrile, 4'-amino-	EtOH-HCl	267(4.43)	56-0429-66
	EtOH	231(4.09),325(4.29)	56-0429-66
Methane, diazodiphenyl-	dioxan	288(4.29)	5-0001-66H
$C_{13}H_{10}N_2O$			
Benzimidazole, 1-hydroxy-2-phenyl-	H_2O	239(4.31),296(4.27)	4-0051-66
	EtOH	239(4.29),298(4.33)	4-0051-66
Benzimidazole, 2-(o-hydroxyphenyl)-	EtOH	232(4.05),247s(3.86), 273s(3.99),289(4.15), 315(4.33),328(4.28)	4-0444-66
Diimide, benzoylphenyl-	dioxan	288(4.12),444(2.08)	24-3337-66
	dioxan	288(4.13),445(2.08)	89-0372-66
Phthalimidine, 2-(2-pyridyl)-	MeOH	290(4.22)	95-0001-66
$C_{13}H_{10}N_2OS$			
Benzo[b]cyclohepta[e][1,4]thiazin-10(11H)-one, 9-amino-	MeOH	238(4.36),349(4.58), 440(3.79)	18-1988-66
$C_{13}H_{10}N_2O_2$			
Formic acid, (phenylazo)-, phenyl ester	dioxan	291(4.09),422(2.12)	24-3337-66
1-Phenazinol, 6-methoxy-	pH 1	272(4.75),283(4.75), 430(3.86),480(3.26), 525(3.04)	88-4867-66
$C_{13}H_{10}N_2O_3$			
Benzanilide, 3'-nitro-	EtOH	262(4.40),324s(3.28)	44-0065-66
1-Phenazinol, 6-methoxy-, 10-oxide	pH 1	277(4.99),386(3.43), 478(3.54)	88-4867-66
	pH 13	252(4.39),301(4.30)	39-1799-66C
Pyrazino[1,2-a]indole-1,4-dione, 2,3-dihydro-6-hydroxy-2-methyl-3-methylene-	MeOH	227(4.46),257(4.28), 308(4.12),376(4.04)	39-1799-66C

Compound	Solvent	λ_{max}(log ε)	Ref.
$C_{13}H_{10}N_2O_4$			
Anthranilic acid, N-(o-nitrophenyl)-, anion	buffer	455(3.87)	35-0947-66
dianion	Me_4NOH	546(4.00)	35-0947-66
Anthranilic acid, N-(p-nitrophenyl)-, anion	buffer	440(4.30)	35-0947-66
dianion	Me_4NOH	493(4.44)	35-0947-66
Benzoic acid, p-(o-anitroanilino)-, anion	buffer	483(3.85)	35-0947-66
dianion	Me_4NOH	550(3.99)	35-0947-66
Benzoic acid, p-nitro-, p-aminophenyl ester	EtOH	250(4.3)	65-1198-66
Myxin	pH 1	283(4.99),340(3.73), 505(3.81)	88-4867-66
4-Pyridazineacetic acid, -benzylidene-1,6-dihydro-3-hydroxy-6-oxo-	THF	322(3.59)	49-1494-66
$C_{13}H_{10}N_2S$			
Diimide, phenyl(thiobenzoyl)-	dioxan	302(4.28),324(4.34), 394(4.19)	24-3337-66
$C_{13}H_{10}N_2S_2$			
3H-Thiapyran-3-carbonitrile, 2-imino-4-mercapto-5-p-tolyl-	MeOH	248(4.08),331(4.36), 472(4.30)	44-2389-66
$C_{13}H_{10}N_4$			
Pyrazole, 1-phenyl-4-(2-pyrazinyl)-	EtOH	273(4.20),295(4.22), 314s(4.13)	44-3845-66
$C_{13}H_{10}N_4O$			
Ketone, 7-methyl-v-triazolo[1,5-a]pyridin-3-yl 3-pyridyl	MeOH	227(4.19),269(3.73), 278(3.73),291(3.81), 329(4.34)	24-2918-66
$C_{13}H_{10}N_4O_4$			
Benzaldehyde, 2,4-dinitrophenylhydrazone	EtOH	463(--)	65-1927-66
Benzaldehyde, 2-nitro-, p-nitrophenylhydrazone	EtOH	530(--)	65-1927-66
Benzaldehyde, 3-nitro-, p-nitrophenylhydrazone	EtOH	540(--)	65-1927-66
Benzaldehyde, 4-nitro-, m-nitrophenylhydrazone	EtOH	590(--)	65-1927-66
Benzaldehyde, 4-nitro-, p-nitrophenylhydrazone	EtOH	560(--)	65-1927-66
$C_{13}H_{10}N_4O_4S$			
8(7H)-Pyrazino[2,3-d]pyridazinone, 5-tosyl-	EtOH	204(4.20),228(4.08), 251(3.93),319(3.18)	4-0512-66
$C_{13}H_{10}N_4O_5$			
Benzaldehyde, 3-hydroxy-, 2,4-dinitrophenylhydrazone	EtOH	462(--)	65-1927-66
Salicylaldehyde, 2,4-dinitrophenylhydrazone	EtOH	460(--)	65-1927-66
$C_{13}H_{10}N_6$			
Bis-s-triazolo[4,3-b:3',4'-f]pyridazine, 1-methyl-8-phenyl-	EtOH	296(3.89)	78-2073-66

Compound	Solvent	$\lambda_{max}(\log \epsilon)$	Ref.
$C_{13}H_{10}O$			
3H-Benz[e]inden-3-one, 1,2-dihydro-	EtOH	244(4.68),251(4.80), 274(3.93),284(4.00), 294(3.78),330(3.46), 344(3.52)	12-1909-66
Benzophenone	MeOH	207(4.26),253(4.25), 288s(3.39),331(2.29)	73-0835-66
protonated	FSO_3H-SbF_5	291(3.06),344(4.89)	35-1488-66
	H_2SO_4	344(4.38)	39-0650-66B
Naphtho[1,2-b]furan, 3-methyl-	EtOH	249(4.74),275(3.80), 283(3.74),321(2.99), 337(2.87)	39-0725-66C
Naphtho[2,1-b]furan, 1-methyl-	EtOH	220(4.54),236(4.45), 243(4.44),294(3.97), 312(3.78),319(3.58), 326(3.71)	39-0725-66C
$C_{13}H_{10}OS_3$			
[2,2';5',2"-Terthiophene]-5-methanol	MeOH	250(3.79),355(4.19)	88-4227-66
$C_{13}H_{10}O_2$			
Benzoic acid, phenyl ester	EtOH	235(4.2)	65-1198-66
Benzophenone, 4-hydroxy-, cation	H_2SO_4	373(4.53)	39-0650-66B
2,4-Dodecadiene-6,8,10-triynoic acid, methyl ester, trans-trans	ether	254s(4.24),265(4.67), 277(4.93),295(4.10), 312(4.44),333(4.66), 358(4.60)	39-1617-66C
$C_{13}H_{10}O_3$			
Benzophenone, 2,4-dihydroxy-, cation	H_2SO_4	362(4.36)	39-0650-66B
p-Benzoquinone, (α-hydroxybenzyl)-	EtOH	210(4.16),246(4.20), 442(1.56)	39-1627-66C
p-Benzoquinone, (phenoxymethyl)-	C_6H_{12}	221(4.11),246(4.29), 269s(--),275s(--), 450(1.40)	39-1627-66C
$C_{13}H_{10}O_3S$			
p-Benzoquinone, 2-acetyl-3-(3-methyl-2-thienyl)-	EtOH	245(4.32),430(3.21)	33-1794-66
p-Benzoquinone, 2-acetyl-3-(5-methyl-2-thienyl)-	EtOH	247(4.19),280s(3.95), 462(3.62)	33-1794-66
$C_{13}H_{10}O_4$			
p-Benzoquinone, 2-acetyl-3-(3-methyl-2-furyl)-	EtOH	243(4.23),457(3.60)	33-1794-66
p-Benzoquinone, 2-acetyl-3-(5-methyl-2-furyl)-	EtOH	229(4.04),261(4.18), 471(3.76)	33-1794-66
3-Buten-2-one, 4-(3-methyl-4,7-dioxo-benzoisofuran-1-yl)-	ether	226(4.46),298s(4.02), 308(4.08),319s(3.97), 392(3.98)	33-1806-66
Crotonaldehyde, β-(3-methyl-4,7-di-oxobenzoisofuran-1-yl)-	ether	229(4.28),292(3.83), 382(3.72)	33-1806-66
Xanthotoxin, 4-methyl-	EtOH	222(4.36),248(4.49), 300(4.24)	4-0042-66
Xanthotoxin, 5'-methyl-	EtOH	219(4.35),251(4.42), 302(4.05)	4-0042-66
$C_{13}H_{10}O_5$			
p-Benzoquinone, 2-carbomethoxy-3-(3-methyl-2-furyl)-	EtOH	242(4.18),260s(4.06), 452(3.70)	33-1794-66

Compound	Solvent	λ_{max}(log ε)	Ref.
p-Benzoquinone, 2-carbomethoxy-3-(5-methyl-2-furyl)-	EtOH	224(4.08),267(4.10), 470(3.76)	33-1794-66
$C_{13}H_{11}^{+}$			
Diphenylcarbonium ion	FSO_3H-SbF_5	292(3.46),440(4.58)	35-1488-66
$C_{13}H_{11}BrN_2O$			
3-Methyl-2-azabenzo[h]quinolizinium bromide, 2-oxide	EtOH	245(4.26),251(4.27), 290(3.99),352(3.78), 367(3.97),388(4.04)	44-0941-66
$C_{13}H_{11}BrO$			
Anisole, 2-bromo-3-(2,4-hexadiynyl)-	ether	275(3.36),282(3.38)	24-2822-66
$C_{13}H_{11}BrO_2$			
2,4-Hexadiyn-1-ol, 1-(2-bromo-3-methoxyphenyl)-	ether	256(2.95),279(3.42), 283(3.43)	24-2822-66
$C_{13}H_{11}ClFeO$			
Ferrocene, (2-formyl-1-chlorovinyl)-	EtOH	272(3.93),309(3.91), 380(3.18),503(3.15)	101-0173-66B
$C_{13}H_{11}ClN_2$			
Heptafulvene, 3-chloro-8,8-dicyano-4-isopropyl-	MeOH	240(4.07),257(4.09), 397(4.40)	18-2444-66
$C_{13}H_{11}ClO$			
3,11-Tridecadiene-5,7,9-triyn-2-ol, 1-chloro-	ether	234(4.68),244(4.82), 254(4.85),268(4.79), 289(3.98),309(4.25), 330(4.37),355(4.21)	24-3433-66
$C_{13}H_{11}N$			
Carbazole, 1-methyl-	EtOH	240(4.64),250(4.52), 260(4.33),286(4.03), 295(4.20),318s(3.52), 328(3.61),340(3.56)	77-0272-66
Cyclopenta[c]quinolizine, 4-methyl-	EtOH	242s(4.4),252(4.46), 267s(4.1),348(4.55), 375s(4.0),445s(3.4)	39-0324-66C
$C_{13}H_{11}NO$			
Benz[h]quinolin-4(1H)-one, 2,3-dihydro-	EtOH	220(4.53),263(4.33), 281(4.13),325(3.87), 395(4.04)	54-0671-66
Phenol, o-benzylideneamino-	MeOH	238s(4.00),265(4.16), 345(3.98)	46-2245-66
Phenol, p-benzylideneamino-	MeOH	264(4.13),336(4.15)	46-2245-66
2-Picoline, 3-benzoyl-	MeOH	202(4.42),251(4.17), 316(2.55)	44-3206-66
$C_{13}H_{11}NO_2$			
Nicotinic acid, 5-phenyl-, methyl ester	n.s.g.	255(4.20)	22-2387-66
Phenol, m-[N-(o-hydroxybenzylidene)-amino]-	MeOH	230s(4.06),268(3.98), 341(3.98),435(2.38)	46-2245-66
Phenol, o-[N-(o-hydroxybenzylidene)-amino]-	MeOH	228s(4.14),269(3.93), 349(3.99),444(3.20)	46-2245-66
Phenol, p-[N-(o-hydroxybenzylidene)-amino]-	MeOH	230(4.24),269(3.96), 349(4.23),435(2.60)	46-2245-66

Compound	Solvent	λ_{max}(log ϵ)	Ref.
Phenol, p-[N-(p-hydroxybenzylidene)-amino]-	C_6H_{12}	333(4.16)	46-2245-66
	MeOH	230s(4.06),284(4.16), 333(4.22),420(2.45)	46-2245-66
	EtOH	334(4.24),423s(2.15)	46-2245-66
	iso-PrOH	334(4.23),428s(1.68)	46-2245-66
	tert-BuOH	332(4.24),435s(1.38)	46-2245-66
	dioxan	334(4.19)	46-2245-66
	EtOAc	330(4.20)	46-2245-66
	CCl_4	333(4.15)	46-2245-66
Salicylanilide	hexane	314(3.98)	59-0281-66
	MeOH-acid	302(3.95)	59-0281-66
	MeOH-base	338(3.95)	59-0281-66
	THF	310(4.04)	59-0281-66
$C_{13}H_{11}NO_3$			
Nicotinic acid, 2-hydroxy-6-methyl-5-phenyl-	n.s.g.	248(3.98),345(3.69)	22-2387-66
Picolinic acid, 3-(benzyloxy)-, hydrochloride	EtOH	217(4.10),288(3.61)	44-0636-66
$C_{13}H_{11}N_3$			
Benzimidazole, 2-(o-aminophenyl)-	EtOH	249(4.26),287(4.16), 296(4.11),339(4.08)	4-0289-66
$C_{13}H_{11}N_3O$			
s-Triazolo[1,5-a]pyridine, 1-benzyl-2-oxo-	EtOH	270(3.86),306(3.69)	32-1020-66
$C_{13}H_{11}N_3O_2$			
Benzaldehyde, p-nitrophenylhydrazone	EtOH	520(--)	65-1927-66
Uracil, 5-cyano-6-ethyl-1-phenyl-	0.08N HCl	279(4.20)	25-0381-66
	0.08N NaOH	278(4.12)(anom.)	25-0381-66
$C_{13}H_{11}N_3O_2S$			
Benzenesulfonamide, p-2-benzimidazolyl-	EtOH	222(4.02),314(4.21)	95-0600-66
$C_{13}H_{11}N_3O_3$			
Benzaldehyde, 3-hydroxy-, o-nitrophenylhydrazone	EtOH	472(--)	65-1927-66
Benzaldehyde, 3-hydroxy-, p-nitrophenylhydrazone	EtOH	520(--)	65-1927-66
Salicylaldehyde, o-nitrophenylhydrazone	EtOH	463(--)	65-1927-66
Salicylaldehyde, p-nitrophenylhydrazone	EtOH	510(--)	65-1927-66
$C_{13}H_{11}N_3O_3S_2$			
2-Thiazoline-4-carboxylic acid, 2-(6-acetamido-2-benzothiazolyl)-	EtOH	225(4.44),327(4.33)	35-2015-66
$C_{13}H_{11}N_3O_5$			
Aspartic acid, N-(2-quinoxalinylcarbonyl)-	EtOH	206(3.46),243(3.41), 318(2.96),326(2.99)	87-0266-66
$C_{13}H_{12}$			
1,4-Pentadiyne, 1-methyl-5-p-tolyl-	EtOH	222(4.61),225(4.73), 233(4.12),246(4.18), 258(4.37),273(4.50), 289(4.37)	78-0867-66

Compound	Solvent	λ_{max}(log ε)	Ref.
2,4,6,8-Tridecatetrayne	EtOH	214(4.87),225(5.31), 237(5.48),273(2.51), 290(2.63),298(2.18), 308(2.60),318(2.15), 330(2.52),341(2.00), 357s(2.00)	39-0139-66C
$C_{13}H_{12}BrNO_4$			
Indole-3-carboxaldehyde, 5-bromo-1-ethyl-6-methoxy-2-methyl-4,7-dioxo-	MeOH	218(4.30),244(4.06), 268(3.98),302(3.98), 340(3.77),450(2.90)	35-0804-66
Indole-3-carboxaldehyde, 6-bromo-1-ethyl-5-methoxy-2-methyl-4,7-dioxo-	MeOH	218(4.30),245s(4.07), 265(3.98),304(4.03), 340(3.77),450(2.81)	35-0804-66
$C_{13}H_{12}ClNO$			
3-Pyridinemethanol, 2-chloro-6-methyl-5-phenyl-	n.s.g.	238(4.02),280(3.84)	22-2387-66
$C_{13}H_{12}ClNO_2$			
Cinnamic acid, p-chloro-α-cyano-β-methyl-, ethyl ester, cis	EtOH	282.5(3.95)	22-1033-66
trans	EtOH	287.5(4.05)	22-1033-66
1-Pyrrolecarboxylic acid, 4-chloro-3-phenyl-, ethyl ester	EtOH	233(4.22),255(4.06)	95-0158-66
$C_{13}H_{12}ClNO_3$			
Picolinic acid, 3-(benzyloxy)-, hydrochloride	EtOH	217(4.10),288(3.61)	44-0636-66
$C_{13}H_{12}ClNO_5$			
Furo[2,3-b]pyridine-2,5-dicarboxylic acid, 4-chloro-, diethyl ester	5% EtOH	230(4.36)	4-0202-66
$C_{13}H_{12}Cl_2N_6S$			
Acetone, 3-(4,6-dichloro-s-triazin-2-yl)-, 4-phenyl-3-thiosemicarbazone	dioxan	254(4.02),318(4.36)	23-0829-66
$C_{13}H_{12}Cl_4$			
Indene, 4,5,6,7-tetrachloro-1,1,2,3-tetramethyl-	EtOH	222(4.28),228(4.29), 234(4.30),242(4.26), 277(4.20),284(4.19)	44-4260-66
Spiro[4.4]nona-1,3,6,8-tetraene, 1,2,3,4-tetrachloro-6,7,8,9-tetramethyl-	EtOH	272(3.93),320(2.75)	44-4260-66
$C_{13}H_{12}NO_2P$			
Diphenylformylphosphinic amide	EtOH	213(4.29),223(4.29), 253(2.93),260(3.13), 266(3.24),273(3.17)	78-0987-66
	EtOH-KOH	227.5(4.247)	78-0987-66
$C_{13}H_{12}N_2$			
Heptafulvene, 8,8-dicyano-3-isopropyl-	MeOH	232(4.06),255(4.10), 397(4.42)	18-2444-66
1H-Pyrrolo[2,3-f]isoquinoline, 2,3-dimethyl-	neutral	275(4.49),315(4.13)	2-0118-66

Compound	Solvent	λ_{max}(log ε)	Ref.
$C_{13}H_{12}N_2O$			
7-Azaindoline, 6-hydroxy-4-methyl-1-phenyl-	EtOH	281(4.1),340(4.2),360s(3.4)	78-3233-66
Azobenzene, 2-hydroxy-5-methyl-	hexane	218s(4.13),246(3.92),322(4.31),395(3.91)	19-0029-66
	MeOH	218s(4.10),235s(3.92),244(3.94),322(4.30),385(3.92)	
	50% MeOH	220s(4.04),244s(3.87),324(4.29),380(3.91)	19-0029-66
	MeOH-KOH	241(4.10),322(4.11),466(3.92)	19-0029-66
Azobenzene, p-methoxy-	$C_2H_4Cl_2$	345(4.41)	49-0171-66
	MeCN	345(4.41)	49-0171-66
$AlBr_3$ complex	$C_2H_4Cl_2$	325s(5.21),485(4.65)	49-0171-66
	MeCN	320(3.31),460(4.64)	49-0171-66
$AlCl_3$ complex	$C_2H_4Cl_2$	325s(5.21),485(4.65)	49-0171-66
	MeCN	320(3.31),460(4.64)	49-0171-66
$GaBr_3$ complex	$C_2H_4Cl_2$	325s(5.21),485(4.65)	49-0171-66
	MeCN	320(3.31),460(4.64)	49-0171-66
$GaCl_3$ complex	$C_2H_4Cl_2$	325s(5.21),485(4.65)	49-0171-66
	MeCN	320(3.31),460(4.64)	49-0171-66
$SbCl_5$ complex	$C_2H_4Cl_2$	325s(5.21),485(4.65)	49-0171-66
	MeCN	320(3.31),460(4.64)	49-0171-66
$SnCl_4$ complex	$C_2H_4Cl_2$	325s(5.21),485(4.65)	49-0171-66
	MeCN	320(3.31),460(4.64)	49-0171-66
Benzanilide, 2-amino-	pH 1	264(4.00)	35-4001-66
Benzophenone, 2-amino-, oxime, anti	iso-PrOH	220(4.5),340(3.7)	9-0363-66
Benzophenone, 2-amino-, oxime, syn	iso-PrOH	230s(4.3),300s(3.0)	9-0363-66
Heptafulvene, 8,8-dicyano-3-hydroxy-4-isopropyl-	MeOH	243(4.19),295(3.64),440(4.60)	18-2444-66
4-Pyrimidinol, 5-allyl-2-phenyl-	pH 13	231(4.34),283(4.05)	44-0406-66
	50% EtOH	242(4.12),291(4.03)	44-0406-66
Urea, 1,1-diphenyl-	EtOH	242(4.06)	94-0001-66
Urea, 1,3-diphenyl-	EtOH	256(4.57)	94-0001-66
$C_{13}H_{12}N_2OS$			
o-Picolinanisidide, thio-	heptane	239(4.20),285(4.09),364(4.03)	65-1499-66
	EtOH	242(4.10),290(4.08),362(4.01)	65-1499-66
	$CHCl_3$	292(4.08),366(4.01)	65-1499-66
	H_2SO_4	218(4.30),252(4.06),326(3.88)	65-1499-66
	DMF	292(4.08),363(4.00)	65-1499-66
p-Picolinanisidide, thio-	heptane	238(4.18),288(4.17),368(3.55)	65-1499-66
	EtOH	235(4.23),290(4.16),357(3.93)	65-1499-66
	$CHCl_3$	292(4.17),365(3.94)	65-1499-66
	H_2SO_4	267(4.16),365(3.83)	65-1499-66
	DMF	292(4.16),356(3.90)	65-1499-66
$C_{13}H_{12}N_2O_2$			
1,3-Isoindolinediol, 2-(2-pyridyl)-	MeOH	246(4.21),293(3.76)	95-0001-66
9H-Pyrido[3,4-b]indole-3-carboxylic acid, 3,4-dihydro-1-methyl-	EtOH-acid	240(3.96),353(2.29)	94-0856-66
	EtOH	237(4.02),350(4.22)	94-0856-66
	EtOH-base	237(4.3),315(4.16)	94-0856-66

Compound	Solvent	$\lambda_{max}(\log \epsilon)$	Ref.
$C_{13}H_{12}N_2O_2S$			
Thiazolin-2-imine, N-acetyl-3-(benzoylmethyl)-	iso-PrOH-acid	245(4.25),285(4.02)	20-0380-66
$C_{13}H_{12}N_2O_3$			
1H,5H-Azocin[4,5,6-cd]indole-4-carboxylic acid, 3,4,6,7-tetrahydro-6-oxo-	n.s.g.	280(3.73)	35-3941-66
1,2-Benzisoxazole, 4,5,6,7-tetrahydro-3-(m-nitrophenyl)-	EtOH	230(4.30)	32-0559-66
Carbanilic acid, p-hydroxy-, p-aminophenyl ester	EtOH	245(4.47),288(3.55)	87-0444B-66
Pyrazino[1,2-a]indole-1,4-dione, 2,3-dihydro-6-hydroxy-2,3-dimethyl-	pH 13	297(4.11)	39-1799-66C
	MeOH	292(4.13),341(3.88)	39-1799-66C
$C_{13}H_{12}N_2O_4$			
Cinnamic acid, α-cyano-β-methyl-m-nitro-, ethyl ester	EtOH	257.5(4.33)	22-1033-66
Cinnamic acid, α-cyano-β-methyl-o-nitro-, ethyl ester	EtOH	271.5(4.08)	22-1033-66
Cinnamic acid, α-cyano-β-methyl-p-nitro-, ethyl ester, cis	EtOH	277.5(4.13)	22-1033-66
trans	EtOH	278.5(4.17)	22-1033-66
4-Pyridazineacetic acid, 1,6-dihydro-3-hydroxy-6-oxo-1-phenyl-, methyl ester	THF	336(3.76)	49-0644-66
3(2H)-Pyridazinone, 2-(benzoyloxy)-6-ethoxy-	EtOH	232(4.25),287(3.43), 311(3.49)	95-0082-66
3(2H)-Pyridazinone, 2-(benzoyloxy)-6-methoxy-4-methyl-	EtOH	232(4.30),286(3.56), 301(3.55)	95-0082-66
3(2H)-Pyridazinone, 6-methoxy-2-(phenacyloxy)-	EtOH	249(4.13),294(3.42)	95-0082-66
$C_{13}H_{12}N_2O_4S$			
Benzenesulfonic acid, p-[(p-methoxyphenyl)azo]-, anion	H_2O	350(4.37)	35-2240-66
	53% H_2SO_4	462(4.73)	35-2240-66
$C_{13}H_{12}N_2O_5$			
Pyrazino[1,2-a]indole, 1,2,3,4-tetrahydro-3,6-dihydroxy-3-(hydroxymethylene)-2-methyl-1,4-dioxo-	pH 13	250(4.28),296(4.14)	39-1799-66C
	MeOH	238(4.23),290(3.89), 347(3.84)	39-1799-66C
$C_{13}H_{12}N_2O_6$			
Pyrrolo[1,2-b]pyridazine-5,6,7-tricarboxylic acid, trimethyl ester	MeOH	222(4.27),249(4.43), 256s(4.36),263s(4.26), 284(3.64),295(3.76), 305(3.78),307(3.78), 343(3.59)	39-2218-66C
$C_{13}H_{12}N_2S$			
Thiocarbanilide, iodine complex	$CHCl_3$	312(--)	60-0029-66
$C_{13}H_{12}N_4O$			
Diphenylcarbazone	dioxan	285(3.92)	24-3337-66
Lumiflavine, 2-deoxy-	pH 7.0	396(3.95),456(4.08)	33-2365-66
Purine, 6-methoxy-3-methyl-8-phenyl-	pH 8	310(4.43)	39-0010-66C
$C_{13}H_{12}N_4OS$			
Lumiflavine, 2-thio-	pH 7.0	312(4.36),390(3.70), 498(4.11)	33-2365-66

Compound	Solvent	λ max(log ε)	Ref.
$C_{13}H_{12}N_4O_2$			
Isoalloxazine, 7,8,10-trimethyl-	EtOH	260(4.44),330(3.86), 444(3.64)	30-1101-66F
Lumiflavine	pH 7.0	370(4.00),446(4.08)	33-2365-66
$C_{13}H_{12}N_4O_3$			
Theophylline, 7-hydroxy-8-phenyl-	pH 1	237(4.35),300(4.31)	5-0142-66A
	pH 13	248(4.32),314(4.13)	5-0142-66A
$C_{13}H_{12}N_4O_6$			
Pyrazole-3-carboxylic acid, 1-(2,4-dinitrophenyl)-4-methyl-, ethyl ester	EtOH	308(4.07)	22-0619-66
Pyrazole-3-carboxylic acid, 1-(2,4-dinitrophenyl)-5-methyl-, ethyl ester	EtOH	275s(3.95)	22-0619-66
$C_{13}H_{12}N_4S$			
Diphenylthiocarbazone	pH 9.3	505(4.50)	28C-0346-66A
	dioxan	253(4.44)	24-3337-66
	DMF	450(4.33),620(4.50)	28C-0346-66A
Purine, 3-methyl-6-(methylthio)-8-phenyl-	pH 5	250(4.18),290s(--), 338(4.49),350s(--)	39-0010-66C
$C_{13}H_{12}N_6$			
Benzaldehyde, (3-methyl-s-triazolo-[4,3-b]pyridazin-6-yl)hydrazone	EtOH	230(4.29),308(4.42)	78-2073-66
$C_{13}H_{12}N_6O_2$			
Adenine, N-methyl-9-(p-nitrobenzyl)-	pH 1	268(4.36)	87-0576-66
	pH 7	271(4.34)	87-0576-66
	pH 13	271(4.33)	87-0576-66
$C_{13}H_{12}O$			
Anisole, p-phenyl-	EtOH	261.0(4.34)	56-0429-66
Fluoren-1(2H)-one, 3,4-dihydro-	EtOH	232(3.80),238(3.93), 302(4.31)	87-0719-66
Fluoren-3(2H)-one, 1,9a-dihydro-	EtOH	227(3.66),234(3.78), 288(4.24),312(4.25)	87-0719-66
3,4-Heptadien-6-yn-1-ol, 7-phenyl-	n.s.g.	205(4.62),218s(4.46), 257(4.13),274(4.29), 288(4.17)	77-0585-66
2,4,6-Heptatrienal, 7-phenyl-	EtOH	247(3.91),253(3.91), 355(4.70)	12-1215-66
Naphtho[1,2-b]furan, 2,3-dihydro-3-methyl-	EtOH	244(4.62),306(3.56), 319(3.53),334(3.47)	39-0725-66C
Naphtho[2,1-b]furan, 1,2-dihydro-1-methyl-	EtOH	235(4.60),283(3.66), 294(3.69),312(3.38), 327(3.41),346(3.06)	39-0725-66C
1H-Naphtho[2,1-b]pyran, 2,3-dihydro-	EtOH	233(4.83),258s(3.45), 267(3.61),277(3.70), 288(3.58),320(3.29), 334(3.37)	22-3644-66
4H-Naphtho[1,2-b]pyran, 2,3-dihydro-	EtOH	236(4.49),285s(3.62), 296(3.64),310(3.54), 317s(3.38),324(3.46)	22-3644-66
$C_{13}H_{12}OS$			
Sulfoxide, ethyl 5-phenyl-2,4-pentadiynyl	EtOH	247(3.91),260(4.12), 275(4.28),292(4.19)	22-3024-66

Compound	Solvent	λ_{max}(log ε)	Ref.
$C_{13}H_{12}O_2$			
1-Azulenecarboxylic acid, 4-methyl-, methyl ester	C_6H_{12}	235(4.34),289(4.61), 294(4.61),300(4.65), 340(3.70),351(3.72), 369(3.92),535(2.75), 572(2.69),626(2.27)	24-2669-66
	C_6H_{12}	236(4.34),289(4.53), 294(4.53),300(4.56), 340(3.66),356(3.69), 372(3.93),533(2.71), 575(2.65),626(2.28)	24-2669-66
2,4-Heptadiyne-1,6-diol, 1-phenyl-	EtOH	245(3.55),258(3.58)	70-0833-66
2,4,6-Heptatrienoic acid, 7-phenyl-	EtOH	241(3.89),249s(3.80), 321s(4.62),335(4.73), 346s(4.69)	12-1215-66
2-Naphthaleneacetic acid, methyl ester	EtOH	224(5.00),268(3.75), 277(3.76)	77-0485-66
2,4-Pentadienoic acid, 4-hydroxy-2,3-dimethyl-5-phenyl-, γ-lactone	n.s.g.	322(4.3)	54-0043-66
5-Tridecene-7,9,11-triyn-4-one, 1-hydroxy-	ether	250(4.58),260(4.68), 286(3.81),304(4.44), 325(4.32),349(4.22)	24-0990-66
$C_{13}H_{12}O_2S$			
2-Thiophenecarboxylic acid, 3-phenyl-, ethyl ester	EtOH	273(4.55)	88-1953-66
$C_{13}H_{12}O_2S_2$			
1-Propanone, 1-[5-(5-acetyl-2-thienyl)-2-thienyl]-	ether	358(4.44)	24-0984-66
$C_{13}H_{12}O_3$			
p-Benzoquinone, 2-methyl-5-(3,4-dimethyl-2-furyl)-	EtOH	265(4.15),310(3.40), 488(3.77)	33-1794-66
p-Benzoquinone, 2-methyl-5-(3,5-dimethyl-2-furyl)-	EtOH	250(4.13),311(3.45), 483(3.70)	33-1794-66
1-Naphthoic acid, 2-methoxy-, methyl ester	hexane	270(3.62),280(3.73), 293(3.38),336(3.42)	59-1537-66
2-Naphthoic acid, 3-methoxy-, methyl ester	hexane	233(4.76),269(3.74), 278(3.70),290(3.38), 340(3.23)	59-1537-66
$C_{13}H_{12}O_3S$			
Hydroquinone, 2-acetyl-3-(3-methyl-2-thienyl)-	EtOH	227(4.19),314(3.50), 372(3.47)	33-1794-66
$C_{13}H_{12}O_4$			
Hydroquinone, 2-acetyl-3-(5-methyl-2-furyl)-	EtOH	240(4.07),276(3.99), 335(3.89)	33-1794-66
1,4-Naphthoquinone, 5,6-dimethoxy-2(3)-methyl-	EtOH	214(4.65),260(4.47), 395(3.82)	5-0172-66A
$C_{13}H_{12}O_5$			
2-Benzopyran-1-one, 4-acetyl-6-hydroxy-8-(hydroxymethyl)-5-methyl-	EtOH	238(4.41),245s(--), 267s(--),332(3.77)	39-0126-66C
Fumaric acid, benzoyl-, dimethyl ester	ether	238(4.30)	24-1558-66
Maleic acid, (α-hydroxybenzyl)-methoxy-, γ-lactone, methyl ester	ether	251(4.03)	24-1558-66

Compound	Solvent	$\lambda_{max}(\log \epsilon)$	Ref.
1,4-Naphthoquinone, 2,5,7-trimethoxy-	EtOH	214(4.53),260(4.15), 296(3.99),416(3.43)	44-1496-66
Phthalic acid, 3-(2-carboxyethyl)-4-methyl-, methyl ester	EtOH	221(4.52),260(3.69), 301(3.50),310(3.52)	33-0858-66
Phthalic anhydride, 3-(3-carboxypropyl)-4-methyl-	ether	222(4.60),260(3.82), 301(3.59),310(3.57)	33-0858-66
$C_{13}H_{12}O_7$			
L-Ascorbic acid, 2-benzoate	EtOH	234(4.29)	94-1039-66
$C_{13}H_{12}S$			
1H-Naphtho[1,2-c]thiopyran, 3,4-dihydro-	EtOH	230(4.96),274(3.76), 280(3.80),291(3.67), 307(2.93),313(2.72), 322(2.78)	22-0236-66
1H-Naphtho[2,1-b]thiopyran, 2,3-dihydro-	EtOH	214(4.61),260(4.73), 278(3.85),288(3.94), 300(3.84),332(3.16), 344(3.14)	22-0228-66 +22-3644-66
4H-Naphtho[1,2-b]thiopyran, 2,3-dihydro-	EtOH	220(4.67),243(4.32), 310(3.85),318s(3.81)	22-0228-66 +22-3644-66
1,3-Pentadiyne, 5-(ethylthio)-1-phenyl-	EtOH	247(3.93),260(4.21), 275(4.34),292(4.26)	22-3024-66
Sulfide, benzyl phenyl	EtOH	255(3.83),280(3.32)	22-0228-66
$C_{13}H_{13}BrN_2O_2S$			
Thiazolin-2-imine, N-acetyl-3-[2-(m-bromophenyl)-2-hydroxyethyl]-	iso-PrOH-base	299.0(4.12)	20-0380-66
$C_{13}H_{13}ClN_2OS$			
Thiazole, 5-(p-chlorophenyl)-3-morpholino-	EtOH	238s(4.01),270(4.27), 280s(4.23),294s(3.94), 310s(3.31),318(3.32)	32-1009-66
$C_{13}H_{13}ClN_2O_2S$			
Thiazolin-2-imine, N-acetyl-3-[2-(m-chlorophenyl)-2-hydroxyethyl]-	iso-PrOH	299.5(4.14)	20-0380-66
$C_{13}H_{13}ClN_6O_3S$			
Sulfanilamide, N -(2-chloro-6-methoxy-9-methyl-8-purinyl)-	pH 1.2	286(4.45)	87-0373-66
	pH 12.9	260(4.31),291(4.39)	87-0373-66
$C_{13}H_{13}Cl_2NOS_2$			
5-(p-Chlorophenyl)-3-morpholino-1,2-dithiolane chloride	n.s.g.	238(4.45),320(4.61), 360(4.41)	32-1000-66
$C_{13}H_{13}Cl_2N_3$			
Pyrimidine, 2,6-dichloro-5-[(phenethylamino)methyl]-	EtOH	212(4.11),252(4.18), 286(3.66)	5-0119-66B
$C_{13}H_{13}IN_4O$			
3-Benzyl-1-methylhypoxanthinium	pH 1	254(3.99),280s(3.52)	44-2202-66
	pH 7	256s(3.82),295s(3.24)	44-2202-66
	pH 13	unstable	44-2202-66
$C_{13}H_{13}N$			
Acridine, 1,2,3,4-tetrahydro-	C_6H_{12}	276f(3.64)	5-0149-66H
	MeOH-acid	325(4.07)	5-0149-66H

Compound	Solvent	$\lambda_{max}(\log \epsilon)$	Ref.
Acridine, 1,2,3,4-tetrahydro- (cont.)	MeOH	284f(3.54)	5-0149-66H
Indene-3-butyronitrile	EtOH	222(3.88),250(3.90)	87-0719-66
Pyrrolo[2,1-a]isoquinoline, 5,6-dihydro-2-methyl-	MeOH	250(3.62),303(4.26), 313s(4.23)	95-0856-66
$C_{13}H_{13}NO$			
4-Biphenylamine, 4'-methoxy-	EtOH-HCl	266(4.32)	56-0429-66
	EtOH	276(4.34)	56-0429-66
Cycloheptatriene, nitrosobenzene adduct	n.s.g.	237(4.06),260s(3.81)	77-0634-66
1-Cyclopentanone, 2-indol-2-yl-	MeOH	271(3.86),280(3.87), 289(3.77)	24-2504-66
Fluoren-1(2H)-one, 3,4-dihydro-, oxime	EtOH	237(4.06),244(3.95), 299(4.38)	87-0719-66
1-Naphthol, 2-(N-methylacetimidoyl)-	MeOH	407(4.06),422(4.03)	35-2407-66
	CH_2Cl_2	412(4.06),429(4.03)	35-2407-66
	$CHCl_3$	412(4.05),431(4.03)	35-2407-66
	CCl_4	416(3.96),433(3.91)	35-2407-66
	MeCN	407(4.06),423(4.03)	35-2407-66
2-Naphthol, 3-(N-methylacetimidoyl)-	$CHCl_3$	374(3.27)	35-2407-66
	MeCN	370(3.26)	35-2407-66
2-Picoline, 3-benzoyl-1,4-dihydro-	MeOH	240(3.95),332(4.28)	44-3206-66
$C_{13}H_{13}NO_2$			
Cinnamic acid, α-cyano-β-methyl-, ethyl ester, trans	EtOH	282.5(4.05)	22-1033-66
Cinnamic acid, α-cyano-p-methyl-, ethyl ester	EtOH	317.5(4.39)	22-1033-66
2-Pyridinemethanol, 3-(benzyloxy)-	EtOH	218(4.29),278(3.85)	44-0636-66
2(1H)-Pyridone, 1-(phenethyloxy)-	EtOH	227(3.78),300(3.74)	78-0025-66
Pyrrole-2-carboxaldehyde, 5-methoxy-3-methyl-4-phenyl-	EtOH	220(4.06),255(4.03), 330(4.40)	44-0048-66
	EtOH-base	345(4.45)	44-0048-66
$C_{13}H_{13}NO_2S$			
o-Toluidine, 4-(phenylsulfonyl)-	EtOH	294(4.26)	78-2177-66
$C_{13}H_{13}NO_3$			
Cinnamic acid, α-cyano-p-hydroxy-β-methyl-, ethyl ester	EtOH	327.5(4.14)	22-1033-66
1-Cyclohexene-1-carboxamide, 2-hydroxy-6-oxo-4-phenyl-	pH 2	258(4.25)	44-2429-66
	pH 12	269(4.28)	44-2429-66
Isoquinoline, 1-acetyl-6,7-dimethoxy-	EtOH	229(4.71),253s(4.59), 345(4.08)	95-0973-66
2(1H)-Pyridone, 4-acetoxy-6-methyl-1-phenyl-	EtOH	230s(3.95),305(3.85)	70-1546-66
2(1H)-Pyridone, 3-acetyl-4-hydroxy-6-methyl-1-phenyl-	EtOH	235s(4.02),268(3.52), 330(4.18)	70-1546-66
2-Pyrroline-1-carboxylic acid, 4-oxo-3-phenyl-, ethyl ester	EtOH	244(4.21),322(4.06)	95-0158-66
$C_{13}H_{13}NO_3S_2$			
5-Thiazolidineacetic acid, N-benzyl-4-oxo-2-thioxo-, methyl ester	EtOH	261(4.11),296(4.22)	95-0095-66
$C_{13}H_{13}NO_5$			
Furo[2,3-b]pyridine-2,5-dicarboxylic acid, diethyl ester	5% EtOH	225(4.36)	4-0202-66
Furo[2,3-b]pyridine-3,5-dicarboxylic acid, 2,6-dimethyl-, dimethyl ester	EtOH	217(4.54),254s(3.88), 291(3.89)	39-1950-66C

Compound	Solvent	λ max(log ε)	Ref.
Tartronic acid, (2-indol-3-ylethyl)-indolyl-, disodium salt	EtOH	222(4.50),280(3.70)	30-0611-66A
free acid	EtOH	222(4.50),282(3.70)	30-0611-66A
$C_{13}H_{13}NO_5S$			
Furo[2,3-b]pyridine-2,5-dicarboxylic acid, 4-mercapto-, diethyl ester	5% EtOH	275(4.39)	4-0202-66
$C_{13}H_{13}NO_6$			
Furo[2,3-b]pyridine-2,5-dicarboxylic acid, 4-hydroxy-, diethyl ester	5% EtOH	240(4.38)	4-0202-66
Quinaldic acid, 4,5,8-trihydroxy-6-methoxy-, ethyl ester	EtOH	219(4.36),269(4.36), 349(3.82),426(3.53)	24-0160-66
$C_{13}H_{13}N_3O_2$			
2-Furaldehyde, 3-methyl-4-phenyl-, semicarbazone	EtOH	306(4.61)	44-0052-66
Imidazole, 4-[2-(piperonylidene-amino)ethyl]-	EtOH	223(4.25),267(4.02), 307(3.97)	44-2380-66
1H-Imidazo[4,5-c]pyridine, 4,5,6,7-tetrahydro-4-[3,4-(methylene-dioxy)phenyl]-	EtOH	286(3.18)	44-2380-66
Triazene, 3-hydroxy-1-(p-methoxy-phenyl)-3-phenyl-	1% acetone	264(5.16)	65-2117-66
$C_{13}H_{13}N_3O_3$			
Butyric acid, 2-(2-quinoxaline-carboxamido)-	EtOH	207(3.23),244(3.52), 319(2.72),327(2.75)	87-0266-66
$C_{13}H_{13}N_3O_4$			
Threonine, N-(2-quinoxalinyl-carbonyl)-	EtOH	207(3.31),244(3.52), 319(2.77),327(2.74)	87-0266-66
$C_{13}H_{13}N_3O_4S$			
Thiazolin-2-imine, N-acetyl-3-[2-(m-nitrophenyl)-2-hy-droxyethyl]-	iso-PrOH	263(3.98),299(4.02)	20-0380-66
Thiazolin-2-imine, N-acetyl-3-[2-(p-nitrophenyl)-2-hy-droxyethyl]-	iso-PrOH	293.0(4.24)	20-0380-66
$C_{13}H_{13}N_3S$			
5H-Thiopyrano[2,3-d]pyrimidine, 2-amino-6,7-dihydro-4-phenyl-	MeOH	210(4.46),240(4.38), 322(4.12)	24-0872-66
$C_{13}H_{13}N_5$			
Adenine, N,3-dimethyl-8-phenyl-	pH 8.0	236(4.22),323(4.37)	39-0010-66C
Tetramethylammonium salt of cyclo-pentadiene-1,2,3,4-tetracarbo-nitrile (anion)	n.s.g.	232s(4.53),235(4.58), 243(4.65),288(4.12), 298(4.15)	39-1641-66C
$C_{13}H_{14}$			
Benzene, (6-methylene-1-cyclohexen-1-yl)-	EtOH	242(4.19)	94-1418-66
Naphthalene, 2-propyl-	EtOH	225(4.23),276(3.90), 319(2.73)	23-1283-66
$C_{13}H_{14}BN$			
Borane, (N-methylanilino)-di(1-propynyl)-	C_6H_{12}	225(4.51),250(4.38)	22-3850-66
	C_6H_{12}	225(4.51),250(4.38)	28C-0376-66A

Compound	Solvent	λ_{max}(log ε)	Ref.
$C_{13}H_{14}ClNOS_2$			
3-Morpholino-5-phenyl-1,2-dithiolane chloride	n.s.g.	242s(4.13),313(4.27), 361(4.11)	32-1000-66
$C_{13}H_{14}ClN_3O_2$			
3-Pyrazoline-3-carboxamide, 4-chloro-N,N,1-trimethyl-5-oxo-2-phenyl-	EtOH	288(3.95)	95-0867-66
$C_{13}H_{14}Cl_2N_2O$			
1-Pyrrolin-5-one, 3,3-dichloro-2-(ethylamino)-4-methyl-4-phenyl-	EtOH	248(4.21)	28C-1276-66A
$C_{13}H_{14}N_2O$			
Cinnamonitrile, β-morpholino-	n.s.g.	231(4.05),279(4.07)	32-0986-66
p-Phenylenediamine, N'-furfurylidene-N,N-dimethyl-	EtOH	284(4.03),388(4.19)	12-1747-66
2-Pyrrolidinone, 5-imino-4-isopropylidene-3-phenyl-	EtOH	259(4.26)	22-3895-66
$C_{13}H_{14}N_2OS$			
β-Carboline, 1,2,3,4-tetrahydro-3-methyl-6-(methylthio)-1-oxo-	EtOH	231(4.15)	23-0307-66
β-Carboline, 1,2,3,4-tetrahydro-4-methyl-6-(methylthio)-1-oxo-	EtOH	248(4.20)	23-0307-66
Morpholine, 4-(5-phenyl-3-isothiazolyl)-	EtOH	231(3.96),245s(3.77), 265(3.99),276s(3.91), 289s(3.59),308(3.12)	32-1009-66
$C_{13}H_{14}N_2O_2$			
Cinnamic acid, p-amino-α-cyano-β-methyl-, ethyl ester	EtOH	377(4.16)	22-1033-66
$C_{13}H_{14}N_2O_2S$			
Thiazolin-2-imine, N-acetyl-3-(2-hydroxy-2-phenylethyl)-	iso-PrOH-acid	282.0(3.88)	20-0380-66
	iso-PrOH	299.0(4.13)	20-0380-66
	iso-PrOH-base	298.5(4.13)	20-0380-66
$C_{13}H_{14}N_2O_2S_2$			
Thiazolin-2-imine, N-butyryl-3-(2-thenoylmethyl)-	iso-PrOH-acid	265(4.18),289(4.25)	20-0380-66
$C_{13}H_{14}N_2O_3$			
Indole-3-carboxaldehyde, 1-ethyl-4,7-dihydro-2,6-dimethyl-4,7-dioxo-, oxime	MeOH	264(4.30),358(3.48), 485(3.60)	44-1012-66
3(2H)-Pyridazinone, 2-(benzyloxy)-6-ethoxy-	EtOH	316(3.510)	95-0082-66
3(2H)-Pyridazinone, 2-(benzyloxy)-6-methoxy-5-methyl-	EtOH	308(3.513)	95-0082-66
$C_{13}H_{14}N_2O_4$			
Urea, 1-acetyl-3-(2-hydroxy-1-phenylvinyl) acetate (ester)	EtOH	260(4.07)	44-3917-66
$C_{13}H_{14}N_2O_5$			
2-Furoic acid, 5-[(2-carbethoxy-2-cyanovinyl)amino]-, ethyl ester	5% EtOH	345(4.48)	4-0202-66

Compound	Solvent	λ max(log ε)	Ref.
$C_{13}H_{14}N_2O_6$			
Uracil, 1-(2,3-di-O-acetyl-5-deoxy-β-D-erythropent-4-enofuranosyl)-	MeOH	258(3.97)	35-5684-66
$C_{13}H_{14}N_2S$			
2-Thenylideneimine, N-(p-dimethylaminophenyl)-	EtOH	263(4.09),288s(3.95), 390(4.18)	12-1747-66
$C_{13}H_{14}N_4$			
Heptafulvene, 2,3-diamino-8,8-dicyano-5-isopropyl-	MeOH	245(4.11),335(4.45), 465(4.45)	18-2444-66
Heptafulvene, 3-hydrazino-8,8-dicyano-5-isopropyl-	MeOH	244(4.20),330(3.93), 465(4.50)	18-2444-66
$C_{13}H_{14}N_4O_4$			
2-Cyclopentylideneacetaldehyde, 2,4-dinitrophenylhydrazone	$CHCl_3$	392(4.42)	22-1400-66
4,5-Hexadien-2-one, 4-methyl-, 2,4-dinitrophenylhydrazone	EtOH	350(4.35)	28C-1601-66A
Pyrazole, 3-tert-butyl-1-(2,4-dinitrophenyl)-	EtOH	328(4.19)	22-3744-66
Pyrazole, 3-ethyl-4,5-dimethyl-1-(2,4-dinitrophenyl)-	EtOH	333(3.84)	22-2977-66
Pyrazole, 5-ethyl-3,4-dimethyl-1-(2,4-dinitrophenyl)-	EtOH	328(3.80)	22-2977-66
Pyrazole, 3-isopropyl-5-methyl-1-(2,4-dinitrophenyl)-	EtOH	320(3.87)	28C-0782-66A
Pyrazole, 5-isopropyl-3-methyl-1-(2,4-dinitrophenyl)-	EtOH	303(3.81)	28C-0782-66A
Pyrazole, 3-methyl-5-propyl-1-(2,4-dinitrophenyl)-	EtOH	303(3.80)	28C-0782-66A
$C_{13}H_{14}N_4O_6$			
2-Pyrazoline-3-carboxylic acid, 4-methyl-1-(2,4-dinitrophenyl)-, ethyl ester	$CHCl_3$	379(4.39)	22-0619-66
$C_{13}H_{14}N_6O_3S$			
Sulfanilamide, N^1-(2-methoxy-7-methyl-6-purinyl)-	pH 1.2	247(3.85),292(4.37)	87-0373-66
	pH 12.9	252(4.23),295(4.36)	87-0373-66
$C_{13}H_{14}O$			
Cyclohexanone, 2-benzylidene-	EtOH	292(4.24)	35-1419-66
	EtOH	288.0(4.28)	39-0075-66B
2,4-Heptadienal, 1-phenyl-, trans	EtOH	274.5(4.51)	12-1215-66
2-Tridecene-4,6,8-triyn-1-ol	EtOH	210(4.62),224s(4.57), 232(4.93),243(5.13), 258(3.56),273(3.86), 290(4.14),309(4.28), 331(4.10)	39-0139-66C
Tropone, cyclopentadiene adduct	n.s.g.	257(3.89),267(3.90)	77-0015-66
$C_{13}H_{14}OS$			
4,6-Nonadien-8-yn-3-ol, 9-(2-thienyl)-	ether	317(4.43),339(4.34)	24-1642-66
$C_{13}H_{14}O_2$			
o-Benzoquinone, 4,5-dimethyl-, cyclopentadiene adduct	MeOH	228s(3.27),302(2.53)	88-6175-66

Compound	Solvent	λ_{max}(log ε)	Ref.
3-Butenoic acid, 2-methylene-3-phenyl-, ethyl ester	EtOH	240(3.98)	88-1787-66
Furan, 3-acetyl-4,5-dihydro-2-methyl-4-phenyl-	hexane	193(4.55),270(4.06)	78-0407-66
	EtOH	205(4.22),280(4.12)	78-0407-66
2,4-Heptadienoic acid, 7-phenyl-, 2,4-di-trans	EtOH	258(4.49)	12-1215-66
5-Hexen-2-one, 1-benzoyl-	C_6H_{12}	247(3.86),308(4.13)	78-3607-66
	MeOH	246(3.76),313(4.13)	78-3607-66
1(2H)-Naphthalenone, 3,4-dihydro-2-(1-methoxyethylidene)-	n.s.g.	261(3.86),316(4.07)	88-2963-66
5-Tridecene-7,9,11-triyne-1,4-diol	ether	232(4.87),243(5.07), 257(3.61),272(3.90), 289(4.18),308(4.29), 330(4.13)	24-0990-66
$C_{13}H_{14}O_3$			
2-Pentenoic acid, 4,4-dihydroxy-2,3-dimethyl-5-phenyl-, γ-lactone	n.s.g.	217(4.1)	54-0043-66
$C_{13}H_{14}O_4$			
2-Benzopyran-1-one, 6,8-dimethoxy-3,5-dimethyl-	EtOH	241(4.57),248(4.63), 264(4.03),280(3.81), 294(3.54),334(3.83)	39-0126-66C
1-Buten-3-one, 1-(p-acetoxyphenyl)-2-methoxy-	MeOH	223(4.04),296(4.39)	5-0053-66I
1-Naphthol, 3,6,8-trimethoxy-	$CHCl_3$	246(4.78),292(3.64), 302(3.66),314(3.59), 329(3.49)	39-2234-66C
2-Naphthaleneethanol, α,3-dihydroxy-8-methoxy-	EtOH	222(4.59),244(4.41), 248(4.39),279(3.49), 287(3.56),296(3.48), 318(3.16),332(3.18)	44-1747-66
Sclerin	EtOH	216(4.36),263(3.90), 333(3.54)	88-5205-66
	EtOH-KOH	246s(3.79),313(3.62)	88-5205-66
2-Tetralone, 3-carbomethoxy-5-methoxy-	EtOH	214(4.11),262(3.81), 278(3.59)	44-1747-66
	base	225(3.88),279(4.05), 288(4.06)	44-1747-66
Tricyclo[4.2.1.$0^{2,5}$]nona-3,7-diene-3,4-dicarboxylic acid, dimethyl ester	EtOH	268(3.28)	35-4273-66
$C_{13}H_{14}O_5$			
2-Indancarboxylic acid, 5,7-dimethoxy-1-oxo-, methyl ester	MeOH	230(4.23),277(4.26), 300s(3.84)	44-1451-66
$C_{13}H_{14}O_8$			
1,2,4-Benzenetricarboxylic acid, 4-ethyl ester	EtOH	251(3.25),301(2.65)	78-0031-66
$C_{13}H_{15}^+$			
Dicyclopropylphenylcarbonium ion	FSO_3H-SbF_5	285(4.22),322(4.10), 340(4.10)	35-1488-66
$C_{13}H_{15}ClN_2O_2$			
Morpholine, 4-[2-(p-chlorobenzoyl)acetimidoyl]-	EtOH	243(4.15),273s(3.19), 330(4.21)	32-1009-66

Compound	Solvent	λ_{max}(log ϵ)	Ref.
$C_{13}H_{15}IN_2O_7$			
Uridine, 5'-deoxy-5'-iodo-, 2',3'-diacetate	MeOH	258(4.02)	35-5684-66
$C_{13}H_{15}N$			
Carbazole, 1,2,3,4-tetrahydro-9-methyl-	EtOH	231(4.55),287(3.84), 294(3.82)	94-0934-66
Cyclohept[b]indole, 5,6,7,8,9,10-hexahydro-	EtOH	229(4.57),285(3.86), 292(3.83)	94-0934-66
$C_{13}H_{15}NO$			
Acrylophenone, 3-(tert-butylimino)-	C_6H_{12}	242(4.00),289(4.13)	35-3169-66
Carbazole, 1,2,3,4-tetrahydro-6-methoxy-	MeOH	229(4.39),288(3.94), 297(3.88)	20-0181-66
Carbazole, 1,2,3,4-tetrahydro-7-methoxy-	MeOH	229(4.51),270(3.65), 300(3.68)	20-0181-66
Carbazole, 1,2,3,4-tetrahydro-8-methoxy-	MeOH	226(4.47),271(3.74), 290s(--)	20-0181-66
	EtOH	227(4.61),274(3.88), 280s(3.84),290(3.65)	94-0934-66
Isocarbostyril, 3-butyl-	EtOH	227(4.30),238(4.06), 247(3.96),278(4.05), 287(4.05),320(3.67), 332(3.74),347(3.54)	44-2090-66
$C_{13}H_{15}NO_2$			
1-Aza-8,9-benzocyclononene-2,7-dione, N-methyl-	EtOH	275(2.30),277(3.95)	35-1049-66
Indole-3-carboxaldehyde, 1-ethyl-4-hydroxy-2,6-dimethyl-	MeOH	221(4.52),252(4.18), 274s(3.94),344(3.92)	44-1012-66
Indol-4-ol, 1-ethyl-2-methyl-, acetate	MeOH	223(4.49),270(3.83), 287(3.70),296(3.77)	35-0804-66
Isocarbostyril, 3-butyl-2-hydroxy-	EtOH	226(4.30),242(3.96), 248(3.97),292(3.90), 327(3.68),343(3.53)	44-2090-66
Isocarbostyril, 3-tert-butyl-2-hydroxy-	EtOH	227(4.00),242(3.77), 249(3.72),291(3.57), 328(3.39),340(3.26)	44-2090-66
Spiro[indoline-3,2'-oxiran]-2-one, 3'-isopropyl-3'-methyl-	EtOH	218(4.45),248s(3.73), 270s(3.38),302(3.18)	44-0077-66
$C_{13}H_{15}NO_2S$			
L-Proline, (N-phenylthioacetyl)-	H_2O	272(4.22),327s(1.99)	1-2781-66
	MeOH	275(4.13),340s(1.85)	1-2781-66
	dioxan	276(4.14),359s(1.69)	1-2781-66
L-Proline, N-thiobenzoyl-, methyl ester	isooctane	287(3.94),395(2.39)	1-2781-66
	MeOH	280(4.02),363(2.10)	1-2781-66
	dioxan	286(4.07),385(1.71)	1-2781-66
$C_{13}H_{15}NO_3$			
Indole-3-lactic acid, β-methyl-, methyl ester	EtOH	224(4.50),283(3.75), 290(3.68)	30-0611-66A
$C_{13}H_{15}NO_4$			
Benzoic acid, p-[N-(2-formylethyl)-formamido]-, ethyl ester	EtOH	266(4.19)	87-0590-66

Compound	Solvent	λ_{max}(log ϵ)	Ref.
$C_{13}H_{15}NO_5$			
1,4-Cyclohexadiene-1-acetic acid, α-(1-aminoethylidene)-4-methoxy-2,5-dioxo-, ethyl ester, trans	MeOH	210(4.05),273(4.22), 345(3.21),465(3.16)	35-2536-66
Fumaric acid, p-anisidino-, dimethyl ester	EtOH	239(3.98),324(4.09)	24-2526-66
2-Furoic acid, 5-[(2-acetyl-3-oxo-1-butenyl)amino]-, ethyl ester	5% EtOH	360(4.47)	4-0202-66
$C_{13}H_{15}N_3$			
2-Benzimidazoleacetonitrile, α-butyl-	$CHCl_3$	276(3.86),282(3.86)	22-3989-66
2-Benzimidazoleacetonitrile, α-isobutyl-	$CHCl_3$	276(3.98),282(3.87)	22-3989-66
Imidazole, 4-[2-[(p-methylbenzylidene)amino]ethyl]-	EtOH	254(4.12)	44-2380-66
1H-Imidazo[4,5-c]pyridine, 4,5,6,7-tetrahydro-4-p-tolyl-	EtOH	216(4.27)	44-2380-66
$C_{13}H_{15}N_3O$			
Benzobicyclo[4.2.0]oct-3-en-2-one, semicarbazone	EtOH	281(4.24)	22-0147-66
Imidazole, 4-[2-(p-methoxybenzylidene)amino]ethyl]-	EtOH	213s(4.26),268(4.22)	44-2380-66
1H-Imidazo[4,5-c]pyridine, 4,5,6,7-tetrahydro-4-(p-methoxyphenyl)-	EtOH	225(4.22)	44-2380-66
s-Triazolo[1,5-a]pyridine, 1-benzyl-5,6,7,8-tetrahydro-2-oxo-	EtOH	244(3.64)	32-1020-66
s-Triazolo[4,3-a]pyridine, 1-benzyl-5,6,7,8-tetrahydro-3-oxo-	EtOH	246(3.64)	7-0199-66
$C_{13}H_{15}N_3OS$			
4(5H)-Pyrimidinone, 2-amino-5-(3-mercaptopropyl)-6-phenyl-	pH 13.5	230(4.30),287(3.89)	24-0872-66
	MeOH	258(3.65),307(3.99)	24-0872-66
1,3,4-Thiadiazole, 2-(N-acetylisopropylamino)-5-phenyl-	EtOH	224(3.69),290(4.11)	83-0921-66
$C_{13}H_{15}N_3O_2$			
Indole, 5-(dimethylamino)-3-(2-nitropropenyl)-	EtOH	238(4.40),415(4.12)	23-0387-66
3-Pyrazoline-3-carboxaldehyde, 4-(dimethylamino)-2-methyl-5-oxo-1-phenyl-	pH 1	265(3.96)	88-4849-66
	H_2O	315(3.65),389(4.03)	88-4849-66
3,6-Pyridazinedione, 4-(dimethylamino)-1-methyl-2-phenyl-	EtOH	304(4.13)	95-1082-66
3(2H)-Pyridazinone, 4-(dimethylamino)-6-methoxy-2-phenyl-	EtOH	318(4.05)	95-1082-66
$C_{13}H_{15}N_5O_6$			
2-Pyrazoline, 3-tert-butyl-1-picryl-	EtOH	394(4.23)	22-0618-66
$C_{13}H_{16}$			
Cyclohexane, benzylidene-	EtOH	245(4.13)	35-1419-66
Indene, 1,2-diethyl-	EtOH	260(4.05)	11-0109-66
Indene, 1,1,2,3-tetramethyl-	heptane	207(4.32),262(4.06)	44-0081-66
3-Nortricyclylcyclohexenone	hexane	229(4.19),338(1.53), 352(1.38),368(0.95)	35-5935-66
$C_{13}H_{16}ClNO_5$			
N-tert-Butyl-5-phenylisoxazolium perchlorate	CH_2Cl_2	296(4.34)	44-2039-66

Compound	Solvent	$\lambda_{max}(\log \epsilon)$	Ref.
$C_{13}H_{16}INO_2$			
6,7-Dimethoxy-1,2-dimethyl-isoquinolinium iodide	MeOH	252(4.81),322(3.95)	83-0159-66
$C_{13}H_{16}N_2$			
Benzimidazole, 2-(1-butylvinyl)-	$CHCl_3$	294(4.17)	22-3989-66
Benzimidazole, 2-(1-isobutylvinyl)-	$CHCl_3$	294(4.17)	22-3989-66
4-Piperidineacetonitrile, α-phenyl-, hydrobromide	EtOH	252(2.18),258(2.48), 264(2.20)	78-2721-66
Quinazoline, 2-tert-butyl-4-methyl-	pH 0.0	236(4.56),270(3.46), 320(3.39)	39-0234-66C
	pH 8.0	225(4.65),259(3.50), 308(3.43),318s(3.38)	39-0234-66C
$C_{13}H_{16}N_2O$			
Indole, N-(2-aminoethyl)-3-methyl-, acetate	EtOH	228(4.34),292(3.79)	94-0555-66
$C_{13}H_{16}N_2OS$			
L-Prolineamide, N-phenylthioacetyl-	MeOH	276(4.15),344(1.69)	1-2781-66
	dioxan	279(4.12),358(1.69)	1-2781-66
$C_{13}H_{16}N_2O_2$			
Benzimidazoleacetic acid, α-ethyl-, ethyl ester	$CHCl_3$	276(3.85),284(3.81)	22-3989-66
3a,6a-Diazapentalene, 4,6-diacetyl-1,2,3-trimethyl-	n.s.g.	225(4.12),263(3.79), 335(4.19),410(4.39)	35-5588-66
Indole-2-acetic acid, 3-(3-amino-1-propyl)-	MeOH	221(4.68),270(4.00), 279(4.04),289(3.93)	5-0116-66F
Morpholine, 4-(2-benzoyl-acetimidoyl)-	EtOH	232(4.14),267s(3.48), 324(4.21)	32-1009-66
2,4-Pyrrolidinedione, 5-(3,4-dimethyl-pyrrol-2-ylmethylene)-3,3-dimethyl-	EtOH	225s(3.75),275(4.00), 288s(3.91),447(4.33)	39-1155-66C
isomer	EtOH	225(3.80),277(3.79), 379s(4.31),397(4.45), 454s(3.48)	39-1155-66C
$C_{13}H_{16}N_2O_2S$			
Pyrazole, 5-ethyl-3-methyl-1-p-toluenesulfonyl-	EtOH	240(4.22)	28C-0782-66A
$C_{13}H_{16}N_2O_2S_2$			
Thiazolin-2-imine, N-butyryl-3-(2-hydroxy-2-thienylethyl)-	iso-PrOH-acid	233(4.11),289(3.89)	20-0380-66
$C_{13}H_{16}N_2O_4$			
Benzimidazole, 1-(4-deoxy-β-D-xylohexopyranosyl)-	pH 2	253(3.73),262(3.75), 268(3.82),275(3.75)	39-1549-66C
	EtOH	245(3.86),274(3.61), 281(3.64)	39-1549-66C
Mesoxalic acid, diethyl ester, phenylhydrazone	EtOH	335(4.29)	22-2981-66
$C_{13}H_{16}N_2O_4S$			
Cephalosporanic acid, 7-methoxy-carboxamido-, methyl ester	EtOH	261(3.88)	44-3409-66

Compound	Solvent	$\lambda_{max}(\log \epsilon)$	Ref.
$C_{13}H_{16}N_2O_7S$			
5-Thia-1-azabicyclo[4.2.0]oct-2-ene-2-carboxylic acid, 7-(carboxy-amino)-3-(hydroxymethyl)-8-oxo-, N-ethyl ester, acetate	EtOH	230(3.85),247(3.87)	44-3409-66
$C_{13}H_{16}N_4O_4$			
3-Pentenal, 2,2-dimethyl-, 2,4-dinitrophenylhydrazone	n.s.g.	360(4.19)	22-0734-66
4-Pentenal, 3-ethyl-, 2,4-dinitrophenylhydrazone	n.s.g.	358(4.32)	54-0757-66
2-Pyrazoline, 3-tert-butyl-1-(2,4-dinitrophenyl)-	EtOH	400(4.27)	22-0618-66
2-Pyrazoline, 3,4,5,5-tetramethyl-1-(2,4-dinitrophenyl)-	$CHCl_3$	400(4.32)	22-0610-66
$C_{13}H_{16}N_4O_6$			
Chromose A, 2,4-dinitro-phenylhydrazone	EtOH	374(4.49)	78-2785-66
	EtOH-NaOH	460(4.48)	78-2785-66
Inosine, 2',3'-O-ethoxymethylene-	pH 7	251(4.08)	24-3778-66
$C_{13}H_{16}N_4O_8$			
1-Pyrroline, 3,5,5-trimethyl-, 1-oxide, picrate	n.s.g.	235(3.94)	39-2295-66C
$C_{13}H_{16}O$			
1-Indanone, 2,3-diethyl-	EtOH	246(4.10),290(3.40)	11-0109-66
1-Indanone, 4,7-diethyl-	EtOH	254(4.03),306(3.45)	35-1205-66
1-Indanone, 2,2,5,7-tetramethyl-	MeOH	260(4.15),292(3.34)	35-2838-66
2-Indanone, 1,1,3,3-tetramethyl-	EtOH	252s(2.54),259(2.84), 265(3.06),272(3.13), 295s(1.29)	44-1393-66
Ketone, 1-benzocyclobutenyl butyl	EtOH	254s(3.00),260(3.09), 266(3.18),273(3.13)	87-0656-66
Ketone, 5-hexenyl phenyl	EtOH	241(3.91),279(2.93)	94-1418-66
Naphtho[2,1-b]furan, 1,2,6,7,8,9-hexahydro-1-methyl-	EtOH	290(3.32)	39-0725-66C
[4.4.3]Propella-3,8-dien-12-one	MeCN	285s(1.36),297(1.40), 307s(1.32),318s(1.02)	78-0279-66B
$C_{13}H_{16}OS$			
6-Nonen-8-yn-3-ol, 9-(2-thienyl)-	ether	277s(4.16),292(4.29), 309(4.21)	24-1642-66
$C_{13}H_{16}O_2$			
3-Butenoic acid, 2-methyl-3-phenyl-, ethyl ester	EtOH	238.0(3.99)	35-1959-66
6,9-Ethano-4-chromone, 5,6,9,10-tetrahydro-2,6-dimethyl-	EtOH	266(4.00)	39-2324-66C
6,9-Ethano-4-chromone, 5,6,9,10-tetrahydro-2,7-dimethyl-	EtOH	266(3.93)	39-2324-66C
2-Furanol, 2,5-dihydro-4,5,5-tri-methyl-2-phenyl-	N H_2SO_4	265(3.34),322(4.46)	65-2064-66
1(2H)-Naphthalenone, 3,4-dihydro-6-hydroxy-4-isopropyl-	EtOH	228(4.12),286(4.16), 338(3.63)	78-1977-66
$C_{13}H_{16}O_3$			
Cinnamic acid, α-ethoxy-, ethyl ester	MeOH	217s(4.06),223s(3.97), 279(4.21)	5-0053-66I

Compound	Solvent	λ max(log ε)	Ref.
Isocoumarin, 3,4-dihydro-8-hydroxy-4,5,6,7-tetramethyl-	EtOH	217(4.44),260(4.00), 330(3.59)	88-5205-66
$C_{13}H_{16}O_4$			
Cinnamic acid, 4-ethoxy-3-methoxy-, methyl ester	EtOH	217(4.10),235(4.08), 296(4.15),324(4.27)	39-1775-66C
1-Naphthoic acid, 1,2,3,5,6,8-hexa-hydro-4,7-dioxo-, ethyl ester	EtOH	325(4.37),394(4.14)	88-0927-66
	NaOEt	394(5.24)	88-0927-66
1-Naphthoic acid, 2,3,4,4a,5,6,7,8-octahydro-4a-methyl-2,5-dioxo-, methyl ester	EtOH	250(4.00)	88-3489-66
$C_{13}H_{16}O_5$			
Isocoumarin, 3,4-dihydro-3-hydroxy-6,8-dimethoxy-3,5-dimethyl-	EtOH	222(4.26),264(3.95), 304(3.73)	39-0126-66C
$C_{13}H_{16}O_6$			
Harpagenone, 6,8-di-O-acetyl-5:9-anhydro-dihydro-	MeOH	212(4.01)	33-1552-66
2H-Oxecin-4,5-dicarboxylic acid, 3,4,9,10-tetrahydro-3-oxo-, dimethyl ester, cis-cis	C_6H_{12}	248f(4.19)	35-4685-66
	EtOH	248(4.15),272s(4.01)	35-4685-66
	MeCN	248(4.09),268s(3.97)	35-4685-66
$C_{13}H_{16}S$			
Benzo[b]thiophene, 5-tert-butyl-3-methyl-	n.s.g.	232(4.51),238s(4.41), 262(3.73),292(3.41), 302(3.47)	22-3674-66
Benzo[b]thiophene, 6-tert-butyl-3-methyl-	n.s.g.	232(4.53),262(3.74), 289(3.35),300(3.25)	22-3674-66
$C_{13}H_{16}S_2$			
Pentane, 3,3-di-2-thienyl-	n.s.g.	236.5(4.29)	22-2253-66
$C_{13}H_{17}BrN_2$			
4-Piperidineacetonitrile, α-phenyl-, hydrobromide	EtOH	252(2.18),258(2.48), 264(2.20)	78-2721-66
$C_{13}H_{17}BrO$			
5-Indanol, 2-bromo-1,1,3,3-tetra-methyl-	heptane	281(3.55),287(3.54)	44-0081-66
$C_{13}H_{17}ClO_3$			
Lactic acid, 3-(p-chlorophenyl)-, tert-butyl ester	EtOH	221(4.05),247s(2.59), 254(2.66),260(2.70), 267(2.70),275(2.59)	22-3243-66
$C_{13}H_{17}ClO_5$			
1,2,3,4,5,6,7,8-Octahydroxanthylium perchlorate	1% $HClO_4$	233s(3.43),315(4.20)	80-0109-66
$C_{13}H_{17}N$			
2,3-Benzoquinuclidine, 6,6-dimethyl-	pentane	255(2.68),261(2.77), 268(2.74)	5-0128-66D
$C_{13}H_{17}NO$			
3-Buten-2-one, 4-p-cumenyl-, α-oxime	n.s.g.	288.5(4.46)	83-0626-66
3-Buten-2-one, 4-p-cumenyl-, β-oxime	n.s.g.	293(4.39)	83-0626-66
Carbostyril, 5,8-diethyl-3,4-dihydro-	EtOH	253(3.95)	35-1205-66
Isocarbostyril, 5,8-diethyl-3,4-dihydro-	EtOH	239(3.85),290(3.30)	35-1205-66

Compound	Solvent	λ_{max}(log ϵ)	Ref.
Ketone, 1-methyl-2-piperidyl phenyl, hydrochloride	MeOH	246(4.10),286(3.09)	44-2149-66
$C_{13}H_{17}NO_2$			
Benzocyclobutene, O-(2-hydroxypropyl)-1-acetyl-, oxime	EtOH	260(3.15),266(3.33), 272(3.32)	87-0656-66
$C_{13}H_{17}NO_2S$			
Pyrrolidine, 1-[β-(methylsulfonyl)-styryl]-	EtOH	249(4.09)	87-0489-66
L-Valine, N-phenylthioacetyl-	H_2O	267(4.26),323(1.84)	1-2781-66
	MeOH	270(4.02),340(1.74)	1-2781-66
	dioxan	269(4.05),356(1.67)	1-2781-66
L-Valine, N-(thiobenzoyl)-, methyl ester	isooctane	289(3.71),401(2.33)	1-2781-66
	MeOH	289(3.84),374(2.38)	1-2781-66
	dioxan	287(3.82),387(2.46)	1-2781-66
$C_{13}H_{17}NO_3$			
10bH-Oxazolo[2,3-a]isoquinoline, 2,3,5,6-tetrahydro-8,9-dimethoxy-	H_2O	247(4.25),309(4.02), 361(4.00)	83-0817-66
	MeOH	249(4.06),312(3.78), 371(3.72)	83-0817-66
$C_{13}H_{17}NO_4$			
2-Azabicyclo[3.2.0]hepta-3,6-diene-4,7-dicarboxylic acid, 1,2,3-trimethyl-, dimethyl ester	EtOH	221(3.74),241(3.64), 319(4.26)	39-1950-66C
1H-Azepine-3,6-dicarboxylic acid, 1,2,7-trimethyl-, dimethyl ester	EtOH	222(4.37),301(3.75)	39-1950-66C
Crotonic acid, 3-amino-2-(2,5-dihydroxy-p-tolyl)-, ethyl ester	MeOH	285(4.18)	35-2536-66
Cyclopentylideneacetic acid, 2-carboxy-α-cyano-, diethyl ester	EtOH	238(4.07),356(2.43)	78-2575-66
	EtOH-base	248(3.97),356(4.11)	78-2575-66
$C_{13}H_{17}NO_5$			
Malonic acid, [3-(butylamino)-3-hydroxy-1-(1-hydroxy-1-methylethoxy)allylidene]-, di-δ-lactone	EtOH	330(4.15)	78-0455-66
Malonic acid, [3-(isobutylamino)-3-hydroxy-1-(1-hydroxy-1-methylethoxy)allylidene]-, di-δ-lactone	EtOH	330(4.65)	78-0455-66
$C_{13}H_{17}N_3O$			
4(3H)-Quinazolinone, 2-methylamino-tert-butyl-	EtOH	230(4.5),280(4.2), 340(3.5)	83-0914-66
$C_{13}H_{17}N_3O_2$			
2-Pyrazoline, 3-tert-butyl-1-(p-nitrophenyl)-	EtOH	410(4.32)	22-0618-66
Tryptophan, 5-(dimethylamino)-	pH 13	232(4.42)	23-0387-66
$C_{13}H_{17}N_5O_5S_2$			
Glucopyranuronamide, 1-deoxy-1-(6-ethylthio-8-mercapto-9-purinyl)-	H_2O	257(4.19),330(4.44)	94-1360-66
$C_{13}H_{18}$			
Indan, 1,1,2,3-tetramethyl-	heptane	258(2.83),264(3.03), 271(3.06)	44-0081-66

Compound	Solvent	$\lambda_{max}(\log \epsilon)$	Ref.
Indan, 2,2,4,5-tetramethyl-	MeOH	269(3.08),273(3.06), 278(3.06)	35-2838-66
Indan, 2,2,5,7-tetramethyl-	MeOH	265(2.81),269(2.91), 279(2.93)	35-2838-66
$C_{13}H_{18}BrNO_3$			
3,4-Dihydro-2-(2-hydroxyethyl)-6,7-dimethoxyisoquinolinium bromide	H_2O	246(4.33),309(4.07), 360(4.04)	83-0817-66
$C_{13}H_{18}BrNS$			
3,4-Dihydro-2-(2-ethylthioethyl)-isoquinolinium bromide	MeOH	285(4.23),288(4.22)	83-0846-66
$C_{13}H_{18}ClNO$			
Ketone, 1-methyl-2-piperidyl phenyl, hydrochloride	MeOH	246(4.10),286(3.09)	44-2149-66
$C_{13}H_{18}ClNO_4$			
3,5-Pyridinedicarboxylic acid, 4-(chloromethyl)-1,4-dihydro-1,2,6-trimethyl-, dimethyl ester	EtOH	231(4.22),252(4.01), 340(3.89)	39-1950-66C
$C_{13}H_{18}N_2$			
Benzimidazole, 2-(1,3-dimethylbutyl)-	$CHCl_3$	276(3.85),282(3.83)	22-3989-66
Benzimidazole, 2-(1-methylpentyl)-	$CHCl_3$	276(3.83),282(3.83)	22-3989-66
$C_{13}H_{18}N_2O$			
2-Benzimidazoleethanol, β-isobutyl-	$CHCl_3$	276(3.81),282(3.82)	22-3989-66
Cyclodec[c]isoxazole, 3-amino-4,5,8,9,12,13-hexahydro-, 6,10-di-trans	n.s.g.	256(3.90)	18-1125-66
isomer	n.s.g.	244(3.90)	18-1125-66
Tryptamine, 5-methoxy-N,N-dimethyl-	EtOH	224(4.46),277(3.84), 296(3.76)	44-2284-66
$C_{13}H_{18}N_2OS$			
L-Leucinamide, N-(thiobenzoyl)-	MeOH	284(3.86),374(2.42)	1-2781-66
	dioxan	289(3.87),384(2.53)	1-2781-66
L-Valineamide, N-phenylthioacetyl-	MeOH	271(4.04),344(1.73)	1-2781-66
	dioxan	274(3.93),356(1.61)	1-2781-66
$C_{13}H_{18}N_2O_2$			
2,4-Pyrrolidinedione, 5-(3,4-dimethyl-pyrrol-2-ylmethyl)-3,3-dimethyl-	EtOH	220(3.79)	39-1155-66C
$C_{13}H_{18}N_2O_5S$			
p-Toluenesulfonic acid, hydrazide, hydrazone with 2,3-didehydro-2,3-dideoxy-D-threo-hexose	n.s.g.	229(4.09)	39-0441-66C
$C_{13}H_{18}N_2O_6$			
Uridine, 2',3'-O-isopropylidene-5-methyl-	pH 7	266(3.98)	24-3884-66
	NaOH	266(3.98)	24-3884-66
$C_{13}H_{18}N_2O_7$			
Uridine, 5-(hydroxymethyl)-2',3'-O-isopropylidene-	pH 7	263(4.01)	24-3884-66

Compound	Solvent	$\lambda_{max}(\log \varepsilon)$	Ref.
$C_{13}H_{18}N_4O$			
2(8H)-Pteridinone, 6,7-diiso-propyl-8-methyl-	pH 0.2	206s(4.24),241(3.70), 290(4.08),308s(4.00)	39-1065-66C
	pH 4.5	224(4.42),292(4.11), 305s(3.98)	39-1065-66C
	pH 13.5	225(4.38),313(4.14)	39-1065-66C
$C_{13}H_{18}N_4O_2$			
Lumazine, 6,7-diisopropyl-8-methyl-, hydrate	pH -2.7	244(4.02),361(4.19)	24-3503-66
	pH 3	277(4.25),316(3.95)	24-3503-66
	pH 5	256(4.19),275(4.06), 404(4.12)	24-3503-66
	pH 12	230(4.40),282(4.14), 322(3.93)	24-3503-66
	pH 13	263(4.26),313(3.51), 415(4.03)	24-3503-66
$C_{13}H_{18}N_4O_6$			
Lumazine, 6,7-dimethyl-8-ribityl-	pH -2.7	245(4.08),361(4.10)	24-3503-66
	pH 3	274(4.20),306(3.95)	24-3503-66
	pH 6	258(4.13),275(3.95), 407(4.01)	24-3503-66
	pH 13	230(4.13),280(4.06), 313(3.91)	24-3503-66
$C_{13}H_{18}N_6O_6$			
Glycine, acetylglycylglycyl-L-histidyl-, copper complex	pH 11	555(1.95)	37-1072-66
nickel complex	pH 11	425(2.00),485(1.93)	37-1022-66
$C_{13}H_{18}O$			
Cyclopentanone, 2-allyl-2-(1-cyclopenten-1-yl)-	isooctane	294(2.24),301(2.37), 312(2.31),324(2.04)	22-0281-66
3,5-Hexadienal, 6-(1-cyclohexen-1-yl)-4-methyl-	n.s.g.	262s(4.37),270(4.46), 280(4.36)	22-0728-66
2,4-Hexadienal, 6-cyclohexylidene-4-methyl-	n.s.g.	240(3.60),260(3.65), 338(4.63)	22-0728-66
2-Indanol, 1,1,3,3-tetramethyl-	EtOH	251s(2.53),257(2.81), 264(3.03),271(3.09)	44-1393-66
β-Ionone, dehydro-	n.s.g.	222(--),345(4.06)	22-3874-66
$C_{13}H_{18}O_2$			
Benzoisopyran-8-ol, 4,5,6,7-tetra-methyl-	EtOH	212(4.12),280(2.85)	88-5205-66
6,9-Ethano-4-chromone, 5,6,7,8,9,10-hexahydro-2,6-dimethyl-	EtOH	270(3.92)	39-2324-66C
2(3H)-Naphthalenone, 4,4a,5,6,7,8-hexahydro-6α-acetyl-8a-methyl-	EtOH	237(4.20)	44-4188-66
6β-isomer	EtOH	241(4.14)	44-4188-66
2,4-Pentadienoic acid, 5-cyclohexyl-4-hydroxy-2,3-dimethyl-, γ-lactone	n.s.g.	273(4.3)	54-0043-66
$C_{13}H_{18}O_3$			
2-Cyclohexen-1-one, 4-hydroxy-3,5,5-trimethyl-4-(3-oxo-1-butenyl)-	n.s.g.	240(4.08)	28C-1397-66A
Lactic acid, 3-phenyl-, tert-butyl ester	EtOH	207(3.94),242(2.06), 247(2.13),252(2.24), 258(2.32),264(2.22), 267s(1.98)	22-3243-66
Unknown compound from croton species	isooctane	220(4.44),265(4.06)	102-0921-66

Compound	Solvent	λ max(log ε)	Ref.
$C_{13}H_{18}O_4$			
Acrylic acid, β-(1-hydroxy-2,6,6-trimethyl-4-oxocyclohex-2-en-1-yl)-, methyl ester	n.s.g.	230(4.06)	28C-1397-66A
$C_{13}H_{18}O_5$			
9-Cyclononene-1,2-dicarboxylic acid, 3-oxo-, dimethyl ester, cis	C_6H_{12}	254f(3.81)	35-4685-66
	EtOH	230(3.77),255(3.54)	35-4685-66
	MeCN	255(3.44)	35-4685-66
$C_{13}H_{18}O_6$			
2H-Oxecin-4,5-dicarboxylic acid, 3,4,7,8,9,10-hexahydro-3-oxo-, dimethyl ester	C_6H_{12}	253(3.99)	35-4685-66
	EtOH	238(3.99)	35-4685-66
	MeCN	255(3.78)	35-4685-66
$C_{13}H_{18}S$			
Cyclohexane, 1-methyl-1-(phenylthio)-	EtOH	270(3.17)	22-0228-66
Cyclohexane, 1-methyl-2-(phenylthio)-	EtOH	259(3.84)	22-0228-66
$C_{13}H_{19}Cl_2NO_6$			
4-[5-Chloro-3,3-dimethyl-1,5-cyclohexadien-1-yl)hydroxymethylene]-morpholinium perchlorate	EtOH	280(3.76)	44-2073-66
$C_{13}H_{19}IN_2S_2$			
3-Ethyl-2-[(3-ethyl-4-methyl-4-thiazolin-2-ylidene)methyl]-4-methylthiazolium iodide	MeOH	409(4.63)	9-0150-66
$C_{13}H_{19}N$			
3-Benzazecine, 1,2,3,4,5,6,7,8-octahydro-	EtOH	266(2.92),274(2.93)	87-0864-66
2H-Benzo[b]quinolizine, 1,3,4,6,7,10,11,11a-octahydro-	EtOH	264(1.81)	44-1707-66
Piperidine, N-benzyl-3-methyl-	MeOH	251(1.8),258(1.8), 264(1.6)	83-0295-66
Quinoline, 1,2,3,4-tetrahydro-2,2,4,4-tetramethyl-	MeOH	248(3.99),300(3.43)	30-0341-66E
$C_{13}H_{19}NO$			
6H-Benzo[c]quinolizin-6-one, 1,2,3,4,4a,5,7,8,9,10-decahydro-	EtOH-HCl	340(3.94)	94-1399-66
	EtOH	334(4.13)	94-1399-66
$C_{13}H_{19}NO_2$			
Isoquinoline, 1,2,3,4-tetrahydro-2-(2-hydroxyethyl)-6,7-dimethoxy-, hydrochloride	H_2O	279(3.52),284(3.53), 288(3.48)	83-0817-66
Pyridine, 3,5-diacetyl-1,4-dihydro-1,2,4,6-tetramethyl-	n.s.g.	258(4.04),283(3.92), 375(3.80)	39-1083-66B
$C_{13}H_{19}NO_4$			
Muconic acid, α-(diethylamino)-, dimethyl ester	ether	338(4.51)	24-0450-66
3,5-Pyridinedicarboxylic acid, 1,4-dihydro-1,4-dimethyl-, diethyl ester	n.s.g.	225(4.12),254(3.90), 372(3.96)	39-1083-66B
3,5-Pyridinedicarboxylic acid, 1,4-dihydro-2,6-dimethyl-, diethyl ester	n.s.g.	232(4.20),252s(3.90), 371(3.88)	39-1083-66B
Pyrrolidine, 2,5-bis(ethoxycarbonylmethylene)-N-methyl-	$CHCl_3$	305(4.72),315(4.80)	24-3444-66

Compound	Solvent	λ max(log ε)	Ref.
$C_{13}H_{19}NO_5$			
3,5-Pyridinedicarboxylic acid, 1,4-dihydro-4-(methoxymethyl)-2,6-dimethyl-, dimethyl ester	EtOH	234(4.24),350(3.88)	39-1075-66C
$C_{13}H_{19}N_3O_5$			
Cytidine, 2',3'-O-isopropylidene-N-methyl-	pH 1	217(3.99),280(4.18)	39-0588-66C
$C_{13}H_{19}N_3O_6$			
Cytidine, N,N-dimethyl-, 5'-acetate	pH 1	219(3.97),286(4.21)	39-0588-66C
$C_{13}H_{19}N_5$			
Pteridine, 2-sec-butylamino-4,6,7-trimethyl-	pH 1.0	227(4.45),236s(4.29), 363(4.09)	39-0226-66C
	pH 7.0	235(4.44),274(4.06), 386(3.92)	39-0226-66C
Pteridine, 2-isobutylamino-4,6,7-trimethyl-	pH 1.0	224(4.45),237s(4.29), 362(4.09)	39-0226-66C
	pH 7.0	235(4.44),274(4.04), 385(3.92)	39-0226-66C
$C_{13}H_{19}N_5O_5$			
Purine, 2-(dimethylamino)-9-β-D-glucopyranosyl-	pH -0.89	228(4.51),252(4.00), 336(3.42)	24-3022-66
	pH 7	256(4.12),331(3.77)	24-3022-66
$C_{13}H_{19}N_7O_2$			
Proline, N-(5-amino-3-butyl-3H-v-triazolo[4,5-d]pyrimidin-7-yl)-	pH 1	263(4.26)	87-0417-66
	pH 7	235(4.24),294(4.18)	87-0417-66
	pH 13	235(4.24),294(4.18)	87-0417-66
$C_{13}H_{20}ClNO_2$			
Isoquinoline, 1,2,3,4-tetrahydro-2-(2-hydroxyethyl)-6,7-dimethoxy-, hydrochloride	H_2O	279(3.52),284(3.53), 288(3.48)	83-0817-66
$C_{13}H_{20}Cl_2O$			
Cyclododecanone, 2-(dichloromethylene)-	EtOH	248(3.56)	77-0567-66
$C_{13}H_{20}N_2$			
Benzopyrazole, 3-cyclohexyl-4,5,6,7-tetrahydro-	EtOH	227(3.85)	22-2971-66
$C_{13}H_{20}N_2O$			
7-Azaindoline, 1-butyl-6-methoxy-4-methyl-	EtOH	254(3.87),314(3.94)	78-3233-66
7(4H)-Benzimidazolone, 5,6-dihydro-2-isobutyl-5,5-dimethyl-	EtOH	268(4.10)	25-0731-66
$C_{13}H_{20}N_2O_3$			
Benzoic acid, p-[[2-[(2-hydroxyethyl)amino]ethyl]amino]-, ethyl ester	EtOH	304(4.34)	87-0868-66
$C_{13}H_{20}N_4O_3$			
Isobarbituric acid, 6-(cyclohexylimino)-1,3-dimethyl-, 5-N-methyloxime	pH 1	261(4.08)	5-0134-66B

Compound	Solvent	$\lambda_{max}(\log \epsilon)$	Ref.
$C_{13}H_{20}O$			
2,4-Cyclopentadien-1-one, 2,4-di-tert-butyl-	gas	209(--),385(--)	35-3433-66
	isooctane	210(3.86),390(2.43)	35-3433-66
	EtOH	217(3.87),395(2.36)	35-3433-66
Cyclopropenone, 1-tert-butyl-2-(3,3-dimethyl-2-buten-2-yl)-	ether	239(4.23)	88-3763-66
2(3H)-Naphthalenone, 7-ethyl-4,4a,5,6,7,8-hexahydro-4a-methyl-, trans	EtOH	239(4.15)	44-3109-66
$C_{13}H_{20}O_2$			
Bicyclo[3.3.1]nonane-2,9-dione, 6,6,8,8-tetramethyl-	MeOH	285(2.04)	88-4173-66
Bicyclo[3.2.1]octane-6,8-dione, 3-tert-butyl-1-methyl-	C_6H_{12}	304(2.36)	22-3888-66
1,6-Decalindione, 4-isopropyl-, trans	EtOH	286(1.72)	78-1977-66
Methane, bis(2-oxocyclohexyl)-, meso	EtOH	289(1.67)	28C-1543-66A
threo	EtOH	290(1.64)	28C-1543-66A
Resorcinol, 5-heptyl-, hydrate	EtOH	276(3.29),281(3.28)	39-1608-66C
2,4-Undecadienoic acid, 4-hydroxy-2,3-dimethyl-, lactone	n.s.g.	270(4.3)	54-0043-66
$C_{13}H_{20}O_3$			
2-Heptanol, 1-(3,5-dihydroxyphenyl)-	EtOH	276(3.20),282(3.20)	39-1608-66C
2-Pentenoic acid, 5-cyclohexyl-4,4-dihydroxy-2,3-dimethyl-, γ-lactone	n.s.g.	216(4.1)	54-0043-66
$C_{13}H_{20}O_4$			
Aconic acid, γ-octyl-	EtOH	216(4.10)	44-0616-66
1,5-Cycloheptenedicarboxylic acid, diethyl ester	EtOH	223(3.90)	78-1521-66
Cyclohexanecarboxylic acid, 2-acetoxy-1,3-dimethyl-2-vinyl-	EtOH	217.5(2.04)	39-0764-66C
Malonic acid, (3-methyl-2,4-pentadienyl)-, diethyl ester	EtOH	229(4.62)	33-0858-66
$C_{13}H_{21}NO$			
Morpholine, 4-(3,5,5-trimethyl-1,3-cyclohexadien-1-yl)-	hexane	277(4.11)	44-2073-66
$C_{13}H_{21}NO_3$			
Cyclooctanone, 2-(diacetylmethyl)-, oxime	EtOH	240(4.03)	18-1129-66
$C_{13}H_{21}NO_3S_2$			
5-Thiazolidineacetic acid, N-butyl-4-oxo-2-thioxo-, butyl ester	EtOH	263(4.09),296(4.19)	95-0095-66
5-Thiazolidineacetic acid, 4-oxo-N-pentyl-2-thioxo-, propyl ester	EtOH	262(4.11),295(4.21)	95-0095-66
5-Thiazolidineacetic acid, 4-oxo-N-propyl-2-thioxo-, pentyl ester	EtOH	262(4.11),296(4.21)	95-0095-66
$C_{13}H_{21}NO_4S$			
3-Pentene-1-sulfonic acid, 2-morpholino-2,3-tetramethylene-4-hydroxy-, sultone	MeOH	210(3.41)	5-0074-66I
$C_{13}H_{21}N_3O$			
Ketone, 3a,4,5,6,7,7a-hexahydro-3a-methylinden-1-yl methyl, semicarbazone	EtOH	266(4.42)	78-0541-66B

Compound	Solvent	λ_{max}(log ε)	Ref.
$C_{13}H_{21}N_5$			
Purine, 2-(hexylamino)-6,9-dimethyl-	pH 1.5	230(4.60),254s(3.81), 325(3.63)	39-0226-66C
	pH 7.5	225(4.46),250(3.91), 257s(3.81),315(3.81)	39-0226-66C
$C_{13}H_{21}N_5O_8$			
Pyrimidine, 2-(dimethylamino)-4-(β-D-glucopyranosylamino)-6-methoxy-5-nitro-	pH 6.0	250(3.70),360(4.32)	24-3022-66
$C_{13}H_{21}N_6OP$			
Phosphonic diamide, P-(3-anilino-5-methyl-1H-1,2,4-triazol-1-yl)-N,N,N',N'-tetramethyl-	EtOH	246(4.13),260(4.20)	54-0429-66
Phosphonic diamide, P-(5-anilino-3-methyl-1H-1,2,4-triazol-1-yl)-N,N,N',N'-tetramethyl-	EtOH	260(4.40)	54-0429-66
$C_{13}H_{21}N_6O_7P$			
5'-Adenylic acid, 2-aminopropyl ester	n.s.g.	259(4.12)	33-0076-66
$C_{13}H_{22}ClNO_4$			
N-(3,5,5-Trimethyl-2-cyclohexenylidene)pyrrolidinium perchlorate	MeCN	268(4.30)	44-0639-66
$C_{13}H_{22}Cl_3NO_3$			
N-(3,5,5-Trimethyl-2-cyclohexen-1-ylidene)morpholinium trichloroacetate	EtOH	278(4.29)	44-2073-66
$C_{13}H_{22}NO_3P$			
Phosphorimidic acid, N-o-tolyl-, triethyl ester	EtOH	238(3.97),280(3.08)	35-3781-66
$C_{13}H_{22}N_2O$			
Cyclododec[c]isoxazole, 3-amino-4,5,6,7,8,9,10,11,12,13-decahydro-	n.s.g.	253(3.90)	18-1125-66
$C_{13}H_{22}N_2O_2S_2$			
5-Thiazolidineacetamide, 4-oxo-N-pentyl-3-propyl-2-thioxo-	EtOH	263(4.10),297(4.21)	95-0101-66
$C_{13}H_{22}N_2O_{11}$			
β-D-Glucopyranuronic acid, 1-deoxy-1-(5-methyl-2,4-dioxo-1,2,3,4-tetrahydro-1-pyrimidinyl)-2,3,4-tri-O-acetyl-, methyl ester	MeOH	261.5(4.01)	94-1354-66
$C_{13}H_{22}O$			
3-Buten-2-one, 4-(2,2,6-trimethylcyclohexyl)-	n.s.g.	232.0(4.02)	22-2212-66
3-Buten-2-one, 4-(3,3,5-trimethylcyclohexyl)-	n.s.g.	232.0(4.07)	22-2212-66
Cyclohexanone, 2-allyl-4-tert-butyl-	C_6H_{12}	289(1.57)	22-3881-66
2-Hydrindanone, 1,1,3,3-tetramethyl-, cis	C_6H_{12}	265s(0.87),273(1.00), 300(1.23),310(1.20), 321s(0.97)	44-1393-66

Compound	Solvent	$\lambda_{max}(\log \epsilon)$	Ref.
$C_{13}H_{22}OS$			
Cyclohexanone, 2-(butylthio-methylene)-4,4-dimethyl-	EtOH	315(4.18)	23-2003-66
$C_{13}H_{22}O_3$			
2-Undecenoic acid, 4,4-dihydroxy-2,3-dimethyl-, γ-lactone	n.s.g.	215(4.1)	54-0043-66
$C_{13}H_{22}O_6S_2$			
Glucopyranoside, methyl 2,3-di-O-methyl-4,6-dithio-, diacetate	EtOH	231(3.90)	39-1291-66C
$C_{13}H_{23}ClN_2O_4$			
1-(3-Piperidinoallylidene)piperidinium perchlorate	EtOH-100°K	320(4.79)	61-0052-66
phototropic form	EtOH-100°K	351(4.56)	61-0052-66
$C_{13}H_{23}NO$			
1-Octen-3-one, 1-piperidino-	EtOH	308(4.45)	70-1739-66
$C_{13}H_{23}N_3OS$			
Glycine, N,N-dimethyl-, [3-[(3-methyl-2-butenyl)thio]-2-butenylidene]hydrazide	C_6H_{12}	308(4.43)	78-0259-66
$C_{13}H_{25}NO$			
Azacyclotridecan-2-one, N-methyl-	MeOH	210(3.82)	24-0724-66
$C_{13}H_{26}N_2$			
1,6-Diazaspiro[4.4]nonane, 1,6-dipropyl-	MeOH	208(3.9)	24-0724-66
picrate	MeOH	213(5.3),354(5.3)	24-0724-66
Pyrrolidine, 1-isopropyl-2-[(3-isopropylamino)propylidene]-	MeOH	217(3.8)	24-1918-66
$C_{13}H_{26}O$			
3-Heptanone, 4,5-diethyl-2,2-dimethyl-	hexane	298(1.64)	22-3578-66
$C_{13}H_{28}N_2$			
Pyrrolidine, 1-isopropyl-2-(3-isopropylaminopropyl)-	MeOH	217(2.8)	24-1918-66
Pyrrolidine, 1-propyl-2-(3-propylaminopropyl)-	MeOH	207(3.2)	24-0737-66

Compound	Solvent	λ_{max}(log ε)	Ref.
$C_{14}H_2F_{10}N_2O_6$			
Hydrobenzoin, decafluoro-, dinitrate	$CHCl_3$	266(3.27)	70-0337-66
$C_{14}H_4Br_2O_4$			
1,4,9,10-Anthracenetetrone, 2,3-dibromo-	$CHCl_3$	242(4.48),385(3.35), 450s(2.60)	22-0580-66
$C_{14}H_4F_4O_8$			
1,4,5,8-Naphthalenetetracarboxylic acid, 2,3,6,7-tetrafluoro-	EtOH	224(4.56),300(3.91)	30-0855-66D
$C_{14}H_4F_{10}$			
Ethane, 1,2-bis(pentafluorophenyl)-	EtOH	282(4.318)	70-0337-66
$C_{14}H_4F_{10}O_2$			
Hydrobenzoin, decafluoro-	EtOH	262(3.114)	70-0337-66
$C_{14}H_5Br_4N$			
4H-Benzo[def]carbazole, 1,3,5,7-tetrabromo-	dioxan	249(4.72),272(4.30), 287(4.30),298(4.14), 370(4.34)	24-1279-66
$C_{14}H_6Br_2O_2$			
Pyracene-1,2-dione, 5,6-dibromo-	EtOH	225(4.52),231(4.53), 243s(4.27),324(3.89), 332(3.89)	35-0853-66
$C_{14}H_6Cl_4O_5$			
Xanthen-9-one, 2,4,5,7-tetrachloro-1,3,6-trihydroxy-8-methyl-	n.s.g.	248(3.91),272(3.72), 364(3.79)	88-3547-66
$C_{14}H_6Cl_{10}Si$			
Silane, dimethylbis-(pentachlorophenyl)-	C_6H_{12}	217(4.93)	101-0102-66B
$C_{14}H_6F_4N_2O_4$			
Phenanthrene, 9,9,10,10-tetrafluoro-9,10-dihydro-2,7-dinitro-	EtOH	313(4.484)	65-1815-66
$C_{14}H_6N_2O_4S$			
p-Benzoquinone, 2,3-dicyano-5-phenylsulfonyl-	MeCN	235(4.13),260(4.02)	44-3671-66
$C_{14}H_6O_2$			
Pyracyloquinone	EtOH	230(4.46),245(4.25), 307(4.24),314(4.23), 346(3.85)	35-0853-66
$C_{14}H_6O_4$			
1,2,5,6-Anthracenetetrone	DMSO	277(4.59),349(3.65), 471(3.66)	24-2322-66
$C_{14}H_6O_5$			
1,2,5,6-Anthracenetetrone, 9-hydroxy-	DMSO	523(3.92)	24-2322-66
$C_{14}H_7Br_2NO_3$			
Anthraquinone, 1-amino-2,3-dibromo-4-hydroxy-	$CHCl_3$	261(4.66),308(3.94), 534(4.06)	22-0580-66

Compound	Solvent	λ_{max}(log ε)	Ref.
$C_{14}H_7ClN_2O_4S$			
Phthalonitrile, 4-chloro-3,6-dihydroxy-5-(phenylsulfonyl)-	MeCN	222(4.44),270(3.93), 377(4.00)	44-3671-66
$C_{14}H_7ClO_3S$			
2-Anthracenesulfinyl chloride, 9,10-dihydro-9,10-dioxo-	$CHCl_3$	258(4.52),320(3.63)	22-2942-66
$C_{14}H_7NO_2$			
3H-Fluoreno[9,1-cd][1,2]oxazin-3-one	EtOH	242(3.34),261(2.97), 310(2.54)	28C-0293-66A
$C_{14}H_7N_3$			
1-Naphthalenemethanetricarbonitrile	C_6H_{12}	228(4.7),277(3.9), 287(3.8),314(2.8), 319(2.9)	44-0919-66
$C_{14}H_7N_5O_2$			
2-Quinoxalinecarbonitrile, 5-nitro-3-(2-pyridyl)-	EtOH	244(4.53),283(4.26), 350(3.91)	44-3384-66
$C_{14}H_8$			
Cyclopentundecene, 4,5,10,11-tetradehydro-	pentane	300(4.70),312(4.68), 324(4.65),340s(4.49), 393s(3.51),415s(3.39), 428(3.17),457(2.91), 558(2.53),578(2.58), 605(2.62),630(2.63), 656(2.60),695(2.51), 732(2.42),773(1.94)	35-0602-66
1,3,7,11-Cyclotetradecatetraene-5,9,13-triyne	pentane	246(3.73),251(3.77), 258(3.79),305(5.14), 348(3.50),375(3.73), 393(3.93),406(4.42), 448(2.18),463(2.18), 497(2.57),517(2.51), 523(2.53),535(2.46), 548(2.57),558(3.45)	35-0602-66
$C_{14}H_8BrNO_3$			
Anthraquinone, 1-amino-2-bromo-4-hydroxy-	$CHCl_3$	254(4.51),292(3.84), 529(3.95),558s(3.83)	22-0580-66
$C_{14}H_8Br_2N_4$			
Benzimidazo[1,2-c][1,2,3]benzotriazine, dibromo-2-methyl-	EtOH	234(4.46),289(4.60), 296(4.61),325s(4.03)	4-0444-66
$C_{14}H_8ClFN_2S$			
Benzimidic acid, p-chloro-N-(p-fluorophenyl)-, anhydride with isothiocyanic acid	C_6H_{12}	222(4.33),270(4.38), 276s(4.35),346(3.62)	44-0722-66
$C_{14}H_8Cl_2F_4N_2$			
Anthracene, 9,10-dichloro-9,10-bis-(difluoroamino)-9,10-dihydro-	EtOH	273(3.20),280(3.17), 298(2.19),309(2.00), 332(1.63)	44-3686-66
$C_{14}H_8Cl_2N_2O_2$			
Diimide, bis(p-chlorobenzoyl)-	dioxan	253(4.53),480(1.70)	24-2039-66 +89-0372-66

Compound	Solvent	λ max(log ε)	Ref.
$C_{14}H_8Cl_2O$			
Fluorene-9-carbonyl chloride, 9-chloro-	CH_2Cl_2	241(4.47),277(3.92)	24-0205-66
$C_{14}H_8Cl_2O_2$			
Phenol, 4-chloro-2-(5-chloro-3-benzofuranyl)-	MeOH	242(4.20),248(4.17), 254(4.18),260(4.11), 285(3.87),293(3.91)	23-1092-66
$C_{14}H_8F_2$			
Anthracene, 9,10-difluoro-	EtOH	245(4.94),253(5.32), 320s(3.03),352(3.72), 368s(3.83),370(3.92), 375(3.33),385(2.64), 391(3.79)	44-3686-66
$C_{14}H_8F_2N_2$			
9,10-Anthracenediimine, N,N'-difluoro-	EtOH	243(4.46),265s(4.14), 300s(3.54)	44-3686-66
$C_{14}H_8F_3NO_2S$			
Phenothiazine-1-carboxylic acid, 2-(trifluoromethyl)-	EtOH	258(4.43),320(3.54)	87-0835-66
	$EtOH-NH_3$	258(4.54),318(3.62)	87-0835-66
$C_{14}H_8F_4$			
Phenanthrene, 9,9,10,10-tetra-fluoro-9,10-dihydro-	EtOH	266(4.323)	65-1807-66
$C_{14}H_8F_4N_2$			
Anthracene, 9-(difluoroamino)-9-fluoro-10-fluorimino-	EtOH	248(4.07)	44-3686-66
$C_{14}H_8F_6N_2$			
9,10-Anthracenediamine, N,N,N',N',9,10-hexafluoro-9,10-dihydro-, cis	EtOH	262s(2.90),267(3.02), 274(3.05)	44-3686-66
trans	EtOH	270(3.12),278(3.17)	44-3686-66
$C_{14}H_8N_2$			
Benz[f]isoquinoline-6-carbonitrile	EtOH	209(4.40),222(4.31), 228(4.42),244s(4.54), 250(4.58),270s(4.25), 287(3.90),298(4.03), 310(4.03),332(3.20), 340(2.81),349(3.18)	39-1078-66C
$C_{14}H_8N_2O$			
Canthin-6-one	EtOH	241s(4.29),258(4.21), 265(4.19),296(3.99), 346s(3.99),357(4.17), 378(4.14)	5-0159-66A
$C_{14}H_8N_2O_4$			
3H-Indol-3-one, 5-nitro-2-phenyl-, 1-oxide	EtOH	244s(4.12),287(4.20), 346(3.91)	44-0065-66
$C_{14}H_8N_2O_4S$			
Phthalonitrile, 3,6-dihydroxy-4-(phenylsulfonyl)-	MeCN	235(4.30),260(4.03), 364(3.95)	44-3671-66

Compound	Solvent	λ_{max}(log ε)	Ref.
$C_{14}H_8N_4O_2$			
4,4'-Diazaindigo	EtOH	249(4.61),314(4.51), 600(4.22)	24-2146-66
5,5'-Diazaindigo	EtOH	257(4.59),549(4.26)	24-2146-66
6,6'-Diazaindigo	EtOH	270(4.57),629(4.21)	24-2146-66
7,7'-Diazaindigo	EtOH	238(4.63),312(4.51), 556(4.26)	24-2146-66
$C_{14}H_8O_2$			
1,2-Anthracenedione	MeOH	235(4.39),299(4.54), 443(3.71)	24-2322-66
1,4-Anthracenedione	dioxan	253(4.64),275s(4.05), 323(3.69)	95-0357-66
1,2-Pyracenedione	EtOH	213(4.63),238(4.72), 246(4.68),318s(3.69), 332(3.81),354(3.83), 364s(3.75)	35-0853-66
$C_{14}H_8O_4$			
1,2-Anthraquinone, 5,6-dihydroxy-	DMSO	310(4.49),515(3.90)	24-2322-66
Anthraquinone, 1,8-dihydroxy-	MeOH	250(4.1),290(4.0), 360(4.0)	83-0783-66
Naphtho[2,3-b]furan-3-carboxaldehyde, 4,9-dihydro-8-methyl-4,9-dioxo-	EtOH	252(4.36)	78-0685-66
4H-Naphtho[1,2-b]pyran-4,5,6-trione, 2-methyl-	EtOH	233(4.32),241(4.37), 253s(4.31),270s(3.92), 305(3.78)	88-4855-66
$C_{14}H_8O_5$			
1,2-Anthracenedione, 5,6,9-trihydroxy-	DMSO	540(4.04)	24-2322-66
Xanthen-9-one, 4-hydroxy-2,3-(methylenedioxy)-	EtOH	244(4.22),287(3.46), 322(4.13)	78-1777-66
	NaOH	242(4.21),275(2.63), 349(4.13)	78-1777-66
	NaOAc	241(4.22),275(4.16), 348(4.13)	78-1777-66
$C_{14}H_8O_5S$			
1-Anthracenesulfonic acid, 9,10-dihydro-9,10-dioxo-	dioxan	258(4.67),280(4.10), 333(3.67)	99-0357-66
2-Anthracenesulfonic acid, 9,10-dihydro-9,10-dioxo-	dioxan	256(4.31),280(4.12), 328(3.35)	99-0357-66
$C_{14}H_8O_8S_2$			
1,4-Anthracenedisulfonic acid, 9,10-dihydro-9,10-dioxo-	dioxan	265(4.59),328(3.47)	99-0357-66
1,5-Anthracenedisulfonic acid, 9,10-dihydro-9,10-dioxo-	dioxan	258(4.70),325(3.52)	99-0357-66
1,6-Anthracenedisulfonic acid, 9,10-dihydro-9,10-dioxo-	dioxan	260(4.50),280(3.90), 328(3.56)	99-0357-66
1,7-Anthracenedisulfonic acid, 9,10-dihydro-9,10-dioxo-	dioxan	260(4.50),278(4.00), 328(3.53)	99-0357-66
1,8-Anthracenedisulfonic acid, 9,10-dihydro-9,10-dioxo-	dioxan	258(4.52),328(3.50)	99-0357-66
2,6-Anthracenedisulfonic acid, 9,10-dihydro-9,10-dioxo-	dioxan	259(4.66),278(4.10), 333(3.61)	99-0357-66
2,7-Anthracenedisulfonic acid, 9,10-dihydro-9,10-dioxo-	dioxan	258(4.64),278(4.00), 328(3.68)	99-0357-66

Compound	Solvent	λ_{max}(log ε)	Ref.
$C_{14}H_9BrN_4$			
Benzimidazo[1,2-c][1,2,3]benzo-triazine, bromo-9(10)-methyl-	EtOH	230(4.36),288(4.54), 295(4.53),326s(3.91)	4-0444-66
$C_{14}H_9ClN_2O_2$			
Diimide, benzoyl(p-chlorobenzoyl)-	dioxan	245(4.38)	24-2039-66
	dioxan	245(4.38)	89-0372-66
$C_{14}H_9ClN_2O_2$			
Indole, 2-(p-chlorophenyl)-4-nitro-	EtOH	274(4.10),405(3.85)	78-3131-66
Indole, 2-(p-chlorophenyl)-6-nitro-	EtOH	254(4.20),382(4.00)	78-3131-66
$C_{14}H_9ClN_4$			
Pyrimidine, 2-(p-chlorophenyl)-5-(2-pyrazinyl)-	EtOH	300(4.56)	44-3845-66
$C_{14}H_9ClO$			
Benzofuran, 3-chloro-2-phenyl-	EtOH	200(4.38),302(4.47), 317(4.32)	39-0533-66C
$C_{14}H_9Cl_3$			
Ethylene, 2-chloro-1,1-bis-(o-chlorophenyl)-	MeOH	238s(4.1)	24-0680-66
Ethylene, 2-chloro-1,1-bis-(p-chlorophenyl)-	MeOH	241(4.4),260s(4.2)	24-0680-66
$C_{14}H_9Cl_5S$			
2,5-Norbornadiene, 1,3(2),4,7,7-pentachloro-2(3)-(methylthio)-5-phenyl-	EtOH	228(4.07),268(3.91)	44-1863-66
$C_{14}H_9F_3N_2O_4S$			
Benzoic acid, 2-(2-amino-4-trifluoro-methylphenylthio)-3-nitro-	EtOH	259(4.46),366(3.67)	87-0835-66
	EtOH-NH_3	263(4.50),341(3.77)	87-0835-66
$C_{14}H_9F_5O_3$			
3-Furoic acid, 2-methyl-5-(penta-fluorophenyl)-, ethyl ester	EtOH	276(4.32)	30-1343-66A
$C_{14}H_9N$			
Cycloprop[a]acenaphthylene-7-carbo-nitrile, 6b,7a-dihydro-, endo	$CHCl_3$	285(3.29),296(3.39), 310(3.20)	78-0189-66A
exo	CCl_4	287(2.99),299(3.10), 312(3.93)	78-0189-66A
Phenaline, 2-cyano-	$CHCl_3$	330(2.27),347(2.43), 363(2.45)	78-0189-66A
Phenanthrene, 4,5-imino-	EtOH	234(4.78),273(4.24), 284(4.14),333(4.12), 345(4.03)	24-1279-66
$C_{14}H_9NO_2$			
4H-3,1-Benzoxazin-4-one, 2-phenyl-	EtOH	241(4.35),275s(4.22), 285(4.33),298(4.29), 321(4.06)	18-1942-66
2H-Fluoreno[2,3-d]oxazol-2-one, 3,5-dihydro-	EtOH	274(4.17),287(4.00), 320(4.17)	87-0719-66
Phthalisoimide, N-phenyl-	MeCN	325(3.83)	35-5001-66
	dioxan	326(--)	35-5001-66

Compound	Solvent	λ_{max}(log ε)	Ref.
$C_{14}H_9NO_3$			
Fluorene-1-carboxylic acid, 9-oximino-	EtOH	243(3.10),258(3.09), 308(2.49)	28C-0293-66A
$C_{14}H_9NO_4$			
3H-Phenoxazin-3-one, 8-acetoxy-	C_6H_{12}	247(4.30),254s(4.26), 265s(4.05)	24-1470-66
	MeOH-HCl	327(4.14),339s(4.11)	24-1470-66
	MeOH-KOH	358s(3.86),444(3.99)	24-1470-66
$C_{14}H_9NO_5$			
Resazurin, acetate	EtOH	204(4.33),234(4.45), 488(4.06),512(4.03), 560s(3.76)	49-0129-66
$C_{14}H_9NO_6$			
3,4,5-Pyridinetricarboxylic acid, 2-phenyl-	MeOH-NaOMe	253s(4.08),288(3.98)	39-1121-66C
hydrogen sulfate	MeOH	260(3.97),304(3.99)	39-1121-66C
$C_{14}H_9N_3$			
2H-Benz[h]pyrazolo[4,3-c]quinoline	EtOH	211(4.32),255(4.88), 292(3.93),302(4.01), 313(3.98),330(3.23), 341(2.90)	54-0681-66
3H-Indole, 3-diazo-2-phenyl-	n.s.g.	348(4.05)	5-0017-66G
$C_{14}H_9N_3O_2$			
1,3-Propanedione, 2-diazo-1-phenyl-3-(2-pyridyl)-	CH_2Cl_2	246s(4.25),260(4.28)	24-3128-66
$C_{14}H_9N_3O_4$			
Diimide, benzoyl(p-nitrobenzoyl)-	dioxan	256(4.40)	24-2039-66
	dioxan	256(4.40),476(1.70)	89-0372-66
Indole, 3,5-dinitro-2-phenyl-	EtOH	263(4.49),333(4.10)	44-0065-66
Indole, 3,6-dinitro-2-phenyl-	EtOH	225(4.2),302(4.32), 338(4.24),414s(3.40)	44-0065-66
$C_{14}H_{10}$			
Acetylene, diphenyl-	THF	280(4.34),290(4.23), 299(4.36)	35-3027-66
Phenanthrene, tetracyanoethylene complex	$CHCl_3$	535(3.24)	60-0018-66
$C_{14}H_{10}BrN_3O_2$			
1H-Indazole, 6-bromo-3-methyl-1-(p-nitrophenyl)-	EtOH	231(4.52),268(3.88), 344(4.31)	78-3131-66
$C_{14}H_{10}ClNO_4S$			
10H-Indeno[2,1-d]pyrido[2,1-b]thiazolium perchlorate	MeOH	214(4.39),237s(4.04), 261(3.87),341(4.36)	39-0686-66C
$C_{14}H_{10}ClN_3O_2$			
1H-Indazole, 4-chloro-3-methyl-1-(p-nitrophenyl)-	EtOH	227(4.19),267(3.70), 346(4.17)	78-3131-66
1H-Indazole, 6-chloro-3-methyl-1-(m-nitrophenyl)-	EtOH	256(4.50),306(4.00)	78-3131-66
1H-Indazole, 6-chloro-3-methyl-1-(p-nitrophenyl)-	EtOH	230(4.25),269(3.74), 343(4.22)	78-3131-66

Compound	Solvent	λ_{max}(log ε)	Ref.
$C_{14}H_{10}Cl_2N_2O_2$			
Aziridine, 2,2-dichloro-1-(p-nitrophenyl)-3-phenyl-	dioxan	304(4.16)	78-1279-66
Aziridine, 2,2-dichloro-3-(p-nitrophenyl)-1-phenyl-	dioxan	227(4.04)	78-1279-66
$C_{14}H_{10}Cl_3N$			
Aziridine, 2,2-dichloro-1-(p-chlorophenyl)-3-phenyl-	dioxan	243(4.27)	78-1279-66
Aziridine, 2,2-dichloro-3-(p-chlorophenyl)-1-phenyl-	dioxan	226(4.31)	78-1279-66
$C_{14}H_{10}Cl_4S$			
2,5-Norbornadiene, 1,4,7,7-tetrachloro-2(7)-(methylthio)-5-phenyl-	EtOH	214(3.83),257(3.70)	44-1863-66
$C_{14}H_{10}CuN_2O_2$			
Salicylaldimine, copper chelate	MeOH	253(2.66),268(3.07)	65-1590-66
$C_{14}H_{10}F_3^+$			
(Trifluoromethyl)diphenylcarbonium ion	FSO_3H-SbF_5	331(3.63),485(4.33)	35-1488-66
$C_{14}H_{10}F_4N_2$			
Anthracene, 9,10-bis(difluoroamino)-9,10-dihydro-, cis	EtOH	256s(2.60),261(2.69), 267(2.66),273(2.51)	44-3686-66
trans	EtOH	255(3.42)	44-3686-66
$C_{14}H_{10}Fe$			
Butadiyne, 1-ferrocenyl-	$CHCl_3$	260(3.51),285(3.37), 311(3.29),448(2.43)	101-0399-66B
$C_{14}H_{10}N_2O$			
Cinnoline, 3-(p-hydroxyphenyl)-	MeOH	266(4.47),298(4.27)	87-0784-66
4-Cinnolinol, 3-phenyl-	MeOH	262(4.22),310(4.07), 352(4.08)	87-0784-66
$C_{14}H_{10}N_2O_2$			
4H-3,1-Benzoxazin-4-one, 1,2-dihydro-2-phenylimino-	EtOH	244(3.91),282(3.96), 292(3.93),332(3.43)	18-1942-66
1,3-Butanedione, 1-(2-naphthyl)-2-diazo-	CH_2Cl_2	249(4.57),280s(4.06), 320(3.33)	24-3128-66
Diimide, dibenzoyl-	dioxan	229(4.26),242(4.30)	24-2039-66
Indole, 4-nitro-2-phenyl-	EtOH	230(4.24),275(4.19), 405(4.00)	78-3131-66
Indole, 5-nitro-2-phenyl-	EtOH	223s(4.15),297(4.60), 341s(4.01)	44-0065-66
2,4(1H,3H)-Quinazolinedione, 3-phenyl-	EtOH	243(4.05),312(3.54)	18-1942-66
$C_{14}H_{10}N_2O_3S$			
2H-1,2-Benzothiazin-4(3H)-one, 3-(2-pyridylmethylene)-, 1,1-dioxide	EtOH	282(4.11),462(4.09)	44-0162-66
$C_{14}H_{10}N_2O_4$			
Azodicarboxylic acid, diphenyl ester	MeOH	273(3.56)	24-2039-66
	dioxan	250s(3.32)	24-2039-66
	dioxan	250s(3.30),409(1.60)	89-0372-66
3H-Phenoxazine-1-carboxylic acid, 2-amino-3-oxo-, methyl ester	EtOH	234(3.91),428(3.78)	69-3824-66

Compound	Solvent	λ_{max}(log ε)	Ref.
$C_{14}H_{10}N_2O_5$			
Anthranilic acid, N-benzoyl-4-nitro-	EtOH	272(4.44),344(3.50)	44-0065-66
Anthranilic acid, N-benzoyl-5-nitro-	EtOH	219s(4.36),241s(4.23), 328(4.40)(anom.)	44-0065-66
Cinnabarin	EtOH	230(4.20),270s(4.04), 434(3.89)	39-0072-66C
Styrene oxide, p,p'-dinitro-, cis	EtOH	215s(4.16),273(4.28)	44-3976-66
Styrene oxide, p,p'-dinitro-, trans	EtOH	216s(4.18),286(4.40)	44-3976-66
$C_{14}H_{10}N_2O_6$			
Biphenylene, 3,7-dimethoxy- 1,5-dinitro-	dioxan	303(4.07),454(4.01)	39-1769-66C
	MeCN	238(--),300(--), 451(--)	39-1769-66C
$C_{14}H_{10}N_4$			
Benzimidazo[1,2-c][1,2,3]benzo- triazine, 3-methyl-	EtOH	230(4.42),282(4.64), 289(4.61),322s(3.90)	4-0289-66
Benzimidazo[1,2-c][1,2,3]benzo- triazine, 10-methyl-	EtOH	227(4.48),286(4.54), 292(4.52),326s(3.83)	4-0289-66
$C_{14}H_{10}N_4O_4$			
1H-Indazole, 3-methyl-5-nitro- 1-(p-nitrophenyl)-	EtOH	206(--),216s(--), 261(--),303(--), 325s(--),349(--)	39-1781-66C
$C_{14}H_{10}O$			
Benzofuran, 2-phenyl-	EtOH	302(4.36),314(4.22)	88-5225-66
$C_{14}H_{10}OS_3$			
Terthienyl, 5-acetyl-	ether	379(4.59)	24-0984-66
$C_{14}H_{10}O_2$			
Benzofuran, 2-(o-hydroxyphenyl)-	EtOH	289(4.09),314(4.25), 328(4.22)	18-1535-66
	EtOH	232(4.12),288(4.24), 314(4.45),328(4.43)	39-0544-66C
1,6:8,13-Diepoxy[14]annulene	C_6H_{12}	306(5.23),345(4.16), 382(3.93),555f(2.89)	89-0734-66
1H-Naphtho[2,1-b]pyran-1-one, 3-methyl-	99% EtOH	225s(4.48),230(4.50), 253s(4.38),261(4.39), 305(4.10),320s(3.85), 335(3.73)	94-0129-66
$C_{14}H_{10}O_3$			
Anthrone, 1,8-dihydroxy-	MeOH	250(4.4),275(4.2), 285(4.2),430(4.2)	83-0783-66
Maturinone	EtOH	246(4.33),288(4.00)	78-0685-66
Naphtho[2,3-b]furan-4,9-dione, 3,8-dimethyl-	EtOH	210(4.21),250(4.37), 290(3.75),352(3.64)	78-0301-66
1H-Naphtho[2,1-b]pyran-1-one, 5-hydroxy-3-methyl-	99% EtOH	236(4.41),272(4.15), 314s(3.85),327(3.93), 337(3.93)	94-0129-66
1H-Naphtho[2,1-b]pyran-1-one, 6-hydroxy-3-methyl-	99% EtOH	212(4.43),225s(4.35), 234(4.38),258(4.27), 265(4.26),295s(3.82), 308(3.98),325(3.98), 340(4.06),367s(3.32)	94-0129-66
4H-Naphtho[2,3-b]pyran-4-one, 5-hydroxy-2-methyl-	99% EtOH	219(4.19),257(4.59), 265(4.60),294(3.42), 307(3.45),319s(2.94), 390(3.57)	94-0121-66

Compound	Solvent	$\lambda_{max}(\log \epsilon)$	Ref.
$C_{14}H_{10}O_4$			
4,4'-Biphenyldicarboxylic acid	n.s.g.	270(3.68)	70-1378-66
Maturone	EtOH	250(4.46),298(3.81), 354(3.51)	78-0685-66
1H-Naphtho[2,1-b]pyran-1-one, 5,6-dihydroxy-3-methyl-	99% EtOH	218s(4.44),225(4.45), 246s(4.22),266s(4.04), 344(3.61)	94-0129-66
$C_{14}H_{10}O_5$			
Alternariol	EtOH	258(4.58)	102-0719-66
Xanthone, 1,3-dihydroxy-7-methoxy-	EtOH	235(4.45),259(4.50), 311(4.14),369(3.80)	39-2265-66C
Xanthone, 1,5-dihydroxy-3-methoxy-	EtOH	248(4.21),315(4.14), 355(3.25)	78-1777-66
	NaOH	240s(4.17),263(4.22), 278s(4.13),344(4.14)	78-1777-66
	NaOAc	253(4.18),287(4.14), 314(4.13)	78-1777-66
	$AlCl_3$	240(4.15),267(4.20), 335(4.14)	78-1777-66
Xanthone, 2,8-dihydroxy-1-methoxy-	EtOH	238(4.43),262(4.52), 290(3.72),322(3.64)	78-1785-66
	NaOH	254(4.50),275(4.40), 350(3.60)	78-1785-66
	$AlCl_3$	238(4.46),277(4.45), 311(3.79),350(3.60)	78-1785-66
Xanthone, 3,4-dihydroxy-2-methoxy-	EtOH	239(4.19),255(4.19), 285(3.36),333(4.13)	78-1777-66
	NaOAc	232(4.18),276(4.14), 364(4.14)	78-1777-66
$C_{14}H_{10}O_8S_2$			
Benzoic acid, 4,4'-disulfonyldi-	n.s.g.	270(3.73)	44-3418-66
$C_{14}H_{11}Br$			
Stilbene, p-bromo-, trans	hexane	227(4.28),301(4.60), 315(4.55),322(4.33)	99-0089-66
	EtOH	228(4.28),305(4.64)	99-0089-66
$C_{14}H_{11}BrN_2$			
Benzimidazole, 2-(o-bromophenyl)-5-methyl-	EtOH	240s(3.83),288(4.05)	4-0444-66
$C_{14}H_{11}BrO$			
Acetophenone, p-(p-bromophenyl)-	hexane	280(4.36)	99-0204-66
	EtOH	285(4.30)	99-0204-66
$C_{14}H_{11}Cl$			
Stilbene, p-chloro-, trans	hexane	228(4.43),300(4.53), 312(4.50),320(4.33)	99-0089-66
	EtOH	228(4.26),299(4.57), 310(4.55),320(4.42)	99-0089-66
$C_{14}H_{11}ClN_2$			
Benzimidazole, 2-(o-chlorophenyl)-5-methyl-	EtOH	294(4.00)	4-0444-66
Benzimidazole, 2-(6-chloro-m-tolyl)-	EtOH	287(4.17)	4-0444-66

Compound	Solvent	λ_{max}(log ε)	Ref.
$C_{14}H_{11}ClN_2O_2$			
7-Methyl-3-nitrobenzo[c]quinolizinium chloride	EtOH	255(4.57),291(4.02), 305s(3.92),320(3.79), 346(3.72),361(4.06), 379(4.12)	44-2346-66
$C_{14}H_{11}ClN_4O_6S$			
Acetaldehyde, [(p-chlorophenyl)sulfonyl]-, 2,4-dinitrophenylhydrazone	EtOH	223(4.44),350(4.36)	24-1580-66
$C_{14}H_{11}ClO$			
Acetophenone, p-(p-chlorophenyl)-	hexane	279(4.41),322(2.55)	99-0204-66
	EtOH	285(4.36),341(2.35)	99-0204-66
$C_{14}H_{11}ClO_2$			
Benzophenone, 3-chloro-4-methoxy-, cation	H_2SO_4	392(4.45)	39-0650-66B
Benzophenone, 4-chloro-4'-methoxy-, cation	H_2SO_4	394(4.54)	39-0650-66B
$C_{14}H_{11}ClO_4S$			
1-Methylnaphtho[2,1-b]thiopyrylium perchlorate	HOAc	288s(4.14),305(4.16), 423(3.91)	78-0007-66
$C_{14}H_{11}Cl_2N$			
Acetimidoyl chloride, 2-chloro-	dioxan	224.7(4.12)	78-1279-66
Aziridine, 2,2-dichloro-1,3-diphenyl-	dioxan	220(4.18)	78-1279-66
3-Chloro-7-methylbenzo[c]quinolizinium chloride	EtOH	234(4.17),258(4.46), 303s(3.59),313(3.65), 345s(3.65),361(4.09), 379(4.22)	44-2346-66
$C_{14}H_{11}F$			
Stilbene, p-fluoro-, trans	hexane	226(4.24),295(4.49), 310(4.41)	99-0089-66
	EtOH	226(4.19),295(4.46), 324(4.15)	99-0089-66
$C_{14}H_{11}FO$			
Acetophenone, p-(p-fluorophenyl)-	hexane	276(4.32),325(2.60)	99-0204-66
	EtOH	280(4.20)	99-0204-66
$C_{14}H_{11}I$			
Stilbene, p-iodo-, trans	hexane	228(4.17),315(4.57), 325(4.40)	99-0089-66
	EtOH	230(4.10),310(4.51)	99-0089-66
$C_{14}H_{11}IO$			
Acetophenone, p-(p-iodophenyl)-	hexane	218(4.30),286(4.40), 327(2.52)	99-0204-66
	EtOH	292(4.27)	99-0204-66
$C_{14}H_{11}MnO_4$			
Manganese, (5-methyl-2-phenyl-π-1-oxapentadienyl)tricarbonyl-	dioxan	231(3.78),266(3.97), 307(3.76)	39-0194-66A

Compound	Solvent	$\lambda_{max}(\log \epsilon)$	Ref.
$C_{14}H_{11}N$			
Benz[f]isoquinoline, 9-methyl-	hexane	211(4.45),220(4.38), 245(4.71),251(4.76), 273(4.09),283(4.05), 295(4.14),313(2.70), 320(3.03),327(2.89), 336(3.40),344(2.92), 348(3.09),352(3.53)	39-1078-66C
Benz[h]isoquinoline, 5-methyl-	hexane	214(4.26),224(4.20), 247(4.67),289(3.98), 300(3.97),314(3.08), 328(3.16),343(3.13)	39-1078-66C
Benzo[f]quinoline, 5-methyl- (inflections not listed)	hexane	212(4.49),234(4.62), 252(4.37),269(4.37), 318(3.13),325(2.97), 333(3.38),341(2.94), 350(3.42)	39-1078-66C
Indole, 3-phenyl-	EtOH	224(4.51),271(4.21)	94-0934-66
Phenanthrene, 9,10-dihydro- 4,5-imino-	EtOH	246(4.50),256(4.36), 294(4.08)	24-1279-66
$C_{14}H_{11}NO$			
Benzamide, N-benzylidene-	n.s.g.	268(4.25)	88-4569-66
Carbazole-2-carboxaldehyde, 1-methyl-	EtOH	207(4.27),252(4.59), 279(3.86),322(4.38)	87-0237-66
$C_{14}H_{11}NOS$			
Phenothiazine, 2-acetyl-	EtOH	212(4.94),245(4.38), 274(4.38),284(4.34), 334s(3.49),405(3.17)	7-0182-66
$C_{14}H_{11}NOSe$			
Phenoselenazine, 2-acetyl-	EtOH	217s(4.21),249(4.32), 285(4.37),341s(3.43), 403(3.21)	7-0182-66
$C_{14}H_{11}NO_2$			
1H-Benz[f]indole-3-acetic acid	EtOH	222(4.29),251(4.84), 317(3.45),330(3.59), 348(3.71),363(3.70)	39-0477-66C
Carbazole-3-carboxaldehyde, 1-methoxy-	MeOH	237(4.59),248(4.42), 272(4.65),285(4.46), 335(4.24)	77-0417-66
Phenoxazine, 2-acetyl-	EtOH	225(4.31),273(4.47), 325(3.84),397(3.22)	7-0182-66
Stilbene, p-nitro-, trans	hexane	234(4.26),260s(4.00), 310(4.26),333(4.46)	99-0089-66
	EtOH	243(4.22),339(4.28)	99-0089-66
$C_{14}H_{11}NO_3$			
p-Benzoquinone, 2-cyano- 3-(5-isopropyl-2-furyl)-	EtOH	226(4.25),255s(3.88), 288(3.87),363(3.88), 495(3.57)	33-1794-66
p-Benzoquinone, 2-cyano- 3-(3,4,5-trimethyl-2-furyl)-	EtOH	230(4.10),300s(3.80), 360s(3.11),522(3.79)	33-1794-66
Fluorene, 1-methoxy-2-nitro-	EtOH	314(4.16)	87-0719-66
2-Naphthonitrile, 3-acetoxy- 8-methoxy-	EtOH	215(4.59),252(4.61), 290(3.45),299(3.54), 310(3.47),352(3.71)	44-1747-66

Compound	Solvent	$\lambda_{max}(\log \epsilon)$	Ref.
3H-Phenoxazin-3-one, 7-hydroxy-2,9-dimethyl-	MeOH-HCl	223(4.51),261(4.35), 510(4.68)	24-1470-66
	MeOH-HOAc	473(4.40)	24-1470-66
	MeOH	255(4.17),281s(3.78)	24-1470-66
	MeOH-KOH	240(4.42),296(3.93), 580(4.87)	24-1470-66
3H-Phenoxazin-3-one, 8-hydroxy-1,7-dimethyl-	MeOH-HCl	278(4.21),445(4.27), 563(4.05)	24-1470-66
	MeOH	263(4.25),370(4.15), 512(4.06)	24-1470-66
	MeOH-KOH	277(4.24),395(4.16), 623(3.92)	24-1470-66
Resorufin, ethyl ether	EtOH	225(4.41),252(4.15), 280s(3.76),400s(4.01), 465(4.37)	49-0129-66
Styrene oxide, p-nitro-, cis	hexane	268(4.03)	44-3976-66
Styrene oxide, p-nitro-, trans	hexane	216(4.26),274(4.19)	44-3976-66
$C_{14}H_{11}NO_4$			
Resazurin, ethyl ether	EtOH	204(4.45),226(4.55), 494(4.20),525(4.23), 560s(4.00)	49-0129-66
$C_{14}H_{11}N_3$			
1,2,3-Triazole, 4,5-diphenyl-	EtOH	249(4.10)	44-3914-66
$C_{14}H_{11}N_3O$			
4(3H)-Quinazolinone, 2-(m-aminophenyl)-	EtOH	232(4.57),288(4.17), 323s(3.95)	32-0264-66
4(3H)-Quinazolinone, 2-(p-aminophenyl)-	EtOH	236(4.39),335(4.45)	32-0264-66
$C_{14}H_{11}N_3O_2$			
Benzimidazole, 5-methyl-2-(o-nitrophenyl)-	EtOH	240s(4.05),284(4.01)	4-0289-66
Indazole, 3-methyl-1-(m-nitrophenyl)-	EtOH	253(4.41),305(4.08)	78-3131-66
Indazole, 3-methyl-1-(p-nitrophenyl)-	EtOH	232(4.30),355(4.25)	78-3131-66
$C_{14}H_{11}N_3O_3$			
Indazole, 6-methoxy-1-(p-nitrophenyl)-	EtOH	228(4.34),280(4.04), 342(4.21)	78-3131-66
$C_{14}H_{11}N_3O_4$			
Indazol-6-ol, 5-methoxy-1-(p-nitrophenyl)-	EtOH	237(4.08),262(3.95), 300(3.87),355(4.12)	78-3131-66
$C_{14}H_{12}$			
Biphenylene, 2,7-dimethyl-	EtOH	243(4.80),252(5.07), 333s(3.51),347(3.83), 366(3.97)	39-1767-66C
Stilbene	hexane	222(4.21),228(4.22), 232(4.11),306(4.44), 317(4.28)	99-0085-66
	$C_2H_4Cl_2$	297(4.58),308(4.54), 322(4.35)	99-0085-66
Stilbene, cis	benzene	none	62-0095-66D
	EtI	none	62-0095-66D

Compound	Solvent	$\lambda_{max}(\log \epsilon)$	Ref.
$C_{14}H_{12}BrNO_5$			
Nicotinic acid, 1-(m-bromophenyl)-1,2-dihydro-4,6-dihydroxy-2-oxo-, ethyl ester	EtOH	305(4.50)	78-0455-66
$C_{14}H_{12}BrN_3$			
Indazole, 1-(p-aminophenyl)-6-bromo-3-methyl-	EtOH	220(4.40),264(4.30),293(3.83)	78-3131-66
$C_{14}H_{12}ClN$			
6-Methylbenzo[c]quinolizinium chloride	EtOH	230(4.13),258(4.36),284(3.88),304s(3.46),340(3.46),358(3.88),374(3.98)	44-2346-66
7-Methylbenzo[c]quinolizinium chloride	EtOH	231(4.02),256(4.40),285(3.83),305s(3.46),333s(3.62),354(4.02),371(4.16)	44-2346-66
2-Picoline, 6-(o-chlorostyryl)-, cis	EtOH	290(4.07)	44-2346-66
$C_{14}H_{12}ClNO$			
Benzophenone, p-chloro-, O-methyloxime, cis	decane	238(4.27),264(4.03)	35-2775-66
trans	decane	238(4.18),267(4.13)	35-2775-66
$C_{14}H_{12}ClNO_4$			
6-Methylbenzo[c]quinolizinium perchlorate	EtOH	263(4.18),275s(--),285(4.31),329(3.66),344(3.98),361(4.13)	44-2616-66
8-Methylbenzo[c]quinolizinium perchlorate	EtOH	272s(--),285(4.47),325(3.79),341(3.98),357(4.08)	44-2616-66
$C_{14}H_{12}ClNO_4S$			
2-Methyl-3-phenylthiazolo[3,2-a]-pyridinium perchlorate	EtOH	239(4.21),314(4.17)	4-0027-66
$C_{14}H_{12}ClNO_5$			
Nicotinic acid, 1-(m-chlorophenyl)-1,2-dihydro-4,6-dihydroxy-2-oxo-, methyl ester	EtOH	304(4.46)	78-0455-66
Nicotinic acid, 1-(p-chlorophenyl)-1,2-dihydro-4,6-dihydroxy-2-oxo-, methyl ester	EtOH	305(4.50)	78-0455-66
$C_{14}H_{12}N_2$			
Benzimidazole, 2-benzyl-	$MeOCH_2CH_2OH$	278(4.0)	44-1498-66
Benzimidazole, 2-o-tolyl-	$CHCl_3$	290(4.20)	44-1498-66
Morphanthridine, 6-amino-, hydrochloride	EtOH	224s(4.2),247(4.02),284(3.86)	4-0206-66
$C_{14}H_{12}N_2O$			
Benzimidazole, 2-(2-hydroxy-5-methylphenyl)-	EtOH	234(4.07),275s(3.97),287(4.09),294(4.17),322(4.22),331(4.19)	4-0444-66
Benzimidazole, 2-(2-hydroxyphenyl)-5-methyl-	EtOH	235(3.92),250s(3.67),275s(3.86),292(4.03),318(4.35),332(4.31)	4-0444-66
Benzimidazole, 1-methoxy-2-phenyl-	H_2O	239(4.24),296(4.34)	4-0051-66
	EtOH	238(4.27),297(4.33)	4-0051-66

Compound	Solvent	λ_{max}(log ε)	Ref.
Benzimidazole, 1-methoxy-2-phenyl-, hydrochloride	EtOH	240(4.28),310(4.40)	4-0051-66
Benzimidazole, 3-methyl-2-phenyl-, 1-oxide	H_2O	237(4.49),287(4.19)	4-0051-66
	EtOH	237(4.48),291(4.18)	4-0051-66
hydrochloride	H_2O	239(4.44),287(4.21)	4-0051-66
	EtOH	239(4.47),286(4.22)	4-0051-66
$C_{14}H_{12}N_2OS$			
Urea, N-phenyl-N'-(thiobenzoyl)-	dioxan	226(4.2),257(4.1), 275(4.1),408(2.5)	24-0782-66
$C_{14}H_{12}N_2O_2$			
Isoxazole, 5-methyl-3'-phenyl-3,5'-methylenedi-	EtOH	240(4.23)	32-1064-66
$C_{14}H_{12}N_2O_2S$			
Dibenzo[b,f][1,4,5]thiadiazepine, 3,8-dimethyl-, 11,11-dioxide	EtOH	243(4.49),320(3.83), 430(2.79)	39-0255-66B
$C_{14}H_{12}N_2O_3$			
Anthranilic acid, N-(phenylcarbamoyl)-	pH 1	226(4.32),260(4.29), 312(3.75)	35-4001-66
	EtOH	265(4.58),306(3.96)	18-1942-66
o-Benzotoluidide, 5'-nitro-	EtOH	242(4.23)	44-0065-66
5-Isoxazolemethanol, 3-[(3-phenyl-5-isoxazolyl)methyl]-	EtOH	240(4.22)	32-1064-66
Pyrazino[1,2-a]indole-1,4-dione, 1,2,3,4-tetrahydro-6-methoxy-2-methyl-3-methylene-	MeOH	223(4.35),228(4.35), 253(4.26),322(4.14)	39-1799-66C
	pH 13	230(4.34),291(4.22)	39-1799-66C
$C_{14}H_{12}N_2O_3S$			
Dibenzo[b,f][1,4,5]thiadiazepine, 3,8-dimethyl-, 5,11,11-trioxide	EtOH	240(4.54),334(3.84)	39-0255-66B
$C_{14}H_{12}N_2O_3S_2$			
3H-Phenothiazine-8-sulfonamide, N,N-dimethyl-3-oxo-	MeOH	226(4.27),240s(--), 295(4.50),359(3.89), 375s(--),492(3.85)	95-0541-66
$C_{14}H_{12}N_2O_4$			
Bibenzyl, 4,4'-dinitro-	EtOH	214s(4.15),276(4.31)	44-3976-66
$C_{14}H_{12}N_2O_4S_2$			
5-Thia-1-azabicyclo[4.2.0]oct-2-ene-2-carboxylic acid, 3-(hydroxy-methyl)-8-oxo-7-[2-(2-thienyl)-acetamido]-, γ-lactone	EtOH	236(4.09),260s(3.83)	87-0741-66
$C_{14}H_{12}N_2S_2$			
Thiazole, 2,2'-p-phenylenebis-[4-methyl-	C_6H_{12}	236(4.16),341(4.57)	22-2857-66
$C_{14}H_{12}N_4$			
13,14-Diazatricyclo[6.4.1.1^{2,7}]tetra-deca-3,5,9,11-tetraene-13,14-di-carbonitrile	MeCN	230(4.18),237(4.21)	35-2591-66
$C_{14}H_{12}N_4O$			
Pyrazolo[5,1-c]-as-triazine, 6-benzoyl-4,6-dihydro-3-methyl-4-methylene-	EtOH	232s(4.16),268s(4.09), 306(4.24)	39-1127-66C

Compound	Solvent	$\lambda_{max}(\log \epsilon)$	Ref.
$C_{14}H_{12}N_4O_2$			
Lumazine, 8-benzyl-3-methyl-	pH -2.7	244(4.05),345(4.06)	24-3503-66
	pH 6	261(4.30),400(3.98)	24-3503-66
	pH 13	235(4.34),282(4.13), 308(3.93)	24-3503-66
Pyrazolo[5,1-c]-as-triazine-3-carboxylic acid, 4-phenyl-, ethyl ester	EtOH	243(4.37),285s(3.65), 372(3.46)	39-1127-66C
$C_{14}H_{12}N_4O_4$			
Acetophenone, 2,4-dinitrophenylhydrazone	$CHCl_3$	380(4.39)	22-0717-66
$C_{14}H_{12}N_4O_6S$			
Acetaldehyde, (phenylsulfonyl)-, 2,4-dinitrophenylhydrazone	EtOH	217(4.34),351(4.36)	24-1580-66
$C_{14}H_{12}O$			
Acetophenone, p-phenyl-	hexane	276(4.39),315s(2.55)	99-0089-66
	EtOH	282(4.38),343s(2.64)	99-0089-66
Naphtho[2,3-b]furan, 3,9-dimethyl-	EtOH	245(4.74),314(3.96), 323(3.88),339(3.76)	39-0725-66C
Stilbene oxide, cis	hexane	218s(4.09),261(3.67)	44-3976-66
Stilbene oxide, trans	hexane	228(4.37),267(3.95)	44-3976-66
4-Stilbenol, trans	hexane	230(4.26),293(4.52), 305(4.56),320(4.55)	99-0089-66
	EtOH-NaOEt	242(4.20),276(3.65), 325(4.30),355(4.44), 397s(3.40)	99-0089-66
$C_{14}H_{12}O_2$			
Acetophenone, p-(p-hydroxyphenyl)-	hexane	228(4.11),320(4.28)	99-0204-66
1H-Benz[cd]azulene-9-carboxylic acid, 2,8-dihydro-	EtOH	212(4.27),240(3.77), 322(3.88)	35-3950-66
1H-Benz[e]inden-1-one, 2,3-dihydro-8-methoxy-	EtOH	231(4.63),263(4.10), 316(3.85),342(3.86)	39-0734-66C
3-Benzofuranol, 2,3-dihydro-2-phenyl-, cis	MeOH	280(3.49),286(3.43)	88-1171-66
Benzophenone, 4-methoxy-, cation	H_2SO_4	383(4.52)	39-0650-66B
2,4,6-Heptatrienoic acid, 5-hydroxy-3-methyl-7-phenyl-, δ-lactone	EtOH	232(4.20),240(4.18), 300(4.08),348(4.30)	35-0624-66
Tricyclo[6.4.1.1^{2,7}]tetradeca-3,5,9,11-tetraene-13,14-dione	MeOH	254(3.94),273s(--)	35-5025-66
Tropone dimer, m. 118°	MeOH	224(3.95),291(3.65)	35-5026-66
Tropone dimer, m. 124°	MeOH	267(3.81),292(3.72)	35-5026-66
Tropone dimer, m. 141°	MeOH	225(3.97),256(3.62), 266(3.64),285s(3.28)	35-5026-66
$C_{14}H_{12}O_2S$			
Naphtho[1,2-c]thiophene-1-carboxylic acid, 4,5-dihydro-, methyl ester	EtOH	259(4.00)	88-1953-66
Thiirane, 2,3-diphenyl-, 1,1-dioxide	EtOH	268(3.65),274(3.61)	44-0349-66
$C_{14}H_{12}O_3$			
1(2H)-Anthracenone, 3,4-dihydro-8,9-dihydroxy-	EtOH	275(4.7),298(3.7), 326s(3.6),405(3.7)	78-2761-66
Benzo-1,3-dioxan-6-ol, 2-phenyl-	EtOH	208(4.27),228s(3.87), 295(3.48)	39-1627-66C
	EtOH-NaOH	240(3.94),316(3.52)	39-1627-66C

Compound	Solvent	$\lambda_{max}(\log \epsilon)$	Ref.
p-Benzoquinone, [(benzyloxy)methyl]-	C_6H_{12}	210(4.10),245(4.31), 442(1.30),455s(--)	39-1627-66C
Kawain, 5,6-dehydro-	EtOH	233(4.10),257(4.08), 345(4.35)	95-1184-66
Xanthyletin	n.s.g.	225(4.42),266(4.34), 302(3.8),347(4.14)	1-2497-66
$C_{14}H_{12}O_4$			
p-Benzoquinone, 2-acetyl-3-(3,5-dimethyl-2-furyl)-	EtOH	244(4.21),280s(3.91), 320s(3.26),492(3.69)	33-1794-66
p-Benzoquinone, 2-acetyl-3-(4,5-dimethyl-2-furyl)-	EtOH	252(4.06),310(3.24), 495(3.80)	33-1794-66
Biphenyl-x,x-dicarboxylic acid, dihydro-	n.s.g.	270(3.27)	70-1378-66
isomer	n.s.g.	270(3.19)	70-1378-66
2-Naphthoic acid, 3-acetoxy-, methyl ester	EtOH	236(4.76),271(3.73), 280(3.80),291s(3.59), 333(3.08)	44-1747-66
Xanthotoxin, 4,5'-dimethyl-	99% EtOH	214(4.28),215(4.46), 301(4.02)	4-0042-66
Xanthyletin, 3',4'-dihydro-3'-oxo-	n.s.g.	296(4.02),327(4.19)	1-2497-66
$C_{14}H_{12}O_5$			
p-Benzoquinone, 2-carbomethoxy-3-(3,4-dimethyl-2-furyl)-	EtOH	263(4.06),312(3.19), 496(3.88)	33-1794-66
p-Benzoquinone, 2-carbomethoxy-3-(3,5-dimethyl-2-furyl)-	EtOH	240(4.14),260s(4.10), 280s(3.99),490(3.77)	33-1794-66
Clausenin	EtOH	216(4.14),279(4.44), 320(4.08)	88-5767-66
1,4-Naphthoquinone, 2(3)-methyl-5-methoxy-6-hydroxy-, acetate	EtOH	205(4.47),251(4.25), 355(3.72)	5-0172-66A
$C_{14}H_{12}O_5S$			
p-Benzoquinone, 2-acetyl-3-(3,4-dimethoxy-2-thienyl)-	EtOH	227(4.06),270(4.10), 500(3.65)	33-1794-66
$C_{14}H_{12}O_6$			
p-Benzoquinone, 2-acetyl-3-(3,4-dimethoxy-2-furyl)-	EtOH	245(4.23),472(3.62)	33-1794-66
$C_{14}H_{12}S$			
Naphtho[c]thiophene, 1,3-dimethyl-	EtOH	229(4.3),254(4.43), 265(4.5),272(4.45), 280s(4.4),335(3.9), 350s(3.75)	88-1953-66
4-Stilbenethiol, trans	$C_2H_4Cl_2$	324(4.43)	99-0089-66
Styrene sulfide, cis	EtOH	226(4.03),268(2.90)	44-3976-66
Styrene sulfide, trans	EtOH	238(4.23),270(3.15)	44-3976-66
$C_{14}H_{13}^+$			
Methyldiphenylcarbonium ion	FSO_3H-SbF_5	312(4.04),422(4.57)	35-1488-66
$C_{14}H_{13}BrN_2O_2$			
2-Formyl-1-phenacylpyridinium bromide, 2-oxime	EtOH	237(4.05),301(3.89)	44-0941-66
$C_{14}H_{13}BrN_4O$			
3-Amino-2-phenacyl-s-triazolo-[4,3-a]pyridinium bromide	MeOH	245(4.29),275(3.66)	44-0265-66

Compound	Solvent	λ_{max}(log ε)	Ref.
$C_{14}H_{13}BrN_4O_8$			
1-(3-Bromopropyl)-3,4-dihydro-4-oxopyridinium picrate	H_2O	260(4.42),353(4.12)	87-0638-66
$C_{14}H_{13}BrN_6O$			
Adenine, 9-p-bromoacetamidobenzyl-	EtOH-N HCl	260(4.40)	87-0576-66
	EtOH	260(4.39)	87-0576-66
	EtOH-N NaOH	261(4.44)	87-0576-66
$C_{14}H_{13}Cl$			
1,4-Methanonaphthalene, 6-chloro-1,4-dihydro-9-isopropylidene-	heptane	186(4.45),201(4.52), 219(4.42),251(3.24), 260(3.22),266(3.23), 274(3.32),280(3.46), 288(3.47)	44-1988-66
$C_{14}H_{13}ClN_2$			
6-Aminomorphanthridine hydrochloride	EtOH	224s(4.2),247(4.02), 284(3.86)	4-0206-66
$C_{14}H_{13}ClN_2O$			
Benzimidazole, 1-methoxy-2-phenyl-, hydrochloride	EtOH	240(4.28),310(4.40)	4-0051-66
Benzimidazole, 3-methyl-2-phenyl-, 1-oxide, hydrochloride	H_2O	239(4.44),287(4.21)	4-0051-66
	EtOH	239(4.47),286(4.22)	4-0051-66
$C_{14}H_{13}ClO_3$			
2-Cyclohexen-1-one, 4-(p-chlorobenzoyl)-3-methoxy-	EtOH	255(4.52)	70-2097-66
2-Cyclohexen-1-one, 6-(p-chlorobenzoyl)-3-methoxy-	EtOH	254.5(4.42)	70-2097-66
$C_{14}H_{13}Cl_2N$			
Aniline, N-[4-(dichloromethyl)-4-methyl-2,5-cyclohexadien-1-ylidene]-	MeOH	251(4.37),343(3.66)	44-3178-66
$C_{14}H_{13}F$			
1,4-Methanonaphthalene, 6-fluoro-1,4-dihydro-9-isopropylidene-	heptane	180(4.46),198(4.48), 210(4.36),216s(4.31), 248(2.98),278(3.54), 285(3.53)	44-1988-66
$C_{14}H_{13}N$			
Methylamine, N-(diphenylmethylene)-	iso-PrOH	244(4.03)	88-5273-66
4-Stilbazole, 3-methyl-	EtOH	303(4.37)	44-4322-66
4-Stilbenamine, trans	hexane	230(4.24),320(4.53)	99-0089-66
	EtOH	234(4.20),318(4.39)	99-0089-66
$C_{14}H_{13}NO$			
Acetophenone, p-(4-aminophenyl)-	EtOH-HCl	277(4.40)	56-0429-66
	EtOH	237(4.10),332(4.29)	56-0429-66
Benzophenone, 2-methyl-, oxime, anti	EtOH	240(4.1)	9-0363-66
Benzophenone, 2-methyl-, oxime, syn	EtOH	250(4.1)	9-0363-66
Carbazole, 3-methoxy-6-methyl-	EtOH	229(4.48),244s(4.25), 255(4.10),266(3.97), 300s(4.14),308(4.28), 349(3.51),362(3.48)	77-0272-66
Glycozoline	n.s.g.	227(4.52),252(4.16), 264(4.60),304(4.17)	88-0661-66

Compound	Solvent	λ_{max}(log ϵ)	Ref.
2-Pyrrolidinone, 1-(1-naphthyl)-	MeOH	206(5.6),223(5.7), 280(4.4)	24-1918-66
$C_{14}H_{13}NO_2$			
Benzo[h]quinolin-4(1H)-one, 2,3-dihydro-8-methoxy-	EtOH	220(4.43),240s(4.18), 283(4.56),386(3.82)	54-0671-66
$C_{14}H_{13}NO_2S$			
2-Thenylidene imine, N-(p-carb-ethoxyphenyl)-	EtOH	273(4.22),332(4.27)	12-1747-66
$C_{14}H_{13}NO_3$			
2-Naphthonitrile, 3-acetoxy-1,4-dihydro-8-methoxy-	EtOH	202(4.46),275(3.32), 281(3.37)	44-1747-66
Nicotinic acid, 2-hydroxy-6-methyl-5-phenyl-, methyl ester	n.s.g.	251(4.26),347(4.03)	22-2387-66
$C_{14}H_{13}NO_4$			
Skimmianine	EtOH	250(4.86),320(3.98), 335(3.97)	78-3245-66
$C_{14}H_{13}NO_5$			
Nicotinic acid, 1,2-dihydro-4,6-dihydroxy-2-oxo-1-o-tolyl-, methyl ester	EtOH	303(4.30)	78-0455-66
Nicotinic acid, 1,2-dihydro-4,6-dihydroxy-2-oxo-1-p-tolyl-, methyl ester	EtOH	304(4.32)	78-0455-66
$C_{14}H_{13}NO_6$			
Nicotinic acid, 1,2-dihydro-4,6-dihydroxy-1-(p-methoxyphenyl)-2-oxo-, methyl ester	EtOH	305(4.33)	78-0455-66
$C_{14}H_{13}N_3$			
Benzimidazole, 2-(2-amino-4-methylphenyl)-	EtOH	250(4.20),290(4.14), 298(4.14),344(4.03)	4-0289-66
Benzimidazole, 2-(o-aminophenyl)-5-methyl-	EtOH	251(4.16),292(4.04), 299(4.08),339(4.11)	4-0289-66
Indazole, 1-(p-aminophenyl)-3-methyl-	EtOH	216(4.22),258(4.08), 303(3.75)	78-3131-66
$C_{14}H_{13}N_3O$			
Acetanilide, 2'-(phenylazo)-	hexane	224(4.18),272(4.18), 325(4.07),380(3.82), 460s(2.97)	19-0029-66
	MeOH	231(4.16),269(4.10), 324(3.98),380s(3.62), 460s(2.88)	19-0029-66
	70% MeOH	230(4.15),271(4.10), 327(4.03),380s(3.63), 460s(2.86)	19-0029-66
1-Butanone, 1-(v-triazolo[1,5-a]-quinolin-3-yl)-	MeOH	229(4.18),253(4.23), 262s(4.05),281s(3.87), 291(4.08),302(4.18), 323(4.21),336(4.14)	24-2918-66
1-Propanone, 2-methyl-1-(v-triazolo-[1,5-a]quinolin-3-yl)-	MeOH	229(4.16),253(4.22), 262s(4.04),281s(3.85), 290(4.07),302(4.17), 323(4.20),336(4.13)	24-2918-66

Compound	Solvent	λ max(log ε)	Ref.
1H-Pyrazolo[4,3-c]pyridin-4(5H)-one, 3,6-dimethyl-1-phenyl-	EtOH	222(4.36),239(4.39), 285(4.08)	70-1723-66
$C_{14}H_{13}N_3O_2S$			
Benzimidazole, 2-[p-(methyl-sulfamoyl)phenyl]-	EtOH	248(4.08),311(4.45), 314(4.45)	95-0600-66
$C_{14}H_{13}N_3O_5$			
Glutamic acid, N-(2-quinoxalinyl-carbonyl)-	EtOH	207(3.33),244(3.57), 320(2.74),328(2.73)	87-0266-66
$C_{14}H_{13}N_3O_5S$			
Benzenesulfonamide, 4-acetamido-N-(p-nitrophenyl)-	EtOH	262(4.41),312(4.13)	30-0114-66A
$C_{14}H_{13}N_5O$			
Propiophenone, β-(9-adenyl)-	EtOH	212(4.28),249(4.32)	23-1872-66
$C_{14}H_{13}OP$			
9-Phosphabicyclo[4.2.1]nonatriene, 9-phenyl-, oxide	EtOH	267(3.30),275(3.35), 286(3.30)	35-3832-66
isomer	EtOH	259(3.51),267(3.51), 273(3.50)	35-3832-66
$C_{14}H_{13}P$			
9-Phosphabicyclo[4.2.1]nonatriene, 9-phenyl-	EtOH	286(3.46)	35-3832-66
stereoisomer	EtOH	240(3.54),270s(3.23)	35-3832-66
9-Phosphabicyclo[6.1.0]nonatriene, 9-phenyl-	EtOH	235(3.56),274s(3.30)	35-3832-66
$C_{14}H_{14}$			
Bibenzyl	EtOH	215s(4.09),259(3.65)	44-3976-66
1,4-Methanonaphthalene, 1,4-dihydro-9-isopropylidene-	heptane	182(4.52),197(4.48), 209(4.41),216(4.47), 221s(4.38),267(3.23), 273(3.39),281(3.42)	44-1988-66
Octalene, 3,4-dihydro-	n.s.g.	240(4.09)	35-3677-66
Phenanthrene, 1,2,3,4-tetrahydro-	EtOH	226s(4.82),230(4.96), 274(3.73),280(3.80), 292(3.70),307(2.96), 314(2.77),322(2.93)	22-0228-66
$C_{14}H_{14}Br_4$			
Cyclopentadiene, 1,2,3,4-tetrabromo-5-(2,3-dipropylcyclopropenylidene)-	C_6H_{12}	320(4.70)	77-0180-66
	MeCN	318(4.60)	77-0180-66
$C_{14}H_{14}Cl_4$			
Cyclopentadiene, 1,2,3,4-tetrachloro-5-(2,3-dipropylcyclopropenylidene)-	C_6H_{12}	316.5(4.42)	77-0180-66
	MeCN	314(4.64)	35-3359-66
	MeCN	314(4.56)	77-0180-66
$C_{14}H_{14}Fe$			
Ferrocene, 1,1'-divinyl-	EtOH	212(4.43),232(4.45), 275(4.11)	39-0251-66C
$C_{14}H_{14}HgN_8O_4$			
Mercury, bis(theophyllin-7-yl)-	neutral	274(4.26)	44-1411-66

Compound	Solvent	λ_{max}(log ε)	Ref.
$C_{14}H_{14}N_2$			
1H-Benz[f]indole, 3-(2-aminoethyl)-, hydrochloride	EtOH	218(4.25),248(4.67), 332(3.52),349(3.59), 364(3.58)	39-0477-66C
1,5-Biphenylenediamine, 3,7-dimethyl-, dihydrochloride	EtOH-HCl	246(4.81),255(4.95), 341(3.72),355(3.90)	39-1769-66C
Pyridine, 2-(N-o-tolylacetimidyl)-	EtOH	230(4.22),267(3.91), 319(3.09)	35-3781-66
$C_{14}H_{14}N_2O$			
7-Azaindoline, 6-methoxy-4-methyl-1-phenyl-	EtOH	276(4.20),330(4.34)	78-3233-66
Benzoic acid, 2,6-dimethyl-, hydrazide	MeOH	247(3.73)	88-1063-66
Nicotinamide, N,N-dimethyl-5-phenyl-	n.s.g.	249(4.17)	22-2387-66
Nitrone, N-o-tolyl-α-methyl-α-2-pyridyl-	EtOH	302(4.00)	35-3781-66
$C_{14}H_{14}N_2O_2$			
Acetic acid, (3-methyl-4-phenylfurfurylidene)hydrazide	EtOH	306(4.44)	44-0052-66
2,3-Diaza-8-oxabicyclo[3.2.1]octa-3,6-diene, 2-acetyl-5-methyl-6-phenyl-	EtOH	278(3.74)	44-0052-66
m-Isonicotinophenetidide	EtOH	280.5(3.95)	56-0205-66
	7.1% HCl	269(3.89),317(3.60)	56-0205-66
o-Isonicotinophenetidide	EtOH	280(3.84)	56-0205-66
	7.1% HCl	269(3.91),316(3.42)	56-0205-66
p-Isonicotinophenetidide	EtOH	229(4.05),289(3.92)	56-0205-66
	7.1% HCl	232(4.12),269(3.86), 346(3.61)	56-0205-66
m-Nicotinophenetidide	EtOH	277.5(3.99)	56-0205-66
	7.1% HCl	264(4.01)	56-0205-66
o-Nicotinophenetidide	EtOH	297(3.94)	56-0205-66
	7.1% HCl	265(3.94)	56-0205-66
p-Nicotinophenetidide	EtOH	291(4.02)	56-0205-66
	7.1% HCl	263.5(3.97)	56-0205-66
m-Picolinophenetidide	EtOH	222(4.27),293(3.98)	56-0205-66
	7.1% HCl	269(3.95),308(3.80)	56-0205-66
o-Picolinophenetidide	EtOH	223(4.18),311(3.99)	56-0205-66
	7.1% HCl	270(3.97)	56-0205-66
p-Picolinophenetidide	EtOH	225(4.12),302(4.00)	56-0205-66
	7.1% HCl	226(4.09),267(3.94), 340(3.77)	56-0205-66
Pyridine, 2-(m-ethoxybenzamido)-	EtOH	288(4.18)	56-0205-66
	7.1% HCl	243(4.10),303(4.32)	56-0205-66
Pyridine, 2-(o-ethoxybenzamido)-	EtOH	249(4.09),291(4.22)	56-0205-66
	7.1% HCl	248(4.06),303(4.28)	56-0205-66
Pyridine, 2-(p-ethoxybenzamido)-	EtOH	291(4.40)	56-0205-66
	7.1% HCl	230(4.04),246(3.99), 314(4.50)	56-0205-66
Pyridine, 3-(m-ethoxybenzamido)-	EtOH	212(4.45),265(4.15)	56-0205-66
	7.1% HCl	215(4.44),262(4.24)	56-0205-66
Pyridine, 3-(o-ethoxybenzamido)-	EtOH	271(4.13)	56-0205-66
	7.1% HCl	265(4.20)	56-0205-66
Pyridine, 3-(p-ethoxybenzamido)-	EtOH	280.5(4.39)	56-0205-66
	7.1% HCl	215(4.27),288(4.35)	56-0205-66
Pyridine, 4-(m-ethoxybenzamido)-	EtOH	213(4.48),267(4.32)	56-0205-66
	7.1% HCl	281(4.43)	56-0205-66

Compound	Solvent	λ_{max}(log ε)	Ref.
Pyridine, 4-(o-ethoxybenzamido)-	EtOH	267.5(4.27)	56-0205-66
	7.1% HCl	283(4.39)	56-0205-66
Pyridine, 4-(p-ethoxybenzamido)-	EtOH	282(4.46)	56-0205-66
	7.1% HCl	302(4.42)	56-0205-66
$C_{14}H_{14}N_2O_3$			
1H,5H-Azocin[4,5,6-c,d]indole-4-carboxylic acid, 3,4,6,7-tetrahydro-6-oxo-, methyl ester	n.s.g.	280(3.74)	35-3941-66
$C_{14}H_{14}N_2O_3S$			
Benzenesulfonamide, 4-acetamido-N-phenyl-	EtOH	227(4.00),265(4.33)	30-0114-66A
$C_{14}H_{14}N_2O_4$			
4-Pyridazineacetic acid, 1,6-dihydro-3-hydroxy-6-oxo-1-phenyl-, ethyl ester	EtOH	229(4.13),290(3.39)	49-0644-66
3(2H)-Pyridazinone, 2-(benzoyloxy)-6-ethoxy-5-methyl-	EtOH	232(4.30),286(3.53), 303(3.52)	95-0082-66
3(2H)-Pyridazinone, 6-ethoxy-2-(phenacyloxy)-	EtOH	244(4.119),293s(--), 316(3.455)	95-0082-66
3(2H)-Pyridazinone, 6-methoxy-5-methyl-2-(phenacyloxy)-	EtOH	248(4.125),291(3.462)	95-0082-66
$C_{14}H_{14}N_2O_5$			
4-Oxazoleacetic acid, 2-anilino-5-carbomethoxy-, methyl ester	EtOH	245(3.96),305(4.44)	95-0300-66
$C_{14}H_{14}N_2O_6$			
Glutamic acid, 5-ester with N-(hydroxymethyl)phthalimide	H_2O	299(3.43)	10-0742-66A
Indan, 1-acetoxyimino-2-acetoxymethyl-2-nitro-	EtOH	260(4.15),298(3.67), 302(3.65)	94-1408-66
Pyrrolo[1,2-a]pyrazine-6,7,8-tricarboxylic acid, 3-methyl-, trimethyl ester	MeOH	224(3.97),250(4.17), 258(4.13),265s(3.95), 331(3.97)	39-2218-66C
	MeOH-$HClO_4$	221(4.03),257s(4.20), 261(4.21),330s(3.26), 371(3.48)	39-2218-66C
Pyrrolo[1,2-b]pyridazine-5,6,7-tricarboxylic acid, 2-methyl-, trimethyl ester	MeOH	223(4.07),250(4.18), 255s(4.12),265(4.05), 285(3.71),291s(3.77), 296(3.81),307(3.83), 336(3.68)	39-2218-66C
$C_{14}H_{14}N_4O_2$			
Lumiflavine, 2-iminol-methyl ester	pH 7.0	378(4.00),442(4.08)	33-2365-66
$C_{14}H_{14}N_4O_3$			
Theophylline, 7-methoxy-8-phenyl-	pH 1	237(4.35),300(4.31)	5-0142-66A
	pH 13	237(4.35),300(4.31)	5-0142-66A
$C_{14}H_{14}N_4O_4$			
Glutamine, N^2-(2-quinoxalinylcarbonyl)-	EtOH	207(3.36),244(3.58), 318(2.85),328(2.87)	87-0266-66
$C_{14}H_{14}N_6O$			
7(8H)-Pteridinone, 2-(dimethylamino)-6-methyl-8-(2-pyridyl)-	pH -0.89	242(4.37),258(4.31), 306(3.94),344(3.89)	24-2997-66

Compound	Solvent	λ_{max}(log ϵ)	Ref.
7(8H)-Pteridinone, 2-(dimethylamino)-6-methyl-8-(2-pyridyl)- (cont.)	pH 5.0	223(4.17),251(4.26), 312(3.77),369(4.20)	24-2997-66
Purine, 9-(p-acetamidobenzyl)-6-amino-	EtOH-N HCl	254(4.41)	87-0576-66
	EtOH	254(4.40)	87-0576-66
	EtOH-N NaOH	254(4.47)	87-0576-66
$C_{14}H_{14}N_6O_2$			
Adenine, N,N-dimethyl-9-(p-nitrobenzyl)-	pH 1	271(4.45)	87-0576-66
	pH 7	277(4.45)	87-0576-66
	pH 13	276(4.45)	87-0576-66
$C_{14}H_{14}O$			
Benzhydrol, α-methyl-	hexane	253(2.54),258(2.61), 264(2.49)	22-0269-66
	C_6H_{12}	253(2.63),258(2.69), 264(2.58)	22-0269-66
5,5-Benzotricyclo[5.3.0.0^{8,10}]decan-1-one	EtOH	235(3.94)	44-1390-66
Naphtho[2,3-b]furan, 2,3-dihydro-3,9-dimethyl-	EtOH	236(4.75),270(3.60), 280(3.71),292(3.62), 324(3.47),338(3.56)	39-0725-66C
Naphtho[1,2-b]oxepin, 2,3,4,5-tetrahydro-	EtOH	229(4.70),278s(3.64), 288(3.70),293s(3.63), 308s(3.21),313s(2.94), 322(2.94)	22-3644-66
Naphtho[2,1-b]oxepin, 1,2,3,4-tetrahydro-	EtOH	231(4.89),272s(3.65), 277s(3.70),280(3.73), 290(3.63),310(3.03), 316s(2.99),324(3.14)	22-3644-66
Naphtho[2,3-b]oxepin, 2,3,4,5-tetrahydro-	EtOH	225(4.96),269(3.71), 278(3.75),289(3.56), 306(2.88),315s(2.76), 321(2.94)	22-3644-66
Phenol, 2-benzyl-6-methyl-	MeOH	274(3.28)	35-5855-66
	MeOH-base	279(3.30),293(3.04)	35-5855-66
$C_{14}H_{14}OS$			
3,5-Heptadiyn-2-ol, 2-methyl-7-(phenylthio)-	EtOH	236(3.82),247(3.82), 262(3.54)	22-3024-66
Phenol, 2-methyl-6-(phenylthiomethyl)-	MeOH	257(3.88)	35-5855-66
	MeOH-base	254(4.01)	35-5855-66
m-Tolyl sulfoxide	EtOH	202(4.63),235(4.25), 270(3.17),279(3.21)	39-0239-66A
o-Tolyl sulfoxide	EtOH	202(4.63),235(4.22), 268(3.21),275(3.22)	39-0239-66A
p-Tolyl sulfoxide	EtOH	202(4.63),240(4.30)	39-0239-66A
$C_{14}H_{14}O_2$			
Bibenzyl-4,4'-diol	50% dioxan	278(3.56)	71-0550-66A
3,5-Heptadiyn-2-ol, 2-methyl-7-phenoxy-	EtOH	244(3.00),263(3.08), 270(3.18),276(3.11)	22-3024-66
2-Naphthaleneacetic acid, α-methyl-, methyl ester	EtOH	225(5.06),270(3.74), 275(3.76)	77-0485-66
2,4-Octadiyne-1,6-diol, 1-phenyl-	EtOH	252(3.58),258(3.66), 264(3.64)	70-0833-66
1-Oxaspiro[4.4]non-2-en-4-one, 2-phenyl-	EtOH	220(3.90),244(3.88), 304(4.18)	88-0233-66
[8]Paracyclo-4-phene, 3,6-dioxo-, cis	EtOH	219(4.06),250s(3.48), 280s(2.85)	35-0515-66

Compound	Solvent	$\lambda_{max}(\log \epsilon)$	Ref.
[8]Paracyclo-4-phene, 3,6-dioxo-, trans	EtOH	223(3.87),249(3.52), 300s(2.80)	35-0515-66
Sorbic acid, 4-hydroxy-2,3-dimethyl-6-phenyl-, γ-lactone	n.s.g.	274(4.3)	54-0043-66
6-Tetradecene-8,10,12-triyn-5-one, 1-hydroxy-	ether	250(4.56),260(4.66), 285(3.83),303(4.15), 324(4.34),347(4.24)	24-0990-66
$C_{14}H_{14}O_2S$			
m-Tolyl sulfone	EtOH	204(4.60),240(4.19), 273(3.44),281(3.30)	39-0239-66A
o-Tolyl sulfone	EtOH	202(4.62),233(4.19), 272(3.40),280(3.38)	39-0239-66A
p-Tolyl sulfone	EtOH	202(4.60),222(4.07), 245(4.32),275(3.04)	39-0239-66A
$C_{14}H_{14}O_3$			
p-Benzoquinone, 2-methyl-5-(3,4,5-trimethyl-2-furyl)-	EtOH	264(4.13),310s(3.52), 500(3.68)	33-1794-66
2-Cyclohexen-1-one, 4-benzoyl-3-methoxy-	EtOH	250(4.46)	70-2097-66
2-Cyclohexen-1-one, 6-benzoyl-3-methoxy-	EtOH	251(4.47)	70-2097-66
Kawain, dihydro-5,6-dehydro-	EtOH	284(3.85)	95-1184-66
$C_{14}H_{14}O_4$			
5H-Benzocyclohepten-5-one, 3,4,6-trimethoxy-	EtOH	240(4.08),278(4.62), 325(3.82),340(3.83), 385(3.78)	39-2072-66C
2,3-Ethenoindan-1-one, 2,5,6-trimethoxy-	EtOH	242(4.17),282(3.85), 328(3.85)	39-2072-66C
Obliquin, dihydro-	MeOH	211(4.26),229(4.03), 295(3.57),342(3.74)	39-0114-66C
$C_{14}H_{14}O_4S_2$			
p-Tolyl disulfone	n.s.g.	250(4.36)	44-3418-66
$C_{14}H_{14}O_5$			
2-Benzopyran-1-one, 4-acetyl-6,8-dimethoxy-5-methyl-	EtOH	237(4.42),335(3.72)	39-0126-66C
2-Naphthaleneacetic acid, α,3-dihydroxy-8-methoxy-, methyl ester	EtOH	223(4.69),250(4.51), 279(3.58),288(3.63), 298(3.54),326(3.27), 336(3.32)	44-1747-66
$C_{14}H_{14}O_6$			
[2,2'-Bifuran]-5,5'-dicarboxylic acid, 4,4'-dimethyl-, dimethyl ester	$CHCl_3$	239(3.85),326(4.61), 341(4.51)	39-0976-66C
Hydroquinone, 2-acetyl-3-(4,5-dimethoxy-2-furyl)-	EtOH	327(3.62),370(3.61)	33-1794-66
1-Phthalanacrylic acid, 5,6-dimethoxy-3-oxo-, methyl ester	EtOH	225(4.30),255(4.01), 296(3.86)	39-2072-66C
$\Delta^{1,\beta}$-Phthalanpropionic acid, 5,6-dimethoxy-3-oxo-, methyl ester	EtOH	255(4.23),298(3.77)	39-2072-66C
	base	225(--),275(--)	39-2072-66C
$C_{14}H_{14}O_6S_2$			
Anisole, 4,4'-disulfonyldi-	n.s.g.	277(4.45)	44-3418-66

Compound	Solvent	λ_{max}(log ε)	Ref.
$C_{14}H_{14}O_8$			
1,4-Naphthoquinone, 2,3-dihydroxy-5,6,7,8-tetramethoxy-	EtOH	270(4.00),307(3.36),376(3.15)	39-0426-66C
$C_{14}H_{14}S$			
Benzo[f]thiachroman, 4-methyl-	EtOH	254(4.75),278(3.84),288(3.94),299(3.84),332(3.18),347(3.07)	78-0007-66
Benzo[h]thiachroman, 4-methyl-	EtOH	244(4.34),308(3.85),318s(3.80)	78-0007-66
Benzo[b]thiophene, 2-(1-cyclohexen-1-yl)-	EtOH	234(4.36),290(4.43)	87-0551-66
Benzyl sulfide	EtOH	260(2.85)	22-0228-66
Naphtho[1,2-b]thiepin, 2,3,4,5-tetrahydro-	EtOH	222s(4.65),228(4.70),233s(4.68),302(3.87)	22-0228-66
	EtOH	222s(4.65),228(4.70),233s(4.68),302(3.87)	22-3644-66
Naphtho[2,1-b]thiepin, 1,2,3,4-tetrahydro-	EtOH	223(4.64),255(4.23),280(3.79),290(3.83),302(3.71),331(2.84)	22-0228-66 +22-3644-66
Naphtho[2,3-b]thiepin, 2,3,4,5-tetrahydro-	EtOH	226(4.95),262(4.24),280(3.82),290(3.87),302(3.73),331(2.75)	22-0228-66
	EtOH	226(4.25),262(4.24),280(3.82),290(3.87),302(3.73),326(2.75)	22-3644-66
2H-Naphtho[1,2-b]thiopyran, 3,4-dihydro-6-methyl-	EtOH	220(4.67),250(4.24),318(3.82),335s(3.76)	22-0228-66
m-Tolyl sulfide	EtOH	201(4.64),253(4.17),277(3.72)	39-0239-66A
o-Tolyl sulfide	EtOH	201(4.61),250(4.09),280(3.67)	39-0239-66A
p-Tolyl sulfide	EtOH	201(4.64),233s(4.3),252(4.29),276(3.98)	39-0239-66A
$C_{14}H_{15}^+$			
Methyl(2-methylpropenyl)phenylethynylcarbonium ion	FSO_3H-SbF_5	312(4.12),389(4.25),461(4.45)	35-1488-66
$C_{14}H_{15}Cl$			
1,4-Methanonaphthalene, 6-chloro-1,2,3,4-tetrahydro-9-isopropylidene-	heptane	199(4.77),269(3.25),274(3.40),281(3.46)	44-1988-66
$C_{14}H_{15}ClN_2$			
1H-Benz[f]indole, 3-(2-aminoethyl)-, hydrochloride	EtOH	218(4.25),248(4.67),332(3.52),349(3.59),364(3.58)	39-0477-66C
$C_{14}H_{15}ClN_4O$			
Acetamide, N-[(4-amino-2-methyl-5-pyrimidinyl)methyl]-2-chloro-2-phenyl-	EtOH	229(4.09),274(3.73)	44-2951-66
$C_{14}H_{15}ClN_6O$			
Purine, 9-(p-acetamidobenzyl)-6-amino-, hydrochloride	EtOH	254(4.40)	87-0576-66
$C_{14}H_{15}ClO$			
Cyclohexanone, 2-(o-chlorobenzylidene)-5-methyl-	EtOH	278(4.09)	35-1419-66

Compound	Solvent	λ_{max}(log ε)	Ref.
Cyclohexanone, 2-(p-chlorobenzylidene)-5-methyl-	EtOH	292(4.25)	35-1419-66
$C_{14}H_{15}F$			
1,4-Methanonaphthalene, 6-fluoro-1,2,3,4-tetrahydro-9-isopropylidene-	heptane	192(4.77),265s(3.38),269(3.49),273(3.54),278(3.52)	44-1988-66
$C_{14}H_{15}N$			
6H-Cyclohepta[b]quinoline, 7,8,9,10-tetrahydro-	C_6H_{12}	272f(3.68)	5-0149-66H
	MeOH	271f(3.60)	5-0149-66H
	MeOH-H_2SO_4	321(4.02)	5-0149-66H
$C_{14}H_{15}NO$			
1,2-Benzisoxazole, 4,5,6,7-tetrahydro-5-methyl-3-phenyl-	EtOH	240(4.09)	32-0559-66
2H-Benzo[a]quinolizin-2-one, 1,6,7,11b-tetrahydro-4-methyl-	EtOH	319(4.24)	44-0797-66
Cyclohexanone, 2-indol-2-yl-	MeOH	273s(3.9),281(3.9),289(3.8)	24-2504-66
10H-Dibenz[b,f]azepin-10-one, 5,6,7,8,9,11-hexahydro-	MeOH	218(4.08),235(4.12),304(3.68),336(3.95)	24-2504-66
$C_{14}H_{15}NO_2$			
Acetamide, N-(3-methyl-4-phenylfurfuryl)-	EtOH	245(4.18)	44-0052-66
Carbazole, 1,2,3,4-tetrahydro-6-methoxy-3-methyl-1-oxo-	n.s.g.	232(4.23),315(4.39)	25-1684-66
Cinnamic acid, α-cyano-p,β-dimethyl-, ethyl ester, cis	EtOH	292.5(3.93)	22-1033-66
Cinnamic acid, α-cyano-β-ethyl-, ethyl ester, cis	EtOH	276.5(3.86)	22-1033-66
2H-Pyrano[2,3-b]quinolin-5-one, 3,4,5,10-tetrahydro-2,2-dimethyl-	MeOH	234(4.57),310(4.02)	39-1084-66C
2H-Pyrano[3,2-c]quinolin-5-one, 3,4,5,6-tetrahydro-2,2-dimethyl-	MeOH	228(4.75),282(4.00),313(3.98)	39-1084-66C
Pyrrolo[3,2,1-hi]indole-1-carboxylic acid, 4,5-dihydro-2-methyl-, ethyl ester	MeOH	220(4.58),240(4.35),297(4.15)	35-4061-66
$C_{14}H_{15}NO_3$			
Azepino[3,2,1-hi]indole-7-carboxylic acid, 1,2,4,5,6,7-hexahydro-6-oxo-, methyl ester	EtOH	218(4.22),253(4.03)	35-4061-66
	EtOH-NaOH	245(4.20),285(4.07)	35-4061-66
Cinnamic acid, α-cyano-p-methoxy-β-methyl-, ethyl ester, cis	EtOH	316.5(4.02)	22-1033-66
Indole-3-carboxaldehyde, 1-ethyl-4,7-dihydro-2,5,6-trimethyl-4,7-dioxo-	MeOH	216(4.10),247(3.87),257(3.86),278(3.82),281(3.77),338(3.43)	44-1012-66
Indole-3-carboxaldehyde, 4-acetoxy-1-ethyl-2-methyl-	MeOH	252(4.36),270(4.18),343(3.94)	35-0804-66
Isoquinoline, 3-acetyl-6,7-dimethoxy-1-methyl-	EtOH	259(4.56),311(4.02),322(3.97)	95-0072-66
4-Piperidone, 3-acetyl-N-benzoyl-	EtOH	283(3.91)	44-0797-66
$C_{14}H_{15}NO_4$			
1,3-Cyclohexanedione, 5,5-dimethyl-2-(p-nitrophenyl)-	EtOH-base	245s(3.88),290(4.24),400(3.88)	30-0827-66B
4,7-Indoloquinone, 3-(acetoxymethyl)-1-ethyl-2-methyl-	MeOH	230(4.19),255(4.07),345(3.42),437(3.01)	35-0804-66

Compound	Solvent	$\lambda_{max}(\log \epsilon)$	Ref.
1-Isoquinolinecarboxylic acid, 6,7-dimethoxy-, ethyl ester, picrate	EtOH	230(4.45),247(4.48), 340(3.96),400(4.01)	95-0973-66
$C_{14}H_{15}NO_5$			
Quinaldic acid, 4-hydroxy-5,8-dimethoxy-, ethyl ester	EtOH	235(4.48),257(3.86), 329(3.90),359(3.75)	24-0160-66
$C_{14}H_{15}NO_6$			
Furo[2,3-b]pyridine-2,5-dicarboxylic acid, 4-methoxy-, diethyl ester	5% EtOH	240(4.46)	4-0202-66
Quinaldic acid, 4,5-dihydroxy-6,7-dimethoxy-, ethyl ester	EtOH	224(4.46),247(4.19), 268(4.27),328(3.60), 362(3.57)	24-0160-66
Quinaldic acid, 4,5-dihydroxy-6,8-dimethoxy-, ethyl ester	EtOH	217(4.32),238(4.29), 265(4.30),344(3.81), 415(3.57)	24-0160-66
Quinaldic acid, 4,5-dihydroxy-7,8-dimethoxy-, ethyl ester	EtOH	215(4.52),237(4.28), 264(4.29),293(4.21), 372(3.51)	24-0160-66
3-Quinolinecarboxylic acid, 4,5-dihydroxy-6,8-dimethoxy-, ethyl ester	EtOH	226(4.53),353(4.12)	24-0160-66
$C_{14}H_{15}NO_8$			
Malonic acid, [3-(5-nitro-2-furyl)-acryloyl]-, diethyl ester	EtOH	246(3.95),304(4.19) 377(4.38)	95-0617-66
$C_{14}H_{15}NS$			
4H-Indeno[2,1-d]thiazole, 2-tert-butyl-	MeOH	227s(3.77),241s(3.56), 295(4.20)	39-0686-66C
$C_{14}H_{15}N_3$			
7-Azaindoline, 6-amino-4-methyl-1-phenyl-	EtOH	280(4.15),338(4.24)	78-3233-66
Azobenzene, p-(dimethylamino)-	MeCN	410(4.46)	49-0171-66
	$C_2H_4Cl_2$	410(4.46)	49-0171-66
$AlBr_3$ complex	MeCN	325(3.57),515(4.75)	49-0171-66
	$C_2H_4Cl_2$	330(3.36),530(4.76)	49-0171-66
$AlCl_3$ complex	MeCN	325(3.57),515(4.75)	49-0171-66
	$C_2H_4Cl_2$	330(3.36),530(4.76)	49-0171-66
$GaBr_3$ complex	MeCN	325(3.57),515(4.75)	49-0171-66
	$C_2H_4Cl_2$	330(3.36),530(4.76)	49-0171-66
$GaCl_3$ complex	MeCN	325(3.57),515(4.75)	49-0171-66
	$C_2H_4Cl_2$	330(3.36),530(4.76)	49-0171-66
$SbCl_5$ complex	MeCN	325(3.57),515(4.75)	49-0171-66
	$C_2H_4Cl_2$	330(3.36),530(4.76)	49-0171-66
$2SbCl_5$ complex	MeCN	405(4.46)	49-0171-66
	$C_2H_4Cl_2$	430(4.48)	49-0171-66
$SnCl_4$ complex	MeCN	325(3.57),515(4.75)	49-0171-66
	$C_2H_4Cl_2$	330(3.36),530(4.76)	49-0171-66
$C_{14}H_{15}N_3O_2$			
Benzoic acid, p-(2-pyrazinylmethyl)-amino-, ethyl ester	EtOH	275s(4.08),299(4.35)	87-0868-66
$C_{14}H_{15}N_3O_2S$			
Phenothiazine, 2,8-bis(methylamino)-, 5,5-dioxide	EtOH	267(4.97),311(4.00)	87-0960-66
$C_{14}H_{15}N_3O_3$			
Norvaline, N-(2-quinoxalinylcarbonyl)-	EtOH	207(3.23),244(3.46), 319(2.64),327(2.67)	87-0266-66

Compound	Solvent	$\lambda_{max}(\log \epsilon)$	Ref.
1H-Pyrrolo[1,2-c]imidazole-1,3,5(2H,6H)-trione, 2-(anilinomethyl)-7,7a-dihydro-7a-methyl-	MeOH-HCl	240(3.90),282(2.90)	87-0142-66
	MeOH	241(4.06),287(3.24)	87-0142-66
	MeOH-KOH	238(4.10),287(3.25)	87-0142-66
Valine, N-(2-quinoxalinylcarbonyl)-	EtOH	207(3.37),244(3.70), 317(2.95),327(2.97)	87-0266-66
$C_{14}H_{15}N_3O_3S$			
p-Azobenzenesulfonic acid, p-(dimethylaminoamino)-, anion	0.1N H_2SO_4	316(3.18),508(4.61)	35-2240-66
	H_2O	463(4.34)	35-2240-66
	86.5% H_2SO_4	407(4.50)	35-2240-66
Methionine, N-(2-quinoxalinylcarbonyl)-	EtOH	206(3.36),244(3.57), 320(2.81),327(2.84)	87-0266-66
$C_{14}H_{15}N_3S$			
Picolinamide, N-(p-dimethylamino)-phenyl-thio-	heptane	252(4.22),302(4.15), 406(4.04)	65-1499-66
	EtOH	258(4.23),304(4.10), 402(3.97)	65-1499-66
	$CHCl_3$	272(4.21),306(4.10), 414(3.96)	65-1499-66
	DMF	270(4.22),308(4.10), 410(4.02)	65-1499-66
	H_2SO_4	278(4.13),312(3.86)	65-1499-66
$C_{14}H_{15}N_5$			
Adenine, 3-benzyl-N,N-dimethyl-	neutral	277s(4.10),295(4.19)	44-1411-66
Adenine, 7-benzyl-N,N-dimethyl-	neutral	293(4.07)	44-1411-66
Adenine, 9-benzyl-N,N-dimethyl-	neutral	277(4.29)	44-1411-66
$C_{14}H_{15}N_5O$			
4(3H)-Pteridinone, 2-amino-7,8-dihydro-7,7-dimethyl-6-phenyl-	pH 1.0	261(4.22),290s(3.81), 381(4.06)	24-3008-66
	pH 7.0	233(4.36),282(4.04), 337(3.98)	24-3008-66
	pH 14.0	240(4.23),280(3.81), 342(4.01)	24-3008-66
9H-Purine-9-propanol, 6-amino-α-phenyl-	EtOH	215(--),262(4.28)	23-1872-66
$C_{14}H_{15}N_5O_6$			
1H-Pyrrolo[2,3-d]pyrimidine-2,4,6(3H)-trione, 5,7-dihydro-5-(1,2,3,4-tetrahydro-6-hydroxy-1,3-dimethyl-2,4-dioxo-5-pyrimidinyl)-1,3-dimethyl-	pH -0.89	289(3.90)	24-3524-66
	pH 6	262(4.30),287(3.94)	24-3524-66
	pH 12	222(4.40),264(4.25), 309(4.08)	24-3524-66
$C_{14}H_{15}OP$			
Phosphine oxide, ethyldiphenyl-	EtOH	265.0(3.14)	77-0425-66
$C_{14}H_{15}P$			
Phosphine, ethyldiphenyl-	EtOH	251(3.91),278(2.98)	77-0425-66
$C_{14}H_{16}$			
1,4-Methanonaphthalene, 1,2,3,4-tetrahydro-9-isopropylidene-	heptane	195(4.74),262(3.14), 268(3.32),274(3.37)	44-1988-66
Naphthalene, 1,2,3,5-tetramethyl-	EtOH	233(4.98),280(3.77), 290(3.84)	39-1892-66C
Naphthalene, 1,2,3,8-tetramethyl-	EtOH	234(4.92),282(3.73), 290(3.77)	39-1892-66C

Compound	Solvent	λ_{max}(log ε)	Ref.
$C_{14}H_{16}BN$			
Borane, benzylmethylamino-di(1-propynyl)-	C_6H_{12}	210(4.54),250(4.51)	22-3850-66
$C_{14}H_{16}BrNO$			
1-(β-Hydroxyphenethyl)-2-picolinium bromide	MeOH	270(3.8),350(1.8)	83-0385-66
$C_{14}H_{16}ClN_3O_4$			
Imidazo[4,5-c]pyridine, 4-chloro-1-(2,3-O-isopropylidene-α-D-ribofuranosyl)-	pH 1	257s(3.63),267(3.81), 273(3.81)	69-0756-66
	pH 11	251(3.84),265(3.74), 272(3.66)	69-0756-66
β-isomer	pH 1	255(3.77),267(3.81), 272(3.81)	69-0756-66
	pH 11	253(3.89),264(3.81), 272(3.72)	69-0756-66
$C_{14}H_{16}Cl_2N_2$			
1,5-Biphenylenediamine, 3,7-dimethyl-, dihydrochloride	EtOH	246(4.81),255(4.95), 341(3.72),355(3.90)	39-1769-66C
$C_{14}H_{16}Cl_2N_2Ni$			
Nickel, dichloro(4,4',6,6'-tetramethyl-2,2'-bipyridine)-	$CHCl_3$	489(2.16),610(1.08), 830(1.51),975(1.88)	12-0201-66
$C_{14}H_{16}Cl_2N_2O$			
5-Pyrrolin-2-one, 4,4-dichloro-3-ethyl-5-(ethylamino)-3-phenyl-	EtOH	250(4.20)	28C-1276-66A
$C_{14}H_{16}N_2$			
Pyridine, N-[(o-tolyl)-2-aminoethyl]-	EtOH	241(4.06),285(3.33)	35-3781-66
$C_{14}H_{16}N_2O$			
4-Isoquinolinecarboxamide, N,N-diethyl-	EtOH	273(3.64),308(3.51), 321(3.65)	22-1763-66
2-Pyrazolin-5-one, 4-allyl-3,4-dimethyl-1-phenyl-	EtOH	242(4.21)	88-6413-66
3-Pyrazolin-5-one, 2-allyl-3,4-dimethyl-1-phenyl-	EtOH	247(4.01),275(3.96)	88-6413-66
2-Pyrrolidinone, 5-imino-4-isopropylidene-3-methyl-3-phenyl-	EtOH	260(4.23)	22-3895-66
$C_{14}H_{16}N_2OS$			
β-Carboline, 1,2,3,4-tetrahydro-4,4-dimethyl-6-(methylthio)-1-oxo-	EtOH	233(4.20)	23-0307-66
Isothiazole, 3-morpholino-5-p-tolyl-	EtOH	244s(4.06),289(4.41), 390s(3.51)	32-1009-66
$C_{14}H_{16}N_2O_2$			
Acetic acid, 2-(3-methyl-4-phenylfurfuryl)hydrazide	EtOH	221(4.32)	44-0052-66
1,2-Diazabicyclo[3.2.0]hept-3-en-6-ol, 2-acetyl-5-methyl-4-phenyl-	EtOH	310(4.34)	44-0052-66
1H-1,2-Diazepin-6-ol, 1-acetyl-6,7-dihydro-5-methyl-4-phenyl-	EtOH	308(3.85)	44-0052-66
8-Oxa-2,3-diazabicyclo[3.2.1]oct-6-ene, 2-acetyl-6-methyl-7-phenyl-	EtOH	253(4.16)	44-0052-66
m-Phenylenediamine, N,N'-bis(2-acetylvinyl)-	EtOH	352(4.61)	18-1079-66

Compound	Solvent	$\lambda_{max}(\log \epsilon)$	Ref.
p-Phenylenediamine, N,N'-bis-(2-acetylvinyl)-	EtOH	386(4.67)	18-1079-66
$C_{14}H_{16}N_2O_2S$			
Isothiazole, 5-(p-methoxyphenyl)-3-morpholino-	EtOH	248s(4.07),284(4.39)	32-1009-66
Sulfanilamide, N^4,N^4-dimethyl-N^1-phenyl-	EtOH	282(4.39)	30-0114-66A
$C_{14}H_{16}N_2O_2S_2$			
5-Thiazolidineacetamide, N-benzyl-3-ethyl-4-oxo-2-thioxo-	EtOH	262(4.09),297(4.19)	95-0101-66
5-Thiazolidineacetamide, 3-benzyl-N-ethyl-4-oxo-2-thioxo-	EtOH	261(4.04),297(4.15)	95-0101-66
5-Thiazolidineacetamide, 3-ethyl-4-oxo-2-thioxo-N-o-tolyl-	EtOH	261(4.14),296(4.23)	95-0101-66
$C_{14}H_{16}N_2O_3$			
3(2H)-Pyridazinone, 2-(benzyloxy)-6-ethoxy-5-methyl-	EtOH	310(3.643)	95-0082-66
$C_{14}H_{16}N_2O_4$			
1H-Pyrrolo[2,3-b]pyridine-3,5-dicarboxylic acid, 1,2,6-trimethyl-, dimethyl ester	EtOH	240(4.55),299(3.96)	39-1950-66C
$C_{14}H_{16}N_4O$			
Acetamide, N-[(4-amino-2-methyl-5-pyrimidinyl)methyl]-2-phenyl-	EtOH	236(3.92),276(3.74)	44-2951-66
Acetanilide, 4'-[N-[2-(imidazol-4-yl)ethyl]formimidoyl]-	EtOH	246(3.62),280(4.37), 295(4.25)	44-2380-66
$C_{14}H_{16}N_4O_2$			
13,14-Diazatricyclo[6.4.1.1^{2,7}]tetradeca-3,5,9,11-tetraene-13,14-dicarboxamide	MeCN	234(4.21),240(4.20)	35-2591-66
$C_{14}H_{16}N_4O_4$			
2-Cycloheptenone, 3-methyl-, 2,4-dinitrophenylhydrazone	EtOH	259(4.26),383(4.50)	78-0105-66B
Cyclohexanone, 2-ethylidene-, 2,4-dinitrophenylhydrazone	n.s.g.	215(4.26),257(4.17), 380(4.36)	39-0397-66C
2-Cyclohexen-1-one, 3,5-dimethyl-, 2,4-dinitrophenylhydrazone	EtOH	390(4.37)	39-0336-66C
2-Cyclohexylideneacetaldehyde, 2,4-dinitrophenylhydrazone	$CHCl_3$	385(4.54)	22-1400-66
3,5-Heptadien-2-one, 4-methyl-, 2,4-dinitrophenylhydrazone	$CHCl_3$	345(4.32)	78-0443-66B
Pyrazole, 1-(2,4-dinitrophenyl)-3,5-diethyl-4-methyl-	EtOH	333(3.79)	22-2977-66
$C_{14}H_{16}N_6O_4S$			
Sulfanilamide, N -(2,6-dimethoxy-9-methylpurin-8-yl)-	pH 1.2	239s(3.98),287(4.36)	87-0373-66
	pH 12.9	254(4.36),288(4.30)	87-0373-66
$C_{14}H_{16}N_6O_5$			
1H-Pyrrolo[2,3-d]pyrimidine-2,4,6(3H)-trione, 5-(6-amino-1,2,3,4-tetrahydro-1,3-dimethyl-2,4-dioxopyrimidinyl)-5,7-dihydro-1,3-dimethyl-	pH 5	268(4.23),330(3.56)	24-3524-66
	pH 10	223(4.45),270(4.18), 310(3.87)	24-3524-66

Compound	Solvent	$\lambda_{max}(\log \epsilon)$	Ref.
(cont.)	MeOH	248(4.15),254(4.15), 268(4.23),328(3.65)	24-3524-66
$C_{14}H_{16}O$			
Cyclohexanone, 2-benzylidene-5-methyl-	EtOH	286(4.14)	35-1419-66
2-Cyclohexen-1-one, 3-(2-ethylphenyl)-	EtOH	229(4.13),266(4.03)	22-1012-66
2-Cyclohexen-1-one, 3-(3-ethylphenyl)-	EtOH	225(4.14),285(4.28)	22-1012-66
2-Cyclohexen-1-one, 3-(4-ethylphenyl)-	EtOH	226(4.00),295(4.30)	22-1012-66
1-Naphthalenemethanol, 2,3,4-trimethyl-	MeOH	228s(4.15),233(5.12), 272s(3.61),283s(3.78), 291(3.84),300s(3.72)	44-2009-66
$C_{14}H_{16}OS_2$			
1(2H)-Naphthalenone, 2,2-bis-(methylthiomethyl)-	MeOH	234(4.52)	35-5855-66
	MeOH-KOH	234(4.52)	35-5855-66
2(1H)-Naphthalenone, 1,1-bis-(methylthiomethyl)-	MeOH	235(3.97),310(3.81)	35-5855-66
	MeOH-KOH	235(3.97),310(3.81)	35-5855-66
$C_{14}H_{16}O_2$			
1,3-Cyclohexanedione, 5,5-dimethyl-2-phenyl-	EtOH-base	228(3.45),290(4.25)	30-0827-66B
6-Heptene-1,3-dione, 4-methyl-1-phenyl-	C_6H_{12}	247(3.79),308(4.18)	78-3607-66
	MeOH	247(3.78),309(4.22)	78-3607-66
2-Naphthaleneacetic acid, 3,4-dihydro-α-methyl-, methyl ester	EtOH	265(4.01)	77-0485-66
[8]Paracyclophane, 3,6-dioxo-	EtOH	224(3.68),265s(2.41), 270(2.50),279(2.41), 300s(1.40)	35-0515-66
4,6,12-Tetradecatriene-8,10-diyne-1,3-diol	ether	250(4.53),266(4.46), 279(4.21),296(4.47), 315(4.60),337(4.47)	24-3433-66
8,10,12-Tetradecatriynoic acid	ether	209(5.19),239(2.03), 253(2.23),267(2.37), 278(2.23),286(2.34), 296(2.05),306(2.10), 308(2.10)	24-2091-66
6-Tetradecene-8,10,12-triyne-1,5-diol	ether	231(4.81),243(5.03), 256(3.59),271(3.89), 288(4.16),307(4.28), 329(4.11)	24-0990-66
$C_{14}H_{16}O_3$			
2,4,8-Decatrien-6-ynal, 10-acetoxy-4,8-dimethyl-	pet ether	310s(4.50),325(4.57), 342(4.50)	33-0369-66
2-Hexenoic acid, 4,4-dihydroxy-2,3-dimethyl-6-phenyl-, γ-lactone	n.s.g.	216(4.1)	54-0043-66
2-Naphthaleneacetic acid, 5,6,7,8-tetrahydro-3,4-dihydroxy-1,5-dimethyl-, γ-lactone	EtOH	287(3.33)	78-0301-66
$C_{14}H_{16}O_4$			
7H-Cyclobut[a]inden-7-one, 1,2,2a,7a-tetrahydro-4,5,7a-trimethoxy-	EtOH	230(4.17),268(4.03), 312(3.98)	39-2072-66C
1-Naphthol, 6,7,8-trimethoxy-3-methyl-	n.s.g.	228(4.57),247(4.66), 305(3.64)	44-1883-66

Compound	Solvent	λ_{max}(log ε)	Ref.
Phthalic anhydride, 4-(3-ethoxy-propyl)-3-methyl-	ether	222(4.59),262(3.74), 302(3.57),311(3.59)	33-0858-66
$C_{14}H_{16}O_5$			
2-Benzopyran-1-one, 4-acetyl-6,8-di-methoxy-5-methyl-3,4-dihydro-	EtOH	266(4.03),308(3.87)	39-0126-66C
$C_{14}H_{16}O_6$			
α-D-arabino-Hexopyranosid-3-ulose, methyl 4,6-O-benzylidene-	EtOH	257(2.50)	39-0695-66C
1-Phthalanpropionic acid, 5,6-di-methoxy-3-oxo-, methyl ester	EtOH	223(4.28),258(3.91), 294(3.78)	39-2072-66C
m-Toluic acid, 2-[1-(hydroxymethylene)-acetonyl]-4,6-dimethoxy-	EtOH	225s(4.13),265(3.90), 286(3.84)	39-0126-66C
$C_{14}H_{16}S$			
Benzo[b]thiophene, 2-cyclohexyl-	EtOH	230(4.50),262(3.97)	87-0551-66
$C_{14}H_{17}Cl$			
Cyclohexane, 1-(p-chlorobenzylidene)-4-methyl-, endo	EtOH	280(3.45)	35-1419-66
exo	EtOH	252(4.23)	35-1419-66
$C_{14}H_{17}ClN_6S$			
4-Pyrimidinethiol, 2-amino-6-(butyl-amino)-5-(p-chlorophenylazo)-	EtOH	226(4.30),261(3.94), 306(4.40),427(4.26)	87-0977-66
$C_{14}H_{17}ClO_3S$			
Dimethylsulfonium 3-carboxy-2-(m-chlorophenyl)allylide, ethyl ester, oxide	EtOH	234s(3.91),272(3.78), 347(4.23)	88-1787-66
Dimethylsulfonium 3-carboxy-2-(o-chlorophenyl)allylide, ethyl ester, oxide	EtOH	344(4.37)	88-1787-66
Dimethylsulfonium 3-carboxy-2-(p-chlorophenyl)allylide, ethyl ester, oxide	EtOH	240(4.00),277s(3.98), 347(4.25)	88-1787-66
$C_{14}H_{17}N$			
Carbazole, 1-ethyl-1,2,3,4-tetrahydro-	EtOH	229(4.48),285(3.86), 292(3.83)	94-0934-66
$C_{14}H_{17}NO$			
4H-Pyrrolo[3,2,1-ij]quinolin-2-one, 1,2,5,6-tetrahydro-4,4,6-trimethyl-	MeOH	255(4.03),283(3.27)	30-0341-66E
$C_{14}H_{17}NO_2$			
Carbazole, 1,2,3,4-tetrahydro-6,7-dimethoxy-	MeOH	229(4.5),305(3.80), 343(3.08)	20-0181-66
Indole-3-carboxaldehyde, 1-ethyl-4-hydroxy-2,5,6-trimethyl-	MeOH	218(4.46),253(4.19), 278(3.92),345(3.88)	44-1012-66
Indol-4-ol, 1-ethyl-2,6-dimethyl-, acetate	MeOH	225(4.65),276(3.93)	44-1012-66
4-Piperidinone, 3-acetyl-1-benzyl-	MeOH	286(3.92)	24-3750-66
2-Pyrrolidinone, 1-benzoyl-4,5,5-trimethyl-	EtOH	231(3.97),244(3.95)	78-0015-66B
$C_{14}H_{17}NO_2S$			
L-Proline, N-phenylthioacetyl-, methyl ester	isooctane	276(4.12),325(2.64)	1-2781-66
	MeOH	275(4.12),319s(2.71)	1-2781-66
	dioxan	278(4.17),330(2.71)	1-2781-66

Compound	Solvent	$\lambda_{max}(\log \epsilon)$	Ref.
$C_{14}H_{17}NO_3$			
1,3(2H,4H)-Isoquinolinedione, 8-methoxy-4,5,6,7-tetramethyl-	EtOH	210(4.40),224s(4.12), 270(3.95),339(3.56)	88-5205-66
$C_{14}H_{17}N_3$			
Benzimidazole, 2-(2-piperidinovinyl)-	pH 2	238(3.78),268(3.83), 343(4.61)	33-0889-66
	EtOH	236(3.98),331(4.57)	33-0889-66
$C_{14}H_{17}N_3O_9$			
6-Azauracil, 1-(2,3,5-tri-O-acetyl-β-D-arabinofuranosyl)-	EtOH	262(3.803)	73-4002-66
$C_{14}H_{17}N_5O_2$			
Pteridine, 2-amino-3,4-dihydro-4-(4,4-dimethyl-2,6-dioxocyclohexyl)-	pH 6	284(4.26),310(3.89)	39-1117-66C
	pH 11	284(4.28),340(3.84)	39-1117-66C
$C_{14}H_{17}N_5O_3$			
Benzoic acid, p-[[3-[(2-amino-4-hydroxy-5-pyrimidinyl)-amino]propyl]amino]-	pH 1	226(4.18),268(4.02), 302(3.96)	87-0590-66
	pH 13	285(4.33)	87-0590-66
$C_{14}H_{17}N_5O_6$			
2-Pyrazoline, 5-isopropyl-4,4-dimethyl-1-picryl-	EtOH	395(4.40)	22-0618-66
$C_{14}H_{18}$			
Cyclohexane, 1-benzylidene-4-methyl-, exo-	EtOH	246(4.08)	35-1419-66
Indene, 2,3-dihydro-1,6-pentamethylene-	EtOH	210(4.39),226s(3.87), 278s(3.95),291(4.03)	35-0515-66
$C_{14}H_{18}ClNO$			
Quinoline, 1-(chloroacetyl)-1,2,3,4-tetrahydro-2,2,4-trimethyl-	MeOH	247(4.08)	30-0341-66E
$C_{14}H_{18}N_2$			
Azepine, 1-methyl-, dimer	MeOH	242(3.65)	89-0839-66
13,14-Diazatricyclo[6.4.1.1^{2,7}]tetradeca-3,5,9,11-tetraene, 13,14-dimethyl-	hexane	230(4.20),238(4.22)	89-0839-66
$C_{14}H_{18}N_2OS$			
Urea, N-cyclohexyl-N'-benzoyl-2-thio-	dioxan	226(4.2),255(4.0), 265(4.0),408(2.4)	24-0782-66
$C_{14}H_{18}N_2O_2$			
Morpholine, 4-(2-p-toluoyl-acetimidoyl)-	EtOH	244(4.16),271s(3.42), 326(4.26)	32-1009-66
2,4-Pyrrolidinedione, 3,3-dimethyl-5-(3,4,5-trimethylpyrrol-2-yl-methylene)-	EtOH	227(3.85),283(3.83), 289(3.82),388s(4.24), 412(4.46),428s(4.39), 468s(3.87)	39-1155-66C
$C_{14}H_{18}N_2O_2S$			
Pyrazole, 3,5-diethyl-1-p-toluenesulfonyl-	EtOH	242(4.18)	28C-0782-66A
Pyrazole, 5-isopropyl-3-methyl-1-p-toluenesulfonyl-	EtOH	243(4.15)	28C-0782-66A

Compound	Solvent	λ_{max}(log ε)	Ref.
Pyrazole, 3-methyl-5-propyl-1-p-toluenesulfonyl-	EtOH	243(4.15)	28C-0782-66A
$C_{14}H_{18}N_2O_3$			
Morpholine, 4-(2-p-anisoyl-acetimidoyl)-	EtOH	227s(3.68),248(4.11), 278s(3.48),328(4.76)	32-1009-66
2,4-Pyrrolidinedione, 5-(5-formyl-3,4-dimethylpyrrol-2-ylmethyl)-3,3-dimethyl-	EtOH	281s(3.92),311(4.26)	39-1155-66C
$C_{14}H_{18}N_2O_4$			
3-Azabicyclo[4.1.0]hept-4-ene-1,5-dicarboxylic acid, 2-cyano-2,3,4-trimethyl-, dimethyl ester	EtOH	229s(3.48),298(4.09)	39-1950-66C
Mesoxalic acid, diethyl ester, methylphenylhydrazone	EtOH	325(4.16)	22-2981-66
$C_{14}H_{18}N_4$			
Imidazole, 4-[2-[[p-(dimethylamino)-benzylidene]amino]ethyl]-	EtOH	327(4.34)	44-2380-66
$C_{14}H_{18}N_4O$			
5-Pyrazolinone, 3-tert-butyl-1-phenyl-4-phenylazo-	$CHCl_3$	394(4.38)	22-2990-66
$C_{14}H_{18}N_4O_2$			
Oxindole, 3-(hydrazinomorpholino-methylene)-1-methyl-	EtOH	255(4.14),335(3.85)	95-1152-66
$C_{14}H_{18}N_4O_4$			
2-Hexenal, 4-ethyl-, 2,4-dinitro-phenylhydrazone	n.s.g.	375(4.45)	22-0717-66
3-Hexenal, 4-ethyl-, 2,4-dinitro-phenylhydrazone	n.s.g.	360(4.34)	22-0717-66
3-Pentenal, 2,2,4-trimethyl-, 2,4-dinitrophenylhydrazone	n.s.g.	360(4.20)	22-0734-66
2-Pyrazoline, 1-(2,4-dinitrophenyl)-5-isopropyl-4,4-dimethyl-	EtOH	394(4.24)	22-0618-66
$C_{14}H_{18}N_4O_7S$			
7H-Azeto[2,1-b]pyrazolo[3,4-d][1,3]-thiazine-6-carbamic acid, 8a-carb-oxy-1,3a,4,5a,6,8a-hexahydro-3a-(hydroxymethyl)-7-oxo-, dimethyl ester, acetate (ester)	EtOH	285(3.96)	44-3409-66
$C_{14}H_{18}N_6O_6$			
Acetic acid, bis(6-amino-1,2,3,4-tetrahydro-1,3-dimethyl-2,4-di-oxo-5-pyrimidinyl)-	pH 1	242s(3.96),260(4.37)	24-3524-66
	pH 7	240s(3.96),269(4.37)	24-3524-66
	MeOH	242(4.03),267(4.40)	24-3524-66
$C_{14}H_{18}O$			
Chamecynone	MeOH	230(4.00)	88-3663-66
Crotonophenone, 2',3,4',5'-tetra-methyl-	EtOH	260(5.92)	44-2716-66
1-Indanone, 2,2,4,6,7-pentamethyl-	MeOH	262(4.11),304(3.28)	35-2838-66
1-Indanone, 2,2,5,6,7-pentamethyl-	MeOH	260(4.22),306(3.55)	35-2838-66
Naphthalene, 1,2-dihydro-1-isopropyl-7-methoxy-	EtOH	271(4.06)	78-1977-66

Compound	Solvent	λ_{max}(log ε)	Ref.
[8]Paracyclophane, 3-oxo-	EtOH	232(3.17),272s(2.41), 278(2.44),286(2.46)	35-0515-66
6,12-Tetradecadiene-8,10-diyn-3-ol	ether	242(4.36),260(3.87), 275(4.13),292(4.29), 310(4.20)	24-1642-66
4,6,10,12-Tetradecatetraene-8-yn-1-ol	ether	234(4.09),304s(4.54), 317(4.68),321(4.67), 339(4.70)	24-2828-66
$C_{14}H_{18}O_2$			
3H-Benz[e]indene-3,7-dione, 1,2,3a,4,-5,7,8,9,9a,9b-decahydro-3a-methyl-	EtOH	237(4.28)	30-0880-66F
Δ^8-Bicyclo[5.4.0]undecen-2-one, 9-ethynyl-7-hydroxy-1-methyl-	EtOH	217(3.80),290(2.32)	78-0949-66
6,9-Ethano-4-chromone, 5,6,9,10-tetrahydro-2,6,7-trimethyl-	EtOH	267(3.94)	39-2324-66C
Isochamecynone, hydroxy-	EtOH	228(3.96)	88-3663-66
4,6-Tetradecadiene-8,10-diyne-1,12-diol	ether	226(4.38),236(4.59), 294(4.52),309s(4.44)	24-2828-66
$C_{14}H_{18}O_3$			
2-Benzopyran-1-one, 3,4-dihydro-8-methoxy-4,5,6,7-tetramethyl-	EtOH	213(4.56),253(3.96), 305(3.38)	88-5205-66
Cinnamic acid, 2-ethoxy-3-methyl-, ethyl ester	MeOH	258(3.88)	5-0053-66I
2,4,6,8-Decatetraenal, 10-acetoxy-4,8-dimethyl-	pet ether	318(4.62),333(4.83), 350(4.80)	33-0369-66
1,5(6H)-Indandione, 4-(3-oxobutyl)-7,7a-dihydro-7a-methyl-	EtOH	250(4.16)	30-0880-66F
2(3H)-Naphthalenone, 4,4a,5,6-tetrahydro-6α-hydroxy-1,4a-dimethyl-, acetate	n.s.g.	282(4.43)	88-2615-66
1,7(2H,6H)-Phenanthrenedione, 3,4,4a,4b,5,9,10,10a-octahydro-4-hydroxy-	EtOH	238(4.19)	65-0836-66
2,5(3H,6H)-Phenanthrenedione, 4,4a,4b,7,8,8a,9,10-octahydro-8-hydroxy-	EtOH	241(4.26),295s(2.31)	65-0843-66
$C_{14}H_{18}O_3S$			
Dimethylsulfoxonium 3-carboxy-2-phenylallylide, ethyl ester	EtOH	230(3.91),270(3.83), 346(4.19)	88-1787-66
$C_{14}H_{18}O_4$			
4a(2H)-Naphthalenecarboxylic acid, 3,4,5,6-tetrahydro-7-methoxy-2-oxo-, ethyl ester	EtOH	310(4.48)	88-0927-66
1(2H)-Naphthalenone, 3,4-dihydro-6,7,8-trimethoxy-3-methyl-	n.s.g.	219s(4.27),229(4.28), 277(4.20)	44-1883-66
1-Propanone, 1,1'-(2,5-dimethoxy-p-phenylene)di-	EtOH	224(4.05),260(3.99), 355(3.60)	44-3321-66
$C_{14}H_{18}O_5$			
2,4-Cyclodecadiene-1,2-dicarboxylic acid, 10-oxo-, dimethyl ester, cis-cis	C_6H_{12}	254(4.08)	35-4685-66
	EtOH	265(3.97),290s(3.73)	35-4685-66
	MeCN	264(3.84),290s(3.54)	35-4685-66
Norpsilostachyin, anhydrodihydro-	MeOH	295(1.62)	78-1139-66

Compound	Solvent	$\lambda_{max}(\log \epsilon)$	Ref.
$C_{14}H_{18}S$			
Benzo[b]thiophene, 5-tert-butyl-2,3-dimethyl-	n.s.g.	233(4.53),263(3.85), 292(3.39),302(3.34)	22-3674-66
Benzo[b]thiophene, 6-tert-butyl-2,3-dimethyl-	n.s.g.	233(4.52),265s(3.87), 289s(3.36),301(3.11)	22-3674-66
Thiophene, 2-(1-adamantyl)-	isooctane	235(3.95)	54-1045-66
Thiophene, 2-(2-adamantyl)-	isooctane	237(3.91)	54-1045-66
Thiophene, 3-(1-adamantyl)-	isooctane	236(3.77)	54-1045-66
Thiophene, 3-(2-adamantyl)-	isooctane	236(3.76)	54-1045-66
$C_{14}H_{19}^+$			
Cyclohexylmethylphenylcarbonium ion	FSO_3H-SbF_5	348(4.35),394(3.00)	35-1488-66
$C_{14}H_{19}ClO_5$			
1,2,3,4,5,6,7,8-Octahydro-9-methyl-xanthylium perchlorate	1% $HClO_4$	247(3.56),306(4.23)	80-0109-66
	HOAc	306(4.14)	80-0109-66
$C_{14}H_{19}Cl_2N$			
2,5-Cyclohexadiene, 1-(cyclohexyl-imino)-4-(dichloromethyl)-4-methyl-	C_6H_{12}	242(4.32),311(2.39)	44-3178-66
$C_{14}H_{19}N$			
1-Cyclohexene, 3-(methylamino-methyl)-2-phenyl-	EtOH	241(4.00)	44-3482-66
1,5-Ethano-1H-1-benzazepine, 2,3,4,5-tetrahydro-11,11-dimethyl-	hexane	252(2.94),262(2.95), 269(2.95)	5-0128-66D
$C_{14}H_{19}NO$			
3-Buten-2-one, 4-p-cumenyl-3-methyl-, oxime	n.s.g.	275(4.44)	83-0626-66
1-Penten-3-one, 1-p-cumenyl-, α-oxime	n.s.g.	289(4.51)	83-0626-66
β-oxime	n.s.g.	295(4.51)	83-0626-66
$C_{14}H_{19}NO_2$			
Cinnamic acid propylamide, 2-ethoxy-	MeOH	216(4.17),222s(4.09), 271(4.28),276s(4.27)	5-0053-66I
2H-Furo[2,3-b]indole, 5-ethoxy-3,3a,8,8a-tetrahydro-3a,8-di-methyl-	HCl	250(3.63),323(3.62)	25-1638-66
	EtOH	250(4.02),324(3.49)	25-1638-66
$C_{14}H_{19}NO_2S$			
L-Valine, N-phenylthioacetyl-, methyl ester	isooctane	268(4.08),364(1.62)	1-2781-66
	MeOH	267(4.10),338s(1.82)	1-2781-66
	dioxan	267(4.06),380s(1.51)	1-2781-66
$C_{14}H_{19}NO_3$			
2H,11bH-[1,3]Oxazino[2,3-a]isoquino-line, 3,4,6,7-tetrahydro-9,10-dimethoxy-	H_2O	233(3.91),309(3.55), 359(3.52)	83-0817-66
	MeOH	232(3.92),312(3.17), 367(3.15)	83-0817-66
$C_{14}H_{19}NO_7S_2$			
5H-Pyrano[3,2-d]thiazole-5-methanol, 3a,6,7,7a-tetrahydro-6,7-dihydroxy-2-(methylthio)-, triacetate	MeOH	221(3.98)	23-0094-66
$C_{14}H_{19}N_3$			
Benzimidazole, 2-(2-piperidinoethyl)-	pH 2	241(3.66),270(3.99), 276(4.01),344(3.09)	33-0889-66

Compound	Solvent	λ_{max}(log ϵ)	Ref.
Benzimidazole, 2-(2-piperidinoethyl)- (cont.)	EtOH	244(3.79),274(3.87), 281(3.91),329(3.01)	33-0889-66
$C_{14}H_{19}N_3O$			
3-Pentenal, 2,2-dimethyl-4-phenyl-, semicarbazone	n.s.g.	242(4.34)	22-0734-66
4-Quinazolinone, 3-isopropyl-2-(isopropylamino)-	EtOH	230(4.5),280(4.2), 340(3.5)	83-0914-66
$C_{14}H_{19}N_3O_4$			
Glycine, N-[(4,5,6,7-tetrahydro-5,5-dimethyl-7-oxo-2-benzimidazolyl)carbamoyl]-, ethyl ester	EtOH	243(4.11),277(4.27)	25-0731-66
$C_{14}H_{19}N_5O_4$			
β-D-Ribofuranoside, methyl 5-(6-aminopurin-1-yl)-5-deoxy-2,3-O-isopropylidene-	EtOH-HCl	259(4.07)	4-0485-66
	EtOH	227(4.26),271(4.05)	4-0485-66
	EtOH-NaOH	268(4.10),273(4.11)	4-0485-66
β-D-Ribofuranoside, methyl 5-deoxy-2,3-O-isopropylidene-5-(purin-6-ylamino)-	EtOH-HCl	280(4.24)	4-0485-66
	EtOH	268(4.28)	4-0485-66
	EtOH-NaOH	275.5(4.14)	4-0485-66
$C_{14}H_{19}N_5O_5$			
Adenine, 9-(2,3-O-isopropylidene-D-mannofuranosyl)-	H_2O	260(4.15)	44-0339-66
$C_{14}H_{20}$			
1,3-Cyclohexadiene, 1,2,3,4,5-pentamethyl-6-methylene-5-vinyl-	MeOH	254(4.28)	24-1097-66
Indan, 2,2,4,5,6-pentamethyl-	MeOH	271(3.02),276(2.96), 281(3.05)	35-2838-66
Indan, 2,2,4,5,7-pentamethyl-	MeOH	270(2.85)	35-2838-66
[8]Paracyclophane	MeCN	230(3.83),268s(2.43), 276(2.55),282(2.48)	35-0515-66
$C_{14}H_{20}BN$			
Borane, (diethylamino)bis-(3-methyl-3-buten-1-ynyl)-	C_6H_{12}	225(4.15),260(4.26)	22-3850-66
	C_6H_{12}	225(4.15),260(4.26)	28C-0376-66A
$C_{14}H_{20}BrNS$			
3,4-Dihydro-2-(β-propyl-thioethyl)-isoquinolinium bromide	MeOH	285(4.24),287(4.24)	83-0846-66
$C_{14}H_{20}N_2$			
Pyridine, 4-(p-dimethylaminophenyl)-1,2,5,6-tetrahydro-N-methyl-	EtOH	286.5(4.25)	78-2735-66
dihydrochloride	pH 1	246(4.11)	78-2735-66
	EtOH	298(4.28)	78-2735-66
$C_{14}H_{20}N_2OS$			
L-Leucineamide, N-phenylthioacetyl-	MeOH	265(4.02),341(1.76)	1-2781-66
	dioxan	265(4.05),352(1.74)	1-2781-66
$C_{14}H_{20}N_2O_2$			
Butyramide, N,N'-o-phenylenebis-	MeOH	240s(4.10)	78-1175-66
1-Pyridinecarboxylic acid, 1,2-dihydro-2-cyclohexylimino-, ethyl ester	EtOH	260(3.55)	32-1020-66
2-Pyrrolidinone, 5-(5-formyl-3,4-dimethylpyrrol-2-ylmethyl)-3,4-dimethyl-	EtOH	280s(3.86),313(4.30)	39-1155-66C

Compound	Solvent	$\lambda_{max}(\log \epsilon)$	Ref.
$C_{14}H_{20}N_2O_3$			
Benzoic acid, p-[3-(2-hydroxyethyl)-1-imidazolidinyl]-, ethyl ester	EtOH	307(4.17)	87-0868-66
$C_{14}H_{20}N_2O_4$			
Benzene, 1,2-di-tert-butyl-3,5-dinitro-	isooctane	266(3.98)	54-0619-66
Benzene, 1,2-di-tert-butyl-4,5-dinitro-	isooctane	266(3.88)	54-0619-66
$C_{14}H_{20}N_2O_5S$			
Cephalosporanic acid, 7-amino-, tert-butyl ester	EtOH	246s(3.77),266(3.84)	87-0444-66
$C_{14}H_{20}N_2O_6$			
Uridine, 2',3'-O-isopropylidene-3,5-dimethyl-	H_2O	267(--)	24-3884-66
	MeOH	267(3.97)	24-3884-66
$C_{14}H_{20}N_6O$			
Urea, 1-[bis(isopropylidenehydrazino)-methylene]-3-phenyl-	EtOH	250(4.52),270s(4.28)	39-0006-66C
$C_{14}H_{20}O$			
2'-Acetonaphthone, 1',4',4'a,5',6',7'-hexahydro-4'a,8'-dimethyl-	EtOH	232(4.00)	88-4989-66
Bicyclo[8.2.2]tetradeca-10,12,13-trien-4-ol	EtOH	230(3.84),270s(2.44), 276(2.53),284s(2.44)	35-0515-66
2,5-Cyclohexadienylideneacetaldehyde, 2,3,4,4,5,6-hexamethyl-	hexane	312(4.38)	24-2479-66
2(3H)-Naphthalenone, 4,4a,5,6,7,8-hexahydro-7α-isopropenyl-4a-methyl-	EtOH	245(4.10)	44-1016-66
isomer	MeOH	243(4.15)	44-4188-66
Naphtho[2,3-b]furan, 4,4a,5,6,7,8,8a,9-octahydro-3,8a-dimethyl-	heptane	222.5(3.84)	39-0377-66C
4-Nor-α-cyperone	MeOH	243(4.13)	44-4188-66
14-Noreudesma-4,6-dien-3-one	EtOH	295(4.39)	78-2869-66
14-Nor-7β(H)-eudesma-4,11-dien-3-one	EtOH	241(4.21)	78-2869-66
2,4-Pentadienal, 5-(2,6,6-trimethyl-1-cyclohexen-1-yl)-, trans	n.s.g.	265(4.13),327(4.29)	78-0293-66
$C_{14}H_{20}OS$			
4-Thiaionylideneacetaldehyde, trans	C_6H_{12}	266(4.16),360(4.11)	78-0265-66
$C_{14}H_{20}O_2$			
1(2H)-Naphthalenone, 3,4,5,8-tetrahydro-4-isopropyl-6-methoxy-	EtOH	238(3.84),241(3.85)	78-1977-66
1(2H)-Naphthalenone, 3,4,7,8-tetrahydro-4-isopropyl-6-methoxy-	EtOH	339(3.96)	78-1977-66
14-Noreudesm-4-ene-3,6-dione	EtOH	250(4.11)	78-2869-66
14-Nor-7β(H)-eudesm-4-ene-3,6-dione	EtOH	294(4.08)	78-2869-66
$C_{14}H_{20}O_2S$			
4-Thiaionylideneacetic acid, trans	C_6H_{12}	255(4.19),349(4.12)	78-0265-66
$C_{14}H_{20}O_3$			
p-Toluenepropionic acid, β-hydroxy-, tert-butyl ester	EtOH	212(3.96),216s(3.92), 252s(2.80),258(2.85), 264(2.88),272(2.82)	22-3243-66

Compound	Solvent	$\lambda_{max}(\log \epsilon)$	Ref.
$C_{14}H_{20}O_4S_8$			
Formic acid, dithiobis[thio-, cyclic O,O':O,O'-bis(2,2-dimethyltrimethylene) ester	MeOH-1% dioxan	240(4.27),285(3.81)	50-0965-66C
$C_{14}H_{20}O_5$			
2-Cyclodecene-1,2-dicarboxylic acid, 10-oxo-, dimethyl ester, cis	C_6H_{12}	255f(3.73)	35-4685-66
	EtOH	225(3.88),256(3.71)	35-4685-66
	MeCN	245(3.22),263(3.40)	35-4685-66
$C_{14}H_{20}O_8$			
Ascorbic acid, 2,6-dibutyrate	neutral	231(4.0)	95-0376-66
	anion	254(4.2)	95-0376-66
$C_{14}H_{21}Br$			
Benzene, 2-bromo-1,4-di-tert-butyl-	C_6H_{12}	261(2.25),267(2.35), 276(2.24)	54-0457-66
$C_{14}H_{21}NO_2$			
Benzene, 1,2-di-tert-butyl-4-nitro-	isooctane	271.5(4.04)	54-0619-66
$C_{14}H_{21}NO_3$			
Phenol, 3,5-di-tert-butyl-2-nitro-	C_6H_{12}	256(3.34),280(3.48)	23-0961-66
	EtOH	273.5(3.36)	23-0961-66
$C_{14}H_{21}NO_3S$			
p-Toluenesulfonamide, N-[(2-hydroxycyclohexyl)methyl]-	EtOH	227(3.98)	44-1372-66
$C_{14}H_{21}NO_4$			
Benzene, 1-heptyloxy-4-methoxy-2-nitro-	MeOH	357(3.38)	78-0093-66B
Benzene, 1-heptyloxy-4-methoxy-3-nitro-	MeOH	357(3.38)	78-0093-66B
3,5-Pyridinedicarboxylic acid, 1,4-dihydro-1,2,6-trimethyl-, diethyl ester	n.s.g.	232(4.21),258(4.16), 345(3.86)	39-1083-66B
3,5-Pyridinedicarboxylic acid, 1,4-dihydro-2,4,6-trimethyl-, diethyl ester	n.s.g.	232(4.28),252s(3.92), 349(3.91)	39-1083-66B
$C_{14}H_{21}N_3O$			
3,5-Hexadienal, 6-(1-cyclohexenyl)-4-methyl-, semicarbazone	n.s.g.	262s(4.62),272(4.69), 282s(4.61)	22-0728-66
2,4-Hexadienal, 6-cyclohexylidene-4-methyl-, semicarbazone	n.s.g.	332(4.82),348(4.80)	22-0728-66
$C_{14}H_{21}N_3O_5$			
Cytidine, 2',3'-O-isopropylidene-N,N-dimethyl-	pH 1	219(3.77),285(4.18)	39-0588-66C
$C_{14}H_{21}N_5$			
Pteridine, 2,8-dihydro-6,7-diisopropyl-8-methyl-2-(methylimino)-	pH 4.5	239(4.45),295(4.05)	39-1065-66C
	pH 9.5	232(4.31),318(4.19), 328s(4.11)	39-1065-66C
$C_{14}H_{21}N_5OS_2$			
Adenine, 9-(2-deoxy-3,5-di-S-ethyl-3,5-dithio-β-D-threopentofuranosyl)-	MeOH	259.5(4.20)	69-0224-66

Compound	Solvent	λ_{max}(log ε)	Ref.
$C_{14}H_{21}N_5O_2$			
Pyrazolo[5,1-c]-as-triazin-4-ol, 6-acetyl-4,6-dihydro-3-methyl-4-(piperidinomethyl)-	EtOH	247(4.18)	39-1214-66C
$C_{14}H_{22}$			
Benzene, pentamethylpropyl-	C_6H_{12}	270(2.40)	24-1097-66
1,3,5-Cycloheptatriene, 1,2,3,4,5,6,7-heptamethyl-	hexane	250s(3.78)	24-1097-66
Phenanthrene,$\Delta^{4a(4b)}$-dodecahydro-, cis	EtOH	210(4.00)	88-6471-66
$C_{14}H_{22}Cl_2N_2$			
Pyridine, 4-(p-dimethylaminophenyl)-1,2,5,6-tetrahydro-N-methyl-, dihydrochloride	pH 1	246(4.11)	78-2735-66
	EtOH	298(4.28)	78-2735-66
$C_{14}H_{22}N_2$			
Malononitrile, (1-pentylhexylidene)-	EtOH	238(4.11)	44-2784-66
$C_{14}H_{22}N_2O$			
3(2H)-Pyridone, 4,5-dihydro-5,5-dimethyl-2-(4,5,5-trimethylpyrrolidin-2-ylidene)-	EtOH	287(4.13),332(4.12)	78-0015-66B
	EtOH-HCl	254(4.02),305(4.42)	78-0015-66B
$C_{14}H_{22}N_2O_2$			
Unknown oxidation product	EtOH	259(3.91),341(4.18)	78-0015-66B
$C_{14}H_{22}N_2O_2S_2$			
5-Thiazolidineacetamide, N-cyclohexyl-4-oxo-3-propyl-2-thioxo-	EtOH	263(4.10),297(4.22)	95-0101-66
$C_{14}H_{22}O$			
Bicyclo[4.3.0]non-6-en-8-one, 3-tert-butyl-1-methyl-, cis	C_6H_{12}	224(4.27),316(1.66), 327(1.70),339(1.60)	22-3888-66
trans	C_6H_{12}	226(4.14),316(1.52), 327(1.56),339(1.45)	22-3888-66
Cyclobutanone, 2-tert-butyl-4-(1-methyl-4-pentylidene)-, trans	isooctane	245(4.13),251(4.15), 259s(4.00),341(1.69), 356(1.83),370(1.78), 385s(1.43)	22-0281-66
Ether, o-tert-butylphenyl tert-butyl	EtOH	227(3.89),275(3.28), 281(3.26)	77-0172-66
2(3H)-Naphthalenone, 4,4a,5,6,7,8-hexahydro-7α-isopropyl-4aβ-methyl-	EtOH	242(4.14)	44-1016-66
trans isomer	EtOH	240(4.13)	44-3109-66
Tetradecatrienal, 5,8-di-cis-2-trans	EtOH	223(4.10)	54-0117-66
$C_{14}H_{22}OS$			
4-Thiaionylideneethanol, trans	C_6H_{12}	233(4.19),300(3.98)	78-0265-66
$C_{14}H_{22}O_2$			
Benzene, 1-heptyloxy-4-methoxy-	MeOH	226(3.98),287(3.40)	78-0093-66B
Bicyclohexyl, 1,1'-dimethyl-3,3'-dioxo-	C_6H_{12}	291(1.58)	22-0381-66
2,4-Dodecadienoic acid, 4-hydroxy-2,3-dimethyl-, γ-lactone	n.s.g.	270(4.3)	54-0043-66
14-Nor-5β,7β(H)-eudesm-11-en-3-one, 5-hydroxy-	EtOH	295(1.85)	78-2869-66

Compound	Solvent	λ_{max}(log ε)	Ref.
$C_{14}H_{22}O_2S_4$			
Xanthic acid, 2,5-bornylenedi-, dimethyl ester	C_6H_{12}	216(4.22),227(4.25), 276(4.38),357(1.95)	39-0885-66B
$C_{14}H_{22}O_3$			
1-Cyclohexene-1-propionic acid, 5-isopropyl-2-methyl-6-oxo-, methyl ester	EtOH	245(4.01)	78-0219-66A
$C_{14}H_{22}O_4$			
2,4-Hexadienedioic acid, di-tert-butyl ester, trans-trans	C_6H_{12}	263(4.51)	39-0387-66C
$C_{14}H_{22}O_7$			
Roridinic acid	EtOH	256(4.32)	33-2527-66
$C_{14}H_{23}NO_3$			
Morpholine, 4-(cyclohexylidene-ethoxyacetyl)-	MeOH	214s(4.10)	5-0053-66I
$C_{14}H_{23}NO_3S_2$			
5-Thiazolidineacetic acid, N-butyl-4-oxo-2-thioxo-, pentyl ester	EtOH	262(4.22),296(4.32)	95-0095-66
$C_{14}H_{23}N_3O_6$			
Uridine, 5-(diethylaminomethyl)-	pH 2	267(--)	70-1718-66
	pH 12	267(3.81)	70-1718-66
$C_{14}H_{23}N_6OP$			
1,2,4-Triazole, 3-anilino-5-ethyl-1-bis(dimethylamino)phosphoryl-	EtOH	245(4.18),261(4.19)	54-0429-66
1,2,4-Triazole, 5-anilino-3-ethyl-1-bis(dimethylamino)phosphoryl-	EtOH	261(4.39)	54-0429-66
$C_{14}H_{23}O_3PSe$			
Butyl phenyl phosphoroselenoate	MeOH	248(3.45)	65-0254-66
$C_{14}H_{24}NO_3P$			
Phosphorimidic acid, N-(o-ethyl-phenyl)-, triethyl ester	EtOH	240(3.81),275(3.19)	35-3781-66
$C_{14}H_{24}N_2O$			
[$\Delta^{2,3'}$(2'H,3H)-Bi-1H-azepin]-2'-one, octahydro-1,1'-dimethyl-	MeOH	212(4.4),302(3.7)	24-0724-66
[$\Delta^{2,3'}$-Bipyrrolidin]-2'-one, 1,1'-diisopropyl-	MeOH	217(3.7),296(4.5)	24-1918-66
[$\Delta^{2,3'}$-Bipyrrolidin]-2'-one, 1,1'-dipropyl-	MeOH	218(3.7),295(4.5)	24-0724-66
$C_{14}H_{24}O$			
Cyclohexanone, 2-allyl-4-tert-butyl-2-methyl-, cis	C_6H_{12}	298(1.56)	22-3881-66
trans	C_6H_{12}	292(1.49)	22-3881-66
Cyclohexanone, 4-tert-butyl-2-methallyl-	C_6H_{12}	282(1.54)	22-3881-66
$C_{14}H_{24}O_3$			
2-Dodecenoic acid, 4,4-dihydroxy-2,3-dimethyl-, γ-lactone	n.s.g.	208(4.1)	54-0043-66
Geranic acid, 2-ethoxy-, ethyl ester	MeOH	233(4.08)	5-0053-66I

Compound	Solvent	$\lambda_{max}(\log \epsilon)$	Ref.
$C_{14}H_{26}N_2$			
$\Delta^{2,3'}$-Bipyrrolidine, 1,1'-dipropyl-	MeOH	207(4.0)	24-1923-66
$C_{14}H_{26}O$			
Cyclohexanone, 4-tert-butyl-2-isobutyl-	C_6H_{12}	276(1.53),283(1.52)	22-3881-66
7-Tridecen-6-one, 8-methyl-	n.s.g.	242(1.99)	54-0753-66
$C_{14}H_{27}FN_2$			
Diimide, α-fluorocyclohexyl-tert-octyl-	n.s.g.	365(1.11)	35-4677-66
$C_{14}H_{28}N_2O$			
Piperidine, 3-hydroxy-5,5-dimethyl-2-(4,5,5-trimethyl-2-pyrrolidinyl)-, dihydrochloride	EtOH	none	78-0015-66B
$C_{14}H_{28}N_2O_2$			
3-Heptanone, 4-(N-leucylamino)-6-methyl-	pH 9.4	220(3.97),330(3.71)	33-1131-66

Compound	Solvent	λ_{max}(log ϵ)	Ref.
$C_{15}H_8BrN_5O_6$			
Pyrazole, 4-bromo-3-phenyl-1-picryl-	EtOH	355(3.86)	22-3744-66
$C_{15}H_8Br_2N_4O_4$			
Pyrazole, 3,4-dibromo-1-(2,4-dinitrophenyl)-5-phenyl-	EtOH	234(4.39),298(3.94)	22-3744-66
$C_{15}H_8Br_2O_3$			
2(3H)-Benzofuranone, 3-(3,5-dibromosalicylidene)-	EtOH	248(4.01),316(3.83), 368(3.98)	44-3854-66
Coumarin, 6,8-dibromo-3-(o-hydroxyphenyl)-	EtOH	276(4.04),292(3.98), 338(3.91)	44-3854-66
$C_{15}H_8Cl_2N_4O_4$			
Pyrazole, 3,4-dichloro-1-(2,4-dinitrophenyl)-5-phenyl-	EtOH	291(3.96)	22-3744-66
$C_{15}H_8Cl_2O_3$			
2(3H)-Benzofuranone, 3-(3,5-dichlorosalicylidene)-	EtOH	247(4.00),310(3.88), 350(3.97)	44-3854-66
Coumarin, 6,8-dichloro-3-(o-hydroxyphenyl)-	EtOH	275(4.10),292(4.02), 338(3.95)	44-3854-66
$C_{15}H_9BrN_4O_4$			
Pyrazole, 4-bromo-1-(2,4-dinitrophenyl)-3-phenyl-	EtOH	346(4.15)	22-3744-66
Pyrazole, 4-bromo-1-(2,4-dinitrophenyl)-5-phenyl-	EtOH	293(3.95)	22-3744-66
$C_{15}H_9BrO_3$			
2(3H)-Benzofuranone, 3-(5-bromosalicylidene)-	EtOH	246(4.14),312(3.89), 378(4.07)	44-3854-66
Coumarin, 6-bromo-3-(o-hydroxyphenyl)-	EtOH	273(4.11),286(4.01), 334(4.04)	44-3854-66
$C_{15}H_9ClO_3$			
2(3H)-Benzofuranone, 3-(5-chlorosalicylidene)-	EtOH	245(4.01),336(3.95), 380(3.79)	44-3854-66
Coumarin, 6-chloro-3-(o-hydroxyphenyl)-	EtOH	272(4.08),284(3.99), 332(3.99)	44-3854-66
$C_{15}H_9Cl_2N_3O_2$			
2-Quinoxalinecarboxylic acid, 3-(2,4-dichloroanilino)-, sodium salt	EtOH	294(4.48),390(3.85)	87-0770-66
$C_{15}H_9NO_5$			
2(3H)-Benzofuranone, 3-(5-nitrosalicylidene)-	EtOH	234(4.14),312(4.14), 360(4.09)	44-3854-66
Coumarin, 3-(o-hydroxyphenyl)-6-nitro-	EtOH	225(4.15),268(4.38), 325(4.02)	44-3854-66
$C_{15}H_9N_3$			
Pyrido[2,3-a]phenazine	EtOH	268(4.43),294(4.53), 370(3.90)	44-3384-66
$C_{15}H_9N_3OS$			
Ketone, 2-thienyl v-triazolo[1,5-a]quinolin-3-yl	DMF	301s(4.00),314(4.17), 337(4.51),353(4.48)	24-2918-66

Compound	Solvent	λ_{max}(log ε)	Ref.
$C_{15}H_9N_3O_2$			
Ketone, 2-furyl v-triazolo[1,5-a]-quinolin-3-yl	MeOH	240(4.21),249(4.21), 299s(4.05),312(4.21), 337(4.50),347s(4.44)	24-2918-66
$C_{15}H_9N_3O_4$			
1,3-Propanedione, 2-diazo-1-(o-nitrophenyl)-3-phenyl-	CH_2Cl_2	251(4.45)	24-3128-66
1,3-Propanedione, 2-diazo-1-(p-nitrophenyl)-3-phenyl-	CH_2Cl_2	263(4.38)	24-3128-66
Stilbene, α-cyano-2',4-dinitro-	acetone-base	550(4.55)	83-0131-66
Stilbene, α-cyano-3',4-dinitro-	acetone-base	550(4.60)	83-0131-66
$C_{15}H_{10}BrN_3O_2$			
2-Quinoxalinecarboxylic acid, 3-(2-bromoanilino)-, sodium salt	EtOH	288(4.43),390(3.81)	87-0770-66
$C_{15}H_{10}ClN_3O_2$			
2-Quinoxalinecarboxylic acid, 3-(p-chloroanilino)-, sodium salt	EtOH	290(4.43),390(3.78)	87-0770-66
$C_{15}H_{10}ClN_7O_4$			
s-Triazine, 2-chloro-4,6-bis-(p-nitroanilino)-	EtOH	225(4.23),323(4.49)	80-0139-66
$C_{15}H_{10}N_2O$			
Spiro[2H-indole-2,3'[3H]indole]-2'-ol	EtOH	226(4.59),249(4.68), 323(4.20),338(4.34)	78-3337-66
$C_{15}H_{10}N_2O_2$			
2-Oxazolin-5-one, 2-phenyl-4-(3-pyridylmethylene)-	HOAc	257(4.21),359(4.56)	78-0035-66
1,3-Propanedione, 2-diazo-1,3-diphenyl-	CH_2Cl_2	257(4.37)	24-3128-66
4-Pyridinecarboxaldehyde, azlactone	HOAc	260(4.19),366(4.46)	78-0035-66
4(3H)-Quinazolinone, 2-benzoyl-	EtOH	209(4.54),220s(4.43), 253(4.13),263(4.10), 305(3.98)	39-2290-66C
Stilbene, α-cyano-4-nitro-	acetone-base	550(4.62)	83-0131-66
$C_{15}H_{10}N_2O_2S$			
4,5-Thiazolidinedione, 2-(diphenylamino)-	dioxan	325(3.9)	24-3572-66
$C_{15}H_{10}N_2O_3$			
Isoxazole, 3-(o-nitrophenyl)-5-phenyl-	EtOH	260(4.35)	32-1073-66
$C_{15}H_{10}N_2O_5S$			
2-Oxazolin-5-one, 2-phenyl-4-(3-pyridylmethylene)-, compound with SO_3	HOAc	256(4.20),362(4.53)	78-0035-66
Pyridine-4-carboxaldehyde, azlactone, compound with SO_3	HOAc	256(4.17),365(4.47)	78-0035-66
$C_{15}H_{10}N_4O$			
1,2,3-Triazolo[1,5-a]quinazolin-5(4H)-one, 3-phenyl-	EtOH	208(4.50),222(4.47), 251(4.35),270s(4.14)	39-2290-66C

Compound	Solvent	$\lambda_{max}(\log \epsilon)$	Ref.
$C_{15}H_{10}N_4O_4$			
Pyrazole, 1-(2,4-dinitrophenyl)-3-phenyl-	EtOH	253(4.36),342(4.05)	22-3744-66
Pyrazole, 1-(2,4-dinitrophenyl)-4-phenyl-	EtOH	240(4.28),343(4.16)	22-3744-66
Pyrazole, 1-(2,4-dinitrophenyl)-5-phenyl-	EtOH	234(4.36),290(3.95)	22-3744-66
2-Quinoxalinecarboxylic acid, 3-(m-nitroanilino)-, sodium salt	EtOH	221(4.56),288(4.50), 387(3.85)	87-0770-66
$C_{15}H_{10}N_4O_9$			
Ketone, bis(2,4-dinitrobenzyl)	MeOH	245(4.47)	32-0311-66
$C_{15}H_{10}N_6O_2$			
5H-s-Triazolo[4,3-b]-s-triazole, 6-(p-nitrophenyl)-3-phenyl-	EtOH	300(4.43)	4-0119-66
$C_{15}H_{10}N_8O_9$			
Formazan, 3-acetyl-1,5-bis-(2,4-dinitrophenyl)-	benzene	340(4.48),430(4.32)	70-1326-66
	MeOH	355(4.26),400(4.20), 620(4.27)	70-1326-66
$C_{15}H_{10}O$			
9-Anthraldehyde	EtOH	247s(4.66),251(4.73), 280(4.06),295(4.04), 320(2.88),328(2.88)	39-1729-66C
$C_{15}H_{10}OSe$			
Isoselenoflavone	n.s.g.	258(4.70),352(4.43)	20-0260-66
Selenoflavone	n.s.g.	274(5.00),354(4.54)	20-0260-66
$C_{15}H_{10}O_2$			
Benzofuran, 2-benzoyl-	MeOH	227(4.05),269s(3.96), 313(4.39)	73-0835-66
1,3-Indandione, 2-phenyl-	DMF	290(4.2),350(4.1), 450(3.3)	30-0631-66A
$C_{15}H_{10}O_3$			
2(3H)-Benzofuranone, 3-salicylidene-	EtOH	243(4.08),316(3.84), 372(4.13)	44-3854-66
Coumarin, 3-(o-hydroxyphenyl)-	EtOH	272(3.87),290(3.88), 326(3.98)	44-3854-66
Isoflavone, 2'-hydroxy-	EtOH	240(4.39),285(4.00)	18-1535-66
$C_{15}H_{10}O_4$			
1,4-Phenanthrenequinone, 2-hydroxy-7-methoxy-	EtOH	237(4.495),302(4.289), 353(3.975)	23-1086-66
	EtOH-base	238(4.484),304(4.275), 358(3.964),500(2.820)	23-1086-66
$C_{15}H_{10}O_5$			
Xanthone, 2-methoxy-3,4-(methylenedioxy)-	EtOH	237(4.24),255(4.20), 291(3.38),326(4.14)	78-1777-66
Xanthone, 4-methoxy-2,3-(methylenedioxy)-	EtOH	245(4.23),280(3.38), 310(4.13)	78-1777-66
$C_{15}H_{10}O_7$			
Norvogeletin	MeOH	275(4.25),353(4.37)	24-2430-66

Compound	Solvent	$\lambda_{max}(\log \epsilon)$	Ref.
$C_{15}H_{11}BrN_2$			
Pyrazole, 4-bromo-3,5-diphenyl-	EtOH	251(4.36)	22-3744-66
$C_{15}H_{11}BrN_4O_4$			
3-Pyrazoline, 3-bromo-1-(2,4-dinitrophenyl)-5-phenyl-	$CHCl_3$	381.5(4.32)	22-0610-66
$C_{15}H_{11}BrO$			
2-Propyn-1-ol, 1-(m-bromophenyl)-1-phenyl-	EtOH	219(4.18),260(3.60)	22-2885-66
2-Propyn-1-ol, 1-(o-bromophenyl)-1-phenyl-	EtOH	215(4.03),258(3.70)	22-2885-66
2-Propyn-1-ol, 1-(p-bromophenyl)-1-phenyl-	EtOH	220(4.23),262(3.67)	22-2885-66
$C_{15}H_{11}ClFeO$			
Ferrocene, 1,1'-(1-chloro-5-oxo-1,3-pentadienyl)-	EtOH	230(4.07),307(3.32), 475(2.53)	101-0173-66B
Ferrocene, 1-(1-chloro-2-formylvinyl)-1'-ethynyl-	EtOH	255(4.03),313(4.16), 380(3.27),497(3.25)	101-0173-66B
$C_{15}H_{11}ClN_2O$			
Azirino[1,2-a]quinazoline, 5-chloro-1,1a-dihydro-3-phenyl-, 2-oxide	n.s.g.	224(4.20),254(4.28), 308(3.90),319(3.88), 342s(3.70)	88-2609-66
5H-1,4-Benzodiazepine, 7-chloro-5-phenyl-, 4-oxide	n.s.g.	239(4.20),285s(3.64), 322(3.87),350s(3.68)	88-2609-66
4(3H)-Quinazolinone, 2-(α-chlorobenzyl)-	EtOH	209(4.35),227(4.41), 272(3.91),303(3.63), 315(3.49)	39-2290-66C
2(3H)-Quinazolinone, 6-chloro-3-methyl-4-phenyl-	iso-PrOH	242(4.70),425(3.78)	44-1007-66
hydrochloride	THF	235(4.53),296(3.90), 412(3.65)	44-1007-66
$C_{15}H_{11}ClN_2S$			
2(3H)-Quinazolinethione, 6-chloro-3-methyl-4-phenyl-	iso-PrOH	225(4.46),276(4.38), 326(4.47),485(3.30)	44-1007-66
hydrochloride	HOAc	289(4.52),312(4.49), 470(3.43)	44-1007-66
$C_{15}H_{11}ClO$			
2-Propyn-1-ol, 1-(o-chlorophenyl)-1-phenyl-	EtOH	225(4.23),259(3.95)	22-2885-66
2-Propyn-1-ol, 1-(p-chlorophenyl)-1-phenyl-	EtOH	225(4.03),256(3.66)	22-2885-66
$C_{15}H_{11}ClO_4Se$			
4-Phenyl-1-selenanaphthalenium perchlorate	HOAc-1% $HClO_4$	265(5.13),359(4.19), 415(3.89)	20-0169-66
$C_{15}H_{11}ClO_6$			
4'-Hydroxyflavylium perchlorate	MeOH-acid	438(4.64)	61-0530-66
7-Hydroxyflavylium perchlorate	MeOH-acid	430(4.43)	61-0530-66
$C_{15}H_{11}ClO_7$			
4',7-Dihydroxyflavylium perchlorate	MeOH-acid	459(4.71)	61-0530-66

Compound	Solvent	λ_{max}(log ε)	Ref.
$C_{15}H_{11}Cl_3N_2$			
2-Pyrazoline, 3-(p-chlorophenyl)-1-(3,4-dichlorophenyl)-	MeOH	363(4.10)	89-0699-66
$C_{15}H_{11}FN_2OS$			
p-Anisimidic acid, N-(p-fluorophenyl)-, anhydride with isothiocyanic acid	C_6H_{12}	226(4.36),290(4.32), 336(3.74)	44-0722-66
$C_{15}H_{11}FN_2S$			
p-Toluimidic acid, N-(p-fluorophenyl)-, anhydride with isothiocyanic acid	C_6H_{12}	222(4.33),270(4.33), 342(3.62)	44-0722-66
$C_{15}H_{11}MnO_5$			
Manganese, (3-benzoyl-1-methyl-π-allyl)tetracarbonyl-	dioxan	231(3.76),268(4.03)	39-0194-66A
$C_{15}H_{11}N$			
4-Stilbenecarbonitrile, trans	hexane	228(4.35),305(4.60), 317(4.53)	99-0089-66
	EtOH	228(4.28),310(4.55), 317(4.50)	99-0089-66
$C_{15}H_{11}NO$			
2H-Azirine, 2-benzoyl-3-phenyl-	ether	247(4.39)	35-1844-66
2-Indolinone, 3-benzylidene-	EtOH	255(4.07),323(4.13)	33-0985-66
Isocarbostyril, 3-phenyl-	EtOH	232(4.35),305(4.16), 337(3.95)	44-2090-66
Isoxazole, 3,5-diphenyl-	ether	245(4.34),265(4.38)	35-1844-66
1aH-Oxazirino[2,3-a]quinoline, 1a-phenyl-	EtOH	251(4.50),317(3.83)	88-2145-66
	EtOH	252(4.50),317(3.83)	94-0555-66
Oxazole, 2,5-diphenyl-	ether	302(4.48),315(4.44)	35-1844-66
$C_{15}H_{11}NO_2$			
Cyclopenta[c]quinolizine-1,3-dicarboxaldehyde, 4-methyl-	EtOH	241(4.56),308(4.50), 320s(4.5),395(3.81)	39-0324-66C
Flavone, 3-amino-	C_6H_{12}	240(4.26),302(3.68), 359(4.03)	5-0225-66C
	EtOH	241(4.31),305(3.74), 362(4.03)	5-0225-66C
hydrochloride	EtOH	241(4.30),305(3.74), 362(4.03)	5-0225-66C
Isocarbostyril, 2-hydroxy-3-phenyl-	EtOH	230(4.07),252(3.84), 302(3.72),332(3.60)	44-2090-66
$C_{15}H_{11}NO_4$			
Cinnamic acid, p-nitro-, phenyl ester	EtOH	300(4.5)	65-1202-66
$C_{15}H_{11}NO_5$			
Cinnamic acid, p-nitro-, p-hydroxyphenyl ester	EtOH	300(4.4)	65-1202-66
Naphthostyril-4,5-dicarboxylic acid, dimethyl ester	EtOH	222(4.66),271(4.14), 332s(3.28),350(3.38), 383(3.50)	39-1028-66C
$C_{15}H_{11}NO_6$			
3,4,5-Pyridinetricarboxylic acid, 2-phenyl-, monomethyl ester	MeOH	259(4.11),292(4.06)	39-1121-66C

Compound	Solvent	$\lambda_{max}(\log \epsilon)$	Ref.
$C_{15}H_{11}NS$			
Isothiazole, 3,5-diphenyl-	EtOH	207(4.40),251(4.42), 277(4.34)	78-2119-66
Thiazole, 2,4-diphenyl-	C_6H_{12}	251(4.43),311(3.96)	22-2857-66
$C_{15}H_{11}N_3$			
2-Benzimidazoleacetonitrile, 1-phenyl-	EtOH	246(3.85),274(3.52), 282(3.46)	4-0278-66
$C_{15}H_{11}N_3O$			
2H-Benz[h]pyrazolo[4,3-c]quinoline, 7-methoxy-	EtOH	212(4.29),263(4.83), 291(4.05),301(4.01), 315(3.92),336(3.28), 352(3.23)	54-0681-66
$C_{15}H_{11}N_3O_2$			
1,3-Propanedione, 2-diazo-3-phenyl-1-(p-aminophenyl)-	CH_2Cl_2	242(4.28),285(4.17), 319(4.27)	24-3128-66
Pyrazole, 1-(p-nitrophenyl)-3-phenyl-	EtOH	257(4.09),341(4.39)	22-3744-66
2-Quinoxalinecarboxylic acid, 3-anilino-, sodium salt	EtOH	285(4.47),395(3.80)	87-0770-66
as-Triazin-3(4H)-one, 5,6-diphenyl-, 2-oxide	EtOH	255(4.17),354(3.78)	44-3914-66
$C_{15}H_{11}N_3O_3$			
2-Oxazolin-5-one, 4-[(1-acetyl-4-imidazolyl)methylene]-2-phenyl-	CH_2Cl_2	247s(4.06),271(4.25), 359s(4.37),374(4.46), 392(4.32)	87-0766-66
5-Quinazolinecarboxylic acid, 2-(m-aminophenyl)-3,4-dihydro-4-oxo-	pH 13	246s(4.50),305(4.11)	32-0264-66
5-Quinazolinecarboxylic acid, 2-(p-aminophenyl)-3,4-dihydro-4-oxo-	pH 13	225s(4.26),290(4.39)	32-0264-66
$C_{15}H_{11}N_3O_4S$			
Indole, 2-[(2,4-dinitrophenyl)-thio]-3-methyl-	MeOH	300f(4.2),330s(4.1)	32-1301-66
$C_{15}H_{11}N_3O_4S_2$			
4-Thiazolecarboxylic acid, 2,2'-(2,6-pyridinediyl)bis-, dimethyl ester	MeOH	222(4.70),278s(4.25), 285(4.25),295s(4.21), 321(4.15)	39-1354-66C
	10N HCl	237s(4.22),283(4.17), 294(4.15),334(4.26)	39-1354-66C
$C_{15}H_{11}N_3O_5$			
Propionitrile, 3-hydroxy-2-(p-nitrophenyl)-3-(m-nitrophenyl)-	EtOH-piperidine	550(2.39)	83-0131-66
$C_{15}H_{11}N_5$			
5H-s-Triazolo[4,3-b]-s-triazole, 3,6-diphenyl-	EtOH	255(4.44)	4-0119-66
$C_{15}H_{11}N_5O_6$			
Pyrazoline, 3-phenyl-1-picryl-	$CHCl_3$	416(4.42)	22-0610-66
Pyrazoline, 5-phenyl-1-picryl-	$CHCl_3$	388(4.18)	22-0610-66
$C_{15}H_{12}$			
5H-Dibenzo[a,d]cycloheptene	EtOH	284(4.06)	5-0058-66B
Indene, 3-phenyl-	heptane	230(4.28)	44-0081-66
9,10-Methanoanthracene, 9,10-dihydro-	EtOH	252(2.97),264s(3.03), 273(3.30),278(3.41)	5-0058-66B

Compound	Solvent	$\lambda_{max}(\log \epsilon)$	Ref.
$C_{15}H_{12}BF_4N_3OS$			
2-(N^{ω}-benzoylazo)-3-methylbenzo-thiazolium tetrafluoroborate	MeCN	368(4.12)	5-0065-66J
$C_{15}H_{12}BrN_3O_2S$			
1H-1,2,3-Triazole, 5-(p-bromophenyl)-1-(p-tolylsulfonyl)-	EtOH	232(4.30)	44-1587-66
$C_{15}H_{12}ClN$			
Indole, 1-benzyl-2-chloro-	MeOH	270(3.99),280(3.94), 290(3.79)	44-2627-66
Indole, 2-(p-chlorophenyl)-1-methyl-	EtOH	242(4.27),303(4.29)	44-1423-66
$C_{15}H_{12}ClNO_2$			
Flavone, 3-amino-, hydrochloride	EtOH	241(4.30),305(3.74), 362(4.03)	5-0225-66C
$C_{15}H_{12}ClNO_5$			
N,5-Diphenylisoxazolium perchlorate	pH 1	222(3.87),320(4.36)	44-2039-66
$C_{15}H_{12}ClN_5$			
s-Triazine, 2,4-dianilino-6-chloro-	EtOH	271(4.43)	80-0139-66
$C_{15}H_{12}Cl_2FeO$			
Ferrocene, 1-(1-chlorovinyl)-1'-(2-formyl-1-chlorovinyl)-	EtOH	230(4.25),275(3.98), 310(3.79),483(2.96)	101-0173-66B
$C_{15}H_{12}Cl_2N_2$			
2-Pyrazoline, 1,3-bis(p-chlorophenyl)-	MeOH	363(4.07)	89-0699-66
2-Pyrazoline, 1-(3,4-dichloro-phenyl)-3-phenyl-	MeOH	354(4.05)	89-0699-66
$C_{15}H_{12}Cl_2N_2O$			
2(3H)-Quinazolinone, 6-chloro-3-methyl-4-phenyl-, hydrochloride	THF	235(4.53),296(3.90), 412(3.65)	44-1007-66
$C_{15}H_{12}Cl_2N_2O_2S$			
1,2,4-Benzothiadiazine, 2-p-chloro-benzyl-6-chloro-7-methyl-, 1,1-dioxide	EtOH	229(4.59),271(4.04), 301(3.71)	7-0286-66
1,2,4-Benzothiadiazine, 4-p-chloro-benzyl-6-chloro-7-methyl-, 1,1-dioxide	EtOH	275(3.91)	7-0286-66
$C_{15}H_{12}Cl_2N_2S$			
2(3H)-Quinazolinethione, 6-chloro-3-methyl-4-phenyl-, hydrochloride	HOAc	289(4.52),312(4.49), 470(3.43)	44-1007-66
$C_{15}H_{12}Cl_4O_2$			
Bicyclo[2.2.1]hepta-2,5-diene, 1,2,3,4-tetrachloro-7,7-di-methoxy-5-phenyl-	isooctane	218(4.01),247(3.96), 275(3.47)	35-0582-66
$C_{15}H_{12}Cl_4S_2$			
2,5-Norbornadiene, 1,4,7,7-tetra-chloro-2,3-bis(methylthio)-5-phenyl-	EtOH	220(4.04),257(3.96)	44-1863-66
$C_{15}H_{12}N_2$			
Quinoline, 4-amino-2-phenyl-	EtOH	260(4.52),322(4.89)	32-1073-66

Compound	Solvent	λ_{max}(log ε)	Ref.
$C_{15}H_{12}N_2O$			
Cinnoline, 3-(p-methoxyphenyl)-	MeOH	265(4.50),297(4.27)	87-0784-66
Cinnoline, 4-methoxy-3-phenyl-	MeOH	248(4.59),288(3.83), 332(3.54)	87-0784-66
4(1H)-Cinnolinone, 1-methyl-3-phenyl-	MeOH	264(4.17),312(4.06), 360(4.11)	87-0784-66
4-Cinnolinone, 2-methyl-3-phenyl-	MeOH	255(3.93),356(4.18), 371(4.21)	87-0784-66
4-Imidazolin-2-one, 4,5-diphenyl-	EtOH	219s(4.23),244(3.83), 260(3.79),308(4.03)	44-3914-66
Pyrazole, 3-phenoxy-1-phenyl-	EtOH	272(4.28)	44-1538-66
4(3H)-Quinazolinone, 2-benzyl-	EtOH	206(4.40),226(4.47), 266(3.95),304(3.61), 315(3.53)	39-2290-66C
$C_{15}H_{12}N_2OSe$			
Selenazolidin-4-one, 5-phenyl-2-phenylimino-	EtOH	273(4.11)	28C-0778-66A
$C_{15}H_{12}N_2O_2$			
Benzo[g]quinoxalin-2(1H)-one, 3-acetonyl-	$CHCl_3$	294(3.99),308(4.04), 322(4.22),403(4.43)	78-3253-66
	DMSO	295(4.06),309(4.13), 322(4.19),400(4.37), 418(4.25)	78-3253-66
	CF_3COOH	317(3.93),419(4.15), 445(4.17)	78-3253-66
Carbazole, 1-methyl-2-(2-nitrovinyl)-	EtOH	211(4.30),239s(--), 249(4.50),295(3.86), 380(4.40)	87-0237-66
Indole, 4-nitro-2-p-tolyl-	EtOH	216(4.18),277(4.20), 411(3.98)	78-3131-66
Indole, 6-nitro-2-p-tolyl-	EtOH	217(4.20),258(4.22), 394(4.20)	78-3131-66
4(3H)-Quinazolinone, 2-(α-hydroxybenzyl)-	EtOH	207(4.42),226(4.49), 267(3.99),274s(3.97), 304(3.70),315(3.59)	39-2290-66C
$C_{15}H_{12}N_2O_2S$			
Acetamide, N-(10,11-dihydro-10-oxobenzo[b]cyclohepta[e][1,4]thiazin-7-yl)-	MeOH	235(4.31),286(4.23), 318(4.19),477(3.84)	18-1980-66
Indole, 2-methyl-3-(o-nitrophenylthio)-	MeOH	270f(4.1),360(3.5)	32-1301-66
Indole, 3-methyl-2-(o-nitrophenylthio)-	MeOH	290(4.2),360(3.6)	32-1301-66
Urea, N-benzoyl-N'-(thiobenzoyl)-	dioxan	226(4.4),270(4.2), 310s(3.9),498(2.2)	24-0782-66
$C_{15}H_{12}N_2O_4$			
Cinnamic acid, 4-nitro-, p-aminophenyl ester	EtOH	240(4.2),300(4.4)	65-1202-66
$C_{15}H_{12}N_2O_5$			
Acetanilide, 4'-hydroxy-, p-nitrobenzoate	EtOH	250(4.4)	65-1198-66
Ketone, bis(o-nitrobenzyl)	MeOH	260(4.03)	32-0311-66
Ketone, bis(p-nitrobenzyl)	MeOH	274(4.31)	32-0311-66
Ketone, (o-nitrobenzyl) (p-nitrobenzyl)	MeOH	266(4.21)	32-0311-66

Compound	Solvent	λ_{max}(log ε)	Ref.
$C_{15}H_{12}N_4$			
Benzimidazo[1,2-c][1,2,3]benzo-triazine, 9,10-dimethyl-	EtOH	227(4.36),295(4.50), 328s(3.82)	4-0289-66
$C_{15}H_{12}N_4O$			
Pyrazol-3-ol, 1-phenyl-4-(phenylazo)-	MeOH	258(4.10),364(4.34)	44-1538-66
5-Pyrazolone, 3-phenyl-4-(phenylazo)-	$CHCl_3$	400(4.18)	22-2990-66
as-Triazine, 3-amino-5,6-diphenyl-, 2-oxide	EtOH	222s(4.25),257(4.31),	44-3914-66
$C_{15}H_{12}N_4O_4$			
Pyrazoline, 1-(2,4-dinitrophenyl)-3-phenyl-	$CHCl_3$	412(4.46)	22-0610-66
Pyrazoline, 1-(2,4-dinitrophenyl)-5-phenyl-	$CHCl_3$	380(4.15)	22-0610-66
$C_{15}H_{12}N_4O_4S$			
1H-1,2,3-Triazole, 5-(p-nitrophenyl)-1-(p-tolylsulfonyl)-	EtOH	274(4.00)	44-1587-66
$C_{15}H_{12}N_4S$			
1,3,4-Thiadiazole, 2-anilino-5-(N-phenylformimidoyl)-	MeOH	240(4.64)	44-3528-66
$C_{15}H_{12}O$			
Chalcone	EtOH	230(3.95),312(4.43)	49-0896-66
$C_{15}H_{12}O_2$			
Chalcone, 4-hydroxy-	EtOH	248(4.08),350(4.39)	49-0896-66
Chalcone, 4'-hydroxy-	EtOH	230(--),320(--)	49-0896-66
Naphtho[2,3-b]furan-9(4H)-one, 3,8-dimethyl-4-methylene-	EtOH	211(4.25),234(4.13), 250(4.12),290(4.12), 340(4.00)	78-0301-66
2H-Oxeto[2,3-b]benzofuran, 2a,7a-dihydro-2-phenyl-	MeOH	276(3.49),282(3.43)	24-1723-66
4-Stilbenecarboxylic acid, trans	EtOH	230(4.12),318(4.56)	99-0089-66
	ether	227(4.36),313(4.62)	99-0089-66
$C_{15}H_{12}O_2S_3$			
α-Terthienylmethanol, acetate	MeOH	252(3.89),350(4.27)	88-4227-66
$C_{15}H_{12}O_3$			
Benzofuran, 2-(p-hydroxyphenyl)-7-methoxy-	EtOH	250(4.02),300(4.54), 308(4.54)	78-2913-66
	NaOEt	328(4.60)	78-2913-66
Benzoic acid, 2-benzoyl-, methyl ester	$CHCl_3$	325(2.09)	35-0781-66
Chalcone, 2',4-dihydroxy-	EtOH	244(4.07),274s(3.81), 373(4.42),453s(3.30)	49-0896-66
	pH 13	236(4.16),394(4.36)	49-0896-66
Chalcone, 4,4'-dihydroxy-	EtOH	240(4.15),348(4.49)	49-0896-66
Freelingyne	EtOH	365(4.65)	88-1939-66
1H-Naphtho[2,1-b]pyran-1-one, 5-methoxy-3-methyl-	99% EtOH	211s(4.64),235(4.46), 269(4.21),312s(3.91), 321(3.95),334(3.91)	94-0129-66
1H-Naphtho[2,1-b]pyran-1-one, 6-methoxy-3-methyl-	99% EtOH	212(4.49),226(4.37), 232(4.37),257(4.30), 265(4.25),290s(3.86), 304(3.99),320(3.91), 334(3.97)	94-0129-66

Compound	Solvent	λ max(log ε)	Ref.
1-Oxaphenalene, 3-acetoxy-2-methyl-	EtOH	241(4.42),248s(4.26), 292s(3.59),314s(3.81), 335(4.94),343s(3.91), 368(3.71),387s(3.52)	39-0523-66C
$C_{15}H_{12}O_4$			
Xanthone, 1,7-dimethoxy-	EtOH	240(4.46),255(4.54), 280(3.90),375(3.94)	39-0430-66C
$C_{15}H_{12}O_5$			
Xanthone, 1-hydroxy-3,4-dimethoxy-	EtOH	234(4.43),259(4.42), 305(4.05),365(3.57)	78-1785-66
	NaOH	231(4.43),270(4.23), 295(3.57)	78-1785-66
	AlCl3	266(4.41),274(4.34)	78-1785-66
Xanthone, 1-hydroxy-3,5-dimethoxy-	EtOH	245(4.25),306(4.16), 355(3.36)	78-1777-66
	NaOH	240(4.25),271(4.16), 303(4.14),313(4.18)	78-1777-66
	AlCl3	240s(4.17),267(4.23), 278s(4.21),308s(4.14), 331(4.20)	78-1777-66
Xanthone, 1-hydroxy-3,7-dimethoxy-	MeOH	231(4.54),259(4.59), 292(4.39),369(3.83)	39-2265-66C
Xanthone, 1-hydroxy-5,6-dimethoxy-	EtOH	235s(4.66),243(4.69), 268(4.04),310(4.30), 354(3.88)	78-1785-66
Xanthone, 4-hydroxy-2,3-dimethoxy-	EtOH	237(4.17),256(4.21), 291(3.72),308(3.89)	78-1777-66
	NaOH	237(4.16),275(4.17), 299(3.96),333(3.49)	78-1777-66
	NaOAc	237(4.17),257(4.16), 276(4.16),305(3.91)	78-1777-66
Xanthone, 5-hydroxy-1,3-dimethoxy-	EtOH	248(4.26),304(4.15), 344(3.29)	78-1777-66
	NaOH	241(4.23),265(4.18), 283(4.18),308(4.16)	78-1777-66
	NaOAc	247(4.23),285(4.14), 303(4.15)	78-1777-66
Xanthone, 8-hydroxy-1,2-dimethoxy-	EtOH	238(4.47),260(4.55), 290(3.75),322(3.70)	78-1785-66
	NaOH	238(4.58),263(4.25), 326(3.85)	78-1785-66
	AlCl3	238(4.50),276(4.49), 310(3.83),351(3.64)	78-1785-66
$C_{15}H_{13}BrN_2$			
2-Pyrazoline, 3-(p-bromophenyl)-1-phenyl-	MeOH	364(4.02)	89-0699-66
$C_{15}H_{13}ClN_2$			
Benzimidazole, 2-(o-chlorophenyl)-5,6-dimethyl-	EtOH	250s(3.82),299(4.17)	4-0444-66
2-Pyrazoline, 1-(p-chlorophenyl)-3-phenyl-	MeOH	356(4.02)	89-0699-66
2-Pyrazoline, 3-(p-chlorophenyl)-1-phenyl-	MeOH	362(4.03)	89-0699-66
$C_{15}H_{13}ClN_2O$			
2(1H)-Quinazolinone, 6-chloro-3,4-dihydro-3-methyl-4-phenyl-	iso-PrOH	260(4.04),300(3.30)	44-1007-66

Compound	Solvent	$\lambda_{max}(\log \epsilon)$	Ref.
$C_{15}H_{13}ClN_2OS$			
2(1H)-Quinazolinethione, 6-chloro-3,4-dihydro-4-hydroxy-3-methyl-4-phenyl-	iso-PrOH	255s(3.98),287(4.43)	44-1007-66
$C_{15}H_{13}ClN_2O_2$			
2(1H)-Quinazolinone, 6-chloro-3,4-dihydro-4-hydroxy-3-methyl-4-phenyl-	iso-PrOH	255(4.13),300(3.18)	44-1007-66
$C_{15}H_{13}ClN_2O_2S$			
1,2,4-Benzothiadiazine-1,1-dioxide, 2-benzyl-6-chloro-7-methyl-	EtOH	229(4.51),271(4.03), 301(3.71)	7-0286-66
1,2,4-Benzothiadiazine-1,1-dioxide, 4-benzyl-6-chloro-7-methyl-	EtOH	276(3.91)	7-0286-66
$C_{15}H_{13}ClN_2S$			
2(1H)-Quinazolinethione, 6-chloro-3,4-dihydro-3-methyl-4-phenyl-	iso-PrOH	290(4.36)	44-1007-66
$C_{15}H_{13}Cl_2N$			
Aziridine, 3,3-dichloro-1-phenyl-2-p-tolyl-	dioxan	223(4.29)	78-1279-66
Aziridine, 3,3-dichloro-2-phenyl-1-p-tolyl-	dioxan	221(4.19)	78-1279-66
$C_{15}H_{13}Cl_2NO$			
Acetimidoyl chloride, 2-chloro-2-(p-methoxyphenyl)-N-phenyl-	dioxan	221.3(4.18)	78-1279-66
Aziridine, 3,3-dichloro-1-(p-methoxyphenyl)-2-phenyl-	dioxan	222(4.18)	78-1279-66
Aziridine, 3,3-dichloro-2-(p-methoxyphenyl)-1-phenyl-	dioxan	248(4.29)	78-1279-66
$C_{15}H_{13}IOS$			
Methylphenylsulfonium p-iodophenacylide	EtOH	251(4.35),315(4.30)	78-2145-66
$C_{15}H_{13}N$			
Indole, 3-methyl-2-phenyl-	EtOH	229(4.37),310(4.29)	94-0934-66
Isoindole, 2-p-tolyl-	MeOH	243(4.50),288(3.94), 298(3.94),337(3.62)	89-0967-66
$C_{15}H_{13}NO$			
Cyclopenta[c]quinolizine, 1-acetyl-4-methyl-	EtOH	236(4.57),241s(4.5), 258s(4.1),299(4.12), 348(4.47),410s(3.6)	39-0324-66C
Formamide, N-(2,2-diphenylvinyl)-	MeOH	280(4.4)	83-0493-66
Indole, 2-(p-methoxyphenyl)-	EtOH	248(4.25),310(4.42)	87-0527-66
Indole, 3-(p-methoxyphenyl)-	EtOH	264.5(4.24)	87-0527-66
Isoindole, 2-(p-methoxyphenyl)-	MeOH	245(4.45),288(4.02), 298(4.03),337(3.66)	89-0967-66
2-Propyn-1-ol, 1-(p-aminophenyl)-1-phenyl-	EtOH	212(4.32),290(4.28)	22-2885-66
4-Stilbenecarboxamide, trans	EtOH	227(4.20),315(4.60)	99-0089-66
$C_{15}H_{13}NOS$			
1-Thiaflavanone, 3-amino-, hydrochloride	C_6H_{12}	242(4.46),352(3.50)	5-0225-66C
	EtOH	240(4.48),355(3.57)	5-0225-66C
1-Thiaflavanone, oxime	EtOH	240(4.25),320(3.37)	5-0225-66C

Compound	Solvent	λ_{max}(log ε)	Ref.
$C_{15}H_{13}NO_2$			
Azepino[3,2,1-hi]indole, 6-(hydroxy-methyl)-7-carboxymethylene-1,2,4,7-tetrahydro-, δ-lactone	EtOH	260(4.00),300(4.04), 428(3.70)	35-4061-66
Carbazole, 6-acetoxy-3-methyl-	n.s.g.	230(4.58),237(4.59), 260(4.20),296(4.22), 330(3.60)	88-0661-66
Flavanone, 3-amino-	C_6H_{12}	214(3.51),248(3.85), 318(3.51)	5-0225-66C
hydrochloride	C_6H_{12}	214(4.52),248(3.85), 318(3.52)	5-0225-66C
	EtOH	252(3.85),320(3.50)	5-0225-66C
$C_{15}H_{13}NO_3$			
Ketone, benzyl 2-nitrobenzyl	MeOH	260(3.81)	32-0311-66
Phenacyl carbanilate	EtOH	250(4.32)	44-1959-66
$C_{15}H_{13}NO_4$			
Styrene oxide, p-methoxy-p'-nitro-, trans	hexane	230(4.25),286(4.16)	44-3976-66
$C_{15}H_{13}NO_5S$			
N,5-Diphenylisoxazolium bisulfate	pH 1	220(3.87),320(4.35)	44-2039-66
$C_{15}H_{13}N_3$			
Pyrido[2,3-a]phenazine, 8,9,10,11-tetrahydro-	EtOH	225(4.72),279(4.47), 336(3.77),352(3.76)	44-3384-66
$C_{15}H_{13}N_3O$			
4(3H)-Quinazolinone, 2-anilino-3-methyl-	EtOH	225(4.39),244s(4.12), 284(4.23),327(3.66)	24-1532-66
as-Triazin-3(4H)-one, 5,6-diphenyl-	EtOH	220s(4.11),285(4.06)	44-3914-66
$C_{15}H_{13}N_3O_2$			
Benzimidazole, 5,6-dimethyl-2-(o-nitrophenyl)-	EtOH	244(4.04),288(4.06)	4-0289-66
Indazole, 3,6-dimethyl-1-(m-nitrophenyl)-	EtOH	215(4.28),258(4.36), 306(4.06)	78-3131-66
Indazole, 3,6-dimethyl-1-(p-nitrophenyl)-	EtOH	216(4.27),236(4.26), 266(3.87),356(4.32)	78-3131-66
Pyrazoline, 1-(p-nitrophenyl)-3-phenyl-	$CHCl_3$	425(4.56)	22-0610-66
$C_{15}H_{13}N_3O_2S$			
1,2,3-Triazole, 5-phenyl-1-(p-tolylsulfonyl)-	EtOH	237(4.39)	44-1587-66
$C_{15}H_{13}N_3O_3$			
Indazole, 6-methoxy-3-methyl-1-(m-nitrophenyl)-	EtOH	260(4.44),292s(4.01)	78-3131-66
Indazole, 6-methoxy-3-methyl-1-(p-nitrophenyl)-	EtOH	231(4.24),278(3.82), 352(4.09)	78-3131-66
7-Oxa-1,2,4-triazabicyclo[4.1.0]hep-tan-3-one, 5-hydroxy-5,6-diphenyl-	EtOH	236(4.37),275s(3.53), 310s(2.91)	94-1448-66
4(3H)-Quinazolinone, 3-amino-2-(2,5-dihydroxybenzyl)-	EtOH	222(4.61),248(4.16), 254(4.13),260(4.10), 278(4.13),295(4.14), 364(2.62),382(2.75), 398(2.71)	39-2190-66C

Compound	Solvent	$\lambda_{max}(\log \epsilon)$	Ref.
4(3H)-Quinazolinone, 3-amino-2-(2,5-dihydroxybenzyl)- (cont.)	HCl	227(4.36),282(3.95), 480(2.39)	39-2190-66
	NaOH	238(4.32),260(4.30), 298(4.48),366(3.97), 546(2.58)	39-2190-66
as-Triazin-3(2H)-one, 4,5-dihydro-5,6-diphenyl-	EtOH	220s(4.11),285(4.06)	94-1448-66
$(C_{15}H_{13}N_3O_3)_n$			
Poly-L-p-(phenylazo)phenylalanine	CF_3COOH	306s(3.63),425(4.42)	35-5010-66
$C_{15}H_{13}N_3O_4$			
Indazole, 5,6-dimethoxy-1-(p-nitrophenyl)-	EtOH	236(4.16),299(3.93), 350(4.15)	78-3131-66
$C_{15}H_{13}P$			
Phosphine, diphenyl-1-propynyl-	EtOH	225(4.26),252(3.82)	22-1002-66
$C_{15}H_{14}$			
1H-Cycloprop[a]anthracene, 1a,2,3,9b-tetrahydro-	EtOH	231(5.0),272(3.73), 281(3.74),292(3.56), 310(2.89),318(2.67), 325s(2.89)	5-0058-66B
Stilbene, p-methyl-, trans	hexane	227(4.30),298(4.57), 310(4.55)	99-0089-66
	EtOH	230(4.10),298(4.38), 310(4.36),321(4.20)	99-0089-66
$C_{15}H_{14}Br_2N_4$			
1,1'-Trimethylenebis[4-cyano-pyridinium bromide]	H_2O	231(4.42),237(4.37), 278(3.95)	87-0638-66
$C_{15}H_{14}Br_2Si$			
Dibenzosilole, 5-(1,2-dibromo-ethyl)-5-methyl-	n.s.g.	270f(4.0)	65-0114-66
$C_{15}H_{14}ClNOS$			
1-Thiaflavanone, 3-amino-, hydro-chloride	C_6H_{12}	242(4.46),352(3.50)	5-0225-66C
	EtOH	240(4.48),355(3.57)	5-0225-66C
$C_{15}H_{14}ClNO_2$			
Acetanilide, 2-chloro-2-(p-methoxyphenyl)-	dioxan	247.3(4.31)	78-1279-66
Flavanone, 3-amino-, hydrochloride	C_6H_{12}	214(4.52),248(3.85), 318(3.52)	5-0225-66C
	EtOH	252(3.85),320(3.50)	5-0225-66C
$C_{15}H_{14}ClNO_4$			
1,3-Dimethylbenzo[a]quinolizinium perchlorate	EtOH	265s(--),273(4.42), 280(4.43),332(3.79), 346(4.09),363(4.24)	44-2616-66
1,7-Dimethylbenzo[a]quinolizinium perchlorate	EtOH	238(4.48),274s(4.37), 279(4.42),332s(3.71), 348(4.05),364(4.18)	44-0978-66
$C_{15}H_{14}ClNO_6$			
7,8-Dihydro-10,11-(methylenedioxy)-6H-benzo[c]pyrid[1,2-a]azepinium perchlorate	EtOH	267(4.00),352(3.96)	94-0622-66

Compound	Solvent	λ max(log ε)	Ref.
$C_{15}H_{14}INO_2$			
7,8-Dihydro-10,11-(methylenedioxy)-6H-benzo[c]pyrid[1,2-a]azepinium iodide	EtOH	267(4.21),352(3.97)	94-0622-66
$C_{15}H_{14}NP$			
Phosphine, diphenyl(2-cyanoethyl)-	MeOH	246(4.0)	51-0476-66
$C_{15}H_{14}N_2$			
Benzimidazole, 5,6-dimethyl-2-phenyl-	EtOH	245(4.00),310(4.42), 324s(4.22)	4-0444-66
Benzimidazole, 2-(o-methylbenzyl)-	$MeOCH_2CH_2OH$	274(4.0)	44-1498-66
Malondialdehyde, dianil	$CHCl_3$	380(4.38)	65-1383-66
2-Pyrazoline, 1,3-diphenyl-	MeOH	354(3.97)	89-0699-66
$C_{15}H_{14}N_2O$			
Benzimidazole, 2-(o-hydroxyphenyl)-5,6-dimethyl-	EtOH	234(3.91),294(4.02), 320(4.39),335(4.34)	4-0444-66
$C_{15}H_{14}N_2O_2$			
Cocculolidine, BrCN degradation product	MeOH	255s(3.27),275(2.73)	88-5179-66
$C_{15}H_{14}N_2O_3$			
Acetanilide, 4'-(p-nitrobenzyl)-	EtOH	255(4.39),270s(4.19), 330s(3.18)	60-0214-66
N-Ethylacridinium nitrate	EtOH	262(5.0),360f(4.3), 430f(3.7)	62-0179-66A
Methane, (p-nitrophenyl)-(p-acetamidophenyl)-	EtOH	255(4.39),270s(4.19), 330s(3.19)	65-0223-66
$C_{15}H_{14}N_2O_4$			
Benzoic acid, p-nitro-, (p-dimethylamino)phenyl ester	EtOH	260(4.5)	65-1198-66
$C_{15}H_{14}N_4O_2$			
Lumazine, 8-benzyl-6,7-dimethyl-	pH -2.7	247(4.05),362(4.14)	24-3503-66
	pH 6	258(4.18),276(4.05), 408(4.09)	24-3503-66
	pH 13	242(4.34),313(4.39), 364(3.80)	24-3503-66
$C_{15}H_{14}N_4O_6S$			
Acetaldehyde, (p-tolylsulfonyl)-, 2,4-dinitrophenylhydrazone	EtOH	222(4.42),352(4.39)	24-1580-66
$C_{15}H_{14}N_6S_2$			
Semicarbazide, 1-(3-mercapto-4-phenyl-1,2,4-triazol-5-yl)-4-phenyl-3-thio-	EtOH	268(4.36)	39-0001-66C
$C_{15}H_{14}O$			
Anisole, p-styryl-, trans	hexane	230(4.30),302(4.52), 320(4.48)	99-0089-66
	EtOH	232(4.33),310(4.55)	99-0089-66
Benzofuran, 2-benzyl-2,3-dihydro-	EtOH	248(4.20),269(3.43), 270s(3.43),273(3.42), 277(3.55),284(3.57), 310s(2.09)	1-1561-66
Chroman, 2-phenyl-	n.s.g.	258s(3.17),277(3.36), 283(3.34)	22-1974-66

Compound	Solvent	$\lambda_{max}(\log \epsilon)$	Ref.
Flavan	EtOH	227s(3.82),245s(3.52), 274(3.21),283(3.18)	1-1561-66
2H-Naphtho[1,2-b]pyran, 3,4-dihydro-2-vinyl-	n.s.g.	214(4.58),237(4.61), 241s(4.58),299(3.67), 312(3.57),318s(3.44), 327(3.50)	22-1974-66
Phenol, o-(3-phenylpropenyl)-	EtOH	252(4.00),256s(3.99), 264s(3.85),268s(3.66), 283(3.45),294s(3.49), 313(3.51),318s(3.26)	1-1561-66
$C_{15}H_{14}OS$			
Methylphenylsulfonium phenacylide	EtOH	310(4.18)	78-2145-66
	MeCN	320(4.01)	78-2145-66
	$CHCl_3$	320(3.95)	78-2145-66
	THF	326(4.06)	78-2145-66
$C_{15}H_{14}O_2$			
Acetophenone, p-(p-methoxyphenyl)-	hexane	220(4.19),298(4.34), 340(2.55)	99-0204-66
	EtOH	222(4.14),310(4.28)	99-0204-66
1H-Benz[e]indene-9-carboxaldehyde, 2,3-dihydro-8-methoxy-	EtOH	228(4.62),256s(4.26), 332(3.76)	39-0734-66C
2,4,6-Cycloheptatriene-1-acetic acid, α-phenyl-	EtOH	250(3.56)	44-0912-66
Stilbene oxide, p-methoxy-, trans	hexane	238(4.38),280(3.32), 287(3.25)	44-3976-66
2,4,6,12-Tetradecatetraene-8,10-diyn-oic acid, methyl ester	ether	252(4.18),275(4.35), 281(4.35),348(4.75), 368s(4.65)	24-3194-66
$C_{15}H_{14}O_2S$			
Thioxanthene, 2,4-dimethyl-, 10,10-dioxide	$CHCl_3$	270(3.30)	30-0107-66D
$C_{15}H_{14}O_3$			
Benzophenone, 4,4'-dimethoxy-, cation	H_2SO_4	406(4.61)	39-0650-66B
Glutaric anhydride, 2-(3,4-di-hydro-2-naphthyl)-	$CHCl_3$	266(4.00)	77-0485-66
$C_{15}H_{14}O_3S$			
1,3-Oxathiolane, 2,2-diphenyl-, 3,3-dioxide	ether	252(2.89),258(2.90), 262(2.86),269(2.67)	87-0860-66
$C_{15}H_{14}O_3S_2$			
2H-Thiopyran-3-carboxylic acid, 4-mercapto-2-oxo-5-p-tolyl-, ethyl ester	MeOH	270(3.72),313(3.60), 455(3.99)	44-2389-66
$C_{15}H_{14}O_4$			
Allopteroxylin	MeOH	222(4.30),263(4.71), 342(3.53)	39-0114-66C
p-Benzoquinone, 2-acetyl-3-(5-iso-propyl-2-furyl)-	EtOH	224(4.12),266(4.28), 474(3.86)	33-1794-66
p-Benzoquinone, 2-acetyl-3-(3,4,5-tri-methyl-2-furyl)-	EtOH	250(4.15),515(3.72)	33-1794-66
Glutaric acid, 2-(2-naphthyl)-	EtOH	226(5.07),268(3.77), 277(3.78)	77-0485-66
Phebalosin	EtOH	235s(3.71),246(3.66), 256(3.70),322(4.18)	12-0483-66

Compound	Solvent	$\lambda_{max}(\log \epsilon)$	Ref.
Pteroxylin	MeOH	206(4.21),230(4.17), 255(4.15),320(2.60)	39-0114-66C
$C_{15}H_{14}O_5$			
p-Benzoquinone, 2-carbomethoxy-3-(5-isopropyl-2-furyl)-	EtOH	225(4.07),260(4.18), 472(3.78)	33-1794-66
p-Benzoquinone, 2-carbomethoxy-3-(3,4,5-trimethyl-2-furyl)-	EtOH	267(4.24),512(3.77)	33-1794-66
Clausenin methyl ether	EtOH	220(4.16),262(4.39), 315(4.11),340(4.04)	88-5767-66
2-Naphthoic acid, 3-hydroxy-8-methoxy-, methyl ester, acetate	EtOH	214(4.60),249(4.68), 288(3.69),297(3.75), 307(3.66),347(3.81)	44-1747-66
12-Noryangonin, 11-methoxy-	EtOH	220(4.33),250(4.12), 373(4.47)	83-0507-66
$C_{15}H_{14}Si$			
Dibenzosilole, 5-methyl-5-vinyl-	n.s.g.	270f(4.0)	65-0118-66
$C_{15}H_{15}^+$			
Ethyldiphenylcarbonium ion	FSO_3H-SbF_5	316(4.07),427(4.44)	35-1488-66
$C_{15}H_{15}ClO_4$			
1,4-Methanonaphthalene, 5,8-diacetoxy-6-chloro-7-methyl-	n.s.g.	213(4.58)	16-0054-66
$C_{15}H_{15}N$			
2-Stilbazole, β,3-dimethyl-	EtOH	272(4.05),312(4.06)	44-4322-66
4-Stilbazole, β,3-dimethyl-	EtOH	268(4.11)	44-4322-66
Stilbazole, β,5-dimethyl-	EtOH	271(4.17),296(4.19)	44-4322-66
$C_{15}H_{15}NO$			
2-Biphenylcarboxamide, N,N-dimethyl-	EtOH	227(4.24),296(1.74), 307(1.88)	44-3482-66
$C_{15}H_{15}NO_2$			
4-Flavanol, 3-amino-	C_6H_{12}	276(3.30),283(3.29)	5-0225-66C
4-Flavanol, 3-amino-, hydrochloride	C_6H_{12}	276(3.33),283(3.30)	5-0225-66C
Furo[3,4-b]pyridin-5-ol, 5,7-dihydro-7,7-dimethyl-5-phenyl-	isooctane	259(3.56),265(3.56)	65-0410-66
$C_{15}H_{15}NO_2S_2$			
2H-Thiopyran-3-carboxylic acid, 2-imino-4-mercapto-5-p-tolyl-, ethyl ester	MeOH	334(3.95),390(3.34)	44-2389-66
$C_{15}H_{15}NO_3$			
Azepino[3,2,1-hi]indole, 7-carboxymethylene-1,2,4,5,6,7-hexahydro-6-hydroxy-6-methoxy-, γ-lactone	MeOH	255(4.21),290(3.95), 402(3.74)	35-4061-66
Ethane, 1-(p-methoxyphenyl)-2-(p-nitrophenyl)-	hexane	214(4.18),218(4.18), 269(4.10)	44-3976-66
$C_{15}H_{15}NO_5$			
Acronycidine	EtOH	214(4.26),253(4.81), 344(3.90)	100-0206-66
1H-1-Benzazepine-3,4-dicarboxylic acid, 2,3-dihydro-1-methyl-2-oxo-, dimethyl ester	MeOH	237(4.48),281(4.01)	24-3070-66

Compound	Solvent	λ_{max}(log ε)	Ref.
Nicotinic acid, 1,2-dihydro-4,6-dihydroxy-2-oxo-1-p-tolyl-, ethyl ester	EtOH	306(4.68)	78-0455-66
$C_{15}H_{15}NO_6$			
Nicotinic acid, 1-p-anisyl-1,2-dihydro-4,6-dihydroxy-2-oxo-, ethyl ester	EtOH	304(4.50)	78-0455-66
$C_{15}H_{15}NS$			
1-Thiaflavan, 4-amino-, hydrochloride	EtOH	258(4.086)	5-0225-66C
$C_{15}H_{15}N_3$			
Benzimidazole, 2-(o-aminophenyl)-5,6-dimethyl-	EtOH	252(4.09),294(4.05), 302(4.07),340(4.17)	4-0289-66
Indazole, 1-(p-aminophenyl)-3,6-dimethyl-	EtOH	214(4.50),262(4.30), 305(3.91)	78-3131-66
$C_{15}H_{15}N_3O$			
Indazole, 1-(p-aminophenyl)-6-methoxy-3-methyl-	EtOH	218(4.39),264(4.37), 290(3.96)	78-3131-66
4H-Pyrazolo[4,3-c]pyridin-4-one, 1,5-dihydro-3,5,6-trimethyl-1-phenyl-	EtOH	223(4.38),239(4.37), 292(4.13)	70-1723-66
$C_{15}H_{15}N_3O_2$			
L-Alanine, p-(phenylazo)phenyl-	3N HCl	229(3.98),324(4.21), 418(3.20)	35-5010-66
Indazole, 1-(p-aminophenyl)-5,6-dimethoxy-	EtOH	265(4.30),301(3.75)	78-3131-66
Triazen-3-ol, 1-(p-ethoxyphenyl)-3-phenyl-	1% acetone	264(5.05)	65-2117-66
$C_{15}H_{15}N_3O_2S$			
Benzenesulfonamide, p-(2-benzimidazolyl)-N,N-dimethyl-	EtOH	223(4.21),225(4.23), 247(4.05),255(4.05), 315(4.49)	95-0600-66
$C_{15}H_{15}N_3S$			
1,3,4-Thiadiazole, 2-(isopropylamino)-5-(1-naphthyl)-	EtOH	225(4.50),323(4.06)	83-0921-66
$C_{15}H_{16}$			
Biphenyl, 2,3',4-trimethyl-	EtOH	242.5(4.04)	39-1737-66C
7H-Cyclohepta[a]naphthalene, 8,9,10,11-tetrahydro-	EtOH	226s(4.79),231(4.91), 276s(3.74),286(3.73), 296s(3.63),322(2.59)	22-0228-66 +22-3644-66
1,4-Methanonaphthalene, 1,4-dihydro-9-isopropylidene-6-methyl-	heptane	186(4.46),201(4.47), 211(4.45),217(4.50), 222s(4.43),272(3.30), 277s(3.42),279(3.43), 286(3.48)	44-1988-66
1,4-Nonadiyne, 1-phenyl-	EtOH	238(4.37),249(4.36), 265(3.05),272(3.02), 279(2.93)	78-0867-66
$C_{15}H_{16}Br_2N_4O$			
4-Carbamoyl-4'-cyano-1,1'-trimethylenebis[pyridinium bromide]	H_2O	224(4.31),272(3.88)	87-0638-66
4-Cyano-4'-formyl-1,1'-trimethylenebis[pyridinium bromide], oxime	H_2O	226(4.14),279(4.31)	87-0638-66

Compound	Solvent	λmax(log ε)	Ref.
$C_{15}H_{16}ClNO_2$			
4-Flavanol, 3-amino-, hydrochloride	C_6H_{12}	276(3.33),283(3.30)	5-0225-66C
$C_{15}H_{16}ClNS$			
1-Thiaflavan, 4-amino-, hydrochloride	EtOH	306(4.68)	78-0455-66
$C_{15}H_{16}INO_2$			
1-[3-(3,4-Methylenedioxyphenyl)-propyl]pyridinium iodide	EtOH	260(3.67),266(3.62), 287(3.58)	94-0622-66
$C_{15}H_{16}IP$			
9-Phosphoniabicyclo[4.2.1]nona-2,4,7-triene, 9-methyl-9-phenyl-, iodide	EtOH	267(3.61),273s(3.55)	35-3832-66
isomer	EtOH	268(3.66),273s(3.62)	35-3832-66
$C_{15}H_{16}N_2$			
Carbazole, 2-(2-aminoethyl)-1-methyl-	EtOH	216(4.40),238(4.66), 248(4.51),258(4.30), 296(4.25),323(3.61), 336(3.53)	87-0237-66
Indolo[2,3-a]quinolizine, 2,3,4,6,7,12-hexahydro-	CH_2Cl_2	242(4.28),307(4.25), 315s(4.23)	4-0101-66
$C_{15}H_{16}N_2O_4$			
3(2H)-Pyridazinone, 6-ethoxy-5-methyl-2-(phenacyloxy)-	EtOH	244(4.137),296(3.530)	95-0082-66
$C_{15}H_{16}N_2O_6$			
Pyrrolo[1,2-b]pyrazine-6,7,8-tricarboxylic acid, 1,3-dimethyl-, trimethyl ester	MeOH	230(4.23),249(4.42), 251s(4.42),331(3.94)	39-2218-66C
	MeOH-acid	224(4.27),300s(3.32), 313(3.52),328(3.58), 364(3.80)	39-2218-66C
$C_{15}H_{16}N_4O_2$			
Lumiflavine, 2-imino ethyl ester	pH 7.0	378(4.00),442(4.08)	33-2365-66
$C_{15}H_{16}N_4O_4$			
3-Hexen-5-ynal, 2,2,4-trimethyl-2,4-dinitrophenylhydrazone	n.s.g.	360(4.40)	22-0734-66
5-Indanone, 4,5,8,9-tetrahydro-, 2,4-dinitrophenylhydrazone, trans	n.s.g.	380(4.45)	28C-1598-66A
5-Indanone, 5,6,7,8-tetrahydro-, 2,4-dinitrophenylhydrazone, cis	n.s.g.	386(4.46)	28C-1598-66A
Pyrazole, 3-cyclopropyl-5-ethyl-4-methyl-1-(2,4-dinitrophenyl)-	EtOH	345(3.85)	22-2977-66
Pyrazole, 5-cyclopropyl-3-ethyl-4-methyl-1-(2,4-dinitrophenyl)-	EtOH	338(3.94)	22-2977-66
$C_{15}H_{16}N_4O_6$			
Pyrrole-3-carboxylic acid, 1-(2,4-dinitroanilino)-2,5-dimethyl-, ethyl ester	EtOH	258(4.14),323(4.11)	39-0341-66C
$C_{15}H_{16}O$			
Bibenzyl, p-methoxy-	hexane	225(4.16),278(3.30), 285(3.28)	44-3976-66
Bicyclo[3.1.0]hexan-2-one, 3-benzylidene-6,6-dimethyl-	EtOH	301(4.31)	22-3490-66

Compound	Solvent	$\lambda_{max}(\log \epsilon)$	Ref.
1H-Cyclopenta[2,3]cyclopropa[1,2-a]-naphthalen-1-one, 2,3,4,5,9b,9c-hexahydro-9b-methyl-	EtOH	238(3.97)	35-0161-66
1,4:5,8-Dimethanofluoren-9-one, 1,4,4a,4b,5,8,8a,9a-octahydro-, endo-trans-endo	EtOH	302(1.52)	35-1059-66
endo-trans-exo-	EtOH	221(3.15),308(1.92)	35-1059-66
exo-trans-exo-	EtOH	225(3.41),309(2.14)	35-1059-66
Ethanol, 2-phenyl-2-tropyl-	C_6H_{12}	258(3.59)	44-0912-66
1,4-Methanonaphthalene, 1,4-dihydro-9-isopropylidene-6-methoxy-	heptane	189(4.44),211(4.45), 256(3.37),263(3.37), 287(3.55),294(3.52)	44-1988-66
2(3H)-Phenanthrone, 4,4a,9,10-tetra-hydro-4a-methyl-	EtOH	237(4.15),273(2.69)	35-1965-66
	tert-BuOH	320(1.69)	35-1965-66
Phenol, o-(3-phenylpropyl)-	EtOH	228s(3.54),268s(3.18), 274(3.25),283s(3.19)	1-1561-66
Propanol, 1,1-diphenyl-	hexane	253(2.60),258(2.66), 264(2.55)	22-0269-66
	C_6H_{12}	253(2.65),259(2.72), 265(2.56)	22-0269-66
Spiro[3-cyclohexene-1,2'(1'H)-naphthalen]-1'-one, 3',4'-dihydro-	EtOH	249(4.08),291(3.27)	22-1693-66
Spiro[cyclopentane-1,2'(1'H)-naphthalen]-3-one, 3',4'-dihydro-1'-methylene-	EtOH	245(4.04),282(3.08)	35-0161-66
$C_{15}H_{16}O_2$			
Cycloprop[a]indene-$\Delta^{4(1H)}$,α-acetic acid, 1a,1b,6,6a-tetrahydro-3-hy-droxy-α,1bβ,6α-trimethyl-, γ-lactone	EtOH	300(4.36)	78-1159-66
1H-Naphtho[2,1-b]pyran, 3-ethoxy-2,3-dihydro-	n.s.g.	213(4.58),235(4.66), 240(4.55),296(3.67), 311(3.51),318s(3.35), 326(3.64)	22-1974-66
1-Oxaspiro[4.5]dec-2-en-4-one, 2-phenyl-	EtOH	220(3.88),245(3.94), 305(4.23)	88-0233-66 +32-1073-66
6-Oxaspiro[4.5]dec-7-en-9-one, 7-phenyl-	EtOH	245(3.47),303(3.84)	88-0233-66 +32-1073-66
3(2H)-Phenanthrenone, 1,9,10,10a-tetrahydro-7-methoxy-	EtOH	243(4.00),330(4.39)	87-0304-66
Phenol, p-(p-methoxyphenethyl)-	50% dioxan	276(3.56)	71-0550-66A
$C_{15}H_{16}O_2S$			
p-Benzoquinone, 3-(5-ethyl-2-thienyl)-2,5,6-trimethyl-	EtOH	257(4.28),314(3.46), 450(3.33)	33-1794-66
Thioxanthene, 8a,10a-dihydro-2,4-dimethyl-, 10,10-dioxide	$CHCl_3$	261(3.44)	30-0107-66D
$C_{15}H_{16}O_3$			
$\Delta^{5(3H)}$,α-Azuleneacetic acid, 3a,4,6,7-tetrahydro-6α-hydroxy-α,3aα,8-tri-methyl-3-oxo-, γ-lactone	EtOH	215(4.31),295(4.14)	78-1723-66
1,3-Cyclohexanedione, 2-benzoyl-5,5-dimethyl-	EtOH-NaOH	260s(4.27),270(4.31), 325s(3.58)	99-0230-66
Cycloprop[2,3]indeno[5,6-b]furan-2(4H)-one, 4a,5,5a,6,6a,6b-hexahydro-4α-hydroxy-3,6β-dimethyl-5-methylene-	EtOH	280(4.41)	78-1159-66
Ivaxillarin, anhydro-	EtOH	234(4.22)	44-3232-66

Compound	Solvent	λ_{max}(log ϵ)	Ref.
$C_{15}H_{16}O_4$			
Coumarin, 8-(2-formyl-2-methylpropyl)-7-methoxy-	EtOH	245(3.56),256(3.57), 320(4.10)	78-1489-66
2-Cyclohexen-1-one, 3-methoxy-6-(m-methoxybenzoyl)-	EtOH	253(4.40)	70-2097-66
2-Cyclohexen-1-one, 3-methoxy-6-(p-methoxybenzoyl)-	EtOH	261(4.04)	70-2097-66
1-Epiallohelenalin, dehydro-	EtOH	225(4.26)	78-3279-66
Hydroquinone, 2-acetyl-3-(3,4,5-trimethyl-2-furyl)-	EtOH	219(4.23),238s(4.11), 330(3.61),381(3.67)	33-1794-66
Isoheteropeucenin	MeOH	206(4.36),258(4.30), 296(3.85)	39-0114-66C
Pteroxylin, dihydro-	MeOH	208(4.20),253(4.11), 258(4.15),290(4.15), 327(3.60)	39-0114-66C
$C_{15}H_{16}O_5$			
2-Naphthoic acid, 1,4-dihydro-3-hydroxy-8-methyl-, methyl ester, acetate	EtOH	203(4.67),273(3.38), 280(3.38)	44-1747-66
Sclerin, acetate	EtOH	216(4.41),260(3.93), 313(3.42)	88-5205-66
$C_{15}H_{16}O_6$			
Elephantol	MeOH	209(4.26)	35-3674-66
$C_{15}H_{16}O_8$			
Benzyl alcohol, 2,3,5-trihydroxy-, tetraacetate	EtOH	267(2.89),273s(2.87)	33-0204-66
$C_{15}H_{16}S$			
Naphtho[1,2-b]thiepin, 2,3,4,5-tetrahydro-7-methyl-	EtOH	222s(4.62),228(4.67), 233s(4.61),305(3.89)	22-0228-66
1H-Naphtho[2,1-b]thiopyran, 8-ethyl-2,3-dihydro-	EtOH	220(4.54),258(4.71), 278(3.85),288(3.93), 298(3.88),334(3.24), 345(3.13)	22-0228-66
2H-Naphtho[1,2-b]thiopyran, 6-ethyl-3,4-dihydro-	EtOH	218(4.67),250(4.27), 315(3.82),322s(3.81)	22-0228-66
$C_{15}H_{16}Si$			
Dibenzosilole, 5-ethyl-5-methyl-	n.s.g.	270f(4.0)	65-0114-66
$C_{15}H_{17}ClN_2$			
2,3,4,6,7,12-Hexahydro-1H-indolo-[2,3-a]quinolzin-5-ium chloride	MeOH	223(3.88),248(4.02), 353(4.35)	4-0101-66
$C_{15}H_{17}ClN_6O_4S$			
Sulfanilamide, N^1-[2-chloro-6-(2-methoxyethoxy)-9-methylpurin-8-yl]-	pH 1.2	287(4.45)	87-0373-66
	pH 12.9	261(4.32),292(4.38)	87-0373-66
$C_{15}H_{17}ClO_3$			
Marasmic acid, $SOCl_2$ product	MeOH	239(4.01)	35-2838-66
$C_{15}H_{17}N$			
3-Picoline, 2-(β-methylphenethyl)-	EtOH	267(3.67)	44-4322-66
3-Picoline, 4-(β-methylphenethyl)-	EtOH	264(3.76)	44-4322-66
Quinoline, 2,3-hexamethylene-	MeOH	277f(3.60)	5-0149-66H
	MeOH-H_2SO_4	323(4.05)	5-0149-66H
	C_6H_{12}	270f(3.58)	5-0149-66H

Compound	Solvent	$\lambda_{max}(\log \varepsilon)$	Ref.
$C_{15}H_{17}NO_2$			
2H-Benzo[a]quinolizin-2-one, 1,6,7,11b-tetrahydro-1-methoxy-4-methyl-	EtOH	323(4.17)	44-0797-66
2H-Benzo[a]quinolizin-2-one, 1,6,7,11b-tetrahydro-9-methoxy-4-methyl-	EtOH	287s(3.72),319(4.20)	44-0797-66
Cinnamic acid, α-cyano-β-isopropyl-, ethyl ester, cis	EtOH	272s(3.45)	22-1033-66
Cinnamic acid, α-cyano-β-propyl-, ethyl ester	EtOH	280(3.99)	22-1033-66
$C_{15}H_{17}NO_3$			
Indole-3-carboxaldehyde, 4-acetoxy-1-ethyl-2,6-dimethyl-	MeOH	216(4.43),248(4.18), 269(4.00),310(4.08)	44-1012-66
$C_{15}H_{17}NO_3S_2$			
5-Thiazolidineacetic acid, N-benzyl-4-oxo-2-thioxo-, propyl ester	EtOH	261(4.07),296(4.18)	95-0095-66
$C_{15}H_{17}NO_4$			
3-Isoquinolinecarboxylic acid, 6,7-dimethoxy-1-methyl-, ethyl ester	EtOH	254(4.60),300(3.89), 316(3.77),330(3.37)	95-0072-66
$C_{15}H_{17}NO_5$			
Hydracrylonitrile, 2-(2,4-dimethoxybenzoyl)-2-methyl-, acetate	EtOH	234(3.99),273(3.93), 305(3.79)	78-1027-66
$C_{15}H_{17}NO_6$			
Quinaldic acid, 4-hydroxy-5,6,7-trimethoxy-, ethyl ester	EtOH	229(4.60),264(4.35), 351(3.93)	24-0160-66
Quinaldic acid, 4-hydroxy-5,6,8-trimethoxy-, ethyl ester	EtOH	223(4.32),258(4.39), 332(3.78),380(3.79)	24-0160-66
Quinaldic acid, 4-hydroxy-5,7,8-trimethoxy-, ethyl ester	EtOH	218(4.46),245(4.27), 259(4.28),292(4.16), 330(3.70)	24-0160-66
Quinoline-3-carboxylic acid, 4-hydroxy-5,6,8-trimethoxy-, ethyl ester	EtOH	225(4.44),330(4.15)	24-0160-66
$C_{15}H_{17}N_3$			
7-Azaindoline, 6-(methylamino)-4-methyl-1-phenyl-	EtOH	266(4.13),334(4.03)	78-3233-66
$C_{15}H_{17}N_3O_2$			
3,6-Pyridazinedione, 1-allyl-4-(dimethylamino)-2-phenyl-	EtOH	314(4.19)	95-1082-66
3(2H)-Pyridazinone, 6-(allyloxy)-4-(dimethylamino)-2-phenyl-	EtOH	319(4.03)	95-1082-66
Pyrimidine-4-carboxaldehyde, 2-amino-6-hydroxy-5-(4-phenylbutyl)-	pH 1	269(3.76)	4-0315-66
	pH 7	299(3.66)	4-0315-66
	pH 13	288(3.46),354(3.36)	4-0315-66
Triazen-3-ol, 3-benzyl-1-(p-ethoxyphenyl)-	1% acetone	254(5.65)	65-2117-66*
$C_{15}H_{17}N_3O_3$			
Isoleucine, N-(2-quinoxalinylcarbonyl)-	EtOH	206(3.19),244(3.55), 318(2.79),328(2.58)	87-0266-66
Leucine, N-(2-quinoxalinylcarbonyl)-	EtOH	207(3.45),244(3.69), 320(2.95),327(2.95)	87-0266-66
Norleucine, N-(2-quinoxalinylcarbonyl)-	EtOH	206(3.46),244(3.59), 318(2.79),326(2.84)	87-0266-66

Compound	Solvent	λ_{max}(log ε)	Ref.
$C_{15}H_{17}N_3O_3S$			
Azobenzene-p-sulfonic acid, p'-(trimethylammonium) hydroxide, inner salt	41.1% H_2SO_4	315(4.40)	35-2240-66
	79.7% H_2SO_4	408(4.51)	35-2240-66
Ethionine, N-(2-quinoxalinyl-carbonyl)-	EtOH	206(3.42),245(3.60), 278(2.91),318(2.81), 328(2.84)	87-0266-66
$C_{15}H_{17}N_5O_4$			
Benzoic acid, p-[N-[3-[N-(2-amino-4-hydroxy-5-pyrimidinyl)formamido]-propyl]amino]-	pH 1	224(4.27),270(4.08), 302s(3.86)	87-0590-66
	pH 13	281(4.37)	87-0590-66
$C_{15}H_{18}$			
1,4-Methanonaphthalene, 1,2,3,4-tetra-hydro-9-isopropylidene-6-methyl-	heptane	198(4.74),266(3.23), 271(3.37),274(3.41), 280(3.48)	44-1988-66
Naphthalene, 2-isopropyl-4,8-dimethyl-	C_6H_{12}	278(3.78),288(3.86)	88-5881-66
$C_{15}H_{18}BrNO_5$			
Indole-3-carboxaldehyde, 6-bromo-1-ethyl-4,7-dihydro-5-methoxy-2-methyl-4,7-dioxo-, 3-(dimethyl acetal)	MeOH	233(4.20),298(4.11), 350(3.43),465(3.15)	35-0804-66
$C_{15}H_{18}BrN_3O$			
4-Pyrimidinol, 2-amino-6-(bromo-methyl)-5-(4-phenylbutyl)-, hydro-bromide	pH 1	278(3.94)	4-0324-66
	pH 7	306(3.90)	4-0324-66
	pH 13	300(3.80)	4-0324-66
$C_{15}H_{18}Br_2N_4O_2$			
4-Carbamoyl-4'-formyl-1,1'-trimeth-ylenebis[pyridinium bromide], oxime	H_2O	275(4.31)	87-0638-66
$C_{15}H_{18}ClNO_4S$			
2-tert-Butyl-3-methyl-4H-indeno-[2,1-d]thiazolium perchlorate	MeOH	229(3.73),253s(3.50), 312(4.15)	39-0686-66C
$C_{15}H_{18}ClNO_6$			
N-(2-Methyl-4-oxopent-2-yl)-5-phenyl-isoxazolium perchlorate	CH_2Cl_2	298(4.35)	44-2039-66
$C_{15}H_{18}N_2$			
1H-Indolo[3,2-g]indolizine, 2,3,5,6,11,11b-hexahydro-11-methyl-	EtOH-HCl	226(4.58),284(3.87), 292s(3.82)	87-0864-66
	EtOH	229(4.53),286(3.85), 293s(3.81)	87-0864-66
12H-Indolo[2,3-a]quinolizine, 1,2,3,4,6,7-hexahydro-	EtOH	282(3.85),289s(3.78)	12-1951-66
$C_{15}H_{18}N_2O$			
2-Pyrazolin-5-one, 4-crotyl-3,4-dimethyl-1-phenyl-	EtOH	242(4.23)	88-6413-66
3-Pyrazolin-5-one, 2-crotyl-3,4-dimethyl-1-phenyl-	EtOH	247(4.04),275(3.99)	88-6413-66
2-Pyrazolin-5-one, 4-(1-methylallyl)-3,4-dimethyl-1-phenyl-	EtOH	242(4.24)	88-6413-66
2-Pyrrolidinone, 4-(-ethylpropyli-dene)-5-imino-3-phenyl-	EtOH	260(4.25)	22-2895-66

Compound	Solvent	λ_{max}(log ε)	Ref.
$C_{15}H_{18}N_2O_2$			
Propionamide, 3-(1,4,5,6-tetrahydro-3-pyridyl)-3-oxo-N-benzyl-	EtOH	304(4.43)	44-2487-66
$C_{15}H_{18}N_2O_3$			
4H-1,2-Diazepin-4-one, 1-acetyl-1,2,3,7-tetrahydro-7-methoxy-5-methyl-6-phenyl-	EtOH	263(3.98)	44-0034-66
4-Oxa-1,2-diazaspiro[4.5]dec-1-en-3-ol, 3-phenyl-, acetate	C_6H_{12}	257(2.72),263(2.63), 270(2.45),319(2.46)	88-0411-66
$C_{15}H_{18}N_2O_4$			
Carbamic acid, methyl-, ester with 1-ethyl-3-(hydroxymethyl)-2,6-dimethylindole-4,7-dione	MeOH	228(4.16),264(4.13), 346(3.42),441(3.30)	44-1012-66
$C_{15}H_{18}N_2O_6$			
Quinoxaline, 2-(β-D-glucopyranosyl-oxy)-3-methyl-	MeOH	219s(4.16),240(4.27), 244(4.26),307s(3.73), 317(3.83),328(3.78)	24-0547-66
$C_{15}H_{18}N_2O_9$			
Uracil, 5-β-D-arabinofuranosyl-, 2',3',5'-triacetate	pH 7	262(3.91)	44-2215-66
	pH 12	293(4.13)	44-2215-66
$C_{15}H_{18}N_4$			
Heptafulvene, 8,8-dicyano-5-iso-propyl-2,3-bis(methylamino)-	MeOH	258(4.14),330(4.36), 425(4.42)	18-2444-66
$C_{15}H_{18}N_4O_3$			
2-Quinoxalinecarboxylic acid, 3-(2-morpholinoethylamino)-, sodium salt	EtOH	256(4.49),298(3.52), 384(3.75)	87-0770-66
$C_{15}H_{18}N_4O_4$			
Cyclohexylideneacetaldehyde, α-methyl-, 2,4-dinitrophenyl-hydrazone	$CHCl_3$	387(4.53)	22-1400-66
Cyclohexylideneacetaldehyde, 2-methyl-, 2,4-dinitrophenyl-hydrazone	$CHCl_3$	391(4.40)	22-1400-66
Cyclohexylideneacetaldehyde, 4-methyl-, 2,4-dinitrophenyl-hydrazone	$CHCl_3$	390(4.51)	22-1400-66
3,5-Hexadienal, 2,2,4-trimethyl-, 2,4-dinitrophenylhydrazone	n.s.g.	360(4.39)	22-0734-66
$C_{15}H_{18}N_4S$			
10H-Dipyrido[2,3-b:2',3'-e][1,4]thia-zine, 10-(3-dimethylamino-1-propyl)-	EtOH-base	242(3.94),332(4.36)	87-0116-66
$C_{15}H_{18}N_6$			
s-Triazine derivative	MeOH	252.5(4.06)	89-0959-66
isomer A	MeOH	256(4.03)	89-0959-66
isomer B	MeOH	262.5(3.92)	89-0959-66
$C_{15}H_{18}O$			
Cyclohexanone, 5-methyl-2-(p-methylbenzylidene)-	EtOH	292(4.27)	35-1419-66

Compound	Solvent	$\lambda_{max}(\log \epsilon)$	Ref.
1,4:5,8-Dimethanofluoren-9-one, 1,2,3,4,4a,4b,5,8,8a,9a-decahydro-, endo-trans-endo	EtOH	303(1.44)	35-1059-66
endo-trans-exo	EtOH	302.5(1.57)	35-1059-66
exo-trans-endo	EtOH	224(3.12),307(1.90)	35-1059-66
exo-trans-exo	EtOH	224(3.12),309(1.92)	35-1059-66
9-Fluorenone, 1,2,3,4,4a,9a-hexahydro-1,4a-dimethyl-	EtOH	248(4.12),291(3.3)	88-2449-66
1,6-Methano[10]annulene, 2-tert-butoxy-	C_6H_{12}	260(4.69),316(3.78), 403(2.67)	89-0733-66
1,6-Methano[10]annulene, 3-tert-butoxy-	C_6H_{12}	262(4.85),304(3.84), 383(2.52)	89-0733-66
5,9-Methanobenzocycloocten-10(5H)-one, 6,7,8,9-tetrahydro-5,7-dimethyl-	EtOH	254(4.07),292(2.95)	88-2449-66
5,9-Methanobenzocycloocten-10(5H)-one, 6,7,8,9-tetrahydro-8,11-dimethyl-	EtOH	254(4.12),292(3.3)	88-2449-66
Pentatriafulvalene-1-carboxaldehyde, 5,6-dipropyl-	MeOH	237(4.12),319(3.98), 373(3.91)	35-3359-66
Spiro[cyclohexane-1,2'(1'H)-naphthalen]-1-one, 3',4'-dihydro-	EtOH	249(4.08),292(3.21)	22-1693-66
$C_{15}H_{18}O_2$			
Cyclohexanone, 2-(p-methoxybenzylidene)-5-methyl-	EtOH	318(4.26)	35-1419-66
$\Delta^{2(1H),\alpha}$-Naphthaleneacetic acid, 4a,5,6,7,8,8aβ-hexahydro-3-hydroxy-α,4aα-dimethyl-8-methylene-	EtOH	275(4.42)	78-1159-66
3(2H)-Phenanthrone, 1,4,4aα,9,10,10aβ-hexahydro-6-methoxy-	EtOH	225s(3.78),281(3.34), 289(3.28)	87-0304-66
3(2H)-Phenanthrone, 1,4,4aα,9,10,10aβ-hexahydro-7-methoxy-	EtOH	220s(3.92),278(3.31), 287(3.28)	87-0304-66
3(2H)-Phenanthrone, 1,4,4aβ,9,10,10aβ-hexahydro-7-methoxy-	EtOH	225s(3.88),279(3.28), 287(3.25)	87-0304-66
Toluene, α-(4-methylcyclohexylidene)-3,4-(methylenedioxy)-, endo	EtOH	236(3.72),288(3.76)	35-1419-66
exo	EtOH	258(4.06),295(3.77)	35-1419-66
Warburgiadione	EtOH	292(4.33)	77-0701-66
	EtOH	292(4.33)	77-0393-66
$C_{15}H_{18}O_3$			
$\Delta^{5(3H),\alpha}$-Azuleneacetic acid, 2,3a,4,6,7,8-hexahydro-6α-hydroxy-α,3aα,8α-trimethyl-3-oxo-, γ-lactone	EtOH	296(4.08)	78-1723-66
5-Azuleneacetic acid, 3,3a,4,5,6,7-hexahydro-6α-hydroxy-α,3aα,8-trimethyl-3-oxo-, γ-lactone	EtOH	217(4.19)	78-1723-66
Cyclohexanone, 2-(3,4-methylenedioxybenzylidene)-5-methyl-	EtOH	325(4.01)	35-1419-66
Cycloprop[2,3]indeno[5,6-b]furan-2(4H)-one, 4aα,5,5a,6,6a,6b-hexahydro-4α-hydroxy-3,5α,6bβ-trimethyl-	EtOH	284(4.39)	78-1159-66
5β-isomer	EtOH	284(4.34)	78-1159-66
Isoaromatin	EtOH	220(4.44),322(1.82)	78-3279-66
Isoperuvin, anhydro-	EtOH	216(4.19)	78-1723-66
Ivaxillarin, anhydrodihydro-	EtOH	236(4.23)	44-3232-66
1(2H)-Naphthalenone, 3,4-dihydro-6-methoxy-2-(3-oxobutyl)-	EtOH	224(4.12),274(4.25)	87-0304-66
Naphtho[2,3-b]furan-3-carboxaldehyde, 4,4a,5,6,7,8,8a,9-octahydro-4a,5-dimethyl-6-oxo-	EtOH	271(3.42)	77-0393-66

Compound	Solvent	λ_{max}(log ε)	Ref.
2,4-Pentadienoic acid, 2-ethoxy-5-phenyl-, ethyl ester	MeOH	231(4.00),318(4.70)	5-0053-66I
Perezinone	EtOH	204(4.30),324(4.42)	78-2387-66
3(2H)-Phenanthrone, 1,4,4aα,9,10,10aβ-hexahydro-7-hydroxy-6-methoxy-	EtOH	225s(3.86),288(3.59), 295s(3.49)	87-0304-66
$C_{15}H_{18}O_4$			
1-Epiallohelenalin	EtOH	220(4.19),331(1.85)	78-3279-66
Hydroquinone, (3-methyl-2-butenyl)-, diacetate	ether	209.5(4.22)	24-0885-66
1,2-Indandicarboxylic acid, 4,5,6,7-tetramethyl-	MeCN	272(2.53),277s(--), 281s(--)	39-1147-66B
Isomexicanin H	n.s.g.	219(3.90)	88-1029-66
Ivaxillarin	EtOH	207(4.00)	44-3232-66
Jacquinelin	EtOH	208(3.91),255(4.14)	39-1298-66C
Marasmic acid	MeOH	241(3.99)	35-2838-66
2α-Phenanthrenecarboxylic acid, 1,2,3,4,4aα,4bβ,5,6,7,9,10,10aβ-dodecahydro-4,7-dioxo-	EtOH	239(4.22)	44-0713-66
2β-Phenanthrenecarboxylic acid, 1,2,3,4,4aβ,5,6,7,8,9,10,10aβ-dodecahydro-4,7-dioxo-	EtOH	287s(1.95)	44-0713-66
Psilostachyin, anhydro-	MeOH	210(4.01)	78-1139-66
Psilostachyin B	MeOH	210(4.02)	78-1943-66
$C_{15}H_{18}O_5$			
Pulvilloric acid	C_6H_{12}	215(3.11),317(3.14), 387(2.37)	39-1608-66C
ethanol adduct	EtOH	252(3.93),307(3.55)	39-1608-66C
$C_{15}H_{18}O_6$			
Elephantol, dihydro-	MeOH	211(3.98)	35-3674-66
Elephantolide, dihydro-	MeOH	211(3.94)	35-3674-66
$C_{15}H_{19}BrO$			
Laurinterol	EtOH	225(3.85),283(3.34), 289(3.32)	88-1837-66
$C_{15}H_{19}Br_2N_3O$			
4-Pyrimidinol, 2-amino-6-(bromomethyl)-5-(4-phenylbutyl)-, hydrobromide	pH 1	278(3.94)	4-0324-66
	pH 7	306(3.90)	4-0324-66
	pH 13	300(3.80)	4-0324-66
$C_{15}H_{19}ClN_2$			
1H-Indolo[3,2-g]indolizine, 2,3,5,6,11,11b-hexahydro-11-methyl-, hydrochloride	EtOH	226(4.58),284(3.87), 292s(3.82)	87-0864-66
$C_{15}H_{19}ClN_6O$			
Pyrimidine, 2-amino-4-(butylamino)-5-(p-chlorophenylazo)-6-methoxy-	EtOH	255(4.13),304(3.71), 388(4.51)	87-0977-66
$C_{15}H_{19}N$			
9H-Indeno[2,1-b]pyridine, 2,3,4,4a-tetrahydro-1,4,4-trimethyl-	EtOH	233(3.94),305(3.97)	44-0903-66
$C_{15}H_{19}NO$			
Naphthalene, 1,2-dihydro-8-methoxy-3-pyrrolidino-	EtOH	205(4.03),223(4.08), 245(4.02),316(4.16)	44-1747-66

Compound	Solvent	λ max(log ε)	Ref.
2,4-Undecadiene-8,10-diynamide, N-isobutyl-, 2-cis-4-trans	ether	254(4.48)	24-3194-66
$C_{15}H_{19}NO_2$			
1-Cyclopentene-1-carboxylic acid, 2-(benzylamino)-, ethyl ester	EtOH	295(4.31)	78-2575-66
Indol-4-ol, 1-ethyl-2,5,6-trimethyl-, acetate	MeOH	225(4.58),277(3.94)	44-1012-66
$C_{15}H_{19}NO_2S$			
4-Thiaionylidenecyanoacetic acid, trans	EtOH	265(4.11),395(4.06)	78-0265-66
$C_{15}H_{19}NO_3$			
Cocculolidine	MeOH	215(4.10)	88-5179-66
$C_{15}H_{19}NO_7$			
Malonic acid, [(5-carboxy-2-furyl)-amino]methylene-, triethyl ester	5% EtOH	344(4.48)	4-0202-66
$C_{15}H_{19}N_3O_2$			
4-Pyrimidinemethanol, 2-amino-6-hydroxy-5-(4-phenylbutyl)-, hydrochloride	pH 1	269(3.96)	4-0324-66
	pH 13	286(3.91)	4-0324-66
$C_{15}H_{19}N_3O_8$			
Cytosine, 1-(2,3,5-tri-O-acetyl-β-D-arabinofuranosyl)-	pH 1	276(4.14)	87-0268-66
	pH 7	233(3.91),270(3.96)	87-0268-66
	pH 13	274(4.00)	87-0268-66
Cytosine, N^4,$O^{3'}$,$O^{5'}$-triacetyl-β-D-arabinofuranosyl-	EtOH	248(4.20),300(3.89)	88-3499-66
$C_{15}H_{19}N_4O_{11}P$			
5'-Uridylic acid, p-nitrophenyl ester, ammonium salt	pH 7	261(4.18),292s(3.91)	44-3241-66
	pH 12	261(4.09),292s(3.91)	
$C_{15}H_{20}$			
α-Calocorene	EtOH	214(4.06),220(4.10), 227(3.96),265(3.68), 297s(2.74)	1-2841-66
Cyclohexane, 1-(p-methoxybenzylidene)-4-methyl-, endo	EtOH	235(3.45)	35-1419-66
exo	EtOH	248(4.16)	35-1419-66
Fluorene, 1,2,3,4,4a,9a-hexahydro-1,4a-dimethyl-	EtOH	260(2.95),266(3.08), 273(3.08)	88-2449-66
Indene, 1,1,2,3,5,6-hexamethyl-	heptane	215(4.38),268(4.05)	44-0081-66
5,9-Methanobenzocyclooctene, 5,6,7,8,9,10-hexahydro-1,4a-dimethyl-	EtOH	260(3.03),266(3.02), 274(2.92)	88-2449-66
$C_{15}H_{20}ClNO_2$			
Benzo[c]pyrid[1,2-a]azepine, 1,2,3,4-6,7,8,12b-octahydro-10,11-(methylenedioxy)-, hydrochloride	EtOH	244(3.62),290(3.66)	94-0622-66
$C_{15}H_{20}ClNO_4$			
2,3,4,4a-Tetrahydro-1,4,4-trimethyl-9H-indeno[2,1-b]pyridine perchlorate	EtOH	232(3.60),305(3.78)	44-0903-66

Compound	Solvent	$\lambda_{max}(\log \epsilon)$	Ref.
$C_{15}H_{20}ClN_3O_2$			
4-Pyrimidinemethanol, 2-amino-6-hydroxy-5-(4-phenylbutyl)-, hydrochloride	pH 1	269(3.96)	4-0324-66
	pH 13	286(3.91)	4-0324-66
$C_{15}H_{20}Cl_2N_8S$			
Pyrimidine-6-isothiouronium chloride, 2-amino-4-(butylamino)-5-(p-chlorophenylazo)-	EtOH	225(4.47),261(4.24), 306(4.58)	87-0977-66
$C_{15}H_{20}N_2O_2$			
Benzimidazoleacetic acid, α-butyl-, ethyl ester	$CHCl_3$	276(3.86),284(3.84)	22-3989-66
Benzimidazoleacetic acid, α-isobutyl-, ethyl ester	$CHCl_3$	276(3.87),284(3.94)	22-3989-66
Glutarimide, 2-(diethylamino)-2-phenyl-	EtOH	204(4.43)	87-0615-66
	EtOH-KOH	237(3.28)	87-0615-66
$C_{15}H_{20}N_4O_3$			
Hypoxanthine, 1,7-bis(tetrahydropyran-2-yl)-	pH 1	250(4.00)	44-2685-66
	pH 11	257(3.85)	44-2685-66
	EtOH	258(3.82)	44-2685-66
Hypoxanthine, 1,9-bis(tetrahydropyran-2-yl)-	pH 1	250(4.04)	44-2685-66
	pH 11	251.5(4.00)	44-2685-66
	EtOH	246(3.97),251(3.96)	44-2685-66
$C_{15}H_{20}N_4O_5$			
Cyclooctanone, 3-methoxy-2,4-dinitrophenylhydrazone	EtOH	370(4.27)	39-0164-66B
$C_{15}H_{20}N_5O_9P$			
Cytidine, 2'-deoxy-, 5'-(p-nitrophenyl hydrogen phosphate), ammonium salt	pH 7.0	275(4.21)	44-3241-66
	pH 12.0	280(4.37)	44-3241-66
$C_{15}H_{20}N_5O_{10}P$			
5'-Cytidylic acid, p-nitrophenyl ester, ammonium salt	pH 2.0	280(4.32)	44-3241-66
	pH 7	275(4.22)	44-3241-66
$C_{15}H_{20}O$			
Benz[f]azulene-9-methanol, 1,2,3,3a,4,9,10,10a-octahydro-	pentane	258(2.43),263(2.48), 271(2.42)	35-0603-66
Cyclohexane, 1-(p-methoxybenzylidene)-4-methyl-, endo	EtOH	227(4.01),277(3.32), 285(3.28)	35-1419-66
exo	EtOH	252(4.20)	35-1419-66
Cyclopentanone, 2,5-dicyclopentylidene-	EtOH	300(4.28)	39-0075-66B
$C_{15}H_{20}O_2$			
5-Azuleneacetic acid, 1,3a,4,5,6,7,-8,8a-octahydro-6α-hydroxy-3aα,8α-dimethyl-α-methylene, lactone	EtOH	211(4.00)	78-1499-66
2,4,6,8,10-Dodecapentaen-1-al, 12-hydroxy-2,6,10-trimethyl-	pet ether	266(3.98),347(4.73), 365(4.93),386(4.91)	33-0369-66
Pentadienoic acid, 3-methyl-5-(2,6,6-trimethylcyclohexa-1,3-dien-1-yl)-, 2-cis-4-trans	n.s.g.	256(3.91),347(4.08)	28C-1397-66A
2-trans-4-trans	n.s.g.	259(3.96),343(4.08)	28C-1397-66A

Compound	Solvent	$\lambda_{max}(\log \epsilon)$	Ref.
$C_{15}H_{20}O_3$			
5-Azuleneacetic acid, 1,2,3,3a,4,-5,6,7,8,8aβ-decahydro-6α-hydroxy-4aα,8α-dimethyl-α-methylene-3-oxo-, lactone	EtOH	212(4.03)	78-1499-66
5-Azuleneacetic acid, 1,2,3,3a,4,5,-6,7,8,8a-decahydro-6α-hydroxy-4aα,8α-dimethyl-α-methylene-3-oxo-, γ-lactone	EtOH	295(1.59)	78-1709-66
5-Azuleneacetic acid, 2β,3β-epoxy-1,2,3,3a,4,5,6,7,8,8aβ-decahydro-6α-hydroxy-3aα,8α-dimethyl-α-methylene-, lactone	EtOH	213(3.99)	78-1499-66
Bisabolangelone	EtOH	254(4.56)	88-3541-66
2-Naphthoic acid, 2,3,4,6,7,8-hexa-hydro-1,5-dimethyl-6-oxo-, ethyl ester	EtOH	295(4.28)	35-1766-66
Peroxycinnamic acid, α,β-dimethyl-, tert-butyl ester, cis	EtOH	232(3.86)	35-1959-66
trans	EtOH	238(3.98)	35-1959-66
α-Pipitzol	EtOH	240(3.60),280(3.89)	78-2387-66
β-Pipitzol	EtOH	244(3.58),282(3.85)	78-2387-66
Reduction product of marasmic acid	MeOH	203(3.81)	35-2838-66
$C_{15}H_{20}O_3S$			
Dimethylsulfonium 3-carboxy-2-p-tolylallylide, ethyl ester, oxide	EtOH	237(3.90),256(3.88), 346(4.20)	88-1787-66
$C_{15}H_{20}O_4$			
Abscisin	n.s.g.	254(4.23)	22-3874-66
	0.002N HCl	262(4.33)	50-0627-66B
Abscisin II	n.s.g.	254(4.25)	28C-1397-66A
5-Azuleneacetic acid, decahydro-6α,8aβ-dihydroxy-3aα,5α-dimethyl-α-methylene-3-oxo-, γ-lactone	EtOH	205(3.74)	78-1723-66
1-Epiallohelenalin, dihydro-	EtOH	222(3.98),329(1.85)	78-3279-66
Isoperuvin	EtOH	219(4.20)	78-1723-66
Jacquinelin, dihydro-	EtOH	230(4.10)	39-1298-66C
Peruvin	EtOH	214(3.96)	78-1723-66
2β-Phenanthrenecarboxylic acid, 1,2,3,4,4aβ,5,6,7,8,9,10,10aβ-dodecahydro-4α-hydroxy-7-oxo-	EtOH	277(1.40),287(1.30)	44-0713-66
2β-Phenanthrenecarboxylic acid, 1,2,3,4,4aα,4bβ,5,6,7,9,10,10aβ-dodecahydro-4,7-dioxo-, methyl ester?	EtOH	239(4.21)	44-0713-66
Psilostachyin C	MeOH	210(4.01)	44-1629-66
$C_{15}H_{20}O_4S$			
Dimethylsulfonium 3-carboxy-2-p-anis-ylallylide, ethyl ester, oxide	EtOH	247(3.87),269(3.90), 341(4.27)	88-1787-66
$C_{15}H_{20}O_5$			
Hydratropic acid, 2-carboxy-3-methoxy-4,5,6-trimethyl-	EtOH	211(4.29),280(3.04)	88-5205-66
Psilostachyin	MeOH	212(4.10)	78-1139-66
Spiro[1,3-dioxolane-2,1'(2'H)-naphtha-lene]-5'-carboxylic acid, 3',4',6',7',8',8'a-hexahydro-8'a-methyl-6'-oxo-, methyl ester	EtOH	247(4.01)	88-3489-66

Compound	Solvent	$\lambda_{max}(\log \epsilon)$	Ref.
$C_{15}H_{20}O_5S$			
Cumanin, cyclic sulfite	EtOH	213(4.02)	78-1499-66
Epicumanin, cyclic sulfite	EtOH	212(4.00)	78-1499-66
$C_{15}H_{20}S_2$			
Thiophene, 2,2'-(1-propylbutylidene)di-	n.s.g.	237(4.27)	22-2253-66
$C_{15}H_{21}ClN_2O_4$			
[5-(Dimethylamino)-5-phenyl-2,4-pentadienylidene]dimethylammonium perchlorate	MeCN	432(4.99)	24-2479-66
$C_{15}H_{21}N$			
2-Cyclohexene-1-methylamine, N,N-dimethyl-2-phenyl-	EtOH	242.5(4.03)	44-3482-66
	EtOH	242(4.03)	94-1418-66
Epiguaipyridine	EtOH	215s(3.58),270(3.70), 273s(3.71),278s(3.90)	35-3109-66
Patchoulipyridine	EtOH	217(3.81),270s(3.74), 273(3.80),277s(3.77), 283(3.67)	35-3109-66
$C_{15}H_{21}NO_2$			
2H-Benzo[b]quinolizin-1(6H)-one, 3,4,7,10,11,11a-hexahydro-, ethylene ketal	EtOH	267(2.08)	44-1716-66
$C_{15}H_{21}NO_2S$			
L-Leucine, N-thiobenzoyl-, ethyl ester	isooctane	289(3.70),400(2.26)	1-2781-66
	MeOH	286(3.83),377(2.27)	1-2781-66
	dioxan	289(3.80),387(2.44)	1-2781-66
$C_{15}H_{21}NO_3$			
Anhydrocycloheximide	EtOH	240(3.91)	44-0425-66
Azepino[3,2,1-hi]indole-7-methanol, 1,2,4,5,6,7-hexahydro-6,6-dimethoxy-	EtOH	252(3.86),293(3.38)	35-4061-66
α-Pipitzol, oxime	EtOH	278(3.87)	78-2387-66
β-Pipitzol, oxime	EtOH	278(3.87)	78-2387-66
$C_{15}H_{21}NO_4$			
o-Anisic acid, 6-(1-carbamoylethyl)-3,4,5-trimethyl-, methyl ester	EtOH	213(4.30),283(3.32)	88-5205-66
9-Oxa-1,3,5-cyclodecatriene-2,3-dicarboxylic acid, 1-(dimethylamino)-, dimethyl ester, cis	EtOH	257(4.05),270(4.08), 326(4.01),356(3.97)	35-4685-66
$C_{15}H_{21}N_3O_2$			
Imidazo[1,5-a]pyrazine, 2-(p-carbethoxyphenyl)octahydro-	EtOH	307(4.49)	87-0868-66
$C_{15}H_{21}N_5O$			
7(8H)-Pteridinone, 8-cyclohexyl-2-(dimethylamino)-6-methyl-	pH -0.89	239(4.40),272(4.36), 280s(4.33),343(3.80)	24-3022-66
	pH 6	223(4.32),248(4.26), 300(3.87),364(4.19)	24-3022-66
	MeOH	223(4.29),247(4.21), 303(3.81),358(4.24)	24-3022-66
$C_{15}H_{21}N_5O_5$			
Zeatin, 9-β-D-ribofuranosyl-	pH 1	208(4.30),266(4.27)	31-0515-66

Compound	Solvent	λ_{max}(log ϵ)	Ref.
Zeatin, 9-β-D-ribofuranosyl- (cont.)	pH 1	208(4.30),266(4.27)	39-0921-66C
	pH 7.2	211(4.29),270(4.25)	39-0921-66C
	pH 13	215(4.26),270(4.26)	39-0921-66C
$C_{15}H_{21}N_5O_6$			
7(8H)-Pteridinone, 2-(dimethylamino)-8-(β-D-galactopyranosyl)-6-methyl-	pH -0.89	239(4.36),264(4.16), 297(3.97),346(3.80)	24-3022-66
	pH 6	249(4.21),318s(3.76), 368(4.15)	24-3022-66
7(8H)-Pteridinone, 2-(dimethylamino)-8-(β-D-glucopyranosyl)-6-methyl-	pH -0.89	239(4.35),264(4.15), 297(3.96),344(3.81)	24-3022-66
	pH 5	248(4.23),320s(3.76), 368(4.15)	24-3022-66
$C_{15}H_{22}$			
Calamenene	EtOH	263s(2.59),269(2.74), 272s(2.68),278(2.76)	1-2841-66
Indan, 1,1,2,3,5,6-hexamethyl-	heptane	271(3.22),274(3.22), 279(3.28)	44-0081-66
Naphthalene, 5,6,7,8-tetrahydro-2-isopropyl-4,8-dimethyl-	C_6H_{12}	268(2.53),276(2.45)	88-5881-66
$C_{15}H_{22}INO_3$			
Oxazolo[2,3-a]isoquinoline, 2,3,5,6-tetrahydro-8,9-dimethoxy-10b-methyl-, methiodide	H_2O	277(3.51),281(3.52), 286(3.49)	83-0817-66
	MeOH	279(3.31),283(3.32), 288(3.29)	83-0817-66
$C_{15}H_{22}N_2$			
2-Pyrazoline, 1-benzyl-4-ethyl-5-propyl-	EtOH	246(3.56)	4-0413-66
$C_{15}H_{22}N_2O$			
4-Isoquinolinecarboxamide, N,N-diethyl-1,2,3,4-tetrahydro-2-methyl-	EtOH	264(2.56),380(1.43)	22-1763-66
$C_{15}H_{22}N_4O_3$			
2H-Oxazolo[2,3-h]pteridine-2,4(3H)-dione, 1,6a,8,9-tetrahydro-6,6a-diisopropyl-3-methyl-	pH 3	275(4.17),310(3.80)	24-3503-66
	pH 10	232(4.34),283(4.05), 315(3.92)	24-3503-66
$C_{15}H_{22}N_6O$			
Urea, 1-[bis(isopropylidenehydrazino)-methylene]-3-p-tolyl-	EtOH	253(4.54),270s(4.35)	39-0006-66C
$C_{15}H_{22}O$			
Azuleno[5,6-b]furan, 4,5,5a,6,8,-8a,9-octahydro-3,5,8-trimethyl-	heptane	225(3.84)	39-0377-66C
Cyclohexanone, 2-allyl-2-(1-cyclohex-1-enyl)-	C_6H_{12}	300(1.32)	22-0281-66
α-Cyperone	EtOH	248.5(4.18)	78-2311-66
Cyperotundone	EtOH	245(3.96)	94-0890-66
2,4,6,10-Dodecatetraenal, 3,7,11-trimethyl-	EtOH	341(4.56)	39-2154-66C
6,8,10,12-Pentadecatetraenal	ether	279(4.24),289(4.54), 303(4.74),318(4.70)	24-3544-66
Spiro[cyclohexane-1,1'-indan]-2-one, 4',5',6',7'-tetrahydro-2'-methyl-	C_6H_{12}	286(1.79)	22-0281-66

Compound	Solvent	$\lambda_{max}(\log \epsilon)$	Ref.
$C_{15}H_{22}O_2$			
$\Delta^{6(1H)},\alpha$-Azuleneacetic acid, 2,3,3a,4,5,7,8,8a-octahydro-7-hydroxy-α,1,4-trimethyl-, γ-lactone	EtOH	220(4.10)	39-0377-66C
7αH-Eremophila-10,11-dien-9-one, 8α-hydroxy-	EtOH	246(3.90)	12-0303-66
Pivalophenone, 3-tert-butyl-2-hydroxy-	EtOH	262(3.91),343(3.54)	77-0172-66
	EtOH-NaOH	239(3.90),310(3.25), 349(3.36)	77-0172-66
Pivalophenone, 3-tert-butyl-4-hydroxy-	EtOH	227(4.13),286(4.16)	77-0172-66
	EtOH-NaOH	248(3.87),350(4.45)	77-0172-66
$C_{15}H_{22}O_2S$			
4-Thiaionylideneacetic acid, methyl ester, trans	C_6H_{12}	253(4.23),342(4.15)	78-0265-66
$C_{15}H_{22}O_3$			
Jacquinelin, deoxyhexahydro-	EtOH	280(1.64)	39-1298-66C
Pentanoic acid, 3-hydroxy-3-phenyl-, tert-butyl ester	EtOH	242(2.35),247(2.41), 257(2.52),264(2.43)	22-0131-66
Teleocidic anhydride	EtOH	258(3.85)	88-2515-66
α,β-Unsaturated lactone	EtOH	250(3.99)	78-2387-66
$C_{15}H_{22}O_4$			
Ambrosiol	MeOH	213(3.92)	44-0681-66
5-Azuleneacetic acid, 1,2,3,3a,4,-5,6,7,8,8aβ-decahydro-2α,3α,6α-trihydroxy-3aα,8α-dimethyl-α-methylene-, lactone	EtOH	219(4.19)	78-1499-66
Cumanin	EtOH	213(3.97)	78-1499-66
2-Cyclohexene-1-carboxylic acid, 2-methyl-4-oxo-3-(3-oxopentyl)-, ethyl ester	EtOH	241(4.03)	35-1766-66
1-Epiallohelenalin, tetrahydro-	EtOH	229-302(1.56)	78-3279-66
Epicumanin	EtOH	214(4.00)	78-1499-66
Unknown acid	EtOH	277(3.95)	78-2387-66
$C_{15}H_{22}O_8S$			
β-L-Idose, 3,6-di-O-acetyl-5-(acetylthio)-5-deoxy-1,2-O-isopropylidene-	EtOH	230(3.62)	39-1287-66C
$C_{15}H_{23}IN_2$			
4-(p-Dimethylaminophenyl)-1,2,5,6-tetrahydro-N-methylpyridinium methiodide	EtOH	294(4.3)	78-2735-66
$C_{15}H_{23}IN_2S_2$			
3-Ethyl-2-[5-(3-ethyl-2-thiazolidinylidene)-1,3-pentadienyl]-2-thiazolinium iodide	MeOH	442(4.97)	9-0150-66
$C_{15}H_{23}NO$			
Cyclopropanemethanol, α,α-dipropyl-2-(4-pyridyl)-, cis	EtOH-HCl	261(4.05)	44-0399-66
	EtOH-NaOH	259(3.48)	44-0399-66
trans	EtOH-HCl	258(4.14)	44-0399-66
	EtOH-NaOH	258(3.30)	44-0399-66
$C_{15}H_{23}NO_3$			
m-Cresol, 4,6-di-tert-butyl-2-nitro-	C_6H_{12}	251(3.34),289(3.53)	23-0961-66
	EtOH	271(3.26)	23-0961-66

Compound	Solvent	λ_{max}(log ϵ)	Ref.
$C_{15}H_{23}NO_4$			
3,5-Pyridinedicarboxylic acid, 1,4-dihydro-1,2,4,6-tetramethyl-, diethyl ester	n.s.g.	232(4.24),256(4.09), 342(3.93)	39-1083-66B
$C_{15}H_{23}NO_5$			
2H-Oxecin-4,5-dicarboxylic acid, 3-(dimethylamino)-7,8,9,10-tetrahydro-, dimethyl ester, cis-cis	EtOH	329(4.08)	35-4685-66
$C_{15}H_{23}N_3$			
Pyridine, 1-(dimethylamino)-4-(p-dimethylamino)-1,2,3,6-tetrahydro-	EtOH	287(4.31)	78-2735-66
dihydrochloride	pH 1	249(4.11)	78-2735-66
	EtOH	294(4.31)	78-2735-66
$C_{15}H_{23}N_3O_4S$			
Morpholine, 2-benzyloxy-3,5-dihydroxy-4-(propylthioureido)-	EtOH	243(4.16)	39-2121-66C
$C_{15}H_{23}N_6O_7P$			
Adenosine-5'-phosphoric acid, ester with L-prolinol	n.s.g.	259(4.09)	33-0076-66
ester with L-valinol	n.s.g.	251(4.08)	33-0076-66
$C_{15}H_{23}O_4P$			
Phosphonic acid, (2,4,6-trimethylphenacyl)-, diethyl ester	C_6H_{12}	271(4.06)	44-1304-66
$C_{15}H_{24}$			
α-Cubebene	n.s.g.	208(3.63)	88-6365-66
β-Cubebene	n.s.g.	210(3.64)	88-6365-66
α-Curcumene, dihydro-	isooctane	266(2.54),273(2.57)	33-1029-66
1,3-Cyclohexadiene, 1-methyl-4-(1,2,2-trimethylcyclopentyl)-	EtOH	271(3.81)	44-0315-66
β-Farnesene, trans	isooctane	224(4.20)	33-1029-66
β-Sesquiphellandrene	EtOH	232(3.93)	12-0283-66
$C_{15}H_{24}ClN_3O$			
5-Hydroxyiminosparteinium chloride	EtOH	260.5(5.13)	94-0147-66
$C_{15}H_{24}O$			
1(2H)-Azulenone, 3,4,5,6,7,8-hexahydro-2-isopropyl-4,8-dimethyl-	EtOH	243(4.10)	94-1310-66
2,4,6,10-Dodecatetraen-1-ol, 3,7,11-trimethyl-	EtOH	281(4.55)	39-2154-66C
2(3H)-Naphthalenone, 7α-tert-butyl-4,4a,5,6,7,8-hexahydro-4aβ-methyl-	EtOH	242(4.11)	44-3109-66
1-Naphthol, 1,2,3,4,4aα,8aβ-hexahydro-4α-isopropyl-1α,6-dimethyl-	n.s.g.	260(3.43)	78-1641-66
6,8,10,12-Pentadecatetraen-1-ol	ether	280(4.30),291(4.59), 304(4.78),319(4.74)	24-3544-66
$C_{15}H_{24}O_2$			
Curdione	EtOH	299(2.40)	94-1310-66
Diosphenol A	EtOH	277(4.03)	78-0337-66
Diosphenol B	EtOH	277(3.78)	78-0337-66
2(1H)-Naphthalenone, 4a,5,6,7,8,8a-hexahydro-8-hydroxy-5-isopropyl-3,8-dimethyl-	n.s.g.	236(4.16)	78-1641-66

Compound	Solvent	$\lambda_{max}(\log \epsilon)$	Ref.
$C_{15}H_{24}O_3$			
Ilicic acid	EtOH	206(3.96)	44-1632-66
Todomatuic acid	EtOH	222(4.13)	57-1020-66D
$C_{15}H_{25}ClN_2O_4$			
1-(5-Piperidino-2,4-pentadienylidene)-piperidinium perchlorate	EtOH-100°K	417(5.23)	61-0052-66
phototropic form	EtOH-100°K	245(4.08),446(4.90)	61-0052-66
$C_{15}H_{25}N_3O_{15}P_2$			
Cytidine, 4-oxo-6-deoxy-D-glucosyl-, diphosphate	HCl	280(4.13)	37-1376-66
	pH 13	270(4.11),320(3.65)	37-1376-66
$C_{15}H_{25}N_6OP$			
Phosphonic diamide, P-(3-anilino-5-isopropyl-1H-1,2,4-triazol-1-yl)-N,N,N',N'-tetramethyl-	EtOH	245(4.15),261(4.19)	54-0429-66
Phosphonic diamide, P-(5-anilino-3-isopropyl-1H-1,2,4-triazol-1-yl)-N,N,N',N'-tetramethyl-	EtOH	261(4.40)	54-0429-66
Phosphonic diamide, P-(3-anilino-5-propyl-1H-1,2,4-triazol-1-yl)-N,N,N',N'-tetramethyl-	EtOH	245(4.18),261(4.19)	54-0429-66
Phosphonic diamide, P-(5-anilino-3-propyl-1H-1,2,4-triazol-1-yl)-N,N,N',N'-tetramethyl-	EtOH	261(4.40)	54-0429-66
$C_{15}H_{25}N_6O_7PS$			
Adenosine-5'-phosphoric acid, ester with methioninol	n.s.g.	260(4.12)	33-0076-66
$C_{15}H_{26}NO_3P$			
Phosphorimidic acid, N-(o-propylphenyl)-, triethyl ester	EtOH	240(3.99),282(3.22)	35-3781-66
$C_{15}H_{26}N_6O_4$			
1,4-Dicyclohexyl-4-methyl-1-(2-nitro-2-aci-nitroethylidene)-2-tetrazenium hydroxide, inner salt	EtOH	255(3.84),393(4.37)	44-0923-66
$C_{15}H_{26}O$			
Cyclohexanone, 2-methallyl-2-methyl-4-tert-butyl-, cis	C_6H_{12}	296(1.53)	22-3881-66
trans	C_6H_{12}	294(1.48)	22-3881-66
β-Simensal	EtOH	231(4.57)	88-0295-66
$C_{15}H_{26}O_3$			
Nonanoic acid, 9-(2-acetylcyclopropyl)-, methyl ester	isooctane	192(3.91)	22-2223-66
$C_{15}H_{28}N_4O_{11}$			
2(1H)-Pyrimidinone, 1-(tetra-O-acetyl-β-D-glucopyranosyl)-4-(p-aminobenzamido)-	pH 1	260(4.31),306(4.09)	44-4014-66
	pH 7	250(4.15),323(4.38)	44-4014-66
	pH 13	330(4.47)	44-4014-66
$C_{15}H_{28}O$			
Cyclohexanone, 4-tert-butyl-2-isobutyl-2-methyl-, cis	C_6H_{12}	295(1.43)	22-3881-66
trans	C_6H_{12}	295(1.48)	22-3881-66

Compound	Solvent	$\lambda_{max}(\log \epsilon)$	Ref.
$C_{15}H_{28}O_2$			
2,5-Undecadienal, diethyl acetal cis-cis	EtOH-HCl	222(4.16)	54-0117-66
	EtOH	none	54-0117-66
$C_{15}H_{28}O_3$			
Tridecanoic acid, 10-methyl-12-oxo-, methyl ester	isooctane	186(3.23)	22-2223-66
$C_{15}H_{30}N_2$			
2-Pyrazoline, 5-hexyl-1-methyl-4-pentyl-	EtOH	245(3.56)	4-0413-66
$C_{15}H_{32}GeO$			
2-Propanone, (tributylgermyl)-	n.s.g.	280(2.07)	65-0158-66*

Compound	Solvent	$\lambda_{max}(\log \epsilon)$	Ref.
$C_{16}F_{10}$ Pyrene, decafluoro-	EtOH	242(4.52),262(4.02), 274(4.23),294(3.24), 306(3.65),322(4.07), 338(4.25)	30-0855-66D
$C_{16}H_2F_{10}O$ Furan, 2,5-bis(pentafluorophenyl)-	EtOH	310(4.48)	30-1343-66A
$C_{16}H_3F_{10}N$ Pyrrole, 2,5-bis(pentafluorophenyl)-	EtOH	314(4.488)	30-1343-66A
$C_{16}H_4F_6O_8$ 2,2',6,6'-Biphenyltetracarboxylic acid, 3,3',4,4',5,5'-hexafluoro-	EtOH	270(3.17)	30-0855-66D
$C_{16}H_4F_{10}O_2$ 1,4-Butanedione, 1,4-bis-(pentafluorophenyl)-	EtOH	230(4.18),276(3.31)	30-1343-66A
$C_{16}H_6Cl_4$ Pyrene, 1,3,6,8-tetrachloro-	$C_2H_4Cl_2$- 15% SO_3	520(5.0)	65-1722-66
$C_{16}H_6F_5NO_2$ 2-Oxazolin-5-one, 4-(2,3,4,5,6-pentafluorobenzylidene)-2-phenyl-	HOAc	340(4.498)	65-0660-66
$C_{16}H_8F_5NO_3$ Cinnamic acid, α-benzamido-2,3,4,5,6-pentafluoro-	EtOH	228(4.24),290(4.10)	65-0660-66
$C_{16}H_8N_2S_2$ 2,5-Cyclohexadiene-$\Delta^{1,\alpha}$-malononitrile, 4-(1,3-benzodithiol-2-ylidene)-	$CHCl_3$	547(4.57),590(4.79), 638(4.75)	89-0517-66
	CF_3COOH	540s(1.18),570(1.49), 612(1.70)	89-0517-66
$C_{16}H_8O_4$ [1]Benzopyrano[4,3-c][1]benzopyran-5,11-dione	EtOH	253(3.98),268(3.92), 380(4.10)	88-4671-66
$C_{16}H_9ClN_4$ Pyrido[2,3-f]quinoxaline, 2-chloro-3-(2-pyridyl)-	EtOH	228(4.51),274(4.26), 301(4.39),355(3.95)	44-3384-66
$C_{16}H_9NO$ 11H-Indeno[1,2-c]isoquinolin-11-one	MeOH	225(4.49),285(4.67), 300(4.73),335(4.13)	88-1991-66
$C_{16}H_9NO_2$ 9H-Benzo[a]phenoxazin-9-one	dioxan	470(4.3)	65-1938-66
$C_{16}H_{10}$ Pyrene	dioxan	240(4.76),263(4.34), 274(4.62),308(4.02), 321(4.40),337(4.61), 353(2.85),364(2.56), 373(2.50)	24-1279-66

Compound	Solvent	λ_{max}(log ε)	Ref.
$C_{16}H_{10}BClO_2$			
Benzeneboronic acid, m-chloro-, cyclic 2,3-naphthylene ester	C_6H_{12}	305(3.74),311(3.71), 319(4.00)	39-0314-66B
Benzeneboronic acid, o-chloro-, cyclic 2,3-naphthylene ester	C_6H_{12}	306(3.73),312(3.69), 320(3.96)	39-0314-66B
Benzeneboronic acid, p-chloro-, cyclic 2,3-naphthylene ester	C_6H_{12}	305(3.86),312(3.81), 319(4.12)	39-0314-66B
$C_{16}H_{10}BNO_4$			
Benzeneboronic acid, m-nitro-, cyclic 2,3-naphthylene ester	C_6H_{12}	306(3.92),312(3.86), 320(4.12)	39-0314-66B
Benzeneboronic acid, o-nitro-, cyclic 2,3-naphthylene ester	C_6H_{12}	303(3.57),309(3.53), 317(3.57),323(3.42)	39-0314-66B
Benzeneboronic acid, p-nitro-, cyclic 2,3-naphthylene ester	C_6H_{12}	310(3.94),315(3.93), 322(4.04)	39-0314-66B
$C_{16}H_{10}BrN_3O$			
Spiro[2-oxindoline-3,5'-pyrazole], 3'-(p-bromophenyl)-	EtOH	261(4.70),311(4.22), 324(4.21)	95-1156-66
$C_{16}H_{10}Br_2$			
Fluorene, 9-(3,3-dibromoallylidene)-	CH_2Cl_2	240(4.47),251(4.37), 261(4.47),271(4.55), 334(4.34),348(4.60), 366(4.59)	24-0658-66
$C_{16}H_{10}F_3N_3O_2$			
2-Quinoxalinecarboxylic acid, 3-(m-trifluoroanilino)-, sodium salt	EtOH	290(4.49),390(3.84)	87-0770-66
$C_{16}H_{10}F_5NO_3$			
Alanine, N-benzoyl-3-(pentafluorophenyl)-	EtOH	224(4.127)	65-0660-66
$C_{16}H_{10}F_6NiO_2S_4$			
Nickel, bis[1,1,1-trifluoro-4-(2-thienyl)-4-thiolobut-3-en-2-one]-	$CHCl_3$	565s(3.23),658s(2.43)	12-1401-66
$C_{16}H_{10}Fe$			
Hexatriyne, 1-ferrocenyl-	$CHCl_3$	264s(3.92),293(3.68), 310(3.75),330(3.64), 455(2.86)	101-0399-66B
$C_{16}H_{10}N_2$			
6,12-Diazachrysene (or calycanine)	EtOH	267(4.75),332s(3.82), 334(3.83),353(3.67), 370(3.63)	78-3421-66
Quino[8,7-f]quinoline	C_6H_{12}	224(4.52),268(4.61), 298(3.83),310(3.79), 333(3.55),342(3.29), 349(3.85),359(3.29), 367(4.01)	78-0181-66B
$C_{16}H_{10}N_2O$			
Dibenzo[c,f][2,7]naphthyridin-6-ol	HOAc	262(4.57),325(3.71), 360(3.85),372(3.84)	87-0161-66
$C_{16}H_{10}N_2O_2$			
5,12-Diazabenz[a]anthracene-6,7-diol	EtOH	264(4.67),272(4.60), 330(3.94),344(3.98)	39-1245-66C

Compound	Solvent	λ_{max}(log ε)	Ref.
Dibenzo[c,f][2,7]naphthyridine-6,7-diol	HOAc	269(4.20),324(3.87),356(3.82),378(3.96),397(3.94)	87-0161-66
$C_{16}H_{10}N_4$			
Pyrido[3,2-h]quinoxaline, 2-(2-pyridyl)-	EtOH	223(4.56),296(4.49),345(4.00),363(3.95)	44-3384-66
$C_{16}H_{10}N_4O$			
Ketone, 3-pyridyl v-triazolo[1,5-a]-quinolin-3-yl	MeOH	239(4.35),296s(3.97),309s(4.16),332(4.42),344(4.36)	24-2918-66
1(2H)-Phthalazinone, 2-(1-phthalazyl)-	EtOH	208(4.62),243(4.79)	4-0381-66
Pyrido[2,3-f]quinoxalin-2-ol, 3-(2-pyridyl)-	EtOH	242(4.61),303(4.27),381(4.20)	44-3384-66
$C_{16}H_{10}O$			
Benzo[b]naphtho[1,2-d]furan	EtOH	246(4.67),280(3.73),313(4.14)	44-2646-66
Phenanthro[9,10-b]furan	EtOH	237(4.48),249(4.72),254(4.81),280(4.17),290(4.03),302(4.09),320(2.90),335(3.08),352(3.11)	35-3759-66
$C_{16}H_{10}O_2$			
1,4-Naphthoquinone, 2-phenyl-	EtOH	249(4.39),303(3.83),337(3.66)	88-2277-66
	EtOH	249(4.39),303(3.83),337(3.66)	35-3759-66
$C_{16}H_{10}O_4$			
Naphtho[2,3-c]furan-1-acrolein, 4,9-dihydro-3-methyl-4,9-dioxo-	ether	247(4.52),316(4.03),325s(4.00),378(4.17)	33-1806-66
$C_{16}H_{10}O_5$			
Glutaconic acid, 4-(1,3-dihydroxy-5-phenyl-2,4-pentadienylidene)-3-hydroxy-, di-δ-lactone	EtOH	232(4.10),272(4.01),392(4.35)	35-0624-66
Pseudobaptigenin	MeOH	226(4.4),248s(--),294(4.4)	39-0509-66C
	MeOH-base	225(4.4),258(4.4),333(3.9)	39-0509-66C
$C_{16}H_{10}O_8$			
2,2',6,6'-Biphenyltetracarboxylic acid	EtOH	272(3.88)	30-0855-66D
Bisphloroglucyl dioxotetra-cyclotrimethine	EtOH	311(4.00),355(3.12),521s(4.44),567(5.23)	5-0153-66I
Ceroalbolinic acid	EtOH	294(4.50),418(3.76)	78-1507-66
$C_{16}H_{11}BO_2$			
Benzeneboronic acid, cyclic 2,3-naphthylene ester	C_6H_{12}	304(3.76),311(3.73),318(4.04)	39-0314-66B
$C_{16}H_{11}Br$			
Fluorene, 9-(3-bromoallylidene)-	CH_2Cl_2	239(4.53),250(4.42),259(4.47),270(4.55),323(4.34),333(4.61),352(4.56)	24-0658-66

Compound	Solvent	λ_{max}(log ϵ)	Ref.
$C_{16}H_{11}BrN_4O_4$			
Pyrazole, 4-bromo-1-(2,4-dinitrophenyl)-3-methyl-5-phenyl-	EtOH	306(3.91)	22-3744-66
Pyrazole, 4-bromo-1-(2,4-dinitrophenyl)-5-methyl-3-phenyl-	EtOH	240(4.33),337(3.88)	22-3744-66
$C_{16}H_{11}ClN_2O_4S$			
1-Naphthalenesulfonic acid, 4-hydroxy-3-(p-chlorophenylazo)-	H_2O	363(3.99),494(4.35)	35-2240-66
	96% H_2SO_4	440(4.06),527(4.53), 555(4.51)	35-2240-66
$C_{16}H_{11}ClO$			
1-Indanone, 2-(p-chlorobenzylidene)-, cis	MeOH	228(4.23),321(4.56)	35-4489-66
trans	MeOH	230(4.15),322(4.55)	35-4489-66
$C_{16}H_{11}ClO_2$			
Phenanthro[9,10-b]-p-dioxin, 2-chloro-2,3-dihydro-	benzene	361(3.10)	24-1881-66
Spiro[oxetane-2,9'(10'H)-phenanthren]-10'-one, 3-chloro-	dioxan	244(4.48),280(3.83), 332(3.47)	24-1881-66
$C_{16}H_{11}NS_3$			
2-Thiopheneacrylonitrile, α-[5-(2-thenyl)-2-thienyl]-, trans	EtOH	229(4.15),265(3.92), 376(4.46)	12-1243-66
$C_{16}H_{11}N_3O_2$			
Imidazo[2,1-b]quinazoline-2,5(1H,3H)-dione, 1-phenyl-	EtOH	228(4.50),258s(3.90), 267(4.06),275s(4.00), 298s(3.30),311(3.45), 322s(3.36)	24-1532-66
$C_{16}H_{11}N_3O_3$			
1-Naphthol, 4-[(p-nitrophenyl)azo]-	EtOH	238(4.48),262s(3.91), 284(3.86),347(3.86), 469(4.60)	19-0037-66
2-Naphthol, 1-[(p-nitrophenyl)azo]-	EtOH	226(4.52),255s(3.85), 297(3.75),327(4.01), 410s(4.02),487(4.40), 495s(4.39)	19-0037-66
$C_{16}H_{11}P$			
Phosphine, butadiynyldiphenyl-	EtOH	228(4.34),250(4.00), 265(3.81),281(3.56)	22-1002-66
$C_{16}H_{12}BClN_2$			
1H-Naphtho[2,3-d]-1,3,2-diazaborole, 2-(m-chlorophenyl)-2,3-dihydro-	C_6H_{12}	322(4.22),329(4.17), 337(4.43)	39-0314-66B
1H-Naphtho[2,3-d]-1,3,2-diazaborole, 2-(o-chlorophenyl)-2,3-dihydro-	C_6H_{12}	325(4.10),332(4.09), 340(4.26)	39-0314-66B
1H-Naphtho[2,3-d]-1,3,2-diazaborole, 2-(p-chlorophenyl)-2,3-dihydro-	C_6H_{12}	322(4.28),329(4.22), 337(4.52)	39-0314-66B
$C_{16}H_{12}BN_3O_2$			
1H-Naphtho[2,3-d]-1,3,2-diazaborole, 2,3-dihydro-2-(m-nitrophenyl)-	C_6H_{12}	325(4.09),336(4.16)	39-0314-66B
1H-Naphtho[2,3-d]-1,3,2-diazaborole, 2,3-dihydro-2-(o-nitrophenyl)-	C_6H_{12}	315(3.78),322(3.79), 329(3.96)	39-0314-66B

Compound	Solvent	λ_{max}(log ϵ)	Ref.
$C_{16}H_{12}BrClO$			
1-Indanone, 2-bromo-2-(p-chlorobenzyl)-	MeOH	255(4.13),298(3.42)	35-4489-66
$C_{16}H_{12}BrClO_2$			
Stilbene, α-acetoxy-α'-bromo-m'-chloro-	EtOH	209(4.45),228(4.30),285(3.88)	39-0533-66C
Stilbene, α-acetoxy-α'-bromo-o'-chloro-	EtOH	208(4.40),233(4.18)	39-0533-66C
Stilbene, α-acetoxy-α'-bromo-p'-chloro-	EtOH	206(4.32),232(4.30),290(3.91)	39-0533-66C
$C_{16}H_{12}BrNO_4$			
Stilbene, α-acetoxy-α'-bromo-m'-nitro-	EtOH	207(4.40),219(4.42),265(4.10)	39-0533-66C
Stilbene, α-acetoxy-α'-bromo-p'-nitro-	EtOH	204(4.36),252(4.20),321(3.88)	39-0533-66C
$C_{16}H_{12}BrN_3$			
2-Naphthylamine, 1-[(p-bromophenyl)-azo]-	benzene	460(4.22)	44-2571-66
	$C_6H_{11}Me$	450(4.20)	44-2571-66
$C_{16}H_{12}BrN_3O$			
1-Naphthol, 7-amino-8-[(m-bromophenyl)azo]-	benzene	515(4.20)	44-2571-66
	$C_6H_{11}Me$	515(4.17)	44-2571-66
1-Naphthol, 7-amino-8-[(p-bromophenyl)azo]-	benzene	510(4.26)	44-2571-66
	$C_6H_{11}Me$	512(4.26)	44-2571-66
$C_{16}H_{12}Br_2N_6O_2$			
Pyrimidine, 2(6)-amino-4,6(2)-bis-(p-bromoanilino)-5-nitro-	pH 1	284(4.34),295(4.34),325s(4.14),410(4.22)	87-0573-66
	pH 11	284(4.38),293s(4.35),336(4.14),407(4.24)	87-0573-66
$C_{16}H_{12}ClNO_4S$			
3-Phenyl-4H-indeno[2,1-d]thiazolium perchlorate	MeOH	233(4.00),319(4.12)	39-0686-66C
$C_{16}H_{12}ClN_3$			
2-Naphthylamine, 1-[(m-chlorophenyl)-azo]-	benzene	460(4.15)	44-2571-66
	$C_6H_{11}Me$	449(4.14)	44-2571-66
2-Naphthylamine, 1-[(p-chlorophenyl)-azo]-	benzene	457(4.16)	44-2571-66
	$C_6H_{11}Me$	450(4.17)	44-2571-66
$C_{16}H_{12}ClN_3O$			
1-Naphthol, 7-amino-8-[(m-chlorophenyl)azo]-	benzene	511(4.21)	44-2571-66
	$C_6H_{11}Me$	513(4.16)	44-2571-66
1-Naphthol, 7-amino-8-[(p-chlorophenyl)azo]-	benzene	510(4.25)	44-2571-66
	$C_6H_{11}Me$	512(4.26)	44-2571-66
$C_{16}H_{12}ClN_3O_2$			
2-Quinoxalinecarboxylic acid, 3-(3-chloro-o-toluidino)-, sodium salt	EtOH	286(4.40),393(3.79)	87-0770-66
$C_{16}H_{12}Cl_2FeO_2$			
Ferrocene, 1,1'-bis(1-chloro-2-formylvinyl)-	EtOH	285(4.32),313(4.26),422(3.33),490(3.25)	101-0173-66B
$C_{16}H_{12}Cl_2O_2$			
Stilbene, α-acetoxy-α',o'-dichloro-	EtOH	207(4.40),263(4.01)	39-0533-66C

Compound	Solvent	λ_{max}(log ϵ)	Ref.
Stilbene, α-acetoxy-α',p'-dichloro-	EtOH	204(4.29),229(4.26), 280(3.99)	39-0533-66C
$C_{16}H_{12}Cl_4O_2$			
1,4-Benzodioxan, 5,6,7,8-tetrachloro-2-methyl-2-p-tolyl-	EtOH	301(3.40)	78-0745-66
$C_{16}H_{12}Cl_{10}Si_2$			
Disilane, 1,1,2,2-tetramethyl-1,2-bis(pentachlorophenyl)-	C_6H_{12}	216(4.96)	101-0102-66B
$C_{16}H_{12}FN_3$			
2-Naphthylamine, 1-[(m-fluorophenyl)-azo]-	benzene	456(4.13)	44-2571-66
	$C_6H_{11}Me$	450(4.13)	44-2571-66
2-Naphthylamine, 1-[(p-fluorophenyl)-azo]-	benzene	450(4.15)	44-2571-66
	$C_6H_{11}Me$	437(4.11)	44-2571-66
$C_{16}H_{12}FN_3O$			
1-Naphthol, 7-amino-8-[(m-fluoro-phenyl)azo]-	benzene	512(4.20)	44-2571-66
	$C_6H_{11}Me$	512(4.19)	44-2571-66
1-Naphthol, 7-amino-8-[(p-fluoro-phenyl)azo]-	benzene	506(4.16)	44-2571-66
	$C_6H_{11}Me$	502(4.12)	44-2571-66
$C_{16}H_{12}F_5N_3OS$			
1-Naphthol, 7-amino-8-(phenylazo)-, m-pentafluorosulfur derivative	benzene	516(4.20)	44-2571-66
	$C_6H_{11}Me$	518(4.18)	44-2571-66
1-Naphthol, 7-amino-8-(phenylazo)-, p-pentafluorosulfur derivative	benzene	524(4.23)	44-2571-66
	$C_6H_{11}Me$	526(4.20)	44-2571-66
$C_{16}H_{12}F_5N_3S$			
C.I. Solvent Yellow 5, m-penta-fluorosulfur derivative	benzene	465(4.15)	44-2571-66
	$C_6H_{11}Me$	458(4.15)	44-2571-66
C.I. Solvent Yellow 5, p-penta-fluorosulfur derivative	benzene	468(4.17)	44-2571-66
	$C_6H_{11}Me$	462(4.17)	44-2571-66
$C_{16}H_{12}IN_3$			
2-Naphthylamine, 1-[(m-iodophenyl)-azo]-	benzene	460(4.18)	44-2571-66
	$C_6H_{11}Me$	453(4.16)	44-2571-66
2-Naphthylamine, 1-[(p-iodophenyl)-azo]-	benzene	463(4.22)	44-2571-66
	$C_6H_{11}Me$	456(4.20)	44-2571-66
$C_{16}H_{12}IN_3O$			
1-Naphthol, 7-amino-8-[(m-iodophenyl)-azo]-	benzene	515(4.24)	44-2571-66
	$C_6H_{11}Me$	513(4.17)	44-2571-66
1-Naphthol, 7-amino-8-[(p-iodophenyl)-azo]-	benzene	517(4.28)	44-2571-66
	$C_6H_{11}Me$	517(4.27)	44-2571-66
$C_{16}H_{12}N_2$			
Naphthalene, 1-(phenylazo)-	heptane	216s(4.67),249(4.05), 255(4.06),262(4.06), 269(4.04),274(4.04), 289(3.92),371(4.09), 453(2.89)	19-0037-66
	EtOH	217(4.62),255(4.02), 262(4.03),268s(4.04), 273(4.04),290s(3.95), 372(4.11),460s(2.97)	19-0037-66
6H-Pyrido[4,3-b]carbazole, 5-methyl-	85% EtOH-HCl	238(4.43),246s(4.40), 271s(4.35),306(4.86), 350(3.84)	87-0237-66

Compound	Solvent	$\lambda_{max}(\log \varepsilon)$	Ref.
6H-Pyrido[4,3-b]carbazole, 5-methyl- (cont.)	EtOH	224(4.43),238(4.32), 273(4.72),284(4.84), 293(4.88),313(3.72), 328(3.80),373(3.63), 390(3.55)	87-0237-66
$C_{16}H_{12}N_2O$			
1H-Benz[f]indole-3-acetonitrile, N-acetyl-	EtOH	213(4.48),228(4.54), 267(4.47),273(4.42), 305(3.81),317(3.89), 331(3.86),374(3.94)	39-0477-66C
Canthin-6-one, 4-ethyl-	EtOH	241s(4.27),257(4.25), 265(4.26),296(3.95), 344(4.01),357(4.19), 377(4.14)	5-0159-66A
1-Naphthol, 4-(phenylazo)-	benzene	277(4.14),410(4.14), 448(4.13)	19-0037-66
	EtOH	238(4.24),272(4.07), 331s(3.62),409(4.18), 464s(4.04)	19-0037-66
2-Naphthol, 1-(phenylazo)-	C_6H_{12}	230(4.59),265s(4.02), 280(3.82),304(3.84), 428s(4.10),468s(4.12)	19-0037-66
	EtOH	229(4.55),256(4.05), 263s(4.00),280(3.71), 313(3.85),421(4.03), 480(4.19),505s(4.14)	19-0037-66
$C_{16}H_{12}N_2O_2$			
Acrylonitrile, 3-amino-2-benzoyl- 3-phenoxy-	MeCN	207(4.25),230(4.21), 298(4.30)	24-2302-66
	dioxan	234(4.05),303(4.25)	24-2302-66
Naphtho[2,3-f]quinoxaline-7,12-dione, 1,2,3,4-tetrahydro-	EtOH	536s(--),567(4.06), 616s(--)	27-0281-66
Pyrazine, 3,6-bis(2-hydroxyphenyl)-	dioxan	292(4.32),373(4.39)	49-0077-66
Quinazoline, 2-benzoyl-4-methoxy-	EtOH	208(4.61),216s(4.54), 250(4.29),290s(3.85), 310s(3.70)	39-2290-66C
4(3H)-Quinazolinone, 2-benzoyl- 3-methyl-	EtOH	212(4.41),222s(4.36), 257(4.21),303(3.87), 316(3.73)	39-2290-66C
$C_{16}H_{12}N_2O_3$			
Acetic acid, 4-(3-cinnolinyl)phenoxy-	MeOH	265(4.50),296(4.27)	87-0784-66
4-Cinnolone-1-acetic acid, 3-phenyl-	MeOH	264(4.14),311(4.02), 358(4.09)	87-0784-66
$C_{16}H_{12}N_4O$			
1,2,3-Triazolo[1,5-a]quinazoline, 5-methoxy-3-phenyl-	EtOH	212(4.58),220(4.57), 254(4.45),262(4.40), 272(4.34),340(3.75)	39-2290-66C
1,2,3-Triazolo[1,5-a]quinazolin- 5(4H)-one, 4-methyl-3-phenyl-	EtOH	210(4.51),224(4.54), 253s(4.28),270s(4.13)	39-2290-66C
$C_{16}H_{12}N_4O_2$			
1-Naphthylamine, 4-[(p-nitrophenyl)- azo]-	benzene	292(4.17),482(4.40)	19-0037-66
	EtOH	225s(4.02),249(4.17), 297(4.11),531(4.47)	19-0037-66
2-Naphthylamine, 1-[(m-nitrophenyl)- azo]-	benzene	470(4.15)	44-2571-66
	$C_6H_{11}Me$	461(4.18)	44-2571-66

Compound	Solvent	$\lambda_{max}(\log \epsilon)$	Ref.
2-Naphthylamine, 1-[(p-nitrophenyl)-azo]-	benzene	496(4.27)	44-2571-66
	$C_6H_{11}Me$	486(4.27)	44-2571-66
	heptane	230(4.54),240(4.55), 276s(4.09),286(4.13), 368(3.95),485(4.27)	19-0037-66
	EtOH	232(4.52),242(4.54), 278(4.09),287(4.07), 312s(3.88),370(3.92), 514(4.34)	19-0037-66
$C_{16}H_{12}N_4O_3$			
1-Naphthol, 7-amino-8-[(m-nitrophenyl)-azo]-	benzene	520(4.18)	44-2571-66
1-Naphthol, 7-amino-8-[(p-nitrophenyl)-azo]-	benzene	547(4.33)	44-2571-66
$C_{16}H_{12}N_4O_4$			
Pyrazole, 1-(2,4-dinitrophenyl)-3-methyl-4-phenyl-	EtOH	235(4.32),348(4.13)	22-3744-66
Pyrazole, 1-(2,4-dinitrophenyl)-4-methyl-3-phenyl-	EtOH	353(4.17)	22-3744-66
Pyrazole, 1-(2,4-dinitrophenyl)-4-methyl-5-phenyl-	EtOH	304(3.93)	22-3744-66
Pyrazole, 1-(2,4-dinitrophenyl)-5-methyl-4-phenyl-	EtOH	239(4.40),317(3.79)	22-3744-66
$C_{16}H_{12}O$			
Anthracene, 2-acetyl-	EtOH	242(4.57),258(4.74), 266(4.77),280(4.70), 324(3.52),338(3.65), 359(3.62),395(3.60)	39-1729-66C
Cyclopropenone, benzylphenyl-	ether	253(4.29)	25-1379-66
Cyclopropenone, phenyl-p-tolyl-	EtOH	223(4.25),228s(4.22), 234s(4.16),293s(4.40), 302(4.47),317s(4.29)	44-1660-66
Furan, 3,4-diphenyl-	EtOH	225(4.26)	88-2277-66
$C_{16}H_{12}OSe$			
Selenoflavone, 6?-methyl-	n.s.g.	280(4.79),359(4.34)	20-0260-66
Selenoflavone, 7-methyl-	n.s.g.	276(4.98),353(4.47)	20-0260-66
Selenoflavone, 8-methyl-	n.s.g.	276(4.69),356(4.36)	20-0260-66
$C_{16}H_{12}O_2$			
Cyclobuta[1,2-b:4,3-b']bisbenzofuran, 5a,5b,10b,10c-tetrahydro-, cis	C_6H_{12}	282(3.9),290(3.89)	24-1723-66
trans	C_6H_{12}	286(3.89),294(4.01)	24-1723-66
1,3-Indandione, 2-benzyl-	isooctane	222(4.11),233(3.46), 240(3.49)	88-1499-66
Rotoxen, 6,6a-dihydro-	EtOH	245(4.61),295(4.41), 310(4.35),345(4.69), 360(4.56)	39-0544-66C
$C_{16}H_{12}O_3$			
6H-Benz[cd]azulene-9-carboxylic acid, 6-oxo-, ethyl ester	EtOH	227(4.52),274(4.06), 335(3.79),374(3.79)	35-3950-66
8H-Benz[cd]azulene-9-carboxylic acid, 8-oxo-, ethyl ester	EtOH	227s(4.26),247(4.45), 287(4.04),308(4.02), 359(3.55),374(3.67), 394(3.64)	35-3950-66

Compound	Solvent	λ_{max}(log ε)	Ref.
1H-Naphtho[2,1-b]pyran-1-one, 2-acetyl-3-methyl-	99% EtOH	213(4.60),264(4.30), 272s(4.25),305(3.97), 320s(3.76),333(3.51)	94-0129-66
$C_{16}H_{12}O_4$			
2-Benzofurancarboxylic acid, 3-(o-methoxyphenyl)-	EtOH	280(4.2)	24-0680-66
3-Benzofurancarboxylic acid, 6-methoxy-2-phenyl-	EtOH	229(4.05),318(4.26)	78-3209-66
1H-Naphtho[2,1-b]pyran-1-one, 2-acetyl-5-hydroxy-3-methyl-	H_2O	216(4.71),278(4.24), 330(3.93),341(3.93)	94-0129-66
$C_{16}H_{12}O_5$			
Flavone, 4',5,7-trihydroxy-6-methyl-	EtOH	275(4.18),335(4.20)	28C-0662-66A
Flavone, 4',5,7-trihydroxy-8-methyl-	EtOH	274(4.33),300(4.22), 330(4.27)	28C-0662-66A
Glutaconic acid, 4-(1,3-dihydroxy-5-phenyl-2-pentenylidene)-3-hydroxy-, di-δ-lactone	EtOH	270(4.10),331(3.94)	35-0624-66
Maturone, acetate	EtOH	252(4.23),297(3.78)	78-0685-66
4H-Naphtho[2,3-b]pyran-4-one, 3-acetyl-5,10-dihydroxy-2-methyl-	EtOH	256s(4.31),276(4.42), 353(3.64),425(3.50)	94-0121-66
$C_{16}H_{12}O_6$			
Flavone, 3,5,7-trihydroxy-8-methoxy-	EtOH	254s(3.88),276(4.17), 375(3.76)	78-0941-66
	EtOH-NaOAc	271s(3.79),283(3.70)	78-0941-66
	EtOH-$AlCl_3$	430(3.88)	78-0941-66
Xanthone, 5-acetoxy-1-hydroxy-3-methoxy-	EtOH	238(4.40),251(4.34), 306(4.12),345(3.60)	78-1777-66
	$AlCl_3$	266(4.11),327(4.00)	78-1777-66
$C_{16}H_{12}O_7$			
Vogeletin	MeOH	277(4.21),349(4.34)	24-2430-66
$C_{16}H_{13}^+$			
Diphenylpropynylcarbonium ion	FSO_3H-SbF_5	372(4.21),420s(--), 452(4.20)	35-1488-66
Methylphenyl(phenylethynyl)-carbonium ion	FSO_3H-SbF_5	334(3.40),406(3.38), 493(4.57)	35-1488-66
$C_{16}H_{13}BN_2$			
1H-Naphtho[2,3-d]-1,3,2-diazaborole, 2,3-dihydro-2-phenyl-	C_6H_{12}	321(4.13),327(4.07), 336(4.40)	39-0314-66B
$C_{16}H_{13}BrO$			
Anthrone, 7-bromo-2,3-dimethyl-	EtOH	277(4.17),306(4.00)	78-0349-66A
$C_{16}H_{13}BrO_2$			
Stilbene, α-acetoxy-α'-bromo-	EtOH	205(4.33),226(4.25), 283(3.86)	39-0533-66C
$C_{16}H_{13}Cl$			
1,3-Butadiene, 1-chloro-1,4-diphenyl-	hexane	233(4.21),328(4.77), 350s(4.55)	44-2175-66
$C_{16}H_{13}ClN_2O$			
5H-1,4-Benzodiazepine, 7-chloro-3-methyl-5-phenyl-, 4-oxide	n.s.g.	236(4.25),280(3.64), 324(3.83)	88-2609-66

Compound	Solvent	λ_{max}(log ϵ)	Ref.
3-Indolizine-7-carboxamide, N-(p-chlorophenyl)-2-methyl-	EtOH	226(4.48),280(4.50), 357(4.29)	56-0061-66
$C_{16}H_{13}ClN_6O_3$			
s-Triazine, 2-(p-anisidino)-4-chloro-6-(p-nitroanilino)-	EtOH	270(4.26),330(4.30)	80-0139-66
$C_{16}H_{13}ClO$			
1-Indanone, 2-(p-chlorobenzyl)-	MeOH	244(4.18),277(3.33), 295(3.46)	35-4489-66
$C_{16}H_{13}ClO_2$			
Stilbene, α-acetoxy-α'-chloro-	EtOH	205(4.39),223(4.26), 276(3.99)	39-0533-66C
$C_{16}H_{13}ClO_4Se$			
3-Methyl-4-phenyl-1-selenanaphthalenium perchlorate	HOAc-1% $HClO_4$	265(4.80),355(3.98), 420(4.04)	20-0169-66
6-Methyl-4-phenyl-1-selenanaphthalenium perchlorate	HOAc-1% $HClO_4$	272(4.42),365(3.98), 437(3.87)	20-0169-66
7-Methyl-4-phenyl-1-selenanaphthalenium perchlorate	HOAc-1% $HClO_4$	270(4.52),370(4.09), 437(3.92)	20-0169-66
8-Methyl-4-phenyl-1-selenanaphthalenium perchlorate	HOAc-1% $HClO_4$	271(4.53),359(4.25), 443(3.96)	20-0169-66
$C_{16}H_{13}ClO_6$			
4'-Methoxyflavylium perchlorate	MeOH-acid	439(4.65)	61-0530-66
7-Methoxyflavylium perchlorate	MeOH-acid	430(4.53)	61-0530-66
$C_{16}H_{13}ClO_7$			
4'-Hydroxy-7-methoxyflavylium perchlorate	MeOH-acid	460(4.73)	61-0530-66
7-Hydroxy-4'-methoxyflavylium perchlorate	MeOH-acid	462(4.69)	61-0530-66
$C_{16}H_{13}ClS$			
Butadiene, 1-chloro-2-phenyl-1-(phenylthio)-	EtOH	226(4.19),254(4.21), 280s(4.02)	44-1694-66
$C_{16}H_{13}Cl_2N$			
Indolenine, 3-(o-chlorobenzylidene)-2-methyl-, hydrochloride	EtOH	278(3.89),289(3.80)	35-1077B-66
$C_{16}H_{13}N$			
Aporphane, dehydro-	n.s.g.	243(4.34),257(4.44), 325(3.82)	88-2941-66
$C_{16}H_{13}NO$			
3H-Pyrido[3,2,1-jk]carbazol-3-one, 1,2-dihydro-4-methyl-	MeOH	237(4.33),263(4.51), 308s(3.92),322(3.94), 339(3.80),510(3.22)	5-0116-66F
$C_{16}H_{13}NO_2S_2$			
Anthraquinone, 1-amino-2,4-bis-(methylthio)-	$CHCl_3$	259(4.60),341(3.83), 540(4.02)	22-0580-66
$C_{16}H_{13}NO_4$			
Acridone, 4-formyl-3-hydroxy-1-methoxy-10-methyl-	EtOH	254(4.39),282(4.40), 304s(4.27),376(4.00)	12-0275-66
	EtOH-base	246(4.39),308(4.44)	12-0275-66

Compound	Solvent	λ max(log ε)	Ref.
3H-Phenoxazin-3-one, 7-acetoxy-1,8-dimethyl-	C_6H_{12}	248(4.24),255s(4.22), 264s(4.00)	24-1470-66
	MeOH-HCl	274s(3.63),345(4.13)	24-1470-66
	MeOH-KOH	360s(4.07),440(4.02)	24-1470-66
3H-Phenoxazin-3-one, 7-acetoxy-2,9-dimethyl-	C_6H_{12}	248(4.24),255s(4.12), 264s(3.86)	24-1470-66
	MeOH-KOH	362(4.19),378(4.18), 429(4.10)	24-1470-66
3H-Phenoxazin-3-one, 8-acetoxy-1,7-dimethyl-	C_6H_{12}	248(4.19),254s(4.18), 263s(3.99)	24-1470-66
	MeOH-HCl	274s(3.62),345(4.09)	24-1470-66
	MeOH-KOH	360s(3.99),442(3.95)	24-1470-66
$C_{16}H_{13}NO_5$			
2-Acridancarboxylic acid, 1-hydroxy-3-methoxy-9-oxo-, methyl ester	EtOH	223(4.15),245(4.25), 270(4.73),375(3.75)	12-0275-66
Cinnamic acid, p-nitro-, p-anisyl ester	EtOH	300(4.4)	65-1202-66
$C_{16}H_{13}NO_6$			
Resazurin, carbethoxy-, methyl ether	EtOH	202(4.29),232(4.45), 493(4.16),525(4.17), 560s(3.93)	49-0129-66
$C_{16}H_{13}N_3$			
1-Naphthylamine, 4-(phenylazo)-	C_6H_{12}	216(4.36),246(4.30), 274(4.27),418(4.40)	19-0037-66
	EtOH	249(4.16),281(4.16), 340s(3.49),352(3.54), 440(4.34)	19-0037-66
2-Naphthylamine, 1-(phenylazo)-	benzene	439(4.13)	44-2571-66
	$C_6H_{11}Me$	430(4.13)	44-2571-66
	C_6H_{12}	230(4.56),242(4.55), 274(4.25),283(4.19), 348(3.97),450(4.16)	19-0037-66
	EtOH	230(4.48),245(4.51), 274(4.23),283s(4.16), 345(3.95),450(4.16)	19-0037-66
$C_{16}H_{13}N_3O$			
Imidazo[2,1-b]quinazolin-5-one, 1-phenyl-	EtOH	224(4.56),286(4.48), 328(3.54)	24-1532-66
1-Naphthol, 7-amino-8-(phenylazo)-	benzene	509(4.39)	44-2571-66
as-Triazine, 3-methoxy-5,6-diphenyl-	EtOH	257(4.11),328(3.9)	44-3914-66
$C_{16}H_{13}N_3O_2$			
2-Quinoxalinecarboxylic acid, 3-(o-toluidino)-, sodium salt	EtOH	285(4.41),400(3.77)	87-0770-66
as-Triazine, 3-methoxy-5,6-diphenyl-, 2-oxide	EtOH	252(4.21),346(3.89)	44-3914-66
as-Triazin-3(4H)-one, 4-methyl-5,6-diphenyl-, 2-oxide	EtOH	252(4.11),365(3.79)	44-3914-66
$C_{16}H_{13}N_3O_3$			
Indolizine-7-carboxamide, 2-methyl-N-(m-nitrophenyl)-	EtOH	227(4.50),279(4.52), 359(4.28)	56-0061-66
6-Oxa-4-azaspiro[2.4]hept-4-en-7-one, 1-(1-acetyl-4-imidazolyl)-5-phenyl-	CH_2Cl_2	263(4.26)	87-0766-66
2-Oxazolin-5-one, 4-[1-(1-acetyl-4-imidazolyl)ethylidene]-2-phenyl-	CH_2Cl_2	247s(4.07),272(4.30), 359s(4.34),375(4.40), 392s(4.25)	87-0766-66

Compound	Solvent	$\lambda_{max}(\log \epsilon)$	Ref.
$C_{16}H_{13}N_3O_3S$			
Oxazole, 5-amino-4-(benzylthio)-2-(p-nitrophenyl)-	EtOH	247(4.06),412(4.17)	44-3612-66
$C_{16}H_{13}N_3O_4$			
Indazole, 6-acetoxy-3-methyl-1-(p-nitrophenyl)-	EtOH	231(4.23),258(3.71), 345(4.12)	78-3131-66
$C_{16}H_{14}$			
1,3-Butadiene, 1,1-diphenyl-	C_6H_{12}	236s(3.71),297(4.26)	44-0589-66
Indan, 1-benzylidene-	EtOH	229(4.10),236(4.04), 244(3.75),284(4.31), 296(4.34),304s(4.34), 318(4.47),332(4.31)	5-0057-66H
Inden, 3-(2,4,6-cycloheptatrien-1-yl)-	EtOH	250(4.17),315(3.05), 330(2.90)	5-0057-66H
Methane, cyclopropylidenediphenyl-	EtOH	226(4.30),234(4.23), 258(4.29)	88-3267-66
Naphthalene, 1,4-dihydro-1-phenyl-	EtOH	266(3.31),272(3.33), 285(3.26),288(3.24)	39-0861-66B
Phenanthrene, 9,10-dimethyl-	EtOH	255(4.75),270(4.47), 280(4.28),287(4.05), 300(4.11)	35-0611-66
$C_{16}H_{14}BrNO_2S$			
Benzo[b]cyclohepta[e][1,4]thiazin-10(11H)-one, 9-bromo-8-isopropyl-, 5-oxide	MeOH	255(4.33),368(4.13), 391(4.11)	18-1988-66
$C_{16}H_{14}BrNO_3S$			
Benzo[b]cyclohepta[e][1,4]thiazin-10(11H)-one, 9-bromo-8-isopropyl-, 5,5-dioxide	MeOH	252(4.30),363(4.25)	18-1988-66
$C_{16}H_{14}Br_2O_2$			
Ethylene, 1,1-dibromo-2,2-bis-(p-methoxyphenyl)-	FSO_3H-SbF_5	350(4.1),500s(4.5), 540(4.8)	27-0324-66
	H_2SO_4	350(4.2),420s(3.9), 550(4.8)	27-0324-66
$C_{16}H_{14}ClN$			
Indole, 3-(o-chlorobenzyl)-2-methyl-	EtOH	275(3.82),279(3.82), 290(3.75)	35-1077B-66
$C_{16}H_{14}ClNO_6$			
6-Carboxybenzo[c]quinolizinium perchlorate, ethyl ester	EtOH	233(4.32),262(4.62), 310s(3.82),335(3.82), 350(4.12),367(4.23)	44-2346-66
$C_{16}H_{14}ClN_3O_4S_2$			
Benzenesulfonamide, p-chloro-N-[1-(p-tolylsulfonyl)pyrazol-5-yl]-	MeOH	214s(4.0),228(4.3), 249(4.2),275s(3.6)	24-0183-66
p-Toluenesulfonamide, N-[1-[(p-chlorophenyl)sulfonyl]pyrazol-5-yl]-	MeOH	211s(4.0),231(4.4), 246s(4.2),274s(3.6)	24-0183-66
$C_{16}H_{14}Cl_2N_4$			
1,4-Dihydro-1,4-bis(2-pyridylmethylene)pyrazinium dichloride	H_2O	218(4.0),230(3.9), 235(3.9),243s(3.8), 278(4.3),288(4.4)	39-1209-66C

Compound	Solvent	$\lambda_{max}(\log \epsilon)$	Ref.
$C_{16}H_{14}Cl_2O$			
2-Cyclopropen-1-one, 2-[1,2-(dichloromethano)cyclohexyl]-3-phenyl-	n.s.g.	264(4.18)	88-3763-66
$C_{16}H_{14}Cl_3NO$			
Mesito-2,4,6-trichloroanilide	EtOH	240s(4.0),268s(3.2)	28C-0369-66A
$C_{16}H_{14}Cl_3NO_4$			
Benzanilide, 2',4',6'-trichloro-3,4,5-trimethoxy-	EtOH	210(4.75),265(4.2)	28C-0584-66A
$C_{16}H_{14}Cl_3N_3O$			
7-Azaindoline, 4-methyl-1-phenyl-6-trichloroacetamido-	EtOH	302(4.17),338(4.08)	78-3233-66
$C_{16}H_{14}CuN_2O_2$			
Copper, [α,α'-(ethylenedinitrilo)-di-o-cresolato(2-)]-	MeOH	355(3.43),373(3.81)	65-1590-66
$C_{16}H_{14}N_2$			
3H-Dibenz[c,f]imidazo[1,2-a]azepine, 2,9-dihydro-	EtOH	229(4.14),256(3.91), 295(3.83)	4-0206-66
3H-Pyrido[4,3-b]carbazole, 4,6-dihydro-5-methyl-	EtOH-HCl	236(4.35),272s(4.42), 283(4.65),302(4.14), 314(4.29),375(4.36)	87-0237-66
	EtOH	238(4.48),248(4.50), 276(4.70),292s(4.46), 316(4.13),328(4.07), 342(3.72)	87-0237-66
$C_{16}H_{14}N_2O$			
Canthin-6-one, 4-ethyl-4,5-dihydro-	EtOH	266(4.15),274(4.08), 284(4.15),315(3.91), 329(4.00)	5-0159-66A
2-Imidazolin-5-one, 1-benzyl-2-phenyl-, hydrochloride	MeOH	252(4.10),320(3.94)	69-2468-66
Indole, 5-acetamido-2-phenyl-	EtOH	249(4.45),259s(4.41), 319(4.42)	44-0065-66
Indolizine-7-carboxamide, 2-methyl-N-phenyl-	EtOH	225(4.46),275(4.40), 352(4.22)	56-0061-66
1H-Isoindole-1-carboxamide, 1'-methyl-3-phenyl-	iso-PrOH	253(4.14)	35-3173-66
2H-Isoindole-1-carboxamide, N'-methyl-3-phenyl-	iso-PrOH	221(4.16),230(4.18), 258(4.52),295(3.77), 369(4.35)	35-3173-66
4(3H)-Quinazolinone, 2-benzyl-3-methyl-	EtOH	208(4.50),226(4.48), 270(3.99),305(3.67), 318(3.57)	39-2290-66
$C_{16}H_{14}N_2OSe$			
Selenazolidin-4-one, 3-methyl-5-phenyl-2-phenylimino-	EtOH	225s(4.7),273(3.81)	28C-0778-66A
Selenazolin-4-one, 2-(N-methylanilino)-5-phenyl-	EtOH	228(4.34),246(4.30)	28C-0778-66A
$C_{16}H_{14}N_2O_2$			
Diimide, di-p-toluoyl-	dioxan	253s(4.46)	24-2039-66
	dioxan	254s(4.44),474(1.70)	89-0372-66
2,5-Pyrazinediol, 3,6-bis-(2-hydroxyphenyl)-	dioxan	255(4.32),323(4.08)	49-0077-66

Compound	Solvent	$\lambda_{max}(\log \epsilon)$	Ref.
Quinazoline, 2-(α-hydroxybenzyl)-4-methoxy-	EtOH	211s(4.43),222(4.70), 255(3.98),299(3.43), 311(3.43)	39-2290-66C
4(3H)-Quinazolinone, 2-(α-hydroxybenzyl)-3-methyl-	EtOH	210(4.50),226(4.48), 269(3.99),277(3.94), 304(3.62),316(3.49)	39-2290-66C
$C_{16}H_{14}N_2O_3S$			
Benzo[b]cyclohepta[e][1,4]thiazin-10(11H)-one, 8-isopropyl-7-nitro-	MeOH	238(4.36),290(4.32), 444(3.74),515(3.79)	18-1980-66
$C_{16}H_{14}N_2O_4S$			
Benzo[b]cyclohepta[e][1,4]thiazin-10(11H)-one, 8-isopropyl-9-nitro-, 5-oxide	MeOH	223(4.37),247(4.39), 370(4.15),395(4.21)	18-1988-66
$C_{16}H_{14}N_2O_5$			
2'-Uridenine, 5'-O-benzoyl-2',3'-dideoxy-	EtOH	230(4.16),261(3.97)	44-0205-66
$C_{16}H_{14}N_2O_5S$			
Benzo[b]cyclohepta[e][1,4]thiazin-10(11H)-one, 8-isopropyl-9-nitro-, 5,5-dioxide	MeOH	243(4.27),365(4.05), 390(4.07)	18-1988-66
$C_{16}H_{14}N_4$			
2-Naphthylamine, 1-[(m-aminophenyl)-azo]-	benzene	440(4.15)	44-2571-66
	$C_6H_{11}Me$	432(4.17)	44-2571-66
2-Naphthylamine, 1-[(p-aminophenyl)-azo]-	benzene	458(4.29)	44-2571-66
	$C_6H_{11}Me$	450(4.25)	44-2571-66
Pyrazole, 5(3)-methyl-3(5)-phenyl-4-(phenylazo)-	$CHCl_3$	339(4.31),420(3.08)	22-2990-66
$C_{16}H_{14}N_4O$			
1-Naphthol, 7-amino-8-[(m-aminophenyl)-azo]-	benzene	508(4.24)	44-2571-66
1-Naphthol, 7-amino-8-[(p-aminophenyl)-azo]-	benzene	509(4.39)	44-2571-66
Pyrazole-4,5-dione, 3-methyl-1-phenyl-, 4-phenylhydrazone	EtOH	251(4.35),394(4.38)	94-0770-66
Pyrazole-4,5-dione, 3-phenyl-, 4-(methylphenylhydrazone)	$CHCl_3$	414(4.06)	22-2990-66
5-Pyrazolone, 3-methyl-1-phenyl-4-(phenylazo)-	$CHCl_3$	393(4.35)	22-2990-66
as-Triazine, 3-(methylamino)-5,6-diphenyl-, 2-oxide	EtOH	226(4.29),262(4.41), 374(3.72)	44-3914-66
$C_{16}H_{14}N_4O_2$			
2-Quinoxalinecarboxylic acid, 5-amino-3-(2-pyridyl)-, ethyl ester	EtOH	229(4.33),299(4.61)	44-3384-66
$C_{16}H_{14}N_4O_4$			
Pyrazoline, 1-(2,4-dinitrophenyl)-3-methyl-4-phenyl-	$CHCl_3$	395(4.51)	22-0610-66
Pyrazoline, 1-(2,4-dinitrophenyl)-3-methyl-5-phenyl-	$CHCl_3$	390(4.30)	22-0610-66
Pyrazoline, 1-(2,4-dinitrophenyl)-4-methyl-3-phenyl-	$CHCl_3$	416(4.46)	22-0610-66
Pyrazoline, 1-(2,4-dinitrophenyl)-5-methyl-3-phenyl-	$CHCl_3$	415.5(4.47)	22-0610-66

Compound	Solvent	λ_{max}(log ε)	Ref.
$C_{16}H_{14}N_4O_5$			
Pyrazoline, 1-(2,4-dinitrophenyl)-3-(p-methoxyphenyl)-	$CHCl_3$	428(4.50)	22-0610-66
$C_{16}H_{14}N_8O_8$			
2,3-Butanedione, bis(2,4-dinitrophenylhydrazone)	EtOH-KOH	556(4.73)	96-0297-66
	$CHCl_3$	393(4.67),435s(4.60)	96-0297-66
$C_{16}H_{14}O$			
Benzo[b]naphtho[1,2-d]furan, 8,9,10,11-tetrahydro-	EtOH	249(4.41),302(3.98), 315(3.95),329(3.93)	44-2646-66
4,5-Benzo-3-oxa-6-phenylbicyclo[4.1.0]-heptane	EtOH	230s(3.87),273(2.19), 280(3.16)	44-1332-66
3-Buten-1-one, 1,4-diphenyl-	C_6H_{12}	248(4.43),285s(3.54), 292s(3.22),327(2.43)	35-4905-66
Crotonaldehyde, 2,3-diphenyl-	EtOH	277(4.00)	35-1518-66
1H-Cyclopenta[7,8]naphtho[2,3-b]-furan, 2,3-dihydro-6-methyl-	EtOH	243(4.77),250(4.88), 305(3.77),318(3.90), 334(3.85)	39-0725-66C
Fluorene, 7-acetyl-1-methyl-	n.s.g.	312(4.42)	39-0481-66C
[2.2]Paracyclophane, 1-oxo-	EtOH	223(4.03),334(2.38)	35-3515-66
Styryl ether, cis	EtOH	279(4.19)	24-0642-66
Styryl ether, trans	EtOH	281(4.50)	24-0642-66
Tricyclo[8.4.1.1^{3,8}]hexadeca-3,5,7,10,12,14-hexaen-2-one	C_6H_{12}	213(4.48),217(4.49), 268(4.08),315s(3.65)	89-0603-66
$C_{16}H_{14}O_2$			
Chalcone, 4-hydroxy-3-methyl-	pH 13	269(4.14),431(4.49)	49-0896-66
	EtOH	252(4.16),360(4.38), 440s(2.55)	49-0896-66
Chalcone, 4'-hydroxy-3'-methyl-	pH 13	273(4.08),303(4.17), 393(4.36)	49-0896-66
	EtOH	224(4.05),325(4.40)	49-0896-66
1H-Cyclopenta[7,8]naphtho[2,3-b]-furan-1-one, 2,3,7,8-tetrahydro-6-methyl-	EtOH	223(4.60),229(4.62), 238(4.62),266(4.03), 350(3.99)	39-0734-66C
Fluorene, 2-acetyl-7-methoxy-	n.s.g.	230(4.16),327(4.50)	39-0481-66C
2-Propyn-1-ol, 1-(m-methoxyphenyl)-1-phenyl-	EtOH	227(4.28),274(3.60)	22-2885-66
2-Propyn-1-ol, 1-(p-methoxyphenyl)-1-phenyl-	EtOH	230(4.08),275(3.48)	22-2885-66
Rotoxen, 6,6a,12,12a-tetrahydro-	EtOH	276(3.59),283(3.61)	39-0544-66C
Spiro[benzofuran-2(3H),4'-chroman]	EtOH	281(3.82),286(3.79)	39-0544-66C
4-Stilbenecarboxylic acid, methyl ester, trans	hexane	228(4.16),308(4.55), 320(4.61),332(4.39)	99-0089-66
	EtOH	230(4.19),320(4.60)	99-0089-66
4-Stilbenol, acetate, trans	hexane	227(4.20),300(4.47), 319(4.30)	99-0089-66
	EtOH	227(4.24),298(4.54), 310(4.52)	99-0089-66
$C_{16}H_{14}O_3$			
Chalcone, 2',4-dihydroxy-3-methyl-	pH 13	235(4.15),260s(4.03), 415(4.38)	49-0896-66
	EtOH	249(4.08),258s(4.06), 381(4.46),470s(2.95)	49-0896-66
Macluran, 3-methoxycyano-	EtOH	280(3.19)	78-0139-66A
Maturinin	EtOH	250(4.45),372(3.95)	78-0685-66
Naphth[2,1-b]oxonine-8,13-dione, 8,9,10,11-tetrahydro-	EtOH	218(4.67),236s(--), 301(3.67)	44-2646-66

Compound	Solvent	λ_{max}(log ε)	Ref.
$C_{16}H_{14}O_4$			
Alloimperatorin	EtOH	221(4.34),244(3.96), 252(3.97),268(4.07), 273(4.07),317(3.89)	102-0367-66
1(2H)-Anthracenone, 3,4-dihydro-8,9-dihydroxy-, 8-acetate	EtOH	266(4.42),289(3.46), 300(3.41),384(3.56)	78-2761-66
2,3-Benzocyclooctenedicarboxylic acid, dimethyl ester	EtOH	242(4.36)	77-0508-66
Maturin	EtOH	251(4.43),353(3.93)	78-0685-66
1H-Naphtho[2,1-b]pyran-1-one, 5,6-dimethoxy-3-methyl-	EtOH	214(4.50),233(4.53), 263(4.20),272(4.17), 316(3.96),334(3.80)	94-0129-66
1-Oxaphenalene, 3-acetoxy-5-methoxy-2-methyl-	EtOH	243(4.45),252(4.38), 262(4.34),290s(3.59), 303(3.70),336s(3.80), 345s(3.83),351(3.85), 364s(3.66),381s(3.34)	39-0523-66C
$C_{16}H_{14}O_5$			
Acetophenone, 2'-hydroxy-4'-methoxy-2-[3,4-(methylenedioxy)phenyl]-	MeOH	206(4.9),229(4.6), 276(4.8),315(4.5)	39-0509-66C
	MeOH-base	212(5.0),229(4.9), 277(4.7),350(4.5)	39-0509-66C
2-Furanacrylic acid, ester with 3-phenyllactic acid	pH 7.5	308.5(4.37)	69-1425-66
Heraclenin	EtOH	218(4.26),246s(4.30), 248(4.32),263s(4.05), 300(4.02)	12-0483-66
Isoheraclenin	EtOH	220(4.50),250(4.55), 262(4.39),300(4.32)	93-0616-66
Xanthone, 1,2,8-trimethoxy-	EtOH	241(4.42),265(4.56), 290(3.89),338(3.94)	78-1785-66
Xanthone, 1,3,5-trimethoxy-	EtOH	246(4.74),300(4.34), 336(3.86)	78-1777-66
Xanthone, 2,3,4-trimethoxy-	EtOH	246(4.21),278(3.96), 304(4.13)	78-1777-66
$C_{16}H_{14}O_5S$			
Hydroquinone, 2-acetyl-3-(2-thienyl)-, diacetate	EtOH	228(4.12),260s(3.85), 284s(3.70)	33-1794-66
$C_{16}H_{14}O_6$			
p-Benzoquinone, 2-acetyl-3-(5-carbethoxymethyl-2-furyl)-	EtOH	260(4.19),454(3.73)	33-1794-66
1,2,4-Naphthalenetriol, triacetate	EtOH	223(5.7),281(3.6)	88-4057-66
$C_{16}H_{15}^+$			
Cyclopropyldiphenylcarbonium ion	FSO_3H-SbF_5	319(4.02),358(3.92), 435(4.48)	35-1488-66
$C_{16}H_{15}BF_4N_4O$			
2-(N^{ω}-Benzoylazo)-1,3-dimethylbenzimidazolium tetrafluoroborate	MeCN	359(4.22)	5-0065-66J
$C_{16}H_{15}Br$			
Ethylene, 2-bromo-1,1-di-p-tolyl-	H_2SO_4	350(3.9),480(4.6)	27-0324-66
$C_{16}H_{15}BrO_2$			
Ethylene, 2-bromo-1,1-bis-(o-methoxyphenyl)-	CF_3COOH	360(4.0),540(4.8)	27-0324-66

Compound	Solvent	λ_{max}(log ε)	Ref.
$C_{16}H_{15}Cl$			
Ethylene, 1-chloro-2,2-di-o-tolyl-	MeOH	234(4.2)	24-0680-66
Ethylene, 1-chloro-2,2-di-p-tolyl-	MeOH	238(4.2),260s(4.1)	24-0680-66
$C_{16}H_{15}ClN_2O$			
2-Imidazolin-5-one, 1-benzyl-2-phenyl-, hydrochloride	MeOH	252(4.10),320(3.94)	69-2468-66
$C_{16}H_{15}ClN_2OS$			
2(1H)-Quinazolinethione, 6-chloro-3,4-dihydro-4-methoxy-3-methyl-4-phenyl-	iso-PrOH	252(3.95),288(4.50)	44-1007-66
$C_{16}H_{15}ClN_2O_2$			
2(1H)-Quinazolinone, 6-chloro-3,4-dihydro-4-methoxy-3-methyl-4-phenyl-	iso-PrOH	253(4.20),300(3.30)	44-1007-66
$C_{16}H_{15}ClN_2O_5$			
5-Imidazolone, 1-benzyl-2-methyl-, perchorate	MeOH	252(4.17),320(3.89)	69-2468-66
$C_{16}H_{15}IN_2O_6$			
Uracil, 1-(3-deoxy-3-iodo-β-D-arabinofuranosyl)-, 5'-benzoate	EtOH	229(4.29),261(3.94)	44-0205-66
$C_{16}H_{15}N$			
Indole, N-benzyl-3-methyl-	n.s.g.	225(4.56),289(3.73)	16-0064-66
$C_{16}H_{15}NO$			
Aziridine, 3-benzoyl-1-methyl-2-phenyl-, cis	C_6H_{12}	326(2.15)	44-1244-66
trans	C_6H_{12}	348(2.46)	44-1244-66
Indole, 2-(p-methoxyphenyl)-3-methyl-	EtOH	245(4.32),308(4.36)	87-0527-66
Indole, 3-(p-methoxyphenyl)-2-methyl-	EtOH	270(4.20)	87-0527-66
Indolinone, 7-phenacyl-	MeOH	252(4.29),290(3.52)	35-4061-66
Pyridine, 4-[(5-methoxy-1-indanylidene)methyl]-	EtOH	238(4.01),296s(4.12), 307(4.24),340(4.54)	44-3337-66
$C_{16}H_{15}NO_2$			
Benzamide, 2-benzoyl-N,N-dimethyl-	MeOH	330(2.12)	35-0781-66
Indole, 3-(3,4-dimethoxyphenyl)-	EtOH	268(4.29)	94-0934-66
$C_{16}H_{15}NO_2S$			
Benzo[b]cyclohepta[e][1,4]thiazin-10(11H)-one, 8-isopropyl-, 5-oxide	pH 13	235(4.34),250(4.30), 393(4.21),460(4.06)	18-1988-66
	MeOH	237(4.43),363(4.20), 382(4.23)	18-1988-66
L-Tyrosine, N-(thiobenzoyl)-	H_2O	282(4.03),360s(2.53)	1-2781-66
	MeOH	275(3.95),372(2.38)	1-2781-66
	dioxan	280(3.93),388(2.42)	1-2781-66
$C_{16}H_{15}NO_3$			
Fluorene-8a(6H)-carbonitrile, 7,8-dihydro-2,4-dimethoxy-6-oxo-	MeOH	222(3.85),227(3.85), 250(4.03),256(4.03), 338(4.40)	44-1451-66
$C_{16}H_{15}NO_3S$			
Benzo[b]cyclohepta[e][1,4]thiazin-10(11H)-one, 8-isopropyl-, 5,5-dioxide	MeOH	234(4.56),360(4.23), 377(4.25)	18-1988-66

Compound	Solvent	$\lambda_{max}(\log \epsilon)$	Ref.
$C_{16}H_{15}NO_4$			
Alanine, N-[3-(2-furyl)acryloyl]-	pH 7.5	303(4.41)	69-1425-66
3-phenyl-, trans			
$C_{16}H_{15}NO_5S$			
Tropone, 5-acetamido-2-tosyloxy-	MeOH	228(4.42),262s(4.03), 344(4.16)	18-1310-66
$C_{16}H_{15}N_3O$			
as-Triazin-3(4H)-one, 2-methyl-	EtOH	220s(4.10),295(4.10)	94-1448-66
5,6-diphenyl-			
as-Triazin-3(4H)-one, 4-methyl-	EtOH	220s(4.22),290(4.13)	44-3914-66
5,6-diphenyl-	EtOH	220s(4.22),290(4.13)	94-1448-66
$C_{16}H_{15}N_3OS$			
p-Benzoquinone, 2,6-dimethyl-, 4-azine	MeCN	475(4.67)	5-0065-66J
with 3-methyl-2-benzothiazolinone			
$C_{16}H_{15}N_3O_2$			
Indazole, 6-ethyl-3-methyl-	EtOH	234(4.27),258(3.87), 355(4.22)	78-3131-66
1-(p-nitrophenyl)-			
Indazole, 3,4,6-trimethyl-	EtOH	274(4.31)	78-3131-66
1-(m-nitrophenyl)-			
Indazole, 3,4,6-trimethyl-	EtOH	215(4.37),236(4.26), 266(3.96),361(4.25)	78-3131-66
1-(p-nitrophenyl)-			
$C_{16}H_{15}N_3O_3$			
Indazole, 6-ethoxy-3-methyl-	EtOH	232(4.35),278(3.94), 353(4.19)	78-3131-66
1-(p-nitrophenyl)-			
Oxazirino[2,3-f]hexahydro-as-triazin-	EtOH	232(4.41),272s(3.44)	94-1448-66
4-one, 6-hydroxy-3-methyl-			
6,7-diphenyl-			
Oxazirino[2,3-f]hexahydro-as-triazin-	EtOH	231(4.28),266s(2.15)	94-1448-66
4-one, 6-hydroxy-5-methyl-			
6,7-diphenyl-			
$C_{16}H_{15}N_3O_4$			
Indazole, 4,6-dimethoxy-3-methyl-	EtOH	228(4.19),265(3.98), 361(4.01)	78-3131-66
1-(p-nitrophenyl)-			
Indazole, 5,6-dimethoxy-3-methyl-	EtOH	215(4.17),240(4.23), 265(3.96),305(3.89), 365(4.21)	78-3131-66
1-(p-nitrophenyl)-			
Morphanthridine, 6-hydrazino-,	EtOH	244(4.02),284(3.9)	4-0206-66
oxalate			
$C_{16}H_{16}$			
Biphenylene, 2,3,6,7-tetramethyl-	EtOH	246(4.7),256(4.9), 352(3.8),372(4.0)	39-1767-66C
1-Butene, 1,1-diphenyl-	98% H_2SO_4	321(3.95),431(4.59)	77-0625-66
1-Butene, 1,3-diphenyl-, cis	hexane	241(4.15),269s(3.20)	35-5777-66
1-Butene, 1,3-diphenyl-, trans	hexane	252(4.33),284(3.26), 293(3.11)	35-5777-66
2-Butene, 1,2-diphenyl-	EtOH	245(4.03)	78-1001-66
2-Butene, 1,3-diphenyl-, cis	hexane	245(4.19),269s(3.67)	35-5777-66
2-Butene, 1,3-diphenyl-, trans	hexane	230s(3.98),262(3.28), 268(3.00)	35-5777-66
Ethylene, 1,1-di-p-tolyl-	H_2SO_4	320(3.9),450(4.6)	27-0324-66
Indan, 1-methyl-3-phenyl-, cation	98% H_2SO_4	304(3.72),415(4.18)	77-0625-66
Naphthalene, 1,2,3,4-tetrahydro-	EtOH	264(3.44),268(3.42), 277(3.33)	39-0861-66B
1-phenyl-			

Compound	Solvent	λ_{max}(log ε)	Ref.
[2.2]Metaparacyclophane	n.s.g.	240s(3.5),275f(2.5)	35-1324-66
1-Propene, 2-methyl-1,1-diphenyl-	EtOH	230(4.08),244(4.13)	88-3267-66
Tricyclo[8.4.1.1^{3,8}]hexadeca-3,5,7,10,12,14-hexaene	C_6H_{12}	221(4.62),258(3.92)	89-0603-66
$C_{16}H_{16}BF_4N_3O_3S_2$			
2-[N-p-Toluenesulfonylazo)-3-methyl-6-methoxybenzothiazolium tetrafluoroborate	MeCN-HBF_4	460(4.18)	5-0116-66G
$C_{16}H_{16}ClFN_4O_7$			
Purine, 6-chloro-2-fluoro-9-β-D-ribofuranosyl-, triacetate	EtOH	269(4.09)	44-3258-66
$C_{16}H_{16}ClNO$			
Mesito-m-chloroanilide	EtOH	210(4.3),249(4.3), 260s(4.2),286s(3.3)	28C-0369-66A
Mesito-o-chloroanilide	EtOH	238(4.1)	28C-0369-66A
Mesito-p-chloroanilide	EtOH	252(4.3)	28C-0369-66A
$C_{16}H_{16}ClNO_4$			
Benzanilide, m'-chloro-3,4,5-trimethoxy-	EtOH	215(4.6),275(4.2)	28C-0584-66A
Benzanilide, o'-chloro-3,4,5-trimethoxy-	EtOH	212(4.7),268(4.2), 300s(3.8)	28C-0584-66A
Benzanilide, p'-chloro-3,4,5-trimethoxy-	EtOH	210(4.6),279(4.3)	28C-0584-66A
$C_{16}H_{16}ClNO_6$			
7,8-Dihydro-10,11-(methylenedioxy)-2-methyl-6H-benzo[c]pyrid[1,2-a]-azepinium perchlorate	EtOH	263(3.83),289(3.57), 345(3.74)	94-0622-66
$C_{16}H_{16}ClN_2O_2S$			
1,2,4-Benzothiadiazine, 4-(p-chlorobenzyl)-6-chloro-3,4-dihydro-2,7-dimethyl-, 1,1-dioxide	EtOH	220(4.57),259(4.20), 327(3.65)	7-0687-66
$C_{16}H_{16}ClN_3$			
as-Triazine, 3-(o-chlorophenyl)-1,4,5,6-tetrahydro-1-methyl-5-phenyl-	MeOH	260(3.63)	87-0881-66
as-Triazine, 3-(p-chlorophenyl)-1,4,5,6-tetrahydro-1-methyl-5-phenyl-	MeOH	230(4.20),300(3.69)	87-0881-66
$C_{16}H_{16}Cl_2N_4O_7$			
Purine, 2,6-dichloro-9-β-D-ribofuranosyl-, triacetate	EtOH	273(4.05)	44-3258-66
$C_{16}H_{16}CuN_2O_4$			
Benzylideneimine, 2-hydroxy-3-methoxy-, copper chelate	MeOH	292(4.47),365(3.77)	65-1590-66
$C_{16}H_{16}N_2$			
Canthine, 4-ethyl-4,5-dihydro-	EtOH	235(4.63),253(4.40), 260s(4.25),283s(3.92), 290(4.22),347(3.71), 363(3.77)	5-0159-66A
	EtOH-HCl	253(4.48),305(4.24), 386(3.71)	5-0159-66A

Compound	Solvent	$\lambda_{max}(\log \epsilon)$	Ref.
Canthine, 5-ethyl-4,5-dihydro-	EtOH	236(4.58),254(4.35), 261s(4.24),284(3.96), 292(4.14),348(3.66), 364(3.72)	5-0159-66A
Malondialdehyde, methyl-, dianil	$CHCl_3$	400s(4.0)	65-1383-66
2-Pyrazoline, 5-methyl-1,3-diphenyl-	MeOH	357(3.99)	89-0699-66
$C_{16}H_{16}N_2O$			
Formamide, N-[2-(1-methylcarbazol-2-yl)-ethyl]-	EtOH	216(4.40),238(4.64), 248(4.51),259(4.31), 286s(4.03),297(4.24), 323(3.64),337(3.54)	87-0237-66
Morphanthridine, 6-(2-hydroxy-ethylamino)-	EtOH	226s(4.23),249(4.06)	4-0206-66
$C_{16}H_{16}N_2O_2$			
Cyclobutenedione, bis(3,5-dimethyl-2-pyrryl)-	EtOH	316(4.32),357(4.17), 385(4.20),440(4.51)	5-0153-66I
2-Cyclobuten-1-one, 2-(3,5-dimethyl-pyrrol-2-yl)-4-(3,5-dimethyl-2H-pyrrol-2-ylidene)-3-hydroxy-	EtOH	246(3.88),302(3.91), 349(3.57),551(5.43)	5-0153-66I
	$CHCl_3$	554(4.36)	5-0153-66I
$C_{16}H_{16}N_2O_2S$			
L-Tyrosineamide, N-thiobenzoyl-	MeOH	280(3.95),380(2.25)	1-2781-66
	dioxan	287(3.95),387(2.49)	1-2781-66
$C_{16}H_{16}N_2O_3$			
Acetanilide, 4'-(p-nitrophenethyl)-	EtOH	247(4.35),265s(4.21), 330s(2.98)	65-0214-66
Mesito-m-nitroanilide	EtOH	244(4.4),286s(3.8), 306(3.2)	28C-0369-66A
Mesito-o-nitroanilide	EtOH	231(4.3),338(3.5)	28C-0369-66A
Mesito-p-nitroanilide	EtOH	242s(3.8),314(4.3)	28C-0369-66A
$C_{16}H_{16}N_2O_4$			
Biphenyl, 4,4',5,5'-tetramethyl-2,2'-dinitro-	CH_2Cl_2	280(4.15)	44-4204-66
$C_{16}H_{16}N_2O_4S$			
Tropolone, 4-acetyl-, p-toluene-sulfonylhydrazone	MeOH	228(4.36),292(4.19), 384(3.68)	18-0253-66
$C_{16}H_{16}N_2O_6$			
Benzanilide, 3,4,5-trimethoxy-m'-nitro-	EtOH	217(4.4),271(4.4), 292s(4.3),341s(3.2)	28C-0584-66A
Benzanilide, 3,4,5-trimethoxy-o'-nitro-	EtOH	216(4.5),268(4.3), 351s(3.8)	28C-0584-66A
Benzanilide, 3,4,5-trimethoxy-p'-nitro-	EtOH	213(4.6),323(4.4)	28C-0584-66A
Uracil, 1-(5-O-benzoyl-3-deoxy-β-D-threo-pentofuranosyl)-	EtOH	229(4.14),262(3.99)	44-0205-66
$C_{16}H_{16}N_2O_6S_2$			
5-Thia-1-azabicyclo[4.2.0]oct-2-ene-2-carboxylic acid, 3-(hydroxy-methyl)-8-oxo-7-[2-(2-thienyl)-acetamido]-, acetate, sodium salt	n.s.g.	234(4.19),260(3.39)	39-1142-66C
$C_{16}H_{16}N_2O_8$			
8H-Pyrido[1,2-b]pyridazine-5,6,7,8-tetracarboxylic acid, tetramethyl ester	MeOH	219(4.01),275(3.82), 303(4.03),369(3.79), 443(3.72)	39-2218-66C

Compound	Solvent	λ_{max}(log ε)	Ref.
$C_{16}H_{16}N_4O_2$			
Lumazine, 8-benzyl-3,6,7-trimethyl-	pH -2.7	250(4.13),363(4.17)	24-3503-66
	pH 6	263(4.31),280s(3.86), 410(4.07)	24-3503-66
	pH 13	241(4.33),310(4.37), 364(3.84)	24-3503-66
$C_{16}H_{16}O$			
1H-Cyclopenta[7,8]naphtho[2,3-b]furan, 2,3,7,8-tetrahydro-6-methyl-	EtOH	241(4.85),247(4.81), 288(3.86),294(3.83), 300(3.81),319(3.35), 334(3.44)	39-0725-66C
Spiro[naphthalene-2(1H),2'-[5]nor-bornene]-1-one, 3,4-dihydro-	EtOH	248(4.02),290(3.22)	22-1693-66
Stilbene, α-ethoxy-	MeOH	287(4.19)	5-0053-66I
$C_{16}H_{16}O_2$			
8H-Benz[cd]azulene-9-carboxylic acid, 1,2-dihydro-, ethyl ester	EtOH	214(4.48),238(3.98), 327(4.07)	35-3950-66
Benzo[b]naphtho[1,2-d]furan-7a-ol, 7a,8,9,10,11,11a-hexahydro-	EtOH	233(4.83),282(3.63)	44-2646-66
1-Butanone, 3-hydroxy-1,3-diphenyl-	EtOH	246(4.13)	28C-1429-66A
Ethylene, 1,1-bis(p-methoxyphenyl)-	CF_3COOH	330(3.8),490(4.8)	27-0324-66
2,6,8,12-Hexadecatetraynoic acid	EtOH	204(3.95),236(2.71), 254(2.48)	12-1207-66
2-trans	EtOH	208(3.98),241s(3.97), 253(2.72)	12-1207-66
12-trans	EtOH	208(4.21),239(3.04), 253(2.70)	12-1207-66
2,12-di-trans	EtOH	207(4.29),241s(3.11), 253(2.76)	12-1207-66
6a,7-Secorotoxen, 6,6a,12,12a-tetra-hydro-	EtOH	220s(4.15),278(3.74), 284(3.73)	39-0544-66C
	EtOH-KOH	243(3.84),280s(3.60), 285(3.68),296(3.60)	39-0544-66C
Tropylphenylacetic acid, methyl ester	C_6H_{12}	258(3.57)	44-0912-66
$C_{16}H_{16}O_3$			
1H-Benz[e]indene-3-carboxylic acid, 2,4,5,9b-tetrahydro-9b-methyl-2-oxo-, methyl ester	EtOH	233(4.01)	35-0161-66
Glutaric anhydride, 2-(3,4-dihydro-2-naphthyl)-2-methyl-	$CHCl_3$	266(4.01)	77-0485-66
Styrene, m-(3-hydroxy-4,5-dimethoxyphenyl)-	EtOH	222(4.52),250(4.41)	44-0189-66
	EtOH-NaOH	241(4.49),298(3.67)	44-0189-66
Styrene oxide, p,p'-dimethoxy-, trans	hexane	242(4.50),279(3.73)	44-3976-66
6-Tetradecene-8,10,12-triyn-1-ol, 4,5-epoxy-, acetate	ether	234(4.87),238(4.86), 249(5.03),259(3.76), 274(4.01),291(4.29), 311(4.42),333(4.26)	24-1830-66
$C_{16}H_{16}O_4$			
Benzophenone, 2,4,4'-trimethoxy- (cation)	H_2SO_4	412(4.56)	39-0650-66B
p-Benzoquinone, 2-acetyl-3-(5-iso-propyl-3-methyl-2-furyl)-	EtOH	235s(4.13),248(4.17), 276s(3.97),495(3.69)	33-1794-66
Glutaric acid, 2-methyl-2-(2-naphthyl)-	EtOH	226(5.01),268(3.78)	77-0485-66
Maturin, dihydro-	EtOH	249(4.82),332(3.96), 348(3.91)	78-0685-66

Compound	Solvent	λ_{max}(log ϵ)	Ref.
Valeric acid, 5-(2-hydroxy-1-naphthoyl)-	EtOH	225(4.74)	44-2646-66
Warburgin	EtOH	370(4.30)	77-0393-66
$C_{16}H_{16}O_5$			
Benzophenone, 4-hydroxy-3,3',4'-trimethoxy-, cation	H_2SO_4	437(4.36)	39-0650-66B
Biphenyl-4-carboxylic acid, 3',4',5'-trimethoxy-	EtOH	217(4.53),265(4.11)	44-0189-66
Fluorene-8a(6H)-carboxylic acid, 7,8-dihydro-2,4-dimethoxy-6-oxo-	MeOH	224(3.90),250(4.04), 334(4.38)	44-1451-66
Yangonin, 11-methoxy-	EtOH	221(4.34),250(4.15), 365(4.41)	83-0503-66
$C_{16}H_{16}O_6$			
Heraclenine hydrate	EtOH	220(4.30),250(4.35), 262(4.15),300(4.10)	93-0616-66
$C_{16}H_{16}O_7$			
L-Ascorbic acid, 5,6-O-isopropylidene-, 3-benzoate	EtOH	230(4.36)	94-1039-66
$C_{16}H_{17}^+$			
Isopropyldiphenylcarbonium ion	FSO_3H-SbF_5	322(4.27),422(4.47)	35-1488-66
$C_{16}H_{17}BrN_4O_6$			
Isoalloxazine, 7-bromo-8-methyl-10-(1-D-ribityl)-	H_2O	270(4.68),360(4.10), 450(4.17)	87-0495-66
$C_{16}H_{17}ClN_2O_2S$			
2H-1,2,4-Benzothiadiazine, 4-benzyl-6-chloro-3,4-dihydro-2,7-dimethyl-, 1,1-dioxide	EtOH	221(4.49),259(4.20), 329(3.64)	7-0687-66
$C_{16}H_{17}N$			
4-Stilbenamine, N,N-dimethyl-, trans	hexane	239(4.18),345(4.53)	99-0089-66
	EtOH	234(4.06),350(4.49)	99-0089-66
$C_{16}H_{17}NO$			
Cyclohexanone, 2-(2-methyl-4-quinolyl)-	EtOH-HCl	240(4.63),310s(3.99), 318(4.05)	94-0762-66
	EtOH	230(4.57),233s(4.53), 280(3.66),305(3.57), 308s(3.45),318(3.60)	94-0762-66
Mesitanilide	EtOH	250(4.0)	28C-0369-66A
$C_{16}H_{17}NO_2$			
Biphenyl, 3,3',4,4'-tetramethyl-2-nitro-	isooctane	241(4.22)	44-4204-66
Biphenyl, 3',4,4',5-tetramethyl-2-nitro-	isooctane	240(4.28),305s(3.31)	44-4204-66
Pyrrolo[2,1-a]isoquinoline-3-carboxylic acid, 5,6-dihydro-2-methyl-, ethyl ester	MeOH	231(4.22),312(4.51), 326(4.46)	95-0856-66
$C_{16}H_{17}NO_3$			
Acetophenone, 2-amino-4'-methoxy-2-(p-methoxyphenyl)-	EtOH-HCl	224(4.26),276s(4.20), 286(4.22),292s(4.21)	87-0527-66

Compound	Solvent	$\lambda_{max}(\log \epsilon)$	Ref.
6,7-Dihydro-2-hydroxy-9,10-dimethoxy-4-methylbenzo[a]quinolizinium hydroxide, inner salt	EtOH	217(4.27),237(4.38), 257(4.37),313(4.24)	44-0797-66
$C_{16}H_{17}NO_4$			
Benzanilide, 3,4,5-trimethoxy-	EtOH	211(4.6),275(4.3)	28C-0584-66A
Cyclohexanecarbonitrile, 1-(3,5-dimethoxybenzyl)-2,4-dioxo-	MeOH	224(4.00),260(4.18)	44-1451-66
Lunine	EtOH	222(4.35),247(4.60), 267s(4.06),313(4.08), 323(4.07)	12-2185-66
$C_{16}H_{17}NO_5$			
3H-1-Benzazepine-3,4-dicarboxylic acid, 2-ethoxy-, dimethyl ester	MeOH	240(4.4),275(4.1)	24-3070-66
Fumaric acid, (2-ethoxyindol-3-yl)-, dimethyl ester	MeOH	223(4.57),267(3.92), 283(3.81),376(3.66)	24-3070-66
Indole-3-carboxaldehyde, 4,7-diacetoxy-1-ethyl-2-methyl-	MeOH	253(4.34),272(4.18), 343(3.94)	35-0804-66
Maleic acid, (2-ethoxyindol-3-yl)-, dimethyl ester	MeOH	225(4.56),267(4.08), 280(4.00),356(4.27)	24-3070-66
$C_{16}H_{17}NO_5S$			
4H-1,4-Benzothiazine-2,3-dicarboxylic acid, 4-acetyl-, diethyl ester	EtOH	208(4.25),228(4.18), 260s(--),320(3.23)	32-1147-66
$C_{16}H_{17}NO_6$			
4,7-Indoloquinone, 3-(diacetoxymethyl)-1-ethyl-2-methyl-	MeOH	230(4.15),255(3.98), 328(3.42),425(3.00)	35-0804-66
$C_{16}H_{17}N_3$			
as-Triazine, 1,4,5,6-tetrahydro-1-methyl-3,5-diphenyl-	MeOH	292(3.62)	87-0881-66
$C_{16}H_{17}N_3O$			
7-Azaindoline, 6-acetamido-4-methyl-1-phenyl-	EtOH	292(4.20),334(4.20)	78-3233-66
1H-Pyrazolo[4,3-c]pyridin-4(5H)-one, 6-methyl-1-phenyl-3-propyl-	EtOH	223(4.33),239(4.35), 284(4.03)	70-1723-66
$C_{16}H_{17}N_3O_2$			
Indazole, 1-(p-aminophenyl)-5,6-dimethoxy-3-methyl-	EtOH	262(4.27),297(3.93)	78-3131-66
$C_{16}H_{17}N_5$			
s-Triazolo[4,3-b]pyridazine, 3-phenyl-6-piperidino-	EtOH	272(4.31)	78-2073-66
$C_{16}H_{18}$			
Benz[f]azulene, 5,6,7,8-tetrahydro-4,10-dimethyl-	EtOH	249(4.45),292(4.66), 557(2.65),597(2.59), 654(2.17)	44-3013-66
7H-Cyclohepta[a]naphthalene, 8,9,10,11-tetrahydro-5-methyl-	EtOH	234(4.90),287s(3.78), 292(3.82),304(3.68), 324(2.93)	22-0228-66
1,4-Pentadiyne, 1-benzyl-5-butyl-	EtOH	233(3.23),242(3.27), 244(3.27),258(3.32), 264(3.11),272(3.27), 288(3.10)	78-0867-66

Compound	Solvent	λ_{max}(log ϵ)	Ref.
$C_{16}H_{18}BrNO_5S$			
Fumaric acid, [(o-acetamidophenyl)-thio]-bromo-, diethyl ester	C_6H_{12}	214(4.37),246(4.18), 264s(--),287(4.16)	32-1147-66
	EtOH	210(4.34),243(4.02), 264s(--),290(4.10)	32-1147-66
$C_{16}H_{18}ClN_3S$			
Methylene blue	BuOH	610s(4.6),650(4.9)	50-0296-66B
	BuOH-20% C_6H_{12}	610s(4.7),655(4.9)	60-1439-66
	EtOH-1% NaOEt	538(4.12)	50-0296-66B
	$BuNH_2$	508(4.39)	50-0296-66B
	$AmNH_2$	504(4.43)	50-0296-66B
	$C_6H_{13}NH_2$	505(4.42)	50-0296-66B
	$C_7H_{15}NH_2$	508(4.39)	50-0296-66B
	aniline	674(--)	50-0296-66B
	$PhCH_2NH_2$	519(4.28)	50-0296-66B
	piperidine	510(3.71)	50-0296-66B
	pyridine	664(4.92)	50-0296-66B
	quinoline	669(--)	50-0296-66B
	Bu_2NH	499(--)	50-0296-66B
	Et_3N	496(--)	50-0296-66B
			50-0296-66B
$C_{16}H_{18}N_2O$			
Benzoic acid hydrazide, 2,6-diisopropyl-	MeOH	246(3.66)	88-1063-66
Ketone, 3-indolyl 2-quinuclidinyl	EtOH	244(4.09),260s(3.98), 302(4.05)	4-0090-66
2-Naphthonitrile, 1,4-dihydro-8-methoxy-3-pyrrolidino-	EtOH	214(4.07),280(4.10), 289(4.12)	44-1747-66
Nicotinamide, diethyl-5-phenyl-	n.s.g.	249(4.16)	22-2387-66
2-Pyrazolin-5-one, 4,4-diallyl-3-methyl-1-phenyl-	EtOH	243(4.22)	88-6413-66
3-Pyrazolin-5-one, 2,4-diallyl-3-methyl-1-phenyl-	EtOH	248(4.00),276(4.01)	88-6413-66
$C_{16}H_{18}N_2O_3$			
Hydrazine, 1,2-diacetyl-1-(3-methyl-4-phenylfurfuryl)-	EtOH	220(--)	44-0052-66
8-Oxa-2,3-diazabicyclo[3.2.0]oct-6-ene, 2,3-diacetyl-6-methyl-7-phenyl-	EtOH	262(4.20)	44-0052-66
$C_{16}H_{18}N_4$			
as-Triazine, 1,4,5,6-tetrahydro-1,6-dimethyl-5-phenyl-3-(3-pyridyl)-	MeOH	314(3.60)	87-0881-66
$C_{16}H_{18}N_4O$			
4-Pyrimidineacetonitrile, 2-amino-6-hydroxy-5-(4-phenylbutyl)-	pH 1	266(3.92)	4-0324-66
	pH 7	295(4.00)	4-0324-66
	pH 13	284(3.94)	4-0324-66
$C_{16}H_{18}N_4O_4$			
4-Cyclohexylidene-2-butenal, 2,4-dinitrophenylhydrazone	$CHCl_3$	385(4.46)	22-1400-66
Pyrazole, 3-cyclobutyl-1-(2,4-dinitrophenyl)-5-ethyl-4-methyl-	EtOH	340(3.76)	22-2977-66

Compound	Solvent	$\lambda_{max}(\log \epsilon)$	Ref.
$C_{16}H_{18}O$			
Benzhydrol, α-propyl-	hexane	253(2.63),259(2.69), 265(2.59)	22-0269-66
	C_6H_{12}	253(2.60),259(2.67), 265(2.56)	22-0269-66
6,8-Hexadecadiene-10,12,14-triyn-1-ol	ether	259(4.78),269(5.04), 288(4.20),305(4.45), 325(4.62),348(4.53)	24-3544-66
Spiro[6H-benzocycloheptene-6,1'-[3]-cyclohexen]-5(7H)-one, 8,9-dihydro-	EtOH	243(3.67),281(3.17)	22-1693-66
$C_{16}H_{18}O_2$			
Bibenzyl, 4,4'-dimethoxy-	hexane	225(4.23),279(3.59)	44-3976-66
	50% dioxan	278(3.56)	71-0550-66A
Hexadec-2-ene-6,8,12-triynoic acid	EtOH	208(3.98),241s(2.97), 253(2.72)	12-1207-66
Hexadec-12-ene-2,6,8-triynoic acid	EtOH	208(4.21),239(3.04), 253(2.70)	12-1207-66
1-Oxaspiro[5.5]undec-2-en-4-one, 2-phenyl-	EtOH	245(3.79),305(4.20)	88-0233-66
2(3H)-Phenanthrone, 4,4a,9,10-tetra-hydro-7-methoxy-4a-methyl-	EtOH	231(4.40)	35-1766-66
2(3H)-Phenanthrone, 4,4a,9,10-tetra-hydro-8-methoxy-4a-methyl-	MeOH	228(4.29)	35-1776-66
$C_{16}H_{18}O_3$			
2-Benzofuranacrylic acid, 3-ethyl-β,4,6-trimethyl-	n.s.g.	248(3.75),339(4.48)	78-2015-66
5,9-Methanobenzocyclooctene-8-carb-oxylic acid, 5,6,7,8,9,10-hexa-hydro-8,11-dimethyl-10-oxo-	EtOH	255(4.20),294(3.24)	77-0184-66
3(2H)-Phenanthrone, 1,9,10,10a-tetra-hydro-6,7-dimethoxy-	EtOH	242(4.02),310(4.11), 346(4.31)	87-0304-66
4,6,12-Tetradecatriene-8,10-diyn-1-ol, 3-acetoxy-	ether	248(4.53),265(4.47), 278(4.22),295(4.46), 314(4.60),336(4.48)	24-3433-66
4,6,12-Tetradecatriene-8,10-diyn-3-ol, 1-acetoxy-	ether	249(4.54),265(4.47), 278(4.22),295(4.46), 314(4.60),336(4.47)	24-3433-66
$C_{16}H_{18}O_4$			
Cyclohexa[f]-2H,3H-benzofuran, 9-acet-oxy-4,8-dimethyl-2-oxo-	EtOH	272(2.99),280(2.99)	78-0301-66
Heteropeucenin, 7-methyl ether	MeOH	205(4.38),259(4.35), 296(3.76),330(3.70)	39-0114-66C
$C_{16}H_{18}O_5$			
2-Benzofuranpropionic acid, 2-acetyl-2,3-dihydro-4,6-dimethyl-3-oxo-, methyl ester	n.s.g.	272(4.14),330(3.81)	78-2015-66
$C_{16}H_{18}O_6$			
Cyclohexanecarboxylic acid, 1-(3,5-di-methoxybenzyl)-2,4-dioxo-	MeOH	225(4.06),258(4.16)	44-1451-66
$C_{16}H_{18}S$			
Naphtho[1,2-b]thiepin, 7-ethyl-2,3,4,5-tetrahydro-	EtOH	222s(4.61),228(4.66), 233s(4.60),306(3.89)	22-0228-66

Compound	Solvent	$\lambda_{max}(\log \epsilon)$	Ref.
$C_{16}H_{18}Si$			
Dibenzosilole, 5,5-diethyl-	n.s.g.	275(4.0)	65-0114-66
$C_{16}H_{19}N$			
1,6-Methano[10]annulene, 2-piperidino-	C_6H_{12}	255(4.49),275s(4.29), 353(3.91),410s(3.15)	89-0733-66
$C_{16}H_{19}NO$			
Carbazol-4(1H)-one, 1,3-diethyl-2,3-dihydro-	MeOH	212(4.61),242(4.35), 265(4.25),295(4.15)	5-0116-66F
$C_{16}H_{19}NO_3$			
2H-Benzo[a]quinolizin-2-one, 1,6,7,11b-tetrahydro-9,10-dimethoxy-4-methyl-	EtOH	294s(3.90),319(4.22)	44-0797-66
Indole-3-carboxaldehyde, 4-acetoxy-1-ethyl-2,5,6-trimethyl-	MeOH	252(4.27),311(4.06)	44-1012-66
Norpluviine	EtOH	284(3.77)	39-0676-66C
Piperlonguminine	EtOH	245(3.99),256(4.02), 307(4.33),340(4.52)	88-1797-66
Pyrrole-2-carboxylic acid, 4-ethyl-5-(hydroxymethyl)-3-methyl-, benzyl ester	EtOH	280(4.30)	78-0241-66A
$C_{16}H_{19}NO_5$			
Cyclohexanecarboxamide, 1-(3,5-dimethoxybenzyl)-2,4-dioxo-	MeOH	222(4.00),264(4.15)	44-1451-66
$C_{16}H_{19}NO_9$			
Pyrrole-2,5-dicarboxylic acid, 3-[(1,2-dicarboxyvinyl)oxy]-1-ethyl-, tetramethyl ester	ether	271(4.28),278s(4.22)	24-1558-66
$C_{16}H_{19}N_3O_2S$			
Phenothiazine, 2,8-bis(dimethylamino)-, 5,5-dioxide	EtOH	272(4.89),315(4.08)	87-0960-66
Phenothiazine, 2,8-bis(ethylamino)-, 5,5-dioxide	EtOH	268(4.95),312(4.04)	87-0960-66
$C_{16}H_{19}N_7$			
Etioluciferin	MeOH-HCl	223(4.42),306(4.30), 410(3.70)	88-3427-66
	MeOH-NaOH	227(4.38),273(4.26), 365(3.88)	88-3427-66
$C_{16}H_{20}BrN_3O_4$			
4,7-Indoloquinone, 5-amino-6-bromo-1-ethyl-3-(hydroxymethyl)-2-methyl-, propyl carbamate	MeOH	241(4.31),316(4.05), 358(3.88),520(3.26)	35-0804-66
4,7-Indoloquinone, 6-amino-5-bromo-1-ethyl-3-(hydroxymethyl)-2-methyl-, propyl carbamate	MeOH	238(4.31),317(4.02), 360(3.96),520(2.78)	35-0804-66
$C_{16}H_{20}Br_2Si_2$			
Disilane, 1,2-bis(p-bromophenyl)-1,1,2,2-tetramethyl-	C_6H_{12}	244(4.51)	101-0102-66B
$C_{16}H_{20}ClNO_4$			
[2-(4,4-Dimethyl-1(4H)-naphthylidene)ethylidene]dimethylammonium perchlorate	MeCN	370(4.60)	24-2479-66

Compound	Solvent	λ_{max}(log ε)	Ref.
$C_{16}H_{20}Cl_2Si_2$			
Disilane, 1,2-bis(p-chlorophenyl)-1,1,2,2-tetramethyl-	C_6H_{12}	241.5(4.45)	101-0102-66B
$C_{16}H_{20}N_2$			
Aniline, N-(p-dimethylaminobenzyl)-N-methyl-	EtOH	258(4.53)	39-1605-66C
Indolo[2,3-a]quinolizine, 1,2,3,4,6,7,12,12b-octahydro-12-methyl-	EtOH-HCl	226(4.59),285(3.86), 293s(3.81)	87-0864-66
	EtOH	229(4.55),286(3.83), 293s(3.80)	87-0864-66
$C_{16}H_{20}N_2O$			
3,4-Diazabicyclo[4.2.0]oct-2-en-5-one, 1,6,7,8-tetramethyl-2-phenyl-	MeOH	277.5(4.08)	24-1236-66
$C_{16}H_{20}N_2O_2$			
2,2'-Dipyrromethene-4'-carboxylic acid, 4-ethyl-3,3',5-trimethyl-, methyl ester, hydrobromide	$CHCl_3$	258(365),263(3.64), 284s(3.06),397(3.72), 477(4.96)	39-1436-66C
2,2'-Dipyrromethene-3-carboxylic acid, 3',4,4',5-tetramethyl-, hydrobromide, ethyl ester	$CHCl_3$	301s(3.19),361(3.66), 413(3.56),509(4.78)	39-1436-66C
2,2'-Dipyrromethene-4-carboxylic acid, 3,3',4',5-tetramethyl-, hydrobromide, ethyl ester	$CHCl_3$	262(3.71),395(3.72), 476(4.89)	39-0098-66C
2,2'-Dipyrromethene-4'-carboxylic acid, 3,3',5,5'-tetramethyl-, ethyl ester	EtOH-HCl	210(4.27),223s(4.10), 260(3.66),351(3.69), 461(4.99)	12-1871-66
	EtOH	213(4.15),226s(4.08), 254s(3.78),415(4.46), 467s(4.13)	12-1871-66
$C_{16}H_{20}N_2O_2S_2$			
5-Thiazolidineacetamide, 3-benzyl-N-butyl-4-oxo-2-thioxo-	EtOH	261(4.07),297(4.18)	95-0101-66
$C_{16}H_{20}N_2O_3$			
3-Pyrrolin-2-one, 5-(5-carbethoxy-3,4-dimethylpyrrol-2-ylmethylene)-3,4-dimethyl- (also inflections)	EtOH	216(3.89),259(4.32), 275(3.90),387(4.48), 407(4.42)	39-1155-66C
$C_{16}H_{20}N_2O_4$			
3-Azabicyclo[3.3.1]nonan-9-ol, 3-methyl-, p-nitrobenzoate	EtOH	259(4.13)	44-3482-66
isomer	EtOH	259(4.14)	44-3482-66
4,7-Indoloquinone, 1-ethyl-3-(hydroxymethyl)-2,5,6-trimethyl-, methyl carbamate	MeOH	228(4.23),271(4.22), 278(4.22),342(4.11), 445(3.32)	44-1012-66
$C_{16}H_{20}N_2O_4S_4$			
3-Thiophenecarboxylic acid, 5,5-dithiobis[2-amino-4-methyl-, diethyl ester	MeOH	225(4.58),308(4.10), 373(4.01)	24-2712-66
$C_{16}H_{20}N_2O_5$			
Serine, (7-methoxyindole-2-carbonyl)-O,N-dimethyl-, methyl ester	MeOH	233(4.38),288(4.26)	39-1799-66C

Compound	Solvent	$\lambda_{max}(\log \epsilon)$	Ref.
$C_{16}H_{20}N_2O_6$			
5H,11H-Benzo[1,2-e:4,5-e']bis[1,4]-oxazepine-5,11-dione, 1,7-bis-(2-hydroxyethyl)-	H_2O	247(4.28),283(3.98), 407(3.40)	24-3910-66
$C_{16}H_{20}N_4$			
Piperazine, 1,4-bis(2-pyridylmethyl)-	H_2O	257s(3.8),262(3.8), 268(3.4)	39-1209-66C
$C_{16}H_{20}N_4O_4$			
1-Cyclohexene-1-carboxaldehyde, 3,5,5-trimethyl-, 2,4-dinitrophenylhydrazone	$CHCl_3$	382.0(4.44)	28C-0848-66A
Cyclooctylideneacetaldehyde, 2,4-dinitrophenylhydrazone	$CHCl_3$	391(4.48)	22-1400-66
5-Hexen-2-one, 3-isopropenyl-3-methyl-, 2,4-dinitrophenylhydrazone	$CHCl_3$	361(4.32)	22-0273-66
3,7-Octadien-2-one, 3,4-dimethyl-, 2,4-dinitrophenylhydrazone, cis	$CHCl_3$	371(4.32)	22-0273-66
trans	$CHCl_3$	382(4.35)	22-0273-66
4,7-Octadien-2-one, 3,4-dimethyl-, 2,4-dinitrophenylhydrazone	$CHCl_3$	365(4.18)	22-0273-66
$C_{16}H_{20}N_4O_5$			
2-Cyclohexen-1-one, 2-(3-hydroxypropyl)-3-methyl-, 2,4-dinitrophenylhydrazone	EtOH	390(4.47)	44-2171-66
$C_{16}H_{20}N_4O_6$			
Inosine, 2',3'-O-isopropylidene-1-methyl-, 5'-acetate	pH 1	254.5(3.98)	44-2685-66
	pH 11	251(3.96)	44-2685-66
	EtOH	246(3.90),251(3.89), 270s(--)	44-2685-66
$C_{16}H_{20}N_4S$			
10H-Dipyrido[2,3-b:2',3'-e][1,4]thiazine, 10-(3-dimethylamino-2-methylpropyl)-	EtOH	244(4.32),329(3.94)	87-0116-66
$C_{16}H_{20}N_7O_8P$			
Adenosine, 2'-deoxy-, 5'-(p-nitrophenyl hydrogen phosphate), ammonium salt	pH 7.0	259(4.22),288s(3.86)	44-3241-66
	pH 12	258(4.18),292s(3.86)	44-3241-66
$C_{16}H_{20}N_7O_9P$			
5'-Adenylic acid, p-nitrophenyl ester, ammonium salt	pH 7	259(4.24),290s(3.86)	44-3241-66
	pH 12	257(4.22),292s(3.89)	44-3241-66
Guanosine, 2'-deoxy-, 5'-(p-nitrophenyl hydrogen phosphate), ammonium salt	pH 2.0	258(4.16),275(4.14)	44-3241-66
	pH 7.0	256(4.18),275(4.14), 301s(3.83)	44-3241-66
	pH 12.0	268(4.22),302(3.86)	44-3241-66
$C_{16}H_{20}N_7O_{10}P$			
5'-Guanylic acid, p-nitrophenyl ester, ammonium salt	pH 2.0	258(4.17),275(4.16)	44-3241-66
	pH 7.0	258(4.18),275(4.15), 301s(3.86)	44-3241-66
	pH 12.0	268(4.22),300s(3.88)	44-3241-66

Compound	Solvent	$\lambda_{max}(\log \epsilon)$	Ref.
$C_{16}H_{20}O$			
6,8,14-Hexadecatriene-10,12-diyn-1-ol, all trans	ether	249(4.48),265(4.42), 279s(4.20),296(4.44), 315(4.57),336(4.44)	24-3544-66
Spiro[6H-benzocycloheptene-6,1'-cyclohexan]-5(7H)-one, 8,9-dihydro-	EtOH	243(3.67),283(3.16)	22-1693-66
$C_{16}H_{20}O_2$			
2,12-Hexadecadiene-6,8-diynoic acid	EtOH	207(4.29),241s(3.11), 253(2.76)	12-1207-66
10,12,14-Hexadecatriynoic acid	ether	210(5.21),239(1.98), 252(2.19),263(2.18), 267(2.33),278(2.18), 286(2.33),296(2.03), 306(2.08),308(2.08)	24-2091-66
Spiro[1,3-dioxolane-2,2'(1'H)-phenanthrene], 3',4',5',6',7',8-hexahydro-	EtOH	268(2.73)	65-0843-66
4,6,10,12-Tetradecatetraen-8-yn-1-ol, acetate	ether	233(4.10),304(4.55), 316(4.71),321(4.69), 339(4.74)	24-2828-66
$C_{16}H_{20}O_3$			
Cyclobutane-1-carboxylic acid, 2-benzoyl-1,2,3,4-tetramethyl-, cis	MeOH	243(3.15)	24-1236-66
Cyclohexanone, 2-(2,3-dimethoxybenzylidene)-5-methyl-	EtOH	288(4.18)	35-1419-66
2,5-Phenanthrenedione, 7-acetyl-3,4,4a,4b,6,7,8,9a,9,10-decahydro-	EtOH	239(4.16)	44-1360-66
3(2H)-Phenanthrone, 1,4,4aα,9,10,10aβ-hexahydro-6,7-dimethoxy-	EtOH	230s(3.86),283(3.59), 287(3.59),292s(3.51)	87-0304-66
3(2H)-Phenanthrone, 1,4,4aβ,9,10,10aβ-hexahydro-6,7-dimethoxy-	EtOH	230s(3.93),282(3.66), 286(3.66),292s(3.58)	87-0304-66
$C_{16}H_{20}O_4$			
4,7-Azulenedicarboxylic acid, 1,2,3,3a,4,5-hexahydro-6-(hydroxymethyl)-2,2-dimethyl-, γ-lactone, methyl ester	MeOH	274(3.88)	35-2838-66
1,3-Cyclopentanedione, 4-hydroxy-2-(3-methylbutyryl)-4-(4-methyl-3-penten-1-ynyl)-	EtOH-acid EtOH-base	263(4.32) 271(4.38)	39-1615-66C 39-1615-66C
Marasmic acid, methyl ester	MeOH	243(4.01)	35-2838-66
3H,6H-3a,8b-Methano-1H-indeno[4,5-c]-furan-4-carboxylic acid, 5a,7,8,8a-tetrahydro-7,7-dimethyl-1-oxo-, methyl ester	MeOH	233(3.80)	35-2838-66
1(2H)-Naphthalenone, 3,4-dihydro-6,7-dimethoxy-2-(3-oxobutyl)-	EtOH	233(4.27),275(4.09), 314(3.92)	87-0304-66
2α-Phenanthrenecarboxylic acid, 1,2,3,4,4aα,4bβ,5,6,7,9,10,10aβ-dodecahydro-4,7-dioxo-, methyl ester	EtOH	239(4.23)	44-0713-66
Spiro[1,3-dioxolane-2,2'(1'H)-phenanthrene]-5',8'-dione, 3',4',4'b,-6',7',8'a,9',10'-octahydro-	EtOH	291(1.90)	65-0843-66
Warburgin, tetrahydro-	EtOH	255(3.42)	77-0393-66
$C_{16}H_{20}O_4S$			
Naphtho[1,2-c]thiophene-1-carboxylic acid, 3-ethoxy-1,3,3a,4,5,9b-hexahydro-9b-hydroxy-, methyl ester	EtOH	228(4.03),268s(3.31)	88-1953-66

Compound	Solvent	$\lambda_{max}(\log \epsilon)$	Ref.
$C_{16}H_{20}O_6$			
m-Toluic acid, 4,6-dimethoxy-2-[1-(methoxymethylene)acetonyl]-, methyl ester	EtOH	254(4.06),285s(3.79)	39-0126-66C
$C_{16}H_{21}NO$			
2,5-Cyclohexadiene-$\Delta^{1,\alpha}$-acetonitrile, 3,5-di-tert-butyl-4-oxo-	hexane	300(4.43),312(4.54)	70-0888-66
Cyclohexanone, 2-(p-dimethylamino-benzylidene)-5-methyl-	EtOH	254(4.02),372(4.13)	35-1419-66
2,4-Dodecadienamide, N-isobutyl-, 2-cis-4-trans	ether	254(4.51)	24-3197-66
$C_{16}H_{21}NO_2$			
Serratinine derivative	n.s.g.	238(4.04)	88-1537-66
$C_{16}H_{21}NO_3$			
3-Indolecarboxylic acid, 5-hydroxy-1-isopropyl-2,6-dimethyl-, ethyl ester	MeOH	298(4.35)	35-2536-66
$C_{16}H_{21}NO_4$			
Azepino[3,2,1-hi]indole, 7-carbo-methoxy-1,2,4,5,6,7-hexahydro-6,6-dimethoxy-	EtOH	257(3.74),302(3.42)	35-4061-66
$C_{16}H_{21}N_3O$			
4-Quinazolone, 2-(N-acetyl)isopropyl-amino-3-isopropyl-	EtOH	230(4.3),280(4.0), 308(3.6),320(3.5)	83-0914-66
$C_{16}H_{21}N_3O_8$			
An aminocarboxylic acid	EtOH	216(4.23),236(4.26), 335(3.30)	78-0171-66B
$C_{16}H_{22}$			
Naphthalene, 1,2-dihydro-5-isopropyl-1,3,8-trimethyl-	EtOH	218(4.50),226(4.56), 233(4.40),271(4.10)	78-1949-66
Naphthalene, 1,2-dihydro-5-isopropyl-2,3,8-trimethyl-	EtOH	219(4.55),225(4.56), 233s(4.40),271(4.09)	78-1949-66
Naphthalene, 1,2-dihydro-5-isopropyl-3,4,8-trimethyl-	EtOH	219(4.47),224(4.46), 262(3.99)	78-1949-66
Naphthalene, 1,2-dihydro-5-isopropyl-3,6,8-trimethyl-	EtOH	223(4.70),263(4.18)	78-1949-66
Naphthalene, 1,2-dihydro-8-isopropyl-2,5,6-trimethyl-	EtOH	219(4.45),225(4.56), 231s(4.62),267(4.06)	78-1949-66
Naphthalene, 1,2-dihydro-8-isopropyl-3,5,6-trimethyl-	EtOH	221(4.52),226(4.53), 234s(4.30),269(4.56)	78-1949-66
$C_{16}H_{22}Cl_2N_4$			
1,4-Dihydro-1,4-bis(3,4,5,6-tetra-hydro-2-pyridylmethylene) pyrazinium dichloride	EtOH	225.0(3.95)	39-1209-66C
$C_{16}H_{22}INO_2$			
1-Butyl-6,7-dimethoxy-2-methyl-isoquinolinium iodide	MeOH	260(4.39),325(3.54)	83-0159-66
$C_{16}H_{22}N_2O_2$			
3-Pyrrolin-2-one, 4-ethyl-5-(3-ethyl-5-formyl-4-methylpyrrol-2-yl-methyl)-3-methyl-	EtOH	211(4.19),278(3.80), 311(4.29)	39-1155-66C

Compound	Solvent	$\lambda_{max}(\log \epsilon)$	Ref.
$C_{16}H_{22}N_2O_3$			
2-Pyrroline, 2-ethoxy-5-(5-formyl-3,4-dimethylpyrrol-2-ylmethyl)-3,3-dimethyl-4-oxo-	EtOH	281s(3.87),313(4.27)	39-1155-66C
$C_{16}H_{22}N_2O_4$			
Peruvin, pyrazoline of	EtOH	322(2.28)	78-1723-66
$C_{16}H_{22}N_2O_4S$			
1-[5-(2-Carboxy-3-oxobutylidene)-3-methyl-4-oxo-2-thiazolidinylidene]piperidinium hydroxide, inner salt, ethyl ester	dioxan	248(4.11),442(4.11)	24-0307-66
	CH_2Cl_2	247(4.14),445(4.59)	24-0307-66
$C_{16}H_{22}N_2O_7$			
4,1-Benzoxazepine-8-carboxylic acid, 7-[bis(2-hydroxyethyl)amino]-1,2,3,5-tetrahydro-1-(2-hydroxyethyl)-5-oxo-	pH 4.62	241(4.28),279(4.06), 385(3.53)	24-3910-66
$C_{16}H_{22}N_4$			
Tetraethylammonium 1,2,3-tricyanocyclopentadienide	MeCN	220(--),288(--), 298(--)	35-3046-66
$C_{16}H_{22}N_6O_5$			
Adenosine, N-(dimethylaminomethylene)-, cyclic 2',3'-(ethyl orthoformate)	EtOH	232(4.06),312(4.47)	73-3198-66
$C_{16}H_{22}N_6O_6$			
Acetic acid, bis(6-amino-1,2,3,4-tetrahydro-1,3-dimethyl-2,4-dioxopyrimidin-5-yl)-, ethyl ester	MeOH	242s(4.04),267(4.45)	24-3524-66
Guanosine, N-(dimethylaminomethylene)-, cyclic 2',3'-(ethyl orthoformate)	EtOH	237(4.05),305(4.18)	73-3198-66
$C_{16}H_{22}O$			
1(2H)-Naphthalenone, 3,4-dihydro-5-isopropyl-2,7,8-trimethyl-	EtOH	215(4.48),256(3.99), 308(3.38)	78-1949-66
1(2H)-Naphthalenone, 3,4-dihydro-5-isopropyl-3,7,8-trimethyl-	EtOH	214(4.42),258(4.00), 305(3.36)	78-1949-66
1(2H)-Naphthalenone, 3,4-dihydro-8-isopropyl-2,3,5-trimethyl-	EtOH	217(4.29),252(3.88), 300(3.21)	78-1949-66
1(2H)-Naphthalenone, 3,4-dihydro-8-isopropyl-2,4,5-trimethyl-	EtOH	219(4.32),254(3.96), 301(3.38)	78-1949-66
1(2H)-Naphthalenone, 3,4-dihydro-8-isopropyl-2,5,7-trimethyl-	EtOH	220(4.38),257(3.91), 304(3.31)	78-1949-66
$C_{16}H_{22}O_2$			
2-Furanol, 2-tert-butyl-2,5-dihydro-5,5-dimethyl-4-phenyl-	2N H_2SO_4	232(3.49),357(4.35)	65-2064-66
2,4-Pentadienoic acid, 3-methyl-5-(2,6,6-trimethylcyclohexa-1,3-dien-1-yl)-, methyl ester, 2-cis-4-trans	n.s.g.	259(3.85),353(4.16)	28C-1397-66A
2-trans-4-trans	n.s.g.	259(3.93),343(4.08)	28C-1397-66A

Compound	Solvent	λ_{max}(log ε)	Ref.
$C_{16}H_{22}O_3$			
2,5-Cyclohexadiene-$\Delta^{1,\alpha}$-acetic acid, 3,5-di-tert-butyl-4-oxo-	hexane	302(4.41),315(4.44)	70-0888-66
$C_{16}H_{22}O_4$			
1H-3a,6-Epoxyazulene-7-acetic acid, octahydro-4β,8aβ-dimethyl-α-methylene-1-oxo-, methyl ester	EtOH	202(3.87)	78-1723-66
2,4-Pentadienoic acid, 5-(1-hydroxy-2,6,6-trimethyl-4-oxocyclohex-2-en-1-yl)-3-methyl-, methyl ester, 2-cis-4-trans	n.s.g.	266(4.27)	28C-1397-66A
2β-Phenanthrenecarboxylic acid, 1,2,3,4,4aβ,5,8,9,10,10aβ-decahydro-4α-hydroxy-7-methoxy-	EtOH	279(1.30),287(1.28)	44-0713-66
Spiro[1,3-dioxolane-2,2'(1'H)-phenanthren]-8'(5'H)-one, 4',4'a,4'b,5'-6',7',8'a,9'-octahydro-5'-hydroxy-	EtOH	284(1.46)	65-0835-66
Succinic acid, (2,3,4,5,6-pentamethylbenzyl)-	MeCN	270(2.53)	39-1147-66B
Unknown dihydro Cannizzaro ester	MeOH	231(4.00)	35-2838-66
$C_{16}H_{22}O_6$			
5,6-Azulenedione, 1,2,3,3a,4,7,8,8aβ-octahydro-2α,3α-dihydroxy-3α,8α-dimethyl-, diacetate	EtOH	282(1.90)	78-1499-66
$C_{16}H_{22}O_7$			
Roridinic acid, 2',3'-epoxy-2'-anhydro-, dimethyl ester	EtOH	258(4.2)	33-2547-66
$C_{16}H_{22}O_8$			
1,3-Butadiene-1,1,4,4-tetracarboxylic acid, tetraethyl ester	MeOH	275(4.70)	44-2641-66
1,2,4-Cyclopentanetrione, 3-hydroxy-3-(3,4-dihydroxy-4-methylpentanoyl)-5-isovaleryl-	MeOH-acid	271(4.07)	39-0675-66C
	MeOH-base	255(4.12)	39-0675-66C
$C_{16}H_{22}O_9$			
Sweroside	n.s.g.	246(3.92)	88-5229-66
$C_{16}H_{22}Si_2$			
Disilane, 1,1,2,2-tetramethyl-1,2-diphenyl-	C_6H_{12}	236.0(4.26)	101-0102-66B
$C_{16}H_{23}N$			
Indene, 1-(3-dimethylamino-1-methylpropyl)-1-methyl-	EtOH	259(3.85),283(3.95), 294(2.57)	44-0903-66
Indene, 3-(1,1-dimethyl-3-dimethylaminopropyl)- (* one logε missing)	EtOH	224(3.90),246(3.92), 249(3.94),263(3.71), 280(2.92),290(*)	44-0903-66
$C_{16}H_{23}NO_2S$			
L-Leucine, N-phenylthioacetyl-, ethyl ester	isooctane	268(4.09),365s(1.75)	1-2781-66
	MeOH	268(4.30),335(2.20)	1-2781-66
	dioxan	270(4.28),362s(2.02)	1-2781-66
$C_{16}H_{23}NO_3$			
Piperlonguminine, tetrahydro-	EtOH	234(3.60),290(3.59)	88-1797-66

Compound	Solvent	λ_{max}(log ϵ)	Ref.
$C_{16}H_{23}NO_4$			
Crotonic acid, 3-(ethylamino)-2-(3-ethyl-2,5-dihydroxyphenyl)-, ethyl ester	MeOH	295(4.35)	35-2536-66
2,4,10-Cyclodecatriene-1,2-dicarboxylic acid, 10-(dimethylamino)-, dimethyl ester, all cis	EtOH	250(4.07),267(4.03), 320(4.04),356(3.94)	35-4685-66
$C_{16}H_{23}NO_8S$			
α-D-Glucopyranoside, methyl 3-acetamido-3-deoxy-6-p-toluenesulfonate	MeOH	225(4.09),256(2.67), 263(2.76),268(2.76), 273(2.72)	94-0902-66
α-D-Mannopyranoside, methyl 3-acetamido-3-deoxy-6-p-toluenesulfonate	MeOH	225(4.06),257(2.62), 263(2.74),268(2.73), 273(2.67)	94-0902-66
$C_{16}H_{23}NO_9S_2$			
β-D-Glucopyranose, 2-deoxy-2-[(dithiocarboxy)amino]-, methyl ester, 1,3,4,6-tetraacetate	MeOH	223(3.61),251(4.12), 275(3.71)	23-0094-66
$C_{16}H_{24}Cl_2CuN_2O_9S_2$			
Bis[2-(2-methylthioethyl)pyridine]-copper perchlorate	acetone	570(2.35)	12-1835-66
$C_{16}H_{24}INO_3$			
3,4,6,7-Tetrahydro-9,10-dimethoxy-2H-11b-methyl-1,3-oxazino[2,3-a]isoquinolinium methiodide	H_2O	278(3.51),282(3.52), 287(3.49)	83-0817-66
	MeOH	280(3.58),284(3.60), 288(3.56)	83-0817-66
$C_{16}H_{24}NO_2$			
Phenoxy, 2,6-di-tert-butyl-4-formyl-, O-methyloxime	C_6H_{12}	297s(--),320s(--), 340(4.34),511(2.88), 547(3.00),597(3.08)	35-5284-66
$C_{16}H_{24}N_2$			
3-Picoline, 1-ethyl-1,2,3,6-tetrahydro-4-(p-dimethylaminophenyl)-	EtOH	280.5(4.28)	78-2745-66
dihydrochloride	EtOH	290(4.26)	78-2745-66
3-Picoline, 1-ethyl-1,2,5,6-tetrahydro-4-(p-dimethylaminophenyl)-	EtOH	269(4.20)	78-2745-66
dihydrochloride	EtOH	277.5(4.18)	78-2745-66
Quinoxaline, 2,3-di-tert-butyl-1,2-dihydro-	EtOH	213(4.20),234(4.37), 270(3.60),355(3.54)	44-3954-66
$C_{16}H_{24}N_2O_2$			
2-Pyrrolidone, 4-ethyl-5-(3-ethyl-5-formyl-4-methylpyrrol-2-yl-methyl)-3-methyl-	EtOH	278(3.87),313(4.34)	39-1155-66C
Pyridine, 2,3,4,5-tetrahydro-5,5-dimethyl-3-oxo-2-(1-acetyl-4,5,5-trimethylpyrrolidin-2-ylidene)-	EtOH-HCl	253(4.04)	78-0015-66B
	EtOH	265(3.70),330(4.09)	78-0015-66B
$C_{16}H_{24}N_4O_4$			
2-Octanone, 3,4-dimethyl-, 2,4-dinitrophenylhydrazone	$CHCl_3$	366(4.37)	22-0273-66

Compound	Solvent	$\lambda_{max}(\log \epsilon)$	Ref.
$C_{16}H_{24}O$			
2,5-Cyclohexadien-1-one, 2,6-di-tert-butyl-4-ethylidene-	hexane	300(4.29)	70-0888-66
1-Cyclohexenesorbaldehyde, γ,2,6,6-tetramethyl-	pet ether	274(4.27)	22-0728-66
isomer	pet ether	240(3.98),263(4.02)	22-0728-66
$C_{16}H_{24}O_2$			
1,3,5-Cycloheptatriene-1-carboxylic acid, 2,3,4,5,6,7-hexamethyl-, ethyl ester	MeOH	263(3.70)	24-1097-66
2,4,6-Cycloheptatriene-1-carboxylic acid, 2,3,4,5,6,7-hexamethyl-, ethyl ester	MeOH	247(3.70)	24-1097-66
2,5-Cyclohexadien-1-one, 4-(2-hydroxy-ethylidene)-2,6-di-tert-butyl-	hexane	298(4.44)	70-0888-66
2,4,6,10-Dodecatetraenoic acid, 3,7,11-trimethyl-, methyl ester	EtOH	314(4.48)	39-2154-66C
2,6,8,12-Hexadecatetraenoic acid, all cis	EtOH	231.5(4.23)	12-1207-66
2,6,8-cis-12-trans	EtOH	232.5(4.25)	12-1207-66
2-trans	EtOH	231.5(4.14)	12-1207-66
2,12-trans	EtOH	231(4.27)	12-1207-66
Hydrocinnamic acid, 2,3,4,5,6-penta-methyl-, ethyl ester	MeOH	270(2.45)	24-1097-66
Longifolene-ω-carboxylic acid	EtOH	235(4.14)	22-1325-66
isomer	EtOH	235(4.12)	22-1325-66
14-Nor-7βH-eudesma-3,5-diene, 3-acetoxy-	EtOH	235(4.24)	78-2869-66
2,4-Pentadienoic acid, 5-(2,6,6-tri-methyl-1-cyclohexen-1-yl)-, ethyl ester, trans	n.s.g.	257(4.07),309(4.25)	78-0293-66
$C_{16}H_{24}O_3$			
Propionic acid, 3-hydroxy-3-(p-iso-propylphenyl)-, tert-butyl ester	EtOH	212(3.97),217s(3.95), 251(2.31),256(2.41), 263(2.48),267s(2.35), 271(2.34)	22-3243-66
$C_{16}H_{24}O_4$			
A methyl ester	EtOH	275(3.75)	78-2387-66
Tetrahydro Cannizzaro ester	MeOH	200(2.58)	35-2838-66
$C_{16}H_{25}NO$			
Longifolene-ω-carboxamide	EtOH	231(4.16)	22-1325-66
isomer	EtOH	230.5(4.21)	22-1325-66
$C_{16}H_{25}NO_2$			
Benzaldoxime, 3,5-di-tert-butyl-O-methyl-4-hydroxy-	C_6H_{12}	227(4.18),294s(--), 303s(--)	35-5284-66
$C_{16}H_{25}NO_4$			
Benzoic acid, p-(3,3-diethoxypropyl-amino)-, ethyl ester	EtOH	225(3.91),307(4.43)	87-0590-66
2,10-Cyclodecadiene-1,2-dicarboxylic acid, 3-(dimethylamino)-, methyl ester, cis-cis	EtOH	319(4.13)	35-4685-66
$C_{16}H_{26}$			
3,5-Hexadiene, 4-methyl-6-(2,6,6-tri-methylcyclohexenyl)-	EtOH	247(4.13)	39-2154-66C

Compound	Solvent	λ max(log ε)	Ref.
2,6,8,10-Tridecatetraene, 2,6,10-trimethyl-	EtOH	269s(--),279(4.66),	39-2154-66C
$C_{16}H_{26}ClNO_4$			
[2-(2,3,4,4,5,6-Hexamethyl-2,5-cyclohexadien-1-ylidene)ethylidene]-dimethylammonium perchlorate	MeCN	390(4.68)	24-2479-66
$C_{16}H_{26}IN_3$			
Pyridine, 1,2,3,6-tetrahydro-1-(dimethylamino)-4-(p-dimethylaminophenyl)-, methiodide	EtOH	292(4.33)	78-2745-66
$C_{16}H_{26}N_2$			
Quinoxaline, 2,3-di-tert-butyl-5,6,7,8-tetrahydro-	EtOH	219(3.96),283(3.98), 300s(3.40)	44-3954-66
$C_{16}H_{26}O_2$			
[Bicyclohexyl]-3,3'-dione, 6,6,6',6'-tetramethyl-	C_6H_{12}	286(1.67)	22-0381-66
Diosphenol A methyl ether	EtOH	254(3.93)	78-0337-66
2,6-Dimethyl farnesoate, 2,6-di-cis	EtOH	210(4.10)	39-2144-66C
6-cis-2-trans	EtOH	210(4.16)	39-2144-66C
2-trans-6-trans	EtOH	210(4.15)	39-2144-66C
$C_{16}H_{26}O_3$			
2,4,8-Decatrienoic acid, 2-ethoxy-5,9-dimethyl-, ethyl ester	MeOH	290(4.32)	5-0053-66I
$C_{16}H_{26}O_6S_2$			
L-Gulitol, 5,6-bisacetylthio-5,6-dideoxy-1,2:3,4-di-O-isopropylidene-	EtOH	230(3.99)	39-1287-66C
$C_{16}H_{26}O_7$			
Roridinic acid, dimethyl ester	C_6H_{12}	254(4.33)	33-2527-66
	EtOH	258(4.32)	33-2527-66
$C_{16}H_{27}NO_3S_2$			
5-Thiazolidineacetic acid, N-hexyl-4-oxo-2-thioxo-, pentyl ester	EtOH	262(4.07),296(4.22)	95-0095-66
5-Thiazolidineacetic acid, 4-oxo-N-propyl-2-thioxo-, octyl ester	EtOH	262(4.09),296(4.19)	95-0095-66
$C_{16}H_{27}N_3OS$			
Glycine, N,N-dimethyl-, [3-(3,4-dihydro-4,4,6-trimethyl-2H-thiopyran-5-yl)-1-methylallylidene]hydrazide	C_6H_{12}	257(4.15),322(4.23)	78-0259-66
Glycine, N,N-dimethyl-, [1-methyl-5-[(3-methyl-2-butenyl)thio]-2,4-hexadienylidene]hydrazide	C_6H_{12}	328(4.46)	78-0259-66
$C_{16}H_{27}N_6OP$			
1,2,4-Triazole, 3-anilino-5-butyl-1-bis(dimethylamido)phosphoryl-	EtOH	245(4.18),261(4.19)	54-0429-66
1,2,4-Triazole, 5-anilino-3-butyl-1-bis(dimethylamido)phosphoryl-	EtOH	261(4.37)	54-0429-66
$C_{16}H_{27}N_6O_7P$			
L-Isoleucinol, adenosine-5'-phosphoric acid ester	n.s.g.	259(4.09)	33-0076-66

Compound	Solvent	$\lambda_{max}(\log \epsilon)$	Ref.
L-Leucinol, adenosine-5'-phosphoric acid ester	n.s.g.	259(4.08)	33-0076-66
$C_{16}H_{27}O_5P$			
2,4,6-Octatrienoic acid, 2,6-dimethyl-8-phosphono-, triethyl ester	EtOH	302(4.59)	33-0369-66
$C_{16}H_{28}NO_3P$			
Phosphorimidic acid, N-(o-butylphenyl)-, triethyl ester	EtOH	238(3.90),281(3.14)	35-3781-66
$C_{16}H_{29}NO$			
4,6-Undecadiene, 7-methyl-5-morpholino-	n.s.g.	214(1.94)	54-0753-66
$C_{16}H_{30}N_2O_2$			
Diimide, α-acetoxycyclohexyl-tert-octyl-	n.s.g.	249s(--),360(1.47), 283s(--)	35-4677-66
$C_{16}H_{30}O_2$			
2,6-Dodecadienal, diethyl acetal, di-cis	EtOH-HCl	226(4.21)	54-0117-66
	EtOH	none	54-0117-66
2-cis-6-trans	EtOH-HCl	229(4.39)	54-0117-66
	EtOH	none	54-0117-66
$C_{16}H_{32}N_2$			
2-Pyrazoline, 4-ethyl-1-octyl-5-propyl-	EtOH	247(3.55)	4-0413-66
$C_{16}H_{36}Sn$			
Tin, tetrabutyl-	hexane	202.0(3.99)	39-0242-66A

Compound	Solvent	λ_{max}(log ϵ)	Ref.
$C_{17}H_9NO_3$ Liriodenine	EtOH	252(4.60),274(4.53), 312(4.05),432(4.19)	95-0124-66
$C_{17}H_{10}BrNOS$ 5H-Thiazolo[2,3-a]isoquinolin-5-one, 3-(p-bromophenyl)-	benzene	288(4.42),316s(4.02), 470s(3.59),496(3.86), 528(3.78)	4-0282-66
	EtOH	226(4.36),250(4.27), 279(4.41),310s(3.92), 466(3.84)	4-0282-66
$C_{17}H_{10}F_6$ Cyclopentene, 3,3,4,4,5,5-hexafluoro-1,2-diphenyl-	EtOH	271.0(3.98)	78-1755-66
$C_{17}H_{10}N_2O_3$ Dibenzo[cf][2,7]naphthyridine-6-carboxylic acid, 7-hydroxy-	HOAc	263(4.59),325(3.73), 360(3.84),375(3.83)	87-0161-66
$C_{17}H_{10}O$ 2H-Dibenz[cd,h]azulen-2-one	MeOH	225(4.32),266(4.63), 303(3.95),328(3.78), 370(3.52),384s(3.48), 390(3.38),504(3.41), 523(3.37)	35-3875-66
	CH_2Cl_2	268(4.65),303(3.98), 315(3.89),330(3.82), 359(3.94),372(3.50), 495(3.28),508(3.39)	35-3875-66
	96% H_2SO_4	235(4.28),245s(4.21), 306s(4.73),314(4.82), 384(3.94),402(3.97), 450s(2.36),486(3.32), 511(3.60),545(3.62)	35-3875-66
$C_{17}H_{10}OS$ 5H-Naphtho[1,8-bc]thiophen-5-one, 2-phenyl-	EtOH	242(4.13),286(4.29), 390(4.45)	12-1908-66
	H_2SO_4	229(4.37),261s(3.78), 266(3.80),288(3.44), 367(4.03),454(4.55)	12-1908-66
$C_{17}H_{10}O_2$ Phenanthro[9,10-b]furan-2-carboxaldehyde	EtOH	244(4.61),249(4.62), 257(4.61),333(4.19), 342(4.17),358(4.09)	35-3759-66
$C_{17}H_{11}^+$ Dibenz[cd,h]azulenium cation	96% H_2SO_4	240(4.20),275s(4.07), 310(4.72),384(3.80), 399(3.73),407s(3.13), 524s(3.50),553(3.36)	35-3875-66
$C_{17}H_{11}ClN_2O_2$ Quinoline, 2-(2-chloro-5-nitrostyryl)-	EtOH	205(4.46),220s(4.32), 275(4.52),330(4.33)	44-3683-66
$C_{17}H_{11}ClO$ Cyclopropenone, (α-chlorostyryl)phenyl-	ether	337.5(4.42)	25-1379-66

Compound	Solvent	λ_{max}(log ε)	Ref.
Cyclopropenone, (α-chlorostyryl)-phenyl-, isomer	ether	325(4.62)	25-1379-66
$C_{17}H_{11}ClO_3$			
2-Naphthoic acid, 6-chloro-4-hydroxy-1-phenyl-	n.s.g.	252(4.60),309(3.81)	39-0603-66C
$C_{17}H_{11}N$			
Benz[a]acridine	MeOH	277f(4.7),360f(4.0)	41-0949-66
Benz[b]acridine	MeOH	275(5.2),360f(4.0), 460(3.5)	41-0949-66
Benz[c]acridine	MeOH	280f(4.9),360f(4.0)	41-0949-66
Naphtho[1,2-f]quinoline	C_6H_{12}	212(4.45),217(4.46), 226(4.39),280(4.69), 316(3.87),354(3.22), 372(3.29)	78-0181-66B
$C_{17}H_{11}NO$			
Naphth[1,2-d]isoxazole, 3-phenyl-	EtOH	277(4.00),288(4.01), 313(3.54),320(3.46), 327(3.63)	32-0559-66
Naphth[2,1-d]isoxazole, 3-phenyl-	EtOH	232(4.59),321(3.58), 335(3.72)	32-0559-66
$C_{17}H_{11}NOS$			
5H-Thiazolo[2,3-a]isoquinolin-5-one, 3-phenyl-	benzene	284(4.38),315s(3.91), 467s(3.55),494(3.83), 525(3.76)	4-0282-66
	EtOH	223(4.40),248(4.23), 280(4.45),311s(3.96), 465(3.87)	4-0282-66
$C_{17}H_{11}N_3O$			
v-Triazolo[1,5-a]quinoline, 3-benzoyl-	MeOH	249(4.55),295s(4.07), 307(4.20),330(4.37), 340s(4.30)	24-2918-66
$C_{17}H_{11}N_3O_2S$			
1H-Thieno[3,2-c]pyrazole, 1-(p-nitrophenyl)-3-phenyl-	EtOH	245(4.30),280s(--), 360(4.51)	39-1527-66C
$C_{17}H_{11}N_3O_4$			
9,10-Ethanoanthracene-11-carbonitrile, 9,10-dihydro-9,10-dinitro-	EtOH	253(3.37)	28C-0841-66A
1,4-Ethanoanthracene-2-carbonitrile, 1,2,3,4-tetrahydro-9,10-dinitro-	EtOH	330(3.39)	28C-0841-66A
$C_{17}H_{12}$			
Benzanthrene	EtOH	228(4.66),250(4.20), 312(4.08),329(4.25), 344(4.13)	24-3298-66
15H-Cyclopenta[a]phenanthrene	EtOH	220(4.45),269(4.80), 273(4.81),292(4.23), 302(4.17),314(3.96)	39-0963-66C
17H-Cyclopenta[a]phenanthrene	EtOH	224(4.41),239(4.24), 269(4.72),274(4.70), 292(4.17),303(4.16), 315(4.04)	39-0963-66C

Compound	Solvent	λ_{max}(log ε)	Ref.
1,3-Pentadiyne, 1,5-diphenyl-	EtOH	223(4.66),233(3.54), 245(3.84),258(4.19), 272(4.38),288(4.29)	78-0867-66
1,4-Pentadiyne, 1,5-diphenyl-	EtOH	239(4.95),251(5.05), 264(3.23),272(3.12), 279(2.96),282s(2.55)	78-0867-66
$C_{17}H_{12}BrNO_2S$			
Acetophenone, 4'-bromo-2-[(3-hydroxy-1-isoquinolyl)thio]-	EtOH	256(4.34),288(4.53), 280s(4.03),295(3.89), 357(3.87)	4-0282-66
$C_{17}H_{12}ClN$			
Dibenzo[a,f]quinolizinium chloride	H_2O	220(4.45),241(4.20), 278(4.32),350(3.92), 368(4.16),387(4.29)	44-3683-66
Isoquinoline, 1-(o-chlorostyryl)-	EtOH	232(4.45),276(4.06), 297(4.14),348(4.25)	44-3683-66
Isoquinoline, 3-(o-chlorostyryl)-	EtOH	220(4.55),272(4.26), 316(4.46)	44-3683-66
Naphtho[1,2-c]quinolizinium chloride	EtOH	224(4.54),242(4.21), 262(4.37),278(4.36), 333(3.62),370(4.08), 388(4.21)	44-3683-66
Naphtho[2,1-c]quinolizinium chloride	H_2O	214(4.51),249(4.44), 273(4.16),308(4.26), 380(3.75),398(3.81)	44-3683-66
Naphtho[2,3-c]quinolizinium chloride	H_2O	225(4.42),239(4.51), 246(4.57),285s(4.31), 299(4.38),313(4.58), 365s(3.69),384(3.75), 400(3.75)	44-3683-66
Pyridine, 2-[2-(1-chloro-2-naphthyl)-vinyl]-	EtOH	205(4.24),224(4.41), 253(4.24),274(4.40), 286(4.36),325(4.50)	44-3683-66
Pyridine, 2-[2-(2-chloro-1-naphthyl)-vinyl]-	EtOH	205(4.24),225(4.40), 253(4.25),274(4.40), 287(4.36),325(4.50)	44-3683-66
Pyridine, 2-[2-(3-chloro-2-naphthyl)-vinyl]-	EtOH	222(4.50),274(4.43), 285s(4.34),326(4.35)	44-3683-66
Quinoline, 2-(o-chlorostyryl)-	EtOH	222(4.38),282(4.40), 327(4.36),339(4.34)	44-3683-66
$C_{17}H_{12}ClN_3O$			
Spiro[1-methyl-2-oxindoline-3,5'-pyrazole], 3'-(p-chlorophenyl)-	EtOH	261(--),313(--), 326(--)	95-1156-66
$C_{17}H_{12}F_3N_3$			
2-Naphthylamine, 1-(α,α,α-trifluoro-m-tolylazo)-	benzene	460(4.14)	44-2571-66
	$C_6H_{11}Me$	452(4.15)	44-2571-66
2-Naphthylamine, 1-(α,α,α-trifluoro-p-tolylazo)-	benzene	462(4.15)	44-2571-66
	$C_6H_{11}Me$	455(4.14)	44-2571-66
$C_{17}H_{12}F_3N_3O$			
1-Naphthol, 7-amino-8-(α,α,α-trifluoro-m-tolylazo)-	benzene	515(4.20)	44-2571-66
	$C_6H_{11}Me$	516(4.13)	44-2571-66
1-Naphthol, 7-amino-8-(α,α,α-trifluoro-p-tolylazo)-	benzene	519(4.21)	44-2571-66
	$C_6H_{11}Me$	521(4.19)	44-2571-66

Compound	Solvent	λ_{max}(log ε)	Ref.
$C_{17}H_{12}N_2$			
1H-Pyrrolo[2,3-f]isoquinoline, 2-phenyl-	neutral	285(4.67),330(4.44)	2-0118-66
Pyrrolo[1,2-a]quinoxaline, 4-phenyl-	EtOH	229(4.37),247(4.47), 271(4.37),339(3.87), 351(3.87)	39-0852-66C
$C_{17}H_{12}N_2O$			
Dibenzo[c,f][2,7]naphthyridin-6-ol, 7-methyl-	HOAc	265(4.52),319(3.71), 362(3.78),384(3.82)	87-0161-66
Dibenzo[c,f][2,7]naphthyridin-6(5H)-one, 5-methyl-	$CHCl_3$	259(4.61),316(3.71), 329(3.73),372(3.91)	87-0161-66
Naphth[1,2-d]imidazole, 2-(o-hydroxyphenyl)-	EtOH	231(4.68),264s(4.33), 273(4.41),306(4.10), 320(4.21),334(4.51), 350(4.62)	4-0444-66
$C_{17}H_{12}N_2S$			
Imidazo[2,1-b]thiazole, 6-(4-biphenylyl)-	n.s.g.	208(4.83),294(4.72)	22-1277-66
1H-Thieno[3,2-c]pyrazole, 1,3-diphenyl-	EtOH	226(4.39),255(4.32), 312(4.47)	39-1527-66C
$C_{17}H_{12}N_4$			
Benzonitrile, m-[(2-amino-1-naphthyl)azo]-	benzene	466(4.15)	44-2571-66
	$C_6H_{11}Me$	458(4.17)	44-2571-66
Benzonitrile, p-[(2-amino-1-naphthyl)azo]-	benzene	480(4.25)	44-2571-66
	$C_6H_{11}Me$	472(4.23)	44-2571-66
Pyrazolo[5,1-c]-as-triazine, 3,4-diphenyl-	EtOH	247(4.40),283(3.56)	39-1127-66C
$C_{17}H_{12}N_4O$			
Benzonitrile, m-[(2-amino-8-hydroxy-1-naphthyl)azo]-	benzene	518(4.22)	44-2571-66
Benzonitrile, p-[(2-amino-8-hydroxy-1-naphthyl)azo]-	benzene	529(4.27)	44-2571-66
s-Triazolo[1,5-a]pyrazin-2-ol, 5,6-diphenyl-	EtOH	203(4.43),241(4.09), 288(3.91)	39-2038-66C
s-Triazolo[4,3-a]pyrazin- -ol, 5,6-diphenyl- (only four absorbancies given)	EtOH	207(4.34),232s(3.99), 266(3.89),288s(3.56?), 365(?)	39-2038-66C
$C_{17}H_{12}N_4O_6$			
Pyrazole-3-carboxylic acid, 1-(2,4-dinitrophenyl)-4-phenyl-, methyl ester	EtOH	320(3.96)	22-0619-66
$C_{17}H_{12}N_4S$			
s-Triazolo[4,3-b]pyridazine, 3-phenyl-6-(phenylthio)-	EtOH	272(4.31)	78-2073-66
$C_{17}H_{12}O$			
17H-Cyclopenta[a]phenanthren-17-one, 15,16-dihydro-	EtOH	218(4.18),265(4.89), 284(4.52),296(4.38), 334(3.24),350(3.40), 367(3.44)	39-0955-66C
Cyclopropenone, phenyl-β-styryl-, trans	ether	322(4.56)	88-3763-66
2H-Dibenz[cd,h]azulen-2-ol	EtOH	231(4.59),238(4.58), 255s(4.51),267(4.31), 278(4.05),295(3.83), 308(4.07),321(4.19), 348(3.60),365(3.57)	35-3875-66

Compound	Solvent	$\lambda_{max}(\log \epsilon)$	Ref.
Naphthalene, 1-benzoyl-	MeOH	208s(4.53),221(4.70), 251(4.22),280(3.82), 289(3.77),307(3.71)	73-0835-66
Naphthalene, 2-benzoyl-	MeOH	218(4.61),254(4.50), 285(4.06),292(4.04), 335(3.38)	73-0835-66
$C_{17}H_{12}OS$			
5H-Naphtho[1,8-bc]thiophen-5-one, 3,4-dihydro-2-phenyl-	EtOH	233(4.09),274(4.34), 333s(3.86),342(3.86)	12-1908-66
	H_2SO_4	244s(3.71),272(4.01), 298(4.00),405(3.36), 480(3.57)	12-1908-66
$C_{17}H_{12}O_2$			
2-Cyclopenten-1-one, 4,5-epoxy-3,4-diphenyl-	EtOH	307(4.01)	35-1518-66
2-Furaldehyde, 4,5-diphenyl-	EtOH	234(4.32),256(4.14), 329(4.29)	35-1518-66
2-Pyrone, 4,5-diphenyl-	EtOH	237(4.24),282(3.85)	35-1518-66
$C_{17}H_{12}O_2S_2$			
2,5-Cyclohexadiene-$\Delta^{1,\alpha}$-acetic acid, 4-(1,3-dithiolan-2-ylidene)-2-hydroxy-α-phenyl-, γ-lactone	$CHCl_3$	505(4.58)	89-0517-66
	CF_3COOH	540(3.54)	89-0517-66
$C_{17}H_{12}O_3$			
2-Pyrone, 5,6-epoxy-4,5-diphenyl-	EtOH	281(4.10)	35-3759-66
$C_{17}H_{12}O_4$			
Naphtho[2,3-c]furan-4,9-dione, 1-methyl-3-(3-oxo-1-butenyl)-	ether	249(4.52),321(3.99), 332s(3.95),385(4.10)	33-1806-66
$C_{17}H_{12}O_5$			
1H-Cyclopenta[7,8]naphtho[2,3-b]furan-7-carboxylic acid, 2,3-dihydro-10-hydroxy-6-methyl-3-oxo-	EtOH	228(4.37),278(4.72), 325(3.73),340(3.70), 399(3.93)	39-0725-66C
6H-[1,3]Dioxolo[5,6]benzofuro[3,2-c]-[1]benzopyran, 3-methoxy-	EtOH	215(4.49),232s(4.22), 243s(4.13),251(4.09), 294(3.82),340(4.54), 359(4.57)	88-1805-66
Isoflavone, 7-methoxy-4',5'-(methylenedioxy)-	MeOH	225(4.6),248s(--), 261s(--),294(4.2)	39-0509-66C
$C_{17}H_{12}O_6$			
Aflatoxin B_1	pH 7.3	367(4.3)	57-1539-66A
$C_{17}H_{12}S$			
2H-Naphtho[1,8-bc]thiophene, 2-phenyl-	EtOH	246(4.08),323(3.75), 335(3.77)	12-1908-66
$C_{17}H_{13}ClN_2O_2$			
Pyrimidine, 2-chloro-6-phenoxy-5-phenoxymethyl-	EtOH	210(4.30),219(4.31), 259(3.83)	5-0119-66B
$C_{17}H_{13}ClO_4S$			
2,6-Diphenylthiopyrylium perchlorate	n.s.g.	251(4.26),404(4.15)	33-2046-66

Compound	Solvent	$\lambda_{max}(\log \epsilon)$	Ref.
$C_{17}H_{13}N$			
Benzo[6,7]cyclohept(1,2,3-de]isoquinoline, 7,8-dihydro-, hydrochloride	EtOH	232(4.59),306(3.95), 340(3.96)	4-0247-66
$C_{17}H_{13}NO$			
Naphth[1,2-d]isoxazole, 4,5-dihydro-1-phenyl-	EtOH	257(4.04)	32-0559-66
Naphth[2,1-d]isoxazole, 4,5-dihydro-3-phenyl-	EtOH	222(4.39),285(4.22), 300(4.12)	32-0559-66
Protoberberine, dehydro-	n.s.g.	233(4.32),322(4.17), 343(4.08),358(3.90)	35-5369-66
Quinoline, 2-phenacyl-	pH 1	236(4.68),317(4.05)	7-0767-66
	EtOH	218(4.63),279(4.06), 293(4.09),326(4.00), 436(4.08)	7-0767-66
$C_{17}H_{13}NO_2$			
Isocarbostyril, 4-benzoyl-N-methyl-	EtOH	226(4.32),296(4.04), 307(4.05),320(3.92)	78-2445-66
$C_{17}H_{13}NO_2S$			
Acetophenone, 2-[(3-hydroxy-1-isoquinolyl)thio]-	EtOH	230(4.56),245s(4.34), 284(3.88),295(3.83), 357(3.86)	4-0282-66
$C_{17}H_{13}NO_3$			
1,4-Pentadien-3-one, 5-(p-nitrophenyl)-1-phenyl-	benzene	335(--)	65-0336-66
	EtOH	328(--)	65-0336-66
	HOAc	333(--)	65-0336-66
2,4-Pentadien-1-one, 1-(p-nitrophenyl)-5-phenyl-	benzene	360(--)	65-0336-66
	EtOH	355(--)	65-0336-66
	HOAc	274(4.29),360(4.39)	65-0336-66
$C_{17}H_{13}NO_5$			
Nicotinic acid, 1,2-dihydro-4,6-dihydroxy-1-(2-naphthyl)-2-oxo-, methyl ester	EtOH	306(4.37)	78-0455-66
$C_{17}H_{13}NO_6$			
Benz[cd]indole-4,5-dicarboxylic acid, N-acetyl-1,2-dihydro-2-oxo-, dimethyl ester	EtOH	229s(4.58),233(4.62), 255(4.25),274s(4.13), 330s(3.48),346(3.59), 363s(3.56)	39-1028-66C
$C_{17}H_{13}N_3O$			
Spiro[1-methyl-2-oxindoline-3,5'-pyrazole], 3'-phenyl-	EtOH	243(4.62),312(4.15), 325(4.15)	95-1156-66
$C_{17}H_{13}N_3O_2$			
Spiro[2-oxindoline-3,5'-pyrazole], 3'-(p-methoxyphenyl)-	EtOH	261(4.34),272(4.33), 310(3.94),323(3.93)	95-1156-66
$C_{17}H_{13}N_3O_2S$			
Benzamide, N-(2-benzoyl-4-methyl-Δ^3-1,2,3-thiadiazolin-5-ylidene)-	MeOH	226(4.28),287(3.90), 389(4.32)	24-1618-66
$C_{17}H_{13}N_3O_3$			
Acetanilide, 4'-(benzoyldiazoacetyl)-	CH_2Cl_2	288(4.36)	24-3128-66

Compound	Solvent	λ_{max}(log ε)	Ref.
Benzoic acid, o-(2,3-dihydro-5-oxo-imidazo[2,1-b]quinazolin-1(5H)-yl)-	EtOH	229(4.66),273(4.14), 320s(3.52)	24-1532-66
5H-Quinazolino[3,2-a]quinazoline-5,12(6H)-dione, 6-(2-hydroxyethyl)-	EtOH	235(4.68),284(4.37), 325(3.75)	24-1532-66
$C_{17}H_{13}N_5$			
s-Triazolo[4,3-a]pyrazine, 3-amino-5,6-diphenyl-	EtOH	210(4.34),234s(4.16), 263(4.19),355(3.50)	39-2038-66C
$C_{17}H_{13}OP$			
Phosphine oxide, diphenyl-1,3-pentadiynyl-	EtOH	225(4.45),247(3.62), 261(3.40),274(3.23)	22-1002-66
$C_{17}H_{13}P$			
Phosphine, diphenyl-1,3-pentadiynyl-	EtOH	228(4.43),250(4.00), 266(3.78),283(3.48)	22-1002-66
$C_{17}H_{14}$			
15H-Cyclopenta[a]phenanthrene, 16,17-dihydro-	EtOH	216(4.48),259(4.77), 280(4.14),288(4.07), 300(4.17)	39-0963-66C
2H-Dibenz[cd,h]azulene, 6,7-dihydro-	EtOH	240(3.37),269(3.55), 291(3.79),302(4.37)	4-0247-66
Naphthalene, 1-methyl-2-phenyl-	EtOH	274s(--),282(--), 296s(--)	22-0645-66
Naphthalene, 1-methyl-3-phenyl-	EtOH	251(4.74),290(4.06)	22-0645-66
$C_{17}H_{14}BrN_3O_3$			
Acetanilide, 2-bromo-2'-(3,4-dihydro-1-isoquinolyl)-4'-nitro-, hydrobromide	EtOH	233(4.43)	33-1222-66
$C_{17}H_{14}ClN$			
Benzo[6,7]cyclohept[1,2,3-de]iso-quinoline, 7,8-dihydro-, hydrochloride	EtOH	232(4.59),306(3.95), 340(3.96)	4-0247-66
$C_{17}H_{14}ClNO_4$			
2-Acridinecarboxylic acid, 9-chloro-1,3-dimethoxy-, methyl ester	EtOH	237(4.25),264(4.99), 327s(3.37),340(3.67), 356(3.86),393(3.67)	12-0275-66
$C_{17}H_{14}ClNO_4S$			
5,6-Dihydro-3-phenylthiazolo[2,3-a]-isoquinolinium perchlorate	EtOH	325(4.13)	4-0282-66
2-Methyl-3-phenyl-4H-indeno[2,1-d]-thiazolium perchlorate	MeOH	241(3.99),318(4.11)	39-0686-66C
3-Methyl-2-phenyl-4H-indeno[2,1-d]-thiazolium perchlorate	MeOH	226s(3.83),254s(3.71), 341(4.28)	39-0686-66C
$C_{17}H_{14}ClNO_5$			
4-Benzoyl-2-methylisoquinolinium perchlorate	EtOH	221(4.62),323(3.78)	78-2445-66
$C_{17}H_{14}ClN_3O_4$			
2-Quinoxalinecarboxylic acid, 3-(5-chloro-2,4-dimethoxyanilino)-, sodium salt	EtOH	288(4.43),321(4.28), 408(3.78)	87-0770-66

Compound	Solvent	$\lambda_{max}(\log \epsilon)$	Ref.
$C_{17}H_{14}Cl_2N_2O$			
2-Pyrrolidinone, 4,4-dichloro-3-methyl-3-phenyl-5-(phenylimino)-	EtOH	228(4.21),285(3.90)	28C-1276-66A
1-Pyrrolin-5-one, 3,3-dichloro-2-(methylamino)-4,4-diphenyl-	EtOH	249(4.20)	28C-1276-66A
$C_{17}H_{14}N_2$			
Olivacine	EtOH-HCl	242(4.48),306(4.90),348(3.82)	87-0237-66
	EtOH	224(4.39),238(4.31),267(4.57),276(4.73),286(4.90),292(4.87),314(3.65),328(3.75)	87-0237-66
9H-Pyrido[3,4-b]indole, 3,4-dihydro-1-phenyl-	EtOH-acid	253(4.04),276(3.94),287(3.92),368(4.38)	94-0856-66
	EtOH	231(4.33),247(4.38),323(4.12)	94-0856-66
$C_{17}H_{14}N_2O$			
5-Pyrazolone, 4,4-ethylene-1,3-diphenyl-	MeOH	208(4.40),253(4.29),300(4.02)	24-2962-66
4-Pyrimidinol, 5-methyl-2,6-diphenyl-	EtOH	305(3.95)	22-1819-66
2-Pyrrolidinone, 4-(diphenylmethylene)-5-imino-	EtOH	279(4.20)	22-3895-66
$C_{17}H_{14}N_2OS$			
Oxazole, 5-(benzylideneamino)-4-(methylthio)-2-phenyl-	EtOH	274(4.16),388(4.37)	44-3612-66
$C_{17}H_{14}N_2O_2$			
Quinazoline, 2-benzoyl-4-ethoxy-	EtOH	210(4.59),220s(4.55),250(4.29),300s(3.85),310s(3.71)	39-2290-66C
$C_{17}H_{14}N_2O_3$			
2-Indolinone, 3-(5-carbethoxy-2-pyridyl)-	EtOH	220(4.31),262(4.15),265(4.15),335s(4.20),346(4.26),359s(4.16),421(3.56)	39-1028-66C
4(3H)-Quinazolinone, 2-(α-acetoxybenzyl)-	EtOH	210(4.33),227(4.43),268(3.86),303(3.53),316(3.42)	39-2290-66C
$C_{17}H_{14}N_2O_4S$			
2H-1,2-Benzothiazin-4(3H)-one, 3-[(p-acetamido)benzylidene]-, 1,1-dioxide	EtOH	265(4.18),374(4.36)	44-0162-66
$C_{17}H_{14}N_2O_5$			
Cinnamic acid, p-nitro-, p-acetamidophenyl ester	EtOH	250(4.3),300(4.4)	65-1202-66
$C_{17}H_{14}N_4O$			
1,2,3-Triazolo[1,5-a]quinazoline, 5-ethoxy-3-phenyl-	EtOH	212(4.58),220(4.56),254(4.44),263(4.38),272(4.32),340(3.76)	39-2290-66C
$C_{17}H_{14}N_4O_2$			
4-Pyridazinecarboxaldehyde, 1,6-dihydro-3-hydroxy-6-oxo-1-phenyl-, 4-(phenylhydrazone)	THF	258(4.11),350(4.08)	49-1494-66

Compound	Solvent	λ_{max}(log ε)	Ref.
as-Triazine, 3-acetamido-5,6-diphenyl-, 2-oxide	EtOH	223(4.48),250(4.46), 349(3.90)	44-3914-66
$C_{17}H_{14}N_4O_4$			
Pyrazole, 1-(2,4-dinitrophenyl)-3,4-dimethyl-5-phenyl-	EtOH	327(3.90)	22-3744-66
Pyrazole, 1-(2,4-dinitrophenyl)-4,5-dimethyl-3-phenyl-	EtOH	241(4.35),347(3.92)	22-3744-66
$C_{17}H_{14}N_4O_6$			
2-Pyrazoline-3-carboxylic acid, 1-(2,4-dinitrophenyl)-4-phenyl-, methyl ester	$CHCl_3$	385(4.39)	22-0619-66
$C_{17}H_{14}O$			
15H-Cyclopenta[a]phenanthren-17-ol, 16,17-dihydro-	EtOH	259(4.77),280(4.09), 288(3.98),300(4.12)	39-0963-66C
2H-Dibenz[cd,h]azulen-2-one, 1,6,7,11b-tetrahydro-	EtOH	247(4.00),297(3.43)	4-0247-66
1-Indanone, 2-(p-methylbenzylidene)-, cis	MeOH	232(4.01),275s(3.86), 334(4.33)	35-4489-66
trans	MeOH	234(3.99),278s(3.89), 331(4.45)	35-4489-66
1,4-Pentadien-3-one, 1,5-diphenyl-	benzene	327(--)	65-0336-66
	EtOH	329(--)	65-0336-66
	HOAc	332(4.43)	65-0336-66
	H_2SO_4	447(--)	65-0336-66
ferric chloride added	HOAc	460(4.55)	65-0336-66
2,4-Pentadien-1-one, 1,5-diphenyl-	benzene	342(4.56)	65-0336-66
	EtOH	343(4.52)	65-0336-66
	HOAc	269(3.97),345(4.54)	65-0336-66
ferric chloride added	HOAc	485(4.18)	65-0336-66
$C_{17}H_{14}O_2$			
Cyclopenta[a]phenanthrene-11,17-dione, 11,12,13,14,15,16-hexahydro-	EtOH	216(4.61),248(4.33), 315(3.82)	39-0955-66C
3(2H)-Furanone, 2-methyl-2,5-diphenyl-	EtOH	220(4.15),244(3.99), 306(4.27)	32-1073-66
	EtOH	220s(4.15),244(3.99), 306(4.27)	88-0233-66
$C_{17}H_{14}O_3$			
Benzofuran, 3-acetonyl-2-(o-hydroxyphenyl)-	EtOH	280(4.07),307(4.17)	39-0544-66C
3(2H)-Benzofuranone, 2-allyl-2-(o-hydroxyphenyl)-	EtOH	251(3.97),279(3.47), 331(3.62)	39-0544-66C
	EtOH-KOH	246(4.14),308(3.86)	39-0544-66C
Crotonic acid, 3-methyl-, ester with 6-hydroxy-2,4-hexadiynophenone	ether	213(4.40),265(4.08), 277(4.17),293(4.03)	24-2413-66
Isoaurone, 4'-methoxy-5-methyl-, cis	EtOH	380(4.42)	78-1789-66
Isoaurone, 4'-methoxy-5-methyl-, trans	EtOH	380(4.40)	78-1789-66
4-Pentynophenone, 2'-hydroxy-2-(o-hydroxyphenyl)-	EtOH	256(4.02),285s(3.41), 328(3.66)	39-0544-66C
	EtOH-KOH	266s(3.82),367(3.58)	39-0544-66C
Phenanthro[9,10-b]-p-dioxin, 2,3-dihydro-2-methoxy-	benzene	367(3.16)	24-1881-66
$C_{17}H_{14}O_4$			
2-Heptene-4,6-diyn-1-ol, 7-(m-acetoxyphenyl)-, acetate	ether	210(4.65),229(4.52), 239(4.53),252(4.51),	24-1223-66

Compound	Solvent	λ_{max}(log ε)	Ref.
2-Heptene-4,6-diyn-1-ol, 7-(m-acetoxyphenyl)-, acetate (cont.)		264(4.10),280(4.41), 298(4.27),319(4.31)	24-1223-66
1H-Naphtho[2,1-b]pyran-1-one, 2-acetyl-5-methoxy-3-methyl-	EtOH	216(4.69),275(4.24), 312s(3.85),324(3.91), 337(3.88)	94-0129-66
1H-Naphtho[2,1-b]pyran-1-one, 2-acetyl-6-methoxy-3-methyl-	EtOH	220(4.47),240(4.30), 258(4.32),265(4.31), 306(4.01),320(3.93), 334(3.95)	94-0129-66
Valeric acid, 3,5-dihydroxy-3,4-diphenyl-4,5-epoxy-, lactone	EtOH	258(2.53),261(2.46), 264(2.46)	35-3759-66
$C_{17}H_{14}O_4S_2$			
Spiro[5H-dibenzo[a,d]cycloheptene-5,2'-[1,3]dithiolane] , 1',1',3',3'-tetroxide	EtOH	223(4.58),293(3.99)	87-0860-66
$C_{17}H_{14}O_5$			
Pterocarpin	EtOH	281(3.60),287(3.66), 311(3.89)	88-1805-66
$C_{17}H_{14}O_6$			
Benzofuran, 2-(2-hydroxy-4-methoxyphenyl)-3-(hydroxymethyl)-5,6-(methylenedioxy)-	EtOH	271(4.11),322(4.36)	88-1805-66
Flavone, 3,5-dihydroxy-4',7-dimethoxy-	EtOH	267(4.37),322s(3.25), 363(4.40)	78-0071-66B
$AlCl_3$ complex	EtOH	270(4.35),348s(3.16), 414(3.35)	78-0071-66B
Xanthone, 4-acetoxy-2,3-dimethoxy-	EtOH	241(4.23),269(4.13), 302(4.13)	78-1777-66
Xanthone, 5-acetoxy-1,3-dimethoxy-	EtOH	239(4.47),299(4.12), 330(3.75)	78-1777-66
$C_{17}H_{14}O_8$			
Axillarin	EtOH	260(4.36),270s(4.33), 295s(4.04),358(4.33)	24-3539-66
$C_{17}H_{14}S_2$			
Spiro[5H-dibenzo[a,d]cycloheptene-5,2'[1,3]dithiolane]	EtOH	214(4.49),293(4.05)	87-0860-66
$C_{17}H_{15}BrN_2O$			
6,7,9,10-Tetrahydro-6-oxo-5H-isoquino-[2,1-d][1,4]benzodiazepin-8-ium bromide	EtOH	240(4.34),292(4.17)	33-1222-66
$C_{17}H_{15}BrO$			
1-Indanone, 2-bromo-2-(p-methylbenzyl)-	MeOH	257(4.07),300(3.44)	35-4489-66
$C_{17}H_{15}BrO_2$			
1-Indanone, 2-bromo-2-(p-methoxybenzyl)-	MeOH	225(4.22),256(4.09), 300(3.42)	35-4489-66
Stilbene, α-acetoxy-α'-bromo-2'-methyl-	EtOH	214(4.13),230(4.12), 263(3.92)	39-0533-66C
$C_{17}H_{15}Cl$			
1,3-Butadiene, 1-chloro-4-phenyl-1-p-tolyl-, trans, trans	hexane	237(4.09),328(4.74), 352s(4.49)	44-2175-66

Compound	Solvent	λ_{max}(log ε)	Ref.
$C_{17}H_{15}ClN_2O$			
Indolizine-7-carboxamide, N-(p-chlorophenyl)-2,5-dimethyl-	EtOH	228(4.48),280(4.42), 361(4.37)	56-0061-66
$C_{17}H_{15}ClO_3$			
Acetic acid, (p-chlorophenoxy)-, 1-phenylallyl ester	n.s.g.	279(3.16),287(3.07)	28C-1099-66A
Acetic acid, (p-chlorophenoxy)-, 3-phenylallyl ester	n.s.g.	248(4.27),252(4.27)	28C-1099-66A
$C_{17}H_{15}ClO_5$			
3(2H)-Benzofuranone, 7-chloro-6-methoxy-4-(p-methoxybenzyloxy)-	EtOH	232(4.42),285(4.24), 322(3.70)	87-0242-66
	dioxan	236(4.38),283(4.28), 317(3.77)	87-0242-66
$C_{17}H_{15}ClO_7$			
4',7-Dimethoxyflavylium perchlorate	MeOH-acid	459(4.73)	61-0530-66
$C_{17}H_{15}IN_2S_2$			
3-Methyl-2-[(3-methyl-2-benzothiazolinylidene)methyl]benzothiazolium iodide	MeOH	420(4.89)	9-0289-66
	n.s.g.	423(4.90)	27-0330-66
$C_{17}H_{15}N$			
Aporphane, N-methyldehydro-	n.s.g.	243(4.47),257(4.48), 324(3.93)	88-2941-66
Benzo[6,7]cyclohept[1,2,3-de]isoquinoline, 1,7,8,12b-tetrahydro-	EtOH	253(4.01)	4-0247-66
hydrochloride	EtOH	298(4.00)	4-0247-66
$C_{17}H_{15}NO$			
7H-Benzo[c]carbazole, 5,6-dihydro-2-methoxy-	EtOH	230(4.53),284(4.18), 314(4.13)	94-0934-66
Benzo[6,7]cyclohept[1,2,3-de]isoquinolin-3(2H)-one, 1,7,8,12b-tetrahydro-	EtOH	235(4.01)	4-0247-66
Isoquinoline, 4-benzoyl-1,2-dihydro-2-methyl-	EtOH	288(4.11),295(4.12), 323(4.09)	78-2445-66
6H-Naphtho[2,3-c]quinolizin-6-one, 1,2,3,4-tetrahydro-	EtOH-HCl	224(4.32),229(4.34), 263(5.09)	94-1399-66
	EtOH	226(4.25),253(4.85), 262(4.80)	94-1399-66
$C_{17}H_{15}NO_2$			
Anonaine	EtOH	234(4.19),272(4.26), 315(3.56)	95-0129-66
hydrochloride	EtOH	260(4.09),307(3.93)	95-0124-66
1-Anthracenecarbamic acid, ethyl ester	EtOH	254(5.15),332s(3.42), 349s(3.69),356(3.72), 365(3.82),380(3.76)	12-1859-66
2-Anthracenecarbamic acid, ethyl ester	EtOH	261(4.92),271(4.88), 322s(3.28),338(3.51), 355(3.59),374(3.52), 392(3.39)	12-1859-66
9-Anthracenecarbamic acid, ethyl ester	EtOH	249s(4.98),255(5.21), 318s(3.06),332(3.45), 347(3.75),365(3.92), 384(3.87)	12-1859-66
Cyclopenta[c]quinazoline, 1,3-diacetyl-4-methyl-	EtOH	242(4.50),254s(4.4), 310(4.44),332s(4.4), 405(3.73)	39-0324-66C

Compound	Solvent	λ_{max}(log ε)	Ref.
Isocarbostyril, 2-(β-hydroxyphenethyl)-	MeOH	278(3.97),286(3.97), 326(3.61)	83-0715-66
1-Naphthaleneacrylic acid, α-cyano-β-methyl-, ethyl ester, cis	EtOH	269(3.79),276(3.79), 301(3.43)	22-1033-66
2-Naphthaleneacrylic acid, α-cyano-β-methyl-, ethyl ester, cis	EtOH	265(4.22),308(3.98)	22-1033-66
trans	EtOH	266(4.26),274(4.30), 309(4.07)	22-1033-66
1-Phenanthrenecarbamic acid, ethyl ester	EtOH	256(4.71),281s(4.00), 298(4.07),339(2.72), 355(2.64)	12-1859-66
2-Phenanthrenecarbamic acid, ethyl ester	EtOH	264(4.88),293(4.30), 301s(3.86),323(2.70), 339(2.74),355(2.59)	12-1859-66
3-Phenanthrenecarbamic acid, ethyl ester	EtOH	253s(4.71),259(4.75), 280(4.25),298(4.16), 308(4.17),340(3.04), 357(3.01)	12-1859-66
4-Phenanthrenecarbamic acid, ethyl ester	EtOH	255(4.72),286s(4.00), 297(4.06),339(2.72), 355(2.59)	12-1859-66
9-Phenanthrenecarbamic acid, ethyl ester	EtOH	255(4.71),276(4.13), 298(4.02),341(2.61), 357(2.47)	12-1959-66
$C_{17}H_{15}NO_4$			
Acridone, 4-formyl-1,3-dimethoxy-10-methyl-	EtOH	227(4.14),247(4.40), 256(4.40),282(4.40), 384(4.01)	12-0275-66
Benzoic acid, 2-benzoyl-, anhydride with dimethylcarbamic acid	$CHCl_3$	325(2.09)	35-0781-66
Cyclopenta[c]quinolizine-1,2-dicarboxylic acid, 4-methyl-, dimethyl ester	EtOH	236(4.46),250s(4.4), 270s(4.05),288(3.78), 342(4.50),370s(4.15), 445(3.64)	39-0324-66C
Cyclopenta[c]quinolizine-2,3-dicarboxylic acid, 4-methyl-, dimethyl ester	EtOH	253(4.04),270(4.23), 288(3.97),342(4.52), 370s(4.1),453(3.58)	39-0324-66C
$C_{17}H_{15}NO_5$			
Acridone, 4-carboxy-1,3-dimethoxy-10-methyl-	EtOH	225(4.19),271(4.57), 293s(4.08),381(3.90)	12-0275-66
Chalcone, 2'-hydroxy-2-methoxy-5'-methyl-3'-nitro-, cis	EtOH	241(4.32),276(3.82), 343(3.79)	78-1789-66
trans	EtOH	264(4.08),310(4.04), 375(4.28)	78-1789-66
$C_{17}H_{15}N_3$			
2-Naphthylamine, 1-(m-tolylazo)-	benzene	439(4.11)	44-2571-66
	$C_6H_{11}Me$	430(4.14)	44-2571-66
2-Naphthylamine, 1-(p-tolylazo)-	benzene	440(4.15)	44-2571-66
	$C_6H_{11}Me$	429(4.17)	44-2571-66
$C_{17}H_{15}N_3O$			
1-Naphthol, 7-amino-8-(m-tolylazo)-	benzene	504(4.20)	44-2571-66
	$C_6H_{11}Me$	502(4.18)	44-2571-66
1-Naphthol, 7-amino-8-(p-tolylazo)-	benzene	503(4.22)	44-2571-66
	$C_6H_{11}Me$	501(4.21)	44-2571-66
2-Naphthylamine, 1-[(m-methoxyphenyl)azo]-	benzene	446(4.13)	44-2571-66
	$C_6H_{11}Me$	432(4.14)	44-2571-66

Compound	Solvent	λ max(log ε)	Ref.
2-Naphthylamine, 1-[(p-methoxy-phenyl)azo]-	benzene	449(4.18)	44-2571-66
	$C_6H_{11}Me$	443(4.18)	44-2571-66
$C_{17}H_{15}N_3OS$			
1-Naphthol, 7-amino-8-[[m-(methyl-thio)phenyl]azo]-	benzene	510(4.23)	44-2571-66
	$C_6H_{11}Me$	507(4.19)	44-2571-66
1-Naphthol, 7-amino-8-[[p-(methyl-thio)phenyl]azo]-	benzene	519(4.34)	44-2571-66
$C_{17}H_{15}N_3O_2$			
1-Naphthol, 7-amino-8-[(m-methoxy-phenyl)azo]-	benzene	507(4.22)	44-2571-66
	$C_6H_{11}Me$	503(4.16)	44-2571-66
1-Naphthol, 7-amino-8-[(p-methoxy-phenyl)azo]-	benzene	503(4.28)	44-2571-66
	$C_6H_{11}Me$	500(4.25)	44-2571-66
2-Quinoxalinecarboxylic acid, 3-(2,3-xylidino)-, sodium salt	EtOH	283(4.38),400(3.75)	87-0770-66
2-Quinoxalinecarboxylic acid, 3-(2,6-xylidino)-, sodium salt	EtOH	262(4.40),385(3.72)	87-0770-66
Stilbene, α-cyano-4'-(dimethylamino)-4-nitro-	acetone	446(4.52)	83-0131-66
	+ base	550(4.53)	83-0131-66
$C_{17}H_{15}N_3O_2S$			
2-Naphthylamine, 1-[[m-(methylsulfon-yl)phenyl]azo]-	benzene	466(4.16)	44-2571-66
2H-Pyrazolo[4,3-c]quinoline, 4,5-dihydro-5-(p-tolylsulfonyl)-	EtOH	224(4.35),237(4.36), 269(4.07)	54-0681-66
$C_{17}H_{15}N_3O_3$			
Anthranilic acid, N-(3,4-dihydro-3-methyl-4-oxo-2-quinazolinyl)-, methyl ester	EtOH	218(4.62),242s(4.26), 281s(4.07),291(4.27), 329(4.22)	24-1532-66
Indolizine-7-carboxamide, 2,5-dimethyl-N-(m-nitrophenyl)-	EtOH	230(4.51),279(4.48), 367(4.34)	56-0061-66
Isoquino[2,1-d][1,4]benzodiazepin-6(7H)-one, 5,9,10,14b-tetrahydro-2-nitro-, hydrochloride	EtOH	272(4.04)	33-1222-66
$C_{17}H_{15}N_3O_3S$			
1-Naphthol, 7-amino-8-[[m-(methyl-sulfonyl)phenyl]azo]-	benzene	518(4.21)	44-2571-66
$C_{17}H_{15}N_3O_4$			
Anthranilic acid, N-[3,4-dihydro-3-(2-hydroxyethyl)-4-oxo-2-quinazolinyl]-	EtOH	219(4.65),240s(4.35), 280s(4.16),291(4.30), 326(4.29)	24-1532-66
4-Quinazolinone, 2-hydroxyethylamino-3-(2-carboxyphenyl)-	EtOH	218(4.95),290(4.27), 330(4.22),382(4.14)	24-1532-66
$C_{17}H_{15}N_3S$			
2-Naphthylamine, 1-[[m-(methylthio)-phenyl]azo]-	benzene	450(4.14)	44-2571-66
	$C_6H_{11}Me$	439(4.15)	44-2571-66
2-Naphthylamine, 1-[[p-(methylthio)-phenyl]azo]-	benzene	467(4.28)	44-2571-66
	$C_6H_{11}Me$	458(4.27)	44-2571-66
$C_{17}H_{15}N_5O$			
Semicarbazide, N-(5,6-diphenyl-pyrazin-2-yl)-	EtOH	226(4.31),283(4.16), 339(4.02)	39-2038-66C

Compound	Solvent	$\lambda_{max}(\log \epsilon)$	Ref.
$C_{17}H_{16}$			
Cyclopenta[a]phenanthrene, 13,14,15,16-tetrahydro-	EtOH	239(4.76),303(3.84), 315(3.91),337(3.63)	39-0955-66C
1H-Cyclopropa[a]naphthalene, 1a,2,3,7b-tetrahydro-1-phenyl-	n.s.g.	237(4.31)	39-1004-66C
Naphthalene, 1,2-dihydro-4-methyl-1-phenyl-	EtOH	260(3.93),265s(3.92), 272s(3.86),285s(3.53), 297s(3.13)	22-0645-66
Naphthalene, 1,2-dihydro-4-methyl-3-phenyl-	EtOH	235s(3.90),279(4.15)	22-0645-66
$C_{17}H_{16}BrCl_2N$			
2-Fluorenamine, 7-bromo-N,N-bis-(2-chloroethyl)-	0.2N HCl	306(4.45)	87-0593-66
	pH 1	306(4.47),342s(--)	87-0593-66
	EtOH	232s(--),307(4.54), 342s(--)	87-0593-66
$C_{17}H_{16}BrNOS$			
1-(Benzoylmethylthio)-3,4-dihydro-isoquinolinium bromide	EtOH	282(4.22)	4-0282-66
$C_{17}H_{16}Br_2ClN$			
2-Fluorenamine, N,N-bis-2-(bromoethyl)-7-chloro-	pH 1	272(4.30),306(4.32)	87-0593-66
	EtOH	240s(--),308(4.51), 343s(--)	87-0593-66
$C_{17}H_{16}Br_2N_2O_2$			
2-Fluorenamine, N,N-bis-(2-bromoethyl)-7-nitro-	2N HCl	246(3.96),319(4.21), 420(3.89)	87-0593-66
	0.2N HCl	250s(--),275(4.14), 410(4.32)	87-0593-66
	pH 1	250s(--),274(4.15), 410(4.34)	87-0593-66
	EtOH	272(4.17),410(4.40)	87-0593-66
$C_{17}H_{16}Br_2N_2O_6$			
Propane, 2,2-bis(bromomethyl)-1,3-bis(o-nitrophenoxy)-	EtOH	213(4.78),263(4.18), 329(3.98)	78-0199-66
$C_{17}H_{16}ClN$			
Benzo[6,7]cyclohept[1,2,3-de]isoquinoline, 1,7,8,12b-tetrahydro-, hydrochloride	EtOH	298(4.00)	4-0247-66
$C_{17}H_{16}ClNO$			
N-(2-Hydroxy-2-phenylethyl)isoquinolinium chloride	MeOH	266(3.63),275(3.63), 335(3.57)	83-0715-66
$C_{17}H_{16}ClNO_2$			
Anonaine, hydrochloride	EtOH	260(4.09),307(3.93)	95-0124-66
$C_{17}H_{16}ClN_3O_4S_2$			
Pyrazole, 5-[(p-chlorophenylsulfonyl)-(p-tolylsulfonyl)amino]-1-methyl-	EtOH	208s(3.9),239(4.4), 267s(3.6),277s(3.5), 354s(2.7)	24-0178-66
$C_{17}H_{16}ClN_3O_6$			
2,3-Dihydro-6-nitro-1,4-diphenyl-1,4-diazepinium perchlorate	MeOH	249(4.13),366(4.12)	39-0093-66C

Compound	Solvent	$\lambda_{max}(\log \epsilon)$	Ref.
$C_{17}H_{16}ClN_5O_2$			
s-Triazine, 2,4-di-p-anisidino-6-chloro-	EtOH	279(4.49)	80-0139-66
$C_{17}H_{16}ClN_5O_5S_2$			
3-(N-Methylbenzothiazol-2-ylidene)-triazen-1-yl-(N-methyl-6-methoxy-benzothiazolidinium) perchlorate	MeCN	498(4.60)	5-0116-66G
$C_{17}H_{16}Cl_2FN$			
2-Fluorenamine, N,N-bis-(2-chloroethyl)-7-fluoro-	0.2N HCl	269(4.20),298(4.26), 304(4.27)	87-0593-66
	pH 1	271s(--),297(4.32), 303(4.31),330s(--)	87-0593-66
	EtOH	298(4.49),328s(--)	87-0593-66
$C_{17}H_{16}Cl_3N$			
2-Fluorenamine, 7-chloro-N,N-bis-(2-chloroethyl)-	pH 1	228s(--),306(4.43)	87-0593-66
	EtOH	307(4.51),340s(--)	87-0593-66
$C_{17}H_{16}CuN_2O_2$			
Copper, [α,α'-(propylenedinitrilo)-di-o-cresolato(2-)]-	MeOH	257(3.93),270(4.35), 360(2.42)	65-1590-66
$C_{17}H_{16}F_3N_3$			
as-Triazine, 1,4,5,6-tetrahydro-1-methyl-5-phenyl-3-(p-tri-fluoromethylphenyl)-	MeOH	317(3.59)	87-0881-66
$C_{17}H_{16}N_2O$			
Indolizine-7-carboxamide, 2,5-dimethyl-N-phenyl-	EtOH	228(4.47),277(4.38), 357(4.30)	56-0061-66
1-Isoindolinecarboxamide, 2,N-di-methyl-3-phenyl-	iso-PrOH	222(4.37),260(4.41), 298(3.70),306(3.70), 356(4.28)	35-3173-66
Isoquino[2,1-d][1,4]benzodiazepin-6(7H)-one, 5,9,10,14b-tetrahydro-	EtOH	240(3.96)	33-1183-66
$C_{17}H_{16}N_2O_2$			
Formamide, N-[2-(9-formyl-1-methyl-carbazol-2-yl)ethyl]-	EtOH	229(4.62),235s(4.58), 249(4.27),260(4.26), 270(4.21),283(4.18), 295s(4.05),312(3.78), 325(3.18),337(3.08)	87-0237-66
Indolizine-7-carboxamide, 2-methyl-N-(p-methoxyphenyl)-	EtOH	225(4.48),274(4.33), 348(4.23)	56-0061-66
2-Pyrazolin-5-one, 4-(2-hydroxyethyl)-1,3-diphenyl-	MeOH	208(4.42),265(4.12)	24-2962-66
Quinazoline, 2-(α-hydroxybenzyl)-4-ethoxy-	EtOH	208s(4.50),223(4.64), 255(3.94),300(3.46), 311(3.46)	39-2290-66C
$C_{17}H_{16}N_2O_3$			
Acetic acid, (1-methyl-2-oxo-1,2-di-hydro-6,7-benzoquinoxalin-3-yl)-, ethyl ester	$CHCl_3$	295(4.13),305(4.06), 319(4.16),388(4.27)	78-3253-66
	DMSO	305(4.18),319(4.26), 386(4.33)	78-3253-66
	CF_3COOH	287(4.31),378s(4.19), 392(4.22)	78-3253-66
Flavanone, 3-acetamido-, oxime	EtOH	256(3.84),306(3.31)	5-0225-66C

Compound	Solvent	λ_{max}(log ε)	Ref.
$C_{17}H_{16}N_2O_4$			
Cinnamic acid, p-nitro-, p-(dimethylamino)phenyl ester	EtOH	260(4.2),300(4.5)	65-1202-66
$C_{17}H_{16}N_4$			
2-Naphthylamine, 1-[[p-(methylamino)-phenyl]azo]-	benzene	467(4.39)	44-2571-66
	$C_6H_{11}Me$	455(4.33)	44-2571-66
Pyrazole, 3,5-dimethyl-1-phenyl-4-(phenylazo)-	$CHCl_3$	334(4.42),428(3.15)	22-2990-66
$C_{17}H_{16}N_4O$			
1-Naphthol, 7-amino-8-[[p-(methyl-amino)phenyl]azo]-	benzene	513(4.46)	44-2571-66
	$C_6H_{11}Me$	507(4.35)	44-2571-66
Pyrazole, 5-methoxy-3-methyl-1-phenyl-4-(phenylazo)-	$CHCl_3$	399(4.32),400(3.72)	22-2990-66
Pyrazole-4,5-dione, 3-methyl-1-phenyl-, 4-(methylphenylhydrazone)	$CHCl_3$	385(4.22)	22-2990-66
5-Pyrazolone, 2,3-dimethyl-1-phenyl-4-(phenylazo)-	$CHCl_3$	355(4.31),420(3.45)	22-2990-66
5-Pyrazolone, 3,4-dimethyl-1-phenyl-4-(phenylazo)-	$CHCl_3$	277(4.11),400(2.30)	22-2990-66
Δ^1-1,2,4-Triazolin-3-one, 5-methyl-4-[(α-methylbenzylidene)amino]-5-phenyl-	EtOH	245(3.88),314(3.91)	88-5815-66
$C_{17}H_{16}N_4O_2$			
as-Triazine, 3-(2-hydroxyethylamino)-5,6-diphenyl-, 2-oxide	EtOH	229(4.35),262(4.45), 374(3.75)	44-3914-66
$C_{17}H_{16}N_4O_4$			
Pyrazoline, 1-(2,4-dinitrophenyl)-3,4-dimethyl-5-phenyl-	$CHCl_3$	395(4.32)	22-0610-66
$C_{17}H_{16}N_4O_6$			
Ketone, o-carbomethoxybenzyl methyl, 2,4-dinitrophenylhydrazone	EtOH	360(4.33)	12-1265-66
Pyrazoline, 3-(3,4-dimethoxyphenyl)-1-(2,4-dinitrophenyl)-	$CHCl_3$	426(4.48)	22-0610-66
$C_{17}H_{16}O$			
1H-Cyclopenta[5,6]naphtho[2,3-b]furan, 2,3-dihydro-6,9-dimethyl-	EtOH	243(4.88),250(4.90), 305(4.00),318(4.14), 335(3.96),351(4.01)	39-0734-66C
1H-Cyclopenta[7,8]naphtho[2,3-b]furan, 2,3-dihydro-7,10-dimethyl-	EtOH	250(4.75),292s(3.52), 309(3.73),320(3.87), 338(3.80)	39-0734-66C
Cyclopenta[a]phenanthrene, 11,12,13,-14,15,16-hexahydro-17-oxo-	EtOH	228(5.92),280(3.82)	39-0955-66C
2-Cyclopenten-1-ol, 3,4-diphenyl-	EtOH	253(4.11)	22-3774-66
3-Cyclopenten-1-ol, 3,4-diphenyl-	EtOH	228(4.23),273(4.01)	22-3774-66
1-Indanone, 2-(p-methylbenzyl)-	MeOH	246(4.13),276s(3.34), 291(3.47)	35-4489-66
1-Indanone, 4-methyl-3-o-tolyl-	EtOH	251(4.09),298(3.43)	4-0247-66
[2.3]Paracyclophane, 1-oxo-	EtOH	229(4.00),260s(3.66), 295s(2.45),330s(2.60)	35-3515-66
[2.3]Paracyclophane, 2-oxo-	EtOH	222(4.19),275(2.58), 307(2.29)	35-3515-66
$C_{17}H_{16}O_2$			
Acrylic acid, 3,3-di-o-tolyl-	EtOH	270(3.99)	4-0247-66

Compound	Solvent	λ_{max}(log ε)	Ref.
Chalcone, 4'-hydroxy-3',5'-dimethyl-	EtOH	220s(4.10),246s(3.84), 328(4.37)	49-0896-66
1H-Cyclopenta[7,8]naphtho[2,3-b]furan, 2,3-dihydro-7-(hydroxymethyl)-6-methyl-	EtOH	245(4.83),252(4.96), 309(3.84),322(3.99), 339(3.94)	39-0725-66C
2-Cyclopentene-1,4-diol, 1,2-diphenyl-	EtOH	250(4.13)	22-3774-66
5H-Dibenzo[a,d]cycloheptene-5-acetic acid, 10,11-dihydro-	EtOH	266(2.79)	4-0247-66
14β-Gona-1,3,5(10),8-tetraene-11,17-dione	dioxan	236(4.06),292(3.87)	19-0611-66
1-Indanone, 2-(m-methoxybenzyl)-	EtOH	248(4.13),283(3.58)	88-6489-66
1-Indanone, 2-(p-methoxybenzyl)-	MeOH	225(4.18),245(4.17), 285(3.62)	35-4489-66
$C_{17}H_{16}O_2S$			
Methylphenylsulfonium α-acetylphenacylide	EtOH	279(4.06)	78-2145-66
$C_{17}H_{16}O_3$			
3(2H)-Benzofuranone, 2-(o-hydroxyphenyl)-2-propyl-	EtOH	251(4.00),279(3.45), 331(3.65)	39-0544-66C
	EtOH-KOH	245(4.15),307(3.87)	39-0544-66C
Chalcone, 2',4-dihydroxy-3,5-dimethyl-	pH 13	233(4.18),262(4.05), 437(4.41)	49-0896-66
	EtOH	255(4.09),384(4.40), 500(2.82)	49-0896-66
Chalcone, 4,4'-dihydroxy-3,5-dimethyl-	EtOH	242(4.04),360(4.46)	49-0896-66
Chalcone, 4,4'-dihydroxy-3',5'-dimethyl-	EtOH	240(4.15),348(4.49)	49-0896-66
4-Pentenophenone, 2'-hydroxy-2-(o-hydroxyphenyl)-	EtOH	256(4.03),285s(3.44), 329(3.67)	39-0544-66C
	EtOH-KOH	265s(3.80),367(3.61)	39-0544-66C
1-Phenanthrenecarboxylic acid, 1,2,3,4-tetrahydro-1-methyl-4-oxo-, methyl ester	EtOH	250(4.4),303(3.8), 317(3.89)	88-4997-66
$C_{17}H_{16}O_4$			
Acrylic acid, 2-(6-hydroxy-m-tolyl)-3-(p-methoxyphenyl)-, cis	EtOH	292(4.40)	78-1789-66
trans	EtOH	280(4.41)	78-1789-66
Alloimperatorin methyl ether	EtOH	220(4.31),244s(4.10), 250(4.13),266(4.08), 308(3.92)	102-0367-66
Macluran, 1,3-dimethoxycyano-	EtOH	275(2.67)	78-0139-66A
Macluran, 3,9-dimethoxycyano-	EtOH	282(3.03)	78-0139-66A
Mandelic acid, 2-hydroxy-α-(p-methoxybenzyl)-5-methyl-, γ-lactone	EtOH	276s(3.71),281(3.75)	78-1789-66
Maturin, cyclic methyl acetal	EtOH	218(4.21),250(4.81), 335(4.00),349(3.94)	78-0685-66
$C_{17}H_{16}O_4S_2$			
Spiro[5H-dibenzo[a,d]cycloheptene-5,2'-[1,3]dithiolane], 10,11-dihydro-, 1',1',3',3'-tetraoxide	EtOH-DMF	268(3.18)	87-0860-66
$C_{17}H_{16}O_5$			
Alloimperatorin methyl ether oxide	EtOH	220(4.43),244s(4.27), 251(4.32),266(4.24), 308(4.11)	78-2923-66

Compound	Solvent	$\lambda_{max}(\log \epsilon)$	Ref.
Benzoic acid, p-hydroxy-, ester with (2-hydroxy-3-methoxyphenyl)-2-propanone	EtOH	262(4.28)	78-2913-66
	EtOH-NaOEt	309(4.52)	78-2913-66
Benzoic acid, p-hydroxy-, ester with 2-hydroxy-3-methoxy-α-methylstyrene	EtOH	257(4.38),305s(--)	78-2913-66
	EtOH-NaOEt	309(4.52)	78-2913-66
$C_{17}H_{16}O_5S$			
Hydroquinone, 2-acetyl-3-(3-methyl-2-thienyl)-, diacetate	EtOH	238s(3.93),290s(3.32)	33-1794-66
Hydroquinone, 2-acetyl-3-(5-methyl-2-thienyl)-, diacetate	EtOH	228s(4.07),278(3.90)	33-1794-66
$C_{17}H_{16}O_6$			
Acenaphthenequinone, 3,4,5,6-tetramethoxy-8-methyl-	n.s.g.	222(4.39),255(5.13), 278(4.13),335(3.81)	44-1883-66
Flavanone, 3',5'-dihydroxy-4',7-dimethoxy-	EtOH	286(4.28)	88-1293-66
	EtOH-KOH	286(3.70)	88-1293-66
	EtOH-NaOAc	286(4.28)	88-1293-66
	EtOH-$AlCl_3$	306(4.56)	88-1293-66
Hydroquinone, 2-acetyl-3-(5-methyl-2-furyl)-, diacetate	EtOH	205(4.22),282(4.08), 309(3.91)	33-1794-66
$C_{17}H_{16}S_2$			
Spiro[5H-dibenzo[a,d]cycloheptene-5,2'-[1,3]dithiolane, 10,11-dihydro-	EtOH	265(3.16)	87-0860-66
$C_{17}H_{17}^+$			
Cyclobutyldiphenylcarbonium ion	FSO_3H-SbF_5	335(4.36),388(3.83)	35-1488-66
$C_{17}H_{17}ClN_2O_5$			
2,3-Dihydro-6-hydroxy-1,4-diphenyl-1,4-diazepinium perchlorate	MeOH	243(3.98),419(4.47)	39-0093-66C
$C_{17}H_{17}ClO_5$			
Spiro[cyclohexane-1,2'-indan]-1',2,4-trione, 4'-chloro-5',7'-dimethoxy-6-methyl-	MeOH	235(4.45),273(4.40), 315(3.96)	44-1462-66
$C_{17}H_{17}Cl_2HgNO_2$			
Isoquinoline, 3,4-dihydro-6,7-dimethoxy-1-phenyl-, $HgCl_2$ complex	MeCN	237(4.43),314(3.95), 373(3.87)	94-0842-66
$C_{17}H_{17}IN_2O_8S$			
Uracil, 1-(3-deoxy-3-iodo-β-D-arabinosyl)-, 5'-benzoate 2'-methanesulfonate	EtOH	232(4.16),260(4.00)	44-0205-66
$C_{17}H_{17}N$			
Benzo[6,7]cyclohept[1,2,3-de]isoquinoline, 1,2,3,7,8,12b-hexahydro-, hydrochloride	EtOH	266(2.89),273(2.84)	4-0247-66
Indole, N-benzyl-2,3-dimethyl-	n.s.g.	225(4.91),285(3.79), 291(3.76)	16-0064-66
Indole, N-benzyl-3-ethyl-	n.s.g.	225(4.23),288(3.66)	16-0064-66
Isoquinoline, 1-benzylidene-1,2,3,4-tetrahydro-2-methyl-	n.s.g.	346(3.50)	88-2941-66
$C_{17}H_{17}NO$			
Formamide, N-[(10,11-dihydro-5H-dibenzo[a,d]cyclohepten-5-yl)methyl]-	EtOH	264(2.77)	4-0247-66

Compound	Solvent	λ_{max}(log ε)	Ref.
6H-Naphtho[2,3-c]quinolizin-6-one, 1,2,3,4,4a,5-hexahydro-	EtOH-HCl	241(4.45),252(4.43), 272(4.59),437(3.88)	94-1399-66
	EtOH	241(4.44),272(4.69), 437(3.97)	94-1399-66
12H-Naphtho[1,2-c]quinolizin-12-one, 1,2,3,4,13,13a-hexahydro-	EtOH-HCl	218(4.44),262(4.51), 322(3.72),407(3.73)	94-1399-66
	EtOH	218(4.43),262(4.52), 322(3.70),407(3.72)	94-1399-66
5H-Oxazolo[2,3-a]isoquinoline, 2,3,6,10b-tetrahydro-2-phenyl-	C_6H_{12}	263(2.79),269(2.72)	83-0715-66
	MeOH	264(2.92),271(2.92), 289(2.77)	83-0715-66
$C_{17}H_{17}NOS$			
Phenothiazine, 10-(tetrahydro-2-pyranyl)-	MeCN	251(4.49),298(3.51)	44-1982-66
$C_{17}H_{17}NO_2$			
Indole, 1-benzoyl-, dimethyl acetal	EtOH	218(4.57),268(3.89), 290(3.61)	88-4701-66
Isocarbostyril, 3,4-dihydro-2-(β-hydroxyphenethyl)-	MeOH	227(3.94),250(3.83)	83-0715-66
Pyridine, 4-[(5,6-dimethoxy-1-indanylidene)methyl]-	EtOH	241(4.06),295(4.03), 305(4.09),350(4.49)	44-3337-66
$C_{17}H_{17}NO_2S$			
L-Tyrosine, N-thiobenzoyl-, methyl ester	MeOH	280(3.86),384(2.29)	1-2781-66
	dioxan	287(3.58),380(1.76)	1-2781-66
$C_{17}H_{17}NO_3$			
Fluorene-8a(6H)-carbonitrile, 7,8-dihydro-2,4-dimethoxy-8-methyl-6-oxo-	MeOH	222(3.88),227(3.88), 246(4.06),254(4.06), 339(4.43)	44-1456-66
$C_{17}H_{17}NO_3S$			
Crotonanilide, N-(p-tolylsulfonyl)-	EtOH	213(4.48),227(4.53), 266s(3.56),274s(3.40)	54-0671-66
$C_{17}H_{17}NO_5$			
Normelicopicine	EtOH	249(4.36),271(4.20), 273(4.61),308(4.02), 422(3.80)	100-0206-66
Styrene, 3-(benzyloxy)-4,5-dimethoxy-β-nitro-, trans	iso-PrOH	239(4.03),349(4.20)	33-0403-66
$C_{17}H_{17}NO_6$			
Diphenylamine-2-carboxylic acid, 4'-carbomethoxy-3',5'-dimethoxy-	EtOH	300(4.25),339(4.02)	12-0275-66
$C_{17}H_{17}N_3$			
Cyanazomethine, 3-methyl-4'-(dimethylamino)-α-diphenyl-	EtOH	459(4.23)	20-0426-66
$C_{17}H_{17}N_3OS$			
Acetamide, N-isopropyl-N-[5-(1-naphthyl)-1,3,4-thiadiazol-2-yl]-	EtOH	224(4.47),307(3.94)	83-0921-66
$C_{17}H_{17}N_3O_4S_2$			
p-Toluenesulfonamide, N-(1-methylpyrazol-5-yl)-N-(phenylsulfonyl)-	EtOH	207s(4.0),234(4.4), 265s(3.5),276s(3.4), 340s(2.3)	24-0178-66

Compound	Solvent	$\lambda_{max}(\log \epsilon)$	Ref.
p-Toluenesulfonamide, N-[1-(p-tolyl-sulfonyl)pyrazol-5-yl]-	MeOH	213s(4.0),228(4.3), 251(4.2),275s(3.6)	24-0183-66
$C_{17}H_{17}N_5O$			
Adenine, N-benzoyl-9-(3-methyl-2-butenyl)-	pH 1	288(4.41)	94-0087-66
	pH 7	280(4.31)	94-0087-66
	pH 13	300(4.11)	94-0087-66
Triacanthine, N'-benzoyl-, hydrobromide	pH 1	299.5(4.45)	94-0087-66
	pH 7	299(4.18)	94-0087-66
	pH 13	329.5(4.23)	94-0087-66
$C_{17}H_{17}N_5O_6$			
Benzoic acid, p-[N-[3-(2-formamido-4-hydroxy-5-pyrimidinyl)formamido]-propyl]formamido]-	pH 1	262(4.22)	87-0590-66
	pH 13	238(4.25),275s(4.03)	87-0590-66
$C_{17}H_{18}$			
2-Butene, 2-phenyl-1-o-tolyl-	EtOH	244(4.03)	78-1001-66
Cyclopenta[a]phenanthrene, 11,12,13,14,15,16-hexahydro-	EtOH	230(4.28),275(3.63), 285(3.76),292(3.54)	39-0955-66C
$C_{17}H_{18}BN$			
Borane, (N-methylanilino)bis-3-methyl-3-buten-1-ynyl-	C_6H_{12}	230(4.34),270(4.23)	22-3850-66
	C_6H_{12}	230(4.34),270(4.23)	28C-0376-66A
$C_{17}H_{18}BrNO_2$			
Ethanol, 2,2'-[(7-bromofluoren-2-yl)imino]di-	pH 1	266(4.43),272(4.43), 294(4.09),299(4.06), 306(4.09)	87-0593-66
	EtOH	233s(--),315(4.47), 345s(--)	87-0593-66
$C_{17}H_{18}ClN$			
Benzo[6,7]cyclohept[1,2,3-de]isoquinoline, 1,2,3,7,8,12b-hexahydro-, hydrochloride	EtOH	266(2.89),273(2.84)	4-0247-66
$C_{17}H_{18}ClNO_2$			
Ethanol, 2,2'-[(7-chlorofluoren-2-yl)imino]di-	pH 1	272(4.45),294(4.07), 306(4.16)	87-0593-66
	EtOH	232s(--),313(4.48), 345s(--)	87-0593-66
$C_{17}H_{18}ClNO_5$			
3,4-Dihydro-N-(2-hydroxy-2-phenyl-ethyl)isoquinolinium perchlorate	MeOH	289.5(3.98)	83-0715-66
$C_{17}H_{18}ClN_3$			
as-Triazine, 3-(p-chlorophenyl)-1,4,5,6-tetrahydro-1,6-di-methyl-5-phenyl-	MeOH	230(4.24),304(3.67)	87-0881-66
$C_{17}H_{18}ClN_3O_7$			
1H-Imidazo[4,5-c]pyridine, 4-chloro-1-β-D-ribofuranosyl-, triacetate	pH 1	255(3.72),265(3.79), 272(3.80)	69-0756-66
	pH 11	254(3.83),265(3.79), 272(3.62)	69-0756-66

Compound	Solvent	λ_{max}(log ε)	Ref.
$C_{17}H_{18}Cl_2N_4O_7$			
Purine, 2,6-dichloro-9-(5-deoxy-β-D-ribohexofuranosyl)-, triacetate	pH 1, 7	252(3.86),273(4.12), 280s(--)	87-0234-66
	pH 13	255s(--),258(4.18), 265s(--),280s(--)	87-0234-66
$C_{17}H_{18}Cl_3HgNO_2$			
Isoquinoline, 3,4-dihydro-6,7-dimethoxy-1-phenyl-, hydrochloride, mercuric chloride adduct	MeCN	240(4.47),283(4.08), 313(4.0),371(3.94)	94-0842-66
$C_{17}H_{18}FNO_2$			
Ethanol, 2,2'-[(7-fluorofluoren-2-yl)imino]di-	pH 1	264(4.38),292(3.92), 304(4.01)	87-0593-66
	EtOH	303(4.50),335s(--)	87-0593-66
$C_{17}H_{18}FN_3$			
as-Triazine, 3-(p-fluorophenyl)-1,4,5,6-tetrahydro-1,6-dimethyl-5-phenyl-	MeOH	287(3.63)	87-0881-66
$C_{17}H_{18}FN_3O_5$			
Cytidine, N^4-p-toluoyl-5-fluoro-2'-deoxy-	EtOH	265(4.19),330(4.27)	87-0566-66
$C_{17}H_{18}N_2$			
Tryptamine, 2-benzyl-	EtOH	224(4.62),278s(3.97), 282(4.00),291(3.94)	39-0425-66C
$C_{17}H_{18}N_2O$			
p-Cresol, α-[(2-indol-3-ylethyl)-amino]-	50% dioxan	278(3.65)	71-0550-66A
7H-1-Pyrindine, 6-benzamido-7,7-dimethyl-	MeOH	225(4.38)	78-0035-66
7H-2-Pyrindine, 6-benzamido-7,7-dimethyl-	MeOH	222(4.35)	78-0035-66
$C_{17}H_{18}N_2OS$			
Thioxanthone, 3,6-bis(dimethylamino)-	EtOH	268(4.66),384(4.61)	4-0228-66
$C_{17}H_{18}N_2OSe$			
Selenoxanthone, 3,6-bis(dimethylamino)-	EtOH	260(4.48),277(4.53), 388(4.62)	4-0228-66
$C_{17}H_{18}N_2O_2$			
Indolo[2,3-a]quinolizin-2(1H)-one, 1-acetyl-3,4,6,7,12,12b-hexahydro-	MeOH	222(4.54),281(4.15), 289(4.15)	24-3750-66
Indolo[2,3-a]quinolizin-2(1H)-one, 12-acetyl-3,4,6,7,12,12b-hexahydro-	MeOH	240(4.22),263(4.08), 289(3.74),298(3.72)	24-3750-66
4(1H)-Pyridone, 1-(2-indol-3-ylethyl)-5-acetyl-2,3-dihydro-	MeOH	219(4.53),265(4.22), 290(4.46),316(4.22)	24-3750-66
$C_{17}H_{18}N_2O_3$			
2-Benzimidazolecarboxylic acid, 4,5,6,7-tetrahydro-6,6-dimethyl-4-oxo-, benzyl ester	EtOH	244(3.92),278(4.25)	25-0731-66
$C_{17}H_{18}N_2O_5$			
Butyric acid, 4-(4-hydroxy-3-methoxyphenyl)-4-(6-methyl-2-pyridylamino)-2-oxo-	H_2O	230(4.03),290(3.77), 350(3.59)	50-0504-66D

Compound	Solvent	$\lambda_{max}(\log \epsilon)$	Ref.
$C_{17}H_{18}N_2O_5S$			
5-Thia-1-azabicyclo[4.2.0]oct-3-ene-2-carboxylic acid, 3-(hydroxymethyl)-8-oxo-7-(2-phenylacetamido)-, methyl ester	n.s.g.	254(3.90)	39-1142-66C
5-Thia-1-azabicyclo[4.2.0]oct-2-en-8-one, 3-(hydroxymethyl)-7-(2-phenylacetamido)-, acetate, 5-oxide	n.s.g.	252(4.03)	39-1142-66C
$C_{17}H_{18}N_2O_8$			
4aH-Pyrido[1,2-b]pyridazine-5,6,7,8-tetracarboxylic acid, 4a-methyl-, tetramethyl ester	MeOH	267(4.13),399(3.86)	39-2218-66C
$C_{17}H_{18}N_4O_2$			
Lumazine, 8-benzyl-3-ethyl-6,7-dimethyl-	pH -2.7	250(4.11),365(4.17)	24-3503-66
	pH 6	264(4.33),280s(3.90), 412(4.09)	24-3503-66
	pH 12	242(4.30),312(4.35), 365(3.80)	24-3503-66
as-Triazine, 1,4,5,6-tetrahydro-1,6-dimethyl-3-(p-nitrophenyl)-5-phenyl-	MeOH	265(4.13),395(3.62)	87-0881-66
$C_{17}H_{18}N_4O_7$			
Butanoic acid, 3-(2-furoyl)-2-methyl-, methyl ester, 2,4-dinitrophenylhydrazone	EtOH	262(4.24),380(4.43)	12-0683-66
$C_{17}H_{18}O$			
1H-Cyclopenta[5,6]naphtho[2,3-b]furan, 2,3,8,9-tetrahydro-6,9-dimethyl-	EtOH	237(4.79),272(3.75), 283(3.83),295(3.72), 333(3.54),347(3.64)	39-0734-66C
1H-Cyclopenta[7,8]naphtho[2,3-b]furan, 2,3,7,8-tetrahydro-6,7-dimethyl-	EtOH	242(4.82),248(4.79), 290(3.82),302(3.77), 320(3.37),336(3.44)	39-0725-66C
1H-Cyclopenta[7,8]naphtho[2,3-b]furan, 2,3,7,8-tetrahydro-7,10-dimethyl-	EtOH	245(4.78),290(3.78), 302(3.68),323(3.19), 338(3.28)	39-0734-66C
Tricyclo[8.2.2.2^{4,7}]hexadeca-4,6,10,-12,13,15-hexaene, 5-methoxy-	EtOH	225(4.27),248s(3.54), 281(2.83),290(2.89), 309(2.80)	44-1227-66
$C_{17}H_{18}O_2$			
Cyclohexanone, 2-(2-methoxy-1-naphthyl)-	EtOH	231(4.85)	44-2646-66
1H-Cyclopenta[7,8]naphtho[2,3-b]furan-7-methanol, 2,3,7,8-tetrahydro-6-methyl-	EtOH	242(4.87),248(4.88), 289(3.87),302(3.83), 321(3.38),336(3.49)	39-0725-66C
Propionic acid, 3,3-di-o-tolyl-	EtOH	263(2.84)	4-0247-66
$C_{17}H_{18}O_3$			
Bibenzyl, 4-acetoxy-4'-methoxy-	50% dioxan	272(3.27)	71-0550-66A
1H-Cyclopenta[7,8]naphtho[2,3-b]furan-7-methanol, 2,3,7,8-tetrahydro-10-hydroxy-6-methyl-	EtOH	247(4.73),312(3.82)	39-0725-66C
Ether, 3-methoxyphenyl 2-methoxycinnamyl	EtOH	255(4.23),285(3.90), 300(3.82)	78-3491-66
Styrene, 4-(3,4,5-trimethoxyphenyl)-	EtOH	224(4.43),254(4.36)	44-0189-66
Valerophenone, 2'-hydroxy-2-(o-hydroxyphenyl)-	EtOH	255(4.03),285s(3.42), 329(3.68)	39-0544-66C

Compound	Solvent	λ max(log ε)	Ref.
Valerophenone, 2'-hydroxy-2-(o-hydroxyphenyl)- (cont.)	EtOH-KOH	264s(3.81),364(3.64)	39-0544-66C
$C_{17}H_{18}O_4$			
Agatharesinol	n.s.g.	266(4.42)	88-2395-66
o-Benzoquinone, 4-(α-ethyl-o-methoxybenzyl)-5-methoxy-	EtOH	274(4.01)	88-3767-66
p-Benzoquinone, 2-(α-ethyl-o-methoxybenzyl)-5-methoxy-	EtOH	268(4.32)	88-3767-66
Cyclohexa[f]benzofuran, 9-acetoxy-3,4,8-trimethyl-5-oxo-	EtOH	212(3.72),254(3.99),316(3.87)	78-0301-66
4-Stilbenol, 3,3',4'-trimethoxy-	EtOH	332(4.51)	24-2638-66
$C_{17}H_{18}O_5$			
Benzophenone, 2,2',4,4'-tetramethoxy-, cation	H_2SO_4	428(4.63)	39-0650-66B
p-Benzoquinone, 2-carboxy-3-(3,4,5-trimethyl-2-furyl)-, isopropyl ester	EtOH	244(4.11),265s(4.05),330s(2.92),513(3.75)	33-1794-66
Spiro[cyclohexane-1,2'-indan]-1',2,4-trione, 5',7'-dimethoxy-6-methyl-	MeOH	228(4.33),273(4.47),303s(4.03)	44-1456-66
$C_{17}H_{18}O_6$			
Mycochromenic acid	n.s.g.	246(4.31),280(3.51),322(3.54),333(3.48)	88-5107-66
Psoralen, 8-methoxy-5-(3-methoxy-2,3-dihydroxybutyl)-	EtOH	220(4.38),244s(4.19),251(4.23),266(4.19),309(4.06)	78-2923-66
$C_{17}H_{18}O_8$			
Ascorbic acid, 6-benzoate 2-butyrate	neutral	232(4.4)	95-0376-66
anion	n.s.g.	253(4.2)	95-0376-66
$C_{17}H_{19}ClN_2S$			
Chlorpromazine, hydrochloride	EtOH	256(4.54),310(3.60)	36-0144-66
$C_{17}H_{19}Cl_3N_2SZn$			
3,6-Bis(dimethylamino)thioxanthylium trichlorozincate	EtOH	284(4.88),564(5.12)	4-0228-66
$C_{17}H_{19}Cl_3N_2SeZn$			
3,6-Bis(dimethylamino)selenoxanthylium trichlorozincate	EtOH	285(4.72),570(5.05)	4-0228-66
$C_{17}H_{19}N$			
2-Aza-3,4-benzo-5,6-dipropylpentatriafulvalene, HI salt	MeOH	222(4.34),245(4.12),265(3.98),319(4.27)	35-3359-66
Isoquinoline, 1,2,3,4-tetrahydro-5-methyl-4-o-tolyl-, hydrochloride	EtOH	264(2.72)	4-0247-66
$C_{17}H_{19}NO$			
Benzanilide, N,2,4,6-tetramethyl-	EtOH	246s(3.8)	28C-0369-66A
m-Benzotoluidide, 2,4,6-trimethyl-	EtOH	252(4.2),272s(3.9),285s(3.4)	28C-0369-66A
o-Benzotoluidide, 2,4,6-trimethyl-	EtOH	250s(3.8)	28C-0369-66A
p-Benzotoluidide, 2,4,6-trimethyl-	EtOH	252(4.2),270s(3.9),285s(3.3)	28C-0369-66A
13H-Dibenzo[a,f]quinolizin-13-one, 1,2,3,4,6,7,11b,12-octahydro-	EtOH	333(4.16)	44-0797-66
2(1H)-Isoquinolineethanol, 3,4-dihydro-α-phenyl-	MeOH	250(2.61),256(2.71),263(2.73),272(2.67)	83-0715-66

Compound	Solvent	λ_{max}(log ε)	Ref.
12H-Naphtho[1,2-c]quinolizin-12-one, 1,2,3,4,6,7,13,13a-octahydro-	EtOH-HCl	260(4.34),362(3.81)	94-1399-66
	EtOH	277(4.16),360(4.03)	94-1399-66
$C_{17}H_{19}NO_2$			
m-Benzanisidide, 2,4,6-trimethyl-	EtOH	247(4.2),279(3.9), 286(3.7)	28C-0369-66A
o-Benzanisidide, 2,4,6-trimethyl-	EtOH	246(4.2),281(3.9), 291s(3.8)	28C-0369-66A
p-Benzanisidide, 2,4,6-trimethyl-	EtOH	253(4.2)	28C-0369-66A
Pyridine, 4-[(5,6-dimethoxy-1-indanyl)methyl]-	EtOH	230s(3.90),264(3.44), 288(3.74),290s(3.73)	44-3337-66
$C_{17}H_{19}NO_3$			
Coclaurine	EtOH	227(3.57),286(3.78)	95-0296-66
$C_{17}H_{19}NO_4$			
Benzanilide, 3,4,5-trimethoxy-N-methyl-	EtOH	216(4.3),251(3.8), 272(3.8)	28C-0584-66A
m-Benzotoluidide, 3,4,5-trimethoxy-	EtOH	213(4.6),278(4.3)	28C-0584-66A
o-Benzotoluidide, 3,4,5-trimethoxy-	EtOH	213(4.6),265(4.1)	28C-0584-66A
p-Benzotoluidide, 3,4,5-trimethoxy-	EtOH	214(4.5),278(4.3)	28C-0584-66A
2,4-Cyclohexanedione, 1-cyano-1-(3,5-dimethoxybenzyl)-6-methyl-	MeOH	225(4.00),263(4.18)	44-1456-66
Isoindoline, 1,3-bis(carbethoxy-methylene)-N-methyl-	$CHCl_3$	241(4.37),285(3.86), 297(4.02),375(4.54), 392(4.55)	24-3444-66
$C_{17}H_{19}NO_5$			
m-Anisidide, 3,4,5-trimethoxy-	EtOH	216(4.6),278(4.2), 296(4.1)	28C-0584-66A
o-Anisidide, 3,4,5-trimethoxy-	EtOH	213(4.6),278(4.1), 292(4.1)	28C-0584-66A
p-Anisidide, 3,4,5-trimethoxy-	EtOH	212(4.6),283(4.3)	28C-0584-66A
1H-1-Benzazepine-3,4-dicarboxylic acid, 2-ethoxy-1-methyl-, dimethyl ester	MeOH	250(4.3),275(4.1), 310(3.8)	24-3070-66
$C_{17}H_{19}N_3$			
Benzimidazole, 2-(2-dimethylamino-ethyl)-1-phenyl-	EtOH	248(3.79),267(3.49), 276(3.52),283(3.50)	4-0278-66
dihydrochloride	EtOH	246(4.10),275(3.79), 282(3.76)	4-0278-66
$C_{17}H_{19}N_3O$			
4H-Pyrazolo[4,3-c]pyridin-4-one, 3-butyl-1,5-dihydro-6-methyl-1-phenyl-	EtOH	223(4.40),239(4.40), 284(4.09)	70-1723-66
4H-Pyrazolo[4,3-c]pyridin-4-one, 1,4-dihydro-3-isobutyl-6-methyl-1-phenyl-	EtOH	223(4.35),239(4.37), 285(4.06)	70-1723-66
as-Triazine, 3-p-anisyl-1,4,5,6-tetra-hydro-1-methyl-5-phenyl-	MeOH	238(4.17),280(3.83)	87-0881-66
$C_{17}H_{19}N_3O_5$			
Morpholine, 2-(benzyloxy)-3,5-dihydroxy-4-isonicotinamido-, hydrate	H_2O	262(3.55)	39-2121-66

Compound	Solvent	$\lambda_{max}(\log \epsilon)$	Ref.
$C_{17}H_{19}N_5O_3S$			
Adenine, 9-(2-S-benzyl-2-thio-β-D-arabinofuranosyl)-	pH 1	258(4.10)	44-3263-66
	pH 7	261(4.11)	44-3263-66
	pH 13	261(4.12)	44-3263-66
Adenine, 9-(3-S-benzyl-3-thio-β-D-arabinofuranosyl)-	pH 1	258(4.19)	44-3263-66
	pH 7	260(4.20)	44-3263-66
	pH 13	260(4.21)	44-3263-66
$C_{17}H_{19}N_5O_4$			
Adenine, 7-α-D-arabinofuranosyl-3-benzyl-	pH 1	225s(4.19),276(4.17)	44-1413-66
	pH 13	227s(4.07),277(4.04)	44-1413-66
$C_{17}H_{19}N_5O_4S$			
Adenosine, 2',5'-dideoxy-, 3'-p-toluenesulfonate	pH 11	230(4.14),259(4.22)	35-1549-66
	MeOH	227(4.15),259(4.19)	35-1549-66
$C_{17}H_{19}N_5O_5S$			
Adenosine, 2'-deoxy-, 3'-p-toluenesulfonate	MeOH	228(4.13),259(4.20)	69-0224-66
Adenosine, 2'-deoxy-, 5'-p-toluenesulfonate	MeOH	260(4.16)	69-0224-66
$C_{17}H_{20}ClN$			
Isoquinoline, 1,2,3,4-tetrahydro-5-methyl-4-o-tolyl-, hydrochloride	EtOH	264(2.72)	4-0247-66
$C_{17}H_{20}Cl_4$			
Indene, 4,5,6,7-tetrachloro-1,1,2,3-tetraethyl-	EtOH	224(4.22),231(4.24),237(4.25),246(4.22),284(4.17)	44-4260-66
Spiro[4.4]nona-1,3,6,8-tetraene, 1,2,3,4-tetrachloro-6,7,8,9-tetraethyl-	EtOH	276(3.88),325(2.70)	44-4260-66
$C_{17}H_{20}INO_4$			
O-Methylluninium iodide	EtOH	218(4.65),260(4.67),336(3.80)	12-2185-66
$C_{17}H_{20}IN_5$			
Tetraethylammonium salt of 5-iodocyclopentadiene-1,2,3,4-tetracarbonitrile	MeCN	243(4.63),250(4.61),290(3.99),300(4.00)	35-4055-66
$C_{17}H_{20}N_2$			
Quinoline, 3-(N-benzylaminomethyl)-1,2,3,4-tetrahydro-	pH 1	257(2.75),262(2.75),268(2.63)	94-0566-66
	EtOH	252(3.92),304(3.32)	94-0566-66
$C_{17}H_{20}N_2O_2$			
4-Piperidone, 3-acetyl-1-(2-indol-3-ylethyl)-	MeOH	220(4.62),282(4.11),289(4.14)	24-3750-66
$C_{17}H_{20}N_2O_6S$			
2-Hydrazinoethyl benzhydryl sulfone, oxalate	EtOH	220(4.25),258(2.90)	87-0860-66
$C_{17}H_{20}N_2Se$			
Selenoxanthene, 3,6-bis(dimethylamino)-	EtOH	252(4.50),310(3.59)	4-0228-66

Compound	Solvent	$\lambda_{max}(\log \epsilon)$	Ref.
$C_{17}H_{20}N_4$			
Glutaraldehyde, bis(phenylhydrazone)	EtOH	286(4.24)	39-1989-66C
$C_{17}H_{20}N_4O_4$			
2-Butenal, 2-methyl-4-cyclohexylidene-, 2,4-dinitrophenylhydrazone	$CHCl_3$	410(4.51)	22-1400-66
Cyclohexanone, 5-methyl-2-(1-methylpropadienyl)-, 2,4-dinitrophenylhydrazone	EtOH	364(4.35)	28C-1601-66A
Cyclopentanone, 2-(1-methyl-4-pentenylidene)-, 2,4-dinitrophenylhydrazone, cis	$CHCl_3$	391(4.37)	22-0281-66
trans	$CHCl_3$	393(4.38)	22-0281-66
Indone, 3a,4,5,6,7,7a-hexahydro-3,3a-dimethyl-, 2,4-dinitrophenylhydrazone	EtOH	390(4.41)	44-3063-66
2-Pentenal, 4-cyclohexylidene-, 2,4-dinitrophenylhydrazone	$CHCl_3$	390(4.42)	22-1400-66
Pyrazole, 5-cyclopentyl-3-ethyl-4-methyl-1-(2,4-dinitrophenyl)-	EtOH	325(3.76)	22-2977-66
$C_{17}H_{20}N_4O_6$			
Riboflavine	EtOH	223(4.36),267(4.40), 348(3.86),445(3.92)	30-1101-66F
$C_{17}H_{20}N_6O_2$			
Tetraethylammonium salt of 5-nitrocyclopentadiene-1,2,3,4-tetracarbonitrile	MeCN	211(4.40),265(4.46), 273(4.49),349(3.56)	35-4055-66
$C_{17}H_{20}N_8$			
Tetraethylammonium salt of 5-azidocyclopentadiene-1,2,3,4-tetracarbonitrile	MeCN	257(4.60),270(4.54), 300(3.98)	35-4055-66
$C_{17}H_{20}O$			
Benzhydrol, α-butyl-	hexane	253(2.83),259(2.86), 265(2.76)	22-0269-66
	C_6H_{12}	253(2.85),259(2.87), 265(2.76)	22-0269-66
1,9,15-Hexadecatriene-11,13-diyn-8-one, cis-cis	hexane	233(4.27),249(4.29), 260(4.26),307(4.10), 315(4.13),336(4.05)	24-0590-66
Spiro[benzocyclooctene-6(5H),1'-[3]-cyclohexen]-5-one, 7,8,9,10-tetrahydro-	EtOH	260(3.33)	22-1693-66
$C_{17}H_{20}O_2$			
Indene, 2,5-diacetyl-7-isopropyl-3-methyl-	EtOH	261(4.29),302(4.22)	78-2845-66
3(2H)-Phenanthrone, 2-ethyl-1,9,10,10a-tetrahydro-7-methoxy-	EtOH	242(4.06),328(4.47)	87-0304-66
$C_{17}H_{20}O_3$			
Cacalol, 2-acetyl-	EtOH	252(4.29),316(4.28)	78-0301-66
1H-Cyclopenta[7,8]naphtho[2,3-b]furan-7-carboxylic acid, 2,3,6,6a,7,8,9a,-10-octahydro-6-methyl-	EtOH	293(3.60)	39-0725-66C
2,4,6,10-Dodecatetraen-8-ynal, 12-acetoxy-2,6,10-trimethyl-	pet ether	352(4.80),371(4.69)	33-0369-66

Compound	Solvent	λ_{max}(log ε)	Ref.
1-Phenanthrenecarboxylic acid, 1,2,3,4,4aα,9,10,10aα-octahydro-1α-methyl-4-oxo-, methyl ester	EtOH	266(2.6),273(2.61)	88-4997-66
Spiro[1,3-dioxolane-2,3'(2'H)-phenanthrene], 1',4',9',10'-tetrahydro-7'-methoxy-	EtOH	273(4.20)	87-0304-66
$C_{17}H_{20}O_4$			
1H-Cyclopenta[7,8]naphtho[2,3-b]furan-7-carboxylic acid, 2,3,3a,4,5,7,8,-10b-octahydro-10-hydroxy-6-methyl-	EtOH	290(3.29)	39-0725-66C
1aH-Cyclopropa[a]naphthalene-1a-acetic acid, 1-carboxy-1,2,3,7b-tetrahydro-7b-methyl-, dimethyl ester	MeOH	228(3.94)	35-0161-66
Glutaric acid, 2-(3,4-dihydro-2-naphthyl)-, dimethyl ester	EtOH	266(4.1)	77-0485-66
Glutaric acid, 2-(3,4-dihydro-2-naphthyl)-2-methyl-, 1-methyl ester	EtOH	265(4.14)	88-4997-66
Naphthalene, 1-carbomethoxymethyl-2-carbomethoxymethylenetetrahydro-1-methyl-	EtOH	216(4.17),265s(2.96), 273(2.50)	35-1965-66
2-Phenanthrenecarboxylic acid, 1,2,3,4,4aα,9,10,10aβ-octahydro-7-methoxy-3-oxo-, methyl ester	pH 13	284(4.17)	87-0304-66
	EtOH	220s(3.95),258(3.97), 285s(3.36)(anom.)	87-0304-66
4aβ-isomer	EtOH	225(3.89),276(3.40), 285(3.30)(anom.)	87-0304-66
Unidentified lactone	EtOH	274(3.06),281(3.07)	78-0301-66
$C_{17}H_{20}O_5$			
1-Epiallohelenalin, acetyl-	EtOH	215(4.17),329(1.83)	78-3279-66
2-Naphthoic acid, 1,2,3,4-tetrahydro-8-methoxy-3-oxo-4-(3-oxobutyl)-, methyl ester	EtOH	260(3.75),278s(3.48), 282(3.31)	44-1747-66
$C_{17}H_{20}O_6$			
Cyclohexanecarboxylic acid, 1-(3,5-dimethoxybenzyl)-2,4-dioxo-, methyl ester	MeOH	225(4.05),260(4.20)	44-1451-66
$C_{17}H_{20}O_9$			
3-Isoferuloylquinic acid	EtOH	326(4.32)	102-0767-66
$C_{17}H_{21}BrN_2O_5$			
4,7-Indoloquinone, 5-bromo-1-ethyl-3-(hydroxymethyl)-6-methoxy-2-methyl-, propyl carbamate	MeOH	229(4.21),298(4.11), 350(3.43),465(3.16)	35-0804-66
4,7-Indoloquinone, 6-bromo-1-ethyl-3-(hydroxymethyl)-5-methoxy-2-methyl-, propyl carbamate	MeOH	233(4.21),300(4.11), 360(3.54),475(2.94)	35-0804-66
$C_{17}H_{21}BrO_2$			
Laurinterol, acetate	EtOH	270(2.83),278(2.82)	88-1837-66
$C_{17}H_{21}NO$			
2,4,6-Heptatrienoic acid amide, 7-phenyl-, N-isobutyl-, trans	EtOH	242(3.98),249(3.93), 321s(4.69),334(4.81), 348s(4.71)	12-1215-66

Compound	Solvent	λmax(log ε)	Ref.
$C_{17}H_{21}NO_3$			
2H-Benzo[a]quinolizin-2-one, 1,6,7,11b-tetrahydro-9,10-dimethoxy-3,4-dimethyl-	EtOH	285(3.73),335(4.20)	44-0797-66
2-Naphthoic acid, 3-(1-pyrrolidinyl)-8-methoxy-1,2-dihydro-, methyl ester	EtOH	225(3.92),245(3.93), 314(4.05),327(4.06)	44-1747-66
Phenethylamine, 3-(benzyloxy)-4,5-dimethoxy-, as oxalate	pH 13	271(2.92)	33-0403-66
	iso-PrOH	268(2.93)	33-0403-66
Pluviine	EtOH	282.5(3.60)	39-0676-66C
$C_{17}H_{21}NO_4$			
2H-Benzo[a]quinolizin-2-one, 1,6,7,11b-tetrahydro-1α,9,10-trimethoxy-4-methyl-	EtOH	292s(3.86),323(4.20)	44-0797-66
1β-isomer	EtOH	290(3.79),324(4.15)	44-0797-66
Carbostyril, 4,8-dimethoxy-3-(4-methyl-2-oxopentyl)-	MeOH	237(4.29),253(4.43), 282(3.93),292s(3.90), 332(3.57),346s(3.47)	44-1276-66
Carbostyril, 3-(2,3-epoxy-3-methylbutyl)-4,6-dimethoxy-1-methyl-	EtOH	222(4.65),246(4.61), 287(4.00),296(3.99), 330(3.87),340(3.88)	78-1153-66
$C_{17}H_{21}NO_5$			
1-Cyclohexanecarboxamide, 1-(3,5-dimethoxybenzyl)-6-methyl-2,4-dioxo-	MeOH	224(3.97),263(4.15)	44-1456-66
Lunidine	EtOH	216s(4.31),228(4.41), 237s(4.34),258(4.30), 266s(4.33),317(3.82), 331(3.78)	12-2185-66
$C_{17}H_{21}N_3O_3$			
1,2-Benzisoxazole, 3a,4,5,6,7,7a-hexahydro-3-(m-nitrophenyl)-7a-(1-pyrrolidinyl)-	EtOH	260(4.26)	32-0559-66
4-Pyrimidinepropionic acid, 2-amino-6-hydroxy-5-(4-phenylbutyl)-	pH 1	267(3.92)	4-0315-66
	pH 13	282(3.87)	4-0315-66
$C_{17}H_{21}N_3O_9$			
Cytosine, N-acetyl-1-β-D-arabinofuranosyl-, 2',3',5'-triacetate	pH 1	247(4.05),302(3.95)	87-0268-66
	pH 7	247(4.22),298(3.93)	87-0268-66
	pH 13	275(4.02)	87-0268-66
$C_{17}H_{21}N_5$			
Tetraethylammonium salt of cyclopentadiene-1,2,3,4-tetracarbonitrile	MeCN	237(4.63),244(4.75), 287(4.13),298(4.16)	35-3046-66 +35-4055-66
$C_{17}H_{21}N_5S$			
Tetraethylammonium salt of 5-mercaptocyclopentadiene-1,2,3,4-tetracarbonitrile	MeCN	233(4.36),271(4.35), 340(3.60)	35-4055-66
$C_{17}H_{21}N_7O_2$			
Alanine, N-(5-amino-3-butyl-3H-v-triazolo[4,5-d]pyrimidin-7-yl)-3-phenyl-	pH 1	258(4.22),273s(--)	87-0417-66
	pH 7, 13	233s(--),265s(--), 292(4.11)	87-0417-66
$C_{17}H_{22}$			
2,9,16-Heptadecatriene-4,6-diyne, 2,9-cis	ether	226(3.18),238(3.63), 251(3.99),265(4.20), 281(4.09)	24-0586-66

Compound	Solvent	λ_{max}(log ε)	Ref.
$C_{17}H_{22}BrN_3O_4$			
4,7-Indoloquinone, 5-bromo-1-ethyl-3-(hydroxymethyl)-2-methyl-6-(methylamino)-, propyl carbamate	MeOH	249(4.24),316(4.08), 350(3.78),550(3.18)	35-0804-66
$C_{17}H_{22}ClNO$			
3-Pyrrolidinol, 4-(p-chloro-benzylidene)-N-cyclohexyl-	EtOH	263(4.42)	44-0001-66
$C_{17}H_{22}N_2O_2$			
1,2-Benzisoxazole, 3a,4,5,6,7,7a-hexahydro-7a-morpholino-3-phenyl-	EtOH	264(4.10)	32-0559-66
$C_{17}H_{22}N_2O_5$			
4,7-Indoloquinone, 1-ethyl-3-(hydroxymethyl)-5-methoxy-2-methyl-, propyl carbamate	MeOH	229(4.18),299(4.18), 343(3.40),450(3.07)	35-0804-66
4,7-Indoloquinone, 1-ethyl-3-(hydroxymethyl)-6-methoxy-2-methyl-, propyl carbamate	MeOH	228(4.17),287(4.10), 340(3.64),445(3.13)	35-0804-66
$C_{17}H_{22}N_2O_9$			
Thymine, 1-(3-deoxy-β-D-xylohexo-pyranosyl)-, 2',4',6'-triacetate	EtOH	263(3.92)	39-1549-66C
Thymine, 1-(4-deoxy-β-D-xylohexo-pyranosyl)-, 2',3',6'-triacetate	pH 7	262.5(3.99)	39-1549-66C
$C_{17}H_{22}N_4$			
Heptafulvene, 8,8-dicyano-5-isopropyl-2,3-bis(dimethylamino)-	MeOH	285(4.01),340(4.17), 455(4.24)	18-2444-66
$C_{17}H_{22}N_4O_4$			
2,5-Undecadienal, 2,4-dinitrophenyl-hydrazone, 5-cis-2-trans	$CHCl_3$	375(4.47)	54-0117-66
$C_{17}H_{22}N_4O_8$			
Oxazole, 4,5-di-tert-butyl-, picrate	EtOH	214(4.36)	44-3954-66
$C_{17}H_{22}O$			
Cyclohexanone, 2-(p-isopropyl-benzylidene)-5-methyl-	EtOH	292(4.26)	35-1419-66
1,6,8,10-Heptadecatetraen-4-yn-3-one, all trans	ether	277(4.16),293(4.14), 342(4.43)	24-3552-66
1,8,15-Heptadecatriene-11,13-diyn-10-ol	ether	247(3.60),254(3.85), 269(4.01),284(3.90)	24-3201-66
Spiro[benzocyclooctene-6(5H),1'-cyclo-hexan]-5-one, 7,8,9,10-tetrahydro-	EtOH	236(3.18),271(2.72), 308(2.33)	22-1693-66
$C_{17}H_{22}O_2$			
1,15-Heptadecadiene-11,13-diyn-10-ol, 8,9-epoxy-, cis	ether	213(4.61),227(3.76), 240(3.83),253(4.01), 267(4.14),282(4.02)	24-2096-66
$C_{17}H_{22}O_3$			
Cyclobutanecarboxylic acid, 2-benzoyl-1,2,3,4-tetramethyl-, methyl ester, cis	MeOH	243.5(4.04)	24-1236-66
1H-Cyclopenta[7,8]naphtho[2,3-b]furan, 2,3,3a,4,5,7,8,10b-octahydro-10-hydroxy-6-methyl-7-(hydroxymethyl)-	EtOH	289(3.31)	39-0725-66C

Compound	Solvent	$\lambda_{max}(\log \epsilon)$	Ref.
2,4,6,8,10-Dodecapentaenal, 12-acetoxy-2,6,10-trimethyl-	pet ether	264(4.06),346(4.74), 364(4.93),384(4.90)	33-0369-66
Nonanoic acid, 9-(p-hydroxymethylbenzoyl)-, lactone	MeOH	246(4.14)	44-4277-66
Spiro[1,3-dioxolane-2,3'(2'H)-phenanthrene], 1',4',4'aα,9',10',10aβ-hexahydro-7'-methoxy-	EtOH	218(3.95),278(3.34), 287(3.31)	87-0304-66
4'aβ-isomer	EtOH	220(3.92),278(3.30), 287(3.27)	87-0304-66
$C_{17}H_{22}O_4$			
1,2-Indandicarboxylic acid, 4,5,6,7-tetramethyl-, dimethyl ester	$C_6H_{11}Me$	272(2.52),278s(--), 282s(--)	39-1147-66B
α-Pipitzol, acetate	EtOH	232(3.83),263(3.65)	78-2387-66
β-Pipitzol, acetate	EtOH	232(3.82),266(3.62)	78-2387-66
$C_{17}H_{22}O_5$			
Axivalin	EtOH	209(3.97)	44-3232-66
Azuleno[6,5-b]furan-2,5-dione, 3,3aα,4,4a,6,8,9,9aβ-octahydro-4β-hydroxy-3α,4aβ,8α-trimethyl-, acetate	EtOH	278(2.42)	78-1709-66
1-Epiisotenulin	EtOH	225(3.98),320(1.91)	78-1709-66
Gaillardin	EtOH	209(end absorption)	35-5292-66
Hysterin, dehydro-	EtOH	213(4.00)	44-0673-66
Isogaillardin	EtOH	211(4.12)(end absorption)	35-5292-66
$C_{17}H_{22}O_6$			
Cumanin, diformate	EtOH	213(3.99)	78-1499-66
$C_{17}H_{23}IN_2S_2$			
3-Ethyl-2-[5-(3-ethyl-4-methyl-4-thiazolin-2-ylidene)-1,3-pentadienyl]-4-methylthiazolium iodide	MeOH	556(4.90)	9-0150-66
$C_{17}H_{23}NO$			
2,4-Heptadienoic acid amide, N-isobutyl-7-phenyl-	EtOH	259.5(4.55)	12-1215-66
Propionamide, N-methyl-N-[(2-phenyl-2-cyclohexen-1-yl)methyl]-	EtOH	242(3.98)	94-1418-66
3-Pyrrolidinol, 4-benzylidene-N-cyclohexyl-	EtOH	257(4.28)	44-0001-66
$C_{17}H_{23}NO_2$			
Bicyclo[2.2.2]oct-2-ene, 4-hydroxy-1-methyl-5-(1-oxo-3-pyrrolidino-2-butenyl)-	EtOH	314(4.53)	39-2324-66C
$C_{17}H_{23}NO_4$			
Carbostyril, 3-(2-hydroxy-4-methylpentyl)-4,8-dimethoxy-	MeOH	237(4.36),254(4.55), 282(4.03),292s(3.99), 332(3.72),346s(3.55)	44-1276-66
α-Pipitzol, oxime, acetate	EtOH	247(3.80)	78-2387-66
$C_{17}H_{23}NO_5$			
Ribalone, O^6-methyl-	EtOH	215(4.49),234(4.62), 276(3.95),283s(3.88), 344(3.88)	78-1153-66

Compound	Solvent	λ_{max}(log ε)	Ref.
$C_{17}H_{23}NO_8S_2$			
4-Thiazoline-2-thione, 3-ethyl-5-(D-arabino-tetracetoxybutyl)-	MeOH	320(4.29)	12-0445-66
$C_{17}H_{23}N_3$			
6H-Pyrido[2,1-c][1,2,4]benzotriazine, 6-cyclohexyl-7,8,9,10-tetrahydro-	EtOH	242(3.61),288(4.25)	30-0361-66D
$C_{17}H_{23}N_3O_4$			
Tetramethylammonium salt of 2,3-dicyanocyclopentadiene-1,4-dicarboxylic acid, diethyl ester	n.s.g.	255s(4.65),262(4.72), 302(4.16),313(4.16)	39-1641-66C
$C_{17}H_{24}$			
Cyclohexane, 1-(p-isopropylbenzylidene)-4-methyl-, endo	EtOH	270(3.38)	35-1419-66
exo	EtOH	249(4.16)	35-1419-66
$C_{17}H_{24}N_2$			
Dipyrromethene, 3,3',4,4'-tetraethyl-,	EtOH	467s(4.79),483(4.85)	39-0098-66C
hydrobromide	$CHCl_3$	472s(4.75),483(4.86)	39-0098-66C
$C_{17}H_{24}N_4O_4$			
Cyclopentanone, 2-(1-methylpentyl)-, 2,4-dinitrophenylhydrazone	$CHCl_3$	367(4.35)	22-0281-66
isomer	$CHCl_3$	367(4.36)	22-0281-66
$C_{17}H_{24}O$			
Crotonaldehyde, β-ionylidene-	EtOH	349(4.52)	54-0343-66
Cyclohexanone, 2-benzyl-4-tert-butyl-	C_6H_{12}	282(1.67)	22-3881-66
1,6,8,10-Heptadecatetraen-4-yn-3-ol	ether	287(4.47),298(4.67), 313(4.61)	24-3552-66
8-Phenanthrol, 1,2,3,4,9,10,11,12-octahydro-1,1,12-trimethyl-, trans	EtOH	274(3.08)	35-1776-66
	EtOH-NaOH	243(3.86),293(3.56)	35-1776-66
$C_{17}H_{24}O_2$			
9-Heptadecene-4,6-diyn-3-one, 1-hydroxy-	ether	227(3.38),240(3.50), 253(3.77),267(3.93), 283(3.81)	24-3552-66
$C_{17}H_{24}O_3$			
7αH-Eremophila-10,11-dien-9-one, 8α-acetoxy-	EtOH	247(3.84)	12-0303-66
Nonanoic acid, 9-p-toluoyl-	MeOH	251(4.16)	44-4277-66
2-Oxa-A-norestr-5(10)-en-1-one, 17β-hydroxy-3ξ-methyl-	MeOH	219(4.00)	44-4255-66
1-Phenanthrenemethanol, 1,2,3,4,4a,9,-10,10a-octahydro-4α-hydroxy-7-methoxy-α-methyl-	EtOH	220(3.94),278(3.26), 283(3.21)	44-1360-66
isomer	EtOH	220(3.93),278(3.29), 286(3.20)	44-1360-66
$C_{17}H_{24}O_4$			
Isohysterin, deoxy-	EtOH	222(4.20)	44-0673-66
Nonanoic acid, 9-(α-hydroxy-p-toluoyl)-	MeOH	251(4.31)	44-4277-66
$C_{17}H_{24}O_5$			
Hysterin	EtOH	213(4.00)	44-0673-66
Isohysterin	EtOH	220(4.19)	44-0673-66
Unknown acid acetate	EtOH	246(4.03)	78-2387-66

Compound	Solvent	λ_{max}(log ε)	Ref.
$C_{17}H_{24}S_2$			
Nonane, 5,5-di-2-thienyl-	n.s.g.	237(4.14)	22-2253-66
$C_{17}H_{25}BrO$			
Benzofuran, 2-(bromomethyl)-4,7-di-tert-butyl-2,3-dihydro-	MeOH	277(2.9),286(2.9)	77-0327-66
$C_{17}H_{25}NO$			
Piperidine, 4-[(1,2,3,4-tetrahydro-6-methoxy-1-naphthyl)methyl]-	EtOH	220(3.89),280(3.29), 287(3.26)	44-3337-66
$C_{17}H_{25}NO_2$			
1-Benzazepine, 2,3,4,5-tetrahydro-5-(p-hydroxyphenyl)-7-methoxy-1-methyl-	EtOH-HCl	225(4.02),280(3.19)	94-1033-66
	EtOH	254(3.83),299(3.38)	94-1033-66
Piperidine, 4-[(5,6-dimethoxy-1-indanyl)methyl]-	EtOH	220(3.86),228s(3.85), 287(3.71),291s(3.69)	44-3337-66
Spiro[5H-2-benzazepine-5,1'-cyclohexan]-4'-ol, 1,2,3,4-tetrahydro-7-methoxy-2-methyl-	EtOH-HCl	232(3.99),280s(3.21)	94-1033-66
	EtOH	230(3.91),280s(3.29)	94-1033-66
$C_{17}H_{25}NO_3$			
Hydrocinnamonitrile, α-(tert-butoxymethyl)-2,4-dimethoxy-α-methyl-	EtOH	230(3.94),278(3.42), 284(3.38)	78-1027-66
2,4-Pentanedione, 3-(12-oxo-4,8-cyclododecadien-1-yl)-, oxime	EtOH	241(4.11)	18-1129-66
$C_{17}H_{25}NO_4$			
Hydracrylonitrile, 2-(tert-butoxymethyl)-3-(2,4-dimethoxyphenyl)-2-methyl-	EtOH	230(3.91),278(3.38)	78-1027-66
$C_{17}H_{25}NO_5$			
Benzoic acid, p-[N-(2-formylethyl)-formamido]-, ethyl ester, diethyl acetal	EtOH	266(4.21)	87-0590-66
$C_{17}H_{25}NO_9S_2$			
2-Thiazolidinethione, 3-ethyl-5-hydroxy-5-(D-arabino-tetraacetoxybutyl)-	MeOH	247(3.88),274(4.19)	12-0445-66
$C_{17}H_{25}N_3O_6$			
D-Valine, N-(2-amino-3,4-cresotoyl)-L-threonyl-	pH 1	221(4.44),285(3.81)	23-0799-66
$C_{17}H_{26}NO_2$			
Phenoxy, 4-acetyl-2,6-di-tert-butyl-, O-methyloxime	C_6H_{12}	306s(--),319s(--), 336(4.20),342(4.20), 535(2.95),591(2.91)	35-5284-66
$C_{17}H_{26}N_2$			
Pyridine, 4-[p-(dimethylamino)phenyl]-1,2,3,6-tetrahydro-2,2,6,6-tetramethyl-	EtOH	286(4.31)	78-2735-66
dihydrochloride	pH 1	246.5(4.20)	78-2735-66
	EtOH	296.5(4.33)	78-2735-66
$C_{17}H_{26}O$			
2,5-Cyclohexadien-1-one, 2,6-di-tert-butyl-4-propylidene-	hexane	300(4.21)	70-0888-66

Compound	Solvent	λ_{max}(log ε)	Ref.
1,7,9-Heptadecatrien-4-yn-3-ol	EtOH	227(4.35)	95-1051-66
$C_{17}H_{26}OS$			
4-Thiaionylidene ethylidene acetone	C_6H_{12}	299(4.18),368(4.20)	78-0265-66
$C_{17}H_{26}O_2$			
9-Phenanthrenecarboxylic acid, 1,2,3,4-5,6,7,8,8a,9,10,10a-dodecahydro-9-methyl-, methyl ester, cis-cis	EtOH	210(4.02)	78-2523-66
cis-trans	EtOH	210(4.03)	78-2523-66
Podocarp-8(14)-en-7-one, 2β-hydroxy-	EtOH	241(4.19)	35-1766-66
$C_{17}H_{26}O_3$			
Diosphenol A, acetate	EtOH	243(4.04)	78-0337-66
Diosphenol B, acetate	EtOH	243(4.01)	78-0337-66
2-Phenanthrenemethanol, 1,2α,3,4,4a,-5,8,9,10,10a-decahydro-4-hydroxy-7-methoxy-α-methyl-	EtOH	278(1.88)	44-1360-66
$C_{17}H_{26}O_4$			
p-Benzoquinone, 2,5-dihydroxy-3-undecyl-	EtOH	291(4.28),420(2.50)	12-0169-66
$C_{17}H_{26}O_5$			
Cyclohepta[c]pyran-8-acetic acid, 1,3,4,4aα,5,6,7,8,9,9a-decahydro-7β-hydroxy-1,3-dimethoxy-5β,9aβ-dimethyl-α-methylene-, lactone	EtOH	213(3.97)	78-1499-66
$C_{17}H_{26}O_6$			
$\Delta^{1,\alpha}$,2-Cyclopentanediacetic acid, 2-carboxy-α^2,5-dimethyl-, diethyl methyl ester	EtOH	222(3.92)	78-2319-66
$C_{17}H_{26}O_8S_2$			
Cumanin, dimethanesulfonate	EtOH	213(3.99)	78-1499-66
$C_{17}H_{27}ClN_2O_4$			
1-(7-Piperidino-2,4,6-heptatrienylidene)piperidinium perchlorate	EtOH-100°K	520(5.45)	61-0052-66
phototropic form	EtOH-100°K	550(4.57)	61-0052-66
$C_{17}H_{27}NO_2$			
Acetophenone, 3',5'-di-tert-butyl-4'-hydroxy-, O-methyloxime	C_6H_{12}	266(4.13)	35-5284-66
$C_{17}H_{27}NO_4$			
Benzene, 1-decyloxy-4-methoxy-2-nitro-	MeOH	357(3.38)	78-0093-66B
Benzene, 1-decyloxy-4-methoxy-3-nitro-	MeOH	357(3.38)	78-0093-66B
$C_{17}H_{28}N_2$			
Piperidine, 4-(p-dimethylaminophenyl)-2,2,6,6-tetramethyl-	EtOH	254.5(4.2)	78-2735-66
dihydrochloride	EtOH	256(4.25)	78-2735-66
$C_{17}H_{28}O$			
8,10,12,14-Heptadecatetraen-1-ol	ether	279(4.38),291(4.68), 304(4.87),319(4.83)	24-3544-66

Compound	Solvent	λ_{max}(log ϵ)	Ref.
$C_{17}H_{28}O_2$			
Benzene, 1-decyloxy-4-methoxy-	MeOH	226(3.93),287(3.36)	78-0093-66B
2-Cyclohexen-1-one, 2,4-dimethyl-3-(3-methyl-7-oxooctyl)-	EtOH	248(4.15)	78-0175-66A
$C_{17}H_{29}NO_3S_2$			
5-Thiazolidineacetic acid, N-butyl-4-oxo-2-thioxo-, octyl ester	EtOH	262(4.10),296(4.20)	95-0095-66
$C_{17}H_{29}N_6OP$			
Phosphonic diamide, P-(5-anilino-3-pentyl-1H-1,2,4-triazol-1-yl)-N,N,N',N'-tetramethyl-	EtOH	261(4.36)	54-0429-66
$C_{17}H_{30}O_2$			
2-Cyclohexen-1-one, 2,4-dimethyl-3-(3-methyl-7-hydroxyoctyl)-	EtOH	248(4.10)	78-0175-66A
$C_{17}H_{36}ClN_5NiO_4$			
5,7,7,12,14,14-Hexamethyl-1,4,8,11-tetraazacyclotetradecane nickel cyanide perchlorate	MeOH	366(1.85),549(1.65)	12-0609-66
	pyridine	366(1.89),549(1.73)	12-0609-66
	DMSO	366(1.91),549(1.68)	12-0609-66

Compound	Solvent	λ_{max}(log ϵ)	Ref.
$C_{18}H_8F_{10}O_4$			
Hydrobenzoin, decafluoro-, diacetate	EtOH	264(3.158)	70-0337-66
$C_{18}H_9BrO_4$			
5,12-Naphthacenedione, 6-bromo-1,11-dihydroxy-	C_6H_{12}	449(4.22),476(4.15)	5-0145-66F
	H_2SO_4	564(4.30),614(4.38)	5-0145-66F
5,12-Naphthacenedione, 11-bromo-1,6-dihydroxy-	C_6H_{12}	452(4.20),469(4.12),479(4.10)	5-0145-66F
	H_2SO_4	545(4.46),589(4.70)	5-0145-66F
$C_{18}H_{10}Fe$			
Octatetrayne, 1-ferrocenyl-	$CHCl_3$	287(4.32),328(3.60),348(3.59),374(3.55),465(3.17)	101-0399-66B
$C_{18}H_{10}N_2S_3$			
2,5-Thiophenediacetonitrile, α,α'-di-2-thenylidene-	n.s.g.	298(4.01),346(4.28),436(4.57)	12-1243-66
$C_{18}H_{10}N_4$			
Dipyrido[2,3-a:2',3'-h]phenazine	EtOH	234(4.52),301(4.96),380(3.96),402(4.00)	44-3384-66
Dipyrido[2,3-a:3',2'-j]phenazine	EtOH	219(4.70),249(4.37),305(4.92),372(3.98),392(4.05)	44-3384-66
$C_{18}H_{10}N_4O_6$			
4,4'-Biisoxazole, 3,3'-dinitro-5,5'-diphenyl-	EtOH	249(4.40)	32-0454-66
$C_{18}H_{10}O$			
Benzanthrone	EtOH	230(4.54),236(4.45),256(4.30),286(3.88),310(3.90),394(4.00)	24-3298-66
$C_{18}H_{10}O_3$			
2,3-Naphthalenedicarboxylic anhydride, 1-phenyl-	EtOH	277s(4.06),286(4.15),295s(4.06),324(3.22),337(3.31)	44-2376-66
$C_{18}H_{10}O_4$			
[2,2'-Biindan]-1,1',3,3'-tetrone	EtOH	220(4.67),288(4.29)	35-2007-66
	H_2SO_4	216(3.26),309(3.45)	35-2007-66
5,12-Naphthacenedione, 1,6-dihydroxy-	C_6H_{12}	450(4.13),467(4.00),479(3.98)	5-0145-66F
	H_2SO_4	550(4.36),595(4.65)	5-0145-66F
5,12-Naphthacenedione, 1,11-dihydroxy-	C_6H_{12}	448(4.01),476(3.93)	5-0145-66F
	H_2SO_4	559(3.90),607(4.57)	5-0145-66F
Naphthacenequinone, 6,11-dihydroxy-	EtOH	261(4.59)	35-2007-66
	H_2SO_4	298(3.83)	35-2007-66
$C_{18}H_{10}O_5$			
5,12-Naphthacenedione, 1,6,11-trihydroxy-	C_6H_{12}	461(4.16),492(4.43),518(4.19),528(4.48)	5-0145-66F
	H_2SO_4	531(4.45),573(4.67)	5-0145-66F
1,4-Naphthoquinone, 3-(1,3-dioxo-indan-2-yl)-2-hydroxy-	MeCN	248(4.50),253(4.44),276(4.23),333(3.53)	44-0062-66
$C_{18}H_{11}N_3O_3S$			
Oxazolo[5,4-d]pyrimidin-7-one-5-thione, 4-benzoyl-2-phenyl-	dioxan	319(3.97),332s(3.87)	50-0737-66C

Compound	Solvent	λ_{max}(log ϵ)	Ref.
$C_{18}H_{12}$			
Biphenylene, 1-phenyl-	EtOH	229(4.30),267(4.38), 342s(3.36),358(3.56), 376(3.53)	39-1767-66C
Biphenylene, 2-phenyl-	EtOH	238(4.39),265(4.75), 335s(3.75),351(3.96), 368(4.03)	39-1767-66C
Chrysene	EtOH	220(4.58),241(4.35), 259(4.93),268(5.17), 283(4.11),294(4.08), 306(4.11),320(4.11), 343(2.65),351(2.20), 361(2.71)	31-0793-66
sym-Dibenzofulvalene	hexane	248(4.29),285(4.35), 291s(4.32),360s(4.26), 380(4.43),401(4.48)	44-2705-66
$C_{18}H_{12}Br_2Cl_2$			
7,8;9,10-Dibenzotricyclo[4.2.2.0^{2,5}]-deca-7,9-diene, 1,6-dibromo-3,4-dichloro-, cis	EtOH	256(2.53),261(2.55), 270(2.41)	5-0009-66A
$C_{18}H_{12}ClN_5O$			
s-Triazolo[4,3-b]pyridazine, 8-benzamido-6-chloro-3-phenyl-	EtOH	207(4.38),273(4.38), 321(4.06)	4-0218-66
$C_{18}H_{12}FN_3O_3$			
Cytosine, 5-fluoro-, dibenzoyl deriv.	EtOH	236(4.33),276(4.11), 331(3.45)	87-0566-66
$C_{18}H_{12}N_2O_4$			
p-Terphenyl, 4,4'-dinitro-	dioxan	333(4.50)	99-0204-66
	$C_2H_4Cl_2$	337(4.45)	99-0204-66
$C_{18}H_{12}N_4S_2$			
Pyrazino[2,3-d]pyridazine, 5,6-bis(phenylthio)-	EtOH	206(4.77),220s(4.68), 243(4.43),260s(4.35), 300(3.92)	4-0512-66
$C_{18}H_{12}N_5O_6$			
Hydrazyl, 2,2-diphenyl-1-picryl-	MeCN	561(4)	35-1923-66
$C_{18}H_{12}N_6$			
Bis-s-triazolo[4,3-b:3',4'-f]-pyridazine, 1,8-diphenyl-	C_6H_{12}	245(4.48)	78-2073-66
$C_{18}H_{12}OS_2$			
[18]Annulene-1,4-oxide-7,10:13,16-disulfide	EtOH	229(4.19),281(4.46), 292(4.47),405(3.99), 425(3.98)	12-0257-66
$C_{18}H_{12}O_2$			
Coumarin, 3-(2-indenyl)-	EtOH	257(3.94),370(4.24)	24-0088-66
2-Naphthoic acid, 3-(hydroxymethyl)-1-phenyl-, lactone	EtOH	242(4.65),281s(3.69), 291(3.88),302(3.87), 333s(3.53),343(3.70)	44-2376-66
α-Truxone	$CHCl_3$	274(3.30),280s(3.28), 293(2.65)	12-1455-66

Compound	Solvent	$\lambda_{max}(\log \epsilon)$	Ref.
$C_{18}H_{12}O_2S$			
[18]Annulene-1,4;7,10-dioxide-13,16-sulfide	EtOH	224(4.16),244(4.07),320(4.26),333(4.63),343(4.98),376(3.71),391(3.85),413(4.22),431(3.32)	12-1461-66
$C_{18}H_{12}O_2S_2$			
1,3-Indandione, 2-[4-(1,3-dithiolan-2-ylidene)-2,5-cyclohexadien-1-ylidene]-	$CHCl_3$	512s(4.22),545(4.45),585(4.89),638(4.93)	89-0517-66
$C_{18}H_{12}O_3$			
[18]Annulene-1,4;7,10;13,16-trioxide	EtOH	220(3.89),239(3.87),309(4.26),323(4.81),332(5.46),370(3.51),391(3.74),405(4.24),426(3.08)	12-1221-66
Furo[2,3-c]coumarin, 2-methyl-3-phenyl-	MeOH	293(3.82),321(4.04)	83-0798-66
4H-Pyrano[2,3-d]coumarin, 2-phenyl-	MeOH	236(4.32),311(3.71)	83-0798-66
$C_{18}H_{12}O_6$			
1,2-Anthraquinone, 5,6-diacetoxy-	MeOH	296(4.45),433(3.61)	24-2322-66
Pseudobaptigenin, acetate	MeOH	228(4.7),258s(--),281s(--)	39-0509-66C
$C_{18}H_{12}S$			
Aceanthra[7,8-b]thiophene	dioxan	235(4.36),285(4.63),295(4.63),328s(3.06),342s(3.38),354(3.63),378(3.76),402(3.60)	22-3667-66
Aceanthra[8,7-b]thiophene	dioxan	240(4.42),265(4.77),289(4.64),292(4.64),326s(3.34),345s(3.58),360(3.78),379(3.86),399(3.85)	22-3667-66
Aceanthra[9,10-b]thiophene	dioxan	240(4.20),263(4.63),278(4.61),290(4.55),337(3.42),354(3.73),372(3.73),393(3.84)	22-3667-66
Acenaphtho[3,4-d]benzo[b]thiophene	EtOH	227(4.34),246(4.50),267(4.67),277(4.96),285(4.23),299(3.83),301(3.84),316(3.91),330(3.77),354(3.63),374(3.68)	22-3618-66
Acenaphtho[4,3-d]benzo[b]thiophene	EtOH	230(4.40),245(4.60),265(4.73),277(4.91),284(4.24),298(3.76),302(3.71),316(3.80),330(3.64),356(3.60),375(3.67)	22-3618-66
Acenaphtho[4,5-d]benzo[b]thiophene	EtOH	248(4.52),254(4.59),261(4.59),266(4.56),275(4.56),294(4.09),305(4.27),321(3.80),338(3.67),355(3.84)	22-3618-66
$C_{18}H_{13}BrO_3$			
Coumarin, 4-hydroxy-3-(3-bromo-3-phenylpropen-1-yl)-	MeOH	311(3.93)	83-0798-66

Compound	Solvent	λ_{max}(log ε)	Ref.
$C_{18}H_{13}ClO_2$			
Coumarin, 3-(1-chloro-2-indanyl)-	EtOH	275(4.17),304(3.86)	24-0088-66
1H-Cyclobut[a]indene-1-carboxylic acid, 7a-chloro-2,2a,7,7a-tetrahydro-2-(o-hydroxyphenyl)-, δ-lactone	EtOH	262(3.13),268(3.25), 274(3.30),279s(3.07)	24-0088-66
$C_{18}H_{13}ClO_3$			
2-Naphthoic acid, 1-(o-chlorophenyl)-4-methoxy-	n.s.g.	252(4.49),296(3.79), 306(3.76),340(3.57)	39-0603-66C
2-Naphthoic acid, 1-(p-chlorophenyl)-4-methoxy-	n.s.g.	251(4.40),297(3.81), 306(3.80),340(3.58)	39-0603-66C
$C_{18}H_{13}Cl_2NO_2$			
Acrylic acid, 3,3-bis(p-chlorophenyl)-2-cyano-, ethyl ester	EtOH	230(4.22),305(4.17)	22-1033-66
$C_{18}H_{13}N$			
Cyclopenta[c]quinolizine, 4-phenyl-	EtOH	258(4.41),278s(4.4), 358(4.37),390s(3.8), 488(3.24)	39-0324-66C
$C_{18}H_{13}NO$			
Benz[f]isoindolin-1-one, 2-phenyl-	MeOH	237(4.66),276(4.07), 306(4.25)	95-0001-66
11H-Benzo[a]carbazole-9-carboxaldehyde, 10-methyl-	EtOH	208(4.40),232(4.36), 254(4.51),298(4.59), 336(4.34)	87-0237-66
$C_{18}H_{13}NO_2$			
Methane, 2,4-cyclopentadien-1-ylidene(p-nitrophenyl)phenyl-	MeOH	280(4.29),334(4.30)	44-2149-66
$C_{18}H_{13}NO_3$			
Lysicamine	EtOH	235(4.47),270(4.41), 307(3.76),400(3.94)	88-6257-66
$C_{18}H_{13}N_3O_6$			
Cinnamic acid, α-cyano-p-nitro-β-(p-nitrophenyl)-, ethyl ester	EtOH	280(4.39)	22-1033-66
$C_{18}H_{14}$			
Benzo[1,2:4,5]dicyclooctene	EtOH	254(4.39)	77-0509-66
Benzo[c]octalene	n.s.g.	245(4.23)	35-3677-66
3,3'-Biindene	EtOH	227(4.48),234(4.47), 255(4.20)	44-2705-66
15H-Cyclopenta[a]phenanthrene, 16,17-dihydro-17-methylene-	EtOH	215(4.22),274(4.69), 287(4.44),299(4.34), 307(4.09),333(3.08), 349(3.28),367(3.34)	39-0963-66C
15H-Cyclopenta[a]phenanthrene, 17-methyl-	EtOH	223(4.30),270(4.70), 276(4.73),294(4.14), 316(3.74),331(2.71), 348(2.69),366(2.46)	39-0963-66C
7,8;9,10-Dibenzotricyclo[4.2.2.0^{2,5}]-deca-3,7,9-triene	EtOH	266(3.12),273(2.23)	5-0009-66A
Naphthalene, 2-styryl-	hexane	226(4.60),251(4.30), 272(4.54),281(4.54), 304(--),315s(4.62), 328(4.48)	99-0085-66

Compound	Solvent	$\lambda_{max}(\log \epsilon)$	Ref.
Naphthalene, 2-styryl- (cont.)	$C_2H_4Cl_2$	253(4.21),273(4.53), 284(4.56),330(4.65), 332(4.53)	99-0085-66
1,4-Pentadiyne, 5-phenyl-1-p-tolyl-	EtOH	243(4.42),250(4.42), 254(4.47),271(2.98), 275(2.88),278(2.82), 282(2.68),286(2.46)	78-0867-66
m-Terphenyl	EtOH	247(4.54)	22-1012-66
$C_{18}H_{14}ClN$			
o-Terphenyl-4'-amine, 4-chloro-	EtOH-HCl	238(4.47),260s(4.11)	56-0429-66
	EtOH	234(4.38),284(4.17)	56-0429-66
$C_{18}H_{14}ClNO_2$			
Cinnamic acid, p-chloro-α-cyano-β-phenyl-, ethyl ester	EtOH	225(4.17),304(4.16)	22-1033-66
isomer	EtOH	226(4.24),304(4.15)	22-1033-66
$C_{18}H_{14}Cl_2N_2O_2S_2$			
5-Thiazolidineacetamide, 3-benzyl-N-(2,4-dichlorophenyl)-4-oxo-2-thioxo-	EtOH	258(4.26),296(4.24)	95-0101-66
$C_{18}H_{14}Fe$			
Acetylene, 1-ferrocenyl-2-phenyl-	EtOH	252(4.28),298(4.23), 453(2.70)	101-0173-66B
	$CHCl_3$	252(4.26),297s(4.17), 340s(3.23),444(2.70)	101-0399-66B
$C_{18}H_{14}N_2$			
1H-Pyrrolo[2,3-f]isoquinoline, 3-methyl-2-phenyl-	neutral	290(4.76),335(4.41)	2-0118-66
$C_{18}H_{14}N_2O_2$			
o-Terphenyl-4'-amine, 4-nitro-	EtOH-HCl	254s(4.16),294(4.04)	56-0429-66
	EtOH	235(4.41),373(4.00)	56-0429-66
$C_{18}H_{14}N_2O_3$			
Furo[2,3-d:4,5-d']diisoxazole, 3a,4a,7a,7b-tetrahydro-3,7-diphenyl-	EtOH	259(4.45)	88-2911-66
$C_{18}H_{14}N_2O_4$			
Cinnamic acid, α-cyano-p-nitro-β-phenyl-, ethyl ester	EtOH	274(4.22)	22-1033-66
isomer	EtOH	276(4.23)	22-1033-66
$C_{18}H_{14}N_2O_5$			
Hydantoin, 5-(4-acetoxy-1-naphthylmethylene)-, acetyl derivative	MeOH	230(4.43)	87-0057-66
$C_{18}H_{14}N_4O$			
Hypoxanthine, 3-(diphenylmethyl)-	pH 1	253(4.06)	44-1413-66
	pH 7	264(4.13)	44-1413-66
	pH 13	264(4.02),280s(3.93)	44-1413-66
$C_{18}H_{14}N_4O_5$			
2-Furaldehyde, 3-methyl-4-phenyl-, 2,4-dinitrophenylhydrazone	EtOH	398(4.73)	44-0052-66
$C_{18}H_{14}N_4O_6$			
Pyrazole-3-carboxylic acid, 1-(2,4-dinitrophenyl)-5-phenyl-, ethyl ester	EtOH	none	22-0619-66

Compound	Solvent	λ_{max}(log ϵ)	Ref.
$C_{18}H_{14}N_4S_6$			
1,3,4-Thiadiazole, 2,2'-dithiobis-[5-(benzylthio)-	EtOH	304(4.10)	83-0648-66
$C_{18}H_{14}N_6$			
Pyrazino[2,3-d]pyridazine, 5,8-dianilino-	EtOH	206(4.47),227s(4.23), 281(4.54),291s(4.50)	4-0512-66
$C_{18}H_{14}O$			
17H-Cyclopenta[a]phenanthren-17-one, 15,16-dihydro-11-methyl-	EtOH	222(4.11),264(4.83), 288(4.49),301(4.32), 342(3.11),358(3.38), 376(3.43)	39-0955-66C
17H-Cyclopenta[a]phenanthren-17-one, 15,16-dihydro-12-methyl-	EtOH	219(4.18),268(4.85), 286(4.42),299(4.36), 340(2.96),355(3.22), 373(3.28)	39-0955-66C
Phenol, p-4-biphenylyl-	dioxan	290(4.45)	99-0089-66
	$C_2H_4Cl_2$	288(4.44)	99-0089-66
$C_{18}H_{14}O_2$			
Anthracene, 1,5-diacetyl-	MeOH	211(4.40),228s(4.30), 260(4.75),369(3.71), 390(3.81),407s(3.74)	39-1729-66C
Anthracene, 1,6-diacetyl-	MeOH	213(4.18),262s(4.72), 266(4.74),302(4.15), 350(3.45),368(3.56), 390(3.57)	39-1729-66C
Anthracene, 1,8-diacetyl-	MeOH	203(4.28),211(4.36), 257(4.78),320s(3.54), 351s(3.64),369s(3.81), 387(3.87)	39-1729-66C
Coumarin, 3-(2-indanyl)-	EtOH	273(4.10),308(3.91)	24-0088-66
17H-Cyclopenta[a]phenanthren-17-one, 15,16-dihydro-3-methoxy-	EtOH	219(4.21),269s(4.78), 277(4.83),319(4.22), 347s(3.62),366(3.28)	39-0955-66C
17H-Cyclopenta[a]phenanthren-17-one, 15,16-dihydro-11-methoxy-	EtOH	214(4.21),261(4.86), 292(4.42),303(4.29), 346(3.42),363(3.72), 381(3.82)	39-0955-66C
2-Naphthoic acid, 3,4-dihydro-3-(hydroxymethyl)-1-phenyl-, lactone	EtOH	230(4.15),235(4.14), 300(4.01)	44-2376-66
p-Terphenyl-4,4"-diol	dioxan	294(4.48)	99-0204-66
$C_{18}H_{14}O_3$			
2,4-Pentadienoic acid, 5-hydroxy-3-methoxy-4,5-diphenyl-, δ-lactone	n.s.g.	226(4.24),236(4.23), 312(3.95)	35-3759-66
Rotoxen-12(6H)-one, 6,6-dimethyl-	EtOH	255(4.16),266s(4.11), 278s(3.87),311(3.79)	39-0542-66C
$C_{18}H_{14}O_4$			
p-Dioxino[2,3-b]phenanthro[9,10-e]-p-dioxin, 9a,11,12,13a-tetrahydro-	benzene	364(3.13)	24-1881-66
Muconic acid, diphenyl ester, trans-trans	EtOH	267(4.55)	39-0387-66C
Spiro[phenanthrene-9(10H),8'-[2,5,7]-trioxabicyclo[4.2.0]octan-10-one	dioxan	243(4.44),280(3.85), 326(3.47)	24-1881-66
$C_{18}H_{14}O_5$			
Coumarin, 5-(benzyloxy)-4-formyl-7-methoxy-	MeCN	244s(3.93),341(3.98)	35-4534-66

Compound	Solvent	λ max(log ε)	Ref.
1H-Cyclopenta[7,8]naphtho[2,3-b]furan-7-carboxylic acid, 2,3-dihydro-10-hydroxy-6-methyl-3-oxo-, methyl ester	EtOH	227(4.35),273(4.71), 343(3.75),390(3.92)	39-0725-66C
Maleic anhydride, bis(o-methoxyphenyl)-	EtOH	300(3.98)	88-4671-66
$C_{18}H_{14}O_6$			
1H-Naphtho[2,1-b]pyran-1-one, 5,6-dihydroxy-3-methyl-, diacetate	EtOH	211s(4.56),231s(4.32), 256(4.24),262(4.24), 307(3.96),333(3.61)	94-0129-66
$C_{18}H_{14}O_7$			
Xanthone, 1,5-diacetoxy-3-methoxy-	EtOH	238(4.45),273(3.95), 296(4.05),325(3.92)	78-1777-66
Xanthone, 2,8-diacetoxy-1-methoxy-	EtOH	236(4.72),243(4.74), 272(4.15),295(3.64), 347(3.95)	78-1785-66
$C_{18}H_{15}B$			
Borane, triphenyl-	isooctane	200s(5.00),238(4.28), 276(4.54),287(4.59)	46-0611-66
$C_{18}H_{15}BCl_2F_4N_5S_2$			
Bis(3-ethyl-6-chlorobenzothiazol-2-yl)triazamethinecyanine perchlorate	MeOH	487(4.62)	27-0318-66
$C_{18}H_{15}ClO$			
1-Indanone, 2-(p-chlorobenzylidene)-3,3-dimethyl-, cis	MeOH	233(4.04),276s(3.93), 325(3.35)	35-4489-66
trans	MeOH	224(4.05),278s(4.12), 322(4.29)	35-4489-66
$C_{18}H_{15}ClO_4S$			
4-Methyl-2,6-diphenylthiopyrylium perchlorate	n.s.g.	263(4.39),395(4.25)	33-2046-66
$C_{18}H_{15}ClO_5$			
2-Methyl-4,6-diphenylpyrylium perchlorate	MeOH-$HClO_4$	255(4.18),335(4.38), 370(4.46)	65-1728-66
$C_{18}H_{15}ClSe$			
Triphenylselenonium chloride	MeOH	230(4.10),238(4.06), 257(4.14),265(4.17), 271(4.1)	78-0653-66
$C_{18}H_{15}N$			
Pyrrolo[2,1-a]isoquinoline, 5,6-dihydro-2-phenyl-	MeOH	253(4.46),285(4.32), 314(4.39)	95-0856-66
o-Terphenyl, 4'-amino-	EtOH-HCl	237(4.42),260s(4.00)	56-0429-66
	EtOH	237(4.37),275(4.10)	56-0429-66
p-Terphenyl, 4-amino-	dioxan	308(4.40)	99-0089-66
$C_{18}H_{15}NO$			
11H-Benzo[a]carbazole-9-methanol, 10-methyl-	EtOH	238s(--),246(4.62), 254(4.66),281(4.69), 298s(4.29),308(4.35), 322(3.85),338(3.83), 354(3.85)	87-0237-66
1-Naphthol, 2-(N-phenylacetimidoyl)-, enol	C_6H_{12}	368(3.97)	35-2407-66
	MeOH	374(3.80)	35-2407-66

Compound	Solvent	λ_{max}(log ϵ)	Ref.
1-Naphthol, 2-(N-phenylacetimidoyl)-, enol (cont.)	$CHCl_3$	372(3.83)	35-2407-66
	CCl_4	371(3.94)	35-2407-66
	MeCN	368(3.85)	35-2407-66
keto form	C_6H_{12}	425(3.15)	35-2407-66
	MeOH	425(3.85)	35-2407-66
	$CHCl_3$	428(3.73)	35-2407-66
	CCl_4	425(3.30)	35-2407-66
	MeCN	421(3.66)	35-2407-66
Quinaldine, 3-acetyl-4-phenyl-	EtOH	237(4.50),285(3.78), 320s(3.54)	44-3852-66
o-Terphenyl, 4'-amino-4-hydroxy-	EtOH-HCl	244(4.26),274s(3.99)	56-0429-66
	EtOH	240(4.31),276(4.10)	56-0429-66
	EtOH-NaOH	256s(4.18),292(4.17)	56-0429-66
$C_{18}H_{15}NO_2$			
11H-Benzo[a]carbazole-9-carboxylic acid, 5,6-dihydro-10-methyl-	EtOH	220(4.36),254(4.44), 259(4.44),331(4.38), 350(4.37)	87-0237-66
6,7-Benzomorphan-3,8-dione, 5-phenyl-	EtOH	249(4.02),294(3.37)	44-1905-66
Cinnamic acid, α-cyano-β-phenyl-, ethyl ester	EtOH	303(4.11)	22-1033-66
$C_{18}H_{15}NO_3$			
Noracronycine, de-N-methyl-	EtOH	252s(4.37),275(4.63), 295(4.48),410(3.66)	78-3245-66
$C_{18}H_{15}NO_4$			
Hernovine	EtOH	233(4.34),271(4.16), 278s(4.12),313(3.77)	95-0763-66
Ovigerine, hydrochloride	EtOH	234(4.29),270(4.10), 317(3.77)	88-1577-66
$C_{18}H_{15}NO_5$			
Nicotinic acid, 1,2-dihydro-4,6-dihydroxy-1-(2-naphthyl)-2-oxo-, ethyl ester	EtOH	308(4.12)	78-0455-66
$C_{18}H_{15}NO_6$			
Cyclopenta[c]quinolizine-1,2,4-tricarboxylic acid, trimethyl ester	EtOH	242(4.43),265s(4.2), 280s(4.1),360(4.31), 400s(3.9),480s(3.5)	39-0324-66C
$C_{18}H_{15}N_3$			
Carbazole, 3-(2-phenylhydrazino)-	MeOH	232(4.72),266(4.27), 302(4.40),350(3.63)	5-0116-66F
$C_{18}H_{15}N_3O$			
Acetamide, N-[1-(phenylazo)-2-naphthyl]-	C_6H_{12}	225(4.46),247(4.56), 268s(4.16),287(4.23), 384(4.00),428(4.13)	19-0037-66
	EtOH	223(4.46),245(4.50), 287(4.21),380s(3.94), 418(4.02)	19-0037-66
Acetamide, N-[4-(phenylazo)-1-naphthyl]-	EtOH	217(4.49),272(4.08), 390(4.20),470s(3.25)	19-0037-66
Acetophenone, 3'-[(2-amino-1-naphthyl)azo]-	benzene	456(4.15)	44-2571-66
	$C_6H_{11}Me$	446(4.17)	44-2571-66
Acetophenone, 4'-[(2-amino-1-naphthyl)azo]-	benzene	477(4.25)	44-2571-66
	$C_6H_{11}Me$	469(4.24)	44-2571-66

Compound	Solvent	λ_{max}(log ϵ)	Ref.
$C_{18}H_{15}N_3O_2$			
Acetophenone, 3'-[(2-amino-8-hydroxy-1-naphthyl)azo]-	benzene	511(4.20)	44-2571-66
Acetophenone, 4'-[(2-amino-8-hydroxy-1-naphthyl)azo]-	benzene	530(4.28)	44-2571-66
Spiro[1-methyl-2-oxindoline-3,5'-pyrazole], 3'-(p-methoxyphenyl)-	EtOH	264(4.62),275(4.59), 313(4.23),326(4.24)	95-1156-66
$C_{18}H_{15}N_3O_3$			
Alanine, 3-phenyl-N-(2-quinoxalinylcarbonyl)-	EtOH	207(3.41),244(3.65), 319(2.91),327(2.92)	87-0266-66
$C_{18}H_{15}N_3O_6$			
2-Pyrazoline-4-carboxylic acid, 3-[2-(5-nitro-2-furyl)vinyl]-5-oxo-1-phenyl-, ethyl ester	EtOH	240(4.47),393(4.28)	95-1187-66
$C_{18}H_{15}N_5O$			
Pterin, 7,8-dihydro-6,7-diphenyl-, hydrochloride	pH 7.0	248(4.30),286(4.00), 372(4.08)	24-3008-66
	pH 14.0	252(4.22),290s(3.81), 384(4.12)	24-3008-66
$C_{18}H_{15}OP$			
Phosphine oxide, triphenyl-, $MoOBr_3$ complex	CH_2Cl_2	725(1.61)	39-0630-66A
$MoOCl_3$ complex	CH_2Cl_2	443(1.66),730(1.57)	39-0630-66A
$C_{18}H_{15}P$			
Phosphine, triphenyl-	MeOH	256(4.0)	51-0476-66
	EtOH	261(4.04),282(3.10)	77-0425-66
tropylium perchlorate	MeCN	262s(--),268(3.69), 275(3.60)	27-0172-66
$C_{18}H_{16}$			
1,3-Cyclohexadiene, 1,3-diphenyl-	EtOH	251(4.42),315(3.97)	22-1012-66
15H-Cyclopenta[a]phenanthrene, 16,17-dihydro-17-methyl-	EtOH	218(4.39),259(4.69), 280(4.07),288(3.99), 300(4.05),320(2.76), 336(2.86),352(2.87)	39-0963-66C
7,8;9,10-Dibenzotricyclo[4.2.2.0^{2,5}]-deca-7,9-diene	EtOH	265(3.13),272(3.23)	5-0009-66A
α-Truxane	EtOH	265s(2.18),268(2.51), 276(2.56)	12-1455-66
$C_{18}H_{16}BCl_2F_4N_3S_2$			
Bis(6-chloro-3-ethylbenzothiazol-2-yl)azamethinecyanine tetrafluoroborate	MeOH	370(4.72),382(4.74)	27-0318-66
$C_{18}H_{16}BCl_2F_4N_5S_2$			
Bis(6-chloro-3-ethylbenzothiazol-2-yl)triazatrimethinecyanine tetrafluoroborate	MeOH	487(4.61)	27-0318-66
$C_{18}H_{16}BrClO$			
1-Indanone, 2-bromo-2-(p-chlorobenzyl)-3,3-dimethyl-	MeOH	253(4.09),296(3.53)	35-4489-66

Compound	Solvent	λ_{max}(log ϵ)	Ref.
$C_{18}H_{16}BrNO_2$			
1-Naphthaleneacetamide, 3-bromo-1,2,3,4-tetrahydro-4-oxo-1-phenyl-	EtOH	252(4.06),294(3.31)	44-1905-66
isomer	EtOH	253(4.03),293(3.33)	44-1905-66
$C_{18}H_{16}Br_2ClN_2O_4PS_2$			
6-Bromo-2-[(6-bromo-N-ethylbenzothiazolinylidene)phosphino]-3-ethylbenzothiazolium perchlorate	MeOH	480(4.71)	24-1325-66
	$C_2H_4Cl_2$	489(4.68)	24-1325-66
$C_{18}H_{16}Cl_2N_2O$			
2-Pyrrolidone, 4,4-dichloro-1,3-dimethyl-3-phenyl-5-(phenylimino)-	EtOH	233(4.41),306(3.42)	28C-1276-66A
2-Pyrrolidone, 4,4-dichloro-3-ethyl-3-phenyl-5-(phenylimino)-	EtOH	289(3.94)	28C-1276-66A
1-Pyrrolin-5-one, 3,3-dichloro-2-(ethylamino)-4,4-diphenyl-	EtOH	251(4.19)	28C-1276-66A
$C_{18}H_{16}Cl_2N_6O_5$			
1H-Pyrimido[4,5-b]indole-2,4(3H,9H)-dione, 7-(6-amino-1,3-dimethyl-2,4-dioxo-1,2,3,4-tetrahydro-5-pyrimidinyl)-5,8-dichloro-6-hydroxy-1,3-dimethyl-	pH 13	226(4.37),278(4.69), 344(3.75)	24-3524-66
$C_{18}H_{16}FeO_4$			
Ferrocene, maleic anhydride adduct	EtOH	327(3.94),412(2.91), 557(3.27)	44-0610-66
$C_{18}H_{16}IN$			
7,8-Dihydro-2-methylbenzo[6,7]cyclohepta[1,2,3-de]isoquinolinium iodide	EtOH	290(4.15)	4-0247-66
$C_{18}H_{16}N_2$			
o-Terphenyl-4,4'-diamine	EtOH	240(4.34),284(4.20)	56-0429-66
p-Terphenyl-4,4"-diamine	dioxan	316(4.57)	99-0204-66
	$C_2H_4Cl_2$	314(4.48)	99-0204-66
$C_{18}H_{16}N_2O$			
2,3'-Biindole, 2'-ethoxy-	MeOH	292(4.23),314(4.40)	24-3063-66
2-Indolinone, 3-methyl-3-(3-methyl-2-indolyl)-	MeOH	228(4.51),252s(3.90), 285(3.90),292(3.87)	44-2627-66
2-Pyrrolidinone, 5-imino-3-methyl-4-(diphenylmethylene)-	EtOH	283(4.19)	22-3895-66
$C_{18}H_{16}N_2OS$			
Benzo-1,4-thiazino[3,2-b]tropone, 9-amino-8-isopropyl-	MeOH	239(4.35),348(4.60), 435(3.77)	18-1988-66
$C_{18}H_{16}N_2O_3$			
2-Quinazolinemethanol, 4-methoxy-α-phenyl-, acetate	EtOH	211s(4.43),225(4.55), 265(3.76),300(3.38), 310(3.38)	39-2290-66C
4(3H)-Quinazolinone, 2-(α-acetoxybenzyl)-3-methyl-	EtOH	208(4.50),225(4.44), 270(3.92),278s(3.91), 305(3.59),317s(3.47)	39-2290-66C
$C_{18}H_{16}N_2O_4S$			
Benzenesulfonamide, p-(1,3-dihydroxybenz[f]isoindolin-2-yl)-	MeOH	254(3.46),283(3.63), 320s(--)	95-0001-66

Compound	Solvent	λ_{max}(log ϵ)	Ref.
$C_{18}H_{16}N_2O_6$			
Pyrrolo[2,1-a]phthalazine-1,2,3-tricarboxylic acid, 6-methyl-, trimethyl ester	MeOH	225(4.42),235s(4.37), 266(4.67),274(4.68), 282s(4.42),310(4.18)	39-2218-66C
$C_{18}H_{16}N_4$			
N-Iminoquinoline dimer	EtOH	318(3.82)	94-0512-66
Indole, 3,3'-azobis[2-methyl-	EtOH	281(4.21),373(4.46), 389(4.43),411(4.43)	39-1345-66C
$C_{18}H_{16}N_4O$			
Acetanilide, 3'-[2-amino-1-naphthyl)-azo]-	benzene	450(4.14)	44-2571-66
	$C_6H_{11}Me$	440(4.15)	44-2571-66
Acetanilide, 4'-[2-amino-1-naphthyl)-azo]-	benzene	460(4.27)	44-2571-66
Imidazole-4(or 5)-carboxamide, N-benzyl-5(or 4)-(benzylidene-amino)-	pH 1	249(4.35)	44-2202-66
	pH 7	244(4.32),333(4.17)	44-2202-66
	pH 13	248(4.35),368(4.17)	44-2202-66
$C_{18}H_{16}N_4O_2$			
Acetanilide, 3'-[(2-amino-8-hydroxy-1-naphthyl)azo]-	benzene	507(4.21)	44-2571-66
Acetanilide, 4'-[(2-amino-8-hydroxy-1-naphthyl)azo]-	benzene	511(4.32)	44-2571-66
1-Naphthylamine, N,N-dimethyl-4-(p-nitrophenylazo)-	EtOH	252s(4.07),292(4.17), 478(4.28)	19-0037-66
$C_{18}H_{16}N_4O_4$			
Benzobicyclo[3.2.1]oct-2-en-8-one, 2,4-dinitrophenylhydrazone	$CHCl_3$	362(4.39)	22-0147-66
Benzobicyclo[3.3.0]oct-3-en-6-one, 2,4-dinitrophenylhydrazone	$CHCl_3$	367(4.33)	22-0147-66
Benzobicyclo[4.2.0]oct-3-en-2-one, 2,4-dinitrophenylhydrazone	$CHCl_3$	396(4.42)	22-0147-66
Pyrazole, 1-(2,4-dinitrophenyl)-3-ethyl-4-methyl-5-phenyl-	EtOH	331(3.91)	22-2977-66
$C_{18}H_{16}N_6$			
Benzaldehyde, 3,6-pyridazinediyl-dihydrazone	THF	358(4.63)	78-2073-66
$C_{18}H_{16}N_8O_9$			
1-Propanone, 1-[1,5-bis(2,4-dinitro-phenyl)formazanyl)-2,2-dimethyl-	benzene	328(4.14),460(4.24), 650(2.60)	70-1326-66
	MeOH	350(3.85),430(3.82), 630(4.42)	70-1326-66
	90% MeOH	340(3.88),430(3.83), 630(4.23)	70-1326-66
	50% MeOH	338(3.99),430(3.95), 630(3.86)	70-1326-66
	10% MeOH	320(4.00),360(3.99), 410(3.93),450(3.87), 620(2.98)	70-1326-66
	50% MeOH-50% benzene	340(3.99),445(4.08), 632(4.07)	70-1326-66
	33.3% MeOH-benzene	332(4.06),453(4.18), 638(3.53)	70-1326-66
	25% MeOH-benzene	330(4.10),455(4.18), 640(2.88)	70-1326-66

Compound	Solvent	$\lambda_{max}(\log \epsilon)$	Ref.
$C_{18}H_{16}O$			
Bicyclo[3.1.0]hexan-2-one, 4,6-diphenyl-	EtOH	225s(4.00),266s(2.92), 273s(2.79)	35-4905-66
Cyclobutanone, 3-phenyl-2-styryl-, cis	C_6H_{12}	247(4.07),292s(2.93), 310s(2.72)	35-4905-66
2-Cyclohexen-1-one, 4,5-diphenyl-	EtOH	222s(4.18),235s(4.04), 264s(3.42),269s(3.25), 322(1.56)	35-4905-66
$C_{18}H_{16}O_2$			
3,5-Hexadienoic acid, 6,6-diphenyl-, cis	C_6H_{12}	231(4.19),236(4.19), 294(4.36)	35-4895-66
trans	EtOH	227(4.15),290(4.39)	35-4895-66
5-Hexen-1,4-olide, 6,6-diphenyl-	EtOH	230(4.20),254(4.22)	35-4895-66
[3.3]Paracyclophane, 1,11-dioxo-	EtOH	221(4.01),236(4.01), 285(3.53),302(3.36)	35-3515-66
[3.3]Paracyclophane, 1,10-dioxo-	EtOH	226(4.02),270(4.02), 322s(2.40)	35-3515-66
[3.3]Paracyclophane, 2,11-dioxo-	EtOH	220(4.04),240s(3.30), 299(2.56),312(2.49)	35-3515-66
1,4-Pentadien-3-one, 5-p-anisyl-1-phenyl-	HOAc	359(4.43)	65-0336-66
ferric chloride added	HOAc	504(4.44)	65-0336-66
2,4-Pentadien-1-one, 1-p-anisyl-5-phenyl-	benzene	345(--)	65-0336-66
	EtOH	350(--)	65-0336-66
	HOAc	348(4.60)	65-0336-66
	H_2SO_4	488(--)	65-0336-66
ferric chloride added	HOAc	468(4.45),502(4.45)	65-0336-66
2,4-Pentadien-1-one, 5-p-anisyl-1-phenyl-	benzene	362(--)	65-0336-66
	EtOH	369(--)	65-0336-66
	HOAc	268(4.14),372(4.38)	65-0336-66
	H_2SO_4	500(--)	65-0336-66
ferric chloride added	HOAc	548(4.43)	65-0336-66
4H-Pyran-4-one, 2,3-dihydro-2-methyl-2,6-diphenyl-	EtOH	244(3.85),304(4.25)	32-1073-66
	EtOH	244(3.85),304(4.25)	88-0233-66
Triphenylene, 1,2,3,4,9,10,11,12-octahydro-1,12-dioxo-	EtOH	222(4.25),262(4.71), 282(3.93),292(3.86), 304(3.62),348(3.59), 360(3.59)	24-1805-66
$C_{18}H_{16}O_3$			
1H-Cyclopenta[7,8]naphtho[2,3-b]furan-7-carboxylic acid, 2,3-dihydro-6-methyl-, methyl ester	EtOH	249(4.93),323(3.87), 325(3.88),328(3.90), 337(3.86)	39-0725-66C
Rotoxen-12(6H)-one, 6a,12a-dihydro-6,6-dimethyl-	EtOH	221(4.37),254(4.01), 283s(3.33),324(3.52)	39-0542-66C
$C_{18}H_{16}O_4$			
Benzoic acid, o-benzoyl-, 1-ethoxyvinyl ester	$CHCl_3$	330(2.13)	35-0781-66
Cinnamic acid, ester with 3-phenyllactic acid	pH 7.5	281(4.34)	69-1425-66
1-Propanone, 1-[2-(p-hydroxyphenyl)-7-methoxy-3-benzofuranyl]-	EtOH	312(4.14)	78-2913-66
	EtOH-NaOEt	365(4.22)	78-2913-66
$C_{18}H_{16}O_4S_2$			
Spiro[5H-dibenzo[a,d]cycloheptene-5,2'-m-dithiane], 1',1',3',3'-tetraoxide	EtOH-DMF	297(3.97)	87-0860-66

Compound	Solvent	λ_{max}(log ε)	Ref.
$C_{18}H_{16}O_5$			
1(2H)-Anthracenone, 3,4-dihydro-8,9-dihydroxy-, diacetate	EtOH	253(4.75),295(3.80),358(3.40)	78-2761-66
Flavone, 5-hydroxy-4',7-dimethoxy-6-methyl-	EtOH	277(4.24),332(4.32)	28C-0662-66A
Globuxanthone	MeOH	251(4.56),267(4.57),310s(4.05),406(3.96)	39-2186-66C
Isoaurone, 3',4',6'-trimethoxy-, cis	EtOH	390(4.34)	78-1789-66
Isoaurone, 3',4',6'-trimethoxy-, trans	EtOH	394(4.32)	78-1789-66
Naphtho[2,1-b]furan-1,2-dicarboxylic acid, diethyl ester	EtOH	241(4.19),250(4.16),293(4.18),303(4.54),324(4.11),337(4.16)	39-0463-66C
Naphtho[2,3-b]furan-4,9-diol, 3,8-dimethyl-, diacetate	EtOH	246(4.75),319(3.93),328(3.86)	78-0301-66
1H-Naphtho[2,1-b]pyran-1-one, 2-acetyl-5,6-dimethoxy-3-methyl-	EtOH	213(4.33),260(4.65),311(3.53),325(3.45),376(3.68)	94-0121-66
4H-Naphtho[2,3-b]pyran-4-one, 3-acetyl-5,10-dimethoxy-2-methyl-	EtOH	219(4.55),270(4.24),318(3.99),334s(3.84)	94-0129-66
$C_{18}H_{16}O_6$			
Flavone, 5-hydroxy-3,4',7-trimethoxy-	EtOH	268(4.18),320s(4.14),346(4.16)	78-0071-66B
aluminum chloride complex	EtOH	278(4.17),348(3.16),399(3.11)	78-0071-66B
Symphoxanthone	MeOH	238(4.64),260(4.64),323s(3.89),389(3.73)	39-2186-66C
Trimethyl ether "IIIb"	EtOH	280(4.21),310(3.47),413(3.43)	78-1507-66
Xanthone, 1,3,5,6-tetrahydroxy-2-(3,3-dimethylallyl)-	MeOH	251(4.42),283(3.81),324(4.21)	39-2265-66C +39-0178-66C
$C_{18}H_{16}O_7$			
Isoflavanone, 2-hydroxy-2',7-dimethoxy-4',5'-(methylenedioxy)-	EtOH	230(4.28),274(4.25),303(4.21)	18-1525-66
Penduletin	EtOH	214(4.57),271(4.37),344(4.39)	24-3222-66
$C_{18}H_{16}S_2$			
Spiro[5H-dibenzo[a,d]cycloheptene-5,2'-m-dithiane]	EtOH	217(4.59),285(4.00)	87-0860-66
$C_{18}H_{17}ClFN_3OS$			
4-Morpholinecarboxamide, N-[p-chloro-N-(p-fluorophenyl)benzimidoyl]thio-	MeOH	246(4.39),284s(4.25)	44-0722-66
$C_{18}H_{17}ClO$			
1-Indanone, 2-(p-chlorobenzyl)-3,3-dimethyl-	MeOH	243(4.11),295(3.66),325(3.67)	35-4489-66
$C_{18}H_{17}ClO_5$			
Spiro[cyclohexa-2,5-diene-1,2'-indan]-1',4-dione, 4'-chloro-2,5',7-trimethoxy-6-methyl-	MeOH	236(4.51),280(4.47),318(4.08)	44-2291-66
$C_{18}H_{17}ClO_6$			
Radicicol	EtOH	265(4.25)	12-1265-66
	EtOH-KOH	254(4.35),274(4.35),319(4.18)	12-1265-66

Compound	Solvent	$\lambda_{max}(\log \varepsilon)$	Ref.
$C_{18}H_{17}N$			
Cyclopent[b]indole, N-benzyl-	n.s.g.	228(4.68),288(3.86), 295(3.81)	16-0064-66
Phenalene-$\Delta^{1,\gamma}$-propenylamine, N,N-dimethyl-	MeCN	496(4.53)	89-0516-66
1-Pyrroline, 3-(diphenylmethylene)-2-methyl-	EtOH	231(4.09),287(4.17)	94-0187-66
$C_{18}H_{17}NO$			
Benzomorphan-3-one, 5-phenyl-	EtOH	252(2.57),258(2.66), 265(2.64),269(2.67), 273(2.40)	44-1905-66
8-chloro derivative	EtOH	260(2.75),265(2.79), 270(2.83),277(2.73)	44-1905-66
$C_{18}H_{17}NO_2$			
Anthraquinone, 1-[2-(dimethylamino)-ethyl]-	EtOH	505(3.84)	27-0281-66
Benzomorphan-3-one, 8-hydroxy-5-phenyl-	EtOH	252s(--),262(2.65), 268s(--),272(2.49)	44-1905-66
	EtOH	258(2.62),268(2.51)	44-1905-66
Indole-2-acetic acid, N-benzyl-, methyl ester	n.s.g.	218(4.71),282(3.94), 292(3.82)	16-0064-66
Indole-2-carboxylic acid, 1-benzyl-, ethyl ester	n.s.g.	206(4.52),225(4.39), 294(4.33)	16-0064-66
Indoline, 1-acetyl-7-phenacyl-	MeOH	243(4.34)	35-4061-66
1-Naphthaleneacetamide, 1,2,3,4-tetra-hydro-4-oxo-1-phenyl-	EtOH	249(4.06),295(3.32)	44-1905-66
Pyrrole, 2,3-bis(p-methoxyphenyl)-	EtOH	247(4.30),295(4.18)	87-0527-66
$C_{18}H_{17}NO_3$			
Alanine, N-cinnamoyl-3-phenyl-	pH 7.5	277(4.41)	69-1425-66
$C_{18}H_{17}NO_4$			
Actinodaphnine	EtOH	282(4.19),307(4.25)	95-0129-66
Hernangerine	EtOH	225(4.43),270(4.11), 310(3.75)	95-1143-66
Nandigerine	EtOH	225(4.40),271(4.13), 314(3.74)	88-1577-66
$C_{18}H_{17}NO_5$			
2-Acridancarboxylic acid, 1,3-dimeth-oxy-10-methyl-9-oxo-, methyl ester	EtOH	225(4.22),269(4.73), 274(4.75),298(3.79), 368s(3.83),381(3.88)	12-0275-66
4-Acridancarboxylic acid, 1,3-dimeth-oxy-10-methyl-9-oxo-, methyl ester	EtOH	225(4.18),256(4.53), 271(4.53),380(3.92)	12-0275-66
2-Acridinecarboxylic acid, 1,3,9-trimethoxy-, methyl ester	EtOH	235(4.29),262(5.03), 320s(3.38),334(3.65), 351(3.82),372s(3.61), 388(3.67)	12-0275-66
$C_{18}H_{17}N_3$			
1-Naphthylamine, N,N-dimethyl-4-(phenylazo)-	EtOH	208(4.65),244s(4.10), 277(4.14),419(4.23)	19-0037-66
2-Naphthylamine, 1-[(m-ethylphenyl)-azo]-	benzene	438(4.14)	44-2571-66
	$C_6H_{11}Me$	431(4.12)	44-2571-66
2-Naphthylamine, 1-[(p-ethylphenyl)-azo]-	benzene	440(4.14)	44-2571-66
	$C_6H_{11}Me$	431(4.19)	44-2571-66

Compound	Solvent	$\lambda_{max}(\log \epsilon)$	Ref.
$C_{18}H_{17}N_3O$			
11-Morphanthridinone, 6-(1-piperazinyl)-	EtOH	238(4.40),252s(4.35), 321(3.93)	33-1433-66
1-Naphthol, 7-amino-8-[(m-ethylphenyl)-azo]-	benzene	502(4.22)	44-2571-66
	$C_6H_{11}Me$	500(4.19)	44-2571-66
1-Naphthol, 7-amino-8-[(p-ethylphenyl)-azo]-	benzene	504(4.25)	44-2571-66
	$C_6H_{11}Me$	501(4.22)	44-2571-66
2-Naphthylamine, 1-[(m-ethoxyphenyl)-azo]-	benzene	449(4.14)	44-2571-66
	$C_6H_{11}Me$	434(4.16)	44-2571-66
2-Naphthylamine, 1-[(p-ethoxyphenyl)-azo]-	benzene	450(4.21)	44-2571-66
	$C_6H_{11}Me$	444(4.23)	44-2571-66
$C_{18}H_{17}N_3O_2$			
1-Naphthol, 7-amino-8-[(m-ethoxyphen-yl)azo]-	benzene	506(4.23)	44-2571-66
	$C_6H_{11}Me$	502(4.19)	44-2571-66
1-Naphthol, 7-amino-8-[(p-ethoxyphen-yl)azo]-	benzene	504(4.29)	44-2571-66
	$C_6H_{11}Me$	502(4.29)	44-2571-66
$C_{18}H_{17}N_3O_3S$			
2H-Pyrazolo[4,3-c]quinoline, 4,5-dihydro-7-methoxy-5-(p-tolylsulfonyl)-	EtOH	223(4.41),240(4.40), 260(4.27)	54-0681-66
$C_{18}H_{17}N_3O_4S$			
1-Naphthalenesulfonic acid, 4-hydroxy-3-(p-dimethylaminophenylazo)-	H_2O	520(4.34)	35-2240-66
	0.1N H_2SO_4	485(4.33)	35-2240-66
	74% H_2SO_4	518(4.40)	35-2240-66
$C_{18}H_{17}N_5O_2$			
Pyridine, 2,6-diamino-4-[(diphenyl-methyl)amino]-3-nitro-	pH 1	225(4.55),275(3.91), 345(4.05)	44-1890-66
$C_{18}H_{17}N_5O_7$			
1,3-Diazaazulene, 7-isopropyl-5-methyl-, picrate	MeOH	225(4.10),257(4.18), 358(3.86),370(3.61)	18-2450-66
$C_{18}H_{17}O_5P$			
Phosphonic acid, [α-(1,3-dioxo-2-indanyl)benzyl]-, dimethyl ester	isooctane	240(4.57),295(2.9)	30-1097-66F
	MeOH	225(4.34),245(4.12), 295(2.9)	30-1097-66F
$C_{18}H_{18}$			
Anthracene, 9-tert-butyl-	EtOH	249(4.81),256(5.26), 304s(--),318(3.03), 332(3.42),348(3.75), 362s(--),367(3.96), 382s(--),386(3.95)	44-4265-66
Benzene, hexavinyl-	hexane	224(4.30),258(4.60)	44-1310-66
1,3-Butadiene, 2,3-dimethyl-1,4-diphenyl-, cis-cis	C_6H_{12}	290(4.65)	35-2213-66
$C_{18}H_{18}BF_4N_3S_2$			
Bis(3-ethyl-2-benzothiazolyl)aza-methinecyanine tetrafluoroborate	MeOH	365(4.68),375(4.70)	27-0318-66
$C_{18}H_{18}BrClO_5$			
Benzophenone, 2-(bromomethyl)-3-chloro-2',4,6-trimethoxy-4'-hydroxy-6'-methyl-	MeOH	281(3.91),315(3.98)	44-2291-66

Compound	Solvent	$\lambda_{max}(\log \epsilon)$	Ref.
$C_{18}H_{18}BrNO_5$			
Azulene-1,3-dicarboxylic acid, 2-acetamido-6-bromo-, diethyl ester	MeOH	234(4.23),245s(4.22), 272(4.13),314s(4.69), 324(4.78),360s(3.87), 379(3.89),496(2.39)	18-1954-66
$C_{18}H_{18}ClN_2O_4PS_2$			
3-Ethyl-2-[(3-ethyl-2-benzothiazolin-ylidene)phosphino]benzothiazolium perchlorate	MeOH	472(4.68)	24-1325-66
	$C_2H_4Cl_2$	478(4.71)	24-1325-66
$C_{18}H_{18}ClN_3O_5$			
2-Indolinone, 5-chloro-3-[(3,4-di-hydro-6-hydroxy-4,5-dimethyl-3-oxopyrazinyl)methylene]-6,7-dimethoxy-1-methyl-	pH 13	253(4.27),300(4.03), 430(4.14)	39-1803-66C
	MeOH	204(4.06),325(3.49)	39-1803-66C
$C_{18}H_{18}ClN_5O_4S_2$			
Bis(3-ethyl-2-benzothiazolyl)triaza-trimethinecyanine perchlorate	MeOH	485(4.62)	27-0318-66
$C_{18}H_{18}Cl_{10}Si_3$			
Trisilane, 1,1,2,2,3,3-hexamethyl-1,3-bis(pentachlorophenyl)-	C_6H_{12}	213(5.02)	101-0102-66B
$C_{18}H_{18}CuN_2O_4$			
Copper, [[α,α'-(ethylenedinitrilo)-bis(6-methoxy-o-cresolato)](2-)]-	MeOH	288(4.14),370(3.63), 380(2.29)	65-1590-66
$C_{18}H_{18}F_3N_3$			
as-Triazine, 1,4,5,6-tetrahydro-1,6-dimethyl-5-phenyl-3-(p-tri-fluoromethylphenyl)-	MeOH	224(3.82),322(3.53)	87-0881-66
$C_{18}H_{18}INO_2$			
6,7-Dimethoxy-2-methyl-1-phenyl-isoquinolinium iodide	MeOH	255(4.62),325(3.99)	83-0159-66
$C_{18}H_{18}N_2$			
Dipyrromethene, 3,5,5'-trimethyl-3'-phenyl-	EtOH-HCl	222(4.26),260(3.53), 365(3.68),474(4.92)	12-1871-66
	EtOH	219(4.16),277(3.63), 443(4.45)	12-1871-66
	CCl_4	448(4.47)	12-1871-66
Dipyrromethene, 4,5,5'-trimethyl-3'-phenyl-	EtOH-N HCl	227(3.82),489(4.80)	12-1871-66
	EtOH	225(3.80),439(4.40)	12-1871-66
	CCl_4	444(4.41)	12-1871-66
$C_{18}H_{18}N_2O$			
Isoquino[2,1-d][1,4]benzodiazepin-6(7H)-one, 5,9,10,14b-tetra-hydro-5-methyl-	EtOH	237(3.98)	33-1222-66
2-Pyrazolin-5-one, 4-benzyl-3,4-dimethyl-1-phenyl-	EtOH	243(4.15)	88-6413-66
3-Pyrazolin-5-one, 2-benzyl-3,4-dimethyl-1-phenyl-	EtOH	250(4.03),277(4.03)	88-6413-66
$C_{18}H_{18}N_2O_2$			
Indolizine-7-carboxamide, N-(p-meth-oxyphenyl)-2,5-dimethyl-	EtOH	227(4.51),274(4.35), 353(4.33)	56-0061-66

Compound	Solvent	$\lambda_{max}(\log \epsilon)$	Ref.
$C_{18}H_{18}N_2O_3$			
Imidazole-2-methanol, α,4-diphenyl-, acetate	H_2SO_4	261(4.31),396(2.39), 592(3.12)	24-1815-66
$C_{18}H_{18}N_2O_7$			
Ether, 3-hydroxy-3-methylbutyl p-nitrophenyl, p-nitrobenzoate	EtOH	263(4.18),302(4.13)	94-0939-66
$C_{18}H_{18}N_4O$			
Imidazole-4-carboxamide, 5-amino-N,1-dibenzyl-	pH 1	243(4.03),268(4.12)	44-2202-66
	pH 7	268(4.22)	44-2202-66
	pH 13	267.5(4.22)	44-2202-66
Imidazole-4-carboxamide, 1-benzyl-5-(benzylamino)-	pH 1	254(3.86)	44-2202-66
	pH 7	269(3.94)	44-2202-66
	pH 13	269(3.94)	44-2202-66
Imidazole-4(or 5)-carboxamide, N-benzyl-5(or 4)-(benzylamino)-, hydrochloride	pH 1	247(4.08),282(4.14)	44-2202-66
	pH 7	281(4.19)	44-2202-66
	pH 13	288(4.19)	44-2202-66
$C_{18}H_{18}N_6O_5$			
1H-Pyrimido[4,5-b]indole-2,4(3H,9H)-dione, 7-(6-amino-1,2,3,4-tetrahydro-1,3-dimethyl-2,4-dioxo-5-pyrimidinyl)-6-hydroxy-1,3-dimethyl-	pH 13	224(4.32),276(4.63), 341s(3.41)	24-3524-66
$C_{18}H_{18}N_8O_8$			
Succindialdehyde, bis(2,4-dinitrophenylhydrazone)	n.s.g.	355(4.52)	22-0734-66
$C_{18}H_{18}O$			
Ether, bis(p-methylstyryl)	EtOH	283(4.53)	24-0642-66
Ether, bis(β-methylstyryl)	EtOH	279(4.19)	24-0642-66
[2.4]Paracyclophane, 2-oxo-	EtOH	217(4.12),285(2.85)	35-3515-66
1(2H)-Triphenylenone, 3,4,9,10,11,12-hexahydro-	EtOH	220(4.40),254(4.60), 260(4.64),282(3.89), 292(3.93),303(3.80), 350(3.41)	24-1805-66
$C_{18}H_{18}O_2$			
2-Furanol, 2,5-dihydro-5,5-dimethyl-2,4-diphenyl-	N H_2SO_4	241(3.53),304(3.87), 395(4.67)	65-2064-66
$C_{18}H_{18}O_3$			
Benzo[3,4]cyclobuta[1,2-b]benzofuran-1,4-dione, 4a,4b,9b,9c-tetrahydro-2,3,4a,9c-tetramethyl-	CH_2Cl_2	284(3.53),288(3.53), 374s(2.29)	24-1723-66
1H-Cyclopenta[7,8]naphtho[2,3-b]furan-7-carboxylic acid, 2,3,7,8-tetrahydro-6-methyl-, methyl ester	EtOH	242(4.78),249(4.80), 290(3.79),302(3.74), 322(3.27),337(3.37)	39-0725-66C
Ether, bis(p-methoxystyryl), cis-trans	EtOH	284(4.57)	24-0642-66
trans-trans	EtOH	285(4.58)	24-0642-66
14β-Gona-1,3,5(10),8-tetraene-11,17-dione, 3-methoxy-	MeOH	244(4.15),292(3.67)	19-0611-66
9-Phenanthrenebutyric acid, 1,2,3,4-tetrahydro-1-oxo-	EtOH	244(4.39),251(4.61), 257(3.68),281(3.90), 291(4.02),302(3.90), 347(3.47)	24-1805-66
Propionic acid, o-benzoylphenyl-α-methyl-, methyl ester	$CHCl_3$	250(4.14)	35-4523-66
pseudo methyl ester	$CHCl_3$	262(2.73)	35-4523-66

Compound	Solvent	$\lambda_{max}(\log \epsilon)$	Ref.
Rotoxen, 6,6a,12,12a-tetrahydro-12-hydroxy-6,6-dimethyl-	EtOH	220s(4.07),276(3.85), 284(3.82)	39-0542-66C
$C_{18}H_{18}O_4$			
Acrylic acid, 3-(p-methoxyphenyl)-2-(6-methoxy-m-tolyl)-, cis	EtOH	285(4.40)	78-1789-66
trans	EtOH	289(4.46)	78-1789-66
3-Flavene, 4'-hydroxy-2,8-dimethoxy-3-methyl-	EtOH	263(4.13),274(4.11)	78-2913-66
Maturin, cyclic ethyl acetal	EtOH	218(4.14),250(4.74), 335(3.92),350(3.95)	78-0685-66
$C_{18}H_{18}O_5$			
Benzoic acid, p-hydroxy-, ester with 1-(2-hydroxy-3-methoxyphenyl)-2-butanone	EtOH	262(4.32)	78-2913-66
	EtOH-NaOEt	309(4.54)	78-2913-66
Benzoic acid, p-hydroxy-, α-ethyl-2-hydroxy-3-methoxystyryl ester	EtOH	257(4.38)	78-2913-66
	EtOH-NaOEt	309(4.50)	78-2913-66
Isoflavan, 2',7-dimethoxy-4',5'-(methylenedioxy)-	EtOH	291(3.86),301(3.82)	88-1805-66
1,4-Naphthalenedibutyric acid, γ-oxo-	EtOH	215(4.84),225(4.89), 236(4.60),304(4.18)	24-1805-66
$C_{18}H_{18}O_5S$			
Ethenol, 2-methoxy-2-phenyl-1-(p-tolylsulfonyl)-, acetate	MeOH	229(4.2),262(4.2)	24-0048-66
$C_{18}H_{18}O_6$			
Acrylic acid, 3-(3,4-dimethoxyphenyl)-2-(2-hydroxy-4-methoxyphenyl)-, cis	EtOH	315(4.26)	78-1789-66
trans	EtOH	309(4.38)	78-1789-66
Frenolicin, deoxy-	MeOH	246(3.96),274(4.06), 420(3.63)	35-4109-66
$C_{18}H_{18}O_7$			
Frenolicin	MeOH	234(4.26),284s(3.54), 362(3.72)	35-4109-66
	MeOH-NaOH	280(3.81),425(3.79)	35-4109-66
1,3-Propanedione, 1-(3,6-dihydroxy-2,4-dimethoxyphenyl)-3-(p-methoxyphenyl)-	EtOH	223(4.24),283(4.26), 348s(3.82)	44-3228-66
$C_{18}H_{18}O_7S$			
Hydroquinone, 2-acetyl-3-(3,4-dimethoxy-2-furyl)-, diacetate	EtOH	250(3.96),282(3.90), 306(3.78)	33-1794-66
$C_{18}H_{19}BrO_4$			
Ethylene, 1-bromo-2,2-bis-(3,4-dimethoxyphenyl)-	H_2SO_4	370(4.1),450s(3.9), 580(4.7)	27-0324-66
	FSO_3H-SbF_5	320(3.9),490(4.8)	27-0324-66
$C_{18}H_{19}ClN_4O_4$			
1-Methyl-2-(p-dimethylaminophenylazo)-quinolinium perchlorate	MeOH	585(5.00)	5-0116-66G
$C_{18}H_{19}ClO_5$			
Spiro[cyclohex-2-ene-1,2'-indan]-1',4-dione, 4'-chloro-2,5',7'-trimethoxy-6-methyl-	MeOH	235(4.44),275(4.29), 315(3.98)	44-1462-66

Compound	Solvent	$\lambda_{max}(\log \epsilon)$	Ref.
Spiro[cyclohex-3-ene-1,2'-indan]-1',2-dione, 4'-chloro-4,5',7'-trimethoxy-6-methyl-	MeOH	234(4.44),266(4.31), 318(3.95)	44-1462-66
$C_{18}H_{19}ClO_6$			
Spiro[benzofuran-2(3H),1'-[2]cyclohexene]-3,4'-dione, 7-chloro-4-ethoxy-2',6-dimethoxy-6'-methyl-	EtOH	218(4.36),235(4.36), 291(4.37),328(3.75)	87-0242-66
$C_{18}H_{19}Cl_2NO$			
Fluoren-2-amine, N,N-bis-(2-chloroethyl)-7-methoxy-	0.2N HCl	284(4.43)	87-0593-66
	pH 1	285(4.40)	87-0593-66
	EtOH	299(4.55),334(3.94), 348s(--)	87-0593-66
$C_{18}H_{19}N$			
Benzo[6,7]cyclohept[1,2,3-de]isoquinoline, 1,2,3,7,8,12b-hexahydro-2-methyl-	EtOH	262(2.76)	4-0247-66
Indole, N-benzyl-3-ethyl-2-methyl-	n.s.g.	226(4.54),285(3.83), 292(3.79)	16-0064-66
$C_{18}H_{19}NO$			
Acrylophenone, 3-[(α-methylbenzyl)-amino]-4'-methyl-, cis	PrOH	347(4.38)(anom.)	39-1221-66B
	$CHCl_3$	347(4.38)	39-1221-66B
trans	PrOH	346.5(4.40)(anom.)	39-1221-66B
	$CHCl_3$	333(4.32)	39-1221-66B
Benzomorphan, 8-hydroxy-5-phenyl-	EtOH	252s(--),258(2.71), 264(2.69),268s(--), 273(2.43)	44-1905-66
Pyridine, 4-[1-(3,4-dihydro-6-methoxy-1-naphthyl)ethyl]-	EtOH	265(4.09)	44-3337-66
$C_{18}H_{19}NO_2$			
1-Pyrroline, 2,3-bis(p-methoxyphenyl)-	EtOH	268(4.26),286s(4.02), 295s(3.72)	87-0527-66
$C_{18}H_{19}NO_3$			
Stepharine	EtOH	233(4.40),283(3.50)	95-0460-66
$C_{18}H_{19}NO_4$			
Hernovine	EtOH	221(4.41),272(4.01), 306(3.64)	88-1577-66
3H-Indolo[7a,1-a]isoquinolin-3-one, 5,6,8,9-tetrahydro-11-hydroxy-2,12-dimethoxy-	MeOH	241(4.12),283(3.47)	88-2557-66
Laurelliptine	EtOH	281(4.11),305(4.15)	12-0135-66
$C_{18}H_{19}NO_5$			
Azulene-1,3-dicarboxylic acid, 2-acetamido-, diethyl ester	MeOH	243(4.37),268(4.32), 306s(4.80),315(4.88), 351(3.95),366(3.98), 480(2.77)	18-1954-66
Melicopicine	EtOH	215(4.20),268(4.72), 294(3.84),400(3.89)	100-0206-66
Tyrosine, benzyloxycarbonyl-, methyl ester	MeOH	277(3.31)	5-0226-66F
	MeOH-NaOH	292(3.37)	5-0226-66F

Compound	Solvent	$\lambda_{max}(\log \epsilon)$	Ref.
$C_{18}H_{19}NO_6$			
Azulene-1,3-dicarboxylic acid, 2-acetamido-6-hydroxy-, diethyl ester	MeOH	234(4.51),275(4.46), 304(4.70),337(3.87), 366(3.92),505(2.93)	18-1954-66
Diphenylamine-2,4'-dicarboxylic acid, 3',5'-dimethoxy-	EtOH	298(4.24),348(4.02)	12-0275-66
$C_{18}H_{19}NO_8$			
9aH-Quinolizine-1,2,3,4-tetracarboxylic acid, 8-methyl-, tetramethyl ester	MeOH	236(4.21),294(4.05),	39-1121-66C
$C_{18}H_{19}N_3O$			
4(1H)-Cinnolinone, 1-(2-dimethylaminoethyl)-3-phenyl-	MeOH	264(4.11),312(4.00), 361(4.09)	87-0784-66
11-Morphanthridinol, 6-(1-piperazinyl)-	EtOH	231s(4.19),246s(4.03), 305(3.93)	33-1433-66
$C_{18}H_{19}N_3OS$			
p-Benzoquinone, 2,6-diethyl-, azine with N-methylbenzothiazol-2-one	MeOH	484(4.66)	5-0116-66G
4-Morpholinecarboxamide, N-(N-phenylbenzimidoyl)thio-	EtOH	200(4.57),238(4.29), 286(4.19)	44-0722-66
$C_{18}H_{19}N_3O_3$			
Alanine, N-acetyl-3-[p-(phenylazo)-phenyl]-, methyl ester	dioxan	228(4.20),324(4.25), 438(3.23)	35-5010-66
	CF_3COOH	300(3.83),427(4.21)	35-5010-66
$C_{18}H_{19}N_3O_5$			
Isonicotinic acid, [[6-(benzyloxy)-5-hydroxy-p-dioxan-2-yl]methylene]-hydrazide	EtOH	261(4.06)	39-2121-66C
$C_{18}H_{19}N_3O_6S$			
Cephalosporanic acid, 7-(α-amino-phenylacetamido)-, zwitter ion	EtOH	260(3.87)	87-0746-66
$C_{18}H_{20}$			
Anthracene, 9-tert-butyl-9,10-dihydro-	EtOH	260s(2.96),267(3.11), 275(3.13)	39-0861-66B
Stilbene, α-butyl-	EtOH	225(3.72),268(4.21)	35-0476-66
$C_{18}H_{20}BrNO_4$			
2,3,6,7-Tetrahydroxy-N-methyl-aporphinium bromide	MeOH	260(4.16),289(4.28), 325(4.19)	5-0144-66E
$C_{18}H_{20}Br_2O_4$			
Pentatriafulvalene-1,3-dicarboxylic acid, 2,4-dibromo-5,6-dipropyl-, dimethyl ester	n.s.g.	260(4.45),340(4.62)	88-3697-66
$C_{18}H_{20}ClNO_4$			
Dehydro-N-methylisolaudanosolinium chloride	MeOH	260(4.12),289(4.23), 325(4.13)	5-0144-66E
3,5-Pyridinedicarboxylic acid, 4-(chloromethyl)-1,4-dihydro-2,6-dimethyl-1-phenyl-, dimethyl ester	EtOH	235(4.35),347(3.83)	39-1950-66C
$C_{18}H_{20}ClNO_6$			
8-(Benzyloxy)-3,4-dihydro-7-methoxy-2-methylisoquinolinium perchlorate	iso-PrOH	239(4.12),301(4.07), 379(3.43)	88-1599-66

Compound	Solvent	λ_{max}(log ε)	Ref.
$C_{18}H_{20}ClN_3O_5S_2$			
Isosporidesmin B	MeOH	219(4.60),262(4.10)	39-1803-66C
$C_{18}H_{20}N_2$			
2H-1,5-Methanobenzo[d][1,3]diazocine, 3-benzyl-3,4,5,6-tetrahydro-	pH 1	258(2.96),262(2.96), 268(2.86)	94-0566-66
	EtOH	269.5(2.90)	94-0566-66
$C_{18}H_{20}N_2O$			
Indole, 3-[2-(p-methoxybenzyl)amino]-	50% dioxan	277(3.80)	71-0550-66A
$C_{18}H_{20}N_2O_2$			
2H-Pyrrolium, 2-[2-hydroxy-4-oxo-3-(1,3,5-trimethylpyrrol-2-yl)-2-cyclobuten-1-ylidene]-1,3,5-trimethyl-	$CHCl_3$	580(4.16)	5-0153-66I
$C_{18}H_{20}N_2O_4$			
o-Cresol, α,α'-(ethylenedinitrilo)-bis[6-methoxy-	MeOH	225(3.34),265(4.27), 295(3.82),325(3.54)	65-1590-66
$C_{18}H_{20}N_2O_5$			
Azulene-1,3-dicarboxylic acid, 2-acetamido-6-amino-, diethyl ester	MeOH	267(4.20),288(4.27), 337(4.85),393(4.23)	18-1954-66
4,5-Diazaspiro[2.4]hept-5-ene-6,7-dicarboxylic acid, 4-benzoyl-, diethyl ester	MeOH	237(3.84),302(4.15)	35-3798-66
$C_{18}H_{20}N_2O_7S_2$			
Cephalosporanic acid, 7-[2-(2-thienyl)-acetamido]-, methoxymethyl ester	EtOH	236(4.07),261(3.88)	87-0741-66
$C_{18}H_{20}N_2O_8$			
Pyridazino[1,6-a]azepine-6,7,8,9-tetracarboxylic acid, 8,9-dihydro-2-methyl-, tetramethyl ester	MeOH	234(3.93),318(4.26), 325s(4.24),393(3.78), 448(3.75)	39-2218-66C
	MeOH-$HClO_4$	258s(3.79),298s(3.51)	39-2218-66C
4H-Pyrido[1,2-b]pyridazine-5,6,7,8-tetracarboxylic acid, 2,4a-dimethyl-, tetramethyl ester	MeOH	266(4.14),351s(3.71), 400(3.89)	39-2218-66C
$C_{18}H_{20}N_4$			
Cyclohexane, 1,2-bis(phenylazo)-, trans	hexane	264(4.27),409(2.45)	39-1989-66C
	EtOH	268(4.22),405(2.52)	39-1989-66C
$C_{18}H_{20}N_4O_6$			
Purine, 6-(benzyloxy)-2-methoxy-9-β-D-ribofuranosyl-	pH 1	236(3.79),267(4.06)	44-3258-66
	pH 11	245(3.83),265(4.07)	44-3258-66
$C_{18}H_{20}N_4O_8$			
Pyrrole-3,4-dicarboxylic acid, N-(2,4-dinitroanilino)-2,5-dimethyl-, diethyl ester	EtOH	235(4.35),258(4.25), 320(4.19)	39-0341-66C
$C_{18}H_{20}N_6$			
Tetraethylammonium salt of cyclopentadiene-1,2,3,4,5-pentacarbonitrile	MeCN	246(4.60),255(4.98), 281(4.04),291(4.01)	35-3046-66
	MeCN	246(4.78),255(4.98), 281(4.04),291(4.01)	35-4055-66

Compound	Solvent	λ_{max}(log ϵ)	Ref.
$C_{18}H_{20}N_8$			
Tetraethylammonium salt of 5-(cyano-azo)cyclopentadiene-1,2,3,4-tetracarbonitrile	MeCN	254(4.12),276(4.29), 405(4.33)	35-4055-66
$C_{18}H_{20}O$			
9-Anthrol, 9-tert-butyl-9,10-dihydro-	$CHCl_3$	258s(2.84),264(2.94), 272(2.90)	44-4265-66
14β-Gona-1,3,5(10),6,8-pentaen-3-ol, 17-methyl-	MeOH	228(4.83),258(3.59), 268(3.71),278(3.74), 288(3.58),326(3.30), 339(3.38)	27-0251-66
$C_{18}H_{20}O_2$			
Butyrophenone, 4-ethoxy-4-phenyl-	n.s.g.	242(4.09)	22-0717-66
Butyrophenone, 3-hydroxy-4'-methyl-3-p-tolyl-	EtOH	256(4.20)	28C-1429-66A
2,5-Cyclohexadien-1-one, 4,4'-(di-methylvinylene)bis[4-methyl-	MeOH	245.5(4.47)	5-0031-66D
Equilin	EtOH	223s(3.83),279(3.40), 286(3.22)	88-5015-66
8,11-Ethenobenzocyclododecene-5,14-dione, 1,4,4a,6,7,12,13,14a-octahydro-, cis	EtOH	226(3.66),274(2.42), 282s(2.38),300s(1.86)	35-0515-66
trans	EtOH	225s(3.76),264s(2.37), 272(2.40),280s(2.38), 296s(1.47)	35-0515-66
9-Phenanthrenebutyric acid, 1,2,3,4-tetrahydro-	EtOH	215(4.55),233(4.85), 279(3.77),284(3.79), 286(3.81),309(3.30), 325(3.15)	24-1805-66
Spiro[benzofuran-2(3H),1'-[2]cyclo-hexen]-4'-one, 3-ethylidene-2',4,6-trimethyl-	n.s.g.	230(4.52),238(4.57), 256(4.23),264(4.11), 268(4.15),306(4.03), 314(4.06),318(4.09)	78-2015-66
$C_{18}H_{20}O_3$			
Estra-1,3,5(10),8(9)-tetraen-17-one, 3,14-dihydroxy-	n.s.g.	275(4.18)	88-3585-66
$C_{18}H_{20}O_4$			
Butyrophenone, 3-hydroxy-4'-methoxy-3-(p-methoxyphenyl)-	EtOH	221(4.27),278(4.30)	28C-1429-66A
Ethylene, 1,1-bis(3,4-dimethoxy-phenyl)-	H_2SO_4	350(4.0),400(3.9), 530(4.7)	27-0324-66
1,4-Naphthalenedibutyric acid	EtOH	228(4.81),269(3.62), 279(3.83),300(3.77), 316(2.88),320(2.53)	24-1805-66
$C_{18}H_{20}O_5$			
Mandelic acid, 2-methoxy-α-(p-methoxybenzyl)-5-methyl-	EtOH	280(3.80)	78-1789-66
$C_{18}H_{20}S_2$			
[1]Benzothiopyran[8,7-h][1]benzothio-pyran, 1,2,3,7,8,9-hexahydro-1,7-dimethyl-	C_6H_{12}	252(4.79),320(4.05), 330(4.05),345(4.01)	78-0007-66
$C_{18}H_{21}ClO_6$			
Radicicol, tetrahydro-	EtOH	215(4.40),265(3.90), 310(3.68)	12-1265-66

Compound	Solvent	λ max(log ε)	Ref.
$C_{18}H_{21}Li$			
Lithium, 1,1-diphenylhexyl-	pyridine	484(--)	35-2109-66
	$BuNH_2$	445(--)	35-2109-66
	tetrahydro-pyran	436(--)	35-2109-66
$C_{18}H_{21}N$			
Quinoline, 1,2,3,4-tetrahydro-2,2,4-trimethyl-4-phenyl-	MeOH	248(3.99),300(3.43)	30-0877-66*
$C_{18}H_{21}NO$			
2',6'-Benzoxylide, 2,4,6-trimethyl-	EtOH	244s(3.8),272s(3.0)	28C-0369-66A
$C_{18}H_{21}NO_2$			
Biphenyl, 3,3',4,4',5,5'-hexamethyl-2-nitro-	isooctane	243(4.19)	44-4204-66
13H-Dibenzo[a,f]quinolizin-13-one, 1,2,3,4,6,7,11b,12-octahydro-9-methoxy-	EtOH	285(3.57),333(4.16)	44-0797-66
$C_{18}H_{21}NO_3$			
Azepino[3,2,1-hi]indole-Δ7(4H),α-acetic acid, 1,2,5,6-tetrahydro-6-oxo-, tert-butyl ester	EtOH	251(4.18),378(3.72)	35-4061-66
Ethanol, 2,2'-[(7-methoxyfluoren-2-yl)imino]di-	0.2N HCl	219s(--),283(4.39), 320s(--),315(4.08)	87-0593-66
	pH 1	182(4.39),220s(--), 302s(--),315(4.09)	87-0593-66
	EtOH	304(4.54),338s(--)	87-0593-66
Indano[5,4-f]quinoline, 2-carboxy-4b,5,6,6a,9a,9b,10,11-octahydro-6a-methyl-7-oxo-	EtOH	231(3.90),278(3.86)	37-1587-66
Litsericinone, N-methyl-	EtOH	240s(3.66),290(3.60)	95-1205-66
$C_{18}H_{21}NO_4$			
2',6'-Benzoxylidide, 3,4,5-trimethoxy-	EtOH	212(4.7),261(4.1)	28C-0584-66A
7-Isoquinolinol, 1,2,3,4-tetrahydro-1-(p-hydroxyphenyl)-6,8-dimethoxy-	MeOH-HCl	225s(4.32),278(3.54)	35-4212-66
	MeOH	225s(4.39),280(3.60)	35-4212-66
$C_{18}H_{21}NO_5$			
Azepino[3,2,1-hi]indole-7-acetic acid, 1,2,4,5,6,7-hexahydro-6-oxo-7-carboxy-, α-ethyl 7-methyl ester	EtOH	257(3.81),312(3.40)	35-4061-66
$C_{18}H_{21}NO_6$			
2,5-Cyclopentadiene-1,2-dicarboxylic acid, 4-(dipropyl-2-cyclopropen-1-ylidene)-3-nitro-, dimethyl ester	n.s.g.	225(4.00),279(4.38), 340(3.87),380(3.93)	88-3697-66
$C_{18}H_{21}N_3$			
Benzimidazole, 1-benzyl-2-(2-dimethylaminoethyl)-	n.s.g.	273(3.82),282(3.81), 292(3.67)	4-0278-66
as-Triazine, 1,4,5,6-tetrahydro-1,6-dimethyl-5-phenyl-3-m-tolyl-	MeOH	289(3.62)	87-0881-66
as-Triazine, 1,4,5,6-tetrahydro-1,6-dimethyl-5-phenyl-3-p-tolyl-	MeOH	227(4.21),289(3.68)	87-0881-66
$C_{18}H_{21}N_3O$			
1H-Pyrazolo[4,3-c]pyridin-4(5H)-one, 6-methyl-3-pentyl-1-phenyl-	EtOH	223(4.36),239(4.40), 286(4.08)	70-1723-66

Compound	Solvent	$\lambda_{max}(\log \epsilon)$	Ref.
as-Triazine, 1,4,5,6-tetrahydro-1,6-dimethyl-3-(p-methoxyphenyl)-5-phenyl-	MeOH	239(4.17),279(3.81)	87-0881-66
$C_{18}H_{21}N_3O_5$			
Azulene-1,3-dicarboxylic acid, 2-acetamido-6-hydrazino-, diethyl ester	MeOH	222(4.27),271(4.26), 288s(4.19),341(4.77), 399(4.26)	18-1954-66
$C_{18}H_{21}N_3O_6$			
Pyrrole-3,4-dicarboxylic acid, 2,5-dimethyl-N-(4-nitroanilino)-, diethyl ester	EtOH	335(4.15)	39-0341-66C
$C_{18}H_{21}N_5O_2$			
Tetraethylammonium salt of 2,3,4,5-tetracyanocyclopentadiene-carboxylic acid	MeCN	250(4.71),258(4.90), 285(4.02),295(4.02)	35-4055-66
$C_{18}H_{21}N_5O_5S_2$			
Adenine, 9-(3-S-benzyl-3-thio-β-D-arabinofuranosyl)-, 2'-methanesulfonate	pH 1	257(4.19)	44-3263-66
	pH 7	258(4.19)	44-3263-66
	pH 13	258(4.18)	44-3263-66
isomer	pH 1	293(4.12)	44-3263-66
	pH 7	293(4.12)	44-3263-66
	pH 13	275(3.95),334(3.51)	44-3263-66
$C_{18}H_{21}N_7O_2$			
s-Triazolo[4,3-a]-s-triazine, 5,7-dimorpholino-3-phenyl-	EtOH	264(4.46)	49-1713-66
$C_{18}H_{22}BrNO_4$			
N-Methylisolaudanosolinium bromide	MeOH	285(3.95)	5-0144-66E
$C_{18}H_{22}ClNO$			
6,7-Dihydro-2-(p-hydroxyphenethyl)-4,7-dimethylpyrindinium chloride	acid or neutral	222(4.08),267(3.74)	88-0445-66
	base	242(4.20),292(3.48)	88-0445-66
$C_{18}H_{22}ClNO_4$			
7-Isoquinolinol, 1,2,3,4-tetrahydro-1-(p-hydroxyphenyl)-1,2,3,4-tetrahydro-, hydrochloride	MeOH	225s(4.32),278(3.54)	35-4212-66
$C_{18}H_{22}N_2$			
Azocumene	C_6H_{12}	252(2.82),258(2.78), 264(2.57),367(1.64)	35-0137-66
$C_{18}H_{22}N_2O_2$			
Urea, 1-[p-(benzyloxy)phenyl]-3-butyl-	EtOH	245(4.38),293(3.29)	87-0444B-66
$C_{18}H_{22}N_2O_3$			
1H-Isoquino[2,1-a][1,6]naphthyridin-13-one, 2,3,4,6,7,11b,12,13-octahydro-9,10-dimethoxy-	EtOH	290(3.75),330(4.15)	44-0797-66
$C_{18}H_{22}N_2O_4$			
Azepine-N-carboxylic acid, ethyl ester, dimer	EtOH	232(4.15),238(4.17)	35-2590-66

Compound	Solvent	$\lambda_{max}(\log \epsilon)$	Ref.
Azepine-N-carboxylic acid, ethyl ester, dimer, exo	EtOH	213(4.14),241(3.97)	35-2590-66
$C_{18}H_{22}N_2O_5$			
Azulene-1,3-dicarboxylic acid, 2-amino-6-(2-hydroxyethylamino)-, diethyl ester	MeOH	252(4.09),297s(3.91), 343(4.76),409(4.24)	18-1954-66
$C_{18}H_{22}N_2O_{10}$			
Pyrazine, β-D-glucopyranosyloxy-, tetraacetate	pH 7.0	200(4.13),272(3.73), 285s(3.64)	24-0542-66
	MeOH	205(4.04),274(3.75), 290s(3.48)	24-0542-66
$C_{18}H_{22}N_2O_{11}$			
β-D-Glucopyranuronic acid, 1-deoxy-1-(1,2-dihydro-4-methoxy-2-oxo-1-pyrimidinyl)-2,3,4-tri-O-acetyl-, methyl ester	EtOH	276.5(3.75)	94-1354-66
$C_{18}H_{22}N_4$			
Adipaldehyde, bis(phenylhydrazone)	EtOH	285(4.33)	39-1989-66C
Cyclohexanone, 2-(β-phenylhydrazino)-, phenylhydrazone	n.s.g.	208(4.36),252(4.28), 276(4.30)	35-3865-66
Hexane, 1,6-bis(phenylazo)-	EtOH	263(4.31),408(2.38)	39-1989-66C
$C_{18}H_{22}N_4O_4$			
Pyrazole, 3-cyclohexyl-5-ethyl-4-methyl-1-(2,4-dinitrophenyl)-	EtOH	339(3.76)	22-2977-66
Pyrazole, 5-cyclohexyl-3-ethyl-4-methyl-1-(2,4-dinitrophenyl)-	EtOH	326(3.76)	22-2977-66
$C_{18}H_{22}N_4O_{10}$			
β-D-Glucopyranuronamide, 1-deoxy-1-(4-amino-1,2-dihydro-2-oxo-1-pyrimidinyl)-, tetraacetate	50% EtOH	251(4.23),300(3.85)	94-1354-66
$C_{18}H_{22}N_4O_{13}$			
Uracil, 6-(2,3,4,6-tetra-O-acetyl-β-D-glucopyranosylamino)-5-nitro-	pH 0	227(4.32),244s(3.93), 318(4.09)	24-3022-66
	pH 5	234(4.12),284s(3.18), 332(4.20)	24-3022-66
$C_{18}H_{22}O$			
Benzhydrol, α-pentyl-	C_6H_{12}	253(2.92),259(2.92), 265(2.80)	22-0269-66
	hexane	253(2.92),259(2.93), 265(2.81)	22-0269-66
8α-Estra-1,3,5(10)-trien-17-one	EtOH	258s(2.38),266(2.45), 274(2.48)	39-0054-66C
2(3H)-Naphthalenone, 4,4a,5,6,7,8-hexahydro-4aα,8-dimethyl-8-phenyl-4aβ-isomer	EtOH	242(4.18)	44-2543-66
	MeOH	248(3.94)	44-2543-66
9,17-Octadecadiene-12,14-diynal, 16-oxo-, cis	hexane	210(4.24),247(3.54), 260(3.78),274(4.00), 291(3.95)	39-1220-66C
Phenol, 4-sec-butyl-2-(α-methylbenzyl)-	octane	278(3.41),285(3.36)	3-0950-66
	CCl_4	279(3.42),285(3.39)	3-0950-66

Compound	Solvent	λ_{max}(log ε)	Ref.
$C_{18}H_{22}O_2$			
2,5-Cyclohexadien-1-one, 4-methyl-4-[1-methyl-2-(1-methyl-4-oxo-2-cyclohexen-1-yl)propenyl]-	MeOH	238.5(4.37)	5-0031-66D
Estra-4,8(14)-diene-3,17-dione	MeOH	233(4.23)	44-3780-66
Estra-4,9(10)-diene-3,7-dione	EtOH	300(4.31)	30-0880-66F
Estra-1,3,5(10),7-tetraene-3,17β-diol	EtOH	227s(4.10),277(3.57), 286(3.45)	88-5015-66
Estra-1,3,5(10),9(11)-tetraene-3,17β-diol	EtOH	263(4.22)	87-0510-66
Estrone	EtOH	280(3.36)	30-0880-66F
19-Norandrosta-4,8(14)-diene-3,17-dione	MeOH	234(4.20)	13-0365-66B
2(1H)-Phenanthrone, 3,4,9,12-tetrahydro-7-methoxy-1,1,12-trimethyl-	EtOH	277(3.26),284(3.23)	35-1766-66
$C_{18}H_{22}O_3$			
1H-Cyclopenta[7,8]naphtho[2,3-b]furan-7-carboxylic acid, 2,3,6,6a,7,8,-9a,10-octahydro-6-methyl-, methyl ester	EtOH	293(3.57)	39-0725-66C
Estra-1,3,5-trien-17-one, 1,3-dihydroxy-	EtOH	279(3.32)	5-0158-66E
Estra-1,3,5(10)-trien-17-one, 2,3-dihydroxy-	EtOH	290(3.61)	87-0027-66
Estra-1,3,5(10)-trien-17-one, 3,9α-dihydroxy-	EtOH	280(3.27)	37-0540-66
3(2H)-Phenanthrone, 2-ethyl-1,9,10,10a-tetrahydro-6,7-dimethoxy-	EtOH	224(3.98),242(3.99), 309(4.12),344(4.31)	87-0304-66
$C_{18}H_{22}O_3S$			
Isovaleric acid, ester with 4-hydroxy-9-(2-thienyl)-6-nonen-8-yn-3-one	ether	294(4.14),310(4.08)	24-1642-66
Isovaleric acid, ester with 5-hydroxy-9-(2-thienyl)-6-nonen-8-yn-3-one	ether	293(4.05),310(3.96)	24-1642-66
$C_{18}H_{22}O_4$			
1H-5,11-Dioxadibenzo[a,d]pentalene, 2,3,4,4a,6,6a-hexahydro-6aα-hydroxy-4aβ,6β,7,9-tetramethyl-3-oxo-	n.s.g.	284(3.50)	44-1725-66
Glutaric acid, 2-(3,4-dihydro-2-naphthyl)-2-methyl-, dimethyl ester	EtOH	265(4.16)	77-0485-66
Pentatriafulvalene-1,2-dicarboxylic acid, 5,6-dipropyl-, dimethyl ester	MeOH	223(4.24),330(4.53)	35-3359-66
Pentatriafulvalene-1,3-dicarboxylic acid, 5,6-dipropyl-, dimethyl ester	MeOH	265(4.54),336(4.42)	35-3359-66
Pentatriafulvalene-1,4-dicarboxylic acid, 5,6-dipropyl-, dimethyl ester	MeOH	330s(3.83),364(4.12)	35-3359-66
Pentatriafulvalene-2,3-dicarboxylic acid, 5,6-dipropyl-, dimethyl ester	MeOH	267(4.39),319(4.62)	35-3359-66
Spiro[cyclohexane-1,1'(2'H)-naphthalen]-4'(3'H)-one, 4-acetoxy-7'-methoxy-	EtOH	227(4.12),279(4.13)	94-1033-66
Spiro[1,3-dioxolane-2,3'(2'H)-phenanthrene], 1',4',9',10'-tetrahydro-6',7'-dimethoxy-	EtOH	217(4.40),276(4.01), 300(3.88),310s(3.79)	87-0304-66
4,6-Tetradecadiene-8,10-diyne-1,12-diol, diacetate	ether	227(4.35),236(4.53), 293(4.48),309s(4.44)	24-2828-66

Compound	Solvent	λ_{max}(log ε)	Ref.
$C_{18}H_{22}O_5$			
2-Phenanthrenecarboxylic acid, 1,2,3,4,4aα,9,10,10aβ-octahydro-6,7-dimethoxy-3-oxo-, methyl ester	EtOH	232(3.98),257(3.95), 281s(3.75),292s(3.56)	87-0304-66
4aβ-isomer (anom.)	EtOH	230s(3.99),282(3.73), 286(3.73),292s(3.63)	87-0304-66
	pH 13	287(4.21)	87-0304-66
Zearalenone	MeOH	236(4.47),274(4.14), 316(3.78)	88-3109-66
$C_{18}H_{23}FO$			
Estra-1,4-dien-3-one, 10β-fluoro-	EtOH	241(4.09)	5-0174-66B
$C_{18}H_{23}NO_3$			
2H-Benzo[a]quinolizin-2-one, 3-ethyl-1,6,7,11b-tetrahydro-9,10-dimethoxy-4-methyl-	EtOH	286(3.69),333(4.15)	44-0797-66
Isonicotinic acid, 3-(2,6-dihydroxy-4-pentylphenyl)-1,2,5,6-tetrahydro-1-methyl-, δ-lactone	EtOH	255(4.00),305(4.12)	35-3664-66
Litsericine, N-methyl-	EtOH	240s(3.50),290(3.54)	95-0134-66
Mecambrine A, 7a,8,9,10,11,12-hexahydro-	EtOH	240s(3.63),290(3.57)	95-1205-66
Mecambrine B, 7a,8,9,10,11,12-hexahydro-	EtOH	240s(3.49),290(3.47)	95-1205-66
$C_{18}H_{23}NO_4$			
1-Benzazepin-2-one, 5-(p-acetoxyphenyl)-7-methoxy-2,3,4,5-tetrahydro-	EtOH	250(4.14),290s(3.42)	94-1033-66
2-Benzazepin-1-one, 5-(p-acetoxyphenyl)-7-methoxy-1,3,4,5-tetrahydro-	EtOH	251(3.98)	94-1033-66
$C_{18}H_{23}NO_5$			
3H-Indole-3,3-dipropionic acid, 2-ethoxy-, dimethyl ester	n.s.g.	254(3.79)	24-3063-66
$C_{18}H_{23}N_7O_2$			
Benzaldehyde, (4,6-dimorpholino-s-triazin-2-yl)hydrazone	EtOH	228(4.66),308(4.49)	49-1713-66
$C_{18}H_{24}$			
Naphthalene, 1,4-di-tert-butyl-	isooctane	221(4.90),286(3.96), 316(2.72)	88-2905-66
$C_{18}H_{24}BrNO_3$			
Litsericine, N-methyl-, hydrobromide	EtOH	238(3.69),290(3.51)	95-0134-66
$C_{18}H_{24}N_2O$			
1,2-Benzisoxazole, 3a,4,5,6,7,7a-hexahydro-5-methyl-3-phenyl-7a-(1-pyrrolidinyl)-	EtOH	264(4.08)	32-0559-66
$C_{18}H_{23}N_2O_3$			
3-Pyrrolin-2-one, 5-(5-carbethoxy-3-ethyl-4-methylpyrrol-2-ylmethylene)-4-ethyl-3-methyl-	$CHCl_3$	255(4.30),260(4.33), 280(3.90),388(4.39), 406(4.36)	39-1155-66C

Compound	Solvent	$\lambda_{max}(\log \epsilon)$	Ref.
$C_{18}H_{24}N_4O_4$			
2,6-Dodecadienal, 2,4-dinitrophenyl-hydrazone, 6-cis-2-trans	$CHCl_3$	378(4.47)	54-0117-66
trans-trans	$CHCl_3$	376(4.46)	54-0117-66
$C_{18}H_{24}N_4O_5$			
Imidazolidine, 1-(p-carbethoxyphenyl)-3-(2-hydroxyethyl)-2-(5-uracilyl)-	EtOH	303(4.17)	87-0868-66
$C_{18}H_{24}N_4O_6$			
Inosine, 2',3'-O-isopropylidene-5'-O-(tetrahydro-2-pyranyl)-	pH 1	252(3.93)	44-2685-66
	pH 11	254.5(4.10)	44-2685-66
	EtOH	246(4.05),250(4.04), 267s(--)	44-2685-66
$C_{18}H_{24}O$			
Bakuchiol	EtOH	260(4.27)	88-4561-66
	EtOH-KOH	285(4.32)	88-4561-66
$C_{18}H_{24}O_2$			
2-Butene, 2,3-bis(4-oxo-1-methyl-cyclohex-2-en-1-yl)-	MeOH	232(4.37)	5-0019-66D
Estra-4,8(14)-dien-3-one, 17β-hydroxy-	MeOH	233(4.22)	44-3780-66
Estra-p-quinol(10β)	EtOH	240(4.06)	5-0174-66B
4,7-Methanoindene-1,8-dione, 3,5-di-tert-butyl-3a,4,7,7a-tetrahydro-	isooctane	197(4.14),223(4.00), 343(1.71)	35-3433-66
19-Norandrosta-5(10),9(11)-dien-3-one, 17β-hydroxy-	MeOH	240(4.25)	13-0087-66B
$C_{18}H_{24}O_3$			
Biphenyl, 6-acetyl-1,4,5,6-tetrahydro-2',6'-dimethoxy-3,4'-dimethyl-	MeOH	210(4.69),272(3.05), 280(3.04)	5-0165-66C
Estra-1,3,5(10)-triene-3,9α,17β-triol	EtOH	280(3.29)	37-0540-66
Estr-4-ene-3,17-dione, 11α-hydroxy-	EtOH	240(4.20)	35-3120-66
Estr-4-en-3-one, 9α,10α-epoxy-17β-hydroxy-	EtOH	243(4.13)	87-0510-66
19-Norandrosta-4,9(10)-dien-3-one, 11β,17β-dihydroxy-	MeOH	296(4.25)	13-0087-66B
$C_{18}H_{24}O_4$			
1(2H)-Naphthalenone, 2-(2-ethyl-3-oxo-butyl)-3,4-dihydro-6,7-dimethoxy-	EtOH	233(4.25),275(4.08), 314(3.91)	87-0304-66
Spiro[1,3-dioxolane-2,3'(2'H)-phenan-threne], 1',4',4'aα,9',10',10'aβ-hexanhydro-6',7'-dimethoxy-	EtOH	223(3.88),282(3.57), 286(3.56),291s(3.49)	87-0304-66
Spiro[1,3-dioxolane-2,3'(2'H)-phenan-threne], 1',4',4'aβ,9',10',10'aβ-hexahydro-6',7'-dimethoxy-	EtOH	223(3.91),282(3.60), 286(3.60),291s(3.52)	87-0304-66
$C_{18}H_{24}O_6$			
1H-2-Benzopyran-7-carboxylic acid, 3,4-dihydro-6,8-dimethoxy-1-oxo-3-pentyl-, methyl ester	EtOH	262(3.30)	39-1608-66C
$C_{18}H_{25}N$			
Estra-1,3,5(10)-trien-3-amine	EtOH	238(4.01),292(3.26)	5-0174-66B
hydrochloride	EtOH	212(3.98),260s(2.72), 265(2.83),274(2.88)	5-0174-66B
Naphthalen-1,4-imine, 9-butyl-1,4-dihydro-1,2,3,4-tetramethyl-	C_6H_{12}	270s(3.35),276(3.42), 283(3.36)	44-2009-66

Compound	Solvent	λ_{max}(log ϵ)	Ref.
$C_{18}H_{25}NO_2$			
3-Pyrrolidinol, N-cyclohexyl-4-(p-methoxybenzylidene)-	EtOH	265(4.41)	44-0001-66
$C_{18}H_{25}N_3$			
6H-Pyrido[2,1-c][1,2,4]benzotriazine, 6-cyclohexyl-7,8,9,10-tetrahydro-1-methyl-	EtOH	242(3.61),288(4.25)	30-0361-66D
$C_{18}H_{25}N_3O$			
Cyclohexylideneacetaldehyde, 2,2,6-trimethyl-, phenylsemicarbazone	MeOH	253(4.36)	22-1400-66
$C_{18}H_{26}ClN$			
Estra-1,3,5(10)-trien-3-amine, hydrochloride	EtOH	212(3.98),260s(2.72), 265(2.83),274(2.88)	5-0174-66B
$C_{18}H_{26}N_2O_2S$			
L-Proline, N-thiobenzoyl-, cyclohexylammonium salt	H_2O	279(4.04),345s(2.60)	1-2781-66
	MeOH	281(4.04),360(2.46)	1-2781-66
	dioxan	285(3.99),380(2.50)	1-2781-66
$C_{18}H_{26}O_2$			
Cyclohexanone, 2-benzyl-4-tert-butyl-2-methyl-, cis	C_6H_{12}	299(1.85)	22-3881-66
trans	C_6H_{12}	282(1.67)	22-3881-66
$C_{18}H_{26}O$			
Benzene, 1,3-dimethoxy-2-(3-methyl-1,3-butadienyl)-5-pentyl-	hexane	227(4.27),242(3.88), 298(4.37),312s(4.20)	5-0165-66C
2-Cyclohexen-1-one, 4-methyl-4-[1-methyl-2-(1-methyl-4-oxo-cyclohexyl)propenyl]-	MeOH	231.5(4.06)	5-0031-66D
Estra-1(10),5-diene-3β,17β-diol	KBr	242(4.18)	44-1761-66
Hexane, 3,4-bis(4-oxocyclohexenyl)-	EtOH	285(3.13)	39-1213-66C
10α-19-Nortestosterone	EtOH	245(4.20)	88-1023-66
$C_{18}H_{26}O_3$			
Estr-4-en-3-one, 1β,17β-dihydroxy-	EtOH	243(4.22)	35-3120-66
	0.07N NaOH	243(4.03),302(3.44)	35-3120-66
Estr-4-en-3-one, 6β,17β-dihydroxy-	EtOH	238(4.12)	35-3120-66
	0.07N NaOH	244(3.94),276(3.63), 365s(2.95)	35-3120-66
Estr-4-en-3-one, 9α,17β-dihydroxy-	EtOH	242(4.14)	87-0510-66
Estr-4-en-3-one, 10β,17β-dihydroxy-	EtOH	236(4.16)	35-3120-66
	EtOH	235(4.12)	87-0510-66
Estr-4-en-3-one, 11α,17β-dihydroxy-	EtOH	242(4.21)	35-3120-66
2,6-Octanedione, 4-(p-methoxyphenyl)-5,5,7-trimethyl-	EtOH	227(4.04),276(3.23), 282(3.16)	88-2243-66
$C_{18}H_{26}O_4$			
Cumanin, acetonide	EtOH	213(3.96)	78-1499-66
Epicumanin, acetonide	EtOH	214(3.98)	78-1499-66
2-Oxaandrost-4-en-3-one, 1ξ,17β-dihydroxy-	EtOH	227.5(4.15)	94-1370-66
4-Oxaandrost-1-en-3-one, 5ξ,17β-dihydroxy-	EtOH	217(3.89)	94-1370-66
Succinic acid, pentamethylbenzyl-, dimethyl ester	$C_6H_{11}Me$	270(2.40)	39-1147-66B

Compound	Solvent	λ_{max}(log ε)	Ref.
$C_{18}H_{26}O_7S$			
Cumanin, acetate, methanesulfonate	EtOH	213(3.99)	78-1499-66
$C_{18}H_{27}Cl_2N_3O_8$			
[4-[(Dimethylamino)methylene]-3-phenyl-2-pentenediylidene]-bis[dimethylammonium perchlorate	MeCN	432(3.85)	24-2479-66
$C_{18}H_{27}NO_4$			
2,10-Cyclodecadiene-1,2-dicarboxylic acid, 3-(N-pyrrolidinyl)-, dimethyl ester, cis-cis	EtOH	315(4.18)	35-4685-66
$C_{18}H_{27}NO_6$			
Domoic acid, trimethyl ester	EtOH	245(4.41)	95-0874-66
$C_{18}H_{28}ClNO_4$			
[4-(2,3,4,4,5,6-Hexamethyl-2,5-cyclohexadien-1-ylidene)-2-butenylidene]-dimethylammonium perchlorate	MeCN	455(4.77)	24-2479-66
$C_{18}H_{28}N_2$			
1-Propanone, 1-(3-cyclohexen-1-yl)-, azine	n.s.g.	235(3.54)	5-0019-66D
$C_{18}H_{28}N_2O_2S$			
Δ^3-1,3,4-Thiadiazoline, 2,5-bis-(3-cyclohexen-1-yl)-2,5-diethyl-, 1,1-dioxide	CH_2Cl_2	369(2.22)	5-0019-66D
L-Valine, N-thiobenzoyl-, cyclohexylammonium salt	H_2O	247(--),279(3.92), 356s(2.54)	1-2781-66
	MeOH	286(3.92),375s(2.41)	1-2781-66
	dioxan	286(3.85),385(2.47)	1-2781-66
$C_{18}H_{28}N_2O_4S$			
Cyclohexanone, 4,4'-(2,5-dimethyl-Δ^3-1,3,4-thiadiazolin-2,5-diyl)bis-[4-methyl-, S,S-dioxide	CH_2Cl_2	285(1.68),370(2.23)	5-0019-66D
$C_{18}H_{28}O$			
2,5-Cyclohexadien-1-one, 2,6-di-tert-butyl-4-isobutylidene-	hexane	300(4.41)	70-0888-66
$C_{18}H_{28}O_2$			
10,16-Heptadecadien-8-ynoic acid, methyl ester	C_6H_{12}	228(4.22)	69-0625-66
$C_{18}H_{28}Si_3$			
Trisilane, 1,1,2,2,3,3-hexamethyl-1,3-diphenyl-	C_6H_{12}	243.0(4.28)	101-0102-66B
$C_{18}H_{30}NO_3P$			
Phosphorimidic acid, N-(o-cyclohexylphenyl)-, triethyl ester	EtOH	239(3.94),281(3.22)	35-3781-66
$C_{18}H_{30}O$			
Bakuchiol, hexahydro-	EtOH	224(3.69),278(3.09)	88-4561-66
Bicyclo[3.1.0]hex-3-en-2-one, 1,3,4,5,6,6-hexaethyl-	EtOH	239(3.72),270(3.43), 332(2.93)	35-1005-66

Compound	Solvent	$\lambda_{max}(\log \epsilon)$	Ref.
2,4-Cyclohexadien-1-one, 2,3,4,5,6,6-hexaethyl-	EtOH	339(4.04)	35-1005-66
$C_{18}H_{31}NO$			
3-Pyridinol, 6-dodecyl-2-methyl-	EtOH-acid	231(3.76),301(3.98)	44-1275-66
	EtOH	224(3.92),288(3.77)	44-1275-66
	EtOH-KOH	245(4.06),310(3.79)	44-1275-66
$C_{18}H_{31}NO_3S_2$			
5-Thiazolidineacetic acid, 4-oxo-N-pentyl-2-thioxo-, octyl ester	EtOH	262(4.10),296(4.20)	95-0095-66
$C_{18}H_{32}O$			
2,4-Cyclohexadien-1-ol, 2,3,4,5,6,6-hexaethyl-	EtOH	262(3.43)	35-1005-66
$C_{18}H_{32}O_2$			
2,5,8-Tetradecatrienal, diethyl acetal, 2,5,8-tri-cis	EtOH-HCl	223(4.16)	54-0117-66
	EtOH	233(2.21)	54-0117-66
$C_{18}H_{33}NO$			
Morpholine, 4-(3-methyl-1-pentylidene-3-octenyl)-	n.s.g.	218(1.98)	54-0753-66
$C_{18}H_{36}N_6Ni$			
5,7,7,12,14,14-Hexamethyl-1,4,8,11-tetrazacyclotetradecane nickel cyanide	MeOH	362(1.20),532(1.08), 781(0.78)	12-0609-66

Compound	Solvent	λ_{max}(log ϵ)	Ref.
$C_{19}H_9BrO_5$			
1,4-Naphthoquinone, 3-(2-bromo-1,3-dioxoindan-2-yl)-2-hydroxy-	MeCN	232(4.48),248(4.51), 254(4.51),275(4.21), 335(3.60)	44-0062-66
$C_{19}H_9ClO_4$			
1,4-Naphthoquinone, 2-chloro-3-(1,3-dioxoindan-2-yl)-	MeCN	224(4.80),248(4.47), 253(4.50),272(4.15), 302(3.78),340(3.59)	44-0062-66
$C_{19}H_{10}Br_2O_8S$			
Bromopyrogallol Red	acid	475(4.51),558(3.48)	73-3127-66
monoanion	n.s.g.	475(4.10),558(4.02)	73-3127-66
dianion	n.s.g.	558(4.74)	73-3127-66
trianion	n.s.g.	558(4.24),598(4.20)	73-3127-66
tetraanion	n.s.g.	598(4.44)	73-3127-66
$C_{19}H_{10}Cl_4$			
Spiro[2,4]hepta-1,4,6-triene, 4,5,6,7-tetrachloro-1,2-diphenyl-	EtOH	222(4.44),230(4.47), 287s(4.47),301(4.59), 318(4.49)	44-4260-66
$C_{19}H_{11}N_5$			
Dipyrido[2,3-f:3',2'-h]quinoxaline, 2-(2-pyridyl)-	EtOH	228(4.43),258(4.61), 314(4.35)	44-3384-66
$C_{19}H_{12}N_2$			
Benz[a]indolo[3,2-h]quinolizine	MeOH	277(4.31),300(4.27), 375(4.26),400(4.21)	4-0395-66
$C_{19}H_{12}OS$			
Dibenzothiophene, 2-benzoyl-	MeOH	236(4.49),248(4.51), 269(4.39),291(4.20), 315(4.00)	73-0835-66
Dibenzothiophene, 4-benzoyl-	MeOH	229(4.69),257(4.48), 288(4.04),297(4.14), 360(3.79)	73-0835-66
$C_{19}H_{12}O_2$			
Dibenzofuran, 2-benzoyl-	MeOH	245(4.48),260(4.52), 276s(4.40),310s(3.85)	73-0835-66
$C_{19}H_{12}O_4$			
Tetrangulol	MeOH	225(4.70),250s(4.31), 315(4.39),425(3.85)	44-2920-66
	base	230(4.85),290(4.26), 325s(4.50),380(3.97), 465(4.06)	44-2920-66
$C_{19}H_{12}O_5$			
[18]Annulene-5-carboxylic acid, 1,4;7,10;13,16-triepoxy-, methyl ester	EtOH	343(5.12),418(4.04), 444(3.88)	12-1221-66
$C_{19}H_{12}O_8S$			
Pyrogallol Red	acid	466(4.44),542(3.57)	73-3127-66
monoanion	n.s.g.	466(4.12),542(4.02)	73-3127-66
dianion	n.s.g.	542(4.63)	73-3127-66
trianion	n.s.g.	542(4.19),584(4.18)	73-3127-66
tetraanion	n.s.g.	584(4.39)	73-3127-66

Compound	Solvent	λ_{max}(log ε)	Ref.
$C_{19}H_{13}IN_2$			
14H-Benz[a]indolo[3,2-h]quinolizin-7-ium iodide	MeOH	277(4.39),299(4.33), 375(4.29),404(4.25)	4-0395-66
$C_{19}H_{13}NO_3$			
Fluorescein oxime, decarboxylated	EtOH	231(4.90),251(4.93), 286s(3.97),354(3.84)	1-1631-66
$C_{19}H_{13}N_3O_2$			
1H-Indazole, 1-(p-nitrophenyl)-3-phenyl-	EtOH	231(4.47),307(3.84), 361(4.32)	78-3131-66
2-Quinoxalinecarboxylic acid, 3-(1-naphthylamino)-, sodium salt	EtOH	248(4.32),260(4.28), 328(4.23),405(3.81)	87-0770-66
$C_{19}H_{13}P$			
Phosphine, 1,3,5-heptatriynyldiphenyl-	EtOH	228(4.78),253(4.51), 272(4.18),297(3.40), 316(3.34),338(3.08)	22-1002-66
$C_{19}H_{14}BrNO_4$			
Thalidastine, deoxy-, bromide	EtOH	247(4.55),271(4.50), 278(4.50),308(4.39), 349(4.35),463(3.95)	88-3069-66
$C_{19}H_{14}BrN_3$			
Cyclopenta[c]quinolizine, 1-[(p-bromophenyl)azo]-4-methyl-	EtOH	260(4.47),267(4.42), 336(4.10),343(4.10), 381(4.04),488(4.50)	39-0324-66C
Cyclopenta[c]quinazoline, 3-[(p-bromophenyl)azo]-4-methyl-	EtOH	240s(4.1),253(4.09), 349(4.14),391(4.25), 467(4.40)	39-0324-66C
$C_{19}H_{14}ClN$			
6-Phenylbenzo[c]quinolizinium chloride	EtOH	236(4.27),258(4.47), 310s(3.70),361(3.92), 375(4.00)	44-2346-66
$C_{19}H_{14}ClNO_4S$			
2,3-Diphenylthiazolo[3,2-a]pyridinium perchlorate	EtOH	251s(4.09),318(4.26)	4-0027-66
$C_{19}H_{14}ClNO_6S$			
10-(Phenylsulfonyl)acridizinium perchlorate	EtOH	237(4.11),255s(3.64), 360(3.38),382(3.28), 402(3.06)	44-0565-66
$C_{19}H_{14}F_3N_3O_2$			
1-Naphthol, 7-amino-8-[(α,α,α-trifluoro-m-tolyl)azo]-, acetate	benzene	437(3.97)	44-2571-66
$C_{19}H_{14}N_2$			
Benz[a]indolo[3,2-h]quinolizine, 8,9-dihydro-	MeOH	257(4.07),277(4.32), 329(4.32),398(3.91)	4-0395-66
Carbazole, 3-(benzylideneamino)-	MeOH	210(4.59),239(4.64), 248(4.64),286(4.57), 350(4.33)	5-0116-66F
6H-Dibenzo[a][3,6]phenanthroline, 2,8-dihydro-	EtOH	269(4.62),336(4.55), 354(3.68),372(3.76)	87-0161-66
o-Terphenyl-4-carbonitrile, 4'-amino-	EtOH-HCl	244(4.47),270s(4.12)	56-0429-66
	EtOH	233(4.42),316(4.19)	56-0429-66

Compound	Solvent	$\lambda_{max}(\log \epsilon)$	Ref.
$C_{19}H_{14}N_2O$			
11H-Benzo[a]carbazole, 10-methyl-9-(2-nitrovinyl)-	EtOH	256(4.43),279(4.34),313(4.03),322(4.03)	87-0237-66
$C_{19}H_{14}N_2O_2$			
Isoxazole, 3',5-diphenyl-3,5'-methylenedi-	EtOH	254(4.62)	32-1064-66
Isoxazole, 5,5'-methylenebis[3-phenyl-	EtOH	242(4.32)	32-1064-66
$C_{19}H_{14}N_2O_5$			
Picolinic acid, 3-(benzyloxy)-, p-nitrophenyl ester	EtOH	287(4.08)	44-0636-66
$C_{19}H_{14}N_4O_2$			
Lumazine, 8-methyl-6,7-diphenyl-	pH -2.7	257(4.04),283(4.07),398(4.05)	24-3503-66
	pH 3	228(4.27),283(4.22),345(3.98)	24-3503-66
	pH 7	266(4.19),289(4.27),424(4.18)	24-3503-66
	pH 12	244(4.27),281(4.14),352(4.05),425s(3.42)	24-3503-66
	pH 13	279(4.31),435(4.35)	24-3503-66
$C_{19}H_{14}O$			
Benzophenone, 4-phenyl-	MeOH	209s(4.48),258s(4.15),291(4.40)	73-0835-66
$C_{19}H_{14}OS$			
Benzophenone, 4-(phenylthio)-	MeOH	245s(4.08),253(4.11),314(4.22)	73-0835-66
$C_{19}H_{14}O_2$			
Indan, 1,3-difurfurylidene-	EtOH	295(4.45),308(4.50),322(4.44),341s(4.45),357(4.54),376(4.41)	5-0057-66H
1,4-Naphthoquinone, 2-cinnamyl-	C_6H_{12}	246s(--),251(4.51),264s(--),284s(--),293s(--),330(3.45)	22-1828-66
$C_{19}H_{14}O_4$			
2-Pyrone, 4-hydroxy-5,6-diphenyl-, acetate	EtOH	227(4.26),343(4.21)	35-3759-66
$C_{19}H_{14}O_5$			
Tetrangomycin	MeOH	206(4.43),267(4.50),330(3.46),400(3.72)	44-2920-66
$C_{19}H_{14}O_6$			
1H-Cyclopenta[7,8]naphtho[2,3-b]furan-7-carboxylic acid, 10-acetoxy-2,3-dihydro-6-methyl-3-oxo-	EtOH	229(4.23),272(4.75),315(3.92),328(3.85),379(3.88)	39-0725-66C
$C_{19}H_{14}O_8$			
Xanthone, 1,2,8-triacetoxy-	EtOH	235s(4.63),244(4.67),268(4.09),344(3.88)	78-1785-66
Xanthone, 2,3,4-triacetoxy-	EtOH	242(4.28),263(4.14),287(3.36),337(3.47)	78-1777-66

Compound	Solvent	$\lambda_{max}(\log \epsilon)$	Ref.
$C_{19}H_{14}S$			
Thiobenzophenone, 4-phenyl-	iso-PrOH	345(4.23),603(2.37)	77-0586-66
$C_{19}H_{15}^+$			
Triphenylcarbonium ion	FSO_3H-SbF_5	403(4.59),429(4.59)	35-1488-66
$C_{19}H_{15}BrN_2O_9$			
2,5-Di-O-(p-nitrobenzoyl)-3-deoxy-β-D-ribofuranosyl bromide	CH_2Cl_2	261(4.46)	44-1163-66
$C_{19}H_{15}ClN_2O_2Sn$			
Tin, chlorobis(8-quinolinolato)-methyl-	benzene	383(3.61)	101-0249-66B
	dioxan	381(3.67)	101-0249-66B
	pyridine	383(3.66)	101-0249-66B
$C_{19}H_{15}IN_2$			
5,6-Dihydro-14H-benz[a]indolo[3,2-h]-quinolizin-7-ium iodide	MeOH	246s(4.16),255s(4.11),277(4.06),331(4.03),389(4.16)	4-0395-66
8,9-Dihydro-14H-benz[a]indolo[3,2-h]-quinolizin-7-ium iodide	MeOH	255(4.02),280s(3.95),300(4.00),380(4.26)	4-0395-66
9,14b-Dihydrobenz[a]indolo[3,2-h]-quinolizin-7-ium iodide	MeOH	257(4.20),277(4.41),330(4.46),395(3.94)	4-0395-66
7,8-Dihydro-13H-benz[g]indolo[2,3-a]-quinolizinium iodide	n.s.g.	252(4.5),281(4.0),351(4.5)	28D-1141-66A
$C_{19}H_{15}NO_2$			
Benz[f]isoindolin-1-one, 2-(m-methoxyphenyl)-	MeOH	237(4.70),308(4.28)	95-0001-66
Benz[f]isoindolin-1-one, 2-(p-methoxyphenyl)-	MeOH	239(4.79),310(4.20)	95-0001-66
1-Pyrenecarbamic acid, ethyl ester	EtOH	236s(4.60),243(4.80),268(4.35),278(4.54),313s(3.95),329(4.28),340(4.44),383(3.17)	12-1859-66
2-Pyrenecarbamic acid, ethyl ester	EtOH	254(4.89),284s(4.32),310(3.98),323(4.39),339(4.60),367(2.92),379s(2.55),388(3.02)	12-1859-66
4-Pyrenecarbamic acid, ethyl ester	EtOH	236s(4.58),242(4.81),255s(4.20),265s(4.36),275(4.54),311s(3.93),325(4.24),340(4.39),378(2.78)	12-1859-66
$C_{19}H_{15}N_3$			
Cyclopenta[c]quinazoline, 4-methyl-1-(phenylazo)-	EtOH	256(4.46),266s(4.4),337s(4.1),344(4.17),379(4.17),475(4.45)	39-0324-66C
Cyclopenta[c]quinazoline, 4-methyl-3-(phenylazo)-	EtOH	236s(4.25),256(4.20),355(4.28),390(4.37),456(4.48)	39-0324-66C
$C_{19}H_{15}N_3O$			
Benzanilide, 2'-(phenylazo)-	hexane	224(4.29),274(4.30),328(4.22),386(3.97),460s(3.15)	19-0029-66
	MeOH	228(4.26),273(4.27),330(4.23),380s(3.92),460s(2.98)	19-0029-66

Compound	Solvent	λ_{max}(log ϵ)	Ref.
Benzanilide, 2'-(phenylazo)- (cont.)	50% MeOH	230(4.24),273(4.21), 331(4.21),380s(3.83), 460s(3.00)	19-0029-66
	50% MeOH-pyridine	333(4.19),380s(3.94), 460s(3.08)	19-0029-66
	96% MeOH-4% hexane	228(4.27),273(4.27), 329(4.22),380s(3.92), 460s(2.98)	19-0029-66
	96% HOAc-4% hexane	331(4.19),380s(3.95), 460s(3.05)	19-0029-66
	96% C_5H_5N-4% hexane	335(4.13),380s(3.90), 460s(3.08)	19-0029-66
$C_{19}H_{15}N_3O_2S$			
Benzenesulfonanilide, 4-(2-benzimidazolyl)-	EtOH	222(4.44),247(4.05), 254(4.01),319(4.48)	95-0600-66
$C_{19}H_{16}$			
15H-Cyclopenta[a]phenanthrene, 11,17-dimethyl-	EtOH	222(4.32),270(4.70), 277(4.81),295s(4.13), 306s(4.13),317(3.69), 350(2.93),368(2.83)	39-0963-66C
15H-Cyclopenta[a]phenanthrene, 12,17-dimethyl-	EtOH	223(4.38),272(4.81), 276s(4.81),290(4.27), 318(3.88),349(2.71), 367(2.43)	39-0963-66C
2,5-Hexadiyne, 1-phenyl-6-p-tolyl-	EtOH	223(4.60),227(4.76), 247(4.20),260(4.38), 275(4.51),292(4.39)	78-0867-66
1-Propene, 1-(2-naphthyl)-1-phenyl-	EtOH	239(4.67),248(4.64), 276(4.06),286(3.94), 297(--)	23-1283-66
$C_{19}H_{16}BrNO_2S$			
Acetic acid, [o-[4-(p-bromophenyl)-2-thiazolyl]phenyl]-, ethyl ester	EtOH	250s(3.98),262s(4.09), 275(4.18),309s(3.70)	4-0282-66
$C_{19}H_{16}ClNO_2$			
Isoquinoline, 1-(o-chlorostyryl)-6,7-dimethoxy-	EtOH	209(4.54),242(4.46), 250(4.46),277s(4.32), 355(4.20)	44-3683-66
$C_{19}H_{16}ClNO_6$			
9,10-Dimethoxydibenzo[a,f]quinolizinium perchlorate	EtOH	225(4.38),268(4.66), 280s(4.44),350s(3.82), 390(4.21),407(4.35)	44-3683-66
$C_{19}H_{16}Cl_5OSb$			
Tritylium hydroxypentachloroantimonate(V)	CH_2Cl_2	408(--),431(4.51)	78-0557-66
$C_{19}H_{16}INO_4$			
Escholamine, iodide	MeOH	225(4.6),254(4.8), 292(4.0),314(4.1)	73-3362-66
$C_{19}H_{16}N_2O$			
Benzoic acid hydrazide, 2,6-diphenyl-	MeOH	260(3.95)	88-1063-66

Compound	Solvent	$\lambda_{max}(\log \epsilon)$	Ref.
$C_{19}H_{16}N_2O_2$			
1,2-Diazabicyclo[3.2.0]hept-3-en-6-one, 2-benzoyl-5-methyl-4-phenyl-	EtOH	225(4.15),265(3.94), 305s(--),331(4.22)	44-0034-66
$C_{19}H_{16}N_2O_4$			
Isoxazolo[2,3-a]indole-3-carboxylic acid, 2-amino-3a,4-dihydro-4-oxo-3a-phenyl-, ethyl ester	EtOH	220(4.17),313(3.44), 357(3.97)	88-3857-66
Orotic acid, 1,3-dibenzyl-	N HCl	273(3.91)	44-0201-66
	pH 7	272(3.97)	44-0201-66
	pH 13	272(3.97)	44-0201-66
$C_{19}H_{16}N_4$			
Toluene, α,α-bis(phenylazo)-	n.s.g.	282(4.30),410(2.60)	77-0622-66
$C_{19}H_{16}N_4O$			
Hypoxanthine, 1,3-dibenzyl-, HBr salt	pH 1	254(4.01),280s(3.59)	44-2202-66
	pH 7	245s(3.97),304s(2.71)	44-2202-66
Hypoxanthine, 1,7-dibenzyl-	neutral	257(3.88)	44-1411-66
Hypoxanthine, 1,9-dibenzyl-	neutral	253(4.02)	44-1411-66
Hypoxanthine, 3,7-dibenzyl-	neutral	266(4.07)	44-1411-66
Hypoxanthine, 7,9-dibenzyl-	pH 1	255(4.02)	44-2202-66
	pH 7	267(3.97)	44-2202-66
	pH 13	264(3.88)	44-2202-66
$C_{19}H_{16}N_4O_4$			
Fluoren-3(2H)-one, 1,9a-dihydro-, 2,4-dinitrophenylhydrazone	$CHCl_3$	404(4.51)	87-0719-66
$C_{19}H_{16}N_4S$			
Purine-6(1H)-thione, 1,7-dibenzyl-	neutral	244(4.00),274(4.08)	44-1411-66
9H-Purine-6(1H)-thione, 1,9-dibenzyl-	neutral	323(4.37)	44-1411-66
$C_{19}H_{16}N_6O_2$			
Carbanilic acid, p-[(6-amino-9H-purin-9-yl)methyl]-, phenyl ester	EtOH-HCl	246(4.42)	87-0576-66
	EtOH	245(4.40)	87-0576-66
	EtOH-NaOH	243(4.42)	87-0576-66
$C_{19}H_{16}O$			
15H-Cyclopenta[a]phenanthrene, 3-methoxy-17-methyl-	EtOH	223(4.28),273(4.84), 294(4.22),304s(4.16), 318(4.10),338(2.92), 355(3.04),374(3.03)	39-0963-66C
15H-Cyclopenta[a]phenanthrene, 11-methoxy-17-methyl-	EtOH	221(4.37),272(4.70), 280(4.74),287(4.63), 324(3.69),339(3.29), 356(3.49),374(3.57)	39-0963-66C
17H-Cyclopenta[a]phenanthren-17-one, 15,16-dihydro-11,12-dimethyl-	EtOH	218(4.09),268(4.88), 292(4.45),304(4.32), 349(3.05),366(3.34), 383(3.40)	39-0955-66C
p-Terphenyl, p-methoxy-	$C_2H_4Cl_2$	288(4.51)	99-0089-66
	dioxan	288(4.54)	99-0089-66
$C_{19}H_{16}O_4$			
3(2H)-Benzofuranone, 2-allyl-2-(o-hydroxyphenyl)-, acetate	EtOH	250(3.96),327(3.58)	39-0544-66C

Compound	Solvent	$\lambda_{max}(\log \epsilon)$	Ref.
$C_{19}H_{16}O_5$			
2H-1-Benzopyran-3-propionic acid, 4-hydroxy-2-oxo-β-phenyl-, methyl ester	EtOH	250(4.23),381(4.28)	88-2287-66
1H-Cyclopenta[7,8]naphtho[2,3-b]furan-7-carboxylic acid, 2,3-dihydro-10-methoxy-6-methyl-3-oxo-, methyl ester	EtOH	225(4.30),269(4.77), 315(3.80),330(3.77), 385(3.88)	39-0725-66C
$C_{19}H_{16}O_8$			
Ceroalbolinic acid, trimethyl ether	EtOH	287(4.59),332(3.84), 410(3.63)	78-1507-66
$C_{19}H_{17}Cl$			
15H-Cyclopenta[a]phenanthrene, 3-chloro-16,17-dihydro-17,17-dimethyl-	EtOH	220(4.66),224(4.67), 255s(4.78),261(4.84), 283(4.24),292(4.18), 304(4.26),323(2.99), 331(2.74),338(3.24), 347(2.75),355(3.31)	5-0106-66J
$C_{19}H_{17}ClN_2O_2$			
1-Benzamido-3-hydroxy-4-methyl-5-phenylpyridinium chloride	EtOH	232(4.48),301(4.01)	44-0034-66
	EtOH-NaOH	230(4.48),333(3.93)	44-0034-66
$C_{19}H_{17}NO$			
o-Terphenyl-4'-amine, 4-methoxy-	EtOH-HCl	245(4.38),267s(4.08)	56-0429-66
	EtOH	239(4.39),276(4.19)	56-0429-66
$C_{19}H_{17}NO_2$			
Aporphane, N-carbethoxydehydro-	n.s.g.	252(3.80),259(3.86), 304(3.40)	88-2941-66
Benzomorphan-3,8-dione, 2-methyl-5-phenyl-	EtOH	250(4.00),294(3.38), 341(2.42)	44-1905-66
Cinnamic acid, α-cyano-β-benzyl-, ethyl ester, trans	EtOH	281(4.05)	22-1033-66
Pyrrolo[3,2,1-hi]indole-1-carboxylic acid, 4,5-dihydro-2-phenyl-	MeOH	251(4.26),309(3.98)	35-4061-66
$C_{19}H_{17}NO_2S$			
Acetic acid, [o-(4-phenyl-2-thiazolyl)-phenyl]-, ethyl ester	EtOH	246(4.17),254s(4.11), 268(4.07),310(3.76)	4-0282-66
$C_{19}H_{17}NO_3$			
Acronycine, de-N-methyl-	EtOH	265(4.60),295(4.43), 333s(3.43),395(3.78)	78-3245-66
Cinnamic acid, α-cyano-p-methoxy-β-phenyl-, ethyl ester	EtOH	233(4.12),338(4.16)	22-1033-66
isomer	EtOH	232(4.13),337(4.11)	22-1033-66
Isocarbostyril, 2-(benzoyloxy)-3-isopropyl-	EtOH	232(4.73),277(4.27), 285(4.26),327(3.94)	44-2090-66
Isocarbostyril, 2-(benzoyloxy)-3-propyl-	EtOH	233(4.56),278(4.10), 286(4.04),326(3.69)	44-2090-66
1-Naphthaleneacetamide, 1,2,3,4-tetrahydro-N-methyl-3,4-dioxo-1-phenyl-	EtOH	239(4.02),302(3.55), 360(3.66)	44-1905-66
Noracronycine	EtOH	255(4.50),285(4.75), 293s(4.69),413(3.79)	78-3245-66
$C_{19}H_{17}NO_5$			
Cassythidine	n.s.g.	217(4.59),286(4.10), 309(4.17)	12-0297-66

Compound	Solvent	λ_{max}(log ε)	Ref.
$C_{19}H_{17}NO_8$			
2,3,4,5-Pyridinetetracarboxylic acid, 6-phenyl-, tetramethyl ester	MeOH	272(4.13),291s(3.98)	39-1121-66C
$C_{19}H_{17}N_3O_2$			
Benzoic acid, p-[(2-amino-1-naphthyl)-azo]-, ethyl ester	benzene	471(4.23)	44-2571-66
	$C_6H_{11}Me$	462(4.22)	44-2571-66
1H-Pyrrolo[1,2-c]imidazole-1,3,5(2H,6H)-trione, 2-(anilino-methyl)-7,7a-dihydro-7a-phenyl-	MeOH-HCl	240(3.71),263(2.77), 267(2.67),282(2.74)	87-0142-66
	MeOH	240(4.13),267(3.03), 287(3.27)	87-0142-66
	MeOH-KOH	231(4.33),287(3.27)	87-0142-66
$C_{19}H_{17}N_3O_3$			
Imidazo[2,1-b]quinazolin-5-one, 1-(4-carbethoxyphenyl)-1,2,3,5-tetrahydro-	EtOH	229(4.60),242s(4.27), 256s(4.11),311(4.60)	24-1532-66
$C_{19}H_{17}N_5$			
Tetramethylammonium salt of 5-phenyl-cyclopentadiene-1,2,3,4-tetracarbo-nitrile	MeCN	262(4.66),300(4.10)	35-4055-66
$C_{19}H_{17}N_5O$			
Tetramethylammonium salt of 5-(p-hydroxyphenyl)cyclopenta-diene-1,2,3,4-tetracarbonitrile	MeCN	264(4.58)	35-4055-66
$C_{19}H_{17}OP$			
Phosphine oxide, (1-cyclopenten-1-yl-ethynyl)diphenyl-	EtOH	228(4.24),246(4.16), 255(4.15)	22-1002-66
Phosphine oxide, 1,3-heptadiynyl-diphenyl-	EtOH	226(4.54),248(3.57), 262(3.46),274(4.28)	22-1002-66
$C_{19}H_{17}P$			
Phosphine, (1-cyclopenten-1-yl-ethynyl)diphenyl-	EtOH	225(4.26),231(4.23), 246(4.20),256(4.13)	22-1002-66
Phosphine, (1,3-heptadiynyl)diphenyl-	EtOH	229(4.46),252(4.00), 267(3.78),284(3.51)	22-1002-66
$C_{19}H_{17}PS$			
Phosphine sulfide, (1-cyclopenten-1-ylethynyl)diphenyl-	EtOH	221(4.48),227(4.48), 241(4.40),252(4.34)	22-1002-66
$C_{19}H_{18}$			
15H-Cyclopenta[a]phenanthrene, 16,17-dihydro-11,17-dimethyl-	EtOH	219(4.45),227(4.19), 257(4.80),294(4.02), 306(4.07),324(2.85), 339(3.00),355(3.03)	39-0963-66C
15H-Cyclopenta[a]phenanthrene, 16,17-dihydro-12,17-dimethyl-	EtOH	217(4.63),261(4.83), 290(4.06),303(4.13), 321(2.57),337(2.64), 353(2.41)	39-0963-66C
$C_{19}H_{18}BrNO_2$			
1-Naphthaleneacetamide, 3-bromo-1,2,3,4-tetrahydro-N-methyl-4-oxo-1-phenyl-	EtOH	253(4.02),291(3.32)	44-1905-66

Compound	Solvent	λ_{max}(log ϵ)	Ref.
$C_{19}H_{18}Cl_2N_2O$			
2-Pyrrolidone, 4,4-dichloro-1,3-dimethyl-3-phenyl-5-(o-tolylimino)-	EtOH	232(4.42),308(3.24)	28C-1276-66A
2-Pyrrolidone, 4,4-dichloro-1,3-dimethyl-3-phenyl-5-(p-tolylimino)-	EtOH	234(4.25),312(3.51)	28C-1276-66A
$C_{19}H_{18}Cl_3NO_4S$			
L-Methione, N-carboxy-N-benzyl-2,4,5-trichlorophenyl ester	EtOH	227s(4.10),279(3.01),288(3.05)	44-3398-66
$C_{19}H_{18}N_2$			
11H-Benzo[a]carbazole, 9-(2-aminoethyl)-10-methyl-	EtOH	228(4.43),246(4.58),255(4.64),280(4.65),307(4.34),322s(3.90),337(3.85),353(3.85)	87-0237-66
Isoquinoline, 1,2,3,4-tetrahydro-2-[2-(3-indolyl)ethyl]-	MeOH	227(4.30),276s(3.85),283(3.87),291s(3.80)	4-0395-66
$C_{19}H_{18}N_2O$			
Benzaldehyde, (3-methyl-4-phenylfurfuryl)hydrazone	EtOH	230(3.73),290(3.69)	44-0052-66
2-Pyrrolidinone, 4-(diphenylmethylene)-5-imino-3,3-dimethyl-	EtOH	265(4.16)	22-3895-66
2-Pyrrolidinone, 5-imino-3,3-diphenyl-4-isopropylidene-	EtOH	261(4.15)	22-3895-66
$C_{19}H_{18}N_2O_2S_2$			
5-Thiazolidineacetamide, 3-benzyl-N-o-tolyl-4-oxo-2-thioxo-	EtOH	260(4.13),297(4.21)	95-0101-66
5-Thiazolidineacetamide, 3,N-dibenzyl-4-oxo-2-thioxo-	EtOH	261(4.07),297(4.19)	95-0101-66
$C_{19}H_{18}N_2O_3$			
1,4-Pentadien-3-one, 1-(p-dimethylaminophenyl)-5-(p-nitrophenyl)-	benzene	324(4.38),445(4.37)	65-0336-66
	EtOH	322(4.39),457(4.39)	65-0336-66
	HOAc	323(4.44),466(4.39)	65-0336-66
ferric chloride added	HOAc	625(4.26)	65-0336-66
2,4-Pentadien-1-one, 5-(p-dimethylaminophenyl)-1-(p-nitrophenyl)-	benzene	314(4.13),358(4.09),464(4.55)	65-0336-66
	EtOH	472(--)	65-0336-66
	HOAc	277(4.24),353(4.22),468(4.26)	65-0336-66
2-Quinazolinemethanol, 4-ethoxy-α-phenyl-, acetate	EtOH	213s(4.45),225(4.57),263(3.76),300(3.45),311(3.43)	39-2290-66C
$C_{19}H_{18}N_2O_3S_2$			
5-Thiazolidineacetamide, 3-benzyl-N-(2-hydroxy-5-methylphenyl)-4-oxo-2-thioxo-	EtOH	258(4.21),297(4.33)	95-0101-66
$C_{19}H_{18}N_4O_2$			
Carbazic acid, 3-(5,6-diphenylpyrazinyl)-, ethyl ester	EtOH	226(4.31),283(4.16),339(4.02)	39-2038-66C
Imidazole-4(or 5)-carboxamide, N-benzyl-5-(or 4)-(N-benzylformamido)-	pH 1	244s(3.94)	44-2202-66
	pH 7	241(3.99)	44-2202-66
	pH 13	266(4.09)	44-2202-66

Compound	Solvent	λ max(log ε)	Ref.
$C_{19}H_{18}O$			
15H-Cyclopenta[a]phenanthrene, 16,17-dihydro-3-methoxy-17-methyl-	EtOH	212(4.42),225(4.30), 236(4.36),261(4.82), 282(4.20),291(4.12), 302(3.95),326(2.83), 341(3.16),358(3.28)	39-0963-66C
15H-Cyclopenta[a]phenanthrene, 16,17-dihydro-11-methoxy-17-methyl-	EtOH	219s(4.42),228(4.45), 247(4.60),253(4.59), 276(4.44),298(3.95), 330(3.25),345(3.55), 362(3.65)	39-0963-66C
Cyclopenta[a]phenanthrene, 11,12,15,16-tetrahydro-12,12-dimethyl-17-oxo-	EtOH	219(4.87),241(4.11), 251s(4.16),262s(4.57), 270(4.89),281(5.00), 324(4.48),336(4.51), 362s(4.25)	39-0955-66C
2,4,6-Heptatrien-1-ol, 1,7-diphenyl-	CH_2Cl_2-$HClO_4$	308(4.7),325(4.6), 340(4.5)	24-2491-66
2H-Pyran, 2,4-dimethyl-2,6-diphenyl-	EtOH	227(3.88),250(3.87), 323(3.95)	22-0689-66
$C_{19}H_{18}O_2$			
Cyclopenta[a]phenanthren-17-one, 13,14,15,16-tetrahydro-, ethylene ketal	EtOH	239(4.81),302(3.89), 314(4.00),319(3.84), 336(3.70)	39-0955-66C
[3.4]Paracyclophane, 2,12-dioxo-	EtOH	219(4.09),245s(3.41), 289(3.87)	35-3515-66
[3.4]Paracyclophane, 2,13-dioxo-	EtOH	250(3.85),285(3.57), 320s(2.70)	35-3515-66
$C_{19}H_{18}O_3$			
14β-Gona-1,3,5(10),6,8-pentaen-11-one, 17,17-ethylenedioxy-	EtOH	216(4.58),245(4.31), 313(3.81)	39-0955-66C
	dioxan	228(4.26),248(4.30), 308(3.91)	19-0611-66
2,4-Pentadien-1-one, 5-(2,4-dimethoxy-phenyl)-1-phenyl-	benzene	352(--)	65-0336-66
	EtOH	360(--)	65-0336-66
	HOAc	367(--)	65-0336-66
$C_{19}H_{18}O_5$			
Flavone, 4',5,7-trimethoxy-6-methyl-	n.s.g.	266(4.17),321(4.38)	28C-0662-66A
Flavone, 4',5,7-trimethoxy-8-methyl-	n.s.g.	270(4.34),325(4.24)	28C-0662-66A
2-(2,4,6-Trimethylbenzoyl)benzoic methyl carbonic ahydride	$CHCl_3$	325(2.12)	35-0781-66
$C_{19}H_{18}O_6$			
Hydrocinnamic acid, 3-(α-carboxy-p-methoxystyryl)-4-hydroxy-	EtOH-acid	220(4.29),310(4.29)	88-2287-66
	EtOH	218(4.27),298(4.23)	88-2287-66
	EtOH-NaOH	288(4.37)	88-2287-66
$C_{19}H_{18}O_7$			
Flavone, 3-hydroxy-3',4',5,7-tetra-methoxy-	MeOH	250(4.37),360(4.37)	28D-1141-66A +28D-1144-66A
Flavone, 4'-hydroxy-3,5,6,7-tetra-methoxy-	EtOH	216(4.03),260(4.00), 328(4.06)	24-3222-66
$C_{19}H_{19}BF_4N_4$			
N-Ethyl-2-(p-anilinophenylazo)-pyridinium tetrafluoroborate	MeCN	535(4.68)	5-0116-66G

Compound	Solvent	$\lambda_{max}(\log \epsilon)$	Ref.
$C_{19}H_{19}IN_2S_2$			
3-Ethyl-2-[(3-ethyl-2-benzothiazolinylidene)methyl]benzothiazolium iodide	MeOH	424(4.92)	9-0289-66
$C_{19}H_{19}IN_2Se_2$			
3-Ethyl-2-[(3-ethyl-2-benzoselenazolinylidene)methyl]benzoselenazolium iodide	MeOH	430(4.88)	9-0289-66
$C_{19}H_{19}N$			
Carbazole, N-benzyl-1,2,3,4-tetrahydro-	n.s.g.	225(4.73),286(3.82), 293(3.79)	16-0064-66
1-Pyrroline, 3-(diphenylmethylene)-2,5-dimethyl-, picrate	EtOH	228(4.12),287(4.12)	94-0187-66
1-Pyrroline, 3-(diphenylmethylene)-2-ethyl-	EtOH	231(4.20),287(4.20)	94-0187-66
$C_{19}H_{19}NO$			
1,4-Pentadien-3-one, 1-(p-dimethylaminophenyl)-5-phenyl-	benzene	304(4.25),416(4.48)	65-0336-66
	EtOH	313(4.26),432(4.47)	65-0336-66
	HOAc	316(4.29),444(4.40)	65-0336-66
ferric chloride added	HOAc	620(4.46)	65-0336-66
2,4-Pentadien-1-one, 1-(p-dimethylaminophenyl)-5-phenyl-	benzene	336(4.41),380(4.49)	65-0336-66
	EtOH	238(4.02),338(4.38), 397(4.47)	65-0336-66
	HOAc	343(4.55),403(4.63)	65-0336-66
ferric chloride added	HOAc	527(4.49)	65-0336-66
2,4-Pentadien-1-one, 5-(p-dimethylaminophenyl)-1-phenyl-	benzene	290(4.47),430(4.59)	65-0336-66
	EtOH	292(4.19),439(4.51)	65-0336-66
	HOAc	334(4.32),442(4.26)	65-0336-66
ferric chloride added	HOAc	668(3.26)	65-0336-66
Quinoline, N-benzoyl-1,2-dihydro-2,2,4-trimethyl-	EtOH	320(3.65)	5-0128-66D
$C_{19}H_{19}NO_2$			
Benzomorphan-3-one, 8-hydroxy-2-methyl-5-phenyl-	EtOH	261(2.58),269(2.43)	44-1905-66
2(1H)-Isoquinoline-2-carboxylic acid, 1-benzylidene-3,4-dihydro-, cis	n.s.g.	226(4.14),288(3.94)	88-2941-66
trans	n.s.g.	231(4.05),302(4.23)	88-2941-66
1-Naphthaleneacetamide, 1,2,3,4-tetrahydro-N-methyl-4-oxo-1-phenyl-	EtOH	249(4.05),293(3.27)	44-1905-66
Nuciferine, 6a,7-dehydro-	n.s.g.	253(4.97),264(4.98), 293(4.26),327(4.49)	88-2937-66
$C_{19}H_{19}NO_3$			
1-Naphthaleneacetamide, 1,2,3,4-tetrahydro-3-hydroxy-N-methyl-4-oxo-1-phenyl-	EtOH	250(4.01),294(3.40), 331(2.51)	44-1905-66
$C_{19}H_{19}NO_4$			
Actinodaphnine, N-methyl-	EtOH	282(4.15),307(4.21)	95-0129-66
1-Caryachine	EtOH	291.5(4.02)	95-0177-66
Cassythicine	EtOH	219(4.54),272s(4.08), 283(4.20),305(4.26)	12-2339-66
Escholamine, tetrahydro-	MeOH	237(4.1),290(4.1)	73-3362-66
Protopapaverine	90% EtOH	235(4.41),263(4.43), 281(4.47),362(4.19)	78-0485-66B

Compound	Solvent	λ_{max}(log ϵ)	Ref.
Protopapaverine (cont.)	EtOH-HCl	233(4.44),257(4.69), 325(3.93)	78-0485-66B
	EtOH-NaOH	284(4.63),393(4.25)	78-0485-66B
$C_{19}H_{19}NO_5$			
Cassythine	n.s.g.	218(4.57),283(4.27), 303(4.23)	12-0297-66
$C_{19}H_{19}N_3O$			
2-Cyclohexen-1-one, 3,5-diphenyl-, semicarbazone	EtOH	283(4.21)	22-1012-66
11-Morphanthridinone, 6-(4-methyl-1-piperazinyl)-	EtOH	238(4.39),250s(4.37), 321(3.94)	33-1433-66
$C_{19}H_{19}N_3O_3$			
Anthranilic acid, N-(3-butyl-3,4-dihydro-4-oxo-2-quinazolinyl)-	EtOH	218(4.64),234s(4.32), 292(4.34),325(4.15)	24-1532-66
$C_{19}H_{19}N_3O_6$			
Uracil, 2,2'-anhydro-1-(3-acetamido-3-deoxy-4,6-O-benzylidene-β-D-mannosyl)-	EtOH	225(4.01),245(3.94)	44-0211-66
$C_{19}H_{19}N_5O_4$			
Benzoic acid, p-N-(2-pyrazinylmethyl)-N-thyminylamino-, ethyl ester	EtOH	271(4.26),303(4.40)	87-0868-66
$C_{19}H_{20}BrClN_2O_4$			
6-Bromo-2,3-dihydro-5,7-dimethyl-1,4-diphenyl-1,4-diazepinium perchlorate	MeOH	230(3.98),376(4.09)	39-0093-66C
$C_{19}H_{20}Br_2N_2O_2$			
Fluoren-2-amine, N,N-bis-(2-bromopropyl)-7-nitro-	2N HCl	246(3.95),282s(--), 320(4.19),420(3.85)	87-0593-66
	0.2N HCl	249(3.96),281(4.05), 319(4.07),413(4.09)	87-0593-66
	0.1N HCl	249(3.96),280(4.06), 319(4.05),415(4.13)	87-0593-66
	EtOH	276(4.18),422(4.36)	87-0593-66
$C_{19}H_{20}Cl_2FN$			
Fluoren-2-amine, N,N-bis-(2-chloropropyl)-7-fluoro-	0.2N HCl	298(4.43),302s(--), 330s(--)	87-0593-66
	pH 1	298(4.44),330s(--)	87-0593-66
	EtOH	299(4.49),330s(--)	87-0593-66
$C_{19}H_{20}Cl_2N_2O$			
Acetamide, N-[7-[bis(2-chloroethyl)-	0.2N HCl	242s(--),295s(--), 312(4.33)	87-0593-66
	pH 1	314(4.36),342s(--)	87-0593-66
	EtOH	314(4.45),343s(--)	87-0593-66
$C_{19}H_{20}Cl_2N_2O_2$			
Fluoren-2-amine, N,N-bis-(2-chloropropyl)-7-nitro-	4N HCl	244(3.96),321(4.32), 412s(--)	87-0593-66
	2N HCl	250s(--),276(4.09), 323(3.93),417(4.06)	87-0593-66
	0.2N HCl	271(4.18),410(4.35)	87-0593-66
	0.1N HCl	271(4.17),410(4.35)	87-0593-66

Compound	Solvent	λ_{max}(log ϵ)	Ref.
Fluoren-2-amine, N,N-bis-(2-chloropropyl)-7-nitro- (cont.)	EtOH	270(4.17),410(4.36)	87-0593-66
$C_{19}H_{20}CuN_2O_4$			
1,2-Propanediamine, N,N'-bis-(2-hydroxy-3-methoxybenzylidene)-, copper chelate	MeOH	280(4.70),370(4.40), 590(3.89)	65-1590-66
$C_{19}H_{20}FN_3O_2S$			
4-Morpholinecarboxamide, N-[N-(p-fluoro-phenyl)-p-anisimidoyl]thio-	MeOH	275(4.42)	44-0722-66
$C_{19}H_{20}F_3N_3OS_2$			
Rhodanine, 5-[2-[1,3-diethyl-5-(tri-fluoromethyl)-2-benzimidazolinyli-dene]ethylidene]-3-ethyl-	EtOH	515(--)	65-0828-66
$C_{19}H_{20}INO_2$			
1-Benzyl-6,7-dimethoxy-2-methyl-isoquinolinium iodide	MeOH	257(4.62),320(3.90)	83-0159-66
$C_{19}H_{20}N_2O_4$			
3-Azabicyclo[4.1.0]hept-4-ene-1,5-di-carboxylic acid, 2-cyano-2,4-di-methyl-3-phenyl-, dimethyl ester	EtOH	205s(4.06),236(3.56), 299(4.15)	39-1950-66C
$C_{19}H_{20}N_2O_8S_2$			
Carbonic acid, ethyl ester, anhydride with 3-(hydroxymethyl)-8-oxo-7-[2-(2-thienyl)acetamido]-5-thia-1-aza-bicyclo[4.2.0]oct-3-ene-2-carbox-ylic acid, acetate	EtOH	234(4.03),266(3.79)	87-0741-66
5-Thia-1-azabicyclo[4.2.0]oct-3-ene-2-carboxylic acid, 3-(hydroxymethyl)-8-oxo-7-[2-(2-thienyl)acetamido]-, hydroxymethyl ester, diacetate	EtOH	235(4.12),270(3.48)	87-0741-66
$C_{19}H_{20}N_4OS$			
Pyrimido[4',5':4,5]pyrimido[2,3-c]-[1,4]thiazine-3-ethanol, 1,6-di-hydro-4,9-dimethyl-1-phenyl-	EtOH	373(4.05)	44-2951-66
$C_{19}H_{20}N_4O_4$			
1-Indanone, 2,2,5,7-tetramethyl-, 2,4-dinitrophenylhydrazone	MeOH	384(4.48)	35-2838-66
3-Pentenal, 2,2-dimethyl-4-phenyl-, 2,4-dinitrophenylhydrazone	n.s.g.	360(4.34)	22-0734-66
$C_{19}H_{20}N_4O_5$			
1(2H)-Naphthalenone, 3,4-dihydro-2,5-dimethyl-7-methoxy-, 2,4-dinitrophenylhydrazone	$CHCl_3$	392(4.43)	78-0301-66
$C_{19}H_{20}O_2$			
Estra-1,3,5(10),8,14-pentaen-17-one, 3-methoxy-	EtOH	312.6(4.49)	94-0262-66
Estra-1,3,5(10),8,14-pentaen-17-one, 4-methoxy-	EtOH	230(4.25),247(4.00), 297(4.34),307(4.39), 320(4.22)	13-0309-66B

Compound	Solvent	λ_{max}(log ε)	Ref.
2-Furanol, 2,5-dihydro-5,5-dimethyl-2-phenyl-4-p-tolyl-	N H_2SO_4	245(3.51),306(3.94),410(4.87)	65-2064-66
2-Furanol, 2,5-dihydro-5,5-dimethyl-4-phenyl-2-p-tolyl-	N H_2SO_4	243(3.54),325(3.88),402(4.73)	65-2064-66
2,4-Pentanedione, 3,3-dibenzyl-	hexane	248s(--),254(2.56),259(2.68),265(2.62),269(2.44),301(2.08)	18-0901-66
	EtOH	249(2.47),253(2.57),259(2.68),265(2.60),268s(--),294(2.09)	18-0901-66
$C_{19}H_{20}O_3$			
Acrylic acid, 2-ethoxy-3,3-diphenyl-, ethyl ester	MeOH	225(4.24),282(4.00)	5-0053-66I
Chalcone, 4,4'-dihydroxy-3,3',5,5'-tetramethyl-	pH 13	266(4.09),390s(4.22),454(4.65)	49-0896-66
	EtOH	247(4.13),362(4.48)	49-0896-66
14β-Estra-1,3,5(10),6,8-pentaene-17α-carboxylic acid, 16ξ-hydroxy-	EtOH	230(4.84),280(3.63),321(2.76)	35-0799-66
14β-Gona-1,3,5(10),8-tetraen-11-one, 17,17-ethylenedioxy-	dioxan	237(4.05),288(3.82)	19-0611-66
1-Phenanthrenebutyric acid, 9,10-dihydro-7-methoxy-	MeOH	280(4.27)	44-1285-66
$C_{19}H_{20}O_4$			
Acrylic acid, 3-(p-methoxyphenyl)-2-(6-methoxy-m-tolyl)-, methyl ester, cis	EtOH	308(4.42)	78-1789-66
trans	EtOH	316(4.54)	78-1789-66
Methane, bis(5-acetyl-2-methylphenyl)-	EtOH	230(4.43),275(4.46)	87-0292-66
1,2-Phenanthrenediacetic acid, 1,2,3,4-tetrahydro-2-methyl-, di-cis?	EtOH	229(4.93),280(3.74),320(2.72)	35-0799-66
3,3'-Spirobichroman, 8,8'-dimethoxy-	PrOH	228(4.19),280(3.73)	78-0199-66
	THF	280(3.71)	78-0199-66
$C_{19}H_{20}O_5$			
Clausenidin	EtOH	222(4.23),284(4.53),328(4.11)	88-5767-66
1H-5,11-Dioxadibenzo[a,d]pentalene, 2,3,4,4a,6,6a-hexahydro-6a-hydroxy-4aβ,7,9-trimethyl-6α-(carboxymethyl)-3-oxo-, γ-lactone	n.s.g.	286(3.51)	44-1725-66
A furonaphthalene diacetate	EtOH	214(4.50),227(4.56),236(4.59)	78-2387-66
$C_{19}H_{20}O_6$			
Acrylic acid, 2-(2,4-dimethoxyphenyl)-3-(3,4-dimethoxyphenyl)-, cis	EtOH	309(4.23)	78-1789-66
trans	EtOH	315(4.36)	78-1789-66
$C_{19}H_{20}O_7$			
Elephantol, methacrylate	MeOH	209(4.33)	35-3674-66
Elephantopin	MeOH	210(4.43)	35-3674-66
$C_{19}H_{21}^+$			
Cyclohexyldiphenylcarbonium ion	FSO_3H-SbF_5	338(4.23),427(4.56)	35-1488-66
$C_{19}H_{21}BrO_2$			
Androsta-1,4,6-triene-3,17-dione, 2-bromo-	MeOH	222(4.16),270(4.09),304(4.01)	73-1363-66

Compound	Solvent	$\lambda_{max}(\log \epsilon)$	Ref.
$C_{19}H_{21}Br_2N$			
Fluoren-2-amine, N,N-bis-(2-bromopropyl)-	pH 1	267(4.27),302(4.32)	87-0593-66
	EtOH	304(4.47),341s(--)	87-0593-66
$C_{19}H_{21}ClN_2O_4$			
2,3-Dihydro-5,7-dimethyl-1,4-diphenyl-1,4-diazepinium perchlorate	MeOH	238(3.85),347(4.46)	39-0093-66C
$C_{19}H_{21}ClN_4O_4$			
1-Ethyl-4-(p-dimethylaminophenylazo)-quinolinium perchlorate	MeOH	610(4.79)	5-0116-66G
$C_{19}H_{21}ClO_2$			
Androsta-1,4,6-triene-3,17-dione, 2-chloro-	MeOH	220(4.16),265(4.03), 305(3.99)	73-1363-66
$C_{19}H_{21}Cl_2N$			
Fluoren-2-amine, N,N-bis-(2-chloropropyl)-	0.2N HCl	270s(--),301(3.79), 330s(--)	87-0593-66
	0.1N HCl	270s(--),301(4.41), 330s(--)	87-0593-66
	EtOH	301(4.48),330s(--)	87-0593-66
$C_{19}H_{21}N$			
Benzomorphan, 2-methyl-5-phenyl-	EtOH	259(2.72)	44-1905-66
$C_{19}H_{21}NO$			
Ethylamine, 2-(5H-dibenzo[a,d]cyclo-hepten-10-yloxy)-N,N-dimethyl-	EtOH	283(4.09)	54-0389-66
$C_{19}H_{21}NO_2$			
Carbamic acid, [(10,11-dihydro-5H-dibenzo[a,d]cyclohepten-5-yl)-methyl]-, ethyl ester	EtOH	263(2.77)	4-0247-66
Estra-1,3,5(10)-triene-2-carbonitrile, 3-hydroxy-17-oxo-	EtOH	236(4.04),304(3.60)	88-0205-66
Nuciferine	EtOH	271(4.23),305s(3.62)	95-0129-66
$C_{19}H_{21}NO_4$			
Alkaloid HF-1	MeOH	220(4.3),283(3.8)	73-1355-66
(+)-Coreximine	MeOH	288(3.96)	39-2010-66C
(-)-Coreximine	MeOH	288(3.95)	39-2010-66C
Isoboldine	EtOH	280(4.11),305(4.15)	12-0135-66
	EtOH	219(4.58),268s(4.02), 280(4.16),304(4.21), 313s(4.18)	12-2331-66
	EtOH	280(4.03),302(4.05)	39-0753-66C
Laurotetanine	EtOH	282(4.14),303(4.14)	95-0129-66
$C_{19}H_{21}N_3O$			
3-Butenal, 2,2-dimethyl-4,4-diphenyl-, semicarbazone	n.s.g.	247.5(4.36)	22-0734-66
4(1H)-Cinnolinone, 1-(3-dimethyl-aminopropyl)-3-phenyl-, maleic acid salt	MeOH	264(4.13),312(4.03)	87-0784-66
11-Morphanthridinol, 6-(4-methyl-1-piperazinyl)-	EtOH	230s(4.19),249s(4.03), 304(3.90)	33-1433-66
$C_{19}H_{21}N_3O_2$			
Piperazine, 1-(o-anthraniloylbenzoyl)-4-methyl-	EtOH	234(4.35),382(3.75)	33-1433-66

Compound	Solvent	λ_{max}(log ε)	Ref.
$C_{19}H_{21}N_3O_7S_2$			
Alanine, N-[[3-(hydroxymethyl)-8-oxo-7-[2-(2-thienyl)acetamido]-5-thia-1-azabicyclo[4.2.0]oct-2-en-2-yl]-carbonyl]-, acetate	EtOH	235(4.19),260(3.91)	87-0741-66
$C_{19}H_{21}O_5P$			
Phosphonic acid, α-(3-methoxy-1-oxo-indan-2-yl)benzyl-, dimethyl ester	MeOH	245(4.58),305(3.0)	30-1097-66F
$C_{19}H_{22}FNO_2$			
2-Propanol, 1,1'-[(7-fluorofluoren-2-yl)imino]di-	pH 1	264(4.37),292(3.92), 304(4.02)	87-0593-66
	EtOH	304(4.47),344s(--)	87-0593-66
$C_{19}H_{22}N_2$			
1,6-Diazaspiro[4.4]nonane, 1,6-diphenyl-	MeOH	206(4.7),250(4.8), 293(4.1)	24-0724-66
dipicrate	$CHCl_3$	255(3.7)	24-0724-66
2-Pentene, 2-benzylamino-4-benzylimino-	MeOH	322(4.60)	88-0459-66
cation	MeOH	321(4.62)	88-0459-66
$C_{19}H_{22}N_2O$			
Ellipticine, tetrahydromethoxy-N-methyl-	EtOH	234(4.61),248(4.60), 258s(4.44),269(4.22), 293(4.12),302(4.35), 342(3.66),356(3.66)	12-1947-66
Peraksine	EtOH	226(4.53),278(4.02), 290s(3.19)	78-3293-66
$C_{19}H_{22}N_2O_2S$			
p-Toluenesulfonamide, N-[2-(2-ethyl-indol-3-yl)ethyl]-	EtOH	224(4.66),284(3.88), 291(3.83)	39-0425-66C
$C_{19}H_{22}N_2O_3$			
Acetamide, N-[7-[bis(2-hydroxyethyl)-amino]fluoren-2-yl]-	pH 1	221s(--),293(4.47), 304s(--),317(4.39)	87-0593-66
	EtOH	255s(--),318(4.54)	87-0593-66
$C_{19}H_{22}N_2O_4$			
2-Propanol, 1,1'-[(7-nitrofluoren-2-yl)imino]di-	2N HCl	240(4.11),320(4.36)	87-0593-66
	pH 1	240s(--),318(4.35)	87-0593-66
	EtOH	275(4.17),430(4.37)	87-0593-66
$C_{19}H_{22}N_2O_7$			
Benzimidazole, 1-(4-deoxy-β-D-xylo-hexopyranosyl)-, triacetate	EtOH	245(3.89),274(3.59), 281(3.60)	39-1549-66C
$C_{19}H_{22}N_4O$			
p-Benzoquinone, 2,6-diethyl-, azine with 1,3-dimethyl-2-benzimidazol-inone	MeCN	501(4.80)	5-0065-66J
$C_{19}H_{22}N_4O_4$			
3,5-Heptadien-2-one, 6-(3-cyclohexen-1-yl)-, 2,4-dinitrophenylhydrazone	$CHCl_3$	270(4.31),310(4.37), 405(4.65)	70-0480-66
3,5-Hexadienal, 6-(1-cyclohexenyl)-4-methyl-, 2,4-dinitrophenyl-hydrazone	n.s.g.	268(4.57),360(4.36)	22-0728-66

Compound	Solvent	λ_{max}(log ε)	Ref.
2,4-Hexadienal, 6-cyclohexylidene-4-methyl-, 2,4-dinitrophenylhydrazone	n.s.g.	251(4.12),266(4.16), 340(4.19),408(4.48)	22-0728-66
$C_{19}H_{22}N_4O_6$			
Isocytosine, 1-(3-acetamido-3-deoxy-4,6-O-benzylidene-β-D-mannosyl)-	EtOH-HCl	255(3.93)	44-0211-66
	EtOH	252(3.81)	44-0211-66
$C_{19}H_{22}N_4O_7$			
Furo[2,3-b]pyridine-5-carboxylic acid, 4-hydroxy-2-[N-[5-(morpholinomethyl)-2-oxo-3-oxazolidinyl]formimidoyl]-, ethyl ester	5% EtOH	312.5(4.50)	4-0202-66
$C_{19}H_{22}N_4O_8S_3$			
s-Trithiane-2,4-dione, bis(o-ethoxyphenylhydrazone), 1,1,3,3,5,5-hexaoxide, cis	n.s.g.	394(4.36)	99-0318-66
trans	n.s.g.	401(4.71)	99-0318-66
$C_{19}H_{22}O_2$			
Estra-1,3,5(10),8,14-pentaen-17β-ol, 3-methoxy-	MeOH	310(4.49)	88-2321-66
13α-Estra-1,3,5(10),8,14-pentaen-17β-ol, 3-methoxy-	MeOH	310(4.44)	88-2321-66
Estra-5(10),6,8,14-tetraene-17β-carboxylic acid	EtOH	217(4.40),223(4.37), 230(4.18),265(4.29)	35-0799-66
8α,9α-Estra-1,3,5(10),14(15)-tetraen-17-one, 3-methoxy-	n.s.g.	278(3.4)	88-3585-66
Estra-1,3,5(10)-triene-6,17-dione, 4-methyl-	MeOH	253(4.07)	39-1034-66C
2-Phenanthrone, 12-allyl-2,3,4,9,10,12-hexahydro-8-methoxy-1-methyl-	MeOH	228(4.29)	35-1786-66
$C_{19}H_{22}O_3$			
Androsta-4,6-diene-3,17-dione, 1α,2α-epoxy-	MeOH	288(4.30)	73-1363-66
10α-Androsta-4,9(11)-diene-3,6,17-trione	EtOH	248(4.06)	33-1529-66
8α-Estra-1,3,5(10),9(11)-tetraen-17-one, 14β-hydroxy-3-methoxy-	n.s.g.	260(4.3)	88-3585-66
14β-Gona-1,3,5(10),8-tetraen-11ξ-ol, 17,17-ethylenedioxy-	MeOH	226(4.07),261(4.09), 267(4.07)	19-0611-66
8,14-Secoestra-1,3,5(10),9(11)-tetraene-14,17-dione, 4-methoxy-	EtOH	224(4.44),261(4.10)	13-0309-66B
$C_{19}H_{22}O_4$			
Atractylon	heptane	222.5(3.84)	39-1866-66C
Cacalol, 2-acetyl-, acetate	EtOH	242(4.17),308(4.39)	78-0301-66
3-Cyclohexene-1-carboxylic acid, 2-(hydroxymethyl)-6,6-dimethyl-2-(6-methylene-7-oxobicyclo[3.2.1]-oct-2-yl)-5-oxo-, lactone	EtOH	231(4.18)	78-1659-66
Grandiflorone	EtOH	235(4.03),277(4.02)	39-1496-66C
Isovaleric acid, 2-(2,4-hexadiynylidene)-1,6-dioxaspiro[4.5]dec-3-en-8-yl ester, cis	ether	215(4.10),223(4.08), 236(4.05),249s(3.63), 264(3.61),317(4.28)	24-2416-66

Compound	Solvent	λ_{max}(log ε)	Ref.
Isovaleric acid, 2-(2,4-hexadiyn-ylidene)-1,6-dioxaspiro[4.5]dec-3-en-8-yl ester, trans	ether	213(4.09),221s(4.04), 230s(3.93),235s(3.94), 240(3.94),253(3.49), 269s(3.68),311(4.33), 321(4.33),342s(4.05)	24-2416-66
Odoratin	n.s.g.	213(4.23)	77-0576-66
Propene, 1-(o-methoxyphenyl)-3-(2,4,5-trimethoxyphenyl)-	EtOH	250(4.23),260(4.15), 295(4.03)	78-3491-66
$C_{19}H_{22}O_5$			
Mandelic acid, α-anisyl-2-ethoxy-5-methyl-	EtOH	280(3.80)	78-1789-66
Mandelic acid, α-anisyl-2-methoxy-5-methyl-, methyl ester	EtOH	276(3.64),281(3.63)	78-1789-66
Pentatriafulvalene-2,3-dicarboxylic acid, 1-formyl-5,6-dipropyl-, dimethyl ester	n.s.g.	228(4.02),272(4.55), 325s(--),345(4.16)	88-3697-66
$C_{19}H_{22}O_6$			
2-Benzofuranpropionic acid, 2-(2-carboxy-1-methylvinyl)-2,3-dihydro-4,6-dimethyl-3-oxo-, dimethyl ester	n.s.g.	232(4.11),275(4.25), 320(3.98)	78-2015-66
Marasmic acid enol diacetate	MeOH	248(4.28)	35-2838-66
Strigol	n.s.g.	236(4.26)	57-1189-66D
$C_{19}H_{23}BrO_2$			
Androsta-1,4,6-trien-3-one, 2-bromo-17β-hydroxy-	MeOH	220(4.14),270(4.08), 306(4.02)	73-1363-66
$C_{19}H_{23}ClN_2$			
9-(Dimethylaminopropyl)-10-methyl-acridinium chloride, hydrochloride	n.s.g.	215(4.31),267(4.54), 281(3.92),361(4.05), 425(3.45)	33-0572-66
7H-Yohimban, 7-chloro-	EtOH	225(4.33),260s(3.34), 290(3.38)	44-1765-66
	CH_2Cl_2	283(3.48)	44-1765-66
$C_{19}H_{23}ClO_2$			
Androsta-1,4,6-trien-3-one, 2-chloro-17β-hydroxy-	MeOH	220(4.18),268(4.08), 308(4.06)	73-1363-66
$C_{19}H_{23}ClO_6$			
Cyclohexanecarboxylic acid, 1-(2-chloro-3,5-dimethoxybenzyl)-6-methyl-2,4-dioxo-, ethyl ester	MeOH	230(4.02),265(4.13), 290s(3.85)	44-1462-66
$C_{19}H_{23}N$			
1-Butene, 1,2-diphenyl-4-(dimethyl-amino)-, hydrochloride	H_2O	255(4.13)	78-1001-66
$C_{19}H_{23}NO$			
Benzanilide, 2,2',4,4',6,6'-hexamethyl-	EtOH	270s(3.1)	28C-0369-66A
Stilbene, 2-(2-dimethylaminoethoxy)-methyl-, trans, hydrochloride	H_2O	203(4.38),227(4.16), 297(4.40)	39-0668-66C
$C_{19}H_{23}NO_2$			
Estra-1,3,5(10)-triene-2-carbo-nitrile, 3,17β-dihydroxy-	EtOH	236(4.04),305(3.57)	88-0205-66
	EtOH-NaOH	223(4.39),246(4.00), 337(3.59)	88-0205-66

Compound	Solvent	λ_{max}(log ε)	Ref.
2-Propanol, 1,1'-(fluoren-2-yl-imino)di-	pH 1	266(4.39),289(3.99), 301(4.09)	87-0593-66
	EtOH	230s(--),306(4.47), 343s(--)	87-0593-66
$C_{19}H_{23}NO_3$			
2H-Benzo[a]quinolizin-2-one, 1,6,7,11b-tetrahydro-9,10-dimethoxy-4-methyl-3-propyl-	EtOH	285(3.70),334(4.17)	44-0797-66
6H-Dibenzo[a,f]quinolizin-13(12H)-one, 1,2,3,4,7,11b-hexahydro-9,10-dimethoxy-	EtOH	286(3.71),332(4.16)	44-0797-66
Magnocurarine	H_2O	224(4.11),283(3.55)	39-1340-66C
Metaphanine	EtOH	205(4.65),285(3.34)	88-6169-66
$C_{19}H_{23}NO_4$			
Benzanilide, 3,4,5-trimethoxy-2',4',6'-trimethyl-	EtOH	212(4.7),262(4.1)	28C-0584-66A
Cepharamine	EtOH	259(3.82)	88-6229-66
$C_{19}H_{23}NO_8$			
2,3,4,5-Piperidinetetracarboxylic acid, 6-phenyl-, tetramethyl ester	MeOH	207(3.93),250(2.78), 256(2.78),263(2.78)	39-1121-66C
$C_{19}H_{23}N_3O_2$			
Butyrophenone, 4-ethoxy-4-phenyl-, semicarbazone	EtOH	274(4.31)	22-0717-66
$C_{19}H_{23}N_3O_3$			
4-Pyrimidineacrylic acid, 2-amino-6-hydroxy-5-(4-phenylbutyl)-, ethyl ester	pH 1	233(4.25),315(4.00)	4-0315-66
	pH 7	243(4.30),345(3.85)	4-0315-66
	pH 13	239(4.46),345(3.79)	4-0315-66
$C_{19}H_{23}N_3O_5$			
Isonicotinamide, N-[2-(benzyloxy)-3,5-dimethoxymorpholino]-	EtOH	258(3.57)	39-2121-66C
$C_{19}H_{23}N_5O_4$			
Benzoic acid, p-[octahydro-3-(1,2,3,4-tetrahydro-2,4-dioxo-5-pyrimidinyl)-imidazo[1,5-a]pyrazin-2-yl]-, ethyl ester	EtOH	304(4.47)	87-0868-66
$C_{19}H_{23}N_5O_4S_2$			
Adenosine, 2',5'-dideoxy-5'-S-ethyl-3'-O-p-toluenesulfonyl-5'-thio-	MeOH	226(4.22),258(4.20)	69-0224-66
$C_{19}H_{23}N_5O_5$			
Benzoic acid, p-[N-[3-[N-(2-acetamido-4-hydroxy-5-pyrimidinyl)amino]-propyl]formamido]-, ethyl ester	EtOH	271(4.35),310s(4.00)	87-0590-66
$C_{19}H_{23}N_5O_8$			
Adenine, N-acetyl-9-(4-deoxy-β-D-xylohexopyranosyl)-, 2,3,6-triacetate	pH 7	272.5(4.25)	39-1549-66C
$C_{19}H_{23}N_5O_9$			
Hypoxanthine, 9-(3-acetamido-3-deoxy-β-D-glucopyranosyl)-, 2',4',6'-triacetate	MeOH	247(4.05)	24-0575-66

Compound	Solvent	$\lambda_{max}(\log \epsilon)$	Ref.
$C_{19}H_{23}N_5O_9S$			
β-D-Glucopyranuronic acid, 1-deoxy-1-[7-(ethylthio)-3H-v-triazolo[4,5-d]-pyrimidin-3-yl]-, methyl ester, 2,3,4-triacetate	MeOH	228(4.11),301(4.23)	94-1360-66
$C_{19}H_{24}BrN_5O_4$			
1,1,2,2-Tetramethyl-3,3-bis-(p-nitrobenzyl)guanidinium bromide	MeCN	267(4.35)	44-1430-66
$C_{19}H_{24}Br_3N_5O_4$			
1,1,2,2-Tetramethyl-3,3-bis(p-nitro-benzyl)guanidinium tribromide	MeCN	268(4.80)	44-1430-66
$C_{19}H_{24}ClN$			
1-Butene, 1,2-diphenyl-4-(dimethyl-amino)-, hydrochloride	H_2O	255(4.13)	78-1001-66
$C_{19}H_{24}ClNO$			
Stilbene, 2-(2-dimethylaminoethoxy)-methyl-, trans, hydrochloride	H_2O	203(4.38),227(4.16), 297(4.40)	39-0668-66C
$C_{19}H_{24}Cl_2N_2$			
9-(Dimethylaminopropyl)-10-methyl-acridinium chloride, hydrochloride	n.s.g.	215(4.31),267(4.54), 281(3.92),361(4.05), 425(3.45)	33-0572-66
$C_{19}H_{24}N_2$			
Yohimban	EtOH	225(4.51),273s(3.88), 280(3.90),290s(3.83)	44-2695-66
$C_{19}H_{24}N_2O$			
Aspidoalbine	EtOH	242(3.86),293(3.48)	23-0028-66
3(2H)-Isoquinolinone, octahydro-2-(2-indol-3-ylethyl)-, trans	EtOH	222(4.57),273s(3.76), 282(3.79),290(3.73)	44-2695-66
Pyrazolo[2,3-d]estra-1,3,5(10)-trien-17β-ol	EtOH	260(3.98),266(3.96), 289(3.86),305(3.69)	88-0205-66
$C_{19}H_{24}N_2O_2$			
Spiroxyane A, 17β-hydroxy-	EtOH	251(3.86),282(3.26)	44-1765-66
Strychnosplendine	n.s.g.	245(3.70),297(3.29)	88-2353-66
$C_{19}H_{24}N_2O_{11}$			
β-D-Glucopyranuronic acid, 1-deoxy-1-(4-ethoxy-2-oxo-1(2H)-pyrimidinyl)-, methyl ester, 2,3,4-triacetate	EtOH	276.5(3.74)	94-1354-66
$C_{19}H_{24}N_4O_3S$			
Acetamide, N-[(4-amino-2-methyl-5-pyrimidinyl)methyl]-2-[[1-(2-hydroxy-ethyl)acetonyl]thio]-2-phenyl-, as acetate	EtOH	228(4.09),275(3.77)	44-2951-66
$C_{19}H_{24}N_4O_{11}S$			
β-D-Glucopyranuronic acid, 1-deoxy-1-[[6-(ethylthio)-5-nitro-4-pyrimid-inyl]amino]-, 2,3,4-triacetate, methyl ester	MeOH	256(4.44),338(3.92)	94-1360-66

Compound	Solvent	λ_{max}(log ϵ)	Ref.
$C_{19}H_{24}N_4O_{13}$			
Uracil, 6-(2,3,4,6-tetra-O-acetyl-β-D-glucopyranosylamino)-3-methyl-5-nitro-	pH 0	229(4.30),318(4.07)	24-3022-66
	pH 5	237(4.11),290s(3.64),332(4.21)	24-3022-66
$C_{19}H_{24}N_6O_6$			
L-Glutamic acid, N-[p-[[3-[(2-amino-4-hydroxy-5-pyrimidinyl)amino]-propyl]amino]benzoyl]-	pH 1	222(4.17),259(4.04),295s(3.86)	87-0590-66
$C_{19}H_{24}O$			
Estra-1,3,5(10)-trien-6-one, 4-methyl-	MeOH	252(3.99)	39-1034-66C
$C_{19}H_{24}O_2$			
Androsta-1,4-diene-3,17-dione	n.s.g.	244(4.23)	19-0347-66
10α-Androsta-4,9(11)-diene-3,17-dione	EtOH	237(4.16)	33-1529-66
Androsta-1,4,9(11)-trien-3-one, 17β-hydroxy-	EtOH	239(4.20)	33-1183-66
Estra-1,3,5(10),8(14)-tetraen-17β-ol, 3-methoxy-	MeOH	278(3.34)	44-3780-66
	MeOH	276(4.23)	88-2321-66
Estra-1,3,5(10),9(11)-tetraen-17β-ol, 3-methoxy-	EtOH	263(4.25)	87-0510-66
Estra-1,3,5(10)-trien-6-one, 17β-hydroxy-4-methyl-	EtOH	253(4.04),299(3.37)	5-0180-66B
Estra-4,8(14),9-trien-3-one, 17β-hydroxy-17-methyl-	MeOH	359(4.37)	44-3780-66
Gona-1,3,5(10)-trien-17-one, 13β-ethyl-3-hydroxy-	EtOH	223(3.86),282(3.32),288(3.30)	35-3120-66
d-isomer	EtOH	282(3.34),287(3.32)	44-2512-66
8α-Gona-1,3,5(10)-trien-17-one, 13β-ethyl-3-hydroxy-	EtOH	279(3.36),285(3.32)	87-0338-66
$C_{19}H_{24}O_3$			
9β,10α-Adrenosterone	MeOH	239(4.20)	54-0721-66
Androsta-1,4-diene-3,11-dione, 17β-hydroxy-	EtOH	240(4.22)	33-1183-66
Androsta-1,4-diene-3,17-dione, 4-hydroxy-	EtOH	240(3.93),309(3.81)	5-0180-66B
	0.05N NaOH	227(4.28),368(3.81)	5-0180-66B
Androsta-4,6-dien-3-one, 1α,2α-epoxy-17β-hydroxy-	MeOH	290(4.33)	73-1363-66
5β-Androst-1-ene-3,11,17-trione	MeOH	225(3.97)	37-5313-66
5β-Androst-2-ene-1,11,17-trione	MeOH	225(3.84)	37-5313-66
9β,10α-Androst-4-ene-3,11,17-trione	MeOH	239(4.19)	54-0731-66
	EtOH	238(4.20)	33-1529-66
8α-Estra-1,3,5(10),9(11)-tetraene-8,17β-diol	EtOH	258(4.25)	88-5015-66
Estra-1,3,5(10)-triene-2-carboxaldehyde, 3,17β-dihydroxy-	EtOH	227(4.19),268(4.16),338(3.54)	88-0205-66
	EtOH-NaOH	239(4.25),275(4.02),289(3.77)	88-0205-66
8α-Estra-1,3,5(10)-trien-17-one, 14β-hydroxy-3-methoxy-	n.s.g.	278(3.4)	88-3585-66
8α,9β-Estra-1,3,5(10)-trien-17-one, 14β-hydroxy-3-methoxy-	n.s.g.	278(3.4)	88-3585-66
Gon-4-ene-3,6,17-trione, 13β-ethyl-	EtOH	254(3.99)	35-3120-66
9,10-Secoandrosta-1,3,5(10)-triene-11,17-dione, 3-hydroxy-	MeOH	279(3.39)	54-0731-66
8,14-Seco-1,3,5(10),9(11)-estratetraen-14-one, 17β-hydroxy-3-methoxy-	MeOH	265(4.32)	88-2321-66

Compound	Solvent	λ_{max}(log ϵ)	Ref.
8,14-Seco-13α-estra-1,3,5(10),9(11)-tetraen-14-one, 17β-hydroxy-3-methoxy-	MeOH	264(4.32)	88-2321-66
Testololactone, 1-dehydro-16-oxo-	EtOH	241(4.22)	88-5337-66
	EtOH-base	244(4.19),277(4.27)	88-5337-66
$C_{19}H_{24}O_4$			
Androst-4-ene-3,17-dione, 9α-hydroxy-6,19-epoxy-	EtOH	240(4.08)	37-0540-66
Androst-4-en-3-one, 4,9α,17β-trihydroxy-	EtOH	278(4.10)	37-0540-66
3H-Benzofuro[3a,3-b]benzofuran-3-one, 1,2,4,4a,6,6a-hexahydro-6aα-methoxy-4aβ,6α,7,9-tetramethyl-	n.s.g.	282(3.48),290(3.51)	37-0540-66
3-Cyclohexene-1-carboxylic acid, 2-(hydroxymethyl)-6,6-dimethyl-2-(6-methyl-7-oxobicyclo[3.2.1]oct-2-yl)-5-oxo-, γ-lactone	EtOH	225(4.00)	78-1659-66
9,10-Secoandrosta-1,3,5(10)-triene-9,17-dione, 3,4-dihydroxy-	EtOH	282(3.29)	37-0540-66
9,10-Secoandrosta-1,3,5(10)-triene-9,17-dione, 3,19-dihydroxy-	EtOH	278(3.30)	37-0540-66
$C_{19}H_{24}O_5$			
Perezinone, dihydro-, diacetate	EtOH	206(3.94),332(3.91), 265(3.72)	78-2387-66
$C_{19}H_{24}O_6$			
Cyclohexanecarboxylic acid, 1-(3,5-dimethoxybenzyl)-2-methyl-4,6-dioxo-, ethyl ester	MeOH	235(4.11),277(3.90), 295(3.81)	44-1456-66
$C_{19}H_{24}O_7$			
Elephantol, dihydro-, isobutyrate	MeOH	211(4.01)	35-3674-66
Elephantopin, tetrahydro-	MeOH	210(4.03)	35-3674-66
$C_{19}H_{25}ClO_2$			
5α-Androst-1-ene-3,17-dione, 2-chloro-	MeOH	246(3.96)	73-1363-66
Androst-4-ene-3,17-dione, 19-chloro-	EtOH	239(4.23)	44-0693-66
$C_{19}H_{25}ClO_4$			
3(2H)-Benzofuranone, 7-chloro-6-methoxy-4-[(3,7-dimethyl-6-octenyl)oxy]-	EtOH	209(4.32),235(4.26), 285(4.26),320(3.69)	87-0242-66
$C_{19}H_{25}NO_3S$			
Ethylamine, 2-[2-(diphenylmethyl)-sulfonyl]ethoxy]-N,N-dimethyl-	EtOH	253(2.65),259(2.76), 262(2.71),269(2.55)	87-0860-66
$C_{19}H_{25}NO_4$			
Magnocurarine (iodide)	EtOH	290(3.63)	95-0124-66
Nerinine, deoxy-	EtOH	280(3.09)	44-0189-66
Spiro[5H-1-benzazepine-5,1'-cyclohexan]-2(1H)-one, 3,4-dihydro-5-(p-acetoxyphenyl)-7-methoxy-1-methyl-	EtOH	247(4.44),290s(3.65)	94-1033-66
Spiro[5H-2-benzazepine-5,1'-cyclohexan]-1(2H)-one, 3,4-dihydro-5-(p-acetoxyphenyl)-7-methoxy-2-methyl-	EtOH	252(4.14)	94-1033-66
$C_{19}H_{25}NO_5$			
α-Pipitzol, oxime, diacetate	EtOH	248(3.87)	78-2387-66

Compound	Solvent	$\lambda_{max}(\log \epsilon)$	Ref.
β-Pipitzol, oxime, diacetate	EtOH	248(3.87),309(2.11)	78-2387-66
$C_{19}H_{25}NO_5S_2$			
Methanesulfinic acid, [[2-[2-(dimethylamino)ethoxy]ethyl]-sulfonyl]diphenyl-	EtOH	253(2.61),259(2.72), 269(2.47)	87-0860-66
$C_{19}H_{25}N_3O_3$			
Phthalhydrazide, 3-undecanamido-	DMF	300s(3.83)	18-0932-66
$C_{19}H_{25}N_5O_7$			
Pseudourea, 1,3-dicyclohexyl-2-picryl-	EtOH	215(4.35)	88-6103-66
$C_{19}H_{25}N_6O_8P$			
L-Tyrosinol, adenosine-5'-phosphoric acid ester	n.s.g.	261(4.10)	33-0076-66
$C_{19}H_{26}BrIN_2S_2$			
x-Bromo-bis(3-ethyl-2-thiazolinyl)-pentamethinecyanine iodide	MeOH	537(5.37)	9-0150-66
$C_{19}H_{26}F_2O_2$			
Androst-4-en-3-one, 6,6-difluoro-17β-hydroxy-	EtOH	229(4.08)	44-0991-66
$C_{19}H_{26}N_2O$			
Estra-2,4-diene-2-carbonitrile, 3-amino-17-hydroxy-	EtOH	241(3.97),331(3.68)	32-0152-66
Pyrazolo[2,3-d]est-4-en-17β-ol	EtOH	260(3.95)	88-0205-66
$C_{19}H_{26}N_2O_4$			
Androst-5-ene-2,3,4-trione,	EtOH	263(4.01)	44-2755-66
17β-hydroxy-, 2,4-dioxime	base	221(3.87),263(3.91), 267(3.73)	44-2755-66
Pyrrole-3-carboxylic acid, 5,5'-iso-propylidenebis[4-methyl-, diethyl ester	EtOH	210(4.41),229s(4.30), 267(3.73)	39-0098-66C
$C_{19}H_{26}N_4O_2$			
5α-Androstan-3-one, 2,4-bis(diazo)-17β-hydroxy-	EtOH	259(3.60),322(4.40)	44-2755-66
5β-Androstan-3-one, 2,4-bis(diazo)-17β-hydroxy-	EtOH	259(3.55),322(4.42)	44-2755-66
$C_{19}H_{26}N_4O_6$			
Inosine, 2',3'-O-isopropylidene-	pH 1	254.5(3.96)	44-2685-66
1-methyl-5'-O-(tetrahydro-	pH 11	251.5(3.93)	44-2685-66
2-pyranyl)-	EtOH	246(3.89),251(3.89)	44-2685-66
$C_{19}H_{26}O_2$			
Androsta-3,5-dien-17-one, 11β-hydroxy-	EtOH	228(4.27),234(4.36), 243(4.12)	37-5336-66
5β-Androst-1-ene-3,17-dione	MeOH	230(3.98)	37-5313-66
5β-Androst-2-ene-1,17-dione	MeOH	226(3.88)	37-5313-66
Biphenyl, 1,4,5,6-tetrahydro-6-isopropylene-2',6'-dimethoxy-3,4'-dimethyl-	MeOH	210(4.59),271(2.95), 279(2.91)	5-0165-66C
4β,10β-Cyclo-9(10→5β)-abeo-10αH-estr-2-en-1-one, 17β-hydroxy-4α-methyl-	EtOH	238(3.73),269s(3.36)	33-1049-66

Compound	Solvent	λ_{max}(log ϵ)	Ref.
17-Epiestradiol, 3-methyl ether	MeOH	219(3.89),279(3.27), 287(3.25)	25-1340-66
Estra-4,8(14)-dien-3-one, 17β-hydroxy-17-methyl-	MeOH	233(4.20)	44-3780-66
Estra-1,3,5(10)-triene-2,17β-diol,	EtOH	220s(3.88),285(3.16)	5-0180-66B
1-methyl-	0.05N NaOH	244(3.77),301(3.34)	5-0180-66B
Estra-1,3,5(10)-triene-4,17β-diol,	EtOH	220s(3.88),284(3.26)	5-0180-66B
1-methyl-	0.05N NaOH	246(3.94),299(3.55)	5-0180-66B
8α-Estra-1,3,5(10)-trien-17β-ol, 3-methoxy-	EtOH	278(3.28)	87-0338-66
Gona-1,3,5(10)-triene-3,17β-diol, 13β-ethyl-	EtOH	223(3.88),283(3.32), 289(3.29)	35-3120-66
d-isomer	EtOH	281(3.31),288(3.27)	44-2512-66
l-isomer	EtOH	280(3.33),287(3.29)	44-2512-66
19-Norandrosta-5(10),9(11)-dien-3-one, 17β-hydroxy-17α-methyl-	MeOH	240(4.30)	13-0087-66B
D-Norandrost-4-en-3-one, 16β-formyl-	MeOH	241(4.19)	13-0505-66A
Testosterone, 6-dehydro-	EtOH	238(4.32)	44-3671-66
Vitamin A acid, 9-demethyl-, all trans	n.s.g.	346(4.62)	78-0293-66
Vitamin A acid, 9,13-didemethyl-, methyl ester, all trans	n.s.g.	350(4.63)	78-0293-66
$C_{19}H_{26}O_2S$			
4-Thiavitamin A acid, all trans	C_6H_{12}	310(4.27),367(4.49)	78-0265-66
$C_{19}H_{26}O_3$			
Androsta-1,4-dien-3-one, 4,17β-dihydroxy-	EtOH	244(3.90),304(3.75)	94-1370-66
Androst-1-ene-3,5-dione, 17β-hydroxy-,	EtOH	240(3.39),311(3.38)	33-2218-66
10(5→4)-abeo-	EtOH-NaOH	224(3.40),231(3.39), 249(3.37),344(3.39)	33-2218-66
9β,10α-Androst-4-ene-3,17-dione,	MeOH	241.5(4.23)	54-0731-66
11α-hydroxy-	EtOH	240(4.21)	33-1529-66
Androst-4-ene-3,17-dione, 11β-hydroxy-	EtOH	241(4.15)	83-1011-66
9β,10α-Androst-4-ene-3,17-dione,	MeOH	240(4.20)	54-0731-66
11β-hydroxy-	EtOH	239(4.20)	33-1529-66
Estra-1,3,5(10)-triene-8,17-diol, 3-methoxy-	n.s.g.	220s(3.93),277(3.38), 285(3.38)	88-5015-66
Estr-4-en-3-one, 9α,10α-epoxy-17β-hydroxy-17α-methyl-	EtOH	244(4.12)	87-0510-66
19-Norandrosta-4,9(10)-dien-3-one, 1β,17β-dihydroxy-17α-methyl-	MeOH	296(4.29)	13-0087-66B
A-Nortestosterone, 1-(hydroxy-	EtOH	254(4.25),290(3.99)	88-1057-66
methylene)-	EtOH-KOH	235(4.37),340(4.24)	88-1057-66
Taxininol, anhydrodeformyldehydro-	EtOH	228(3.97)	95-1172-66
9β,10α-Testololactone	MeOH	240(4.23)	54-0701-66
Testosterone, 6β,19-epoxy-	EtOH	240(4.02)	44-0693-66
$C_{19}H_{26}O_4$			
Androst-1-ene-3,4-dione, 5β,17β-dihydroxy-	EtOH	260(3.67)	94-1370-66
Androst-4-ene-3,17-dione, 2β,9α-dihydroxy-	EtOH	243(4.08)	37-0540-66
Androst-4-ene-3,17-dione, 4,9α-dihydroxy-	EtOH	278(4.09)	37-0540-66
Androst-4-ene-3,17-dione, 9α,19-dihydroxy-	EtOH	245(4.17)	37-0540-66
4-Oxaestr-5-en-3-one, 17β-acetoxy-	C_6H_{12}	201(3.86)	78-1317-66
	EtOH	202(3.92)	78-1317-66

Compound	Solvent	λ_{max}(log ε)	Ref.
4-Oxaestr-5(10)-en-3-one, 17β-acetoxy-	C_6H_{12}	222(3.51)	78-1317-66
	EtOH	223(3.52)	78-1317-66
2-Oxa-A-norestr-3(5)-en-1-one, 17β-acetoxy-3-methyl-	MeOH	232(3.45)	44-4255-66
2-Oxa-A-norestr-5(10)-en-1-one, 17β-acetoxy-3ξ-methyl-	MeOH	219(3.99)	44-4255-66
9,10-Secoandrosta-1,3,5(10)-trien-9-one, 3,4,17β-trihydroxy-	EtOH	281(3.28)	37-0540-66
$C_{19}H_{26}O_6$			
Cumanin, diacetate	EtOH	213(4.02)	78-1499-66
Isocumanin, diacetate	EtOH	219(4.21)	78-1499-66
$\Delta^{1(2H),\alpha}$-Naphthaleneacetic acid, octahydro-2,6α-dihydroxy-5α-(hydroxymethyl)-5,8aα-dimethyl-, γ-lactone, diacetate	MeOH	210.5(4.25)	88-3489-66
$C_{19}H_{26}O_{12}$			
Daphylloside	MeOH	235(3.95)	95-0943-66
$C_{19}H_{27}BrO_2$			
5α-Androst-1-en-3-one, 2-bromo-17β-hydroxy-	MeOH	255(3.86)	73-1363-66
$C_{19}H_{27}ClO$			
Androsta-3,5-dien-17β-ol, 3-chloro-	EtOH	237s(4.31),244(4.38), 250s(4.24)	32-1241-66
$C_{19}H_{27}ClO_2$			
5α-Androst-1-en-3-one, 2-chloro-17β-hydroxy-	MeOH	247(3.93)	73-1363-66
Testosterone, 19-chloro-	EtOH	240(4.19)	44-0693-66
$C_{19}H_{27}IN_2S_2$			
2,2'-Bis(3-ethylthiazolinyl)pentamethinecyanine iodide	MeOH	540(5.13)	9-0150-66
$C_{19}H_{27}NO_5$			
Fumaric acid, [benzyl(2-tert-butoxyethyl)amino]-, dimethyl ester	ether	278(4.19)	24-1558-66
$C_{19}H_{27}N_3O_2S$			
Isothiourea, N,N'-dicyclohexyl-S-(p-nitrophenyl)-	dioxan	232(4.06)	35-5855-66
$C_{19}H_{27}N_3O_3$			
4-Pyrimidinecarboxaldehyde, 2-amino-6-hydroxy-5-(4-phenylbutyl)-, diethyl acetal	pH 1	269(3.89)	4-0315-66
	pH 7	299(3.73)	4-0315-66
	pH 13	288(3.85)	4-0315-66
$C_{19}H_{27}N_5O_5$			
β-D-Ribofuranoside, methyl 5-deoxy-2,3-O-isopropylidene-5-[9-(tetrahydro-2-pyranyl)purin-6-yl]amino-	EtOH	210(4.29),266(4.23)	4-0485-66
	EtOH-NaOH	265.5(4.26)	4-0485-66
$C_{19}H_{28}N_2O$			
5α-Estr-2-ene-2-carbonitrile, 3-amino-17β-hydroxy-	EtOH	262(3.98)	32-0152-66
Pyridine, 1-acetyl-4-(p-dimethylaminophenyl)-1,2,3,4-tetrahydro-2,2,6,6-tetramethyl-	EtOH	292(4.36)	78-2735-66

Compound	Solvent	λ_{max}(log ε)	Ref.
3-Pyrrolidinol, N-cyclohexyl-4-(p-dimethylaminobenzylidene)-	EtOH	301(4.45)	44-0001-66
$C_{19}H_{28}N_2O_2$			
3,17a-Diaza-A,D-bishomoandrost-4a-ene-4,17-dione	MeOH	219.5(4.31)	88-0983-66
$C_{19}H_{28}N_2O_2S$			
L-Proline, N-(phenylthioacetyl)-, cyclohexylammonium salt	H_2O	272(4.09),325s(2.10)	1-2781-66
	MeOH	278(4.10),340s(1.89)	1-2781-66
	dioxan	277(4.12),375s(1.83)	1-2781-66
$C_{19}H_{28}N_2O_4$			
5α-Androstane-2,3,4-trione,	EtOH	268(3.91)	44-2755-66
17β-hydroxy-, 2,4-dioxime	base	348(4.04)	44-2755-66
5β-Androstane-2,3,4-trione,	EtOH	264(4.01)	44-2755-66
17β-hydroxy-, 2,4-dioxime	base	333(4.12)	44-2755-66
$C_{19}H_{28}O_2$			
5β-Androst-1-en-3-one, 17β-hydroxy-	EtOH	230(3.85)	88-4113-66
10α-Estr-4-en-3-one, 17β-hydroxy-17α-methyl-	MeOH	243(4.21)	88-1023-66
Gon-4-en-3-one, 13β-ethyl-17β-hydroxy-	EtOH	241(4.24)	35-3120-66
	EtOH	240.5(4.24)	44-2512-66
A-Nor-5β-androst-1-ene-1-carboxaldehyde, 17β-hydroxy-	EtOH	243(3.96)	88-1057-66
D-Norandrost-4-en-3-one, 16β-(hydroxymethyl)-	MeOH	243(4.19)	13-0505-66A
$C_{19}H_{28}O_3$			
Androstane-3,5-dione, 17β-hydroxy-,	EtOH	291(4.01)	33-2218-66
10(5→4)-abeo-	EtOH-NaOH	313(4.08)	33-2218-66
Androst-4-en-3-one, 12α,17β-dihydroxy-	EtOH	241(4.21)	44-2116-66
9β,10α-Androst-4-en-3-one, 11α,17β-dihydroxy-	EtOH	241.5(4.21)	33-1529-66
Estrane-3,5-dione, 17β-hydroxy-17α-	EtOH	292(3.95)	33-2218-66
methyl-, 10(5→4)-abeo-	EtOH-KOH	312(4.18)	33-2218-66
ferric complex	EtOH	551(3.04)	33-2218-66
Estran-3-one, 4β,5β-epoxy-17β-hydroxy-17α-methyl-	EtOH	296(1.72)	33-2218-66
Gon-4-en-3-one, 13β-ethyl-	EtOH	243(4.19)	35-3120-66
1β,17β-dihydroxy-	0.066N NaOH	243(3.98),302(3.50)	35-3120-66
Gon-4-en-3-one, 13β-ethyl-6β,17β-dihydroxy-	EtOH	238(4.15)	35-3120-66
Gon-4-en-3-one, 13β-ethyl-10β,17β-dihydroxy-	EtOH	236(4.17)	35-3120-66
Gon-4-en-3-one, 13β-ethyl-11α,17β-dihydroxy-	EtOH	242(4.18)	35-3120-66
5β-A-Norandrostan-2-one, 17β-hydroxy-	EtOH	311(4.02)	88-1057-66
3-(hydroxymethylene)-	EtOH-KOH	311(4.24)	88-1057-66
A-Nortestosterone, 3,5-dihydro-	EtOH	278(3.99)	88-1057-66
1-(hydroxymethylene)-	EtOH-KOH	310(4.43)	88-1057-66
Podocarp-8(14)-en-7-one, 2β-acetoxy-	EtOH	240(4.20)	35-1766-66
Podocarp-8-en-14-one, 3β-acetoxy-	MeOH	244(3.88)	35-1766-66
Podocarp-12-en-14-one, 3β-acetoxy-	MeOH	229(3.86)	35-1766-66
9β,10α-Testosterone, 11α-hydroxy-	MeOH	242(4.21)	54-0721-66
9β,10α-Testosterone, 16α-hydroxy-	MeOH	242(4.22)	54-0712-66
$C_{19}H_{28}O_4$			
2-Chromanvaleric acid, 6-hydroxy-α,2,5,7,8-pentamethyl-	EtOH	292.0(3.55)	73-2434-66

Compound	Solvent	λ_{max}(log ϵ)	Ref.
$C_{19}H_{28}O_5$			
1,4-Cyclohexadiene-1-octanoic acid, ε-hydroxy-α,ε,2,4,5-pentamethyl-3,6-dioxo-	EtOH	263(--),269(4.17)	73-2434-66
$C_{19}H_{29}ClN_2O$			
Pyridine, 1-acetyl-4-(p-dimethylamino)-phenyl)-1,2,3,4-tetrahydro-2,2,6,6-tetramethyl-, hydrochloride	pH 1	254(4.25)	78-2735-66
	EtOH	292(4.22)	78-2735-66
$C_{19}H_{29}NO_2$			
A-Nor-5β-androst-1-ene-1-carboxaldehyde, 17β-hydroxy-, oxime, anti	EtOH	242(4.14)	88-1057-66
syn	EtOH	242(4.22)	88-1057-66
$C_{19}H_{29}NO_{11}S_2$			
Thiazolidine-2-thione, 5-acetoxy-3-ethyl-5-(D-arabino-tetraacetoxy-butyl)-	MeOH	242(3.87),273(4.19)	12-0445-66
$C_{19}H_{30}$			
5α-Androst-16-ene	C_6H_{12}	191.5(3.84)	39-1266-66C
$C_{19}H_{30}N_2O_2S$			
L-Valine, N-(phenylthioacetyl)-, cyclohexylammonium salt	H_2O	269(4.02),324(1.90)	1-2781-66
	MeOH	271(4.10),336(1.82)	1-2781-66
	dioxan	271(4.03),360s(1.73)	1-2781-66
$C_{19}H_{30}O$			
8β-d_1-5α-Androstan-12-one	EtOH	238(4.10)	35-0536-66
2,5-Cyclohexadien-1-one, 2,6-di-tert-butyl-4-pentylidene-	hexane	300(4.30)	70-0888-66
$C_{19}H_{30}O_2$			
D-Norandrostan-16-one, 3-methoxy-	MeOH	289(1.62)	88-4573-66
isomer	MeOH	293(1.58)	88-4573-66
$C_{19}H_{30}O_3$			
Taxininol, anhydrodeformyldihydro-	EtOH	210(3.83)	95-1172-66
Taxininol, anhydroisodeformyldihydro-	EtOH	215(3.72)	95-1172-66
$C_{19}H_{30}O_4$			
Rapanone	EtOH	292(4.21),428(2.45)	12-0169-66
$C_{19}H_{30}O_5$			
Normutilincarboxylic acid	EtOH	295(1.53)	78-0359-66B
$C_{19}H_{31}N_3$			
Tetrapropylammonium salt of cyclopentadiene-1,2-dicarbonitrile	MeCN	265(4.13),282(4.21)	35-3046-66
Tetrapropylammonium salt of cyclopentadiene-1,3-dicarbonitrile	MeCN	224(4.51),273(4.35), 277(4.36)	35-3046-66
$C_{19}H_{33}NO_3S_2$			
5-Thiazolidineacetic acid, N-hexyl-4-oxo-2-thioxo-, octyl ester	EtOH	263(3.92),297(4.02)	95-0095-66
$C_{19}H_{34}N_2$			
7,9-Diazatrispiro[5.1.1.5.2.2]nonadecane, 7,9-dimethyl-	MeOH	219(3.6)	24-0724-66
	$CHCl_3$	239(3.6)	24-0724-66

Compound	Solvent	λ_{max}(log ϵ)	Ref.
$C_{19}H_{34}O_2$			
2,6,9-Pentadecatrienal, diethyl acetal, tri-cis	EtOH-HCl	229(4.34)	54-0117-66
	EtOH	none	54-0117-66
$C_{19}H_{36}Cl_3N_5O_{12}$			
[4-[(Dimethylamino)methylene]-3-[2-(dimethylamino)-1-[(methylimino)-methyl]vinyl]-2-pentenediylidene]-bis[dimethylammonium perchlorate], methoperchlorate	MeCN	303(4.66),418(4.67)	24-2479-66
$C_{19}H_{38}N_2$			
1,6-Diazaspiro[4.4]nonane, 1,6-dihexyl-	MeOH	209(4.0)	24-0724-66
$C_{19}H_{42}IN$			
Cetyltrimethylammonium iodide	CH_2Cl_2	241(--)	60-0286-66

Compound	Solvent	$\lambda_{max}(\log \epsilon)$	Ref.
$C_{20}H_8Cl_2O_2S$			
Benzo[b]thiophene, 2-(8-chloro-2-oxo-1-acenaphthenylidene)-6-chloro-2,3-dihydro-3-oxo-	PhCl	304(--),474(4.04), 509(3.95)	65-1055-66
	H_2SO_4	645(--)	65-1055-66
$C_{20}H_9ClO_2S$			
Benzo[b]thiophene, 2-(8-chloro-2-oxo-1-acenaphthenylidene)-2,3-dihydro-3-oxo-	PhCl	302(--),484(3.78), 512(3.78)	65-1055-66
	H_2SO_4	660(4.00)	65-1055-66
$C_{20}H_{10}$			
Dibenzo[ghi,mno]fluoranthene	n.s.g.	246(4.78),249(4.74), 253(4.96),288(4.52)	35-0380-66
$C_{20}H_{10}Cl_4$			
Cyclopentadiene, 1,2,3,4-tetrachloro-5-(diphenyl-2-cyclopropen-1-ylidene)-	C_6H_{12}	242(4.42),269(4.36), 349(4.62)	78-1275-66
	MeCN	240(4.42),270(4.36), 345(4.60)	78-1275-66
$C_{20}H_{10}O_2$			
peri-Xanthenoxanthenone	$CHCl_3$	313(3.67),327(3.83), 371(3.38),392(3.71), 416(4.05),443(4.17)	39-1836-66C
$C_{20}H_{11}N_5$			
Cinchoninonitrile, 2-s-triazolo-[1,5-a]quinolin-2-yl-	EtOH	324(3.91),337(4.09), 350(4.17)	94-0512-66
$C_{20}H_{12}$			
Azuleno[5,6,7-cd]phenalene	hexane	316(5.29),426(4.60), 451(5.10),614(2.77), 633(2.87),666(3.08), 702(3.04),741(3.21)	89-0516-66
$C_{20}H_{12}Cl_2N_4$			
Benzimidazo[1,2-c]quinazoline, 6-(2-amino-5-chlorophenyl)-2-chloro-	EtOH	233(4.55),271s(4.65), 280(4.74),310(4.06), 324(4.03),342(3.92)	4-0289-66
$C_{20}H_{12}Cl_4N_4S_2$			
Pyrazino[2,3-d]pyridazine, 5,8-bis-(3,4-dichlorobenzylthio)-	EtOH	205(4.87),228s(4.54), 255(4.32),301(3.70)	4-0512-66
$C_{20}H_{12}CuN_4$			
Copper, [porphinato(2-)]-	dioxan	517(4.1),551(3.92)	83-0001-66
$C_{20}H_{12}F_2N_4$			
Benzimidazo[1,2-c]quinazoline, 6-(2-amino-5-fluorophenyl)-2-fluoro-	EtOH	231(4.42),267s(4.57), 275(4.62),311(4.04), 325(4.05),341(3.93)	4-0289-66
$C_{20}H_{12}F_6N_2O_2S_2$			
p-Benzoquinone, 2,5-dianilino-3,6-bis[(trifluoromethyl)thio]-	EtOH	240(4.26),360(4.10)	44-3671-66
$C_{20}H_{12}N_2$			
Naphtho[2,3-b]phenazine	MeCN	223(4.46),238(4.49), 273(4.56),307(4.65)	44-3734-66
	DMF	409(3.53),433(3.54), 582(2.86)	44-3734-66

Compound	Solvent	λ_{max}(log ε)	Ref.
$C_{20}H_{12}N_2O_6S_2$			
Phthalonitrile, 3,6-dihydroxy-4,5-bis-(phenylsulfonyl)-, sodium salt	EtOH	248(4.42),402(3.98)	44-3671-66
$C_{20}H_{12}N_8O_2S_2$			
1,4-Naphthalenediol, 2,3-bis(s-triazolo-[4,3-b]pyridazin-6-ylthio)-	EtOH-DMF	276(4.35)	49-1523-66
$C_{20}H_{12}N_8O_9$			
Ketone, 1,5-bis(2,4-dinitrophenyl)-formazanyl phenyl	benzene	355(4.30),415(4.27)	70-1326-66
	MeOH	355(4.12),428(4.10), 633(4.50)	70-1326-66
$C_{20}H_{12}O_2S$			
Anthraquinone, 2-(phenylthio)-	$CHCl_3$	240(4.2),300(4.09), 390(3.4)	22-2942-66
$C_{20}H_{12}O_4S$			
Anthraquinone, 2-(phenylsulfonyl)-	$CHCl_3$	260(4.04),325(3.15)	22-2942-66
1,4-Naphthoquinone, 2-(1,3-dioxo-indan-2-yl)-3-(methylthio)-	MeCN	250(4.38),303(3.83), 427(4.36)	44-0062-66
$C_{20}H_{12}O_6$			
1,4-Naphthoquinone, 3-(1,3-dioxo-indan-2-ylidene)-2,3-dihydro-2-hydroxy-3-methoxy-	MeCN	231(4.89),250(4.43), 290(3.40)	44-0062-66
$C_{20}H_{12}O_7$			
Succinic anhydride, dipiperonylidene-, cis-trans	dioxan	262(3.98),310s(3.87), 355(4.01),432(4.41)	44-3342-66
	$CHCl_3$	318(3.93),358s(4.07), 440(4.45)	44-3342-66
trans-trans	dioxan	264(4.12),309(4.15), 425(4.18)	44-3342-66
	$CHCl_3$	318(4.23),428(4.13)	44-3342-66
$C_{20}H_{13}BrN_2O_2$			
Anthraquinone, 1-amino-4-anilino-2-bromo-	$CHCl_3$	257(4.51),285(4.23), 584(4.06),623(4.07)	22-0580-66
1H-Pyrazolo[1,2-a]pyrazol-4-ium, 1,3-dibenzoyl-2-bromo-, hydroxide, inner salt	n.s.g.	230(3.98),260(4.14), 330(4.09),428(4.43)	35-5588-66
$C_{20}H_{13}BrO_4$			
5,12-Naphthacenedione, 6-bromo-8-ethyl-1,11-dihydroxy-	C_6H_{12}	452(4.23),479(4.16)	5-0145-66F
	H_2SO_4	570(4.26),627(4.38)	5-0145-66F
5,12-Naphthacenedione, 6-bromo-9-ethyl-1,11-dihydroxy-	C_6H_{12}	453(4.23),462(4.19), 480(4.15)	5-0145-66F
	H_2SO_4	572(4.36),620(4.60)	5-0145-66F
5,12-Naphthacenedione, 11-bromo-8-ethyl-1,6-dihydroxy-	C_6H_{12}	456(4.23),474(4.13), 484(4.12)	5-0145-66F
	H_2SO_4	555(4.45),601(4.68)	5-0145-66F
5,12-Naphthacenedione, 11-bromo-9-ethyl-1,6-dihydroxy-	C_6H_{12}	454(4.22),482(4.13)	5-0145-66F
	H_2SO_4	549(4.45),593(4.70)	5-0145-66F
$C_{20}H_{13}ClHg$			
Mercury, chloro-(α-fluoren-9-ylidene-benzyl)-	THF	253(3.48),262(4.51), 307(4.04),322(4.11)	35-3027-66

Compound	Solvent	$\lambda_{max}(\log \epsilon)$	Ref.
$C_{20}H_{13}ClN_4$			
Benzimidazo[1,2-c]quinazoline, 6-(o-aminophenyl)-x-chloro-	EtOH	230(4.46),266(4.59), 276(4.64),305(4.01), 318(4.02),334s(3.80)	4-0289-66
$C_{20}H_{13}Cl_2N$			
Indole, 2,3-bis(p-chlorophenyl)-	EtOH	248(4.32),311(4.27)	87-0527-66
$C_{20}H_{13}NOS_3$			
Rhodanine, 5-(2,6-diphenyl-4H-thiopyran-4-ylidene)-	DMF	266(4.38),465(4.52)	33-2046-66
$C_{20}H_{13}NO_4$			
Benzoic acid, o-(3,6-dihydroxy-9-acridinyl)-	EtOH	229(4.69),261(5.28), 380(3.97),410(3.61)	1-1631-66
$C_{20}H_{13}NO_5$			
Fluorescein, oxime	EtOH	241(4.30),282s(3.79), 347(3.72)	1-1631-66
Spiro[(2-hydroxy-3-oxo-2,3-dihydroiso-indole)-1,9'-(2,7-dihydroxyxanthene]	EtOH	219s(4.62),250s(4.29), 320(3.81)	1-1631-66
$C_{20}H_{13}N_3O_4$			
Indole, 2,3-bis(p-nitrophenyl)-	EtOH	264(4.18),360(4.17)	87-0527-66
$C_{20}H_{14}$			
Biphenyl, 2-(phenylethynyl)-	THF	254(4.40),292(4.34), 305(4.28)	35-3027-66
7,8:9,10-Dibenzosesquifulvalene	C_6H_{12}	244(4.67),264(4.32), 373(4.38)	5-0034-66H
	EtOH	244(4.67),263(4.30), 377(4.27)	5-0034-66H
	EtOH-$HClO_4$	223(4.83),230(4.84), 267(4.51),273s(4.43), 299(3.83),320(3.62), 405(3.22)	5-0034-66H
	MeCN	243(4.61),261(4.26), 376(4.29)	5-0034-66H
	dioxan	245(4.72),264(4.39), 378(4.39)	5-0034-66H
	$MeNO_2$	385(4.19)	5-0034-66H
	CF_3COOH	260(4.02),270(3.96), 284s(3.58),338(3.11), 446(2.73)	5-0034-66H
Fluorene, benzylidene-	THF	328(4.18)	35-3027-66
Indeno[1,2-a]fluorene, 7,12-dihydro-	C_6H_{12}	247s(4.33),259(4.66), 265(4.67),282(4.89), 293(4.26),306(3.51), 320(3.45),332(3.17)	33-1850-66
Indeno[2,1-c]fluorene, 5,8-dihydro-	EtOH	235(4.15),244(4.17), 253s(4.34),260(4.49), 275(4.16),286(4.19), 300(4.37),311(4.37)	33-1931-66
$C_{20}H_{14}ClNO_6$			
8-Hydroxybenzo[a]quinolizinium perchlorate, benzoate	EtOH	260s(--),271s(--), 281(4.49),324(3.84), 339(4.05),355(4.13)	44-2616-66

Compound	Solvent	λ_{max}(log ε)	Ref.
$C_{20}H_{14}Cl_2N_4S_2$			
Pyrazino[2,3-d]pyridazine, 5,8-bis-(p-chlorobenzylthio)-	EtOH	204(4.58),222(4.51), 255(4.35),301(3.69)	4-0512-66
$C_{20}H_{14}D_2$			
Bicyclo[4.2.0]octa-1,3,5-triene-7,8-d_2, 7,8-diphenyl-	MeOH	261(3.26),267(3.35), 273(3.28)	5-0044-66C
isomer	MeOH	260(3.27),266(3.37), 272(3.29)	5-0044-66C
$C_{20}H_{14}Fe$			
Butadiyne, 1-ferrocenyl-4-phenyl-	$CHCl_3$	275(4.28),292(4.26), 310(4.25),355s(3.49), 446(3.04)	101-0399-66B
$C_{20}H_{14}N_2$			
Naphtho[2,3-b]phenazine, 5,14-dihydro-	MeCN	222(4.46),240(4.50), 273(4.75),308(4.84), 326(4.58)	44-3734-66
$C_{20}H_{14}N_2O$			
1aH-Oxazirino[2,3-a]quinoxaline, 1a,2-diphenyl-	EtOH	262(4.58),326(3.86)	94-1316-66
$C_{20}H_{14}N_2O_2$			
Anthraquinone, 1-[(2-aminocyclohexyl)amino]-	EtOH	495(3.82)	27-0281-66
1H-Pyrazolo[1,2-a]pyrazol-4-ium, 1,3-dibenzoyl-, hydroxide, inner salt	n.s.g.	228(4.22),250(4.17), 342(4.21),430(4.48)	35-5588-66
$C_{20}H_{14}N_2S_3$			
2,5-Thiophenediacetonitrile, α,α'-bis-(5-methyl-2-thenylidene)-, cis	EtOH	264(4.16),347(4.10)	12-1243-66
trans	EtOH	297(3.93),357(4.31), 449(4.56)	12-1243-66
$C_{20}H_{14}N_4O$			
6-Pyridazone, 1-(3-phenyl-6-pyridazino)-3-phenyl-	EtOH	255(4.60),324(3.87)	1-2637-66
$C_{20}H_{14}O_4$			
Benz[a]anthracene-7,12-dione, 8,11-dimethoxy-	MeOH	236(4.14),280(4.04), 430(3.71)	44-2920-66
$C_{20}H_{14}O_5$			
Cyclopenta[7,8]phenanthro[10,1-bc]-furan-3,6,9(8H)-trione, 7,11b-dihydro-2-methoxy-11b-methyl-	EtOH	239(4.51),287(4.10), 305(4.07)	39-0743-66C
5,12-Naphthacenedione, 8-ethyl-1,6,11-trihydroxy-	C_6H_{12}	463(4.20),494(4.49), 521(4.25),531(4.58)	5-0145-66F
	H_2SO_4	536(4.44),578(4.72)	5-0145-66F
5,12-Naphthacenedione, 9-ethyl-1,6,11-trihydroxy-	C_6H_{12}	463(4.20),494(4.49), 520(4.23),531(4.58)	5-0145-66F
	H_2SO_4	532(4.45),574(4.75)	5-0145-66F
$C_{20}H_{14}O_8$			
Succinic acid, dipiperonylidene-, trans-trans	EtOH	230s(4.40),287(4.29), 327(4.43)	44-3342-66

Compound	Solvent	$\lambda_{max}(\log \epsilon)$	Ref.
$C_{20}H_{15}BrN_2O$			
1,2-Diazetidin-3-one, 4-(o-bromophenyl)-1,2-diphenyl-	CH_2Cl_2	267(4.06)	88-5245-66
1,2-Diazetidin-3-one, 4-(p-bromophenyl)-1,2-diphenyl-	CH_2Cl_2	265(4.05)	88-5245-66
$C_{20}H_{15}ClN_2O$			
1,2-Diazetidin-3-one, 4-(m-chlorophenyl)-1,2-diphenyl-	CH_2Cl_2	266(4.06)	88-5245-66
1,2-Diazetidin-3-one, 4-(o-chlorophenyl)-1,2-diphenyl-	CH_2Cl_2	267(4.05)	88-5245-66
1,2-Diazetidin-3-one, 4-(p-chlorophenyl)-1,2-diphenyl-	CH_2Cl_2	266(4.08)	88-5245-66
$C_{20}H_{15}ClN_2O_3$			
Camptothecin, chloro-	n.s.g.	224(4.46),255(4.39), 290(3.78),370(4.32)	35-3888-66
$C_{20}H_{15}ClN_2O_4S$			
2,3-Naphthalenedicarbonitrile, 4a-chloro-1,4,4a,5,8,8a-hexahydro-6,7-dimethyl-1,4-dioxo-8a-(phenylsulfonyl)-	MeCN	220(4.18),263(4.08), 350(2.30)	44-3671-66
$C_{20}H_{15}N$			
Indole, 2,3-diphenyl-	EtOH	248(4.35),308(4.23)	87-0527-66
Isoindole, 1,3-diphenyl-	EtOH	228s(4.24),237(4.27), 268s(4.22),273(4.27), 322(4.13),335(4.16), 387(4.37)	78-1011-66
$C_{20}H_{15}NO_2$			
Cyclopenta[c]quinolizine-1-carboxylic acid, 4-phenyl-, methyl ester	EtOH	236(4.37),275(4.56), 348(4.42),390s(3.7), 445(3.48)	39-0324-66C
Cyclopenta[c]quinolizine-2-carboxylic acid, 4-phenyl-, methyl ester	EtOH	235(4.39),248(4.38), 294(4.32),362(4.47), 390s(4.1),493(3.51)	39-0324-66C
Indole, 2,3-bis(p-hydroxyphenyl)-	EtOH	253(4.51),309(4.32)	87-0527-66
$C_{20}H_{15}N_3$			
2-Naphthylamine, 1-(2-naphthylazo)-	benzene	460(4.28)	44-2571-66
	$C_6H_{11}Me$	452(4.26)	44-2571-66
$C_{20}H_{15}N_3O$			
1-Naphthol, 7-amino-8-(2-naphthylazo)-	benzene	517(4.35)	44-2571-66
	$C_6H_{11}Me$	513(4.32)	44-2571-66
$C_{20}H_{15}N_3O_2$			
1H-Indazole, 3-methyl-1-(m-nitrophenyl)-	EtOH	245(4.48),263(4.48), 311s(3.98)	78-3131-66
1H-Indazole, 3-methyl-1-(p-nitrophenyl)-	EtOH	216(4.23),247(4.42), 280(4.08),356(4.21)	78-3131-66
5-Norbornene-2,3-dicarboximide, 7-(di-2-pyridylmethylene)-, endo	MeOH	241(4.17),271(4.05)	44-2141-66
$C_{20}H_{15}N_3O_3$			
1,2-Diazetidin-3-one, 4-(m-nitrophenyl)-1,2-diphenyl-	MeOH	264(4.22)	88-5245-66

Compound	Solvent	$\lambda_{max}(\log \epsilon)$	Ref.
1,2-Diazetidin-3-one, 4-(o-nitrophenyl)-1,2-diphenyl-	CH_2Cl_2	264(4.23)	88-5245-66
1,2-Diazetidin-3-one, 4-(p-nitrophenyl)-1,2-diphenyl-	CH_2Cl_2	270(4.36)	88-5245-66
$C_{20}H_{15}N_3O_5$			
Benzanilide, 4-nitro-4'-(p-nitrobenzyl)-	EtOH	272(5.34),330s(3.97)	65-0214-66
$C_{20}H_{15}OP$			
Phosphine oxide, diphenyl(phenylethynyl)-	EtOH	230(4.26),244(4.26), 253(4.36),264(4.28), 286(3.26)	22-1002-66
$C_{20}H_{15}P$			
Phosphine, diphenyl(phenylethynyl)-	EtOH	230(4.20),246(4.20), 254(4.28),265(4.20), 287(3.56)	22-1002-66
$C_{20}H_{15}PS$			
Phosphine sulfide, diphenyl(phenylethynyl)-	EtOH	229(4.32),245(4.32), 252(4.34),264(4.26), 285(3.90)	22-1002-66
$C_{20}H_{16}$			
Bicyclo[4.2.0]octa-1,3,5-triene, 7,8-diphenyl-	MeOH	261(3.27),266(3.37), 273(3.28)	5-0044-66C
isomer	MeOH	261(3.26),267(3.35), 273(3.27)	5-0044-66C
15H-Cyclopenta[a]phenanthrene, 17-isopropenyl-	EtOH	223(4.40),271(4.74), 276(4.75),295(4.30), 316(3.40)	39-0963-66C
17H-Cyclopenta[a]phenanthrene, 17-isopropylidene-	EtOH	214(4.44),228(4.50), 270(4.63),302(4.67), 310(4.77),347s(3.61), 367s(3.50)	39-0963-66C
Fluorene, 9-cyclohepta-2,4,6-trien-1-yl-	EtOH	266(4.37),290(3.91), 302(3.97)	5-0034-66H
Phenanthrene, 9,10-dihydro-9-phenyl-	MeOH	267(4.21),300(3.58)	88-0939-66
Stilbene, 4-phenyl-	$C_2H_4Cl_2$	320(4.55)	99-0085-66
$C_{20}H_{16}ClFN_2S$			
4(1H)-Quinazolinethione, 2-(p-chlorophenyl)-1-(p-fluorophenyl)-5,6,7,8-tetrahydro-	MeOH	202s(4.52),252s(4.02) 343(4.43)	44-0722-66
$C_{20}H_{16}ClN_4$			
Verdazyl, p-chloro-	n.s.g.	416(3.88),710(3.61)	88-4103-66
Verdazyl, p-chloro-, cation	n.s.g.	246(4.33),321(4.05), 550(3.59)	88-4103-66
$C_{20}H_{16}FN_3O_3$			
Cytosine, 5-fluoro-, di-p-toluoyl derivative	EtOH	244(4.47),278(4.34), 329(3.72)	87-0566-66
$C_{20}H_{16}N_2O$			
1,2-Diazetidin-3-one, 1,2,4-triphenyl-	CH_2Cl_2	267(4.08)	88-5245-66
$C_{20}H_{16}N_2O_3$			
Benzanilide, 4'-(p-nitrobenzyl)-	EtOH	272(5.41),330s(3.49)	65-0214-66

Compound	Solvent	$\lambda_{max}(\log \epsilon)$	Ref.
$C_{20}H_{16}N_2O_4$			
4,4'-Biisoxazole, 3,3'-dimethoxy-5,5'-diphenyl-	EtOH	260(4.37)	32-0454-66
Camptothecin	n.s.g.	220(4.57),254(4.47),290(3.70),370(4.30)	35-3888-66
1-Isoquinolinecarbonitrile, 4-(3,4,5-trimethoxybenzoyl)-	MeOH	330(4.05)	88-5731-66
Spiro[bicyclo[4.2.0]octa-1,3,5-triene-7,2'-[1,3]dioxolan]-8-one, azine	EtOH	262(4.19),305(4.30),316(4.25)	44-1866-66
$C_{20}H_{16}N_2O_4S$			
2,3-Naphthalenedicarbonitrile, 1,4,4a,5,8,8a-hexahydro-6,7-dimethyl-1,4-dioxo-4a-(phenylsulfonyl)-	MeCN	220(4.17),261(4.06),367(2.31)	44-3671-66
$C_{20}H_{16}N_2O_5$			
Carbanilic acid, p-(benzyloxy)-, p-nitrophenyl ester	EtOH	238(4.35),310(4.00)	87-0444B-66
$C_{20}H_{16}N_4O_2$			
Lumazine, 3,8-dimethyl-6,7-diphenyl-	pH -1.9	242(3.80),273(4.04),397(3.97)	24-3503-66
	pH 3	283(4.16),345(3.92)	24-3503-66
	pH 7	271(4.23),290s(4.09),425(4.08)	24-3503-66
	pH 13	245(4.24),285s(3.97),355(4.01)	24-3503-66
Pyrazino[2,3-d]pyridazine, 5,8-bis-(benzyloxy)-	EtOH	208(4.46),256(4.25)	4-0512-66
$C_{20}H_{16}N_4O_3$			
Lumazine, 8-(2-hydroxyethyl)-6,7-diphenyl-	pH -2.7	257(4.13),281(4.12),398(4.06)	24-3503-66
	pH 4	268(4.19),289(4.23),426(4.08)	24-3503-66
	pH 11	244(4.29),281s(4.07),368(4.16)	24-3503-66
$C_{20}H_{16}N_4S_2$			
Pyrazino[2,3-d]pyridazine, 5,8-bis(benzylthio)-	EtOH	206(4.35),256(4.11),304(3.22)	4-0512-66
$C_{20}H_{16}N_6O_2$			
Bis-s-triazolo[4,3-b:3',4'-f]pyridazine, 1,8-bis(p-methoxyphenyl)-	EtOH	268(4.39)	78-2073-66
$C_{20}H_{16}O$			
Acetophenone, p-biphenylyl-	$C_2H_4Cl_2$	305(4.52)	99-0089-66
	dioxan	303(4.61)	99-0089-66
$C_{20}H_{16}O_4$			
Dibenzotricyclo[3.3.0.0^{2,8}]octadiene-1,2-dicarboxylic acid, dimethyl ester	MeCN	217s(4.40),265(3.15),271(3.18),279(3.08)	35-2882-66
$C_{20}H_{16}O_6$			
1H-Cyclopenta[7,8]naphtho[2,3-b]furan-7-carboxylic acid, 10-acetoxy-2,3-dihydro-6-methyl-3-oxo-, methyl ester	EtOH	221(4.22),265(4.77),312(3.96),325(3.91),360(3.77),375(3.81)	39-0725-66C

Compound	Solvent	λ max(log ε)	Ref.
$C_{20}H_{16}O_7$ 4H-Naphtho[2,3-b]pyran-4-one, 3-acetyl-5,10-dihydroxy-2-methyl-, diacetate	EtOH	214(4.29),253(4.65), 282s(3.99),307(3.59), 320(3.47),365(3.66), 379(3.61)	94-0121-66
$C_{20}H_{16}O_8$ Ascorbic acid, 2,6-dibenzoate	neutral	234(4.5)	95-0376-66
	anion	234(4.5),255s(4.3)	95-0376-66
L-Ascorbic acid, 2,6-dibenzoate	EtOH	237(4.51)	94-1039-66
L-Ascorbic acid, 3,6-dibenzoate	EtOH	230(4.51)	94-1039-66
$C_{20}H_{16}S$ Aceanthra[10,9-b]thiophene, 2,3-dimethyl-	dioxan	234(4.53),273s(4.58), 282(4.65),297(4.68), 306s(4.65),342s(3.54), 360(3.75),380(3.86), 402(3.72)	22-3667-66
$C_{20}H_{17}BO_2$ Borinic acid, diphenyl-, ester with 2-hydroxy-4-methyl-2,4,6-cyclo-heptatrien-1-one	$C_2H_4Cl_2$	252(4.39),337(4.00), 397(3.60)	80-1409-66
$C_{20}H_{17}ClN_4O_4S$ 2-(p-Anilinophenylazo)-N-methyl-benzothiazolium perchlorate	MeCN	604(4.91)	5-0116-66G
$C_{20}H_{17}NO_2$ 11H-Benzo[a]carbazole-9-carboxylic acid, 10-methyl-, ethyl ester	EtOH	237(4.41),253(4.55), 288(4.75),316(4.35)	87-0237-66
5H-Naphtho[2,3-a]carbazole-5,13(8H)-dione, 7b,9,10,11,11a,12-hexahydro-	EtOH	494(3.87)	27-0281-66
$C_{20}H_{17}NO_6$ Pyrrole-2,3-dicarboxylic acid, 4,5-bis(p-methoxyphenyl)-	EtOH-acid	238(4.39),270s(4.15), 329(4.00)	87-0527-66
	EtOH	238(4.35),281(4.62), 296s(4.11),318s(4.00)	87-0527-66
	EtOH-base	231(4.17),258(4.29), 288(4.25),298(4.27)	87-0527-66
$C_{20}H_{17}N_3O$ 1H-Pyrazolo[4,3-c]pyridin-4(5H)-one, 3,6-dimethyl-1,5-diphenyl-	EtOH	223(4.42),234(4.39), 291(4.12)	70-1723-66
$C_{20}H_{17}N_3O_2S$ p-Toluenesulfonamide, N-[α-(phenylazo)-benzylidene]-	dioxan	290(4.30),440(2.40)	24-3337-66
$C_{20}H_{17}N_4$ Verdazyl	n.s.g.	404(3.89),710(3.61)	88-4103-66
Verdazyl, cation	n.s.g.	240(4.25),318(4.08), 545(3.72)	88-4103-66
nitrate	MeOH	244(4.18),314(4.33), 539(4.05)	49-1280-66
$AuCl_4^-$ salt	HCOOH	317(4.41),547(4.08)	49-1280-66
$PdCl_4^=$ salt	HCOOH	315(4.62),548(4.37)	49-1280-66
$PtCl_6^=$ salt	HCOOH	315(4.62),548(4.36)	49-1280-66
tetranitromethane salt	MeOH	242(4.32),318(4.40), 347(4.37),539(4.06)	49-1280-66

Compound	Solvent	λ_{max}(log ϵ)	Ref.
$C_{20}H_{18}$			
Benzene, 1,4-bis(1-cycloheptatrienyl)-	EtOH	240(3.14),337(3.37)	35-3527-66
Benzene, 1,4-bis(2-cycloheptatrienyl)-	EtOH	258(4.50)	35-3527-66
Benzene, 1,4-bis(3-cycloheptatrienyl)-	EtOH	239(3.38),312(3.47)	35-3527-66
Benzene, 1,4-bis(7-cycloheptatrienyl)-	EtOH	259(3.82),272s(--)	35-3527-66
15H-Cyclopenta[a]phenanthrene, 17-isopropyl-	EtOH	224(4.31),271(4.70), 275(4.71),316(4.02)	39-0963-66C
15H-Cyclopenta[a]phenanthrene, 11,12,17-trimethyl-	EtOH	219(4.35),280(4.93), 311(4.14),355s(2.78)	39-0963-66C
17H-Cyclopenta[a]phenanthrene, 17-isopropyl-	EtOH	221(4.66),240(4.33), 269(4.65),274(4.60), 293(4.07),303(4.08), 317(3.95)	39-0963-66C
7,8:9,10-Dibenzotricyclo[4.2.2.0^{2,5}]-deca-3,7,9-triene, 1,6-dimethyl-	EtOH	265(3.09),273(3.19)	5-0009-66A
m-Terphenyl, 2-ethyl-	EtOH	243(4.49)	22-1012-66
m-Terphenyl, 3-ethyl-	EtOH	248.5(4.54)	22-1012-66
m-Terphenyl, 4-ethyl-	EtOH	251(4.67)	22-1012-66
m-Terphenyl, 4'-ethyl-	EtOH	242(4.43)	22-1012-66
m-Terphenyl, 5'-ethyl-	EtOH	248(4.59)	22-1012-66
$C_{20}H_{18}ClNO_2$			
Anthraquinone, 1-[(2-chloro-cyclohexyl)amino]-	EtOH	504(3.85)	27-0281-66
$C_{20}H_{18}N_2$			
Pyridazine, 4-(1-methylpropenyl)-3,6-diphenyl-	MeOH	266.5(4.44)	24-1232-66
$C_{20}H_{18}N_2O$			
Anthra[1,9-ef][1,5]benzo[b][1,4]dia-zepin-9(1H)-one, 2,3,4,4a,5,14a-hexahydro-	EtOH	465(3.48)	27-0281-66
Quinoline, N-benzoyl-4-(cyano-methylene)-1,2,3,4-tetrahydro-2,2-dimethyl-	MeOH	331(3.92)	5-0128-66D
Urea, 1-methyl-1,3,3-triphenyl-	EtOH	270(4.29)	34-0436-66
$C_{20}H_{18}N_2O_2$			
Naphtho[2,3-a]phenazine-8,13-dione, 1,2,3,4,4a,5,14,14a-octahydro-, cis	EtOH	536s(--),571(4.11), 615s(--)	27-0281-66
trans	EtOH	532s(--),566(4.09), 608s(--)	27-0281-66
$C_{20}H_{18}N_2O_2Sn$			
Tin, bis(8-quinolinolato)dimethyl-	benzene	380(3.69)	101-0249-66B
	dioxan	378(3.70)	101-0249-66B
	pyridine	379(3.72)	101-0249-66B
$C_{20}H_{18}N_2O_3$			
4H-1,2-Diazepin-4-one, 1-benzoyl-1,7-dihydro-7-methoxy-5-methyl-6-phenyl-	MeOH	300(4.08)	44-0048-66
4H-1,2-Diazepin-4-one, 2-benzoyl-2,3-dihydro-7-methoxy-5-methyl-6-phenyl-	MeOH	215(4.34),367(2.95)	44-0048-66
Glutaconimide, 2-benzamido-4-ethyl-4-phenyl-	EtOH	274(4.01)	7-1519-66

Compound	Solvent	λ_{max}(log ϵ)	Ref.
$C_{20}H_{18}N_2O_4$			
Carbostyril, 6-methoxy-, dimer	EtOH	266(4.40),310(3.74)	94-1102-66
1-Indanone, 2-methyl-2-nitroso-, dimer	EtOH	249(4.43),293(3.80)	44-2090-66
$C_{20}H_{18}N_2O_4S_3$			
5-Thia-1-azabicyclo[4.2.0]oct-2-en-8-one, 3-(mercaptomethyl)-7-[2-(2-thienyl)acetamido]-, benzoate, 5-oxide	EtOH	239(4.40),263(4.28)	39-1142-66C
$C_{20}H_{18}N_2O_6$			
3H-Phenoxazine-1,9-dicarboxylic acid, 2-(1-aziridinyl)-4,6-dimethyl-3-oxo-, dimethyl ester	MeOH	242(4.56),418(4.47)	44-2564-66
$C_{20}H_{18}N_2O_8$			
Otobain, 6',8-dinitro-	EtOH	249(4.24),347(4.04)	39-1717-66C
$C_{20}H_{18}N_2O_{10}$			
β-D-Ribofuranoside, methyl 2,5-di-O-p-nitrobenzoyl-3-deoxy-	MeOH	258(4.43)	44-1163-66
$C_{20}H_{18}N_6$			
Pyrazino[2,3-d]pyridazine, 5,8-di-p-toluidino-	EtOH	203(4.36),223(3.89), 280(4.34)	4-0512-66
$C_{20}H_{18}NiO_2S_2$			
Nickel, bis(1-phenyl-3-thio-1,3-butanedionato)-	$CHCl_3$	510s(3.48),641(2.30)	12-1401-66
$C_{20}H_{18}O$			
Cyclohexanone, 2,6-dibenzylidene-	EtOH	330.0(4.40)	39-0075-66B
Ether, bis(4-phenyl-1,3-butadienyl), all trans	EtOH	324(4.79)	24-0642-66
2(1H)-Pentalenone, 4,5,6,6a-tetrahydro-1,3-diphenyl-	EtOH	228(4.32),262(4.00)	44-1336-66
$C_{20}H_{18}O_2$			
1-Cyclobutene, 3,4-dibenzoyl-1,2-dimethyl-, cis	MeOH	244(4.35)	24-1232-66
trans	MeOH	245.5(4.46)	24-1232-66
p-Terphenyl, 4,4"-dimethoxy-	$C_2H_4Cl_2$	295(4.55)	99-0204-66
$C_{20}H_{18}O_4$			
Gona-1,3,5(10),6,8-pentaene-12-carboxaldehyde, 11,17-dioxo-, cyclic 17-(ethylene acetal)	EtOH	218(4.54),255(4.13), 362(3.87)	39-0955-66C
Naphthalene, 3,4-dihydro-2,3-dimethyl-7,8-(methylenedioxy)-1-(3,4-methylenedioxyphenyl)-	EtOH	218(4.37),232(4.36), 275(4.11),284(4.07)	39-1775-66C
$C_{20}H_{18}O_6$			
d-Asarinin	EtOH	238(3.95),288(3.91)	94-0641-66
Robustic acid, dihydrodemethyl-	EtOH	215(4.49),226s(4.43), 252s(4.09),333(4.31)	39-0606-66C
	EtOH-NaOEt	255(4.47),333(4.35)	39-0606-66C
Stilbene, α,2,2'-triacetoxy-	EtOH	267(4.32)	39-0544-66C

Compound	Solvent	$\lambda_{max}(\log \epsilon)$	Ref.
$C_{20}H_{18}O_7$			
Naphtho[2,3-b]furan-3-methanol, 4,9-dihydroxy-8-methyl-, triacetate	EtOH	240(4.83),246(4.86), 318(4.02),344(3.92)	78-0685-66
Norsolorinic acid	EtOH	270(4.32),286(4.35), 312(4.44),453s(3.95), 466(4.00)	39-1727-66C
$C_{20}H_{18}O_8$			
Ceroalbolinic acid, tetramethyl ether	EtOH	285(4.59),312(3.60)	78-1507-66
Ceroalbolinic acid, methyl ester, trimethyl ether	EtOH	281(4.55),312(3.83), 342(3.55),415(3.68)	78-1507-66
$C_{20}H_{19}ClN_2$			
Piperidine, N-benzyl-4-[cyano-(p-chlorophenyl)methylene]-, hydrochloride	EtOH	256(4.12)	78-2721-66
$C_{20}H_{19}FN_2$			
Piperidine, N-benzyl-4-[cyano-(p-fluorophenyl)methylene]-, hydrochloride	EtOH	250(4.02)	78-2721-66
$C_{20}H_{19}F_3O_2$			
Anthracene, 9-(trifluoroacetoxy)-9-tert-butyl-9,10-dihydro-	$CHCl_3$	256(3.24),265(3.17), 273(3.08)	44-4265-66
$C_{20}H_{19}NO_2$			
11H-Benzo[a]carbazole-9-carboxylic acid, 5,6-dihydro-10-methyl-, ethyl ester	EtOH	223(4.34),253(4.45), 260(4.45),289(3.98), 328(4.34)	87-0237-66
Crotonic acid, 3-benzyl-2-cyano-4-phenyl-, ethyl ester	EtOH	226.5(4.34)	22-1033-66
$C_{20}H_{19}NO_3$			
Acronycine	EtOH	223(4.19),279(4.46), 292(4.44),306(4.25), 391(3.82)	12-0275-66
	EtOH	225(4.21),280(4.58), 293(4.54),308s(--), 395(3.82)	78-3245-66
Anthraquinone, 1-[(2-hydroxy-cyclohexyl)amino]-	EtOH	515(3.85)	27-0281-66
Isocarbostyril, 2-(benzoyloxy)-3-butyl-	EtOH	232(3.96),277(3.56), 287(3.51),320(3.20)	44-2090-66
Isocarbostyril, 4-(3,4-dimethoxy-styryl)-2-methyl-	EtOH	278(4.20),333(4.33)	78-2445-66
4-Quinolineacetic acid, N-benzoyl-1,2-dihydro-2,2-dimethyl-	MeOH	323(3.91)	5-0128-66D
$C_{20}H_{19}NO_4$			
Acrylic acid, 2-cyano-3,3-bis(p-methoxyphenyl)-, ethyl ester	EtOH	234(4.21),335(4.31)	22-1033-66
1-Naphthaleneacetic acid, 1,2,3,4-tetrahydro-3,4-dioxo-1-phenyl-, ethyl ester, 3-oxime	EtOH	273(4.14)	44-1905-66
$C_{20}H_{19}NO_5$			
1-Naphthaleneacetic acid, 1,2,3,4-tetrahydro-3-nitro-4-oxo-1-phenyl-, ethyl ester	EtOH	238(4.14),306(3.59), 351(3.62)	44-1905-66

Compound	Solvent	$\lambda_{max}(\log \epsilon)$	Ref.
$C_{20}H_{19}N_2O_2P$			
Tetracyanoethylene adduct of 9-phenyl-9-phosphabicyclo[4.2.1]nonatriene methiodide	EtOH	267(3.63),273s(3.57)	35-3832-66
$C_{20}H_{19}OP$			
Phosphine oxide, (1-cyclohexen-1-ylethynyl)diphenyl-	EtOH	223(3.99),229(4.01), 242(3.96),252(3.94)	22-1002-66
$C_{20}H_{19}P$			
Phosphine, (1-cyclohexen-1-ylethynyl)-diphenyl-	EtOH	225(4.06),231(4.04), 243(3.97),255(3.90)	22-1002-66
$C_{20}H_{19}PS$			
Phosphine sulfide, (1-cyclohexen-1-ylethynyl)diphenyl-	EtOH	222(4.08),228(4.09), 241(4.00),252(3.92)	22-1002-66
$C_{20}H_{20}$			
15H-Cyclopenta[a]phenanthrene, 16,17-dihydro-17-isopropyl-	EtOH	216(4.51),259(4.79), 281(4.17),288(3.86), 300(4.14)	39-0963-66C
15H-Cyclopenta[a]phenanthrene, 16,17-dihydro-11,12,17-trimethyl-	EtOH	218(4.33),225(4.26), 261(4.71),287(3.98), 301(3.97),312(4.01), 343(3.17),360(3.07)	39-0963-66C
7,8:9,10-Dibenzotricyclo[4.2.2.0^{2,5}]-deca-7,9-diene, 1,6-dimethyl-	EtOH	265(3.08),272(3.17)	5-0009-66A
$C_{20}H_{20}BrClN_2$			
Piperidine, N-benzyl-4-[cyano-(m-chlorophenyl)methylene]-, hydrobromide	EtOH	247(4.06)	78-2721-66
$C_{20}H_{20}BrClO_6$			
Benzophenone, 4-acetoxy-2-(bromomethyl)-3-chloro-2',4,6-trimethoxy-6'-methyl-	MeOH	318(3.82)	44-2291-66
$C_{20}H_{20}BrNO_6$			
Azulene-1,3-dicarboxylic acid, 2-diacetylamino-6-bromo-, diethyl ester	MeOH	233(4.39),274(4.28), 316(4.83),350(4.03), 373(3.75),509(2.69)	18-1954-66
$C_{20}H_{20}Br_2OZr_2$			
Zirconium, dibromotetra-π-cyclopentadienyl-μ-oxodi-	C_6H_{12}	280(3.88)	22-3548-66
$C_{20}H_{20}ClFN_2$			
Piperidine, N-benzyl-4-[cyano-(p-fluorophenyl)methylene]-, hydrochloride	EtOH	250(4.02)	78-2721-66
$C_{20}H_{20}ClNO_4S$			
2-tert-Butyl-3-phenyl-4H-indeno-[2,1-d]thiazolium perchlorate	MeOH	232(3.87),318(4.14)	39-0686-66C
$C_{20}H_{20}ClNO_7$			
4-[(3,4-Dimethoxyphenyl)acetyl]-2-methylisoquinolinium perchlorate	EtOH	232(4.41),281(4.14), 333(4.08)	78-2445-66

Compound	Solvent	$\lambda_{max}(\log \epsilon)$	Ref.
$C_{20}H_{20}ClNS$			
Piperidine, 3-(chlorothioxanthen-9-ylidenemethyl)-1-methyl-, hydrochloride	pH 1	231(4.28),314(3.42)	33-1483-66
$C_{20}H_{20}ClN_2O_4PS$			
1-Ethyl-2-[(3-ethyl-2-benzothiazolinylidene)phosphino]quinolinium perchlorate	MeOH	554(4.52)	24-1325-66
	$C_2H_4Cl_2$	562(4.54)	24-1325-66
$C_{20}H_{20}Cl_2FeN_6O_8$			
1,10-Phenanthroline-ferrous perchlorate, acrylonitrile complex	H_2O	225(4.66),270(4.42), 290s(4.00),328(3.68), 365(3.42)	7-0298-66
$C_{20}H_{20}Cl_2Hf_2O$			
Hafnium, dichlorotetra-π-cyclopentadienyl-μ-oxodi-	C_6H_{12}	255(4.14)	22-3548-66
$C_{20}H_{20}Cl_2N_2$			
Piperidine, N-benzyl-4-[cyano-(p-chlorophenyl)methylene]-, hydrochloride	EtOH	256(4.12)	78-2721-66
$C_{20}H_{20}Cl_2OTi_2$			
Titanium, dichlorotetra-π-cyclopentadienyl-μ-oxodi-	$CHCl_3$	240(--),410(4.00)	22-3548-66
$C_{20}H_{20}Cl_2OZr_2$			
Zirconium, dichlorotetra-π-cyclopentadienyl-μ-oxodi-	C_6H_{12}	275(4.07)	22-3548-66
$C_{20}H_{20}F_3IN_2S$			
2-(p-Dimethylaminostyryl)-3-ethyl-6-(trifluoromethyl)benzothiazolium iodide	EtOH	546(--)	65-0828-66
$C_{20}H_{20}I_2OZr_2$			
Zirconium, diiodotetra-π-cyclopentadienyl-μ-oxodi-	$CHCl_3$	282(4.03)	22-3548-66
$C_{20}H_{20}N_2$			
Piperidine, N-benzyl-4-(cyanophenylmethylene)-, hydrochloride	EtOH	250(4.10)	78-2721-66
$C_{20}H_{20}N_2O_2$			
Anthraquinone, 1-[(2-aminocyclohexyl)amino]-, cis	EtOH	509(3.84)	27-0281-66
trans	EtOH	507(3.85)	27-0281-66
Kopsanone, 10-oxo-	EtOH	242(3.91),296(3.55)	33-1237-66
Kopsanone, 10-lactam	EtOH	243(3.89),301(3.55)	39-1260-66C
$C_{20}H_{20}N_2O_6$			
3-Phenoxazone, 4,6-dimethyl-1,9-dicarbomethoxy-2-(dimethylamino)-	MeOH	228(--),254(--), 438(--),458(--)	44-2564-66
$C_{20}H_{20}N_4O_6$			
Diimide, dibenzoyl-, bis(O-carboxyoxime), diethyl ester	$CHCl_3$	247(4.47),422(1.85)	88-0405-66

Compound	Solvent	λ_{max}(log ϵ)	Ref.
$C_{20}H_{20}N_6O_2$			
p-Anisaldehyde, 3,6-pyridazine-diyldihydrazone	EtOH	250(4.11),347(4.61)	78-2073-66
$C_{20}H_{20}O$			
2-Cyclohexen-1-one, 6-ethyl-3,5-diphenyl-	EtOH	285(4.29)	22-1012-66
$C_{20}H_{20}O_3$			
4-Estren-3-one, 9α,10α-epoxy-17α-ethynyl-17β-hydroxy-	EtOH	243(4.12)	87-0510-66
1-Phenanthrenepropionic acid, 7-methoxy-2-methyl-, methyl ester	EtOH	224(4.33),234(4.32), 258(4.86),279(4.29), 290(4.16),302(4.01)	44-1285-66
$C_{20}H_{20}O_4$			
Cinnamic acid, β-butyl-o-carboxy-α-phenyl-	EtOH	209(4.31),221(4.22), 280s(3.32)	35-0476-66
Cinnamic acid, 3-benzoyl-2-ethoxy-, ethyl ester	MeOH	254(4.15),279s(4.02)	5-0053-66I
1,3-Cyclobutanedicarboxylic acid, 2,4-diphenyl-, dimethyl ester, exo-endo	C_6H_{12}	220(4.11),233s(4.00)	88-1127-66
exo-exo	C_6H_{12}	223(4.12)	88-1127-66
14β-Gona-1,3,5(10),6,8-pentaen-11-one,	MeOH	246(4.40),309(3.78)	19-0611-66
17,17-(ethylenedioxy)-3-methoxy-	EtOH	221(4.65),246(4.49), 310(3.76)	39-0955-66C
1,4-Pentadien-3-one, 1-phenyl-5-(2,4,6-trimethoxyphenyl)-	benzene	370(--)	65-0336-66
	EtOH	383(--)	65-0336-66
	HOAc	292(4.25),394(4.22)	65-0336-66
	H_2SO_4	480(--)	65-0336-66
2,4-Pentadien-1-one, 5-phenyl-1-(2,4,6-trimethoxyphenyl)-	benzene	325(--)	65-0336-66
	EtOH	328(--)	65-0336-66
	HOAc	334(4.52)	65-0336-66
$C_{20}H_{20}O_5$			
1H-Cyclopenta[7,8]naphtho[2,3-b]furan-7-carboxylic acid, 10-acetoxy-2,3,7,8-tetrahydro-6-methyl-, methyl ester	EtOH	244(4.76),268(3.86), 291(3.88),338(3.27), 373(2.37)	39-0725-66C
3-Flavene, 4'-hydroxy-2,8-dimethoxy-3-methyl-, acetate	EtOH	264(4.05),273(4.02)	78-2913-66
Otobain, hydroxy-	EtOH	235(3.91),287(3.84)	39-1775-66C
Xanthone, 1-hydroxy-3,7-dimethoxy-2-(1,1-dimethylallyl)-	MeOH	233(4.54),264(4.57), 311(4.26),372(3.84)	39-2265-66C
Xanthone, 1-hydroxy-3,7-dimethoxy-2-(3,3-dimethylallyl)-	MeOH	233(4.55),264(4.64), 299(4.42),370(3.84)	39-2265-66C
Xanthone, 1-hydroxy-3,7-dimethoxy-4-(3,3-dimethylallyl)-	MeOH	233(4.50),265(4.61), 314(4.15),385(3.91)	39-2265-66C
$C_{20}H_{20}O_6$			
Jacareubin, dihydro-	MeOH	251(4.60),286(4.00), 330(4.35)	39-0175-66C
$C_{20}H_{20}O_7$			
Anthraquinone, 1,3,6,8-tetrahydroxy-2-(1-hydroxyhexyl)- (or averantin)	n.s.g.	223(4.53),258s(4.18), 266(4.24),287s(4.47), 294(4.53),325(4.01), 454(4.03)	39-0855-66C

Compound	Solvent	$\lambda_{max}(\log \epsilon)$	Ref.
$C_{20}H_{20}O_8$			
Combretol	EtOH	212(4.63),266(4.26), 345(4.23)	39-0125-66C
	EtOH-NaOH	213(--),288(--), 366(--)	39-0125-66C
Hydroquinone, 2-acetyl-3-(5-carbethoxymethyl-2-furyl)-, diacetate	EtOH	204(4.22),280(4.08), 308(3.91)	33-1794-66
$C_{20}H_{20}O_{10}$			
Cassiaside	EtOH	225(4.41),279(4.67), 329(3.40),410(3.81)	95-1087-66
$C_{20}H_{21}ClN_2$			
Piperidine, N-benzyl-4-(cyanophenylmethylene)-, hydrochloride	EtOH	250(4.10)	78-2721-66
$C_{20}H_{21}ClO_3$			
p-Dioxin, 2-(2-chloroethoxy)-2,3-dihydro-3,6-dimethyl-2,5-diphenyl-	hexane	274(3.99)	87-0762-66
$C_{20}H_{21}ClO_6$			
Radicicol, dimethyl ether	EtOH	279(4.28)	12-1265-66
$C_{20}H_{21}Cl_2NS$			
Piperidine, 3-(chlorothioxanthen-9-ylidenemethyl)-1-methyl-, hydrochloride	pH 1	231(4.28),314(3.42)	33-1483-66
$C_{20}H_{21}N$			
Carbazole, N-benzyl-1,2,3,4-tetrahydro-3-methyl-	n.s.g.	228(4.67),287(3.77), 294(3.73)	16-0064-66
Carbazole, N-benzyl-1,2,3,4-tetrahydro-4-methyl-	n.s.g.	228(4.52),284(3.72), 291(3.65)	16-0064-66
Naphthalen-1,4-imine, 1,4-dihydro-1,2,3,4-tetramethyl-9-phenyl-	MeOH	206(4.55),268(3.14)	44-2009-66
1-Pyrroline, 3-(diphenylmethylene)-2-isopropyl-	EtOH	230(4.23),288(4.22)	94-0187-66
1-Pyrroline, 3-(diphenylmethylene)-2-propyl-	EtOH	231(4.16),287(4.16)	94-0187-66
1-Pyrroline, 3-(diphenylmethylene)-2,5,5-trimethyl-	EtOH	229(4.17),287(4.20)	94-0187-66
$C_{20}H_{21}NO$			
1-Indanone, 2-benzylidene-3-(tert-butylamino)-	isooctane	238(4.16),245s(4.13), 310(4.38)	88-1499-66
Indone, 2-benzyl-3-(tert-butylamino)-	MeOH	218(4.19),255s(4.26), 264(4.30),312s(3.08), 430(3.28)	88-1499-66
Indone, 2-[α-(tert-butylamino)benzyl]-	isooctane	238(4.62),244(4.63)	88-1499-66
4H-Pyrrolo[3,2,1-if]quinoline, 1,2,5,6-tetrahydro-2-oxo-6-phenyl-4,4,6-trimethyl-	MeOH	255(4.03),283(3.27)	30-0341-66E
Quinoline, 1,2-dihydro-1-phenylacetyl-2,2,4-trimethyl-	MeOH	247(4.22),293(3.66)	30-0341-66E
$C_{20}H_{21}NO_2$			
1,4-Pentadien-3-one, 1-(p-methoxyphenyl)-5-(p-dimethylaminophenyl)-	benzene	332(4.24),412(4.53)	65-0336-66
	EtOH	247(4.08),344(4.22), 434(4.59)	65-0336-66
	HOAc	358(4.27),443(4.45)	65-0336-66

Compound	Solvent	λ_{max}(log ϵ)	Ref.
1,4-Pentadien-3-one, 1-(p-methoxyphenyl)-5-(p-dimethylaminophenyl)-, ferric chloride added	HOAc	622(4.68)	65-0336-66
2,4-Pentadien-1-one, 1-(p-methoxyphenyl)-5-(p-dimethylaminophenyl)-	benzene	296(4.31),424(4.46)	65-0336-66
	EtOH	238(4.06),306(4.25), 435(4.56)	65-0336-66
	HOAc	338(3.34),434(4.27)	65-0336-66
ferric chloride added	HOAc	660(4.41)	65-0336-66
2,4-Pentadien-1-one, 5-(p-methoxyphenyl)-1-(p-dimethylaminophenyl)-	benzene	388(4.84)	65-0336-66
	EtOH	242(4.02),268(3.97), 402(4.59)	65-0336-66
	HOAc	412(4.61)	65-0336-66
ferric chloride added	HOAc	545(4.55)	65-0336-66
$C_{20}H_{21}NO_3$			
Isoquinoline, 1,2-dihydro-4-(3,4-dimethoxyphenylacetyl)-2-methyl-	EtOH	230(4.40),287(4.39), 345(4.26)	78-2445-66
$C_{20}H_{21}NO_4$			
Eschscholtzidine	EtOH	291(4.08)	23-1259-66
Isocarbostyril, 4-[1-hydroxy-2-(3,4-dimethoxyphenethyl)]-2-methyl-	EtOH	230(4.41),285(3.99), 330(3.63)	78-2445-66
$C_{20}H_{21}NO_5$			
Cassythine, N-methyl-	n.s.g.	219(4.46),282(4.15), 302(4.13)	12-0297-66
Hunnemannine	MeOH	233(4.1),285(3.9)	73-1355-66
Isoquinoline, 3,4-dihydro-1-(p-methoxybenzoyl)-5,6,7-trimethoxy-	MeOH	296.5(4.35)	44-0516-66
$C_{20}H_{21}NO_6$			
1,3-Azulenedicarboxylic acid, 2-diacetylamino-, diethyl ester	MeOH	234(4.51),275(4.46), 304(4.70),337(3.87), 366(3.92),505(2.93)	18-1954-66
$C_{20}H_{21}NO_8$			
2,3,4,5-Pyridinetetracarboxylic acid, 1,6-dihydro-1-methyl-6-phenyl-, tetramethyl ester	MeOH	235(4.17),284(4.13), 394(3.79)	39-1121-66C
$C_{20}H_{21}NS$			
Piperidine, 1-methyl-3-(thioxanthen-9-ylidenemethyl)-	EtOH	230(4.49),268(4.15), 320(3.52)	33-1483-66
$C_{20}H_{21}N_5O_6$			
Adenosine, N-benzoyl-, cyclic 2',3'-(ethyl orthoformate)	EtOH	246s(3.95),279(4.23)	73-1785-66
$C_{20}H_{22}BF_4N_3O_2S_2$			
Bis(3-ethyl-6-methoxy-2-benzothiazolyl)azamethinecyanine tetrafluoroborate	MeOH	385(4.69)	27-0318-66
$C_{20}H_{22}BF_4N_5O_2S_2$			
Bis(3-ethyl-6-methoxy-2-benzothiazolyl)triazamethinecyanine tetrafluoroborate	MeOH	512(4.62)	27-0318-66

Compound	Solvent	λmax(log ε)	Ref.
$C_{20}H_{22}ClNO_4$			
[2-(10,10-Dimethyl-9(10H)-anthrylidene)ethylidene]dimethylammonium perchlorate	MeCN	350(4.42)	24-2479-66
Protopapaverine methochloride isomer	90% EtOH	251(3.39),260(3.38), 288(3.20),371(3.18)	78-0485-66B
	EtOH-HCl	256(4.65),323(3.95)	78-0485-66B
	EtOH-NaOH	249(4.47),260(4.46), 286(3.27),370(3.20)	78-0485-66B
$C_{20}H_{22}ClNO_5$			
1-Cyclohexene-1-carboxamide, 4-[(8-chloro-1,2,3,4-tetrahydro-5-methoxy-1-methyl-4-oxo-2-naphthyl)methyl]-2-hydroxy-6-oxo-	MeOH-HCl	226(4.43),258(4.37), 328(3.64)	44-2429-66
	MeOH-NaOH	227(4.40),262(4.35), 328(3.64)	44-2429-66
$C_{20}H_{22}ClN_2O_6PS_2$			
3-Ethyl-2-[(3-ethyl-6-methoxy-2-benzothiazolinylidene)phosphino]-6-methoxybenzothiazolium perchlorate	MeOH	486(4.69)	24-1325-66
	$C_2H_4Cl_2$	494(4.59)	24-1325-66
$C_{20}H_{22}INO_4$			
Protopapaverine methiodide	90% EtOH	229(4.47),254(4.65), 276(4.26),315(3.79)	78-0485-66B
	EtOH-HCl	229(4.48),282(3.64), 315(3.68)	78-0485-66B
	EtOH-NaOH	227(4.51),314(3.65), 391(3.68)	78-0485-66B
$C_{20}H_{22}N_2$			
Dipyrromethene, 3,3',4',5,5'-pentamethyl-4-phenyl-	EtOH-HCl	216(4.60),364(3.82), 485(4.85)	12-1871-66
	EtOH	444(4.40)	12-1871-66
4-Piperidineacetonitrile, N-benzyl-α-phenyl-, hydrobromide	EtOH	256(2.57),261(2.30)	78-2721-66
$C_{20}H_{22}N_2O$			
Kopsanone	EtOH	244(3.85),296(3.53)	33-1237-66
	EtOH	244(3.85),295(3.54)	39-1260-66C
	5N HCl	249s(3.14),256s(3.24), 263(3.31),269(3.26)	39-1260-66C
$C_{20}H_{22}N_2O_2$			
Diimide, bis(2,4,6-trimethylbenzoyl)-	EtOH	247(4.07)	78-1309-66
	$CHCl_3$	445s(1.9)	78-1309-66
Epikopsanol-10-lactam	EtOH	239(3.92),289(3.65)	39-1260-66C
Kopsanol-10-lactam	EtOH	245(3.93),297(3.53)	39-1260-66C
$C_{20}H_{22}N_2O_5$			
α-D-Altropyranoside, methyl 4,6-O-benzylidene-3-deoxy-3-(phenylazo)-	EtOH	268(4.01),415(1.19)	39-0695-66C
1,3-Azulenedicarboxylic acid, 2-acetamido-6-ethyleneimino-, diethyl ester	MeOH	238(3.95),268(4.10), 334(4.77),367s(4.13), 399(4.21),445s(3.17)	18-1954-66
α-D-arabino-Hexopyranosid-3-ulose, methyl 4,6-O-benzylidene-, phenylhydrazone	MeOH	285(4.24)	39-0695-66C
β-D-xylo-Hexopyranosid-3-ulose, methyl 4,6-O-benzylidene-, phenylhydrazone	MeOH	275(3.97)	39-0695-66C

Compound	Solvent	$\lambda_{max}(\log \epsilon)$	Ref.
α-D-Mannoside, methyl 4,6-O-benzylidene-3-deoxy-3-(phenylazo)-	MeOH	266(3.98),401(2.10)	39-1508-66C
$C_{20}H_{22}N_2O_6S$			
5-Thia-1-azabicyclo[4.2.0]oct-3-ene-2-carboxylic acid, 3-(acetoxymethyl)-8-oxo-7-(2-phenylacetamido)-, ethyl ester	n.s.g.	248(3.90)	39-1142-66C
$C_{20}H_{22}N_2O_8S_2$			
5-Thia-1-azabicyclo[4.2.0]oct-3-ene-2-carboxylic acid, 3-(hydroxymethyl)-7-[2-(2-thienyl)acetamido]-, ester with ethyl glycolate, acetate	EtOH	236(4.10),260(3.86)	87-0741-66
$C_{20}H_{22}N_4O$			
Pyrazole-4,5-dione, 3-tert-butyl-1-phenyl-, 4-(methylphenylhydrazone)	$CHCl_3$	406(4.18)	22-2990-66
$C_{20}H_{22}N_4O_4$			
1-Indanone, 2,2,5,6,7-pentamethyl-, 2,4-dinitrophenylhydrazone	MeOH	386(4.47)	35-2838-66
$C_{20}H_{22}N_4O_5$			
Imidazole-4-carboxamide, N-benzyl-5-formamido-1-β-D-ribofuranosyl-, anhydronucleoside from	pH 1	250s(4.01)	44-2202-66
	pH 7	246(4.02)	44-2202-66
	pH 13	262(4.05)	44-2202-66
Inosine, 1-benzyl-2',3'-isopropylidene-	pH 1	245s(--),252(3.99)	44-2202-66
	pH 7	245s(--),252(3.98)	44-2202-66
	pH 13	246s(--),252(3.97)	44-2202-66
$C_{20}H_{22}N_8O_8$			
2,3-Heptanedione, 6-methyl-, bis-(2,4-dinitrophenylhydrazone)	$CHCl_3$	395(4.64),435s(4.56)	78-1857-66
$C_{20}H_{22}O$			
Ether, 2-(α-benzylphenethylidene)-cyclopropyl ethyl	C_6H_{12}	248(2.76),253(2.79), 259(2.82),261(2.81), 265(2.75),268(2.72)	78-2621-66
[4.4]Paracyclophane, 1-oxo-	EtOH	263(3.95),278(3.60)	35-3515-66
[4.4]Paracyclophane, 2-oxo-	EtOH	267(2.86),272(2.88), 287(2.78)	35-3515-66
$C_{20}H_{22}O_2$			
Estra-1,3,5(10),8,14-pentaen-17-one, 3-methoxy-6α-methyl-	n.s.g.	316(4.40)	78-1019-66
6β-isomer	n.s.g.	314(4.43)	78-1019-66
Hexane, 3,4-bis(5-oxocyclohepta-1,3,6-trienyl)-	EtOH	233(4.42),306(4.01)	39-1213-66C
3-Hexyne, 1,6-bis(m-methoxyphenyl)-	EtOH	274(3.59),280(3.57)	94-0262-66
$C_{20}H_{22}O_3$			
Estra-1,3,5(10),8,14-pentaen-17-one, 2,3-dimethoxy-	EtOH	243(4.18),328(4.36)	87-0027-66
14β-Estra-3,5(10),6,8-tetraen-17-one, 3-acetoxy-	dioxan	274.5(4.17)	13-0365-66B
Estrone, 8(14)-dehydro-, acetate	MeOH	267(3.06),272(3.06), 275(3.06)	13-0365-66B

Compound	Solvent	$\lambda_{max}(\log \epsilon)$	Ref.
$C_{20}H_{22}O_4$			
4-Benzofuranol, 2-isopropyl-7-(2'-methoxyphenylacetyl)-2,3-dihydro-	EtOH	222(4.32),282(4.22)	39-0749-66C
Estra-4,6,8(14)-trien-3-one, 17-ethynyl-17β-hydroxy-	MeOH	345(4.46)	44-3780-66
14β-Gona-1,3,5(10),8-tetraene-11,17-dione, 3-methoxy-, 17-(ethylene acetal)	MeOH	245(4.15),291(3.71)	19-0611-66
Malonic acid, phenyltropyl-, ethyl ester	hexane	258(3.59)	44-0912-66
$C_{20}H_{22}O_6$			
Benzophenone, 4'-acetoxy-2',4,6-trimethoxy-2,6'-dimethyl-	M-OH	233s(4.18),281(3.94), 302s(3.85)	44-2291-66
$C_{20}H_{22}O_7$			
Elephantin	MeOH	215(4.40)	35-3674-66
$C_{20}H_{23}BrN_2$			
4-Piperidineacetonitrile, N-benzyl-α-phenyl-, hydrobromide	EtOH	256(2.57),261(2.30)	78-2721-66
$C_{20}H_{23}ClN_2O_4$			
2,3-Dihydro-5,6,7-trimethyl-1,4-diphenyl-1H-1,4-diazepinium perchlorate	MeOH	250(4.12),348(4.36)	39-0093-66C
$C_{20}H_{23}ClO_6$			
Radicicol, alcohol derivative	EtOH	238s(4.29),292(3.64)	12-1265-66
$C_{20}H_{23}NO$			
3-Picoline, 1-benzyl-1,2,3,6-tetra-hydro-4-(p-methoxyphenyl)-hydrochloride	EtOH	250(4.16)	78-2745-66
	EtOH	250(4.15)	78-2745-66
3-Picoline, 1-benzyl-1,2,5,6-tetra-hydro-4-(p-methoxyphenyl)-hydrochloride	EtOH	239(4.12)	78-2745-66
	EtOH	241(4.11)	78-2745-66
Quinoline, 1-acetyl-1,2,3,4-tetrahydro-2,2,4-trimethyl-4-phenyl-	MeOH	253(4.03)	30-0341-66E
$C_{20}H_{23}NOS$			
Piperidine, 1-methyl-3-(thioxanthen-2-ylmethyl)-, S-oxide	EtOH	212(4.43),250(3.82), 273(3.42)	33-0042-66
isomer B	EtOH	211(4.42),275(3.10)	33-0042-66
$C_{20}H_{23}NO_2$			
6H-Dibenzo[a,f]quinolizin-13(12H)-one, 1,2,7,11b-tetrahydro-9,10-dimethoxy-3-methyl-	EtOH	235(4.27),265(4.16), 381(3.89)	44-0797-66
$C_{20}H_{23}NO_4$			
Aporphine, 10-hydroxy-1,2,9-trimethoxy-	EtOH	280(4.21),304(4.17)	12-0437-66
	EtOH-KOH	335(3.98)	12-0437-66
Isoboldine, O-methyl-	EtOH	279(4.14),304(4.16)	12-0135-66
Isoquinoline, 3,4-dihydro-5,6,7-tri-methoxy-1-(p-methoxybenzyl)-, hydrochloride	MeOH	272(3.96),315(3.43)	44-0516-66
$C_{20}H_{23}NO_5$			
Norpluviine, diacetate	EtOH	279(3.98)	39-0676-66C

Compound	Solvent	λ_{max}(log ε)	Ref.
$C_{20}H_{23}NO_6$			
1,3-Azulenedicarboxylic acid, 2-acetamido-6-ethoxy-, diethyl ester	MeOH	245(4.34),278(4.24), 327(4.79),372(3.97), 408s(3.77)	18-1954-66
$C_{20}H_{23}N_3$			
3H-Dibenz[c,f]imidazo[1,2-a]azepine, 2,9-dihydro-9-[3-(methylamino)-propyl]-	EtOH	231s(4.12),289(3.78), 260(3.98)	4-0206-66
$C_{20}H_{23}N_3O$			
Cinnoline, 3-(p-diethylaminoethoxy-phenyl)-	MeOH	264(4.52),295(4.30)	87-0784-66
Cinnoline, 4-(2-diethylaminoethoxy)-3-phenyl-	MeOH	248(4.54),288(3.79), 330(3.54)	87-0784-66
4(1H)-Cinnolinone, 1-(2-diethylamino-ethyl)-3-phenyl-	MeOH	264(4.16),312(4.01), 360(4.11)	87-0784-66
$C_{20}H_{23}N_3O_2$			
Dipyrromethene-5-carboxylic acid, 3,3',4-trimethyl-5'-(4-methyl-pyrrol-2-yl)-, ethyl ester	EtOH	210(4.02),261(4.14), 268(4.13),295(4.16), 307s(4.09),354(3.70), 474(4.48),495s(4.42)	39-0098-66C
hydrobromide	EtOH	264s(3.88),271(3.92), 302(4.27),392(3.77), 413(3.75),539(4.70), 570(4.78)	39-0098-66C
$C_{20}H_{23}N_3O_7S_2$			
Glycine, N-[[3-(hydroxymethyl)-8-oxo-7-[2-(2-thienyl)acetamido]-5-thia-1-azabicyclo[4.2.0]oct-2-en-2-yl]-carbonyl]-, ethyl ester, acetate	EtOH	233(4.11)	87-0741-66
$C_{20}H_{23}N_5O_{10}$			
Pteridine, 2-amino-7-(β-D-glucopyran-osyloxy)-, tetraacetate	pH -0.89	245(3.77),307(4.03)	24-0536-66
	pH 5.0	232(4.21),255(3.81), 266(3.79),355(3.99)	24-0536-66
	MeOH	232(4.27),265(3.86), 357(4.03)	24-0536-66
$C_{20}H_{24}ClNO$			
3-Picoline, 1-benzyl-1,2,3,6-tetra-hydro-4-(p-methoxyphenyl)-, hydrochloride	EtOH	250(4.15)	78-2745-66
3-Picoline, 1-benzyl-1,2,5,6-tetra-hydro-4-(p-methoxyphenyl)-, hydrochloride	EtOH	241(4.11)	78-2745-66
$C_{20}H_{24}ClNO_4$			
Isoquinoline, 3,4-dihydro-5,6,7-tri-methoxy-1-(p-methoxybenzyl)-, hydrochloride	MeOH	272(3.96),315(3.43)	44-0516-66
$C_{20}H_{24}ClN_3S$			
Perchlorperazine (dihydrochloride)	EtOH	256(4.54),309(3.62)	36-0144-66
$C_{20}H_{24}Cl_{10}Si_4$			
Tetrasilane, 1,1,2,2,3,3,4,4-octa-methyl-1,4-bis(pentachlorophenyl)-	C_6H_{12}	217(5.03)	101-0102-66B

Compound	Solvent	λ_{max}(log ϵ)	Ref.
$C_{20}H_{24}INO_3$			
3,4-Dihydro-4-(α-hydroxy-3,4-dimethoxyphenethyl)-2-methylisoquinolinium iodide	EtOH	220(4.51),286(4.18)	78-2445-66
$C_{20}H_{24}INO_4$			
Laurifoline iodide	EtOH	227(4.48),281(4.04), 307(4.14)	39-1340-66C
Magnoflorine iodide	EtOH	227(4.65),271(3.93), 310(3.8)	39-1340-66C
	EtOH	276(3.89),318(3.74)	95-0124-66
$C_{20}H_{24}N_2$			
Phenazine, 2,6-di-tert-butyl-	EtOH	255(5.1),365(4.2)	7-1791-66
$C_{20}H_{24}N_2O$			
Epikopsanol	EtOH	227(3.74),262(2.91)	39-1260-66C
	5N HCl	262(--),270(--)	39-1260-66C
Kopsanol	EtOH	246(3.78),297(3.39)	39-1260-66C
	5N HCl	249s(3.24),255s(3.29), 262(3.36),269(3.31)	39-1260-66C
$C_{20}H_{24}N_2O_2$			
Beninine, 1,2-dehydro-	EtOH	226(4.2),252(3.64), 306(3.61)	33-2072-66
Kopsininic acid	EtOH	240(4.11),288(3.71)	33-1237-66
Lochnerine	50% EtOH	227(4.50),278(3.94)	23-1523-66
Tetracyclotrimethine, bis(3,5-dimethyl-4-ethylpyrrol-2-yl)-	$CHCl_3$	560(5.30)	5-0153-66I
Vallesiachotamine, $LiAlH_4$ reduction product	EtOH	225(4.53),281(3.91), 290(3.89)	35-1792-66
$C_{20}H_{24}N_2O_3$			
Geissoschizine	n.s.g.	225(4.6),268(4.2)	102-1065-66
Imidazolidine, 1-(p-carbethoxyphenyl)-3-(2-hydroxyethyl)-2-phenyl-	EtOH	308(4.36)	87-0868-66
Strictamine	EtOH	213(4.37),262(3.80)	44-1641-66
$C_{20}H_{24}N_2O_5$			
1,3-Azulenedicarboxylic acid, 2-acetamido-6-(dimethylamino)-, diethyl ester	MeOH	223(4.27),270(4.28), 344(4.82),402(4.37)	18-1954-66
$C_{20}H_{24}N_2O_6$			
1,3-Azulenedicarboxylic acid, 2-acetamido-6-(2-hydroxyethylamino)-, diethyl ester	MeOH	220(3.99),269(4.17), 290(4.13),342(4.78), 399(4.27)	18-1954-66
$C_{20}H_{24}N_2O_6S_2$			
5-Thia-1-azabicyclo[4.2.0]oct-2-ene-2-carboxylic acid, 3-(hydroxymethyl)-8-oxo-7-[2-(2-thienyl)acetamido]-, isobutyl ester, acetate	EtOH	235(4.16)	87-0741-66
$C_{20}H_{24}N_4O_2$			
Benzimidazole, 2-benzyl-1-(2-diethylaminoethyl)-5-nitro-	EtOH-HCl	241(4.46),302(4.02)	39-1511-66C
Benzimidazole, 2-benzyl-1-(2-diethylaminoethyl)-6-nitro-	EtOH-HCl	236(4.26),307(4.05)	39-1511-66C

Compound	Solvent	λ_{max}(log ε)	Ref.
$C_{20}H_{24}N_4O_4$			
3,5-Heptadien-2-one, 6-(2-methyl-3-cyclohexen-1-yl)-, 2,4-dinitrophenylhydrazone	CHCl3	270(4.34),310(4.41), 405(4.68)	70-0480-66
3,5-Heptadien-2-one, 6-(4-methyl-3-cyclohexen-1-yl)-, 2,4-dinitrophenylhydrazone	CHCl3	270(4.33),315(4.39), 405(4.67)	70-0480-66
2(3H)-Naphthalenone, 4,4a,5,6,7,8-hexahydro-7-isopropenyl-4a-methyl-, 2,4-dinitrophenylhydrazone	CHCl3	395(4.40)	44-4188-66
4-Nor-α-cyperone, 2,4-dinitrophenylhydrazone	CHCl3	398(4.46)	44-4188-66
$C_{20}H_{24}N_4O_8$			
4-Cyclohexene-1,3-dicarboxylic acid, 2,4-dimethyl-6-oxo-, diethyl ester, 2,4-dinitrophenylhydrazone	EtOH	373(4.28)	39-0336-66C
$C_{20}H_{24}N_4O_9S_2$			
β-D-Glucopyranuronic acid, 1-deoxy-1-(6-ethylthio-8-mercapto-9H-purin-9-yl)-, methyl ester, 2,3,4-triacetate	MeOH	258(4.22),329(4.49)	94-1360-66
$C_{20}H_{24}O_2$			
Estra-4,8(14)-dien-3-one, 17-ethynyl-17β-hydroxy-	MeOH	231(4.24)	44-3780-66
Estra-5(10),6,8,14-tetraene-17β-carboxylic acid, methyl ester	EtOH	218(4.39),223(4.37), 230(4.19),265(4.30)	35-0799-66
Estra-1,3,5(10),8-tetraen-17-one, 3-methoxy-6β-methyl-	n.s.g.	281(4.20)	78-1019-66
D-Homoestra-1,3,5(10),14(15)-tetraen-17a-one, 3-methoxy-	n.s.g.	278(3.3)	88-3585-66
9α-D-Homoestra-1,3,5(10),8(14)-tetraen-17a-one, 3-methoxy-	n.s.g.	278(3.3)	88-3585-66
9β-isomer	n.s.g.	278(3.36)	88-3585-66
$C_{20}H_{24}O_3$			
Androsta-1,4,6-triene-3,17-dione, 2-methoxy-	MeOH	219(4.27),282(4.25)	73-1363-66
Estra-1,3,5(10),8-tetraen-17-one, 2,3-dimethoxy-	EtOH	224(4.31),287(4.00), 306(3.97)	87-0027-66
Estra-1,3,5(10),9(11)-tetraen-17-one, 2,3-dimethoxy-	EtOH	263(4.14),271(4.10), 306(3.85)	87-0027-66
Fuerstionone	ether	256(5.20),350s(3.20), 418(3.59)	33-1151-66
D-Homoestra-1,3,5(10)-triene-16,17a-dione, 3-methoxy-	EtOH-HCl	256(4.20)	22-1222-66
	EtOH	256s(--),283(4.21)	22-1222-66
	EtOH-KOH	283(4.38)	22-1222-66
	dioxan	245(4.0)	22-1222-66
$C_{20}H_{24}O_4$			
Estra-1,3,5(10)-trien-6-one, 3,17β-dihydroxy-, 17-acetate	EtOH	222(4.33),256(3.96), 327(3.49)	5-0180-66B
	0.05N NaOH	243(4.39),270s(3.86), 374(3.17)	5-0180-66B
Fuerstionone, 5,6-dehydro-7-oxo-	ether	220s(4.80),294s(3.25), 315(3.27),403(3.33)	33-1151-66
8,14-Secoestra-1,3,5(10),9(11)-tetraene-14,17-dione, 2,3-dimethoxy-	EtOH	267(4.14),310(3.91)	87-0027-66

Compound	Solvent	λ_{max}(log ϵ)	Ref.
$C_{20}H_{24}O_5$			
1H-Cyclopenta[7,8]naphtho[2,3-b]furan-7-carboxylic acid, 10-acetoxy-2,3,3a,4,5,7,8,10b-octahydro-6-methyl-, methyl ester	EtOH	288(3.54)	39-0725-66C
Mandelic acid, 2-methoxy-α-(p-methoxybenzyl)-5-methyl-, ethyl ester	EtOH	276(3.66),281(3.65)	78-1789-66
$C_{20}H_{24}O_6$			
Enmein, tetradehydrodihydro-	EtOH	290(1.85)	78-3423-66
Fastigilin C	EtOH	222.5(4.47)	78-1907-66
$C_{20}H_{25}BrO_2$			
A-Homoestra-2,4,5(10)-trien-17-one, 3-bromo-4-methoxy-	n.s.g.	281(3.90)	78-0391-66A
$C_{20}H_{25}BrO_3$			
A-Norpregn-3(5)-ene-2,16,20-trione, 17α-bromo-	EtOH	232(4.29)	88-5337-66
$C_{20}H_{25}ClN_2O_2$			
Spegatrine, chloride	50% EtOH	272(3.87),296s(3.63)	23-1523-66
	50% EtOH-NaOH	270(3.81),309(3.43), 320s(3.34)	23-1523-66
$C_{20}H_{25}ClO_3$			
A-Norpregn-3(5)-ene-2,16,20-trione, 17α-chloro-	EtOH	233(4.25)	88-5337-66
$C_{20}H_{25}ClO_5$			
Radicicol, olefin from	EtOH	246(3.72),292(3.56)	12-1265-66
$C_{20}H_{25}ClO_6$			
Radicicol, tetrahydro-, dimethyl ether	EtOH	245(3.72),292(3.60)	12-1265-66
$C_{20}H_{25}N$			
3-Butenylamine, N,N,2-trimethyl-3-phenyl-4-o-tolyl-, hydrochloride	H_2O	250(3.95)	78-1001-66
3-Butenylamine, N,N,2-trimethyl-3-phenyl-4-p-tolyl-, hydrochloride	H_2O	259.5(4.19)	78-1001-66
3-Butenylamine, N,N,2-trimethyl-4-phenyl-3-o-tolyl-, hydrochloride	H_2O	255.5(4.16)	78-1001-66
$C_{20}H_{25}NO$			
3-Butenylamine, 4-(p-methoxyphenyl)-N,N,2-trimethyl-3-phenyl-, hydrochloride	H_2O	267(4.23)	78-1001-66
2,6,8,12-Hexadecatetraynoic amide, N-isobutyl-	EtOH	254(3.08)	12-1207-66
$C_{20}H_{25}NO_3$			
6H-Dibenzo[a,f]quinolizin-13(12H)-one, 1,2,3,4,7,11b-hexahydro-9,10-dimethoxy-3-methyl-	EtOH	285(3.69),334(4.11)	44-0797-66
Hetisine, dehydro-	EtOH	290(1.60)	23-0001-66
$C_{20}H_{25}NO_4$			
Codamine	EtOH	282(3.79)	78-0485-66B
Estra-1,3,5(10)-trien-6-one, 3,17β-dihydroxy-, oxime, 17-acetate	EtOH	212(4.38),256(4.02), 307(3.54)	5-0180-66B

Compound	Solvent	λ_{max}(log ε)	Ref.
Estra-1,3,5(10)-trien-6-one, 3,17β-dihydroxy-, oxime, 17-acetate	0.05N NaOH	242(4.17),270s(3.93), 333(3.49)	5-0180-66B
Isoquinoline, 1,2,3,4-tetrahydro-5,6,7-trimethoxy-1-(p-methoxybenzyl)-, hydrochloride	MeOH	278(3.54),284(3.51)	44-0516-66
8-Isoquinolinol, 1,2,3,4-tetrahydro-6,7-dimethoxy-1-(p-methoxybenzyl)-2-methyl-	pH 1	225s(4.37),274(3.38), 281s(3.28)	33-1757-66
	pH 13	216(4.54),277s(3.62), 283(3.63)	33-1757-66
	iso-PrOH	228s(4.36),277(3.40), 285s(3.20)	33-1757-66
hydrochloride	pH 1	226s(4.33),274(3.35), 281(3.28)	33-1757-66
	pH 13	216(4.54),277s(3.43), 284(3.60)	33-1757-66
	iso-PrOH	227s(4.36),276(3.38), 284(3.26)	33-1757-66
Pseudocodamine	EtOH	284(3.78)	24-1764-66
Pseudolaudanine	EtOH	284(3.78)	24-1764-66
	EtOH	226s(4.11),282(3.72)	78-0485-66B
$C_{20}H_{25}NO_5$			
1-Isoquinolinemethanol, 1,2,3,4-tetrahydro-5,6,7-trimethoxy-α-(p-methoxyphenyl)-	MeOH	277(3.52),283(3.51)	44-0516-66
$C_{20}H_{26}ClNO_3$			
Colletine chloride	EtOH	227(4.22),284(3.70)	39-1340-66C
$C_{20}H_{26}ClNO_4$			
Tembetarine chloride	EtOH	284(3.84)	39-1340-66C
$C_{20}H_{26}INO_3$			
Petaline iodide	iso-PrOH	223(4.45),280(3.60), 285(3.60)	88-1599-66
	n.s.g.	206s(4.87),233s(4.35), 278(3.59),284(3.59)	78-1335-66
$C_{20}H_{26}N_2$			
1,1'-Azopropane, 1,1'-dimethyl-1,1'-diphenyl-	isooctane	375(1.57)	88-3125-66
$C_{20}H_{26}N_2O$			
Pyrazolo[2,3-d]estra-1,3,5(10)-trien-17β-ol, 1'-methyl-	EtOH	257(3.81),263(3.85), 272(3.81),293(3.73), 299(4.74),304(3.72), 312(3.64)	88-0205-66
$C_{20}H_{26}N_2O_2$			
Ajmaline	50% EtOH	248(3.96),291(3.50)	23-1523-66
Beninine	EtOH-HCl	248(3.38),270(3.19), 277(3.20)	33-2072-66
	EtOH	246(3.85),291(3.43)	33-2072-66
	EtOH-KOH	246(3.85),291(3.43)	33-2072-66
	12N HCl	270(3.42),277(3.41)	33-2072-66
Ibogaine, 18-hydroxy-	EtOH	228(4.39),294(3.96)	35-2532-66
2,3-Seco-A-norandrost-5-ene-2,3-dicarbonitrile, 17β-acetoxy-	EtOH	215(3.87)	44-2755-66
Spiroxyane, 17α-hydroxy-16α-methyl-	EtOH	252(3.85),283(3.16)	44-1765-66
epimer B	EtOH	252(3.87),280(3.18)	44-1765-66

Compound	Solvent	λ_{max}(log ε)	Ref.
Unknown ketone from macralstonine	EtOH	230(4.61),285(3.90), 292(3.87)	33-0946-66
$C_{20}H_{26}N_2O_3$			
Isopteropodal, methyl-	EtOH	253(3.86),270s(3.28)	39-2245-66C
Strictamine, dihydro-	EtOH-HCl	235(3.90),290(3.51)	44-1641-66
	EtOH	244(3.90),298(3.51)	44-1641-66
$C_{20}H_{26}N_2O_4$			
Isoquinoline, 1-(6-amino-4-hydroxy-3-methoxybenzyl)-1,2,3,4-tetrahydro-6,7-dimethoxy-2-methyl-	EtOH	289(3.75)	12-0437-66
$C_{20}H_{26}N_2O_{11}$			
β-D-Glucopyranuronic acid, 1-deoxy-1-(4-ethoxy-5-methyl-2-oxo-1(2H)-pyrimidinyl)-, methyl ester, 2,3,4-triacetate	MeOH	283(3.78)	94-1354-66
$C_{20}H_{26}N_4O_4$			
2,5,8-Tetradecatrienal, 2,4-dinitrophenylhydrazone, 2-trans-5,8-di-cis	pet ether	354(4.48)	54-0117-66
	$CHCl_3$	375(4.45)	54-0117-66
$C_{20}H_{26}N_4O_9$			
1,3-Cyclohexanedicarboxylic acid, 4-hydroxy-2,4-dimethyl-6-oxo-, diethyl ester, 2,4-dinitrophenylhydrazone	EtOH	360(4.29)	39-0336-66C
$C_{20}H_{26}N_4O_{13}$			
Uracil, 6-(β-D-glucopyranosylamino)-1,3-dimethyl-5-nitro-, 2,3,4,6-tetraacetate	pH 3	238(4.01),326(3.83)	24-3022-66
	pH 8	238s(4.03),338(3.83)	24-3022-66
$C_{20}H_{26}N_6O_9$			
v-Triazolo[4,5-d]pyrimidine, 5-(dimethylamino)-3-(β-D-glucopyranosyl)-, 2',3',4',6'-tetraacetate	pH -1.9	238(4.43),257(3.77), 344(3.45)	24-3022-66
	pH 5	227(4.24),263(3.89), 340(3.90)	24-3022-66
$C_{20}H_{26}O_2$			
Androst-4-ene-3,17-dione, 19-methylene-	EtOH	238(4.20)	44-0693-66
Estra-1,3,5(10),6-tetraen-17β-ol, 3-methoxy-6-methyl-	n.s.g.	262(3.74),305(3.34)	78-1019-66
Estra-1,3,5(10),8-tetraen-17β-ol, 3-methoxy-6β-methyl-	n.s.g.	281(4.19)	78-1019-66
Gona-1,3,5(10)-trien-17-one, 13β-ethyl-3-methoxy-	EtOH	222(3.86),279(3.28), 286(3.26)	35-3120-66
d-isomer	EtOH	278(3.32),286(3.30)	44-2512-66
8α-Gona-1,3,5(10)-trien-17-one, 13β-ethyl-3-methoxy-	EtOH	278(3.34)	87-0338-66
8α,9α,14β-Gona-1,3,5(10)-trien-17-one, 13β-ethyl-3-methoxy-	MeOH	278(3.32),287(3.30)	88-0801-66
8α,9β,14β-isomer	MeOH	278(3.30),287(3.28)	88-0801-66
Gona-1,3,5(10)-trien-17-one, 3-hydroxy-13β-propyl-	EtOH	281(3.22),287(3.21)	44-2512-66
A-Homoestr-1(10),2,4a-trien-4-one, 17β-hydroxy-17α-methyl-	n.s.g.	232(4.34),367(3.99)	78-0391-66A

Compound	Solvent	$\lambda_{max}(\log \epsilon)$	Ref.
1-Indanone, 2-benzylidenehexahydro-4-hydroxy-7-isopropyl-4-methyl-	EtOH	295(4.48)	78-0219-66A
Naphth[1,8-bc]oxocin-11-ol, 2,3,4,5-tetrahydro-10-isopropyl-2,2,6-trimethyl-	ether	240(4.88),290(3.76), 303(3.67),320(3.39), 335(3.44)	33-1151-66
$C_{20}H_{26}O_2S$			
Estrone, 2-(methylthiomethyl)-	MeOH	288(3.48)	35-5855-66
	MeOH-KOH	293(3.43)	35-5855-66
Estrone, 4-(methylthiomethyl)-	MeOH	287(3.38)	35-5855-66
	MeOH-KOH	292(3.36)	35-5855-66
$C_{20}H_{26}O_3$			
Androsta-1,4-diene-3,17-dione, 2-methoxy-	EtOH	253(4.20)	37-0540-66
Androsta-4,6-dien-3-one, 1α,2α-epoxy-17β-hydroxy-17α-methyl-	MeOH	290(4.31)	73-1363-66
Androsta-1,4,6-trien-3-one, 17β-hydroxy-2-methoxy-	MeOH	220(4.25),283(4.25)	73-1363-66
Cinnamic acid, p-[(3,7-dimethyl-2,6-octadienyl)oxy]-, methyl ester	EtOH	229(3.87),301(4.11), 314(4.16)	12-0451-66
	EtOH	227(4.11),300s(4.38), 310(4.43)	100-0206-66
Estra-p-quinol, 10β-acetate	EtOH	251(4.15)	5-0174-66B
Estra-1,3,5(10),8-tetraen-17β-ol, 2,3-dimethoxy-	EtOH	282(4.03),305(3.96)	87-0027-66
Estra-1,3,5(10)-trien-17-one, 2,3-dimethoxy-	EtOH	286(3.51)	87-0027-66
Fuerstion	C_6H_{12}	253(3.65),445(4.07)	33-1151-66
	EtOH	252(3.64),445(4.09)	33-1151-66
Fuerstionone, dihydro-	ether	256(4.43),350(3.20), 418(3.57)	33-1151-66
D-Homoestra-1,3,5(10)-trien-17a-one, 14α-hydroxy-3-methoxy-	n.s.g.	277(3.2)	88-3585-66
8α,9α-D-Homoestratrien-17a-one, 14β-hydroxy-3-methoxy-	n.s.g.	277(2.4)	88-3585-66
8α,9β-D-Homoestra-1,3,5(10)-trien-17a-one, 14β-hydroxy-3-methoxy-	n.s.g.	277(3.3)	88-3585-66
8β,9β-isomer	n.s.g.	278(3.3)	88-3585-66
8,10,14,16-Octadecatetraen-12-yn-3-one, 1-acetoxy-	ether	229(4.16),235(4.16), 317(4.70),339(4.73)	24-0142-66
Royleanone, 9-dehydro-	EtOH	213(4.21),245s(3.89), 329(3.87),455(2.88)	12-0329-66
$C_{20}H_{26}O_4$			
Estr-1-ene-3,5-dione, 17β-acetoxy-, 10(5→4)-abeo-	EtOH	238(3.84),309(3.81)	33-2218-66
Fuerstioquinone, 7-oxo-	ether	220(4.91),278(3.27), 300s(3.10),403(3.27)	33-1151-66
$C_{20}H_{26}O_5$			
Isodocarpin	n.s.g.	232(3.68)	88-3153-66
2-Phenanthrenecarboxylic acid, 2-ethyl-1,2,3,4,4aβ,9,10,10aβ-octahydro-6,7-dimethoxy-3-oxo-	EtOH	230s(3.88),282(3.60), 287(3.62),291s(3.56)	87-0304-66
9,10-Secoandrosta-1,3,5(10)-triene-9,17-dione, 3,4-dihydroxy-2-methoxy-	EtOH	273(2.95)	37-0540-66
$C_{20}H_{26}O_6$			
Enmein	EtOH	233(3.85)	78-3423-66

Compound	Solvent	$\lambda_{max}(\log \epsilon)$	Ref.
Fastigilin A	EtOH	222(4.26),305(--)	78-1907-66
Fastigilin B	EtOH	222.5(4.35)	78-1907-66
Nodosin	n.s.g.	233(3.90)	88-3153-66
$C_{20}H_{26}O_7$			
Cyclohexaneacetic acid, 3-carboxy-4-[6-carboxy-1-(hydroxymethyl)-5,5-dimethyl-4-oxo-2-cyclohexen-1-yl]-α-methyl-, γ-lactone	EtOH	228(3.91)	78-1659-66
$C_{20}H_{27}BrO$			
Totara-8,11,13-trien-7-one, 6α-bromo-	EtOH	217(3.89),261(3.87), 292s(3.46)	78-2845-66
$C_{20}H_{27}ClO_3$			
Androsta-1,4-dien-3-one, 4-chloro-11β,17β-dihydroxy-17-methyl-	EtOH	245(3.93)	83-1011-66
A-Norandrost-3-en-2-one, 17β-acetoxy-3-chloro-	EtOH	246(4.13)	44-3189-66
$C_{20}H_{27}ClO_4$			
2-Oxaandrost-4-en-3-one, 17β-acetoxy-4-chloro-	EtOH	239(4.05)	44-3189-66
$C_{20}H_{27}ClO_7$			
Radicicol, diol from	EtOH	290(3.56)	12-1265-66
Radicicol, isocoumarin from	EtOH	243(4.55),250(4.69), 265(4.13),282(3.80), 293(3.47),336(3.78)	12-1265-66
$C_{20}H_{27}Cl_2N$			
Quinoline, 2,4-dichloro-3-pentyl-6,8-dipropyl-	EtOH	245(4.65),288(3.51), 315(3.45),330(3.34)	23-2525-66
$C_{20}H_{27}N$			
Naphthalen-1,4-imine, 9-cyclohexyl-1,4-dihydro-1,2,3,4-tetramethyl-	MeOH	276(3.08)	44-2009-66
$C_{20}H_{27}NO$			
Estra-1,3,5(10)-triene, 3-acetamido-	EtOH	208(4.51),247(4.28), 278s(3.29),289(3.11)	5-0174-66B
$C_{20}H_{27}NO_3$			
2H-Benzo[a]quinolizin-2-one, 3-butyl-1,6,7,11b-tetrahydro-9,10-dimethoxy-4-methyl-	EtOH	285(3.69),334(4.17)	44-0797-66
2H-Benzo[a]quinolizin-2-one, 1,6,7,11b-tetrahydro-3-isobutyl-9,10-dimethoxy-4-methyl-	EtOH	286(3.70),333(4.16)	44-0797-66
Taxinonitrile, anhydro-	EtOH	211(3.73)	95-1172-66
$C_{20}H_{27}NO_5$			
Serratinine derivative	EtOH	228(4.02)	88-1537-66
$C_{20}H_{27}N_3O$			
4(3H)-Quinolizinone, 3-cyclohexyl-2-(cyclohexylamino)-	EtOH	225(4.4),270(4.2), 340(3.5)	83-0914-66

Compound	Solvent	λ_{max}(log ε)	Ref.
$C_{20}H_{27}N_3O_3$			
Cyclopentanone, 2-[2-(3,4-dihydro-6-methoxy-1(2H)-naphthylidene)ethyl]-3-hydroxy-2-methyl-, semicarbazone	MeOH	214(4.55),266(4.31)	25-1340-66
$C_{20}H_{27}N_5O_{11}$			
β-D-Galctopyranosylamine, N-[2-(dimethylamino)-5-nitro-4-pyrimidinyl]-, 2,3,4,6-tetraacetate	pH -0.89	237(4.35),275(4.10), 323(3.98)	24-3022-66
	pH 5	226(4.24),280(3.69), 368(4.34)	24-3022-66
β-D-Glucopyranosylamine, N-[2-(dimethylamino)-5-nitro-4-pyrimidinyl]-, 2,3,4,6-tetraacetate	pH -0.89	233(4.35),275(4.08), 323(3.97)	24-3022-66
	pH 4	226(4.24),274(3.81), 367(4.26)	24-3022-66
$C_{20}H_{28}N_2O$			
Androsta-2,4-diene-2-carbonitrile, 3-amino-17β-hydroxy-	n.s.g.	238(4.02),331(3.72)	32-0152-66
Pyrazolo[2,3-d]estr-4-en-17β-ol, 1'-methyl-	EtOH	276(4.00)	88-0205-66
Spiroxyane, 2-deoxy-17α-hydroxy-16α-methyl-	EtOH	243(3.83),296(3.46)	44-1765-66
$C_{20}H_{28}N_2O_2$			
Spiroxyane, 2-deoxy-17α-hydroxy-16α-(hydroxymethyl)-	EtOH	243(3.83),296(3.42)	44-1765-66
$C_{20}H_{28}N_2O_3$			
Sitsirikinediol, dihydro-	MeOH	226(4.53),282(3.85), 290(3.78)	78-0321-66
$C_{20}H_{28}N_2O_4$			
Camptothecin, dodecahydro-	n.s.g.	234(3.67),303(3.87)	35-3888-66
$C_{20}H_{28}N_4O_4S$			
4-Thiazolidinone, 5-(β-acetyl-β-carbethoxyethylidene)-2-(dicyanomethylene)-3-methyl-, triethylammonium salt	dioxan	297(4.15),380(4.22), 459(4.39)	24-0307-66
$C_{20}H_{28}O$			
Abietinal, dehydro-	EtOH	268(2.80),276(2.86)	1-2829-66
Furan, 3-methyl-2-[3-methyl-5-(2,6,6-trimethyl-1-cyclohexen-1-yl)-2,4-pentadienyl]-	EtOH	225(4.18),256(4.21)	77-0600-66
γ-Retinal	EtOH	408(4.76)	39-2154-66C
Totara-8,11,13-trien-7-one	EtOH	218(3.96),254(3.92), 296(3.28)	78-2845-66
$C_{20}H_{28}O_2$			
Androst-4-ene-3,17-dione, 19-methyl-	EtOH	242(4.21)	44-0693-66
Androst-4-en-3-one, 17β-hydroxy-6-methylene-	EtOH	260(3.92)	32-1268-66
1α,5β-Cyclo-10α-androst-3-en-2-one, 17β-hydroxy-4-methyl-	EtOH	234(3.84),261s(3.57)	33-1049-66
Estra-1,3,5(10)-triene-1,17β-diol, 3,4-dimethyl-	EtOH	223s(3.93),287(3.31)	33-1049-66
Estra-1,3,5(10)-trien-17β-ol, 3-methoxy-6β-methyl-	n.s.g.	286(3.46)	78-1019-66

Compound	Solvent	$\lambda_{max}(\log \epsilon)$	Ref.
Gona-1,3,5(10)-triene-3,17β-diol,	EtOH	282(3.32),288(3.29)	35-3120-66
13β-propyl-	EtOH	281(3.36),288(3.31)	44-2512-66
benzene solvate	EtOH	223(3.84),282(3.30), 288(3.26)	35-3120-66
Gona-1,3,5(10)-trien-17-ol, 13-ethyl-3-methoxy-, cis-syn-trans	MeOH	278(3.75),287(3.78)	88-0801-66
Testosterone, 19-methylene-	EtOH	239(4.18)	44-0693-66
Vitamin A acid, 9-demethyl-, methyl ester	n.s.g.	348(4.56)	78-0293-66
Vitamin A acid, 13-demethyl-, methyl ester, all trans	n.s.g.	354(4.62)	78-0293-66
$C_{20}H_{28}O_2S$			
2,4,6,8-Nonatetraenoic acid, 9-(3,4-dihydro-4,4,6-trimethyl-2H-thiopyran-5-yl)-3,7-dimethyl-, methyl ester, all trans	C_6H_{12}	310(4.27),363(4.48)	78-0265-66
$C_{20}H_{28}O_3$			
Androsta-5,16-diene-17-carboxylic acid, 3β-hydroxy-	EtOH	226(3.80)	32-1254-66
Androst-1-ene-3,5-dione, 17β-hydroxy-17α-methyl-, 10(5→4)-abeo-	EtOH	239(3.91),311(3.83)	33-2218-66
5α-Androst-1-ene-3,17-dione, 2-methoxy-	MeOH	264(3.93)	73-1363-66
Androst-4-ene-3,17-dione, 6α-methoxy-	EtOH	242(3.93)	22-2277-66
Androst-1-en-3-one, 4,5-epoxy-17β-hydroxy-17α-methyl-	EtOH	232(3.95)	33-2218-66
5β-Androst-1-en-3-one, 2-formyl-17β-hydroxy-	EtOH	249(3.76)	88-4113-66
	EtOH-NaOH	306(4.03)	88-4113-66
Cinerin I	hexane	220(4.32)	39-0332-66C
Estra-1,3,5(10)-trien-17β-ol, 2,3-dimethoxy-	EtOH	287(3.58)	87-0027-66
Fuerstioquinone	ether	255s(3.27),315(3.04), 404(3.29),583(1.83)	33-1151-66
D-Homoandrost-5-ene-16,17a-dione, 3β-hydroxy-	EtOH-HCl	256(4.23)	22-1222-66
	EtOH	256(4.22)	22-1222-66
	EtOH-KOH	284(4.43)	22-1222-66
	dioxan	246(4.07)	22-1222-66
D-Homoandrost-5-ene-17,17a-dione, 3β-hydroxy-	EtOH	267(3.86)	22-1222-66
	EtOH-KOH	312(3.57)	22-1222-66
Kaur-16-en-19-oic acid, 15-oxo-	EtOH	234(3.92)	12-0861-66
Lambertianic acid	EtOH	200(4.00)(end abs.)	78-0679-66
Mutilintrione	EtOH	303(2.19)	78-0359-66B
Royleanone	$CHCl_3$	278(4.20),283s(4.19), 408(2.63)	12-0329-66
Solidagenone	EtOH	223(4.00)	77-0740-66
$C_{20}H_{28}O_4$			
A-Norandrostane-2,5-dione, 17β-acetoxy-10(5→3)-abeo-	EtOH	310(3.26)	33-2218-66
	EtOH-KOH	310(4.17)	33-2218-66
A-Norandrostane-3,6-dione, 17β-acetoxy-	EtOH	281(3.87)	33-2218-66
	EtOH-KOH	314(4.09)	33-2218-66
A-Norandrostan-2-one, 3α,5α-epoxy-17β-hydroxy-, acetate	EtOH	305(1.65)	33-2218-66
B-Norandrostan-3-one, 4β,5β-epoxy-17β-hydroxy-, acetate	EtOH	300(1.43)	33-2218-66
4-Oxaandrost-5-en-3-one, 17β-acetoxy-	C_6H_{12}	197(3.85)	78-1317-66
	EtOH	198(3.81)	78-1317-66

Compound	Solvent	λ max(log ε)	Ref.
Podocarp-8(14)-ene-7,9-dione,	EtOH	266(4.04)	35-1766-66
2β-acetoxy-8-methyl-	EtOH	266.5(4.03)	35-1766-66
2,3-Seco-A-norandrosta-1,3-dien-6-one,	EtOH	293(3.95)	33-2218-66
17β-acetoxy-3-hydroxy-	EtOH-KOH	310(4.24)	33-2218-66
2,3-Seco-A-norandrosta-3,5-dien-6-one,	EtOH	293(3.89)	33-2218-66
17β-acetoxy-3-hydroxy-	EtOH-KOH	313(4.16)	33-2218-66
Taxininol, anhydro-	EtOH	215(3.72)	95-1172-66
$C_{20}H_{28}O_5$			
2-Oxa-A-norestr-5(10)-en-1-one, 17β-acetoxy-3ξ-methoxy-3ξ-methyl-	MeOH	218(3.85)	44-4255-66
$C_{20}H_{28}O_6$			
Butyric acid, 2-methyl-, 4-ester with 3a,5,6,11a-tetrahydro-4,6-dihydroxy-3,6,10-trimethylcyclodeca[b]furan-2,9(3H,4H)-dione	EtOH	250(3.97)	78-3173-66
Enmein, dehydrotetrahydro-	EtOH	205.5(3.56)	78-1659-66
Enmein, dihydro-	EtOH	295(1.59)	78-3423-66
Helianginone, tetrahydro-	EtOH	246(3.77)	78-3173-66
$C_{20}H_{28}O_7$			
Cyclohexaneacetic acid, 3-carboxy-4-(7,7a-dihydro-6-hydroxy-7,7-dimethyl-1-oxo-3a(6H)-phthalanyl)-α-methyl-	EtOH	205(3.47)	78-1659-66
$C_{20}H_{28}O_8$			
Hexanedioic acid, 3-[2-carboxy-4-(1-carboxyethyl)cyclohexyl]-3-(hydroxymethyl)-2-isopropylidene-, γ-lactone	EtOH	236(3.90)	78-1659-66
$C_{20}H_{29}ClO_3$			
Androst-4-en-3-one, 4-chloro-11β,17β-dihydroxy-17α-methyl-	EtOH	257(4.17)	83-1011-66
$C_{20}H_{29}ClO_4$			
Callicarpone, chlorohydrin	EtOH	265.5(3.97)	88-3519-66
$C_{20}H_{29}N$			
Quinoline, 3-pentyl-6,8-dipropyl-	EtOH	237(4.50),298(3.59), 310(3.57),322(3.53)	23-2525-66
$C_{20}H_{29}NO_2$			
2H-[1]Benzopyrano[4,3-c]pyridin-10-ol, 1,3,4,5-tetrahydro-2,5,5-trimethyl-8-pentyl-	EtOH	225(4.32),280(4.00)	35-3664-66
2,4-Quinolinediol, 3-pentyl-6,8-dipropyl-	EtOH	234(4.73),278(4.00), 290s(3.86),330s(--)	23-2525-66
$C_{20}H_{29}NO_3$			
3-Aza-A-homoestr-4a-en-4-one, 17β-acetoxy-	n.s.g.	220(4.24)	25-1378-66
$C_{20}H_{29}N_5O_9$			
β-D-Glucopyranosylamine, N-[5-amino-2-(dimethylamino)-4-pvrimidinyl]-, 2,3,4,6-tetraacetate	n.s.g.	240(4.43),315(3.57)	24-3022-66

Compound	Solvent	λ max(log ε)	Ref.
$C_{20}H_{30}$			
8,11,13-Totaratriene	EtOH	220(3.84),263(3.37)	78-2845-66
$C_{20}H_{30}N_2O$			
5α-Androst-2-ene-2-carbonitrile, 3-amino-17β-hydroxy-	EtOH	263(4.03)	32-0152-66
$C_{20}H_{30}N_4$			
Tetrapropylammonium salt of 1,3,5-cyclopentadienetricarbonitrile	MeCN	236(4.78),265(4.05)	35-3046-66
$C_{20}H_{30}N_4O$			
1H-Cyclopenta[5,6]naphtho[1,2-g]quinazolin-1-ol, 8,10-diamino-2,3,3a,-3b,4,5,5a,6,11,11a,11b,12,13,13a-tetradecahydro-13a-methyl-	EtOH-HCl	272(3.82)	32-0152-66
	EtOH	283(3.82)	32-0152-66
$C_{20}H_{30}O$			
Abietinal	EtOH	241(4.15)	1-2829-66
Hexadeca-2,4,6,10,14-pentaenal, 3,7,11,15-tetramethyl-	EtOH	341(4.55)	39-2154-66C
8,11,13-Totaratrien-12-ol	EtOH	226(3.88),282(3.58)	78-2845-66
$C_{20}H_{30}O_2$			
Gon-4-en-3-one, 13β-ethyl-17β-hydroxy-6α-methyl-	n.s.g.	240(4.22)	78-1019-66
Gon-4-en-3-one, 13β-ethyl-17β-hydroxy-7α-methyl-	EtOH	242(4.22)	87-0782-66
Gon-4-en-3-one, 17β-hydroxy-13β-propyl-	EtOH	241(4.18)	35-3120-66
	EtOH	240.5(4.24)	44-2512-66
Hexane, 3,4-bis(1-methyl-4-oxo-cyclohex-2-enyl)-	EtOH	232(4.24)	39-1213-66C
Hexane, 3,4-bis(2-methyl-4-oxo-cyclohexenyl)-	EtOH	285(3.10)	39-1213-66C
D-Homoandrost-17-en-16-one, 3β-hydroxy-	EtOH	228(4.02)	22-1222-66
A-Homotestosterone	n.s.g.	235(4.22)	78-0391-66A
Neoabietic acid	EtOH	251(4.32)	12-2403-66
18-Norpregn-13(17)-en-20-one, 3α-hydroxy-	EtOH	257.5(4.13)	78-0541-66B
Ozic acid	n.s.g.	237(4.28)	77-0044-66
Testosterone, 18-methyl-	MeOH	241(4.21)	44-1026-66
Testosterone, 19-methyl-	EtOH	243(4.21)	44-0693-66
$C_{20}H_{30}O_3$			
Androstane-3,5-dione, 17β-hydroxy-17α-methyl-, 10(5→4)-abeo-	EtOH	290(3.93)	33-2218-66
	EtOH-KOH	310(4.13)	33-2218-66
5α-Androstan-3-one, 17β-hydroxy-2-(hydroxymethylene)-	EtOH	284(3.92)	32-1241-66
5β-Amdrostan-3-one, 17β-hydroxy-4-(hydroxymethylene)-	EtOH	280(4.06)	32-1241-66
Androstan-5-one, 17β-hydroxy-6-(hydroxymethylene)-, 10(5→4)-abeo-	EtOH	278(3.88)	33-2218-66
	EtOH-KOH	316(4.23)	33-2218-66
Androst-4-en-3-one, 11β,17β-dihydroxy-17α-methyl-	EtOH	243(4.14)	83-1011-66
Androst-4-en-3-one, 12α,17β-dihydroxy-17α-methyl-	EtOH	242(4.20)	44-2119-66
5α-Androst-1-en-3-one, 17β-hydroxy-2-methoxy-	MeOH	264(3.92)	73-1363-66
Estrane-3,5-dione, 17β-hydroxy-7α,17α-dimethyl-, 10(5→4)-abeo-	EtOH	293(3.97)	33-2218-66

Compound	Solvent	λ_{max}(log ε)	Ref.
Estrane-3,5-dione, 17β-hydroxy-7α,17α-dimethyl-, 10(5→4)-abeo-, ferric complex	EtOH	550(3.12)	33-2218-66
Gon-4-en-3-one, 1β,17β-dihydroxy-13β-propyl-	EtOH	243.5(4.18)	35-3120-66
	0.66N NaOH	243(3.93),303(3.43)	35-3120-66
D-Homoandrostane-16,17a-dione, 3β-hydroxy-	EtOH-HCl	256(4.24)	22-1222-66
	EtOH	256(4.19)	22-1222-66
	EtOH-KOH	284(4.42)	22-1222-66
	dioxan	244(4.05)	22-1222-66
D-Homoandrostane-17,17a-dione, 3β-hydroxy-	EtOH	266(3.80)	22-1222-66
	EtOH-KOH	315(3.63)	22-1222-66
Mutilindione A, hydroxy-	EtOH	303(2.50)	78-0359-66B
Podocarp-8(14)-en-7-one, 2β-acetoxy-8-methyl-	EtOH	249(4.24)	35-1766-66
$C_{20}H_{30}O_4$			
Agathic acid	EtOH	218(4.12)	12-2403-66
2-Chromanvaleric acid, 6-hydroxy-α,2,5,7,8-pentamethyl-	EtOH	292.0(3.55)	73-2434-66
$C_{20}H_{30}O_5$			
Prosta-5,8(12),13-trienoic acid, 15,19-dihydroxy-9-oxo-	EtOH	278(4.3)	37-0257-66
Prosta-5,10,13-trienoic acid, 15,19-dihydroxy-9-oxo-	EtOH	217(4.0)	37-0257-66
Taxinolactone, isopropylidenedihydro-, product with EtOH-NaOH	EtOH	232(3.76)	94-0502-66
$C_{20}H_{30}O_6$			
δ-Cesalpin	EtOH	215(3.92)	32-0662-66
$C_{20}H_{30}O_9$			
Roridinic acid, di-O-acetyl-, dimethyl ester	C_6H_{12}	254(4.38)	33-2527-66
	EtOH	256(4.30)	33-2527-66
$C_{20}H_{31}NO$			
2,5-Cyclohexadien-1-one, 2,6-di-tert-butyl-4-(piperidinomethylidene)-	hexane	270(3.96)	70-0888-66
$C_{20}H_{31}NO_3$			
4-Azaandrost-5-en-3-one, 17β-acetoxy-4-methyl-	C_6H_{12}	234(4.18)	78-1317-66
	H_2O	236(4.13)	78-1317-66
	EtOH	235(4.11)	78-1317-66
$C_{20}H_{32}$			
2,6,8,10,14-Hexadecapentaene, 2,6,11,15-tetramethyl-	EtOH	273s(--),286(4.53), 297s(--)	39-2154-66C
Rimua-1(10),5-diene	EtOH	232(4.16),239(4.21), 247s(4.02)	39-1737-66C
$C_{20}H_{32}NO_2$			
Phenoxy, 2,6-di-tert-butyl-4-pivaloyl-, O-methyloxime	C_6H_{12}	265(3.88),305(3.77), 377(3.83),392(2.83), 600(2.68)	35-5284-66
$C_{20}H_{32}N_2O_4$			
Leucine, N-(N,N-dimethyl-D-allo-isoleucyl)-3-phenoxy-	EtOH	225(4.02),275(3.13)	6-0083-66

Compound	Solvent	$\lambda_{max}(\log \epsilon)$	Ref.
$C_{20}H_{32}N_2Si_2$			
Aniline, 4,4'-(tetramethyldisilanylene)bis[N,N-dimethyl-	C_6H_{12}	274(4.70)	101-0102-66B
$C_{20}H_{32}O$			
Abietinol	EtOH	234(4.25),241(4.28)	1-2829-66
2,4,6,10,14-Hexadecapentaen-1-ol, 3,7,11,15-tetramethyl-	EtOH	280(4.58)	39-2154-66C
2,6,10,14-Hexadecatetraenal, 3,7,11,15-tetramethyl-	hexane	233(4.16)	39-2154-66C
Rimu-5(10)-en-6-one	EtOH	249.5(4.06)	39-1737-66C
$C_{20}H_{32}O_3$			
Labd-7-en-15-oic acid, 6-oxo-	n.s.g.	240(4.07)	12-2133-66
$C_{20}H_{32}O_4$			
Prosta-8(12),13-dienoic acid, 15-hydroxy-9-oxo-	EtOH	278(4.43)	37-0257-66
Prosta-10,13-dienoic acid, 15-hydroxy-9-oxo-	EtOH	217(4.01)	37-0257-66
Taxa-4(16),11-diene-5α,9α,10β,13α-tetrol	n.s.g.	227.0(3.85)	77-0923-66
$C_{20}H_{32}O_5$			
Prosta-8(12),13-dienoic acid, 15,19-dihydroxy-9-oxo-	EtOH	278(4.32)	37-0257-66
Prosta-10,13-dienoic acid, 15,19-dihydroxy-9-oxo-	EtOH	217(4.04)	37-0257-66
Taxicin I, 5-deoxydihydro-	EtOH	283(3.76)	39-1933-66C
$C_{20}H_{33}NO$			
2,6,8,12-Hexadecatetraenamide, N-isobutyl-, all cis	EtOH	232(4.34)	12-1207-66
2-trans	EtOH	230.5(4.40)	12-1207-66
12-trans	EtOH	229(4.34)	12-1207-66
2,12-di-trans-6-cis	EtOH	228.5(4.26)	12-1207-66
$C_{20}H_{33}NO_2$			
Pivalophenone, 3',5'-di-tert-butyl-4'-hydroxy-, O-methyloxime	C_6H_{12}	272(3.18),278(3.15)	35-5284-66
$C_{20}H_{34}O$			
Abienol	EtOH	238(4.30)	12-1535-66
Communic acid, methyl ester, trans	EtOH	232(4.44)	12-1535-66
α-Ocimene, cis	EtOH	234.5(4.34)	12-1535-66
α-Ocimene, trans	EtOH	231(4.44)	12-1535-66
β-Ocimene, cis	EtOH	237.5(4.32)	12-1535-66
β-Ocimene, trans	EtOH	232(4.44)	12-1535-66
$C_{20}H_{34}O_2$			
Rimuan-6-one, 5β-hydroxy-	EtOH	305(1.68)	39-1737-66C
$C_{20}H_{34}O_4$			
Podocarp-8-ene-7,11,13-triol, 13-(1-hydroxy-1-methylethyl)-	EtOH	210(3.60)	88-3519-66
$C_{20}H_{34}Si_4$			
Tetrasilane, 1,1,2,2,3,3,4,4-octamethyl-1,4-diphenyl-	C_6H_{12}	250.5(4.33)	101-0102-66B

Compound	Solvent	λ_{max}(log ε)	Ref.
$C_{20}H_{36}N_2O$ [$\Delta^{2,3'}$-Bipyrrolidin]-2'-one, 1,1'-dihexyl-	MeOH	204(4.1),298(4.6)	24-0724-66
$C_{20}H_{37}NO_3$ Acetamide, N-(1-hydroxymethyl)-2-oxo-3-heptadecenyl]-	EtOH	230(4.16)	35-3643-66
$C_{20}H_{38}N_2O$ [2,3'-Bipyrrolidin]-2'-one, 1,1'-dihexyl-	MeOH	207(3.9)	24-1923-66
$C_{20}H_{38}O$ Phytenal	CH_2Cl_2	238(4.05)	39-2144-66C

Compound	Solvent	λ_{max}(log ε)	Ref.
$C_{21}H_3F_{15}O_2$			
1,3-Dioxolane, 2,4,5-tris(pentafluorophenyl)-	EtOH	264(3.188)	70-0337-66
$C_{21}H_{10}Cl_2O_2S$			
Benzo[b]thiophen-3(2H)-one, 6-chloro-2-(8-chloro-2-oxo-1-acenaphthenylidene)-4-methyl-	PhCl	302(--),475(3.88), 508(3.78)	65-1055-66
	H_2SO_4	610(--)	65-1055-66
$C_{21}H_{11}Cl_2N_5$			
1,3,5-Benzenetricarbonitrile, 2,4-dichloro-6-(diphenylhydrazino)-	n.s.g.	653(3.57)	27-0303-66
$C_{21}H_{11}N_5$			
Tripyrido[2,3-a:3',2'-c:2",3"-h]phenazine	EtOH	231(4.48),244(4.51), 266(4.55),307(4.64), 314(4.62),370(4.04), 390(4.07)	44-3384-66
$C_{21}H_{12}N_4$			
4a,5,10,15-Tetraazabenzo[a]naphth[1,2,3-de]anthracene	EtOH-HCl	245(4.70),309(4.44), 360(3.82),377(3.97), 438(4.28),463(4.48)	39-1245-66C
	EtOH	235(4.60),251(4.57), 266(4.58),291(4.40), 412(4.16),434(4.17)	39-1245-66C
$C_{21}H_{12}N_4O_3$			
7H-5,7a,12-Triazabenz[a]anthracen-7-one, 6-(o-nitrophenyl)-	$CHCl_3$	262(4.66),305(4.17), 363(4.24),382(4.36), 398(4.09)	39-1245-66C
$C_{21}H_{12}O_8S$			
[18]Annulene-5,11,18-tricarboxylic acid, 1,4-epithio-7,10;13,16-diepoxy-	EtOH	212(4.27),250(4.26), 333(4.49),440(3.65)	12-1461-66
[18]Annulene-6,11,18-tricarboxylic acid, 1,4-epithio-7,10;13,16-diepoxy-	EtOH	217(4.18),255(4.08), 324(4.45),440(3.85)	12-1461-66
$C_{21}H_{12}O_9$			
[18]Annulene-5,11,18-tricarboxylic acid, 1,4;7,10;13,16-triepoxy-	EtOH	243(4.18),324(4.46)	12-1221-66
$C_{21}H_{13}BrN_4O_4$			
Pyrazole, 4-bromo-1-(2,4-dinitrophenyl)-3,5-diphenyl-	EtOH	236(4.53),338(3.96)	22-3744-66
$C_{21}H_{13}ClO_2$			
5H-Dibenzo[a,d]cycloheptene-5,10(11H)-dione, 11-(p-chlorophenyl)-	$CHCl_3$	275(4.18),310(3.78), 375(2.45)	78-0141-66B
$C_{21}H_{13}N$			
Benzo[h]naphtho[1,2-f]quinoline	C_6H_{12}	212(4.38),232(4.32), 275(4.82),296(4.39), 318(3.99),336(3.88), 354(3.25),369(3.15)	78-0287-66A

Compound	Solvent	λ_{max}(log ϵ)	Ref.
Benzo[h]naphtho[2,1-f]quinoline	C_6H_{12}	217(4.41),237(4.43), 276(4.79),285(4.76), 297(4.31),321(4.02), 355(3.21),370(2.48)	78-0287-66A
Phenanthro[9,10-f]quinoline	C_6H_{12}	257(4.61),278(4.61), 288(4.59),320(3.87), 355(2.97),369(2.74)	78-0287-66A
Phenanthro[9,10-h]quinoline	C_6H_{12}	254(4.66),266(4.63), 278(4.63),288(4.62), 318(3.87),328(3.83), 335(3.82),351(3.63), 369(3.42)	78-0287-66A
$C_{21}H_{13}N_5$ Pyrido[2,3-f]quinoxaline, 2,3-di-2-pyridyl-	EtOH	242(4.48),292(4.56), 346(3.94),363(3.92)	44-3384-66
$C_{21}H_{13}N_5O_4S_4$ [2,4'-Bithiazole]-4-carboxylic acid, 2',2'''-(2,6-pyridinediyl)bis-, dimethyl ester	EtOH	232s(4.73),268s(4.49), 298(4.57),314s(4.44), 335s(4.20),355(3.78)	39-1354-66C
	10N HCl	258s(4.40),320(4.55), 382(4.32)	39-1354-66C
$C_{21}H_{14}$ Naphtho[2',1':1,2]fluorene	EtOH	272(4.80),282(4.88), 297(4.62),316(4.44), 332(3.16),338(2.94), 347(3.29),356(2.88), 366(3.32)	24-3298-66
Perylene, 1-methyl-	n.s.g.	248(4.69),256(4.44), 264s(4.92),382(4.49), 404(4.16),427(4.01)	88-3801-66
$C_{21}H_{14}D_2O$ 2-Indanone-1,3-d_2, 1,3-diphenyl-	CH_2Cl_2	262(4.05),269(4.06), 277(4.02),296(2.56), 305(2.64),316(2.62), 327(2.36)	5-0044-66C
$C_{21}H_{14}N_2O$ 2,5-Cyclohexadienone, 4-(4,5-diphenyl-2H-imidazol-2-ylidene)-	$CHCl_3$	440(4.64)	89-0311-66
$C_{21}H_{14}N_2O_3S$ Barbituric acid, 4-(2,6-diphenyl-4H-thiopyran-4-ylidene)-	DMF	267(4.35),342(4.10), 456(4.53)	33-2046-66
$C_{21}H_{14}N_2O_4$ 6-(3-Dimethylaminopropyl)morphanthridine hydrogen oxalate	EtOH	223s(4.3),244(3.93), 311(3.57)	4-0206-66
$C_{21}H_{14}N_4O$ 7H-5,7a,12-Triazabenz[a]anthracene, 6-(o-aminophenyl)-7-oxo-	$CHCl_3$	290(4.58),356(3.99), 384(4.06),400(4.00)	39-1245-66C
$C_{21}H_{14}N_4O_2$ Cinchoninic acid, 2-s-triazolo[1,5-a]quinolin-2-yl-, methyl ester	EtOH	338(4.07)	94-0512-66

Compound	Solvent	λ_{max}(log ϵ)	Ref.
$C_{21}H_{14}N_4O_9$			
Flavone, 3-amino-, picrate	EtOH	240(4.50),363(4.42)	5-0225-66C
$C_{21}H_{14}O$			
Anthracene, 9-benzoyl-	MeOH	209(4.29),219(4.23), 247(5.00),253(5.12), 311(3.21),317(3.26), 334(3.50),346(3.73), 363(3.89),383(3.85)	73-0835-66
5H-Dibenzo[a,d]cyclohepten-5-one, 10-phenyl-	EtOH	258(4.49),312(4.04), 345s(3.60)	78-0141-66B
Phenanthrene, 1-benzoyl-	MeOH	212(4.50),244s(4.71), 252(4.80),293s(3.97), 308s(3.76),331s(3.47), 340s(3.41),358(3.33), 376(2.84)	73-0835-66
$C_{21}H_{14}O_2$			
4-Hydroxy-1,3-diphenyl-2-benzo-pyrylium hydroxide, inner salt	benzene	390s(4.14),403(4.17), 544(4.43)	35-4942-66
	C_6H_{12}	236(4.32),285(4.47), 392(4.11),406(4.15), 544(4.41)	35-4942-66
	EtOH	378(4.09),525(4.33)	35-4942-66
$C_{21}H_{14}O_2S$			
Spiro[2H-oxeto[2,3-b]benzofuran-2,9'-thioxanthene], 2a,7a-dihydro-	EtOH	265(5.40),339(3.38)	24-1723-66
$C_{21}H_{14}O_2S_2$			
1-Naphthylideneacetic acid, 4-(1,3-dithiolan-2-ylidene)-1,4-dihydro-2-hydroxy-α-phenyl-, γ-lactone	$CHCl_3$	470(4.40)	89-0517-66
	CF_3COOH	482(4.04)	89-0517-66
$C_{21}H_{14}O_4$			
5-Benzofuranacrylic acid, 6-hydroxy-α-inden-2-yl-7-methoxy-, δ-lactone	EtOH	250(4.42),278(4.15), 370(4.43)	24-0088-66
$C_{21}H_{14}O_9$			
Furan, 3-methyl-2,5-bis(3-carbomethoxy-1,4-benzoquinon-2-yl)-	EtOH	236(4.38),270s(4.29), 485(3.94)	33-1794-66
$C_{21}H_{15}^+$			
Diphenyl(phenylethynyl)carbonium ion	FSO_3H-SbF_5	449(4.43),504(4.57)	35-1488-66
$C_{21}H_{15}ClN_4O$			
Pyrazole-4,5-dione, 1,3-diphenyl-, 4-[(o-chlorophenyl)hydrazone]	hexane	266(4.36),404(4.31)	44-1722-66
	EtOH	266(4.61),407(4.56)	44-1722-66
	$CHCl_3$	268(4.42),409(4.39)	44-1722-66
Pyrazole-4,5-dione, 1,3-diphenyl-, 4-[(p-chlorophenyl)hydrazone]	hexane	265(4.42),408(4.42)	44-1722-66
	EtOH	265(4.51),408(4.51)	44-1722-66
	$CHCl_3$	266(4.50),411(4.53)	44-1722-66
$C_{21}H_{15}ClO_4$			
1H-Cyclobut[a]indene-1-carboxylic acid, 7a-chloro-2,2a,7,7a-tetra-hydro-2-(6-hydroxy-7-methoxy-5-benzofuranyl)-, δ-lactone	EtOH	256(4.11),282s(3.34)	24-0088-66

Compound	Solvent	$\lambda_{max}(\log \epsilon)$	Ref.
$C_{21}H_{15}F_6N_2P$			
Phosphorane, triphenyl[[2,2,2-trifluoro-1-(trifluoromethyl)-ethylidene]hydrazono]-	EtOH	221(4.43),268(4.22), 273(4.22),280(4.21)	35-3617-66
$C_{21}H_{15}N$			
Quinoline, 2-(4-biphenylyl)-	C_6H_{12}	275(4.59),297s(4.42), 329(4.31),342s(4.19)	5-0107-66I
	MeOH	275(4.56),295s(4.34), 328(4.30),340s(4.19)	5-0107-66I
	$CHCl_3$	278(4.54),300s(4.33), 343s(4.16)	5-0107-66I
	dioxan	277(4.58),300s(4.41), 314s(4.36),330(4.35), 342s(4.26)	5-0107-66I
Quinoline, 6-(4-biphenylyl)-	C_6H_{12}	268(4.59),292s(4.37), 329s(3.92)	5-0107-66I
	MeOH	272(4.65),300s(4.38), 330s(4.04)	5-0107-66I
	$CHCl_3$	273(4.59),298s(4.34), 330s(3.96)	5-0107-66I
	dioxan	272(4.61),302s(4.40), 330s(4.03)	5-0107-66I
$C_{21}H_{15}NO_4$			
Benz[cd]indole-5-carboxylic acid, N-acetyl-1,2-dihydro-2-oxo-4-phenyl-, methyl ester	EtOH	206(4.62),222s(4.56), 251(4.37),298s(3.79), 329s(3.74),347s(3.78), 358(3.78)	39-1028-66C
$C_{21}H_{15}NS$			
Thiazole, 2-(4-biphenylyl)-4-phenyl-	C_6H_{12}	268(4.39),326(4.2)	22-2857-66
Thiazole, 4-(4-biphenylyl)-2-phenyl-	C_6H_{12}	292.5(4.41)	22-2857-66
$C_{21}H_{15}N_3O$			
as-Triazine, 3-phenoxy-5,6-diphenyl-	EtOH	258(4.19),327(3.9)	44-3914-66
$C_{21}H_{15}N_3O_2$			
as-Triazine, 3-phenoxy-5,6-diphenyl-, 2-oxide	EtOH	250(4.33),343(3.92)	44-3914-66
$C_{21}H_{16}$			
Fluorene, 9-β-phenylethylidene-	EtOH	230(4.64),248(4.51), 257(4.66),283(4.21), 299(4.11),313(4.12)	39-1810-66C
Fluorene, 9-styryl-	EtOH	220(4.38),261(4.53), 294(3.79),304(3.84)	39-1810-66C
Indene, 1,3-diphenyl-	dioxan	268(3.8)	24-0392-66
Indene, 1,3-diphenyl-, K salt	DMSO	375(4.3),443(4.41)	24-0392-66
Spiro[cyclopropane-1,9'-fluorene], 1-phenyl-	EtOH	211(4.77),270(4.25), 293(3.93),304(3.93)	39-1810-66C
$C_{21}H_{16}N_2$			
13H-Benzo[a]pyrido[3,4-h]carbazole, 8,12-dimethyl-	85% EtOH-HCl	233(4.42),246(4.39), 264s(--),274(4.48), 294s(--),309(4.80), 330(4.45),351(4.11)	87-0237-66

Compound	Solvent	λ max(log ε)	Ref.
13H-Benzo[a]pyrido[3,4-h]carbazole, 8,12-dimethyl- (cont.)	EtOH	233(4.42),238s(4.40), 283(4.99),290s(4.91), 314(4.15),326(4.20), 337(4.12),352(3.81), 382(3.77),397(3.74)	87-0237-66
Imidazole, 2,4,5-triphenyl-	EtOH	302(4.42)	35-3825-66
$C_{21}H_{16}N_2O$			
2-Indolinone, 1-methyl-3-(α-4-pyridylbenzylidene)-	EtOH	258(4.31),323(3.91), 410(3.18)	44-0077-66
$C_{21}H_{16}N_2O_2$			
5-Norbornene-2,3-dicarboximide, 7-(α-2-pyridylbenzylidene)-, endo	MeOH	242(4.19),276s(3.82)	44-2141-66
Ourouparine (as sulfate)	n.s.g.	256(4.5),284(4.0), 359(4.5)	28D-1141-66A
2-Pyridinemethanol,α-phenyl-α-[6-phenyl-6-(2-pyridyl)-2-fulvenyl]-, cis	MeOH	240(4.17),262s(4.10), 268s(4.09),324(4.39)	44-2141-66
$C_{21}H_{16}N_2O_3$			
Ourouparine oxide	EtOH	217(4.58),251(4.26), 257s(4.26),298(3.86), 366(4.55),384(4.51)	28D-1141-66A
$C_{21}H_{16}N_2O_4$			
Glutaconimide, 4-ethyl-4-phenyl-2-phthalimido-	EtOH	289s(3.43)	7-1519-66
$C_{21}H_{16}N_4$			
Benzaldehyde, (3-phenyl-2-quinoxalinyl)hydrazone	EtOH	239(4.31),313(4.33), 399(4.06)	95-0622-66
	EtOH-NaOEt	262(4.25),387(4.14), 480(4.05)	95-0622-66
Benzimidazo[1,2-c]quinazoline, 6-(o-aminophenyl)methyl-	EtOH	230(4.50),270s(4.61), 277(4.68),304(4.06), 320(4.04),336s(3.86)	4-0289-66
$C_{21}H_{16}N_4O$			
Pyrazole-4,5-dione, 1,3-diphenyl-, 4-(phenylhydrazone)	$CHCl_3$	406(4.36)	22-2990-66
as-Triazine, 3-anilino-5,6-diphenyl-, 2-oxide	EtOH	237(4.24),293(4.50), 384(3.43)	44-3914-66
$C_{21}H_{16}N_4O_4$			
2-Pyrazoline, 1-(2,4-dinitrophenyl)-3,5-diphenyl-	$CHCl_3$	415(4.44)	22-0610-66
$C_{21}H_{16}O$			
5H-Dibenzo[a,d]cyclohepten-5-one, 10,11-dihydro-10-phenyl-	EtOH	252(4.14),290s(3.36)	78-0141-66B
Indan, 1-benzylidene-3-furfurylidene-	EtOH	295(4.52),335(4.48), 349(4.53),368s(4.36)	5-0057-66H
Indene, 3-benzyl-1-furfurylidene-	EtOH	249(4.02),291(3.99), 345s(4.38),359(4.53), 371(4.52)	5-0057-66H
$C_{21}H_{16}O_2$			
2H-Oxeto[2,3-b]benzofuran, 2a,7a-dihydro-2,2-diphenyl-	EtOH	260(3.31),278(3.53), 286s(3.47)	24-1723-66

Compound	Solvent	λ_{max}(log ϵ)	Ref.
Phenol, o-(2,2-diphenylvinyl)-, formate	n.s.g.	230(4.31),294(4.20)	24-1723-66
Tropone, diphenylketene adduct	EtOH	290(3.34)	78-1809-66
$C_{21}H_{16}O_4$			
Tetrangulol, di-O-methyl-	MeOH	221(4.67),302(4.48), 350(3.79),380(3.78)	44-2920-66
$C_{21}H_{17}ClO_2$			
Phenol, p-[3-(p-chlorophenyl)-4-chromanyl]-, cis	MeOH	201(5.03),224s(4.60), 276(3.72),284(3.65)	87-0516-66
trans	MeOH	201(4.90),220s(4.45), 276(3.66),284s(3.59)	87-0516-66
$C_{21}H_{17}Cl_2N_5O_2$			
Acetanilide, 2,2-bis[(p-chlorophenyl)-azo]-2'-hydroxy-N-methyl-	EtOH	438(4.27)	22-0400-66
	EtOH-KOH	515(4.35)	22-0400-66
$C_{21}H_{17}N$			
Isoindole, 2-methyl-1,3-diphenyl-	EtOH	228(4.47),268s(4.13), 276(4.20),332(4.01), 371(4.27)	78-1011-66
$C_{21}H_{17}NO$			
Indole, 2-(p-methoxyphenyl)-3-phenyl-	EtOH	253(4.45),309(4.30)	87-0527-66
Indole, 3-(p-methoxyphenyl)-2-phenyl-	EtOH	249(4.45),308(4.25)	87-0527-66
2-Indolinone, 5-methyl-3,3-diphenyl-	$CHCl_3$	260(3.75),283(3.27)	44-1637-66
$C_{21}H_{17}NO_2$			
Acridine, 2,7-dimethoxy-9-phenyl-	EtOH	229s(4.27),261(4.74), 358s(4.01),376(4.22)	1-1631-66
$C_{21}H_{17}N_3O$			
Benzaldehyde, o-amino-, bisanhydro trimer	MeOH	241(4.16),285(3.36), 304s(3.15)	23-2323-66
	EtOH	240(4.23),286(3.50), 300s(3.33)	39-0956-66B
$C_{21}H_{17}N_3O_2S$			
2H-Benz[h]pyrazolo[4,3-c]quinoline, 10,11-dihydro-10-(p-tolylsulfonyl)-	EtOH	227(4.48),254(4.64), 305(3.90)	54-0681-66
$C_{21}H_{17}N_3O_4$			
Indazole, 6-methoxy-3-(p-methoxy-phenyl)-1-(p-nitrophenyl)-	EtOH	241(4.28),288(4.08), 367(4.12)	78-3131-66
$C_{21}H_{17}N_3O_5$			
Benzanilide, 4-nitro-4'-(p-nitro-phenethyl)-	EtOH	272(5.55),330s(4.02)	65-0214-66
$C_{21}H_{18}$			
Fluorene, 9-phenethyl-	EtOH	211(4.68),230(3.88), 264(4.25),267(4.25), 291(3.76),301(3.94)	39-1810-66C
$C_{21}H_{18}BN$			
Pyridine, compound with (3-buten-1-ynyl)diphenylborane	EtOH	228(4.23)	22-1981-66

Compound	Solvent	$\lambda_{max}(\log \varepsilon)$	Ref.
$C_{21}H_{18}IP$			
Methyldiphenyl(phenylethynyl)-phosphonium iodide	EtOH	237(4.23),248(4.24), 259(4.34),268(4.26), 290(3.36)	22-1002-66
$C_{21}H_{18}N_2$			
13H-Benzo[a]pyrido[3,4-h]carbazole, 10,11-dihydro-8,12-dimethyl-	85% EtOH-HCl	226(4.29),239(4.24), 246(4.19),273(4.84), 288s(4.43),308(4.30), 319(4.39),332(4.62), 347(4.29),366(4.24)	87-0237-66
	EtOH	231(4.34),249(4.22), 255s(4.28),279(4.89), 313s(4.39),328s(4.24), 353(3.80)	87-0237-66
2-Pyrazoline, 1,3,5-triphenyl-	MeOH	356(3.93)	89-0699-66
$C_{21}H_{18}N_2O$			
1,2-Diazetidin-3-one, 1,2-diphenyl-4-p-tolyl-	CH_2Cl_2	267(4.11)	88-5245-66
$C_{21}H_{18}N_2O_2$			
1,2-Diazetidin-3-one, 4-(m-methoxy-phenyl)-1,2-diphenyl-	CH_2Cl_2	270(4.08)	88-5245-66
1,2-Diazetidin-3-one, 4-(p-methoxy-phenyl)-1,2-diphenyl-	CH_2Cl_2	268(4.16)	88-5245-66
$C_{21}H_{18}N_2O_3$			
Benzanilide, 4'-(p-nitrophenethyl)-	EtOH	272(5.50),330s(3.36)	65-0214-66
$C_{21}H_{18}N_4O_3$			
Lumazine, 8-(2-hydroxyethyl)-3-methyl-6,7-diphenyl-	pH -2.7	260(4.13),273(4.14), 399(4.03)	24-3503-66
	pH 3	236(4.27),285(4.20), 357(4.06)	24-3503-66
	pH 6	272(4.34),429(4.13)	24-3503-66
	pH 7	272(4.35),430(4.15)	24-3503-66
	pH 11	246(4.26),282s(4.04), 369(4.15)	24-3503-66
$C_{21}H_{18}O$			
Chroman, 2,2-diphenyl-	n.s.g.	277(3.44),284(3.46)	22-1974-66
Fluoren-9-ol, 9-phenethyl-	EtOH	228(4.37),236(4.27), 276(4.09),307(3.54)	39-1810-66C
$C_{21}H_{18}O_2$			
Fluorene, 3,6-dimethoxy-9-phenyl-	n.s.g.	222(4.6),243(4.33), 267(3.90),322(4.19)	39-1050-66C
$C_{21}H_{18}O_4$			
4H-Furo[2,3-h]benzopyran-4-one, 8-iso-propyl-3-(o-methoxyphenyl)-	EtOH	225(4.34),275(4.04), 310(3.88)	39-0749-66C
$C_{21}H_{18}O_7$			
Isolisetin	EtOH	260(4.56),285(4.31), 337(4.09)	78-0333-66A
Lisetin	EtOH	258(4.60),284(4.36), 338(4.15)	78-0333-66A

Compound	Solvent	$\lambda_{max}(\log \epsilon)$	Ref.
$C_{21}H_{19}B$			
Borane, tri-p-tolyl-	isooctane	235(4.18),249(4.30), 282(4.28),297(4.52)	46-0611-66
$C_{21}H_{19}ClN_2O_4$			
4,5,7,8-Tetramethyl[1,4]diazepino-[7,1,2-cd:5,4,3-c'd']diindolizinium perchlorate	MeCN	389(4.60),860(3.44)	39-0483-66B
$C_{21}H_{19}ClN_2O_6$			
8,9-Dihydro-2,3-dimethoxy-14H-benz-[a]indolo[3,2-h]quinolizin-7-ium perchlorate	MeOH	254(4.21),266(4.23), 284(4.34),324(4.07), 350(4.09),403s(4.16), 415(4.20)	4-0395-66
$C_{21}H_{19}ClO_5$			
5,6,7,8-Tetrahydro-1,3-diphenyl-2-benzopyrylium perchlorate	$MeOH-HClO_4$	248(4.26),344(4.34)	65-1728-66
$C_{21}H_{19}FN_2OS$			
4(1H)-Quinazolinethione, 1-(p-fluoro-phenyl)-5,6,7,8-tetrahydro-2-(p-methoxyphenyl)-	MeOH	230(4.09),274s(3.91), 288s(4.01),297s(4.06), 342(4.43)	44-0722-66
$C_{21}H_{19}FN_2S$			
4(1H)-Quinazolinethione, 1-(p-fluoro-phenyl)-5,6,7,8-tetrahydro-2-p-tolyl-	MeOH	258(3.92),276s(3.89), 342(4.42)	44-0722-66
$C_{21}H_{19}N$			
Naphtho[1,2-f]quinoline, 10-isopropyl-6-methyl-	C_6H_{12}	215(4.52),231(4.41), 283(4.74),299(4.28), 318(3.94),358(3.12), 376(3.14)	78-0181-66B
$C_{21}H_{19}NO$			
Cyclopropanemethanol, 2-(2-pyridyl)-α,α-diphenyl-, trans	EtOH-HCl	278(4.03)	44-0399-66
	EtOH-NaOH	270(3.74)	44-0399-66
Cyclopropanemethanol, 2-(4-pyridyl)-α,α-diphenyl-, cis	EtOH-HCl	258(4.07)	44-0399-66
	EtOH-NaOH	259(3.29)	44-0399-66
trans	EtOH-HCl	257(4.21)	44-0399-66
	EtOH-NaOH	258(3.44)	44-0399-66
$C_{21}H_{19}NO_2$			
Aniline, N-[bis(p-methoxyphenyl)-methylene]-	C_6H_{12}	222(4.44),276(4.41), 332(3.64)	35-2775-66
$C_{21}H_{19}NO_4$			
Nitidine, dihydro-	EtOH	230(4.59),282(4.52), 315(4.25)	95-0631-66
$C_{21}H_{19}N_4$			
Verdazyl, p-methyl-	n.s.g.	398(3.88),710(3.58)	88-4103-66
Verdazyl, p-methyl-, cation	n.s.g.	255(4.29),323(4.09), 560(3.61)	88-4103-66
$C_{21}H_{19}N_4O$			
Verdazyl, p-methoxy-	n.s.g.	392(3.89),718(3.56)	88-4103-66
Verdazyl, p-methoxy-, cation	n.s.g.	267(4.28),331(4.12), 580(3.48)	88-4103-66

Compound	Solvent	λ_{max}(log ϵ)	Ref.
$C_{21}H_{19}N_5O_2$			
Acetanilide, 2'-hydroxy-N-methyl-2,2-bis(phenylazo)-	EtOH	428(4.35)	22-0400-66
	EtOH-KOH	504(4.60)	22-0400-66
$C_{21}H_{20}AsClPd$			
Palladium, chloro-π-allyl-(triphenylarsine)-	$CHCl_3$	243(4.35),324(2.50)	101-0578-66B
$C_{21}H_{20}ClNO_4$			
Epicryptopirubine chloride	n.s.g.	210s(4.27),225(4.37),242(4.50),275s(4.16),368(4.32)	88-3975-66
$C_{21}H_{20}ClPPd$			
Palladium, chloro-π-allyl-(triphenylphosphine)	$CHCl_3$	243(4.30),323(3.60)	101-0578-66B
$C_{21}H_{20}Cl_2N_2O$			
1-Pyrrolin-5-one, 3,3-dichloro-4,4-diphenyl-2-piperidino-	EtOH	265(4.26)	28C-1276-66A
$C_{21}H_{20}N_2$			
6H-[1,4]Diazepino[7,1,2-cd:5,4,3-c'd']diindolizine, 4,5,7,8-tetramethyl-	MeCN	514(3.82)	39-0483-66B
	HOAc	560(3.81)	39-0483-66B
$C_{21}H_{20}N_2O$			
Acetamide, N-[2-(10-methyl-11H-benzo-[a]carbazol-9-yl)ethyl]-	EtOH	229(4.46),246(4.61),255(4.66),280(4.67),307(4.35),322s(3.88),337(3.84),353(3.84)	87-0237-66
$C_{21}H_{20}N_2O_2$			
Acetic acid, benzylidene(3-methyl-4-phenylfurfuryl)hydrazide	EtOH	234(4.26),288(4.53)	44-0052-66
Benz[g]indolo[2,3-a]quinolizine-1-carboxylic acid, 5,7,8,13,13b,4-hexahydro-, methyl ester	EtOH	221(4.66),273(3.93),281(3.95),289(3.87)	28D-1141-66A
$C_{21}H_{20}N_2O_6$			
3H-Phenoxazine-1,9-dicarboxylic acid, 2-(1-azetidinyl)-4,6-dimethyl-3-oxo-, dimethyl ester	MeOH	230(4.39),252(4.50),434(4.40),454(4.45)	44-2564-66
$C_{21}H_{20}N_2O_7$			
5H-Oxazolo[4,5-b]phenoxazine-4,6-dicarboxylic acid, 2-(methoxy-methyl)-9,11-dimethyl-, dimethyl ester	MeOH	226(4.59),250(4.54)	44-2564-66
$C_{21}H_{20}N_2O_8$			
4H-Pyrido[2,1-a]phthalazine-1,2,3,4-tetracarboxylic acid, 7-methyl-, tetramethyl ester	MeOH	225(4.44),255s(3.92),307(3.95),328(3.92),453(4.04)	39-2218-66C
	$MeOH-HClO_4$	218(4.30),246(4.51),283(3.75),326(3.65)	39-2218-66C
11bH-Pyrido[2,1-a]phthalazine-1,2,3,4-tetracarboxylic acid, 7-methyl-, tetramethyl ester	MeOH	225(4.24),271(4.32),335(3.73),460(3.88)	39-2218-66C
	$MeOH-HClO_4$	235s(4.31),275s(4.33),280(4.33),333(3.91),460(3.94)	39-2218-66C

Compound	Solvent	λ_{max}(log ϵ)	Ref.
$C_{21}H_{20}N_4O_5$			
2-Pyridinecarbamic acid, 4-[(diphenylmethyl)amino]-6-hydroxy-5-nitro-, ethyl ester	pH 1	236s(4.46),266s(3.85), 353(4.24)	44-1890-66
	pH 7	236s(4.43),266s(3.85), 353(4.23)	44-1890-66
	pH 13	255(4.08),281(3.82), 361(4.14)	44-1890-66
$C_{21}H_{20}O$			
Inden-2(4H)-one, 5,6,7,7a-tetrahydro-1,3-diphenyl-	EtOH	218(4.33),258(3.97)	44-1336-66
2,4,6,8-Nonatetraen-1-ol, 1,9-diphenyl-	CH_2Cl_2-$HClO_4$	335(4.7),350(4.9), 367(4.8)	24-2491-66
$C_{21}H_{20}OS$			
2,6-Xylenol, α-(benzylthio)-α-phenyl-	MeOH	278(3.45)	35-5855-66
	MeOH-KOH	283(3.36),306(3.36)	35-5855-66
$C_{21}H_{20}O_3$			
Estra-1,3,5(10),6,8,14-hexaene-17α-carboxylic acid, 16-oxo-, ethyl ester	EtOH	219(4.35),238(4.15), 268(4.47),278(4.56), 318(4.40)	35-0799-66
$C_{21}H_{20}O_4$			
4H-Furo[2,3-h]-1-benzopyran-4-one, 8,9-dihydro-8-isopropyl-3-(o-methoxyphenyl)-	EtOH	218(4.37),248(4.52), 302(4.08)	39-0749-66C
$C_{21}H_{20}O_5$			
4-Pentenophenone, 2'-hydroxy-2-(o-hydroxyphenyl)-, diacetate	EtOH	241(3.98),290s(3.03)	39-0544-66C
$C_{21}H_{20}O_7$			
Isopiscidone	EtOH	262(4.46),300s(3.93)	78-0333-66A
Piscerythrone	EtOH	265(4.37),294(4.17)	78-0333-66A
Solorinic acid	EtOH	269(4.43),281(4.40), 311(4.45),460(3.98)	39-1727-66C
$C_{21}H_{20}O_8$			
Ceroalbolinic acid, tetra-O-methyl-, methyl ester	EtOH	180(4.73),341(3.86)	78-1507-66
$C_{21}H_{20}O_9$			
Bayin	EtOH	257(4.09),331(4.40)	12-1717-66
$C_{21}H_{20}O_{10}$			
4'-Flavanol, 7-glucosyloxy-4'-hydroxy-	EtOH	256(4.31),317(4.21), 360(4.45)	22-3212-66
Pigment B from Tamarindus indica L.	EtOH	270(4.3),334(4.3)	102-0177-66
	EtOH-NaOEt	280(4.2),400(4.3)	102-0177-66
	EtOH-NaOAc	279(4.4),388(4.2)	102-0177-66
	EtOH-NaOAc-H_3BO_3	270(4.1),335(4.1)	102-0177-66
	EtOH-$AlCl_3$	274(4.3),304(4.2), 340(4.3),380(4.1)	102-0177-66
$C_{21}H_{20}O_{11}$			
Epiorientin	EtOH	258(4.10),269s(4.10), 352(4.10)	78-1147-66
	EtOH-NaOEt	275(4.20),407(4.10)	78-1147-66

Compound	Solvent	λ_{max}(log ε)	Ref.
Epiorientin (cont.)	EtOH-NaOAc	279(4.20),325(3.90), 394(3.80)	78-1147-66
	EtOH-NaOAc-H_3BO_3	264(4.20),374(3.90)	78-1147-66
	EtOH-$AlCl_3$	276(4.10),394(4.10)	78-1147-66
Populnin	EtOH	265(4.08),325(3.85), 370(4.11)	22-3212-66
Tamarindus indica L. pigment A	EtOH	258(4.1),270(4.1), 350(4.1)	102-0177-66
	EtOH-NaOEt	272(4.2),402(4.3)	102-0177-66
	EtOH-NaOAc	279(4.1),380(4.1)	102-0177-66
	EtOH-NaOAc-H_3BO_3	262(4.4),371(4.3)	102-0177-66
	EtOH-$AlCl_3$	269(4.1),350(4.1), 380(4.0)	102-0177-66
$C_{21}H_{21}ClN_2$			
Piperidine, 4-[cyano(p-chlorophenyl)-methylene]-N-phenethyl-, hydrochloride	EtOH	255(4.10)	78-2721-66
$C_{21}H_{21}ClN_2O_4$			
5a,6-Dihydro-4,5,7,8-tetramethyl[1,4]-diazepino[7,1,2-cd:5,4,3-c'd']di-indolizinium perchlorate	MeCN	560(3.87)	39-0483-66B
$C_{21}H_{21}IN_2S_2$			
3-Ethyl-2-[3-(3-ethyl-2-benzothiazol-inylidene)propenyl]benzothiazolium iodide	MeOH	556(5.13)	9-0289-66
	EtOH	558(--)	65-0828-66
	n.s.g.	557(5.20)	27-0330-66
$C_{21}H_{21}NO_3$			
2H-Benzo[a]quinolizin-2-one, 1,6,7,11b-tetrahydro-9,10-dimethoxy-4-phenyl-	EtOH	285(3.77),331(4.13)	44-0797-66
$C_{21}H_{21}NO_4$			
N-Nornuciferine, N-carbethoxy-10a,11-dehydro-	n.s.g.	256(4.59),263(4.64), 311(4.06),323(4.06), 356(3.26),375(3.28)	88-2937-66
$C_{21}H_{21}NO_6$			
Coulteropine	EtOH	286(3.85)	78-1095-66
$C_{21}H_{21}NO_7$			
Narcotoline	MeOH	288s(3.3),310(3.5)	83-0196-66
$C_{21}H_{21}N_3O_2S$			
7-Azaindoline, 4-methyl-1-phenyl-6-p-toluenesulfonamido-	EtOH	282(4.21),336(4.19)	78-3233-66
$C_{21}H_{21}N_5O_4$			
2-Pyridinecarbamic acid, 6-amino-4-[(diphenylmethyl)amino]-5-nitro-, ethyl ester	pH 1	216(4.59),243(3.53), 269s(4.01),346(4.22)	44-1890-66
$C_{21}H_{21}OP$			
Phosphine oxide, (1-cyclohepten-1-ylethynyl)diphenyl-	EtOH	223(4.23),228(4.26), 245(4.15),254(4.17)	22-1002-66

Compound	Solvent	$\lambda_{max}(\log \epsilon)$	Ref.
$C_{21}H_{22}BrN_5$			
3,9-Dibenzyl-N,N-dimethyladeninium bromide	pH 1	274s(--),287(4.27)	44-2202-66
	pH 7	274s(--),286(4.28)	44-2202-66
$C_{21}H_{22}ClNO_5$			
m-Toluamide, α-[(7-chloro-2,3-dihydro-6-methoxy-3-oxo-4-benzofuranyl)oxy]-N,N-diethyl-	EtOH	208(4.57),235(4.42), 286(4.30),320(3.84)	87-0242-66
$C_{21}H_{22}ClN_2O_4PS$			
1-Ethyl-2-[(3-ethyl-2-benzothiazolinylidene)phosphino]-4-methylquinolinium perchlorate	MeOH	549(4.54)	24-1325-66
	$C_2H_4Cl_2$	557(4.55)	24-1325-66
$C_{21}H_{22}ClN_2O_5PS$			
1-Ethyl-2-[(3-ethyl-6-methoxy-2-benzothiazolinylidene)phosphino]-quinolinium perchlorate	MeOH	558(4.54)	24-1325-66
	$C_2H_4Cl_2$	565(4.56)	24-1325-66
$C_{21}H_{22}Cl_2N_2$			
Piperidine, 4-[cyano(p-chlorophenyl)-methylene]-N-phenethyl-, hydrochloride	EtOH	255(4.10)	78-2721-66
$C_{21}H_{22}CrO_4$			
Chromium, tricarbonyl(estra-1,3,5(10)-trien-17-one)-	n.s.g.	318(3.92)	39-0054-66C
$C_{21}H_{22}IP$			
(1-Cyclohexen-1-ylethynyl)methyl-diphenylphosphonium iodide	EtOH	225(4.48),238(4.24), 250(4.20),256(4.23)	22-1002-66
$C_{21}H_{22}N_2$			
Piperidine, 4-(cyanophenylmethylene)-N-phenethyl-, hydrochloride	EtOH	250(4.06)	78-2721-66
$C_{21}H_{22}N_2O$			
Naphth[2,1-d]isoxazole, 3a,4,5,9b-tetrahydro-3-phenyl-9b-(1-pyrrolidinyl)-	EtOH	264(4.16)	32-0559-66
Piperidine, N-benzyl-4-[cyano(p-anisyl)methylene]-, hydrobromide	EtOH	272(3.98)	78-2721-66
$C_{21}H_{22}N_2O_2$			
Benz[a]indolo[3,2-h]quinolizine, 5,6,8,9,14,14b-hexahydro-2,3-dimethoxy-	MeOH	225(4.63),275s(4.02), 283(4.09),290s(4.03)	4-0395-66
Dipyrromethene-3'-carboxylic acid, 4',5,5'-trimethyl-3-phenyl-,	EtOH-N HCl	222(4.44),263(3.80), 379(3.87),483(4.90)	12-1871-66
ethyl ester	EtOH	220(4.24),459(4.51)	12-1871-66
Dipyrromethene-4'-carboxylic acid, 3',5,5'-trimethyl-3-phenyl-,	EtOH-N HCl	222(4.33),260(3.91), 472(4.88)	12-1871-66
ethyl ester	EtOH	217s(4.42),263s(4.04), 448(4.40),483s(4.35)	12-1871-66
Naphth[1,2-d]isoxazole, 3a,4,5,9b-tetrahydro-3a-morpholino-1-phenyl-	EtOH	256(4.02)	32-0559-66
$C_{21}H_{22}N_2O_7$			
3H-Phenoxazine-1,9-dicarboxylic acid, 4,6-dimethyl-2-[(2-methoxyethyl)-amino]-3-oxo-, dimethyl ester	MeOH	226(4.29),250(4.45), 436(4.41),442(4.46)	44-2564-66

Compound	Solvent	λ_{max}(log ε)	Ref.
$C_{21}H_{22}N_4OS$			
Acetamide, N-[(4-amino-2-methyl-5-pyrimidinyl)methyl]-2-(benzylthio)-2-phenyl-	EtOH	236s(4.03),275(3.73)	44-2951-66
$C_{21}H_{22}N_4O_2S$			
Pyrimido[4',5':4,5]pyrimido[2,1-c]-[1,4]thiazine-8-ethanol, 5,10-dihydro-2,7-dimethyl-10-phenyl-, acetate	EtOH	372(4.04)	44-2951-66
$C_{21}H_{24}N_4O_5$			
3(2H)-Phenanthrone, 1,4,4aα,9,10,10aβ-hexahydro-7-methoxy-, 2,4-dinitrophenylhydrazone	EtOH	229(4.39)	87-0304-66
4aβ-isomer	EtOH	220(4.38),229(4.38),265s(4.09),287s(3.70),365(4.38)	87-0304-66
$C_{21}H_{22}N_4O_5S$			
Tryptophan, 2-(o-nitrophenylthio)-alanyl-, methyl ester	MeOH	240(4.2),280f(4.0),390(3.6)	32-1301-66
hydrochloride	MeOH	280(4.2),360(3.6)	32-1301-66
$C_{21}H_{22}N_8O_{10}$			
Heptanoic acid, 3,6-dioxo-, ethyl ester, bis(2,4-dinitrophenylhydrazone)	$CHCl_3$	359(4.63)	39-0341-66C
$C_{21}H_{22}N_8O_{11}$			
Inosine, 5',5'''-carbonate	pH 2	248(4.05)	44-3402-66
	pH 12	253(4.07)	44-3402-66
$C_{21}H_{22}O_3$			
Estra-1,3,5(10),8,14,16-hexaene, 17-acetoxy-3-methoxy-	EtOH	357(4.41)	13-0547-66B
$C_{21}H_{22}O_4$			
4H-Furo[2,3-h]-1-benzopyran-4-one, 2,3,8,9-tetrahydro-8-isopropyl-3-(o-methoxyphenyl)-	EtOH	220(4.46),245(4.32),295(4.04)	39-0749-66C
Gona-1,3,5(10),6,8-pentaene-11,17-dione, cyclic bis(ethylene acetal)	EtOH	229(4.92),280(3.75)	39-0955-66C
$C_{21}H_{22}O_4W$			
Tungsten, tricarbonyl(estra-1,3,5(10)-trien-17-one)-	n.s.g.	319(4.45)	39-0054-66C
$C_{21}H_{22}O_5$			
Psoralen, 5-[(3,6-dimethyl-6-formyl-2-heptenyl)oxy]-	EtOH	242(4.35),249(4.40),258(4.35),267(4.33),307(4.28)	78-1489-66
$C_{21}H_{22}O_6$			
2H,12H-Pyrano[2,3-a]xanthen-12-one, 3,4-dihydro-5,8,9-trimethoxy-2,2-dimethyl-	MeOH	246(4.69),308(4.38)	39-0175-66C
Xanthone, 1-hydroxy-3,5,6-trimethoxy-2-(3,3-dimethylallyl)-	MeOH	246(4.59),282(3.90),318(4.28)	39-0175-66C

Compound	Solvent	λ_{max}(log ε)	Ref.
Xanthone, 1-hydroxy-3,5,6-trimethoxy-2-(3,3-dimethylallyl)- (cont.)	MeOH	254(4.62),285(4.01),328(4.31)	39-0178-66C
	MeOH	254(4.62),285(4.01),328(4.31)	39-2265-66C
Xanthone, 1-hydroxy-3,5,6-trimethoxy-4-(3,3-dimethylallyl)-	MeOH	243(4.58),255s(4.51),287s(4.04),319(4.44),377s(3.88)	39-2265-66C
$C_{21}H_{22}O_{10}$			
Digicitrin	n.s.g.	282(4.26),338(4.23)	24-3218-66
$C_{21}H_{23}BrN_2O$			
Piperidine, N-benzyl-4-[cyano(p-anisyl)methylene]-, hydrobromide	EtOH	272(3.98)	78-2721-66
$C_{21}H_{23}ClN_2$			
Piperidine, 4-(cyanophenylmethylene)-N-phenethyl-, hydrochloride	EtOH	250(4.06)	78-2721-66
$C_{21}H_{23}ClN_4O_5S$			
Tryptophan, 2-(o-nitrophenylthio)-alanyl-, methyl ester, hydrochloride	MeOH	280(4.2),360(3.6)	32-1301-66
$C_{21}H_{23}F_3O_2$			
Estra-1,3,5(10)-triene-3,17β-diol, 17α-(trifluoro-1-propynyl)-	EtOH	280(3.32),286(3.28)	78-2829-66
$C_{21}H_{23}N$			
11H-Benzo[a]carbazole, 6a,11a-dihydro-5,6,6a,11,11a-pentamethyl-	MeOH	227(4.40),255(4.41),272s(4.12)	44-2009-66
1-Pyrroline, 2-butyl-3-(diphenylmethylene)-, picrate	EtOH	231(4.23),287(4.23)	94-0187-66
$C_{21}H_{23}NO$			
Aziridine, 1-cyclohexyl-2-phenyl-3-benzoyl-, cis	C_6H_{12}	327(2.18)	44-1244-66
trans	C_6H_{12}	349(2.47)	44-1244-66
$C_{21}H_{23}NO_3$			
1,4-Pentadien-3-one, 1-(p-dimethylaminophenyl)-5-(2,4-dimethoxyphenyl)-	benzene	412(4.55)	65-0336-66
	EtOH	273(4.11),434(4.61)	65-0336-66
	HOAc	262(4.12),435(4.41)	65-0336-66
ferric chloride added	HOAc	627(4.71)	65-0336-66
2,4-Pentadien-1-one, 1-(p-dimethylaminophenyl)-5-(2,4-dimethoxyphenyl)-	benzene	314(3.99),395(4.62)	65-0336-66
	EtOH	259(4.16),408(4.65),	65-0336-66
	HOAc	260(4.18),418(4.65)	65-0336-66
ferric chloride added	HOAc	556(3.54)	65-0336-66
2,4-Pentadien-1-one, 5-(p-dimethylaminophenyl)-1-(2,4-dimethoxyphenyl)-	benzene	286(4.27),419(4.56)	65-0336-66
	EtOH	288(4.15),430(4.48)	65-0336-66
	HOAc	338(4.39),432(4.24)	65-0336-66
ferric chloride added	HOAc	655(4.31)	65-0336-66
4-Quinolineacetic acid, N-benzoyl-2,2-dimethyl-1,2,3,4-tetrahydro-, methyl ester	MeOH	276(3.78)	5-0128-66D
$C_{21}H_{23}NO_4$			
2(1H)-Isoquinolinecarboxylic acid, 1-benzylidene-3,4-dihydro-6,7-dimethoxy-, ethyl ester	n.s.g.	227(4.36),328(4.40)	88-2937-66

Compound	Solvent	$\lambda_{max}(\log \epsilon)$	Ref.
$C_{21}H_{23}NO_5$			
Cassythine, N,O-dimethyl- (or ocoteine)	n.s.g.	206(4.42),283(4.10), 300(4.09)	12-0297-66
$C_{21}H_{23}N_3O$			
Benzoquinone, 2,6-diethyl-, 4-azine with N-ethylcarbostyril	MeOH	515(4.75)	5-0116-66G
2-Cyclohexen-1-one, 4-ethyl-3,5-diphenyl-, semicarbazone	EtOH	283(4.22)	22-1012-66
$C_{21}H_{23}N_3OS$			
Phenothiazine-2-carbonitrile, 10-[3-(4-hydroxypiperidino)propyl]-	$MeSO_3H$	232(4.73),268(4.43), 312s(--)	95-0510-66
$C_{21}H_{23}N_3O_2S$			
Phenothiazine-2-carbonitrile, 10-[3-(4-hydroxypiperidino)propyl]-, 3-oxide	$MeSO_3H$	245(4.56),275(4.16), 308(3.85),364(3.69)	95-0510-66
$C_{21}H_{23}N_3O_5$			
1,3-Azulenedicarboxylic acid, 2-acetamido-6-(2-cyanoethylamino)-, diethyl ester	MeOH	268(4.20),289(4.13), 342(4.74),398(4.24)	18-1954-66
$C_{21}H_{23}N_5O_5S$			
Adenine, 9-(2,5-di-O-acetyl-3-S-benzyl-3-thio-β-D-arabinofuranosyl)-	MeOH	259(4.17)	44-3263-66
$C_{21}H_{24}BrNO_4$			
6,7,8-Trimethoxy-1-(p-methoxybenzyl)-2-methylisoquinolinium bromide	MeOH	224(4.37),262(4.70), 285s(3.70),311(3.77), 322(3.79),353(3.65)	33-1757-66
$C_{21}H_{24}ClNO_4$			
N-Methylcanadinium chloride	EtOH	212(4.3),233s(4.04), 287(3.8)	25-0237-66
$C_{21}H_{24}F_3N_3S$			
Trifluoroperazine (as dihydrochloride)	EtOH	258(4.50),308(3.50)	36-0144-66
$C_{21}H_{24}INO_4$			
5,6,7-Trimethoxy-1-(p-methoxybenzyl)-2-methylisoquinolinium iodide	MeOH	265(4.61),318(3.69)	44-0516-66
$C_{21}H_{24}N_2$			
Acetonitrile, N-β-phenethyl-4-piperidylphenyl-, hydrobromide	EtOH	251(2.51),256(2.46), 261(2.62)	78-2721-66
$C_{21}H_{24}N_2O$			
Kopsanone, N(a)-methyl-	EtOH	252(3.96),298(3.58)	33-1237-66
Kopsanone, methine	EtOH	242(3.91),292(3.52)	33-1237-66
1,4-Pentadien-3-one, 1,5-bis(p-dimethylaminophenyl)-	benzene	430(4.84)	65-0336-66
	EtOH	272(4.13),457(4.66)	65-0336-66
	HOAc	290(4.14),463(4.57)	65-0336-66
ferric chloride added	HOAc	667(4.83)	65-0336-66
2,4-Pentadien-1-one, 1,5-bis(p-dimethylaminophenyl)-	benzene	422(4.63)	65-0336-66
	EtOH	276(4.14),437(4.64)	65-0336-66
	HOAc	334(4.28),425(4.52)	65-0336-66
ferric chloride added	HOAc	660(4.63)	65-0336-66

Compound	Solvent	λ_{max}(log ϵ)	Ref.
$C_{21}H_{24}N_2O_3$			
Perakine, dihydro-	EtOH	220(4.30),258(3.69)	78-3293-66
Quebrachidine	EtOH	243(3.87),293(3.51)	102-1065-66
$C_{21}H_{24}N_2O_4$			
Isopteropodine	EtOH	246(4.22),280s(3.27)	39-2245-66C
	EtOH	225(4.00),246(4.22), 280s(3.27)	88-0931-66
Mitraphylline	EtOH	243(4.53),280s(3.25)	44-1765-66
Pteropodine	EtOH	246(4.2),280s(3.25)	39-2245-66C
	EtOH	225(3.93),246(4.20), 280s(3.25)	88-0931-66
$C_{21}H_{24}N_2O_5$			
1,3-Azulenedicarboxylic acid, 2-acet- amido-6-allylamino-, diethyl ester	MeOH	270(4.23),289(4.17), 342(4.80),399(4.29)	18-1954-66
$C_{21}H_{24}N_2O_8S_2$			
Carbonic acid, isobutyl ester, anhyd- ride with 3-(hydroxymethyl)-8-oxo- 7-[2-(2-thienyl)acetamido]-5-thia- 1-azabicyclo[4.2.0]oct-2-ene- 2-carboxylic acid, acetate	EtOH	234(4.08),266(3.83)	87-0741-66
$C_{21}H_{24}O_2$			
14ξ-Estra-1,3,5(10),6,8-pentaene-17ξ- carboxylic acid, ethyl ester	EtOH	231(4.98),275(3.72), 286(3.76),308(3.07), 313(2.84),321(2.90)	35-0799-66
A-Homoestra-1(10),2,4a-trien-4-one, 17α-ethynyl-17β-hydroxy-	n.s.g.	237(4.42),315(4.01)	78-0391-66A
$C_{21}H_{24}O_3$			
Estra-1,3,5(10),8,14-pentaen-17β-ol,	MeOH	310(4.51)	88-2321-66
3-methoxy-, acetate	MeOH	311(4.49)	25-1340-66
13α-Estra-1,3,5(10),8,14-pentaen- 17β-ol, 3-methoxy-, 17-acetate	MeOH	310(4.51)	88-2321-66
$C_{21}H_{24}O_4$			
7H-Benz[e]inden-7-one, 1,2,3,3a,4,5,8,- 9,9a,9b-decahydro-3,5-dihydroxy- 3a-methyl-, 3-benzoate	EtOH	234(4.45)	22-1537-66
1,2-Dibenzofurandione, 4,6-di-tert- butyl-8-methoxy-	EtOH	217(4.39),252(4.33), 260(4.32),294(3.80)	39-0362-66C
1,4-Dibenzofurandione, 2,6-di-tert- butyl-8-methoxy-	EtOH	215(4.47),243(4.39), 251(4.37),291(3.82)	39-0362-66C
$C_{21}H_{24}O_7$			
Shikonin, β-hydroxyisovaleryl-	EtOH	273(4.13)	88-3677-66
Thujaplicatin, 3-O-methyl-	EtOH	281(3.60)	23-1541-66
	EtONa	293(3.86)	23-1541-66
$C_{21}H_{24}O_8$			
Thujaplicatin, hydroxy-O-methyl-	EtOH	281(3.58)	23-1827-66
	EtONa	292(3.87)	23-1827-66
$C_{21}H_{24}O_9$			
Chromomycinone	EtOH	232(4.38),282(4.60), 326(3.85),340(3.85), 412(4.01)	78-2761-66

Compound	Solvent	λ_{max}(log ε)	Ref.
Chromomyciquinone, 3'-dehydro-	EtOH	256(4.27),300(4.06), 415(3.78)	78-2761-66
Isochromomycinone	EtOH	232(4.36),282(4.55), 328(3.83),341(3.84), 418(3.98)	78-2761-66
$C_{21}H_{25}BrN_2$			
Acetonitrile, N-β-phenethyl-4-piperidylphenyl-, hydrobromide	EtOH	251(2.51),256(2.46), 261(2.62)	78-2721-66
$C_{21}H_{25}Br_3O_2$			
A-Homoestra-2,4-dien-17-one, 3-bromo-5(10)-dibromomethylene-4-methoxy-	n.s.g.	270(3.65)	78-0391-66A
$C_{21}H_{25}F_3O_2$			
Androsta-1,4-dien-3-one, 17β-hydroxy-17α-(trifluorovinyl)-	EtOH	244(4.14)	78-2829-66
Estra-1,3,5(10)-trien-17β-ol, 3-methoxy-17α-(trifluorovinyl)-	EtOH	220(3.92),229s(3.84), 278(3.29),287(3.27)	78-2829-66
$C_{21}H_{25}F_3O_4$			
Androst-4-ene-3,12-dione, 17β-hydroxy-, trifluoroacetate	EtOH	238(4.22)	44-2116-66
$C_{21}H_{25}NO$			
Pyrrolidine, 1-[2-[(o-styrylbenzyl)-oxy]ethyl]-	H_2O	203(4.41),227(4.15), 297(4.38)	39-0668-66C
$C_{21}H_{25}NO_2$			
Morpholine, 4-[2-[(o-styrylbenzyl)-oxy]ethyl]-	H_2O	203(4.39),227(4.16), 297(4.40)	39-0668-66C
$C_{21}H_{25}NO_4$			
Glaucine	EtOH	217(4.60),280(4.18), 301(4.17)	39-0753-66C
Palmatin, tetrahydro-	MeOH	282(3.8)	73-1355-66
	EtOH	282(3.74)	95-0460-66
$C_{21}H_{25}NO_5$			
Stepharotine	EtOH	283(3.72)	95-0460-66
$C_{21}H_{25}NO_7$			
Narcotolinediol	MeOH	280(3.6)	83-0196-66
$C_{21}H_{25}N_3$			
5H-Imidazo[5,6-a]dibenz[b,e]azepine, 11-(3-dimethylaminopropyl)-	EtOH	231s(4.1),260(3.94), 293(3.85)	4-0206-66
$C_{21}H_{25}N_3O$			
Ibogaine, 18-cyano-	EtOH	213(4.54),282(3.98), 295(3.89)	35-2532-66
$C_{21}H_{25}N_3O_2$			
Imidazo[1,5-a]pyrazine, 2-(p-carbethoxyphenyl)-3-phenyloctahydro-	EtOH	307(4.50)	87-0868-66
$C_{21}H_{25}N_5O_4$			
β-D-Ribofuranoside, methyl 5-[7-benzyl-6(1H)-iminopurin-1-yl]-5-deoxy-2,3-O-isopropylidene-	EtOH-HCl	222(4.39),278(3.89)	4-0485-66
	EtOH	265(3.97)	4-0485-66
	EtOH-NaOH	265(4.00)	4-0485-66

Compound	Solvent	λ max(log ε)	Ref.
$C_{21}H_{26}ClNO_4$			
6-Hydroxy-2,3,5-trimethoxy-N-methyl-aporphinium chloride	EtOH	229(4.55),273s(3.98), 283(4.10),310(4.19), 320s(4.09)	25-0237-66
$C_{21}H_{26}F_2O_2$			
Androst-4-en-3-one, 17α-ethynyl-6,6-difluoro-17β-hydroxy-	EtOH	229(4.09),332(1.54)	44-0991-66
$C_{21}H_{26}INO_4$			
Corydine methiodide	EtOH	222(4.71),267(4.09), 273(4.09),306(3.79)	39-1340-66C
N,N-Dimethyllaurotetaninium iodide	EtOH	284(4.10),304(4.14)	95-0129-66
10-Hydroxy-1,2,9-trimethoxy-N-methyl-aporphinium iodide	EtOH	283(4.10),307(4.12)	12-0437-66
Hydroxytrimethoxy-N-methyl-aporphinium iodide	EtOH	222(4.71),267(4.09), 273(4.09),306(3.79)	25-0118-66
Xanthoplanine iodide	EtOH	220(4.75),284(4.15), 304(4.21)	39-1340-66C
$C_{21}H_{26}N_2$			
1,6-Diazaspiro[4.4]nonane, 1,6-dibenzyl-	MeOH	219(4.0)	24-0724-66
$C_{21}H_{26}N_2O_2$			
Pleiocarpininic acid	EtOH	256(3.93),302(3.45)	33-1237-66
Pyridine, 2,3,4,5-tetrahydro-5,5-di-methyl-3-oxo-2-(1-benzoyl-4,5,5-trimethylpyrrolidin-2-ylidene)-	EtOH	225(4.03),260(4.00), 332(3.78),365(3.72)	78-0015-66B
	EtOH-HCl	255(4.16)	78-0015-66B
5H-Pyrrolo[1,2-c]imidazole, 1-(3,3-dimethyl-1,4-dioxobutyl)-6,7-dihydro-5,5,6-trimethyl-3-phenyl-	EtOH	271(4.15)	78-0015-66B
$C_{21}H_{26}N_2O_3$			
Heyneanine	EtOH	225(4.51),284(3.87), 292(3.81)	88-1251-66
Isositsirikine	MeOH	224(4.55),283(3.92), 291(3.84)	78-0321-66
Minovincinine	EtOH	227(4.08),299(4.05), 329(4.23)	88-2483-66
Sitsirikine	MeOH	226(4.56),282(3.90), 290(3.84)	78-0321-66
Strychnosplendine, N-acetyl-	n.s.g.	252(3.70),280s(3.10)	88-2353-66
Tabernemontanine	EtOH	240(4.19)	102-1065-66
$C_{21}H_{26}N_2O_4$			
Aspidoalbidine, N-acetyl-16,17-dihydroxy-	EtOH	225(4.06),259(3.70)	23-0028-66
$C_{21}H_{26}N_2O_5$			
2-Pyrazoline-3,4-dicarboxylic acid, 1-benzoyl-5,5-pentamethylene-, diethyl ester	MeOH	224(4.16),287(4.35)	35-3798-66
$C_{21}H_{26}N_2O_6$			
Propane, 2,2-bis(4-carbethoxy-5-formyl-3-methylpyrrol-2-yl)-	EtOH	221(4.24),267(4.06), 342(4.51)	39-0098-66C

Compound	Solvent	λ max (log ε)	Ref.
$C_{21}H_{26}N_4O$			
p-Benzoquinone, 2,6-diethyl-, 4-azine with 1,3-diethylbenzimidazolin-2-one	MeOH	517(4.80)	5-0116-66G
$C_{21}H_{26}N_4O_2$			
Benzimidazole, 2-benzyl-1-(2-diethylaminopropyl)-5-nitro-	EtOH-HCl	240(4.46),301(4.04)	39-1511-66C
$C_{21}H_{26}N_4O_4$			
Cyperotundone, 2,4-dinitrophenylhydrazone	$CHCl_3$	260(4.12),297(3.78), 396(4.35)	94-0890-66
2,4,6,10-Dodecatetraenal, 3,7,11-trimethyl-, 2,4-dinitrophenylhydrazone	EtOH	418(4.63)	39-2154-66C
3,5-Octadien-2-one, 6-(2-methyl-3-cyclohexen-1-yl)-, 2,4-dinitrophenylhydrazone	$CHCl_3$	270(4.38),310(4.45), 408(4.72)	70-0480-66
$C_{21}H_{26}N_4O_4S$			
Acetamide, N-[(4-amino-2-methyl-5-pyrimidinyl)methyl]-2-[1-(2-hydroxyethyl)acetonyl]thio]-2-phenyl-, acetate	EtOH	229(4.09),275(3.76)	44-2951-66
$C_{21}H_{26}N_6O_9$			
Purine, 2,6-diacetamido-9-(2,3,6-tri-O-acetyl-4-deoxy-β-D-xylohexosyl)-	pH 7	235(4.47),264(4.10), 286(4.11)	39-1549-66C
Purine, 2,6-diacetamido-9-(2,4,6-tri-O-acetyl-3-deoxy-β-D-xylohexosyl)-	n.s.g.	236(4.50),262(4.04), 288(4.12)	39-1549-66C
$C_{21}H_{26}O$			
2,5-Cyclohexadien-1-one, 4-benzylidene-2,6-di-tert-butyl-	hexane	344(4.48)	70-0888-66
$C_{21}H_{26}O_2$			
Pregna-1,4,9(11)-triene-3,20-dione	EtOH	240(4.18)	95-0565-66
$C_{21}H_{26}O_3$			
Androsta-1,4,9(11)-trien-3-one, 17β-acetoxy-	EtOH	238(4.12)	33-1183-66
1α,5β-Cyclo-10α-androsta-3,9(11)-dien-2-one, 17β-acetoxy-	EtOH	280(3.88)	33-1183-66
Estra-1,3,5(10),8-tetraen-17β-ol, 3-methoxy-, acetate	MeOH	278(4.20)	25-1340-66
	MeOH	276(4.24)	88-2321-66
Estra-1,3,5(10)-trien-6-one, 17β-acetoxy-4-methyl-	MeOH	252(4.04)	39-1034-66C
Estra-1(10),3,9(11)-trien-2-one, 17β-acetoxy-5α-methyl-	EtOH	242(4.07),288(3.95)	33-1183-66
Merogedunin, anhydro-	MeOH	228(4.15),275(4.06)	39-0506-66C
10α-Pregna-4,9(11)-diene-3,6,20-trione	EtOH	248(4.03)	33-1529-66
Pregna-4,6,8(14)-triene-3,20-dione, 17α-hydroxy-	EtOH	348(4.39)	78-0325-66A
$C_{21}H_{26}O_4$			
Androsta-1,4-diene-3,17-dione, 4-acetoxy-	EtOH	242(4.15)	5-0180-66B
Androsta-1,4-diene-3,11-dione, 17β-acetoxy-	EtOH	239(4.22)	33-1183-66
Androsta-1,5-diene-3,17-dione, 4α-acetoxy-	EtOH	226(4.03)	5-0180-66B

Compound	Solvent	λ_{max}(log ϵ)	Ref.
Androsta-3,9-diene-2,11-dione, 17β-acetoxy-, 1(10→5β)-abeo-	EtOH	245(4.00)	33-1183-66
	EtOH-KOH	245(3.95)	33-1183-66
1α,5β-Cyclo-10α-androst-3-ene-2,11-dione, 17β-acetoxy-	EtOH	268(3.47)	33-1183-66
2(4aH)-Dibenzofuranone, 4,6-di-tert-butyl-4a-hydroxy-8-methoxy-	EtOH	265(3.0)	39-0362-66C
β-Digiprogenin	EtOH	240(4.06)	94-0552-66
Estra-1(10),3-dien-2-one, 9ξ,11ξ-epoxy-17β-hydroxy-5α-methyl-, acetate	EtOH	242(4.13)	33-1183-66
Estra-1,3,5(10)-triene-4-carboxaldehyde, 3,17β-dihydroxy-, 17-acetate	EtOH	223(4.02),272(3.98), 355(3.42)	27-0110-66
Estra-1,3,5(10)-trien-6-one, 2,17β-dihydroxy-1-methyl-, 17-acetate	EtOH	234(4.00),287(4.08)	5-0180-66B
	0.05N NaOH	254(3.93),343(4.38)	5-0180-66B
$C_{21}H_{26}O_5$			
Androst-4-en-19,11β-olide, 17-acetoxy-3-oxo-	EtOH	241(4.15)	22-1537-66
2-Benzofuranacrylic acid, 2-(2-carboxyethyl)-3-ethylidene-2,3-dihydro-β,4,6-trimethyl-, dimethyl ester	n.s.g.	245(4.12),268(4.24), 325(3.94)	78-2015-66
Odoratin, ethylene ketal	n.s.g.	212(4.17)	77-0576-66
$C_{21}H_{26}O_6$			
1H-Benzofuro[3a,3-b]benzofuran-6-acetic acid, 2,3,4,4a,6,6a-hexahydro-6aα-hydroxy-4aβ,7,9-trimethyl-3-oxo-, ethyl ester	n.s.g.	226(4.13),278(3.51), 290(3.61)	44-1725-66
1H-Benzofuro[3a,3-b]benzofuran-6-acetic acid, 2,3,4,4a,6a-hexahydro-6aα-methoxy-4aβ,7,9-trimethyl-3-oxo-, methyl ester	n.s.g.	284(3.48),290(3.52)	44-1725-66
Bicyclo[3.2.1]octane-1-carboxylic acid, 2-[6-carboxy-1-(hydroxymethyl)-5,5-dimethyl-4-oxo-2-cyclohexen-1-yl]-6-methyl-7-oxo-, γ-lactone	EtOH	224(4.00)	78-1659-66
Enmeinenonoic acid, dihydro-, methyl ester	EtOH	224(3.99)	78-3423-66
Pregn-4-en-19-oic acid, 11β,17α,21-trihydroxy-3,20-dioxo-, γ-lactone	EtOH	242(4.09)	22-1537-66
$C_{21}H_{26}O_8$			
Chromomycinone, 1-deoxo-	EtOH	244(4.36),261(3.87), 298(3.78),340(3.82)	78-2761-66
Isochromomycinone, 1-deoxo-	EtOH	244(4.42),297(3.44), 342(3.59)	78-2761-66
$C_{21}H_{26}O_9$			
Chromomycinone, dihydro-	EtOH	231(4.37),280(4.57), 325(3.81),340(3.80), 410(3.96)	78-2761-66
epimer	EtOH	231(4.36),281(4.58), 326(3.82),340(3.83), 413(3.98)	78-2761-66
Chromomyciquinone	EtOH	256(4.30),305(3.10), 415(3.68)	78-2761-66
$C_{21}H_{26}O_{13}$			
Leucoglycodrin	n.s.g.	201(4.30),226(3.97), 276(3.32),282(3.29)	88-3773-66

Compound	Solvent	$\lambda_{max}(\log \epsilon)$	Ref.
$C_{21}H_{27}BrO_3$			
Pregn-4-ene-3,16,20-trione, 17α-bromo-	EtOH	239(4.29)	88-5337-66
$C_{21}H_{27}ClN_2O_6$			
Spegatrine, O[10]-methyl-, perchlorate (unchanged by acid or base)	50% EtOH	272(3.86),291s(3.63), 303s(3.51)	23-1523-66
$C_{19}H_{27}ClO_2$			
19-Norandrost-4-en-3-one, 17α-chloro-ethynyl-17β-methoxy-	EtOH	240(4.22)	78-2829-66
$C_{21}H_{27}ClO_3$			
Pregn-4-ene-3,16,20-trione, 17α-chloro-	EtOH	238(4.25)	88-5337-66
$C_{21}H_{27}F_3O_2$			
Androst-4-en-3-one, 17β-hydroxy-17α-trifluorovinyl-	EtOH	240(4.20)	78-2829-66
$C_{21}H_{27}F_3O_4$			
Androst-4-en-3-one, 12α,17β-dihydroxy-, 17-trifluoroacetate	EtOH	241(4.22)	44-2116-66
$C_{21}H_{27}IN_2S_2$			
2,2'-Bis(3-ethyl-4-methylthiazolyl)-pentamethinecyanine iodide	MeOH	640(5.08)	9-0150-66
$C_{21}H_{27}NO$			
Stilbene, 2-(2-diethylaminoethoxy)-methyl-, hydrochloride, trans	H_2O	203(4.38),227(4.16), 297(4.40)	39-0668-66C
$C_{21}H_{27}NO_4$			
Isoquinoline, 1,2,3,4-tetrahydro-6,7-dimethoxy-3-(3,4-dimethoxybenzyl)-2-methyl-, perchlorate	MeOH	285(3.86)	24-2873-66
Isoquinoline, 1,2,3,4-tetrahydro-5,6,7-trimethoxy-1-(p-methoxybenzyl)-2-methyl-, hydrochloride	MeOH	281(3.51)	44-0516-66
Laudanosine	n.s.g.	207s(4.89),232(4.17), 282(3.77)	78-1335-66
$C_{21}H_{28}ClNO_4$			
Laudanosine, hydrochloride	n.s.g.	207s(5.03),232(4.30), 281(3.80)	78-1335-66
[2-(2,3,4,4,5,6-Hexamethyl-2,5-cyclo-hexadien-1-ylidene)ethylidene]methyl-phenylammonium perchlorate	MeCN	421(4.67)	24-2479-66
$C_{21}H_{28}Cl_2O_2$			
Androsta-3,5-dien-17β-ol, 3,4-dichloro-, acetate	EtOH	240s(4.15),247(4.21), 255s(4.14)	32-1241-66
$C_{21}H_{28}F_2O_3$			
Androst-4-en-3-one, 6,6-difluoro-17β-hydroxy-, acetate	EtOH	228(4.10),333(1.48)	44-0991-66
$C_{21}H_{28}N_2$			
Pyrrolidine, 1-benzyl-2-(3-benzyl-aminopropyl)-	MeOH	216(4.1)	24-0737-66

Compound	Solvent	$\lambda_{max}(\log \epsilon)$	Ref.
$C_{21}H_{28}N_2O_2$			
Estra-2,4-diene-2-carbonitrile, 3-amino-17β-hydroxy-, acetate	EtOH	331(3.56)	32-0152-66
Ibogaine, 18-methoxy-	EtOH	225(4.45),284(3.99), 298(3.91)	35-2532-66
5H-Pyrrolo[1,2-c]imidazole, 1-(3,3-dimethyl-4-hydroxy-1-oxobutyl)-6,7-dihydro-5,5,6-trimethyl-3-phenyl-	EtOH-HCl	249(4.05)	78-0015-66B
	EtOH	272(4.12)	78-0015-66B
$C_{21}H_{28}N_2O_3$			
Isositsirikine, dihydro-	MeOH	226(4.57),284(3.92), 291(3.84)	78-0321-66
Sitsirikine, dihydro-	MeOH	226(4.61),282(3.95), 290(3.87)	78-0321-66
$C_{21}H_{28}N_4O_2$			
5H-Pyrrolo[1,2-c]imidazole, 1-(3,3-dimethyl-1,4-dioxobutyl)-6,7-dihydro-5,5,6-trimethyl-3-phenyl-, dioxime	EtOH-HCl	257(4.12)	78-0015-66B
	EtOH	264(4.07)	78-0015-66B
$C_{21}H_{28}N_4O_4$			
2,6,9-Pentadecatrienal, 2-trans-6,9-di-cis, 2,4-dinitrophenylhydrazone	$CHCl_3$	377(4.47)	54-0117-66
$C_{21}H_{28}N_4O_8$			
Malonic acid, [[[5-[N-[5-(morpholinomethyl)-2-oxo-3-oxazolidinyl]formimidoyl]-2-furyl]amino]methylene]-, diethyl ester	5% EtOH	370(4.47)	4-0202-66
$C_{21}H_{28}O$			
2(3H)-Naphthalenone, 8α-p-cumenyl-4,4a,5,6,7,8-hexahydro-4aα,8-dimethyl-	EtOH	242(4.16)	44-2543-66
4aβ-isomer	EtOH	243(4.01)	44-2543-66
$C_{21}H_{28}O_2$			
Estra-5(10),6,8-triene-17β-carboxylic acid, ethyl ester	EtOH	269(2.58),278(2.46)	35-0799-66
14β-Estra-5(10),6,8-triene-17α-carboxylic acid, ethyl ester	EtOH	269(2.59),278(2.48)	35-0799-66
Gona-1,3,5(10),6-tetraen-17β-ol, 13-ethyl-3-methoxy-6-methyl-	n.s.g.	262(3.85),305(3.39)	78-1019-66
Gona-1,3,5(10)-trien-17-one, 13β-ethyl-3-methoxy-6β-methyl-	n.s.g.	280(3.34)	78-1019-66
8α-Gona-1,3,5(10)-trien-17-one, 3-methoxy-13β-propyl-	EtOH	280(3.36),287(3.36)	87-0338-66
18-Nor-5α,8α,9β,14β-androsta-12,15-diene-3,17-dione, 4α,8,14-trimethyl-	EtOH	247(3.96)	78-3459-66
Podocarpa-1,8,11,13-tetraen-3-one, 13-isopropyl-12-methoxy-	n.s.g.	227(4.25)	32-0206-66
Pregna-1,4-diene-3,20-dione	n.s.g.	244(4.24)	19-0347-66
5β-Pregna-1,9(11)-diene-3,20-dione	EtOH	226.5(4.03)	95-0565-66
10α-Pregna-4,9(11)-diene-3,20-dione	EtOH	238(4.16)	33-1529-66
	EtOH	238(4.17)	35-4538-66
Pregna-5,14,16-trien-20-one, 3β-hydroxy-	EtOH	310(4.10)	87-0179-66
$C_{21}H_{28}O_3$			
Androsta-1,4,6-trien-3-one, 17β-hydroxy-2-methoxy-17α-methyl-	MeOH	220(4.17),281(4.15)	73-1363-66

Compound	Solvent	λ max(log ε)	Ref.
1α,5β-Cyclo-10α-androst-9(11)-en-2-one, 17β-acetoxy-	EtOH	241(3.92)	33-1183-66
1β,5α-Cycloandrost-3-en-2-one, 17β-acetoxy-	EtOH	242(3.74),270s(3.51)	33-1049-66
Estra-1(10),9(11)-dien-2-one, 17β-acetoxy-5α-methyl-	EtOH	282(4.10)	33-1183-66
Estra-1,3,5(10)-triene-2,17β-diol	EtOH	220s(3.89),286(3.30)	5-0180-66B
1-methyl-, 17-acetate	0.05N NaOH	244(3.89),301(3.53)	5-0180-66B
Estra-1,3,5(10)-triene-4,17β-diol,	EtOH	220s(3.92),285(3.27)	5-0180-66B
1-methyl-, 17-acetate	0.05N NaOH	247(3.86),301(3.52)	5-0180-66B
Gona-4,6-dien-3-one, 17β-acetoxy-13β-ethyl-	EtOH	283(4.39)	87-0782-66
Merogedunin, anhydrodihydro-	MeOH	278(3.9)	39-0506-66C
Pregna-1,4-diene-3,20-dione, 2-hydroxy-	EtOH	255(4.16),290s(3.51)	94-1370-66
5β-Pregna-1,9(11)-diene-3,20-dione, 17α-hydroxy-	EtOH	225.5(4.03)	95-0565-66
Pregna-4,14-diene-3,20-dione, 17α-hydroxy-	EtOH	239(4.19)	78-0325-66A
Pregna-5,16-diene-15,20-dione, 3-hydroxy-	EtOH	242(3.95)	94-0552-66
5β,10α-Pregn-4-ene-3,15,20-trione	MeOH	242(4.32)	54-0701-66
9β,10α-Pregn-4-ene-3,11,20-trione	EtOH	237(4.18)	33-1529-66
9β,10α-Progesterone, 6-dehydro-16α-hydroxy-	MeOH	286(4.43)	54-0712-66
$C_{21}H_{28}O_4$			
Androst-4-en-19-al, 17β-hydroxy-3-oxo-, acetate	EtOH	246(4.19),314(2.50)	27-0110-66
Androst-5-ene-6-carboxaldehyde, 17β-hydroxy-4-(hydroxymethylene)-3-oxo-	EtOH	225(4.06),330(3.65)	32-1284-66
Androst-1-ene-3,5-dione, 17β-acetoxy-, 10(5→4)-abeo-	EtOH	239(3.96),311(3.87)	33-2218-66
ferric complex	EtOH	552(2.97)	33-2218-66
9β,10α-Androst-4-ene-3,17-dione, 11α-acetoxy-	EtOH	238(4.22)	33-1529-66
Androst-1-en-3-one, 4α,5α-epoxy-17β-hydroxy-, acetate	EtOH	228(4.02),330(1.72)	33-2218-66
Androst-1-en-3-one, 4β,5β-epoxy-17β-hydroxy-, acetate	EtOH	232(3.95),333(1.81)	33-2218-66
1α,5β-Cyclo-10α-androstane-2,11-dione, 17β-hydroxy-, acetate	EtOH	none	33-1183-66
1α,5β-Cyclo-10α-androstan-2-one, 17β-acetoxy-9ξ,11ξ-epoxy-	EtOH	285(1.88)	33-1183-66
1,3-Cyclopentanedione, 4-hydroxy-5-(3-methyl-2-butenyl)-2-(3-methylbutyryl)-4-(4-methylpent-3-en-1-ynyl)-	EtOH-acid EtOH-base	245(3.78),283(3.87) 285(4.18)	39-1615-66C 39-1615-66C
p-Geranyloxyferulic acid, methyl ester	EtOH	236(3.90),301(4.00), 325(4.11)	12-0451-66
Merogedunin	MeOH	212(3.97),230(3.98)	39-0506-66C
Taxininol, anhydrodeformyldehydro-, monoacetate	EtOH	228(4.16)	95-1172-66
$C_{21}H_{28}O_5$			
Androst-4-ene-3,17-dione, 2β-acetoxy-9α-hydroxy-	EtOH	243(4.17)	37-0540-66
9β,10α-Pregn-4-ene-3,11,20-trione, 17α,21-dihydroxy-	EtOH	237.5(4.20)	33-1528-66
6,7-Secokaur-1-ene-6,7-dioic acid, 20-hydroxy-3-oxo-, 6→20-lactone, 7-methyl ester	EtOH	233(3.98)	78-3423-66

Compound	Solvent	λ_{max}(log ε)	Ref.
$C_{21}H_{28}O_6$			
2-Oxa-A-norestr-5(10)-en-1-one, 3ξ,17β-diacetoxy-3ξ-methyl-	MeOH	225(3.87)	44-4255-66
$C_{21}H_{29}ClO_2$			
Androsta-5,16-dien-3β-ol, 17-chloro-, acetate	EtOH	210(3.88)	32-1241-66
$C_{21}H_{29}ClO_4$			
Pregn-4-ene-3,20-dione, 21-chloro-14α,17α-dihydroxy-	EtOH	240(4.17)	78-0325-66A
$C_{21}H_{29}FO_4$			
Pregn-4-ene-3,20-dione, 11β-fluoro-17α,21-dihydroxy-	MeOH	239(4.20)	35-3016-66
$C_{21}H_{29}FO_5$			
Pregn-4-ene-3,20-dione, 6α-fluoro-11α,16α,17-trihydroxy-	EtOH	237.5(4.16)	87-0513-66
$C_{21}H_{29}IO_4$			
Pregn-4-ene-3,20-dione, 14α,17α-dihydroxy-21-iodo-	EtOH	240(4.24)	78-0325-66A
$C_{21}H_{29}NO_2$			
Estra-1,3,5(10)-trien-17β-ol, 2-amino-1-methyl-, 17-acetate	EtOH	239(3.89),294(3.31)	5-0180-66B
	0.05N HCl	214(4.00),266(2.48)	5-0180-66B
Estra-1,3,5(10)-trien-17β-ol, 4-amino-1-methyl-, 17-acetate	EtOH	239(3.87),292(3.27)	5-0180-66B
	0.05N HCl	214(3.99),265(2.48)	5-0180-66B
$C_{21}H_{29}NO_3$			
2H-Benzo[a]quinolizin-2-one, 1,6,7,11b-tetrahydro-9,10-dimethoxy-4-methyl-	EtOH	287(3.70),335(4.18)	44-0797-66
Pregna-5,16-diene-15,20-dione, 3-hydroxy-, oxime	EtOH	273(4.26)	94-0552-66
$C_{21}H_{29}N_3O$			
1H-Cyclopenta[5,6]naphtho[1,2-g]quinazolin-1-ol, 10-amino-2,3,3a,3b,4-5,11,11a,11b,12,13,13a-dodecahydro-11a,13a-dimethyl-	HCl	252(4.16)	32-0152-66
	EtOH	244(4.15),310(3.78)	32-0152-66
$C_{21}H_{29}N_7O_{11}P_2$			
Pyridinium, 3-carbamoyl-1-(5-hydroxypentyl)-, hydroxide, ester with adenosine 5'-(trihydrogen pyrophosphate), inner salt	pH 5.7	261(4.29)	5-0180-66J
$C_{21}H_{30}N_2O_2$			
5α-Estr-2-ene-2-carbonitrile, 3-amino-17β-acetoxy-	EtOH	262(3.91)	32-0152-66
5H-Pyrrolo[1,2-c]imidazole, 1-(1,4-dihydroxy-3,3-dimethylbutyl)-6,7-dihydro-5,5,6-trimethyl-3-phenyl-	EtOH-HCl	247(3.94)	78-0015-66B
	EtOH	263(3.85)	78-0015-66B
$C_{21}H_{30}N_2O_3$			
1H-Cyclopenta[5,6]naphtho[1,2-g]quinazoline-8,10(3bH,9H)-dione, tetradecahydro-1-hydroxy-11a,13a-dimethyl-	EtOH	264(3.85)	32-0179-66
	EtOH-NaOH	289(3.88)	32-0179-66

Compound	Solvent	$\lambda_{max}(\log \epsilon)$	Ref.
$C_{21}H_{30}O$			
18-Norandrosta-4,13-dien-3-one, 17α-ethyl-17β-methyl-	MeOH	239(4.19)	24-3836-66
$C_{21}H_{30}O_2$			
Abietic acid, methyl ester	EtOH	241(4.42)	12-2403-66
Androsta-3,5-dien-17-one, 3-ethoxy-	n.s.g.	241(4.21)	25-0025-66
Cannabichromene	EtOH	228(4.40),280(3.95)	77-0020-66
Cannabidiol	MeOH	210(4.56),227s(4.20), 273(3.06),280(3.04)	5-0165-66C
$\Delta^{1(6)}$-Cannabinol, tetrahydro-	EtOH	275(3.10),282(3.12)	78-1481-66
Δ^{6}-Cannabinol, tetrahydro-, trans	EtOH	209(4.61),276(3.12), 283(3.14)	35-1832-66
d-Gona-1,3,5(10)-triene-3,17β-diol, 13β,17α-diethyl-	EtOH	282(3.30),289(3.26)	44-2512-66
dl-isomer	EtOH	281(3.35),288(3.31)	44-2512-66
Gona-1,3,5(10)-trien-17β-ol, 3-methoxy-13β-propyl-	EtOH	278(3.31),287(3.29)	35-3120-66
Hexadeca-2,4,6,8,10,14-hexaenoic acid, 3,7,11,15-tetramethyl-, methyl ester	EtOH	370(4.68)	39-2154-66C
D-Homo-C-norgona-5,13(17a)-dien-11-one, 17-ethyl-3β-hydroxy-10,17a-dimethyl-	EtOH	256(4.01)	27-0114-66
isomer	EtOH	257(4.02)	27-0114-66
18-Nor-5α,8α,9β,14β-androsta-12,15- dien-17-one, 3α-hydroxy-4α,8,14- trimethyl-	EtOH	249(3.88)	78-3459-66
18-Nor-5α,8α,9β,14β-androst-12-ene- 3,17-dione, 4α,8,14-trimethyl-	EtOH	246(3.92)	78-3459-66
$C_{21}H_{30}O_2S$			
2,4,6,8-Nonatetraen-1-ol, 9-(3,4- dihydro-4,4,6-trimethyl-2H-thio- pyran-5-yl)-3,7-dimethyl-, acetate, all trans	C_6H_{12}	296(4.29),310(4.42), 334(4.49)	78-0265-66
$C_{21}H_{30}O_3$			
Androst-3-en-5-one, 17β-acetoxy-, 10(5→4)-abeo-	EtOH	249(3.86)	33-2218-66
Androst-4-en-3-one, 17-acetoxy-	EtOH	243(4.21),314(1.83)	27-0110-66
5β-Androst-1-en-3-one, 17β-acetoxy-	EtOH	231(3.86)	88-4113-66
1β,5α-Cycloandrostan-2-one, 17β-acetoxy-	EtOH	none	33-1049-66
10ξ-Estr-9(11)-en-2-one, 17β-acetoxy- 5α-methyl-	EtOH	220(2.95)(end abs.)	33-1183-66
Gon-4-en-3-one, 17β-acetoxy-13β-ethyl-	EtOH	240(4.23)	35-3120-66
Jasmolin I	hexane	219(4.33)	39-0332-66C
Pregn-4-ene-3,20-dione, 6α-hydroxy-	EtOH	241.5(4.19)	94-1096-66
9β,10α-Pregn-4-ene-3,20-dione, 11α-hydroxy-	EtOH	242(4.23)	33-1529-66
9β,10α-Pregn-4-ene-3,20-dione, 11β-hydroxy-	EtOH	240(4.20)	33-1529-66
9β,10α-Pregn-4-ene-3,20-dione, 15α-hydroxy-	MeOH	242(4.21)	54-0701-66
$C_{21}H_{30}O_4$			
Androstane-3,5-dione, 17β-acetoxy-, 10(5→4)-abeo-	EtOH	291(3.96)	33-2218-66
ferric complex	EtOH	544(3.10)	33-2218-66
Androstan-3-one, 4α,5α-epoxy- 17β-hydroxy-, acetate	EtOH	299(1.71)	33-2218-66

Compound	Solvent	λ max(log ε)	Ref.
Androstan-3-one, 4β,5β-epoxy-17β-hydroxy-, acetate	EtOH	303(1.65)	33-2218-66
Gon-4-en-3-one, 17β-acetoxy-13β-ethyl-	EtOH	237(4.14)	35-3120-66
10β-hydroxy-	EtOH	238.5(4.16)	35-3120-66
A-Norandrost-3-en-2-one, 17β-acetoxy-3-methoxy-	EtOH	250(4.06)	44-3189-66
4-Oxa-A-homoandrost-4'-en-3-one,	C_6H_{12}	202(3.97)	78-1317-66
17β-acetoxy-	EtOH	204(4.00)	78-1317-66
Pregn-4-ene-3,20-dione, 14α,17α-dihydroxy-	EtOH	240(4.21)	78-0325-66A
Pregn-4-ene-3,20-dione, 17α,21-dihydroxy-	MeOH	242(4.21)	54-0731-66
9β,10α-Testosterone, 11α-hydroxy-, 17-acetate	MeOH	242(4.21)	54-0721-66
$C_{21}H_{30}O_5$			
2-Chromanvaleric acid, 6-acetoxy-α,2,5,7,8-pentamethyl-	EtOH	284.0(3.33)	73-2434-66
9β,10α-Pregn-4-ene-3,20-dione, 11α,17α,21-trihydroxy-	EtOH	240(4.22)	33-1529-66
$C_2H_4Cl_2$solvate	EtOH	241(4.23)	33-1529-66
9β,10α-Pregn-4-ene-3,20-dione,	MeOH	241(4.22)	54-0731-66
11β,17α,21-trihydroxy-	EtOH	240(4.22)	33-1529-66
$C_{21}H_{31}NO$			
Cyclopropanemethanol, 2-(4-pyridyl)-	EtOH-HCl	262(4.20)	44-0399-66
α,α-dicyclohexyl-, trans	EtOH-NaOH	258(3.44)	44-0399-66
$C_{21}H_{31}NO_2$			
1H-Cyclopenta[7,8]phenanthro[2,3-c]-isoxazol-1-ol, tetradecahydro-10,10a,12a-trimethyl-	MeOH	223(3.69)	73-1064-66
1H-Cyclopenta[7,8]phenanthro[2,3-d]-isoxazol-1-ol, tetradecahydro-10,10a,12a-trimethyl-	MeOH	227(3.68)	73-1064-66
$C_{21}H_{31}NO_2S$			
5α-Androstan-2-one, 17β-hydroxy-17α-methyl-3α-thiocyano-	MeOH	292(1.78)	87-0693-66
3β-thiocyano isomer	MeOH	327(1.86)	87-0693-66
$C_{21}H_{31}N_3O$			
1H-Cyclopenta[5,6]naphtho[1,2-g]quin-	HCl	260(4.08)	32-0152-66
azolin-1-ol, 10-aminotetradecahydro-11a,13a-dimethyl-	EtOH	236(4.02),268(3.69)	32-0152-66
$C_{21}H_{31}N_3OS$			
8H-Cyclopenta[5,6]naphtho[1,2-g]quin-	EtOH-HCl	229(3.89),283(4.26)	32-0152-66
azoline-8-thione, 10-aminohexadeca-	EtOH	274(4.20)	32-0152-66
hydro-1-hydroxy-11a,13a-dimethyl-	EtOH-NaOH	220(4.07),268(4.12)	32-0152-66
$C_{21}H_{31}N_3O_2$			
8H-Cyclopenta[5,6]naphtho[1,2-g]quin-	EtOH-HCl	288(3.91)	32-0152-66
azolin-8-one, 10-aminohexadecahydro-	EtOH	276(3.81)	32-0152-66
1-hydroxy-11a,13a-dimethyl-	EtOH-NaOH	285(3.69)	32-0152-66
$C_{21}H_{31}Na_2O_7P$			
Chroman, 6-acetoxy-2,5,7,8-tetramethyl-2-(5-hydroxy-4-methylpentyl)-, disodium phosphate	EtOH	285(3.34)	73-4598-66

Compound	Solvent	$\lambda_{max}(\log \epsilon)$	Ref.
$C_{21}H_{32}$			
Androsta-5,7-diene, 4,4-dimethyl-	EtOH	273(4.01),282(4.00)	39-1847-66C
5α-Androsta-6,8(14)-diene, 4,4-dimethyl-	EtOH	245s(--),253(4.35), 261s(--)	39-1847-66C
$C_{21}H_{32}ClN_3O_2$			
[[5-[(5-Carboxy-3-ethyl-4-methylpyrrol-2-yl)methyl]-4-ethyl-3-methylpyrrol-2-yl]methylene]dimethylammonium chloride, ethyl ester	$CHCl_3$	278(4.24),630(4.27)	39-0030-66C
$C_{21}H_{32}N_2O$			
5α-Androst-2-ene-2-carbonitrile, 3-amino-17-hydroxy-17α-methyl-	EtOH	202.5(4.03)	32-0152-66
Cyclopenta[7,8]phenanthro[2,3-c]-pyrazol-1-ol, hexadecahydro-trimethyl-	MeOH	222(3.73)	73-1064-66
$C_{21}H_{32}N_2O_2$			
Urea, 1,3-dicyclohexyl-1-(3-methylsalicyl)-	MeOH	278(3.32)	35-5855-66
	MeOH-KOH	242(3.75),294(3.49)	35-5855-66
$C_{21}H_{32}N_4O$			
1H-Cyclopenta[5,6]naphtho[1,2-g]quinazolin-1-ol, 8,10-diaminotetradecahydro-11a,13a-dimethyl-	MeOH-HCl	220(4.15),274(3.66)	32-0152-66
	EtOH	216(4.11),281(3.72)	32-0152-66
Guanidine, (2-cyano-17β-hydroxy-5α-androst-2-en-3-yl)-	n.s.g.	233(3.97)	32-0152-66
$C_{21}H_{32}O$			
Androst-5-en-7-one, 4,4-dimethyl-	EtOH	242(4.06)	22-0850-66
5α-Pregn-9(11)-en-12-one	EtOH	240(4.12)	35-0536-66
$C_{21}H_{32}O_2$			
Communic acid, methyl ester, cis	EtOH	235(4.30)	1-1074-66
Communic acid, methyl ester, trans	EtOH	232(4.41)	1-1074-66
	EtOH	232(4.42)	12-2403-66
6H-Dibenzo[b,d]pyran-1-ol, 6a,7,8,9,-10,10a-hexahydro-6,6,9-trimethyl-3-pentyl-	EtOH	275(3.07),282(3.07)	78-1481-66
2,4,6,10,14-Hexadecapentaenoic acid, 3,7,11,15-tetramethyl-, methyl ester	EtOH	318(4.52)	39-2154-66C
15-Oxa-D-homo-5α-androst-17(17a)-en-16-one, 4,4-dimethyl-	EtOH	218.5(3.83)	39-1847-66C
5β,17α-Pregnane-3,20-dione	EtOH	282(1.70)	78-1615-66
5α-Pregn-16-en-20-one, 3β-hydroxy-	EtOH	240(3.91),322(1.79)	102-0707-66
5α-Pregn-16-en-20-one, 12β-hydroxy-	EtOH	243(3.91)	35-0536-66
$C_{21}H_{32}O_3$			
5α-Androstan-3-one, 17β-hydroxy-2-(hydroxymethylene)-1α-methyl-	MeOH	285(4.38)	73-1064-66
D-Norandrost-4-ene-16β-carboxaldehyde, 3-oxo-, dimethyl acetal	MeOH	240(4.20)	13-0505-66A
15-Oxa-D-homo-5α-androstane-16,17a-dione, 4,4-dimethyl-	EtOH	244(3.78),270s(--)	39-1847-66C
	EtOH-NaOH	269.5(4.23)	39-1847-66C
5α-Pregn-16-en-20-one, 3β,6β-dihydroxy-	EtOH	240(3.98)	78-0123-66B
5α-Pregn-16-en-20-one, 3β,15α-dihydroxy-	EtOH	241.5(3.80)	5-0169-66D

Compound	Solvent	λ_{max}(log ε)	Ref.
$C_{21}H_{32}O_4$			
Agathic acid, methyl ester	EtOH	220(4.11)	12-2403-66
5α-Androst-3-en-2-one, 17β-acetoxy-3-hydroxy-	EtOH	272(3.75)	44-2395-66
Isotaxininol, anhydrodeformyldihydro-, monoacetate	EtOH	212(3.73)	95-1172-66
Mutilin, hydroxymethylene deriv.	EtOH	268(3.93)	78-0359-66B
	EtOH-NaOH	308(4.31)	78-0359-66B
5α-Pregn-16-en-20-one, 3β,14β,15α-trihydroxy-	EtOH	231.5(3.87)	94-0809-66
$C_{21}H_{32}O_5$			
5α-Etianic acid, 3β,12β-dihydroxy-11-oxo-, methyl ester	EtOH	289(1.60)	33-1632-66
2-Propen-1-one, 1,2-bis[1-methyl-4,4-(ethylenedioxy)cyclohexyl]-	MeOH	225(3.37)	5-0031-66D
$C_{21}H_{33}Li_3N_7O_{17}P_3$			
Oxycoenzyme A, trilithium salt	pH 7.0	259(4.17)	35-2299-66
$C_{21}H_{33}NOSn$			
Tin, (8-quinolinolato)tributyl-	C_6H_{12}	362(3.31)	101-0249-66B
$C_{21}H_{34}$			
5α-Pregn-9(11)-ene	C_6H_{12}	191.5(3.85)	39-1266-66C
5β-Pregn-11-ene	C_6H_{12}	190(4.01)	39-1266-66C
$C_{21}H_{34}Cl_2N_2$			
Piperidine, 3-benzoyloxy-5,5-dimethyl-2-(4,5,5-trimethyl-2-pyrrolidinyl)-, dihydrochloride (as trihydrate)	EtOH	228(4.07)	78-0015-66
$C_{21}H_{34}N_4O_4$			
Pentadecanal, 2,4-dinitrophenylhydrazone	$CHCl_3$	360(4.26)	54-0117-66
$C_{21}H_{34}O_2$			
Cannabichromene, tetrahydro-	EtOH	275(3.09),281(3.10)	77-0020-66
Labd-8(20),13-dien-15-oic acid, methyl ester	EtOH	220(4.20)	78-0203-66B
$C_{21}H_{34}O_4$			
Cassaidic acid, methyl ester	EtOH	224(4.20)	35-5865-66
Drevogenin P, anhydrotetrahydro-	EtOH	283(1.57)	33-1632-66
Isocassaidic acid, methyl ester	EtOH	224(4.22)	35-5856-66
Isodrevogenin P, anhydrotetrahydro-	EtOH	285(1.53)	33-1632-66
$C_{21}H_{34}O_5$			
Androstan-3-one, 17β-acetoxy-4β,5β-dihydroxy-	dioxan	273(1.60)	33-1986-66
4,9-Pentadecadienoic acid, 8-hydroxy-11-isopropyl-4,8-dimethyl-6,14-dioxo-, methyl ester	EtOH	239(4.10)	44-1797-66
Taxicin-I, 5-deoxydihydro-2-O-methyl-	EtOH	281(3.76)	39-1933-66C
Trisporic C acid, tetrahydro-, methyl ester, acetate	n.s.g.	249(4.10)	78-0175-66A
$C_{21}H_{36}O_2$			
Labd-8-en-15-oic acid, methyl ester	C_6H_{12}	198(3.95),228(3.00)	12-2133-66

Compound	Solvent	λ_{max}(log ϵ)	Ref.
$C_{21}H_{36}O_3$			
Labd-13-en-15-oic acid, 8β-hydroxy-, methyl ester	EtOH	220(4.19)	78-0203-66B
$C_{21}H_{36}O_5$			
Humulinone, tetrahydrodeoxy-	acid	254(4.08)	39-2308-66
	base	275(4.2)	39-2308-66
$C_{21}H_{40}O_2$			
2,4-Heneicosanedione	isooctane	298(3.48)	44-0628-66
Phytenoic acid, methyl ester, cis	EtOH	220(4.16)	39-2144-66C
Phytenoic acid, methyl ester, trans	EtOH	219(4.21)	39-2144-66C
$C_{21}H_{41}N_3O$			
Phytenal, semicarbazone	EtOH	274(4.47)	39-2144-66C

Compound	Solvent	λ_{max}(log ϵ)	Ref.
$C_{22}H_6Cl_4N_4O_2$			
Bisbenzimidazo[1,2-a:1',2'-a']benz-[1,2-c:4,5-c']dipyrrole-7,15-dione, 6,14,x,x-tetrachloro-	$C_{10}H_7Cl$	420(3.94)	33-0534-66
$C_{22}H_8Br_2N_4O_2$			
Bisbenzimidazo[1,2-a:1',2'-a']benz-[1,2-c:4,5-c']dipyrrole-7,15-dione, 6,14-dibromo-	$C_{10}H_7Cl$	415(3.91),470s(3.81)	33-0534-66
Bisbenzimidazo[1,2-a:2',1'-e']benz-[1,2-c:4,5-c']dipyrrole-13,15-dione, 6,14-dibromo-	$CHCl_3$	313(4.73),337(4.20), 351(4.25),423(4.14)	33-0534-66
	$C_{10}H_7Cl$	422(4.10)	33-0534-66
$C_{22}H_8Cl_2N_4O_2$			
Bisbenzimidazo[1,2-a:1',2'-a']benz-[1,2-c:4,5-c']dipyrrole-7,15-dione, 6,14-dichloro-	$C_{10}H_7Cl$	415(3.92),480s(3.8)	33-0534-66
Bisbenzimidazo[1,2-a:2',1'-e']benz-[1,2-c:4,5-c']dipyrrole-13,15-dione, 6,14-dichloro-	$CHCl_3$	307(4.75),334(4.28), 348(4.34),421(4.18)	33-0534-66
	$C_{10}H_7Cl$	418(4.12)	33-0534-66
$C_{22}H_{12}ClN_3O_2$			
5,12-Diazabenz[a]anthracene, 7-chloro-6-(o-nitrophenyl)-	$CHCl_3$	286(4.85),366(4.02)	39-1245-66C
$C_{22}H_{12}Cl_2N_2$			
5,12-Diazabenz[a]anthracene, 7-chloro-6-(o-chlorophenyl)-	$CHCl_3$	288(4.78),368(3.95)	39-1245-66C
$C_{22}H_{12}Cl_4O_2$			
Anthracene, 9,10-dimethyl-, tetrachloro-o-benzoquinone adduct	C_6H_{12}	230s(4.59),253s(4.20), 302(3.48)	23-2507-66
$C_{22}H_{12}N_2O$			
5-Oxa-10,15-diazabenzo[a]naphth-[1,2,3-de]anthracene	EtOH-HCl	247(4.57),258(4.54), 278(4.51),308(4.70), 368(3.71),420(3.94)	39-1245-66C
	EtOH	242(4.71),256(4.48), 284(4.66),311(4.29), 325(4.30),341(3.69), 358(3.86),375(4.03), 395(4.10),415(3.92)	39-1245-66C
$C_{22}H_{13}N$			
4H-Benzo[def]naphtho[2,3-b]carbazole	benzene	298(4.94),320(4.42), 334(4.46),345s(4.20), 384s(3.92),400(4.17), 418(4.13),445(4.07)	24-1279-66
$C_{22}H_{13}NO_2S$			
9H-Benzo[a]phenoxazin-9-one, 5-(phenylthio)-	benzene	518(--)	65-1938-66
	dioxan	508(--)	65-1938-66
	pyridine	523(--)	65-1938-66
	PhNO	528(--)	65-1938-66

Compound	Solvent	λmax(log ε)	Ref.
$C_{22}H_{13}N_3$			
5,10,15-Triazabenzo[a]naphth[1,2,3-de]-anthracene	EtOH	238(4.74),246(4.74), 280(4.54),312(4.06), 329(3.71),345(3.78), 374(4.03),382(4.04), 394(4.27),421(4.28), 447(4.22)	39-1245-66C
$C_{22}H_{14}$			
m-Terphenyl, 3,3"-diethynyl-	EtOH	249(4.58)	22-1012-66
$C_{22}H_{14}Br_2$			
Indene, 3-bromo-1-(α-bromo-benzylidene)-2-phenyl-	EtOH	247(4.22),277(4.19), 340(4.12)	35-4525-66
stereoisomer	EtOH	250(4.18),274(4.19), 337(4.10)	35-4525-66
$C_{22}H_{14}Cl_2N_4S_4$			
p-Dithiino[2,3-d:5,6-d']dipyridazine, 1,6-bis[(p-chlorobenzyl)thio]-	EtOH	206(4.71),241(4.56), 262(4.55),320(3.54)	4-0541-66
$C_{22}H_{14}Cl_4N_4S_2$			
Pyridine, 2,2'-[dithiobis(o-phenyl-eneimino)]bis[3,5-dichloro-	EtOH	273(4.56),285(4.54), 347(4.20)	95-0050-66
$C_{22}H_{14}Fe$			
Hexatriyne, 1-ferrocenyl-6-phenyl-	$CHCl_3$	255(4.72),289(4.57), 312s(4.20),334(4.23), 356(4.04),450(3.35)	101-0399-66B
$C_{22}H_{14}N_2O_2$			
9H-Benzo[a]phenoxazin-9-one, 5-p-anilino-	pyridine	430(4.4)	65-1938-66
	40% pyridine	600s(4.3),640(4.4)	65-1938-66
$C_{22}H_{14}N_4$			
1,1,2,2-Cyclobutanetetracarbonitrile, 3-(2,2-diphenylvinyl)-	C_6H_{12}	233s(--),268(4.12)	44-0589-66
$C_{22}H_{14}O_4$			
2,2'-Bifuran, 5,5'-dibenzoyl-	$CHCl_3$	263(4.14),370(4.55), 388(4.53)	39-0976-66C
$C_{22}H_{14}O_6$			
Benzo[1,2-b:4,5-b']bisbenzofuran, 2,8-diacetoxy-	dioxan	262(4.33),305s(4.40), 317(4.77),326(4.65), 334(4.86)	1-2202-66
$C_{22}H_{15}Br$			
Indene, 1-(α-bromobenzylidene)-2-phenyl-	EtOH	252(4.34),274(4.38), 339(3.99)	35-4525-66
$C_{22}H_{15}ClO_2S$			
Dibenzobicyclo[2.2.2]octatriene, 7-chloro-8-(phenylsulfonyl)-	CCl_4	256(3.44),268s(3.48), 274(3.52),280(3.54)	35-3095-66
Dibenzotricyclo[3.2.1.0^{2,8}]octa-3,6-diene, 1-chloro-8-(phenylsulfonyl)-	CCl_4	256(3.41),265(3.38), 273(3.36),281(3.10)	35-3095-66
$C_{22}H_{15}ClO_5$			
9-Methyl-1,2;5,6-dibenzoxanthylium perchlorate	HOAc	468(3.23)	24-1822-66

Compound	Solvent	λ_{max}(log ε)	Ref.
9-Methyl-1,2;6,7-dibenzoxanthylium perchlorate	HOAc	472(3.34)	24-1822-66
9-Methyl-2,3;5,6-dibenzoxanthylium perchlorate	HOAc	420(2.90),560(3.44)	24-1822-66
$C_{22}H_{15}NO_4$			
Naphthostyril, N,2-diacetyl-3-benzoyl-	EtOH	210(4.57),224s(4.44), 261(4.45),286s(4.12), 357s(3.42),386(3.55)	39-1028-66C
$C_{22}H_{15}N_3O$			
Benzo[a][3,6]phenanthroline, 6-anilino-7-hydroxy-	$CHCl_3$	260(4.53),292(4.36), 295(4.35),320(4.45), 430(3.66)	87-0161-66
5,12-Diazabenz[a]anthracene, 7-anilino-6-hydroxy-	$CHCl_3$	273(4.53),364(3.95), 394(3.99)	39-1245-66C
$C_{22}H_{15}OP$			
Phosphine oxide, diphenyl-(phenylbutadiynyl)-	EtOH	237(4.75),248(4.48), 259(4.00),273(4.26), 289(4.40),307(4.32)	22-1002-66
$C_{22}H_{15}P$			
Phosphine, diphenyl(phenylbutadiynyl)-	EtOH	236(4.58),247(4.52), 259(4.43),275(4.30), 291(4.23),307(4.08)	22-1002-66
$C_{22}H_{16}$			
Anthracene, 9-styryl-	$C_2H_4Cl_2$	259(5.13),288(4.02), 292(4.05),372(4.09), 389(4.15)	99-0085-66
Ethylene, 1,2-di-1-naphthyl-	hexane	232(4.73),290(3.85), 335(4.33)	99-0085-66
Ethylene, 1-(1-naphthyl)-2-(2-naphthyl)-	$C_2H_4Cl_2$	295(4.30),330(4.38)	99-0085-66
$C_{22}H_{16}ClNO_4S$			
2,3-Diphenyl-4H-indeno[2,1-d]thiazolium perchlorate	MeOH	245(4.05),352(4.28)	39-0686-66C
$C_{22}H_{16}Cl_2O_2S$			
Dibenzobicyclo[3.2.1]octadiene, 4,5-dichloro-8-(phenylsulfonyl)-, endo-syn	CCl_4	252(2.94),259(3.10), 266(3.25),273(3.17)	35-3095-66
$C_{22}H_{16}Cl_4FeN_4$			
6-Ethynyl-1,3,5-triphenylverdazylium chloroferrate	HCOOH	318(4.21),553(4.08)	49-1280-66
$C_{22}H_{16}N_2$			
6H-Pyrido[4,3-b]carbazole, 5-methyl-1-phenyl-	85% EtOH-HCl	237(4.46),250(4.40), 258(4.41),276(4.45), 314(4.93),355(3.96)	87-0237-66
	EtOH	225(4.45),243(4.39), 290(4.83),295(4.86), 333(3.81),348(3.72), 379(3.69),395(3.67)	87-0237-66

Compound	Solvent	λ max(log ε)	Ref.
$C_{22}H_{16}N_2O_2$			
Naphtho[2,3-b]phenazine-7,12-dione, 5,14-dihydro-5,14-dimethyl-	EtOH	245(4.71),361(4.40), 632(3.09)	44-3734-66
$C_{22}H_{16}N_4S_4$			
p-Dithiino[2,3-d:5,6-d']dipyridazine, 1,6-bis(benzylthio)-	EtOH	206(4.63),244(4.45), 264(4.52),338(3.56)	4-0541-66
$C_{22}H_{16}O$			
Cyclopropene, 3-benzoyl-1,2-diphenyl-	EtOH	228(4.39),235(4.41), 244(4.28),294s(4.31), 308(4.40),321s(4.24)	18-1975-66
$C_{22}H_{16}O_2$			
o-Toluic acid, α-2-anthryl-	50% EtOH	258(5.30),325(3.40), 341(3.63),358(3.76), 378(3.68)	24-0396-66
$C_{22}H_{16}O_6$			
1-Naphthacenecarboxylic acid, 2-ethyl-5,7-dihydroxy-6,11-dihydro-6,11-dioxo-, methyl ester	hexane	242(4.65),255(4.63), 262(4.68),279(4.26), 290(4.28),445(4.28), 462(4.18),475(4.20)	94-0802-66
$C_{22}H_{16}O_{11}$			
Furan, 2,5-bis(2-carbomethoxy-3,6-dioxocyclohexadien-1-yl)-3,4-dimethoxy-	EtOH	216(4.46),336s(4.31), 270s(4.03),341(4.07), 518(3.74)	33-1794-66
$C_{22}H_{17}BrO$			
Chalcone, α-(bromomethyl)-p'-phenyl-	isooctane	285(4.54)	88-4037-66
$C_{22}H_{17}Cl$			
1,3-Butadiene, 1-chloro-1,4,4-triphenyl-, trans	EtOH	249(4.13),338(4.58)	44-2175-66
$C_{22}H_{17}ClN_2$			
6H-Pyrido[4,3-b]carbazole, 1-(p-chlorophenyl)-3,4-dihydro-5-methyl-	85% EtOH-HCl	222(4.32),232(4.35), 238(4.37),245(4.37), 278s(4.41),287(4.49), 306(4.12),319(4.15), 392(4.30)	87-0237-66
	EtOH	240(4.61),248(4.58), 283(4.54),298s(4.35), 320(4.09),331(4.13), 343s(3.93)	87-0237-66
$C_{22}H_{17}ClO_2$			
2H-1-Benzopyran, 3-(p-chlorophenyl)-4-(p-methoxyphenyl)-	MeOH	243(4.39),284s(4.02), 304(4.05),325(4.06)	87-0516-66
$C_{22}H_{17}Cl_2N_3O_6S_3$			
p-Toluenesulfonamide, N-[(p-chlorophenyl)sulfonyl]-N-[1-[(p-chlorophenyl)sulfonyl]pyrazol-5-yl]-	MeOH	216(4.2),240(4.6), 269s(3.5),278s(3.2)	24-0183-66
$C_{22}H_{17}N$			
Pyrrole, 2,3,4-triphenyl-	EtOH	254(--),303s(--)	12-1871-66
Pyrrole, 2,4,5-triphenyl-	EtOH	238(--),257(--), 300s(--),320(--)	12-1871-66

Compound	Solvent	$\lambda_{max}(\log \epsilon)$	Ref.
$C_{22}H_{17}NO$			
2-Indolinone, 3-(diphenylmethylene)-1-methyl-	EtOH	254(4.26),388(4.02)	44-0077-66
2(1H)-Naphthalenone, 1-[(1-methyl-4(1H)-quinolylidene)ethylidene]-, (spectra in other solvents, not listed)	benzene	637(--),690s(--)	61-0817-66
	EtOH-HCl	475(4.3)(anom.)	61-0817-66
	$EtOH-NH_3$	408(3.8),645(4.5)	61-0817-66
	MeCN	435(3.8),667(4.7)	61-0817-66
$C_{22}H_{17}NO_2$			
Spiro[indoline-3,2'-oxiran]-2-one, 1-methyl-3',3'-diphenyl-	EtOH	230(4.38),256s(3.99), 274s(3.58),313(3.28), 344(3.04)	44-0077-66
$C_{22}H_{17}NO_4$			
Cyclopenta[c]quinolizine-1,2-dicarboxylic acid, 4-phenyl-, dimethyl ester	EtOH	238s(4.35),251(4.41), 284(4.37),350(4.45), 380s(4.0),463(3.61)	39-0324-66C
$C_{22}H_{17}N_3$			
2-Naphthylamine, 1-(3-biphenylylazo)-	benzene	450(4.13)	44-2571-66
	$C_6H_{11}Me$	434(4.17)	44-2571-66
2-Naphthylamine, 1-(4-biphenylylazo)-	benzene	457(4.14)	44-2571-66
	$C_6H_{11}Me$	450(4.26)	44-2571-66
$C_{22}H_{17}N_3O$			
1-Naphthol, 7-amino-8-(3-biphenylylazo)-	benzene	508(4.22)	44-2571-66
	$C_6H_{11}Me$	507(4.19)	44-2571-66
1-Naphthol, 7-amino-8-(4-biphenylylazo)-	benzene	517(4.35)	44-2571-66
	$C_6H_{11}Me$	516(4.33)	44-2571-66
$C_{22}H_{17}N_4$			
Verdazyl, 6-ethynyl-1,3,5-triphenyl-	dioxan	243(4.19),270(4.44), 320(4.08),384(3.99), 705(3.70)	49-0846-66
$C_{22}H_{18}$			
Anthracene, 1,4-dimethyl-9-phenyl-	$CHCl_3$	265(4.88),360(3.81), 379(4.00),399(3.95)	44-4082-66
Anthracene, 2,3-dimethyl-9-phenyl-	$CHCl_3$	264(5.06),354(3.76), 371(3.92),393(3.88)	44-4082-66
Benzene, 1,3-distyryl-	dioxan	325(4.37)	18-1547-66
Fluorene, 9-(1,6-dimethyl-2,4-norcaradien-7-ylidene)-	MeCN	231(4.71),249(4.50), 258(4.52),267(4.21), 277(4.20),297(4.02), 313(4.03),325(3.95)	89-0251-66
$C_{22}H_{18}BN$			
Pyridine, compound with (1,3-pentadien-1-yl)diphenylborane	EtOH	240(4.00)	22-1981-66
$C_{22}H_{18}ClNO_2$			
Indole, 5-chloro-2,3-bis-(p-methoxyphenyl)-	EtOH	257(4.48),314(4.23)	87-0527-66
$C_{22}H_{18}CuN_4O_2$			
Copper, bis(8-amino-7-quinolyl methyl ketonato)-	n.s.g.	245(4.54),285(.81), 346(4.14),386(4.05), 400(4.08),505(3.25), 575(2.38)	24-3806-66

Compound	Solvent	λ_{max}(log ε)	Ref.
$C_{22}H_{18}FNO_2$			
Indole, 5-fluoro-2,3-bis-(p-methoxyphenyl)-	EtOH	251(4.46),310(4.35)	87-0527-66
Indole, 7-fluoro-2,3-bis-(p-methoxyphenyl)-	EtOH	253(4.54),305(4.29)	87-0527-66
$C_{22}H_{18}Fe_2$			
Acetylene, diferrocenyl-	EtOH	230(4.29),265(4.06),302(4.04),453(2.91)	101-0173-66B
	$CHCl_3$	258(4.02),300(3.85),345s(3.06),444(2.90)	101-0399-66B
$C_{22}H_{18}N_2O_2$			
5-Norbornene-2,3-dicarboximide, 5-methyl-7-(α-2-pyridylbenzylidene)-	MeOH	243(4.19)	44-2149-66
$C_{22}H_{18}N_2O_2S_3$			
2,5-Thiophenediacetonitrile, α,α'-bis[5-(methoxymethyl)-2-thenylidene]-	n.s.g.	299(3.96),352(4.34),445(4.63)	12-1243-66
$C_{22}H_{18}N_2O_3$			
Anthraquinone imine, N-[p-(dimethylamino)phenyl]-1,8-dihydroxy-	MeOH	227(4.7),250(4.5),280(4.4),410(4.1)	83-0783-66
$C_{22}H_{18}N_2O_5$			
Camptothecin, acetate	n.s.g.	220(4.59),254(4.46),290(3.79),365(4.34)	35-3888-66
$C_{22}H_{18}N_2O_6$			
1-Indanone, 2-acetyl-2-nitroso-, dimer	EtOH	253(4.26),299(3.73)	44-2090-66
$C_{22}H_{18}N_4$			
Benzimidazo[1,2-c]quinazoline, 6-(6-amino-m-tolyl)-2-methyl-	EtOH	230(4.62),267(4.68),276(4.75),304(4.07),320(4.05),339(3.93)	4-0289-66
Formazan, 1,3,5-triphenyl-5-(2-propynyl)-	dioxan	249(4.20),281(4.02),388(4.26)	49-0846-66
$C_{22}H_{18}N_4NiO_2$			
Nickel, bis(8-amino-7-quinolyl methyl ketonato)-	n.s.g.	233(4.49),282(4.64),300(4.30),336(3.78),429(4.10),568(3.75),634(3.71)	24-3806-66
$C_{22}H_{18}N_4O$			
Pyrazole-4,5-dione, 1,3-diphenyl-, 4-(o-tolylhydrazone)	hexane	263(4.45),412(4.43)	44-1722-66
	EtOH	263(4.54),414(4.49)	44-1722-66
	$CHCl_3$	265(4.44),414(4.41)	44-1722-66
Pyrazole-4,5-dione, 1,3-diphenyl-, 4-(p-tolylhydrazone)	hexane	265(4.38),410(4.55)	44-1722-66
	EtOH	266(4.63),414(4.59)	44-1722-66
	$CHCl_3$	268(4.52),417(4.51)	44-1722-66
$C_{22}H_{18}N_4O_2$			
Pyrazole-4,5-dione, 1,3-diphenyl-, 4-[(o-methoxyphenyl)hydrazone]	hexane	267(4.49),417(4.43)	44-1722-66
	EtOH	267(4.43),427(4.37)	44-1722-66
	$CHCl_3$	269(4.41),429(4.37)	44-1722-66
Pyrazole-4,5-dione, 1,3-diphenyl-, 4-[(p-methoxyphenyl)hydrazone]	hexane	270(4.59),424(4.55)	44-1722-66
	EtOH	272(4.53),430(4.40)	44-1722-66

Compound	Solvent	$\lambda_{max}(\log \epsilon)$	Ref.
Pyrazole-4,5-dione, 1,3-diphenyl-, 4-[(p-methoxyphenyl)hydrazone]	$CHCl_3$	274(4.42),433(4.40)	44-1722-66
$C_{22}H_{18}N_4O_2Pd$			
Palladium, bis(8-amino-7-quinolyl methyl ketonato)-	n.s.g.	242(4.74),288(4.52), 298(4.60),419(4.20), 539(3.91),571(4.14)	24-3806-66
$C_{22}H_{18}N_4O_4$			
3-Butenal, 4,4-diphenyl-, 2,4-dinitrophenylhydrazone	n.s.g.	250(4.36),360(4.38)	22-0717-66
Crotonaldehyde, 2,3-diphenyl-, 2,4-dinitrophenylhydrazone	n.s.g.	387(4.51)	35-1518-66
$C_{22}H_{18}O$			
Benzyl alcohol, α-(2,3-diphenyl-2-Cyclopropen-1-yl)-	EtOH	224(4.39),230(4.44), 237(4.35),310(4.30), 320(4.39),337(4.22)	18-1975-66
Chalcone, α-methyl-p'-phenyl-	isooctane	284(4.43)	88-4037-66
$C_{22}H_{18}O_2$			
Indan. 3-furfurylidene-1-(p-methoxybenzylidene)-	EtOH	292s(4.55),303(4.59), 316s(4.51),339s(4.52), 353(4.57),371s(4.40)	5-0057-66H
$C_{22}H_{18}O_2S$			
Methylphenylsulfonium dibenzoyl-methylide	EtOH	288(3.87),317(3.80)	78-2145-66
$C_{22}H_{18}O_3$			
Benzofuran, 2,3-bis(p-methoxyphenyl)-	EtOH	228s(4.28),246(4.31), 310(4.44)	87-0527-66
$C_{22}H_{18}O_5$			
Cyclohexanone, 2,6-dipiperonylidene-	EtOH	370(4.37)	44-0639-66
$C_{22}H_{18}O_6$			
2-Naphthoic acid, 7-ethoxy-3-(hydroxy-methyl)-6-methoxy-1-[(3,4-methylene-dioxy)phenyl]-, lactone	EtOH	258(4.55),289(3.85), 347(3.51)	39-1775-66C
$C_{22}H_{18}O_7$			
Isopaulownin	EtOH	238(3.96),287(3.91)	94-0641-66
$C_{22}H_{18}O_8$			
1,2,5,6-Anthracenetetrol, tetraacetate	dioxan	259(4.98),349(3.44), 366(3.50),386(3.47)	24-2322-66
Succinic acid, dipiperonylidene-, dimethyl ester, cis-trans	dioxan	350(4.47)	44-3342-66
trans-trans	dioxan	235(4.39),291(4.34), 328(4.45)	44-3342-66
Succinic acid, dipiperonylidene-, ethyl ester, cis-trans	EtOH	346(4.43)	44-3342-66
trans-trans	EtOH	218(4.36),230s(4.30), 290(4.27),328(4.38)	44-3342-66
$C_{22}H_{19}BrO_7$			
Elephantol, p-bromobenzoate	MeOH	246(4.41)	35-3674-66

Compound	Solvent	λ max(log ε)	Ref.
$C_{22}H_{19}ClN_2O$			
Benzamide, p-chloro-N-[2-(1-methylcarbazol-2-yl)ethyl]-	EtOH	217(4.55),239(4.79), 248(4.69),258(4.48), 286s(4.14),296(4.34), 323(3.68),336(3.55)	87-0237-66
$C_{22}H_{19}ClN_2O_3$			
Alstoniline chloride	n.s.g.	284(3.9),333s(3.9), 382(4.5)	28D-1141-66A
$C_{22}H_{19}ClO_2$			
Chroman, 3-(p-chlorophenyl)-4-(p-methoxyphenyl)-, cis	MeOH	222(3.96),275(3.66), 283(3.59)	87-0516-66
$C_{22}H_{19}ClO_4S_2$			
4,7-Dimethyl-2-[(7-methyl-4H-1-benzothiopyran-4-ylidene)methyl]-1-benzothiopyrylium perchlorate	CH_2Cl_2	262(4.49),319(3.93), 337(3.86),402(3.75), 575s(4.32),613(4.71), 720s(3.17)	78-0007-66
$C_{22}H_{19}NO$			
Aziridine, 3-benzoyl-1-benzyl-2-phenyl-, cis	C_6H_{12}	324(2.17)	44-1244-66
trans	C_6H_{12}	347(2.45)	44-1244-66
2-Indolinone, 3-(diphenylmethyl)-1-methyl-	EtOH	256(3.90),267s(3.76)	44-0077-66
$C_{22}H_{19}NO_2$			
Indole, 2,3-bis(m-methoxyphenyl)-	EtOH	249(4.38),312(4.25)	87-0527-66
Indole, 2,3-bis(o-methoxyphenyl)-	EtOH	248(4.36),303(4.21)	87-0527-66
Indole, 2,3-bis(p-methoxyphenyl)-	EtOH	253(4.51),308(4.31)	87-0527-66
Indole, 2-(3,4-dimethoxyphenyl)-3-phenyl-	EtOH	257(4.40),315(4.31)	87-0527-66
$C_{22}H_{19}NO_2S$			
Methylphenylsulfonium α-(phenylcarbamoyl)phenacylide	EtOH	241(3.32),286(4.25)	78-2145-66
$C_{22}H_{19}NO_3S_2$			
1-Thiaflavanone, O-(p-tolylsulfonyl)-oxime	C_6H_{12}	226(4.40),265(4.00), 334(3.37)	5-0225-66C
$C_{22}H_{19}NO_4S$			
Flavanone, O-(p-tolylsulfonyl)oxime	C_6H_{12}	216(4.50),258(4.04), 312(3.68)	5-0225-66C
	EtOH	258(4.13),314(3.72)	5-0225-66C
$C_{22}H_{19}N_3O_3S$			
2H-Benz[h]pyrazolo[4,3-c]quinoline, 10,11-dihydro-7-methoxy-10-(p-tolylsulfonyl)-	EtOH	233(4.52),262(4.73), 306(4.17)	54-0681-66
$C_{22}H_{19}N_5O_6$			
Malonic acid, [3-(5-nitro-2-furyl)-acryloyl]-, bisphenylhydrazide	EtOH	236(4.57),276(4.28), 350(4.46)	95-1187-66
$C_{22}H_{20}BN$			
Pyridine, compound with (3-methyl-3-buten-1-ynyl)diphenylborane	EtOH	228(4.32)	22-1981-66

Compound	Solvent	λ max(log ε)	Ref.
$C_{22}H_{20}F_5IN_2S$			
2-[p-(Dimethylamino)styryl]-3-ethyl-6-(pentafluoropropenyl)benzothiazolium iodide	EtOH	556(--)	65-0828-66
$C_{22}H_{20}N_2O$			
[2.2]Paracyclophane, 4-hydroxy-7-(phenylazo)-	EtOH	215s(4.33),234s(4.14), 278(3.75),295(3.75), 393(4.21)	44-1227-66
	EtOH-NaOH	253s(3.93),310(3.75), 493(4.51)	44-1227-66
$C_{22}H_{20}N_2O_3$			
Naphtho[2,3-a]phenazine-8,13-dione, 5-acetyl-1,2,3,4,4a,5,14,14a-octahydro-	EtOH	495(3.90)	27-0281-66
$C_{22}H_{20}N_4O_2S_2$			
[$\Delta^{4,4'}$-Biimidazolidine]-2,2',5,5'-tetrone, 1,1'-diethyl-3,3'-diphenyl-2,2'-dithio-, trans	MeOH	278(3.8),350(3.7), 472(4.46)	24-1851-66
Hydantoin, 3-ethyl-5-[3-ethyl-4-oxo-2-(phenylimino)-5-thiazolidinylidene]-1-phenyl-2-thio-	MeOH	265(3.8),320(3.7), 424(4.53)	24-1851-66
$C_{22}H_{20}N_6O$			
Urea, 1-[bis(benzylidenehydrazino)-methylene]-3-phenyl-	EtOH	230(4.48),279(4.36)	39-0006-66C
$C_{22}H_{20}O_2$			
1(2H)-Naphthalenone, 3,4-dihydro-2-methylene-, dimer	EtOH	230s(--),250(4.22), 282(4.03),300s(--)	22-1693-66
$C_{22}H_{20}O_3$			
Methanol, (2,3-dihydro-2-methoxy-2-benzofuranyl)diphenyl-	MeOH	277(3.58),283(3.52)	24-1723-66
$C_{22}H_{20}O_6$			
2-Naphthoic acid, 7-ethoxy-3,4-dihydro-3-(hydroxymethyl)-6-methoxy-1-(3,4-methylenedioxyphenyl)-, lactone	EtOH	251(4.29),343(4.01)	39-1775-66C
Robustic acid	EtOH	232(4.45),259(4.49), 263s(4.48),347(4.26)	39-0606-66C
	EtOH-NaOEt	245s(4.61),251(4.62), 280(4.43),325s(4.19), 333(4.20)	39-0606-66C
	EtOH-NaOAc	245s(4.48),251(4.59), 280(4.40),326s(4.14), 333(4.20)	39-0606-66C
$C_{22}H_{20}O_7$			
2-Naphthoic acid, 1-(3,4,5-trimethoxyphenyl)-3-(hydroxymethyl)-7,8-(methylenedioxy)-3,4-dihydro-, lactone	EtOH	307(4.00),350(3.97)	78-1797-66
$C_{22}H_{20}O_8$			
Isopaulownin, acetate	EtOH	239(3.96),288(3.90)	94-0641-66

Compound	Solvent	λ_{max}(log ε)	Ref.
$C_{22}H_{20}O_9$			
Ceroalbolinic acid, tri-O-methyl-, methyl ester, acetate	EtOH	280(4.37),340(3.45)	78-1507-66
$C_{22}H_{21}ClO_8$			
Radicicol, diacetate	EtOH	279(4.20)	12-1265-66
$C_{22}H_{21}Cl_3CrN_3$			
Chromium, dichloro(o-chlorobenzyl)-tris(pyridine)-	n.s.g.	377(3.33),445s(--)	101-0542-66B
Chromium, dichloro(p-chlorobenzyl)-tris(pyridine)-	pyridine	378(3.39),440s(--)	101-0542-66B
$C_{22}H_{21}NO$			
Benzamide, 2,4,6-trimethyl-N,N-diphenyl-	EtOH	263s(4.0)	28C-0369-66A
Benzanilide, 2,4,6-trimethyl-2'-phenyl-	EtOH	258s(4.0)	28C-0369-66A
Cyclopropanemethanol, 2-(2-methyl-5-pyridyl)-α,α-diphenyl-, trans	EtOH-HCl	281(3.79)	44-0399-66
	EtOH-NaOH	275.5(3.62)	44-0399-66
$C_{22}H_{21}NO_2$			
Aziridine, 1-cyclohexyl-2,3-dibenzoyl-, cis	C_6H_{12}	324(2.31)	44-1244-66
trans	C_6H_{12}	351(2.72)	44-1244-66
Indoline, 2,3-bis(p-methoxyphenyl)-	EtOH-acid	255s(3.59),261(3.57), 269(3.59),276(3.59), 283s(3.54),298(3.19)	87-0527-66
	EtOH	226(4.39),277(3.71), 284(3.69),299(3.49)	87-0527-66
$C_{22}H_{21}NO_3$			
Acetophenone, 2-anilino-4'-methoxy-2-(p-methoxyphenyl)-	EtOH	223(4.32),248(4.23), 276s(4.27),282(4.27)	87-0527-66
$C_{22}H_{21}NO_4$			
Benzamide, 3,4,5-trimethoxy-N,N-diphenyl-	EtOH	244s(4.2),283(4.1)	28C-0584-66A
Benzanilide, 3,4,5-trimethoxy-2'-phenyl-	EtOH	212(4.6),278s(4.2), 300s(4.9)	28C-0584-66A
$C_{22}H_{21}N_3O_3$			
p-Benzotoluidide, 4-(dimethylamino)-α-(p-nitrophenyl)-	EtOH	302(4.36)	65-0214-66
$C_{22}H_{21}N_4$			
Verdazyl, 6-ethyl-1,3,5-triphenyl-	dioxan	246(4.16),283(4.38), 319(4.19),406(3.95), 429(3.94),720(3.65)	49-0846-66
$C_{22}H_{21}N_5O$			
Benzamide, N-[3-(3-amino-6-indol-3-ylpyrazinyl)propyl]-	MeOH-HCl	223(4.53),270s(4.16), 306(4.26),410(3.61)	88-3445-66
	MeOH-NaOH	226(4.50),271(4.22), 365(3.80)	88-3445-66
$C_{22}H_{21}N_5O_2$			
o-Acetanisidide, N-methyl-2,2-bis-(phenylazo)-	EtOH	430(4.24)	22-0400-66
	EtOH-KOH	503(4.30)	22-0400-66

Compound	Solvent	λ_{max}(log ϵ)	Ref.
$C_{22}H_{21}N_5O_3$			
1H-Imidazo[4,5-b]pyridine-5-carbamic acid, 6-[(diphenylmethyl)amino]-2,3-dihydro-2-oxo-, ethyl ester	pH 1	246(4.59),277(4.25), 305(4.29)	44-1890-66
	pH 13	237(4.52),298(4.25)	44-1890-66
$C_{22}H_{22}$			
m-Terphenyl, 2,2"-diethyl-	EtOH	234(4.35)	22-1012-66
m-Terphenyl, 2,3"-diethyl-	EtOH	243.5(4.43)	22-1012-66
m-Terphenyl, 2,4"-diethyl-	EtOH	247(4.45)	22-1012-66
m-Terphenyl, 3,4"-diethyl-	EtOH	252(4.61)	22-1012-66
m-Terphenyl, 4,4"-diethyl-	EtOH	253(4.68)	22-1012-66
$C_{22}H_{22}AsClPd$			
Palladium, chloro(1-methyl-π-allyl)-(triphenylarsine)-	CHCl	243(4.30),323(2.50)	101-0578-66B
Palladium, chloro(2-methyl-π-allyl)-(triphenylarsine)-	CHCl	246(4.32),324(2.53)	101-0578-66B
$C_{22}H_{22}BN$			
Pyridine, compound with bis(3-methyl-3-buten-1-ynyl)-o-tolylborane	EtOH	215(4.34)	22-1981-66
Pyridine, compound with bis(3-methyl-3-buten-1-ynyl)-p-tolylborane	EtOH	215(4.40)	22-1981-66
$C_{22}H_{22}BrNO$			
4-[2-(Hydroxydiphenylmethyl)cyclopropyl]-1-methylpyridinium bromide	EtOH-HCl	261(4.24)	44-0399-66
	EtOH-NaOH	261(4.25)	44-0399-66
$C_{22}H_{22}ClN_2O_4P$			
1-Ethyl-2-[(1-ethyl-2(1H)-quinolylidene)phosphino]quinolinium perchlorate	MeOH	594(4.65)	24-1325-66
	$C_2H_4Cl_2$	605(4.71)	24-1325-66
$C_{22}H_{22}ClPPd$			
Palladium, chloro(1-methyl-π-allyl)-(triphenylphosphine)-	$CHCl_3$	243(4.28),319(3.64)	101-0578-66B
Palladium, chloro(2-methyl-π-allyl)-(triphenylphosphine)-	$CHCl_3$	248(4.25),322(3.72)	101-0578-66B
$C_{22}H_{22}Cl_2CrN_3$			
Chromium, dichlorobenzyltris-(pyridine)-	M HClO	247(--),275(0.91), 298(0.89),358(0.39)	101-0542-66B
	aq. MeOH	355(3.40)	101-0542-66B
	pyridine	379(3.36),430s(--)	101-0542-66B
$C_{22}H_{22}N_2$			
Vinylenediamine, N,N'-dimethyl-N,N',1-triphenyl-	EtOH	231s(4.16),241(4.10), 287(4.17),314(4.20)	44-1423-66
$C_{22}H_{22}N_2O_3$			
Acetamide, N-[2-(1-anthraquinonylamino)cyclohexyl]-	EtOH	512(3.85)	27-0281-66
Kopsanone, N-acetyl-10-oxo-	EtOH	250(4.14)	33-1237-66
$C_{22}H_{22}N_2O_4$			
1-Indanone, 2-ethyl-2-nitroso-, dimer	EtOH	247(4.39),295(3.76)	44-2090-66
Unnamed alkaloid	EtOH	242(4.15),280(3.42), 286(3.39)	33-2321-66

Compound	Solvent	λ_{max}(log ε)	Ref.
$C_{22}H_{22}N_2O_5$			
Fructicosine, oxo-	EtOH	244(4.21),284(3.41), 292(3.41)	33-2321-66
α-D-Ribohexosid-3-ulose, 4,6-O-benzylidene, phenylhydrazone, methyl ester, anti	EtOH	280(4.36)	39-0695-66C
	dioxan	405(1.98)	39-0695-66C
syn	EtOH	278(4.36)	39-0695-66C
$C_{22}H_{22}N_2O_8$			
5H-Oxazolo[4,5-b]phenoxazine-4,6-dicarboxylic acid, 2-(dimethoxymethyl)-9,11-dimethyl-, dimethyl ester	MeOH	226(4.60),251(4.56), 418(4.25)	44-2564-66
$C_{22}H_{22}NiO_4S_2$			
Nickel, bis[hydrogen (thiobenzoyl)-acetato]-, ethyl ester	CHCl	495s(2.38),675(1.85)	12-1401-66
$C_{22}H_{22}O_3S$			
2,4,6-Cycloheptatriene-1-ethanol, β-phenyl-, p-toluenesulfonate	C_6H_{12}	257(3.60)	44-0912-66
$C_{22}H_{22}O_4$			
Benzofuran-4-ol, 2-isopropyl-5-(2'-methoxyphenylacetyl)-2,3-dihydro-	EtOH	226(4.45),285(4.15)	39-0749-66C
Naphthalene, 7-ethoxy-6-methoxy-2,3-dimethyl-1-(3,4-methylenedioxy)-phenyl]-	EtOH	237(4.69),290(3.94), 313(3.42),328(3.60)	39-1775-66C
$C_{22}H_{22}O_6$			
Robustic acid, dihydro-	EtOH	221s(4.50),236s(4.23), 253s(4.06),333(4.35)	39-0606-66C
	EtOH-NaOAc	221(4.57),258(4.26), 325(4.20)	39-0606-66C
$C_{22}H_{22}O_{10}$			
Swertisin	EtOH	273(4.24),336(4.32)	88-1611-66
$C_{22}H_{22}O_{11}$			
Kaempferol, 7-glucosyl-3-methyl-	EtOH	268(4.30),293(3.99), 353(4.26)	22-0987-66
Mumenine	EtOH	269(4.05),323(3.86), 367(4.03)	22-3212-66
Parkinsonin A	EtOH	258(4.50),271(4.50), 352(4.60)	78-1147-66
	EtOH-NaOEt	270(4.10),412(4.20)	78-1147-66
	EtOH-NaOAc	278(4.60),326(4.30), 395(4.50)	78-1147-66
	+ H_3BO_3	265(4.50),376(4.50)	78-1147-66
	EtOH-$AlCl_3$	267(4.60),279(4.60), 360(4.60)	78-1147-66
Swertiajaponin	EtOH	259s(4.30),271(4.31), 350(4.38)	88-1611-66
$C_{22}H_{22}O_{12}$			
Cacticin	EtOH	255(4.3),358(4.2)	24-1384-66
Isorhamnetin, 3-glucoside	MeOH	256(4.79),356(4.65)	100-0225-66
Quercetin, 7-glucosyl-3-methyl-	EtOH	257(4.39),270s(--), 295(3.92),362(4.34)	22-0987-66
Quercetin, 7-glucosyl-4'-methyl-	EtOH	256(4.43),374(4.32)	22-3212-66

Compound	Solvent	λ_{max}(log ε)	Ref.
$C_{22}H_{23}ClFN_3OS$			
4(1H)-Pyrimidinethione, 2-(p-chloro-phenyl)-1-(p-fluorophenyl)-5,6-di-hydro-5,5-dimethyl-6-morpholino-	MeOH	222s(4.12),281s(3.86), 365(4.22)	44-0722-66
$C_{22}H_{23}ClO_2$			
Estra-1,3,5(10),6,8-pentaene, 17α-chloroethynyl-3,17β-dimethoxy-	EtOH	238(4.72)	78-2829-66
$C_{22}H_{23}ClO_6S_2$			
7-Methoxy-2-[(7-methoxythiochroman-4-yl)methyl]-1-benzothiopyrylium perchlorate	CH_2Cl_2	266(4.25),273s(3.50), 320(3.70),342(3.62), 420(3.57),572s(3.37), 622(4.57)	78-0007-66
$C_{22}H_{23}ClO_9$			
4,6-Bis(3,4-dimethoxyphenyl)-2-methylpyrylium perchlorate	MeOH-$HClO_4$	290(3.93),345(3.83), 455(4.42)	65-1728-66
$C_{22}H_{23}IN_2S_2$			
3-Ethyl-2-[3-(3-ethyl-2-benzothiazol-inylidene)-2-methylpropenyl]-benzothiazolium iodide	MeOH	536(5.06)	9-0289-66
$C_{22}H_{23}NO$			
1,4-Hexadien-3-one, 2,4-diphenyl-1-(1-pyrrolidinyl)-	EtOH	223(4.23),284(4.27)	44-1336-66
$C_{22}H_{23}NO_4$			
Ochotensimine	MeOH	226(4.41),287(4.12)	23-2449-66
$C_{22}H_{23}NO_5$			
Acetamide, N-[2-(9,10-dimethoxy-phenanthro[3,4-d]-1,3-dioxol-5-yl)ethyl]-N-methyl-	EtOH	247s(4.52),257s(4.69), 265(4.76),287s(4.16), 318(4.02),329(4.03), 350(3.58),368(3.54)	12-2339-66
$C_{22}H_{23}NO_6$			
Cassythine, N-methyl-, acetate	n.s.g.	210(4.42),279(4.17), 300(4.11)	12-0297-66
$C_{22}H_{23}NO_7$			
Narcotine	MeOH	292(3.6),310(3.7)	83-0196-66
$C_{22}H_{23}N_3O_2S$			
7-Azaindole, 4-methyl-1-phenyl-6-(N-methyl-p-toluenesulfonamido)-	EtOH	288(4.23),332(4.05)	78-3233-66
$C_{22}H_{24}CrO_5$			
Chromium, tricarbonyl(3-methoxy-estra-1,3,5(10)-trien-17-one)-	n.s.g.	318(3.79)	39-0054-66C
$C_{22}H_{24}N_2$			
3,4-Diazabicyclo[4.2.0]octa-2,4-diene, 1,6,7,8-tetramethyl-2,5-diphenyl-	MeOH	308(4.17)	24-1236-66
$C_{22}H_{24}N_2O$			
[$\Delta^{2,3'}$-Bipyrrolidin]-2'-one, 1,1'-dibenzyl-	MeOH	216(4.4),297(4.7)	24-0724-66
Urea, 1-butyl-1,3,3-triphenyl-	EtOH	264(4.16)	34-0436-66

Compound	Solvent	$\lambda_{max}(\log \epsilon)$	Ref.
$C_{22}H_{24}N_2O_2$			
Anthraquinone, 1-[2-(dimethylamino)-cyclohexylamino]-	EtOH	519(3.84)	27-0281-66
Kopsanone, N-acetyl-	EtOH	252(4.12),281(3.61)	33-1237-66
$C_{22}H_{24}N_2O_3$			
Kopsanol-10-lactam, O-acetate	EtOH	245(3.92),297(3.55)	39-1260-66C
$C_{22}H_{24}N_2O_6$			
2,2'-Biisoquinoline, 1,1',2,2',3,3',-4,4'-octahydro-8,8'-dimethoxy-6,7;6',7'-bis(methylenedioxy)-	dioxan	210(4.62),298(3.64)	24-0273-66
Pyrrole-3-carboxylic acid, 5-[3-(4-carboxy-3,5-dimethyl-2H-pyrrol-2-ylidene)-2-hydroxy-4-oxo-1-cyclobuten-1-yl]-2,4-dimethyl-, diethyl ester	$CHCl_3$	562(5.38)	5-0153-66I
$C_{22}H_{24}N_2O_8$			
3H-Phenoxazine-1,9-dicarboxylic acid, 4,6-dimethyl-2-[(formylmethyl)-amino]-3-oxo-, dimethyl ester, 2-(dimethyl acetal)	MeOH	224(4.29),250(4.39), 422(4.38),444(4.44)	44-2564-66
$C_{22}H_{24}N_2O_{10}$			
Quinoxaline, 2-(β-D-glucopyranosyloxy)-, tetraacetate	MeOH	219s(4.18),236s(4.25), 239(4.32),243(4.30), 288s(3.57),307s(3.68), 319(3.74),331s(3.65)	24-0547-66
$C_{22}H_{24}N_4O_4S$			
2-Cyclohexen-1-one, 2-(p-tolylthio-methyl)-4,4-dimethyl-	EtOH	228(4.23),242(4.17)	23-2003-66
$C_{22}H_{24}N_4O_6$			
3(2H)-Phenanthrone, 1,4,4aβ,9,10,10aβ-hexahydro-6,7-dimethoxy-, 2,4-dinitrophenylhydrazone	EtOH	229(4.32),270(4.03), 365(4.33)	87-0304-66
$C_{22}H_{24}O_3$			
Gona-1,3,5(10),8,14,16-hexaene, 17-acetoxy-13-ethyl-3-methoxy-	EtOH	360(4.42)	13-0547-66B
$C_{22}H_{24}O_4$			
2-Heptenoic acid, 3-(o-carboxyphenyl)-2-phenyl-, dimethyl ester	EtOH	225(4.19),227s(3.61), 312s(3.22)	35-0476-66
α-Pipitzol, benzoate	EtOH	232(4.19),260(3.64)	78-2387-66
β-Pipitzol, benzoate	EtOH	233(4.32),266(3.76)	78-2387-66
$C_{22}H_{24}O_6$			
Xanthone, 2-(3,3-dimethylallyl)-1,3,5,6-tetramethoxy-	MeOH	245(4.69),280(3.98), 431(4.30)	39-0175-66C
$C_{22}H_{24}O_7$			
1H-7,9a-Methanobenz[a]azulene-1,10-dicarboxylic acid, decahydro-4a,7-dihydroxy-1-methyl-8-methylene-2-oxo-, 1,4a-lactone, methyl ester, 7-acetate	n.s.g.	230(3.70)	70-0171-66

Compound	Solvent	λ_{max}(log ε)	Ref.
$C_{22}H_{24}O_8$			
Benzophenone, 2-(acetoxymethyl)-2',4,6-trimethoxy-4'-acetoxy-6'-methyl-	MeOH	235(4.19),277(3.97), 300(3.89)	44-2291-66
Chromomycinone, anhydro-3'-O-methyl-	EtOH	233(4.35),282(4.57), 342(3.86),418(4.01)	78-2761-66
$C_{22}H_{24}O_{10}$			
Flavone, 3'-hydroxy-3,4',5,5',6,7,8-heptamethoxy-	n.s.g.	254s(4.23),272(4.24), 331(4.31)	24-3218-66
$C_{22}H_{25}ClO_2$			
Estra-1,3,5(10),7-tetraene, 17α-chloroethynyl-3,17β-dimethoxy-	EtOH	228(4.26)	78-2829-66
Estra-1,3,5(10),9(11)-tetraene, 17α-chloroethynyl-3,17β-dimethoxy-	EtOH	262(4.31)	78-2829-66
$C_{22}H_{25}ClO_8$			
Radicicol, tetrahydro-, diacetate	EtOH	279(2.96)	12-1265-66
$C_{22}H_{25}F_3O$			
Estra-1,3,5(10)-trien-17β-ol, 4-methyl-17α-(trifluoropropynyl)-	EtOH	219(3.95),258s(2.28), 263(2.35),270(2.23)	78-2829-66
$C_{22}H_{25}F_3O_2$			
Estra-1,3,5(10)-trien-17β-ol, 3-methoxy-17α-(trifluoro-1-propynyl)-	EtOH	219s(3.92),278(3.30), 286(3.21)	78-2829-66
$C_{22}H_{25}NO$			
Aziridine, 1-cyclohexyl-2-phenyl-3-p-tolyl-, cis	C_6H_{12}	323(2.27)	44-1244-66
trans	C_6H_{12}	347(2.51)	44-1244-66
$C_{22}H_{25}NO_4$			
Isoquinoline, 6,7-diethoxy-1-(3,4-dimethoxybenzyl)-, hydrochloride	MeOH	245(4.65),325(3.49), 335(3.52)	24-2873-66
Ochotensimine, dihydro-	MeOH	223s(4.09),282(3.77)	23-2449-66
1,4-Pentadien-3-one, 1-(p-dimethyl-	benzene	410(4.55)	65-0336-66
aminophenyl)-5-(2,4,6-trimethoxy-	EtOH	264(4.07),435(4.60)	65-0336-66
phenyl)-	HOAc	262(4.12),432(4.45)	65-0336-66
ferric chloride added	HOAc	623(4.74)	65-0336-66
2,4-Pentadien-1-one, 5-(p-dimethyl-	benzene	405(4.53)	65-0336-66
aminophenyl)-1-(2,4,6-trimethoxy-	EtOH	275(3.90),420(4.25)	65-0336-66
phenyl)-	HOAc	325(4.35),420(4.27)	65-0336-66
ferric chloride added	HOAc	635(4.21)	65-0336-66
$C_{22}H_{25}N_3$			
3H-Dibenz[c,f]imidazo[1,2-a]azepine, 2,9-dihydro-9-(1-methyl-4-piperidyl)-	EtOH	231s(4.15),260(3.98), 293(3.86)	4-0206-66
$C_{22}H_{25}N_3OS$			
4-Pyrimidinethione, 5,6-dihydro-5,5-dimethyl-6-morpholino-1,2-diphenyl-	MeOH	223s(4.18),276s(3.81), 365(4.18)	44-0722-66
$C_{22}H_{26}Cl_2O_4$			
p-Dioxane, 3,6-bis(2-chloroethoxy)-2,5-dimethyl-3,6-diphenyl-	hexane	257(2.76)	87-0762-66

Compound	Solvent	$\lambda_{max}(\log \epsilon)$	Ref.
$C_{22}H_{26}N_2O$			
[2,3'-Bipyrrolidin]-2'-one, 1,1'-dibenzyl-	MeOH	219(--)	24-1923-66
	$CHCl_3$	240(3.1)	24-1923-66
3,4-Diazabicyclo[4.2.0]oct-4-en-2-ol, 1,6,7,8-tetramethyl-2,5-diphenyl-	MeOH	288(4.20)	24-1236-66
9H-Pyrido[1,2,3-lm]pyrrolo[2,3-d]carbazol-9-one, 1,2,3,11a,11b,12,13,13a-octahydro-1-methyl-12-(1-methylpropenyl)-	EtOH	310(3.85)	78-0217-66B
$C_{22}H_{26}N_2O_2$			
2H,14H-Indolo[3,2,1-ij]oxepino[2,3,4-de]pyrrolo[2,3-h]quinolin-14-one, dodecahydro-6-methyl-4-methylene-	EtOH	260(4.12)	78-0217-66B
9H-Pyrido[1,2,3-lm]pyrrolo[2,3-d]-carbazol-9-one, 1,2,3,11a,11b,12,-13,13a-octahydro-12[1-(2-hydroxyethyl)vinyl]-1-methyl-	EtOH	310(3.85)	78-0217-66B
$C_{22}H_{26}N_2O_3$			
Vincamajine	MeOH	249(3.95),292(3.49)	100-0029-66
$C_{22}H_{26}N_2O_4$			
Akuammine	MeOH	245(3.91),312(3.62)	100-0029-66
Fructicosamine, dihydro-	EtOH	244(4.17),278(3.44), 286(3.41)	33-2321-66
Pleiocarpinic acid	EtOH	248(4.05),283(3.39), 290(3.36)	33-1237-66
Reserpinine	MeOH	219(4.65),297(3.90)	100-0029-66
$C_{22}H_{26}N_2O_5$			
1,3-Azulenedicarboxylic acid, 2-acetamido-6-pyrrolidino-, diethyl ester	MeOH	272(4.26),292s(4.10), 347(4.78),402(4.40)	18-1954-66
$C_{22}H_{26}N_2O_6$			
1,3-Azulenedicarboxylic acid, 2-acetamido-6-morpholino-, diethyl ester	MeOH	233(4.01),271(4.25), 347(4.74),409(4.40)	18-1954-66
$C_{22}H_{26}N_6O_4S_2$			
Tetramethylammonium salt of 5-(2-cyano-1,2-di-p-toluenesulfonylvinyl)tetrazole	CH_2Cl_2	393(4.05)	44-3849-66
$C_{22}H_{26}O_2$			
1,5-Heptanedione, 4,4,6-trimethyl-1,3-diphenyl-	EtOH	242(4.11),278s(3.04)	88-2243-66
$C_{22}H_{26}O_4$			
Estra-1,3,5(10),8,14-pentaen-17-one, 2,3-dimethoxy-, cyclic ethylene ketal	EtOH	243(4.18),326(4.37)	87-0027-66
$C_{22}H_{26}O_6$			
Pentatriafulvalene-2,3-dicarboxylic acid, 1,4-diacetyl-5,6-dipropyl-, dimethyl ester	n.s.g.	275(4.33),348(4.07)	88-3697-66
$C_{22}H_{26}O_9$			
Chromomyciquinone, 3'-dehydro-3'-O-methyl-	EtOH	256(4.24),299(4.04), 415(3.66)	78-2761-66

Compound	Solvent	λ_{max}(log ε)	Ref.
$C_{22}H_{27}BrO_2$			
Estra-1,3,5(10)-triene, 17α-bromoethynyl-3,17β-dimethoxy-	EtOH	218s(3.99),278(3.31), 287(3.29)	78-2829-66
$C_{22}H_{27}ClN_2O_3$			
N^b-Methylakuammidinium chloride	50% EtOH	219(4.70),271(3.87), 277s(3.86),288(3.75)	23-1523-66
$C_{22}H_{27}ClO$			
Estra-1,3,5(10)-triene, 17α-chloroethynyl-17β-methoxy-4-methyl-	EtOH	263(2.36),270(2.24)	78-2829-66
$C_{22}H_{27}ClO_2$			
Estra-1,3,5(10)-triene, 17α-chloroethynyl-3,17β-dimethoxy-	EtOH	278(3.32),287(3.30)	78-2829-66
$C_{22}H_{27}ClO_3$			
Androsta-2,4,6-triene-2-carboxaldehyde, 17β-acetoxy-7-chloro-	MeOH	360(4.37)	24-3057-66
$C_{22}H_{27}ClO_9$			
Radicicol, diformate of diol from	EtOH	291(3.58)	12-1265-66
$C_{22}H_{27}F_3O_2$			
Estra-1,3,5(10)-triene, 3,17β-dimethoxy-17α-trifluorovinyl-	EtOH	220(3.94),229s(3.86), 278(3.30),287(3.28)	78-2829-66
$C_{22}H_{27}IO_2$			
Estra-1,3,5(10)-triene, 3,17β-dimethoxy-17α-iodoethynyl-	EtOH	278(3.32),286(3.30)	78-2829-66
$C_{22}H_{27}LiN_2O_{15}P_2$			
3-Carbamoyl-1-β-ribofuranosylpyridinium hydroxide, 5'-pyrophosphate, 5-ester with phenyl β-D-ribofuranoside, inner salt, lithium salt	pH 7.0	267(3.76)	24-1712-66
$C_{22}H_{27}NO$			
2H-Naphth[2',1':4,5]indeno[1,2-b]pyridin-2-one, 3,4,4a,4b,5,6,6a,11,11a,-11b,12,13-dodecahydro-4a,6a-dimethyl-	EtOH	242(4.25),268(3.91), 276(3.72)	4-0338-66
Piperidine, 1-[2-[(o-styrylbenzyl)oxy]-ethyl]-, hydrochloride, trans	H_2O	203(4.39),227(4.15), 297(4.39)	39-0668-66C
$C_{22}H_{27}NO_5$			
Alkaloid from Croton linearis Jacq., diacetate	EtOH	278s(3.41),285(3.44)	39-1680-66C
Stepharotine, O-methyl-	EtOH	282(3.75)	95-0460-66
Tyrosine, benzyloxycarbonyl-O-tert-butyl-, methyl ester	MeOH	258(2.77),264(2.83), 267(2.84),274(2.77)	5-0226-66F
$C_{22}H_{27}NO_5S$			
Azepino[3,2,1-hi]indole-7-methanol, 1,2,4,5,6,7-hexahydro-6,6-dimethoxy-, p-toluenesulfonate	EtOH	252(3.97),295(3.38)	35-4061-66
$C_{22}H_{27}N_3OS$			
p-Benzoquinone, 2,6-di-tert-butyl-, 4-azine with 3-methyl-2-benzothiazolinone-	MeCN	471(4.68)	5-0065-66J

Compound	Solvent	$\lambda_{max}(\log \epsilon)$	Ref.
$C_{22}H_{27}N_5O_{10}$			
Pteridine, 2-(dimethylamino)-7-(β-D-glucopyranosyloxy)-, tetraacetate	pH -0.89	235(4.25),267(4.03), 305s(3.74),357(3.86)	24-0536-66
	pH 4.0	247(4.32),282(3.89), 390(3.94)	24-0536-66
	MeOH	245(4.31),284(3.92), 385(3.96)	24-0536-66
$C_{22}H_{27}N_7O_2$			
Cypridina oxyluciferin, hydrochloride	MeOH	220(4.45),271(4.15), 302(4.13),347(4.19)	88-3427-66
$C_{22}H_{28}$			
Anthracene, ,4-di-tert-butyl-1,4-dihydro-	EtOH	231(4.75),267s(3.68), 276(3.75),283(3.74), 295s(3.56),312s(2.55), 322s(2.34),328(2.27)	39-0861-66B
Anthracene, 9,10-di-tert-butyl-9,10-dihydro-	EtOH	260s(2.69),268(2.77), 276(2.73)	39-0861-66B
$C_{22}H_{28}ClNO$			
Piperidine, 1-[2-[(o-styrylbenzyl)oxy]-ethyl]-, hydrochloride, trans	H_2O	203(4.39),227(4.15), 297(4.39)	39-0668-66C
$C_{22}H_{28}INO_4$			
O,O'-Dimethylmagnoflorine iodide	EtOH	223(4.71),271(4.18), 297(3.75)	39-1340-66C
$C_{22}H_{28}NO_2$			
o-Tolyloxy , 6-tert-butyl-4-[(3-tert-butyl-5-methyl-4-oxo-2,5-cyclohexa-dien-1-ylidene)amino]-	C_6H_{12}	305(3.88),334(4.18), 550(2.98),601(3.00)	35-5284-66
$C_{22}H_{28}N_2O$			
9H-Pyrido[1,2,3-lm]pyrrolo[2,3-d]carb-azol-9-one, 1,2,3,10,11,11a,11b,12,-13,13a-decahydro-1-methyl-12-(1-methylpropenyl)-	EtOH	257(4.12)	78-0217-66B
$C_{22}H_{28}N_2O_2$			
9H-Pyrido[1,2,3-lm]pyrrolo[2,3-d]carb-azol-9-one, decahydro-12-[1-(2-hydroxyethyl)vinyl]-1-methyl-	EtOH	255(4.11)	78-0217-66B
9H-Pyrido[1,2,3-lm]pyrrolo[2,3-d]carb-azol-9-one, decahydro-11-hydroxy-12-(1-methylallyl)-	EtOH	256(4.12)	78-0217-66B
9H-Pyrido[1,2,3-lm]pyrrolo[2,3-d]carb-azol-9-one, decahydro-11-hydroxy-1-methyl-12-(1-methylpropenyl)-	EtOH	255(4.12)	78-0217-66B
9H-Pyrido[1,2,3-lm]pyrrolo[2,3-d]carb-azol-9-one, decahydro-12-(3-hydroxy-1-methylpropenyl)-1-methyl-	EtOH	255(4.08)	78-0217-66B
$C_{22}H_{28}N_2O_3$			
Beninine, N_a-acetyl-	EtOH	219(4.51),255(4.07), 280s(3.53)	33-2072-66
Ibogaine derivative, methyl ester	EtOH	226(4.40),282(3.91), 295(3.83)	35-2532-66
Unnamed ester	ether	227.5(4.23)	24-3362-66

Compound	Solvent	$\lambda_{max}(\log \epsilon)$	Ref.
$C_{22}H_{28}N_2O_3S$			
L-Tyrosine, N-thiobenzoyl-, cyclohexylammonium salt	H_2O	278(4.00),356s(2.57)	1-2781-66
	MeOH	279(3.94),373s(2.37)	1-2781-66
	dioxan	278(3.95),384(2.42)	1-2781-66
$C_{22}H_{28}N_2O_4$			
Herbaine	EtOH	228(4.63),274(3.85), 297(3.87)	49-0857-66
$C_{22}H_{28}N_2O_5$			
1,3-Azulenedicarboxylic acid, 2-acetamido-6-(diethylamino)-, diethyl ester	MeOH	271(4.28),290s(4.11), 346(4.82),403(4.41)	18-1954-66
Tetraphyllinine	NaOH	225s(4.6),272(4.3)	102-1065-66
	n.s.g.	226(4.64),272(3.64), 298(3.73)	102-1065-66
$C_{22}H_{28}N_2O_6$			
1,3-Azulenedicarboxylic acid, 2-acetamido-6-(3-methoxypropylamino)-, diethyl ester	MeOH	269(4.25),290(4.28), 342(4.81),397(4.32)	18-1954-66
$C_{22}H_{28}N_2O_7$			
1,3-Azulenedicarboxylic acid, 2-acetamido-6-[di(2-hydroxyethyl)amino]-, diethyl ester	MeOH	225(3.97),270(4.25), 347(4.77),404(4.38)	18-1954-66
$C_{22}H_{28}N_4O_3$			
Benzimidazole, 2-(p-ethoxybenzyl)-1-(2-diethylaminoethyl)-5-nitro-	EtOH-HCl	240(4.51),302(4.07)	39-1511-66C
$C_{22}H_{28}N_4O_4$			
2,4-Hexadienal, 6-(2,6,6-trimethylcyclohexylidene)-4-methyl-, 2,4-dinitrophenylhydrazone	n.s.g.	253(4.13),366s(4.38), 418(4.57)	22-0728-66
3,5-Octadien-2-one, 7-methyl-6-(2-methylcyclohexen-1-yl)-, 2,4-dinitrophenylhydrazone	$CHCl_3$	270(4.23),310(4.37), 410(4.52)	70-0480-66
3,5-Octadien-2-one, 7-methyl-6-(4-methyl-3-cyclohexen-1-yl)-, 2,4-dinitrophenylhydrazone	$CHCl_3$	270(4.22),310(4.35), 410(4.54)	70-0480-66
$C_{22}H_{28}O_3$			
Naphth[1,8-bc]oxocin-11-ol, 2,3,4,5-tetrahydro-10-isopropyl-2,2,6-trimethyl-, acetate	hexane	237(4.30),286(3.82), 396(3.85)	33-1151-66
Pregna-4,6,15-triene-3,20-dione, 17α-hydroxy-16-methyl-	EtOH	285(4.44)	73-2768-66
$C_{22}H_{28}O_4$			
Gona-1,3,5(10)-trien-6-one, 17β-acetoxy-13β-ethyl-3-methoxy-	n.s.g.	256(3.89),325(3.43)	78-1019-66
Pregna-4,6-diene-3,20-dione, 16α,17α-epoxy-11α-hydroxy-16β-methyl-	EtOH	284(4.42)	73-4703-66
$C_{22}H_{28}O_5$			
19-Norpregna-4,9(10)-dien-3-one, 17,20;20,21-bis(methylenedioxy)-	MeOH	304(4.30)	13-0087-66B
19-Norpregna-4,9(11)-dien-3-one, 17,20;20,21-bis(methylenedioxy)-	MeOH	238(4.22)	13-0087-66B

Compound	Solvent	λ_{max}(log ε)	Ref.
19-Norpregna-5(10),9(11)-dien-3-one, 17,20;20,21-bis(methylenedioxy)-	MeOH	240(4.29)	13-0087-66B
Pregna-1,4-diene-3,20-dione, 9,11β-epoxy-17α,21-dihydroxy-16β-methyl-	MeOH	249(4.18)	44-0026-66
Pregna-1,4-diene-3,11,20-trione, 17α,21-dihydroxy-16β-methyl-	MeOH	238(4.17)	44-0026-66
Pregn-4-ene-3,11,20-trione, 17α,21-dihydroxy-1,2β-methylene-	MeOH	218(4.06),241(3.95)	24-1118-66
Pregn-4-ene-3,11,20-trione, 16α,17α-epoxy-7β-hydroxy-16β-methyl-	EtOH	235(4.13)	73-4703-66
$C_{22}H_{28}O_6$			
1H-Benzofuro[3a,3-b]benzofuran-6-acetic acid, 2,3,4,4a,6,6a-hexahydro-6aα-methoxy-4aβ,7,9-trimethyl-3-oxo-, ethyl ester	n.s.g.	282(3.42),288(3.49)	44-1725-66
Bicyclo[3.2.1]octane-1-carboxylic acid, 2-[6-carboxy-1-(hydroxymethyl)-5,5-dimethyl-4-oxo-2-cyclohexen-1-yl]-6-methyl-7-oxo-, γ-lactone, ethyl ester	EtOH	225(3.96)	78-1659-66
Bicyclo[3.2.1]octane-1-carboxylic acid, 6-methyl-7-oxo-2-(4,5,6,7-tetrahydro-4,6-dihydroxy-7,7-dimethyl-3a(3H)-isobenzofuranyl)-, δ-lactone, acetate	EtOH	217(3.57)	78-1659-66
19-Norpregna-4,9(10)-dien-3-one, 11 -hydroxy-17,20;20,21-bis(methylene-dioxy)-	MeOH	296(4.30)	13-0087-66B
$C_{22}H_{28}O_7$			
Emmein, 3-acetate	n.s.g.	233(3.96)	77-0297-66
$C_{22}H_{28}S_4$			
2,4,13,15-Tetrathia[5.5]metacyclo-phane, 3,3,14,14-tetramethyl-	$CHCl_3$	267(2.86),273(2.69)	88-5723-66
$C_{22}H_{29}BrO_3$			
Pregn-4-ene-3,20-dione, 6β-bromo-16α,17α-epoxy-16β-methyl-	EtOH	247(4.13)	73-2768-66
Pregn-4-ene-3,20-dione, 6β-bromo-17α-hydroxy-16-methylene-	EtOH	247(4.15)	73-2768-66
$C_{22}H_{29}ClO_2$			
Androst-4-en-3-one, 17α-chloroethynyl-17β-methoxy-	EtOH	241(4.20)	78-2829-66
$C_{22}H_{29}ClO_3$			
Androsta-5,16-diene-16-carboxaldehyde, 17-chloro-3β-hydroxy-, acetate	EtOH	261(4.08)	32-1241-66
Pregn-4-ene-3,20-dione, 6β-chloro-16α,17α-epoxy-16β-methyl-	EtOH	240(4.21)	73-2768-66
Pregn-4-ene-3,20-dione, 6α-chloro-17α-hydroxy-16-methylene-	EtOH	235(4.18)	73-2768-66
Pregn-4-ene-3,20-dione, 6β-chloro-17α-hydroxy-16-methylene-	EtOH	241(4.12)	73-2768-66
$C_{22}H_{29}ClO_7$			
Radicicol, acetate of alcohol from	EtOH	249(3.74),293(3.62)	12-1265-66

Compound	Solvent	λ_{max}(log ε)	Ref.
$C_{22}H_{29}FO_4$			
Pregna-1,4-diene-3,20-dione, 11β-fluoro-17,21-dihydroxy-16α-methyl-	MeOH	241(4.18)	35-3016-66
$C_{22}H_{29}FO_5$			
Pregna-1,4-diene-3,20-dione, 9α-fluoro-11β,17α,21-trihydroxy-16β-methyl-	MeOH	239(4.19)	44-0026-66
$C_{22}H_{29}N$			
Retinylideneacetonitrile	pet ether	390(4.79)	54-0334-66
$C_{22}H_{29}NO$			
Estra-1,3,5(10),9(11)-tetraen-17β-ol, 3-(1-pyrrolidinyl)-	EtOH	295(4.43)	87-0510-66
2H-Naphtho[2',1':4,5]indeno[1,2-b]-pyridin-2-one, 1,3,4,4a,4b,5,6,6a,-11,11a,11b,12,13,13a-tetradeca-hydro-4a,6a-dimethyl-	EtOH	269(3.81),277(3.69)	4-0338-66
$C_{22}H_{29}NO_2$			
Androst-4-en-3-one, 17-(5-isoxazolyl)-	EtOH	235(4.13)	44-3193-66
Benzophenone, 3,5-di-tert-butyl-4-hydroxy-, O-(methyloxime)	C_6H_{12}	242(4.10),271(4.02)	35-5284-66
$C_{22}H_{29}NO_2S$			
Progesterone, 6α-thiocyanato-	EtOH	236.5(4.13)	94-1096-66
Progesterone, 6β-thiocyanato-	EtOH	243(4.13),284s(3.32)	94-1096-66
$C_{22}H_{29}NO_4$			
Isoquinoline, 1-(3,4-diethoxybenzyl)-1,2,3,4-tetrahydro-6,7-dimethoxy-, perchlorate	MeOH	290(3.84)	24-2873-66
Isoquinoline, 6,7-diethoxy-1,2,3,4-tetrahydro-1-(3,4-dimethoxybenzyl)-, perchlorate	MeOH	290(3.83)	24-2873-66
$C_{22}H_{29}N_7O_2$			
Cypridina luciferin, hydrobromide	MeOH-HBr	221(4.46),243s(4.08), 275s(4.18),307(4.30), 320s(3.78)	88-3427-66
	MeOH	218(4.44),270(4.23), 310s(4.06),435(3.95)	88-3427-66
$C_{22}H_{30}$			
Anthracene, 1,4-di-tert-butyl-tetrahydro-	EtOH	228s(4.61),233(4.76), 263(3.33),276(3.43), 284(3.45),294s(3.24), 308s(2.49),320s(2.17), 323(2.23)	39-0861-66B
$C_{22}H_{30}ClNO_2S$			
5'βH-5α-Pregnano[5,6-d]thiazole-3,20-dione, 2'-chloro-	EtOH	206(3.23),237(3.31), 253(3.32)	94-1096-66
$C_{22}H_{30}Cl_2N_2O_8$			
5-Benzylidene-Δ$^{1(6)}$-dehydrosparteinium diperchlorate	EtOH	321.2(4.32)	94-0147-66

Compound	Solvent	λ max(log ε)	Ref.
$C_{22}H_{30}Cl_{10}Si_5$			
Pentasilane, 1,1,2,2,3,3,4,4,5,5-decamethyl-1,5-bis(pentachlorophenyl)-	C_6H_{12}	216.5(5.03)	101-0102-66B
$C_{22}H_{30}Hg$			
Mercury, bis[(2,6,6-trimethyl-1-cyclohexen-1-yl)ethynyl]-	MeOH	257.5(4.43)	24-0689-66
$C_{22}H_{30}INO_4$			
N-Methyllaudanosinium iodide	EtOH	281(3.81)	39-1340-66C
	n.s.g.	207s(4.95),232(4.30), 281(3.76)	78-1335-66
$C_{22}H_{30}N_2O$			
Androsta-5,16-dien-3β-ol, 17-(3 or 5-pyrazolyl)-	EtOH	247(4.04)	44-3193-66
$C_{22}H_{30}N_2O_2$			
Palosin, O-demethyl-	MeOH	221(4.38),259(3.88), 291(3.46)	88-5027-66
$C_{22}H_{30}N_3O_2P$			
1,4-Butanedione, 1,4-diphenyl-2-[tris-(dimethylamino)phosphoranylidene]-	CH_2Cl_2	236(4.13),276(3.68)	78-0567-66
$C_{22}H_{30}N_4O_2S_2$			
Phenothiazine-2-sulfonamide, N,N-dimethyl-10-[3-(4-methyl-1-piperazinyl)propyl]-	$MeSO_3H$	233(4.31),263(4.49), 311(3.51)	95-0510-66
$C_{22}H_{30}N_4O_3S_2$			
Phenothiazine-2-sulfonamide, N,N-dimethyl-10-[3-(4-methyl-1-piperazinyl)propyl]-, S-oxide	$MeSO_3H$	218(4.37),246(4.46), 275(4.20),304(3.92), 355(3.71)	95-0510-66
$C_{22}H_{30}O$			
14'-Apo-β-carotenal	isooctane	397(4.83)	54-0343-66
$C_{22}H_{30}O_2$			
Androst-4-en-3-one, 17α-ethynyl-17β-hydroxy-18-methyl-	MeOH	243(4.20)	44-1026-66
14'-Apo-β-carotenic acid	acetone	386(4.69)	54-0334-66
Estra-1,3,5(10)-triene, 3,17β-dimethoxy-17α-vinyl-	EtOH	219s(3.93),278(3.31), 286(3.29)	78-2829-66
8α-Gona-1,3,5(10)-trien-17-one, 13β-isobutyl-3-methoxy-	EtOH	280(3.34),288(3.33)	87-0338-66
Gon-4-en-3-one, 13β-ethyl-17α-ethynyl-17β-hydroxy-6α-methyl-	n.s.g.	240(4.18)	78-1019-66
Gon-4-en-3-one, 13β-ethyl-17α-ethynyl-17β-hydroxy-7α-methyl-	EtOH	240(4.22)	87-0782-66
A-Homoandrosta-3,4a,6-trien-17β-ol, acetate	MeOH	278(4.29),290(4.42), 305(4.29)	24-3836-66
$C_{22}H_{30}O_3$			
Androsta-1(10),3-dien-2-one, 17β-acetoxy-1-methyl-, 9(10→5β)-abeo-	EtOH	245(4.61),275s(3.85)	33-1049-66
Androsta-1(10),3-dien-2-one, 17β-acetoxy-4-methyl-, 9(10→5)-abeo-	EtOH	250(4.24)	33-1049-66

Compound	Solvent	λ_{max}(log ε)	Ref.
Androsta-2,4-dien-1-one, 17β-acetoxy-3-methyl-	EtOH	322(3.77)	33-1049-66
Androsta-4,9(11)-dien-3-one, 17β-acetoxy-17α-methyl-	EtOH	239(4.24)	33-2218-66
Androst-4-en-3-one, 17β-acetoxy-1,2β-methylene-	MeOH	220(3.93),243(4.06)	24-1118-66
1α,5β-Cyclo-10α-androst-3-en-2-one, 17β-acetoxy-3-methyl-	EtOH	238(3.73),269s(3.40)	33-1049-66
1α,5β-Cyclo-10α-androst-3-en-2-one, 17β-acetoxy-4-methyl-	EtOH	232(3.81),264s(3.58)	33-1049-66
4,10-Cycloandrost-2-en-1-one, 17β-acetoxy-3-methyl-, 9(10→5)-abeo-	EtOH	242(3.67),271s(3.54)	33-1049-66
4α,10α-Cyclo-10βH-estr-2-en-1-one, 17β-acetoxy-2,3-dimethyl-, 9(10→5β)-abeo-	EtOH	231(3.76),276(3.59)	33-1049-66
4β,10β-Cyclo-10αH-estr-2-en-1-one, 17β-acetoxy-3,4α-dimethyl-, 9(10→5β)-abeo-	EtOH	233(3.81),268s(3.47)	33-1049-66
Estra-1(10),2-dien-4-one, 17β-acetoxy-2,5α-dimethyl-	EtOH	317(3.73)	33-1049-66
Estra-1(10),3-dien-2-one, 17β-acetoxy-3,5α-dimethyl-	EtOH	246(4.11)	33-1049-66
Estra-1(10),3-dien-2-one, 17β-acetoxy-4,5α-dimethyl-	EtOH	245(4.27)	33-1049-66
Estra-1,3,5(10)-triene-1,17β-diol, 3,4-dimethyl-, 17-acetate	EtOH	287(3.35)	33-1049-66
Estra-1,3,5(10)-triene-4,17β-diol, 1,2-dimethyl-, 17-acetate	EtOH	286(3.42)	33-1049-66
Pregna-5,16-diene-21-carboxaldehyde, 3β-hydroxy-20-oxo-	EtOH	274(4.00)	32-1254-66
Pregn-4-ene-3,18,20-trione, 18-methyl-	MeOH	240(4.14)	44-1026-66
$C_{22}H_{30}O_4$			
Androsta-1,4-dien-3-one, 2,17β-dihydroxy-, 17-propionate	EtOH	254(4.15),290s(3.51)	94-1370-66
Androst-1-ene-3,5-dione, 17β-acetoxy-17α-methyl-, 10(5→4)-abeo-	EtOH	237(3.86),307(3.76)	33-2218-66
	EtOH-KOH	223(3.92),229(3.90), 247(3.71),342(3.90)	33-2218-66
Androst-9(11)-ene-3,5-dione, 17β-acetoxy-17α-methyl-, 10(5→4)-abeo-	EtOH	289(3.97)	33-2218-66
	EtOH-KOH	314(4.14)	33-2218-66
ferric chloride complex	EtOH	534(3.04)	33-2218-66
Androst-9(11)-en-3-one, 4α,5α-epoxy-17β-hydroxy-17α-methyl-, acetate	EtOH	288(1.92)	33-2218-66
4β,5β-isomer	EtOH	298(1.59)	33-2218-66
Estr-4-en-3-one, 4-acetyl-17β-hydroxy-, acetate	C_6H_{12}	233(4.13)	44-0705-66
Pregn-4-ene-3,20-dione, 17 ,21-dihydroxy-1,2β-methylene-	MeOH	212(3.95),244(4.06)	24-1118-66
$C_{22}H_{30}O_5$			
Estra-1,3,5(10)-trien-3-ol, 17β-acetoxy-2,4-dimethoxy-	MeOH	272(3.01)	35-0856-66
Jasmolin II	hexane	229(4.36)	39-0332-66C
Pregn-4-ene-3,20-dione, 16α,17α-epoxy-7β,11α-dihydroxy-16β-methyl-	EtOH	242(4.17)	73-4703-66
Pregn-5-ene-3,20-dione. 11β,17α,21-trihydroxy-1,2β-methylene-	MeOH	214(3.99),245(4.04)	24-1118-66
Pregn-4-ene-3,11,20-trione, 17α,21-dihydroxy-16β-methyl-	MeOH	238(4.18)	44-0026-66
Taxininol, anhydro-, acetate	EtOH	210(3.74)	95-1172-66

Compound	Solvent	$\lambda_{max}(\log \epsilon)$	Ref.
$C_{22}H_{30}O_6$			
19-Norpregn-4-en-3-one, 11β-hydroxy-17,20;20,21-bis(methylenedioxy)-	MeOH	242(4.24)	13-0087-66B
$C_{22}H_{30}O_7$			
Unsaturated ketodiacid, methyl ester	EtOH	226(3.92)	78-3423-66
$C_{22}H_{30}O_8$			
Isovaleric acid, 3,4-diester with 3a,4-dihydro-3,4-dihydroxyspiro[benzofuran-2(3H),2'-oxirane]-6-methanol, 6-acetate	MeOH	204(3.60),256(4.21)	88-1163-66
$C_{22}H_{31}ClN_4O_4$			
5-Benzylidenesparteine perchlorate	EtOH	242.3(4.08)	94-0147-66
$C_{22}H_{31}ClO_2$			
Androsta-3,5-dien-17β-ol, 3-chloro-, propionate	EtOH	239s(4.26),244(4.36), 253s(4.22)	32-1241-66
$C_{22}H_{31}ClO_3$			
5α-Androst-2-ene-2-carboxaldehyde, 17β-acetoxy-3-chloro-	EtOH	260(4.02)	32-1241-66
5β-Androst-3-ene-4-carboxaldehyde, 17β-acetoxy-3-chloro-	EtOH	255(3.81)	32-1241-66
$C_{22}H_{31}NO$			
1H-Naphth[2',1':4,5]indeno[1,2-b]pyridin-2-ol, tetradecahydro-4a,6a-dimethyl-	EtOH	269(3.81),277(3.69)	4-0339-66
$C_{22}H_{31}NO_2$			
Androst-5-en-3β-ol, 17β-(5-isoxazolyl)-	EtOH	217(3.88)	44-3193-66
	MeOH-NaOMe	264(4.11)	44-3193-66
Δ^4-Conenine-3,11-dione	EtOH	239(4.20)	13-0421-66B
Estra-1,3,5(10)-triene-9α,17β-diol, 3-(1-pyrrolidinyl)-	EtOH	259(4.36)	87-0510-66
$C_{22}H_{31}N_3O_7$			
5H,10H-Benzo[c]furo[3,4-e][1]benzopyran-7-acetic acid, dodecahydro-α,1,1-trimethyl-2,5,12-trioxo-, methyl ester, semicarbazone	EtOH	231(4.16)	78-1659-66
$C_{22}H_{32}N_2O$			
Androst-5-en-3β-ol, 17β-pyrazol-3-yl-	EtOH-HCl	219(3.95)	44-3193-66
$C_{22}H_{32}N_2O_3$			
1H-Cyclopenta[5,6]naphtho[1,2-g]quinazoline-8,10(3bH,9H)-dione, tetradecahydro-1-hydroxy-1,11a,13a-trimethyl-	EtOH	264(3.83)	32-0179-66
	EtOH-NaOH	289(3.90)	32-0179-66
$C_{22}H_{32}O$			
14'-Apo-β-carotenol	isooctane	358(4.82)	54-0343-66
8,11,13-Totaratriene, 12-acetyl-	EtOH	220(4.37),261(4.25)	78-2845-66
$C_{22}H_{32}O_2$			
3,5-Cycloandrostane, 17-ethylenedioxy-6-methylene-	MeOH	205(4.00)	5-0204-66G

Compound	Solvent	λ_{max}(log ϵ)	Ref.
Pregn-4-ene-3,20-dione, 18-methyl-	MeOH	242(4.20)	44-1026-66
8,11,13-Totaratriene, 12-acetoxy-	EtOH	218(4.02),269(2.99), 275(2.98)	78-2845-66
Vitamin A, acetate	n.s.g.	244(4.22)	24-2012-66
$C_{22}H_{32}O_3$			
10ξ-Androstan-1-one, 17β-acetoxy-3ξ-methyl-, 9(10 →5ξ)-abeo-	EtOH	220(3.67)	33-1049-66
Androst-4-ene-3,17-dione, 12-isopropoxy-	EtOH	240(4.21)	44-2116-66
Biphenyl, 2',6'-dimethoxy-4'-pentyl-3-methyl-6-acetyl-1,4,5,6-tetrahydro-	MeOH	209(4.87),273(3.11), 279(3.09)	5-0165-66C
1α,5β-Cyclo-10α-androstan-2-one, 17β-acetoxy-3α-methyl-	EtOH	none	33-1049-66
D-Nor-4-androsten-3-one, 16β-(hydroxymethyl)-, propionate	MeOH	240(4.23)	13-0505-66A
Pregn-4-ene-3,18-dione, 20β-hydroxy-18-methyl-	MeOH	242(4.18)	44-1026-66
Pregn-4-ene-3,20-dione, 17α-hydroxy-18-methyl-	MeOH	241(4.22)	44-1026-66
Pregnenolone, 21-hydroxymethylene-	EtOH	204(3.81),270(3.91)	78-1625-66
Testosterone, 2β-hydroxy-, 17-propionate	EtOH	244(4.16)	94-1370-66
$C_{22}H_{32}O_3S$			
Androst-3-en-5-one, 17β-acetoxy-3-(methylthio)-, 10(5→4)-abeo-	EtOH	319(4.02)	33-2218-66
$C_{22}H_{32}O_4$			
Androst-4-en-3-one, 12α,17β-dihydroxy-17α-methyl-, 12-acetate	EtOH	240(4.19)	44-2119-66
Androst-5-en-19-one, 3-ethylenedioxy-17β-hydroxy-17-methyl-	EtOH	222(3.24),307(2.12)	33-0292-66
Mutilindione B, acetate	EtOH	300(1.73)	78-0359-66B
18-Nor-5α,8α,9β,14β-androst-12-ene-11,17-dione, 3α-acetoxy-4α,8,14-trimethyl-	EtOH	260(3.88)	78-3443-66
5α-Pregn-7-ene-20-carboxylic acid, 3β-hydroxy-6-oxo-	EtOH	243(4.14)	33-1591-66
Pregn-5-ene-16,20-dione, 3β-acetoxy-	EtOH	289(3.60)	33-2218-66
	EtOH-KOH	309(4.22)	33-2218-66
$C_{22}H_{32}O_5$			
Androstane-3,5-dione, 11α-hydroxy-17β-acetoxy-17α-methyl-, 10(5→4)-abeo-	EtOH	291(3.90)	33-2218-66
	EtOH-KOH	311(4.11)	33-2218-66
ferric complex	EtOH	542(3.10)	33-2218-66
Androstan-3-one, 17β-acetoxy-4β,5β-epoxy-11α-hydroxy-17α-methyl-	EtOH	301(1.79)	33-2218-66
Androst-1-ene-3,4-dione, 5β,17β-dihydroxy-, 17-propionate	EtOH	262(3.68)	94-1370-66
2-Chromanvaleric acid, 6-acetoxy-α,2,5,7,8-pentamethyl-, methyl ester	EtOH	284.0(3.32)	73-2434-66
$C_{22}H_{32}O_7S$			
Pregn-4-ene-3,20-dione, 14α,17α,21-trihydroxy-, 21-methanesulfonate	EtOH	240(4.19)	78-0325-66A
$C_{22}H_{33}NO$			
Con-4-enin-3-one	EtOH	241(4.20)	78-0541-66B

Compound	Solvent	$\lambda_{max}(\log \epsilon)$	Ref.
$C_{22}H_{33}NO_2$			
5α-Androst-2-eno[2,3-d]isoxazol-17β-ol, 1α,17-dimethyl-	MeOH	226(3.71)	73-1064-66
5α-Androst-2-eno[3,2-c]isoxazol-17β-ol, 1α,17-dimethyl-	MeOH	222(3.74)	73-1064-66
$C_{22}H_{33}NO_3$			
Cyclopenta[5,6]naphth[1,2-d]azepin-2(3H)-one, dodecahydro-8-hydroxy-5a,7a,8-trimethyl-, acetate	n.s.g.	220(4.23)	25-1378-66
$C_{22}H_{33}N_3O$			
1H-Cyclopenta[5,6]naphtho[1,2-g]quinazolin-1-ol, 10-aminotetradecahydro-1,11a,13a-trimethyl-	HCl	260(4.08)	32-0152-66
	EtOH	236(4.02),268(3.69)	32-0152-66
$C_{22}H_{33}N_3OS$			
Pseudourea, 1-(2-cyano-17 -hydroxy-5 -androst-2-en-3-yl)-2-methyl-2-thio-	EtOH	234(3.98)	32-0152-66
$C_{22}H_{33}N_3O_4S$			
Urea, 1,3-dicyclohexyl-1-[(2-hydroxy-5-nitro-3-(methylthiomethyl)benzyl]-	MeOH	324(3.96)	35-5855-66
	MeOH-KOH	414(4.30)	35-5855-66
$C_{22}H_{34}N_2O$			
5α-Androst-2-eno[2,3-d]pyrazol-17β-ol, 1α,17-dimethyl-	MeOH	223(3.68)	73-1064-66
$C_{22}H_{34}O_2$			
Gon-4-en-3-one, 13β,17α-diethyl-17β-hydroxy-7α-methyl-	EtOH	244(4.22)	87-0782-66
Pregnane-3,20-dione, 16β-methyl-	EtOH	286(1.81)	78-1615-66
Pregn-4-en-3-one, 20β-hydroxy-18-methyl-	MeOH	242(4.25)	44-1026-66
$C_{22}H_{34}O_3$			
5α-Androstan-3-one, 17β-hydroxy-2-(hydroxymethylene)-1α,17α-dimethyl-	MeOH	288(3.90)	73-1064-66
5α-Androstan-3-one, 17β-hydroxy-2-(methoxymethylene)-1α-methyl-	MeOH	275(4.04)	73-1064-66
Androst-4-en-3-one, 17β-hydroxy-12α-isopropoxy-	EtOH	243(4.17)	44-2116-66
$C_{22}H_{34}O_4$			
Ancepsenolide	n.s.g.	208(4.45)	88-0097-66
1,3-Butadiene, 2,3-bis(1-methyl-4,4-ethylenedioxycyclohexyl)-	hexane	182(4.14)	5-0031-66D
$C_{22}H_{34}O_6$			
Taxicin I, 5-deoxy-4,16-dihydro-, 2-acetate	EtOH	283(3.72)	39-1933-66C
$C_{22}H_{35}N_3O$			
Androst-5-en-3-one, 4,4-dimethyl-, semicarbazone	EtOH	228(4.14)	22-0850-66
$C_{22}H_{36}$			
3,4,5,6,7-Decapentaene, 3,8-di-tert-butyl-2,2,9,9-tetramethyl-	n.s.g.	237(5.27),308(4.46), 336(4.36)	35-3155-66

Compound	Solvent	λ_{max}(log ε)	Ref.
$C_{22}H_{36}N_2O_4$			
Ketone, methyl 8-methyl-1,4-dioxa-spiro[4.5]dec-8-yl, azine	MeOH	230(3.48)	5-0019-66D
1-Propanone, 1-(1,4-dioxaspiro[4.5]-dec-8-yl)-, azine	MeOH	234(3.60)	5-0019-66D
$C_{22}H_{36}O$			
Pregnan-20-one, 16β-methyl-	EtOH	287(1.62)	78-1615-66
$C_{22}H_{36}O_2Si$			
Androst-4-en-3-one, 17β-(trimethyl-siloxy)-	EtOH	240(4.34)	87-0433-66
$C_{22}H_{38}NiO_2S_2$			
Nickel, bis(2,2,6,6-tetramethyl-5-thiolohept-4-en-3-one)-	$CHCl_3$	495s(3.00),641(1.81)	12-1401-66
$C_{22}H_{38}Si_4$			
Disilane, 1,1,2,2-tetramethyl-1,2-bis-[p-(trimethylsilyl)phenyl]-	C_6H_{12}	245(4.48)	101-0102-66B
$C_{22}H_{40}Si_5$			
Pentasilane, 1,1,2,2,3,3,4,4,5,5-decamethyl-1,5-diphenyl-	C_6H_{12}	257.5(4.40)	101-0102-66B

Compound	Solvent	λ_{max}(log ε)	Ref.
$C_{23}H_{12}O$			
Naphtho[1,2,3-cd]fluoranthene, 5-oxo-	EtOH	260(4.39),285(4.46), 334(3.58),380(3.90), 402(4.10),426(4.16)	24-3298-66
$C_{23}H_{14}$			
5H-Naphtho[1,2,3-cd]fluoranthene	EtOH	241(4.56),252(4.48), 258(4.50),268(4.49), 276(4.40),286(4.38), 299(4.33),310(4.28), 342(3.74),365(4.04), 373(4.09),389(4.22), 404(4.12)	24-3298-66
$C_{23}H_{14}O$			
Pyrene, 1-benzoyl-	MeOH	207(4.40),242(4.71), 263(4.35),275(4.36), 324(4.05),340(4.16), 356(4.15),364(4.12), 381(3.96)	73-0835-66
$C_{23}H_{14}O_4$			
[$\Delta^{3,3'(2H,2'H)}$-Bibenzofuran]-2-one, 2'-(α-hydroxybenzylidene)-	EtOH	312(4.03),391(4.30)	88-4671-66
$C_{23}H_{15}F_{10}N_2P$			
Phosphorane, [[2,2,3,3,3-pentafluoro-1-(pentafluoroethyl)propylidene]-hydrazono]triphenyl-	n.s.g.	223(4.40),268(4.18), 273(4.21),284(4.22)	35-3617-66
$C_{23}H_{15}MnO_5Pb$			
Manganese, pentacarbonyl(triphenyl-plumbyl)-	glyme	295(4.34)	35-5124-66
$C_{23}H_{15}NOS$			
5H-Thiazolo[2,3-a]isoquinolin-5-one, 2,3-diphenyl-	benzene	294(4.43),333(3.92), 476s(3.70),504(3.89), 534(3.84)	4-0282-66
	EtOH	255(4.32),285(4.48), 315s(3.99),475(3.94)	4-0282-66
$C_{23}H_{15}NO_2S$			
9H-Benzo[a]phenoxazin-9-one, 5-(p-tolylthio)-	benzene	521(--)	65-1938-66
	dioxan	512(--)	65-1938-66
	pyridine	525(--)	65-1938-66
	$PhNO_2$	530(--)	65-1938-66
$C_{23}H_{15}N_3O_2S$			
Oxazolo[5,4-d]pyrimidine-5,7(4H,6H)-dione, 2,4,6-triphenyl-5-thio-	$CHCl_3$	329(4.41),341s(4.22)	50-0737-66C +94-1425-66
$C_{23}H_{16}$			
Benzo[g]chrysene, 9-methyl-	EtOH	274(4.72),285(4.81), 294(4.82),320(4.02), 332(4.05),345s(3.85), 385(3.09)	77-0548-66
6H-Naphtho[1,2,3-cd]fluoranthene, 7,7a-dihydro-	EtOH	284(4.86),300(4.56), 319(4.24),335(3.33), 352(3.58),369(3.70)	24-3298-66

Compound	Solvent	λ_{max}(log ε)	Ref.
$C_{23}H_{16}N_2O_2$			
9H-Benzo[a]phenoxazin-9-one, 5-(N-methylanilino)-	pyridine	450(4.4)	65-1938-66
	40% pyridine	590(4.4),620(4.3)	65-1938-66
$C_{23}H_{16}O_2$			
4-Cyclopentene-1,3-dione, 2,4,5-triphenyl-	dioxan	233(4.40),329(4.09)	88-0295-66
1H-Cyclopropa[b]naphthalene-2,7-dione, 1a,17a-dihydro-1,1-diphenyl-	EtOH	223(4.55),306(3.51)	77-0894-66
$C_{23}H_{16}O_6$			
Tetrangulol, diacetate	MeOH	220(4.65),242(4.43), 290(4.58),338s(3.74), 360s(3.67),400s(3.54)	44-2920-66
$C_{23}H_{17}Cl$			
Benzofulvene, 3-(cyclohepta-2,4,6-trien-1-yl)-ω-(p-chlorophenyl)-	EtOH	241(4.16),285(4.23), 348(4.24)	5-0057-66H
9,10-Benzosesquifulvalene, 8-(p-chlorobenzyl)-	EtOH	245(4.27),303(3.59), 316(3.61),400(4.36)	5-0057-66H
Indan, 1-benzylidene-3-(p-chlorobenzylidene)-	EtOH	231(4.22),283s(4.45), 292(4.47),329(4.51), 342s(4.47)	5-0057-66H
$C_{23}H_{17}NO_2S$			
Acetophenone, 2-[(3-hydroxy-1-isoquinolyl)thio]-4'-phenyl-	EtOH	226(4.49),285(4.38), 295(4.38),356(3.84)	4-0282-66
$C_{23}H_{17}N_3O$			
Benzophenone, 4-[(2-amino-1-naphthyl)azo]-	benzene	480(4.26)	44-2571-66
	$C_6H_{11}Me$	470(4.25)	44-2571-66
$C_{23}H_{17}N_3O_3$			
Diformylanhydrotri-2-aminobenzaldehyde	EtOH	232(4.46),248s(4.34), 284s(3.48),315(3.28), 350s(2.99)	39-0956-66B
$C_{23}H_{17}P$			
Phosphorin, 2,4,6-triphenyl-	MeOH	278(4.61)	89-0846C-66
$C_{23}H_{18}$			
Benzofulvene, 3-(cyclohepta-2,4,6-trien-1-yl)-ω-phenyl-	EtOH	242(4.17),286(4.23), 346(4.24)	5-0057-66H
9,10-Benzosesquifulvalene, 8-benzyl-	C_6H_{12}	247(4.24),304(3.53), 315(3.54),395(4.34)	5-0057-66H
	EtOH	246(4.21),303(3.48), 314(3.51),397(4.31)	5-0057-66H
	MeCN	245(4.28),304(3.65), 314(3.66),399(4.36)	5-0057-66H
	dioxan	401(4.32)	5-0057-66H
	$MeNO_2$	402(4.34)	5-0057-66H
Indan, 1,3-dibenzylidene-	EtOH	227(4.23),282s(4.45), 290(4.47),326(4.53), 338s(4.49)	5-0057-66H
Indene, 3-benzyl-1-benzylidene-	EtOH	238(4.17),249(4.31), 341(4.35)	5-0057-66H
$C_{23}H_{18}ClNO_4S$			
4-(p-Aminophenyl)-2,6-diphenylthiopyrylium perchlorate	HOAc	264(4.33),394(4.21), 518(4.60)	33-2046-66

Compound	Solvent	λ_{max}(log ε)	Ref.
$C_{23}H_{18}Cl_2N_2O$			
2-Pyrrolidinone, 4,4-dichloro-3-methyl-1,3-diphenyl-5-(phenylimino)-	EtOH	226(4.27),304(3.46)	28C-1276-66A
$C_{23}H_{18}N_2$			
Aniline, 3,5-diphenyl-4-(2-pyridyl)-	MeOH	240(4.50),288(3.88)	44-3206-66
$C_{23}H_{18}N_4O$			
Urea, 1,1-diphenyl-3,3-di-2-pyridyl-	EtOH	255(4.28)	34-0436-66
$C_{23}H_{18}N_4O_6$			
Benzofuran, 3-acetonyl-2-(o-hydroxyphenyl)-, 2,4-dinitrophenylhydrazone	EtOH	270(4.31),305(4.22), 360(4.35)	39-0544-66C
$C_{23}H_{18}N_8O_{13}$			
Inosine, cyclic 2',3'-carbonate, 5',5'''-carbonate	pH 2	248(4.06)	44-3402-66
	pH 12	253(4.07)	44-3402-66
$C_{23}H_{18}O$			
4H-Pyran, 2,4,6-triphenyl-	EtOH	248(4.44)	24-2351-66
$C_{23}H_{18}O_2$			
1-Cyclopropene, 3-(p-methoxybenzoyl)-1,2-diphenyl-	EtOH	223(4.57),234(4.40), 292(4.64),308(4.56), 324(4.35)	18-1975-66
$C_{23}H_{18}O_{10}$			
Anthraquinone, 1,2,3,7-tetrahydroxy-5-methyl-, tetraacetate	EtOH	262(4.11),282(3.64), 324(3.11),425(--)	78-1507-66
$C_{23}H_{18}O_{11}$			
Norvogeletin, tetraacetate	MeOH	274(4.52),328(4.18)	24-2430-66
$C_{23}H_{19}F_6IN_2S_2$			
3,3'-Diethyl-6,6'-bis(trifluoromethyl)-thiacarbocyanine iodide	EtOH	561(--)	65-0828-66
$C_{23}H_{19}NO$			
2-Cyclohexen-1-one, 3,5-diphenyl-4-(2-pyridyl)-	MeOH	205(4.42),228(4.03), 265(4.17),270(4.19), 282(4.17)	44-3206-66
cis	MeOH	228(4.01),265(4.18), 271(4.22),282(4.23)	44-3213-66
trans	MeOH	228(4.02),263(4.17), 270(4.21),282(4.17)	44-3213-66
$C_{23}H_{19}NO_2$			
Aziridine, 2,3-dibenzoyl-1-benzyl-, cis	C_6H_{12}	324(2.41)	44-1244-66
trans	C_6H_{12}	348(2.61)	44-1244-66
Quinoline, 2,3-bis(p-methoxyphenyl)-	EtOH	217(4.08),239(4.64), 271(4.47),341(3.99)	87-0527-66
$C_{23}H_{19}NO_4$			
Cyclopenta[c]quinolizine-2,3-dicarboxylic acid, 5-methyl-4-phenyl-, dimethyl ester	EtOH	234(4.44),272(4.30), 285s(4.2),354(4.57), 385s(4.1),480(3.59)	39-0324-66C

Compound	Solvent	λ_{max}(log ε)	Ref.
$C_{23}H_{19}N_3O_2$			
Monoacetylanhydrotri-2-amino-benzaldehyde	EtOH	237(4.27),282s(3.38)	39-0956-66B
$C_{23}H_{20}ClN_3O_6S_3$			
p-Toluenesulfonamide, N-[(p-chloro-phenyl)sulfonyl]-N-[1-(p-tolyl-sulfonyl)pyrazol-5-yl]-	MeOH	209(4.2),241(4.6), 267s(3.5),276s(3.3)	24-0183-66
$C_{23}H_{20}N_2$			
Dipyrromethene, 3,5-dimethyl-3',5-diphenyl-	EtOH-N HCl	221(4.27),273(4.33), 377(3.95),505(4.98)	12-1871-66
	EtOH	225(4.20),285(4.35), 467(4.52)	12-1871-66
	CCl_4	474(4.51)	12-1871-66
Dipyrromethene, 4,5-dimethyl-3',5'-diphenyl-	EtOH-N HCl	220(4.24),284(4.27), 391(3.85),521(4.87)	12-1871-66
	EtOH	238(4.01),290(4.22), 470(4.40)	12-1871-66
	CCl_4	474(4.45)	12-1871-66
Dipyrromethene, 5,5'-dimethyl-3,3'-diphenyl-	EtOH-HCl	222(4.41),263(3.79), 380(3.82),485(4.90)	12-1871-66
	EtOH	220(4.34),258s(3.95), 276s(3.80),457(4.52)	12-1871-66
	CCl_4	465(4.57)	12-1871-66
Imidazole, 1-ethyl-2,4,5-triphenyl-	EtOH	276(4.30)	35-3826-66
$C_{23}H_{20}N_2O_2$			
Benzoic acid, (phenylazo)-, 8-decene-4,6-diynyl ester	ether	227(4.12),237(4.14), 251(4.25),265(4.40), 281(4.39),322(4.40)	24-0586-66
1H-Pyrazolo[1,2-a]pyrazol-4-ium, 5,7-dibenzoyl-1,2,3-trimethyl-, hydroxide, inner salt	n.s.g.	234(4.23),248s(4.22), 360(4.18),440(4.34)	35-5588-66
$C_{23}H_{20}O$			
1-Indanone, 2,2-dibenzyl-	EtOH	251(4.08),299(3.38)	88-6489-66
$C_{23}H_{20}O_2$			
2-Furanol, 2,5-dihydro-5-methyl-2,4,5-triphenyl-	2N H_2SO_4	240(3.60),310(3.96), 401(4.37)	65-2064-66
$C_{23}H_{20}O_3$			
Cyclohexanone, 2-(2-hydroxy-1-naphthyl)-, benzoate	EtOH	223(4.85)	44-2646-66
1-Indanone, 3-ethoxy-2-hydroxy-2,3-diphenyl-, cis	hexane	240(4.05),265s(3.14), 282(3.15),290(3.12), 350(1.93)	35-4942-66
trans	hexane	240s(4.11),268(3.12), 285(3.19),293(3.17), 335(2.04)	35-4942-66
$C_{23}H_{20}O_5$			
Phthalic acid, 5-methoxy-3,4-diphenyl-	EtOH	275(4.18),300(4.00)	35-3759-66
$C_{23}H_{21}IN_2OS$			
3-Methyl-2-[(5,6,7,8-tetrahydro-2-phenyl-4H-1,3-benzoxazin-4-ylidene)-methyl]benzothiazolium iodide	MeOH	440(4.57)	87-0758-66

Compound	Solvent	λ_{max}(log ε)	Ref.
$C_{23}H_{21}IN_2O_2$			
3-Methyl-2-[(5,6,7,8-tetrahydro-2-phenyl-4H-1,3-benzoxazin-4-ylidene)-methyl]benzoxazolium iodide	MeOH	405(4.59),417(4.59)	87-0758-66
$C_{23}H_{21}NO$			
Cyclohexanone, 3,5-diphenyl-4-(2-pyridyl)-, cis-cis	MeOH	258(3.51),262(3.49), 271(3.36)	44-3213-66
cis-trans	MeOH	259(3.59),263(3.63), 270(3.52)	44-3213-66
trans-trans	MeOH	258(3.53),262(3.56), 269(3.43)	44-3213-66
Ketone, 1-benzyl-3-phenyl-2-aziridinyl p-tolyl, cis	C_6H_{12}	323(2.21)	44-1244-66
trans	C_6H_{12}	344(2.42)	44-1244-66
Ketone, 1-benzyl-3-p-tolyl-2-aziridinyl phenyl, cis	C_6H_{12}	325(2.17)	44-1244-66
trans	C_6H_{12}	346(2.44)	44-1244-66
$C_{23}H_{21}NO_2$			
Cyclohexanone, 3-hydroxy-3,5-diphenyl-4-(2-pyridyl)-	MeOH	258(3.55),263(3.57), 269(3.44)	44-3206-66
1,5-Hexanedione, 1,3-diphenyl-2-(2-pyridyl)-	MeOH	249(4.17),265(3.97), 270(3.80),318(2.52)	44-3213-66
Indole, 2,3-bis(p-methoxyphenyl)-1-methyl-	EtOH	249(4.50),301(4.20)	87-0527-66
Indole, 2,3-bis(p-methoxyphenyl)-5-methyl-	EtOH	254(4.49),313(4.32)	87-0527-66
Indole, 2,3-bis(p-methoxyphenyl)-7-methyl-	EtOH	253(4.49),308(4.29)	87-0527-66
$C_{23}H_{21}NO_3$			
Indole, 4-methoxy-2,3-bis-(p-methoxyphenyl)-	EtOH	254(4.52),310(4.27)	87-0527-66
Indole, 5-methoxy-2,3-bis-(p-methoxyphenyl)-	EtOH	252(4.40),316(4.40)	87-0527-66
Indole, 6-methoxy-2,3-bis-(p-methoxyphenyl)-	EtOH	253(4.50),323(4.30)	87-0527-66
Indole, 7-methoxy-2,3-bis-(p-methoxyphenyl)-	EtOH	256(4.53),306(4.30)	87-0527-66
$C_{23}H_{21}NO_6$			
1,3-Azulenedicarboxylic acid, 2-amino-5-benzoyloxy-, diethyl ester	MeOH	245(4.66),317(4.78), 327(4.88),370(3.94), 390(3.96),450(3.44)	18-1310-66
$C_{23}H_{22}N_2O_7S_2$			
5-Thia-1-azabicyclo[4.2.0]oct-3-ene-2-carboxylic acid, 3-(hydroxymethyl)-8-oxo-7-[2-(2-thienyl)acetamido]-, p-methoxyphenyl ester, acetate	EtOH	223(4.25),237(4.14), 268(4.00)	87-0741-66
$C_{23}H_{22}O_4$			
11H-Cyclopenta[a]phenanthrene-17-carboxylic acid, 12,13-dihydro-16-hydroxy-13-methyl-, ethyl ester, acetate	EtOH	235(4.60),283(4.22), 294(4.25),350(4.30)	35-0799-66

Compound	Solvent	λ_{max}(log ε)	Ref.
$C_{23}H_{22}O_5S$			
Phenol, o-[3-(m-methoxyphenoxy)propenyl]-, p-toluenesulfonate	EtOH	252(3.22)	78-3491-66
$C_{23}H_{22}O_6$			
Robustic acid, methyl ester	EtOH	233(4.47),261(4.46), 269(4.47),351(4.28)	39-0606-66C
$C_{23}H_{22}O_7$			
Isolisetin, di-O-methyl-	EtOH	235s(4.41),259(4.53), 272s(4.47),307(4.16), 327(4.18)	78-0333-66A
$C_{23}H_{23}IN_2S_2$			
3-Ethyl-2-[5-(3-ethyl-2-benzothiazolinylidene)-1,3-pentadienyl]benzothiazolium iodide	MeOH	649(5.25)	9-0289-66
$C_{23}H_{23}IN_2Se_2$			
3-Ethyl-2-[5-(3-ethyl-2-benzoselenazolinylidene)-1,3-pentadienyl]benzoselenazolium iodide	MeOH	660(5.19)	9-0289-66
$C_{23}H_{23}NO$			
Cyclohexanol, 3,5-diphenyl-4-(2-pyridyl)-	MeOH	209(4.32),258(3.55), 262(3.58),269(3.45)	44-3213-66
isomer	MeOH	208(4.36),258(3.50), 261(3.50),268(3.38)	44-3213-66
Cyclopropanemethanol, 2-(4-pyridyl)-α,α-di-o-tolyl-, trans	EtOH-HCl	257.5(4.17)	44-0399-66
	EtOH-NaOH	259(3.58)	44-0399-66
$C_{23}H_{23}NO_5$			
Chelerythrine, 9-ethoxy-	EtOH	228(4.53),284(4.67), 320s(4.18)	88-0181-66
$C_{23}H_{23}NO_6$			
Acetamide, N-[2-(1-hydroxy-2-methoxyphenanthro[2,3-d][1,3]dioxol-4-yl)ethyl]-N-methyl-, acetate	EtOH	245s(4.68),252(4.73), 261(4.42),286s(3.46), 315(4.11),326(4.13), 358(3.60),376(3.61)	12-2339-66
$C_{23}H_{23}N_3O_3$			
Benzanilide, 4-(dimethylamino)-4'-(p-nitrophenethyl)-	EtOH	315(4.48)	65-0214-66
$C_{23}H_{23}N_3O_5$			
1,3-Azulenedicarboxylic acid, 2-acetamido-6-(2-pyridylamino)-, diethyl ester	MeOH	228s(4.27),271(4.29), 303s(4.14),354(4.69), 422(4.57)	18-1954-66
$C_{23}H_{23}N_3O_6S$			
5-Thia-1-azabicyclo[4.2.0]oct-3-ene-2-carboxylic acid, 3-(hydroxymethyl)-8-oxo-7-(2-phenylacetamido)-, acetate, pyridine salt	H_2O	250(4.07),256s(3.90), 263s(3.66)	39-1142-66C
$C_{23}H_{23}N_5O_4$			
Acetanilide, 2'-hydroxy-2,2-bis-[(p-methoxyphenyl)azo]-N-methyl-	EtOH	462.5(4.27)	22-0400-66
	EtOH-KOH	472(4.05)	22-0400-66

Compound	Solvent	λ_{max}(log ϵ)	Ref.
$C_{23}H_{24}N_2O_2$			
p-Phenylenediamine, N'-[bis(p-methoxyphenyl)methylene]-N,N-dimethyl-	C_6H_{12}	275(4.47),379(4.03)	35-2775-66
$C_{23}H_{24}O_6$			
4H,8H-Benzo[1,2-b:3,4-b']dipyran-4-one, 9,10-dihydro-2,5-dimethoxy-3-(p-methoxyphenyl)-8,8-dimethyl-	EtOH	250(4.73),311(4.19)	39-0606-66C
Robustic acid, dihydro-, methyl ester	EtOH	224s(4.36),239s(4.14), 329(4.27)	39-0606-66C
Robustic acid, α-isodihydro-, methyl ester	EtOH	261(4.06),343(4.36)	39-0606-66C
$C_{23}H_{24}O_9$			
Chalcone, α,2'-dihydroxy-3,4,4',6'-tetramethoxy-, diacetate	EtOH	252(4.06),344(4.27)	88-4179-66
$C_{23}H_{24}O_{11}$			
Parkinsonin B	EtOH	246(3.50),270(3.50), 350(3.00)	78-1147-66
	EtOH-NaOEt	262(3.50),312(3.50), 390(3.30)	78-1147-66
	EtOH-NaOAc	250(3.50),275(3.30), 363(4.20)	78-1147-66
	+ H_3BO_3	275(4.30),345(4.20), 370(3.90)	78-1147-66
	EtOH-$AlCl_3$	275(3.10),291(3.20), 350(3.00)	78-1147-66
$C_{23}H_{24}S$			
Sulfide, tert-butyl trityl	EtOH	248s(3.34),256s(3.26), 262s(3.08),267s(2.79)	35-1257-66
	50% H_2SO_4-HOAc	406(4.48),433(4.48)	35-1257-66
$C_{23}H_{25}NO$			
1,4-Hexadien-3-one, 5-methyl-2,4-diphenyl-1-(1-pyrrolidinyl)-	EtOH	234(4.28),295(4.14)	44-1336-66
$C_{23}H_{25}N_3OS_2$			
Rhodanine, 5-[3,3-bis[p-(dimethylamino)phenyl]allylidene]-3-methyl-	CH_2Cl_2	514(4.61)	24-0307-66
$C_{23}H_{26}F_2O_3$			
Androsta-1,4-dien-3-one, 17α-ethynyl-6,6-difluoro-17β-hydroxy-, acetate	EtOH	238(4.16)	44-0991-66
$C_{23}H_{26}N_2O_6$			
Phenoxaz ine-1,9-dicarboxylic acid, 3-hydroxy-4,6-dimethyl-2-piperidino-, dimethyl ester	MeOH	226(4.67),408(4.16)	44-2564-66
$C_{23}H_{26}N_2O_{10}$			
Quinoxaline, 2-(β-D-glucopyranosyloxy)-3-methyl-, tetraacetate	MeOH	239(4.28),243(4.27), 305s(3.70),316(3.81), 328(3.76)	24-0547-66
$C_{23}H_{26}N_2O_{11}$			
2(1H)-Quinoxalinone, 3-(β-D-glucopyranosyloxy)-1-methyl-, tetraacetate	MeOH	227(4.26),250s(3.77), 284s(3.67),315(3.92), 326(3.93),340s(3.65)	24-0547-66

Compound	Solvent	$\lambda_{max}(\log \varepsilon)$	Ref.
$C_{23}H_{26}O_5W$			
Tungsten, tricarbonyl(estra-1,3,5(10)-trien-17-one)-, cyclic ethylene acetal	n.s.g.	320(4.36)	39-0054-66C
$C_{23}H_{26}O_6$			
3-Pentenoic acid, 3,4-dimethyl-, 2-ester with 5,8-dihydroxy-2-(1-hydroxy-4-methyl-3-pentenyl)-1,4-naphthoquinone	EtOH	273(4.66)	88-3677-66
$C_{23}H_{26}O_7$			
Cinnamic acid, α-[α-(hydroxymethyl)-3,4-dimethoxyphenethyl]-3,4,5-trimethoxy-, γ-lactone	EtOH	314(4.26)	23-1541-66
$C_{23}H_{27}FO_6$			
Pregna-1,4,8-triene-3,20-dione, 21-acetoxy-11β,17α-dihydroxy-6α-fluoro-	MeOH	238(4.24)	13-0381-66A
$C_{23}H_{27}F_3O_2$			
Estra-1,3,5(10)-triene, 3,17β-dimethoxy-17α-(trifluoro-1-propynyl)-	EtOH	220(3.94),228s(3.88), 278(3.30),286(3.28)	78-2829-66
$C_{23}H_{27}NO_4$			
Estra-1,3,5(10)-triene-2-carbonitrile, 3,17β-diacetoxy-	EtOH	236(4.05),278(3.27), 288(3.25)	88-0205-66
$C_{23}H_{27}N_3O_8S$			
5-Thia-1-azabicyclo[4.2.0]oct-2-ene-2-carboxylic acid, 7-[2-(carboxyamino)-2-phenylacetamido]-3-(hydroxymethyl)-8-oxo-, N-tert-butyl ester, acetate	EtOH	258(3.86)	87-0746-66
$C_{23}H_{28}F_2O_3$			
Androst-4-en-3-one, 17α-ethynyl-6,6-difluoro-17β-hydroxy-, acetate	EtOH	229(4.11),335(1.46)	44-0991-66
$C_{23}H_{28}F_4O_2$			
Pregn-4-ene-3,20-dione, 16α,17α-tetrafluoroethylene-	n.s.g.	237(4.20)	88-3451-66
$C_{23}H_{28}FeO_4$			
Iron, tricarbonyl(retinal)-	n.s.g.	212(4.46),285(4.36), 358(4.31)	39-2060-66C
$C_{23}H_{28}INO_4$			
1-(3,4-Diethoxybenzyl)-6,7-dimethoxy-2-methylisoquinolinium iodide	MeOH	260(4.71),325(3.98), 360s(3.69)	24-2873-66
6,7-Diethoxy-1-(3,4-dimethoxybenzyl)-2-methylisoquinolinium iodide	MeOH	260(4.74),325(4.03), 360s(3.78)	24-2873-66
$C_{23}H_{28}N_2$			
Retinylideneacetodinitrile	acetone	454(4.79)	54-0334-66
$C_{23}H_{28}N_2O_2S$			
Nortricyclene-1-carboxaldehyde, N-(nortricyclyl-1-carbinyl)-, hydrazone	MeCN	220(4.32),257(3.99), 275(3.45)	44-1543-66

Compound	Solvent	λ_{max}(log ϵ)	Ref.
$C_{23}H_{28}N_2O_3$			
Majoridine	MeOH	247(3.94),310(3.54)	100-0029-66
$C_{23}H_{28}N_2O_5$			
1,3-Azulenedicarboxylic acid, 2-acetamido-6-piperidino-, diethyl ester	MeOH	271(4.32),274(4.32), 348(4.77),410(4.42)	18-1954-66
Isoreserpiline	EtOH	227(4.60),305(4.06)	102-1065-66
$C_{23}H_{28}N_2O_6$			
Majdine	MeOH	224(4.51),248(4.12)	100-0029-66
$C_{23}H_{28}O_3$			
Gona-1,3,5(10),8,14-pentaen-17-one, 13β-ethyl-3-methoxy-6β-methyl-, cyclic ethylene acetal	n.s.g.	312(4.47)	78-1019-66
$C_{23}H_{28}O_4$			
Estra-1,3,5(10),9(11)-tetraene-3,17β-diol, 1-methyl-, diacetate	EtOH	217(4.24),249(4.11)	33-1183-66
$C_{23}H_{28}O_5$			
Androst-4-en-19-oic acid, 17α-ethynyl-17β-hydroxy-3-oxo-, 17-acetate	EtOH	242(4.07)	44-0693-66
Estra-1,3,5(10)-triene-2-carboxaldehyde, 3,17β-diacetoxy-	EtOH	261(4.05),296(3.36)	88-0205-66
$C_{23}H_{28}O_6$			
Pregna-1,4,8-triene-3,20-dione, 21-acetoxy-11β,17α-dihydroxy-	MeOH	240(4.21)	13-0381-66A
$C_{23}H_{28}O_7$			
Hydrocinnamic acid, α-[α-(hydroxymethyl)-3,4-dimethoxyphenethyl]-3,4,5-trimethoxy-, γ-lactone, cis	EtOH	278(3.56)	23-1541-66
Thujaplicatin, tri-O-methyl-	EtOH	278(3.54)	23-1541-66
$C_{23}H_{29}ClO_2$			
Estra-1,3,5(10)-triene, 17α-chloroethynyl-3-ethoxy-17α-methoxy-	EtOH	279(3.26),288(3.23)	78-2829-66
Estra-1,3,5(10)-triene, 17α-chloroethynyl-17β-ethoxy-3-methoxy-	EtOH	279(3.28),288(3.26)	78-2829-66
Estra-1,3,5(10)-trien-17β-ol, 17α-chloroethynyl-3-propoxy-	EtOH	221(3.93),279(3.27), 287(3.25)	78-2829-66
$C_{23}H_{29}ClO_3$			
Androsta-2,4,6-triene-2-carboxaldehyde, 17β-acetoxy-7-chloro-17α-methyl-	MeOH	358(4.38)	24-3057-66
Pregna-2,4,6-trien-20-one, 17α-acetoxy-7-chloro-	MeOH	299(4.17),311(4.26), 326(4.10)	24-3057-66
$C_{23}H_{29}FO_5$			
Pregna-1,4-diene-3,20-dione, 11β-fluoro-17α,21-dihydroxy-, 21-acetate	MeOH	241(4.16)	35-3016-66
$C_{23}H_{29}NO_2$			
14'-Apo-β-carotenoic acid, 15'-cyano-	acetone	433(4.74)	54-0334-66
$C_{23}H_{29}NO_3$			
Serratinine, 6-benzylidene-	EtOH	225(3.81),296(4.18)	88-1537-66

Compound	Solvent	λ_{max}(log ε)	Ref.
$C_{23}H_{29}NO_4$			
Estra-2,4-diene-2-carbonitrile, 3,17β-diacetoxy-	EtOH	296(3.92)	32-0152-66
Estra-1,3,5(10)-trien-6-one, 17β-hydroxy-4-methyl-, O-acetyloxime acetate	EtOH	255(4.16),290s(3.24)	5-0180-66B
$C_{23}H_{29}NO_6$			
Nortaxinonitrile, anhydrooxo-, diacetate	EtOH	210(3.81)	95-1172-66
$C_{23}H_{29}NO_{10}S_2$			
α-D-Glucopyranoside, methyl 3-acetamido-3-deoxy-2,6-di-p-toluenesulfonate	MeOH	226(4.32),256(2.98), 263(3.09),268(3.06), 274(3.02)	94-0902-66
$C_{23}H_{29}N_3O_7S_2$			
Alanine, N-[[3-(hydroxymethyl)-8-oxo-7-[2-(2-thienyl)acetamido]-5-thia-1-azabicyclo[4.2.0]oct-2-en-2-yl]-carbonyl]-, tert-butyl ester, acetate	EtOH	235(4.20),265(3.91)	87-0741-66
$C_{23}H_{29}N_5$			
5H-Pyrimido[4",5":3',4']cyclopenta-[1',2':5,6]naphtho[1,2-g]quinazoline, 2-aminododecahydro-5a,7a-dimethyl-	EtOH	227(4.21),255(3.75), 331(3.71)	32-0152-66
$C_{23}H_{29}N_5O_{10}$			
7(8H)-Pteridinone, 2-(dimethylamino)-8-(β-D-galactopyranosyl)-6-methyl-, tetraacetate	pH -0.89	239(4.36),263(4.15), 301(4.02),347(3.82)	24-3022-66
	pH 6	249(4.21),317s(3.76), 369(4.14)	24-3022-66
	MeOH	219(4.17),248(4.18), 319(3.84),362(4.19)	24-3022-66
7(8H)-Pteridinone, 2-(dimethylamino)-8-(β-D-glucopyranosyl)-6-methyl-, tetraacetate	pH -0.89	238(4.37),262(4.15), 300(4.01),346(3.83)	24-3022-66
	pH 5	248(4.23),316s(3.80), 369(4.17)	24-3022-66
	MeOH	217(4.16),246(4.18), 320s(3.86),360(4.21)	24-3022-66
$C_{23}H_{30}Cl_2O_4$			
Pregna-1,4-diene-3,20-dione, 9α,11β-dichloro-6α,16β-dimethyl-	MeOH	236(4.16)	13-0234-66A
$C_{23}H_{30}F_2O_4$			
Pregn-4-ene-3,20-dione, 17α-acetoxy-6,6-difluoro-	EtOH	228(4.08),285(1.92), 340(1.52)	44-0991-66
$C_{23}H_{30}N_2$			
1,7-Diazaspiro[5.5]undecane, 2,8-dimethyl-1,7-diphenyl-	MeOH	208(4.4),228(3.9)	24-0724-66
$C_{23}H_{30}N_2O_4$			
Aspidoalbidine, 17-hydroxy-16-methoxy-N-propionyl-	EtOH	225(4.28),258(3.55)	23-0028-66
	EtOH-NaOH	220(4.24),300(4.39)	23-0028-66

Compound	Solvent	λ_{max}(log ϵ)	Ref.
$C_{23}H_{30}N_2O_5$			
Epiherbaceine	EtOH	226(4.47),277s(3.90), 306(4.10)	24-1008-66
Herbaceine	EtOH	226(4.53),280s(3.83), 300(4.04)	24-1008-66
$C_{23}H_{30}N_2O_6$			
Herbaline	EtOH	215(4.56),273(4.05), 305(3.99)	24-2052-66
$C_{23}H_{30}N_4O$			
p-Benzoquinone, 2,6-di-tert-butyl-, 4-azine with 1,3-dimethyl-2-benzimidazolinone	MeCN	495(4.79)	5-0065-66J
$C_{23}H_{30}N_4O$			
Benzimidazole, 1-(2-diethylaminopropyl)-2-(p-ethoxybenzyl)-5-nitro-	EtOH-HCl	240(4.45),302(4.07)	39-1511-66C
$C_{23}H_{30}N_4O_3$			
2,5,8,11-Heptadecatetraenal, 2,4-dinitrophenylhydrazone, 2-trans-5,8,11-tri-cis	$CHCl_3$	375(4.45)	54-0117-66
$C_{23}H_{30}O_4$			
Androst-4-en-3-one, 17α-ethynyl-17β,19-dihydroxy-, 17-acetate	EtOH	243(4.20)	44-0693-66
12H-Benz[4,5]indeno[2,1-c]oxepin-12-one, 9-ethylidenedodecahydro-3α-hydroxy-11,12bα-dimethyl-, acetate	EtOH	276(4.13)	27-0114-66
5β-Carda-16,20(22)-dienolide, 14β,15β-epoxy-3β-hydroxy-	EtOH	276(4.18)	94-0496-66
Estra-1,3,5(10)-triene-1,17β-diol, 3-methyl-, diacetate	EtOH	269(2.78)	33-1049-66
Estra-1,3,5(10)-triene-3,17β-diol, 4-methyl-, diacetate	EtOH	268(2.82)	33-1049-66
Pregna-4,14-diene-3,20-dione, 17α-acetoxy-	EtOH	239(4.21)	78-0325-66A
$C_{23}H_{30}O_5$			
Androsta-1,4-dien-3-one, 5,17β-diacetoxy-, 10(5→4)-abeo-	EtOH	248(3.94)	33-2218-66
Estr-5(10)-en-3-one, 17β-hydroxy-4-(hydroxymethylene)-, diacetate	EtOH	230(4.25),297(3.77)	27-0110-66
Merogedunin, acetate	MeOH	227(4)	39-0506-66C
10α-Pregna-4,9(11)-diene-3,20-dione, 17α,21-dihydroxy-, 21-acetate	MeOH	238(4.15)	54-0731-66
9β,10α-Pregn-4-en-3-one, 17α,20:20,21-bis(methylenedioxy)-	MeOH	242(4.22)	54-0731-66
$C_{23}H_{30}O_6$			
Pregn-4-ene-3,11-dione, 17,20:20,21-bis(methylenedioxy)-	MeOH	238(4.19)	24-3051-66
9β,10α-Pregn-4-ene-3,11-dione, 17α,20:20,21-bis(methylenedioxy)-	MeOH	239(4.19)	54-0731-66
9β,10α-Pregn-4-ene-3,11,20-trione, 21-acetoxy-17α-hydroxy-	EtOH	238(4.19)	33-1528-66
	MeOH	239(4.21)	54-0731-66

Compound	Solvent	λ_{max}(log ε)	Ref.
$C_{23}H_{31}ClO_3$			
Pregna-3,5-dien-20-one, 3-chloro-17α-hydroxy-, acetate	EtOH	238(4.19),244(4.25), 254s(--)	32-1268-66
$C_{23}H_{31}DO_5$			
Androst-5-en-19-al-19d, 3,17-dioxo-, cyclic 3,17-bis(ethylene acetal)	EtOH	226(3.14),310(2.05)	33-0292-66
$C_{23}H_{31}FO_5$			
Pregn-4-ene-3,20-dione, 9α-fluoro-11β,17-dihydroxy-, 17-acetate	EtOH	239(4.24)	32-1115-66
Pregn-4-ene-3,20-dione, 11β-fluoro-17α,21-dihydroxy-, 21-acetate	MeOH	239(4.19)	35-3016-66
$C_{23}H_{31}FO_6$			
Pregn-4-ene-3,20-dione, 9-fluoro-11β,17,21-trihydroxy-, 17-acetate	EtOH	240(4.22)	32-1115-66
Pregn-4-en-3-one, 9α-fluoro-11β-hydroxy-17,20:20,21-bis(methylenedioxy)-	MeOH	238(4.24)	24-3051-66
$C_{23}H_{31}NO_2$			
Pregna-3,5-dieno[3,4-d]oxazol-20-one, 2'-methyl-	EtOH	261(4.04)	22-3407-66
$C_{23}H_{31}NO_3$			
Androsta-3,5-diene-6-carbonitrile, 17β-acetoxy-3-methoxy-	MeOH	216(4.00),282(4.35)	78-1069-66
Androsta-3,5-dieno[3,4-d]oxazol-17β-ol, 2'-methyl-, acetate	EtOH	262(4.09)	22-3407-66
Androsta-3,5-dieno[4,3-d]oxazol-17β-ol, 2'-methyl-, acetate	EtOH	236(4.22)	22-3407-66
Pregna-1,4-diene-3,20-dione, 11α-acetamido-	MeOH	250(4.25)	44-1342-66
11β-isomer	MeOH	240(4.16)	44-1342-66
$C_{23}H_{31}NO_4$			
5α-Estr-2-ene-2-carbonitrile, 3,17β-diacetoxy-	EtOH	211(4.05)	32-0152-66
Isoquinoline, 6,7-diethoxy-1,2,3,4-tetrahydro-3-(3,4-dimethoxybenzyl)-2-methyl-, perchlorate	MeOH	290(3.80)	24-2873-66
Isoquinoline, 3-(3,4-diethoxybenzyl)-1,2,3,4-tetrahydro-6,7-dimethoxy-2-methyl-, perchlorate	MeOH	290(3.84)	24-2873-66
$C_{23}H_{31}N_3O$			
3-Pyrrolin-2-one, 4-ethyl-5-(4-ethyl-3,3',4',5-tetramethyl-2,2'-dipyrromethen -5-ylmethyl)-3-methyl-, hydrobromide	$CHCl_3$	374(3.80),413(4.56), 495(5.01)	39-1155-66C
$C_{23}H_{31}N_3O_2$			
1H-Cyclopenta[5,6]naphtho[1,2-g]quinazolin-1-ol, 10-aminododecahydro-11a,13a-dimethyl-, acetate	HCl	252(4.16),313(3.80)	32-0152-66
	EtOH	244(4.15),310(3.78)	32-0152-66
$C_{23}H_{32}ClN_3$			
Acridan, 2-chloro-9,9-bis(dimethylaminopropyl)-, dihydrochloride	n.s.g.	293(4.37)	33-0572-66

Compound	Solvent	λ_{max}(log ε)	Ref.
$C_{23}H_{32}Cl_2O_3$			
Androsta-3,5-dien-17β-ol, 4-chloro-3-(2-chloroethoxy)-, acetate	EtOH	258(4.20)	32-1268-66
$C_{23}H_{32}N_2O_4$			
Uracil-androstanol acetate	EtOH	264(3.87)	32-0179-66
	EtOH-NaOH	289(3.91)	32-0179-66
$C_{23}H_{32}O$			
Retinylideneacetone	pet ether	390(4.76)	54-0334-66
$C_{23}H_{32}O_3$			
14 -Apo-β-carotenoic acid, methyl ester	pet ether	374(4.79)	54-0334-66
Biphenyl, 6-isopropyl-2',6'-dimethoxy-3-methyl-4'-pentyl-	MeOH	210(4.20),272(2.58)	5-0165-66C
Pregn-4-ene-3,20-dione, 16α,17α-ethylene-	n.s.g.	241(4.23)	88-3451-66
$C_{23}H_{32}O_3$			
Gona-4,6-dien-3-one, 17β-acetoxy-13β,17α-diethyl-	EtOH	282(4.46)	87-0782-66
Gona-1,3,5(10)-triene-3,17β-diol, 13β,17α-diethyl-, 3-acetate	EtOH	217s(3.92),269(2.89), 276(2.89)	44-2512-66
18-Nor-5α,8α,9α,14β-androsta-12,15-dien-17-one, 3α-acetoxy-4α,8,14-trimethyl-	EtOH	248(3.97)	78-3459-66
Pregna-1,4-dien-3-one, 20β-acetoxy-	n.s.g.	244(4.23)	19-0347-66
5β-Pregna-14,16-dien-20-one, 3β-acetoxy-	EtOH	310(4.15)	23-0844-66
$C_{23}H_{32}O_4$			
12H-Benz[4,5]indeno[2,1-c]oxepin-12-one, 9-ethyldodecahydro-3α-hydroxy-11,12bα-dimethyl-, acetate	EtOH	272(4.15)	27-0114-66
5β,17α-Card-20(22)-enolide, 14β,15β-epoxy-3β-hydroxy-	EtOH	215(4.14)	94-0496-66
Estr-5(10)-ene-6,17-dione, 3β-[(tetrahydro-2-pyranyl)oxy]-	KBr	249(4.05)	44-1761-66
Gitoxigenin, $\Delta^{8(14)}$-anhydro-	n.s.g.	212(4.31)	83-0679-66
Pregn-4-ene-3,20-dione, 17α-acetoxy-	EtOH	238(4.21),285(1.64)	44-0991-66
9β,10α-Pregn-4-ene-3,20-dione, 11α-acetoxy-	EtOH	238(4.24)	33-1529-66
11β-isomer	EtOH	238(4.22)	33-1529-66
$C_{23}H_{32}O_5$			
Androst-5-ene-3,11,17-trione, cyclic 3,17-bis(ethylene acetal)	EtOH	226(3.10),310(2.05)	33-0292-66
5α-Androst-1-en-3-one, 2,17β-diacetoxy-	EtOH	237(4.00)	44-2395-66
5α-Androst-3-en-2-one, 3,17β-diacetoxy-	EtOH	240(3.89)	44-2395-66
Androst-4-en-3-one, 5,17β-diacetoxy-, 10(5→4)-abeo-	EtOH	254(3.87)	33-2218-66
9β,10α-Androst-4-en-3-one, 11α,17β-diacetoxy-	EtOH	238(4.25)	33-1529-66
Androst-4-en-3-one, 12α,17β-diacetoxy-	EtOH	240(4.24)	44-2116-66
Gon-4-en-3-one, 1β,17β-diacetoxy-13β-ethyl-	EtOH	240(4.23)	35-3120-66
Gon-4-en-3-one, 6β,17β-diacetoxy-13β-ethyl-	EtOH	235(4.16)	35-3120-66

Compound	Solvent	λ_{max}(log ϵ)	Ref.
Pregn-4-ene-3,20-dione, 11β,17α-dihydroxy-, 17-acetate	EtOH	242(4.23)	32-1115-66
9β,10α-Pregn-4-ene-3,20-dione, 17α,21-dihydroxy-, 21-acetate	MeOH	242(4.21)	54-0731-66
5α-Pregn-16-ene-15,20-dione, 3β,14β-dihydroxy-	EtOH	247(3.95)	94-0809-66
Pregn-4-en-3-one, 17,20:20,21-bis-(methylenedioxy)-	MeOH	240(4.22)	24-3051-66
Uzarigenin, 11-oxo-	n.s.g.	217(4.19),290s(2.30)	33-0316-66
$C_{23}H_{32}O_6$			
Allog laucotoxigenin	EtOH	216(4.23),308(1.50)	33-1662-66
9β,10α-Pregn-4-ene-3,20-dione, 21-acetoxy-11α,17α-dihydroxy-	EtOH	241(4.23)	33-1529-66
Pregn-4-en-3-one, 11β-hydroxy-17,20:20,21-bis(methylenedioxy)-	MeOH	241(4.21)	24-3051-66
isomer	MeOH	241(4.19)	24-3051-66
9β,10α-Pregn-4-en-3-one, 11α-hydroxy-17α,20:20,21-bis(methylenedioxy)-	MeOH	242(4.22)	54-0731-66
11β-isomer	MeOH	241(4.21)	54-0731-66
$C_{23}H_{33}BF_2O_4$			
Boron, difluoro(4-acetyl-17β-hydroxy-5β-androstan-3-onato)-, acetate	EtOH	302(3.97)	78-2039-66
$C_{23}H_{33}NO$			
1H-Cyclopenta[5,6]naphtho[1,2-g]quinolin-1-ol, tetradecahydro-1,11a,13a-trimethyl-	EtOH	269(3.73),277(3.63)	4-0339-66
$C_{23}H_{33}NO_2$			
Pregn-4-ene-3,20-dione, 16ξ-aziridino-	CHCl	243(4.29)	13-0553-66B
$C_{23}H_{33}NO_3$			
Androst-4-en-19-amide, N,N-diethyl-3,17-dioxo-	EtOH	252(4.09)	44-0693-66
$\Delta^{20,22}$-Norcholenic acid, 18-cyano-3α-hydroxy-	EtOH	213(3.97)	78-0541-66B
Pregna-1,4-dien-3-one, 11α-acetamido-20β-hydroxy-	MeOH	252(4.24)	44-1342-66
11β-isomer	MeOH	241(4.13)	44-1342-66
Pregn-4-ene-3,20-dione, 11α-acetamido-	MeOH	242(4.16)	44-1342-66
11β-isomer	MeOH	238(4.17)	44-1342-66
$C_{23}H_{33}NO_4$			
Pregn-4-ene-3,20-dione, 4-hydroxy-, methylcarbamate	EtOH	248(4.10)	22-3410-66
$C_{23}H_{33}N_3O_2$			
1H-Cyclopenta[5,6]naphtho[1,2-g]quinazolin-1-ol, 10-aminotetradecahydro-11a,13a-dimethyl-, acetate	HCl	260(4.08)	32-0152-66
	EtOH	236(4.02),268(3.69)	32-0152-66
$C_{23}H_{34}Cl_3N_3$			
Acridan, 2-chloro-9,9-bis(dimethylaminopropyl)-, dihydrochloride	n.s.g.	293(4.37)	33-0572-66

Compound	Solvent	$\lambda_{max}(\log \epsilon)$	Ref.
$C_{23}H_{34}N_2O$			
Androst-5-en-3β-ol, 17β-(1-methyl-3-pyrazolyl)-	EtOH	221(3.90)	44-3193-66
Androst-5-en-3β-ol, 17β-(1-methyl-5-pyrazolyl)-	EtOH	214(3.65)	44-3193-66
$C_{23}H_{34}N_2O_5$			
Androst-4-en-3-one, 4,17β-dihydroxy-, bis(methylcarbamate)	EtOH	249(4.11)	22-3410-66
$C_{23}H_{34}O_2$			
Androsta-5,7-dien-17β-ol, 4,4-dimethyl-, acetate	EtOH	262s(--),273(4.01), 282(4.00),292s(--)	39-1847-66C
5α-Androsta-6,8(14)-dien-17β-ol, 4,4-dimethyl-, acetate	EtOH	245s(--),253(4.34), 261s(--)	39-1847-66C
$C_{23}H_{34}O_3$			
5β,14β-Androstane-3β,14β-diol, 17β-(2-furyl)-	EtOH	212.5(3.66)	39-0377-66C
5β,17α-Card-20(22)-enolide, 3β-hydroxy-	EtOH	217(4.18)	94-0496-66
Estr-5-en-3-one, 4α-ethyl-4β-methyl-17β-hydroxy-, acetate	C_6H_{12}	300f(4.7)	44-0705-66
18-Nor-5α,8α,9β,14β-androst-12-en-17-one, 3α-acetoxy-4α,8,14-trimethyl-	EtOH	245(3.98)	78-3459-66
5α-Pregn-16-en-20-one, 12β-acetoxy-	EtOH	234(3.94)	35-0536-66
$C_{23}H_{34}O_4$			
5β-Androstan-3-one, 4β-acetyl-17β-hydroxy-, acetate	EtOH	289(2.04)	78-2039-66
	EtOH-NaOH	313,5(4.02)	78-2039-66
Cannabidiol, dimethyl ether, cis and trans	MeOH	209(4.78),271(3.01), 279(2.94)	5-0165-66C
trans	MeOH	209(4.78),271(3.01), 279(2.97)	5-0165-66C
3,5-Cyclo-20-ethylenedioxypregnan-6-one, 17α-hydroxy-	MeOH	209(3.70)	5-0204-66G
5α-Pregn-7-ene-20-carboxylic acid, 3β-hydroxy-6-oxo-, methyl ester	EtOH	244(4.13)	33-1591-66
Pregn-4-ene-3,20-dione, 17α,21-dihydroxy-6α,16β-dimethyl-	MeOH	241(4.19)	13-0234-66
5β-Pregn-9(11)-en-12-one, 3α,20β-dihydroxy-, 20-acetate	MeOH	240.5(4.10)	44-1349-66
Uzarigenin	n.s.g.	218(4.20)	33-0316-66
$C_{23}H_{34}O_5$			
5α-Androstan-19-one, 3,17-diethylenedioxy-	EtOH	311(1.51)	33-0292-66
5β-Androstan-19-one, 3,17-diethylenedioxy-	EtOH	304(1.46)	33-0292-66
14-Epicassaic acid, methyl ester, 3-acetate	EtOH	223(4.20)	35-5865-66
Mallogenin	n.s.g.	218(4.22)	33-0316-66
Pregn-4-ene-3,20-dione, 11α,17α,21-trihydroxy-6α,16α-dimethyl-	MeOH	241(4.15)	13-0234-66A
Pregn-4-en-3-one, 16α,17α,20β-trihydroxy-, 16-acetate	EtOH	242(4.17)	13-0537-66B
Pregn-4-en-3-one, 16α,17α,20β-trihydroxy-, 20-acetate	EtOH	242(4.18)	13-0537-66B

Compound	Solvent	λ_{max}(log ε)	Ref.
5α-Pregn-16-en-20-one, 3β,14β,15α-trihydroxy-, 3-acetate	EtOH	231(3.83)	94-0809-66
$C_{23}H_{34}O_6S$			
Androst-4-en-3-one, 17β-acetoxy-11α-methanesulfonyloxy-17α-methyl-	EtOH	239(4.24)	33-2218-66
$C_{23}H_{34}O_8$			
Hexanedioic acid, 3-[2-carboxy-4-[1-carboxyethyl)cyclohexyl]-3-(hydroxymethyl)]-2-isopropylidene-, γ-lactone, trimethyl ester	EtOH	236(3.98)	78-1659-66
$C_{23}H_{35}NO$			
Androst-4-en-3-one, 17α-(1-pyrrolidinyl)-, hydrochloride	EtOH	240.5(4.17)	39-1698-66C
$C_{23}H_{35}NO_3$			
Pregn-4-en-3-one, 11α-acetamido-20β-hydroxy-	MeOH	242(4.11)	44-1342-66
11β-isomer	MeOH	237(4.04)	44-1342-66
$C_{23}H_{35}NO_4$			
13,17-Secoandrost-4-ene-3,17-dione, 17-butyl-13-nitro-	EtOH	238(4.28),285(2.11)	5-0158-66E
$C_{23}H_{35}NO_5$			
Carbamic acid, [(17β-hydroxy-3-oxo-5α-androstan-2-yl)carbonyl]-, ethyl ester	EtOH	301(3.76)	32-0179-66
$C_{23}H_{36}N_2O_4$			
Carbamic acid, [(3-amino-17β-hydroxy-5α-androst-2-en-2-yl)carbonyl]-, ethyl ester	EtOH	309(4.11)	32-0179-66
$C_{23}H_{36}O_3$			
5α-Androstan-3-one, 17β-hydroxy-2-methoxymethylene-1α,17α-dimethyl-	MeOH	276(4.02)	73-1064-66
Cannabinol, 1-ethoxyhexahydro-	EtOH	272(2.94),280(2.94)	78-1481-66
$C_{23}H_{36}O_4$			
Labda-8(20),13-dien-15-oic acid, 18-acetoxy-, methyl ester	EtOH	220(4.17)	78-0203-66B
$C_{23}H_{36}O_5$			
Cassaidic acid, methyl ester, 3-acetate	EtOH	224(4.22)	35-5865-66
7,14-Diepicassaidic acid, methyl ester, 3-acetate	EtOH	225(4.24)	35-5865-66
Isocassaidic acid, methyl ester, 3-acetate	EtOH	223(4.25)	35-5865-66
$C_{23}H_{36}O_6$			
Androstan-3-one, 4β,17β-diacetoxy-5β-hydroxy-	dioxan	280(1.45)	33-1986-66
$C_{23}H_{38}O_2$			
Cannabidiol, tetrahydro-, dimethyl ether, cis and trans	MeOH	210(4.60),271(3.10), 279s(3.00)	5-0165-66C
trans	MeOH	210(4.60),271(3.17), 279s(3.08)	5-0165-66C

Compound	Solvent	λ_{max}(log ϵ)	Ref.
$C_{23}H_{39}NO_4$			
Benzene, 1-hexadecyloxy-4-methoxy-2-nitro-	MeOH	357(3.38)	78-0093-66B
Benzene, 1-hexadecyloxy-4-methoxy-3-nitro-	MeOH	357(3.38)	78-0093-66B
$C_{23}H_{40}O_2$			
Benzene, 1-hexadecyloxy-4-methoxy-	MeOH	225(3.97),287(3.42)	78-0093-66B
$C_{23}H_{46}N_2$			
1,6-Diazaspiro[4,4]nonane, 1,6-dioctyl-	MeOH	205(3.9)	24-0724-66

Compound	Solvent	$\lambda_{max}(\log \epsilon)$	Ref.
$C_{24}H_{11}ClO_2S$			
Naphtho[2,1-b]thiophen-1(2H)-one, 2-(6-chloro-2-oxo-1-acenaphthen-ylidene)-	PhCl	332(--),388(--), 435(--)	65-1055-66
	H_2SO_4	665(--)	65-1055-66
Naphtho[2,1-b]thiophen-1(2H)-one, 2-(8-chloro-2-oxo-1-acenaphthen-ylidene)-	PhCl	330(--),387(--), 438(3.70)	65-1055-66
	H_2SO_4	645(--)	65-1055-66
$C_{24}H_{12}$			
1,2:5,6:9,10-Tribenzocyclododeca-1,5,9-triene-3,7,11-triyne	C_6H_{12}	289f(5.39)	77-0087-66
	C_6H_{12}	290f(5.57)	88-0751-66
$C_{24}H_{12}N_6$			
Tetrapyrido[2,3-a:3',2'-c:2",3"-h:-3''',2'''-j]phenazine	$CHCl_3$	274(4.72),320(4.48), 345(4.11),353(4.09), 363(4.14),383(4.13)	44-3384-66
$C_{24}H_{14}$			
Dibenzo[a,l]pyrene	EtOH	240(4.54),271(4.64), 282s(4.45),294(4.54), 305(4.69),319(4.74), 341(3.85),360(4.02), 375(4.25),397(4.27), 420(3.04)	77-0548-66
$C_{24}H_{14}Cl_2FeN_4O_8$			
Dipyridyliron perchlorate, acrylonitrile adduct	H_2O	242(4.28),285(4.58), 416(3.65)	7-0298-66
$C_{24}H_{14}Fe$			
Octatetrayne, 1-ferrocenyl-8-phenyl-	$CHCl_3$	265s(4.78),280(4.83), 308(4.79),345(4.05), 368(4.05),394(3.92), 464(3.58)	101-0399-66B
$C_{24}H_{14}N_2$			
2,5-Cyclohexadiene-$\Delta^{1,\alpha}$-malononitrile, 4-(diphenyl-2-cyclopropen-1-ylidene)-	$CHCl_3$	282(4.13),294(4.15), 309(4.11),509(4.59)	89-0517-66
2,3-Naphthalenedicarbonitrile, 5,8-diphenyl-	$CHCl_3$	260(4.66),330(3.89), 340(3.87),355(3.78)	78-0491-66B
$C_{24}H_{14}N_6$			
Dipyrido[2,3-f:3',2'-h]quinoxaline, 2,3-di-2-pyridyl-	MeOH	219(4.49),263(4.66), 316(4.42)	44-3384-66
$C_{24}H_{14}O_2$			
Azulenylium, 1-[3-(1-azulenyl)-2-hydroxy-4-oxo-2-cyclobuten-1-ylidene]dihydro-, hydroxide, inner salt	$CHCl_3$	284(4.27),311(4.24), 366(3.98),394(3.88), 460(3.54),680(5.04)	89-0893-66
$C_{24}H_{14}O_3$			
2,3-Naphthalenedicarboxylic anhydride, 5,8-diphenyl-	ether	270(4.72),332(3.97), 350(3.91),365(3.90)	78-0491-66B
$C_{24}H_{16}$			
Acenaphthylene, 1,2-diphenyl-	$CHCl_3$	280s(4.10),336(4.23), 356(3.90),430(3.29)	78-0135-66B
17H-Cyclopenta[a]phenanthrene, 17-benzylidene-	EtOH	249(4.52),328(4.65), 384(4.07)	39-0963-66C

Compound	Solvent	$\lambda_{max}(\log \epsilon)$	Ref.
$C_{24}H_{16}N_2O$			
Benzo[a][3,6]phenanthrolin-6-ol, 7-styryl-	$CHCl_3$	266(4.38),318(4.40)	87-0161-66
$C_{24}H_{16}N_2S_2$			
Thiazole, 2,2'-p-phenylenebis[4-phenyl-	C_6H_{12}	262(4.85),354(4.62)	22-2857-66
Thiazole, 4,4'-p-phenylenebis[2-phenyl-	C_6H_{12}	299(4.63),309(4.67)	22-2857-66
$C_{24}H_{16}N_4O_2$			
4(3H)-Quinazolinone, 2-[m-(2-hydroxy-1-naphthylazo)phenyl]-	dioxan	478(4.23)	32-0279-66
4(3H)-Quinazolinone, 2-[p-(2-hydroxy-1-naphthylazo)phenyl]-	dioxan	494(4.45)	32-0279-66
$C_{24}H_{16}N_6O_4$			
4(3H)-Quinazolinone, 2-[[m-(3-carboxy-5-oxo-1-phenyl-2-pyrazolin-4-yl)-azo]phenyl]-	dioxan	410(4.42)	32-0279-66
4(3H)-Quinazolinone, 2-[[p-(3-carboxy-5-oxo-1-phenyl-2-pyrazolin-4-yl)-azo]phenyl]-	dioxan	425(4.55)	32-0279-66
$C_{24}H_{16}O_2$			
7,18-Dioxapentacycylo[17.3.1.1^{2,6}.-1$^{8.12}$.1^{13,17}]hexacosa-1(23),2,4,-6(26),8,10,12(25),13,15,17(24),19,21-dodecaene	THF	215(4.56),246(4.49),	88-2837-66
$C_{24}H_{16}O_4$			
[$\Delta^{3,3'}$(2H,2'H)-Bibenzofuran-2-one, 2'-(α-methoxybenzylidene)-	EtOH	237(4.50),243s(--), 312(4.29),389(4.45)	88-4671-66
2,3-Naphthalenedicarboxylic acid, 5,8-diphenyl-	EtOH	251(4.72),310(4.02), 342(3.70)	78-0491-66B
2,8,14,20-Tetraoxapentacyclo-[19.3.1.1^{3,7}.1^{9,13}.1^{15,19}]octacosa-1(25),3,5,7(28),9,11,13(27),15,17,-19(26),21,23-dodecaene	THF	215(4.77),275(3.76)	88-2837-66
$C_{24}H_{17}Cl_2N_5O_3$			
Acetoacetanilide, 2',3'-dichloro-2-[[p-(3,4-dihydro-4-oxo-2-quinazol-inyl)phenyl]azo]-	dioxan	395(4.68)	32-0279-66
Acetoacetanilide, 2',6'-dichloro-2-[[m-(3,4-dihydro-4-oxo-2-quinazol-inyl)phenyl]azo]-	dioxan	375(4.52)	32-0279-66
$C_{24}H_{17}NO_3$			
Naphthalen-1,4-imine-2,3-dicarboxylic anhydride, 1,2,3,4-tetrahydro-1,4-diphenyl-	MeCN	236(4.32),270(4.30), 320(4.12),333(4.13), 387(4.34)	78-1011-66
$C_{24}H_{18}$			
Naphthalene, 1-(p-phenylstyryl)-	hexane	234(4.62),283(4.32), 335(4.64)	99-0085-66
	$C_2H_4Cl_2$	280(4.34),332(4.39)	99-0085-66
Naphthalene, 2-(p-phenylstyryl)-	$C_2H_4Cl_2$	245(4.20),255(4.15), 278(4.32),290(4.36), 337(4.72)	99-0085-66

Compound	Solvent	λ_{max}(log ε)	Ref.
$C_{24}H_{18}Br_2N_2O_2$			
Benzo[1",2":4,5:4",5":4',5']difuro-[2,3-b:2',3'-b']diindole, 2.9-dibromo-5,5a,7b,12,12a,14b-hexahydro-7b,14b-dimethyl-	THF	234(4.42),258(4.37), 303(4.04),328(4.18)	44-3321-66
$C_{24}H_{18}Br_2O$			
2-Oxabicyclo[4.1.0]hept-3-ene, 7,7-dibromo-1,3,5-triphenyl-	EtOH	254(4.07),258(4.07), 265(4.05)	24-2351-66
$C_{24}H_{18}Br_4N_4$			
Porphine, 2,3,6,7-tetrabromo-1,4,5,8-tetramethyl-	$PhNO_2$	509(--),541(--), 580(--),634(--)	65-0808-66
$C_{24}H_{18}ClN_5O_3$			
Acetoacetanilide, 2'-chloro-2-[[m-(3,4-dihydro-4-oxo-2-quinazolinyl)-phenyl]azo]-	dioxan	375(4.56)	32-0279-66
Acetoacetanilide, 2'-chloro-2-[[p-(3,4-dihydro-4-oxo-2-quinazolinyl)-phenyl]azo]-	dioxan	395(4.61)	32-0279-66
Acetoacetanilide, 4'-chloro-2-[[m-(3,4-dihydro-4-oxo-2-quinazolinyl)-phenyl]azo]-	dioxan	375(4.53)	32-0279-66
Acetoacetanilide, 4'-chloro-2-[[p-(3,4-dihydro-4-oxo-2-quinazolinyl)-phenyl]azo]-	dioxan	395(4.60)	32-0279-66
$C_{24}H_{18}Cl_2O$			
2-Oxabicyclo[4.1.0]hept-3-ene, 7,7-dichloro-1,3,5-triphenyl-	EtOH	259(4.06),267(4.04)	24-2351-66
$C_{24}H_{18}CuN_4S_2$			
Copper, bis(thiopicolinanilidato)-	benzene	368(3.99),440(3.90)	65-1499-66
	heptane	276(4.31),370(4.06), 430(3.90)	65-1499-66
	EtOH	265(4.29),360(4.02), 432(3.86)	65-1499-66
	$CHCl_3$	276(4.31),360(3.97), 436(3.91)	65-1499-66
$C_{24}H_{18}Fe_2$			
Butadiyne, diferrocenyl-	EtOH	224(4.65),285(4.33), 453(3.22)	101-0173-66B
	$CHCl_3$	287(4.25),330s(3.99), 355s(3.15),450(3.23)	101-0399-66B
$C_{24}H_{18}N_2O$			
1,2-Diazetidin-3-one, 4-(1-naphthyl)-	CH_2Cl_2	268(4.18),276(4.23), 286(4.22),297(4.00)	88-5245-66
$C_{24}H_{18}N_2O_2$			
p-Benzoquinone, 2,5-bis(2-methylindol-3-yl)-	THF	230(4.82),281(4.57), 289s(4.53),500(3.99)	44-3321-66
after two days in daylight	THF	230(4.63),280(4.50), 289s(4.46),487(3.82)	44-3321-66
p-Benzoquinone, 2,5-bis(3-methylindol-2-yl)-	THF	251(4.60),324(4.37), 449(4.04)	44-3321-66
after two days in the dark	THF	251(4.50),324(4.32), 450(4.03)	44-3321-66

Compound	Solvent	λmax(log ε)	Ref.
p-Benzoquinone, 2,5-bis(3-methylindol-2-yl)-, after two days in daylight	THF	251(4.42),306(4.23), 324(4.24),450(3.99)	44-3321-66
$C_{24}H_{18}N_2O_5$			
5(4H)-Oxazolinone, 4-(p-benzyloxy-benzylidene)-2-(o-nitrobenzyl)-	$CHCl_3$	250(4.50)	7-1192-66
$C_{24}H_{18}O$			
Ether, bis[2-(1-naphthyl)vinyl], cis-cis	EtOH	314(4.27)	24-0642-66
$C_{24}H_{18}O_2$			
Benzene, 1,4-bis(1-hydroxy-1-phenyl-2-propyn-1-yl)-	EtOH	258(3.03)	22-2877-66
Cyclopent-4-ene-1,3-dione, 2-trityl-	50% EtOH	225s(4.30)	99-0115-66
	75% EtOH-KOH	238(3.72),485(3.76)	99-0115-66
1H-Cyclopropa[b]naphthalene-2,7-dione, 1a,7a-dihydro-1a-methyl-1,1-diphenyl-	EtOH	224(4.63),306(3.47)	77-0894-66
$C_{24}H_{18}O_4$			
Pentatriafulvalene-2,3-dicarboxylic acid, 5,6-diphenyl-, dimethyl ester	MeOH	269(4.68),351(4.85)	35-3359-66
$C_{24}H_{18}O_7S_2$			
1,4:7,10-Diepithio-13,16-epoxy[18]-annulene-5,11,18-tricarboxylic acid, trimethyl ester	EtOH	207(4.20),252(4.17), 327(4.56),450(4.01)	12-0257-66
$C_{24}H_{18}O_8S$			
1,4-Epithio-7,10:13,16-diepoxy[18]-annulene-5,11,18-tricarboxylic acid, trimethyl ester	EtOH	216(4.28),253(4.28), 335(4.50)	12-1461-66
1,4-Epithio-7,10:13,16-diepoxy[18]-annulene-6,11,18-tricarboxylic acid, trimethyl ester	EtOH	217(4.16),257(4.12), 328(4.53)	12-1461-66
$C_{24}H_{18}O_9$			
1,4:7,10:13,16-Triepoxy[18]annulene-5,11,18-tricarboxylic acid, trimethyl ester	EtOH	328(4.55),420(3.88)	12-1221-66
$C_{24}H_{18}O_{12}$			
Ceroalbolinic acid, tetraacetate	EtOH	270(4.38),345(3.23)	78-1507-66
$C_{24}H_{19}NO_3$			
5(4H)-Oxazolinone, 2-benzyl-4-(p-benzyloxybenzylidene)-	$CHCl_3$	247(4.11),368(4.56)	7-1192-66
$C_{24}H_{19}NO_4$			
Indole, 2,3-bis(p-acetoxyphenyl)-	EtOH	248(4.35),309(4.27)	87-0527-66
1,4-Naphthoquinone, 2-(1,3-dioxo-indan-2-yl)-3-piperidino-	MeCN	222(3.70),241(4.48), 285(4.15),510(3.68)	44-0062-66
$C_{24}H_{19}N_3O_8S_2$			
Phthalimide, N-[[[3-(hydroxymethyl)-8-oxo-7-[2-(2-thienyl)acetamido]-5-thia-1-azabicyclo[4.2.0]oct-2-en-2-yl]carbonyl]oxy]-, acetate	EtOH	219(4.62),268(3.84)	87-0741-66

Compound	Solvent	λ_{max}(log ε)	Ref.
$C_{24}H_{19}N_5O_3$			
Acetoacetanilide, 2-[m-(3,4-dihydro-4-oxo-2-quinazolinyl)phenylazo]-	dioxan	375(4.52)	32-0279-66
Acetoacetanilide, 2-[p-(3,4-dihydro-4-oxo-2-quinazolinyl)phenylazo]-	dioxan	395(4.67)	32-0279-66
$C_{24}H_{20}$			
Cyclopentadiene, 1-methyl-2,3,4-triphenyl-	EtOH	235(4.54),318(4.14)	22-3775-66
15H-Cyclopenta[a]phenanthrene, 17-benzyl-16,17-dihydro-	EtOH	216(4.46),260(4.81), 281(4.20),289(4.08), 301(4.14),320(2.77), 336(2.98),352(3.00)	39-0963-66C
$C_{24}H_{20}Cl_2N_2O$			
2-Pyrrolidinone, 4,4-dichloro-3-methyl-1,3-diphenyl-5-(phenylimino)-	EtOH	225(4.29)	28C-1276-66A
$C_{24}H_{20}Cl_2O$			
2-Oxabicyclo[4.1.0]heptane, 7,7-dichloro-1,3,5-triphenyl-	n.s.g.	252(2.87),258(2.9), 264(2.81)	24-2351-66
$C_{24}H_{20}Cl_2O_2$			
Benzene, 1,4-bis(p-chlorostyryl)-2,5-dimethoxy-	dioxan	395(4.66)	18-1547-66
$C_{24}H_{20}N_2O_2$			
Benzo[1",2":4,5:4",5":4',5']difuro-[2,3-b:2',3'-b']diindole, 5,5a,7b,-12,12a,14b-hexahydro-7b,14b-dimethyl-	EtOH	225s(4.32),244s(4.00), 299s(3.83),319(4.08)	44-3321-66
2-Furaldehyde, 3-methyl-4-phenyl-, azine	EtOH	306(4.48)	44-0052-66
Hydroquinone, 2,5-bis(2-methylindol-3-yl)-	EtOH	228(4.78),281(4.27), 289s(4.22),315(4.20)	44-3321-66
Hydroquinone, 2,5-bis(3-methylindol-2-yl)-	EtOH	227(4.74),250s(4.35), 309(4.26),353(4.49)	44-3321-66
$C_{24}H_{20}N_2O_6$			
Benzene, 2,5-dimethoxy-1,4-bis-(p-nitrostyryl)-	dioxan	431(4.55)	18-1547-66
$C_{24}H_{20}N_4$			
Naphthalene, 1,8-dibenzoyl-, dihydrazone	$CHCl_3$	270(4.36)	78-0135-66B
$C_{24}H_{20}N_4O_{11}$			
2(1H)-Pyrimidinone, 1-(3-deoxy-α-D-ribofuranosyl)-4-methoxy-, di-p-nitrobenzoate	MeOH	261(4.45)	44-1163-66
β-isomer	MeOH	262(4.48)	44-1163-66
$C_{24}H_{20}N_6O_6S_2$			
Cystine, N,N'-bis(2-quinoxalinyl-carbonyl)-	EtOH	207(3.66),244(3.85), 320(3.12),327(3.08)	87-0266-66
$C_{24}H_{20}O$			
Benzofulvene, 3-(cyclohepta-2,4,6-trien-1-yl)-ω-(p-methoxyphenyl)-	EtOH	245(4.26),292(4.08), 357(4.38)	5-0057-66H
9,10-Benzosesquifulvalene, 8-(p-methoxybenzyl)-	EtOH	246(4.29),303(3.66), 315(3.68),398(4.39)	5-0057-66H

Compound	Solvent	$\lambda_{max}(\log \epsilon)$	Ref.
2-Cyclohexen-1-one, 3,4,5-triphenyl-, trans	MeOH	282(4.21)	44-3213-66
$C_{24}H_{20}O_3S$			
15H-Cyclopenta[a]phenanthren-17-ol, 16,17-dihydro-, p-toluenesulfonate	EtOH	220(4.48),259(4.83), 279(4.23),288(4.09), 300(4.15)	39-0963-66C
$C_{24}H_{20}O_4$			
1,4-Naphthoquinone, 2,3-dimethyl-, dimer	EtOH	225(4.75),244(4.65), 310(3.82)	39-0123-66C
$C_{24}H_{20}O_5$			
1,4-Naphthoquinone, 2,3-dihydro-3-hydroxy-2,3,3'-trimethyl-2,2'-methylenedi-	EtOH	224(4.58),250(4.38), 264(4.33),307(3.55), 330(3.44)	39-0123-66C
$C_{24}H_{20}O_5S$			
Benzofuran, 3-acetonyl-2-(o-hydroxyphenyl)-, p-toluenesulfonate	EtOH	226(4.39),296(4.17)	39-0544-66C
$C_{24}H_{20}O_9$			
Lactucopicrin oxidation product B'	70% MeOH	216(4.2),286(3.5)	83-0303-66
$C_{24}H_{20}O_{10}$			
1,2,5,6,9-Anthracenepentol, pentaacetate	dioxan	261(5.13),301(3.61), 312(3.55),340s(3.48), 355(3.69),373(3.78), 393(3.73)	24-2322-66
$C_{24}H_{20}O_{11}$			
Vogeletin, tetraacetate	MeOH	254(4.36),306(4.43)	24-2430-66
$C_{24}H_{21}NO_2$			
Anonaine, N-benzyl-	EtOH	211(4.48),270(4.13), 318(3.49)	44-1281-66
Indole, 2-(p-methoxyphenyl)-3-(p-methoxystyryl)-	EtOH	248(4.40),265s(4.34), 299(4.41),338(4.35)	87-0527-66
$C_{24}H_{21}NO_3$			
Indole, 1-acetyl-2,3-bis(p-methoxyphenyl)-	EtOH	241(4.45),298(4.20)	87-0527-66
Ketone, p-methoxybenzyl 2-(p-methoxyphenyl)indol-3-yl	EtOH	211(4.59),258(4.39), 284s(4.11),308(4.17)	87-0527-66
$C_{24}H_{22}$			
Anthracene, 2,3,6-trimethyl-10-p-tolyl-	$CHCl_3$	266(5.06),358(3.72), 374(3.85),394(3.82)	44-4082-66
$C_{24}H_{22}BN$			
Pyridine, compound with (1-cyclopentenylethynyl)diphenylborane	EtOH	233(4.24)	22-1981-66
$C_{24}H_{22}N_2$			
Benzo[f]quinoline, 1,2,3,4-tetrahydro-3-methyl-1-(2-naphthylamino)-	MeOH	210(4.7),250(5.0), 280(4.3),350(3.8)	30-0110-66A
Dipyrromethene, 3,4,5-trimethyl-3',5'-diphenyl-	EtOH-N HCl	224(4.30),283(4.42), 384(3.88),513(4.93)	12-1871-66
	EtOH	226s(4.17),290(4.29), 466(4.44)	12-1871-66

Compound	Solvent	λ_{max}(log ϵ)	Ref.
$C_{24}H_{22}N_2O_8$			
Fumaric acid, 2,2'-(5,10-phenazine-diyl)di-, tetramethyl ester	MeOH	245(4.65),336(3.69), 540(3.00)	39-2218-66C
$C_{24}H_{22}N_4$			
Porphine, 1,4,5,8-tetramethyl-	$PhNO_2$	498(--),531(--), 570(--),624(--)	65-0808-66
$C_{24}H_{22}N_4O$			
1H-1,2,3-Triazole, 4,5-diphenyl-1-(2,4,6-trimethylbenzamido)-	EtOH	242(4.34)	78-1309-66
$C_{24}H_{22}N_4O_4$			
3-Butenal, 2,2-dimethyl-4,4-diphenyl-, 2,4-dinitrophenylhydrazone	n.s.g.	365(4.19)	22-0734-66
$C_{24}H_{22}O_2$			
Benzene, 2,5-dimethoxy-1,4-distyryl-	dioxan	390(4.69)	18-1547-66
Naphthoquinone, 2-cinnamyl-3-(dimethyl-allyl)- (shoulders, not listed)	C_6H_{12}	251(4.54),330(3.54)	22-1828-66
1,5-Pentanedione, 2-methyl-1,3,5-triphenyl-	EtOH	244(4.43),278s(3.34), 318(2.22)	88-2243-66
$C_{24}H_{23}ClO_7$			
Spiro[benzofuran-2(3H),1'-[2]cyclo-hexene]-3,4'-dione, 7-chloro-2',6-dimethoxy-4-(p-methoxybenzyloxy)-6'-methyl-	EtOH	231(4.57),292(4.34), 335(3.81)	87-0242-66
$C_{24}H_{23}IN_2OS$			
3-Ethyl-2-[(5,6,7,8-tetrahydro-2-phenyl-4H-1,3-benzoxazin-4-yli-dene)methyl]benzothiazolium iodide	MeOH	440(4.60)	87-0758-66
$C_{24}H_{23}NO_2$			
Indole, 2,3-bis(p-ethoxyphenyl)-	EtOH	253(4.52),309(4.32)	87-0527-66
Indole, 1-ethyl-2,3-bis(p-methoxy-phenyl)-	EtOH	240(4.49),299(4.17)	87-0527-66
Indole, 3-(p-methoxyphenethyl)-2-(p-methoxyphenyl)-	EtOH	228(4.49),245s(4.38), 280s(4.03),288s(4.14), 306(4.31)	87-0527-66
Isoquinoline, 1-benzyl-1,2,3,4-tetra-hydro-6,7-methylenedioxy-	EtOH	214(4.52),295(4.02)	44-1281-66
$C_{24}H_{23}NO_3$			
Indole-3-methanol, α-(p-methoxybenzyl)-2-(p-methoxyphenyl)-	EtOH	213s(4.52),224(4.52), 243s(4.38),287s(4.19), 300(4.27)	87-0527-66
$C_{24}H_{23}N_3O_7$			
as-Triazine-3,5(2H,4H)-dione, 2-(2-deoxy-β-D-ribofuranosyl)-, di-p-toluenesulfonate	EtOH	241(4.56),270s(3.91)	4-0226-66
$C_{24}H_{23}N_5O_5$			
Adenine, 7-α-D-arabinofuranosyl-N-benzoyl-3-benzyl-	pH 1	223(4.28),301(4.39)	44-1413-66
	pH 7, 13	236(4.14),336(4.32)	44-1413-66
$C_{24}H_{24}NO_2P$			
Crotonic acid, 3-[(triphenylphosphoran-ylidene)amino]-, ethyl ester	EtOH	268(4.37),275(4.41), 297(4.47)	44-3907-66

Compound	Solvent	λ_{max}(log ε)	Ref.
$C_{24}H_{24}N_2$			
2H-1,5-Methano-1,3-benzodiazocine, 3-benzyl-3,4,5,6-tetrahydro-2-phenyl-	pH 1	251(4.07)	94-0566-66
$C_{24}H_{24}N_2O_5$			
1,3-Azulenedicarboxylic acid, 2-acetamido-6-anilino-, diethyl ester	MeOH	270(4.21),274s(4.20), 348(4.58),417(4.30)	18-1954-66
$C_{24}H_{24}N_4O_3$			
4-Pyrimidinol, 2-amino-6-[4-(p-nitrophenyl)-1,3-butadienyl]-5-(4-phenylbutyl)-	$MeOCH_2CH_2OH$	377(4.51)	4-0324-66
	+ base	396(4.40)	4-0324-66
$C_{24}H_{24}N_4O_4$			
3,5-Hexadien-2-one, 6-(3-cyclohexen-1-yl)-6-phenyl-	$CHCl_3$	270(4.26),310(4.42), 405(4.66)	70-0480-66
$C_{24}H_{24}N_4O_5$			
Butyrophenone, 4-ethoxy-4-phenyl-, 2,4-dinitrophenylhydrazone	$CHCl_3$	380(4.39)	22-0717-66
$C_{24}H_{24}N_5$			
Verdazyl, 1,3,5-triphenyl-6-pyrrolidino-	dioxan	244(4.17),276(4.33), 319(4.18),407(3.99), 420s(3.99),688(3.66)	49-0846-66
$C_{24}H_{24}O_6$			
Oxepino[4,5-b:2,3-e']bisbenzofuran, 2,3,5,6-tetrahydro-2-isopropenyl-5,8,9-trimethoxy-	EtOH	284s(4.08),293(4.14), 328(4.58),340(4.56)	39-0550-66C
$C_{24}H_{24}O_7$			
Lisetin, tri-O-methyl-	EtOH	259(4.51),277(4.25), 320(4.18)	78-0333-66A
$C_{24}H_{25}NO$			
2,7-Cycloctadienone, 2,8-diphenyl-3-(1-pyrrolidinyl)-	EtOH	231(4.22),287(4.11)	44-1336-66
$C_{24}H_{25}NO_7$			
Isoquinoline, 7-acetoxy-1-(3-acetoxy-4-methoxybenzylidene)-2-acetyl-1,2,3,4-tetrahydro-6-methoxy-	EtOH	229(4.29),308s(4.19), 325(4.28)	39-2061-66C
Isoquinoline, 7-acetoxy-1-(4-acetoxy-3-methoxybenzylidene)-2-acetyl-1,2,3,4-tetrahydro-6-methoxy-	EtOH	228(4.34),308s(4.43), 320(4.48)	39-2061-66C
$C_{24}H_{25}N_3O$			
4-Pyrimidinol, 2-amino-5-(4-phenylbutyl)-6-(4-phenyl-1,3-butadienyl)-	pH 1	358(4.41)	4-0315-66
	pH 7	362(4.43)	4-0315-66
	pH 13	324(4.59)	4-0315-66
$C_{24}H_{25}N_3O_5$			
1,3-Azulenedicarboxylic acid, 2-acetamido-6-phenylhydrazino-, diethyl ester	MeOH	270s(4.21),285(4.22), 342(4.75),402(4.31)	18-1954-66
$C_{24}H_{25}N_3O_6$			
Pentatriafulvalene-2,3-dicarboxylic acid, 1-(p-nitrophenylazo)-5,6-dipropyl-, dimethyl ester	n.s.g.	292(4.39),450(4.20)	88-3697-66

Compound	Solvent	λ_{max}(log ϵ)	Ref.
$C_{24}H_{25}N_5O_2$			
Pyrido[2,3-b]pyrazine-6-carbamic acid, 8-[(diphenylmethyl)amino]-3,4-dihydro-2-methyl-, ethyl ester	pH 1	239(4.40),269(4.33), 325(4.07)	44-1890-66
$C_{24}H_{25}N_5O_5$			
2-Pyridinecarbamic acid, 6-acetonylamino-4-[(diphenylmethyl)amino]-5-nitro-, ethyl ester	pH 1	222(--),245s(--)	44-1890-66
$C_{24}H_{25}N_5O_6$			
2,6-Pyridinedicarbamic acid, 4-(diphenylmethyl)amino]-3-nitro-, diethyl ester	pH 1	224(4.41),256(4.42), 360(4.12)	44-1890-66
	pH 7	233(4.45),258(4.41), 364(4.23)	44-1890-66
$C_{24}H_{25}N_7O_4$			
Aspartic acid, N-(5-amino-3-ethyl-3H-v-triazolo[4,5-d]pyrimidin-7-yl)-, dibenzyl ester	EtOH	227(4.34),260(3.86), 289(4.07)	87-0417-66
$C_{24}H_{26}BrNO_7$			
Phenoxazine-1,9-dicarboxylic acid, 7-bromo-2-hydroxy-4,6-dimethyl-3-oxo-, dibutyl ester	MeOH	229(4.51),363(3.97), 419(3.97)	44-3694-66
$C_{24}H_{26}ClN_2O_4P$			
1-Ethyl-2-[(1-ethyl-4-methyl-2(1H)-quinolylidene)phosphino]-4-methylquinolinium perchlorate	MeOH	587(4.62)	24-1325-66
	$C_2H_4Cl_2$	595(4.56)	24-1325-66
$C_{24}H_{26}N_2O_3$			
Majorine	MeOH	223(2.94),289(3.10)	100-0029-66
$C_{24}H_{26}N_2O_4$			
Dipyrromethene-4,4'-dicarboxylic acid, 3',5,5'-trimethyl-3-phenyl-, diethyl ester	EtOH-N HCl	249(4.00),370(3.54), 474(5.03)	12-1871-66
	EtOH	216s(4.54),262(4.10), 463(4.48)	12-1871-66
Epikopsanol-10-lactam, N^2,O-diacetate	EtOH	211(4.37),254(4.11), 280s(3.65),289s(3.59)	39-1260-66C
1-Indanone, 2-isopropyl-2-nitroso-, dimer	EtOH	252(4.41),295(3.81)	44-2090-66
1-Indanone, 2-nitroso-2-propyl-, dimer	EtOH	252(4.46),295(3.85)	44-2090-66
1-Indanone, 2,3,3-trimethyl-2-nitroso-, dimer	EtOH	247(4.11),292(3.42)	44-2090-66
Pyrrole-2-carboxylic acid, 4-ethyl-5-[(4-ethyl-5-formyl-3-methylpyrrol-2-yl)carbonyl]-3-methyl-, benzyl ester	CH_2Cl_2	350(4.39)	78-0241-66A
	CH_2Cl_2 +2% CF_3COOH	370(4.38)	78-0241-66A
$C_{24}H_{26}N_2O_6$			
15H-Azepino[2',1':2,3]oxazolo[4,5-b]-phenoxazine-1,14-dicarboxylic acid, 7a,8,9,10,11,12-hexahydro-4,6-dimethyl-, dimethyl ester	MeOH	234(4.57),440(4.05)	44-2564-66
$C_{24}H_{26}N_4O_5$			
1,3-Azulenedicarboxylic acid, 2-acetamido-6-[di(2-cyanoethyl)amino]-, diethyl ester	MeOH	270(4.25),288s(4.16), 341(4.83),400(4.32)	18-1954-66

Compound	Solvent	λ_{max}(log ε)	Ref.
$C_{24}H_{26}N_6$			
Tetraethylammonium salt of 5-(benzylideneamino)cyclopentadiene-1,2,3,4-tetracarbonitrile	MeCN	212(4.34),222(4.38), 280(4.57),365(4.13)	35-4055-66
$C_{24}H_{26}O_3$			
2(1H)-Phenanthrone, 4aβ-allyl-3-furfurylidene-3,4,4a,9,10,10aα-hexahydro-8-methoxy-1α-methyl-	EtOH	320(4.28)	35-1786-66
$C_{24}H_{26}O_6$			
A conjugated dienone from 1α,2α-epoxyscillirosidine	EtOH	285(4.24)	78-3213-66
$C_{24}H_{26}O_8$			
Coumarin, 4-hydroxy-3-[2-hydroxy-4,5-dimethoxy-3-(3-methyl-2-butenyl)-phenyl]-5,7-dimethoxy-	EtOH	320(4.21)	78-0333-66A
$C_{24}H_{26}O_9$			
Chromomycinone, anhydro-3'-O-methyl-, acetate	EtOH	225(4.42),273(4.66), 415(3.92)	78-2761-66
$C_{24}H_{26}O_{12}$			
Pendulin	EtOH	212(4.36),274(4.24), 328(4.25)	24-3222-66
$C_{24}H_{26}O_{13}$			
Isoflavone, 3',5,7-trihydroxy-4,5',6-trimethoxy-, 7-β-D-glucoside	MeOH	267(4.59),323s(4.71)	24-0865-66
$C_{24}H_{27}ClF_2O_4$			
Pregna-1,4-diene-3,20-dione, 6β-chloro-6α,7α-(difluoromethylene)-17α-hydroxy-, acetate	MeOH	245(4.22)	88-3287-66
$C_{24}H_{27}ClO_4$			
Pregna-1,4,6-triene-3,20-dione, 17α-acetoxy-6-chloro-16-methylene-	EtOH	229(4.01),258(4.00), 297(4.03)	73-2768-66
$C_{24}H_{27}NO_3$			
19-Nor-17α-pregn-4-en-20-yno[4,5,6-cd]-pyridin-3-one, 17-acetoxy-	EtOH	217(4.18),244(3.88), 280(3.35)	32-1284-66
$C_{24}H_{27}NO_7$			
Phenoxazine-1,9-dicarboxylic acid, 2-hydroxy-4,6-dimethyl-3-oxo-, dibutyl ester	MeOH	231(4.39),438(4.13)	44-3694-66
$C_{24}H_{27}N_3O_8$			
Phenoxazine-1,9-dicarboxylic acid, 2-amino-7-nitro-3-oxo-, dibutyl ester	MeOH	228(4.42),305(3.88), 434(4.36)	44-3694-66
$C_{24}H_{27}N_5O_4$			
2,6-Pyridinedicarbamic acid, 3-amino-4-[(diphenylmethyl)amino]-, diethyl ester	pH 1	242(4.58),271(4.17), 300(3.99)	44-1890-66
	pH 7	235(4.54),258s(4.11), 293(3.86)	44-1890-66
	pH 13	235(4.50),258s(4.17), 295(3.92)	44-1890-66

Compound	Solvent	λ_{max}(log ε)	Ref.
$C_{24}H_{28}N_2O$			
[$\Delta^{2,3'}$-Bipiperidin]-2'-one, 6,6'-dimethyl-1,1'-diphenyl-	MeOH	209(4.5),300(4.6)	24-0724-66
$C_{24}H_{28}N_2O_2$			
Pyrrole-2-carboxylic acid, 4-ethyl-5-[(4-ethyl-3,5-dimethyl-2H-pyrrol-2-ylidene)methyl]-3-methyl-, benzyl ester	EtOH-HCl	485(4.72)	78-0241-66A
	EtOH	409(4.50)	78-0241-66A
$C_{24}H_{28}N_2O_3$			
Epikopsanol, N-acetyl-, acetate	EtOH	209(4.51),254(4.21), 284(3.59),290s(3.54)	39-1260-66C
Pyrrole-2-carboxylic acid, 4-ethyl-5-[(4-ethyl-3,5-dimethylpyrrol-2-yl)-carbonyl]-3-methyl-, benzyl ester	CH_2Cl_2	348(4.33)	78-0241-66A
	CH_2Cl_2 +2% CF_3COOH	422(4.51)	78-0241-66A
$C_{24}H_{28}N_2O_4$			
Pyrrole-2,5-dicarboxylic acid, 3-ethyl-4-methyl-5-benzyl-, 2-(4-ethyl-3,5-dimethylpyrrol-2-yl) ester	CH_2Cl_2	238(4.29),303(4.30)	78-0241-66A
Vincamedine	MeOH	248(3.94),292(3.48)	100-0029-66
$C_{24}H_{28}N_2O_5$			
1,3-Azulenedicarboxylic acid, 2-acetamido-6-diallylamino-, diethyl ester	MeOH	234s(3.99),248(4.15), 257(4.15),275(4.11), 324(4.68),342(4.55), 389(4.25)	18-1954-66
$C_{24}H_{28}N_2O_6$			
Phenoxazine-1,9-dicarboxylic acid, 2-amino-4,6-dimethyl-3-oxo-, dibutyl ester	MeOH	238(4.58),433(4.48)	44-3694-66
Phenoxazine-1,9-dicarboxylic acid, 2-(hexahydro-1H-azepin-1-yl)-3-hydroxy-4,6-dimethyl-, dimethyl ester	MeOH	226(4.55),410(4.06)	44-2564-66
$C_{24}H_{28}O_2$			
Bixindial	benzene	458(4.93),486(5.10), 520(5.08)	39-2166-66C
Propiophenone, 3-phenyl-3-[2,2,4-trimethyl-6-oxocyclohexyl]-	EtOH	242(4.13),278s(3.04)	88-2243-66
$C_{24}H_{28}O_4$			
Pregna-1,4,6,14-tetraene-3,20-dione, 17α-acetoxy-6-methyl-	EtOH	226(4.06),252(3.93), 296(4.10)	78-0325-66A
$C_{24}H_{28}O_7$			
Vogeletin, tetraethyl-	MeOH	263(4.25),328(4.41)	24-2430-66
$C_{24}H_{28}O_9$			
Chromomycinone, isopropylidene-	EtOH	232(4.37),280(4.59), 322(3.79),410(3.97)	78-2761-66
$C_{24}H_{29}ClF_2O_4$			
Pregn-4-ene-3,20-dione, 6-chloro-6α,7α-(difluoromethylene)-17α-hydroxy-, acetate	EtOH	247(4.02)	88-3287-66

Compound	Solvent	$\lambda_{max}(\log \epsilon)$	Ref.
$C_{24}H_{29}ClN_2O_8$			
Spegatrine, perchlorate, diacetate	50% EtOH	223(4.38),268(3.91), 291s(3.73)	23-1523-66
$C_{24}H_{29}ClO_4$			
Pregna-4,6-diene-3,20-dione, 17α-acetoxy-6-chloro-16-methylene-	EtOH	285(4.32)	73-2768-66
Pregna-2,4,6-triene-2-carboxaldehyde, 17α-acetoxy-7-chloro-20-oxo-	MeOH	355(4.39)	24-3057-66
$C_{24}H_{29}N$			
11H-Benzo[a]carbazole, 11-butyl-6a,11a-dihydro-5,6,6a,11a-tetramethyl-	MeOH	220s(4.48),226s(4.34), 257(4.42),310s(3.45)	44-2009-66
$C_{24}H_{29}N_3O$			
4-Pyrimidinol, 2-amino-5,6-bis-(4-phenylbutyl)-	pH 1	267(3.96)	4-0315-66
	pH 13	281(3.93)	4-0315-66
$C_{24}H_{29}N_3O_5S$			
5-Thia-1-azabicyclo[4.2.0]oct-2-ene-2-carboxamide, N-cyclohexyl-3-(hydroxymethyl)-8-oxo-7-(2-phenylacetamido)-, acetate	EtOH	264(3.96)	39-1142-66C
$C_{24}H_{30}F_2O_4$			
Pregn-4-ene-3,20-dione, 6,7-(difluoromethylene)-17-hydroxy-, acetate	MeOH	247(4.18)	88-3287-66
$C_{24}H_{30}N_2O_5$			
Poweridine	n.s.g.	226(4.6),272(3.7), 298(3.7)	102-1065-66
$C_{24}H_{30}N_4O$			
4-Pyrimidinol, 2-amino-6-(p-aminophenylbutyl)-5-(4-phenylbutyl)-	pH 1	268(3.97)	4-0324-66
	pH 13	282(3.97)	4-0324-66
$C_{24}H_{30}O_2$			
1,10-Decanedione, 1,10-di-p-tolyl-	MeOH	251(4.50)	44-4277-66
1,5-Octanedione, 4-isopropyl-7-methyl-1,3-diphenyl-	EtOH	242(4.10),278s(3.04)	88-2243-66
$C_{24}H_{30}O_3$			
Androstenolone, 16-furfurylidene-	n.s.g.	327(4.27)	88-3891-66
$C_{24}H_{30}O_4$			
Estra-1,3,5(10),6-tetraene-1,17β-diol, 3,4-dimethyl-, diacetate	EtOH	223(4.42),269(3.95)	33-1049-66
Pregna-4,14-diene-3,20-dione, 17α-acetoxy-6-methylene-	EtOH	260(4.01)	78-0325-66A
Pregna-1,4,14-triene-3,20-dione, 17α-acetoxy-6α-methyl-	EtOH	242(4.16)	78-0325-66A
Pregna-4,6,14-triene-3,20-dione, 17α-acetoxy-6-methyl-	EtOH	285(4.34)	78-0325-66A
$C_{24}H_{30}O_5$			
Pregna-4,6-diene-3,20-dione, 11α-acetoxy-16α,17α-epoxy-16β-methyl-	EtOH	282(4.44)	73-4703-66

Compound	Solvent	λ_{max}(log ϵ)	Ref.
Pregna-1,4,9(11)-triene-3,20-dione, 17α,21-dihydroxy-16α-methyl-, 21-acetate	EtOH	240(4.20)	95-0639-66
Pregna-1,4,9(11)-triene-3,20-dione, 17α,21-dihydroxy-16β-methyl-, 21-acetate	MeOH	239(4.19)	44-0026-66
Scillirubrosidin, 3-dehydro-	MeOH	241(4.21),298(3.76)	33-0030-66
$C_{24}H_{30}O_6$			
Pregna-1,4-diene-3,11,20-trione, 17α,21-dihydroxy-16β-methyl-, 21-acetate	MeOH	238(4.18)	44-0026-66
Pregna-1,4,8-triene-3,20-dione, 21-acetoxy-11β,17α-dihydroxy-16α-methyl-	MeOH	240(4.15)	13-0381-66A
$C_{24}H_{30}O_7$			
5,9-Cyclopregn-1-ene-3,20-dione, 11β,17α,21-trihydroxy-, 11-formate 21-acetate	MeOH	272(3.70)	44-2749-66
Hypophyllanthin	EtOH	231(4.56),280(2.23)	78-2899-66
$C_{24}H_{30}O_{11}$			
Harpagoside	MeOH	216(4.19),222(4.12), 276(4.36)	33-1552-66
$C_{24}H_{30}O_{13}$			
Sweroside, tetraacetate	n.s.g.	243(3.96)	88-5229-66
$C_{24}H_{31}BrO_4$			
Pregna-4,6-diene-3,20-dione, 17α-acetoxy-6-(bromomethyl)-	EtOH	282(4.35)	25-0497-66
Pregn-4-ene-3,20-dione, 17α-acetoxy-6β-bromo-16-methylene-	EtOH	248(4.17)	73-2768-66
$C_{24}H_{31}ClO_2$			
Estra-1,3,5(10)-triene, 17α-chloroethynyl-17β-methoxy-3-propoxy-	EtOH	279(3.29),287(3.25)	78-2829-66
$C_{24}H_{31}ClO_4$			
Pregna-4,6-diene-3,20-dione, 17α-acetoxy-6-(chloromethyl)-	EtOH	280(4.37)	25-0497-66
Pregn-4-ene-3,20-dione, 17α-acetoxy-6α-chloro-16-methylene-	EtOH	236(4.17)	73-2768-66
Pregn-4-ene-3,20-dione, 17α-acetoxy-6β-chloro-16-methylene-	EtOH	240(4.18)	73-2768-66
Spiro[androst-5-ene-17,2'(3'H)-furan]-3'-one, 16β-chloro-3-hydroxy-, acetate	EtOH	262(3.91)	32-1268-66
$C_{24}H_{31}FO_4$			
Pregna-4,6-diene-3,20-dione, 17α-acetoxy-6-(fluoromethyl)-	EtOH	278(4.39)	25-0497-66
$C_{24}H_{31}FO_5$			
Pregn-4-ene-3,11,20-trione, 6-fluoro-16α,17-isopropylidenedioxy-	EtOH	233.5(4.17)	87-0513-66

Compound	Solvent	λ_{max}(log ε)	Ref.
$C_{24}H_{31}FO_6$			
Pregna-1,4-diene-3,20-dione, 9α-fluoro-11β,17α,21-trihydroxy-16β-methyl-, 21-acetate	MeOH	238(4.20)	44-0026-66
$C_{24}H_{31}NO_2$			
14'-Apo-β-carotenoic acid, 15'-cyano-, methyl ester	MeOH	435(4.77)	54-0334-66
$C_{24}H_{31}NO_3$			
Androst-4-eno[4,5,6-cd]pyridin-3-one, 17β-hydroxy-, propionate	EtOH	218(4.15),249(3.86), 285(3.45)	32-1284-66
$C_{24}H_{31}N_3O_4$			
Pregna-4,6-diene-3,20-dione, 6-(azido-methyl)-17-hydroxy-, acetate	EtOH	279(4.37)	25-0497-66
$C_{24}H_{32}F_2O_5$			
Pregn-4-ene-3,20-dione, 6α,9α-di-fluoro-11β-hydroxy-16α,17-iso-propylidenedioxy-	EtOH	234(4.21)	87-0513-66
$C_{24}H_{32}N_4O_4S$			
Thiomandelamide, S-[1-acetyl-3-(tetrahydro-2-pyranyloxy)propyl]-N-(2-methyl-4-amino-5-pyrimidyl-methyl)-	EtOH	228(4.10),276(3.76)	44-2951-66
$C_{24}H_{32}O_2$			
Pregn-4-ene-3,20-dione, 17α-propadienyl-	EtOH	242(4.20)	32-1125-66
$C_{24}H_{32}O_3$			
Fauronyl acetate, benzylidene derivative	EtOH	290.5(4.47)	94-0735-66
Pregna-5,16-dien-20-one, 3β-acetoxy-21-methylene-	EtOH	260(3.98)	32-1254-66
$C_{24}H_{32}O_4$			
Crocetin, diethyl ester	$CHCl_3$	257(4.20),412(4.92), 434(5.11),462(5.09)	33-0369-66
Estra-1,3,5(10)-triene-1,17β-diol, 3,4-dimethyl-, diacetate	EtOH	268(1.56)	33-1049-66
Estra-1,3,5(10)-triene-2,17β-diol,	EtOH	274(1.71)	33-1049-66
Heptanoic acid, 3,5-dihydroxy-4-methyl-3,5-diphenyl-, tert-butyl ester	EtOH	242(2.51),247(2.54), 253(2.62),258(2.67), 264(2.59),269(2.33)	22-0131-66
Ketene, phenyl-, diethyl acetal, dimer	EtOH	220(3.49),254(2.65), 259(2.81),265(2.71)	35-3769-66
Pregna-5,16-diene-21-carboxaldehyde, 3β-acetoxy-20-oxo-	EtOH	271(4.00)	32-1254-66
Pregna-4,14-diene-3,20-dione, 17α-acetoxy-6α-methyl-	EtOH	238(4.19)	78-0325-66A
Pregna-3,5,14-trien-20-one, 17α-acetoxy-3-methoxy-	EtOH	239(4.26)	78-0325-66A
Spiroandrost-5-ene-17,2'(3'H)-furan-3'-one, 3β-acetoxy-	EtOH	262(4.01)	32-1268-66

Compound	Solvent	λ_{max}(log ϵ)	Ref.
$C_{24}H_{32}O_5$			
5,9-Cycloandrost-1-en-3-one, 11β,17β-dihydroxy-3-oxo-, 11-acetate 17-propionate	MeOH	272(3.69)	44-2749-66
Pregna-4,6-diene-3,20-dione, 17α-acetoxy-6-(hydroxymethyl)-	EtOH	284(4.37)	78-0365-66
5β-Pregna-1,9(11)-diene-3,20-dione, 17α,21-dihydroxy-16α-methyl-, 21-acetate	EtOH	226(4.05)	95-0639-66
5β-Pregna-1,9(11)-diene-3,20-dione, 17β,21-dihydroxy-16β-methyl-, 21-acetate	EtOH	227(4.01)	95-0639-66
Pregn-4-ene-3,20-dione, 11α-acetoxy-16α,17α-epoxy-16β-methyl-	EtOH	240(4.12)	73-4703-66
Scillirubrosidin	MeOH	298(3.72)	33-0030-66
$C_{24}H_{32}O_6$			
Androst-1-ene-3,5-dione, 11α,17β-diacetoxy-17α-methyl-, 10(5→4)-	EtOH	240(3.87),305(3.73)	33-2218-66
	EtOH-KOH	230(3.90),325(3.76)	33-2218-66
Androst-1-en-3-one, 4α,5α-epoxy-11α,17β-diacetoxy-17α-methyl-	EtOH	224(4.08)	33-2218-66
Androst-1-en-3-one, 4β,5β-epoxy-11α,17β-diacetoxy-17α-methyl-	EtOH	236(3.90)	33-2218-66
Pregna-1,4-diene-3,20-dione, 21-acetoxy-11α,17α-dihydroxy-16α-methyl-	MeOH	248(4.21)	13-0381-66A
Pregna-1,4-diene-3,20-dione, 21-acetoxy-11α,17α-dihydroxy-16β-methyl-	MeOH	247(4.24)	44-0026-66
Pregn-4-ene-3,20-dione, 17α-acetoxy-6β-hydroxy-6α-(hydroxymethyl)-	EtOH	237(4.12)	78-0369-66
Pregn-4-ene-3,11,20-trione, 17α,21-dihydroxy-16β-methyl-, 21-acetate	MeOH	238(4.17)	44-0026-66
Pregn-4-en-3-one, 11β-hydroxy-1,2β-methylene-17,20:20,21-bis-(methylenedioxy)-	MeOH	214(3.98),246(4.02)	24-1118-66
Taxininol, anhydro-, diacetate	EtOH	208(3.79)	95-1172-66
$C_{24}H_{32}O_{13}$			
Sweroside, dihydro-, tetraacetate	n.s.g.	244(3.98)	88-5229-66
$C_{24}H_{33}BrO_5$			
Androst-4-en-3-one, 9α-bromo-11β,17β-dihydroxy-, 11-acetate 17-propionate	MeOH	240(4.16)	44-2749-66
$C_{24}H_{33}ClO_3$			
Pregna-3,5-dien-20-one, 3-chloro-17α-hydroxy-6-methyl-, acetate	EtOH	252(4.31)	32-1241-66
Pregna-3,5-dien-20-one, 6-chloro-16α,17α-epoxy-3-ethoxy-16β-methyl-	EtOH	251(4.29)	73-2768-66
Pregna-3,5-dien-20-one, 6-chloro-3-ethoxy-17α-hydroxy-16-methylene-	EtOH	251(4.31)	73-2768-66
Pregna-5,16,20-triene-21-carboxaldehyde, 20-(2-chloroethoxy)-3β-hydroxy-	EtOH	271(4.15)	32-1254-66
$C_{24}H_{33}ClO_4$			
Androsta-5,16-diene-16-carboxaldehyde, 17-(2-chloroethoxy)-3β-hydroxy-, acetate	EtOH	273(4.18)	32-1268-66

Compound	Solvent	λ_{max}(log ϵ)	Ref.
$C_{24}H_{33}FO_5$			
Pregn-4-ene-3,20-dione, 6α-fluoro-11β-hydroxy-16α,17-isopropylidenedioxy-	EtOH	237.5(4.18)	87-0513-66
$C_{24}H_{33}FO_6$			
Pregn-4-ene-3,20-dione, 9-fluoro-11β,17,21-trihydroxy-, cyclic 17,21-(methyl orthoacetate)	EtOH	240(4.23)	32-1115-66
$C_{24}H_{33}NO_2$			
1H-Cyclopenta[5,6]naphtho[1,2-g]quinolin-1-ol, 2,3,3a,3b,4,5,5a,6,11,-11a,11b,12,13,13a-tetradecahydro-11a,13a imethyl-, acetate	EtOH	269(3.75),277(3.66)	4-0339-66
$C_{24}H_{33}NO_2S$			
Pregna-3,5-dien-20-one, 3-ethoxy-6-thiocyanato-	EtOH	264.5(4.31)	94-1096-66
$C_{24}H_{33}NO_3$			
Androst-5-en-3β-ol, 17β-(5-isoxazolyl)-, acetate	EtOH	217(3.95)	44-3193-66
$C_{24}H_{33}NO_4$			
5α-Androst-2-ene-2-carbonitrile, 3,17β-diacetoxy-	EtOH	211(4.04)	32-0152-66
$C_{24}H_{33}NO_5$			
Taxinonitrile, anhydrodihydro-, diacetate	EtOH	206(3.80)	95-1172-66
$C_{24}H_{34}N_2O_2$			
Androst-5-en-3β-ol, 17β-pyrazol-3-yl-, acetate	EtOH-HCl	219(3.99)	44-3193-66
$C_{24}H_{34}O_3$			
Pregna-3,5-dien-20-one, 16α,17α-epoxy-3-ethoxy-16β-methyl-	EtOH	241(4.27)	73-2768-66
Pregna-3,5-dien-20-one, 3-ethoxy-17α-hydroxy-16-methylene-	EtOH	241(4.27)	73-2768-66
Pregn-5-en-20-one, 3β-acetoxy-21-methylene-	EtOH	209.5(4.05)	78-1625-66
$C_{24}H_{34}O_4$			
5α-Androst-9-en-17-one, 16α-acetyl-3β-hydroxy-16α-methyl-, acetate	$CHCl_3$	306(2.10)	39-2210-66C
5α-Androst-9-en-17-one, 16α-acetyl-3β-hydroxy-16β-methyl-, acetate	$CHCl_3$	307(2.16)	39-2210-66C
D-Homo-5α-androst-9-en-17a-one, 3β-acetoxy-17α-hydroxy-17β-methyl-16-methylene-	EtOH	288(2.09)	39-2210-66C
Pregna-4,16-diene-21-carboxaldehyde, 3,20-dioxo-, 21-(dimethyl acetal)	EtOH	242(4.46)	32-1254-66
Pregn-4-ene-3,20-dione, 17α-acetoxy-18-methyl-	MeOH	240(4.20)	44-1026-66
Pregn-4-ene-3,20-dione, 21-acetoxy-18-methyl-	MeOH	240(4.23)	44-1026-66
Pregn-5-en-20-one, 3β-acetoxy-17α-hydroxy-21-(dimethylaminomethyl)-	EtOH	207(3.89),217(3.87)	78-1625-66

Compound	Solvent	λ_{max}(log ϵ)	Ref.
$C_{24}H_{34}O_5$			
Pregn-4-ene-21-carboxaldehyde, 16α,17α-epoxy-3,20-dioxo-, dimethyl acetal	EtOH	242(4.23)	32-1254-66
Pregn-4-ene-3,20-dione, 17α,21-dihydroxy-18-methyl-, 21-acetate	MeOH	242(4.20)	44-1026-66
5α-Pregn-7-ene-20-carboxylic acid, 3β-acetoxy-6-oxo-	EtOH	243(4.14)	33-1591-66
$C_{24}H_{34}O_6$			
19-Norpregna-5(10),9(11)-diene, 3,3-dimethoxy-17,20:20,21-bis-(methylenedioxy)-	MeOH	235(4.30),242(4.32), 250(4.13)	13-0087-66B
Phyllanthin	EtOH	230(4.33),280(1.89)	78-2899-66
5α-Pregn-7-ene-20-carboxylic acid, 3β-acetoxy-5-hydroxy-6-oxo-	EtOH	248(4.13)	33-1591-66
Pregn-4-ene-3,20-dione, 17α-acetoxy-6α-hydroxy-6β-(hydroxymethyl)-	EtOH	243.5(4.14)	78-0365-66
$C_{24}H_{35}NO$			
Cyclosuffrobuxinine	EtOH	243(3.95)	39-1412-66C
$C_{24}H_{35}NO_3$			
Pregna-5,16-dien-20-one, 3β-hydroxy-, O-methyloxime, acetate	MeOH	243(5.17)	24-0386-66
$C_{24}H_{35}NO_4$			
Androst-5-ene-6-carbonitrile, 17β-acetoxy-3,3-dimethoxy-	MeOH	220(4.39)	78-1069-66
$C_{24}H_{35}N_3O_4$			
Androsta-3,5-diene-6-carbonitrile, 17β-acetoxy-3-methoxy-, semicarbazone, anti	MeOH	230(4.18),284(3.88)	78-1069-66
Androsta-3,5-diene-6-carbonitrile, 17β-acetoxy-3-methoxy-, semicarbazone, syn	MeOH	228(3.93),313(4.45)	78-1069-66
$C_{24}H_{36}Cl_{10}Si_6$			
Hexasilane, 1,1,2,2,3,3,4,4,5,5,6,6-dodecamethyl-1,6-bis(pentachlorophenyl)-	C_6H_{12}	216(5.04)	101-0102-66B
$C_{24}H_{36}F_6PRe$			
Rhenium, bis(hexamethylbenzene)-, hexafluorophosphate	MeCN	414(2.18)	24-2206-66
$C_{24}H_{36}N_2O_2$			
Formimidic acid, N-(2-cyano-17β-hydroxy-17-methyl-5α-androst-2-en-3-yl)-, ethyl ester	EtOH	250(3.91)	32-0152-66
$C_{24}H_{36}O_2$			
Pregna-3,5-dien-20-one, 3-ethoxy-18-methyl-	MeOH	241(4.29)	44-1026-66
$C_{24}H_{36}O_3$			
3,5-Cyclopregnan-17α-ol, 20-ethylenedioxy-6-methylene-	MeOH	206(3.99)	5-0204-66G
Gon-4-en-3-one, 17β-acetoxy-13β,17α-diethyl-7α-methyl-	EtOH	242(4.26)	87-0782-66

Compound	Solvent	$\lambda_{max}(\log \epsilon)$	Ref.
19-Nortestosterone, 4,4-diethyl-, acetate	C_6H_{12}	300f(4.6)	44-0705-66
Pregn-4-en-3-one, 20β-acetoxy-18-methyl-	MeOH	243(4.15)	44-1026-66
$C_{24}H_{36}O_4$			
Pregna-5,16-diene-21-carboxaldehyde, 3β-hydroxy-20-oxo-, 21-(dimethyl acetal)	EtOH	243(3.94)	32-1254-66
Pregna-3,5-dien-20-one, 3-ethoxy-17α-hydroperoxy-18-methyl-	MeOH	243(4.24)	44-1026-66
Pregn-4-ene-3,20-dione, 17α-acetoxy-6α-methyl-	MeOH	240(4.19)	5-0204-66G
$C_{24}H_{36}O_5$			
Androst-5-ene-6-carboxaldehyde, 17β-acetoxy-3,3-dimethoxy-	MeOH	250(4.10)	78-1063-66
$C_{24}H_{37}NO_5$			
Androst-5-ene-6-carboxaldehyde, 17β-acetoxy-3,3-dimethoxy-, oxime, syn	n.s.g.	247(4.25)	78-1069-66
$C_{24}H_{37}Re$			
Rhenium, (hexamethylbenzene)-(hexamethylcyclohexadienyl)-	hexane	406(2.30),484(1.85)	24-2206-66
$C_{24}H_{38}N_2$			
3,5-Conadiene, 3-(dimethylamino)-	ether	271(4.26)	78-0541-66B
$C_{24}H_{38}N_2O$			
Conessine, 7-oxo-	MeOH	237(4.08)	13-0421-66B
Neoconessine, 7-oxo-	n.s.g.	252(3.92)	88-4375-66
Neoconessine, 11-oxo-	n.s.g.	250(3.79)	88-4375-66
$C_{24}H_{38}O_3$			
D-Homoandrost-16-en-17a-one, 3β-hydroxy-16-isobutoxy-	EtOH	250(4.21)	22-1222-66
$C_{24}H_{38}O_4$			
6-Chromanol, 2,5,7,8-tetramethyl-2-(8-carboxy-4,8-dimethyloctyl)-	EtOH	292.0(3.53)	73-2434-66
$C_{24}H_{39}NO_4$			
Cassaine	EtOH	224(4.26)	35-5865-66
$C_{24}H_{41}NO_4$			
Cassaidine	EtOH	224(4.27)	35-5865-66
$C_{24}H_{44}N_2O$			
[$\Delta^{2,3'}$-Bipyrrolidin]-2'-one, 1,1'-dioctyl-	MeOH	205(3.95),298(4.5)	24-0724-66
$C_{24}H_{46}N_4$			
1,2,4,5-Tetrazine, 3,6-undecyl-	EtOH	276(3.49),542(2.73)	88-5067-66
$C_{24}H_{46}Si_6$			
Hexasilane, 1,1,2,2,3,3,4,4,5,5,6,6-dodecamethyl-1,6-diphenyl-	C_6H_{12}	265.0(4.48)	101-0102-66B

Compound	Solvent	λ_{max}(log ε)	Ref.
$C_{24}H_{48}N_4$ 1,2,4,5-Tetrazine, 1,6-dihydro-3,6-diundecyl-	EtOH	310(3.48),426(2.62)	88-5067-66
$C_{24}H_{54}OSn_2$ Distannoxane, hexabutyl-	hexane	206.5(4.19)	39-0242-66A
$C_{24}H_{54}SSn_2$ Distannthiane, hexabutyl-	hexane	208(4.32),247(3.67)	39-0242-66A

Compound	Solvent	λ max(log ε)	Ref.
$C_{25}H_{13}NO_3S$			
Benzo[a]phenothiazin-5-one, 6-(1,3-dioxoindan-2-yl)-	MeCN	248(4.63),255(4.63), 317(4.25),374(4.00), 388(4.00),479(4.05)	44-0062-66
$C_{25}H_{15}IN_2Se_2$			
2,2'-Bis(3-ethylbenzoselenazolyl)-heptamethinecyanine iodide	MeOH	761(5.19)	9-0289-66
$C_{25}H_{15}N_3O_3$			
Oxazolo[5,4-d]pyrimidine-5,7-dione, 4,6-dibenzoyl-2-phenyl-	$CHCl_3$	293s(4.38),301(4.40)	50-0737-66C
$C_{25}H_{16}FN_3O_4$			
Cytosine, 5-fluoro-tribenzoyl-	EtOH	240(4.55)	87-0566-66
$C_{25}H_{16}N_4O_4$			
5-Quinazolinecarboxylic acid, 3,4-dihydro-2-[m-[(2-hydroxy-1-naphthyl)-azo]phenyl]-4-oxo-	pH 13	450(4.21),500(4.16)	7-1394-66
5-Quinazolinecarboxylic acid, 3,4-dihydro-2-[p-[(2-hydroxy-1-naphthyl)-azo]phenyl]-4-oxo-	pH 13	450(4.23),510(4.25)	7-1394-66
$C_{25}H_{16}O$			
Chrysene, 6-benzoyl-	MeOH	222(4.56),250s(4.77), 260(4.90),266(4.91), 280s(4.49),314(3.96), 326(4.04)	73-0835-66
$C_{25}H_{17}Br_4ClN_4O_4$			
1,1-Bis(p-bromophenyl)-2-formyldiazenium perchlorate, bis[(p-bromophenyl)hydrazone]	HCOOH	268(4.16),385(3.78), 566(4.69)	49-0554-66
$C_{25}H_{18}ClNO_4$			
1,3-Diphenylbenzo[a]quinolizinium perchlorate	EtOH	255s(--),275s(--), 293(4.49),360(4.08), 375(4.17)	44-2616-66
$C_{25}H_{18}F_3N_2O_2P$			
Phosphine imide, N-(p-nitrophenyl)-P,P-diphenyl-P-(α,α,α-trifluoro-p-tolyl)-	EtOH	380(4.2)	65-1248-66
$C_{25}H_{18}N_2O$			
Carbazole-9-carboxamide, N,N-diphenyl-	EtOH	265(4.30)	34-0436-66
$C_{25}H_{18}N_4O_2S$			
2-Naphthanilide, 1,4-dihydro-1,4-dioxo-, 4-azine with N-methylbenzothiazol-2-one	MeCN	540(4.66)	5-0116-66G
$C_{25}H_{18}O_2$			
1,4-Naphthoquinone, 2-(3,3-diphenylallyl)- (shoulders, not listed)	C_6H_{12}	251(4.52),330(3.47)	22-1828-66

Compound	Solvent	$\lambda_{max}(\log \epsilon)$	Ref.
$C_{25}H_{19}BBrN$			
Pyridine, compound with [(p-bromo-phenyl)ethynyl]diphenylborane	EtOH	245(4.46),265(4.36)	22-1981-66
$C_{25}H_{19}FN_2O$			
Urea, 1-(p-fluorophenyl)-1,3,3-triphenyl-	EtOH	260(4.29)	34-0436-66
$C_{25}H_{19}F_3NP$			
Phosphine imide, N,P,P-triphenyl-P-(α,α,α-trifluoro-p-tolyl)-	EtOH	270s(3.7)	65-1248-66
Phosphine imide, P,P,P-triphenyl-N-(α,α,α-trifluoro-p-tolyl)-	EtOH	260(4.3)	65-1248-66
$C_{25}H_{19}N_2O_4P$			
Phosphorane, (2,4-dinitrobenzylidene)-triphenyl-	$CHCl_3$	455(4.56)	44-1287-66
$C_{25}H_{19}N_3O_2S$			
Benzenesulfonamide, p-2-benzimidazolyl-N,N-diphenyl-	EtOH	316(4.57),322(4.58)	95-0600-66
$C_{25}H_{19}N_5O_5$			
5-Quinazolinecarboxylic acid, 3,4-dihydro-4-oxo-2-[m-[[1-(phenyl-carbamoyl)acetonyl]azo]phenyl]-	pH 13	380(4.48)	7-1394-66
5-Quinazolinecarboxylic acid, 3,4-dihydro-4-oxo-2-[p-[[1-(phenyl-carbamoyl)acetonyl]azo]phenyl]-	pH 13	415(4.65)	7-1394-66
$C_{25}H_{20}BN$			
Pyridine, compound with (phenyl-ethynyl)diphenylborane	EtOH	235(4.46),245(4.40)	22-1981-66
$C_{25}H_{20}BrP$			
9,10-Dihydro-9,9-diphenyl-9-phosphonia-phenanthrene bromide	EtOH	260(4.1),275(3.7)	39-2003-66C
$C_{25}H_{20}NO_2P$			
Phosphorane, (p-nitrobenzylidene)-triphenyl-	$C_2H_4Cl_2$	237(3.96),266(3.89), 511(4.58)	97-0068-66
$C_{25}H_{20}N_2O$			
Urea, 1,1,3,3-tetraphenyl-	EtOH	267(4.30)	34-0436-66
$C_{25}H_{20}O_{12}$			
Ceroalbolinic acid, methyl ester, tetraacetate	EtOH	262(4.38),330(3.64)	78-1507-66
Norvogeletin, pentaacetate	MeOH	274(4.32),320(4.15)	24-2430-66
$C_{25}H_{21}AuCl_4N_4$			
2-Formyl-1,1-diphenyldiazenium tetra-chloroaurate, diphenylhydrazone	HCOOH	315(3.93),538(4.64)	49-0554-66
$C_{25}H_{21}ClN_4O_4$			
2-Formyl-1,1-diphenyldiazenium perchlorate, diphenylhydrazone	HCOOH	264(4.08),342(3.67), 537(4.68)	49-0554-66

Compound	Solvent	λ_{max}(log ε)	Ref.
$C_{25}H_{21}Cl_4FeN_4$			
2-Formyl-1,1-diphenyldiazenium tetrachloroferrate, diphenylhydrazone	HCOOH	345(3.80),538(4.63)	49-0554-66
$C_{25}H_{21}NO_2S$			
Acetic acid, [o-(4,5-diphenyl-2-thiazolyl)phenyl]-, ethyl ester	EtOH	241(4.33),319(4.12)	4-0282-66
$C_{25}H_{21}N_3O_3$			
Diacetylanhydrotri-2-aminobenzaldehyde	EtOH	237(4.26),274s(3.45), 282s(3.38)	39-0956-66B
$C_{25}H_{21}N_4$			
1,2,4,5-Tetraazapentenyl, 1,1,5,5-tetraphenyl-	dioxan	251(4.11),298(4.02), 381(4.25),631(4.19)	49-0554-66
$C_{25}H_{21}N_5O_3$			
Acetoacetanilide, p'-methyl-2-[m-(3,4-dihydro-4-oxoquinazolin-2-yl)phenylazo]-	dioxan	375(4.48)	32-0279-66
Acetoacetanilide, p'-methyl-2-[p-(3,4-dihydro-4-oxoquinazolin-2-yl)phenylazo]-	dioxan	395(4.68)	32-0279-66
$C_{25}H_{22}ClNO_4S$			
4-(p-Dimethylaminophenyl)-2,6-diphenylthiopyrylium perchlorate	HOAc	259(4.27),383(4.18), 582(4.76)	33-2046-66
$C_{25}H_{22}N_2$			
5H-1,2-Diazepine, 3,7-dimethyl-4,5,6-triphenyl-	n.s.g.	212(4.46),253(3.78)	88-4979-66
$C_{25}H_{22}N_2O_6$			
Cinnamic acid, p-(benzyloxy)-α-[2-(o-nitrophenyl)acetamido]-, methyl ester	EtOH	309(4.43)	7-1192-66
$C_{25}H_{22}N_4$			
Formic acid, 2,2-diphenylhydrazide, diphenylhydrazone	dioxan	255(4.20),284(4.35)	49-0554-66
$C_{25}H_{22}N_4O_{11}$			
2(1H)-Pyrimidinone, 1-(3-deoxy-α-D-ribofuranosyl)-4-methoxy-5-methyl-, bis(p-nitrobenzoate)	MeOH	260(4.42)	44-1163-66
2(1H)-Pyrimidinone, 1-(3-deoxy-β-D-ribofuranosyl)-4-methoxy-5-methyl-, bis(p-nitrobenzoate)	MeOH	261(4.45)	44-1163-66
$C_{25}H_{22}O_6$			
3(2H)-Benzofuranone, 2-[4-(benzyloxy)-benzylidene]-4,5,6-trimethoxy-	EtOH	230(4.07),254(4.06), 358(4.23),391(4.34)	24-3222-66
$C_{25}H_{22}O_7$			
Flavone, 3-hydroxy-5,6,7-trimethoxy-4'-(benzyloxy)-	EtOH	231(4.22),259(4.20), 346(4.24)	24-3222-66
Flavone, 5-hydroxy-3,6,7-trimethoxy-4'-(benzyloxy)-	EtOH	231(4.33),261(4.30), 329(4.33)	24-3222-66

Compound	Solvent	$\lambda_{max}(\log \epsilon)$	Ref.
$C_{25}H_{23}IN_2O$			
1-Methyl-2-[(5,6,7,8-tetrahydro-2-phenyl-4H-1,3-benzoxazin-4-ylidene)-methyl]quinolinium iodide	MeOH	444(4.63)	87-0758-66
$C_{25}H_{23}N$			
Benzofulvene, 3-(cyclohepta-2,4,6-trien-1-yl)-ω-(p-dimethylamino-phenyl)-	EtOH	251(4.28),407(4.43)	5-0057-66H
9,10-Benzosesquifulvalene, 8-(p-dimethylaminobenzyl)-	EtOH	250(4.83),302(3.83), 395(4.41)	5-0057-66H
$C_{25}H_{23}NO_4$			
Cinnamic acid, p-(benzyloxy)-α-(2-phenylacetamido)-, methyl ester	EtOH	309(4.60)	7-1192-66
$C_{25}H_{24}BN$			
Pyridine, compound with [(1-cyclo-hexenyl)ethynyl]diphenylborane	EtOH	232(4.34)	22-1981-66
$C_{25}H_{24}BrOP$			
(3-Methyl-6-oxo-2,4-hexadienyl)tri-phenylphosphonium bromide	EtOH	268s(4.44),274(4.45)	33-0369-66
$C_{25}H_{24}N_2$			
Dipyrromethene, 4-ethyl-3,5-dimethyl-3',5'-diphenyl-	EtOH-N HCl	222(4.41),283(4.34), 379(3.91),514(4.91)	12-1871-66
	EtOH	224s(4.19),289(4.34), 465(4.46)	12-1871-66
Dipyrromethene, 3,3',5,5'-tetramethyl-4,4'-diphenyl-	EtOH-N HCl	226(4.64),249(3.94), 365(3.85),470(4.83)	12-1871-66
	EtOH	225(4.15),279(3.65), 446(4.38)	12-1871-66
$C_{25}H_{24}O_3$			
1-Indanone, 2,2-bis(m-methoxybenzyl)-	EtOH	250(4.11),283(3.74)	88-6489-66
$C_{25}H_{24}O_5$			
Osajin	EtOH	225(4.32),275(4.62), 354(4.45)	39-0701-66C
Scandenone	EtOH	220(4.42),276(4.46), 348s(3.63)	39-0701-66C
$C_{25}H_{24}O_6$			
Chalcone, 4-(benzyloxy)-2'-hydroxy-4',5',6'-trimethoxy-	EtOH	236(4.32),271(4.22), 308(4.31),363(4.36)	24-3222-66
Flavanone, 4'-(benzyloxy)-5,6,7-trimethoxy-	EtOH	231(4.35),274(4.11), 318(3.61)	24-3222-66
$C_{25}H_{25}IN_2S_2$			
3-Ethyl-2-[7-(3-ethyl-2-benzothiazolin-ylidene)-1,3,5-heptatrienyl]benzo-thiazolium iodide	MeOH	747(5.33)	9-0289-66
$C_{25}H_{25}N$			
Piperidine, N-benzyl-4-(diphenyl-methylene)-	EtOH	243(3.54)	78-2721-66

Compound	Solvent	λ_{max}(log ϵ)	Ref.
$C_{25}H_{26}IN_4$			
Verdazyl, 1,3,5-triphenyl-6-(5-iodopentyl)-	dioxan	244(4.11),283(4.40), 319(4.19),405(3.96), 427(3.95),720(3.6)	49-0846-66
$C_{25}H_{26}N_2O_4$			
1H-Isoquino[2,1-a][1,6]naphthyridin-13-one, 2,3,4,6,7,11b,12,13-octahydro-9,10-dimethoxy-N-benzoyl-	EtOH	287(3.75),329(4.14)	44-0797-66
$C_{25}H_{26}N_4O_4$			
3,5-Hexadien-2-one, 6-(4-methyl-3-cyclohexen-1-yl)-6-phenyl-, 2,4-dinitrophenylhydrazone	$CHCl_3$	270(4.26),310(4.41), 405(4.66)	70-0480-66
$C_{25}H_{26}N_4O_7$			
Cacalol, 2-acetyl-, 2,4-dinitrophenylhydrazone, acetate	$CHCl_3$	316(4.20),411(4.39)	78-0301-66
$C_{25}H_{26}N_4O_{13}$			
2(1H)-Pyrimidinone, 1-β-D-glucopyranosyl-4-(p-nitrobenzamido)-, tetraacetate	pH 1	269(4.40)	44-4014-66
	pH 7	269(4.39)	44-4014-66
	pH 13	217(4.24)	44-4014-66
	EtOH	268(4.48)	44-4014-66
$C_{25}H_{26}N_5$			
Verdazyl, 1,3,5-triphenyl-6-piperidino-	dioxan	243(4.12),276(4.33), 319(4.18),405(3.98), 420s(3.96),688(3.64)	49-0846-66
$C_{25}H_{26}N_6O_5$			
Isonicotinic acid hydrazide, dihydrazone with glyoxal benzyl 2-formyl-2-methoxyethyl acetal	EtOH	260(4.36)	39-2121-66C
$C_{25}H_{26}O_5$			
Isonorscandenin, 4'-deoxydihydro-	EtOH	229(4.38),247(4.19), 328(4.37)	39-0192-66C
Munetol	EtOH	227(4.48),259(4.50), 348(3.71)	39-0749-66C
Thamnosin	EtOH	203(4.61),228(4.36), 256(4.35),298(4.06), 335(4.27)	78-2923-66
$C_{25}H_{26}O_6$			
Isonorscandenin, dihydro-	EtOH	256s(4.13),333(4.34)	39-0192-66C
	NaOEt	253s(4.45),327(4.33)	39-0192-66C
$C_{25}H_{26}O_8$			
Agatharesinol, tetraacetate	n.s.g.	253(4.34)	88-2395-66
$C_{25}H_{26}O_9$			
Isoderrisic acid, acetate	EtOH	238(4.63),260s(4.07), 290(3.78)	39-0550-66C
	EtOH-KOH	242(4.44),257s(4.26), 294s(3.96),368(3.53)	39-0550-66C
$C_{25}H_{27}NO$			
2,8-Cyclononadien-1-one, 2,9-diphenyl-3-(1-pyrrolidinyl)-	EtOH	228(4.30),283(4.03)	44-1336-66

Compound	Solvent	λ_{max}(log ε)	Ref.
$C_{25}H_{27}NO_5S$			
Benzoic acid, p-[N-benzyl-N-(2-hydroxyethyl)amino]-, ethyl ester, p-toluenesulfonate	$CHCl_3$	290(4.22),305(4.25)	87-0868-66
$C_{25}H_{27}N_3O_9$			
Phenoxazine-1,9-dicarboxylic acid, 2-formamido-4,6-dimethyl-7-nitro-3-oxo-, dibutyl ester	MeOH	238s(4.44),402(4.08)	44-3694-66
$C_{25}H_{27}N_5O_{13}$			
2(1H)-Pyrimidinone, 1-(2-deoxy-2-carbomethoxyamino-β-D-glucopyranosyl)-4-(p-nitrobenzamido)-, triacetate	pH 1	269(4.35)	44-4014-66
	pH 7	269(4.35)	44-4014-66
	pH 13	271(4.19),322(4.11)	44-4014-66
$C_{25}H_{28}N_2O_6$			
9H,16H-Azocino[2',1':2,3]oxazolo-[4,5-b]phenoxazine-1,15-dicarboxylic acid, 7a,8,10,11,12,13-hexahydro-4,6-dimethyl-, dimethyl ester	MeOH	234(4.51),440(3.96)	44-2564-66
$C_{25}H_{28}N_2O_{12}$			
2-Quinoxalinecarboxylic acid, 3-β-D-glucopyranosyloxy-, ethyl ester, tetraacetate	MeOH	221s(4.19),241(4.39), 244(4.39),305(3.76), 325(3.77),335s(3.74)	24-0547-66
$C_{25}H_{28}N_8O_5$			
2-Pyridinecarbamic acid, 6-(acetonylamino)-4-[(diphenylmethyl)amino]-5-nitro-, ethyl ester, semicarbazone	pH 1	222(4.37),244(4.28), 356(4.02)	44-1890-66
	pH 7	224(4.46),262s(4.06), 290s(3.80),362(3.96)	44-1890-66
	pH 13	225(4.45),262s(4.06), 364(4.00)	44-1890-66
$C_{25}H_{28}O_2$			
2-Bornanone, 3-(α-phenacylbenzyl)-	EtOH	241(4.11),278s(3.02)	88-2243-66
$C_{25}H_{28}O_7$			
Piscerythrone, tetra-O-methyl-	EtOH	255(4.49),285(4.14)	78-0333-66A
Piscidone, tetra-O-methyl-	EtOH	251(4.44),280s(4.01)	78-0333-66A
$C_{25}H_{28}O_{11}$			
Epiorientin, tetra-O-methyl-	EtOH	246(4.50),266(4.40), 334(4.40)	78-1147-66
Parkinsonin A, tri-O-methyl-	EtOH	245(4.40),266(4.30), 334(4.40)	78-1147-66
$C_{25}H_{29}FO_7$			
Pregna-1,4,9(11)-triene-3,20-dione, 16α,21-diacetoxy-6α-fluoro-17α-hydroxy-	MeOH	239(4.18)	13-0381-66A
$C_{25}H_{29}N_5O_{11}$			
2(1H)-Pyrimidinone, 1-(2-deoxy-2-carbomethoxyamino-β-D-glucopyranosyl)-4-(p-aminobenzamido)-, triacetate	pH 1	260(4.31),309(4.05)	44-4014-66
	pH 7	252(4.16),323(4.42)	44-4014-66
	pH 13	331(4.45)	44-4014-66

Compound	Solvent	λ_{max}(log ε)	Ref.
$C_{25}H_{30}F_2O_4$			
Androsta-2,4-diene-3,17β-diol, 17-ethynyl-6,6-difluoro-, diacetate	EtOH	265(3.68)	44-0991-66
$C_{25}H_{30}N_2O_6$			
Aspidoalbine, 21-oxo-	n.s.g.	225(4.27),265(3.98)	35-4984-66
Phenoxazine-1,9-dicarboxylic acid, 2-(hexahydro-1(2H)-azocinyl)-3-hydroxy-4,6-dimethyl-, dimethyl ester	MeOH	226(4.48),410(3.94)	44-2564-66
Spiroxyane A, 17α-acetoxy-1-acetyl-16α-carbomethoxy-	EtOH	229(3.99),270(3.10)	44-1765-66
$C_{25}H_{30}N_2O_7$			
Alanine, N-benzyloxycarbonyl-β-tert-butylaspartylphenyl-, methyl ester	EtOH	248(2.44),252(2.56), 257(2.62),264(2.53), 267(2.36)	44-3400-66
$C_{25}H_{30}N_8$			
Tetraethylammonium salt of 5-[[p-(dimethylamino)phenyl]azo]cyclopentadiene-1,2,3,4-tetracarbonitrile	MeCN	218(4.36),278(4.43), 318(4.05),472(4.52)	35-4055-66
$C_{25}H_{30}N_8O_8$			
Cyclohexanone, 2-methyl-2-(4-methyl-3-oxopentyl)-, bis(2,4-dinitrophenylhydrazone)	$CHCl_3$	365(4.64)	39-0224-66C
$C_{25}H_{30}O_5$			
Unnamed lactone	MeOH	220(4.11)	39-0506-66C
$C_{25}H_{31}ClO_4$			
Pregna-4,6-diene-3,20-dione, 6-chloro-1α,2α-methylene-16α,17α-dihydroxy-, 16,17-cyclic acetal with acetone	EtOH	280(4.22)	44-3467-66
$C_{25}H_{31}IN_4$			
2-[3-(1,3-Diethyl-2-benzimidazolinylidene)propenyl]-1,3-diethylbenzimidazolium iodide	EtOH	498(--)	65-0828-66
$C_{25}H_{31}NO_4$			
Pregn-4-eno[4,5,6-cd]pyridine-3,20-dione, 17-acetoxy-	EtOH	218(4.11),250(3.80), 285(3.45)	32-1284-66
$C_{25}H_{31}N_3O_5$			
Leucine, N-benzyloxycarbonyl-L-tryptophanyl-, methyl ester	EtOH	275s(3.77),282(3.80), 291(3.75)	44-3400-66
$C_{25}H_{32}$			
Retinylidenecyclopentadiene	EtOH	427(4.81)	54-0339-66
$C_{25}H_{32}Cl_2O_4$			
Pregna-4,6-diene-3,20-dione, 6-chloro-1-chloromethyl-16,17-dihydroxy-, cyclic acetal with acetone	EtOH	287(4.35)	44-3467-66
$C_{25}H_{32}N_2O_4$			
5'H-Pregna-1,4,6-trieno[2,1-c]pyrazole-3,20-dione, 1 2-dihydro-16α,17 -dihydroxy-, cyclic acetal with acetone	EtOH	235(3.60),293(4.34)	44-3467-66

Compound	Solvent	λ_{max}(log ϵ)	Ref.
$C_{25}H_{32}N_2O_5$			
Piperidine, 1-acetyl-4-[2-hydroxy-2-[3-(2-hydroxyethyl)indol-2-yl]-ethyl]-3-vinyl-, diacetate	n.s.g.	222(4.66),285(4.02), 294(3.99)	22-2207-66
$C_{25}H_{32}N_2O_6$			
Aspidoalbine, O-methyl-21-oxo-	MeOH	220(4.92),258(4.13), 290(3.70)	35-4984-66
$C_{25}H_{32}O_4$			
Benz[4,5,6]androst-4-en-3-one, 5',17β-dihydroxy-, 17-propionate	EtOH	270(3.97),340(3.61)	32-1284-66
3'H-Cyclopropa[1,2]pregna-1,4,6-triene-3,20-dione, 1,2-dihydro-16,17-dihydroxy-, cyclic acetal with acetone	EtOH	282(4.32)	44-3467-66
$C_{25}H_{32}O_5$			
5β-Carda-16,20(22)-dienolide, 3β-acetoxy-14α,15α-epoxy-	EtOH	275(4.21)	94-0496-66
5β-Carda-16,20(22)-dienolide, 3β-acetoxy-14β,15β-epoxy-	EtOH	276(4.19)	94-0496-66
Pregna-3,5,14-triene-6-carboxaldehyde, 17α-acetoxy-3-methoxy-20-oxo-	EtOH	219(4.04),320(4.16)	78-0325-66A
Pregna-3,5,14-trien-20-one, 3,17α-diacetoxy-	EtOH	234(4.36)	78-0325-66A
$C_{25}H_{32}O_6$			
Pregna-4,14-diene-3,20-dione, 17α,21-diacetoxy-	EtOH	238(4.25)	78-0325-66A
$C_{25}H_{32}O_7$			
5,9-Cyclopregn-1-ene-3,20-dione, 11β,17,21-trihydroxy-, 11β,21-diacetate	MeOH	272(3.69)	44-2749-66
$C_{25}H_{33}BrO_7$			
Pregn-4-ene-3,20-dione, 9α-bromo-11β,17α,21-trihydroxy-, 11β,21-diacetate	MeOH	240(4.21)	44-2749-66
$C_{25}H_{33}ClO_4$			
Androsta-3,5,7-triene-6-carboxaldehyde, 3-(2-chloroethoxy)-17β-hydroxy-, propionate	EtOH	238(3.95),278(3.95), 387(4.04)	32-1268-66
$C_{25}H_{33}ClO_5$			
3'H-Cyclopropa[1,2]pregna-1,4-diene-3,20-dione, 6β-chloro-1β,2β-dihydro-7α,16α,17-trihydroxy-, cyclic 16,17-acetal with acetone	EtOH	236(4.21)	44-3467-66
$C_{25}H_{33}FO_5$			
Cyclobuta[2]cyclopenta[a]phenanthren-3(1H)-one, 5b-fluoro-8-glycoloyl-tetradecahydro-6-hydroxy-5a,7a-dimethyl-, 8-acetate	MeOH	241(3.95)	88-3451-66
16α,21-Cyclo-17-pregn-4-ene-3,20-dione, 9-fluoro-17-glycoloyl-11β-hydroxy-, 17-acetate	MeOH	237(4.20)	88-3451-66

Compound	Solvent	λ_{max}(log ε)	Ref.
$C_{25}H_{33}FO_6$			
Pregna-1,4-diene-3,20-dione, 11β-fluoro-17α,21-dihydroxy-16α-methyl-, 21-propionate	MeOH	242(4.17)	35-3016-66
$C_{25}H_{33}NO_2$			
14'-Apo-β-carotenoic acid, 15'-cyano-, ethyl ester	EtOH	433(4.76)	54-0334-66
$C_{25}H_{33}NO_3$			
Benz[4,5,6]androst-4-en-3-one, 5'-amino-17β-hydroxy-, propionate	EtOH	240(4.26),273(3.78), 360(3.38)	32-1284-66
$C_{25}H_{33}NO_6$			
Pregna-1,4-diene-3,20-dione, 11α-acetamido-17α,21-dihydroxy-, 21-acetate	MeOH	252(4.20)	44-1346-66
Pregna-1,4-diene-3,20-dione, 11β-acetamido-17α,21-dihydroxy-, 21-acetate	MeOH	240(4.11)	44-1346-66
$C_{25}H_{33}N_3O_6$			
1,3-Azulenedicarboxylic acid, 2-acetamido-6-(3-morpholinopropylamino)-, diethyl ester	MeOH	269(4.10),286s(4.03), 343(4.69),399(4.19)	18-1954-66
$C_{25}H_{34}N_2O_2$			
Androsta-5,16-dien-3β-ol, 17-(1-methyl-5-pyrazolyl)-, acetate	EtOH	241(3.93)	44-3193-66
$C_{25}H_{34}O_2$			
Retinylideneacetylacetone	pet ether	413(4.75)	54-0334-66
$C_{25}H_{34}O_3$			
Cochliobolin, 3-anhydro-	MeOH	232(4.32)	88-1211-66
$C_{25}H_{34}O_4$			
3'H-Cyclopropa[1,2]pregna-1,4-diene-3,20-dione, 1β,2β-dihydro-16α,17-dihydroxy-, cyclic acetal with acetone	EtOH	240(4.09)	44-3467-66
3'H-Cyclopropa[6,7]pregna-4,6-diene-3,20-dione, 6β,7β-dihydro-16α,17-dihydroxy-, cyclic acetal with acetone	EtOH	267(4.22)	44-3467-66
14β-Gona-1,3,5(10),8-tetraen-15β-ol, 13-ethyl-3-methoxy-17β-[(tetrahydropyran-2-yl)oxy]-	MeOH	273(4.23)	88-0801-66
Pregna-5,16-dien-20-one, 3β-acetoxy-21-methoxymethylene-	EtOH	279(4.15)	32-1254-66
Pregna-3,5,14-trien-20-one, 17α-acetoxy-3-ethoxy-	EtOH	240(4.26)	78-0325-66A
Pregna-3,5,14-trien-20-one, 17α-acetoxy-6-methyl-3-methoxy-	EtOH	245(4.26)	78-0325-66A
$C_{25}H_{34}O_4S$			
5α-Estran-17-one, 3β-hydroxy-, p-toluenesulfonate	MeOH	225(4.10)	87-0685-66

Compound	Solvent	λ_{max}(log ε)	Ref.
$C_{25}H_{34}O_5$			
Pregna-3,5,14-trien-20-one, 17α-acetoxy-6-(hydroxymethyl)-3-methoxy-	EtOH	249(4.24)	78-0325-66A
Unnamed hemiacetal	MeOH	225(4)	39-0506-66C
$C_{25}H_{34}O_6$			
5β-Carda-16,20(22)-dienolide, 3β,14ξ,15ξ-trihydroxy-, 3-acetate	EtOH	269(4.25)	94-0496-66
Pregn-4-ene-3,20-dione, 17α-hydroxy-, 17-hydrogen succinate	EtOH	241(4.21)	32-1115-66
$C_{25}H_{34}O_7$			
9β,10α-Pregn-4-ene-3,20-dione, 11α,21-diacetoxy-17α-hydroxy-	MeOH	240(4.22)	54-0731-66
	EtOH	239(4.24)	33-1528-66
9β,10α-Pregn-4-ene-3,20-dione, 11β,21-diacetoxy-17α-hydroxy-	EtOH	240(4.23)	33-1528-66
$C_{25}H_{34}O_8S$			
Pregna-1,4-diene-3,20-dione, 11α,17α,21-trihydroxy-16β-methyl-, 21-acetate, 11-methanesulfonate	MeOH	243(4.22)	44-0026-66
$C_{25}H_{34}O_{13}$			
Glutaric acid, 3-oxo-, 2,4-bis(4,5-O-isopropylidene-D-xylofuranosylidene)-, diethyl ester	EtOH	234(3.7),247(3.8)	23-1841-66
$C_{25}H_{35}ClO_4$			
Androsta-3,5-diene-6-carboxaldehyde, 3-(2-chloroethoxy)-17β-hydroxy-, propionate	EtOH	221(3.99),322(4.25)	32-1268-66
$C_{25}H_{35}NO_4$			
Barbonine, 1,2,3,4-tetrahydro-2-methyl-	MeOH	290(3.83)	24-2873-66
Isoquinoline, 6,7-diethoxy-3-(3,4-diethoxybenzyl-1,2,3,4-tetrahydro-2-methyl-, perchlorate	MeOH	290(3.87)	24-2873-66
Pregna-1,4-dien-3-one, 11α-acetamido-20β-acetoxy-	MeOH	252(4.22)	44-1342-66
Pregna-1,4-dien-3-one, 11β-acetamido-20β-acetoxy-	MeOH	242(4.05)	44-1342-66
$C_{25}H_{35}NO_6$			
Pregn-4-ene-3,20-dione, 11α-acetamido-21-acetoxy-17α-hydroxy-	MeOH	242(4.18)	44-1346-66
$C_{25}H_{35}N_2P$			
Phosphorane, tributyl(fluoren-9-ylidenehydrazono)-	dioxan	215(4.23),248(4.78), 295(3.79),306(3.82), 366(4.32),369(4.33)	5-0001-66H
$C_{25}H_{35}N_5O_{11}$			
Propionic acid, 2-[[2-(dimethylamino)-4-(β-D-galactopyranosylamino)-5-pyrimidinyl]imino]-, ethyl ester, tetraacetate	MeOH	232(4.28),304(3.90), 369(3.98)	24-3022-66
Propionic acid, 2-[[2-(dimethylamino)-4-(β-D-glucopyranosylamino)- (cont.)	MeOH	232(4.28),303(3.89), 368(3.95)	24-3022-66

Compound	Solvent	$\lambda_{max}(\log \epsilon)$	Ref.
5-pyrimidinyl]imino]-, ethyl ester, acetate (cont.)			
$C_{25}H_{36}N_2O_2$			
Androst-5-en-3β-ol, 17β-(1-methyl-3-pyrazolyl)-, acetate	EtOH	221(3.83)	44-3193-66
Androst-5-en-3β-ol, 17β-(1-methyl-5-pyrazolyl)-, acetate	EtOH	214(3.83)	44-3193-66
$C_{25}H_{36}O_2$			
Zizanin A, anhydro-	n.s.g.	229(4.45)	88-2211-66
$C_{25}H_{36}O_3$			
Cochiobolin B, 3-anhydro-	MeOH	231(4.30)	88-1329-66
Zizanin B, anhydro-	n.s.g.	228(4.44)	88-2211-66
$C_{25}H_{36}O_4$			
Cochliobolin	MeOH	236(4.00)	88-1211-66
Gona-3,5-diene-3,17β-diol, 13β,17α-diethyl-, diacetate	EtOH	238(4.26)	87-0782-66
Gona-1,3,5(10)-trien-15β-ol, 13-ethyl-3-methoxy-17β-[(tetrahydropyran-2-yl)oxy]-	MeOH	278(3.35),287(3.32)	88-0801-66
isomer	MeOH	278(3.35),287(3.33)	88-0801-66
5α-Pregna-9,16-dien-20-one, 3β-acetoxy-15ξ-methoxy-16-methyl-	EtOH	245(3.88)	39-2201-66C
$C_{25}H_{36}O_5$			
18-Nor-5α,8α,9β,14β-androst-15-en-17-one, 3α,11α-diacetoxy-4α,8,14-trimethyl-	EtOH	226(3.92)	78-3459-66
Pregn-4-ene-20-carboxylic acid, 2β-acetoxy-3-oxo-, methyl ester	EtOH	243(4.20)	33-1581-66
Pregn-4-ene-3,20-dione, 11β,17α-dihydroxy-, 17-butyrate	EtOH	242(4.23)	32-1115-66
5β-Pregn-9(11)-en-12-one, 3α,20β-diacetoxy-	MeOH	239.5(4.12)	44-1349-66
5α-Pregn-16-en-20-one, 3β,15α-diacetoxy-	EtOH	236.5(3.71)	5-0169-66D
$C_{25}H_{36}O_6$			
Pregn-4-ene-3,20-dione, 11,17,21-trihydroxy-, 17-butyrate	EtOH	243(4.21)	32-1115-66
$C_{25}H_{36}O_7$			
5α-Etianic acid, 3β,12β-diacetoxy-11-oxo-, methyl ester	EtOH	295(1.48)	33-1632-66
9β,10α-Pregn-4-en-3-one, 17α,20:20,21-bis(methylenedioxy)-11α-(methoxy-methyl)-	MeOH	241(4.20)	54-0731-66
$C_{25}H_{37}ClO_4$			
Androsta-3,5-diene-6-methanol, 3-(2-chloroethoxy)-17β-hydroxy-, propionate	EtOH	253(4.32)	32-1268-66
$C_{25}H_{37}NO$			
Cyclosuffrobuxine	EtOH	243(3.94)	39-1412-66C

Compound	Solvent	λ_{max}(log ε)	Ref.
$C_{25}H_{37}NO_4$			
19-Norandrost-4-en-3-one, 17β-hydroxy-10β-(N,N-diethylcarboxamido)-, acetate	EtOH	253(4.12)	44-0693-66
Pregn-4-en-3-one, 11α-acetamido-20β-acetoxy-	MeOH	241(4.00)	44-1342-66
$C_{25}H_{37}N_5O_8$			
Sarcosine, N-(2-amino-3,4-cresotoyl)-L-threonyl-D-valyl-L-prolyl-	pH 1	216(4.49),222s(4.46), 330(3.34)	23-0799-66
$C_{25}H_{38}N_2O_3$			
1H-Cyclopenta[5,6]naphtho[1,2-g]quinazoline-8,10(3bH,9H)-dione, 7-butyltetradecahydro-1-hydroxy-11a,13a-dimethyl-	EtOH	273(3.98)	32-0179-66
$C_{25}H_{38}O_3$			
A-Nortestosterone, 1-cyclohexyloxymethylene-3,5-dihydro-	EtOH	279(4.67)	88-1057-66
Zizanin A	n.s.g.	240(4.10)	88-2211-66
$C_{25}H_{38}O_4$			
Cochliobolin B	MeOH	237(4.00)	88-1329-66
Zizanin B	MeOH	237(4.29)	88-1329-66
	n.s.g.	239(4.16)	88-2211-66
$C_{25}H_{38}O_5$			
Pregn-7-en-6-one, 2β,3β,14,21-tetrahydroxy-20β-methyl-, cyclic 2,3-acetal with acetone	MeOH	241(4.04)	88-3457-66
$C_{25}H_{38}O_6$			
Cassaidic acid, methyl ester, diacetate	EtOH	222(4.22)	35-5865-66
Isocassaidic acid, methyl ester, diacetate	EtOH	221(4.24)	35-5865-66
$C_{25}H_{38}O_7$			
Taxicin I, 5-deoxy-2-O-methyldihydro-, diacetate	EtOH	279(3.75)	39-1933-66C
$C_{25}H_{39}NO$			
Cyclobuxomicreine	EtOH	243(3.90)	39-1412-66C
Cyclobuxosuffrine	EtOH	243(3.95)	39-1412-66C
$C_{25}H_{39}NO_7$			
Browniine, dehydro-	EtOH	288(1.48)	23-0001-66
$C_{25}H_{39}N_3O_5$			
Androst-5-ene-6-carboxaldehyde, 17β-acetoxy-, semicarbazone, 3-(dimethyl acetal), anti	MeOH	272(4.46)	78-1069-66
syn	MeOH	233(3.92)	78-1069-66
$C_{25}H_{39}O_5P$			
Phosphonic acid, (20,21-epoxy-3-oxopregn-4-en-20-yl)-, diethyl ester	EtOH	240(4.23)	78-1625-66

Compound	Solvent	$\lambda_{max}(\log \epsilon)$	Ref.
$C_{25}H_{40}N_8O_3S$			
Bis(tetraethylammonium) salt of 5-(sulfoazo)cyclopentadiene-1,2,3,4-tetracarbonitrile	MeCN	210(3.96),271(4.51), 279(4.54),342(4.02), 436(2.39)	35-4055-66
$C_{25}H_{40}O_4$			
2-Chromannonanoic acid, 6-hydroxy-α,ε,2,5,7,8-hexamethyl-, methyl ester	EtOH	292.5(3.53)	73-2434-66
$C_{25}H_{40}O_7$			
Unnamed alcohol	EtOH	224(3.81)	39-1933-66C

Compound	Solvent	λ_{max}(log ε)	Ref.
$C_{26}H_2F_{20}$			
Ethane, 1,1,2,2-tetrakis(pentafluoro-phenyl)-	$CHCl_3$	264(3.42)	30-0598-66A
$C_{26}H_2F_{20}O$			
Ether, bis[bis(pentafluorophenyl)-methyl]	$CHCl_3$	264(3.41)	30-0598-66A
$C_{26}H_{14}$			
1,12:4,5-Dibenzoperylene	benzene	305(4.55),318(4.55), 333(4.60),346(3.91), 364(4.04),384(4.35), 407(4.45),429(3.02)	24-1275-66
	EtOH	240(4.70),262(4.50), 276(4.50)	24-1275-66
$C_{26}H_{14}Cl_2O_2$			
$\Delta^{9,9}$-Bixanthene, 2,2'-dichloro-	$CHCl_3$	255s(4.13),286(4.00), 374(4.23)	78-0349-66A
$\Delta^{9,9}$-Bixanthene, 3,3'-dichloro-	$CHCl_3$	274(4.45),295s(4.22), 332(3.96),345(4.06), 397s(4.04),417(4.15), 437(4.11)	78-0349-66A
$C_{26}H_{14}N_4$			
Spiro[azulene-1(2H),9'-fluorene]-2,2,3,3(3aH)-tetracarbonitrile	C_6H_{12}	277(4.17)	5-0034-66H
$C_{26}H_{16}Cl_2N_2$			
Dibenzo[b,f][1,5]diazocine, 2,8-dichloro-6,12-diphenyl-	iso-PrOH	260(4.58),320s(3.78)	35-1077-66
$C_{26}H_{16}N_2$			
1,4-Ethenonaphthalene-2,3-dicarbo-nitrile, 1,4-dihydro-5,8-diphenyl-	MeCN	235(4.48),315(2.93)	78-0491-66B
$C_{26}H_{16}N_4$			
2,2,3,3-Naphthalenetetracarbonitrile, 1,4-dihydro-1,4-diphenyl-	CH_2Cl_2	253(2.81),259(2.91), 265(2.89),270(2.73)	5-0044-66C
isomer	CH_2Cl_2	255(2.87),260(2.94), 265(2.95)	5-0044-66C
$C_{26}H_{16}N_4O_6$			
5-Quinazolinecarboxylic acid, 3,4-dihydro-4-oxo-2-[[m-(3-carboxy-2-hydroxy-1-naphthyl)azo]phenyl]-	pH 13	450(4.26),500(4.28)	7-1394-66
5-Quinazolinecarboxylic acid, 3,4-dihydro-4-oxo-2-[[p-(3-carboxy-2-hydroxy-1-naphthyl)azo]phenyl]-	pH 13	460(4.39),515(4.41)	7-1394-66
$C_{26}H_{16}O_2S$			
1,3-Indandione, 2-(2,6-diphenyl-4H-thiopyran-4-ylidene)-	HOAc	251(4.47),345(4.13), 479(4.78)	33-2046-66
$C_{26}H_{17}NO_2S_2$			
Anthraquinone, 1-amino-2,4-bis-(phenylthio)-	$CHCl_3$	255(4.64),341(3.83), 540(4.04)	22-0580-66

Compound	Solvent	$\lambda_{max}(\log \epsilon)$	Ref.
$C_{26}H_{17}N_4O_6P$			
Acetonitrile, picryl(triphenylphosphoranylidene)-	$CHCl_3$	465(4.09)	44-1287-66
$C_{26}H_{18}$			
Anthracene, 9-[2-(1-naphthyl)vinyl]-	$C_2H_4Cl_2$	260(5.04),310(4.00), 315(3.97),389(4.12)	99-0085-66
$C_{26}H_{18}Cl_2N_2$			
Dibenzo[b,f][1,5]diazocine, 2,8-dichloro-5,6-dihydro-6,12-diphenyl-	iso-PrOH	257(4.48),320(3.70)	35-1077-66
	iso-PrOH	257(4.48),320(3.70)	44-3356-66
Indolo[3,2-b]indole, 3,8-dichloro-4b,5,9b,10-tetrahydro-4b,9b-diphenyl-	iso-PrOH	248(4.40),315(3.74)	35-1077-66
	iso-PrOH	248(4.40),260s(4.32), 315(3.74)	44-3356-66
$C_{26}H_{18}Fe$			
Ferrocene, 1,1'-bis(phenylethynyl)-	EtOH	253(4.50),304(4.42), 453(2.94)	101-0173-66B
$C_{26}H_{18}N_2$			
1,4-Ethanonaphthalene-2,3-dicarbonitrile, 1,4-dihydro-5,8-diphenyl-	MeCN	250(4.46),307s(2.85)	78-0491-66B
Quinoline, 6,7-diphenyl-5-(2-pyridyl)-	MeOH	207(4.66),238(4.68), 321(3.70)	44-3206-66
$C_{26}H_{18}N_2O_3S$			
Benzoic acid, p-(phenylazo)-, 5-[4-(2-furyl)-3-buten-1-ynyl]-2-thenyl ester	ether	328(4.68),338(4.68)	24-0135-66
$C_{26}H_{18}N_3O_4P$			
Acetonitrile, (2,4-dinitrophenyl)-(triphenylphosphoranylidene)-	$CHCl_3$	437(4.32)	44-1287-66
$C_{26}H_{18}N_4O_4$			
Benzophenone, p-nitro-, azine	dioxan	336(4.20)	5-0001-66H
$C_{26}H_{18}O_2$			
Benzo[ghi]perylene-5-butyric acid	benzene	294(4.64),305(4.73), 332(3.82),352(4.04), 372(4.36),393(4.44)	24-1275-66
$C_{26}H_{18}O_6$			
Biphenyl-3,3'-dicarboxylic acid, 2,4'-dihydroxy-, diphenyl ester	EtOH	233(4.68),325(4.07)	39-0321-66C
$C_{26}H_{18}O_{10}$			
Benzo[1,2-b:4,5-b']bisbenzofuran, 1,3,7,9-tetraacetoxy-	dioxan	231(4.75),263(4.28), 300s(4.40),311(4.73), 321s(4.76),325(4.80), 339(4.80)	1-2202-66
Benzo[1,2-b:4,5-b']bisbenzofuran, 2,3,8,9-tetraacetoxy-	dioxan	229s(4.63),262(4.29), 309s(4.43),322(4.73), 339(4.88)	1-2202-66
$C_{26}H_{19}ClO_4S_3$			
Thiopyrylium, 4-[(5-methyl-1,3-benzodithiol-2-ylidene)methyl]-2,6-diphenyl-, perchlorate	HOAc	258(4.31),392(4.16), 577(4.92)	33-2046-66

Compound	Solvent	λ_{max}(log ϵ)	Ref.
$C_{26}H_{19}F_3N_2O$			
Urea, 1,1,3-triphenyl-3-(α,α,α-trifluoro-m-tolyl)-	EtOH	265(4.29)	34-0436-66
$C_{26}H_{19}N$			
Anthracen-9,10-imine, 9,10-dihydro-9,10-diphenyl-	EtOH	257s(3.26),277(3.17), 284(3.30)	78-1011-66
$C_{26}H_{19}N_5O_2$			
Benzeneazoethene, 4-nitro-1',2'-diphenyl-2'-(phenylazo)-	EtOH	240(4.50),297(4.24), 420(4.31)	78-1309-66
$C_{26}H_{20}$			
Naphthalene, 1,8-distyryl-	dioxan	250(4.72),342(4.34)	44-2407-66
$C_{26}H_{20}BP$			
Borane, [(diphenylphosphino)ethynyl]-diphenyl-	EtOH	213(4.40),223(4.30)	22-1002-66
$C_{26}H_{20}ClNO_4SSe$			
Thiopyrylium, 4-[(3-methyl-2-benzoselenazolinylidene)methyl]-2,6-diphenyl-, perchlorate	HOAc	272(4.00),352(4.15), 518(4.68),542(4.68)	33-2046-66
$C_{26}H_{20}ClNO_4S_2$			
Thiopyrylium, 4-[(3-methyl-2-benzothiazolinylidene)methyl]-2,6-diphenyl-, perchlorate	HOAc	273(3.92),296(3.95), 348(4.12),508(4.71), 532(4.69)	33-2046-66
$C_{26}H_{20}ClNO_5S$			
Thiopyrylium, 4-[(3-methyl-2-benzoxazolinylidene)methyl]-2,6-diphenyl-, perchlorate	HOAc	285(4.01),343(4.08), 484s(4.71),509(4.66)	33-2046-66
$C_{26}H_{20}Cl_2N_2$			
Dibenzo[b,f][1,5]diazocine, 2,8-dichloro-5,6,11,12-tetrahydro-6,12-diphenyl-, cis	iso-PrOH	255s(3.90),283s(3.42)	44-3356-66
trans	iso-PrOH	256(4.34),305(3.60)	44-3356-66
$C_{26}H_{20}MoO_3Sn$			
Molybdenum, tricarbonyl(cyclopentadienyl)(triphenylstannyl)-	n.s.g.	282(4.09)	35-5124-66
$C_{26}H_{20}N_2$			
Benzophenone, azine	dioxan	324(4.01)	5-0001-66H
$C_{26}H_{20}N_4O_2$			
Lumazine, 8-benzyl-3-methyl-6,7-diphenyl-	pH -2.7	271(4.18),376(3.99)	24-3503-66
	pH 5	272(4.35),365(3.82), 430(4.02)	24-3503-66
	pH 12	247(4.31),285(4.18), 364(4.13)	24-3503-66
$C_{26}H_{20}O_4$			
2,3-Naphthalenedicarboxylic acid, 5,8-diphenyl-, dimethyl ester	EtOH	252(4.77),311(4.04), 342s(3.70)	78-0491-66B
$C_{26}H_{21}Cl_4FeN_4$			
1,3,5,6-Tetraphenylverdazylium tetrachloroferrate	HCOOH	323(4.62),550(4.09)	49-1280-66

Compound	Solvent	λ_{max}(log ϵ)	Ref.
$C_{26}H_{21}N$			
p-Toluidine, α-17H-cyclopenta[a]phenanthren-17-ylidene-N,N-dimethyl-	EtOH	371(4.18),434(4.58)	39-0963-66C
$C_{26}H_{21}N_9O_{16}$			
4'-Formyl-3,4-dihydro-4-oxo-1,1'-trimethylenebis[pyridinium picrate], 4'-oxime	H_2O	260(4.59),353(4.46)	87-0638-66
$C_{26}H_{22}$			
Benzene, 1,3-bis(4-phenylbutadienyl)-	dioxan	356(4.66)	18-1547-66
$C_{26}H_{22}BN$			
Pyridine, compound with (phenylethynyl)phenyl-m-tolyl-	EtOH	235(4.46),245(4.43)	22-1981-66
Pyridine, compound with (phenylethynyl)phenyl-o-tolyl-	EtOH	235(4.52),245(4.54)	22-1981-66
Pyridine, compound with (phenylethynyl)phenyl-p-tolyl-	EtOH	235(4.40),245(4.44)	22-1981-66
$C_{26}H_{22}CuN_4O_2S_2$			
Copper, bis(thio-o-picolinanisididato)-	benzene	356(4.02),440(4.05)	65-1499-66
	heptane	274(4.34),360(4.07), 420(3.99)	65-1499-66
	EtOH	274(4.32),355(3.99), 432(3.87)	65-1499-66
	$CHCl_3$	276(4.27),350(3.84), 432(3.87)	65-1499-66
Copper, bis(thio-p-picolinanisididato)-	benzene	415(4.16)	65-1499-66
	heptane	278(4.27),400(4.18)	65-1499-66
	EtOH	225(4.26),278(4.34), 396(4.13)	65-1499-66
	$CHCl_3$	277(4.32),420(4.16)	65-1499-66
$C_{26}H_{22}INO_2$			
1-Ethyl-2-(α-hydroxystyryl)quinolinium iodide, benzoate	EtOH	245(4.59),298s(3.95), 361(4.42)	44-2384-66
$C_{26}H_{22}NO_3P$			
Phosphorane, (p-methoxyphenyl)(p-nitrobenzylidene)diphenyl-	$C_2H_4Cl_2$	249(4.27),350s(3.20), 513(4.58)	97-0068-66
$C_{26}H_{22}N_2O$			
Urea, 1-benzyl-1,3,3-triphenyl-	EtOH	260(4.15)	34-0436-66
Urea, 1,1,3-triphenyl-3-p-tolyl-	EtOH	265(4.32)	34-0436-66
$C_{26}H_{22}N_2O_2$			
Urea, 1-(p-methoxyphenyl)-1,3,3-triphenyl-	EtOH	265(4.31)	34-0436-66
$C_{26}H_{22}N_6$			
Formaldehyde, azodi-, bis-(diphenylhydrazone)	dioxan	253(4.23),470(4.73)	49-0554-66
$C_{26}H_{22}O_2$			
Fulvene, 1,3-bis(α-hydroxybenzyl)-6-phenyl-	MeOH	316(4.40)	44-2149-66
$C_{26}H_{22}O_{10}$			
Bis-3,3'-(tri-O-methylflaviolin)	$CHCl_3$	265(4.49),303(4.25), 381(3.87),402s(3.79)	39-2234-66C

Compound	Solvent	λ max(log ε)	Ref.
$C_{26}H_{22}O_{12}$			
1,2,5,6,9,10-Anthracenehexol, hexaacetate	dioxan	264(5.16),348s(3.67), 362(3.87),381(3.97), 402(3.90)	24-2322-66
$C_{26}H_{23}BrN_4O$			
1,7,9-Tribenzyl-1,6-dihydro-6-oxo-purinium bromide	pH 1	257(3.99)	44-2202-66
	pH 7	257(3.96)	44-2202-66
	pH 13	262(3.82),290s(3.62)	44-2202-66
$C_{26}H_{24}N_2O_2$			
Bisindolo[2,3-b:2',3'-b']benzo-[1,2-d:4,5-d']difuran, 7b,14b-diethyl-5a,7b,12a,14b-tetrahydro-	EtOH	227s(4.31),244s(4.00), 286s(3.85),320(4.08)	44-3321-66
Bisindolo[2,3-b:2',3'-b']benzo-[1,2-d:4,5-d']difuran, 5a,7b,12a,14b-tetrahydro-5,7b,12,14b-tetramethyl-	EtOH	232(4.24),251(4.03), 293s(3.86),320(4.03)	44-3321-66
Bisindolo[2,3-b:2',3'-b']benzo-[1,2-d:4,5-d']difuran, 5a-7b,12a,14b-tetrahydro-5a,7b,12a,14b-tetramethyl-	EtOH	226s(4.35),247s(3.98), 297s(3.81),320(4.03)	44-3321-66
Dipyrromethene-3'-carboxylic acid, 4',5'-dimethyl-3,5-diphenyl-, ethyl ester	EtOH-N HCl	223(4.32),283s(4.21), 291s(4.26),382(4.01), 544(4.98)	12-1871-66
	EtOH	223(4.37),289(4.33), 355(3.81),508(4.49)	12-1871-66
Dipyrromethene-4'-carboxylic acid, 3',5'-dimethyl-3,5-diphenyl-, ethyl ester	EtOH-N HCl	278(4.23),338(3.68), 389(3.75),513(5.04)	12-1871-66
	EtOH	229(4.32),280(4.35), 475(4.43)	12-1871-66
Hydroquinone, 2,5-bis(1,3-dimethyl-indol-2-yl)-	EtOH	230(4.82),298(4.28), 323(4.26)	44-3321-66
Hydroquinone, 2,5-bis(3-ethylindol-2-yl)-	EtOH	228(4.55),247s(4.21), 304(4.04),347(4.17)	44-3321-66
Indole, 2,2'-(2,5-dimethoxy-p-phenyl-ene)bis[3-methyl-	EtOH	225(4.69),251s(4.27), 308s(4.15),349(4.47)	44-3321-66
$C_{26}H_{24}N_2O_2S$			
Benzoic acid, (phenylazo)-, 1-ethyl-7-(2-thienyl)-4-hepten-6-ynyl ester	ether	296s(4.52),310(4.59)	24-1642-66
$C_{26}H_{24}N_3O_2P$			
Phosphine imide, P-[p-(dimethylamino)-phenyl]-N-(p-nitrophenyl)-P,P-diphenyl-	EtOH	290(4.3),390(4.3)	65-1248-66
$C_{26}H_{24}N_4O_2$			
Formamide, N-benzyl-N-[1-benzyl-4-(benzylamino)-1,6-dihydro-6-oxo-5-pyrimidinyl]-	pH 1	229(4.45),260(3.84)	44-2202-66
	pH 7	229(4.45),260(3.84)	44-2202-66
	pH 13	229(4.46),260(3.86)	44-2202-66
Imidazole-5-carboxamide, 4-(N-benzyl-formamido)-N,1-dibenzyl-	pH 1	250s(3.75),257s(3.68)	44-2202-66
	pH 7	250s(3.80),257s(3.74)	44-2202-66
	pH 13	250s(3.81),257s(3.75)	44-2202-66
$C_{26}H_{24}N_4O_2S$			
Pyrimido[4',5':4,5]pyrimido[2,1-c]-[1,4]thiazine-8-ethanol, 5,10-dihydro-2,7-dimethyl-10-phenyl-, benzoate	EtOH	372(4.08)	44-2951-66

Compound	Solvent	$\lambda_{max}(\log \epsilon)$	Ref.
$C_{26}H_{24}O_4$			
Benzo[1",2":3,4:4",5":3',4']dicyclobuta[1,2-b:1',2'-b']bisbenzofuran-6,12-dione, octahydro-5b,6a,11b,12a-tetramethyl-	CH_2Cl_2	282(3.91),290(3.89)	24-1723-66
isomer	CH_2Cl_2	284(3.92),290(3.91)	24-1723-66
$C_{26}H_{24}O_5$			
Munetone	EtOH	230(4.49),265(4.58), 328(4.06)	39-0749-66C
$C_{26}H_{24}O_7$			
Flavone, 4'-(benzyloxy)-3,5,6,7-tetramethoxy-	EtOH	234(4.36),268(4.19), 324(4.29)	24-3222-66
$C_{26}H_{24}O_9$			
Dibenzoylmethane, 2-hydroxy-5-(4-methoxybenzoyloxy)-4,4',6-trimethoxy-	EtOH	264(4.35),392(4.36)	44-3228-66
$C_{26}H_{25}IN_2O$			
1-Ethyl-2-[(5,6,7,8-tetrahydro-2-phenyl-4H-1,3-benzoxazin-4-ylidene)-methyl]quinolinium iodide	MeOH	444(4.65)	87-0758-66
$C_{26}H_{25}NO_7$			
Tyrosine, N,O-dibenzyloxycarbonyl-, methyl ester	MeOH	251(2.73),257(2.86), 261(2.83),263(2.84), 267(2.69)	5-0226-66F
	MeOH-NaOH	292(3.38)	5-0226-66F
$C_{26}H_{25}N_2P$			
Phosphorane, (p-dimethylaminophenylimino)triphenyl-	EtOH	270(4.2)	65-1248-66
Phosphorane, (p-dimethylaminophenyl)-phenyliminodiphenyl-	EtOH	290(4.3)	65-1248-66
$C_{26}H_{25}N_3O_2$			
2H-Imidazole, 2-(p-dimethylamino)-α-(p-methoxyphenyl)benzylidene]-4-(p-methoxyphenyl)-	C_6H_{12}	400(4.21),494(4.55)	24-1815-66
	benzene	410(4.22),515(4.57)	24-1815-66
	EtOH	430(4.30),560(4.60)	24-1815-66
	HOAc	482(4.38),615(4.68)	24-1815-66
	H_2SO_4	260(4.2),345(4.0), 445(4.3),587(4.4)	24-1815-66
	$CHCl_3$	420(4.26),533(4.58)	24-1815-66
	CCl_4	405(4.23),505(4.57)	24-1815-66
	THF	406(4.24),515(4.58)	24-1815-66
	DMF	418(4.23),535(4.57)	24-1815-66
$C_{26}H_{25}N_3O_4$			
Iminobis[2-(4-phenyl-4-ethyl-lutaconimide)]	EtOH	322(3.76)	7-1519-66
$C_{26}H_{26}Cl_4O_4$			
Podocarpa-8,11,13-trien-15-oic acid, 13-(5,6,7,8-tetrachloro-2-methyl-1,4-benzodioxan-2-yl)-	EtOH	291(3.63),301(3.69)	78-0745-66
$C_{26}H_{26}N_2O_2$			
Spiroxyane A, 18-benzylidene-17-oxo-	EtOH	265(4.11),288(4.25)	44-1765-66

Compound	Solvent	λ_{max}(log ε)	Ref.
$C_{26}H_{26}N_2O_3$			
Peraksine, benzoate	EtOH	226(4.61),276(3.72), 282s(3.70),290s(3.53)	78-3293-66
$C_{26}H_{26}N_4O_3S$			
2H-1,4-Thiazin-3(4H)-one, 6-[(2-benzoyl-oxy)ethyl]-5-methyl-4-[(2-methyl-4-amino-5-pyrimidyl)methyl]-2-phenyl-	EtOH	230(4.47),277(3.90)	44-2951-66
$C_{26}H_{26}O_2$			
Benzene, 2,5-dimethoxy-1,4-bis-(p-methylstyryl)-	dioxan	392(4.57)	18-1547-66
2a,6b-Ethenodicyclopenta[fg,op]naph-thacene, 1,2,7,8,8a,12b,12c,12d-octahydro-8a,12b-dimethoxy-	EtOH	265(2.83)	35-4522-66
$C_{26}H_{26}O_4$			
Benzene, 2,5-dimethoxy-1,4-bis-(o-methoxystyryl)-	dioxan	396(4.65)	18-1547-66
Benzene, 2,5-dimethoxy-1,4-bis-(p-methoxystyryl)-	dioxan	396(4.56)	18-1547-66
$C_{26}H_{26}O_5$			
Scandinone	EtOH	268(4.66),325s(3.27)	39-0701-66C
$C_{26}H_{26}O_6$			
Lonchocarpic acid	EtOH	235s(4.28),259(4.46), 286(4.25),339(4.12)	39-0192-66C
	EtOH-NaOAc	256(4.52),278(4.36), 334(4.02)	39-0192-66C
$C_{26}H_{26}O_8$			
[2,2'-Binaphthalene]-1,1'-diol, 3,3',6,6',8,8'-hexamethoxy-	$CHCl_3$	253(4.99),287(4.14), 300(4.06),314(3.98), 331(3.87)	39-2234-66C
$C_{26}H_{26}O_{11}$			
2-Anthroic acid, 5,9,10-trihydroxy-3,6,7-trimethoxy-1-methyl-, methyl ester, triacetate	EtOH	277(4.95),349(3.51), 368(3.65),386(3.62), 407(3.55)	78-1507-66
$C_{26}H_{27}ClN_2O$			
7H-Yohimban, 18-benzylidene-7-chloro-18β-hydroxy-	CH_2Cl_2	252s(4.19),289s(3.40)	44-1765-66
$C_{26}H_{27}NO$			
Acrylophenone, 2-[α-(tert-butylamino)-benzyl]-4'-phenyl-	isooctane	284(4.47)	88-4037-66
Chalcone, α-[(tert-butylamino)methyl]-4'-phenyl-	isooctane	284(4.31)	88-4037-66
Ketone, 4-biphenylyl 1-tert-butyl-2-phenyl-3-azetidinyl, cis	isooctane	282(4.36)	88-4037-66
Ketone, 4-biphenylyl 1-tert-butyl-2-phenyl-3-azetidinyl, trans	isooctane	282(4.42)	88-4037-66
deuterated trans	isooctane	282(4.42)	88-4037-66
$C_{26}H_{27}N_3O_3$			
Imidazole-2-methanol, α-[p-(dimethyl-amino)phenyl]-α,4-bis(p-methoxy-phenyl)-	EtOH	265(4.7)	24-1815-66
	dil. H_2SO_4	260(4.3),510(4.2)	24-1815-66
	H_2SO_4	260(4.2),345(4.0), 445(4.3),587(4.4)	24-1815-66

Compound	Solvent	$\lambda_{max}(\log \epsilon)$	Ref.
$C_{26}H_{27}O_2P$			
1,4-Butanedione, 2-(diethylphenylphosphoranylidene)-1,4-diphenyl-	CH_2Cl_2	240(4.33),280(3.90)	78-0567-66
$C_{26}H_{28}Cl_4O_4$			
Podocarpa-7,13-dien-15-oic acid, 13-(5,6,7,8-tetrachloro-2-methyl-1,4-benzodioxan-2-yl)-	EtOH	236(4.60),302(3.51)	78-0745-66
$C_{26}H_{28}N_2O_2$			
Spiroxyane, 18-benzylidene-17β-hydroxy-	EtOH	245(4.32),280s(3.34)	44-1765-66
epimer	EtOH	244(4.31),280s(3.30)	44-1765-66
$C_{26}H_{28}N_2O_5$			
1,3-Azulenedicarboxylic acid, 2-acetamido-6-(2-phenylethylamino)-, diethyl ester	MeOH	270(4.22),290(4.15), 343(4.83),399(4.32)	18-1954-66
$C_{26}H_{28}N_4Pd$			
Palladium, bis[2-[(3,4-dimethyl-2H-pyrrol-2-ylidene)methyl-]-5-[2-[(3,4-dimethylpyrrol-2-yl)methylene]-3,4-dimethyl-2H-pyrrol-5-yl]-3,5-dimethylpyrrolato]-	$CHCl_3$	272s(4.07),289s(3.98), 345s(3.97),394s(4.52), 420s(4.66),430(4.68), 526(4.29),562s(4.17), 732s(3.59),786(3.89), 859(4.18)	39-0098-66C
$C_{26}H_{28}N_5$			
Verdazyl, 1,3,5-triphenyl-6-(4-methylpiperidino)-	dioxan	243(4.15),275(4.34), 318(4.19),404(3.99), 420s(3.97),688(3.65)	49-0846-66
$C_{26}H_{28}O_6$			
Isonorscandenin, dihydro-4'-O-methyl-	EtOH	228(4.39),255(4.06), 333(4.29)	39-0192-66C
$C_{26}H_{28}O_9$			
Isoderrisic acid, methyl ester, acetate	EtOH	238(4.62),260s(4.13), 288(3.85)	39-0550-66C
	EtOH-KOH	240(4.48),258s(4.31), 292s(4.03),370(3.67)	39-0550-66C
$C_{26}H_{29}NO_2$			
2H-Naphth[2',1':4,5]indeno[1,2-b]pyridin-2-one, 8-(2-furyl)-3,4,-4a,4b,5,6,6a,11,11a,11b,12,13-dodecahydro-4a,6a-dimethyl-	EtOH	242(4.29),265s(4.18), 309(4.27)	4-0339-66
$C_{26}H_{29}N_3O_9$			
Phenoxazine-1,9-dicarboxylic acid, 2-acetamido-4,6-dimethyl-7-nitro-3-oxo-, dibutyl ester	MeOH	227(4.44),404(4.38)	44-3694-66
$C_{26}H_{29}N_7O_4$			
Aspartic acid, N-(5-amino-3-butyl-3H-v-triazolo[4,5-d]pyrimidin-7-yl)-, dibenzyl ester	EtOH	227(4.34),260(3.87), 290(4.07)	87-0417-66
$C_{26}H_{30}N_2O_4$			
1-Indanone, 2-butyl-2-nitroso-, dimer	EtOH	252(4.40),294(3.81)	44-2090-66

Compound	Solvent	$\lambda_{max}(\log \epsilon)$	Ref.
1-Indanone, 2-tert-butyl-2-nitroso-, dimer	EtOH	247(4.40),294(3.57)	44-2090-66
$C_{26}H_{30}N_4$			
Pyrrole, 2-[3,4-dimethyl-2H-pyrrol-2-ylidene)methyl]-5-[2-[(3,4-dimethyl-pyrrol-2-yl)methylene]-3,4-dimethyl-2H-pyrrol-5-yl]-3,5-dimethyl-	EtOH	253(4.22),338(4.06), 582(4.77)	39-0098-66C
	$CHCl_3$-HBr	272(5.14),451(4.90), 693(4.52)	39-0098-66C
$C_{26}H_{30}N_4O_2$			
p-Benzenediacetaldehyde, 2,5-dimethoxy-α,α'-dimethyl-, bis(phenylhydrazone)	EtOH	240(4.17),280(4.53), 302s(4.36)	44-3321-66
1-Propanone, 1,1'-(2,5-dimethoxy-p-phenylene)di-, bis(phenyl-hydrazone)	EtOH	267(4.83),297(4.52)	44-3321-66
$C_{26}H_{30}N_8O_{12}$			
Glutaric acid, 2,4-diacetyl-3-methyl-, diethyl ester, bis(2,4-dinitro-phenylhydrazone)	EtOH	357(4.37)	39-0336-66C
$C_{26}H_{30}O_5$			
Gedunin, 7-deacetoxy-14(15)-deoxy-7-oxo-	MeOH	219(4.30)	39-0944-66C
$C_{26}H_{30}O_6$			
Isogedunin, deacetoxy-7-oxo-	MeOH	218(4.05)	39-0506-66C
Lonchocarpic acid, tetrahydro-	EtOH	215(4.58),239s(4.23), 257s(4.07),335(4.32)	39-0192-66C
	EtOH-NaOAc	222(4.59),254(4.30), 325(4.18)	39-0192-66C
Scandenin, tetrahydro-	EtOH	227(4.52),255(4.08), 335(4.30)	39-0192-66C
$C_{26}H_{30}O_8$			
Scillirosidine, 3-dehydro-1α,2α-epoxy-	EtOH	223(4.15),300(3.77)	78-3213-66
p-Terphenyl, 2,2',2",4,4",5',6,6"-octamethoxy-	dioxan	259(4.29),302(4.16)	1-2202-66
$C_{26}H_{31}ClO_4$			
Benzo[4,5,6]pregn-4-ene-3,20-dione, 5'-chloro-17-hydroxy-, acetate	EtOH	260(4.03),312(3.50)	32-1284-66
$C_{26}H_{31}FO_6$			
Pregna-1,4,9(11)-triene-3,20-dione, 21-acetoxy-6α-fluoro-16α,17α-iso-propylidenedioxy-	MeOH	238(4.20)	13-0381-66A
$C_{26}H_{31}FO_7$			
Pregna-1,4-diene-3,20-dione, 21-acet-oxy-9β,11β-epoxy-6α-fluoro-16α,17α-isopropylidenedioxy-	MeOH	246(4.18)	13-0381-66A
Pregna-1,4,8-triene-3,20-dione, 21-acetoxy-6α-fluoro-11β-hydroxy-16α,17α-isopropylidenedioxy-	MeOH	239(4.20)	13-0381-66A
$C_{26}H_{31}NO_2$			
1H-Naphth[2',1':4,5]indeno[1,2-b]-pyridin-2-ol, 8-(2-furyl)-dodeca-hydro-4a,6a-dimethyl-	EtOH	268(4.15),309(4.26)	4-0338-66

Compound	Solvent	λ max (log ε)	Ref.
$C_{26}H_{31}NO_4$			
16,17-Seco-16-nor-5-androsten-17-oic acid, 3β-hydroxy-15-(3-oxo-2-indolinylidene)-	MeOH	238(4.22),262(4.40), 275s(4.19),300s(3.88), 455(3.70)	44-1363-66
$C_{26}H_{31}NO_6$			
Benzo[4,5,6]pregn-4-ene-3,20-dione, 17-hydroxy-5'-nitro-, acetate	EtOH	249(4.35)	32-1284-66
$C_{26}H_{32}Cl_4O_4$			
15-Podocarpanoic acid, 13-(5,6,7,8-tetrachloro-2-methyl-1,4-benzodioxan-2-yl)-	EtOH	302(3.35)	78-0745-66
$C_{26}H_{32}N_2$			
Lobinaline, demethyl-	EtOH	251(2.79),257(2.79), 264(2.67),268(2.52)	44-3206-66
$C_{26}H_{32}N_2O_4$			
Echitovenidine	EtOH	219(4.42),302(4.13), 327(4.28)	88-2483-66
$C_{26}H_{32}N_2O_7$			
Phenoxazine-1,9-dicarboxylic acid, 2-acetamido-3-hydroxy-4,6-dimethyl-, dibutyl ester	MeOH	234(4.46),408(4.07)	44-3694-66
$C_{26}H_{32}N_6O_9$			
L-Glutamic acid, N-[p-[N-[3-[N-(2-formamino-4-hydroxy-5-pyrimidinyl)-formamido]propyl]formamido]-benzoyl]-, diethyl ester	EtOH	262(4.26),300s(3.96)	87-0590-66
$C_{26}H_{32}O_3$			
Androst-5-en-17-one, 16-[3-(2-furyl)-allylidene]-3β-hydroxy-	EtOH	238(3.51),359(4.58)	4-0339-66
$C_{26}H_{32}O_4$			
Androst-5-en-17-one, 3β-acetoxy-16-furfurylidene-	n.s.g.	325(4.44)	88-3891-66
$C_{26}H_{32}O_6$			
Pregna-4,14-diene-3,20-dione, 17α,21-diacetoxy-6-methylene-	EtOH	260(3.85)	78-0325-66A
Pregna-4,6,14-triene-3,20-dione, 17α,21-diacetoxy-6-methyl-	EtOH	284(4.22)	78-0325-66A
$C_{26}H_{32}O_7$			
3-Cyclohexene-1-acetic acid, 2-[1(3-furyl)-4a-hydroxy-8a-methyl-5-methylene-3-oxo-6-isochromanyl]-2,6,6-trimethyl-5-oxo-	MeOH	212(4.00),237(4.00)	39-0944-66C
Pregna-1,4,8-triene-3,20-dione, 21-acetoxy-11β-hydroxy-16α,17α-isopropylidenedioxy-	MeOH	240(4.19)	13-0381-66A
$C_{26}H_{32}O_8$			
Scillirosidin, 1α,2α-epoxy-	EtOH	300(3.76)	78-3213-66

Compound	Solvent	λ_{max}(log ε)	Ref.
$C_{26}H_{33}BrO_7$			
Pregna-1,4-diene-3,20-dione, 21-acetoxy-9α-bromo-11β-hydroxy-16α,17α-isopropylidenedioxy-	MeOH	241(4.16)	13-0381-66A
$C_{26}H_{33}ClO_6$			
Spiro[benzofuran-2(3H),1'-[2]cyclohexene]-3,4'-dione, 7-chloro-4-(3,7-dimethyl-6-octen-1-yloxy)-2',6-dimethoxy-6'-methyl-	EtOH	218(4.39),235(4.36), 291(4.36),327(3.78)	87-0242-66
$C_{26}H_{33}NO_2$			
1H-Cyclopenta[5,6]naphtho[1,2-g]quinolin-1-ol, 8-(2-furyl)-tetradecahydro-11a,13a-dimethyl-	EtOH	266(4.20),309(4.27)	4-0339-66
1H-Naphth[2',1':4,5]indeno[1,2-b]pyridin-2-ol, 8-(2-furyl)-tetradecahydro-4a,6a-dimethyl-	EtOH	268(4.17),309(4.28)	4-0339-66
$C_{26}H_{33}NO_3$			
Androst-5-en-17-one, 16-[3-(2-furyl)-allylidene]-3β-hydroxy-, oxime	EtOH	338(4.65)	4-0339-66
$C_{26}H_{33}NO_4$			
Benzo[4,5,6]pregn-4-ene-3,20-dione, 5'-amino-17-hydroxy-, acetate	EtOH	240(4.24),273(3.80), 362(3.38)	32-1284-66
16,17-Seco-16-norandrostan-17-oic acid, 3β-hydroxy-15-(3-oxo-2-indolinylidene)-	MeOH	238(4.22),262(4.40), 275s(4.19),300s(3.85), 450(3.72)	44-1363-66
$C_{26}H_{33}N_3O_7$			
1-Ornithine, colchicidyl-	pH 1	255(4.51),374(4.26)	65-0033-66
	EtOH	252(4.03),356(4.34), 412(4.08)	65-0033-66
	50% EtOH-HCl	255(4.51),358(4.28), 365(4.24)	65-0033-66
$C_{26}H_{34}Cl_2O_6$			
Pregna-1,4-diene-21-carboxylic acid, 9α,11β-chloro-6α,16α-dimethyl-17α,21-dihydroxy-3,20-dioxo-, ethyl ester	MeOH	238(4.17)	13-0234-66A
$C_{26}H_{34}Cl_2O_7$			
Pregna-3,5-diene-6-carboxaldehyde, 3-(2-chloroethoxy)-9α-chloro-11β,17α,-21-trihydroxy-20-oxo-, 21-acetate	EtOH	216(4.13),318(4.18)	31-0468-66
$C_{26}H_{34}N_2O$			
5α-Androst-2-ene-2-carbonitrile, 3-anilino-17α-hydroxy-	EtOH	290(4.10)	32-0152-66
$C_{26}H_{34}N_2O_4$			
Isoquinoline-1-carbonitrile, 3-(3,4-diethoxybenzyl)-6,7-diethoxy-1,2,3,4-tetrahydro-2-methyl-	MeOH	255(4.30),320(4.01), 380(3.94)	24-2873-66
5'H-Pregna-1,4,6-trieno[2,1-c]pyrazole-3,20-dione, 1β,2β-dihydro-16α,17-dihydroxy-5'-methyl-, cyclic acetal with acetone	EtOH	289(4.32)	44-3467-66

Compound	Solvent	λ_{max}(log ε)	Ref.
$C_{26}H_{34}N_4O_4$			
2,4,6,10,14-Hexapentaenal, 3,7,11,15-tetramethyl-, 2,4-dinitrophenylhydrazone	EtOH	419(4.63)	39-2154-66C
$C_{26}H_{34}N_4O_5$			
A-Homoandrost-4a-en-4-one, 17-hydroxy-, 2,4-dinitrophenylhydrazone	n.s.g.	381(4.22)	78-0391-66A
$C_{26}H_{34}O_3$			
5α-Androstan-3-one, 2-[3-(2-furyl)-allylidene]-17β-hydroxy-	EtOH	361(4.56)	4-0339-66
5α-Androstan-17-one, 16-[3-(2-furyl)-allylidene]-3β-hydroxy-	EtOH	361(4.60)	4-0339-66
$C_{26}H_{34}O_4$			
Androst-5-en-17β-ol, 3β-acetoxy-16-furfurylidene-	n.s.g.	271(4.39)	88-3891-66
$C_{26}H_{34}O_4S$			
5α-Androst-2-en-17-one, 19-hydroxy-, p-toluenesulfonate	MeOH	225(4.09)	87-0685-66
$C_{26}H_{34}O_5$			
Pregna-3,5,14-triene-6-carboxaldehyde, 17α-acetoxy-3-ethoxy-20-oxo-	EtOH	219(4.04),321(4.17)	78-0325-66A
$C_{26}H_{34}O_6$			
Pregna-4,6-diene-3,20-dione, 17α-acetoxy-6-(acetoxymethyl)-	EtOH	280(4.38)	78-0365-66
Pregna-4,6-diene-3,20-dione, 2α,17α-diacetoxy-6-methyl-	EtOH	289(4.38)	78-0377-66
Pregna-4,6-diene-3,20-dione, 2β,17α-diacetoxy-6-methyl-	EtOH	295(4.36)	78-0377-66
Pregna-4,14-diene-3,20-dione, 17α,21-diacetoxy-6α-methyl-	EtOH	239(4.18)	78-0325-66A
Pregna-1,4,9(11)-triene-21-carboxylic acid, 17α,21-dihydroxy-6α,16α-dimethyl-3,20-dioxo-, ethyl ester	MeOH	238(4.20)	13-0234-66A
Pregna-3,5,14-trien-20-one, 17α,21-diacetoxy-3-methoxy-	EtOH	240(4.33)	78-0325-66A
Royleanone, 9-dehydro-, reductive acetylation product	EtOH	219(4.42),267(4.02)	12-0329-66
$C_{26}H_{34}O_7$			
Pregn-4-ene-3,20-dione, 7β,11α-diacetoxy-16α,17α-epoxy-16β-methyl-	EtOH	235(4.15)	73-4703-66
$C_{26}H_{35}ClO_4$			
Pregna-3,5-dien-20-one, 17α-acetoxy-6-chloro-3-ethoxy-16-methylene-	EtOH	251(4.55)	73-2768-66
Pregna-5,16,20-triene-21-carboxaldehyde, 20-(2-chloroethoxy)-3β-hydroxy-, acetate	EtOH	270(4.15)	32-1254-66
$C_{26}H_{35}ClO_5$			
Pregna-3,5-diene-6-carboxaldehyde, 3-(2-chloroethoxy)-17α-hydroxy-, acetate	EtOH	224(3.92),324(4.18)	32-1268-66

Compound	Solvent	λ_{max}(log ε)	Ref.
$C_{26}H_{35}FO_7$			
Pregna-1,4-diene-3,20-dione, 9α-fluoro-11β,16α,17α,21-tetrahydroxy-, 21-valerate	EtOH	239(4.17)	13-0537-66B
$C_{26}H_{35}NO_3$			
5α-Androstan-3-one, 2-[3-(2-furyl)-allylidene]-17β-hydroxy-, oxime	EtOH	335(4.56)	4-0339-66
16,17-Seco-16-norandrostan-17-oic acid, 3β-hydroxy-15-(2-indolyl)-	MeOH	222(4.64),280(3.94), 290(3.84)	44-1363-66
$C_{26}H_{36}N_4O_4$			
2,6,10,14-Hexadecatetraenal, 3,7,11,15-tetramethyl-, 2,4-dinitrophenyl-hydrazone	EtOH	386(4.30)	39-2154-66C
$C_{26}H_{36}O_3$			
Pregn-5-en-20-one, 21-allylidene-, 3-acetoxy-	EtOH	265(4.38)	32-1125-66
$C_{26}H_{36}O_4$			
Pregna-3,5-dien-20-one, 17α-acetoxy-3-ethoxy-16-methylene-	EtOH	241(4.29)	73-2768-66
$C_{26}H_{36}O_5$			
Pregna-1,4-diene-3,20-dione, 11β,17α-dihydroxy-, 17-valerate	EtOH	244(4.18)	32-1115-66
Pregn-5-en-20-one, 3β,17α-diacetoxy-21-methylene-	EtOH	208(4.06),214(4.04)	78-1625-66
Unnamed C_{22}-ketone derivative	EtOH	236s(4.18),242(4.22), 250s(4.00)	78-1857-66
$C_{26}H_{36}O_6$			
5β-Pregn-7-ene-20-carboxaldehyde, 2β,3β-diacetoxy-6-oxo-	EtOH	246(4.15)	33-1601-66
Pregn-4-ene-3,20-dione, 17α-hydroxy-, methyl succinate	EtOH	242(4.23)	32-1115-66
$C_{26}H_{36}O_7$			
5β-Pregn-7-ene-20-carboxylic acid, 2β,3β-diacetoxy-6-oxo-	EtOH	246(4.17)	33-1601-66
Pregn-4-ene-3,20-dione, 17α-acetoxy-6β-(acetoxymethyl)-6α-hydroxy-	EtOH	241(4.15)	78-0365-66
$C_{26}H_{37}ClO_4$			
Pregna-5,20-diene-21-carboxaldehyde, 3β-acetoxy-20-(2-chloroethoxy)-	EtOH	265(4.30)	32-1254-66
Pregna-5,16,20-triene-21-methanol, 20-(2-chloroethoxy)-3β-hydroxy-, acetate	EtOH	240(3.72)	32-1254-66
$C_{26}H_{38}ClNO_2$			
19-Norandrost-4-en-3-one, 17α-chloro-ethynyl-17β-(diethylaminoethoxy)-	EtOH	238(4.22)	78-2829-66
$C_{26}H_{38}O_4S$			
5α-Estrane-3β,17β-diol, 17α-methyl-, 3-p-toluenesulfonate	MeOH	224.5(4.11)	87-0685-66

Compound	Solvent	$\lambda_{max}(\log \epsilon)$	Ref.
$C_{26}H_{38}O_5$			
Pregna-5,16-diene-21-carboxaldehyde, 3β-acetoxy-20-oxo-, 20-(dimethyl acetal)	EtOH	243(3.95)	32-1254-66
$C_{26}H_{38}O_6$			
Hydrocortisone, 17α,21-(methyl orthobutyrate)	EtOH	242(4.20)	32-1115-66
$C_{26}H_{38}O_8$			
Taxicin I, 5-deoxydihydro-, triacetate	EtOH	279(3.76)	39-1933-66C
$C_{26}H_{39}NO$			
Buxpsiine	n.s.g.	240(4.73),247(4.62), 255s(4.43)	88-0915-66
$C_{26}H_{40}N_2O_3$			
1H-Cyclopenta[5,6]naphtho[1,2-g]quinazoline-8,10(3bH,9H)-dione, tetradecahydro-1-hydroxy-11a,13a-dimethyl-7-pentyl-	EtOH	273(3.95)	32-0179-66
$C_{26}H_{40}O_5$			
2-Chromannonanoic acid, α,ε,2,5,7,8-hexamethyl-6-acetoxy-	EtOH	284.0(3.34)	73-2434-66
$C_{26}H_{41}NO$			
Cyclobuxoviridine	EtOH	269(3.97)	39-1412-66C
9,19-Cyclopregn-16-en-20-one, 3β-(dimethylamino)-4,4,14α-trimethyl-	EtOH	242(4.02)	44-0608-66
$C_{26}H_{41}N_3O_2$			
8H-Cyclopenta[5,6]naphtho[1,2-g]quinazolin-8-one, 10-amino-hexadecahydro-1-hydroxy-11a,13a-dimethyl-7-pentyl-	EtOH-HCl	220(4.11),297(4.11)	32-0179-66
	EtOH	219(4.15),285(3.90)	32-0179-66
$C_{26}H_{45}NO$			
6-Chromanamine, 2-methyl-2-(4,8,12-trimethyltridecyl)-	EtOH	304(3.46)	33-2297-66

Compound	Solvent	$\lambda_{max}(\log \epsilon)$	Ref.
$C_{27}H_{15}N_6P$			
Phosphorane, triphenyl[(tetracyano-2,4-cyclopentadien-1-ylidene)-hydrazono]-	MeCN	273(4.43),412(4.47)	35-4055-66
$C_{27}H_{16}$			
5H-Benzo[b]indeno[2,1-l]fluoranthene	C_6H_{12}	225(4.72),238s(4.63), 258(4.79),266(4.79), 274(4.73),292(4.46), 305(4.61),316(4.47), 352(4.13)	33-1850-66
$C_{27}H_{16}F_9N_2O_2P$			
Phosphine imide, N-(p-nitrophenyl)-P,P,P-tris(α,α,α-trifluoro-p-tolyl)-	EtOH	380(4.2)	65-1248-66
$C_{27}H_{17}F_9NP$			
Phosphine imide, N-phenyl-P,P,P-tris(α,α,α-trifluoro-p-tolyl)-	EtOH	270s(3.6)	65-1248-66
$C_{27}H_{18}$			
5H-Diindeno[1,2-a:1',2'-h]fluorene, 13,15-dihydro-	C_6H_{12}	262(4.87),310(4.13), 320(4.09),327(4.02), 335(4.20)	33-0997-66
5H-Diindeno[1,2-a:2',1'-i]fluorene, 10,15-dihydro-	C_6H_{12}	222s(4.54),243s(4.47), 253s(4.67),262s(4.85), 268(4.98),285s(4.67), 293(4.60),310s(3.51), 320s(3.35),326s(3.18)	33-1850-66
$C_{27}H_{18}ClNO_8$			
8,9-Dihydroxybenzo[a]quinolizinium perchlorate, dibenzoate	EtOH	275s(--),281(4.61), 310s(--),323(3.96), 338(4.16),355(4.23)	44-2616-66
8,10-Dihydroxybenzo[a]quinolizinium perchlorate, dibenzoate	EtOH	283(4.34),303(4.18), 327(4.09),345(3.82), 361(3.88)	44-2616-66
$C_{27}H_{18}Cl_2N_2O$			
7H-1-Pyrindine, 6-benzamido-7,7-bis-(o-chlorophenyl)-	MeOH	235(4.43)	78-0035-66
7H-1-Pyrindine, 6-benzamido-7,7-bis-(p-chlorophenyl)-	MeOH	235(4.47)	78-0035-66
7H-2-Pyrindine, 6-benzamido-7,7-bis-(o-chlorophenyl)-	MeOH	236(4.44)	78-0035-66
7H-2-Pyrindine, 6-benzamido-7,7-bis-(p-chlorophenyl)-	MeOH	232(4.46)	78-0035-66
$C_{27}H_{19}NO_4$			
Aporphine, 10-(benzyloxy)-1,2,9-trimethoxy-	EtOH	281(3.99),302(3.90)	12-0437-66
$C_{27}H_{19}NS$			
Thiazole, 2,4-bis(4-biphenylyl)-	C_6H_{12}	291(4.61)	22-2857-66
$C_{27}H_{20}$			
Fluorene, 9-(2,2-diphenylethylidene)-	EtOH	223(4.59),229(4.61), 248(4.49),257(4.64), 277(4.20),285(4.23),	39-1810-66C

(cont. on next page)

Compound	Solvent	λ_{max}(log ϵ)	Ref.
		301(4.16),314(4.18)	39-1810-66C
Fluorene, 9-(2,2-diphenylvinyl)-	EtOH	212(4.75),261(4.49), 291(3.83),303(3.90)	39-1810-66C
$C_{27}H_{20}BF_4NS$			
Dibenzotropylium tetrafluoroborate, phenothiazine complex	MeCN	542(3.75),665(2.91), 750(2.92),830(2.83)	88-0797-66
$C_{27}H_{20}BN$			
Pyridine, compound with phenylbis-(phenylethynyl)borane	EtOH	235(4.49),245(4.43)	22-1981-66
$C_{27}H_{20}Cl_2N_2$			
Dibenzo[b,f][1,5]diazocine, 2,8-dichloro-5,6-dihydro-5-methyl-6,12-diphenyl-	iso-PrOH	259(4.46),315(3.60)	44-3356-66
Indolo[3,2-b]indole, 3,8-dichloro-4b,5,9b,10-tetrahydro-5-methyl-4b,9b-diphenyl-	iso-PrOH	250(4.36),273s(4.04), 320(3.74)	44-3356-66
5,11(6H,12H)-Methanodibenzo[b,f][1,5]-diazocine, 2,8-dichloro-6,12-diphenyl-, cis-endo	iso-PrOH	258s(3.88),295s(3.30)	44-3356-66
cis-exo	iso-PrOH	255s(4.00),296(3.30)	44-3356-66
trans	iso-PrOH	255s(4.00),295(3.30)	44-3356-66
$C_{27}H_{20}N_2$			
2H-Imidazole, 2,2,4,5-tetraphenyl-	EtOH	260(4.09)	35-3826-66
4H-Imidazole, 2,4,4,5-tetraphenyl-	EtOH	246(4.30),257(4.28), 269(4.27)	35-3826-66
$C_{27}H_{20}N_2O$			
7H-1-Pyrindine, 6-benzamido-7,7-diphenyl-	MeOH	236(4.39)	78-0035-66
7H-2-Pyrindine, 6-benzamido-7,7-diphenyl-	MeOH	234(4.28)	78-0035-66
$C_{27}H_{20}N_2OS$			
2-Pyrazolin-5-one, 4-(2,6-diphenyl-4H-thiopyran-4-ylidene)-3-methyl-1-phenyl-	HOAc	250(4.40),345(4.13), 421(4.33),485s(4.20)	33-2046-66
$C_{27}H_{20}N_3O_8P$			
Acetic acid, picryl(triphenylphos-phoranylidene)-, methyl ester	$CHCl_3$	493(3.59)	44-1287-66
$C_{27}H_{20}N_4O$			
1,4-Cyclopentadiene-1-methanol, 3-(di-2-pyridylmethylene)-α,α-di-2-pyridyl-	MeOH	318(4.39)	44-2149-66
1,4-Cyclopentadiene-1-methanol, 3-(di-3-pyridylmethylene)-α,α-di-3-pyridyl-	MeOH	322(4.40)	44-2149-66
1,4-Cyclopentadiene-1-methanol, 3-(di-4-pyridylmethylene)-α,α-di-4-pyridyl-	MeOH	308(4.40)	44-2149-66
4-Pyridinemethanol, α-2-pyridyl-α-[6-(2-pyridyl)-6-(4-pyridyl)-2-fulvenyl]-	MeOH	313(4.38)	44-2149-66
Δ^1-1,2,4-Triazolin-3-one, 4-[diphenyl-methylene)amino]-5,5-diphenyl-	EtOH	252(4.08),322(3.96)	88-5815-66

Compound	Solvent	λ_{max}(log ε)	Ref.
$C_{27}H_{20}N_6O_6$			
5-Quinazolinecarboxylic acid, 2-[[m-(3-carbethoxy-5-hydroxy-1-phenyl-4-pyrazolyl)azo]phenyl]-3,4-dihydro-4-oxo-	pH 13	388(4.45)	7-1394-66
5-Quinazolinecarboxylic acid, 2-[[p-(3-carbethoxy-5-hydroxy-1-phenyl-4-pyrazolyl)azo]phenyl]-3,4-dihydro-4-oxo-	pH 13	415(4.53)	7-1394-66
$C_{27}H_{20}O_9$			
L-Ascorbic acid, 3,5,6-tribenzoate	EtOH	230(4.77)	94-1039-66
$C_{27}H_{21}NOSn$			
Tin, (8-quinolinolato)triphenyl-	C_6H_{12}	368(3.39)	101-0249-66B
	dioxan	361(3.42)	101-0249-66B
	95% dioxan	318(3.39)	101-0249-66B
$C_{27}H_{21}N_2O_6P$			
Acetic acid, (2,4-dinitrophenyl)-(triphenylphosphoranylidene)-, methyl ester	$CHCl_3$	453(3.99)	44-1287-66
$C_{27}H_{21}N_5O_2$			
Benzeneazoethene, 4-nitro-1',2'-diphenyl-2'-(p-tolylazo)-	EtOH	241(4.52),297(4.33), 420(4.35)	78-1309-66
$C_{27}H_{22}Cl_2N_2$			
Dibenzo[b,f][1,5]diazocine, 2,8-dichloro-5,6,11,12-tetrahydro-5-methyl-6,12-diphenyl-, cis	iso-PrOH	260s(3.78),290s(3.30)	44-3356-66
trans	iso-PrOH	258(4.20),310s(3.48)	44-3356-66
$C_{27}H_{22}N_4O_2$			
1,3-Cyclopentadiene-1,4-dimethanol, α,α,α',α'-tetra-2-pyridyl-	MeOH	262(4.32),267s(4.26)	44-2149-66
$C_{27}H_{24}AsClPd$			
Palladium, chloro(1-phenyl-π-allyl)-(triphenylarsine)-	$CHCl_3$	248(4.17),294(4.17), 352(3.90)	101-0578-66B
$C_{27}H_{24}BNO$			
Pyridine, compound with (p-anisyl)-(phenylethynyl)-p-tolylborane	EtOH	235(4.46),245(4.52)	22-1981-66
$C_{27}H_{24}ClPPd$			
Palladium, chloro(1-phenyl-π-allyl)-(triphenylphosphine)-	$CHCl_3$	242(4.24),290(4.18), 350(4.06)	101-0578-66B
$C_{27}H_{24}F_3N_2P$			
Phosphine imide, N-[p-(dimethylamino)-phenyl]-P,P-diphenyl-P-(α,α,α-trifluoro-p-tolyl)-	EtOH	270(4.2)	65-1248-66
Phosphine imide, P-[p-(dimethylamino)-phenyl]-P,P-diphenyl-N-(α,α,α-trifluoro-p-tolyl)-	EtOH	290(4.1)	65-1248-66
$C_{27}H_{24}NO_4P$			
Phosphorane, bis(p-methoxyphenyl)-(p-nitrobenzylidene)phenyl-	$C_2H_4Cl_2$	248(4.39),350(3.66), 515(4.37)	97-0068-66

Compound	Solvent	$\lambda_{max}(\log \epsilon)$	Ref.
$C_{27}H_{24}N_2O$			
Urea, 1,1,3-triphenyl-3-(2,4-xylyl)-	EtOH	262(4.29)	34-0436-66
$C_{27}H_{24}N_2O_3$			
Carbanilide, 4,4'-bis(benzyloxy)-	EtOH	260(3.83)	87-0444B-66
Urea, 1,1-dimethyl-3-(m-phenoxyphenyl)-3-(p-phenoxyphenyl)-	EtOH	259(4.29)	34-0436-66
$C_{27}H_{24}O_{10}$			
Lisetin, triacetate	dioxan	246(4.45),273s(4.27), 298(3.95),317(4.05)	78-0333-66A
$C_{27}H_{25}NO_8$			
2,3,4,5-Pyridinetetracarboxylic acid, 1,6-dihydro-1-phenyl-6-styryl-, tetramethyl ester	MeOH	251(4.48),287s(4.21), 395(4.03)	39-1121-66C
$C_{27}H_{25}N_2O_2P$			
Aniline, N,N-dimethyl-p-[(p-nitrobenzylidene)diphenylphosphoranyl]-	$C_2H_4Cl_2$	237s(3.88),275s(4.16), 295(4.32),520(4.49)	97-0068-66
$C_{27}H_{25}N_3O_5$			
1,3-Azulenedicarboxylic acid, 2-acetamido-6-(2-quinolylamino)-, diethyl ester	MeOH	230s(4.38),270(4.33), 351(4.66),427(4.51)	18-1954-66
$C_{27}H_{26}O_6$			
Scandenone, 4'-acetyl-	EtOH	273(4.21),350(3.12)	39-0701-66C
$C_{27}H_{26}O_{10}$			
Isopiscidone, triacetate	EtOH	285s(3.88)	78-0333-66A
$C_{27}H_{27}NO$			
Aziridine, 1-cyclohexyl-2-phenyl-3-p-phenylbenzoyl-, cis	C_6H_{12}	331(2.27)	44-1244-66
trans	C_6H_{12}	351(2.87)	44-1244-66
$C_{27}H_{27}N_7O$			
Tetraethylammonium salt of 5-[(2-hydroxy-1-naphthyl)azo]cyclopentadiene-1,2,3,4-tetracarbonitrile	MeCN	275(4.39),283(4.39), 313(3.98),476(4.35), 498(4.36)	35-4055-66
$C_{27}H_{28}BN$			
Pyridine, compound with bis(1-cyclohexen-1-ylethynyl)phenylborane	EtOH	225(4.30)	22-1981-66
$C_{27}H_{28}ClNO_2$			
Triethylamine, 2-[p-[3-(p-chlorophenyl)-2H-1-benzopyran-4-yl]-phenoxy]-	MeOH	242(4.40),284s(4.04), 304(4.06),322(4.07)	87-0516-66
$C_{27}H_{28}Cl_4O_4$			
Podocarpa-8,11,13-trien-15-oic acid, 13-(5,6,7,8-tetrachloro-2-methyl-1,4-benzodioxan-2-yl)-, methyl ester	EtOH	291(3.62),301(3.66)	78-0745-66
$C_{27}H_{28}N_2$			
Dipyrromethene, 4,4'-diethyl-5,5'-dimethyl-3,3'-diphenyl-	EtOH-N HCl	225(4.38),261(3.57), 385(3.86),490(5.04)	12-1871-66

Compound	Solvent	λ_{max}(log ε)	Ref.
Dipyrromethene, 4,4'-diethyl-5,5'-dimethyl-3,3'-diphenyl- (cont.)	EtOH	223(4.36),236(4.20), 457(4.56)	12-1871-66
$C_{27}H_{28}N_2O_4$			
Morphanthridine, 6-(3-N-benzyl-N-methylaminopropyl)-, hydrogen oxalate	EtOH	224s(4.31),244(3.95), 310(3.57)	4-0206-66
$C_{27}H_{28}N_4O_6$			
Dibenzofuran, 4,6-di-tert-butyl-2-(2,4-dinitrophenylazo)-8-methoxy-	C_6H_{12}	229(4.47),252(4.35), 274(4.34),289(4.32), 328(4.25),400(4.35)	39-0362-66C
$C_{27}H_{28}N_4O_{11}$			
10-Hydroxy-1,2,9-trimethoxy-N-methylaporphinium picrate	EtOH	284(4.16),315(4.18), 360(4.15)	12-0437-66
$C_{27}H_{28}O_5$			
Scandenone, dimethyl-	EtOH	225(4.25),276(4.63), 325(3.69)	39-0701-66C
Scandinone, methyl- (or dimethyl-osajin)	EtOH	222s(4.28),267(4.68), 320s(3.83)	39-0701-66
	EtOH	221s(4.36),267(4.72), 320(3.84)	39-0701-66C
$C_{27}H_{30}$			
17H-Cyclopenta[a]phenanthrene, 17-methyl-17-(1,4,5-trimethylhex-2-enyl)-	EtOH	222(4.84),243(4.58), 268(4.67),295(4.06), 307(4.19),320(4.18), 346(3.08)	5-0106-66J
$C_{27}H_{30}ClNO_2$			
Triethylamine, 2-[p-[3-(p-chloro-phenyl)-4-chromanyl]phenoxy]-, cis	MeOH	222(4.46),276(3.68), 284(3.61)	87-0516-66
$C_{27}H_{30}Cl_4O_4$			
Podocarpa-7,13-dien-15-oic acid, 13-(5,6,7,8-tetrachloro-2-methyl-1,4-benzodioxan-2-yl)-, methyl ester	EtOH	236(4.55),302(3.46)	78-0745-66
$C_{27}H_{30}N_2$			
Lobinaline, dehydro-	EtOH	258(3.62),263(3.65), 269(3.53)	44-3206-66
Pyridine, 2-[2,6-diphenyl-4-(1-pyrrolidinyl)cyclohexyl]-	MeOH	256(3.57),262(3.60), 269(3.47)	44-3213-66
$C_{27}H_{30}O_6$			
Isonorscandenin, dihydrodi-O-methyl-	EtOH	263(3.98),344(4.21)	39-0192-66C
$C_{27}H_{30}O_{15}$			
Kaempferol, 7-rhamnoglucosyl-	EtOH	268(4.25),325(4.04), 368(4.31)	22-3212-66
$C_{27}H_{30}O_{17}$			
3-Quercitol, gentiobioside	EtOH	257(4.31),360(4.25)	28D-1144-66A +28D-1141-66A
$C_{27}H_{31}BrN_4$			
b-Norbiladiene-ac, 1,19-dideoxy-1-bromo-2,18-diethyl-3,7,13,17,19-pentamethyl-	$CHCl_3$	269(4.15),418(4.57), 611(4.51),636s(4.45)	39-0098-66C

Compound	Solvent	λ max(log ε)	Ref.
$C_{27}H_{31}Cl_2N_3O_5$			
Formamidine, N-(3,7-dichloro-1,9-dibutoxycarbonyl)-4,6-dimethylphenoxazin-2-yl)-N',N'-dimethyl-	MeOH	235s(4.54),407(4.12)	44-3694-66
$C_{27}H_{31}IN_4S$			
5,6,7,8-Tetrahydro-2-phenyl-4-[(3-methyl-2-benzothiazolinylidene)-methyl]quinazolinium iodide	MeOH	447(4.89)	87-0758-66
$C_{27}H_{31}NO_4$			
Isoquinoline, 8-(benzyloxy)-6,7-dimethoxy-1-(4-methoxybenzyl)-2-methyl-1,2,3,4-tetrahydro-	pH 1	225s(4.36),275(3.38), 281(3.33)	33-1757-66
	iso-PrOH	228s(4.41),275(3.48), 283(3.44)	33-1757-66
Isoquinoline, 1-[4-(benzyloxy)-3-methoxybenzyl]-1,2,3,4-tetrahydro-6,7-dimethoxy-2-methyl-	EtOH	283(3.73)	12-0437-66
$C_{27}H_{32}N_2O$			
Urea, 1-octyl-1,3,3-triphenyl-	EtOH	263(4.18)	34-0436-66
$C_{27}H_{32}N_2O_4$			
Isoquinoline, 1-(6-amino-4-benzyloxy-3-methoxybenzyl)-1,2,3,4-tetrahydro-6,7-dimethoxy-2-methyl-	EtOH	290(3.68)	12-0437-66
$C_{27}H_{32}O_4$			
1-Anthroic acid, 9,10-dihydro-9,10-dioxo-5,6,7,8-tetrapropyl-	C_6H_{12}	241(4.27),269(4.49), 352(3.64)	33-0644-66
$C_{27}H_{32}O_5$			
Osajetin, tri-O-methyl-	EtOH	228(4.53),255(4.28), 259s(4.28),285s(3.91), 325s(3.27)	39-0192-66C
$C_{27}H_{32}O_6$			
Andirobin, deoxy-	MeOH	217(4.18),236(4.23)	39-0944-66C
$C_{27}H_{32}O_7$			
Cedrela odorata B	MeOH	212(4.15)	39-2127-66C
Cedrela odorata D	MeOH	209(4.08),288(4.42)	39-2127-66C
$C_{27}H_{32}O_8$			
Physodic acid, 4-O-methyl-	EtOH	211(4.67),260(4.15)	102-0815-66
$C_{27}H_{32}O_{13}$			
Flavanone, 5,7-dihydroxy-, 7-neohesperidoside	EtOH	286(4.25),330(3.45)	28C-1712-66A
$C_{27}H_{32}O_{15}$			
Flavanone, 3',4',5,7-tetrahydroxy-, 7-neohesperidoside	EtOH	286(4.29)	28C-1712-66A
$C_{27}H_{33}NO_4$			
16,17-Seco-16-nor-5-androsten-17-oic acid, 3β-hydroxy-15-(3-oxo-2-indolinylidene)-, methyl ester	MeOH	236(4.20),238(4.21), 262(4.43),275s(4.19), 295s(3.88),453(3.68)	44-1363-66

Compound	Solvent	λ_{max}(log ε)	Ref.
$C_{27}H_{34}$			
19-Norergosta-1,3,5(10),6,8,14,22-heptaene	EtOH	239(4.65),247(4.71), 256(4.85),265(4.82), 284(4.34),294(4.42), 306(4.35),330(2.96), 347(2.56)	5-0106-66J
$C_{27}H_{34}BrNO_4$			
16,17-Seco-16-norandrostan-17-oic acid, 3β-acetoxy-15-(5-bromo-3-oxo-2-indolinylidene)-, methyl ester	MeOH	245(4.31),263(4.39), 284(4.28),310s(--), 460(3.67)	44-1363-66
$C_{27}H_{34}Cl_4O_4$			
Podocarpan-15-oic acid, 13-(5,6,7,8-tetrachloro-2-methyl-1,4-benzodioxan-2-yl)-, methyl ester	EtOH	302(3.40)	78-0745-66
$C_{27}H_{34}N_2O_2$			
1H-Cyclopenta[5,6]naphtho[1,2-b]phenazin-1β-ol, 2,3,3aα,3bβ,4,5,5aα,6,-13,13α,13bα,14,15,15a-tetradecahydro-13aβ,15aβ-dimethyl-, acetate	EtOH	239(4.5),263(3.3), 321(4.07)	44-2395-66
$C_{27}H_{34}O_3$			
Mutilindione A, benzylidenehydroxy-	EtOH	295(4.32)	78-0359-66B
Mutilindione B, benzylidenehydroxy-	EtOH	295(4.31)	78-0359-66B
$C_{27}H_{34}O_7$			
Pregna-3,5,14-triene-6-carboxaldehyde, 17α,21-diacetoxy-3-methoxy-20-oxo-	EtOH	219(4.06),321(4.18)	78-0325-66A
Pregna-3,5,14-trien-20-one, 3,17α,21-triacetoxy-	EtOH	234(4.29)	78-0325-66A
$C_{27}H_{35}BF_2O_6$			
Boron, difluoro[2-acetyl-17β-hydroxy-6-(1-hydroxyethylidene)androst-4-en-3-onato]-, diacetate	EtOH	250(3.81),271(3.70), 377(4.06)	78-2039-66
$C_{27}H_{35}FO_7$			
Pregna-1,4-diene-3,20-dione, 9α-fluoro-11β,16α,17α,21-tetrahydroxy-, 16-cyclopentanecarboxylate	EtOH	239(4.19)	13-0537-66B
Pregna-1,4-diene-3,20-dione, 9α-fluoro-11β,16α,17α,21-tetrahydroxy-. 21-cyclopentanecarboxylate	EtOH	239(4.16)	13-0537-66B
$C_{27}H_{35}NO_2$			
1H-Cyclopenta[5,6]naphtho[1,2-g]quinolin-1-ol, 8-(2-furyl)-tetradecahydro-1,11a,13a-trimethyl-	EtOH	266(4.18),309(4.25)	4-0338-66
2,4,6,8,10,12-Tridecahexaenoic acid, 2-cyano-3,7,11-trimethyl-13-(2,6,6-trimethyl-1-cyclohexen-1-yl)-, methyl ester	EtOH	448(4.78)	54-0334-66
$C_{27}H_{35}NO_3$			
16,17-Seco-16-norandrostan-15-(2'-indoxylidene)-17-oic acid, methyl ester	EtOH	238(4.25),262(4.43), 275s(4.23),300s(3.90), 455(3.69)	44-1363-66

Compound	Solvent	λ_{max}(log ε)	Ref.
$C_{27}H_{35}N_3O_7$			
1-Lysine, colchicidyl-	pH 1	256(4.51),380(4.29)	65-0033-66
	EtOH	253(4.55),356(4.42), 410(4.13)	65-0033-66
$C_{27}H_{35}N_5O_7$			
1-Arginine, colchicidyl-	pH 1	256(4.40),375(4.28)	65-0033-66
	EtOH	255(4.38),356(4.25), 411(4.01)	65-0033-66
$C_{27}H_{36}ClFO_6$			
Pregna-3,5-diene-6-carboxaldehyde, 3-(2-chloroethoxy)-9α-fluoro-11β,21-dihydroxy-16α-methyl-20-oxo-, 21-acetate	EtOH	216(4.09),324(4.20)	31-0468-66
Pregna-3,5-diene-6-carboxaldehyde, 3-(2-chloroethoxy)-9α-fluoro-11β,16α,17α-trihydroxy-20-oxo-, 16α,17α-acetonide	EtOH	216(4.02),323(4.21)	31-0468-66
$C_{27}H_{36}Cl_2O_6$			
Pregna-3,5-diene-6-carboxaldehyde, 9α-chloro-3-(2-chloroethoxy)-11β,21-dihydroxy-16-methyl-20-oxo-, 21-acetate	EtOH	217(4.10),325(4.19)	31-0468-66
$C_{27}H_{36}N_4O_4$			
Androst-5-en-17-one, 4,4-dimethyl-, 2,4-dinitrophenylhydrazone	$CHCl_3$	371(4.14)	22-0859-66
$C_{27}H_{36}O_3$			
5α-Androstan-3-one, 2-[3-(2-furyl)-allylidene]-17β-hydroxy-17-methyl-	EtOH	360(4.55)	4-0339-66
$C_{27}H_{36}O_6$			
Testosterone, 2,6-diacetyl-, enol acetate	EtOH	243(3.91),281(3.74), 351(3.92)	78-2039-66
$C_{27}H_{36}O_9$			
Nigrescigenin, di-O-acetyl-	EtOH	217(4.21),303(1.4)	33-1844-66
$C_{27}H_{37}NO_3$			
5α-Androstan-3-one, 2-[3-(2-furyl)-allylidene]-17β-hydroxy-17-methyl-, oxime	EtOH	335(4.59)	4-0339-66
$C_{27}H_{38}ClNO_2$			
Triethylamine, 2-[(21-chloro-3-methoxy-19-nor-17-pregna-1,3,5(10)-trien-20-yn-17-yl)oxy]-, hydrochloride	EtOH	278(3.31),287(3.28)	78-2829-66
$C_{27}H_{38}N_4O_2$			
Bilene b, 2,3,7,8,12,13,17,18-octamethyl-2,3,17,18-tetrahydro-, hydrobromide	EtOH	232(4.08),288(3.30), 376(3.81),494(4.94)	39-1155-66C
free base	EtOH	227(4.07),346(3.55), 454(4.48)	39-1155-66C
$C_{27}H_{38}O_3$			
5β,25D-Spirosta-1,9(11)-dien-3-one	EtOH	229(3.95)	95-0558-66

Compound	Solvent	λ_{max}(log ϵ)	Ref.
$C_{27}H_{38}O_4$			
18-Norchola-7,9(11),13-trienoic acid, 3α-acetoxy-12β-methyl-, methyl ester	n.s.g.	225s(4.01),231(4.09), 238s(4.01),284(3.64)	88-3063-66
18-Norchola-8,11,13-trienoic acid, 3α-acetoxy-12-methyl-, methyl ester	isooctane	220(4.10),261(2.48), 268(2.51),275s(2.45)	88-3063-66
$C_{27}H_{38}O_6$			
Pregn-4-ene-3,20-dione, 17α-acetoxy-6α-hydroxy-6β-(hydroxymethyl)-, isopropylidenedioxy derivative	EtOH	242(4.13)	78-0365-66
$C_{27}H_{38}O_7$			
Alloglaucotoxigenin, deoxo-, diacetate	EtOH	215.2(4.145)	33-1662-66
5α-Pregn-7-ene-20-carboxylic acid, 2β,3β-diacetoxy-6-oxo-, methyl ester	EtOH	243(4.11)	33-1591-66
5β-Pregn-7-ene-20-carboxylic acid, 2β,3β-diacetoxy-6-oxo-, methyl ester	EtOH	247(4.16)	33-1591-66
14α-Uzarigenin, 3,15-di-O-acetyl-15α-hydroxy-	EtOH	217(4.175)	33-1662-66
$C_{27}H_{38}O_8$			
5α,14α-Pregn-7-ene-20-carboxylic acid, 2β,3β-diacetoxy-14-hydroxy-6-oxo-, methyl ester	EtOH	239(4.08)	33-1591-66
$C_{27}H_{40}O$			
19-Norcholesta-1,3,5(10),6-tetraen-3-ol, 4-methyl-	EtOH	261s(--),268(4.08), 278s(--),309(3.42)	94-0866-66
$C_{27}H_{40}O_3$			
A-Homo-B-nor-5β,25D-spirost-3-en-4a-one	EtOH	228(3.82)	78-1053-66
5β,25D-Spirost-9(11)-en-3-one	EtOH	205.7(3.60)	95-0558-66
$C_{27}H_{40}O_4$			
5β,25D-Spirost-1-en-3-one, 11α-hydroxy-	EtOH	240(3.99)	44-0734-66
$C_{27}H_{41}NO_2$			
Cyclobuxophyllinine, N-acetyl-	EtOH	243(3.93)	39-1412-66C
$C_{27}H_{41}O_6P$			
Phosphonic acid, (3β-hydroxy-17,22-dioxo-D-homo-24-norchola-5,17a(20)-dien-21-yl)-, isopropyl ester	pH 13	280.5(3.94)	78-1625-66
	EtOH	273(3.92)	78-1625-66
$C_{27}H_{42}$			
19-Norcholesta-1,3,5(10)-triene, 4-methyl-	hexane	263(2.45)	94-0866-66
$C_{27}H_{42}N_2O_4$			
Jervine, tetrahydro-N-nitroso-	n.s.g.	380(2.07)	78-3103-66
$C_{27}H_{42}O$			
2,4-Cholestadien-1-one	EtOH	324(3.76)	94-0873-66
19-Norcholesta-1,3,5(10)-trien-3-ol, 4-methyl-	EtOH	281(3.26),286s(--)	94-0866-66
$C_{27}H_{42}O_2$			
19-Nor-5β-cholest-8(9)-ene-4,7-dione, 5-methyl-	n.s.g.	254(3.99)	39-1010-66C

Compound	Solvent	$\lambda_{max}(\log \epsilon)$	Ref.
$C_{27}H_{42}O_3$			
2-Oxa-A-norcholest-5-ene-1,7-dione, 3,3-dimethyl-	EtOH	230(4.03)	39-1443-66C
$C_{27}H_{42}O_5$			
2-Chromannonanoic acid, 6-acetoxy-α,ε,2,5,7,8-hexamethyl-, methyl ester	EtOH	284.0(3.33)	73-2434-66
$C_{27}H_{43}BrO$			
Cholest-2-en-1-one, 4α-bromo-	hexane	351(1.71)	94-0873-66
	EtOH	223(3.94)	94-0873-66
Cholest-2-en-1-one, 4β-bromo-	hexane	350(1.79)	94-0873-66
	EtOH	224(3.98)	94-0873-66
$C_{27}H_{43}NO$			
5α,22αH,25βH-Solanid-16-en-3β-ol,	EtOH	225.5(3.40)	24-3183-66
perchlorate	EtOH	223(3.08)	24-3183-66
5α,25βH-Solanid-22-en-3β-ol,	EtOH	219(3.27),325(1.05)	24-3183-66
perchlorate			
$C_{27}H_{43}NO_2$			
6-Aza-B-homocholesta-2,4-dien-7-one, 6-hydroxy-	n.s.g.	237(3.70)	56-1911-66
$C_{27}H_{43}N_5O_5$			
Phytenal, 2,4-dinitrophenylhydrazone	EtOH	267(4.44),320(4.28)	39-2144-66C
$C_{27}H_{44}BrFO$			
3-Cholestanone, 2α-bromo-5α-fluoro-	C_6H_{12}	284(1.40)	22-1884-66
3-Cholestanone, 2β-bromo-5α-fluoro-	dioxan	310(1.99)	22-1884-66
$C_{27}H_{44}N_2O_3$			
Cholestane-2,3,4-trione, 2,4-dioxime	EtOH	249(3.87)	44-2755-66
	base	349(4.16)	44-2755-66
$C_{27}H_{44}N_2O_4$			
Soladulcidine, 15α-hydroxy-N-nitroso-	MeOH	235(3.53),371(1.73)	5-0169-66D
Tomatidine, 15α-hydroxy-N-nitroso-	MeOH	234(3.90),367(1.83)	5-0169-66D
$C_{27}H_{44}O$			
Cholest-2-en-4-one	EtOH	231(4.00)	5-0180-66B
$C_{27}H_{44}O_3$			
5α-Cholest-7-en-6-one, 2β,3β-dihydroxy-	EtOH	244(4.15)	33-1591-66
5β-Cholest-7-en-6-one, 2β,3β-dihydroxy-	MeOH	248(4.16)	33-1591-66
2-Oxa-A-nor-5β-cholestane-1,7-dione, 3,3-dimethyl-	EtOH	290(1.65)	39-1443-66C
$C_{27}H_{44}O_4$			
5α,14α-Cholest-7-en-6-one, 2β,3β,14-trihydroxy-	EtOH	240(4.07)	33-1591-66
5β,14α-Cholest-7-en-6-one, 2β,3β,14-trihydroxy-	EtOH	242(4.09)	33-1591-66
$C_{27}H_{44}O_6$			
Ecdysone	EtOH	242(4.08)	33-1601-66
Ecdysterone	n.s.g.	240(4.13)	89-0248-66
22-Isoecdysone	EtOH	244(4.07)	88-3457-66
Ponasterone D	MeOH	244(4.09),326(2.11)	77-0915-66

Compound	Solvent	λ_{max}(log ϵ)	Ref.
$C_{27}H_{44}O_7$			
Ecdysterone	EtOH	240(4.13)	88-4017-66
$C_{27}H_{45}FO$			
3-Cholestanone, 5α-fluoro-	dioxan	284(1.28)	22-1884-66
$C_{27}H_{45}NO$			
Cholest-2-en-4-one, oxime, anti	EtOH	227(3.88)	5-0180-66B
Cholest-2-en-4-one, oxime, syn	EtOH	230(4.00)	5-0180-66B
$C_{27}H_{45}NO_2$			
2,3-Cholestanedione, 2-oxime	dioxan	235(3.91)	44-2015-66
	dioxan-NaOH	310(4.20)	44-2015-66
$C_{27}H_{45}NO_3$			
Alkaloid from Lycopersicon pimpinellifolium	n.s.g.	none	102-0707-66
$C_{27}H_{46}$			
4-Cholestene	C_6H_{12}	192(4.04)	39-1266-66C
5-Cholestene	C_6H_{12}	190(3.92)	39-1266-66C
5α-Cholest-1-ene	C_6H_{12}	none	39-1266-66C
5α-Cholest-2-ene	C_6H_{12}	none	39-1266-66C
5α-Cholest-3-ene	C_6H_{12}	none	39-1266-66C
5α-Cholest-6-ene	C_6H_{12}	none	39-1266-66C
5α-Cholest-7-ene	C_6H_{12}	204.5(3.74)	39-1266-66C
5α-Cholest-14-ene	C_6H_{12}	193.5(3.89)	39-1266-66C
$C_{27}H_{46}N_2O$			
Cyclokoreanine B	EtOH	213(4.17)	39-1805-66C
$C_{27}H_{46}N_2O_2$			
3,4-Cholestanedione, dioxime	EtOH	226(3.80)	5-0180-66B
	0.05N NaOH	276.5(4.02)	5-0180-66B
$C_{27}H_{47}NO$			
δ-Tocopheramine	EtOH	236(3.86),303(3.47)	33-2297-66

Compound	Solvent	λ_{max}(log ε)	Ref.
$C_{28}H_{16}Cl_2O_2$			
Glyoxal, bis(9-chlorofluoren-9-yl)-	CH_2Cl_2	238(4.70),269s(4.30)	24-0205-66
$C_{28}H_{16}F_{12}NP$			
Phosphine imide, tetrakis(α,α,α-trifluoro-p-tolyl)-	EtOH	270(3.8)	65-1248-66
$C_{28}H_{16}N_2$			
Dibenzo[f,h]phenanthro[9,10-c]-cinnoline	$CHCl_3$	264(4.80),271(4.78), 333(4.18),353(4.28), 370(4.26),410(2.52)	78-2053-66
$C_{28}H_{17}N$			
Dinaphtho[2',3':3,4;2",3":5,6]carbazole	benzene	295(4.52),308(4.53), 323(4.06),338(4.09), 358(4.24),386(3.98), 408(4.26),432(4.44)	24-2449-66
$C_{28}H_{17}N_3$			
5,10,15-Triazabenzo[a]naphth[1,2,3-de]-anthracene, 5-phenyl-	EtOH	239(4.75),248(4.75), 284(4.55),316(4.04), 331(3.79),348(3.85), 376(3.97),396(4.21), 407(4.14),430(4.30), 456(4.22)	39-1245-66C
$C_{28}H_{18}$			
Biphenyl, 2,2'-bis(phenylethynyl)-	THF	287(4.69),296s(4.64), 305(4.65)	35-3027-66
isomer	THF	242(4.41),250(4.45), 262(4.46),281(4.72), 291(4.80),328(3.75), 343(3.71)	35-3027-66
$C_{28}H_{18}BrCl$			
Butatriene, 1-(m-bromophenyl)-4-(o-chlorophenyl)-1,4-diphenyl-	$CHCl_3$	405(4.55)	22-2885-66
Butatriene, 1-(m-bromophenyl)-4-(p-chlorophenyl)-1,4-diphenyl-	$CHCl_3$	278(--),427(4.59)	22-2885-66
Butatriene, 1-(o-bromophenyl)-4-(o-chlorophenyl)-1,4-diphenyl-	$CHCl_3$	398(4.52)	22-2885-66
Butatriene, 1-(o-bromophenyl)-4-(p-chlorophenyl)-1,4-diphenyl-	$CHCl_3$	276(--),426(4.41)	22-2885-66
Butatriene, 1-(p-bromophenyl)-4-(o-chlorophenyl)-1,4-diphenyl-	$CHCl_3$	270(--),408(4.54)	22-2885-66
Butatriene, 1-(p-bromophenyl)-4-(p-chlorophenyl)-1,4-diphenyl-	$CHCl_3$	282(--),429(4.56)	22-2885-66
$C_{28}H_{18}Fe_2$			
Octatetrayne, diferrocenyl-	$CHCl_3$	254(4.82),306(4.75), 345(4.13),375s(4.10), 470(3.85)	101-0399-66B
$C_{28}H_{18}N_2$			
Benzo[a][3,6]phenanthroline, 6,7-diphenyl-	EtOH	238(4.54),283(4.54), 385(3.23)	87-0161-66
$C_{28}H_{18}O$			
Anthrone, 10-(diphenylethenylidene)-	$CHCl_3$	275(4.25),305(3.88)	22-1999-66

Compound	Solvent	λ_{max}(log ϵ)	Ref.
$C_{28}H_{18}O_2$			
Glyoxal, difluoren-9-yl-	CH_2Cl_2	265(4.56),288s(3.89), 300(3.83),326s(3.12)	24-0205-66
$C_{28}H_{18}O_5$			
Benzoic acid, 2-benzoyl-, anhydride	$CHCl_3$	325(2.11)	35-0781-66
Benzoic acid, 4-benzoyl-, anhydride	$CHCl_3$	325(2.16)	35-0781-66
$C_{28}H_{19}Br$			
Butatriene, (o-bromophenyl)triphenyl-	$CHCl_3$	272(--),415(4.41)	22-2885-66
Butatriene, (p-bromophenyl)triphenyl-	$CHCl_3$	277(--),424(4.04)	22-2885-66
$C_{28}H_{19}BrN_2O_3$			
1,3-Propanedione, 2-benzoyl-2-[(p-bromophenyl)azo]-1,3-diphenyl-	CH_2Cl_2	245(4.52),359(4.42)	35-4637-66
$C_{28}H_{19}Cl$			
Anthracene, 9-benzylidene-10-(o-chlorophenyl)-9,10-dihydro-	EtOH	220(4.45),242(4.32), 330(4.40)	78-0141-66B
Anthracene, 9-benzylidene-10-(p-chlorophenyl)-9,10-dihydro-	EtOH	250(4.35),330(4.40)	78-0141-66B
	$CHCl_3$	340(4.40)	78-0141-66B
Butatriene, (o-chlorophenyl)-triphenyl-	$CHCl_3$	270(--),414(4.43)	22-2885-66
Butatriene, (p-chlorophenyl)-triphenyl-	$CHCl_3$	277(--),425(4.42)	22-2885-66
$C_{28}H_{19}ClO_5$			
9-Benzyl-1,2:5,6-dibenzoxanthylium perchlorate	HOAc	484(3.31)	24-1822-66
9-Benzyl-1,2:6,7-dibenzoxanthylium perchlorate	HOAc	486(3.38)	24-1822-66
9-Benzyl-2,3:5,6-dibenzoxanthylium perchlorate	HOAc	480(3.16),570(3.48)	24-1822-66
9-Benzyl-3,4:5,6-dibenzoxanthylium perchlorate	HOAc	478(3.22)	24-1822-66
$C_{28}H_{19}ClO_6$			
9-(p-Methoxyphenyl)-1,2:5,6-dibenzo-xanthylium perchlorate	HOAc	474(3.48)	24-1822-66
9-(p-Methoxyphenyl)-1,2:6,7-dibenzo-xanthylium perchlorate	HOAc	478(3.51)	24-1822-66
9-(p-Methoxyphenyl)-2,3:5,6-dibenzo-xanthylium perchlorate	HOAc	460(3.21),530(3.69)	24-1822-66
9-(p-Methoxyphenyl)-3,4:5,6-dibenzo-xanthylium perchlorate	HOAc	470(3.42)	24-1822-66
$C_{28}H_{19}NO_2$			
Butatriene, (p-nitrophenyl)triphenyl-	$CHCl_3$	274(--),453(4.32)	22-2885-66
$C_{28}H_{20}$			
Anthracene, 9,10-dibenzylidene-9,10-dihydro-	EtOH	220(4.78),250s(4.38), 380(4.35)	78-0141-66B
Butatriene, tetraphenyl-	$CHCl_3$	270(--),420(4.60)	22-2885-66
$C_{28}H_{20}BrClO_2$			
2-Butyne-1,4-diol, 1-(m-bromophenyl)-4-(p-chlorophenyl)-1,4-diphenyl-	EtOH	223(4.36),253(--), 259(3.85)	22-2885-66
2-Butyne-1,4-diol, 1-(p-bromophenyl)-4-(p-chlorophenyl)-1,4-diphenyl-	EtOH	226(4.34),252(--), 258(3.90)	22-2885-66

Compound	Solvent	λ_{max}(log ε)	Ref.
$C_{28}H_{20}Br_2$			
Anthracene, 9,10-bis(α-bromobenzyl)-	dioxan	415(4.00)	18-1551-66
$C_{28}H_{20}Br_2O_2$			
2-Butyne-1,4-diol, 1,4-bis(m-bromophenyl)-1,4-diphenyl-	EtOH	220(4.32),252(--), 259(3.85)	22-2885-66
$C_{28}H_{20}Cl_2N_2O$			
Dibenzo[b,f][1,5]diazocine, 2,8-dichloro-5,6-dihydro-6,12-diphenyl-, acetyl derivative	iso-PrOH	263(4.37),331(3.30)	44-3356-66
$C_{28}H_{20}Cl_2O_2$			
2-Butyne-1,4-diol, 1,4-bis(p-chlorophenyl)-1,4-diphenyl-	EtOH	225(4.26),253(--), 259(3.85)	22-2885-66
$C_{28}H_{20}N_2$			
Dinaphtho[2,3-b:2',3'-i]phenazine, 5?,7,16,18?-tetrahydro-	MeCN	262(4.85),269(4.72), 318(3.83)	44-3734-66
$C_{28}H_{20}O$			
1-Naphthol, 2,3,4-triphenyl-	EtOH	243(4.73),260(4.58), 310(3.95)	18-1975-66
$C_{28}H_{20}O_2$			
Cyclobuta[1,2-b:4,3-b']bisbenzofuran, 5a,5b,10b,10c-tetrahydro-5a,5b-diphenyl-	dioxan	282(3.89),290(3.80)	24-1723-66
1-Naphthol, 4-phenoxy-2,3-diphenyl-	EtOH	246(4.51),325(3.79)	35-3064-66
	EtOH	245(4.51),317(3.78)	35-3064-66
$C_{28}H_{21}BrO_2$			
2-Butyne-1,4-diol, 1-(o-bromophenyl)-1,4,4-triphenyl-	EtOH	213(4.43),253(--), 259(4.00)	22-2885-66
$C_{28}H_{21}ClO_2$			
2-Butyne-1,4-diol, 1-(o-chlorophenyl)-1,4,4-triphenyl-	EtOH	219(4.42),253(--), 259(3.90)	22-2885-66
$C_{28}H_{22}BN$			
Pyridine, compound with bis(phenylethynyl)-o-tolylborane	EtOH	235(4.58),245(4.48)	22-1981-66
Pyridine, compound with bis(phenylethynyl)-p-tolylborane	EtOH	235(4.44),245(4.48)	22-1981-66
$C_{28}H_{22}BNO$			
Pyridine, compound with p-anisylbis-(phenylethynyl)borane	EtOH	235(3.57),245(3.51)	22-1981-66
$C_{28}H_{22}Cl_2FeN_6O_8$			
Phenanthrolineferrous perchlorate, acrylonitrile complex	H_2O	225(4.82),263(4.86), 290s(4.42),380(3.76), 415(3.74)	7-0298-66
$C_{28}H_{22}Cl_2N_2$			
Indolo[3,2-b]indole, 3,8-dichloro-4b,5,9b,10-tetrahydro-5,10-dimethyl-4b,9b-diphenyl-	iso-PrOH	254(4.34),280s(4.04), 337(3.70)	44-3356-66

Compound	Solvent	$\lambda_{max}(\log \epsilon)$	Ref.
$C_{28}H_{22}N_4O$			
o-Aminobenzaldehyde, trisanhydro-, tetramer	MeOH	210(4.64),237(4.51), 267(4.04),286(3.43), 369(3.84)	23-2323-66
	EtOH	230s(4.57),236(4.59), 259s(4.19),268s(4.10), 286(3.57),370(3.82)	39-0956-66B
	5N HCl	220(4.26),246s(4.04), 307(3.93),467(3.26)	39-0956-66B
$C_{28}H_{22}N_6$			
1H-Indazole, 1,1'-(azodi-p-phenylene)-bis[3-methyl-	EtOH	241(4.07),392(4.17)	78-3131-66
$C_{28}H_{22}N_6O$			
1H-Indazole, 1,1'-(azoxy-di-p-phenylene)bis[3-methyl-	EtOH	308(3.94),400(4.62)	78-3131-66
$C_{28}H_{22}O_2$			
2-Butyne-1,4-diol, 1,1,4,4-tetraphenyl-	EtOH	218(4.34),252(--), 258(3.85)	22-2885-66
$C_{28}H_{22}O_3$			
3-Butenoic acid, 4-phenoxy-2,3,4-triphenyl-	EtOH	252(4.23),277s(3.94)	35-3064-66
$C_{28}H_{22}Sn$			
Tin, 1-naphthyltriphenyl-	THF	274(3.88),285(3.93), 296(3.83)	35-3027-66
$C_{28}H_{23}ClN_2O_5S$			
4-Antipyryl-2,6-diphenylthiopyrylium perchlorate	HOAc	260(4.39),385(4.29), 462(4.46)	33-2046-66
$C_{28}H_{23}N$			
Divinylamine, 2,2,2',2'-tetraphenyl-	dioxan	360(4.6)	83-0493-66
Photocyclization products of 9,9',-10,10'-tetrahydrodi-2-anthrylamine	EtOH	244(4.53),268(4.26), 292(4.18),300(4.26), 323(3.80),342(3.60)	24-2449-66
$C_{28}H_{23}OP$			
Phosphonium, [4-(p-hydroxyphenyl)-1,3-butadienyl]triphenyl-, hydroxide, inner salt	$CHCl_3$	267(3.97),375(3.90), 510(4.25+)	30-1042-66
$C_{28}H_{24}Cl_2N_2$			
Dibenzo[b,f][1,5]diazocine, 2,8-dichloro-5,6,11,12-tetrahydro-5,11-dimethyl-6,12-diphenyl-, cis	iso-PrOH	225s(4.45),275(3.78), 305s(3.40)	44-3356-66
trans	iso-PrOH	262(4.08),287s(3.88)	44-3356-66
$C_{28}H_{24}N_2O_4$			
Hydroquinone, 2,5-bis(2-methylindol-3-yl)-, diacetate	EtOH	225(4.80),283(4.39), 290(4.40),301s(4.30)	44-3321-66
Unnamed heterocycle	EtOH	234(4.37),251s(4.19), 285(3.68),311(3.90)	44-3321-66
Uridine, 2',3'-didehydro-2',3'-dideoxy-5'-O-trityl-	EtOH	261(3.99)	44-0205-66

Compound	Solvent	λ max (log ε)	Ref.
$C_{28}H_{24}O_6$			
1,4-Ethanonaphthalene-2,5(1H,3H)-dione, hexahydro-3,6-dipiperonylidene-	EtOH	331(4.56)	44-0639-66
$C_{28}H_{24}O_9$			
α-D-arabino-Hexopyranosidulose, methyl, tribenzoate	n.s.g.	231(4.57),273(3.48)	39-1131-66C
$C_{28}H_{25}IN_4S$			
5,6,7,8-Tetrahydro-4-[(3-methyl-2-benzothiazolinylidene)methyl]-2-phenyl-1-(3-pyridyl)quinazolinium iodide	MeOH	452(4.91)	87-0758-66
$C_{28}H_{26}F_6N_2O_8$			
5H-Oxazolo[4,5-b]phenoxazine-4,6-dicarboxylic acid, 8-hydroxy-2-(trifluoromethyl)-9,11-dimethyl-, dibutyl ester, 8-(trifluoroacetate)	MeOH	228(4.55),252(4.49), 434(4.24)	44-3694-66
$C_{28}H_{26}NO_5P$			
Phosphorane, tris(p-methoxyphenyl)-(p-nitrobenzylidene)-	$C_2H_4Cl_2$	250(4.47),350(3.68), 517(4.39)	97-0068-66
$C_{28}H_{26}O_2$			
Benzene, 2,5-dimethoxy-1,4-bis-(4-phenyl-1,3-butadienyl)-	dioxan	419(4.51)	18-1547-66
Pentatriafulvalene, 1,2-dibenzoyl-5,6-dipropyl-	MeOH	260(4.32),361(4.35)	35-3359-66
Pentatriafulvalene, 2,3-dibenzoyl-5,6-dipropyl-	MeOH	252(4.32),310(4.48), 331(4.50)	35-3359-66
$C_{28}H_{27}N_3O_7$			
Alanine, N-[p-(benzyloxy)-α-[2-(o-nitrophenyl)acetamido]cinnamoyl]-, methyl ester	EtOH	301(4.46)	7-1192-66
$C_{28}H_{28}BrOP$			
(3,7-Dimethyl-8-oxo-2,4,6-octatrienyl)-triphenylphosphonium bromide	EtOH	315(4.68)	33-0369-66
$C_{28}H_{28}CuN_6S_2$			
Copper, bis[4'-(dimethylamino)thiopicolinanilidato]-	benzene	364(4.03),456(4.43)	65-1499-66
	heptane	260(4.52),440(4.34)	65-1499-66
	EtOH	258(4.44),436(4.13)	65-1499-66
	$CHCl_3$	316(4.01),452(4.37)	65-1499-66
$C_{28}H_{28}Fe_2$			
Ferrocene, 1,1'-divinyl-, dimer?	EtOH	213(4.44),273(3.88)	39-0251-66C
$C_{28}H_{28}N_2OSi$			
Urea, 1,1,3-triphenyl-3-[(p-trimethylsilyl)phenyl]-	EtOH	270(4.35)	34-0436-66
$C_{28}H_{28}N_2O_2$			
Bisindolo[2,3-b:2',3'-b']benzo[1,2-d:-4,5-d']difuran, 5a,7b,12a,14b-tetrahydro-5,5a,7b,12,12a,14b-hexamethyl-	EtOH	231s(4.17),250s(3.95), 297s(3.71),323(3.94)	44-3321-66
Indole, 2,2'-(2,5-dimethoxy-p-phenylene)bis[1,3-dimethyl-	EtOH	230(4.86),252s(4.33), 303s(4.27),327(4.39)	44-3321-66

Compound	Solvent	λ_{max}(log ε)	Ref.
Indole, 3,3'-(2,5-dimethoxy-p-phenylene)bis[1,2-dimethyl-	EtOH	231(4.45),284(3.87), 315(3.94)	44-3321-66
3H-Indolium, 2-[[2-hydroxy-4-oxo-3-[(1,3,3-trimethyl-2-indolinylidene)-methyl]-2-cyclobuten-1-ylidene]-methyl]-1,3,3-trimethyl-, hydroxide, inner salt	EtOH	277(4.34),287s(4.27), 305(3.88),332(3.79), 376(3.56),402s(3.09), 588s(4.98),625(5.66)	5-0153-66I
$C_{28}H_{28}N_2O_5$			
Alanine, N-[p-(benzyloxy)-α-(2-phenylacetamido)cinnamoyl]-, methyl ester	EtOH	301(4.70)	7-1192-66
$C_{28}H_{28}N_2O_{10}$			
Quinoxaline, 2-(β-D-glucopyranosyloxy)-3-phenyl-, tetraacetate	MeOH	244(4.31),255s(4.22), 305s(3.90),335(4.10)	24-0547-66
$C_{28}H_{28}O_3S_2$			
D-Ribose, 2-deoxy-5-O-trityl-, ethylene mercaptal	EtOH	230(3.91),253(2.95), 258(2.94),264s(--), 269s(--)	87-0791-66
$C_{28}H_{28}Si_2$			
9,18-Disilatetrabenzo[a,c,f,h]cyclodecene, 9,18-diethyl-9,18-dihydro-	n.s.g.	270f(3.9)	65-0114-66
$C_{28}H_{30}BN$			
Pyridine, compound with bis(1-cyclohexen-1-ylethynyl)-p-tolylborane	EtOH	225(4.51)	22-1981-66
Pyridine, compound with diphenyl-(2,6,6-trimethyl-1-cyclohexen-1-yl)-ethynylborane	EtOH	233(4.18)	22-1981-66
$C_{28}H_{30}ClNO_7$			
m-Toluamide, α-[(7-chloro-2',6-dimethoxy-6'-methyl-3,4'-dioxospiro[benzofuran-2(3H),1'-[2]cyclohexen]-4-yl)-oxy]-N,N-diethyl-	EtOH	214(4.55),235(4.47), 292(4.34),330(3.76)	87-0242-66
$C_{28}H_{30}N_2$			
Benzo[f]quinoline, 1,2,3,4-tetrahydro-3-isopropyl-2,2-dimethyl-1-(2-naphthylamino)-	MeOH	210(4.6),250(4.9), 280(4.3),350(3.7)	30-0110-66A
$C_{28}H_{30}N_2O$			
5,16-Androstadien-3β-ol, 17-[3(5)-pyrazolyl]-	EtOH	247(4.04)	44-3193-66
$C_{28}H_{30}O_6$			
Lonchocarpic acid, di-O-methyl-	EtOH	233(4.24),270(4.29), 349(4.06)	39-0192-66C
Scandenin, di-O-methyl-	EtOH	237(4.49),290(4.11), 343(4.24)	39-0192-66C
$C_{28}H_{30}O_{10}$			
Isoderrisic acid, methyl ester, enol, diacetate	EtOH	229(4.41),243(4.41), 259s(4.36),280(4.17), 315(4.12)	39-0550-66C
	EtOH-KOH	242s(4.46),260s(4.34), 281s(4.15),320(4.05)	39-0550-66C

Compound	Solvent	$\lambda_{max}(\log \epsilon)$	Ref.
$C_{28}H_{30}O_{12}$			
Granaticin B	EtOH	223(4.42),285(3.68), 498s(3.71),527(3.76), 566(3.57)	33-1736-66
$C_{28}H_{31}BrN_2O_8$			
3H-Phenoxazine-1,9-dicarboxylic acid, 2-(diacetylamino)-7-bromo-4,6-dimethyl-3-oxo-, dibutyl ester	MeOH	225(4.25),258(4.13), 373(4.12),493(3.98)	44-3694-66
$C_{28}H_{31}ClO_2$			
Estra-1,3,5(10)-triene, 17β-(benzyloxy)-17α-chloroethynyl-3-methoxy-	EtOH	278(3.30),287(3.28)	78-2829-66
$C_{28}H_{31}NO$			
2H-Naphtho[2',1':4,5]indeno[1,2-b]pyridin-2-one, dodecahydro-4a,6a-dimethyl-8-phenyl-	EtOH	245(4.45),288(4.20)	4-0339-66
$C_{28}H_{32}B_2N_4O_8$			
4,4'-Dihydroxyvinylenebis[1-methylpyridinium]dihydroxy[1,2-bis(1-methyl-4(1H)-pyridylidene)-1,2-ethanediolato(2-)]borate, hydroxide, dihydrogen borate, inner salt	EtOH	221(4.9),267(4.0), 379(2.4),400(2.4), 421(2.5),452(2.4), 470s(2.2)	83-0531-66
$C_{28}H_{32}N_2O$			
5,16-Androstadien-3β-ol, 17-(1-methyl-5-pyrazolyl)-	EtOH	241(4.02)	44-3193-66
$C_{28}H_{32}N_2O_2$			
Aniline, [(2,5-dimethoxy-p-phenylene)-divinylene]bis[N,N-dimethyl-	dioxan	420(4.62)	18-1547-66
Hydroquinone, 2,5-bis(1-ethyl-3-methylindolin-3-yl)-	EtOH	254(4.15),301(3.96)	44-3321-66
Indole, 1-[2-(diethylamino)ethyl]-2,3-bis(p-methoxyphenyl)-	EtOH	239(4.49),299(4.19)	87-0527-66
$C_{28}H_{32}N_2O_8$			
Phenoxazine-1,9-dicarboxylic acid, 2-(diacetylamino)-4,6-dimethyl-3-oxo-, dibutyl ester	MeOH	226s(4.30),260s(4.18), 383(4.15)	44-3694-66
$C_{28}H_{32}N_6NiO_6$			
Nitratobis(6,6'-bi-2,4-lutidine)-nickel nitrate	EtOH-CHCl	603(1.30),795(0.60)	12-0201-66
$C_{28}H_{32}O_{16}$			
Isorhamnetin-3-β-rutinoside	MeOH	255(4.30),359(4.29)	100-0225-66
Narcissin	MeOH	255(4.3),356(4.2)	24-1384-66
Quercetin, 4'-methyl-7-rhamnoglucosyl-	EtOH	255(4.35),375(4.23)	22-3212-66
$C_{28}H_{32}O_{17}$			
Flavone-3-β-D-(4-O-β-D-glucopyranosyl-D-glucopyranose), 3,4',5,7-tetrahydroxy-3'-methoxy-	MeOH	255(4.3),356(4.2)	24-1384-66
Isorhamnetin, 3,7-diglucosyl-	EtOH	256(4.36),270s(--), 300(3.85),363(4.24)	22-0987-66

Compound	Solvent	λ_{max}(log ε)	Ref.
$C_{28}H_{33}BrN_2O_8$			
Phenoxazine-1,9-dicarboxylic acid, 2-(diacetylamino)-7-bromo-4,6-dimethyl-3-hydroxy-, dibutyl ester	MeOH	237s(4.52),403(4.15)	44-3694-66
$C_{28}H_{33}BrO_7$			
Gedunin, 2-bromo-	MeOH	210(3.90),255(3.86)	39-0506-66C
$C_{28}H_{33}ClN_2O_8$			
Phenoxazine-1,9-dicarboxylic acid, 2-(diacetylamino)-7-chloro-4,6-dimethyl-3-hydroxy-, dibutyl ester	MeOH	235(4.46),400(4.04)	44-3694-66
$C_{28}H_{33}NO$			
1H-Naphth[2',1':4,5]indeno[1,2-b]-pyridin-2-ol, 2,3,4,4a,4b,5,6,6a,-11,11a,11b,12-dodecahydro-4a,6a-dimethyl-8-phenyl-	EtOH	249(4.08),289(4.15)	4-0339-66
$C_{28}H_{33}NO_3$			
15,17-Seco-16-norandrost-5-en-17-oic acid, 3β-acetoxy-15-(2-indolyl)-, lactam	MeOH	243(4.45),265(4.15), 290(3.83),300(3.81)	44-1363-66
$C_{28}H_{33}NO_5$			
Androst-5-en-17-one, 3β-acetoxy-16-(o-nitrobenzylidene)-	MeOH	260(4.23),320s(3.70)	44-1363-66
$C_{28}H_{34}ClNO_2$			
Androsta-5,16-dien-3β-ol, 17-chloro-16-(N-phenylformimidoyl)-, acetate	EtOH	248(3.93),384(4.31)	32-1241-66
$C_{28}H_{34}N_2O$			
Androsta-5,16-dien-3β-ol, 17-(1-phenyl-5-pyrazolyl)-	EtOH	246(3.99)	44-3193-66
Androst-4-en-3-one, 17β-(1-phenyl-5-pyrazolyl)-	EtOH	237(4.38)	44-3193-66
$C_{28}H_{34}N_2O_2$			
Androsta[17,16-c]pyrazol-3β-ol, 2'-phenyl-, acetate	EtOH	272(4.32)	32-1241-66
$C_{28}H_{34}N_2O_8$			
Phenoxazine-1,9-dicarboxylic acid, 2-(diacetylamino)-3-hydroxy-4,6-dimethyl-, dibutyl ester	MeOH	213s(4.48),230s(4.52), 397(4.08)	44-3694-66
$C_{28}H_{34}N_4$			
2H-Pyrrole, 3,3',4,4'-tetramethyl-2,2'-bis[(3,4,5-trimethylpyrrol-2-yl)-methylene]-	EtOH	262(4.15),333(3.87), 610(4.60)	39-0098-66C
	$CHCl_3$-HBr	403s(4.63),438(4.78), 663(4.50)	39-0098-66C
$C_{28}H_{34}N_4Pd$			
Dipyrromethene, 3,3',4,4',5-pentamethyl-, palladium complex	$CHCl_3$	401(4.32),499(4.77)	39-0098-66C
$C_{28}H_{34}O_2$			
Androst-5-en-17-one, 16-cinnamylidene-3β-hydroxy-	EtOH	235(3.79),333(4.60)	4-0339-66

Compound	Solvent	λ_{max}(log ϵ)	Ref.
$C_{28}H_{34}O_4$			
Pregn-4-ene-3,20-dione, 17α-hydroxy-, benzoate	EtOH	236(4.44)	32-1115-66
$C_{28}H_{34}O_5$			
Pregn-4-ene-3,20-dione, 17α,21-dihydroxy-, 17-benzoate	EtOH	236(4.45)	32-1115-66
$C_{28}H_{34}O_7$			
Isogedunin	MeOH	218(4.05)	39-0506-66C
$C_{28}H_{34}O_8$			
Gedunin, epoxy-	MeOH	211(3.78)	39-0506-66C
$C_{28}H_{34}O_{15}$			
Flavanone, 3',5,7-trihydroxy-4'-methoxy-, 7-neohesperidoside	EtOH	286(4.27)	28C-1712-66A
Flavanone, 4',5,7-trihydroxy-3'-methoxy-, 7-neohesperidoside	EtOH	286(4.31)	28C-1712-66A
$C_{28}H_{35}NO$			
1H-Cyclopenta[5,6]naphtho[1,2-g]quinolin-1-ol, tetradecahydro-11a,13a-dimethyl-8-phenyl-	EtOH	248(4.15),287(4.12)	4-0339-66
$C_{28}H_{35}NO_2$			
Androst-5-en-17-one, 16-cinnamylidene-3β-hydroxy-, oxime	EtOH	233(3.91),329(4.61)	4-0339-66
$C_{28}H_{35}NO_3$			
1H-Cyclopenta[5,6]naphtho[1,2-g]quinolin-1-ol, 8-(2-furyl)tetradecahydro-11a,13a-dimethyl-, acetate	EtOH	266(4.18),309(4.26)	4-0339-66
15,17-Seco-16-norandrostan-17-oic acid, 3β-acetoxy-15-(2-indolyl)-, lactam	MeOH	243(4.41),266(4.08), 293(3.74),320(3.74)	44-1363-66
$C_{28}H_{35}NO_4$			
Pregn-4-ene-3,20-dione, 4-hydroxy-, carbanilate	EtOH	239(4.18)	22-3410-66
$C_{28}H_{35}N_3O_3$			
Tubulosine, demethyl-	pH 1	275(4.07)	88-1081-66
	pH 13	283(4.02),305s(3.93), 322s(3.65)	88-1081-66
	MeOH	279(4.10)	88-1081-66
$C_{28}H_{35}N_3O_7$			
Ostreogrycin A	EtOH-HCl	303(4.20)	39-1653-66C
	EtOH	228(4.51),272s(4.00)	39-1653-66C
	EtOH-NaOH	293(4.34)	39-1653-66C
$C_{28}H_{36}N_2O$			
Androst-5-en-3β-ol, 17β-(1-phenyl-5-pyrazolyl)-	EtOH	230(3.96)	44-3193-66
$C_{28}H_{36}N_2O_4$			
Psychotrine	pH 1	240(4.15),288(3.76), 306(3.80),356(3.83)	35-4068-66
Psychotrine, 6'-O-methyl-7'-demethyl-	pH 1	240(4.23),288(3.88), 304(3.96),357(3.90)	35-4068-66

Compound	Solvent	$\lambda_{max}(\log \epsilon)$	Ref.
$C_{28}H_{36}O_2$			
5α-Androstan-3-one, 2-cinnamylidene-17β-hydroxy-	EtOH	233(3.95),337(4.55)	4-0339-66
$C_{28}H_{36}O_4$			
5α-Androstan-17-one, 16-benzoyl-3β-hydroxy-, acetate	C_6H_{12}	230(3.83),313(4.20)	78-3607-66
	MeOH	246(3.92),315(4.06)	78-3607-66
$C_{28}H_{36}O_6S$			
Androst-5-en-17-one, 3β,19-dihydroxy-, 3-acetate, 19-p-toluenesulfonate	EtOH	226(4.09)	44-0693-66
$C_{28}H_{36}O_8$			
Pregna-4,6-diene-3,20-dione, 2α,17α-diacetoxy-6-(acetoxymethyl)-	EtOH	280(4.33)	78-0365-66
Pregna-4,6-diene-3,20-dione, 2β,17α-diacetoxy-6-(acetoxymethyl)-	EtOH	286(4.36)	78-0365-66
$C_{28}H_{37}BF_2O_6$			
Boron, difluoro[2-acetyl-17-hydroxy-6-(1-hydroxyethylidene)-4-methyl-androst-4-en-3-onato]-, diacetate	EtOH	257(4.05),272(4.04), 364(4.04)	78-2039-66
$C_{28}H_{37}FO_8$			
Pregna-1,4-diene-3,20-dione, 9α-fluoro-11β,16α,17α,21-tetrahydroxy-, 16-acetate 21-valerate	EtOH	239(4.16)	13-0537-66B
$C_{28}H_{37}NO_2$			
5α-Androstan-3-one, 2-cinnamylidene-17β-hydroxy-, oxime	EtOH	231(4.00),322(4.56)	4-0339-66
$C_{28}H_{37}NO_3$			
D-Homo-18-noretiocholan-17a-one, 17-furfurylidene-3α-hydroxy-13α-(2-cyanopropyl)-	EtOH	328(4.35)	78-0541-66B
$C_{28}H_{37}N_3O_7$			
Ostreogrycin G	EtOH	215(4.53)	39-1856-66C
	EtOH-NaOH	235(4.18),295(4.06)	39-1856-66C
$C_{28}H_{38}N_2O_2$			
Androsta-2,4-diene[4,5,6-cd]pyridin-17β-ol, 3-(1-pyrrolidinyl)-, propionate	EtOH	250(3.92),295(3.48)	32-1284-66
$C_{28}H_{38}O_5$			
Jaborosalactone A	MeOH	219(4.30),333(2.03)	78-1121-66
Jaborosalactone B	MeOH	312(3.85)	78-1121-66
$C_{28}H_{38}O_6$			
Withaferin A, 27-deoxy-14-hydroxy-	EtOH	216(4.20)	39-1765-66C
$C_{28}H_{38}O_8$			
Khivorin, 1,3-dideacetyl-	MeOH	212(3.72)	39-0506-66C
$C_{28}H_{39}NO_2$			
Paspalin	EtOH	230(4.4),282(3.9), 291s(3.8)	33-1907-66

Compound	Solvent	$\lambda_{max}(\log \epsilon)$	Ref.
$C_{28}H_{39}NO_4$			
Androsta-3,5-diene-4,6-dicarboxaldehyde, 17β-hydroxy-3-(1-pyrrolidinyl)-, propionate	EtOH	230(4.22),343(4.00), 392(4.08)	32-1284-66
$C_{28}H_{40}NO_2$			
Indophenoxyl, tetra-tert-butyl-	isooctane	312s(4.27),325(4.35), 454(4.34)	35-3303-66
$C_{28}H_{40}O_3$			
25D-Spirost-4-en-3-one, 6-methylene-	EtOH	262(4.09)	78-1053-66
$C_{28}H_{40}O_4$			
Jaborosalactone A, deoxydihydro-	MeOH	227(4.01)	78-1121-66
$C_{28}H_{40}O_7$			
Pregn-16-en-20-one, 2β,3β,11α-trihydroxy-16-methyl-, triacetate	EtOH	249.5(4.01)	95-0639-66
$C_{28}H_{41}BrO_7$			
Colletotrichin, bromo-, acetate	EtOH	254(3.79),289(3.80)	39-0230-66C
$C_{28}H_{41}NO$			
Cholesta-1,4-diene-1-carbonitrile, 3-oxo-	EtOH	214(4.08),250(4.13)	39-0661-66C
$C_{28}H_{41}NO_2$			
Indophenol, tetra-tert-butyl-	isooctane	270(4.32),315(4.13), 480(3.91)	35-3303-66
$C_{28}H_{41}NO_4$			
Pregna-4,6-diene-3,20-dione, 6-[(diethylamino)methyl]-17-hydroxy-, acetate	EtOH	283.5(4.32)	25-0497-66
$C_{28}H_{41}N_2P$			
Phosphorane, (fluoren-9-ylidenehydrazono)tripentyl-	dioxan	215(4.16),248(4.77), 295(3.80),306(3.81), 370(4.32)	5-0001-66H
$C_{28}H_{42}N_2O_3S$			
1H-Cyclopenta[5,6]naphtho[1,2-g]quinazoline-8,10(3bH,9H)-dione, tetradecahydro-1-hydroxy-11a,13a-dimethyl-7-pentyl-10-thio-, acetate	EtOH	245(3.64),341(4.26)	32-0179-66
	EtOH-NaOH	331(4.18)	32-0179-66
$C_{28}H_{42}N_2O_4$			
Androsta-2,4-diene-2-carbonitrile, 3-(N-ethoxymethylidene)amino-17β-hydroxy-, 17-(orthodiethoxyformate)	EtOH	236(4.09),310(3.85)	32-0152-66
1H-Cyclopenta[5,6]naphtho[1,2-g]quinazoline-8,10(3bH,9H)-dione, tetradecahydro-1-hydroxy-11a,13a-dimethyl-7-pentyl-, acetate	EtOH	274(3.97)	32-0179-66
$C_{28}H_{42}O$			
Cholesta-1,4,6-trien-3-one, 2-methyl-	EtOH	267(4.13),304(4.04)	94-0866-66
Cholesta-1,4,6-trien-3-one, 4-methyl-	MeOH	227(3.08),309(3.95)	94-0866-66

Compound	Solvent	$\lambda_{max}(\log \epsilon)$	Ref.
$C_{28}H_{42}O_2$			
5α-Cholesta-1,7-diene-2-carboxaldehyde, 3-oxo-	EtOH	239(3.91)	88-0049-66
	EtOH-NaOH	306(4.23)	88-0049-66
$C_{28}H_{42}O_4$			
Jaborosalactone A, tetrahydrodeoxy-	MeOH	227.5(3.95)	78-1121-66
$C_{28}H_{42}O_6$			
Adigenin	n.s.g.	215(4.13)	33-1855-66
$C_{28}H_{42}O_7$			
Colletotrichin, acetate	EtOH	261(3.88)	39-0230-66C
$C_{28}H_{43}NO$			
Cholest-4-ene-1α-carbonitrile, 3-oxo-	EtOH	242(4.18)	39-0661-66C
5α-Cholest-1-ene-1-carbonitrile, 3-oxo-	EtOH	236(4.04)	39-0661-66C
$C_{28}H_{43}N_3O_2$			
Teleocidin B, dihydro-	EtOH	232(4.53),276s(--), 287(3.98),297s(--)	88-2515-66
$C_{28}H_{43}N_3O_3$			
8H-Cyclopenta[5,6]naphtho[1,2-g]quinazolin-8-one, 10-aminohexadecahydro-1-hydroxy-11a,13a-dimethyl-7-pentyl-, acetate	EtOH-HCl	221(4.09),297(4.12)	32-0179-66
	EtOH	220(4.13),284(3.90)	32-0179-66
$C_{28}H_{43}N_3O_7$			
Ostreogrycin "hydro A"	EtOH-6N HCl	292(3.58)	39-1653-66C
	EtOH	none	39-1653-66C
	EtOH-NaOH	298(4.31)	39-1653-66C
	H_2SO_4	304(4.00),359(4.01)	39-1653-66C
Ostreogrycin G, hexahydro-	EtOH	none	39-1856-66C
	EtOH-NaOH	298(4.26)	39-1856-66C
	H_2SO_4	304(4.00),357(3.94)	39-1856-66C
$C_{28}H_{44}$			
19-Norcholesta-1,3,5(10)-triene, 2,4-dimethyl-	EtOH	269(2.65),278(3.60)	94-0866-66
$C_{28}H_{44}N_2O$			
5α-Androst-2-ene-2-carbonitrile, 17β-hydroxy-3-(N-ethoxymethylidene)-amino-, 17-(orthodiethoxyformate)	EtOH	260(4.03)	32-0152-66
$C_{28}H_{44}O$			
Cholesta-1,4-dien-3-one, 2-methyl-	EtOH	250(4.28)	94-0866-66
Ergosterol	EtOH	262(2.48),271(2.60), 281(2.70),293(2.60)	39-0072-66C
$C_{28}H_{45}ClO$			
A-Homo-5α-cholest-4-en-3-one, 4-chloro-	EtOH	254(4.09)	78-0105-66B
$C_{28}H_{46}$			
Cholesta-3,5-diene, 6-methyl-	EtOH	236(4.26),242(4.30), 249(4.14)	23-2587-66
$C_{28}H_{46}O$			
4α,5-Methanocholestan-3-one	hexane	194(3.92)	44-3869-66
	EtOH	204(3.86),275(1.76)	44-3869-66

Compound	Solvent	λ_{max}(log ε)	Ref.
4β,5-Methanocholestan-3-one	hexane	187(3.91)	44-3869-66
	EtOH	200(3.81),277(1.65)	44-3869-66
A-Norcholest-5-ene-3α-carboxaldehyde, 6-methyl-	EtOH	214(3.40)	23-2837-66
$C_{28}H_{46}O_2$			
Cholest-4-en-3-one, 6α-methoxy-	EtOH	242(4.12)	22-2277-66
Cholest-4-en-3-one, 6β-methoxy-	EtOH	237(4.12)	22-2277-66
Cholest-4-en-6-one, 3β-methoxy-	EtOH	248.5(3.91)	22-2277-66
A-Norcholest-3-ene-3-carboxylic acid, methyl ester	MeOH	238(4.04)	22-1287-66
$C_{28}H_{48}$			
5α-Ergost-8-ene	C_6H_{12}	198(3.89)	39-1266-66C
5α-Ergost-8(14)-ene	C_6H_{12}	206(4.04)	39-1266-66C
$C_{28}H_{48}N_2O$			
Cyclokoreanine B, N-methyl-	EtOH	213(4.21)	39-1805-66C
$C_{28}H_{48}O_4$			
β-Tocopherol-p-quinone, 7-hydroxy-	EtOH	265(4.50),455(2.82)	39-1431-66C
$C_{28}H_{49}NO$			
β-Tocopheramine	EtOH	234(3.88),300(3.51)	33-2297-66
γ-Tocopheramine	EtOH	301(3.53)	33-2297-66
δ-Tocopheramine, N-methyl-	EtOH	242(3.97),309(3.44)	33-2297-66
ζ_2-Tocopheramine	EtOH	233(3.90),300(3.55)	33-2297-66

Compound	Solvent	λ max(log ε)	Ref.
$C_{29}H_{20}Cl_2N_2O$			
Benzophenone, 5-chloro-2-[(6-chloro-3-methyl-4-phenyl-2-quinolyl)-amino]-	EtOH	224(4.78),256(4.55), 283(4.44),334(3.75), 349(3.80),398(4.12)	4-0359-66
	EtOH-HCl	252(4.58),344(3.91), 352(3.91)	4-0339-66
$C_{29}H_{20}O$			
Anthrone, 10-(p-tolylphenethenylidene)-	$CHCl_3$	273(4.32),305(4.01)	22-1999-66
$C_{29}H_{22}$			
Anthracene, 9-benzylidene-9,10-dihydro-10-(p-methylbenzylidene)-	EtOH	233(4.76),247(4.28), 334(4.40)	78-0141-66B
$C_{29}H_{22}ClNO_3$			
Indole, 1-(p-chlorobenzoyl)-2,3-bis-(p-methoxyphenyl)-	EtOH	246(4.58),300(4.30)	87-0527-66
$C_{29}H_{22}F_9N_2P$			
Phosphine imide, N-[p-(dimethylamino)-phenyl]-P,P,P-tris(α,α,α-trifluoro-p-tolyl)-	EtOH	260(4.4)	65-1248-66
$C_{29}H_{22}N_2$			
Dipyrromethene, 5-methyl-3,3',5'-triphenyl-	EtOH-HCl	283(4.15),388(3.81), 521(4.89)	12-1871-66
	EtOH	287(3.55),489(4.53)	12-1871-66
$C_{29}H_{22}N_2O$			
2-Pyridinemethanol, α-phenyl-α-[6-phenyl-6-(2-pyridyl)-2-fulvenyl]-	MeOH-HCl	343(4.35)	44-2149-66
	MeOH	240(4.16),262s(4.10), 267s(4.09),323(4.39)	44-2149-66
3-Pyridinemethanol, α-phenyl-α-[6-phenyl-6-(3-pyridyl)-2-fulvenyl]-	MeOH	327(4.38)	44-2149-66
4-Pyridinemethanol, α-phenyl-α-[6-phenyl-6-(4-pyridyl)-2-fulvenyl]-	MeOH	325(4.37)	44-2149-66
Urea, 1-(1-naphthyl)-1,3,3-triphenyl-	EtOH	260s(4.21)	34-0436-66
Urea, 1-(2-naphthyl)-1,3,3-triphenyl-	EtOH	250(4.53)	34-0436-66
$C_{29}H_{22}N_2O_3$			
2-Pyridinemethanol, α-phenyl-α-[6-phenyl-6-(2-pyridyl)-2-fulvenyl]-, di-N-oxide	MeOH	322(4.29)	44-2149-66
$C_{29}H_{22}N_4O_2$			
Monoformylanhydrotetra-2-amino-benzaldehyde	EtOH	229(4.60),235s(4.59), 258s(4.24),268s(4.13), 282s(3.48),370(3.82)	39-0956-66B
$C_{29}H_{22}O$			
Anthracene, 9-benzylidene-9,10-dihydro-10-(o-methoxybenzylidene)-	$CHCl_3$	345(4.36)	78-0141-66B
2-Cyclopenten-1-one, 2,3,4,5-tetraphenyl-	MeOH	222(4.42),297(4.09)	44-1336-66
cis	EtOH	299(4.16)	22-3775-66
trans	EtOH	299(4.10)	22-3775-66
$C_{29}H_{22}O_6$			
Flavone, 3,7-bis(benzyloxy)-5,8-dihydroxy-	EtOH	279(4.49),369(3.68)	78-0941-66

Compound	Solvent	λ max (log ε)	Ref.
$C_{29}H_{22}O_{12}$ Cryptoclauxin	n.s.g.	233(4.62),325(3.79)	88-2867-66
$C_{29}H_{23}ClO_4$ 2-Butyne-1,4-diol, 1-(p-chlorophenyl)-4-(p-methoxyphenyl)-1,4-diphenyl-	EtOH	228(3.32),269(--), 277(3.00)	22-2885-66
$C_{29}H_{23}NO_2$ Anthracen-9,10-imine-11-carboxylic acid, 9,10-dihydro-9,10-diphenyl-, ethyl ester	n.s.g.	255s(3.30),287(3.10)	78-1011-66
$C_{29}H_{23}NO_5$ 2-Butyne-1,4-diol, 1-(p-methoxyphenyl)-4-(p-nitrophenyl)-1,4-diphenyl-	EtOH	232(4.30),278(3.95)	22-2885-66
$C_{29}H_{23}OP$ Phosphorin, 1,1-dihydro-1-hydroxy-1,2,4,6-tetraphenyl- (hydrate)	EtOH	210(4.83),252s(4.72)	35-1034-66
$C_{29}H_{24}O_{12}$ Theaflavin	EtOH	216(4.55),229s(4.40), 271(4.29),290s(4.24), 384(3.94),470(3.56)	88-1193-66
$C_{29}H_{25}FIN_3O$ 1-(p-Fluorophenyl)-5,6,7,8-tetrahydro-4-[(3-methyl-2-benzoxazolinylidene)-methyl]-2-phenylquinazolinium iodide	MeOH	427(4.87)	87-0758-66
$C_{29}H_{25}FIN_3S$ 1-(p-Fluorophenyl)-5,6,7,8-tetrahydro-4-[(3-methyl-2-benzothiazolinylidene)methyl]-2-phenylquinazolinium iodide	MeOH	450(4.89)	87-0758-66
$C_{29}H_{24}N_2O_3$ 7H-1-Pyrindine, 6-benzamido-7,7-bis-(o-anisyl)-	MeOH	223(4.52)	78-0035-66
7H-1-Pyrindine, 6-benzamido-7,7-bis-(p-anisyl)-	MeOH	225(4.49)	78-0035-66
7H-2-Pyrindine, 6-benzamido-7,7-bis-(o-anisyl)-	MeOH	224(4.49)	78-0035-66
7H-2-Pyrindine, 6-benzamido-7,7-bis-(p-anisyl)-	MeOH	224(4.46)	78-0035-66
$C_{29}H_{25}O_2P$ 2-Phosphorinol, 1,2,5,6-tetrahydro-1,2,4,6-tetraphenyl-, 1-oxide	EtOH	205(4.56),250s(4.08)	35-1034-66
$C_{29}H_{26}ClNO_4S$ Thiopyrylium, 2,6-diphenyl-4-[1,3,3-triphenyl-2-indolinylidene)-methyl]-, perchlorate	HOAc	249(4.31),348(4.05), 517(4.60)	33-2046-66
$C_{29}H_{26}IN_3S$ 5,6,7,8-Tetrahydro-4-[(3-methyl-2-benzothiazolinylidene)methyl]-1,2-diphenylquinazolinium iodide	MeOH	448(4.94)	87-0758-66

Compound	Solvent	λ_{max}(log ε)	Ref.
$C_{29}H_{26}N_2O_4$			
Thymine, 1-(2,3-dideoxy-5-O-trityl-β-D-glycero-pent-2-enofuranosyl)-	EtOH	264(4.04)	44-0205-66
$C_{29}H_{26}N_2O_8$			
Glutamic acid, N-carboxy-N,1-dibenzyl ester, 5-ester with N-(hydroxymethyl)phthalimide	DMF	294.5(3.30)	10-0742-66A
$C_{29}H_{27}BrN_4O_2$			
Antipyrine, 4,4'-(p-bromo-α-hydroxybenzylidene)di-, perchlorate	H_2O	490(4.11)	65-1595-66
$C_{29}H_{27}NO$			
7H-Benzocyclononen-7-one, 10,11-dihydro-6,8-diphenyl-9-(1-pyrrolidinyl)-	EtOH	227(4.40),275(4.15), 315(4.24)	44-1336-66
$C_{29}H_{28}N_2O_2$			
Urea, 3-(p-phenoxyphenyl)-1-methyl-1-(p-tolyl)-3-(2,4-xylyl)-	EtOH	260(4.37)	34-0436-66
$C_{29}H_{28}N_2O_4$			
Dipyrromethene-4,4'-dicarboxylic acid, 5,5'-dimethyl-3,3'-diphenyl-, diethyl ester	EtOH-N HCl	220s(4.45),251(4.01), 388(3.76),483(5.01)	12-1871-66
	EtOH	215(4.32),266(4.03), 472(4.56)	12-1871-66
$C_{29}H_{28}N_2O_7S$			
Uracil, 1-(2-deoxy-5-O-trityl-β-D-lyxosyl)-, 3'-methanesulfonate	EtOH	229(4.23),262(4.14)	44-0205-66
$C_{29}H_{28}N_4O_3S_2$			
3-Ethyl-2-[5-(3-ethyl-2-benzothiazolinylidene)-2-(1,2,3,4-tetrahydro-1,3-dimethyl-6-hydroxy-4-oxo-2-thioxo-5-pyrimidinyl)-1,3-pentadienyl]benzothiazolium hydroxide, inner salt	MeOH	661(5.08)	9-0289-66
$C_{29}H_{28}O_{11}$			
Piscerythrone, tetraacetate	EtOH	243(4.44),286(4.07)	78-0333-66A
Piscidone, tetraacetate	EtOH	245s(4.38),282(3.93)	78-0333-66A
$C_{29}H_{28}S$			
Sulfide, 4-biphenylyldiphenylmethyl tert-butyl	EtOH	261(4.37)	35-1257-66
	50% H_2SO_4	423(4.20),506(4.46)	35-1257-66
$C_{29}H_{29}NO_3$			
2-Bornanone, 3-(trityl-aci-nitro)-	EtOH	288(4.14)	77-0230-66
$C_{29}H_{29}NO_7$			
Benzaldehyde, 2-[(5,6-dihydro-1,2,10-trimethoxy-6-methyl-4H-dibenzo[de,g]quinolin-9-yl)oxy]-4,5-dimethoxy-	n.s.g.	268(5.13),334(4.72)	25-0770-66
$C_{29}H_{29}N_5O_3$			
Antipyrine, 4,4'-(m-amino-α-hydroxybenzylidene)di-, perchlorate	H_2O	483(4.18)	65-1595-66

Compound	Solvent	λ_{max}(log ε)	Ref.
$C_{29}H_{30}N_3O_2P$			
Aniline, 4,4'-[(p-nitrobenzylidene)-phenylphosphoranylidene]bis-[N,N-dimethyl-	$C_2H_4Cl_2$	246(4.31),275s(4.26), 294(4.38),520(4.56)	97-0068-66
$C_{29}H_{30}O_6$			
Acetophenone, 3',5'-bis(5-acetyl-2-methoxybenzyl)-4'-methoxy-	EtOH	229(4.61),270(4.56)	87-0838-66
$C_{29}H_{30}O_{10}$			
Flavone, 3'-(benzyloxy)-3,4',5,5',-6,7,8-heptamethoxy-	n.s.g.	254(4.26),270(4.23), 330(4.34)	24-3218-66
$C_{29}H_{31}NO_7$			
Hernandaline	EtOH	216(4.36),278(4.40), 304(4.20)	88-4279-66
$C_{29}H_{32}O_{13}$			
Chromomycinone, tetraacetate	EtOH	273(4.71),420(3.99)	78-2761-66
Chromomyciquinone, 3'-dehydro-, tetraacetate	EtOH	260(4.37),276(4.10), 352(3.56)	78-2761-66
$C_{29}H_{34}$			
Indene, retinylidene-	EtOH	440(4.86)	54-0339-66
$C_{29}H_{34}N_2O_2$			
Indole, 1-[3-(diethylamino)propyl]-2,3-bis(p-methoxyphenyl)-	EtOH	241(4.48),299(4.18)	87-0527-66
$C_{29}H_{35}IN_4S$			
1-(2-Diethylaminoethyl)-5,6,7,8-tetrahydro-4-[(3-methyl-2-benzothiazolinylidene)methyl]-2-phenylquinazolinium iodide	MeOH	449(4.90)	87-0758-66
$C_{29}H_{36}$			
Cyclopentadiene, retinylidene-isopropylidene-	EtOH	440(4.84)	54-0339-66
$C_{29}H_{36}N_2O$			
Lobinaline, acetyl-	EtOH	232-243(3.80)	44-3206-66
$C_{29}H_{36}O_4$			
Pregn-4-ene-3,20-dione, 17α-hydroxy-, 17-phenylacetate	EtOH	240(4.24)	32-1115-66
$C_{29}H_{36}O_5$			
Pregn-4-ene-3,20-dione, 17,21-dihydroxy-, cyclic methyl orthobenzoate	EtOH	242(4.19)	32-1115-66
Pregn-4-ene-3,20-dione, 17,21-dihydroxy-, 17-phenylacetate	EtOH	241(4.23)	32-1115-66
$C_{29}H_{36}O_8$			
Fissinolide	EtOH	210(4.04)	88-6441-66
$C_{29}H_{36}O_{15}$			
Flavanone, 5,7-dihydroxy-3',4'-dimethoxy-, 7-neohesperidoside	EtOH	238(3.47),285(4.26)	28C-1712-66A

Compound	Solvent	λ_{max}(log ε)	Ref.
$C_{29}H_{37}NO$			
1H-Cyclopenta[5,6]naphtho[1,2-g]quinolin-1-ol, tetradecahydro-1,11a,13a-trimethyl-8-phenyl-	EtOH	248(4.16),284(4.13)	4-0339-66
$C_{29}H_{37}NO_5$			
Phomin	EtOH	213(4.42),219(4.32), 258s(2.69),264(2.48), 267(2.32)	31-0750-66
$C_{29}H_{37}N_3O_3$			
Isotubulosine	pH 1	277(4.06)	88-5077-66
	pH 13	279(4.06),319(3.62)	88-5077-66
	MeOH	279(4.08)	88-5077-66
Tubulosine	pH 1	275(4.06)	88-1081-66
	pH 13	278(4.08),320(3.63)	88-1081-66
	pH 13	279(4.08),320(3.63)	88-5077-66
	MeOH	279(4.10)	88-5077-66
$C_{29}H_{38}ClFO_7$			
Pregna-3,5-diene-6-carboxaldehyde, 3-(2-chloroethoxy)-9α-fluoro-17α,21-dihydroxy-11,20-dioxo-, 21-valerate	EtOH	216(4.04),317(4.01)	31-0468-66
$C_{29}H_{38}ClFO_8$			
Pregna-3,5-diene-6-carboxaldehyde, 3-(2-chloroethoxy)-9α-fluoro-11β,16α,-17α,21-tetrahydroxy-20-oxo-, 16α,17α-acetonide, 21-acetate	EtOH	216(4.08),324(4.23)	31-0468-66
$C_{29}H_{38}N_2O_3$			
Pregna-2,4-dieno[4,5,6-cd]pyridin-20-one, 17-hydroxy-3-(1-pyrrolidinyl)-, acetate	EtOH	249(3.86),295(3.45)	32-1284-66
$C_{29}H_{38}N_2O_4$			
Benz[4,5,6]androsta-2,4-dien-17β-ol, 5'-nitro-3-(1-pyrrolidinyl)-, propionate	EtOH	283(4.48)	32-1284-66
$C_{29}H_{38}N_4O_6S_2$			
Urea, 1,3-dicyclohexyl-1-[7-(thiophene-2-acetamido)cephalosporanoyl]-	EtOH	234(4.17),266(3.81)	87-0741-66
$C_{29}H_{38}O_2$			
1-Anthroic acid, 10-ethyl-9,10-dihydro-9-hydroxy-5,6,7,8-tetrapropyl-, lactone	C_6H_{12}	278(3.54),287(3.50)	33-0644-66
$C_{29}H_{38}O_3$			
9-Furanhexanol, tetrahydro-γ-methyl-α-(2-methyl-2,3-diphenylallyl)-, acetate	EtOH	242(4.43)	88-1939-66
$C_{29}H_{38}O_8$			
Cedrela odorata C	MeOH	212(4.08)	39-2127-66C
$C_{29}H_{38}O_9$			
Roridin A, dehydro-	C_6H_{12}	278.5(4.25)	33-2527-66
	EtOH	262(4.18)	33-2527-66

Compound	Solvent	λ_{max}(log ε)	Ref.
Roridin D	EtOH	260(4.33)	33-2547-66
$C_{29}H_{39}NO_5$			
Pregna-3,5-diene-4,6-dicarboxaldehyde, 17α-hydroxy-20-oxo-3-(1-pyrrolidinyl)-, acetate	EtOH	229(4.17),340(3.99), 390(4.06)	32-1284-66
$C_{29}H_{40}O_2$			
1-Anthroic acid, 10-ethyl-9,10-dihydro-5,6,7,8-tetrapropyl-	C_6H_{12}	286.5(3.46)	33-0644-66
$C_{29}H_{40}O_7$			
Gitoxigenin, dianhydro-, rhamnoside	n.s.g.	222(4.06),336(4.25)	83-0679-66
$C_{29}H_{40}O_9$			
Roridin A	EtOH	263(4.31)	33-2527-66
$C_{29}H_{41}NO_4$			
Pregna-4,6-diene-3,20-dione, 17-hydroxy-6-(piperidinomethyl)-, acetate	EtOH	283(4.32)	25-0497-66
$C_{29}H_{42}O_6$			
18-Norchola-9(11),13-dienic acid, 3α,7α-diacetoxy-12β-methyl-, methyl ester	n.s.g.	210(3.86)(end abs.)	88-3063-66
$C_{29}H_{42}O_8$			
Gitoxigenin, $\Delta^{8(14)}$-anhydro-, rhamnoside	n.s.g.	212(4.30)	83-0679-66
$C_{29}H_{43}NO_5$			
Phomin, hexahydro-	EtOH	258.5(2.37)	31-0750-66
$C_{29}H_{44}O_2$			
Saikogenin A derivative	EtOH	319(4.28)	88-0701-66
$C_{29}H_{44}O_9$			
Roridin A, tetrahydro-	EtOH	none	33-2527-66
$C_{29}H_{45}NO$			
Cholesta-3,5-dieno[3,4-d]oxazole, 2'-methyl-	EtOH	262(4.18)	22-3407-66
Cholesta-3,5-dieno[4,3-d]oxazole, 2'-methyl-	EtOH	236(4.15)	22-3407-66
$C_{29}H_{45}NO_6$			
Cyclopentaneheptadecanoic acid, 2-carboxy-λ-cyano-π-hydroxy-γ,ε,η,ι-tetramethyl-β,κ-dioxo-, π-lactone, methyl ester	EtOH	235(3.7)	31-0355-66
$C_{29}H_{46}N_2O_2$			
1H-Cyclopenta[5,6]naphtho[1,2-g]quinazoline-8,10(3bH,9H)-dione, 1-(1,5-dimethylhexyl)tetradecahydro-11a,13-dimethyl-	EtOH	264(3.74)	32-0179-66
$C_{29}H_{46}O_2$			
Cholest-5-ene-3,7-dione, 4,4-dimethyl-	EtOH	237(3.93)	39-1443-66C

Compound	Solvent	λ_{max}(log ε)	Ref.
5α-Cholest-8-ene-3,11-dione, 4,4-dimethyl-	EtOH	249(--)	39-2359-66C
$C_{29}H_{46}O_4$			
5α-Cholest-7-en-6-one, 2β-acetoxy-3β-hydroxy-	MeOH	245(4.11)	33-1591-66
5β,25D-Spirost-9(11)-en-3-one, dimethyl acetal	EtOH	205(3.48)	95-0558-66
$C_{29}H_{47}NO_2$			
Cholest-2-en-4-one, O-acetyloxime, anti	EtOH	235(3.90)	5-0180-66B
Cholest-2-en-4-one, O-acetyloxime, syn	EtOH	228(4.25)	5-0180-66B
$C_{29}H_{47}NO_3$			
Cholest-4-en-3-one, 4-hydroxy-, methylcarbamate	EtOH	251(4.03)	22-3410-66
$C_{29}H_{47}NO_4$			
B-Homo-6-azacholest-4-en-7-one, 3β-acetoxy-6-hydroxy-	n.s.g.	240(4.20)	56-1911-66
$C_{29}H_{47}N_3$			
1H-Cyclopenta[5,6]naphtho[1,2-g]quinazoline, 10-amino-1-(1,5-dimethylhexyl)tetradecahydro-11a,13a-dimethyl-	HCl	260(3.95)	32-0152-66
	EtOH	236(3.90),268(3.58)	32-0152-66
$C_{29}H_{48}O$			
Cholestan-3-one, 2-ethylidene-	EtOH	246(3.85)	25-1497-66
Cholest-5-en-3-one, 4,4-dimethyl-	C_6H_{12}	292(1.62)	22-0897-66
$C_{29}H_{48}O_2$			
5α-Cholest-8-en-11-one, 3β-hydroxy-4,4-dimethyl-	EtOH	250(3.84)	39-2359-66C
	n.s.g.	250(3.84)	88-5235-66
Spiro[7H-benz[e]indene-7,1'-cyclopentan]-3'-one, 3α-(1,5-dimethylhexyl)-1,2,3,3a,4,5,8,9,9aα,9bβ-decahydro-8β-hydroxy-2',2',3aα,6-tetramethyl-	hexane	199.0(4.08)	39-1374-66C
$C_{29}H_{48}O_4$			
5α-Cholestan-4-one, 5,7β-dihydroxy-, 7-acetate	n.s.g.	305(1.62)	39-1012-66C
5β-Cholestan-4-one, 5,7β-dihydroxy-, 7-acetate	n.s.g.	305(1.60)	39-1012-66C
2-Chromantridecanoic acid, 6-hydroxy-α,ε,ι,2,5,7,8-heptamethyl-3,6-dioxo-	EtOH	292.0(3.48)	73-2434-66
$C_{29}H_{48}O_5$			
1,4-Benzoquinone, 3-(3-hydroxy-3,7,11,15-tetramethyl-15-carboxypentadecyl)-2,5,6-trimethyl-	EtOH	263(--),269(4.27)	73-2434-66
Lumifusidic acid, 16-deacetyl-24,25-dihydro-	EtOH	204(4.00)	87-0015-66
$C_{29}H_{49}N$			
Cholesta-3,5-dien-3-amine, N,N-dimethyl-	ether	271(4.28)	78-0541-66B
$C_{29}H_{49}NO_5$			
Phomin, dodecahydro-	EtOH	none	31-0750-66

Compound	Solvent	$\lambda_{max}(\log \epsilon)$	Ref.
$C_{29}H_{49}N_3O$			
5α-Solanidane, N-nitroso-3α-ethylamino-	n.s.g.	364(1.92)	78-3103-66
5α-Solanidane, N-nitroso-3β-ethylamino-	n.s.g.	360(1.98)	78-3103-66
5β-Solanidane, N-nitroso-3α-ethylamino-	n.s.g.	360(1.95)	78-3103-66
5β-Solanidane, N-nitroso-3β-ethylamino-	n.s.g.	362(1.89)	78-3103-66
$C_{29}H_{50}O$			
5α-Cholestan-2-one, 4,4-dimethyl-	C_6H_{12}	299(1.42)	22-0897-66
5α-Cholestan-3-one, 4,4-dimethyl-	C_6H_{12}	287(1.48)	22-0897-66
$C_{29}H_{50}O_2$			
5α-Cholestan-2-one, 3β-hydroxy-4,4-dimethyl-	C_6H_{12}	282(1.68)	22-0897-66
$C_{29}H_{51}NO$			
α-Tocopheramine	EtOH	280(3.52)	33-2297-66
β-Tocopheramine, N-methyl-	EtOH	239(3.99),304(3.49)	33-2297-66
γ-Tocopheramine, N-methyl-	EtOH	238(3.94),306(3.52)	33-2297-66
δ-Tocopheramine, N,N-dimethyl-	EtOH	244(3.89),290(3.23)	33-2297-66
ζ_2-Tocopheramine, N-methyl-	EtOH	236(3.87),292(3.40)	33-2297-66

Compound	Solvent	λ_{max}(log ϵ)	Ref.
$C_{30}H_{12}Br_2N_4O_2$ 9H,19H-Diperimidino[1,2-a:1',2'-a']-benz[1,2-c:4,5-c']dipyrrole, 8,18-dibromo-9,19-dioxo-	$C_{10}H_7Cl$	570(4.29),615s(4.12)	33-0534-66
$C_{30}H_{12}Cl_2N_4O_2$ 9H,19H-Diperimidino[1,2-a:1',2'-a']-benz[1,2-c:4,5-c']dipyrrole, 8,18-dichloro-9,19-dioxo-	$C_{10}H_7Cl$	574(4.29),620s(4.20)	33-0534-66
$C_{30}H_{16}$ 1,12-Benzo[naphtho-2",3":4,5-perylene]	benzene	292(4.56),306(4.58), 322(4.65),333(4.39), 350(4.73),369(4.94), 402(4.00),426(4.30), 450(4.40)	24-1275-66
	EtOH	254(4.88),262(4.81), 271(4.96)	24-1275-66
$C_{30}H_{18}O_3$ Flavanonol red	THF	457(4.18),487(4.58), 525(4.74)	88-4671-66
$C_{30}H_{18}O_4$ Phenanthro[9,10-b]phenanthro[9',10':-5,6]-p-dioxino[2,3-e]-p-dioxin, 9a,19a-dihydro-	benzene	361(4.2)	24-1881-66
Spiro[11H-oxeto[2,3-b]phenanthro-(9,10-e]-p-dioxin-11,9'(10'H)-phenanthren]-10'-one, 9a,11a-dihydro-	dioxan	272(4.20)	24-1881-66
$C_{30}H_{18}O_5$ Photoflavanonol red	EtOH	234(4.65),315(4.23), 391(4.45)	88-4671-66
$C_{30}H_{18}O_{10}$ Amentoflavone	EtOH	272(4.52),342(4.51)	28D-0707-66A
$C_{30}H_{18}O_{12}$ Aurofusarin	$CHCl_3$	248(4.69),269(4.53), 381(3.99),422s(3.93)	39-2234-66C
	dioxan	243(4.69),267(4.52), 372(4.05)	88-4855-66
	n.s.g.	248(4.62),268s(4.48), 388(3.94)	39-2237-66C
$C_{30}H_{20}$ Hexapentaene, 1,1,6,6-tetraphenyl-	$CHCl_3$	370(--),440(--), 488(4.88)	22-2885-66
$C_{30}H_{20}Br_2O_2$ 2,4-Hexadiyne-1,6-diol, 1,6-bis-(p-bromophenyl)-1,6-diphenyl-	EtOH	222(4.53),262(3.23)	22-2885-66
$C_{30}H_{20}N_4$ 1,4,5,8-Tetraazafulvalene, 2,3,6,7-tetraphenyl-	n.s.g.	494(4.75)	88-5221-66

Compound	Solvent	λ max(log ε)	Ref.
$C_{30}H_{20}O_2$			
Cyclopropenone, diphenyl-, dimer	EtOH	227(4.58),285(4.58), 296(4.62),312(4.48)	44-1336-66
$C_{30}H_{21}BrO_2$			
2,4-Hexadiyne-1,6-diol, 1-(m-bromophenyl)-1,6,6-triphenyl-	EtOH	220(4.69),248(--), 260(3.30)	22-2885-66
2,4-Hexadiyne-1,6-diol, 1-(o-bromophenyl)-1,6,6-triphenyl-	EtOH	220(4.67),248(--), 260(--)	22-2885-66
2,4-Hexadiyne-1,6-diol, 1-(p-bromophenyl)-1,6,6-triphenyl-	EtOH	220(4.67),262(3.30)	22-2885-66
$C_{30}H_{21}ClO_2$			
2,4-Hexadiyne-1,6-diol, 1-(o-chlorophenyl)-1,6,6-triphenyl-	EtOH	219(4.61),248(--), 260(3.18)	22-2885-66
$C_{30}H_{22}$			
Fulvene, 1,2,3,4-tetraphenyl-	EtOH	259(4.47),315s(3.83), 430(3.04)	22-3775-66
$C_{30}H_{22}Cl_2N_2O_2$			
Indolo[3,2-b]indole, 3,8-dichloro-4b,5,9b,10-tetrahydro-4b,9b-diphenyl-, diacetyl deriv.	iso-PrOH	262(4.40),290s(3.74)	44-3356-66
$C_{30}H_{22}N_2$			
Dinaphtho[2,3-b:2',3'-i]phenazine, 7,16-dihydro-7,16-dimethyl-	MeCN	262(5.22),269(5.17), 330(5.15)	44-3734-66
$C_{30}H_{22}O_5$			
3,6(2H,7H)-Oxepindione, 4,5-bis-(o-hydroxyphenyl)-2,7-diphenyl-	EtOH	263(4.40),300(4.31)	88-4671-66
$C_{30}H_{23}N$			
2-Cyclopentene, 1-(diphenylmethylene)-4-[α-(2-pyridyl)benzylidene]-	MeOH	257(4.21),364(4.49)	44-2149-66
Fulvene, 2-(diphenylmethyl)-6-phenyl-6-(2-pyridyl)-	MeOH	323(4.28)	44-2149-66
$C_{30}H_{23}NO_2$			
2,3-Naphthalenedicarboximide, 1,2,3,4-tetrahydro-N,1,4-triphenyl-	CH_2Cl_2	258(3.12),263s(3.02)	5-0044-66C
isomer	CH_2Cl_2	253s(3.12),259(3.09), 264s(3.01),270s(2.81)	5-0044-66C
$C_{30}H_{23}N_5O_3$			
13H-6,12-o-Benzeno-6H-quinazolino-[3,4-a]quinazoline, 7-acetyl-13-(o-formyl-N-nitrosoanilino)-7,11b-dihydro-	EtOH	209(4.79),240s(4.46), 275s(3.78),283(3.72)	23-2323-66
	EtOH	239s(4.22),274s(3.49), 283s(3.42)	39-0956-66B
$C_{30}H_{24}$			
Cyclopentadiene, 2-methyl-1,3,4,5-tetraphenyl-	EtOH	240(4.31),331(4.20)	22-3775-66
$C_{30}H_{24}ClNO_4S$			
Thiopyrylium, 4-[p-(N-methylanilino)-phenyl]-2,6-diphenyl-, perchlorate	HOAc	270(4.22),292(4.19), 390(4.26),576(4.77)	33-2046-66

Compound	Solvent	λ_{max}(log ϵ)	Ref.
$C_{30}H_{24}Cl_2N_2O_2$			
Dibenzo[b,f][1,5]diazocine, 2,8-dichloro-5,11-diacetyl-5,6,11,12-tetrahydro-6,12-diphenyl-	iso-PrOH	260s(3.18)	44-3356-66
$C_{30}H_{24}N_2O$			
Pyridine, 2-[α-[3-(α-methoxy-α-2-pyridylbenzyl)-2,4-cyclopentadien-1-ylidene]benzyl]-	MeOH	243s(4.13),267s(4.05), 325(4.36)	44-2149-66
2-Pyridinemethanol, α-[3-methyl-6-phenyl-6-(2-pyridyl)-2-fulvenyl]-α-phenyl-, isomer A	MeOH	245(4.16),325(4.40)	44-2149-66
isomer B	MeOH	245(4.17),325(4.40)	44-2149-66
$C_{30}H_{24}N_4O_2$			
13H-6,12-o-Benzeno-6H-quinazolino-[3,4-a]quinazoline, 7-acetyl-13-(o-formylanilino)-7,11b-dihydro-	EtOH	229(4.61),235s(4.59), 260s(4.19),268s(4.10), 282s(3.46),372(3.82)	39-0956-66B
$C_{30}H_{24}N_4O_6$			
Inosine, 5'-O-trityl-, 2',3'-carbonate	EtOH-HCl	243(4.17),249(4.14)	69-2076-66
	EtOH-NaOH	244(4.18),251(4.18)	69-2076-66
phenolate	pH 2	243(4.07)	69-2076-66
	EtOH-NaOH	252(4.09)	69-2076-66
$C_{30}H_{24}O$			
4H-Pyran, 2,6-dimethyl-4-(2,3,4-triphenyl-2,4-cyclopentadien-1-ylidene)-	n.s.g.	260(--),417(--)	39-1086-66C
$C_{30}H_{24}O_2$			
Cyclopentadiene, 2-methyl-1,3,4,5-tetraphenyl-, photooxide from	EtOH	250(4.59)	22-3775-66
$C_{30}H_{24}O_7$			
Flavone, 3,7-bis(benzyloxy)-4'-methoxy-5,8-dihydroxy-	EtOH	280(4.37),300s(4.19), 369(3.84)	78-0941-66
$C_{30}H_{24}O_{10}$			
Xylaphin	$C_2H_2Cl_4$	274(3.90),286(3.93), 325(4.20),397(4.00), 478(3.30),590(4.39), 636(4.56)	39-1836-66C
$C_{30}H_{24}Sn$			
Tin, o-biphenylyltriphenyl-	THF	246(3.95)	35-3027-66
$C_{30}H_{25}NO_4$			
5-Norbornene-2,3-dicarboximide, 7-benzylidene-1,5-bis(α-hydroxybenzyl)-	MeOH	248(4.18)	44-2149-66
$C_{30}H_{26}CoN_4$			
Cobalt, bis(N,N'-propanediylidene-dianilinato)(1-)]-	$CHCl_3$	395(4.59),510s(3.4)	65-1372-66
$C_{30}H_{26}CuN_4$			
Copper, bis(N,N'-propanediylidene-dianilinato(1-)]-	$CHCl_3$	337(4.19),435(4.3), 580(3.07),720(3.00)	65-1372-66

Compound	Solvent	λ_{max}(log ε)	Ref.
$C_{30}H_{26}N_2$			
Dipyrromethine, 3,5,5'-trimethyl-3',4,4'-triphenyl-	EtOH-N HCl	221(4.30),413(3.93), 504(4.82)	12-1871-66
	EtOH	470(4.46)	12-1871-66
$C_{30}H_{26}N_4Ni$			
Nickel, bis[N,N'-propanediylidene-dianilinato)(1-)]-	$CHCl_3$	398(4.55),510s(3.54)	65-1372-66
$C_{30}H_{26}NiO_2S_2$			
Nickel, bis(3-thiolo-1,3-diphenyl-prop-2-en-1-one)-	$CHCl_3$	555s(3.38),645s(2.52)	12-1401-66
$C_{30}H_{26}O$			
2-Cyclopenten-1-ol, 1-methyl-2,3,4,5-tetraphenyl-	EtOH	244(2.07)	22-3775-66
$C_{30}H_{26}O_2$			
Anthracene, 9,10-bis(p-methoxybenzyl)-	dioxan	403(4.09)	18-1551-66
Naphthoquinone, 2-(diphenylallyl)-3-(dimethylallyl)- (shoulders, not listed)	C_6H_{12}	251(4.46),329(3.45)	22-1828-66
$C_{30}H_{26}O_3$			
Phenol, 2,2'-(2,3,6,7-tetrahydro-2,7-diphenyl-4,5-oxepindiyl)di-	EtOH	282(3.85)	88-4671-66
$C_{30}H_{26}O_4$			
2-Butyne-1,4-diol, 1,4-bis(m-methoxy-phenyl)-1,4-diphenyl-	EtOH	268(--),275(3.48)	22-2885-66
2-Butyne-1,4-diol, 1,4-bis(p-methoxy-phenyl)-1,4-diphenyl-	EtOH	232(4.32),276(3.30)	22-2885-66
2-Butyne-1,4-diol, 1-(m-methoxyphenyl)-4-(p-methoxyphenyl)-1,4-diphenyl-	EtOH	269(--),275(3.48)	22-2885-66
$C_{30}H_{26}O_{10}$			
Vioxanthin	EtOH	268(4.70),380(4.15)	23-2873-66
$C_{30}H_{26}O_{14}$			
Ergoxanthin, demethyl-	EtOH	258(4.28),369(3.74)	78-2359-66
$C_{30}H_{27}FIN_3S$			
4-[(3-Ethyl-2-benzothiazolinylidene)-methyl]-1-(p-fluorophenyl)-5,6,7,8-tetrahydro-2-phenylquinazolinium iodide	MeOH	450(4.91)	87-0758-66
$C_{30}H_{28}CuN_2O_2$			
Benzylidenimine, N-(α-methylbenzyli-dene)-2-hydroxy-, copper chleate	MeOH	257(4.17),316(3.60), 403(2.85)	65-1590-66
$C_{30}H_{28}N_2O_4$			
Hydroquinone, 2,5-bis(1,3-dimethyl-indol-2-yl)-, diacetate	EtOH	229(4.87),248s(4.37), 308(4.37)	44-3321-66
$C_{30}H_{29}ClO_5$			
3-Phenyl-5,6,7,8-tetrahydro-1-(4,6,8-trimethyl-1-azulyl)vinylene-2-benzopyrylium perchlorate	$MeOH-HClO_4$	245(4.46),345(4.11), 630(4.36)	65-1728-66

Compound	Solvent	λ_{max}(log ϵ)	Ref.
$C_{30}H_{29}N_5$			
Benzimidazoline, 1-(2-dimethylamino-ethyl)-3-phenyl-2-(1-phenyl-benzimidazol-2-yl)-	EtOH	246(4.22),278(3.95), 286(3.94)	4-0278-66
$C_{30}H_{30}BrOP$			
(3,7-Dimethyl-10-oxo-2,4,6,8-decatetra-enyl)triphenylphosphonium bromide	EtOH	354(4.74)	33-0369-66
$C_{30}H_{30}N_4O_2$			
Antipyrine, 4,4'-(m-methylbenzylidene)-di-, perchlorate	H_2O	476(4.00)	65-1595-66
Antipyrine, 4,4'-(p-methylbenzylidene)-di-, perchlorate	H_2O	483(4.25)	65-1595-66
$C_{30}H_{30}N_4O_3$			
Antipyrine, 4,4'-(p-methoxybenzylidene)di-, perchlorate	H_2O	441(3.87)	65-1595-66
$C_{30}H_{30}O_{15}$			
Flavone, 4',5,7-trihydroxy-3-methoxy-, 7-β-D-glucopyranoside, tetraacetate	EtOH	266(4.27),352(4.31)	22-0987-66
$C_{30}H_{30}O_{16}$			
Flavone, 3',4',5,7-tetrahydroxy-3-methoxy-7-β-D-glucopyranoside, tetraacetate	EtOH	257(4.36),270s(--), 363(4.32)	22-0987-66
$C_{30}H_{32}BrO_2P$			
(8-Carbethoxy-3-methyl-2,4,6-octatrien-yl)triphenylphosphonium bromide	EtOH	303(4.70)	33-0369-66
$C_{30}H_{32}CrN_5O_7S$			
Chromium, [1-[[3-hydroxy-4-[(5-hydroxy-3-methyl-1-phenyl-2-pyrazolin-4-yl)-azo]-1-naphthyl]sulfonyl]piperidin-ato(2')](2,4-pentanedionato)-aquo-	n.s.g.	256(4.50),348(4.11), 524(4.33),556(4.34)	35-0186-66
$C_{30}H_{32}CuN_4O_4$			
5,5'-Bi(4'-carbethoxy-3,3',5'-tri-methyldipyrromethenyl), copper complex	$CHCl_3$	281(4.18),347s(4.04), 402(4.40),472(4.75), 741(3.98),806(2.91)	39-0098-66C
$C_{30}H_{34}N_2$			
Pyridine, 2,2'-[(3a,4,7,7a-tetrahydro-4,7-methanoindene-1,8-diylidene)-dineopentylidyne]di-	MeOH	256(4.26),261(4.29)	44-2149-66
$C_{30}H_{34}N_4$			
Porphine, 8,18-diethyl-2,3,7,12,13,17-hexamethyl-	$CHCl_3$	336s(4.26),379s(4.96), 402(5.21),475s(3.55), 499(4.11),534(3.98), 568(3.80),592s(3.20), 623(3.69)	39-0022-66C
$C_{30}H_{34}N_4O_4$			
5,5'-Bi(4'-carbethoxy-3,3',5'-tri-methyldipyrromethenyl)	$CHCl_3$	407(4.59),556s(4.45), 588(4.56),620s(4.45)	39-0098-66C

Compound	Solvent	λ_{max}(log ϵ)	Ref.
$C_{30}H_{34}O_2$			
Benzene, 1,4-bis(p-isopropylstyryl)-2,5-dimethoxy-	dioxan	390(4.81)	18-1547-66
Indene-3-carboxylic acid, 1-retinylidene-	EtOH	466(4.80)	54-0339-66
$C_{30}H_{34}O_{12}$			
Chromomycinone, isopropylidene-, triacetate	EtOH	264(4.59),272(4.62), 293(3.83),304(3.90), 390(3.77)	78-2761-66
Chromomyciquinone, 3'-dehydroisopropylidene-, triacetate	EtOH	259(4.36),275(4.09), 352(3.54)	78-2761-66
Isochromomycinone, isopropylidene-, triacetate	EtOH	226(4.48),272(4.62), 293(3.89),304(3.91), 390(3.76)	78-2761-66
$C_{30}H_{34}O_{13}$			
Chromomyciquinone, 3'-dehydro-3'-O-methyl-, tetraacetate	EtOH	259(4.39),276(4.13), 352(3.89)	78-2761-66
$C_{30}H_{35}NO_2$			
14'-Apo-β-carotenoic acid, 15'-cyano-, benzyl ester	EtOH-C_6H_6	443(4.77)	54-0334-66
$C_{30}H_{35}NO_5$			
15,17-Seco-16-nor-14-androsten-17-oic acid, 3β-acetoxy-15-(3-acetoxy-2-indolyl)-, lactam	MeOH	222(4.35),240(4.28), 270(4.02),322(4.41), 335s(4.19)	44-1363-66
$C_{30}H_{36}N_2O_2$			
Hydroquinone, 2,5-bis(1-ethyl-1,3-dimethylindolin-3-yl)-, inner double salt	EtOH	319(3.95)	44-3321-66
	EtOH-0.012 N HCl	318(3.92)	44-3321-66
	EtOH-0.051 N HCl	252s(3.1),269s(3.0), 316(3.89)	44-3321-66
	EtOH-4N HCl	252s(2.8),269s(2.8), 310(3.72)	44-3321-66
$C_{30}H_{36}N_4Pd$			
5,5'-Bi(4'-ethyl-3,3',4,5'-tetramethyl-2,2'-dipyrromethenyl)-, palladium complex	$CHCl_3$	347s(3.88),402s(4.40), 430(4.50),507(4.27), 520(4.27),865(3.91)	39-0098-66C
$C_{30}H_{36}N_6O_9S_3$			
s-Trithianetrione, tris[(o-isopropoxyphenyl)hydrazone], 1,1,3,3,5,5-hexaoxide, cis	n.s.g.	395(4.53)	99-0318-66
s-Trithianetrione, tris[(o-isopropoxyphenyl)hydrazone], 1,1,3,3,5,5-hexaoxide, trans	n.s.g.	405(4.76)	99-0318-66
$C_{30}H_{36}O_2$			
8'-Apo-β-carotenal, 15,15'-dehydro-4-oxo-	benzene	440(4.94)	1-1195-66
$C_{30}H_{36}O_5$			
Pregna-4,6-diene-3,20-dione, 17-hydroxy-6-(phenoxymethyl)-, acetate	EtOH	278(4.36)	25-0497-66

Compound	Solvent	λ_{max}(log ϵ)	Ref.
$C_{30}H_{36}O_9$ Gedunin, 11β-acetoxy-	n.s.g.	218(4.00)	78-0891-66
$C_{30}H_{37}BrN_4$ Biladiene-ac, 1-bromo-1,19-dideoxy-8,18-diethyl-2,3,7,12,13,17,19-hexamethyl-, dihydrobromide	$CHCl_3$	293(3.49),377(4.25), 439(4.19),463(4.49), 536(5.24)	39-0022-66C
$C_{30}H_{37}FO_8S$ Pregn-4-ene-3,20-dione, 9α-fluoro-11β,17,21-trihydroxy-, 17-acetate 21-p-toluenesulfonate	EtOH	230(4.40)	32-1115-66
$C_{30}H_{38}I_2N_2O_2$ 3,3'-(2,5-Dihydroxy-p-phenylene)bis-[1-ethyl-1,3-dimethylindolinium iodide]-	EtOH	254s(3.56),259s(3.53), 267(3.42),301(3.84)	44-3321-66
$C_{30}H_{38}N_2O_2$ Androst-5-en-3β-ol, 17β-(1-phenyl-5-pyrazolyl)-, acetate	EtOH	230(3.96)	44-3193-66
$C_{30}H_{38}N_2O_5$ Benzo[4,5,6]pregna-2,4-dien-20-one, 17-hydroxy-5'-nitro-3-(1-pyrrolidinyl)-, acetate	EtOH	244(4.22),287(4.11)	32-1284-66
$C_{30}H_{38}O$ Apo-3-lycopenal, 15,15'-dehydro-	benzene	464(4.96),493(4.87)	1-1195-66
$C_{30}H_{38}O_5$ Pregn-4-ene-3,20-dione, 17,21-di-hydroxy-, cyclic methyl phenyl-orthoacetate	EtOH	243(4.20)	32-1115-66
$C_{30}H_{38}O_8S$ Cortisol, 17-acetate 21-p-toluene-sulfonate	EtOH	230(4.34)	32-1115-66
$C_{30}H_{39}Br_3N_4$ Biladiene-ac, 1-bromo-1,19-dideoxy-8,18-diethyl-2,3,7,12,13,17,19-hexamethyl-, dihydrobromide	$CHCl_3$	293(3.49),377(4.25), 439(4.19),463(4.49), 536(5.24)	39-0022-66C
$C_{30}H_{40}N_4O_3$ Bilene b, 1-carbethoxy-2,3,7,8,12,-13,17,18-octamethyl-1-deoxy-17,18-dihydro-	EtOH	227(4.14),283(4.30), 319s(3.61),454(4.46)	39-1155-66C
$C_{30}H_{40}O_6$ Jaborosalactone A, acetate	MeOH	219(4.11)	78-1121-66
$C_{30}H_{40}O_7$ Taxicin I, 2-O-methyl-5-hydro-cinnamate	EtOH	275(3.73)	39-1933-66C
$C_{30}H_{41}NO_3$ Paspaline, O-acetyl-	EtOH	234(4.2),284(3.7), 292s(3.6)	33-1907-66

Compound	Solvent	$\lambda_{max}(\log \epsilon)$	Ref.
$C_{30}H_{42}O$			
8-Apo-β-carotenal, 7',8'-dihydro-	hexane	408(4.80),431(4.90), 458(4.82)	39-2166-66C
$C_{30}H_{42}O_4$			
Holothurinogenone, 17-deoxy-22,25-epoxy-	EtOH	237s(4.10),243(4.15), 252s(4.00)	78-1857-66
$C_{30}H_{42}O_5$			
Holothurinogenone, 22,25-epoxy-	EtOH	236s(4.13),244(4.16), 252s(4.01)	78-1857-66
$C_{30}H_{42}O_7$			
Taxicin I, 4,16-dihydro-2-O-methyl-5-hydrocinnamate	EtOH	281(3.79)	39-1933-66C
$C_{30}H_{42}O_7S_2$			
5α-Card-20(22)-enolide, 3β,15,16β-trihydroxy-19-oxo-, cyclic trimethylene mercaptal, 3,15-diacetate	EtOH	214(4.13)	33-1662-66
$C_{30}H_{42}O_{10}$			
Scillirubroside	MeOH	298(3.72)	33-0030-66
$C_{30}H_{44}$			
2,6,8,10,12,14,16,18,22-Tetracosanonaene, 2,6,10,15,19,23-hexamethyl-	hexane	379(4.92),400(5.12), 424(5.12)	39-2154-66C
4,6,8,10,12,14,16,18,20-Tetracosanonaene, 2,6,10,15,19,23-hexamethyl-	hexane	414(4.92),435(5.16), 465(5.16)	39-2154-66C
$C_{30}H_{44}BrN_3O_3$			
Teleocidin B, dihydro-, bromoacetate	MeOH	232(4.48),287(3.91)	88-2523-66
$C_{30}H_{44}O_2$			
Clerodolone diosphenol	EtOH	272(3.90)	78-2377-66
$C_{30}H_{44}O_3$			
Bourjotinolone diketone, anhydro-derivative	EtOH	280(3.8)	12-0455-66
$C_{30}H_{44}O_3S_2$			
Withaferin A, thioketal derivative	EtOH	226(3.86)	39-1753-66C
$C_{30}H_{44}O_4$			
Glabrolide	EtOH	243(4.06)	32-0772-66
Holothurinogenin, 17-deoxy-22,25-epoxy-	EtOH	237s(4.18),244(4.23), 252s(4.07)	78-1857-66
Isoglabrolide	MeOH	235(3.94)	32-0843-66
$C_{30}H_{44}O_5$			
Holothurigenin	EtOH	244(4.15)	95-0637-66
Holothurinogenin, 22,25-epoxy-	EtOH	237s(4.12),244(4.16), 252s(4.01)	78-1857-66
A-Homo-B-nor-25D-spirost-3-en-4a-one, 5β-(acetoxymethyl)-	EtOH	226(3.74)	78-1053-66
$C_{30}H_{44}O_6$			
Cyclosenegenin	EtOH	209(3.83),230(3.43)	35-1544-66

Compound	Solvent	λ_{max}(log ϵ)	Ref.
$C_{30}H_{44}O_8$			
Drevogenin A, 3-acetate	EtOH	275(2.00)	33-1632-66
$C_{30}H_{46}D_2$			
Oleana-11,13(18)-diene-28,28-d_2	EtOH	242(4.36),250(4.41), 259(4.24)	44-1945-66
$C_{30}H_{46}N_2O$			
Urea, 1-hexadecyl-1-methyl-3,3-diphenyl-	EtOH	255(4.17)	34-0436-66
$C_{30}H_{46}O$			
Glochidone	n.s.g.	228(4.00)	78-1513-66
$C_{30}H_{46}O_2$			
2,5-Cyclohexadienone, 2,6-di-tert-butyl-4-(3,5-di-tert-butyl-4-hydroxybenzyl)-4-methyl-	EtOH	232(4.1)	1-2211-66
$C_{30}H_{46}O_4$			
Cholest-8-ene-7,11-dione, 3β-acetoxy-14α-methyl-	n.s.g.	271(3.92)	35-0790-66
Liquiritic acid	MeOH	249(4.04)	32-0833-66
Momordic acid	EtOH	203(3.79),295(1.39)	88-5137-66
$C_{30}H_{46}O_6$			
A-Nor-5β-ergost-24-en-26-oic acid, 5-formyl-6β,22-dihydroxy-1-oxo-, δ-lactone, 5-(dimethyl acetal)	EtOH	226(3.90)	39-1757-66C
$C_{30}H_{46}O_8$			
Drevogenin A, dihydro-, 3-acetate	C_6H_{12}	284(1.49)	33-1632-66
$C_{30}H_{47}O_6P$			
Phosphonic acid, (3β-hydroxy-17,22-dioxo-D-homo-24-norchola-5,17a(20)-dien-21-yl)-, diisopropyl ester	EtOH	273(3.91)	78-1625-66
$C_{30}H_{47}O_8P$			
Phosphonic acid, [2-[(6β,17-dihydroxy-3-oxoandrost-4-en-17β-yl)carbonyl]-ethyl]-, diisopropyl ester, 17-acetate	EtOH	238(4.23)	78-1625-66
$C_{30}H_{48}$			
Hopa-15,17(21)-diene	EtOH	244(4.44),252(4.49), 261(4.31)	39-1564-66C
A-Neoadianene	EtOH	240(4.30),250(4.31), 255s(4.10)	39-1251-66C
2,6,10,14,16,18,22-Tetracosaheptaene, 2,6,10,14,19,23-hexamethyl-	EtOH	273s(--),285(4.59), 297s(--)	39-2154-66C
Unnamed diene A	hexane	244(4.30)	78-3469-66
$C_{30}H_{48}O$			
E:B-Friedohop-1(10),5-dien-3β-ol	EtOH	234(4.09),241(4.10)	39-1251-66C
Hop-5-en-7-one	EtOH	241(4.17)	39-1556-66C
Lup-1-en-3-one	EtOH	228(4.11)	32-0220-66
Olean-12-en-11-one	EtOH	250(4.10)	44-1945-66

Compound	Solvent	$\lambda_{max}(\log \epsilon)$	Ref.
$C_{30}H_{48}O_2$			
Cholest-5-en-3-one, 2-(hydroxy-methylene)-4,4-dimethyl-	EtOH	280(3.87)	39-1277-66C
Lupane-1,3-dione	EtOH-HCl	257(4.28)	32-0220-66
	EtOH-NaOH	288(4.37)	32-0220-66
$C_{30}H_{48}O_3$			
A-Homo-B-nor-5β-cholest-3-en-4a-one, 3-hydroxy-5-methyl-, acetate	EtOH	230(3.70)	78-1421-66
Saikogenin B	EtOH	282(3.95)	88-0701-66
Saikogenin C	EtOH	242(4.42),251(4.49), 260(4.29)	88-0701-66
6,7-Seco-6-norhopan-7-oic acid, 22-hydroxy-5-oxo-	EtOH	284(4.00)	39-1556-66C
	EtOH-base	284(4.00)	39-1556-66C
$C_{30}H_{48}O_4$			
Saikogenin A	EtOH	242(4.43),251(4.48), 260(4.29)	88-0701-66
	EtOH	242(4.43),250(4.48), 260(4.29)	94-1023-66
Saikogenin D	EtOH	242(4.40),252(4.46), 261(4.28)	88-0701-66
$C_{30}H_{48}O_5$			
Isoescigenin	EtOH	204(3.90)	78-0351-66
$C_{30}H_{48}O_6$			
Isoescigenin epoxide	EtOH	204(3.40)	78-0351-66
$C_{30}H_{49}ClO_3$			
Bourjotinolone C	EtOH	210(4.1)(end abs.)	12-0455-66
$C_{30}H_{49}NO_3$			
Cholest-2-en-4-one, 3-methoxy-, O-acetyloxime	EtOH	256(3.95)	5-0180-66B
$C_{30}H_{50}O$			
Lanost-8-en-2-one	C_6H_{12}	287(1.40),295(1.40)	22-0897-66
Lanost-8-en-3-one	C_6H_{12}	290(1.60)	22-0897-66
$C_{30}H_{50}O_2$			
Lanost-8-en-2-one, 3β-hydroxy-	C_6H_{12}	280(1.60)	22-0897-66
$C_{30}H_{50}O_3$			
1H-Benz[e]inden-6-propionic acid, 3α-(1,5-dimethylhexyl)-2,3,3a,4,5,5aβ,-6,7,9aα,9bβ-decahydro-3aα,6α,8-trimethyl-7-methylene-, 2-hydroxy-ethyl ester	EtOH	233(4.31),239(4.36),	78-1421-66
$C_{30}H_{50}O_4$			
Bryodulcosigenin	EtOH	295(1.8)	5-0162-66D
2-Chromantridecanoic acid, 6-hydroxy-α,ε,ι,2,5,7,8-heptamethyl-, methyl ester	EtOH	292.0(3.53)	73-2434-66
$C_{30}H_{51}BrO$			
Shionan-3-one, 4α-bromo-	n.s.g.	310(2.00)	22-1670-66

Compound	Solvent	$\lambda_{max}(\log \epsilon)$	Ref.
$C_{30}H_{53}NO$			
α-Tocopheramine, N-methyl-	EtOH	290(3.37)	33-2297-66
γ-Tocopheramine, N,N-dimethyl-	EtOH	293(3.44)	33-2297-66
γ-Tocopheramine, N-ethyl-	EtOH	238(3.94),305(3.49)	33-2297-66

Compound	Solvent	λ max(log ε)	Ref.
$C_{31}H_{19}ClO_5$			
9-(1-Naphthyl)-1,2:5,6-dibenzo-xanthylium perchlorate	HOAc	480(3.38)	24-1822-66
9-(1-Naphthyl)-1,2:6,7-dibenzo-xanthylium perchlorate	HOAc	490(3.41)	24-1822-66
9-(1-Naphthyl)-2,3:5,6-dibenzo-xanthylium perchlorate	HOAc	580(3.62)	24-1822-66
9-(1-Naphthyl)-3,4:5,6-dibenzo-xanthylium perchlorate	HOAc	480(3.32)	24-1822-66
$C_{31}H_{20}Cl_3N_2P$			
Phosphorane, tris(p-chlorophenyl)-(fluoren-9-ylidenehydrazono)-	dioxan	212(4.72),240(4.87), 298(3.92),310(3.93), 366(4.29),372(4.29), 379(4.30)	5-0001-66H
$C_{31}H_{20}N_4O_{14}$			
[3,3'-Spirobichroman]-8,8'-diol, bis(3,5-dinitrobenzoate)	PrOH	214(4.93),304(3.42)	78-0199-66
$C_{31}H_{21}N_5O_3$			
2-Naphthanilide, 3-hydroxy-4-[m-(3,4-dihydro-4-oxo-2-quinazolinyl)-phenylazo]-	dioxan	495(4.36),518s(4.35)	32-0279-66
2-Naphthanilide, 3-hydroxy-4-[p-(3,4-dihydro-4-oxo-2-quinazolinyl)-phenylazo]-	dioxan	510(4.49),532s(4.48)	32-0279-66
$C_{31}H_{22}BrN_2P$			
Phosphorane, (p-bromophenyl)(fluoren-9-ylidenehydrazono)diphenyl-	dioxan	214(4.66),246(4.80), 297(3.90),308(3.90), 377(4.33)	5-0001-66H
$C_{31}H_{22}ClN_2P$			
Phosphorane, (p-chlorophenyl)(fluoren-9-ylidenehydrazono)diphenyl-	dioxan	240(4.76),247(4.77), 297(3.90),308(3.90), 372(4.34)	5-0001-66H
$C_{31}H_{22}Cl_2N_4O_7$			
9H-Purine, 2,6-dichloro-9-β-D-ribo-furanosyl-, 2,3,5-tribenzoate	EtOH	231(4.60),275(4.06)	94-1377-66
$C_{31}H_{23}ClN_3O_2P$			
Phosphorane, (p-chlorophenyl)[(p-nitro-α-phenylbenzylidene)hydrazono]-diphenyl-	dioxan	308s(3.9),416(4.22)	5-0001-66H
$C_{31}H_{23}Cl_2N_2P$			
Phosphorane, [[bis(p-chlorophenyl)-methylene]hydrazono]triphenyl-	dioxan	240(4.42),245(4.48), 350(4.12)	5-0001-66H
$C_{31}H_{23}N_2P$			
Phosphorane, (fluoren-9-ylidene-hydrazono)triphenyl-	dioxan	227(4.64),247(4.83), 296(3.91),308(3.93), 375(4.47),388(4.47)	5-0001-66H
$C_{31}H_{23}N_5O_3$			
5-Norbornene-2,3-dicarboximide, 5-[hydroxydi(2-pyridyl)methyl]-7-[di(2-pyridyl)methylene]-	MeOH	249(4.27),254(4.27), 260(4.27),267(4.24)	87-0537-66

Compound	Solvent	λ_{max}(log ϵ)	Ref.
$C_{31}H_{24}N_2O$			
Urea, 1-biphenyl-1,3,3-triphenyl-	EtOH	280(4.46)	34-0436-66
$C_{31}H_{24}N_2O_2$			
Urea, 1-(p-phenoxyphenyl)-1,3,3-triphenyl-	EtOH	265(4.40)	34-0436-66
$C_{31}H_{24}N_3O_2P$			
Phosphorane, [(p-nitro-α-phenylbenzylidene)hydrazono]triphenyl-	dioxan	310s(3.8),421(4.22)	5-0001-66H
$C_{31}H_{24}O$			
2-Fulvenemethanol, α,α,6,6-tetraphenyl-	MeOH	328(4.40)	44-2149-66
$C_{31}H_{25}BO_2$			
5,7,6-Dioxaboraazulene, 4,6,6,8-tetraphenyl-	$C_2H_4Cl_2$	304(4.32),370(4.03), 467(4.15)	80-1409-66
$C_{31}H_{25}N_5O_4$			
5-Norbornene-2,3-dicarboximide, 1,4-bis(α-hydroxy-α,α-di-2-pyridylmethyl)-	MeOH	255s(4.11),260(4.15), 266s(4.04)	44-2149-66
$C_{31}H_{26}$			
Cyclopentadiene, 5-ethyl-1,2,3,4-tetraphenyl-	EtOH	240(4.26),332(4.05)	22-3775-66
$C_{31}H_{27}FIN_3$			
1-(p-Fluorophenyl)-5,6,7,8-tetrahydro-4-[(1-methyl-2(1H)-quinolylidene)-methyl]-2-phenylquinazolinium iodide	MeOH	324(3.93),465(4.78), 492(4.99)	87-0758-66
$C_{31}H_{28}N_4O$			
2-Fulvenemethanol, α,α,6,6-tetra-(6-methyl-2-pyridyl)-	MeOH	325(4.35)	44-2149-66
$C_{31}H_{28}O$			
2-Cyclopenten-1-ol, 1-ethyl-2,3,4,5-tetraphenyl-	EtOH	258(4.03)	22-3775-66
$C_{31}H_{28}O_{14}$			
Ergochrysin A	MeOH	240(4.31),267(4.29), 336(4.29)	24-3863-66
Ergochrysin B	MeOH	242(4.28),260(4.26), 336(4.21)	24-3863-66
ψ-Ergoxanthin	EtOH	272(4.33),370(3.86)	78-2359-66
$C_{31}H_{29}N_5O_4$			
Adenosine, 2'-deoxy-5'-O-trityl-, 3'-acetate	MeOH	259.5(4.20)	69-0224-66
$C_{31}H_{30}O_{15}$			
Ergochrome CD	MeOH	241(4.34),374(3.79)	24-3875-66
$C_{31}H_{31}IN_4S$			
1-(p-Dimethylaminophenyl)-5,6,7,8-tetrahydro-4-[(3-methyl-2-benzothiazolinylidene)methyl]-2-phenylquinazolinium iodide	MeOH	449(4.94)	87-0758-66

Compound	Solvent	λ_{max}(log ε)	Ref.
$C_{31}H_{31}N_5O_{13}$			
2(1H)-Pyrimidinone, 1-(3,4,6-tri-O-acetyl-2-deoxy-2-carbobenzoxyamino-β-D-glucopyranosyl)-4-(p-nitrobenzamido)-	pH 1	278(4.28)	44-4014-66
	pH 7	270(4.24)	44-4014-66
	pH 13	325(4.18)	44-4014-66
$C_{31}H_{32}O_{13}$			
Chromomycinone, anhydro-, pentaacetate	EtOH	242(4.33),260(4.82), 360(3.45)	78-2761-66
$C_{31}H_{33}N_5O_{11}$			
2(1H)-Pyrimidinone, 1-(3,4,6-tri-O-acetyl-2-deoxy-2-carbobenzoxyamino-β-D-glucopyranosyl)-4-(p-aminobenzamido)-	pH 1	260(4.23),309(4.04)	44-4014-66
	pH 7	252(4.30),324(4.40)	44-4014-66
	pH 13	223(4.04),332(4.49)	44-4014-66
$C_{31}H_{34}O_{13}$			
Chromomycinone, 2-deoxy-, pentaacetate	EtOH	258(--),300(--), 356(--)	78-2761-66
$C_{31}H_{34}O_{14}$			
Chromomycinone, pentaacetate	EtOH	263(4.59),272(4.60), 390(4.07)	78-2761-66
Isochromomycinone, pentaacetate	EtOH	260(4.73),302(3.83), 366(3.43)	78-2761-66
$C_{31}H_{35}NO_4$			
Paspalicin	EtOH	231(4.68),250s(4.25), 275s(4.1)	33-1907-66
$C_{31}H_{35}N_3O_2$			
Zingiberene adduct with p-phenylazophenylmaleic anil	C_6H_{12}	325(4.37)	12-0283-66
$C_{31}H_{35}N_4O_2P$			
Aniline, 4,4',4"-[(p-nitrobenzylidene)phosphoranylidyne]tris-[N,N-dimethyl-	$C_2H_4Cl_2$	245s(4.09),275s(4.38), 305(4.67),531(4.49)	97-0068-66
$C_{31}H_{36}N_4$			
Porphine, 3,8,18-triethyl-2,7,12,13,17-pentamethyl-	$CHCl_3$	269(3.86),333s(4.16), 379s(3.90),400(5.14), 500(4.06),536(3.93), 569(3.74),623(3.63)	39-0022-66C
$C_{31}H_{36}N_4O_2$			
Isocinchophyllamine	acid	275(4.00)	22-2309-66
	neutral	207(4.30),228(4.38), 282(3.95)	22-2309-66
	base	205(4.50),228(4.45), 281(3.98)	22-2309-66
$C_{31}H_{38}N_4O_2$			
Bilatriene-abc, 3,7,13,17-tetraethyl-2,8,12,18-tetramethyl-	EtOH	288(4.22),309(4.29), 367(4.62),656(4.09)	39-1155-66C
Cinchophyllamine, dihydro-	EtOH	208(4.38),228(4.35), 282(4.10)	22-2309-66
Isocinchophyllamine, dihydro-	acid	274(4.00)	22-2309-66
	neutral	207(4.30),228(4.39), 282(3.95)	22-2309-66

Compound	Solvent	λ_{max}(log ε)	Ref.
Isocinchophyllamine, (cont.) dihydro-	base	205(4.50),228(4.45), 281(3.98)	22-2309-66
$C_{31}H_{38}O$			
Indene, 3-(1-hydroxyethyl)-1-retinylidene-	EtOH	438(4.89)	54-0339-66
$C_{31}H_{39}NO_5$			
Pregna-4,6-diene-3,20-dione, 17-hydroxy-6-[(p-methoxyanilino)-methyl]-, acetate	EtOH	248(4.23),286(4.37)	25-0497-66
$C_{31}H_{39}N_5O_7$			
L-Leucine, N-benzyloxycarbonyl-L-glutaminyl-L-tryptophanyl-, methyl ester	EtOH	222(4.65),275s(3.73), 282(3.77),290(3.71)	44-3400-66
$C_{31}H_{40}N_4$			
Biladiene-ac, 1,19-dideoxy-1,2,3,7,8,-10,10,12,13,17,18,19-dodecamethyl-, dihydrobromide	EtOH	459(4.43),516(5.14), 610(3.27)	39-0098-66C
	$CHCl_3$	461(4.60),522(5.32), 628(3.24)	39-0098-66C
$C_{31}H_{40}N_4O_2$			
Biladiene-ab, 3,7,13,17-tetraethyl-2,8,12,18-tetramethyl-	EtOH	328(4.59),566(4.46)	39-1155-66C
$C_{31}H_{40}O_8$			
Khayasin (benzene solvate)	MeOH	209(4.10)	39-2127-66C
$C_{31}H_{41}Br_3N_4$			
Biladiene-ac, 1-bromo-1,19-dideoxy-3,8,18-triethyl-2,7,12,13,17,19-hexamethyl-, dihydrobromide	$CHCl_3$	293(3.54),375(4.18), 442s(4.24),461(4.47), 505s(4.65),532(5.14)	39-0022-66C
$C_{31}H_{42}N_4O_2$			
Bilene-b, 3,7,13,17-tetraethyl-2,8,12,18-tetramethyl-, hydrobromide	EtOH	210(4.53),379(3.78), 493(4.89)	39-1155-66C
$C_{31}H_{42}N_4O_5$			
Pandaminone	EtOH	266(3.87)	6-0083-66
$C_{31}H_{44}N_4O_2$			
Bilene-b, 3,7,13,17-tetraethyl-2,8,12,18-tetramethyl-2,3-dihydro-, hydrobromide	$CHCl_3$	374(3.82),497(4.97)	39-1155-66C
$C_{31}H_{44}N_4O_5$			
Pandamine	EtOH	230(3.59),277(1.80)	6-0083-66
$C_{31}H_{45}NO_2$			
14'-Apo-β-carotenoic acid, 15'-cyano-, octyl ester	C_6H_6-EtOH	435(4.76)	54-0334-66
$C_{31}H_{46}N_4O_2$			
Bilene-b, 3,7,13,17-tetraethyl-2,8,12,18-tetramethyl-2,3,17,18-tetrahydro-	EtOH	227(4.03),333(3.49), 454(4.43)	39-1155-66C

Compound	Solvent	$\lambda_{max}(\log \epsilon)$	Ref.
$C_{31}H_{46}O_8$			
Fusidic acid metabolite E	EtOH	222(3.91)	1-1599-66
$C_{31}H_{48}O_2$			
Leucotyliddienoic acid, methyl ester	EtOH	244(4.29),252(4.32), 261(4.15)	88-0607-66
A-Nor-18α-olean-1-ene, 2-acetyl-19β,28-epoxy-	EtOH	239(4.1)	73-3174-66
Stigmasta-7,22,25-trien-3β-ol, acetate	ether	204(4.15)	24-3559-66
$C_{31}H_{48}O_4$			
Liquiritic acid, methyl ester	MeOH	247.5(4.06)	32-0833-66
$C_{31}H_{48}O_5$			
5α-Cholest-7-en-6-one, 2β,3β-diacetoxy-	EtOH	244(4.11)	33-1591-66
Olean-12-ene-30-carboxylic acid, 3β,7β-dihydroxy-, methyl ester	MeOH	248(4.05)	32-0820-66
$C_{31}H_{48}O_6$			
5α,14α-Cholest-7-en-6-one, 2β,3β-diacetoxy-14-hydroxy-	EtOH	240(4.06)	33-1591-66
$C_{31}H_{49}NO_4$			
Cholesterin, 7-acetoximino-, acetate	EtOH	236(4.29)	5-0180-66B
$C_{31}H_{50}N_2O$			
Formimidic acid, N-(2-cyano-3α-cholest-2-en-3-yl)-, ethyl ester	EtOH	254(3.94)	32-0152-66
$C_{31}H_{50}N_2O_3$			
Cyclokoreanine B, N-acetyl-, acetate	EtOH	213(4.20)	39-1805-66C
$C_{31}H_{50}O_2$			
Lupan-3-one, 2-(hydroxymethylene)-	EtOH	292(3.94)	39-1277-66C
Stigmasta-7,25-dien-3β-ol, acetate	ether	202(3.90)	24-3559-66
$C_{31}H_{50}O_3$			
Cholest-5-en-3-one, 2α-acetoxy-4,4-dimethyl-	C_6H_{12}	289(1.80)	22-0897-66
$C_{31}H_{50}O_4$			
Cholestan-7-one, 6-acetyl-3β-hydroxy-, acetate	EtOH	286(1.92)	78-2039-66
19-Nor-5β-cholest-9-ene-4β,7β-diol, 5-methyl-, diacetate	n.s.g.	209(4.07)	39-1010-66C
$C_{31}H_{50}O_5$			
2-Chromantridecanoic acid, 6-acetoxy-α,ε,ι,2,5,7,8-heptamethyl-	EtOH	285.0(3.34)	73-2434-66
$C_{31}H_{50}O_6$			
Lumifusidic acid, 24,25-dihydro-	EtOH	204(4.00)	87-0015-66
$C_{31}H_{51}Na_2O_7P$			
Chroman, 2,5,7,8-tetramethyl-2-(4,8,12-trimethyl-13-hydroxytridecyl)-6-acetoxy-, disodium phosphate	EtOH	285(3.335)	73-4598-66

Compound	Solvent	$\lambda_{max}(\log \epsilon)$	Ref.
$C_{31}H_{52}O_3$			
5α-Cholestan-2-one, 3β-acetoxy-4,4-dimethyl-	C_6H_{12}	293(1.48)	22-0897-66
5α-Cholestan-3-one, 2α-acetoxy-4,4-dimethyl-	C_6H_{12}	290(1.60)	22-0897-66
$C_{31}H_{55}NO$			
α-Tocopheramine, N,N-dimethyl-	EtOH	288(3.22)	33-2297-66
α-Tocopheramine, N-ethyl-	EtOH	291(3.38)	33-2297-66
$C_{31}H_{60}O_2$			
2,6-Hentriacontanedione	$CHCl_3$	278(1.76)	28C-1325-66A

Compound	Solvent	$\lambda_{max}(\log \epsilon)$	Ref.
$C_{32}H_{16}N_8O_2U$			
Phthalocyanine, uranyl derivative	$C_{10}H_7Br$	605(3.57),647(3.87), 666(4.14),700(4.17)	30-1318-66B
	quinoline	647(3.69),700(3.56)	30-1318-66B
$C_{32}H_{18}Fe_2$			
Dodecahexayne, diferrocenyl-	$CHCl_3$	283(4.76),301(4.92), 312(4.97),333(5.03), 350(4.81),398(4.13), 506s(4.16)	101-0399-66B
$C_{32}H_{18}O_5$			
Rotoxen-12(6H)-one, dimer	EtOH	264(4.54),295(4.44), 315(4.56),330(4.46)	39-0544-66C
$C_{32}H_{21}N_5O_5$			
5-Quinazolinecarboxylic acid, 3,4-dihydro-2-[[m-(2-hydroxy-3-N-phenylcarboxamido-1-naphthyl)azo]phenyl]-4-oxo-	pH 13	450(4.27),505(4.29)	7-1394-66
5-Quinazolinecarboxylic acid, 3,4-dihydro-2-[[p-(2-hydroxy-3-N-phenylcarboxamido-1-naphthyl)azo]phenyl]-4-oxo-	pH 13	460(4.30),520(4.34)	7-1394-66
$C_{32}H_{22}$			
Anthracene, 9-benzylidene-9,10-dihydro-10-(1-naphthylmethylene)-	EtOH	220(4.94),335(4.38)	78-0141-66B
Anthracene, 1,4,9-triphenyl-	$CHCl_3$	269(4.86),368(3.85), 385(4.04),404(4.02)	44-4082-66
Anthracene, 2,3,9-triphenyl-	$CHCl_3$	293(4.90),361(3.79), 380(3.93),401(3.83)	44-4082-66
$C_{32}H_{22}Br_2O_2$			
Bianthrone, 7,7'-dibromo-2,2',3,3'-tetramethyl-	CHCl	244(4.62),324(4.06), 412(4.26)	78-0349-66A
$C_{32}H_{22}O$			
Cyclopenta[b]pyran, 2,4,5,6-tetraphenyl-	C_6H_{12}	276(4.34),406(4.36), 524(2.78)	39-0859-66C
	$HOAc-HClO_4$	260(4.10),355(4.17), 445(4.58)	39-0859-66C
$C_{32}H_{22}O_2S$			
Anthracene, 9,10-diphenyl-2-(phenylsulfonyl)-	$CHCl_3$	300(4.52),330(3.51), 370(3.35),380(3.42), 410(3.4)	22-2942-66
$C_{32}H_{22}O_6$			
Rotexen-12(6H)-one, 6a,12a-dihydro-, dehydrodimer	EtOH	257(4.39),327(3.85)	39-0544-66C
	EtOH-KOH	257(4.40),327(3.86)	39-0544-66C
$C_{32}H_{22}O_{12}$			
Aurofusarin, di-O-methyl-	EtOH	227(4.72),269(4.66), 360(4.08)	88-4855-66
$C_{32}H_{22}S$			
Anthracene, 9,10-diphenyl-1-(phenylthio)-	$CHCl_3$	270(4.76),405(4.03), 424(3.96)	22-2939-66

Compound	Solvent	λ_{max}(log ε)	Ref.
Anthracene, 9,10-diphenyl-2-(phenylthio)-	$CHCl_3$	275(4.79),300(4.43), 360(3.79),380(3.94), 400(3.89)	22-2942-66
$C_{32}H_{24}$			
Anthracene, 4a,10-dihydro-9,10,10-triphenyl-	isopentane-$C_6H_{11}Me$	250(4.25),261(4.16), 373s(3.87),386(3.98), 405(3.97),426s(3.71)	88-2193-66
Bicyclo[4.2.0]octa-1,3,5-triene, 7,7,8,8-tetraphenyl-	isopentane-$C_6H_{11}Me$	260s(3.56),266(3.58), 273(3.57),276s(2.73)	88-2193-66
1,3-Cyclohexadiene, 5,6-bis-(diphenylmethylene)-	isopentane-$C_6H_{11}Me$	285(4.48),520(3.96)	88-2193-66
$C_{32}H_{25}N_2OP$			
Phosphorane, (fluoren-9-ylidene-hydrazono)(p-methoxyphenyl)-diphenyl-	dioxan	214(4.65),248(4.85), 296(3.91),307(3.91), 380(4.36),394(4.36)	5-0001-66H
$C_{32}H_{25}N_3$			
Dibenzo[b,h][1,6]naphthyridine, 7,12-dihydro-7-(phenylimino)-6-(5,6,7,8-tetrahydro-2-naphthyl)-	C_6H_{12}	245(4.66),271(4.75), 318(4.32),332(4.35), 350(3.57),368(3.65), 389(3.75),447(3.89)	56-0621-66
$C_{32}H_{26}$			
Fulvene, 6,6-dimethyl-1,2,3,4-tetraphenyl-	EtOH	262(4.34),312(4.13), 405s(3.90)	22-3775-66
$C_{32}H_{26}O_5$			
3,6(2H,7H)-Oxepindione, 4,5-bis-(o-methoxyphenyl)-2,7-diphenyl-	EtOH	262(4.37),300(4.34)	88-4671-66
$C_{32}H_{26}O_7$			
Naphtho[2,3-b]furan-4-carboxaldehyde, 9-methoxy-3-[[(6-methoxy-7-methyl-1H,3H-benzo[h]furo[4,3,2-de]-2-benzopyran-1-yl)oxy]methyl]-8-methyl-	EtOH	250(4.30),338(3.52), 352(3.53)	78-0685-66
$C_{32}H_{27}Cl$			
Cyclopentadiene, 5-chloro-2-isopropyl-1,3,4,5-tetraphenyl-	EtOH	246(4.28),338(3.74)	22-3775-66
$C_{32}H_{27}ClO_5$			
Pyrylium, 2,4-diphenyl-6-[2-(4,6,8-trimethyl-1-azulenyl)vinyl]-, perchlorate	MeOH-$HClO_4$	250(3.81),340(3.87), 370(3.92),465(3.32), 660(3.64)	65-1728-66
$C_{32}H_{27}N_2OP$			
Phosphorane, [(diphenylmethylene)-hydrazono](p-methoxyphenyl)-diphenyl-	dioxan	240(4.68),244(4.83), 342(4.08)	5-0001-66H
$C_{32}H_{28}$			
Cyclopentadiene, 5-isopropyl-1,2,3,4-tetraphenyl-	EtOH	252(4.34),330(4.06)	22-3775-66
2,3,4-Hexatriene, 2,5-dibenzyl-1,6-diphenyl-	C_6H_{12}	274.5(4.40)	78-2621-66

Compound	Solvent	λ_{max}(log ε)	Ref.
$C_{32}H_{28}HgI_2$			
Mercury, bis(1-iodo-2,2-dibenzylvinyl)-	C_6H_{12}	258(4.14)	78-2621-66
$C_{32}H_{28}N_2O_6$			
Indole, 2,2'-(2,5-dihydroxy-p-phenylene)bis[1-acetyl-3-methyl-, diacetate	EtOH	240s(4.54),309(4.43)	44-3321-66
$C_{32}H_{28}O$			
2,4-Cyclopentadien-1-ol, 1-isopropyl-2,3,4,5-tetraphenyl-	EtOH	249(4.64),339(3.76)	22-3775-66
$C_{32}H_{29}FIN_3$			
4-[(1-Ethyl-2-quinolylidene)methyl]-1-(p-fluorophenyl)-5,6,78-tetrahydro-2-phenylquinazolinium iodide	MeOH	322(3.93),465(4.77), 493(4.99)	87-0758-66
$C_{32}H_{30}$			
2,4-Hexadiene, 2,5-dibenzyl-1,6-diphenyl-	C_6H_{12}	256(4.5),338(4.42), 278(4.4)	78-2621-66
$C_{32}H_{30}BNO$			
1-(β-Hydroxyphenethyl)-2-picolinium hydroxide, inner salt, compound with triphenylboron	MeOH	270s(3.7)	83-0385-66
$C_{32}H_{30}CoN_4$			
Methylmalondialdehyde dianil, cobalt chelate	$CHCl_3$	420(4.61),550(3.33)	65-1372-66
$C_{32}H_{30}CuN_4$			
Methylmalondialdehyde dianil, copper chelate	$CHCl_3$	326(4.29),400(4.58), 543(3.05),690(3.01)	65-1372-66
$C_{32}H_{30}N_4Ni$			
Methylmalondialdehyde dianil, nickel chelate	$CHCl_3$	410(4.3),530(3.35)	65-1372-66
$C_{32}H_{30}O_{14}$			
Secalonic acid A	MeOH	247(4.26),340(4.49)	24-3842-66
Secalonic acid B	MeOH	240(4.26),339(4.53)	24-3842-66
Secalonic acid C	MeOH	236(4.27),339(4.52)	24-3842-66
$C_{32}H_{32}Cl_2O_2$			
Naphtho[1',2',3',4':7,8,14,15]-19-nor-cholesta-1,3,5,7,9,14-hexaene-5',8'-dione, 6',7'-dichloro-	$CHCl_3$	265(4.74),313(4.37), 340(4.42),420(3.72), 620(3.49)	5-0106-66J
$C_{32}H_{32}CuN_2O_4$			
Benzylidenimine, 2-hydroxy-3-methoxy-N-(α-methylbenzyl)-, copper chelate	MeOH	270(4.15),373(3.53)	65-1590-66
$C_{32}H_{32}N_2O_6$			
Hydroquinone, 2,5-bis(1-acetyl-3-methylindolin-3-yl)-, diacetate	EtOH	254(4.47),279(4.00), 289(3.92)	44-3321-66
$C_{32}H_{32}N_4O_8$			
meso-Deuteroporphinogen, tetraoxo-	benzene	340(4.60)	5-0092-66J

Compound	Solvent	$\lambda_{max}(\log \epsilon)$	Ref.
$C_{32}H_{32}O_{15}$			
Ergochrome AD	MeOH	246(4.30),338(4.26)	24-3875-66
Ergochrome BD	MeOH	249(4.21),339(4.09)	24-3875-66
$C_{32}H_{33}ClO_5$			
2-Benzopyrylium, 8-[2-(3,8-dimethyl-5-isopropyl-1-azulenyl)vinyl]-5,6,7,8-tetrahydro-3-phenyl-, perchlorate	MeOH-$HClO_4$	245(4.34),320(4.04), 357(4.00),680(4.56)	65-1728-66
$C_{32}H_{34}O_{16}$			
Ergochrome DD	MeOH	251(4.46),361(3.91)	24-3875-66
Flavone, 4',5-dihydroxy-3,6,7-trimethoxy-, 4'-β-D-glucoside, tetraacetate	EtOH	328(4.03),373(4.06)	24-3222-66
$C_{32}H_{36}CuN_4O_4$			
5,5-Bi(4-carbethoxy-3,3',4,5'-tetramethyldipyrromethenyl), copper complex	$CHCl_3$	288(4.15),345s(3.97), 410(4.49),480(4.62), 771s(3.96),825(4.25)	39-0098-66C
Ethane, 1,2-bis(4-carbethoxy-3,3',4'-trimethyldipyrromethen-5-yl)-, copper complex	$CHCl_3$	276(4.17),426(4.29), 476(4.84),520(4.53)	39-0098-66C
$C_{32}H_{36}N_2O_4$			
Hydroquinone, 2,5-bis(1-ethyl-3-methylindolin-3-yl)-, diacetate	EtOH	257(4.24),308(3.78)	44-3321-66
$C_{32}H_{36}O_{11}$			
Hirtin	n.s.g.	217(4.01),277(3.87)	77-0206-66
	base	326(3.65)	77-0206-66
$C_{32}H_{36}O_{13}$			
Chromomycinone, 3',4'-O-isopropylidene-, tetraacetate	EtOH	260(4.84),301(3.80), 360(3.47)	78-2761-66
Isochromomycinone, 3',4'-O-isopropylidene-, tetraacetate	EtOH	260(4.78),302(3.91), 362(3.45)	78-2761-66
$C_{32}H_{37}N_3O_3$			
2H-Pyrrole-5-carboxylic acid, 3-ethyl-2-[[4-ethyl-5-[(4-ethyl-3,5-dimethyl-2H-pyrrol-2-ylidene)methyl]-3-methylpyrrol-2-yl]hydroxymethylene]-4-methyl-, benzyl ester, dihydriodide	EtOH	364(4.20),498(4.51)	78-0241-66A
$C_{32}H_{37}N_5O_7$			
15,17-Seco-16-norandrostan-17-oic acid, 3-hydroxy-15-(3-oxo-2-indolinylidene)-, 2,4-dinitrophenylhydrazone	MeOH	220(3.51),270(3.18), 440(3.31)	44-1363-66
$C_{32}H_{38}N_4$			
Porphin, 8,12,13,18-tetraethyl-2,3,7,17-tetramethyl-	$CHCl_3$	268(3.96),336s(4.32), 378s(5.03),400(5.28), 474s(3.59),498(4.19), 534(4.05),567(3.88), 590s(3.30),622(3.76)	39-0022-66C
$C_{32}H_{38}N_4NiO_4$			
2-Pyrroline, 2-ethoxy-5-(5-formyl-3,4-dimethylpyrrol-2-ylmethyl)-3,3-dimethyl-4-oxo-, nickel complex	EtOH	259(4.27),312s(4.32), 327(4.33),418(4.33), 500s(3.86),726(4.12)	39-1155-66C

Compound	Solvent	λ_{max}(log ε)	Ref.
$C_{32}H_{38}N_4O_6$			
b-Norbilene-a-1,17,19-tricarboxylic acid, 1,19-dideoxy-2,3,7,13,18-pentamethyl-, triethyl ester, hydrochloride	$CHCl_3$	270(4.29),308(4.32), 405(3.83),558s(4.71), 581(4.74)	39-0098-66C
b-Norbilene-a-2,17,19-tricarboxylic acid, 1,19-dideoxy-1,3,7,13,18-pentamethyl-, triethyl ester, hydrochloride	$CHCl_3$	269(4.32),303(4.26), 390(3.91),550s(4.83), 579(4.89)	39-0098-66C
$C_{32}H_{38}O_2$			
Indene-3-carboxylic acid, 1-retinylidene-, ethyl ester	EtOH	486(4.83)	54-0339-66
$C_{32}H_{38}O_{11}$			
Gedunin, 6α,11β-diacetoxy-	n.s.g.	218(4.03)	78-0891-66
Hirtin, dihydro-	n.s.g.	277(3.94)	77-0206-66
$C_{32}H_{38}O_{12}$			
Isochromomycinone, 1-deoxo-3',4'-O-isopropylidene-, tetraacetate	EtOH	235(5.08),283(3.75), 292(3.75)	78-2761-66
$C_{32}H_{38}O_{13}$			
Chromomyciquinone, 3',4'-O-isopropylidene-, tetraacetate	EtOH	258(4.40),275(4.02), 351(3.57)	78-2761-66
$C_{32}H_{40}$			
Indene, 3-isopropyl-1-retinylidene-	EtOH	435(4.86)	54-0339-66
$C_{32}H_{40}IN_3O$			
4-[(3-Ethyl-2-benzoxazolinylidene)-methyl]-5,6,7,8-tetrahydro-1-octyl-2-phenylquinazolinium iodide	MeOH	424(4.93)	87-0758-66
$C_{32}H_{40}N_4O_4$			
b-Norbilene-a-2,18-dicarboxylic acid, 1,19-dideoxy-1,3,7,8,12,13,17,19-octamethyl-, diethyl ester, hydrobromide	$CHCl_3$	265(4.08),310(3.96),	39-0098-66C
$C_{32}H_{41}N_3O_2$			
Indole, 2,3-bis[p-[2-(diethylamino)-ethoxy]phenyl]-	EtOH	253(4.53),309(4.33)	87-0527-66
$C_{32}H_{42}O_4Se$			
12,19-Seleno-$\Delta^{9(11)12,18}$-coriaceolide, acetoxy-	MeOH	312(3.75),323(3.81), 337(3.68)	20-0734-66
$C_{32}H_{42}O_6$			
$\Delta^{9(11),13,18}$-Coriaceolide, 12,19-di-oxo-, acetoxy-	MeOH	219(3.93),293(4.00)	20-0734-66
$C_{32}H_{43}Br_3N_4$			
Biladiene-ac, 1-bromo-1,19-dideoxy-8,12,13,18-tetraethyl-2,3,7,19,19-pentamethyl-, dihydrobromide	$CHCl_3$	292(3.44),374(4.16), 458(4.39),529(4.28)	39-0022-66C

Compound	Solvent	$\lambda_{max}(\log \epsilon)$	Ref.
$C_{32}H_{44}ClFO_8$			
Pregna-3,5-dien-20-one, 6-formyl-9α-fluoro-3-(2-chloroethoxy)-11β,16α-17α,21-tetrahydroxy-, 16α,17α-acetonide, 21-valerate	EtOH	216(4.01),323(4.07)	31-0468-66
$C_{32}H_{44}Si_5$			
Trisilane, 2,2-bis(dimethylphenylsilyl)-1,1,3,3-tetramethyl-1,3-diphenyl-	C_6H_{12}	242(4.59)	101-0665-66B
$C_{32}H_{46}N_2O_2$			
Inodphenoxyl, cyanoisopropyl adduct	isooctane	267(4.40)	35-3303-66
$C_{32}H_{46}O_5$			
Clerodolone, diosphenol, acetate	EtOH	233(3.90)	78-2377-66
Holothurinogenin, 17-deoxy-22,25-epoxy-, acetate	EtOH	237s(4.14),244(4.17), 255s(4.02)	78-1857-66
Isoglabrolide, acetate	MeOH	235(3.93)	32-0843-66
$C_{32}H_{46}O_6$			
Holothurinogenin, 22,25-epoxy-, acetate	EtOH	237s(4.14),243(4.17), 252s(4.03)	78-1857-66
$C_{32}H_{46}O_7$			
Holothurinogenin derivative	EtOH	265(3.88)	78-1857-66
$C_{32}H_{46}O_9$			
5β-Card-20(22)-enolide, 3β-[(6-deoxy-3-O-methyl-β-D-galactopyranosyl)-oxy]-14β,15β-epoxy-, 2'-acetate	EtOH	215(4.20)	94-1133-66
$C_{32}H_{48}N_2O_7$			
Oleanan-28-oic acid, 3β,12α,13-trihydroxy-27-(nitroimino)-, γ-lactone, 3-acetate	EtOH	206(3.70)	78-0057-66A
$C_{32}H_{48}O_2$			
5α-Cholestan-3-one, 2-furfurylidene-	EtOH	327.5(4.32)	39-1277-66C
$C_{32}H_{48}O_2S_4$			
5α-Ergost-24-en-26-oic acid, 22-hydroxy-1,4-dioxo-, δ-lactone, cyclic bis(ethylene mercaptole)	EtOH	226(3.89)	39-1753-66C
$C_{32}H_{48}O_4$			
Polyporenic acid C	EtOH	236(4.16),243(4.22), 251(4.06),278s(1.76)	39-0072-66C
$C_{32}H_{48}O_5$			
Liquiritic acid, acetate	MeOH	248(4.07)	32-0833-66
18α-Liquiritic acid, acetate	MeOH	244.5(4.03)	32-0833-66
$C_{32}H_{48}O_6$			
Cholest-8-ene-7,11-dione, 3β,6α-dihydroxy-14α-methyl-, diacetate	n.s.g.	270(3.80)	35-0790-66
13,27-Cycloolean-11-ene-23,28-dioic acid, 2β,3β-dihydroxy-, dimethyl ester	EtOH	220(3.72)	35-1544-66

Compound	Solvent	$\lambda_{max}(\log \epsilon)$	Ref.
$C_{32}H_{48}O_8$			
Holothurinogenin, 22,25-epoxy-7,8-dihydroxy-, acetate	EtOH	209(3.32)(end abs.)	78-1857-66
$C_{32}H_{49}NO_3$			
Indophenoxyl, tetra-tert-butyl-, tert-butoxy adduct	isooctane	230(3.85),285(4.18)	35-3303-66
$C_{32}H_{49}NO_4$			
18α-Liquiritic acid amide, 3-acetate	MeOH	245(4.06)	32-0843-66
$C_{32}H_{50}O_2$			
E:B-Friedohop-1(10),5-dien-3β-ol, acetate	EtOH	231(4.07),237(4.06), 246s(3.87)	39-1251-66C
Hopa-15,17(21)-diene, 15-acetoxy-	EtOH	257(4.11)	39-1564-66C
$C_{32}H_{50}O_6$			
Holothurinogenin, 22,25-epoxy-tetrahydro-, acetate	EtOH	210(2.31)(end abs.)	78-1857-66
$C_{32}H_{50}O_9$			
Fusidic acid metabolite E, methyl ester	EtOH	222(3.91)	1-1599-66
$C_{32}H_{51}N_8O_{15}P$			
Guanosine, 2'-O-(1-ethoxyethyl)-uridylyl-(3'→5')-2',3'-O-ethoxymethylene-	pH 2	260(4.34)	73-3198-66
$C_{32}H_{52}O_3$			
Lanost-8-en-2-one, 3β-acetoxy-	C_6H_{12}	291(1.43)	22-0897-66
Lanost-8-en-3-one, 2α-acetoxy-	C_6H_{12}	286(1.70)	22-0897-66
$C_{32}H_{52}O_4$			
5α-Lanostane-7,11-dione, 3β-acetoxy-	EtOH	272(4.0)	39-2359-66C
$C_{32}H_{52}O_5$			
2-Chromantridecanoic acid, 6-acetoxy-α,ε,ι,2,5,7,8-heptamethyl-, methyl ester	EtOH	284.0(3.36)	73-2434-66
Holothurinogenin, 22,25-epoxy-3β-ethoxyperhydro-	EtOH	209(2.46)(end abs.)	78-1857-66
$C_{32}H_{54}O_4$			
5α-Lanostan-7-one, 3β-acetoxy-11β-hydroxy-	EtOH	294(1.69)	39-2359-66C

Compound	Solvent	λ max(log ε)	Ref.
$C_{33}H_{22}ClN_7$			
1,3,5-Benzenetricarbonitrile, 2-chloro-4,6-bis(N,N-diphenylhydrazino)-	n.s.g.	630(3.88)	27-0303-66
$C_{33}H_{23}Cl_2N_3O_3$			
5-Norbornene-2,3-dicarboximide, 5-(α-hydroxy-α-2-pyridyl-p-chlorobenzyl)-7-(α-2-pyridyl)-p-chlorobenzylidene)-	MeOH	224(4.41),253(4.31)	87-0537-66
$C_{33}H_{24}BrN_3O_3$			
5-Norbornene-2,3-dicarboximide, 2-(or 3)-bromo-5-(α-hydroxy-α-2-pyridylbenzyl)-7-(α-2-pyridylbenzylidene)-		248(4.19)	87-0537-66
$C_{33}H_{24}N_2$			
Pyrrole, 2-[(3,5-diphenyl-2H-pyrrol-2-ylidene)methyl]-3,5-diphenyl-	EtOH-HCl	296(4.68),396(4.19), 561(5.19)	12-1871-66
	EtOH	292(4.70),360(3.68), 523(4.64)	12-1871-66
$C_{33}H_{24}O_2$			
Cyclopenta[b]pyran, 2-(p-methoxyphenyl)-4,5,6-triphenyl-	C_6H_{12}	237(4.23),254(4.26), 287(4.38),410(4.45), 520s(2.96)	39-0859-66C
	HOAc-$HClO_4$	265(4.07),368(4.16), 473(4.59)	39-0859-66C
Cyclopenta[b]pyran, 4-(p-methoxyphenyl)-2,5,6-triphenyl-	C_6H_{12}	252(4.26),288(4.32), 406(4.33),520s(2.83)	39-0859-66C
	HOAc-$HClO_4$	255(4.10),325(4.08), 448(4.50)	39-0859-66C
$C_{33}H_{25}N_3O_3$			
5-Norbornene-2,3-dicarboximide, 5-(α-hydroxy-α-2-pyridylbenzyl)-7-(α-2-pyridylbenzylidene)-	MeOH	250(4.23)	44-2141-66
	MeOH	248(4.22)	87-0537-66
5-Norbornene-2,3-dicarboximide, 5-(α-hydroxy-α-4-pyridylbenzyl)-7-(α-4-pyridylbenzylidene)-	MeOH	253(4.13)	87-0537-66
$C_{33}H_{25}N_3O_5$			
5-Norbornene-2,3-dicarboximide, 5-(α-hydroxy-α-2-pyridylbenzyl)-7-(α-2-pyridylbenzylidene)-, di-N-oxide	MeOH	262(4.35)	87-0537-66
$C_{33}H_{25}P$			
Phosphorane, (2-fluoren-9-ylidene-ethylidene)triphenyl-	CH_2Cl_2	250(4.59),258(4.56), 287(4.17),365(3.81), 445(4.34)	24-0658-66
$C_{33}H_{26}ClO_4P$			
(2-Fluoren-9-ylideneethyl)triphenyl-phosphonium perchlorate	CH_2Cl_2	310(4.14),318(4.10)	24-0658-66
$C_{33}H_{27}CoN_6O_3$			
Cobalt, tris(8-amino-7-quinolyl methyl ketonato)-	n.s.g.	237(4.69),266(4.70), 289(4.70),299(4.69), 419(4.42),513(4.15)	24-3806-66

Compound	Solvent	λ_{max}(log ε)	Ref.
$C_{33}H_{27}FN_2O_2$			
Urea, 1-(p-fluorophenyl)-1-phenyl-3-(p-phenoxyphenyl)-3-(2,4-xylyl)-	EtOH	260(4.41)	34-0436-66
$C_{33}H_{28}Cl_2N_3P$			
Phosphorane, [[bis(p-chlorophenyl)-methylene]hydrazono][p-(dimethyl-amino)phenyl]diphenyl-	dioxan	240(4.44),245(4.61), 248(4.61),286(4.48), 355(4.18)	5-0001-66H
$C_{33}H_{28}Cl_2O_4$			
Cyclopentadienone, 2,5-bis(benzyloxy)-3,4-dichloro-, dibenzyl acetal	EtOH	308(3.22)	35-0582-66
$C_{33}H_{29}ClN_3P$			
Phosphorane, [(p-chloro-α-phenylbenzyl-idene)hydrazono][p-(dimethylamino)-phenyl]diphenyl]-	dioxan	245(4.70),247(4.68), 286(4.43),348(4.13)	5-0001-66H
$C_{33}H_{29}Cl_3O_4$			
3-Cyclopentene-1,2-dione, 3,4,5-tri-chloro-, bis(dibenzyl acetal)	EtOH	209(4.63),258(2.94)	35-0582-66
$C_{33}H_{30}FN_3O_7$			
Cytosine, 1-[2-deoxy-3,5-di-(O-p-toluoyl)-α-D-ribofuranosyl]-N^4-p-toluoyl-5-fluoro-	CH_2Cl_2	243(4.57),334(4.45)	87-0566-66
β-isomer	CH_2Cl_2	243(4.57),333(4.42)	87-0566-66
$C_{33}H_{30}N_4O_7$			
Inosine, 5'-O-trityl-, 2',3'-diacetate	EtOH	254(4.04)	69-2076-66
$C_{33}H_{32}Cl_2O_2$			
Naphtho[1',2',3',4':7,8,14,15]-19-nor-ergosta-1,3,5,7,9,14,22-heptaene-5',8'-dione, 6',7'-dichloro-	$CHCl_3$	265(4.74),310(4.38), 339(4.42),420(3.66), 615(3.44)	5-0106-66J
$C_{33}H_{33}AlO_6$			
2,4-Pentanedione, 3-phenyl-, aluminum complex (also other metal complexes)	EtOH	232(4.10),299(4.51), 314s(--)	18-0901-66
	$CHCl_3$	302(4.54),314s(--)	18-0901-66
$C_{33}H_{33}BrN_3OP$			
[[2-Amino-6-hydroxy-5-(4-phenylbutyl)-4-pyrimidinyl]methyl]triphenyl-phosphonium bromide	EtOH	303(4.00)	4-0324-66
$C_{33}H_{34}BrOP$			
11-Formyl-3,7-dimethyl-2,4,6,8,10-dodecapentaenyl)triphenyl-phosphonium bromide	EtOH	378(4.81),392(4.80)	33-0369-66
$C_{33}H_{36}$			
Fluorene, retinylidene-	EtOH	432(4.85)	54-0339-66
$C_{33}H_{36}N_2O_5S_2$			
Picralinol, deformo-, N-(p-toluene-sulfonyl)-, p-toluenesulfonate	n.s.g.	225(4.61),255(4.14)	28C-1169-66A

Compound	Solvent	λ_{max}(log ε)	Ref.
$C_{33}H_{36}N_4NiO$			
meso-Etioporphyrin, formyl-, nickel complex	$CHCl_3$	408(4.86),428(4.91), 532(3.64),566(3.82), 659(3.94)	39-0794-66C
$C_{33}H_{36}O_{15}$			
Chromomycinone, hexaacetate	EtOH	260(4.80),302(3.87)	78-2761-66
$C_{33}H_{37}N_3O_8$			
l-Ornithine, N-benzoyl-, product with colchicine	EtOH-HCl	228(4.50),256(4.46), 378(4.20)	65-0033-66
	EtOH	249(4.48),355(4.22), 410(3.98)	65-0033-66
$C_{33}H_{37}N_5$			
meso-Etioporphyrin I, cyano-	$CHCl_3$	406(5.16),515(4.00), 553(4.13),587(3.76), 641(4.12)	39-0794-66C
	CF_3COOH	409(5.45),567(4.06), 616(4.11)	39-0794-66C
$C_{33}H_{38}N_4O$			
meso-Etioporphyrin, formyl-	$CHCl_3$	407(5.15),509(3.99), 538(3.84),577(3.82), 634(3.62)	39-0794-66C
$C_{33}H_{38}N_4O_2$			
meso-Etioporphyrin I, carboxy-	$CHCl_3$	402(5.19),500(4.11), 535(3.88),572(3.76), 620(3.59)	39-0794-66C
2-Porphinepropionic acid, 8,12-diethyl-3,7,13,17,20-pentamethyl-, methyl ester	$CHCl_3$	273(3.92),333s(4.26), 379s(4.80),407(5.31), 506(4.19),540(3.74), 575(3.79),634(3.19)	39-0022-66C
12-Porphinepropionic acid, 8,18-diethyl-2,3,7,13,17-pentamethyl-, methyl ester	$CHCl_3$	335s(4.23),375s(4.90), 402(5.16),500(4.09), 536(3.95),570(3.76), 624(3.62)	39-0022-66C
$C_{33}H_{38}N_4O_3$			
Cinchophyllamine, N-acetyl-	EtOH	227(4.33),281(4.08)	22-2309-66
Isocinchophyllamine, N-acetyl-	EtOH	227(4.31),281(4.02)	22-2309-66
$C_{33}H_{38}O_{14}$			
Chromomycinone, 1-deoxo-, hexaacetate	EtOH	234(5.11),275(3.77), 283(3.79),290(3.78)	78-2761-66
$C_{33}H_{38}O_{15}$			
Chromomycinone, dihydro-, hexaacetate	EtOH	225(4.46),272(4.65), 304(3.93),392(3.79)	78-2761-66
Chromomyciquinone, hexaacetate	EtOH	253(4.37),275(4.09), 355(3.50)	78-2761-66
$C_{33}H_{39}NO_2$			
12'-Apo-β-carotenoic acid, 13'-cyano-, benzyl ester	C_6H_6-EtOH	452(4.80)	54-0334-66
$C_{33}H_{39}N_5O$			
Etioporphyrin, oximinoformyl-	$CHCl_3$	401(5.10),500(4.02), 535(3.79),569(3.64), 619(3.45)	39-0794-66C

Compound	Solvent	$\lambda_{max}(\log \epsilon)$	Ref.
$C_{33}H_{40}N_4O$			
meso-Etioporphyrin, hydroxymethyl-	$CHCl_3$	402(5.20),504(4.12), 537(3.94),570(3.83), 623(3.62)	39-0794-66C
$C_{33}H_{40}N_6$			
meso-Etioporphyrin I, formyl-, hydrazone	$CHCl_3$	407(5.17),505(4.05), 537(3.80),574(3.70), 628(3.44)	39-0794-66C
$C_{33}H_{40}O_2$			
Indene-3-methanol, α-methyl-1-retinylidene-, acetate	EtOH	440(4.86)	54-0339-66
$C_{33}H_{40}O_5$			
1-Anthroic acid, 10-ethyl-5,6,7,8-tetrapropyl-, maleic acid adduct	EtOH	280.5(3.25)	33-0644-66
$C_{33}H_{42}IN_3O$			
1-Decyl-5,6,7,8-tetrahydro-4-[(3-methyl-2-benzoxazolinylidene)-methyl]-2-phenylquinazolinium iodide	MeOH	424(4.88)	87-0758-66
$C_{33}H_{42}IN_3S$			
1-Decyl-5,6,7,8-tetrahydro-4-[(3-methyl-2-benzothiazolinylidene)-methyl]-2-phenylquinazolinium iodide	MeOH	449(4.93)	87-0758-66
$C_{33}H_{42}N_4O_4$			
Cinchophyllamine, N-acetyldihydro-	EtOH	228(4.31),281(4.10)	22-2309-66
$C_{33}H_{42}O$			
5α-Androstane, 17-(diphenylmethylene)-3β-methoxy-	EtOH	218(3.928)	73-2737-66
$C_{33}H_{42}O_8S$			
Pregna-1,4-diene-3,20-dione, 11α,17,21-trihydroxy-, 21-p-toluenesulfonate 17-valerate	EtOH	229(4.36)	32-1115-66
$C_{33}H_{42}O_9S$			
Pregn-4-ene-3,20-dione, 17,21-di-hydroxy-, 17-(methylsuccinate) 21-p-toluenesulfoante	EtOH	230(4.38)	32-1115-66
$C_{33}H_{43}Br_3N_4O_2$			
Biladiene-ac, 1-bromo-1,19-dideoxy-8,18-diethyl-12-(2-methoxycarbonylethyl)-2,3,7,13,17,19-hexamethyl-, dihydrobromide	$CHCl_3$	371(4.14),438s(4.63), 460(--),496s(--), 525(--)	39-0022-66C
$C_{33}H_{44}O_7$			
Hydrocinnamic acid, ester with 8,11-dihydroxy-9-methoxy-2,2,5,13a,14,14-hexamethyl-10-methylene-4,8-methano-benzo[3,4]cyclodeca[1,2-d]-1,3-di-oxol-6(3aH)-one	EtOH	278(3.78)	39-1933-66C

Compound	Solvent	$\lambda_{max}(\log \epsilon)$	Ref.
$C_{33}H_{44}O_8$			
Taxicin I, 2,9-di-O-methyl-, 10-acetate 5-hydrocinnamate	EtOH	276(3.80)	39-1933-66C
$C_{33}H_{46}O_7$			
Taxicin I, 4,16-dihydro-2-O-methyl-9,10-O-isopropylidene-, 5-hydrocinnamate	EtOH	280(3.72)	39-1933-.6C
$C_{33}H_{46}O_9$			
Khayasin C	MeOH	209(4.00)	39-2127-66C
$C_{33}H_{48}N_4O_4$			
Cholest-4-en-3-one, 2,4-dinitrophenylhydrazone	$CHCl_3$	396(4.51)	94-1185-66
	DMF	404(4.38)	94-1185-66
isomer	$CHCl_3$	390(4.49)	94-1185-66
	DMF	396(4.30)	94-1185-66
$C_{33}H_{48}O_5$			
Liquiritic acid, 18,19-didehydro-, methyl ester, 3-acetate	MeOH	277(4.05)	32-0833-66
$C_{33}H_{50}O_4$			
Lupa-13(18),20(29)-dien-28-oic acid, 3β-acetoxy-, methyl ester	EtOH	205(3.86)	78-0897-66
$C_{33}H_{50}O_5$			
Liquiritic acid, methyl ester, acetate	MeOH	250(4.03)	32-0833-66
Lup-13(18)-en-28-oic acid, 3β-acetoxy-12-oxo-, methyl ester	EtOH	240(4.10)	78-0897-66
$C_{33}H_{50}O_8$			
Fusidic acid metabolite C, methyl ester	EtOH	222(4.29)	1-1599-66
$C_{33}H_{53}NO_7$			
Phomin, dodecahydro-, diacetate	EtOH	none	31-0750-66

Compound	Solvent	λ_{max}(log ϵ)	Ref.
$C_{34}H_{16}O_2$			
Violanthrone	n.s.g.	555(4.64),600(4.78)	27-0300-66
$C_{34}H_{17}NO_2$			
Violanthrone, 16-amino-	n.s.g.	570s(3.82),620(4.17), 680(4.14)	27-0300-66
$C_{34}H_{18}$			
4,5:11,12-Dibenzoperopyrene	benzene	296(4.70),311(4.82), 317(4.82),332(4.85), 348(5.10),398(4.36), 418(4.73),445(4.86)	24-3298-66
$C_{34}H_{18}N_2O_2$			
Violanthrone, 16,17-diamino-	n.s.g.	585s(3.72),640(4.01), 700(3.81)	27-0300-66
$C_{34}H_{20}O_2$			
Dibenz[a,h]anthracene-7,14-dione, 5,12-diphenyl-	dioxan	226(4.71),305(4.72), 340s(4.03),401(3.96)	78-0203-66A
$C_{34}H_{22}$			
Dibenz[a,h]anthracene, 5,12-diphenyl-	EtOH	209(4.54),225(4.66), 296(5.02),308(5.04), 332(4.39),346(4.38), 364(4.36)	78-0203-66A
	dioxan	226(4.65),298(4.98), 310(5.02),332(4.30), 348(4.31),364(4.29)	78-0203-66A
$C_{34}H_{22}O_{14}$			
Aurofusarin, diacetate	EtOH	223(4.74),270(4.71), 350(4.12)	88-4855-66
$C_{34}H_{23}N$			
Aniline, N,N-bis(fluoren-9-ylidenemethyl)-	dioxan	295(4.2),360(4.3), 430(4.5)	83-0493-66
$C_{34}H_{24}$			
Naphthalene, 1,8-bis[2-(2-naphthyl)-vinyl]-	dioxan	252(4.94),280s(4.40), 300s(4.30),350(4.46)	44-2407-66
$C_{34}H_{24}ClN_5O_7$			
5-Quinazolinecarboxylic acid, 2-[m-[[3-[(5-chloro-2,4-dimethoxyphenyl)-carbamoyl]-2-hydroxy-1-naphthyl]-azo]phenyl]-3,4-dihydro-4-oxo-	pH 13	460(4.27),505(4.28)	7-1394-66
5-Quinazolinecarboxylic acid, 2-[p-[[3-[(5-chloro-2,4-dimethoxyphenyl)-carbamoyl]-2-hydroxy-1-naphthyl]-azo]phenyl]-3,4-dihydro-4-oxo-	pH 13	475(4.38),515(4.40)	7-1394-66
$C_{34}H_{24}N_4O_2$			
2-Quinolinecarbonitrile, 1-benzoyl-1,2-dihydro-, dimer	MeOH	264(4.19)	88-5731-66
$C_{34}H_{24}N_6O_2$			
Benzamide, N,N''-m-phenylenebis[3-(2-benzimidazolyl)-	H_2SO_4	243(4.56),303(4.66)	22-0926-66

Compound	Solvent	$\lambda_{max}(\log \epsilon)$	Ref.
$C_{34}H_{26}$			
Calicene, 5,6-dimethyl-1,2,3,4-tetraphenyl-	C_6H_{12}	298(4.42),336(4.38)	89-0602-66
	MeCN	237(4.20),300(4.42),334(4.43)	89-0602-66
Dibenz[a,h]anthracene, 5,6,12,13-tetrahydro-5,12-diphenyl-	C_6H_{12}	225(4.7),298s(4.3),307(4.6),323(4.5),335(4.5)	78-0203-66A
p-Terphenyl, 2',5'-distyryl-	C_6H_{12}	211(4.5),253(4.6),280(4.4),350(4.6)	78-0203-66A
$C_{34}H_{26}Cl_2O_2$			
Bianthrone, 5,5'-dichloro-2,2'-diisopropyl-	$CHCl_3$	273(4.46),300(4.26),392(3.30)	78-0349-66A
$C_{34}H_{26}Fe_3$			
Ferrocene, 1,1'-bis(ferrocenylethynyl)-	C_6H_{12}	264(4.27),303(4.22),447(3.11)	101-0173-66B
$C_{34}H_{26}N_2$			
Pyridine, 2,2'-[(3a,4,7,7a-tetrahydro-4,7-methanoindene-1,8-diylidene)dibenzylidyne]di-	MeOH	245(4.37),280(4.35),311s(4.14)	44-2149-66
$C_{34}H_{26}N_2O_2$			
5-Norbornene-2,3-dicarboximide, 5-(diphenylmethyl)-7-(α-2-pyridylbenzylidene)-	MeOH	245(4.20)	87-0537-66
$C_{34}H_{26}N_8O_2$			
Biurea, 1,6-bis(5,6-diphenylpyrazinyl)-	EtOH	225(4.63),276(4.51),333(4.38)	39-2038-66C
$C_{34}H_{26}N_{10}O_{10}$			
Spiro[benz[cd]indole-5(2H),3'-indoline]-2,2'-dione, 3,4-diacetyl-1,2a,3,4-tetrahydro-, 3,4-bis(2,4-dinitrophenylhydrazone)	EtOH	251s(4.12),274s(3.92),362(4.22),415s(3.73)	39-1028-66C
$C_{34}H_{26}O$			
3-Buten-2-one, 3-methyl-4-(tetraphenyl-2,4-cyclopentadien-1-ylidene)-	MeCN	246(4.57),258s(4.54),308s(4.15),420(3.30)	89-0602-66
$C_{34}H_{27}NO$			
2-Indolinone, 1-(diphenylmethyl)-5-methyl-3,3-diphenyl-	EtOH	262(3.8)	44-1637-66
$C_{34}H_{27}N_3$			
Pyrrole, 3,4-dianilino-1,2,5-triphenyl-	EtOH	250(4.56),280s(4.38)	44-1423-66
$C_{34}H_{27}N_3O_3$			
5-Norbornene-2,3-dicarboximide, 5-(α-hydroxy-α-2-pyridylbenzyl)-N-methyl-7-(α-2-pyridylbenzylidene)-	MeOH	248(4.23)	87-0537-66
5-Norbornene-2,3-dicarboximide, 5-(α-hydroxy-α-2-pyridylbenzyl)-6-methyl-7-(α-2-pyridylbenzylidene)-	MeOH	254(4.22)	87-0537-66
$C_{34}H_{28}Cl_2O_2$			
Bianthronyl, 5,5'-dichloro-2,2'-diisopropyl-	EtOH	270(4.40)	78-0349-66A

Compound	Solvent	$\lambda_{max}(\log \epsilon)$	Ref.
$C_{34}H_{28}N_4O_{12}$			
α-D-arabino-Hexopyranosidulose, methyl, tribenzoate, 2,4-dinitrophenylhydrazone	EtOH	353(4.18)	39-1131-66C
$C_{34}H_{28}N_8$			
1,3-Bis(1,5-diphenylverdazyl-3-yl)-benzene	mixt., 77°K	455(4.3),835(3.9)	49-0524-66
	300°K	425s(4.2),715(3.9)	49-0524-66
$C_{34}H_{28}O_2$			
Tricyclo[3.1.0.0^{2,4}]hexane, 3,6-diacetyl-1,2,4,5-tetraphenyl-	n.s.g.	226(4.21),275s(3.74)	88-1503-66
$C_{34}H_{29}N_5O_8$			
Adenosine, N,1-dibenzoyl-, cyclic 2',3'-(ethyl orthoformate) 5'-benzoate	EtOH	231(4.42),249s(4.32), 274(4.25)	73-1785-66
$C_{34}H_{30}N_2O_3$			
Urea, 1-methyl-1-(m-phenoxyphenyl)-3-(p-phenoxyphenyl)-3-(2,4-xylyl)-	EtOH	262(4.29)	34-0436-66
$C_{34}H_{30}O_5$			
Benzoic acid, o-(2,4,6-trimethylbenzoyl)-, anhydride	$CHCl_3$	332(2.64)	35-0781-66
$C_{34}H_{31}ClO_5$			
Pyrylium, 4,6-diphenyl-2-[(3,8-dimethyl-5-isopropyl-1-azulenyl)-vinyl]-, perchlorate	MeOH-$HClO_4$	245(3.90),285(3.75), 370(3.89),450(3.48), 700(4.18)	65-1728-66
$C_{34}H_{34}N_4O_4$			
Protoporphyrin IX	4N HCl	408(5.38)	37-5276-66
	25% HCl	556(4.15),602(3.72)	95-1138-66
$C_{34}H_{34}N_4O_{13}S_4$			
5-Thia-1-azabicyclo[4.2.0]oct-3-ene-2-carboxylic acid, 3-(hydroxymethyl)-8-oxo-7-[2-(2-thienyl)acetamido]-, oxydimethylene ester, diacetate	EtOH	235(4.35),270(3.67)	87-0741-66
$C_{34}H_{34}O_{14}$			
Ergoxanthin, trimethyl ether	EtOH	216(4.30),250(4.39), 337(3.64)	78-2359-66
$C_{34}H_{36}O_{17}$			
Flavone, 5-acetoxy-4'-hydroxy-3,6,7-trimethoxy-, 4'-β-D-glucoside, tetraacetate	EtOH	260(4.06),318(4.19)	24-3222-66
$C_{34}H_{38}O_4$			
B-Homoandrosta-5,7a-diene, 3β,17β-dibenzoyloxy-	MeOH	227(4.56)	24-3836-66
$C_{34}H_{39}N_3O_8$			
Lysine, N-benzoyl-, product with colchicine	EtOH-HCl	228(4.49),248(4.44), 377(4.17)	65-0033-66
	EtOH	245(4.49),357(4.18), 410(3.92)	65-0033-66

Compound	Solvent	$\lambda_{max}(\log \epsilon)$	Ref.
$C_{34}H_{40}N_{12}$			
Tetraethylammonium salt of 5,5"-azobis-[cyclopentadiene-1,2,3,4-tetra-carbonitrile]	MeCN	217(4.72),256(4.39), 272(4.60),282(4.73), 300(4.37),428(4.37), 455(4.37),484(4.34)	35-4055-66
$C_{34}H_{40}O_{16}$			
Harpagoside, pentaacetate	MeOH	216(4.30),222(4.21), 276(4.41)	33-1552-66
	$CHCl_3$	279(4.34)	33-1552-66
	CH_2Cl_2	276.5(4.37)	33-1552-66
$C_{34}H_{42}F_6O_7$			
Holothurinogenin, 22,25-epoxy-, bis(trifluoroacetate)	EtOH	243(4.18)	78-1857-66
$C_{34}H_{44}IN_3O$			
1-Decyl-4-[(3-ethyl-2-benzoxazolidin-ylidene)methyl]-5,6,7,8-tetrahydro-2-phenylquinazolinium iodide	MeOH	424(4.90)	87-0758-66
$C_{34}H_{44}IN_3S$			
1-Decyl-4-[(3-ethyl-2-benzothiazolin-ylidene)methyl]-5,6,7,8-tetrahydro-2-phenylquinazolinium iodide	MeOH	446(4.93)	87-0758-66
$C_{34}H_{46}N_4Pd$			
Dipyrromethene, 3,3',4,4'-tetraethyl-, palladium complex	$CHCl_3$	407(4.30),484s(4.62), 498(4.79)	39-0098-66C
$C_{34}H_{47}NO_2$			
Cholesta-3,5-dieno[3,4-d]oxazol-2'(3'H)-one, 3'-phenyl-	EtOH	267(4.30)	22-3410-66
$C_{34}H_{48}O$			
Cholesta-3,5-diene-6-carboxaldehyde, 3-phenyl-	hexane	315(4.43)	23-2233-66
$C_{34}H_{48}O_{10}$			
5β-Card-20(22)-enolide, 3β-[(6-deoxy-3-O-methyl-β-D-galactopyranosyl)-oxy]-14β,15β-epoxy-, diacetate	EtOH	213.5(4.23)	94-1133-66
$C_{34}H_{49}NO_3$			
Cholest-4-en-3-one, 4-hydroxy-, carbanilate	EtOH	240(4.33)	22-3410-66
$C_{34}H_{50}N_4O_5$			
Cholest-4-en-3-one, 6β-hydroxy-, 2,4-dinitrophenylhydrazone	$CHCl_3$	385(4.52)	22-2277-66
$C_{34}H_{50}O$			
Cholesta-3,5-diene-6-methanol, 3-phenyl-	MeOH	288(4.42)	23-2233-66
Cholestan-3-one, 2-benzylidene-	EtOH	221(3.86),288(4.19)	25-1497-66
$C_{34}H_{50}O_2$			
Cholestan-3-one, 2-benzoyl-	C_6H_{12}	244(3.61),312(4.00)	78-3607-66
	MeOH	245(3.72),314(4.04)	78-3607-66

Compound	Solvent	λ_{max}(log ϵ)	Ref.
Cholest-5-en-3-one, 2-furfurylidene-4,4-dimethyl-	EtOH	323(4.38)	39-1277-66C
$C_{34}H_{50}O_6$			
Tormentolactone, diacetate	n.s.g.	208(3.89),228(3.65)	22-3458-66
$C_{34}H_{50}O_9$			
8ξ-Holothurinogenin, 22,25-epoxy-7,8-dihydroxy-, diacetate	EtOH	210(3.32)(end abs.)	78-1857-66
$C_{34}H_{52}O_6$			
Lup-13(18)-ene-27,28-dioic acid, 3β-acetoxy-, dimethyl ester	EtOH	210(3.78),215(3.69), 220(3.58)(end abs.)	78-0897-66
$C_{34}H_{54}O_2$			
8α-Gona-1,3,5(10)-trien-17-one, 13β-hexadecyl-3-methoxy-	EtOH	280(3.20)	87-0338-66

Compound	Solvent	λ_{max}(log ε)	Ref.
$C_{35}H_{18}O_3$			
Violanthrone, 16-methoxy-	n.s.g.	598(4.59)	27-0300-66
$C_{35}H_{23}N_5O_3$			
2-Naphthamide, 3-hydroxy-4-[m-(3,4-dihydro-4-oxoquinazolin-2-yl)phenylazo]-N-(1-naphthyl)-	dioxan	495(4.42),518s(4.40)	32-0279-66
2-Naphthamide, 3-hydroxy-4-[m-(3,4-dihydro-4-oxoquinazolin-2-yl)phenylazo]-N-(2-naphthyl)-	dioxan	495(4.39),518s(4.37)	32-0279-66
2-Naphthamide, 3-hydroxy-4-[p-(3,4-dihydro-4-oxoquinazolin-2-yl)phenylazo]-N-(1-naphthyl)-	dioxan	510(4.54),532s(4.53)	32-0279-66
2-Naphthamide, 3-hydroxy-4-[p-(3,4-dihydro-4-oxoquinazolin-2-yl)phenylazo]-N-(2-naphthyl)-	dioxan	510(4.53),532s(4.52)	32-0279-66
$C_{35}H_{26}N_2$			
3,4-Diazabicyclo[4.1.0]hepta-2,4-diene, 1,2,5,6,7-pentaphenyl-	n.s.g.	313(4.24)	88-4979-66
5H-1,2-Diazepine, 3,4,5,6,7-pentaphenyl-	n.s.g.	260(4.46),310s(4.18)	88-4979-66
$C_{35}H_{26}O_{14}$			
Viopurpurin, triacetate	EtOH	217(3.75),270(3.97), 277(3.97)	23-2873-66
$C_{35}H_{27}NO_3$			
5-Norbornene-2,3-dicarboximide, 5-(diphenylmethyl)-7-(α-hydroxydiphenylmethyl)-	MeOH	249(4.23)	87-0537-66
$C_{35}H_{28}ClO_4P$			
(4-Fluoren-9-ylidene-2-butenyl)-triphenylphosphonium perchlorate	CH_2Cl_2	335(4.56),355(4.55)	24-0658-66
	+piperidine	233(4.48),260(4.47), 531(4.47)	24-0658-66
$C_{35}H_{28}N_2$			
Dipyrromethene, 5,5'-dimethyl-3,3',4,4'-tetraphenyl-	EtOH-N HCl	239(4.68),315(4.10), 365(4.28),409(4.79)	12-1871-66
	EtOH	222s(4.36),241s(4.26), 262s(4.13),476(4.53)	12-1871-66
$C_{35}H_{29}N_3O_3$			
5-Norbornene-2,3-dicarboximide, 5-(α-hydroxy-α-2-pyridylbenzyl)-N,2(or 3)-dimethyl-7-(α-2-pyridylbenzylidene)-	MeOH	248(4.23)	87-0537-66
5-Norbornene-2,3-dicarboximide, 5-(α-hydroxy-α-2-pyridylbenzyl)-N-ethyl-7-(α-2-pyridylbenzylidene)-	MeOH	250(4.21)	87-0537-66
$C_{35}H_{31}IN_2S_2$			
3-Ethyl-2-[(3-ethyl-4,5-diphenyl-4-thiazolin-2-ylidene)methyl]-4,5-diphenylthiazolium iodide	MeOH	439(4.52)	9-0150-66
$C_{35}H_{31}N_3$			
Dibenzo[b,h][1,6]naphthyridine, 7,12-dihydro-2,9-dimethyl-6-(5,6,7,8-tetrahydro-2-naphthyl)-7-(p-tolylimino)-	C_6H_{12}	249(4.66),275(4.79), 325(4.28),339(4.33), 376(3.65),397(3.75), 460(3.88)	39-1191-66C

Compound	Solvent	λ_{max}(log ε)	Ref.
$C_{35}H_{34}N_4O_6$			
Deuteroporphyrin-4-acrylic acid, 2-vinyl-	4N HCl	414(5.39)	37-5276-66
$C_{35}H_{36}N_2O_6$			
Chondrofoline, dihydrochloride	H_2O	225(4.5),280(4.0)	65-1764-66
$C_{35}H_{36}N_4O_7$			
Deuteroporphyrin-4-propionic acid, β-hydroxy-2-vinyl-	pH 1	405(5.53)	37-5276-66
$C_{35}H_{37}N_5O_{14}$			
2(1H)-Pyrimidinone, 1-(tetra-O-acetyl-β-D-glucopyranosyl)-4-[p-(carbobenzoxyglycylamino)benzamido]-	EtOH	304(4.54)	44-4014-66
$C_{35}H_{38}CuN_4O_5$			
Isochlorine e_4, 6-acetyl-2-devinyl-, dimethyl ester, copper compound	$CHCl_3$	400(--),500(--), 547(--),591(--), 629(4.58)	5-0112-66E
$C_{35}H_{38}N_6O_{14}$			
2(1H)-Pyrimidinone, 1-(3,4,6-tri-O-acetyl-2-deoxy-2-carbomethoxyamino-β-D-glucopyranosyl)-4-[p-(carbobenzoxyglycylamino)benzamido]-	pH 1	309(4.42)	44-4014-66
	pH 7	303(4.45)	44-4014-66
	pH 13	267(4.27),324(4.30)	44-4014-66
$C_{35}H_{40}N_4O_5$			
Isochlorine e_4, 2-acetyl-, dimethyl ester	$CHCl_3$	421(4.95),675(4.56)	5-0112-66E
$C_{35}H_{40}O_7S$			
Pregn-4-ene-3,20-dione, 17,21-dihydroxy-	EtOH	233(4.56)	32-1115-66
$C_{35}H_{40}O_{16}$			
Chromomycinone, dihydro-, heptaacetate	EtOH	260(4.82),301(3.89), 360(3.45)	78-2761-66
$C_{35}H_{42}CoN_4O_4$			
Biladiene-ac, 1,19-dideoxy-7,13-dicarbethoxy-1,2,3,8,10,10,12,17,18,-19-decamethyl-, cobalt complex	$CHCl_3$	267(4.03),303(4.01), 467(4.63),515s(4.53)	39-0098-66C
$C_{35}H_{42}CuN_4O_4$			
Biladiene-ac, 1,19-dideoxy-7,13-dicarbethoxy-1,2,3,8,10,10,12,17,18,-19-decamethyl-, copper complex	$CHCl_3$	387(4.10),495(4.83), 554s(4.23)	39-0098-66C
$C_{35}H_{42}N_2O_4$			
2H-Benzo[a]quinolizine, 2-(6-benzyloxy-3,4-dihydro-7-methoxy-1-isoquinolinylmethyl)-3-ethyl-1,3,4,6,7,11b-hexahydro-9,10-dimethoxy-, as dioxalate	pH 1	241(4.28),289(3.93), 306(4.00),355(3.98)	35-4068-66
	H_2O	241(4.28),289(3.90), 306(3.98),355(3.98)	35-4068-66
$C_{35}H_{42}N_4NiO_4$			
Biladiene-ac, 1,19-dideoxy-7,13-dicarbethoxy-1,2,3,8,10,10,12,17,18,-19-decamethyl-, nickel complex	$CHCl_3$	319(3.85),515(4.55)	39-0098-66C

Compound	Solvent	λ_{max}(log ε)	Ref.
$C_{35}H_{42}O_4$			
Indene-2,3-dicarboxylic acid, 1-retinylidene-, diethyl ester	EtOH	497(4.80)	54-0339-66
$C_{35}H_{44}IN_3$			
1-Decyl-5,6,7,8-tetrahydro-4-[(1-methyl-2(1H)-quinolylidene)methyl]-2-phenylquinazolinium iodide	MeOH	328(3.86),467(4.77), 493(4.97)	87-0758-66
$C_{35}H_{44}N_4O_4$			
Biladiene-ac, 1,19-dideoxy-2,18-dicarbethoxy-8,12-diethyl-1,3,7,13,-17,19-hexamethyl-, dihydrobromide	$CHCl_3$	269(4.57),371(4.16), 435s(4.12),454(4.44), 524(5.32)	39-0880-66C
	EtOH	246(4.18),304(4.33), 362(4.39),428(4.86), 488(3.76),521(3.72), 756(4.50)	39-0880-66C
Biladiene-ac, 1,19-dideoxy-7,13-dicarbethoxy-1,2,3,8,10,10,12,17,18,-19-decamethyl-, dihydrobromide	$CHCl_3$	239s(3.73),241(4.08), 295s(3.93),345(3.93), 473(4.59),527(5.04)	39-0098-66C
$C_{35}H_{44}O_3$			
5α-Androstan-18-ol, 17-(diphenylmethylene)-3β-methoxy-, acetate	EtOH	215(4.046)	73-2737-66
$C_{35}H_{46}IN_3O$			
1-Decyl-5,6,7,8-tetrahydro-2-phenyl-4-[(3-propyl-2-benzoxazolinylidene)-methyl]quinazolinium iodide	MeOH	425(4.91)	87-0758-66
4-[(3-Ethyl-2-benzoxazolinylidene)-methyl]-1-undecyl-2-phenyl-5,6,7,8-tetrahydroquinazolinium iodide	MeOH	424(4.88)	87-0758-66
$C_{35}H_{48}N_2O$			
Urea, 1-hexadecyl-1,3,3-triphenyl-	EtOH	263(4.19)	34-0436-66
$C_{35}H_{48}N_4$			
Biladiene-ac, 1,19-deoxy-2,7,8,12,13,-18-hexaethyl-1,3,17,19-tetramethyl-, dihydrobromide	$CHCl_3$	376(4.30),443(4.13), 461(4.44),530(4.39)	39-0030-66C
$C_{35}H_{48}O_2$			
4,6,8,10,16,18,20,22-Octacosanonaen-12-yn-3-one, 27-hydroxy-2,6,10,15,-19,23,27-heptamethyl-	benzene	465(4.97)	1-1195-66
$C_{35}H_{50}O$			
Cholesta-3,5-diene, 6-acetyl-	hexane	303(4.27)	23-2233-66
$C_{35}H_{52}O_6$			
13,27-Cycloolean-11-ene-23,28-dioic acid, 2β,3β-isopropylidenedioxy-, dimethyl ester	EtOH	221(3.77)	35-1544-66
$C_{35}H_{52}O_7$			
Olean-12-en-30-oic acid, 3β,7β-diacetoxy-11-oxo-, methyl ester	MeOH	248(4.15)	32-0820-66
$C_{35}H_{52}O_{15}$			
Eryscenoside	n.s.g.	216(4.3),273(2.6)	88-1703-66

Compound	Solvent	$\lambda_{max}(\log \epsilon)$	Ref.
$C_{35}H_{53}N_3O_2S$			
p-Toluenesulfonic acid, (1α-cyano-5α-cholestan-3-ylidene)hydrazide	EtOH	233(4.14)	39-0661-66C
$C_{35}H_{54}O_9$			
Adigoside	n.s.g.	215(4.09)	33-1855-66
$C_{35}H_{54}O_{15}$			
Glucopanoside	n.s.g.	218(4.18)	33-0316-66

Compound	Solvent	$\lambda_{max}(\log \epsilon)$	Ref.
$C_{36}H_{18}$			
Anthra[2,3-a]coronene	benzene	290(5.0),320(4.52), 332(4.63)	24-1272-66
	$C_6H_3Cl_3$	352(4.94),370(5.12), 410(3.66),435(3.86), 464(4.06),495(4.02)	24-1272-66
$C_{36}H_{18}Fe_2$			
Hexadecaoctayne, diferrocenyl-	$CHCl_3$	290(4.73),307(4.83), 327(5.04),347(5.05), 385s(4.68),508s(4.20)	101-0399-66B
$C_{36}H_{20}O_4$			
Violanthrone, 16,17-dimethoxy-	n.s.g.	636(4.61)	27-0300-66
$C_{36}H_{23}N_5O_5$			
5-Quinazolinecarboxylic acid, 3,4-dihydro-2-[m-[[2-hydroxy-3-(1-naphthylcarbamoyl)-1-naphthyl]azo]phenyl]-4-oxo-	pH 13	455(4.30),510(4.33)	7-1394-66
5-Quinazolinecarboxylic acid, 3,4-dihydro-2-[m-[[2-hydroxy-3-(2-naphthylcarbamoyl)-1-naphthyl]azo]phenyl]-4-oxo-	pH 13	450(4.28),510(4.31)	7-1394-66
5-Quinazolinecarboxylic acid, 3,4-dihydro-2-[p-[[2-hydroxy-3-(1-naphthylcarbamoyl)-1-naphthyl]azo]phenyl]-4-oxo-	pH 13	475(4.40),520(4.43)	7-1394-66
5-Quinazolinecarboxylic acid, 3,4-dihydro-2-[p-[[2-hydroxy-3-(2-naphthylcarbamoyl)-1-naphthyl]azo]phenyl]-4-oxo-	pH 13	465(4.38),520(4.41)	7-1394-66
$C_{36}H_{24}N_2O_4$			
N,N'-Bi[9,10-ethanoanthracene-11,12-dicarboximide], 9,9',10,10'-tetrahydro-	n.s.g.	215(3.58),249(3.84), 256(3.85),264(3.88), 271(3.91)	44-1317-66
$C_{36}H_{24}N_4$			
2H-Imidazole, 2,2'-(2,5-cyclohexadiene-1,4-diylidene)bis[4,5-diphenyl-	$CHCl_3$	602(4.95)	89-0311-66
$C_{36}H_{26}$			
Fulvene, 2,3,4,5,6-pentaphenyl-	EtOH	267(4.60),340(4.50), 440(3.64)	5-0034-66H
Sesquifulvalene, 7,8,9,10-tetraphenyl-	C_6H_{12}	268(4.42),459(4.34)	5-0034-66H
	MeCN	268(4.43),460(4.33)	5-0034-66H
	dioxan	269(4.40),464(4.30)	5-0034-66H
	$MeNO_2$	464(4.14)	5-0034-66H
$C_{36}H_{26}Sn$			
Tin, triphenyl[8-(phenylethynyl)-1-naphthyl]-	THF	280(3.93),309(4.21), 332(4.18)	35-3027-66
$C_{36}H_{28}$			
1,3,5-Cycloheptatriene, 7-(2,3,4,5-tetraphenylcyclopenta-2,4-dien-1-yl)-	EtOH	252(3.81),318(4.27)	5-0034-66H
Cyclopentadiene, 5-benzyl-1,2,3,4-tetraphenyl-	EtOH	248(4.30),347(4.08)	22-3775-66

Compound	Solvent	λ_{max}(log ε)	Ref.
$C_{36}H_{28}O$			
4H-Pyran, 2,6-dimethyl-4-(tetraphenyl-2,4-cyclopentadien-1-ylidene)-	n.s.g.	261(4.34),425(4.50)	39-1086-66C
$C_{36}H_{29}ClO_5$			
2,6-Dimethyl-4-(2,3,4,5-tetraphenyl-cyclopentadienyl)pyrylium perchlorate	n.s.g.	256s(4.18),301(4.17), 481(4.47)	39-1086-66C
$C_{36}H_{29}N_3O_3$			
5-Norbornene-2,3-dicarboximide, N-allyl-5-(α-hydroxy-α-2-pyridylbenzyl)-7-(α-2-pyridylbenzylidene)-	MeOH	251(4.27)	87-0537-66
$C_{36}H_{30}$			
Anthracene, 1,4-diethyl 2,3,9-triphenyl-	$CHCl_3$	274(4.88),367(3.78), 385(3.95),406(3.87)	44-4082-66
Sesquifulvalene, tetrahydro-tetraphenyl-	EtOH	245(4.35),345(4.05)	5-0034-66H
$C_{36}H_{30}O$			
Cyclopenta[b]pyran, 7-tert-butyl-2,4,5,6-tetraphenyl-	C_6H_{12}	278(4.07),407(4.02), 524s(2.47)	39-0859-66C
2-Cyclopenten-1-ol, 1-benzyl-2,3,4,5-tetraphenyl-	EtOH	256(4.10)	22-3775-66
$C_{36}H_{31}N_3O_3$			
5-Norbornene-2,3-dicarboximide, 5-(α-hydroxy-α-2-pyridylbenzyl)-N-isopropyl-7-(α-2-pyridylbenzylidene)-	MeOH	248(4.25)	87-0537-66
5-Norbornene-2,3-dicarboximide, 5-(α-hydroxy-α-2-pyridylbenzyl)-N-propyl-7-(α-2-pyridylbenzylidene)-	MeOH	250(4.26)	87-0537-66
$C_{36}H_{31}N_3O_4$			
5-Norbornene-2,3-dicarboximide, 5-(α-hydroxy-α-2-pyridylbenzyl)-N-(2-methoxyethyl)-7-(α-2-pyridyl-benzylidene)-	MeOH	250(4.25)	87-0537-66
$C_{36}H_{33}N_5O_3S$			
Adenine, 9-(3-S-benzyl-3-thio-5-O-trityl-β-D-arabinofuranosyl)-	MeOH	266(4.17)	44-3263-66
$C_{36}H_{33}N_5O_5S$			
Adenosine, 2'-deoxy-5'-O-trityl-, 3'-p-toluenesulfonate	MeOH	226(4.39),259(4.19)	69-0224-66
$C_{36}H_{34}N_2O_6$			
Stebisimine	EtOH	238(4.71),279(4.38), 308s(4.10)	39-2313-66C
methiodide	EtOH	340(4.27)	39-2313-66C
$C_{36}H_{34}N_4O_8$			
Deuteroporphyrin-2,4-diacrylic acid	4N HCl	423(5.40)	37-5276-66

Compound	Solvent	λ_{max}(log ϵ)	Ref.
$C_{36}H_{35}Cl_3O_4$			
6aH-Naphtho[1,2,3-fg]aceanthrylene, 9,12-diacetoxy-2,10,11-trichloro-6a-methyl-7-(1,5-dimethylhexyl)-7,8-dihydro-	EtOH	233s(4.71),238(4.76), 246s(4.58),263s(4.47), 270(4.50),297(4.45), 312(4.62),327(4.68), 378s(3.75),394s(3.86), 413(3.98),439(3.90)	5-0106-66J
$C_{36}H_{36}N_4O_9$			
Deuteroporphyrin-4-propionic acid, β-hydroxy-2-(carboxyvinyl)-	pH 1	412(5.56)	37-5276-66
$C_{36}H_{36}O_{15}$			
Ergoxanthin, trimethyl ether, acetate	EtOH	230(4.39),254(4.48), 339(3.73)	78-2359-66
$C_{36}H_{38}N_4O_4$			
Protoporphyrin IX, dimethyl ester	$CHCl_3$	508(4.20),541(4.09), 576(3.88),631(3.78)	95-1138-66
$C_{36}H_{38}O_{16}$			
Glucoside, 3,4-dihydro-4,7,10-trihydroxy-1,3-dimethyl-6-(3,4,5,10-tetrahydro-4,7,9-trihydroxy-1,3-dimethyl-5,10-dioxo-1H-naphtho-[2,3-c]pyran-6-yl)-1H-naphtho-[2,3-c]pyran-9-yl	EtOH	221(4.70),243(4.79), 274(4.15),295(3.99), 309(3.84),341(3.20), 353(3.79),448(3.65)	39-1825-66C
	Na_2HPO_4	238(4.76),298(4.31), 340(3.75),352(3.77), 530(3.73)	39-1825-66C
Granaticin B, tetraacetate	EtOH	242(4.26),356(3.60)	33-1736-66
$C_{36}H_{39}AlO_6$			
2,4-Pentanedione, 3-benzyl-, aluminum complex (also other metal chelates, not listed)	EtOH	264s(--),270s(--), 302(4.48),318s(--)	18-0901-66
	$CHCl_3$	265s(--),272s(--), 305(4.57),318s(--)	18-0901-66
$C_{36}H_{40}$			
Indene, 3-benzyl-1-retinylidene-	EtOH	438(4.88)	54-0339-66
$C_{36}H_{40}CuN_4O_5$			
α-meso-Oxy-1,2,3,4,5,8-hexamethyl-6,7-bis(2-carbethoxyethyl)porphine, copper complex	$CHCl_3$	404(5.38),435(4.74), 494s(3.80),529(4.20), 563(4.09)	12-1481-66
$C_{36}H_{40}N_2O_7$			
Ether, 4,5-dihydroxy-2,4'-bis(2-methyl-6-methoxy-7-hydroxy-1,2,3,4-tetrahydro-1-isoquinolyl)methyl biphenyl	EtOH	286(4.07)	94-0078-66
$C_{36}H_{40}N_4O_8$			
meso-Tetraoxomesoporphinogen	benzene	340(4.73)	5-0092-66J
$C_{36}H_{42}N_4O_5$			
α-meso-Oxy-1,2,3,4,5,8-hexamethyl-6,7-bis(2-carbethoxyethyl)porphine	5N HCl	408(5.33),557(3.88), 609(3.88)	12-1481-66
	HOAc	411(5.04),536(3.62), 683(4.32)	12-1481-66
	$CHCl_3$	405(4.98),541s(3.42), 588(3.81),636(4.08)	12-1481-66

Compound	Solvent	λ_{max}(log ε)	Ref.
$C_{36}H_{42}O_{17}$			
Harpagoside, hexaacetate	MeOH	216(4.21),223(4.11), 278(4.34)	33-1552-66
$C_{36}H_{44}CoN_4O_4$			
5,5'-Bi(5-carbethoxy-3,3'-diethyl-4,4'-dimethyl-2,2'-dipyrromethene), cobalt complex	$CHCl_3$	267(4.38),435(4.35), 481(4.48),542(4.58), 612(3.99)	39-1155-66C
$C_{36}H_{44}N_4O_4$			
5,10,15,20(22H,24H)-Porphinetetrone, 2,3,7,8,12,13,17,18-octaethyl-	benzene	340(4.70)	5-0092-66J
$C_{36}H_{46}N_2$			
3-Hexenedinitrile, 2,5-bis[3-methyl-5-(2,6,6-trimethyl-1-cyclohexen-1-yl)-2,4-pentadienylidene]-	isooctane	445(4.99)	54-0343-66
$C_{36}H_{46}N_4O_4$			
5,5'-Bi(5-carbethoxy-3,3'-diethyl-4,4'-dimethyl-2,2'-dipyrromethene)	$CHCl_3$	270(4.62),311(3.97), 344(4.05),404(3.76), 554(4.75),569(4.74)	39-1155-66C
$C_{36}H_{48}IN_3O$			
4-[(3-Butyl-2-benzoxazolinylidene)-methyl]-1-decyl-5,6,7,8-tetrahydro-2-phenylquinazolinium iodide	MeOH	425(4.90)	87-0758-66
$C_{36}H_{48}N_6O_8$			
L-Leucine, N-benzyloxycarbonyl-L-valyl-L-glutaminyl-L-tryptophanyl-, methyl ester	EtOH	220s(4.57),276s(3.75), 282(3.78),291(3.72)	44-3400-66
$C_{36}H_{48}O_2$			
3-Hexenedial, 2,5-bis[3-methyl-5-(2,6,6-trimethyl-1-cyclohexen-1-yl)-2,4-pentadienylidene]-	isooctane	448(4.82)	54-0343-66
$C_{36}H_{50}N_4O_4$			
Melianone, 2,4-dinitrophenylhydrazone	EtOH	365(4.33)	88-2049-66
1-Pyrroline, 2-(5-carbethoxy-3,3'-diethyl-4,4'-dimethyl-2,2'-dipyrromethen-5'-yl)-5-(5-carbethoxy-3-ethyl-4-methylpyrrol-2-ylmethyl)-4-ethyl-3-methyl-	$CHCl_3$	226(4.31),264(4.58), 280(4.38),453(4.49)	39-1155-66C
$C_{36}H_{50}O_2$			
Benzoic acid, aegiceradienyl ester	EtOH	235(4.76),243s(4.68), 253s(4.44)	12-0169-66
$C_{36}H_{50}O_8$			
Oleana-9(11),13(18)-diene-12,19-dione, 3,16,28-triacetoxy-	EtOH	278(4.13)	88-0701-66
$C_{36}H_{52}N_2O_3$			
Cyclomalayanine B	n.s.g.	227(4.00),273(--), 318(4.35),337(--)	22-0758-66

Compound	Solvent	$\lambda_{max}(\log \epsilon)$	Ref.
$C_{36}H_{52}N_4O_4$			
1-Pyrroline, 2-(5-carbethoxy-3,3'-diethyl-4,4'-dimethyl-2,2'-dipyrromethan-5'-yl)-5-(5-carbethoxy-3-ethyl-4-methylpyrrol-2-ylmethyl)-4-ethyl-3-methyl-	$CHCl_3$	253(4.22),284(4.62), 302(4.48),356(3.32)	39-1155-66C
$C_{36}H_{52}O_2$			
3-Hexene-1,6-diol, 2,5-bis[3-methyl-5-(2,6,6-trimethyl-1-cyclohexen-1-yl)-2,4-pentadienylidene]-	isooctane	415(5.00)	54-0343-66
$C_{36}H_{60}O_{14}$			
Aldgamycin C	MeOH	217(4.13)	88-0839-66

Compound	Solvent	$\lambda_{max}(\log \epsilon)$	Ref.
$C_{37}H_{26}N_2O$			
2-Quinolinemethanol, α-phenyl-α-[6-phenyl-6-(2-quinolyl)-2-fulvenyl]-	MeOH	338(4.35)	44-2149-66
$C_{37}H_{30}N_2O_2$			
Urea, 1-(1-naphthyl)-3-(p-phenoxyphenyl)-1-phenyl-3-(2,4-xylyl)-	EtOH	245(4.48)	34-0436-66
$C_{37}H_{32}N_4O_{17}$			
Ergoxanthin, 2,4-dinitrophenylhydrazone	EtOH	215(4.51),240(4.45), 380(4.42)	78-2359-66
$C_{37}H_{33}N_3O_3$			
5-Norbornene-2,3-dicarboximide, N-butyl-5-(α-hydroxy-α-2-pyridylbenzyl)-7-(α-2-pyridylbenzylidene)-	MeOH	248(4.26)	87-0537-66
5-Norbornene-2,3-dicarboximide, N-sec-butyl-5-(α-hydroxy-α-2-pyridylbenzyl)-7-(α-2-pyridylbenzylidene)-	MeOH	248(4.25)	87-0537-66
5-Norbornene-2,3-dicarboximide, N-tert-butyl-5-(α-hydroxy-α-2-pyridylbenzyl)-7-(α-2-pyridylbenzylidene)-	MeOH	248(4.25)	87-0537-66
5-Norbornene-2,3-dicarboximide, 5-(α-hydroxy-α-2-pyridylbenzyl)-N-isobutyl-7-(α-2-pyridylbenzylidene)-	MeOH	248(4.25)	87-0537-66
$C_{37}H_{34}N_4O_3$			
5-Norbornene-2,3-dicarboximide, N-dimethylaminomethyl-5-(α-hydroxy-α-2-pyridylbenzyl)-7-(α-2-pyridylbenzylidene)-	MeOH	249(4.26)	87-0537-66
$C_{37}H_{36}O_{15}$			
Flavonol, 4'-(benzyloxy)-7-(2,3,4,6-tetraacetyl-β-D-glucosyloxy)-5-hydroxy-3-methyl-	EtOH	268(4.21),351(4.17)	22-0987-66
$C_{37}H_{38}N_2O_6$			
Epistephanine	EtOH	233(4.53),282(4.16)	39-2313-66C
Epistephanine, hydrochloride	EtOH	284(4.12)	39-2313-66C
$C_{37}H_{42}N_2O_6$			
Liensinine, natural	EtOH	285(4.03)	94-0067-66
synthetic	EtOH	285(4.06)	94-0067-66
$C_{37}H_{42}N_4O_6$			
Isochlorine e_4, 2,6-diacetyl-2-devinyl-, dimethyl ester	$CHCl_3$	414(4.99),684(4.59)	5-0112-66E
$C_{37}H_{42}O_2Si$			
Androst-4-en-3-one, 17β-(triphenylsiloxy)-	EtOH	220(4.59),240(4.19)	87-0433-66
$C_{37}H_{44}N_4O_5$			
Mesoisochlorine e_4, 6-acetyl-, dimethyl ester	$CHCl_3$	653(4.63)	5-0112-66E
$C_{37}H_{44}O_{17}$			
Leucochromomyciquinone, octaacetate	EtOH	235(4.89),285(3.64)	78-2761-66

Compound	Solvent	λ_{max}(log ε)	Ref.
$C_{37}H_{45}N_5$			
α-Porphinecarbonitrile, octaethyl-	benzene	413(5.0),513(3.8), 551(4.0),583(3.3), 636(3.8)	5-0133-66E
$C_{37}H_{46}N_4O$			
α-Porphinecarboxaldehyde, octaethyl-	benzene	408(5.0),507(3.8), 541(3.5),577(3.5), 625(3.3),655s(3.0)	5-0133-66E
$C_{37}H_{47}N_5O$			
α-Porphinecarboxaldehyde, octaethyl-, oxime	benzene	408(5.0),506(3.8), 539(3.5),574(3.4), 624(3.2)	5-0133-66E
$C_{37}H_{48}N_4O$			
α-Porphinemethanol, octaethyl-	benzene	407(5.0),507(3.8), 540(3.5),575(3.4), 627(3.2)	5-0133-66E
$C_{37}H_{50}O_3$			
5α-Cholestan-3-one, 2,4-difurfurylidene-	EtOH	365(4.52)	39-1277-66C
$C_{37}H_{56}O_{12}$			
Fusidic acid, β-D-glucopyranuronosyl ester	EtOH	230(3.96)	1-1599-66

Compound	Solvent	$\lambda_{max}(\log \epsilon)$	Ref.
$C_{38}H_{20}Cl_2$			
Acenaphtho[1,2-j]fluoranthene, 4,5-bis(p-chlorophenyl)-	dioxan	224(5.06),244(4.84), 273(4.43),282(4.42), 325(4.33),337(4.61), 355(4.74),400(4.01)	44-2407-66
$C_{38}H_{24}N_2$			
Acenaphtho[5,6-de][1,2]diazepine, 1,4,7,8-tetraphenyl-	DMF	280(4.37),316s(3.54), 395(4.55)	78-0135-66B
	$C_6H_4Cl_2$	296(4.22),399(4.61)	78-0135-66B
	EtOH-CF_3COOH	260(4.44),284s(4.25), 384(4.49),454(4.05)	78-0135-66B
$C_{38}H_{24}O_4$			
Naphthalene, 1,4,5,8-tetrabenzoyl-	DMF	315(4.18),418(2.35)	78-0135-66B
$C_{38}H_{26}$			
Benzene, m-bis(9-anthrylvinyl)-	dioxan	390(4.39)	18-1547-66
$C_{38}H_{28}N_6O_2$			
2,3-Fulvenedimethanol, α,α,α',α',6,6-hexa-2-pyridyl-	MeOH	262(4.32),327(4.41)	44-2149-66
$C_{38}H_{28}Sn$			
Tin, (α-fluoren-9-ylidenebenzyl)-triphenyl-	THF	264(4.70),292(4.20), 303(4.20),318(4.15)	35-3027-66
$C_{38}H_{29}N_5O_8$			
Adenine, N-benzoyl-3-β-D-ribofuranosyl-, 2',3',5'-tribenzoate	EtOH-HCl	231(4.72),304(4.46)	25-0664-66
	EtOH	231(4.72),307(4.21)	25-0664-66
Adenosine, N-benzoyl-, 2',3',5'-tribenzoate	EtOH-HCl	231(4.68),283(4.36)	25-0664-66
	EtOH	231(4.68),279(4.34)	25-0664-66
$C_{38}H_{32}N_2O_2$			
Urea, 1-(2-naphthyl)-3-(p-phenoxyphenyl)-1-p-tolyl-3-(2,4-xylyl)-	EtOH	255(4.59)	34-0436-66
$C_{38}H_{32}N_2O_6S_2$			
Hydroquinone, 2,5-bis(3-methylindol-2-yl)-, di-p-toluenesulfonate	EtOH	225(4.75),249s(4.20), 330s(4.20),357(4.34)	44-3321-66
$C_{38}H_{34}$			
Anthracene, 2,3,9-triphenyl-1,4-dipropyl-	$CHCl_3$	275(5.01),367(3.88), 385(4.05),406(3.97)	44-4082-66
$C_{38}H_{34}O_{14}$			
Vioxanthin, tetraacetate	EtOH	263(5.05),320(4.43)	23-2873-66
$C_{38}H_{36}ClI_2P_2Rh$			
Rhodium, chloroiodo(iodomethane)-methylbis(triphenylphosphine)-	benzene	335(3.41),525s(2.28), 630(2.16)	39-1733-66A
$C_{38}H_{36}N_4O_3$			
5-Norbornene-2,3-dicarboximide, 5-(α-hydroxy-α-2-pyridylbenzyl)-N-dimethylaminopropyl-7-(α-2-pyridylbenzylidene)-	MeOH	249(4.24)	87-0537-66

Compound	Solvent	λ_{max}(log ε)	Ref.
$C_{38}H_{37}Cl_3O_4$			
6aH-Naphtho[1,2,3-fg]aceanthrylene, 9,12-diacetoxy-2,10,11-trichloro-4,6a-dimethyl-7-(1,4,5-trimethylhex-2-enyl)-7,8-dihydro-	EtOH	233s(4.61),238(4.67), 262s(4.38),270(4.41), 298(4.34),312(4.52), 328(4.59),377s(3.66), 394s(3.76),415(3.90), 441(3.85)	5-0106-66J
$C_{38}H_{44}N_4O_6$			
6,7-Porphinedipropionic acid, α-hydroxy-1,2,3,4,5,8-hexamethyl-, diethyl ester, acetate	5N HCl	404(5.44),514s(3.51), 553(4.24),595(3.66)	12-1481-66
	$CHCl_3$	403(5.23),500(4.22), 533(3.86),572(3.87), 625(3.42)	12-1481-66
$C_{38}H_{44}O_8$			
Gambogic acid	EtOH	280(4.25),288(4.30), 360(4.21)	7-0232-66
$C_{38}H_{44}O_{14}$			
Hirtin, dihydro-, triacetate	n.s.g.	235(4.11),280(4.10)	77-0206-66
$C_{38}H_{46}N_2O_8$			
5β-Chol-11-ene-3α,24-diol, bis-(p-nitrobenzoate)	dioxan	260(4.38)	78-1033-66
$C_{38}H_{46}N_4O_3S_2$			
Bilevomepromazine	MeOH	258(4.77),285(4.31), 301s(--),330s(--)	95-0514-66
$C_{38}H_{46}O_9$			
Garcinolic acid	EtOH	278(4.39),286(4.43), 360(4.21)	7-0232-66
Hydrogambogic acid	EtOH	285(4.53),330(4.28)	39-0772-66C
$C_{38}H_{49}N_7O$			
α-Porphinecarboxaldehyde, semicarbazone	benzene	406(<u>5.0</u>),508(<u>3.8</u>), 535(<u>3.5</u>),578(<u>3.4</u>), 625(<u>3.2</u>)	5-0133-66E
$C_{38}H_{50}O_{12}$			
Olean-12-ene-11,15,16,21-tetrone, 3β,22β,24,28-tetraacetoxy-	EtOH	240(4.04),296s(2.70)	78-0351-66
$C_{38}H_{52}N_2O_{12}$			
Olean-12-ene-11,15,16,21-tetrone, 3β,22β,24,28-tetraacetoxy-, oxime	EtOH	244(4.04)	78-0351-66
$C_{38}H_{53}BrO_{10}$			
13β-Olean-9(11)-en-12-one, 11-bromo-16α,21α-epoxy-3β,22β,24,28-tetra-acetoxy-	EtOH	208(3.78),292(3.90)	78-0351-66
$C_{38}H_{54}O_8$			
Gambogic acid, decahydro-	EtOH	218(4.45),232(4.19), 301(4.36),345(3.68)	39-0772-66C
$C_{38}H_{54}O_9$			
Olean-9(11),12-diene-3β,22β,24,28-tetrol, 16α,21α-epoxy-, tetraacetate	EtOH	280(3.95)	78-0351-66

Compound	Solvent	$\lambda_{max}(\log \epsilon)$	Ref.
$C_{38}H_{54}O_{10}$			
Escigenin-11-one, tetraacetate	EtOH	248(4.04)	78-0351-66
13β-Olean-9(11)-en-12-one, 16α,21α-epoxy-3β,22β,24,28-tetraacetoxy-	EtOH	247(4.03)	78-0351-66
$C_{38}H_{55}BrO_{10}$			
13β-Oleanan-12-one, 16α,21α-epoxy-3β,22β,24,28-tetraacetoxy-11β-bromo-	EtOH	310(2.08)	78-0351-66
$C_{38}H_{55}NO_8$			
Gambogic acid, decahydro-, oxime	EtOH	218(4.43),233s(4.21), 301(4.31),350(3.63)	39-0772-66C
$C_{38}H_{56}O_9$			
13β-Olean-9(11)-ene-3β,22β,24,28-tetrol, 16α,21α-epoxy-, tetraacetate	EtOH	204(3.70)	78-0351-66
$C_{38}H_{56}O_{10}$			
13β-Oleanan-12-one, 16α,21α-epoxy-3β,22β,24,28-tetrahydroxy-, tetraacetate	EtOH	281(1.60)	78-0351-66
$C_{38}H_{57}NO_2$			
12'-Apo-β-carotenoic acid, 13'-cyano-, dodecyl ester	BuOH	446(4.79)	54-0334-66

Compound	Solvent	λ_{max}(log ε)	Ref.
$C_{39}H_{29}N_3O_3$			
5-Norbornene-2,3-dicarboximide, 5-(α-hydroxy-α-2-pyridylbenzyl)-N-phenyl-7-(α-2-pyridylbenzylidene)-	MeOH	251(4.31)	87-0537-66
$C_{39}H_{30}N_4O_3$			
5-Norbornene-2,3-dicarboximide, 5-(α-hydroxy-α-2-pyridylbenzyl)-N-(α-pyridylmethyl)-7-(α-2-pyridylbenzylidene)-	MeOH	251(4.30)	87-0537-66
$C_{39}H_{32}O_{16}$			
Viopurpurin, dihydro-, pentaacetate	EtOH	286(4.48),296(4.60), 365(3.70)	23-2873-66
$C_{39}H_{33}BrP_2$			
1,1,1,5,5,5-Hexaphenyl-1,5-diphospha-2,4-pentadienylium bromide	n.s.g.	267(4.12),272(4.12), 290(4.01),295(4.01)	30-1042-66*
$C_{39}H_{33}IP_2$			
1,1,1,5,5,5-Hexaphenyl-1,5-diphospha-2,4-pentadienylium iodide	n.s.g.	267(4.34),305(4.15), 345(4.11)	30-1042-66*
$C_{39}H_{35}IN_2S_2$			
3-Ethyl-2-[5-(3-ethyl-4,5-diphenyl-4-thiazolin-2-ylidene)-1,3-pentadien-yl]-4,5-diphenylthiazolium iodide	MeOH	589(4.96)	9-0150-66
$C_{39}H_{35}N_3O_3$			
5-Norbornene-2,3-dicarboximide, N-cyclohexyl-5-(α-hydroxy-α-2-pyridylbenzyl)-7-(α-2-pyridylbenzylidene)-	MeOH	248(4.28)	87-0537-66
$C_{39}H_{43}N_5$			
Etioporphyrin I, α-(benzylideneamino)-	$CHCl_3$	414(5.16),512(4.11), 543(3.48),580(3.67), 634(2.89)	39-0794-66C
Etioporphyrin I, α-(N-phenylformimidoyl)-	$CHCl_3$	404(5.17),501(4.09), 534(3.83),569(3.76), 628(3.43)	39-0794-66C
$C_{39}H_{44}N_2O_7$			
Hernandezine	EtOH	205(5.10),282(3.95)	100-0094-66
$C_{39}H_{46}N_2O_8$			
Thalibrunine	EtOH	205(5.04),282(3.93)	100-0094-66
$C_{39}H_{46}O_8$			
Gambogic acid, methyl ester	EtOH	279s(4.19),292(4.26), 358(4.17)	39-0772-66C
$C_{39}H_{52}O_2$			
Naphthoquinone, 2-cinnamyl-3-phytyl- (plus shoulders not listed)	C_6H_{12}	251(4.54),330(3.59)	22-1828-66
$C_{39}H_{54}O_2$			
Naphtho[1,2-b:4,3-b']dipyran, 2,3,4,5,6,7-hexahydro-2-methyl-7-phenyl-2-(4,8,12-trimethyldecyl)-	C_6H_{12}	249s(--),252(4.52), 324(3.87),339(3.89)	22-1828-66

Compound	Solvent	$\lambda_{max}(\log \epsilon)$	Ref.
$C_{39}H_{56}N_{10}O_{13}$			
α-Aminitin, dethio-	MeOH	267s(3.59),275(3.64), 296(3.68)	5-0157-66J
$C_{39}H_{56}O_8$			
Gambogic acid, decahydro-, methyl ester	EtOH	218(4.43),235(4.18), 302(4.31),345(3.64)	39-0772-66C
$C_{39}H_{57}NO_8$			
Gambogic acid, decahydro-, methyl ester, oxime	EtOH	217(4.46),233s(4.22), 301(4.33),350(3.64)	39-0772-66C
$C_{39}H_{61}NO_2$			
14'-Apo-β-carotenoic acid, 15'-cyano-, hexadecyl ester	EtOH-benzene	439(4.76)	54-0334-66

Compound	Solvent	λ_{max}(log ε)	Ref.
$C_{40}H_{22}$			
6,6'-Bibenzo[a]pyrene	n.s.g.	_230(4.9),275(5.0), 296(4.9),302(4.9), 378(4.7),400(4.9), 411(4.9)_	78-2599-66
$C_{40}H_{26}$			
Bitryptycyl	dioxan	266s(3.38),272(3.63), 280(3.69)	44-1970-66
Pyrene, 1,3,6,8-tetraphenyl-	$CHCl_3$	250(4.70),260(4.71), 298(4.76),388(4.61)	78-0135-66B
	$C_6H_4Cl_2$	300(4.62),389(4.49)	78-0135-66B
$C_{40}H_{30}BrN_3O_3$			
5-Norbornene-2,3-dicarboximide, N-(p-bromobenzyl)-5-(α-hydroxy-α-2-pyridylbenzyl)-7-(α-2-pyridylbenzylidene)-	MeOH	247(4.27)	87-0537-66
$C_{40}H_{30}O_2$			
Benzene, 1,4-bis(9-anthrylvinyl)-2,5-dimethoxy-	dioxan	413(4.64)	18-1547-66
$C_{40}H_{31}N_3O_3$			
5-Norbornene-2,3-dicarboximide, N-benzyl-5-(α-hydroxy-α-2-pyridylbenzyl)-7-(α-2-pyridylbenzylidene)-	MeOH	250(4.24)	87-0537-66
$C_{40}H_{32}N_8$			
s-Tetrazin-1(2H)-yl, 2,2'-(4,4'-biphenylene)bis[3,4-dihydro-4,6-diphenyl-	mixt. at 97°K	_450(4.3),570(4.1), 770(4.3)_	49-0525-66
	197°K	_440(4.3),530(3.8), 760(4.2)_	49-0525-66
	298°K	_435(4.3),515(3.7), 755(4.1)_	49-0525-66
$C_{40}H_{44}N_4O_3$			
Vobtusine, anhydrodecarbomethoxy-demethyl-	EtOH	268(4.04),300s(3.62)	33-2072-66
	EtOH-KOH	270(4.00),317(3.68)	33-2072-66
$C_{40}H_{45}N_5O$			
Etioporphyrin I, α-(p-anisylidene-amino)-	$CHCl_3$	415(5.16),512(4.11), 545(3.46),582(3.62), 636(2.82)	39-0794-66C
Etioporphyrin I, α-[N-(p-methoxy-phenyl)formimidoyl]-	$CHCl_3$	405(5.17),502(4.10), 535(3.83),570(3.77), 629(3.45)	39-0794-66C
$C_{40}H_{46}N_4O_{10}$			
Deuteroporphyrin IX-2,4-bis(β-hydroxy-propionic acid), tetramethyl ester	N HCl	402(5.64),550(4.22), 591(3.75)	37-5276-66
	$CHCl_3$	403(5.20),500(4.06), 535(3.95),570(3.82), 623(3.62)	37-5276-66
$C_{40}H_{46}O_9$			
Gambogic acid, acetyl-	EtOH	280(4.32),345(3.92), 390(3.89)	7-0232-66
α-Gambogic acid, acetyl-	EtOH	282(4.33),346(3.96), 378s(3.94)	7-0232-66

Compound	Solvent	$\lambda_{max}(\log \epsilon)$	Ref.
Isogambogic acid, acetyl-	EtOH	276(4.38),332(3.97), 385(3.72)	39-0772-66C
$C_{40}H_{48}O_8$			
Gambogic acid, O-methyl-, methyl ester	EtOH	278(4.32),333(4.00)	39-0772-66C
$C_{40}H_{50}N_2$			
3-Hexenedinitrile, 2,5-bis[5-methyl-7-(2,6,6-trimethyl-1-cyclohexen-1-yl)-2,4,6-heptatrienylidene]-	isooctane	475(5.11)	54-0343-66
$C_{40}H_{50}N_4O$			
Pleiomutinol	EtOH	214(3.5),260(3.1), 297(2.9)	33-0964-66
$C_{40}H_{50}O_4$			
Bacterioruberin-2,2'-dione, 15,15'-didehydro-	$CHCl_3$	304(4.32),359(4.46), 512(5.20),542s(5.10)	33-0992-66
$C_{40}H_{50}O_8$			
Pregn-4-ene-3,11,20,21-tetrone, 17-hydroxy-21-(17-hydroxy-3,11-dioxoandrost-4-en-17β-yl)-	EtOH	238(4.50)	28C-0120-66A
$C_{40}H_{50}O_9$			
Hydrogambogic acid, dimethyl ester	EtOH	280s(4.35),286(4.38), 333(4.18)	39-0772-66C
$C_{40}H_{52}N_4O_3$			
Villalstoninetriol	EtOH	229(4.66),258(4.09), 286(3.97),294s(3.95)	33-1173-66
$C_{40}H_{52}O$			
γ-Carotene-4-one, 15,15'-didehydro-	benzene	456(4.99)	1-1195-66
$C_{40}H_{52}O_2$			
β-Carotene-20,20'-dial	isooctane	467(4.93)	54-0343-66
$C_{40}H_{52}O_{12}$			
Oleana-9(11),13(18),15-triene-12,19-dione, 3β,21α,22β,24,28-pentaacetoxy-	EtOH	271(3.99)	78-0351-66
$C_{40}H_{54}O_2$			
γ-Carotene, 15,15'-didehydro-1',2'-dihydro-1'-hydroxy-4-oxo-	benzene	455(4.99)	1-1195-66
$C_{40}H_{54}O_{10}$			
Oleana-9(11),12,15,18-tetraene-3β,21α,-22β,24,28-pentol, pentaacetate	EtOH	301(4.13)	78-0351-66
$C_{40}H_{56}O_2$			
β-Carotene-20,20'-diol	isooctane	448(4.07)	54-0343-66
γ-Carotene, 15,15'-didehydro-1',2'-dihydro-1',4-dihydroxy-	benzene	453(5.02),481(4.93)	1-1195-66
Lycopene, 5,5',6,6'-tetrahydro-6,6'-dioxo-	benzene	484(5.12),519(5.08)	39-2166-66C
$C_{40}H_{56}O_4$			
Foliachrome	EtOH	401(4.94),424(5.13), 451(5.12)	77-0404-66

Compound	Solvent	$\lambda_{max}(\log \epsilon)$	Ref.
Foliaxanthin	EtOH	416(4.97),439(5.13), 467(5.08)	77-0404-66
$C_{40}H_{58}$			
ζ-Carotene, 15,15'-dehydro-, trans	hexane	366s(--),387(4.92), 409(4.92)	39-2154-66C
$C_{40}H_{58}O_{10}$			
Isoescigenin, pentaacetate	EtOH	204(3.90)	78-0351-66
$C_{40}H_{60}$			
ζ-Carotene, all trans	hexane	360(4.61),380(4.95), 401(5.14),425(5.14)	39-2154-66C
15-cis	hexane	286(4.27),296(4.32), 378(4.70),398(4.77), 422(4.62)	39-2154-66C
$C_{40}H_{60}O_4Si$			
Androst-4-en-3-one, 17β,17'β-[(dimethylsilylene)dioxy]di-	EtOH	240(4.60)	87-0433-66
$C_{40}H_{62}$			
Phytofluene	hexane	318(4.42),331(4.73), 347(4.93),366(4.91)	39-2154-66C
$C_{40}H_{64}$			
Phytoene	hexane	276(4.56),286(4.70), 298(4.58)	39-2154-66C

Compound	Solvent	$\lambda_{max}(\log \epsilon)$	Ref.
$C_{41}H_{29}N_3O_3$			
5-Norbornene-2,3-dicarboximide, 5-(α-hydroxy-α-2-quinolylbenzyl)-7-(α-2-quinolylbenzylidene)-	MeOH	233(4.89),254(4.39), 289-316f(4)	87-0537-66
$C_{41}H_{38}N_2$			
4,4'(2H,2'H)-Spirobi[cyclopenta[c]-pyrrole], 5,5',6,6'-tetrahydro-6,6,6',6'-tetramethyl-1,1',3,3'-tetraphenyl-	n.s.g.	330(4.68)	77-0711-66
$C_{41}H_{38}O_{19}$			
Ergoxanthin, pentaacetate	EtOH	218(4.42),243(4.58), 265s(4.34),325(3.65)	78-2359-66
$C_{41}H_{40}AgNP_2S_2$			
Bis(triphenylphosphine)silver diethyl dithiocarbamate	EtOH	258(4.41)	12-0555-66
$C_{41}H_{40}CuNP_2S_2$			
Bis(triphenylphosphine)copper diethyl dithiocarbamate	EtOH	257(4.24),300s(3.75)	12-0555-66
$C_{41}H_{41}N_3O_3$			
5-Norbornene-2,3-dicarboximide, 5-(α-hydroxy-α-2-pyridylbenzyl)-N-octyl-7-(α-2-pyridylbenzylidene)-	MeOH	250(4.26)	87-0537-66
$C_{41}H_{46}N_2O_8$			
Thalicarpine, dehydro-	n.s.g.	268(4.82),331(4.34)	25-0770-66
$C_{41}H_{46}N_4O_3$			
Apovobtusine	EtOH	220(4.34),271(4.04), 310s(3.48)	78-1075-66
Vobtusine, anhydrodecarbomethoxy-	EtOH	270(4.14),298s(3.85)	33-2072-66
$C_{41}H_{48}N_4O_3$			
Apovobtusine, dihydro-	EtOH	245s(3.92),272(3.98), 300s(3.80)	78-1075-66
Vobtusine, anhydrodecarbomethoxy-dihydro-	EtOH	276(3.98),300s(3.80)	33-2072-66
	12N HCl	245(3.52),250s(3.48), 269(3.24),352(3.67)	33-2072-66
$C_{41}H_{48}N_4O_4$			
Vobtusine, decarbomethoxy-	EtOH	220(4.55),264(4.16), 300s(3.60)	78-1075-66
$C_{41}H_{48}O_9$			
Isoallogambogic acid, methyl ester, acetate	EtOH	276(4.36),329(3.96), 385(3.72)	39-0772-66C
$C_{41}H_{50}N_4O_2$			
Pleiomutine	EtOH	216(4.5),228(4.5), 264(4.1),285(3.5)	33-0964-66
$C_{41}H_{50}N_4O_3$			
Apovobtusine, tetrahydro-	EtOH	220(4.36),250s(4.06), 260(4.10),300(3.76)	78-1075-66
Vobtusine, anhydrotetrahydro-decarbomethoxy-	EtOH	256(4.15),302(3.83)	33-2072-66

Compound	Solvent	$\lambda_{max}(\log \epsilon)$	Ref.
$C_{41}H_{52}N_2O$			
Urea, 1,1-bis(p-octylphenyl)-3,3-diphenyl-	EtOH	270(4.36)	34-0436-66
$C_{41}H_{56}O_2$			
Trinorshion-23-ene, 3β-acetoxy-24,24-diphenyl-	C_6H_{12}	250(4.34)	22-1670-66
$C_{41}H_{58}O_9$			
Gambogic acid, decahydro-, methyl ester, acetate	EtOH	220(4.39),242s(4.08), 290(4.24)	39-0772-66C
$C_{41}H_{62}O_{13}$			
5β-Card-20(22)-enolide, 14β,15β-epoxy-3β-hydroxy-, β-D-tridigitoxoside	EtOH	214(4.23)	94-1133-66
$C_{41}H_{68}N_{10}O_{14}$			
Ferrimycin A_1, desferri-	EtOH	201(4.57),233(4.38), 322(3.22)	78-0171-66B

Compound	Solvent	λ_{max}(log ε)	Ref.
$C_{42}H_{24}N_6O_2$			
9H,19H-Diperimidino[1,2-a:1',2'-a']-benz[1,2-c:4,5-c']dipyrrole, 8,18-dianilino-9,19-dioxo-	$C_{10}H_7Cl$	630s(4.32),678(4.42)	33-0534-66
$C_{42}H_{25}N$			
Dinaphtho[2',3":3,4;2",3":5,6]carbazole, 9-(1-anthryl)-	benzene	298(4.55),310(4.58), 362(4.25),385(4.11), 410(4.15),435(4.26)	24-2449-66
$C_{42}H_{26}$			
Naphtho[2,3-k]fluoranthene, 7,13,14-triphenyl-	$CHCl_3$	288(4.68),295(4.68), 328(4.60),343(4.84), 387(3.80),405(4.09), 427(4.27),453(4.17)	44-4082-66
$C_{42}H_{26}N_4$			
Bicyclo[3.2.2]nona-2,8-diene-6,6,7,7-tetracarbonitrile, 4-(2,3,4,5-tetraphenyl-2,4-cyclopentadien-1-ylidene)-	THF	248(4.51),355(4.08)	5-0034-66H
$C_{42}H_{28}$			
9,9'-Bifluorene, 9-(3-fluoren-9-ylidenepropenyl)-	CH_2Cl_2	243(4.74),258(4.75), 269(4.75),292(4.30), 302(4.28),330(4.40), 342(4.59),358(4.57)	24-0658-66
Propane, 2-fluoren-9-yl-1,3-difluoren-9-ylidene-	CH_2Cl_2	306(4.48),322(4.48)	24-0658-66
$C_{42}H_{28}N_4$			
Tricyclo[3.2.2.0^{2,4}]non-8-ene-6,6,7,7-tetracarbonitrile, 3-(2,3,4,5-tetraphenyl-2,4-cyclopentadien-1-yl)-	THF	242(3.94),325(3.53)	5-0034-66H
$C_{42}H_{31}N_7O_4$			
5-Norbornene-2,3-dicarboximide, 5,6-bis(α-hydroxy-α,α-di-2-pyridylmethyl)-7-(di-2-pyridylmethylene)-	MeOH	262(4.41)	44-2149-66
$C_{42}H_{32}$			
Indene, 1,3-diphenyl-2-(1,3,3-triphenylallyl)-	dioxan	279(4.2)	24-0392-66
carbanion, potassium salt	DMSO	420(4.3)	24-0392-66
$C_{42}H_{34}$			
Indene, 1,3-diphenyl-2-(1,3,3-triphenylpropyl)-	dioxan	277(4.1)	24-0392-66
carbanion, potassium salt	DMSO	423(4.3)	24-0392-66
$C_{42}H_{34}N_2$			
2,6-Lutidine, 1-anilino-1,4-dihydro-4-(tetraphenyl-2,4-cyclopentadien-1-ylidene)-	n.s.g.	262(4.44),289(4.44), 472(4.69)	39-1086-66C
perchlorate	n.s.g.	262(4.23),427(4.27)	39-1086-66C
$C_{42}H_{34}O_4$			
Tricyclo[3.2.2.0^{2,4}]nona-6,8-diene-6,7-dicarboxylic acid, 3-(2,3,4,5-tetraphenyl-2,4-cyclopentadien-1-yl)-, dimethyl ester	EtOH	241(4.36),320(3.59)	5-0034-66H

Compound	Solvent	λ_{max}(log ϵ)	Ref.
$C_{42}H_{34}O_{18}$			
Aurofusarin, hexaacetate	n.s.g.	279(4.89),343s(3.98), 366(3.95)	39-2237-66C
$C_{42}H_{38}O_{16}$			
Xylaphin, dihydro-, hexaacetal	$CHCl_3$	282(4.30),292(4.20), 307(3.85),321(3.91), 337(4.11),385(3.50), 406(3.86),433(4.20), 462(4.33)	39-1836-66C
$C_{42}H_{42}O_{18}$			
Erythroaphin-fb, diglucoside	EtOH	252(4.39),274s(4.31), 297(4.42),306(4.47), 423(4.18),514(3.76), 554(4.16),600(4.34)	39-1832-66C
$C_{42}H_{43}N_5O_7$			
Alanine, carbobenzoxy-phenylalanyl-propyltryptophyl-3-phenyl-	MeOH	281(3.78)	5-0220-66B
$C_{42}H_{54}CuO_8$			
Steroid 39, copper complex	EtOH	246(4.31),332(4.14)	33-2218-66
$C_{42}H_{58}CuO_8$			
Steroid 33, copper complex	EtOH	259(3.95),310(4.33)	33-2218-66
$C_{42}H_{60}O_2$			
Spirilloxanthin	benzene	480(5.01),510(5.15), 546(5.08)	39-2166-66C
$C_{42}H_{60}O_3$			
Trinorshion-23-en-3β-ol, 22-methoxy-24,24-diphenyl-, acetate	C_6H_{12}	252(4.15)	22-1670-66
$C_{42}H_{64}O_2$			
Lycopene, 1,1',2,2'-tetrahydro-1,1'-dimethoxy-	benzene	456(5.09),484(5.28), 519(5.21)	22-1670-66
$C_{42}H_{68}N_2O_2$			
5α-Androstan-3-one, 17β-hydroxy-1α,17α-dimethyl-, azine	MeOH	232(4.33)	73-1064-66

Compound	Solvent	$\lambda_{max}(\log \epsilon)$	Ref.
$C_{43}H_{34}N_2O_3$			
Urea, 1-(2-naphthyl)-3-(m-phenoxyphenyl)-1-(p-phenoxyphenyl)-3-(2,5-xylyl)-	EtOH	255s(4.51)	34-0436-66
$C_{43}H_{34}N_5O_{10}P$			
Adenosine, N-benzoyl-, 2',3'-dibenzoate, 5'-(diphenyl phosphate)	EtOH-HCl	230(4.59),283(4.33)	25-0664-66
	EtOH	231(4.54),279(4.28)	25-0664-66
$C_{43}H_{35}N$			
2,6-Lutidine, 1-benzyl-1,4-dihydro-4-(tetraphenyl-2,4-cyclopentadien-1-ylidene)-	acid	262s(4.22),413(4.18)	39-1086-66C
	n.s.g.	262(4.24),289(4.24), 469(4.49)	39-1086-66C
$C_{43}H_{36}N_2$			
2,6-Lutidine, 1,4-dihydro-1-(N-methylanilino)-4-(tetraphenyl-2,4-cyclopentadien-1-ylidene)-	acid	263(4.36),433(4.47)	39-1086-66C
	n.s.g.	263(4.31),288(4.34), 474(4.62)	39-1086-66C
$C_{43}H_{39}IN_2S_2$			
2,2'-Bis(3-ethyl-4,5-diphenylthiazolyl)pentamethinecyanine iodide	MeOH	663(5.14)	9-0150-66
$C_{43}H_{42}FeN_6O_8$			
Verdohemochrome	$CHCl_3$-1% pyridine	397(4.70),500(3.9), 534(4.0),663(4.4)	69-2845-66
$C_{43}H_{45}N_5O_7$			
Alanine, carbobenzoxy-phenylalanyl-prolyl-tryptophyl-3-phenyl-, methyl ester	MeOH	280(3.77)	5-0220-66B
$C_{43}H_{48}N_4O_5$			
Vobtusine, anhydro-	EtOH	294(4.24),324(4.27)	33-2072-66
	12N HCl	297(3.85),330(3.93)	33-2072-66
$C_{43}H_{49}NO_8$			
Gambogic acid, compound with pyridine	EtOH	279s(4.32),292(4.35), 360(4.26)	39-0772-66C
$C_{43}H_{50}N_4O_6$			
Vobtusine	EtOH	235s(4.08),265(4.07), 299(4.15),325(4.21)	33-2072-66
	EtOH	225(4.49),265(4.02), 300s(4.12),328(4.19)	78-1075-66
	12N HCl	238(3.95),281(3.74), 303s(3.81),328(3.90)	33-2072-66
$C_{43}H_{51}NO_9$			
Hydrogambogic acid, pyridine salt	EtOH	280s(4.40),286(4.43), 334(4.23)	39-0772-66C
$C_{43}H_{52}N_4O_5$			
Macralstonine	EtOH	230(4.84),260(4.21), 294(4.12)	33-0946-66
Vobtusine, anhydrotetrahydro-	EtOH	255(4.15),303(3.79)	33-2072-66
$C_{43}H_{52}N_4O_6$			
Vobtusine, dihydro-	EtOH	255(4.11),304(3.82)	33-2072-66

Compound	Solvent	$\lambda_{max}(\log \epsilon)$	Ref.
$C_{43}H_{68}N_{10}O_{12}S$			
Glycinamide, L-tyrosyl-L-isoluecyl-L-glutaminyl-L-asparaginyl-S-ethylcarbamoyl-L-cysteinyl-L-prolyl-L-leucyl-	n.s.g.	286(3.22)	33-0083-66
$C_{44}H_{26}$			
1,2-Butadiene, 1,4-difluoren-9-ylidene-3-(fluoren-9-ylidenemethyl)-	CH_2Cl_2	278(4.70),386(4.49), 404(4.49),565(4.61)	24-0658-66
$C_{44}H_{28}$			
2-Butene, 1,4-difluoren-9-ylidene-3-(fluoren-9-ylidenemethyl)-	CH_2Cl_2	270(4.76),370(4.30), 385(4.30),459(4.54), 477(4.53)	24-0658-66
$C_{44}H_{42}O_{16}$			
Flavone, 3',4'-bis(benzyloxy)-5,7-dihydroxy-3-methoxy-, 7-β-D-glucopyranoside, tetraacetate	EtOH	255(4.27),270s(--), 355(4.22)	22-0987-66
$C_{44}H_{48}O_{25}$			
Flavone, 3,4',5,7-tetrahydroxy-3'-methoxy-, 3,7-di-β-D-glucopyranoside, octaacetate	EtOH	255(4.45),270s(--), 362(4.35)	22-0987-66
$C_{44}H_{50}Cl_4O_6$			
Hydroquinone, tetrachloro-, complex with 17β-hydroxy-17-methylestra-4,6,8(14)-trien-3-one	MeOH	345(4.27)	44-3780-66
$C_{44}H_{58}Cl_2N_2O_4$			
Androsta-5,16-diene-16-carboxaldehyde, 17-chloro-3-hydroxy-, azine, diacetate	EtOH	310(4.41),323(4.42)	32-1241-66
$C_{44}H_{62}CuO_6$			
Pregn-5-ene-21-carboxaldehyde, 3-hydroxy-20-oxo-, copper derivative	EtOH	244(4.09),302(4.28)	44-3193-66
$C_{44}H_{64}O_4$			
Kitol, diacetate	hexane	190(4.08),286(4.59), 295(4.59)	22-3299-66
$C_{44}H_{72}$			
Cyclobutane, tetrakis(2-tert-butyl-3,3-dimethyl-1-butenylidene)-	n.s.g.	250(4.20),258(4.21), 268(4.12),298(3.24), 316(3.31)	35-3155-66
$C_{45}H_{32}N_2$			
Dipyrromethene, 3,3',4,4',5,5'-hexaphenyl-	EtOH-HCl	305(4.27),415(3.99), 561(4.86)	12-1871-66
	EtOH	299(4.39),524(4.50)	12-1871-66
$C_{45}H_{33}N_9$			
1,3,5-Benzenetricarbonitrile, 2,4,6-tris(N,N-diphenylhydrazino)-	n.s.g.	620(4.17)	27-0303-66

Compound	Solvent	λ_{max}(log ε)	Ref.
$C_{45}H_{35}N_5O_8$			
Adenine, 7-α-D-arabinofuranosyl-N-benzoyl-3-benzyl-, 2',3',5'-tribenzoate	pH 1	233(4.68),306(4.27)	44-1413-66
	pH 7, 13	243(4.68),339(4.35), 355s(4.28)	44-1413-66
$C_{45}H_{36}O_8$			
D-Arabinose, 5-O-trityl-, 2,3,4-tribenzoate (benzene solvate)	EtOH	229(4.57),254(3.47), 260(3.39),266(3.36), 270(3.36),273(3.36), 281(3.27)	87-0791-66
$C_{45}H_{46}N_7O_{15}P_3$			
Riboflavin, 5'-O-trityl-, 2',3',4'-tri(2-cyanoethyl)phosphate	EtOH	268(4.51),365(3.96)	65-1749-66
$C_{45}H_{56}O_2$			
1,4-Naphthoquinone, 2-(diphenylallyl)-3-phytyl-	C_6H_{12}	251(4.44),330(3.49)	22-1828-66
$C_{45}H_{57}N_7O_9$			
L-Leucine, N-benzyloxycarbonyl-L-phenylalanyl-L-valyl-L-glutaminyl-L-tryptophanyl-, methyl ester	MeOH	221s(4.64),275s(3.72), 282(3.76),290(3.71)	44-3400-66
$C_{45}H_{60}N_4O_{11}$			
Gambogic acid, decahydro-, methyl ester, 2,4-dinitrophenylhydrazone	$CHCl_3$	219(4.49),238s(4.37), 303(4.27),361(4.34)	39-0772-66C
$C_{45}H_{72}N_2O_{16}$			
β-Solamarin, N-nitroso-	MeOH	234(3.90),363(1.89)	102-1227-66
$C_{45}H_{72}N_2O_{17}$			
α-Solamarin, N-nitroso-	MeOH	234(3.91),366(1.86)	102-1227-66
$C_{46}H_{22}$			
Dibenzo[a,j]difluoreno[2,1,9-cde: 2',1',9'-lmn]perylene	benzene	308(4.66),322(4.68), 356(4.43),376(4.68), 464(4.43),486(4.54), 521(4.69)	24-3298-66
$C_{46}H_{58}N_2$			
3-Hexenedinitrile, 2,5-bis[3,7-dimethyl-9-(2,6,6-trimethyl-1-cyclohexen-1-yl)-2,4,6,8-nonatetraen-ylidene]-	$CHCl_3$	539(5.15)	54-0343-66
$C_{46}H_{74}N_2O_6$			
5β-Pregnan-3α-ol, 20-nitroso-, acetate, dimer	heptane	298(3.95)	44-0524-66
$C_{47}H_{33}P$			
Phosphorane, [2-fluoren-9-ylidene-1-fluoren-9-ylidenemethyl)-ethylidene]triphenyl-	CH_2Cl_2	258(5.02),333(4.42), 390(3.88),574(4.50)	24-0658-66
perchlorate	CH_2Cl_2	313(4.14),325(4.17)	24-0658-66
$C_{47}H_{66}O_{22}$			
Olivomycin D	EtOH	227(4.30),275(4.70), 319(3.97),406(4.04)	30-0863-66F

Compound	Solvent	λ_{max}(log ε)	Ref.
Olivomycin D (cont.)	EtOH	227(4.30),275(4.70), 308s(--),319(3.97), 330s(--),406(4.04)	88-1643-66
$C_{47}H_{84}N_2O_2$ 5,27-Dioxa-1,31-diazatetracyclo-[29.10.10.0^{2,28}.0^{4,30}]henpenta-conta-2,4(30),28-triene	C_6H_{12}	216(4.22),271(3.84), 315(3.74)	54-0041-66
$C_{48}H_{26}N_8$ Benzonitrile, 4,4',4'',4'''-(5,10,15,-20-porphinetetrayl)tetra-	pyridine	423(5.55),488(3.58), 521(4.29),556(3.91), 597(3.77),655(3.53)	4-0495-66
$C_{48}H_{30}N_4O_8$ Benzoic acid, 4,4',4'',4'''-(5,10,15,-20-porphinetetrayl)tetra-	5% $NaHCO_3$	412(5.56),527(3.90), 566(3.90),593(3.57), 654(3.59)	4-0495-66
	pyridine	422(5.21),486(3.49), 517(4.16),552(3.85), 591(3.63),649(3.46)	4-0495-66
$C_{48}H_{38}N_4$ Porphine, 5,10,15,20-tetra-p-tolyl-	benzene	420(5.68),485(3.62), 516(4.28),550(3.99), 592(3.73),650(3.64)	4-0495-66
	pyridine	422(5.66),485(3.61), 517(4.25),552(4.00), 594(3.71),651(3.17)	4-0495-66
$C_{48}H_{38}N_4O_4$ Benzyl alcohol, 4,4',4'',4'''-(5,10,15,20-porphinetetrayl)tetra-	pyridine	422(5.59),485(3.53), 517(4.18),553(3.95), 594(3.66),651(3.64)	4-0495-66
$C_{49}H_{35}P$ Phosphorane, [4-fluoren-9-ylidene-3-(fluoren-9-ylidenemethyl)-2-butenylidene]triphenyl-	CH_2Cl_2	252(5.12),259(5.30), 554(4.61)	24-0658-66
perchlorate	CH_2Cl_2	234(4.85),254(4.72), 262(4.81),342(4.43)	24-0658-66
$C_{49}H_{36}N_2O_5$ Urea, 1,3-bis(m-phenoxyphenyl)-1,3-bis(p-phenoxyphenyl)-	EtOH	262(4.34)	34-0436-66
$C_{49}H_{56}N_6O_8$ Alanine, carbobenzoxyleucyl-phenyl-alanyl-prolyl-tryptophyl-phenyl-, methyl ester	n.s.g.	281(3.78)	5-0220-66B
$C_{49}H_{84}N_2O$ Urea, 1,1-dioctadecyl-3,3-diphenyl-	EtOH	267(4.58)	34-0436-66
$C_{50}H_{62}N_2$ 3-Hexenedinitrile, 2,5-bis[5,9-di-methyl-11-(2,6,6-trimethyl-1-cyclo-hexen-1-yl)-2,4,6,8,10-undeca-pentaenylidene]-	$CHCl_3$	565(5.21)	54-0343-66

Compound	Solvent	λ max(log ε)	Ref.
$C_{50}H_{64}O_4Si$			
Androst-4-en-3-one, 17β,17'β-[(diphenylsilylane)dioxy]di-	EtOH	220(5.04),240(4.60)	87-0433-66
$C_{51}H_{54}O_{25}$			
Flavone, 4'-(benzyloxy)-3,5,7-trihydroxy-3'-methoxy-, 3,7-di-β-D-glucopyranoside, octaacetate	EtOH	255(4.48),270s(--), 355(4.34)	22-0987-66
$C_{52}H_{34}S_2$			
Disulfide, bis(9,10-diphenyl-1-anthryl)-	$CHCl_3$	273(4.90),375(4.05), 395(4.24),419(4.24)	22-2939-66
Disulfide, bis(9,10-diphenyl-2-anthryl)-	$CHCl_3$	270(4.95),370(4.13), 389(4.19),408(4.16)	22-2939-66
$C_{52}H_{38}N_4O_8$			
Benzoic acid, 4,4',4'',4'''-(5,10,15,-20-porphinetetrayl)tetra-, tetramethyl ester	pyridine	423(5.61),485(3.57), 516(4.25),551(3.92), 592(3.72),650(3.51)	4-0495-66
$C_{52}H_{50}N_8$			
Porphine, 5,10,15,20-tetrakis-(p-dimethylaminophenyl)-	6N HCl	438(5.47),597(3.92), 649(4.45)	4-0495-66
$C_{53}H_{70}N_8O_{12}$			
L-Leucine, N-benzyloxycarbonyl-β-tert-butyl-L-aspartyl-L-phenylalanyl-L-valyl-L-glutaminyl-L-tryptophanyl-, methyl ester	MeOH	217s(4.55),274s(3.68), 282(3.78),291(3.64)	44-3400-66
$C_{53}H_{76}O_{23}$			
Chromomycin A_2	EtOH	229(4.37),279(4.67), 317(3.86),331(3.75), 412(3.89)	88-0545-66
$C_{53}H_{80}N_2O$			
Urea, 1,3-dihexadecyl-1,3-di-2-naphthyl-	EtOH	250s(4.36)	34-0436-66
$C_{54}H_{74}N_8O_4Zn$			
Bilene-b, 2,3,7,8,12,13,17,18-octamethyl-2,3,17,18-tetrahydro-, zinc complex	EtOH	238(4.28),294(3.39), 310(3.38),378(4.01), 488(4.87),498(4.97), 513(5.09)	39-1155-66C
$C_{54}H_{94}N_2O_2$			
5α-Cholestane, 3-nitroso-, dimer	heptane	294(3.93)	44-0524-66
$C_{55}H_{46}N_4O_5$			
Hypoxanthine, 3-benzyl-7-[3(or 2),5-di-O-trityl-α-D-arabinofuranosyl]-	pH 1	256(4.09)	44-1413-66
	pH 7	262(4.28)	44-1413-66
	pH 13	266(4.12)	44-1413-66
$C_{55}H_{72}N_4O_5$			
Protophaeophytin	dioxan	275(4.37),298(4.28), 420(5.30),523(3.93), 567(4.26),588(4.14), 639(3.38)	88-5145-66

Compound	Solvent	λ_{max}(log ε)	Ref.
$C_{55}H_{74}MgN_4O_6$ Bacteriochlorophyll	ether	357(4.87),392(4.67), 573(4.34),770(4.98)	35-4500-66
$C_{56}H_{80}O_{26}$ Olivomycin B	EtOH	228(4.34),276(4.64), 318(3.78),406(4.02)	30-0863-66F
	EtOH	228(4.34),276(4.64), 308s(--),318(3.78), 330s(--),406(4.02)	88-1643-66
$C_{56}H_{82}O_{25}$ Olivomycin C	EtOH	228(4.36),277(4.67), 319(3.87),406(4.04)	30-0863-66F
	EtOH	228(4.36),277(4.67), 308s(--),319(3.87), 330s(--),406(4.04)	88-1643-66
$C_{56}H_{90}O_4$ Cholest-4-en-3-one, 6β-[(3β-hydroxy-5,14-dimethyl-18,19-dinor-5β,8α,9β,-10α,14β-cholest-13(17)-en-6α-yl)-oxy]-, acetate	EtOH	236(4.29)	78-3195-66
$C_{57}H_{84}N_2O$ Urea, 1,1,3,3-tetrakis(p-octylphenyl)-	EtOH	275(4.63)	34-0436-66
$C_{58}H_{84}O_6Si$ Androst-4-en-3-one, 17β,17'β,17"β-[(methylsilylidyne)trioxy]tri-	EtOH	240(4.76)	87-0433-66
$C_{58}H_{84}O_{26}$ Olivomycin A	EtOH	228(4.39),277(4.67), 318(3.81),406(4.05)	30-0863-66F
	EtOH	228(4.39),277(4.67), 308s(--),318(3.81), 330s(--),406(4.05)	88-1643-66
$C_{58}H_{94}O_4$ A-Homo-B-nor-5β-cholest-3-en-4a-one, 3,3'-(ethylenedioxy)bis[5-methyl-	EtOH	255(4.36)	78-1421-66
$C_{58}H_{96}O_6$ 5α-Cholestane-3β,5-diol, 6β-[(3β-hydroxy-5,14-dimethyl-18,19-di-nor-5β,8α,9β,10α,14β-cholest-13(17)-en-6α-yl)oxy]-, diacetate	EtOH	207(3.95),210(3.87), 215(3.63),220(3.18) (end absorptions)	78-3195-66
$C_{59}H_{90}O_4$ Coenzyme Q_{10}	EtOH	275(4.15)	10-0548-66A
$C_{59}H_{92}O_4$ Coenzyme Q_{10}, reduced	EtOH	290(3.62)	10-0548-66A
$C_{60}H_{66}N_8$ Porphine, 5,10,15,20-tetrakis-(p-diethylaminophenyl)-	6N HCl	439(5.51),599(3.94), 652(4.52)	4-0495-66
	pyridine	449(5.09),592(4.40), 682(3.99)	4-0495-66

Compound	Solvent	λ_{max}(log ϵ)	Ref.
$C_{62}H_{86}N_{12}O_{16}$			
Valine, N,N'-[(2-amino-4,6-dimethyl-3-oxo-3H-phenoxazine-1,9-diyl)bis-[carbonylimino(2-hydroxypropylidene)-carbonyliminoisobutylidenecarbonyl-1,2-pyrrolidinediylcarbonyl(methyl-imine)methylenecarbonyl]bis[N-methyl-, dilactone	MeOH	241(4.52),443(4.39)	88-2331-66
$C_{62}H_{90}N_{12}O_{18}$			
Valine, N,N'-[(2-amino-4,6-dimethyl-3-oxo-3H-phenoxazine-1,9-diyl)bis-[carbonylimino(2-hydroxypropylidene)-carbonyliminoisobutylidenecarbonyl-1,2-pyrrolidinediylcarbonyl(methyl-imine)methylenecarbonyl]bis[N-methyl-	MeOH	238(4.63),427(4.40), 445(4.42)	88-2331-66
$C_{63}H_{87}N_{13}O_{18}$			
Actinomycin C_2, 7-nitro-	MeOH	443(4.26)	88-3531-66
$C_{63}H_{89}N_{13}O_{16}$			
Actinomycin C_2, 7-amino-	MeOH	522(4.34)	88-3531-66
$C_{64}H_{87}BrClN_{11}O_{16}$			
Actinomycin C_3, 2-deamino-2-chloro-7-bromo-	$CHCl_3$	265(--),390(4.20), 480(--)	88-3595-66
$C_{64}H_{89}BrN_{12}O_{16}$			
Actinomycin C_3, 7-bromo-	$CHCl_3$	434(--),456(4.41)	88-3595-66
$C_{64}H_{89}ClN_{12}O_{16}$			
Actinomycin C_3, 7-chloro-	$CHCl_3$	434(--),454(4.30)	88-3595-66
$C_{66}H_{81}N_9O_{12}S$			
Alanine, carbobenzoxy-valyl-δ-tosyl-ornithyl-leucyl-phenylalanyl-pro-lyl-tryptophyl-3-phenyl-, methyl ester	MeOH	281(3.77)	5-0220-66B
$C_{67}H_{96}$			
Cholesta-3,5-diene, 3-phenyl-6-[(3-phenylcholesta-3,5-dien-6-yl)-methyl]-	hexane	296(4.72)	23-2233-66
$C_{69}H_{60}O_{16}$			
Harpagoside, penta-O-cinnamoyl-	$CHCl_3$	279(5.12)	33-1552-66
	CH_2Cl_2	277(5.13)	33-1552-66
$C_{76}H_{102}N_{14}O_{14}$			
Gramicidin S, actinocyl-	MeOH	442(4.32)	39-1406-66C
$C_{80}H_{56}$			
1,2-Benzanthracene, 9,10-dimethyl-, tetramer	n.s.g.	227(5.1),274(5.2), 300(5.2),374(4.3), 382(4.2)	78-2599-66
$C_{84}H_{56}$			
Cholanthrene, 20-methyl-, tetramer	n.s.g.	226(5.1),274s(5.0), 295s(5.1),300(5.1), 380(4.5),422s(4.5)	78-2599-66

Compound	Solvent	λ_{max}(log ε)	Ref.
$C_{142}H_{222}N_{42}O_{31}$			
Valinamide, D-seryl-L-tyrosyl-L-seryl-L-norleucyl-L-glutamyl-L-histidyl-L-phenylalanyl-L-arginyl-L-tryptophyl-glycyl-L-lysyl-L-prolyl-L-valyl-glycyl-L-lysyl-L-lysyl-L-arginyl-L-arginyl-L-prolyl-L-valyl-L-lysyl-L-valyl-L-tyrosyl-L-prolyl-	N HOAc	275.5(3.85)	31-0526-66
$C_{214}H_{386}N_{56}O_{93}S$			
β-Corticotropin	0.1N HOAc	280(3.9)	33-0134-66
	pH 13	285(3.9),290(4.0)	33-0134-66

1- -66, Acta Chem. Scand., 20 (1966)
0057 J. Sandström and I. Wennerbeck
0262 O. Bucherdt et al.
0689 U. Berg and J. Sandström
1074 B.R. Thomas
1113 P.H. Nielsen and O. Dahl
1195 A.P. Leftwick and B.C.L. Weedon
1561 J. Gripenberg and T. Hase
1599 W.O. Godtfredsen and S. Vangedal
1631 H. Lund, P. Lunde and F. Kaufmann
1778 M.G. Ettlinger et al.
2202 J. Gripenberg and M. Lounasmaa
2211 R. Magnusson
2467 O. Buchardt, J. Becher and L. Lohse
2497 J. Lemmich, E. Lemmich and B.E. Nielsen
2637 H. Lund and S. Gruhn
2781 E. Bach et al.
2829 L. Westfelt
2841 L. Westfelt

2- -66, Indian J. Chem., 4 (1966)
0118 T.R. Govindachari et al.

3- -66, Anal. Chem., 38 (1966)
0612 M.W. Scoggins and J.W. Miller
0950 B.Z. Egan and W.D. Arnold
1702 P.D. Anderson and D.M. Hercules

4- -66, J. Heterocyclic Chem., 3 (1966)
0005 C.W. Noell and C.C. Cheng
0027 C.K. Bradsher and D.F. Lohr, Jr.
0042 N.J. DeSousa, P.B. Nayak and E. Secco
0051 G.W. Stacy, T.E. Wollner and T.R. Oakes
0090 J.I. Degraw and J.G. Kennedy
0093 F.F. Ebetino
0101 R.N. Schut and T.J. Leipzig
0107 L. Joseph and A.H. Albert
0110 R.K. Robins et al.
0115 D.E. O'Brien, R.H. Springer and C.C. Cheng
0119 H.H. Takimoto, G.C. Denault and S. Hotta
0178 I. Tomita, H.G. Brooks and D.E. Metzler
0188 J.A. Settepani and A.B. Borkovec
0202 H.R. Snyder, Jr., and F.F. Ebetino
0206 A.E. Drukker, C.I. Judd and D.D. Dusterhoft
0218 T. Kuraishi and R.N. Castle
0224 L. Bauer, C.N.V. Nambury and F.M. Hershenson
0226 G.L. Tong, W.W. Lee and L. Goodman
0228 R.H. Nealy and J.S. Driscoll
0241 L.B. Townsend and R.K. Robins
0247 L.G. Humber et al.
0278 C.R. Ganellin, H.F. Ridley and R.G.W. Spickett
0282 H.F. Andrew and C.K. Bradsher
0289 L.L. Zaika and M.M. Joullie
0315 B.R. Baker and J.H. Jordaan
0324 B.R. Baker and J.H. Jordaan
0328 L.R. Caswell and P.C. Atkinson
0338-9 T.C. Miller
0359 A.E. Drukker and C.I. Judd
0381 R.N. Castle and S. Takano
0395 K.T. Potts, S.K. Roy and D.R. Liljegren
0413 N. Rabjohn, H.R. Havens and J.L. Rutter
0422 R.C. Bertelson and W.J. Becker
0435 H. Rutner and P.E. Spoerri
0444 L.L. Zaika and M.M. Joullie
0454 M. Hoffer, V. Toome and A. Brossi
0466 K. Aparajithan, A.C. Thompson and J. Sam
0476 S.-C.J. Fu and E. Chinoporos
0485 N.J. Leonard and K.L. Carraway
0495 N. Datta-Gupta and T.J. Bardos
0512 N.R. Patel and R.N. Castle
0518 L. Huestis, I. Emery and E. Steffensen
0541 R.N. Castle, K. Kaji and D. Wise

5- -66A, Ann. Chem. Liebigs, 691 (1966)
0009 M. Arram, I.G. Dinulescu and C.D. Nenitzescu
0142 H. Goldner, G. Dietz and E. Carstens
0159 H.J. Rosenkranz, G. Botyos and H. Schmid
0172 G.S. Sidhu and A.V.B. Sankaram

5- -66B, Ann. Chem. Liebigs, 692 (1966)
0058 E. Müller and H. Kessler
0119 R. Brossmer and E. Röhm
0134 H. Goldner, G. Dietz and E. Carstens
0174 E. Hecker and M. Hopp
0180 H. Dannenberg and H.J. Gross
0220 H. Zahn and D. Brandenburg

5- -66C, Ann. Chem. Liebigs, 693 (1966)
0044 G. Quinkert et al.
0134 W. Grell and H. Machleidt
0165 F. Korte, E. Dlugosch and U. Claussen
0225 R. Bognar and M. Rakosi
0233 H. Goldner, G. Dietz and E. Carstens

5- -66D, Ann. Chem. Liebigs, 694 (1966)
0019 H.H. Inhoffen et al.
0031 H.H. Inhoffen, K.-D. Müller and O. Brendler
0128 G. Kohl and H. Pracejus
0149 F. Bohlmann, K.-M. Kleine and C. Arndt
0162 P. Tunmann, W. Gerner and G. Stapel
0169 H. Rönsch and K. Schreiber

5- -66E, Ann. Chem. Liebigs, 695 (1966)
0112 H.H. Inhoffen, G. Klotmann and G. Jeckel
0133 H.H. Inhoffen et al.
0144 B. Franck and G. Blaschke
0158 K. Irmscher

5- -66F, Ann. Chem. Liebigs, 696 (1966)
0001 H.H. Perkampus and G. Kassebeer
0116 H.-J. Teuber, D. Cornelius and U. Wölcke
0145 H. Brockmann, R. Zunker and H. Brockmann, Jr.
0226 M. Kinoshita and H. Klostermeyer

5- -66G, Ann. Chem. Liebigs, 697 (1966)
0017 U. Simon, O. Süs and L. Horner
0042 S.H. Schroeter et al.
0062 G. Hesse, H. Broll and W. Rupp
0116 S. Hünig et al.
0204 U. Stache and W. Fritsch

5- -66H, Ann. Chem. Liebigs, 698 (1966)
0001 H. Goetz and H. Juds
0034 H. Prinzbach et al.
0057 H. Prinzbach, D. Seip and G. Englert
0149 M. Wilk, H. Schwab and J. Rochlitz

5- -66I, Ann. Chem. Liebigs, 699 (1966)
0053 W. Grell and H. Machleidt
0068 G. Opitz and E. Tempel
0074 G. Opitz and E. Tempel
0107 M. Wilk and J. Rochlitz
0133 H. Balli and R. Gipp
0153 A, Treibs and K. Jacob

5- -66J, Ann. Chem. Liebigs, 700 (1966)
0065 S. Hünig and G. Kaupp
0092 H.H. Inhoffen et al.
0106 H. Dannenberg and K.-F. Hebenbrock
0157 T. Wieland and U. Gebert
0180 K.H. Göbbeler and C. Woenekhaus

6- -66, Ann. Chim.(Paris), 1 (1966)
0083 M. Pais et al.
0179 J. Wiemann and D. Hourdin

7- -66, Ann. chim.(Rome), 56 (1966)
0087 F. Russo and M. Ghelardoni
0182 G. Cordella and F. Sparatore
0190 G. Palazzo and L. Baiocchi
0199 G. Palazzo and L. Baiocchi
0232 M. Amorosa and G. Giovanninetti
0251 A. Arcoria
0286 M. Ghelardoni, F. Russo and M.G. Salmon
0298 G. Condorelli, L.L. Condorelli and V. Carassiti
0531 M. Marzona and R. Carpignano
0687 M. Ghelardoni, F. Russo and M.G. Salmon
0700 E. Cingolani, A. Cancelliere and A. Sordi
0767 P. Bruni and A. Strocchi
0946 M. Grifantini, M.L. Stein and A. Temperilli
1083 M. Ghelardoni, F. Russo and M.G. Salmon
1103 V. Sprio and E. Ajello
1192 G. Lucente, C. Gallina and A. Romeo
1248 F. Piozzi and C. Fuganti
1259 M.M. Melandri et al.
1379 E. Perrotti et al.
1394 A. Arcoria and G. Scarlata

9- -66, Appl. Spectroscopy, 20 (1966)
0150 A. Leifer et al.
0289 A. Leifer et al.
0363 J.G. Pritchard et al.

10- -66A, Arch. Biochem. Biophys., 113 (1966)
0195 A.K. Prince
0399 J.J. Gavin et al.
0548 C.K.R. Kurup, C.S. Vaidyanathan and T. Ramasarma
0742 M. Wilchek, A. Frensdorff and M. Sela

10- -66C, Arch. Biochem. Biophys., 115 (1966)
0129 G. Ramponi, C. Treves and A. Guerritore

10- -66D, Arch. Biochem. Biophys., 116 (1966)
0332 M. Shinitzky, E. Katchalski, V. Grisaro and N. Sharon

11- -66, Arkiv Kemi, 25 (1966)
0109 F. Plenat and G. Bergson
0135 B.J. Lindberg and H. Agback
0263 C. Frisell and G. Bergson

12- -66, Australian J. Chem., 19 (1966)
0135 P.S. Clezy et al.
0169 O.D. Hensens and K.G. Lewis
0201 J.R. Hall, M.R. Litzow and R.A. Plowman
0257 G.M. Badger, G.E. Lewis and U.P. Singh
0275 P.L. MacDonald and A.V. Robertson
0283 D.W. Connell and M.D. Sutherland
0297 S.R. Johns and J.A. Lamberton
0303 R.A. Massy-Westropp and G.D. Reynolds
0329 J.H. Gough and M.D. Sutherland
0337 J.W. Bunting and D.D. Perrin
0437 P.S. Clezy and D.Y.K. Lau
0445 L.K. Dalton
0451 R.H. Prager and H.M. Thredgold
0455 G.J.W. Breen et al.
0483 P.W. Chow, A.M. Duffield and P.R. Jefferies
0503 R.F.C. Brown et al.
0555 C. Kowala and J.M. Swan
0609 Y.M. Curtis and N.F. Curtis
0617 R.K. Norris and S. Sternhell
0683 N.V. Riggs and J.D. Stevens
0841 R.K. Norris and S. Sternhell
0861 J.R. Cannon et al.
0881 M.M. Ray et al.
0949 I.R. Lantzke and D.W. Watts
1045 R.F.C. Brown and R.K. Solly
1207 A. Meisters and P.C. Wailes
1215 A. Meisters and P.C. Wailes

1221 G.M. Badger, J.A. Elix and G.E. Lewis
1243 G.M. Badger, J.A. Elix and G.E. Lewis
1259 N.K. Hart and J.A. Lamberton
1265 R.N. Mirrington et al.
1401 S.H.H. Chaston et al.
1455 J. Bowyer and Q.N. Porter
1461 G.M. Badger, G.E. Lewis and U.P. Singh
1481 P.S. Clezy et al.
1487 D.J. Brown and R.V. Foster
1535 R.M. Carman
1717 R.A. Eade, I. Salasoo and J.J.H. Simes
1747 R.J. Head and R.A. Jones
1751 D.G. Hawthorne and Q.N. Porter
1835 P.S.K. Chia et al.
1859 A.L.J. Beckwith and J.W. Redmond
1871 R.W. Guy and R.A. Jones
1887 G.E. Lewis and J.A. Reiss
1908 D.G. Hawthorne and Q.N. Porter
1947 J.W. Loder
1951 S.R. Johns et al.
2107 C.J. Moye and S. Sternhell
2127 C.C.J. Culvenor and L.W. Smith
2133 R.M. Dawson et al.
2185 N.K. Hart and J.R. Price
2321 D.J. Brown and R.V. Foster
2331 S.R. Johns et al.
2339 S.R. Johns et al.
2403 R.M. Carman and R.A. Marty

13- -66A, Steroids, 7 (1966)
0234 J.N. Gardner et al.
0381 M. Heller, R.H. Lenhard and S. Bernstein
0505 H. Reimann et al.

13- -66B, Steroids, 8 (1966)
0087 J.J. Brown and S. Bernstein
0309 D.P. Strike et al.
0365 L. Re et al.
0421 A.F. Marx et al.
0537 R. Vitali and R. Gardi
0547 G.A. Hughes and H. Smith
0553 F. Schneider, J. Hamsher and R.E. Beyler

16- -66, Bol. inst. quim. univ. na. auton. Mex., 18 (1966)
0023 M.C. Perezamador, M. Salmon and F. Walls
0034 M. Salmnon and F. Walls
0054 M.C. Perezamador et al.
0064 A.J. Namis et al.

17- -66, Boll. sci. fac. chim. ind. Bologna, 24 (1966)
0075 I. Degani, R. Fochi and G. Spunta

18- -66, Bull. Chem. Soc. Japan., (1966)
0160 A. Fujino, F. Kusuda and T. Sakan
0253 S. Ito et al.
0901 Y. Murakami and K. Nakamura
0932 Y. Omote et al.
1079 Y. Omote et al.
1125 M. Ohno and N. Naruse
1129 M. Ohno et al.
1310 T. Nozoe et al.
1525 H. Suginome
1535 H. Suginome and T. Iwadare
1547 T. Nakaya and M. Imoto
1551 T. Nakaya, T. Tomomoto and M. Imoto
1788 M. Kobayashi and N. Koga
1942 M. Kurihara and N. Yoda
1954 M. Tada
1975 N. Obata and I. Moritani
1980 T. Nozoe, T. Asao and K. Takahashi
1988 T. Nozoe et al.
2444 Y. Kitahara, K. Doi and T. Kato
2450 T. Kato
2538 T. Asao and M. Kobayashi

19- -66, Bull.Acad. Polon. Sci., 14 (1966)
0007 C. Cagara and M. Kocor
0023 L. Skulski
0029 L. Skulski
0037 L. Skulski
0347 M. Kocor and M. Maczka
0505 J. Michalski, C. Piechucki and H. Zajac
0611 S. Mejer and S. Respondek

20- -66, Bull. soc. chim. Belges, 75 (1966)
0169 J.M. Danze and M. Renson
0181 G. Van Binst, C. Danheux and R.H. Martin
0260 A. Ruwet and M. Renson
0349 M. Vandewalle and F. Compernolle
0380 J.L.M. Loomans
0391 N. Schamp and H. de Pooter
0426 E. Feytmans-deMedicis
0734 B. Tursch, D. Daloze and G. Chiurdoglu

22- -66, Bull. soc. chim. France, (1966)
0043 J.-P. Durand
0116 P. Leriverend and J.-M. Conia
0121 P. Leriverend and J.-M. Conia
0131 Y. Maroni-Barnaud and G. Gilard
0147 S. Julia, C. Huynh and J. Olivie
0153 H. Najer et al.
0228 D. Cagniant and P. Cagniant
0236 P. Cagniant, G. Jecko and D. Cagniant
0269 A. Jung and F. Jung
0273 J.-M. Conia and P. LePerchec
0281 J.-M. Conia and P. LePerchec
0287 J.-M. Conia and P. LePerchec
0293 J. Elguero et al.
0381 J. Wiemann, S. Risse and P.-F. Casals
0400 J. Schoenleber and Geoffroy-Jambu
0522 M. Fleury
0527 S. David and H. Hirshfeld
0580 R. Panico and D. Pouchot
0610 J. Elguero and R. Jacquier
0618 J. Elguero et al.

0619 J. Elguero et al.
0645 J.-P. Quillet and J. Dreux
0689 P. Rouillier, D. Gagnaire and J. Dreux
0717 M. Julia and G. Le Thuillier
0728 M. Julia and G. Le Thuillier
0734 M. Julia and M. Baillarge
0758 F. Khuong-Huu-Laine et al.
0850 M. Fetizon and M. Golfier
0859 M. Fetizon and M. Golfier
0897 A. Labalche-Combier et al.
0926 G. Rabilloud et al.
0987 A. Grouiller and H. Pacheco
1002 C. Charrier, W. Chodkiewicz and P. Cadiot
1012 R. Maurel, J.-M. Boquet and J.-E. Germain
1033 H. Le Moal et al.
1222 C. Thal and B. Gastambide
1227 J.-A. Retamar
1277 N.P. Buu-Hoi, N.D. Xuong and T. Thu-Cuc
1287 J. Levisalles and I. Tkatchenko
1325 M. Fisch and G. Ourisson
1400 A. Marcou and H. Normant
1537 R. Bucourt and G. Nomine
1582 M. Saquet and A. Thuillier
1658 J. Armand
1670 Y. Tanahashi et al.
1693 J. Brugidou and H. Christol
1763 G. Thuillier et al.
1819 H.B. Kagan and Y.-H. Suen
1828 R. Azerad and M.-O. Cyrot
1884 R. Jacquesy and J. Levisalles
1974 J. Brugidou and H. Christol
1981 J. Soulie and P. Cadiot
1999 J.-C. Cognacq and W. Chodkiewicz
2207 R. Beugelmans et al.
2212 M. De Botton
2223 D. Lefort, J. Dorbu and A. Pourchez
2253 M. Sy and M. Maillet
2277 S. Julia et al.
2309 P. Potier et al.
2387 M. Julia, H. Pinhas and J. Igolen
2535 H. Christol, A. Douche and F. Plenat
2845 R. Jacquier, C. Petrus and F. Petrus
2857 R. Arnaud et al.
2877 J. Rauss-Godineau et al.
2885 J. Rauss-Godineau et al.
2939 M. Gueunier and R. Panico
2942 R. Panico and J. Klein
2971 R. Jacquier, C. Petrus and F. Petrus
2977 R. Jacquier, C. Petrus and F. Petrus
2981 J. Elguero et al.
2990 J. Elguero et al.
3024 G. Poucelot and P. Cadiot
3055 P. Cagniant, P. Faller and D. Cagniant
3212 H. Pacheco and A. Grouiller
3238 J.J. Godfroid et al.
3243 Y. Maroni-Barnaud et al.
3299 C. Giannotti, B.C. Das and E. Lederer
3328 P. Demerseman et al.
3407 B. Fürer et al.
3410 S. Julia and C.P. Papantoniou
3458 P. Potier et al.
3490 S. Julia and G. Linstrumelle
3548 E. Samuel
3578 J.-E. Dubois et al.
3618 P. Faller
3644 D. Cagniant, C. Charaux and P. Cagniant
3667 P. Faller
3674 P. Cagniant and D. Cagniant
3744 J. Elguero et al.
3774 G. Rio and J. Mion-Coatleven
3775 G. Rio and G. Sanz
3850 J. Soulie and P. Cadiot
3874 M. Mousseron-Canet et al.
3881 J.-M. Conia and P. Briet
3888 J.-M. Conia and P. Briet
3895 J. Le Leduc, D. Danion and R. Carrié
3934 J. Seyden-Penne et al.
3947 H. Christol, D. Lafont and F. Plenat
3989 M. Vinot
4028 J. Bourson

23- -66, <u>Can. J. Chem.</u>, <u>44</u> (1966)
0001 M. H. Benn
0028 R.H. Burnell, J.D. Medina and W.A. Ayer
0052 J.A.F. Gardner et al.
0094 T. Ito
0307 J.K. Homer, J.I. DeGraw and W.A. Skinner
0387 J.I. DeGraw
0759 E.J. Moriconi et al.
0799 J.P. Marsh, Jr., and L. Goodman
0819 J.F. King and T. Durst
0829 A. Pollak, B. Stanovnik and M. Tisler
0844 G.R. Pettit and D.M. Piatek
0961 A. Balasubramanian et al.
0972 V. Krishnan and C.C. Patel
1007 E. Bullock, T.S.Chen and C.E. Loader
1086 T.R. Kasturi and T.A. Arunachalam
1092 E.C.M. Coxworth
1259 R.H.F. Manske and K.H. Shin
1283 G.R. Pettit et al.
1317 J.T. Edward and J.-M. Ferland
1523 O. Orazi et al.
1541 H. MacLean and K. Murakami
1827 H. MacLean and K. Murakami
1831 H.J. Anderson and L.C. Hopkins
1841 P.E. Papadakis and S.M. Philip
1872 C. Podesva and K. Vagi
1961 S. Hoshino et al.
2003 J.-C. Richer and D. Perelman
2115 F.I. Carrol, S.C. Kerbow and M.E. Wall
2233 M. Douek and G. Just

2323 S.G. McGeachin
2449 S. McLean, M.-S. Lin and R.H.F. Manske
2465 C.E. Hall and A. Taurins
2473 C.E. Hall and A. Taurins
2507 J.W. Lown and A.S.K. Aidoo
2525 Z. Valenta et al.
2587 G. Just and E. Lee-Ruff
2837 G. Bauslaugh, G. Just and E. Lee-Ruff
2867 J.Y. Savoie and P. Brassard
2873 F. Blank, A.S. Ng and G. Just

24- -66, Chem. Ber., 99 (1966)
0048 K. Schank
0088 C.H. Krauch and W. Metzner
0094 K. Gewald, E. Schinke and H. Bottcher
0135 F. Bohlmann and C. Arndt
0138 F. Bohlmann and A. Seyberlich
0142 F. Bohlmann, K.M. Kleine and H. Bornowski
0160 W. Schäfer and B. Franck
0178 H. Dorn, G. Hilgetag and A. Zubek
0183 H. Dorn, G. Hilgetag and A. Zubek
0205 A. Schönberg and K. Praefcke
0273 D. Korbonits and K. Harsanyi
0307 H. Behringer and K. Falkenberg
0368 B.S. Thyagarajan et al.
0386 G. Drefahl et al.
0392 D. Rewicki
0396 H.G. Farnk and M. Zander
0450 E. Winterfeld and H. Preuss
0475 R. Huisgen et al.
0536 W. Pfleiderer and F. Reisser
0542 F. Reisser and W. Pfleiderer
0547 F. Reisser and W. Pfleiderer
0575 F.W. Lichtenthaler and H.P. Albrecht
0586 F. Bohlmann, H. Mönch and U. Niedballa
0590 F. Bohlmann and K.M. Kleine
0634 K. Dimroth, G. Pohl and H. Follmann
0642 K. Dimroth, H. Follmann and G. Pohl
0658 H. Fischer and H. Fischer
0680 G. Köbrich and H. Trapp
0689 G. Köbrich et al.
0698 A. Roedig et al.
0712 E. Niwa et al.
0724 K.H. Büchel, A. Kiskeri-Bocz and F. Korte
0737 F. Korte, A. Kiskeri-Bocz and K.H. Buchel
0782 J. Goerdeler and H. Schenk
0865 L. Farkas et al.
0872 H. Wamhoff and F. Korte
0885 F. Bohlmann and K.M. Kleine
0889 S. Swaminathan and K. Narasimhan
0944 H. Brederek et al.
0984 F. Bohlmann et al.
0990 F. Bohlmann and G. Florentz
1008 I. Ognyanov and B. Pyuskyulev
1097 H. Knoche
1118 R. Wiechert et al.
1179 H. Gnichtel
1223 F. Bohlmann and H. Bornowski
1226 F. Bohlmann and C. Zdero
1229 G. Maier
1232 G. Maier
1236 G. Maier and F. Seiler
1272 H.G. Franck and M. Zander
1275 M. Zander and W.H. Franke
1279 M. Zander and W.H. Franke
1325 K. Dimroth and P. Hoffmann
1357 H. Bock and K.L. Kompa
1384 L. Hörhammer et al.
1470 H. Musso and D. Dopp
1532 G. Doleschall and K. Lempert
1558 E. Winterfeldt and A.J. Dillinger
1580 H.W. Wanzlick and H. Ahrens
1618 J. Goerdeler and G. Gnad
1642 F. Bohlmann, K.M. Kleine and C. Arndt
1648 F. Bohlmann, C. Arndt and C. Zdero
1712 C. Woenckhaus and M.H. Volz
1723 C.H. Krauch, W. Metzner and G.O. Schenk
1732 K. Öfele
1764 A. Chatterjee and S.K. Kundu
1805 A.U. Rahman and O.L. Tombesi
1815 H. Behringer and U. Turck
1822 M. Kamel and H. Shoeb
1830 F. Bohlmann and A.G. Kapteyn
1851 J. Götze and H. Kampfer
1881 C.H. Krauch, S. Farid and D. Hess
1899 R.W. Hoffmann and J. Schneider
1918 F. Korte and A.K. Bocz
1923 A.K. Bocz
2012 H. Oediger et al.
2039 H. Bock and J. Kroner
2052 I. Ognyanov
2091 F. Bohlmann et al.
2096 F. Bohlmann and K.M. Kleine
2146 W. Lüttke and D. Hunsdiecker
2206 E.O. Fischer and M.W. Schmidt
2302 D. Martin et al.
2322 P. Boldt
2351 K. Dimroth, W. Kinzebach and M. Soyka
2380 E. Wittenberg
2391 E. Wittenberg
2413 F. Bohlmann and C. Zdero
2416 F. Bohlmann and K.M. Rode
2430 H. Wagner et al.
2449 M. Zander and W.H. Franke
2479 C. Jutz, W. Muller and E. Muller
2491 J. Sondermann and H. Kuhn
2504 P. Rosenmund and W.H. Hase
2526 R. Huisgen et al.
2572 E. Winterfeld et al.
2593 T. Kauffmann
2638 H. Nimz
2669 D. Meuche and S. Huneck
2703 J.N. Chatterjea and H.C. Jha
2712 K. Gewald and E. Schinke
2813 A. Roedig et al.
2818 A. Roedig et al.
2822 F. Bohlmann et al.
2828 F. Bohlmann et al.
2873 J. Knabe and K. Detering

2918 M. Regitz
2962 H. Wamhoff and F. Korte
2984 H. Zondler and W. Pfleiderer
2997 E. Bühler and W. Pfleiderer
3008 W. Pfleiderer and H. Zondler
3022 W. Pfleiderer and E. Buhler
3051 H. Laurent, G. Schulz and R. Wiechert
3057 H. Laurent et al.
3063 H. Plieninger and D. Wild
3070 H. Plieninger and D. Wild
3128 M. Regitz
3183 K. Schreiber and C. Horstmann
3194 F. Bohlmann, C. Zdero and P.H. Bonnet
3197 F. Bohlmann and M. Grenz
3201 F. Bohlmann, S. Kohn and E. Waldau
3218 L. Farkas et al.
3222 L. Farkas et al.
3268 K. Hartke et al.
3298 H.G. Frank and R.K. Erunlu
3337 H. Bock, E. Baltin and J. Kroner
3362 F. Bohlmann and A. Peter
3433 F. Bohlmann, S. Köhn and C. Arndt
3437 F. Bohlmann and R. Reinecke
3441 F. Bohlmann and W. Dornfeldt
3444 W. Flitsch and V. v. Weissenborn
3503 W. Pfleiderer et al.
3524 W. Pfleiderer et al.
3539 W. Herz et al.
3544 F. Bohlmann, K.M. Rohde and C. Zdero
3552 F. Bohlmann et al.
3559 W. Sucrow
3572 J. Goerdeler and K. Jonas
3750 E. Winterfeld, H. Radunz and P. Strehlke
3778 A. Holy and K.H. Scheit
3806 H. Hennig et al.
3836 H. Laurent et al.
3842 B. Franck et al.
3863 B. Franck and G. Baumann
3875 B. Franck and G. Baumann
3884 K.H. Scheit
3910 E. Uhlig and D. Walther
3932 E. Niwa et al.

25- -66, Chem. and Ind.(London) (1966)
0025 J.B. Siddall et al.
0031 J.M.Z. Gladych
0118 A.M. Kuck
0157 M.H. Palmer and E.R.R. Russell
0237 J.M. Calderwood and F. Fish
0340 H.L. Retcofsky et al.
0381 R.N. Warrener
0497 D. Burn et al.
0662 J.A. Weisbach et al.
0664 B. Shimizu and M. Miyaki
0731 B. Halpern et al.
0770 H.B. Dutschewska and N.M. Mollov
1300 C. Gallina et al.
1340 C.H. Kuo, D. Taub and N.L. Wendler
1344 J.A. Vida
1378 F. Kohen
1379 E.V. Dehmlow
1380 F.T. Bond and W.E. Musa
1533 I. Coutts and K. Schofield
1598 T. Fujii et al.
1600 G.E. Risinger and H.H. Hsieh
1637 J.D. Renwick
1638 R.B. Longmore and B. Robinson
1683 L. Jurd
1684 D.P. Chakraborty et al.
1720 W. Lawrie et al.
1963 K.W. Ratts and A.N. Yao
1967 T. Fujii et al.
2057 R. Walter and T.C. Purcell
2167 D.E. Minnikin
1497 P.M. Weintraub

27- -66, Chimia, 20 (1966)
0110 D.E. Poel et al.
0114 G. Bozzato, K. Schaffner and O. Jeger
0172 P.A. Chopard
0251 R.D. Hoffsommer, D. Taub and N.L. Wendler
0281 H.P. Kölliker and P. Caveng
0300 P.M. Nair, T.G. Manjrekar and M.K. Unni
0318 H. Balli and F. Kersting
0324 H.M. Buck et al.
0330 O. Riester

28C- -66A, Compt. rend., 262 (1966). Series C.
0018 S. Champy-Hatem
0120 L. Velluz et al.
0285 J.-F. Giudicelli et al.
0293 A. Resplandy, P. LeRoux and C. Mentzer
0346 M. Breant
0362 A. de Corville and L. Kerisit
0369 P. Grammaticakis
0376 J. Soulie
0584 P. Grammaticakis
0652 J.P. Schirmann and J. Dreux
0662 J. Chopin and M. Chadenson
0778 J. Menin et al.
0782 M. Bertrand et al.
0803 B. Briat and M. Le Liboux
0822 M. Delepine and F. Larèze
0841 J. Rigaudy and K. Trang
0848 M. de Botton
0885 C. Gerey and B. Pouyet
0924 Dang-Quoc-Quan
0927 J. Royer and J. Dreux
0985 P. Souchay et al.
1099 G. Meyer, P. Viout and P. Rumpff
1169 J. Levy et al.
1271 J. Wiemann and M. Bouyer
1276 P. Plusquellec and A. Foucaud
1325 A. Kirrmann and C. Wakselman
1397 M. Mousseron-Canet et al.
1429 P. Maroni et al.
1433 P. Cresson and M. Atlanti
1543 A. Plasky, J. Huet and J. Dreux
1598 R. Granger et al.
1601 M. Bertrand and M. Santelli

1605 P. Brechot et al.
1709 S. Gelin and R. Gelin
1712 J. Chopin and G. Dellamonica
1725 M. Mousseron-Canet et al.

28D- -66A, Compt. rend., 262 (1966). Series D.
0707 B. Voirin and P. Lebreton
1141 W.I. Taylor and Raymond-Hamet
1144 A. Sosa and C. Sosa-Bourdovil

29- -66, Discussions Faraday Soc., 37 (1966)
None

30- -66A, Doklady Akad. Nauk S.S.S.R., 166 (1966)(Asterisk indicates data from English translation)
0110 Y.A. Zhdanov et al.
0114 V.A. Izmail'skii and E.V. Chechegoeva
0598 N.N. Vorozhtsov, Jr., et al.(*107)
0611 N.N. Preobrazhenskaya et al.(*120)
0631 Y. Stradin, I. Tutane, O. Neiland and G. Vanag
0635 G.M. Khefets et al.(*144)
1343 V.A. Barkhash et al.(*238)

30- -66B, Doklady Akad. Nauk S.S.S.R., 167 (1966)
0827 E.Y. Gren and G.Y. Vanag
1318 B.D. Berezin et al.

30- -66C, Doklady Akad. Nauk S.S.S.R., 168 (1966)
0577 V.M. Berezovskii and Z.I. Aksel'rod

30- -66D, Doklady Akad. Nauk S.S.S.R., 169 (1966)
0103 S.P. Gubin et al.
0107 V.N. Drozd (*657)
0361 L.N. Yakhontov et al. (*705)
0855 G.G. Yakobson et al. (*762)

30- -66E, Doklady Akad. Nauk S.S.S.R., 170 (1966)
0341 B.A. Lugovik, L.G. Yudin and A.N. Kost (*877)

30- -66F, Doklady Akad. Nauk S.S.S.R., 171 (1966)
0863 Y.A. Berlin et al.
0877 I.P. Romm and E.N. Gur'yanova(*1151)
0880 O.I. Fedorova et al.
1101 V.M. Berezovskii and Z.I. Aksel'rod
1097 B.A. Arbuzov
*1042 N.A. Nemeyanov and O.A. Reutov

31- -66, Experientia, 22 (1966)
0140 Y.T. Lin et al.
0209 E.L. Patterson, W.W. Andres and N. Bohonos
0355 W. Keller-Schierlein
0468 G. Baldratti et al.
0526 R.A. Boissonnas et al.
0705 G. Ramponi, C.Treves and A. Guerritone
0750 W. Rothweiler and C.H. Tamm

32- -66, Gazz. chim. ital., 96 (1966)
0103 S. Carboni et al.
0152 P. DeRuggieri et al.
0179 P. DeRuggieri et al.
0206 L. Mangoni and M. Belardini
0220 R. Caputo and L. Mangoni
0264 A. Arcoria and G. Scarlata
0279 A. Arcoria and G. Scarlata
0311 G. Settimi et al.
0443 P. Maggioni, G. Gaudiano and P. Barvo
0454 P. Bravo and G. Gaudiano
0483 R. Rossi and E. Benedetti
0559 G. Bianchi and E. Frati
0662 L. Canonica et al.
0772 L. Canonica, G. Russo and A. Bonati
0820 L. Canonica et al.
0833 L. Canonica, G. Russo and E. Bombardelli
0843 L. Canonica et al.
0915 L. Canonica et al.
0941 G. Grandolini, A. Fravolini and I. Montanini
0986 A. Bruno and G. Purrello
1000 G. Purrello
1009 A. Bruno and G. Purrello
1020 G. Palazzo and L. Baiocchi
1064 G. Casnati et al.
1073 G. Casnati et al.
1108 G. Bellomonte, G. Caronna and S. Palazzo
1115 R. Vitali, R. Gardi and A. Ercoli
1125 R. Vitali and R. Gardi
1147 G. Pappalardo et al.
1241 R. Sciaky and U. Pallini
1254 R. Sciaky and U. Pallini
1268 R. Sciaky, U. Pallini and B. Patelli
1284 R. Sciaky, U. Pallini and A. Consonni
1301 A. Fontana et al.
1423 G. Favini and I.R. Bellobono

33- -66, Helv. Chim. Acta, 49 (1966)
0030 A. von Wartburg
0042 W. Michaelis, O. Schindler and R. Signer
0053 A. Hofmann and C.H. Eugster
0076 E. Sandrin and R.A. Boissonnas
0083 St. Guttmann
0134 R. Schwyzer and P. Sieber
0168 A. Gabbai et al.
0204 A. Closse, R. Mauli and H.P. Sigg
0292 J. Hill et al.
0316 K.D. Roberts, E. Weiss and T. Reichstein
0349 E. Fetz and C. Tamm
0359 E. Kyburz et al.
0369 U. Schwieter et al.
0403 A. Brossi et al.

0517 P.A. Straub et al.
0534 B.K. Manukian
0572 A. Marxer
0617 Y.R. Naves and P. Ardizio
0625 F.P. Emmenegger and G. Schwarzenbach
0644 H. Hopff and A. Gati
0858 J.S. Scarpa, M. Ribi and C.H. Eugster
0889 R.W. Balsiger, H. Hänni and O. Schindler
0946 T. Kishi et al.
0964 M. Hesse, F. Bodmer and H. Schmid
0985 O. Schindler, R. Blaser and F. Hunziker
0992 U. Schwieter, R. Rüegg and O. Isler
0997 L. Chardonnens and G. Gamba
1002 C.R. Zanesco
1029 Y.R. Naves
1049 J. Frei et al.
1131 R. Zell and H. Erlenmeyer
1151 D. Karanatsios et al.
1173 M. Hesse et al.
1183 L. Lorenc et al.
1222 M. Müller and P. Zeller
1237 C. Kump, J.J. Dugan and H. Schmid
1278 D. Meuche
1433 F. Hunziker, F. Künzle and J. Schmutz
1483 O. Schindler, W. Michaelis and R. Gauch
1529 G. Saucy, H. Els, F. Miksch and A. Furst
1552 H. Lichti and A. von Wartburg
1581 R. Wiechert et al.
1591 A. Furlenmeier et al.
1601 U. Kerb et al.
1632 H.H. Sauer, E. Weiss and T. Reichstein
1662 R. Brandt et al.
1736 S. Bareza et al.
1757 A. Brossi and S. Teitel
1794 N. Baumann et al.
1806 G. Weisgerber and C.H. Eugster
1815 M. Viscontini and A. Bobst
1844 R. Brandt and H. Kaufmann and T. Reichstein
1850 L. Chardonnens and W. Hammer
1855 S. Hoffmann, E. Weiss and T. Reichstein
1907 T. Fehr and W. Acklin
1931 L. Chardonnens and H. Chardonnens
1986 G. Hüppi et al.
2017 H.-R. Blattmann et al.
2046 R. Wizinger and H.J. Angliker
2072 A.A. Gorman et al.
2218 H. Wehrli et al.
2297 U. Schwieter et al.
2321 A. Guggisberg et al.
2365 F. Müller, W. Walker and P. Hemmerich
2527 B. Böhner and C. Tamm
2547 B. Böhner and C. Tamm

34- -66, J. Chem. Eng. Data, 11 (1966)
0418 W.E. Noland, S.P. Hiremath and S. Siddappa
0436 C.C. Chappelow, Jr, R.N. Clark and F.V. Morriss

35- -66, J. Am. Chem. Soc., 88 (1966)
0137 S.F. Nelsen and P.D. Bartlett
0161 O.L. Chapman et al.
0181 P.A. Waitkus, L.I. Peterson and G.W. Griffin
0183 H.E. Zimmerman and G.L. Grunewald
0185 R. Wolfenden et al.
0186 M. Idelson and I.R. Karady
0204 W.W. Moreau and K. Weiss
0334 E. Smissman et al.
0354 M.J.S. Dewar and A.P. Marchand
0378 H. Kristinsson and G.W. Griffin
0380 W.E. Barth and R.G. Lawton
0476 J.E. Mulvaney et al.
0487 E.P. Blanchard, Jr., and A. Cairncross
0515 D.J. Cram, C.S. Montgomery and G.R. Knox
0536 C. Djerassi and L. Tokes
0582 D.M. Lemal, E.P. Gosselink and S.D. McGregor
0602 J. Mayer and F. Sondheimer
0603 J. Mayer and F. Sondheimer
0611 E.H. White et al.
0618 P.J. Chapman et al.
0624 T. Money, J.L. Douglas and A.I. Scott
0781 M.S. Newman and C. Courduvelis
0790 J.C. Knight, D.I. Wilkinson and C. Djerassi
0799 A.L. Wilds et al.
0804 W.A. Remers and M.J. Weiss
0834 T.E. Acker et al.
0852 R.B. Woodward et al.
0853 B.M. Trost
0856 L.R. Axelrod et al.
0935 W.H. Knoth
0947 K. Bowden, A. Buckley and R. Stewart
1005 H. Hart, P.M. Collins and A.J. Waring
1034 C.C. Price, T. Parasaran and T.V. Lakshminarayan
1049 L.J. Dolby and D.L. Booth
1059 R.C. Cookson, J. Henstock and J. Hudec
1077 W. Metlesics and L.H. Sternbach
1077B K.A. Schellenberg and G.W. McLean
1079 P.S. Forgione et al.
1205 P.T. Lansbury and N.R. Manenso
1257 J.A. Kampmeier et al.
1324 D.J. Cram et al.
1330 R.L. Cargill, J.R. Damewood and M.M. Cooper
1419 J.H. Brewster and J.E. Privett
1488 G.A. Olah et al.

1518 A. Padwa and R. Hartman
1533 S. Cohen and S.G. Cohen
1544 Y. Shimizu and S.W. Pelletier
1549 J.R. McCarthy, Jr., et al.
1559 C.P. Lillya and P. Miller
1560 C.P. Lillya and P. Miller
1562 P. Yates and R.J. Crawford
1732 M.E.H. Howden et al.
1766 R.B. Turner et al.
1776 R.B. Turner et al.
1786 R.B. Turner et al.
1792 C. Djerassi et al.
1825 D.I. Schuster and D.J. Patel
1832 R.L. Hively, W.A. Mosher and F.W. Hoffmann
1844 E.F. Ullman and B. Singh
1923 R.I. Walter
1959 J.A. Kampmeier and R.M. Fantazier
1965 H.E. Zimmerman et al.
1979 E. Liganek
2007 G.A. Russell and M.C. Young
2015 E.H. White et al.
2109 R. Waack, M.A. Doran and P.E. Stevenson
2213 J.J. Eisch and W.C. Kaska
2240 R.L. Reeves
2299 T.L. Miller, G.L. Rowley and C.J. Stewart
2407 G.O. Dudek and E.M. Dudek
2412 K.S. Sidhu et al.
2494 J.A. Berson and M.R. Willcott
2532 G. Buchi and R.E. Manning
2536 G.R. Allen, Jr., et al.
2587 P. Dowd
2590 L.A. Paquette and J.H. Barrett
2591 A.L. Johnson and H.E. Simmons
2602 T. Matsuura and K. Ogura
2775 D.Y. Curtin, E.J. Grubbs and C.G. McCarty
2803 W. Reusch, C.K. Johnson and J.A. Manner
2838 J.J. Dugan et al.
2880 W.B. Hammond and N.J. Turro
2882 E. Ciganek
3016 D.H.R. Barton et al.
3027 S.A. Kandil and R.E. Dessy
3046 O.W. Webster
3051 R.S. Neale, N.L. Marcus and R.G. Schepers
3064 A. Padwa, D. Crumrine and A. Shubber
3075 D.G. Farnum, J. Chickos and P.E. Thurston
3095 S.J. Cristol and B.B. Jarvis
3109 G. Buchi et al.
3120 L.L. Smith et al.
3131 P.F. Beal, J.C. Babcock and F.H. Lincoln
3140 E.M. Arnett and D. Hufford
3155 H.D. Hartzler
3165 M. Ikehara et al.
3169 R.B. Woodward and D.J. Woodman
3173 R.I. Fryer et al.
3180 R.-M. Dupeyre and A. Rassat
3303 P.D. Bartlett and S.T. Purrington
3318 M.J.S. Dewar and A.P. Marehand
3359 A.S. Kende, P.T. Izzo and P.T. MacGregor
3408 J.A. Marshall, N. Cohen and A.R. Hochstetler
3433 E.W. Garbisch, Jr., and R.F. Sprecher
3461 P. Radlick and W. Rosen
3515 D.J. Cram and R.C. Helgeson
3527 R.W. Murray and M.L. Kaplan
3617 D.M. Gale et al.
3640 T.A. Khwaja and R.K. Robins
3643 R.C. Gaver and C.C. Sweeley
3664 H.G. Pars et al.
3672 M.J. Turro and W.B. Hammond
3674 S.M. Kupchan et al.
3677 R. Breslow et al.
3759 A. Padwa and R. Hartman
3769 J.E. Baldwin and L.E. Walker
3781 R.J. Sundberg
3798 W.M. Jones and D.L. Muck
3825 D.M. White and J. Sonnenberg
3829 J.P. Ferris and L.E. Orgel
3832 T.J. Katz, C.R. Nicholson and C.A. Reilly
3842 W.J. Middleton
3865 L. Caglioti et al.
3875 E. Galantay et al.
3888 M.E. Wall et al.
3905 S. Kobinata and S. Nagakura
3941 O. Yonemitsu et al.
3950 V. Boekelheide and C.D. Smith
3959 R.J. Crawford et al.
4001 J.F. Bunnett and M.B. Naff
4055 O.W. Webster
4061 J. Blake et al.
4068 S. Teitel and A. Brossi
4109 G.A. Ellestad et al.
4112 M.P. Cava and N.M. Pollack
4113 G. Buchi et al.
4212 S.M. Kupchan et al.
4273 C.O. Smith
4489 D.N. Kevill, E.D. Weiler and N.H. Cromwell
4500 J.R.L. Smith and M. Calvin
4522 H.H. Wasserman and P.M. Keehn
4523 M.S. Newman and S. Mladenovic
4525 H.W. Whitlock and P.E. Sandvick
4534 G. Buchi et al.
4538 M. Uskokovic et al.
4541 P.T. Kwitowski and R. West
4637 R.T. Puckett, C.E. Pfluger and D.Y. Curtin
4677 W.H. Graham
4685 L.A. Paquette and R.W. Begland
4895 H.E. Zimmerman et al.
4905 H.E. Zimmerman and D.J. Sam
4942 E.F. Ullman and W.A. Henderson, Jr.
4984 K.S. Brown, Jr., et al.
5001 M.L. Ernst and G.L. Schmir
5010 M. Goodman and A. Kossoy
5025 T. Mukai, T. Tezuka and Y. Akasaki
5026 A.S. Kende
5063 I.H. Hillier, L. Glass and S.A. Rice

5124 R.E. Dessy and P.M. Weissman
5284 R. Kreilick
5292 S.M. Kupchan et al.
5369 N.C. Yang, A. Shani and G.R. Lenz
5555 R.C. Fahey and D.J. Lee
5588 S. Trofimenko
5684 J.P.H. Verheyden and J.G. Moffatt
5747 M.C. Caserio, R.E. Pratt and R.J. Holland
5777 S.W. Ela and D.J. Cram
5855 M.G. Burdon and J.G. Moffatt
5865 R.L. Clarke et al.
5934 D.M. Lemal and J.P. Lokensgard
5935 J.J. McCullough and J.M. Kelly

36- -66, J. Pharm. Sci., (1966)
0144 R.J. Warren et al.

37- -66, J. Biol. Chem., 241 (1966)
0122 G.F. Bryce and F.R.N. Gurd
0257 M. Hamberg and B. Samuelsson
0540 C.J. Sih et al.
1072 G.F. Bryce et al.
1376 H. Nikaido and K. Nikaido
1439 G.F. Bryce and F.R.N. Gurd
1587 R.G. Coombe et al.
1807 L. Tsai et al.
3424 S. Shifrin et al.
3468 K.N. Murray and S. Chaykin
3776 L.N. Ornston and R.Y. Stanier
4616 T.E. Creighton and C. Yanofsky
5276 S. Sano
5313 J.J. Schneider and N.S. Bhacca
5336 D.K. Fukushima et al.

38- -66A, J. Chem. Phys., 44 (1966)
1803 M.B. Robinm R.R. Hart and N.A. Kuebler
2664 M.B. Robin and N.A. Kuebler

38- -66B, J. Chem. Phys., 45 (1966)
1367 S.D. Thompson et al.

39- -66A, J. Chem. Soc., Sect. A (1966)
0239 C.W.N. Cumper, J.F. Read and A.I. Vogel
0242 C.W.N. Cumper, A. Melnikoff and A.I. Vogel
0544 M.J. Frazer and Z. Goffer
0580 J.H. Morris and P.G. Perkins
0630 F.J. Kohl, J. Lewis and R. Whyman
0933 R.C. Dobbie and H.J. Emeleus
1733 D.N. Lawson, J.A. Osborn and G. Wilkinson
0073 L. Morpurgo and R.J.P. Williams
0194 W.D. Bannister et al.

39- -66B, J. Chem. Soc., Sect. B (1966)
0073 H.N.A. Al-Jallo and E.S. Waight
0075 H.N.A. Al-Jallo and E.S. Waight
0092 J.L. Garraway
0164 N. Heap and G.H. Whitham
0255 M.F. Grundon and B.T. Johnston
0285 G.B. Barlin
0314 R. Hemming and D.G. Johnston
0424 Y. Iskander, R. Tewfik and S. Wazif
0427 A. Albert
0433 J.W. Bunting and D.D. Perrin
0438 A. Albert
0441 J.W. Lown
0483 M. Fraser, A. Melera and D.H. Reid
0489 S. Patai and Y. Gotshal
0498 M.R. Crampton and V. Gold
0521 P.B.D. de la Mare et al.
0533 F.C.R. de Fabrizio et al.
0562 A.R. Katritzky, F.D. Poff and J.D. Rowe
0565 A.R. Katritzky, F.D. Poff and A.J. Waring
0631 A.R. Katritzky and B. Ternai
0650 T.G. Bonner and J. Phillips
0861 W. Carruthers and G.E. Hall
0870 J. Gleghorn et al.
0885 D.E. Bays, G.W. Cannon and R.C. Cookson
0956 A. Albert and H. Yamamoto
0991 E. Spinner and J.C.B. White
0996 E. Spinner and J.C.B. White
1058 R.M. Johnson
1083 P.J. Brignell, U. Eisner and P.G. Farrell
1147 Z. Raciszewski
1221 C.H. McMullen and C.J.M. Stirling

39- -66C, J. Chem. Soc., Sect. C (1966)
0001 F. Kurzer and K. Douraghi-Zadeh
0006 F. Kurzer and K. Douraghi-Zadeh
0010 F. Bergmann, Z. Neiman and M. Kleiner
0022 R.L.N. Harris, A.W. Johnson and I.T. Kay
0030 D. Dolphin et al.
0040 J.H. Atkinson, R.S. Atkinson and A.W. Johnson
0054 A.J. Birch et al.
0056 A. Calderbank
0072 R.C. Cambie and P.W. LeQuesne
0080 R.K. Smalley
0093 C. Barnett et al.
0098 D. Dolphin et al.
0114 F. McDean and D.A.H. Taylor
0123 K. Chandrasenan and R.H. Thomson
0125 S. Mongkolsuk et al.
0126 D.C. Aldridge et al.
0129 R.E. Bew et al.
0135 R.E. Bew et al.
0139 E.R.H. Jones, G. Lowe and P.V.R. Shannon
0164 D.J. Brown and M.N. Paddon-Row
0168 R.F. Curtis et al.
0175 A. Jefferson and F. Scheinmann
0178 B. Jackson et al.
0192 A.P. Johnson, A. Pelter and P. Stainton
0224 N.B. Haynes and C.J. Timmons
0226 D.J. Brown, B.T. England and J.M. Lyall
0230 J.F. Grove, R.N. Speake and G. Ward
0234 W.L.F. Armarego and J.I.C. Smith
0251 P.L. Pauson et al.

0255 B.F. Burrows and W.B. Turner
0285 A. Stuart, H.C.S. Wood and D. Duncan
0321 R.H. Thomson and A.G. Wylie
0324 W.K. Gibson and D. Leaver
0332 P.J. Godin et al.
0336 T.D. Binns and R. Brettle
0341 T.D. Binns and R. Brettle
0362 F.R. Hewgill and B.R. Kennedy
0377 H. Minato and T. Nagasaki
0385 J.A. Elvidge and P. Sims
0387 J.A. Elvidge and P.D. Ralph
0397 T.G. Halsall and J.M. Mellor
0412 A.K. Qureshi and B. Sklarz
0419 A.J. Birch and J.S. Hill
0425 I. Fleming and J. Harley-Mason
0426 H.A. Anderson and R.H. Thomson
0430 H.D. Locksley et al.
0441 D.M. Ciment and R.J. Ferrier
0463 K.H. Takemura
0464 A.C. Day and M.C. Whiting
0470 D.E. Ames, R.F. Chapman and D. Waite
0477 S.Y. Ambekar and S. Siddappa
0481 E.O. Arene and D.A.H. Taylor
0495 J. Cullen and D. Harrison
0506 M. Akisanya et al.
0509 C.W.L. Bevan et al.
0523 A.J. Birch, M. Salahud-Din and D.C. Smith
0533 D.J. Cooper and L.N. Owen
0542 D.J. Adam, L. Crombie and D.A. Whiting
0544 D.J. Adam et al.
0550 D.J. Adam, L. Crombie and D.A. Whiting
0566 R.A. Bowie and O.C. Musgrave
0578 J.C. Craig and R.J. Young
0588 R.L.C. Brinacombe and C.B. Reese
0597 P.L. Coe, R.G. Plevey and J.C. Tatlow
0603 F.G. Baddar et al.
0606 A.P. Johnson and A. Pelter
0661 A,T. Glen, W. Lawrie and J. McLean
0668 G.G. Booth and A.F. Turner
0675 P.R. Ashurst and J.A. Elvidge
0676 G.W. Kirby and H.P. Tiwari
0686 D.H. Reid and W.G. Salmond
0692 A.S. Jones, A.M. Mian and R.T. Walker
0695 G.J.F. Chittenden and R.D. Guthrie
0701 A. Pelter and P. Stainton
0717 T. Kametani et al.
0725 J.S. Moffatt
0734 J.S. Moffatt
0743 J.F. Grovem P.M. Closkey and J.S. Moffatt
0749 S.F. Dykem W.D. Ollis and M. Sainsbury
0753 W. Chan and P. Maitland
0764 A.C. Riemer and W. Rigby
0772 S.A. Ahmad, W. Rigby and R.B. Taylor
0794 A.W. Johnson and D. Oldfield
0813 D. Mackay and W.A. Waters
0852 G.W.H. Cheeseman and B. Tuck
0855 J.H. Birkinshaw et al.
0859 G.V. Boyd and F.W. Clark
0880 D. Dolphin et al.
0921 G. Shaw, B.M. Smallwood and D.V. Wilson
0944 D.E.V. Ekong and E.O. Olagbemi
0955 M.M. Coombs
0963 M.M. Coombs
0971 A.J. Boulton, P.B. Ghosh and A.R. Katritzky
0976 R. Grigg, J.A. Knight and M.V. Sargent
1004 H.G. Heller and R.A.N. Morris
1010 A.R. Davies and G.H.R. Summers
1012 A.R. Davies and G.H.R. Summers
1028 P. Bamfield, A.W. Johnson and A.S. Katner
1034 E. Caspi, D.M. Piatak and P.K. Grover
1050 N. Campbell and A.H. Scott
1065 N.W. Jacobsen
1075 M. Anderson and A.N. Johnson
1078 C.E. Loader and C.J. Timmons
1084 R.M. Bowman and M.F. Grundon
1086 D. Lloyd and F.I. Wesson
1101 R.E. Atkinson, R.F. Curtis and G.T. Phillips
1112 J. Clark and G. Neath
1117 A. Albert and J.J. McCormack
1121 J.M.F. Gagan
1123 D.K. Black
1127 M.W. Partridge and M.F.G. Stevens
1131 P.J. Beynon et al.
1142 J.D. Cocker et al.
1152 W. Cocker, T. McMurry and D.M. Sainsbury
1155 J.H. Atkinson et al.
1165 D.J. Brown, B.T. England and J.S. Harper
1202 F.L. Scott and R.N. Butler
1209 E.E. Glover, G. Jones and G. Trenholm
1213 A.J. Birch and G.S.R. SubbaRao
1214 G.R. Bedford, M.W. Partridge and M.F.G. Stevens
1220 E.R.H. Jones, S. Safe and V. Thaller
1223 S.R. Landor et al.
1226 L.A. Cort
1236 R.N. Haszeldine and A.E. Tipping
1245 M.W. Partridge et al.
1251 R.T. Aplin, H.R. Arthur and W.H. Hui
1260 J.MF. Filho et al.
1266 G.M.L. Cragg et al.
1277 D.J. Hampson, G.D. Meakins and D.J. Morris
1287 T.J. Adley and L.N. Owen
1291 L.N. Owen and P.L. Ragg
1298 J.B. Barrera et al.
1306 P.F. Holt and R. Oakland
1308 J.A. Knight, J.C. Roberts and P. Roffey
1340 S.M. Albonico et al.
1342 G.M. Blackburn and R.J.H. Davies
1345 A.S. Bailey and J.J. Merer

1354 R.J. Clark and J. Walker
1374 T.G. Halsall et al.
1406 A.B. Manger and R. Wade
1412 T. Nakano, S. Terao and Y. Saeki
1431 S. Marcinkiewicz et al.
1433 W.L.F. Armarego et al.
1436 P. Bamfield et al.
1443 D.T. Cropp, J. Holker and W.R. Jones
1496 R.O. Hellyer and J.T. Pinhey
1508 G.J.F. Chittenden and R.D. Guthrie
1511 A.F. Casy and J. Wright
1522 W.K. Warburton
1527 W.A.F. Gladstone and R.O.C. Norman
1549 A.F. Cook and W.G. Overend
1556 R.E. Corbett and H. Young
1564 R.E. Corbett and H. Young
1605 J.M. Fayadh, D.W. Jessop and G.A. Swan
1608 J.F.W. McOmie et al.
1615 P.R. Ashurst and D.R.J. Laws
1617 P. Hodge
1627 J.M. Bruce and P. Knowles
1641 R.C. Cookson and K.R. Friedrich
1653 G.R. Delpierre et al.
1680 L.J. Haynes, G. Husbands and K.L. Stuart
1698 M. Davis et al.
1717 I. Maclean and R. Stevenson
1719 A.C. Day and M.C. Whiting
1727 H.A. Anderson, R.H. Thomson and J.N. Wells
1729 P.H. Gore and C.K. Thadani
1737 R.E. Corbett and S.G. Wyllie
1753 D. Lavie, S. Greenfield and E. Glotter
1757 D. Lavie et al.
1765 E. Glotter, R. Waiman and D. Lavie
1767 P.R. Constantine et al.
1769 C.R. Harrison et al.
1775 F. Kohen, I. Maclean and R. Stevenson
1781 W.A.F. Gladstone et al.
1784 A.S. Jones, A.M. Mian and R.T. Walker
1799 G. Lowe, A. Taylor and L.C. Vining
1803 R. Hodges, J.S. Shannon and A. Taylor
1805 T. Nakuno et al.
1810 N. Campbell and K.W. Delahunt
1825 D.W. Cameron and H.W.-S. Chan
1832 D.W. Cameron et al.
1836 G.M. Blackburn et al.
1847 P. Johnston et al.
1856 D.G.I. Kingston et al.
1866 H. Minato and T. Nagasaki
1892 D.E.A. Rivett
1924 G.W. Perold and H.K.L. Hundt
1933 J.W. Harrison et al.
1950 R.F. Childs and A.W. Johnson
1976 P.M. Greaves, S.R. Landor and D.R.J. Laws
1989 A.J. Bellamy et al.
2003 E.A. Cookson and P.C. Crofts
2010 T. Kametani and M. Ihara
2026 A.M. Parsons and D.J. Moore
2031 J.A. Bee and F.L. Rose
2038 S.E. Mallett and F.L. Rose
2060 A.J. Birch and H. Fitton
2061 A.H. Jackson and J.A. Martin
2072 E.J. Forbes and J. Griffiths
2095 K. Takeda et al.
2100 B.M. Goldschmidt, B.L. Van Duuren and C. Mercado
2115 M.A. Barton et al.
2121 J.E. McCormick
2127 E.K. Adesogan et al.
2144 J.W.K. Burrell et al.
2154 J.B. Davis et al.
2166 M.S. Barber et al.
2184 A.C. Baillie and R.H. Thomson
2186 H.D. Locksley, I. Moore and F. Scheinmann
2190 M.J. Kort and M. Lamchen
2201 G.I. Gregory et al.
2210 P.J. May, F.A. Nice and G.H. Phillipps
2218 R.M. Acheson and M.W. Foxton
2234 P.M. Baker and J.C. Roberts
2237 G.R. Birchall et al.
2239 G.M. Blackburn and R.J.H. Davies
2245 K.C. Chan, F. Morsingh and G.B. Yeoh
2265 H.D. Locksley, I. Moore and F. Scheinmann
2283 S.R. Landor and E.S. Pepper
2290 G. Tennant
2295 L.S. Kaminsky and M. Lamchen
2300 M. Lamchen and T.W. Mittag
2308 J.F. Corbett
2313 D.H.R. Barton, G.W. Kirby and A. Wiechers
2324 A.J. Birch and J.S. Hill
2328 R.D. Chambers et al.
2359 C.W. Shoppee, N.W. Hughes and R.E. Lack

41- -66, J. chim. phys., 63 (1966)
0949 A. Kellmann

43- -66, J. Opt. Soc. Am. (1966)
none

44- -66, J. Org. Chem., 31 (1966)
0001 P.L. Southwick et al.
0026 R. Rausser et al.
0034 J.A. Moore et al.
0048 R.L. Wineholt, E. Wyss and J.A. Moore
0052 J.A. Moore, R.W. Medeiros and R.L. Williams
0062 J.A. VanAllan et al.
0065 W.E. Noland, K.R. Rush and L.R. Smith
0070 W.E. Noland and K.R. Rush
0077 W.L. Anthony
0081 L. Skattebøl and B. Boulette
0162 H. Zinnes, R.A. Comes and J. Shavel, Jr.
0175 Y. Inoue, N. Furutachi and K. Nakanishi

0178 T.J. Delia and G.B. Brown
0189 P.W. Jeffs and T.P. Toube
0201 W.V. Curran and R.B. Angier
0205 J.P. Horwitz et al.
0211 K.A. Watanabe and J.J. Fox
0223 I. Ziderman and E. Dimant
0260 K.T. Potts, H.R. Burton and J. Bhattacharyya
0265 K.T. Potts, H.R. Burton and S.K. Roy
0315 W.G. Dauben and P. Oberhansli
0339 L.M. Lerner and P. Kohn
0342 E.C. Taylor and A. Abul-Hasn
0349 N. Tokura, T. Nagai and S. Matsumura
0369 H.E. Ungnade and L.W. Kissinger
0394 W.G. Brown and F.H. Greenberg
0399 A.P. Gray and H. Kraus
0406 H.J. Minnemeyer et al.
0425 E.R. Garrett and R.E. Notari
0447 K. Ichikawa et al.
0509 D. Horton and J.S. Jewell
0516 S. Kubota et al.
0524 C.H. Robinson, L. Milewich and P. Hofer
0565 C.K. Bradsher and J.D. Turner
0586 A.T. Bottini and E.F. Bottner
0589 J.J. Eisch and G.R. Husk
0608 S.M. Kupchan and G. Ohta
0610 R.E. Bozak
0616 A. Takeda et al.
0628 E.S. Rothman
0636 J.T. Sheehan
0639 N.J. Leonard and W.J. Musliner
0646 H.O. House et al.
0656 S. Swaminathan et al.
0673 A. Romo de Vivar et al.
0681 T.J. Mabry et al.
0693 O. Halpern et al.
0705 V. Permutti and Y. Mazur
0713 Z.G. Hajos, D.R. Parrish and M.W. Goldberg
0722 H.M. Blatter and H. Lukaszewski
0734 K. Takeda, H. Tanida and K. Horiki
0764 L.A. Carpino and D.E. Barr
0768 E.T. McBee, J.A. Bosoms and C.J. Morton
0781 H.C. Brown, H.J. Gisler, Jr, and M.T. Cheng
0797 M. von Strandtmann et al.
0813 B. Urbas and R.L. Whistler
0870 E.E. Schweizer and K.K. Light
0900 T.K. Liao, F. Baiocchi and C.C. Cheng
0903 R.T. Parfitt, E.M. Fry and E.L. May
0912 W.A. Bonner et al.
0919 J.K. Williams, E.L. Martin and W.A. Sheppard
0923 W.E. Thun, D.W. Moore and W.R. McBride
0941 C.K. Bradsher and S.A. Telang
0977 G. Buchi and H. Wuest
0978 C.K. Bradsher and D.F. Lohr, Jr.
0991 G.A. Boswell, Jr.
1007 W. Metlesics et al.
1012 R.H. Roth, W.A. Remers and M.J. Weiss
1016 J.A. Marshall, W.I. Fanta and H. Roebke
1026 G.V. Baddeley et al.
1042 O.L. Chapman and T.H. Koch
1050 K. Conrow
1131 A.G. Anastassiou
1135 D.N. Matthews and E.I. Becker
1163 E. Walton et al.
1181 K.J. Ryan, E.M. Acton and L. Goodman
1227 D.J. Cram and A.C. Day
1244 A. Padwa, L. Hamilton and L. Norling
1275 R.J. Highet and P.F. Highet
1276 J.W. Huffman, S.P. Garg and J.H. Cecil
1281 M.P. Cava and D.R. Dalton
1285 S.G. Boots and W.S. Johnson
1287 J.J. Pappas and E. Gaucher
1289 F. Keller and A.R. Tyrrill
1303 D.R. Eckroth, T.G. Cochran and E.C. Taylor
1304 K.D. Berlin and D.H. Burpo
1310 J.M. Bollinger et al.
1311 E. Hedaya, R.L. Hinman and S. Theodoropulos
1317 E. Hedaya, R.L. Hinman and S. Theodoropulos
1332 H.E. Zaugg and R.J. Michaels
1336 J. Liabattoni and G.A. Berchtold
1342 R. Rausser et al.
1346 R. Rausser et al.
1349 J.C. Grivas
1360 Z.G. Hajos, C.P. Parios and M.W. Goldberg
1363 A. Hassner et al.
1372 E.J. Moriconi and P.H. Mazzocchi
1390 M.M. Fawzi and C.D. Gutsche
1393 J.E. Starr and R.H. Eastman
1411 J.A. Montgomery and H.J. Thomas
1413 H.J. Thomas and J.A. Montgomery
1417 C. Temple, Jr., and J.A. Montgomery
1423 G. Smolinsky and B.I. Feuer
1430 A.J. Papa
1451 H. Newman and R.B. Angier
1456 H. Newman and R.B. Angier
1462 H. Newman and R.B. Angier
1472 T.P. McGovern and M. Beroza
1484 E.H. White and H. Worther
1496 N.N. Gerber and B. Wieclawek
1498 M.F. Sartori et al.
1503 P. Kohn et al.
1538 D.F. O'Brien and J.W. Gates, Jr.
1543 J.W. Wilt et al.
1587 G.R. Harvey
1629 H.B. Kagan et al.
1632 W. Herz, H. Chikamatsu and L.R. Tether
1637 J.C. Sheehan and J.H. Beeson
1641 H.K. Schnoes et al.
1660 D. Seyferth and R. Damrauer
1689 K.W. Ratts and A.N. Yao
1694 W.E. Parham and S.H. Groen

1707 S.M. Kupchan et al.
1716 S.M. Kupchan and C.G. DeGrazia
1722 H. Yasuda and H. Midorikawa
1725 D.S. Deorha and P. Gupta
1728 C.G. Overberger and S. Altscher
1747 S.W. Pelletier et al.
1761 R. Ginsig and A.D. Cross
1765 H. Zinnes and J. Shavel, Jr.
1797 G.W. Kinzer, T.F. Page, Jr, and R.R. Johnson
1852 R.C. Neuman, Jr.
1863 A.J. Fry
1866 M.P. Cava and R.P. Stein
1883 J. Cason and D.M. Lynch
1890 R.D. Elliott, C. Temple, Jr., and J.A. Montgomery
1905 G.N. Walker and D. Alkalay
1912 E.H. White and M.M. Bursey
1945 J. Karliner and C. Djerassi
1959 M.F. Saettone
1970 C. Koukotas et al.
1982 L.G. Vaughan and D.N. Kramer
1988 R. Muneyuki and H. Tanida
2009 E. Wolthuis et al.
2015 M.P. Cava et al.
2026 J.J. Beereboom
2039 R.B. Woodward and D.J. Woodman
2073 G.H. Alt and A.J. Speziale
2090 E.J. Moriconi and F.J. Creegan
2116 P.E. Shaw
2119 P.E. Shaw
2141 R.J. Morhbacher et al.
2149 R.J. Mohrbacher et al.
2171 M.S. Newman, V. deVries and R. Darlak
2175 M.S. Newman and B.C. Ream
2192 D.P. Brust et al.
2202 J.A. Montgomery et al.
2210 C. Temple, Jr. et al.
2215 W. Asbun and S.B. Binkley
2244 H. Hart et al.
2284 S. Ghosal and B. Mukherjee
2291 H. Newman and A. Durante
2333 L.A. Cohen and J.A. Steele
2346 A. Fozard and C.K. Bradsher
2376 L.H. Klemm et al.
2380 F.B. Stocker et al.
2385 G.H. Alt
2389 J.L. Adelfang
2395 S. Kaufman
2407 E.D. Bergmann and I. Agranat
2429 H. Muxfeldt, G. Grethe and W. Rogalski
2487 P.M. Quan and L.D. Quinn
2491 L. Bauer et al.
2512 G. Greenspan et al.
2543 R.E. Ireland and R.C. Kierstad
2564 M.C. Wani and S.G. Levine
2568 M. E. Hermes and R.A. Brown
2571 D.L. Ross and E. Reissner
2607 A. Rosowsky and E.J. Modest
2616 R.E. Doolittle and C.K. Bradsher
2627 J.C. Powers
2641 J. Tsuji and T. Nogi
2646 T.S. Osdene and R.B. Russell
2672 W.L. Hall
2685 N. Nagasawa et al.
2695 G.C. Morrison, W.A. Cetenko and J. Shavel, Jr.
2700 T. Inoi, T. Okamoto and Y. Koizumi
2705 A.G. Anastassiou et al.
2716 R.C. Bansal et al.
2749 C.H. Robinson et al.
2755 M.P. Cava et al.
2784 H. Hart and Y.C. Kim
2920 N.P. Kuntsmann and L.A. Mitscher
2951 A. Takamizawa et al.
3013 C.W. Muth et al.
3038 E.H. White, M.C. Chen and L.A. Dolak
3055 W.W. Paudler and T.J. Kress
3063 L. Rand and R.J. Dolinski
3109 J.A. Marshall and H. Roebke
3178 T.G. Miller
3189 S.D. Levine
3193 N.J. Doorenbos and L. Milewich
3206 M.M. Robison et al.
3213 L. Farkas et al.
3232 W. Herz, V. Sudarsanam and J.J. Schmid
3241 R.K. Borden and M. Smith
3258 J.F. Gerster and R.K. Robins
3263 A.P. Martinez, W.W. Lee and L. Gordmann
3296 R.A. Moss
3321 W.E. Noland and F.J. Baude
3337 D.M. Lynch and W. Cole
3342 D.P. Chakraborty et al.
3356 W. Metlesics et al.
3363 D.J. Zwanenberg et al.
3369 D.K. Wald and M.M. Joullie
3384 F.R. Pfeiffer and F.H. Case
3398 A.A. Costopanagiotis et al.
3400 T.A. Hylton, J. Preston and B. Weinstein
3402 A. Hampton and A.W. Nichol
3409 R.A. Archer and B.S. Kitchell
3418 G.C. Denzer, Jr., et al.
3439 G.A. Gornowitz and J.W. Ryan
3442 C. Ainsworth and R.E. Hackler
3444 J.C. Stickler and W.H. Pirkle
3467 G.W. Krakower and H.A. Van Dine
3482 H.O. House and M.W. Bryant, III
3489 E. Klingsberg
3510 F. Camps et al.
3528 K.T. Potts and R.M. Huseby
3531 B. Loev et al.
3592 D.W.H. MacDowell and T.B. Patrick
3612 A.R. Martin and R. Ketcham
3671 R.M. Scribner
3683 A. Fozard and C.K. Bradsher
3686 A.L. Logothetis
3694 J.P. Marsh, Jr., and L. Goodman
3731 W.J. Middleton
3734 E. Leete, O. Ekechukwu and P. Delvigs
3736 H. Hart and R.M. Lange
3780 W.F. Johns
3787 P.S. Wharton and B.T. Aw
3838 D.O. Spry and H.S. Aaron

3845 H.V. Hansen, J.A. Caputo and R.I. Meltzer
3849 M. Brown and R.E. Benson
3852 E.A. Fehnel and D.E. Cohn
3854 R. Walter, H. Zimmer and T.C. Purcell
3869 W.G. Dauben et al.
3907 G.R. Harvey and K.W. Ratts
3914 T. Sasaki and K. Minamoto
3917 T. Sasaki and K. Minamoto
3921 G.G. Lyle and L.K. Keefer
3928 N.J. Leonard and R.Y. Ning
3941 H. Tanida, T. Tsaji and T. Irie
3954 A. de Groot and H. Wynberg
3958 D.A. Bak and K. Conrow
3976 L.A. Strait et al.
4014 C.L. Stevens, T.S. Sulkowski and M.E. Munk
4043 D.S. Noyce and E.H. Banitt
4054 N.L. Weinberg and E.A. Brown
4082 J.B. Miller
4133 P.L. Pacini and R.G. Ghirardelli
4143 M. Masaki, Y. Chigira and M. Ohta
4188 D.C. Humber and A.R. Pinder
4204 I. Puskas and E.K. Fields
4237 W.K. Musker and R.W. Ashby
4255 D.M. Piatak and E. Caspi
4260 E.T. McBee et al.
4265 R.C. Parish and L.M. Stock
4277 P.R. Jones et al.
4290 M.M. Kreevoy et al.
4321 R.C. Kerber and H.G. Linde, Jr.
4322 M. Feldman, S. Danishefsky and R. Levine

46- -66, J. Phys. Chem., 70 (1966)
0611 B.G. Ramsey
2245 J.W. Ledbetter, Jr.
4097 B.G. Ramsey

49- -66, Monatsh. Chem., 97 (1966)
0036 G. Zigeuner et al.
0077 T.H. Kappe, K. Burdeska and E. Ziegler
0129 E. Ruzicka and J. Jurina
0171 A. Steininger and V. Gutmann
0409 E. Ziegler, T.H. Kappe and H.G. Foraita
0524 R. Kuhn et al.
0554 F.A. Neugebauer and H. Trischmann
0570 F. Wessely and M. Grossa
0644 A. Krbavcic and M. Tisler
0846 R. Kuhn, F.A. Neugebauer and H. Trischmann
0857 I. Ognyanov et al.
0896 H. Wittmann, H. Uragg and H. Sterk
1029 H. Falk, K. Schlogl and W. Steyrer
1280 R. Kuhn et al.
1365 P. Schuster and O.E. Polansky
1384 M. Grossa and F. Wessely
1494 A. Krbavcic and M. Tisler
1523 A. Pollak, B. Stanovnik and M. Tisler
1713 M. Jelenc et al.

50- -66A, Nature, 209 (1966)
1129 P.W. Ratliff and M.J. Follett

50- -66B, Nature, 210 (1966)
0296 R.B. McKay
0298 R.H. Laby and T.C. Morton
0522 R.L. Martin and I.M. Stewart
0627 J.W. Cornforth et al.
1222 H. Ishihara and S.Y. Wang

50- -66C, Nature, 211 (1966)
0737 T. Nishiwaki
0965 B.S. Shasha et al.

50- -66D, Nature, 212 (1966)
0504 O. Nitidandhaprabhas

51- -66, Naturwiss., 53 (1966)
0476 G.P. Schiemenz
0584 L. Horhammer et al.

52- -66, Proc. Royal Soc., 292 (1966)
0122 A. Charlesby, P.M. Kopp and J.F. Read

54- -66, Rec. trav. chim., 85 (1966)
0041 T. Doornbos and J. Strating
0043 G. Lardelli et al.
0117 J.P. Ward and D.A. Van Dorp
0334 H.H. Haeck et al.
0339 H.H. Haeck and T. Kralt
0343 H.H. Haeck and T. Kralt
0347 H.C. Beyerman et al.
0389 G.J.B. Corts and W.T. Nauta
0429 R.G. van den Bos et al.
0457 J.M.A. Baas and B.M. Wepster
0619 R. Van de Graaf and B.M. Wepster
0671 W.N. Speckamp and H.O. Huisman
0681 W.N. Speckamp and H.O. Huisman
0701 J. de Flines et al.
0712 W.F. Van Der Waard et al.
0721 D. van der Sijde et al.
0731 A. Smit and J. Bakker
0753 P. Madsen and S.-O. Lawesson
0757 P.D. Harkes
0929 H. Vieregge et al.
1045 W. Hoek, J. Strating and H. Wynberg
1101 H.C. Van der Plas et al.
1191 C.L. Habraken et al.
1194 C.L. Habraken et al.

56- -66, Roczniki Chem., 40 (1966)
0061 J. Mirek
0205 J. Mirek
0429 E. Czerwinska-Fejgin and W. Polaczkowa
0463 A. Zabza and H. Kuczynski
0621 J. Moszen, M. Bala and E. Sledziewska
1911 W.J. Rodewald and J. Wicha

57- -66A, Science, 151 (1966)
0068 P.E. Brown et al.

0583 J. Meinwald et al.
1539 M.B. Sporn et al.

57- -66B, Science, 152 (1966)
1372 J.D. Dougherty et al.

57- -66C, Science, 153 (1966)
0379 R.B. Setlow

57- -66D, Science, 154 (1966)
1020 W.S. Bowers, H.M. Fales, M.J. Thompson and E.C. Uebel
1189 C.E. Cook et al.

59- -66, Spectrochim. Acta, 22 (1966)
0281 D. Welti
0399 C. Tosi
1537 M. Oki, M. Hirota and S. Hirofuji
1701 C. Tosi
1869 R.D. Srivastara and G. Prasad
2005 B. Ellis and P.J.F. Griffiths

60- -66, Trans. Faraday Soc., 62 (1966)
0018 A.R. Cooper, C.W.P. Crowne and P.G. Farrell
0029 K.R. Bhaskar, R.K. Gosavi and C.N.R. Rao
0286 M.J. Blandamer et al.
0296 M.J. Blandamer et al.
0301 M.J. Blandamer et al.
0788 K.R. Blaskar et al.
1236 D.N. Hague and M. Eigen
1406 R.N. Dixon and G.H. Kirby
1439 R.B. McKay and P.J. Hillson
3162 A. Reiser, G. Bowes and R.J. Horne

61- -66, Ber. Bunsengesellschaft Phys. Chem., 70 (1966)
0052 G. Scheibe, J. Heiss and K. Feldmann
0530 W. Sperling, F.C. Werner and H. Kuhn
0817 J.E. Lohr and G. Kortüm
0862 K.-D. Asmus et al.

62- -66, Z. physik. Chem.(Frankfurt), 48 (1966)
0179 V. Zanker et al.

65- -66, Zhur. Obshchei Khim., 36 (1966)
0033 V.V. Kiselev and N.A. Serova
0114 I.M. Gvertsiteli et al.
0214 V.A. Glushenkov et al.
0254 E.G. Kataev and T.G. Mannafov
0336 V.P. Dzyuba, V.F. Lavrushin and V.N. Tolmachev
0410 L.A. Pavlova et al.
0660 G.V. Shishkin and V.P. Mamaev
0808 A.A. Khachatur'yan et al.
0828 L.M. Yagupol'skii et al.
0835 K.K. Pivnitskii and I.V. Torgov
0843 K.K. Pivnitskii and I.V. Torgov
0862 A.Y. Yakubovich et al.
1034 E.E. Milliaresi (*1048)
1055 A.P. Karishin et al.
1198 A.V. Belotsvetov et al.
1202 A.V. Belotsvetov et al. (*1212)
1208 V.N. Luzgina et al.
1248 I.N. Zhmurova and A.V. Kirsanov
1383 E.I. Filippovich et al.
1499 G.V. Glazneva, E.K. Mamaeva and A.P. Zeif
1577 A.E. Lutskii et al.
1590 A.P. Terent'ev et al.
1595 V.V. Sinev, E.I. Kvyat and O.F. Ginzburg
1722 E.P. Gendrikov and G.V. Semenyuk
1728 G.N. Dorofeenko et al.
1749 E.D. Khomutova, T.A. Shapiro and V.M. Berezovskii
1764 O.N. Tolkachev et al.
1815 L.N. Nikolenko and A.V. Potapova
1927 R.M. Kuritsina and N.V. Chugreeva
1938 G.B. Afanas'eva and I.Y. Postovskii
2064 V.V. Belogorodskii et al.
2098 G.I. Lavrenova et al.
2141 S.M. Shein et al.

67- -66, J. Structural Chem., 7 (1966) (English translation pagination)
0042 Y.S. Bobovich et al.
69- -66, Biochemistry, 5 (1966)
0224 M.J. Robins, J.R. McCarthy, Jr., and R.K. Robins
0625 R.G. Powell and C.R. Smith, Jr.
0689 M.E. Levitch and P. Rietz
0756 R.J. Rousseau et al.
1425 W.O. McClure and H. Neurath
2076 A. Hampton and A.W. Nichol
2082 G.B. Chheda and R.H. Hall
2468 C. Zioudrou and J.S. Fruton
2845 E.Y. Levin
3057 A. Giner-Sorolla et al.
3385 J.R. Little and H.N. Eisen
3454 T.P. King
3638 B.E. Griffin et al.
3824 N.N. Gerber

70- -66, Izvest. Akad. Nauk S.S.S.R., (1966)(English translation pagination)
0122 L.S. Povarov et al.
0171 I. Gurvich, I.M. Mil'shtein and V.F. Kucherov
0256 L.D. Bergel'son and A.N. Grigoryan
0337 N.N. Vorozhtsov et al.
0468 L.D. Bergel'son et al.
0615 M.S. L'vova et al.
0646 B.I. Kozyrkin et al.
0833 M.V. Mavrov and V.F. Kucherov
0848 N.S. Val'fson et al.
0888 V.V. Ershov et al.
1000 I.L. Knunyants et al.
1167 L.N. Ivanova et al.
1171 G.V. Kondrat'eva et al.
1326 I.P. Beletskaya, K.P. Butin and O.A. Reutov
1378 E.P. Kaplan et al.
1414 L.A. Kazitsyna et al.
1546 N.S. Vul'fson et al.

1549 V.P. Mamaev and E.N. Lyubimova
1718 N. Kochetkov et al.
1723 N.S. Vul'fson and G.M. Sukhotina
1739 E.A. El'perina et al.
1791 F.N. Mazitova and R.R. Shagidullin
2092 A.A. Akhrem and A.M. Moiseenkov
2097 V.I. Gunar et al.
2139 V.A. Mironov et al.

71- -66A, Biochem. Biophys. Acta, 112 (1966)
0550 R.W. Cowgill

73- -66, Coll. Czech. Chem. Comm., 31 (1966)
0835 C. Parkanyi et al.
1064 B. Pele and J. Hodkova
1355 L. Slavikova and J. Slavik
1363 B. Pele, J. Hodkova and J. Holubek
1785 S. Chladek, J. Zemlicka and F. Sorm
1824 A. Cerny et al.
1864 J. Pitha, P. Fiedler and J. Gut
2434 J. Weichet et al.
2737 J. Hora
2768 K. Syhora and R. Mazac
3127 V. Suk
3174 J. Klinot, M. Krumpole and A. Vysttrcil
3198 J. Zemlicka et al.
3362 L. Slavikova and J. Slavik
3744 M. Prochazka and M. Palecek
3800 A. Holy and J. Smrt
4002 J. Farkas, J. Beranek and F. Sorm
4286 F. Santavy, L. Hruban and M. Maturova
4598 J. Weichet, L. Blaha and B. Kakac
4703 V. Schwarz et al.

77- -66, Chemical Communications, (1966)
0015 R.C. Cookson et al.
0016 A.K.M. Anisuzzaman and L.N. Owen
0020 Y. Gaoni and R. Mechoulam
0044 C.W.L. Bevan, D. Ekong and J.I. Okogun
0048 J. Burdon, J.C. Tatlow and D.F. Thomas
0087 I.D. Campbell et al.
0137 G.H. Aylward, J.L. Garnett and J.H. Sharp
0141 J.D. Hobson and J.R. Malpass
0172 R.A. Finnegan and D. Knutson
0180 Y. Kitahara et al.
0184 U.R. Ghatak and J. Chakravarty
0206 W.R. Chan and D.R. Taylor
0230 A. Young et al.
0272 W. Carruthers
0297 E. Fujita, T. Fujita and M. Shibuya
0327 B. Miller
0393 C.J.W. Brooks and G.H. Draffan
0401 S. McKenzie and D.H. Reid
0404 L. Cholnoky et al.
0417 J.D. Crum and P.W. Sprague
0425 G. Shaw et al.
0485 S.K. Dasgupta and P.C. Antony
0498 K. Chambers et al.
0508 J.A. Elix, M.V. Sargent and F. Sondheimer
0509 J.A. Elix, M.V. Sargent and F. Sondheimer
0546 L.A. Summers
0548 W. Carruthers
0567 J. Wolinsky and D. Chan
0576 W.R. Chan, D.R. Taylor and R.T. Aplin
0585 S.R. Landor et al.
0586 A. Ceccon, U. Tonellato and U. Miotti
0600 N. Boegman, F. During and C.F. Garbers
0622 A.F. Hegarty and F.L. Scott
0625 V. Bertoli and P.H. Plesch
0634 J. Hutton and W.A. Waters
0696 G.R. Proctor and A.H. Renfrew
0701 C.J.W. Brooks and G.H. Dreffan
0711 J.D. White
0740 T. Anthonsen et al.
0775 W.M. Horspool, J.M. Tedder and Z.U. Din
0894 P.G. Jones
0915 K. Nakanishi et al.
0923 W.R. Chan et al.

78- -66, Tetrahedron, 22 (1966)
0001 A.T. Balaban, E. Romas and C. Rentia
0007 B.D. Tilak et al.
0025 L.A. Paquette
0031 A.B. Turner and J.F.W. McOmie
0037 G. Slater and A.W. Somerville
0053 L. Chierici and G.P. Gardini
0057 K.J. Morgan and D.P. Morrey
0081 W. vonE. Doering and R.A. Odum
0133 K. Gollnick and G. Schade
0139 K. Gollnick, G. Schade and S. Schroeter
0157 Y. Ogata, A. Kawasaki and K. Nakagawa
0199 S. Smolinski
0209 L.L. Leveson and C.W. Thomas
0259 J.L. Baas et al.
0265 J.L. Baas et al.
0293 P.J. van den Tempel and H.O. Huisman
0301 P. Joseph-Nathan, J.J. Morales and J. Romo
0321 J.P. Kutney and R.T. Brown
0337 L.H. Zalkow et al.
0351 J.B. Thomson
0365 G. Cooley et al.
0369 D. Burn et al.
0377 F.K. Butcher et al.
0407 K. Ichikawa, S. Uemura and T. Sugita
0455 A. Butt, S.M.A. Hai and I.A. Akhtar
0463 G.L. Closs and H. Heyn
0557 W.M. Pasika et al.
0567 F. Ramirez, O.P. Madan and C.P. Smith

0653 M. Arshad, A. Beg and A.R. Shaikh
0679 W.G. Dauben and V.F. German
0685 J. Correa and J. Romo
0745 T.R. Kasturi et al.
0867 H. Taniguchi et al.
0891 J.D. Connolly et al.
0897 C.S. Chopra and D.E. White
0925 A.T. Nielsen and W.G. Finnegan
0931 A.R. Katritzky et al.
0941 M. Krishnamurti et al.
0949 R. Selvarajan et al.
0987 K.D. Berlin and M.A.R. Khayat
0995 A.S. Bailey et al.
1001 A,F, Casy, J.L. Myers and P. Pocha
1011 J.C. Emmett and W. Lwowski
1019 G.H. Douglas et al.
1027 T.R. Kasturi and K.M. Damodaran
1033 H. Mitsuhashi and S. Harada
1053 N.Z. Chow, D.C. Huang and Huang-Minlon
1063 G. Just and C. Pace-Asciak
1069 G. Just and C. Pace-Asciak
1075 J. Poisson et al.
1095 F.R. Stermitz, L. Chen, and J.I. White
1121 R. Tschesche et al.
1139 T.J. Mabry et al.
1147 V.K. Bhatia, S.R. Gupta and T.R. Seshadri
1153 R.A. Corral and O.O. Orazi
1159 K. Takeda et al.
1175 K.J. Morgan and A.M. Turner
1201 D. Misiti, H.W. Moore and K. Folkers
1207 H. Nozaki, T. Mori and R. Noyori
1275 E.D. Bergmann and I. Agranat
1279 R.E. Brooks et al.
1309 N.E. Alexandrou
1317 A. Yogev and Y. Mazur
1335 J.C. Craig et al.
1421 J.W. Blunt et al.
1461 R. Mukherjee and A. Chatterjee
1481 Y. Gaoni and R. Mechoulam
1489 J.F. Fisher and H.E. Nordby
1499 J. Romo, P. Joseph-Nathan and G. Siade
1507 T. Rios
1513 A.K. Ganguly et al.
1521 G.L. Buchanan and G.W. McLay
1587 T. Fujita et al.
1615 N.L. Allinger, P. Crabbe and G. Perez
1625 S. Hirai, R.G. Harvey and E.V. Jensen
1641 G.K. Trivedi et al.
1650 T. Kubota et al.
1709 W. Herz et al.
1723 P. Joseph-Nathan and J. Romo
1755 G. Camaggi et al.
1777 O.R. Gottlieb et al.
1785 O.R. Gottlieb et al.
1789 K.G. Marathe, M.J. Byrne and R.N. Vidwans
1797 L.H. Klemm et al.
1809 C. Jutz et al.
1857 J.D. Chanley, T. Mezzetti and H. Sobotka
1907 W. Herz et al.
1929 R.M. Silverstein et al.
1943 T.J. Mabry et al.
1949 B.A. Nagasampagi et al.
1977 M.V.R. Koteswara Rao, G.S. Rao and S. Dev
2003 S.I. Zavialov et al.
2015 D.S. Deorha and P. Gupta
2021 P.S. Venkataramani et al.
2039 M. Gorodetsky et al.
2053 A.N. Hughes and V. Prankprakma
2073 A. Pollak and M. Tisler
2119 R.A. Olofson et al.
2135 J.M. Landesberg and R.A. Olofson
2145 H. Nozaki, M. Takaku and K. Kondo
2177 H. Nozaki et al.
2311 G.L. Chetty et al.
2319 R.K. Gupta et al.
2359 D.J. Aberhart and P. de Mayo
2377 M. Manzoor-I-Khuda
2387 F. Walls et al.
2401 T. Nishiwaki
2445 M. Sainsbury, S.F. Dyke and A.R. Marshall
2461 P. Bouchet, J. Elguero and R. Jacquier
2523 H. Cristol, Y. Pietrasanta and J.L. Vernet
2555 J. Riera and R. Stephens
2575 T.R. Kasturi and A. Srinivasan
2599 M. Wilk, W. Bez and J. Rochlitz
2621 G. Köbrich and W. Drischel
2721 C.M. Lee, A.H. Beckett and J.R. Sugden
2735 A.H. Beçkett et al.
2745 A.H. Beckett, A.F. Casy and M.A. Iorio
2761 M. Miyamoto et al.
2785 M. Miyamoto et al.
2829 C. Burgess et al.
2845 C.R. Bennett and R.C. Cambie
2869 D.W. Theobald
2899 L.R. Row et al.
2913 L. Jurd
2923 D.L. Dreyer
3103 G. Snatzke et al.
3131 E.B. Dennler and A.R. Frasca
3173 H. Morimoto, Y. Sanno and H. Oshio
3195 J.W. Blunt et al.
3209 M. Geoghegan et al.
3213 P.R. Enslin et al.
3233 L. Jakhontov et al.
3245 T.R. Govindachari et al.
3253 R. Mondelli and L. Merlini
3279 A. Romo de Vivar et al.
3293 A.K. Kiang et al.
3337 R. Kikumoto and T. Kobayashi
3351 M. Kamel, M.I. Ali and M.M. Kamel
3421 V.M. Clark and A. Cox
3423 E. Fujita et al.
3443 P.A. Diassi et al.
3459 P.A. Diassi et al.
3469 P. Sengupta, S. Ghosh and L.J. Durham
3491 D. Kumari, S.K. Mukerjee and T.R. Seshadri

3607 M. Gorodetsky and Y. Mazur

78- -66A, Tetrahedron, Supplement 7 (1966)
0049 Altaf-Ur-Rahman and A. J. Boulton
0057 D.H.R. Barton, P.G. Sammes and M. Silva
0139 G.D. Bhatia, S.R. Mukerjee and T.R. Seshadri
0175 L. Caglioti et al.
0189 G. Wittig et al.
0203 S. Vromen et al.
0219 K.I. Takeda, H. Minato and M. Ishikawa
0241 J.A. Ballantine et al.
0287 F. Geers-Evrard and R.H. Martin
0325 G. Cooley, B. Ellis and V. Petrow
0333 C.P. Falshaw et al.
0349 E.D. Bergmann et al.
0391 A.J. Birch and G.S.R. SubbaRao
0415 R.B. Woodward and R.A. Olofson
0441 H.H. Wasserman and M.B. Floyd

78- -66B, Tetrahedron, Supplement 8 (1966)
0015 R.F.C. Brown, V.M. Clark and Lord Todd
0033 A. Deljac et al.
0071 H. Erdtman et al.
0093 F. Korte, W. Klein and E.D. Schmid
0105 G. Stork, M. Nussim and B. August
0123 K.I. Takeda et al.
0135 E.D. Bergmann and I. Agranat
0141 E.D. Bergmann, M. Rabinovitz and S. Glily
0181 R.H. Martin et al.
0203 G. Hugel, A.C. Oehlschlager and G. Ourisson
0217 O. Achmatowicz and J. Szychowski
0279 J. Altman et al.
0321 R.B. Woodward, R.A. Olofson and H. Mayer
0359 A.J. Birch et al.
0443 M. Julia, C. Descoins and C. Risse
0485 B.K. Cassels and V. Deulofeu
0491 C. Dufraisse, J. Rigaudy and M. Ricard
0541 W.S. Johnson et al.

80- -66, Revue Romaine Chim., 11 (1966)
0109 A.T. Balaban and N.S. Barbulescu
0139 J. Reichel, A. Balint and A. Demian
0263 A. Demian and J. Reichel
0701 R. Ripan, C. Mirel and D. Lupu
1409 I. Bally et al.

82- -66, J. Mol. Spect., 20 (1966)
0226 T. Kubota, M. Yanakawa and I. Tanaka

83- -66, Arch. Pharm., 299 (1966)
0001 K.E. Schulte, J. Reisch and H. Walker
0131 G. Schwenker
0159 J. Knabe and N. Ruppenthal
0196 B. Gober and S. Pfeifer
0295 H. Mohrle
0303 H. Loth, G. Schenk and F. Walter
0385 H.J. Roth and S. Al Sarraj
0493 F. Elden and B.S. Nagar
0503 R. Hansel and L. Klaproth
0507 R. Hansel, H. Saver and H. Rimpler
0531 H.J. Roth and S. Al Sarraj
0626 B. Unterhalt
0646 V. Petelin-Hudnik et al.
0679 H.W. Voigtlander, G. Balsam and B. Hampel
0715 H. Mohrle
0783 H. Auterhoff and G. Kinsky
0798 J. Reisch
0817 W. Schneider and E. Kammerer
0846 W. Schneider and E. Kammerer
0883 G. Wagner and D. Heller
0914 K. Hartke et al.
0921 K. Hartke and A. Birke
1011 R. Huttenrauch and A. Schubert

86- -66, Talanta, 13 (1966)
none

87- -66, J. Med. Chem., 9 (1966)
0015 W.O. Godtfredsen et al.
0027 G.H. Douglas, C.R. Walk and H. Smith
0046 F. Schenker et al.
0057 K.C. Tsou et al.
0097 A. Giner-Sorolla and L. Medrek
0101 J.J. Fox, N. Miller and I. Wempen
0105 J.A. Montgomery and K. Hewson
0108 F.R. Gerns, A. Perrotta and G.H. Hitchings
0116 O.R. Rodig, R.E. Collier and R.K. Schlatzer
0121 D.E. O'Brien et al.
0142 M.B. Winstead et al.
0143 A. Giner-Sorolla, L. Medrek and A. Bendich
0161 R.T. Parfitt
0179 A.J. Solo and B. Singh
0234 J.A. Montgomery and K. Hewson
0237 C.W. Mosher et al.
0242 V. Dev et al.
0266 S. Gerchakov, P.J. Whitman and H.P. Schultz
0268 A.P. Martinez, W.W. Lee and L. Goodman
0292 J.I. DeGraw et al.
0304 K.E. Fahrenholtz et al.
0338 G.C. Buzby, Jr., et al.
0354 J.A. Montgomery and K. Hewson
0366 T.Y. Shen, J.F. McPherson and B.O. Linn
0373 A.G. Beaman et al.
0417 Y.F. Shealy, C.A. O'Dell and J.A. Montgomery
0419 G.J. Durr
0420 H.G. Petering et al.
0433 E. Chang and V.K. Jain
0439 J.D. Benigni and D.E. Dickson
0444 R.J. Stedman
0444B C.W. Mosher, E.M. Acton and L. Goodman

0489 J.N. Wells and F.S. Abbott
0495 R.D. Faulkner and J.P. Lambooy
0510 E. Farkas and J.M. Owen
0513 P. Deghenghi et al.
0516 R.W.J. Carney et al.
0527 J. Szmuszkovicz et al.
0537 G.I. Poos et al.
0551 A.W. Chow et al.
0566 R. Duschinsky et al.
0573 D.E. O'Brien et al.
0576 H.J. Schaeffer and E. Odin
0590 G.L. Tong, W.W. Lee and L. Goodman
0593 T.L. Fletcher et al.
0609 E.G. Podrebarac and C.C. Cheng
0615 V.L. Narayanan and C.F. Martin
0638 C.N. Corder and J.L. Way
0656 J.A. Skorcz et al.
0685 R.E. Counsell et al.
0693 P.D. Klimstra, E.F. Nutting and R.E. Counsell
0715 A. Dipple and C. Heidelberger
0719 H.T. Nagasawa and H.R. Gutmann
0741 R.H. Chauvette and E.H. Flynn
0746 J.L. Spencer et al.
0751 G.L. Dunn, P. Actor and V.J. Di-Pasquo
0758 R.W.J. Carney et al.
0762 H.L. Johnson, A.R. Patel and J.F. Oneto
0766 R.A. Pages and A. Burger
0770 J.G. Lombardino
0782 G.C. Buzby, Jr,, C.R. Walk and H. Smith
0784 H.S. Lowrie
0791 P.F. Wiley and E. L. Caron
0835 B.M. Sutton and J.H. Birnie
0838 J.I. DeGraw et al.
0860 M.A. Davis et al.
0864 D. Herst et al.
0868 M.P. Mertes and N.R. Patel
0876 M.P. Mertes, S.E. Saheb and D. Miller
0881 D,L, Trepanier et al.
0960 P.L. Chien and C.C. Cheng
0977 P.L. Warner, Jr., and T.J. Bardos
0980 V.J. Bauer and S.R. Safir
0981 R. Klemm et al.

88- -66, <u>Tetrahedron Letters</u> (1966)
0049 P. Morand, S. Stavric and D. Godin
0059 H. Nozaki et al.
0065 J. Strating et al.
0097 F.J. Schmitz et al.
0145 L. Crombie, D.E. Games and A. McCormick
0181 F.G. Torto et al.
0201 W.H. Dietsche
0205 P. deRuggieri, C. Gandolfi and U. Guzzi
0233 G. Casnati et al.
0295 R.A. Flath et al.
0341 M. Franck-Neumann
0405 R. Huisgen, H. Blaschke and E. Brunn
0411 R.W. Hoffmann and H.J. Luthardt
0445 K. Torssell and K. Wahlberg
0459 L.C. Dorman
0489 K. Nano et al.
0507 D.A. Nelson and J.J. Worman
0545 M. Miyamoto et al.
0607 I. Yosioka et al.
0643 T. Hashizume and H. Iwamura
0655 E. Vogel and W. Maier
0661 D.P. Chakraborty
0701 T. Kubota, F. Tonami and H. Hinoh
0751 H.A. Staab and F. Graf
0797 M. Feldman and J.A. Jackson
0801 J.P. Turnbull and J.H. Fried
0839 M.P. Kunstmann et al.
0915 J. Tomko et al.
0927 D. Nasipuri et al.
0931 G.B. Yeok, K.C. Chan and F. Morsingh
0939 W.L. Dilling
0983 H. Singh and V.V. Parashar
0999 D.S. Glass, R.S. Boikess and S. Winstein
1023 E. Farkas et al.
1029 J. Romo et al.
1057 S. Hara and K. Oka
1063 T. Mukai, H. Tsuruta and T. Nozoe
1081 A. Popelak et al.
1109 J. Kucera and Z. Arnold
1127 I.A. D'yakonov et al.
1145 J. Seydel
1163 P.W. Thies
1171 S.P. Pappas and J.E. Blackwell, Jr.
1185 C.F. Huebner et al.
1193 A.G. Brown et al.
1205 I. McMillan and R.J. Stoodley
1211 L. Canonica et al.
1245 T.A. Geissman et al.
1251 S.M. Kupchan et al.
1293 K. Christiansen and P.M. Boll
1329 L. Canonica et al.
1343 J.P. Horwitz, J. Chua and M. Noel
1347 J. Streith and C. Sigwalt
1499 N.H. Cromwell and E.-M. Wu
1503 N. Obata and I. Moritani
1537 Y. Inubushi et al.
1577 M.P. Cava et al.
1599 G. Grethe, M. Uskokovic and A. Brossi
1611 M. Komatsu and T. Tomimori
1643 Y.A. Berlin et al.
1649 E.F.L.J. Anet
1703 S. Bauer et al.
1727 L.I. Peterson
1759 T.J. Bardos et al.
1787 J. Ide and Y. Kishida
1797 A. Chatterjee and C.P. Dutta
1805 K. Fukui and M. Nakayama
1837 T. Irie et al.
1939 R.A. Massy-Westropp et al.
1953 B.D. Tilak, H.S. Desai and S.S. Gupte
1969 A.D. Argoudelis and J.F. Zieserl
1991 M. Hassner and D.R. Fitchman
2043 C.H. Heathcock
2049 D. Lavie, M.K. Jain and I Kirson

2145 C. Kaneko, S. Yamada and M. Ishikawa
2193 G. Quinkert et al.
2211 S. Nozoe, K. Hirai and K. Tsuda
2243 Y. Maroni-Barnaud et al.
2257 L. Skattebøl et al.
2277 A. Padwa and R. Hartman
2287 A. Stoessl
2307 R.E. McMahon
2321 H. Gibian et al.
2331 H. Brockmann and W. Schramm
2353 M. Koch, M. Plat, B.C. Das and J. LeMen
2361 R.W. Rickards and R.M. Smith
2395 C.R. Enzell and B.R. Thomas
2449 U.R. Ghatak and J. Chakravarty
2483 B. Das et al.
2515 H. Nakata, H. Harada and Y. Hirata
2523 N. Sakabe et al.
2557 A. Mondon and M. Ehrhardt
2579 A.G. Anderson, Jr., and H.L. Ammon
2609 G.F. Field, W.J. Zally and L.H. Sternbach
2615 M. Nakazaki and K. Naemura
2837 N. Sommer and H.A. Staab
2867 Y. Ogihara et al.
2875 W. Mack
2899 I. Suzuki et al.
2905 R.W. Franck and K. Yanagi
2911 A.C. Coda, P. Grunanger and G. Veronesi
2937 M.P. Cava et al.
2941 N.C. Yang, G.R. Lenz and A. Shani
2947 M.P. Cava et al.
2963 D. Nasipuri and K.K. Biswas
2971 J.J. Bloomfield and J.R.S. Irelan
3031 L. Canonica et al.
3063 D. Levy and R. Stevenson
3069 H.F. Andrew and C.K. Bradsher
3109 W.H. Urry et al.
3125 E.M. Kosower and D.J. Severn
3153 E. Fujita, T. Fujita and M. Shibuya
3225 R.N. Warrener and E.N. Cain
3267 K. Sisido and K. Utimoto
3287 C. Beard et al.
3341 A. Fozard and C.K. Bradsher
3427 Y. Kishi et al.
3445 Y. Kishi et al.
3451 P. Sunder-Plassmann, J. Zderic and J.H. Fried
3457 I.T. Harrison, J.B. Siddall and J.H. Fried
3477 A. Umani-Ronchi, P. Bravo and G. Gaudiano
3489 S.W. Pelletier et al.
3499 H.P.M. Fromageot and C.B. Reese
3519 K. Kawazu and T. Mitsui
3531 H. Brockmann et al.
3537 C.L. Coon, D.L. Ross and M.E. Hill
3541 L. Novotny, Z. Samek and F. Sorm
3547 S. Huneck
3585 A.V. Zakharychev et al.
3595 H. Brockmann et al.
3621 K. Vlentin and A. Taurins
3627 D. Caine et al.
3663 T. Nozoe, Y.S. Cheng and T. Toda
3671 N.C. Yang and D.-M. Thap
3677 I. Morimoto and Y. Hirata
3695 C.W. Plujgers and J. Berg
3697 A.S. Kende, P.T. Izzo and W. Fulmor
3723 A. Ballio, S. Barcellona and B. Santurbano
3737 G.L. Dunn, V.J. DiPasquo and J.R.E. Hoover
3743 W.G. Dauben and D.L. Whalen
3763 E.V. Dehmlow
3767 D. Kumari et al.
3773 A.W. Murray and R.W. Bradshaw
3801 H.E. Zieger and E.M. Laski
3857 J.E. Bunney and M. Hooper
3891 M. Stefanovic et al.
3975 D.W. Brown and S.F. Dyke
4017 H. Hoffmeister and H.F. Grutzmacher
4037 N.H. Cromwell and E. Doomes
4057 U.P. Bhatia and B.L. Mathur
4103 C. Schiele and G. Arnold
4113 Y. Shimizu, H. Mitsuhashi and E. Caspi
4173 J.R. Hargreaves and P.W. Hickmott
4179 J.W. Clark-Lewis and R.W. Jamison
4227 M.R. Krishnaswamy et al.
4279 M.P. Cava et al.
4355 O. Buchardt and C. Lohse
4375 M.-M. Janot
4437 T. Ando and M. Nakagawa
4561 G. Mehta, U.R. Nayak and S. Dev
4569 S.W. Breuer et al.
4573 G. Quinkert, Cimbollek and G. Buhr
4671 T. Kubota et al.
4701 C. Kaneko, S. Yamada and I. Yokoe
4729 C. Kaneko et al.
4849 T. Kametani et al.
4855 S. Shibata et al.
4867 O.E. Edwards and D.C. Gillespie
4979 J. Sauer and G. Heinrichs
4989 J.A. Marshall and M.T. Pike
4993 J.H.P. Tyman and R. Pickles
4997 S.K. Dasgupta and P.C. Antony
5015 R.P. Stein, G.C. Buzby, Jr., and H. Smith
5027 H. Achenbach
5067 W. Skorianetz and E sz. Kovats
5077 A. Popelak, E. Haack and H. Spingler
5081 M. Vidal, C. Dumont and P. Arnaud
5107 I.M. Campbell et al.
5137 T. Murakami et al.
5145 H.H. Inhoffen and H. Biere
5179 K. Wada et al.
5205 T. Kubota et al.
5221 V. Mayer et al.
5225 T. Sheradsky
5229 H. Inouye, S. Ueda and Y. Nakamura
5235 C.W. Shoppee et al.
5245 W. Fischer and E. Fahr
5253 R. Brossmer and D. Ziegler

5273 M. Fischer
5337 S.L. Neidleman et al.
5369 H.H. Takimoto and G.C. Denault
5379 J. Buddrus et al.
5451 A.R. Gagneux and R. Goschke
5455 U. Mazzucato et al.
5525 K. Nagarajan and P. Rajagopalan
5723 R.E. Busby and D. Huckle
5731 P.T. Izzo and A.S. Kende
5767 B.S. Joshi and V.N. Kamat
5811 M.J. Jorgenson
5815 J. Warkentin and P.R. West
5821 H. Balli and R. Low
5875 Y. Hirose et al.
5881 B. Lacoume and L.H. Zalkow
5961 F.A.L. Anet and B. Gregorovich
6037 J.K. Crandall and W.H. Machleder
6087 R.A. Finnegan et al.
6103 A.W. Hawtrey
6157 J. Hancock and A.R. Markert
6163 H. Kiefer and T.G. Traylor
6169 H.L. de Waal and B.J. Prinsloo
6175 D.D. Chapman, H.S. Wilgus and J.W. Gates
6229 M. Tomita and M. Kozuka
6257 N. Katsui et al.
6327 J.W. Loder and G.B. Russell
6365 Y. Ohta, T. Sakai and Y. Hirose
6413 Y. Makisumi
6441 R. Zelnik and C.M. Rosito
6471 H. Christol and Y. Pietrasanta
6489 H.W. Thompson
6495 Z.G. Hajos, D.R. Parrish and E.P. Oliveto

89- -66, Angew. Chem., 5 (1966)(International Edition)
0246 E.H. Gold and D. Ginsburg
0248 H. Hoffmeister
0251 H. Prinzbach, U. Fischer and R. Cruse
0311 U. Mayer. H. Baumgärtel and H. Zimmermann
0372 E. Fahr and H. Lind
0511 R.N. Warrener and E.N. Cain
0514 V. Bertini and P. Pino
0516 C. Jutz and R. Kirchlechner
0517 R. Gompper, E. Kutter and H.-U. Wagner
0579 E. Winterfeldt and G. Giesler
0587 A. Albert and K. Tratt
0590 E. Vogel et al.
0602 H. Prinzbach and U. Fischer
0603 E. Vogel et al.
0605 L.A. Summers
0699 A. Wagner et al.
0724 C. Jutz and E. Müller
0732 E. Vogel, W. Schröck and W.A. Boll
0733 W.A. Böll
0734 E. Vogel et al.
0839 K. Hafner and J. Mondt
0846 G. Wittig and P. Fritze
0846C G. Märkl
0893 W. Ziegenbein and H.-E. Sprenger
0896 W. Mack
0959 G. Maier and T. Sayrac
0967 R. Kreher and J. Senbert
1039 H. Prinzbach, M. Arguëlles and E. Druckrey

90- -66, J. Inorg. Nucl. Chem., 28 (1966)
0147 C.H. Misra, S.S. Parmar and S.N. Shukla
1429 R.D. Gillard et al.

93- -66, J. Applied Chem. S.S.S.R., 39 (1966)(English translation ed.)
0616 S.S. Kerimov
2413 Z.N. Grechukhina and V.V. Nesmelov

94- -66, Chem. Pharm. Bulletin (Japan), 14 (1966)
0001 N. Akagi et al.
0046 M. Ikehara and K. Muneyama
0067 T. Kametani et al.
0078 T. Kametani and H. Yagi
0087 M. Miyaki, K. Iwase and B. Shimizu
0089 K. Sirakawa et al.
0121 A. Ueno
0129 A. Ueno and S. Fukushima
0147 K. Sugimoto et al.
0187 S. Ohki et al.
0238 A. Takamizawa et al.
0262 T. Hiraoka and I. Iwai
0314 S. Okuda et al.
0496 M. Okada et al.
0502 S. Uyeo, Y. Maki and Y. Yamamoto
0512 T. Okamoto, M. Hirobe and T. Yamazaki
0552 D. Satoh et al.
0555 C. Kaneko and S. Yamada
0566 T. Kametani and K. Kigasawa
0622 S. Akaboshi and S. Ikegami
0641 K. Takahashi and T. Nakagawa
0666 T. Ueda et al.
0735 H. Hikino et al.
0762 M. Hamana and H. Noda
0770 Y. Maki, M. Suzuki and T. Yamada
0802 Z. Horii et al.
0809 H. Mitsuhashi and M. Fukuoka
0823 G. Tsukamoto, K. Harada and I. Utsumi
0842 A.M.E. Omar and S. Yamada
0856 A.M.E. Omar and S. Yamada
0866 M. Morisaki, H. Izawa and K. Tsuda
0873 H. Izawa, M. Morisaki and K. Tsuda
0890 H. Hikino, K. Aota and T. Takemoto
0902 S. Inouye
0934 Y. Kanaoka et al.
0939 H. Yoshimura et al.
1023 S. Shibata et al.
1033 S. Uyeo et al.
1039 H. Nomura and K. Sugimoto
1065 T. Nakagome et al.
1090 T. Nakagome, A. Misaki and A. Murano
1096 K. Takeda et al.
1102 M. Ishikawa et al.

1133 D. Satoh and M. Horie
1185 M. Onda et al.
1201 T. Hino et al.
1277 I. Iwai and N. Nakamura
1300 K. Harada and S. Enoto
1310 H. Hikino et al.
1314 M. Hashimoto and K. Hattori
1316 C. Kaneko et al.
1321 S. Mizukami and E. Hirai
1354 T. Kishikawa, T. Yamazaki and H. Yuki
1360 T. Kishikawa and H. Yuki
1370 K. Yoshida and T. Kubota
1377 K. Imai, A. Nohara and M. Honjo
1382 S. Yamada and S. Ikegami
1399 Z. Horii et al.
1408 T. Kametani, H. Sugahara and S. Asagi
1418 S. Oida, M. Kurabayashi and E. Ohki
1425 T. Nishiwaki
1448 T. Sasaki and K. Minamoto

95- -66, J. Pharm. Soc. Japan, 86 (1966)
0001 T. Amano
0050 Y. Maki, K. Yamane and M. Sato
0055 T. Kametani and K. Ogasawara
0072 T. Kametani, L.L. Ling and S. Shibuya
0082 M. Yanai et al.
0095 J. Kinagawa and H. Nagase
0101 J. Kinagawa and H. Nagase
0114 S. Fukushima and K. Shimizu
0124 K. Ito and A. Yoshida
0129 T. Nakasoto, S. Asada and Y. Koezuka
0134 T. Nakasoto and S. Asada
0158 N. Yoshida
0177 S.-T. Lu and P.-K. Lan
0296 S.-T. Lu
0300 I. Ito, S. Murakami and K. Tanabe
0376 H. Tanaka and R. Yamamoto
0460 M. Tomita, M. Kozuka and S. Uyeo
0487 Y. Maki, M. Takaya and M. Suzuki
0510 S. Fujisawa and S. Kawabata
0514 S. Fujisawa and S. Kawabata
0541 S. Fujisawa and M. Kawamura
0544 T. Okano, S. Goya and Y. Tsuda
0558 K. Hamamoto et al.
0565 K. Hamamoto, K. Horiki and N. Maezoni
0600 M. Itaya, Y. Takai and T. Kaiya
0608 M. Nagata
0617 M. Ito and A. Sugihara
0622 S. Tagami et al.
0631 H. Ishii and T. Komaki
0637 T. Matsuno and J. Iba
0639 K. Hamamoto and N. Maezono
0649 T. Okano et al.
0763 M. Tomita, S.-T. Lu and Y.-Y. Chen
0766 T. Okamoto and S. Hayashi
0823 T. Kametani et al.
0854 M. Sekiya and Y. Osaki
0856 S. Sakai et al.
0867 T. Takahashi, M. Furukawa and Y. Maki
0874 T. Takemoto et al.
0943 H. Inouye et al.
0952 S. Goya, T. Takahashi and T. Okano
0973 T. Kametani et al.
1051 M. Takahashi and M. Yoshikura
1082 T. Takahashi et al.
1087 Y. Kimura, M. Takido and S. Takahashi
1099 S. Kamiya et al.
1138 M. Meguro, K. Ishibashi and I. Yosioka
1143 H. Furukawa and S.-T. Lu
1152 G. Kobayashi et al.
1156 G. Kobayashi et al.
1172 S. Uyeo, K. Ueda and Y. Yamamoto
1184 Y. Kimura et al.
1187 M. Ito and A. Sugihara
1205 T. Nakasoto and S. Asada
1213 M. Tomoeda, Y. Tani and H. Okada

96- -66, The Analyst, 91 (1966)
0098 J.A.W. Dalziel and M. Thompson
0297 A.M. Parsons
0714 H.R. Savage et al.

97- -66, Z. Chemie, 6 (1966)
0068 G. Tomaschewski and G. Geissler
0318 K. Issleib and W. Gründler

99- -66, Theor. Exptl. Chem., 2 (1966) (English translation edition)
0085 A.E. Lutskii et al.
0089 A.E. Lutskii et al.
0115 E.Y. Gren and G.Y. Vanag
0201 A.E. Lutskii et al.
0204 A.E. Lutskii et al.
0230 E.Y. Gren and G.Y. Vanag
0318 R.G. Dubenko et al.
0357 Y.A. Kolesnik, V.V. Kozlov and L.A. Kazitsina
0469 A.E. Moskvin et al.

100- -66, Lloydia, 29 (1966)
0029 J.L. Kaul and J. Trojanek
0094 H.H.S. Iong, J.L. Beal and M.P. Cava
0206 G.H. Svoboda
0225 L. Horhammer, H. Wagner and E. Khalil
0343 N.R. Farnsworth et al.

101- -66A, J. Organometallic Chem., 5 (1966)
0181 H.P. Fritz and K.E. Schwarzhans
0341 R.B. King, M.B. Bisnette and A. Fronzaglia
0357 N. Tirosh, A. Modiano and M. Cais
0370 M. Cais and N. Maoz
0392 H. Gilman and D.R. Chapman
0480 J.W. Fitch, III, and J.J. Lagowski

101- -66B, J. Organometallic Chem., 6 (1966)
0102 H. Gilman and P.J. Morris
0173 M. Rosenblum et al.
0249 K. Kawakami and R. Okawara
0399 K. Schlogl and W. Steyrer

0542 R.P.A. Snneden and H.P. Throndsen
0578 B.F. Hegarty and W. Kitching
0598 G. Bianchi et al.
0665 H. Gilman and C.L. Smith

102- -66, Phytochemistry, 5 (L966)
0177 V.K. Bhatia, S.R. Gupta and T.R. Seshadri
0367 D.L. Dreyer
0707 K. Schreiber and O. Aurich
0719 G.G. Freeman
0767 J. Corse et al.
0815 C.F. Culberson
0921 R. Bracho and K.J. Crowley
1065 G. Combes et al.
1227 H. Ronsch and K. Schreiber